THERMODYNAMICS

This book differs from other thermodynamics texts in its objective, which is to provide engineers with the concepts, tools, and experience needed to solve practical real-world energy problems. The presentation integrates computer tools (e.g., EES) with thermodynamic concepts to allow engineering students and practicing engineers to solve problems that they would otherwise not be able to solve. The use of examples, solved and explained in detail and supported with property diagrams that are drawn to scale, is ubiquitous in this textbook. The examples are not trivial drill problems, but rather complex and timely real-world problems that are of interest by themselves. As with the presentation, the solutions to these examples are complete and do not skip steps. Similarly, the book includes numerous end-of-chapter problems, both in the book and online. Most of these problems are more detailed than those found in other thermodynamics textbooks. The supplements include complete solutions to all exercises, software downloads, and additional content on selected topics.

Sanford Klein is currently the Bascom Ouweneel Professor of Mechanical Engineering at the University of Wisconsin, Madison. He has been on the faculty at Wisconsin since 1977. He is the Director of the Solar Energy Laboratory and has been involved in many studies of solar and other types of energy systems. He is the author or co-author of more than 160 publications relating to the analysis of energy systems. Professor Klein's current research interests are in solar energy systems and applied thermodynamics and heat transfer. In addition, he is actively involved in the development of engineering computer tools for both instruction and research. He is the primary author of a modular simulation program (TRNSYS), a solar energy system design program (F-CHART), a finite element heat transfer program (FEHT), and the general engineering equation solving program (EES). Professor Klein is a Fellow of the American Society of Mechanical Engineers (ASME); the American Society of Heating, Refrigeration, and Air-Conditioning Engineers (ASHRAE); and the American Solar Energy Society (ASES).

Gregory Nellis is the Elmer R. and Janet A. Kaiser Professor of Mechanical Engineering at the University of Wisconsin, Madison. He received his M.S. and Ph.D. at the Massachusetts Institute of Technology and is a member of the ASHRAE, the ASME, the International Institute of Refrigeration (IIR), and the Cryogenic Society of America (CSA). Professor Nellis carries out applied research that is related to energy systems with a focus on refrigeration technology, and he has published more than 40 journal papers. Professor Nellis's focus has been on graduate and undergraduate education, and he has received the Polygon, Pi Tau Sigma, and Woodburn awards for excellence in teaching as well as the Boom Award for excellence in cryogenic research. He is the co-author of *Heat Transfer* (2009) with Sanford Klein.

Thermodynamics

SANFORD KLEIN

University of Wisconsin, Madison

GREGORY NELLIS

University of Wisconsin, Madison

CAMBRIDGE
UNIVERSITY PRESS

CAMBRIDGE UNIVERSITY PRESS
Cambridge, New York, Melbourne, Madrid, Cape Town,
Singapore, São Paulo, Delhi, Tokyo, Mexico City

Cambridge University Press
32 Avenue of the Americas, New York, NY 10013-2473, USA

www.cambridge.org
Information on this title: www.cambridge.org/9780521195706

First published 2012

Printed in the United States of America

A catalog record for this publication is available from the British Library.

Library of Congress Cataloging in Publication data

Klein, Sanford A., 1950–
Thermodynamics / Sanford Klein, Gregory Nellis.
p. cm.
Includes bibliographical references and index.
ISBN 978-0-521-19570-6 (hardback)
1. Thermodynamics. 2. Engineering – Problems, exercises, etc. I. Nellis, Gregory. II. Title.
QC311.15.K58 2011
536′.7–dc22 2011001982

ISBN 978-0-521-19570-6 Hardback

Additional resources for this publication at www.cambridge.org/kleinandnellis

CONTENTS

* Section can be found on the Web site that accompanies this book (www.cambridge.org/kleinandnellis).

* Section can be found on the Web site that accompanies this book (www.cambridge.org/kleinandnellis).

* Section can be found on the Web site that accompanies this book (www.cambridge.org/kleinandnellis).

* Section can be found on the Web site that accompanies this book (www.cambridge.org/kleinandnellis).

* Section can be found on the Web site that accompanies this book (www.cambridge.org/kleinandnellis).

* Section can be found on the Web site that accompanies this book (www.cambridge.org/kleinandnellis).

* Section can be found on the Web site that accompanies this book (www.cambridge.org/kleinandnellis).

PREFACE

Thermodynamics is a mature science. Many excellent engineering textbooks have been written on the subject, which leads to the question: Why yet another textbook on classical thermodynamics? There is a simple answer to this question: this book is different. The objective of this book is to provide engineers with the concepts, tools, and experience needed to solve practical real-world energy problems. With this in mind, the focus of this effort has been to integrate a computer tool with thermodynamic concepts in order to allow engineering students and practicing engineers to tackle problems that they would otherwise not be able to solve.

It is generally acknowledged that students need to solve problems in order to integrate concepts and skills. The effort required to solve a thermodynamics problem can be broken into two parts. First, it is necessary to identify the fundamental relationships that describe the problem. The set of equations that leads to a useful solution to a problem results from application of appropriate balances and rate relations, simplified with justified assumptions. Identifying the necessary equations is the conceptual part of the problem, and no computer program can provide this capability in general. Proper application of the First and Second Laws of Thermodynamics is at the heart of this process. The ability to identify the appropriate equations does not come easily to most thermodynamics students. This is an area in which problem-solving experience is helpful. A distinguishing feature of this textbook is that it presents detailed examples and discussion that explain how to apply thermodynamics concepts identify a set of equations that will provide solutions to non-trivial problems.

Once the appropriate equations have been identified, they must be solved. In our experience, much of the time and effort required to solve thermodynamics problems results from looking up property information in tables and solving the appropriate equations. Though necessary for obtaining a solution, these tasks contribute little to the learning process. For example, once the student is familiar with the use of property tables, further use of the tables does not contribute to the student's grasp of the subject – nor does doing the tedious algebra that is required to solve a large set of equations. Practical problems that focus on real engineering issues tend to be more interesting to students, but also more mathematically complex. The time and effort required to do problems without computing tools may actually detract from learning the subject matter by forcing the student to focus on the mathematical complexity of the problem rather than on the underlying concepts.

The motivation for writing this book is a result of our experience in teaching mechanical engineering thermodynamics in a manner that is tightly integrated with the EES (Engineering Equation Solver) program. EES eliminates much of the mathematical complexity involved in solving thermodynamics problems by providing a large bank of high-accuracy property data and the capability to solve large sets of simultaneous algebraic and differential equations. EES also provides the capability to check equations for unit consistency; do parametric studies; produce high-quality plots; and apply numerical integration, optimization, and uncertainty analyses. Using EES, students can easily obtain solutions to interesting practical problems that involve nonlinear and implicit sets of equations. They can quickly display the results of these calculations in plots. They can conduct design studies by varying the inputs or constraints and by applying optimization methods. EES is a powerful tool that can be of great advantage for solving thermodynamics problems. However, like all tools, some training and experience are needed to use it effectively. The presentation in this book teaches readers by example how to use EES most effectively, with more advanced features introduced in a sequential manner throughout the text.

A review of the table of contents shows that the topics and order of presentation are similar to those provided in current mechanical engineering thermodynamics textbooks. Sufficient material is provided for both undergraduate and graduate thermodynamics courses. The book can be used in a single-semester undergraduate course by appropriately selecting from the available topics. For example, we typically do not cover Chapter 7 and Chapters 14–16 in our single-semester undergraduate course. Topics such as absorption cycles (9.4), cryogenic cooling cycles (9.5–6), desiccants (12.4.4), exergy relations for psychrometrics (12.6), and fuels (13.5) are also usually not included in our undergraduate courses. The reason that this book can be used for a first course (despite its expanded content) while remaining an effective graduate-level textbook is that all concepts and methods are presented in detail, starting at the beginning without skipping steps. You will not find many occurrences of the clause, "it can be shown that ... " in this textbook. The use of examples, solved and explained in detail and supported with property diagrams that are drawn to scale, is ubiquitous in this textbook. The examples are not trivial drill problems, but rather complex and timely real-world problems that are of interest by themselves. As with the presentation, the solutions to these examples are complete and do not skip steps.

The book includes a large collection of real-world problems at the end of each chapter. A larger selection of problems is provided on the Web site associated with this textbook (www.cambridge.org/kleinandnellis). Most of the problems provided with this book are more detailed than those provided in currently popular thermodynamics textbooks. It may appear upon first review that these problems are too complex for use in a first course in thermodynamics. Our experience, however, is that the organized approach to problem solving presented in this textbook, combined with the use of EES, allows undergraduates to successfully solve these more detailed problems. Indeed, we have found that students are more interested in the course because the problems are challenging and relevant. Complete solutions to all problems are provided to instructors.

This book is unusual in its linking of thermodynamic concepts with detailed instructions for using a powerful equation-solving computer tool that eliminates much of the tedious effort that is otherwise needed to solve thermodynamics problems. It fills an obvious void that we have encountered in teaching both undergraduate and graduate thermodynamics courses. The text and the EES program were developed over many years from our experiences teaching the undergraduate and graduate thermodynamics courses at the University of Wisconsin. It our hope that this text will be a lifelong resource for practicing engineers.

Sanford Klein
Gregory Nellis
June 2011

ACKNOWLEDGMENTS

The development of this book has taken several years and a substantial effort. This has only been possible due to the collegial and supportive atmosphere that makes the Mechanical Engineering Department at the University of Wisconsin such a unique and impressive place. In particular, we would like to acknowledge Doug Reindl and John Pfotenhauer for their encouragement throughout the process.

Several years of undergraduate and graduate students and faculty have used our initial drafts of this manuscript. These students have had to endure carrying multiple volumes of poorly bound paper with no index and many typographical errors. Their feedback has been invaluable to the development of the book.

More than two decades of students and faculty have contributed to the continuous development of the EES program. This program was initially developed specifically for use in undergraduate thermodynamics classes but it has expanded to the point where it is now a commercial program that is widely used in the HVAC&R and other industries. The suggestions and feedback of users at the University of Wisconsin, other universities, and various companies have driven the development of EES to the useful tool that it is today.

Preparing this book has necessarily reduced the time that we have been able to spend with our families. We are grateful to them for allowing us this indulgence. In particular, we wish to thank Jan Klein and Jill, Jacob, Spencer, and Sharon Nellis. We could not have completed this book without their continuous support.

Finally, we are indebted to Cambridge University Press and in particular Peter Gordon for giving us this opportunity and helping us through the process of bringing our manuscript to this final state.

NOMENCLATURE

a	specific Helmholtz free energy (J/kg)
	parameter in an equation of state
A	area (m^2)
	Helmholtz free energy (J)
	amplitude of wave
A^*	critical area (m^2)
A_c	cross-sectional area (m^2)
A_f	frontal area (m^2)
AF	air-fuel ratio (-)
A_s	surface area (m^2)
b	parameter in an equation of state (m^3/kg)
B	parameter defined in Eq. (10-71)
	parameter defined in Eq. (15-106)
BPR	bypass ratio (-)
bwr	back work ratio (-)
c	specific heat capacity (J/kg-K)
	speed of sound (m/s)
	speed of light in a vacuum (3×10^8 m/s)
$\bar{c}$	molar specific heat of an incompressible substance (J/kmol-K)
C	number of chemical species in a mixture (-)
$\dot{C}$	capacitance rate (product of mass flow rate and specific heat) (W/K)
C_D	drag coefficient (-)
COP	coefficient of performance (-)
c_P	specific heat capacity at constant pressure (J/kg-K)
$\bar{c}_P$	molar specific heat capacity at constant pressure (J/kmol-K)
C_R	capacitance ratio (-)
CR	compression ratio (-)
c_v	specific heat capacity at constant volume (J/kg-K)
$\bar{c}_v$	molar specific heat capacity at constant volume (J/kmol-K)
D	diameter (m)
$d\vec{x}$	differential displacement vector (m)
E	energy (J)
	voltage (Volt)
	number of elements (-)
$\dot{E}$	rate of energy transfer (W)
$\bar{E}$	intensity of radiation (W/m^2)
$E_{b,\lambda}$	blackbody spectral emissive power (W/m^2-μm)
EER	energy efficiency rating (Btu/W-hr)
$e_{i,j}$	number of moles of element *j* per mole of substance *i* (-)
E_j	number of moles of element *j* (kmol)

f	frequency (Hz)
	fugacity (Pa)
	partition function (-)
	fraction of flow (-)
F	force (N)
	number of intensive properties (for Gibbs phase rule)
$\vec{F}$	force vector (N)
f_i	fugacity of pure fluid i (Pa)
$\hat{f}_i$	partial fugacity of component i in a mixture (Pa)
g	gravitational acceleration (m/s^2)
	specific Gibbs free energy (J/kg)
G	Gibbs free energy (J)
$\bar{g}$	molar specific Gibbs free energy (J/kmol)
g_i	degeneracy of energy level i (-)
h	specific enthalpy (J/kg)
	Planck's constant (6.625×10^{-34} J/s)
$\bar{h}$	molar specific enthalpy (J/kmol)
H	enthalpy (J)
$\dot{H}$	enthalpy flow rate (W)
h_{av}	enthalpy of an air-water vapor mixture per kg dry air (J/kg$_a$)
h_{conv}	convective heat transfer coefficient (W/m^2-K)
HC	heat of combustion (J/kg)
$\bar{h}_{dep}$	molar specific enthalpy departure (J/kmol or J/kg)
$\bar{h}_{dep,i}$	molar specific enthalpy departure of component i in a mixture (J/kmol)
h_{fg}	specific enthalpy of vaporization (J/kg)
$\bar{h}_{fg}$	molar specific enthalpy of vaporization (J/kmol)
$\bar{h}_{form}$	molar specific enthalpy of formation (J/kmol)
HHV	higher heating value (J/kmol)
$\bar{h}_i$	molar specific enthalpy of component i in a mixture (J/kmol)
$HSPF$	heating season performance factor (Btu/W-hr)
$\bar{h}_{std}$	standardized molar specific enthalpy (J/kmol)
HV	heating value (J/kmol)
i	current (Amp)
$\bar{i}$	unit vector in the x-direction
I	moment of inertia (kg-m^2)
$\bar{j}$	unit vector in the y-direction
k	thermal conductivity (W/m-K)
	ratio of specific heat capacities, c_P/c_v (-)
	Boltzmann's constant (1.3805×10^{-23} J/K)
$\bar{k}$	unit vector in the z-direction
K	spring constant (N/m)
KE	kinetic energy (J)
k_{ij}	binary mixing parameter (-)
K_j	equilibrium constant for reaction j (-)
K_P	coefficient of pressure recovery (-)
K_T	isothermal compressibility (1/Pa)
L	the dimension length
L	distance or length (m)
LHV	lower heating value (J/kmol or J/kg)
m	mass (kg)
	parameter in the RKS and PR equations of state (-)
$\dot{m}$	mass flow rate (kg/s)

M	the dimension mass
M	Mach number (-)
MEP	mean effective pressure (Pa)
mf_i	mass fraction of component i in a mixture (-)
m_i	mass of component i in a mixture (kg)
MW	molar mass (kg/kmol)
n	number of moles (kmol)
	polytropic exponent (-)
	quantum number (-)
N	number of particles (-)
	number (-)
	engine speed (rpm)
N_A	Avogadro's number (6.022×10^{26} kmol^{-1})
n_i	number of moles of component i in a mixture (kmol)
N_i	number of particles in energy level i (-)
NTU	number of transfer units (-)
p	momentum (N-s)
P	pressure (Pa)
	probability (-)
P^0	standard state pressure or low pressure at which the ideal gas law is valid (Pa)
P_0	dead state pressure (Pa)
	stagnation pressure (Pa)
P_{atm}	atmospheric pressure (Pa)
P_{crit}	critical pressure (Pa)
$P_{crit,eff}$	effective critical (or pseudo-critical) pressure of a mixture (Pa)
PE	potential energy (J)
P_{gage}	gage pressure (Pa)
P_i	partial pressure of component i in a mixture (Pa)
PLF	part load factor (-)
P_r	reduced pressure (-)
PR	pressure ratio (-)
$P_{r,eff}$	effective reduced (or pseudo-reduced) pressure of a mixture (-)
Q	heat transfer (J)
	molar quality (-)
$\dot{Q}$	heat transfer rate (W)
$\dot{Q}''$	heat transfer rate per unit area (W/m^2)
$Q_{1\rightarrow2}$	heat transfer during the process of going from state 1 to state 2 (J)
R	ideal gas constant (J/kg-K)
	radius (m)
	resistance Ω
R_{univ}	universal gas constant (8314.3 J/kmol-K)
s	specific entropy (J/kg-K)
$\bar{s}$	molar specific entropy (J/kmol-K)
S	entropy (J/K)
$\dot{S}$	rate of entropy transfer (W/K)
s_{av}	entropy of an air–water vapor mixture per kg dry air (J/kg$_a$-K)
$\bar{s}_{dep}$	molar specific entropy departure (J/kmol-K)
$\bar{s}_{dep,i}$	molar specific entropy departure of pure gas i in a mixture (J/kmol-K)
S_{gen}	entropy generation (J/K)
$\dot{S}_{gen}$	rate of entropy generation (W/K)
S_m	entropy transfer due to mass transfer (J/K)
$\dot{S}_m$	rate of entropy transfer due to mass transfer (W/K)

$\bar{s}_i$	molar specific entropy of component i in a mixture (J/kmol-K)
S_Q	entropy transfer due to heat (J/K)
$\dot{S}_Q$	rate of entropy transfer due to heat (W/K)
$SEER$	seasonal energy efficiency rating (Btu/W-hr)
SFC	specific fuel consumption (kg/s-N)
SHR	sensible heat ratio (-)
t	the dimension time
t	time (s)
T	the dimension temperature
T	temperature (K)
T_0	dead state temperature (K)
	stagnation temperature (K)
T_B	Boyle temperature (K)
T_{crit}	critical temperature (K)
$T_{crit,eff}$	effective critical (or pseudo-critical) temperature of a mixture (K)
T_{dp}	dew-point temperature (K)
T_r	reduced temperature (-)
$T_{r,eff}$	effective reduced (or pseudo-reduced) temperature of a mixture (-)
$T_{r,B}$	reduced Boyle temperature (-)
T_{wb}	wet bulb temperature (K)
th	thickness (m)
u	specific internal energy (J/kg)
$\bar{u}$	molar specific internal energy (J/kmol)
$\bar{u}_i$	partial molar specific internal energy of component i in a mixture (J/kmol)
$\bar{u}_{std}$	standardized molar internal energy (J/kmol)
U	internal energy (J)
UA	conductance (W/K)
	building heat loss coefficient (W/K)
v	specific volume (m^3/kg)
$\bar{v}$	molar specific volume (m^3/kmol)
$\bar{v}_i$	molar specific volume of component i in a mixture (m^3/kmol)
V	volume (m^3)
$\dot{V}$	volumetric flow rate (m^3/s)
$\tilde{V}$	velocity (m/s)
v_{av}	volume of an air–water vapor mixture per kg dry air (m^3/kg_a)
v_{crit}	critical specific volume (m^3/kg)
$\bar{v}_{crit}$	critical molar specific volume (m^3/kmol)
$v_{crit,eff}$	effective critical (or pseudo-critical) specific volume of a mixture (m^3/kg)
$\dot{V}_{disp}$	displacement rate of a compressor (m^3/s)
V_i	volume of component i in a mixture (m^3)
$\bar{V}_i$	partial molar volume of component i in a mixture(m^3/kmol)
v_r	reduced specific volume (-)
$v_{r,eff}$	effective reduced (or pseudo-reduced) specific volume (-)
W	work (J)
	weight (N)
$W_{1\to2}$	work transfer during the process of going from state 1 to state 2 (J)
W_{lost}	"lost" work or exergy destruction (J)
$\dot{W}$	work transfer rate, power (W)
x	quality
	position (m)
	displacement (m)
X	exergy (J)

X_{des}	exergy destroyed (J)
X_f	exergy associated with a mass transfer (J)
x_i	mole fraction of component i in the liquid phase of a mixture (-)
X_Q	exergy associated with a heat transfer (J)
X_s	exergy of a system (J)
x_f	specific exergy of a flowing substance (J/kg)
x_s	specific exergy of a system (J/kg)
$\dot{X}_{des}$	rate of exergy destruction, also called the irreversibility rate (W)
$\dot{X}_f$	rate of exergy flow with mass flow (W)
$\dot{X}_Q$	rate of exergy flow with heat (W)
y_i	mole fraction of component i in a mixture (-)
	mole fraction of component i in the vapor phase of a mixture (-)
z	elevation in a gravitational field (m)
Z	compressibility factor (-)
Z_{crit}	critical compressibility factor (-)
z_i	total mole fraction of component i in a mixture (-)
Z_i	compressibility factor for component i in a mixture (-)

Greek Symbols

α	reduced Helmholtz free energy (-)
	parameter in the RK or PR equation of state (-)
β	parameter defined in Eq. (15-101) (1/J)
δ	uncertainty in some measurement
	reduced density (-)
	differential amount
Δ	change of some property of a system
ΔG_j^o	standard state Gibbs free energy change of reaction (J)
Δh_{fg}	latent heat of vaporization (J/kg)
$\Delta \bar{h}_{mix}$	molar specific enthalpy change of mixing (J/kmol)
ΔP	pressure drop (Pa)
$\Delta \bar{s}_{mix}$	molar specific entropy change of mixing (J/kmol-K)
ΔT	approach temperature difference (K)
$\Delta \bar{v}_{mix}$	molar specific volume change of mixing (m^3/kmol)
ΔV_{mix}	volume change of mixing (m^3)
ε	Lennard-Jones energy potential (J)
	emissivity (-)
	effectiveness of a heat exchanger (-)
	reaction coordinate or degree of reaction (kmol)
ε_i	energy associated with a energy level i (J)
ε_j	reaction coordinate for reaction j (kmol)
ϕ	fugacity coefficient (-)
	relative humidity (-)
γ	surface tension (N/m)
η	efficiency (-)
η_2	Second Law efficiency (-)
λ	wavelength (μm)
	undetermined multiplier
Π	number of phases (-)
θ	angle (radian)
ρ	density (kg/m^3)

σ	Lennard-Jones length potential (m)
	Stefan-Boltzmann constant (5.67×10^{-8} W/m^2-K^4)
τ	torque (N-m)
	inverse reduced temperature (-)
μ	viscosity (Pa-s)
$\mu_{f,i}$	chemical potential of component i in the liquid phase of a mixture (J/kmol)
$\mu_{g,i}$	chemical potential of component i in the vapor phase of a mixture (J/kmol)
μ_{JT}	Joule-Thomson coefficient (K/Pa)
ν_i	stoichiometric coefficient for component i
$\nu_{i,j}$	stoichiometric coefficient for component i in reaction j
ψ	constraint function that evaluates to zero
ω	angular velocity (rad/s)
	acentric factor (-)
	humidity ratio (kg$_v$/kg$_a$)
ω_{eff}	effective acentric factor of a mixture (-)
Ω	thermodynamic probability (-)

Superscripts

*	quantity evaluated at location of critical area
o	under conditions where fluid behaves as an ideal gas (i.e., at low pressure)

Subscripts

a	dry air
act	actual
amb	ambient
as	adiabatic saturation
atm	atmospheric
av	psychrometric property defined on a per mass of dry air basis
avg	average
b	boundary
	boiler
	blackbody
B	Boyle isotherm
BDC	bottom dead center
BE	Bose-Einstein model
c	compressor
C	cold fluid in a heat exchanger or cold thermal reservoir
comp	compression process
cond	condenser
crit	critical, related to the critical point
CTHB	cold-to-hot blow process
cv	associated with a control volume
cyl	cylinder
d	diffuser
	downstream of a normal shock
	drag
des	destroyed within system
dp	dew point
ec	evaporative cooler

evap	evaporator
exp	expansion process
f	saturated liquid
	fuel
	furnace
FD	Fermi-Dirac model
g	saturated vapor
gage	gage
gen	generated within system
	generator in an absorption cycle
H	hot fluid in a heat exchanger or hot thermal reservoir
	heat pump
HTCB	hot-to-cold blow process
hf	heat transfer fluid
hx	heat exchanger
i	the *i*th component in a mixture
IC	based on incompressible model
in	in, entering a system
ini	initial, at time = 0
load	refrigeration or building load
max	maximum or maximum possible
min	minimum or minimum possible
mix	associated with a mixture
mp	maximum power
MB	Maxwell-Boltzmann model
n	nozzle
net	net output
nom	nominal value
o	overall
	stagnation
	initial
p	pump
	piston
	propulsive
P	related to an isobaric process, at constant pressure
	associated with the products of a reaction
pure	associated with a pure substance
out	out, leaving a system
R	refrigeration cycle
	associated with the reactants of a reaction
Rankine	Rankine cycle
r	reduced
ref	reference
res	residual component
rev	reversible
rh	reheat cycle or reheater
s	associated with a reversible device
sat	saturated
sc	subcool
sh	superheat
sur	surroundings

t	turbine
	at time t
T	related to an isothermal process, at constant temperature
TDC	top dead center
th	thermal
u	upstream of a normal shock
v	valve
	water vapor
vol	volumetric
wb	wet bulb
x	in the x-direction
y	in the y-direction
z	in the z-direction
λ	spectral – as a function of wavelength
∞	free-stream fluid

Other Notes

$\bar{a}$	specific property on molar basis
$f(A)$	function of variable A
ΔA	change in variable A
dA	differential change in the property A
δA	differential amount of the quantity A
	uncertainty in the quantity A
$\mathring{A}$	rate of transfer of quantity A

THERMODYNAMICS

1 Basic Concepts

1.1 Overview

Thermodynamics is unquestionably the most powerful and most elegant of the engineering sciences. Its power arises from the fact that it can be applied to any discipline, technology, application, or process. The origins of thermodynamics can be traced to the development of the steam engine in the 1700's, and thermodynamic principles do govern the performance of these types of machines. However, the power of thermodynamics lies in its generality. Thermodynamics is used to understand the energy exchanges accompanying a wide range of mechanical, chemical, and biological processes that bear little resemblance to the engines that gave birth to the discipline. Thermodynamics has even been used to study the energy exchanges that are involved in nuclear phenomena and it has been helpful in identifying sub-atomic particles. The elegance of thermodynamics is the simplicity of its basic postulates. There are two primary 'laws' of thermodynamics, the First Law and the Second Law, and they always apply with no exceptions. No other engineering science achieves such a broad range of applicability based on such a simple set of postulates.

So, what is thermodynamics? We can begin to answer this question by dissecting the word into its roots: 'thermo' and 'dynamics'. The term 'thermo' originates from a Greek word meaning warm or hot, which is related to temperature. This suggests a concept that is related to temperature and referred to as heat. The concept of heat will receive much attention in this text. 'Dynamics' suggests motion or movement. Thus the term 'thermodynamics' may be loosely interpreted as 'heat motion'. This interpretation of the word reflects the origins of the science. Thermodynamics was developed in order to explain how heat, usually generated from combusting a fuel, can be provided to a machine in order to generate mechanical power or 'motion'. However, as noted above, thermodynamics has since matured into a more general science that can be applied to a wide range of situations, including those for which heat is not involved at all. The term thermodynamics is sometimes criticized because the science of thermodynamics is ordinarily limited to systems that are in *equilibrium*. Systems in equilibrium are not 'dynamic'. This fact has prompted some to suggest that the science would be better named 'thermostatics' (Tribus, 1961).

Perhaps the best definition of thermodynamics is this: *Thermodynamics is the science that studies the conversion of energy from one form to another*. This definition captures the generality of the science. The definition also introduces a new concept – energy. Thermodynamics involves a number of concepts that may be new to you, such as heat and energy, and these terms must each be carefully defined. As you read this, it may seem that heat and energy are familiar words and therefore no further definition of these concepts is necessary. However, the common understanding of these terms differs from the formal definitions that are needed in order to apply the laws of thermodynamics.

The First Law of Thermodynamics states that energy is conserved in all processes (in the absence of nuclear reactions). If energy is conserved (i.e., it is not generated or destroyed) then the amount of energy that is available must be constant. But if the amount of energy is constant then why do we hear on the news that the world is experiencing an energy shortage? How could we be 'running out of energy'? Why do we receive monthly 'energy' bills?

The answer to these questions lies in the difference between the term energy as it is commonly used and the formal, thermodynamic definition of energy. These differences between common vernacular and precise thermodynamic definitions are a source of confusion. The term *energy* that is used in everyday conversations should be thought of as 'the capacity to do work'. This definition is not consistent with the thermodynamic definition of energy, but rather refers to a different thermodynamic concept that is referred to as *exergy* and is studied in Chapter 7. The thermodynamic definition of energy is not as satisfying. Energy is not really 'something'; rather, it is a property of matter. We cannot see, smell, taste, hear or feel energy. We can measure it, but only indirectly. Hopefully, the thermodynamic concept of energy will become clearer as you progress through this book.

The First Law of Thermodynamics is concerned with the conservation of energy. However, energy has both quantity and quality. The quality of energy is not conserved and the Second Law of Thermodynamics can be interpreted as a system for assigning quality to energy. Although energy is conserved, the quality of energy is always reduced during energy transformation processes. Lower quality energy is less useful to us in the sense that its capability for doing work has been diminished. The quality of energy is continuously degraded by all real processes; this observation can be expressed in lay terms as 'running out of energy'.

The Second Law is responsible for the directional nature of all real processes. That is, processes can occur in only one direction and will not spontaneously reverse themselves because doing so would require a spontaneous increase in the quality of energy. The Second Law explains why heat flows from hot to cold and why objects at different temperatures will eventually come to the same temperature. The Second Law explains why gases mix and things break. It can be used to explain why we age and why time moves forward. The Second Law of Thermodynamics is likely the most famous law in all of the physical sciences.

Our society is now facing some very challenging problems. Some of these problems are related to the diminishing supply of petroleum, coal, natural gas, and the other combustible materials that provide the energy (the common definition rather than the thermodynamic definition) that powers our world. Even if these fuels were inexhaustible (which they are not), combustion of carbon-based fuels necessarily produces carbon dioxide, which has been linked to global warming and other climate change phenomena. What alternatives exist to provide the power that we need? Hydrogen-powered fuel cells, biomass, nuclear power plants, solar and wind energy systems have all been mentioned in the popular media as potential solutions. Which one of these alternatives is actually best? What role can each of them play in terms of displacing our current energy supplies? These are huge questions. The solution to our energy problem will likely be one of the biggest challenges facing our species this century. In one sense this is alarming, but it is also very exciting. You are reading this book because you have either a professional or personal interest in the subject of thermodynamics. Thermodynamics plays a major role in addressing these energy-related questions. It is clear that the demand for people who are well-educated in thermodynamics and capable of applying the discipline to a wide range of problems will only increase.

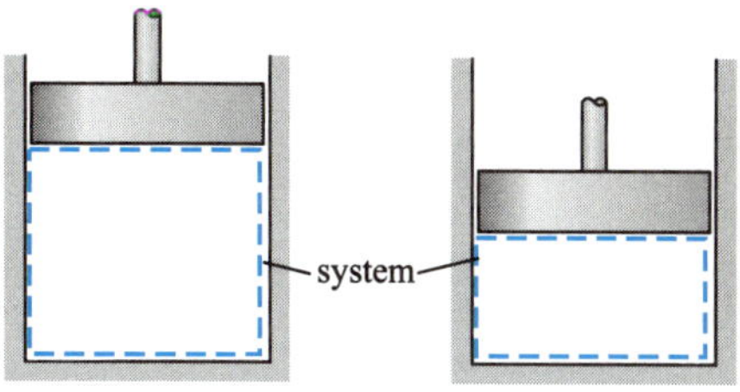

Figure 1-1: A system defined to contain all of the air in a piston-cylinder device.

1.2 Thermodynamic Systems

Every thermodynamic analysis begins with the specification of a *system*. A system is simply any object, quantity of matter, or region of space that has been selected for study. The system provides the precise specification for the focus of the analysis and enables the use of the First and Second Laws of Thermodynamics. Everything that is not part of the system is referred to as the *surroundings*.

The specification of a system requires the identification of its *boundary*, the surface that separates the system from its surroundings. The system boundary may correspond to a real surface. For example, Figure 1-1 illustrates a perfectly-sealed piston-cylinder device that is filled with air. The dashed blue line indicates the boundary of a system that is defined so that it contains all of the air within the cylinder. Notice that the boundary of this system must move as the position of the piston changes.

The laws of thermodynamics can be applied to any system and often there will be more than one logical system choice for a particular problem. For example, Figure 1-2 illustrates an air tank that is being filled from an air line. One choice for a system is the fixed region that corresponds to the internal volume of the air tank. An alternative system is defined by the dashed blue line in Figure 1-2 so that it contains all of the air that was initially in the tank. Notice that the boundary of this system must move as the tank is filled.

We classify systems according to their interactions with the surroundings. If mass does not cross the boundary of a system then it is referred to as a *closed system*. Mass does cross the boundary of an *open system*. An *adiabatic system* is one in which the boundary is impermeable to heat, i.e., the energy transfer that normally occurs when a temperature difference exists between the system and surroundings. A *steady-state* system is one in which all of the properties of the system do not change with time. An *isolated* system has no interaction of any kind with its surroundings.

The First and Second Laws of Thermodynamics apply regardless of the system choice. The definition of a system is dictated by convenience. While there is no wrong choice of system, some system choices simplify the mathematical description of a process whereas others cause the problem to become impossibly complicated. Your ability

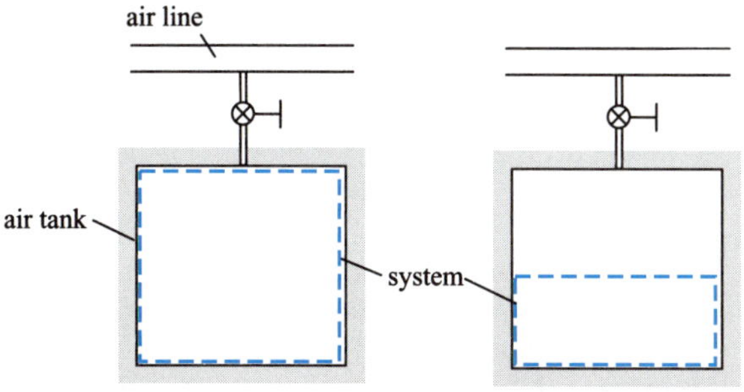

Figure 1-2: A system defined to contain all of the air that is initially in a tank that is being filled.

to select an appropriate system will improve with experience. The first step in every thermodynamics problem is the selection of a system and this is accomplished by clearly indicating its boundaries so that it can be carefully analyzed.

1.3 States and Properties

1.3.1 State of a System

Once a system has been specified, it is next necessary to specify its *state*. The state is a description of the system in terms of quantities that will be helpful in describing its behavior and its interactions with the surroundings.

There are two very different ways to describe a thermodynamic system, referred to as the *microscopic* and *macroscopic* approaches. The microscopic approach recognizes that the system consists of matter that is composed of countless, discrete particles (molecules). These molecules often behave in a manner that may be non-intuitive based on our everyday experience with much larger amounts of matter. The fundamental particles move at high velocities and have kinetic energies in three dimensions. Depending on the complexity of the molecules, they may also store energy due to their rotation and the vibration of the bonds connecting the atoms. The particles interact with each other and with the walls of their container. There are so many particles that it is hopeless to attempt to represent the observed characteristics of a system by describing the behaviors of each of its individual particles. However, we may be able to formulate a molecular model that describes the attractive and repulsive forces between particles and the various ways that a particle can store energy. We cannot directly test the molecular model against the behavior of a single molecule. However, we can apply statistics and probability theory to the molecular model in order to deduce the macroscopic behavior that would result from a large number of particles. Agreement between the calculated statistical behavior and the observed macroscopic behavior lends confidence in the fidelity of the model.

The branch of science that describes the state of a system using this microscopic approach is called Statistical Thermodynamics. Statistical Thermodynamics directly integrates the properties of matter with the conservation of energy. It provides a molecular explanation for the Second Law of Thermodynamics and it allows some physical properties (e.g., the specific heats and entropy of low pressure gases) to be determined more simply than is possible using any alternative method. Chapter 15 provides an introduction to Statistical Thermodynamics.

This text will apply a macroscopic approach to describe thermodynamic systems. In the macroscopic approach, the state of the system is described by a relatively small set of characteristics that are called *properties*. Some of these properties are already familiar to you, such as mass, temperature, pressure and volume. This macroscopic approach works well when the system is sufficiently large such that it contains many molecules. However, the macroscopic approach would not work well for a system that consists of a rarefied gas (i.e., a vacuum with just a few molecules). For example, how would you measure the temperature of such a system that consists almost entirely of vacuum?

1.3.2 Measurable and Derived Properties

Thermodynamic properties are classified as being either *measurable* or *derived*. Measurable properties can be directly measured using an appropriate instrument. Examples of measurable properties include mass, temperature, pressure, volume, velocity, elevation, specific heat capacities, and composition. Derived properties cannot be directly measured. Derived properties include internal energy, enthalpy, entropy, and other related thermodynamic properties that will be defined in this text.

It is not always clear whether a property is measurable or derived. For example, temperature is normally considered to be a measurable property. But how does one actually measure temperature? The common thermometer consists of a precision bore within a transparent glass enclosure that is filled with a liquid that expands when its temperature is increased. By observing the height of the liquid in the bore, we can measure the volume of the fluid and infer the temperature. There are many other ways to measure temperature. For example, thermistors relate the electrical resistance of a material to its temperature. Thermocouples are junctions between two dissimilar metals that generate a voltage potential that is a function of temperature. In each of these instruments, however, something (e.g., volume, resistance, or voltage) is directly measured and temperature is then inferred from this measurement. Although we do not directly measure temperature, it is still considered to be a measurable property.

1.3.3 Intensive and Extensive Properties

Thermodynamic properties are also classified as being either *intensive* or *extensive*. Intensive properties are independent of the amount of mass in the system whereas the values of extensive properties depend directly on the amount of mass. Temperature and pressure, for example, are intensive properties. If you were told the temperature or pressure of a system and nothing more, you would have no idea of the size of the system. Volume and energy are extensive properties. The greater the volume of a system, the more mass it must have. Extensive properties are linearly related to the system mass.

A *specific property* is defined as the ratio of an extensive property of a system to the mass of the system. Thus specific volume, v, is the ratio of volume (an extensive property) to mass:

$$v = \frac{V}{m} \tag{1-1}$$

where V is the volume of the system and m is the mass of the system. The inverse of specific volume is density:

$$\rho = \frac{1}{v} = \frac{m}{V} \tag{1-2}$$

Specific volume and density are both intensive properties. We will encounter several other extensive properties, including internal energy (U), enthalpy (H), and entropy (S). The corresponding specific properties are specific internal energy (u), specific enthalpy (h), and specific entropy (s):

$$u = \frac{U}{m} \tag{1-3}$$

$$h = \frac{H}{m} \tag{1-4}$$

$$s = \frac{S}{m} \tag{1-5}$$

1.3.4 Internal and External Properties

Properties can also be classified as being either *internal* or *external*. The value of an internal property depends on the nature of the matter that composes the system. External properties are independent of the nature of the matter within the system. Examples of external properties include the velocity of the system ($\tilde{V}$) and its elevation in a gravitational field (z). These properties do not depend on whether we are talking about a system composed of helium or one composed of steel. For example, in Section 3.2.2 we will see that a system with a mass $m = 1$ kg that is elevated a distance $z = 1$ m in Earth's gravitational field ($g = 9.81$ m/s^2) will have a potential energy $PE = m\,g\,z = 9.81$ J regardless of the type of matter that the system is composed of. Internal properties depend on the nature of the matter in the system and, as a consequence, they depend upon each other. That is, internal properties are functionally related to one another.

The interdependence of internal properties is of fundamental importance because it allows us to completely fix the state of a system by specifying only a few internal properties. The values of other internal properties can be found by employing the relationships that exist between these properties. It will be shown in Chapter 2 that only two internal intensive properties are required to fix the state of a system containing a pure substance that consists of only one phase (i.e., solid, liquid, or vapor). For example, if the temperature and pressure of water vapor are specified, then the density, specific internal energy, and specific enthalpy all have fixed values. Any other intensive property of the water vapor could also be determined. It is only necessary to know the temperature and pressure of a single phase pure substance in order to determine the specific heat capacity at constant pressure, the magnetic moment, the surface tension, the speed of sound, the electrical resistivity, and many other properties.

You likely have already employed a property relationship in your chemistry class by using the absolute temperature (T) and absolute pressure (P) of a gas in order to calculate its specific volume (v) using the ideal gas law:

$$v = \frac{R\,T}{P} \tag{1-6}$$

where the parameter R is the ideal gas constant. The ideal gas law will be discussed in Section 2.5. It does not apply under all conditions. The accuracy of Eq. (1-6) is reduced as the pressure is increased or as the temperature is decreased. Under some conditions, the ideal gas law may not be sufficiently accurate to be of any use at all. However, this complication does not change the fact that the specific volume is fixed at some value (i.e., it is not an independent variable) when the temperature and pressure are specified. If the ideal gas law is not applicable, a more complicated relation between specific volume, temperature and pressure may be needed, as discussed in Chapter 10. The properties of many substances have been measured and the relationships between internal properties can be expressed using tables, charts, equations, and computer programs, as described in Chapter 2.

1.4 Balances

Balances are the basic tool of engineering. Once a system has been carefully defined, it is possible to apply a balance to the system. A balance is simply a mathematical statement of what we know to be true. Any number of quantities can be balanced for any arbitrary system. The general balance equation, written for a finite period of time and some arbitrary quantity is:

$$In + Generated = Out + Destroyed + Stored \tag{1-7}$$

where *In* is the amount entering the system by crossing its boundary, *Generated* is the amount generated within the system, *Out* is the amount leaving the system by crossing its boundary, *Destroyed* is the amount destroyed within the system, and *Stored* is the amount stored in the system (i.e., the change in the quantity during the period of time). A balance can also be written on a rate basis, in which case the rate of each of the terms in Eq. (1-7) must be balanced at a particular instant in time. It is important to emphasize that the balance provided by Eq. (1-7) makes no sense until you have carefully defined a system and its boundaries.

Every month, most of us define a system that is referred to as our *household* and carry out a money balance on this system:

$$D_{in} + D_{gen} = D_{out} + D_{des} + \Delta D \tag{1-8}$$

where D indicates the amount of money. The variable D_{in} is the amount of money that enters your household (from wages and other forms of income), D_{des} is the amount of money that is destroyed, D_{out} is the amount of money that leaves your household (expenses), and D_{gen} is the amount of money that is created in your household. The term ΔD in Eq. (1-8) is the amount of money stored in your household, i.e., the change in the amount of money contained within your household during the time period of interest. A positive value of ΔD indicates that you managed to save some money during the month while a negative value indicates that you had to dip into your savings. For most of us, D_{des} will be zero every month as we do not often literally burn up or destroy currency. Also, D_{gen} will be zero as we cannot (legally) generate money. Thus, balancing our monthly finances will result in the following equation:

$$D_{in} = D_{out} + \Delta D \tag{1-9}$$

Equation (1-9) shows that, at least on a personal level, money is a *conserved* quantity; that is, it is neither destroyed nor produced. Other quantities are not conserved. For example, we could define a system around the borders of the United States and balance people for a year:

$$P_{in} + P_{gen} = P_{out} + P_{des} + \Delta P \tag{1-10}$$

In Eq. (1-10), P_{in} is the number of people that enter the U.S. by crossing its borders (immigration) and P_{out} is the number of people leaving the U.S. by crossing its borders (emigration). The quantity ΔP is the change in the population of the U.S. during the year (i.e., the number of people at the end of the year less the number of people at the beginning of the year). People are, as you know, not a conserved quantity; they are both destroyed and generated. The quantity P_{des} is the number of people that die and P_{gen} is the number of babies born within the borders of the U.S during the year. Equation (1-10) must be satisfied as it is simply a mathematical statement of what we know to be true.

Mass is a conserved quantity since it cannot be generated or destroyed (in the absence of nuclear reactions). Therefore, a mass balance on a system for a finite period of time leads to:

$$m_{in} = m_{out} + \Delta m \tag{1-11}$$

where m_{in} is the amount of mass that enters the system by crossing its boundary and m_{out} is the amount of mass that leaves the system by crossing its boundary. Note that the quantities m_{in} and m_{out} must be zero for a closed system. The quantity Δm is the amount of mass stored in the system; this is the mass in the system at the end of time period less the mass in the system at the beginning of the time period. A mass balance written on a rate basis at a particular instant in time is:

$$\dot{m}_{in} = \dot{m}_{out} + \frac{dm}{dt} \tag{1-12}$$

where $\dot{m}_{in}$ and $\dot{m}_{out}$ are the rates at which mass is entering and leaving the system, respectively, by crossing its boundaries, and $\frac{dm}{dt}$ is the rate of change in the mass of the system.

The laws of thermodynamics can be expressed most concisely in the form of balances. The First Law of Thermodynamics (which is introduced in Chapter 3) states that energy is a conserved quantity. For any system you care to define, energy cannot be destroyed or generated. Energy can flow into or out of a system (in various ways) and it can be stored in a system (in various forms). However, energy is never generated or destroyed in a system.

The Second Law of Thermodynamics states that the thermodynamic property entropy (which is introduced in Chapter 6) is not a conserved quantity; entropy is always generated and never destroyed. The Second Law suggests that the entropy of the universe is always increasing and it provides directionality to all processes. Any real process will result in the generation of entropy. Therefore, in order for the process to run in reverse (i.e., all inflows become outflows, outflows become inflows, etc.) it would be necessary to destroy entropy. Because it is not possible to destroy entropy, no real process is "reversible". A theoretical process that results in no entropy generation is often referred to as a "reversible" process and provides a useful limit to the behavior of real processes.

1.5 Introduction to EES (Engineering Equation Solver)

Thermodynamics and related thermal science courses (e.g., fluid dynamics and heat transfer) focus on providing a mathematical description of physical phenomena (i.e., an engineering model). An engineering model increases our understanding of the underlying physics. Properly formulated, the model can be used in place of the actual physical system in order to do mathematical experiments that can be conducted more quickly and with less cost than their physical counterparts. A good model is predictive. The model therefore allows the behavior of a system to be explored at conditions that would be difficult or impossible to achieve and it provides resolved information that would be hard to measure in the actual physical system. Coupled with optimization techniques, a model can be used to improve equipment designs and therefore allow us to obtain a desired result at less expense or more quickly or with less effect on the environment.

An engineering model will consist of a set of algebraic and/or differential equations that may be challenging to solve. In general, the more faithfully a mathematical model represents a physical behavior, the more equations are required and the more complicated the model becomes. At one extreme, the model can require so much computational effort that it may be easier to conduct experiments using the actual physical system. At the other extreme, the model can be too simplistic. Although the equations are easy to solve, the model does not accurately represent the physics and therefore is not very useful. A useful model is a compromise between these extremes.

Engineering Equation Solver (EES) is a computer program that has been developed in order to numerically solve the type of algebraic and differential equations that typically appear in models of thermodynamic systems. EES can check the dimensional and unit consistency of the equations in order to catch many common programming errors. In addition, EES provides built-in functions for the thermodynamic and transport properties of many engineering fluids. EES provides the capability to carry out parametric studies and generate high-quality plots; it can do optimization, provide linear and non-linear regression, and automate uncertainty analyses. The combination of equation solving capability with access to engineering property data makes EES a powerful

Equations Window
X*ln(X)=Y^3
sqrt(X)=1/Y

Figure 1-3: Two non-linear equations entered in the EES Equations window.

tool for modeling thermal-fluid systems and it is used extensively in industry. EES will be used throughout this textbook.

An introduction to EES is presented in this section. If you have already become familiar with EES and are comfortable entering and solving equations then you can skip this section. The EES program is probably installed on your departmental computer system. If not, you can use the limited academic version that can be downloaded from www.fchart.com or www.cambridge.org/kleinandnellis. To start EES from the Windows File Manager or from Explorer, double-click on the EES program icon or on any file that was created by EES. EES begins by displaying a splash screen that shows the registration information, the version number and other information. Click the OK button in order to dismiss the splash screen.

You will next see the Equations window. The Equations window is where the mathematical equations that constitute your model are entered. EES is capable of solving large sets of non-linear, coupled algebraic and differential equations. Enter the following two equations on separate lines in the Equations window:

$$x \ln(x) = y^3 \tag{1-13}$$

$$\sqrt{x} = \frac{1}{y} \tag{1-14}$$

Equations (1-13) and (1-14) have no physical significance. However, they are non-linear and coupled (i.e., they must be solved simultaneously) and, as a result, they would be difficult to solve by hand. Note that the equations are entered in the Equations window in the same manner that you would enter text into a word processor. However, there are some rules that must be followed in order for EES to understand your input.

1. Variable names must start with a letter and may consist of any keyboard character except () ‘ | * / + − ^ {}: “ or;. The maximum length of a variable name is 30 characters.
2. EES is not case-sensitive. That is, upper and lower case letters are not distinguished from one another. The variable X and the variable x are identical as far as EES is concerned.
3. Blank lines and spaces are ignored.
4. In general, each equation must be entered on a separate line. However, multiple equations may be entered on one line if they are separated by a semi-colon (;).
5. The caret symbol (^) and ** are both used to indicate the mathematical operation of raising a number to a power. For example, y^3 can be entered as y^3 or y**3.
6. EES uses the standard order of operations that is used by most other computer languages.

After you have entered Eqs. (1-13) and (1-14), the Equations window should appear as shown in Figure 1-3.

It is always important to annotate the equations in the Equations window so that you and others who look at your model (e.g., your thermodynamics professor) can understand what each equation is intended to do. Annotation can be accomplished in EES by adding comments in the Equations window. Comments can be enclosed within curly braces {} or within double quotation marks “ ”, as shown in Figure 1-4. It is good

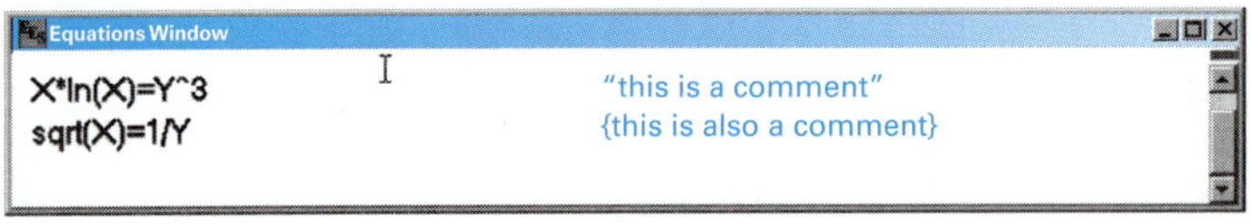

Figure 1-4: Equations window with equations and comments.

practice to enter the comments immediately to the right of each equation; this process is facilitated by pressing the tab key. Information within comments is ignored by EES and comments may span as many lines as needed. EES will display the comments in blue.

It is sometimes difficult to interpret equations that have been entered in text format in the Equations window, particularly when many nested sets of parentheses or operations are employed. Therefore, EES provides a Formatted Equations window that displays the equations that are entered in the Equations window using a mathematical notation. Select Formatted Equations from the Windows menu in order to access the Formatted Equations window (Figure 1-5). Notice that the comments that are entered in the Equations window within quotes are also displayed in the Formatted Equations window whereas comments entered in curly braces are not displayed. Normally, comments within quotes are used to document the equations whereas curly braces are used to "comment out" text that you do not wish EES to use at this time. "Commenting out" a set of equations is a convenient way to remove these equations temporarily. To accomplish this, highlight the equation(s) to be removed and right-click. Select Comment {} from the pop-up menu that appears. To re-instate the equation(s), highlight them again, right-click, and select Undo Comment {}.

The equations in the Formatted Equations window can be copied and pasted, for example into a report documenting the model. Highlight the equation(s) of interest and right-click on the selection. Notice that it is possible to copy the equation as a picture that can be pasted into a word processor. The Professional version of EES can also copy the equation as a LaTeX object or a MathType® equation.

Select Solve from the Calculate menu. A dialog window will appear indicating the progress of the solution. Click the Continue button when the calculations are completed in order to display the Solution window that contains the solution to the set of equations (Figure 1-6). EES can solve thousands of equations very quickly, which makes it a very powerful tool.

The Equations window allows a free form input. As you can see in Figure 1-4, the position of variables within the equation does not matter and it is not necessary to isolate the unknown variable on the left side of an equal sign, as is required in formal programming languages. This capability is convenient because in many problems, it is not possible to isolate the unknown variable. Also, the order in which the equations are entered does not matter. Before the equations are solved, EES will (internally) rearrange them into an order that leads to the most efficient solution process, regardless

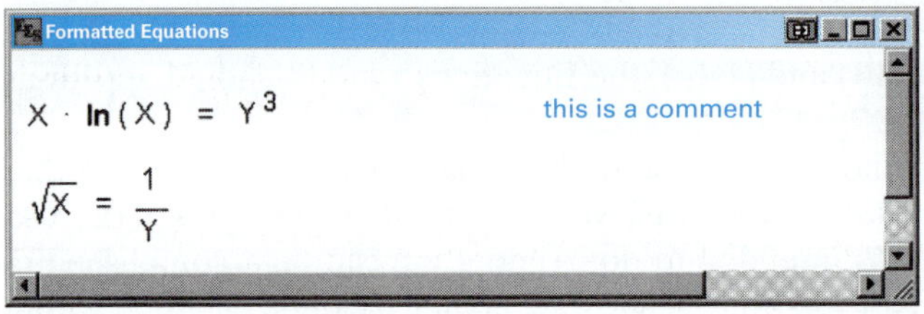

Figure 1-5: Equations displayed in the Formatted Equations window.

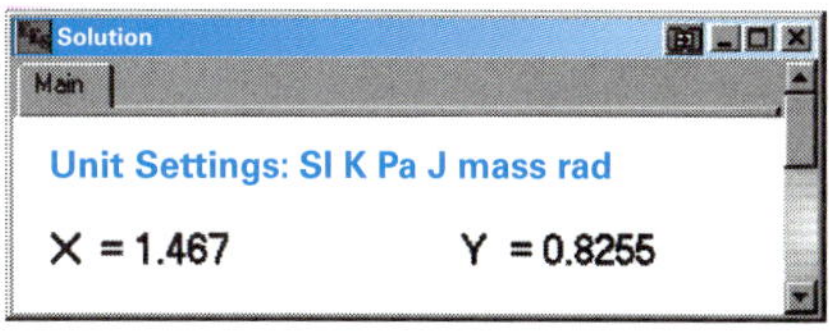

Figure 1-6: Solution window.

of the order that they are entered in the Equations window. Although EES allows you to enter equations in any order, it is still recommended that you enter your equations in an organized manner that progresses logically from the known information to the desired results. Also, it is best to enter and solve a few equations at a time, rather than entering all of the equations needed to solve a problem at once. This strategy allows you to efficiently debug your program because problems are naturally isolated to the last group of equations that were entered. These best practices for using EES will be demonstrated in the example problems presented throughout this textbook.

1.6 Dimensions and Units

Most of the variables used in models of thermodynamic processes and systems represent physical quantities and therefore they are dimensional. It is necessary to know both the value of the variable as well as its associated units. Calculation errors that result from incorrectly converting between units are common and frustrating. Careful attention to units is essential. This section reviews the dimensions and units of the fundamental quantities used in thermodynamic analyses. One of the most basic and powerful features of the EES program is its ability to keep track of units, convert between units, and check equations for unit consistency.

1.6.1 The SI and English Unit Systems

Dimensions are the fundamental measures of a physical quantity. Dimensions are categorized as being either *primary* or *secondary*. Primary dimensions for the quantities encountered in most thermodynamic models are usually chosen to be mass, length, time, and temperature, although other choices are equally valid. Secondary dimensions are combinations of the primary dimensions and result from definitions or physical laws. Examples of secondary dimensions are length/time for velocity and mass-length/time^2 for force. Note that in this text, when dealing with units, the dash symbol (-) will indicate multiplication on one side of the divisor. Therefore, the unit N-m/kg should be read as $\frac{\mathrm{N\,m}}{\mathrm{kg}}$ and the unit J/kg-K should be read as $\frac{\mathrm{J}}{\mathrm{kg\,K}}$. This convention is consistent with how units are entered in the EES program.

The scale for a dimension is its units. Many different units can be used to express a primary dimension. For example, the possible units for the dimension length include inches, feet, yards, miles, meters, centimeters, furlongs, and many others. Secondary dimensions can also be expressed in many different units. Energy, for example, is a secondary dimension that is expressed in terms of primary dimensions as mass-length^2/time^2. The units of energy include ft-$\mathrm{lb_f}$, Btu (British thermal units), calories, Joules, and many others.

Units are commonly categorized as belonging to the English or SI (Systems International) unit system. A list of the standard units used for the primary dimensions in each system is provided in Table 1-1. Some secondary dimensions and their typical units in each system are provided in Table 1-2.

Table 1-1: Standard units for primary dimensions in the SI and English unit systems.

	SI Unit System		English Unit System	
Physical Quantity	Unit	Symbol	Unit	Symbol
mass (M)	kilogram	kg	pound-mass	lb_m
length (L)	meter	m	foot	ft
time (t)	second	s	second	s
temperature (T)	Kelvin	K	Rankine	R
	degree Celsius	°C	degree Fahrenheit	°F

Why is there more than one unit for each dimension? There is no single answer to this question. Convenience certainly provides one explanation. It is possible but not convenient to express the distance between New York and California in inches or feet, but miles is a more convenient measure. Different societies have historically adopted different units for the same dimension and these conventions tend to persist for a long time.

In at least one case, the primary units associated with a secondary dimension remained undiscovered until after the dimension itself had already achieved widespread use. Heat and work were for a long time considered to be unrelated quantities. The science of calorimetry was developed to measure heat and used as its fundamental unit the British Thermal Unit (Btu) or calorie. Work was understood to be a separate quantity that resulted in lifting a weight; therefore, the unit of work was taken to be ft-lb_f. Joule demonstrated in the late 1800's that many of the effects produced by heat could also be produced by work. At that time it became clear that heat and work both refer to the transfer of energy and therefore could share a common unit.

Whatever the reasons, physical quantities can be expressed using many different units and this fact introduces the possibility for unit conversion errors. The practicing engineer must be able to deal with and convert between a variety of units and unit systems. Instruments will report measurements in a variety of units. For example, a vacuum gage may report pressure measurements in units of torr while a water manometer will naturally lead to a pressure measurement in units of inches of water. Engineers must communicate the results of their analyses to a variety of audiences, some of whom are most comfortable thinking in terms of a specific set of units. For example, cooling power

Table 1-2: Units of some secondary dimensions in the SI and English unit systems.

		SI Unit System		English Unit System	
Physical Quantity	Dimensions	Unit	Symbol	Unit	Symbol
force	M-L/t^2	Newton	N	pound-force	lb_f
energy	M-L^2/t^2	Joule	J	British Thermal Unit	Btu
power	M-L^2/t^3	Watt (J/s)	W	horsepower	hp
pressure	M/L-t^2	Pascal (N/m^2)	Pa	pound/inch^2	psi
		bar	bar	atmosphere	atm

should be reported to a refrigeration engineer in the U.S. in units of tons. An automotive engineer in the U.S. would be most comfortable understanding output power in units of horsepower.

In the real world, the inputs to an engineering analysis will be provided in a variety of units (often in mixed units, some SI and some English) and the results of the analysis should be reported in whatever units are most appropriate. However, it is not necessary to carry out the analysis in an arbitrary set of units. In fact, there is a strong argument for working a problem entirely in the SI system of units that is listed in Table 1-1. This unit system is defined so that it is completely self-consistent; that is, no unit conversions are required when working in the standard SI unit system. For example, the SI unit of energy (the Joule) is equal to the product of the SI units of force (the Newton) and distance (the meter):

$$1\text{ J} = 1\text{ N-m} \tag{1-15}$$

The SI unit of force (the Newton) is related to the SI units of mass, distance and time (kg, m, and s, respectively) according to:

$$1\text{ N} = 1\,\frac{\text{kg-m}}{\text{s}^2} \tag{1-16}$$

The same self-consistency is not evident in the English unit system. The unit of energy in the English system (the British Thermal Unit or Btu) is related to the English units of force and distance according to:

$$1\text{ Btu} = 778.17\text{ lb}_f\text{-ft} \tag{1-17}$$

The English unit of force is related to the English units of mass, distance, and time according to:

$$1\text{ lb}_f = 32.174\,\frac{\text{lb}_m\text{-ft}}{\text{s}^2} \tag{1-18}$$

The differences between Eqs. (1-15) and (1-16) and Eqs. (1-17) and (1-18) clearly illustrate the self-consistency of the SI unit system and the need for unit conversions when working in the English unit system. Appendix A contains tables of many common unit conversions.

The units of each variable are self-evident in the SI unit system. It is not necessary to constantly worry about applying the correct unit conversion to each equation during the development of a model. As a result, if you are working in the SI unit system and you check the units of your equations then you are actually carrying out a more powerful and complete check on your equations; you are establishing their dimensional (as well as their unit) consistency.

It is up to the engineer to establish a procedure for dealing with units that works for him or her, and it is not the objective of this text to be prescriptive in this regard. Certainly there are many strategies that work. However, in this book we will consistently adopt the following procedure. Inputs to the problem reported in arbitrary units will be converted to the base SI system listed in Table 1-1 and Table 1-2 (i.e., kg, m, s, K, N, J, etc.). The calculations required to solve the problem will be carried out using the base SI system and unit checking will be rigorously applied in order to establish the dimensional consistency of each equation. The results will be converted from the SI system into whatever units are requested or are logical and convenient. With this approach in mind, the tables in Appendix A are presented so that it is easy to convert from arbitrary units to their SI equivalent and back.

EXAMPLE 1.6-1: WEIGHT ON MARS

The gravitational acceleration on the surface of Mars is $g = 12.5$ ft/s^2, which is about 38% of the gravitational acceleration on earth's surface.

a) What is the weight on Mars of an astronaut with a mass of $m = 175$ lb$_m$?

The inputs to the problem, g and m, are converted to base SI units using the conversions found in Appendix A.

$$g = \frac{12.5\ \text{ft}}{\text{s}^2}\left\| \frac{0.30480\ \text{m}}{\text{ft}} \right. = 3.81\ \frac{\text{m}}{\text{s}^2}$$

$$m = \frac{175\ \text{lb}_m}{}\left\| \frac{0.45359\ \text{kg}}{\text{lb}_m} \right. = 79.38\ \text{kg}$$

The weight of the astronaut is computed according to:

$$W = mg = \frac{79.38\ \text{kg}}{}\left| \frac{3.81\ \text{m}}{\text{s}^2} \right\| \frac{\text{N-s}^2}{\text{kg-m}} = 302.4\ \text{N}$$

Notice that the computed weight of the astronaut is automatically expressed in the SI unit for force, the Newton. No unit conversion is required because the inputs were converted to SI units. If the weight of the astronaut is required in other, non-SI units (e.g., lb$_f$) then the SI result is converted according to:

$$W = \frac{302.4\ \text{N}}{}\left\| \frac{\text{lb}_f}{4.4482\ \text{N}} \right. = 67.98\ \text{lb}_f$$

This example illustrates the strategy of converting inputs to the SI system, carrying out the calculations required for the problem in the SI unit system, and then converting the results to appropriate, non-SI units if necessary. This example was trivial. However, as the problems become more complex, this approach is an effective way of avoiding unit conversion errors.

1.6.2 Working with Units in EES

It is possible to assign both a value and a unit to each of the variables that are used in an EES program. EES has been programmed to check the dimensional and unit consistency of the equations. In order to apply this capability, it is necessary to enter the units of all variables that are used in the analysis. We will demonstrate how this is done in Example 1.6-2.

EXAMPLE 1.6-2: POWER REQUIRED BY A VEHICLE

The two major forces opposing the motion of a vehicle on a level road are the rolling resistance of the tires (F_r) and the aerodynamic drag on the car (F_d). The rolling resistance is the product of the dimensionless rolling resistance coefficient, $f = 0.02$, and the force exerted by the vehicle on the road (i.e., its weight, W).

$$F_r = f\,W \tag{1}$$

The aerodynamic drag is expressed in terms of a dimensionless drag coefficient (C_d) according to:

$$F_d = A_f\,C_d\,\frac{1}{2}\rho\,\tilde{V}^2 \tag{2}$$

where A_f is the frontal area of the vehicle, $\rho = 0.075\ \mathrm{lb_m/ft^3}$ is the density of air and $\tilde{V}$ is the speed of the vehicle. The Toyota Prius has a drag coefficient of $C_d = 0.29$, a frontal area of $A_f = 21.2\ \mathrm{ft^2}$ and a curb weight of $W = 2930\ \mathrm{lb_f}$.

a) Determine the power required by a Prius (in hp) traveling at a velocity of $\tilde{V} =$ 65 mph.

It is good form to put the problem inputs at the top of the Equations window. Each input is entered and immediately annotated. We'll start with the density of air.

```
rho=0.075 [lbm/ft^3]                    "density of air"
```

Note that the units of the numerical constant, 0.075, are entered in square brackets immediately after the constant is typed in EES. If the EES code is solved at this point (select Solve from the Calculate menu) you will see that the units of the variable rho are indicated next to its value in the Solutions window (Figure 1). The ^ symbol in the unit designation for rho is used to raise a unit to a power. EES will also accept the unit lbm/ft3, but the ^ symbol helps EES recognize that the 3 should be a superscript in the formatted output. Also note that the unit for pound-mass is represented in EES as either lb_m or lbm. The underscore causes EES to recognize that the m should be a subscript, but either designation is acceptable.

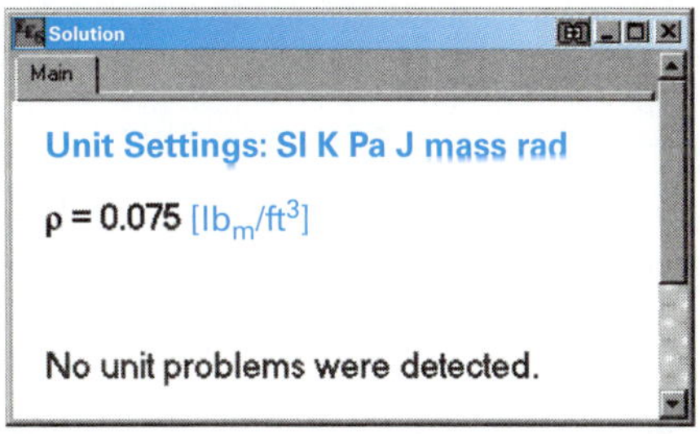

Figure 1: Solution window.

The unit designations that are recognized by EES can be viewed by selecting Unit Conversion Info from the Options menu. Select the dimension from the list on the left and EES will display the defined units associated with that dimension in the list on the right, as shown in Figure 2 for the dimension mass.

As discussed in Section 1.6.1, the inputs to the problem should be converted to SI units. EES provides the function convert in order to easily convert from one

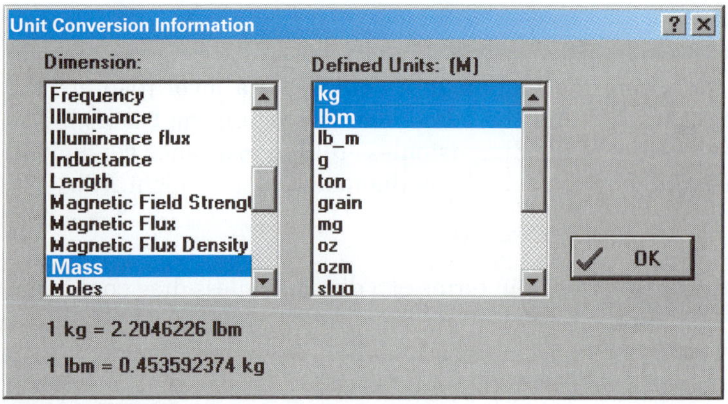

Figure 2: Unit Conversion Information dialog.

unit to another without resorting to the use of tables of unit conversion factors (like those included in Appendix A). The convert function accepts two arguments. The first argument indicates the unit(s) that you wish to convert from, and the second argument is the unit(s) that you wish to convert to. Both sets of units must have the same dimensions (i.e., they must be dimensionally consistent). So, for example, the conversion factor that is required to convert from feet to meters is obtained from the EES code convert(ft,m). According to Appendix A, the function convert(ft,m) will return the conversion factor 0.30480 m/ft. In our case, we wish to convert the density of air from the English units that it was provided in, lb_m/ft^3, to SI units, kg/m^3. We can do this conversion by revising the information in the Equations window as follows:

```
rho=0.075 [lbm/ft^3]*convert(lbm/ft^3,kg/m^3)          "density of air"
```

Select Solve from the Equations menu (or press the F2 key, the shortcut for Solve) and you will see the Solution window (Figure 3).

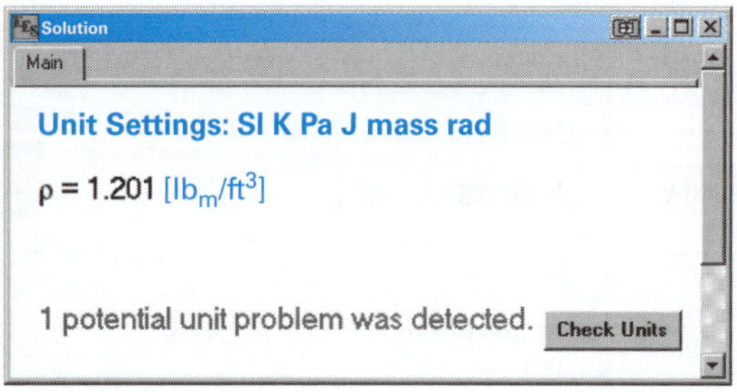

Figure 3: Solution window.

Notice the warning message indicating that EES has detected a unit problem. EES checks the units of your equations each time you solve. You can have EES check units at any time by selecting Check Units from the Calculate menu (or by pressing F8). A window will appear that provides a list of all of the equations that contain unit errors as well as some description of the unit inconsistency that was detected (Figure 4).

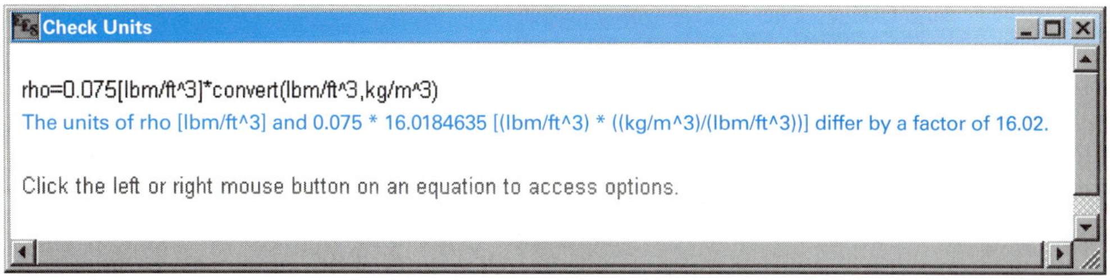

Figure 4: Check Units dialog.

The units of the variable rho were previously set to lb_m/ft^3; these units are no longer consistent with the equation that is used to specify the variable. This is most evident by examining the Formatted Equations window (Figure 5).

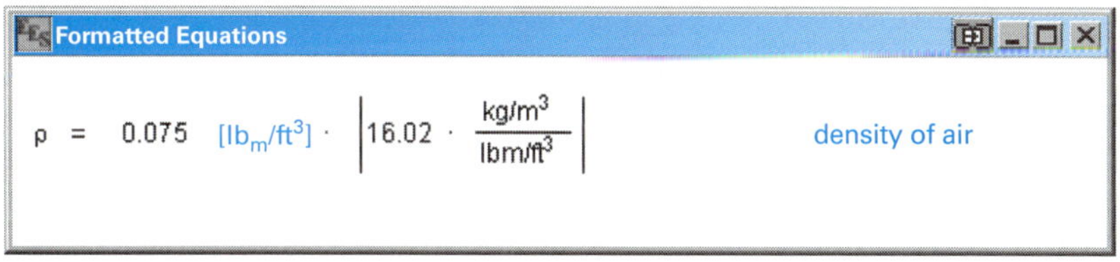

Figure 5: Formatted Equations window.

Figure 5 shows that the units of the variable rho are clearly kg/m^3 and therefore the units assigned to this variable should be updated to reflect the unit conversion that was carried out. There are several ways to set the units of a variable. One way is to highlight the variable in the Equations window and right-click on it; select Variable Info from the pop-up menu that appears (Figure 6).

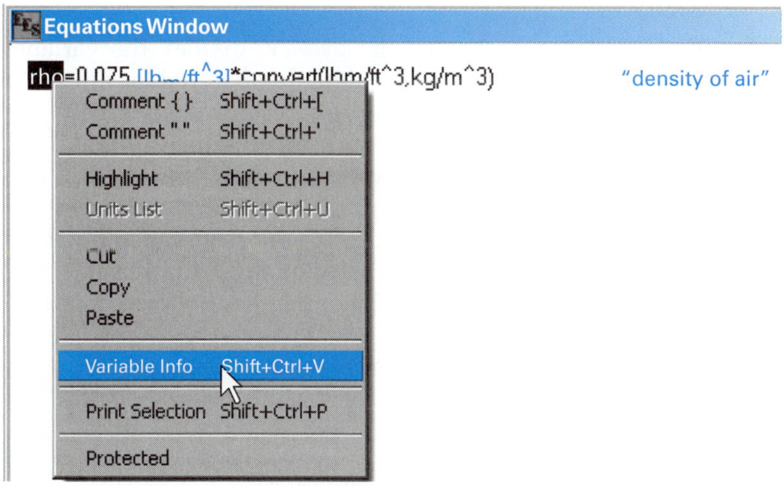

Figure 6: Setting the units of a variable from the Equations window.

A dialog box will appear that allows you to set the units of the variable rho, as shown in Figure 7.

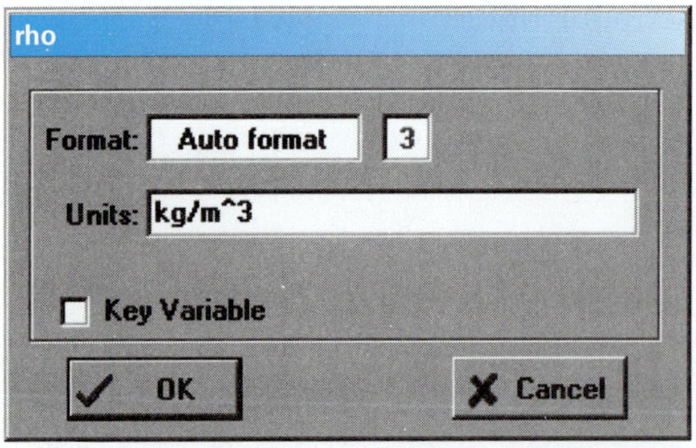

Figure 7: Variable information dialog for the variable rho.

When the EES program is solved, the units of the variable rho are indicated beside its value in the Solution window. Another way of setting the units of a variable is by right-clicking on the variable in the Solution window; a dialog will appear that allows you to set many of the characteristics of the variable including its units, as shown in Figure 8.

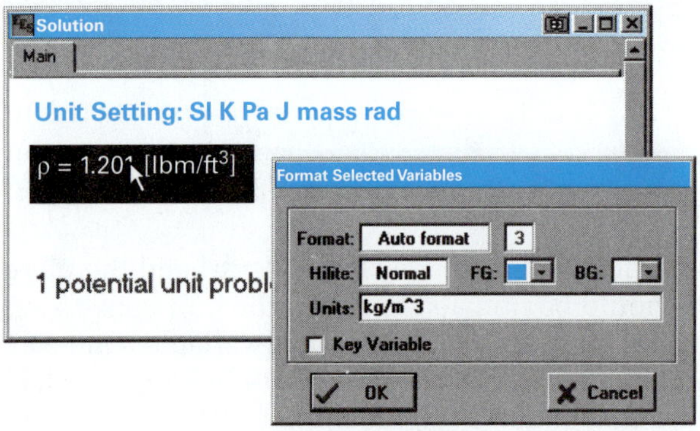

Figure 8: Setting units from the Solution window.

Notice that in addition to the units, the format used to display the variable can be adjusted in the dialog shown in Figure 8. By default, all variables are set to Auto format; Auto format tells EES to automatically decide how many significant figures should be displayed. However, you can change this setting by making a different selection from the pull-down list that appears when you click in the Format box. You can also cause the variable to appear as highlighted or with a specified foreground or background color so that it stands out from other variables in the Solution window. If the Key Variable box is selected, then the variable will also be displayed in a separate Key Variables window along with a comment that can be entered for the variable.

Continue to enter the input variables for the problem, converting each one to SI units.

```
f=0.02 [-]                                  "rolling resistance coefficient"
vel=65 [mph]*convert(mph,m/s)               "velocity of car"
C_d=0.29 [-]                                "drag coefficient of Prius"
A_f=21.2 [ft^2]*convert(ft^2,m^2)           "frontal area of Prius"
W=2930 [lbf]*convert(lbf,N)                 "weight of Prius"
```

Note that the unit [-] entered after the variables f and C_d indicates that there are no units associated with these variables; they are dimensionless. An easy way to set the units of all of the variables in your program at once is to select Variable Info from the Options menu. The Variable Information window that appears allows you to set many of the characteristics of each variable in your problem, including the units (Figure 9).

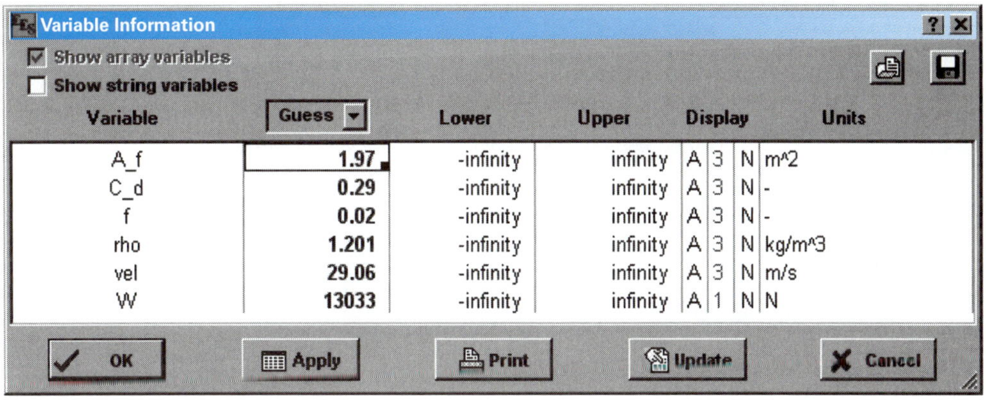

Figure 9: Variable Information window.

If you select Check Units from the Calculate menu (or press F8, the associated shortcut key) you should receive a message indicating that no unit problems were detected. The forces associated with rolling resistance and aerodynamic drag are evaluated using Eqs. (1) and (2).

```
F_r=f*W                          "rolling resistance of the tires"
F_d=A_f*C_d*rho*vel^2/2          "aerodynamic drag"
```

The units of the forces are naturally Newton (N) because all of the inputs have been converted to the SI system; no unit conversions are required. The units of the variables F_r and F_d should be set using any one of the techniques discussed above. The units of the equations should be checked to verify unit consistency. Note that because an SI unit system is used, the unit check actually verifies the dimensional consistency of the equations that have been entered.

The power required to move the car at a steady state velocity is:

$$\dot{W} = (F_r + F_d)\ \tilde{V}$$

```
W_dot=(F_r+F_d)*vel              "power"
```

The units of the variable W_dot must be W (Watts, the SI unit for power). In order to obtain the power in units of hp, it is necessary to use the convert function. Define a new variable W_dot_hp (the power expressed in hp).

```
W_dot_hp=W_dot*convert(W,hp)     "power, in hp"
```

After solving, the Solution window will show that $\dot{W} = 21.45$ hp.

b) Prepare a plot showing the power required by the Prius as a function of the vehicle's velocity.

EES provides the capability of carrying out parametric studies and generating high quality graphs. In order to carry out a parametric study it is necessary to generate a Parametric table. Select New Parametric Table from the Tables menu in order to bring up the New Parametric Table dialog (Figure 10).

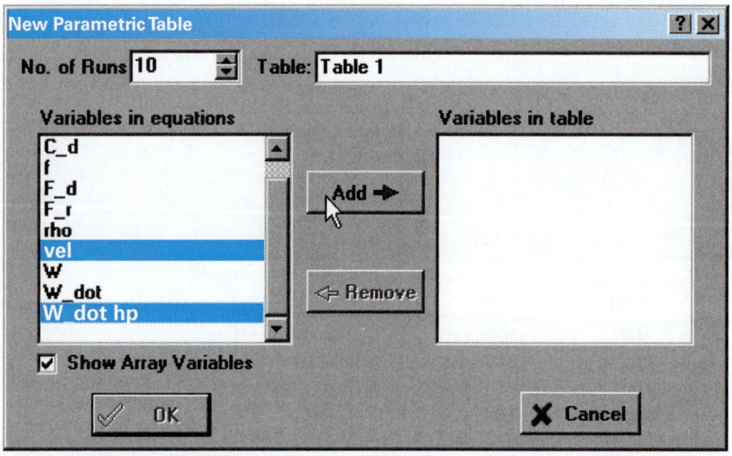

Figure 10: New Parametric Table dialog.

The dialog asks for the number of runs that should be placed in the table and the name of the table. A list of all of the variables used by the EES code is shown on the left. Highlight the variable that will be varied in order to make the plot as well as any variables that will be examined parametrically and select the Add button. In order to make the plot requested by the problem statement, the variables vel and W_dot_hp should be placed in the table. Select OK and a Parametric table will be created with a column for each of the selected variables (Figure 11).

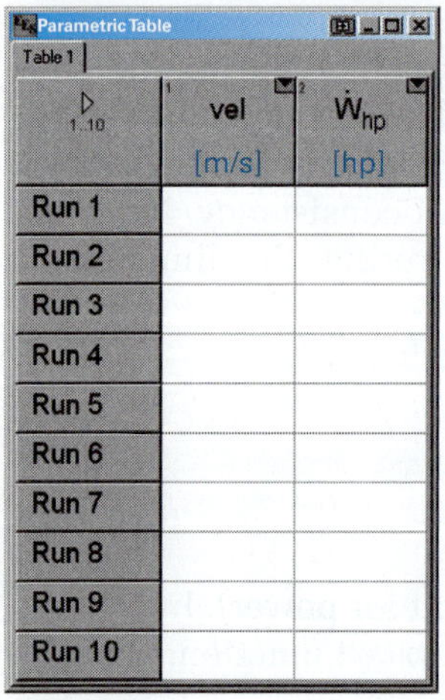

Figure 11: Parametric table, empty.

When Solve Table is selected from the Calculate menu, EES will sequentially work with each row of the table. If a value is set in one of the columns of the table for that row, then EES will assign that value to the corresponding variable in the Equations window and solve the resulting set of equations. The values of the variables in any of the other columns in the table will be calculated. Entered values are, by default, displayed in black type whereas calculated variables are displayed in blue after the calculations are completed. This process is repeated for each row in the table. It is possible to specify the value for velocity in each run manually, but it is much more convenient to do this automatically. Right-click on the heading of the column corresponding to the variable vel and select Alter Values in order to bring up the dialog shown in Figure 12.

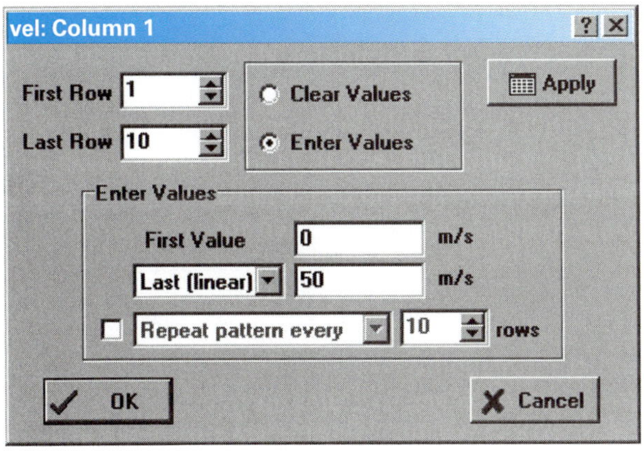

Figure 12: Alter Values dialog.

Fill in the dialog so that the variable vel is varied from $\tilde{V} = 0$ to 50 m/s in equal increments (Figure 13).

Figure 13: Parametric table with velocity values set.

Parametric Table — Table 1

1..10	vel [m/s]	$\dot{W}_{hp}$ [hp]
Run 1	0	
Run 2	5.556	
Run 3	11.11	
Run 4	16.67	
Run 5	22.22	
Run 6	27.78	
Run 7	33.33	
Run 8	38.89	
Run 9	44.44	
Run 10	50	

Select Solve Table from the Calculate menu (or the corresponding shortcut key, F3) and you will receive an error indicating that the value of the variable vel has already been set. It is not possible for EES to set the value of velocity to 0 m/s in Run 1 of the table when the value of the velocity has already been set to 29.06 m/s (corresponding to 65 mph) in the Equations window. In order to proceed, it is necessary to remove the line of code in the Equations window that specifies velocity. One convenient method for temporarily removing the line of code is to highlight the appropriate line in the Equations window and right-click on it. Select Comment {} from the pop-up menu that appears and the line will be "commented out"; that is, it will be surrounded by curly brackets so that it is read as a comment, having no effect on the equation set.

```
{vel=65 [mph]*convert(mph,m/s)            "velocity of car"}
```

It is easy to re-activate (or "uncomment") the line of EES code by highlighting it again, right-clicking, and selecting Undo Comment {}. For now, leave the line "commented out" and solve the Parametric table. The column corresponding to the variable W_dot_hp should be filled in (Figure 14).

Parametric Table — Table 1

1..10	vel [m/s]	$\dot{W}_{hp}$ [hp]
Run 1	0	0
Run 2	5.556	2.021
Run 3	11.11	4.515
Run 4	16.67	7.956
Run 5	22.22	12.82
Run 6	27.78	19.57
Run 7	33.33	28.69
Run 8	38.89	40.65
Run 9	44.44	55.93
Run 10	50	74.99

Figure 14: Parametric table, after it is solved.

Data contained in a Parametric table (or any other type of table in the EES program) can be plotted. Select New Plot Window from the Plots menu and then select X-Y plot in order to bring up the New Plot Setup dialog shown in Figure 15.

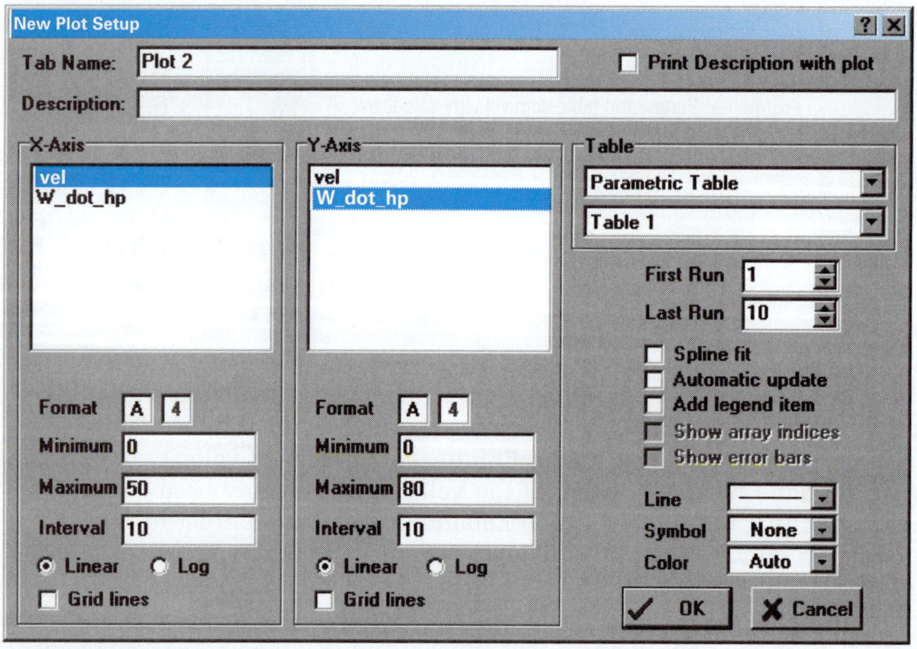

Figure 15: New Plot Setup dialog.

The right side of the dialog is used to specify the source of the data (the table in which the data are stored). There is only one table in our EES code (Table 1, a Parametric table). Select the X- and Y-Axis variables and then hit OK. The plot, with some formatting, is shown in Figure 16.

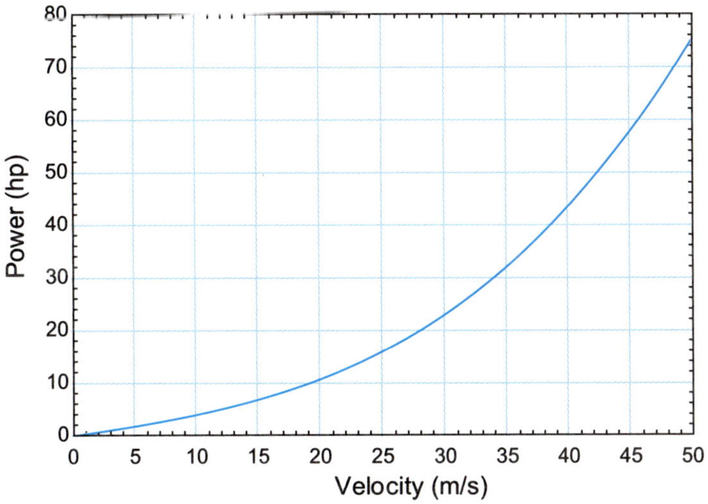

Figure 16: Power required by the Prius as a function of velocity.

c) The Ford Escape has a drag coefficient of $C_d = 0.40$, a frontal area of $A_f = 29.0$ ft^2 and a curb weight of $W = 3272$ lb$_f$. Overlay on your plot from (b) the power required by the Ford Escape as a function of velocity.

The characteristics of the Prius are commented out and those of the Escape are inserted in the Equations window.

```
{C_d=0.29 [-]                                "drag coefficient of Prius"
A_f=21.2 [ft^2]*convert(ft^2,m^2)            "frontal area of Prius"
W=2930 [lbf]*convert(lbf,N)                  "weight of Prius"}
C_d=0.40 [-]                                 "drag coefficient of Escape"
A_f=29.0 [ft^2]*convert(ft^2,m^2)            "frontal area of Escape"
W=3272 [lbf]*convert(lbf,N)                  "weight of Escape"
```

The Parametric table is solved with the new inputs. In order to overlay a new plot onto an existing plot, select Overlay Plot from the Plots menu. The result is shown in Figure 17.

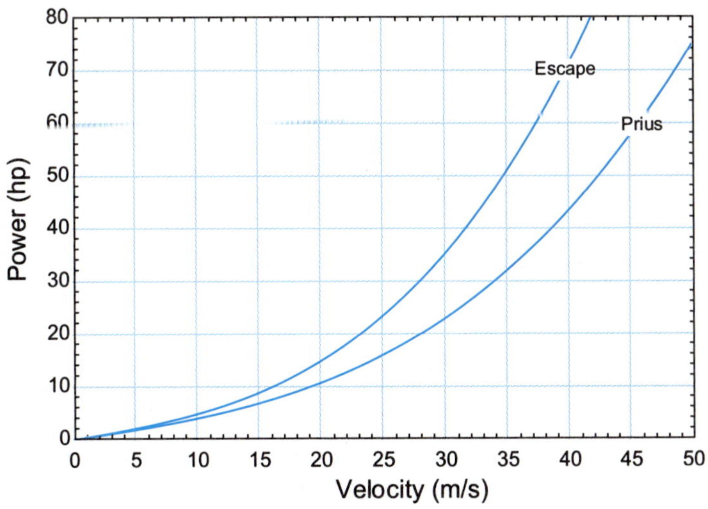

Figure 17: Power required by the Prius and the Escape as a function of velocity.

1.7 Specific Volume, Pressure, and Temperature

Thermodynamic properties and their interrelationships are the subject of Chapter 2. However, it is worth briefly discussing three properties that you are probably already familiar with: temperature, pressure, and specific volume (or its inverse, density).

1.7.1 Specific Volume

Density is a familiar concept. The density of a substance is an intensive property defined as the ratio of the mass of the substance to the volume that it occupies:

$$\rho = \frac{m}{V} \tag{1-19}$$

where m is the mass of the substance and V is its volume. The specific volume of a substance is defined as the ratio of the volume that the substance occupies to its mass:

$$v = \frac{V}{m} = \frac{1}{\rho} \tag{1-20}$$

Sometimes the specific volume defined in Eq. (1-20) is referred to as mass specific volume in order to make it clear that v is the volume per unit mass (as opposed to molar specific volume, which is the volume per unit mole).

1.7.2 Pressure

Pressure is defined as the normal force exerted on a surface per unit area. From a molecular perspective, the pressure exerted on a surface is a manifestation of the change in momentum of individual molecules as they interact with the surface. The momentum, and therefore the pressure, is greater when the molecules are traveling faster (which is related to the temperature of the fluid) or when there are more molecules (which is related to the density of the fluid). Pressure is a relatively easily measured property and therefore we often use it to describe the state of engineering fluids. The dimension of pressure is force/area. There are many different units for pressure because it is such an important property for so many different applications. The most common pressure units are the Pascal (1 Pa $=$ 1 N/m^2) and the psi (1 psi $=$ 1 lb$_f$/in^2). Unit conversions for pressure can be found in Table A-4 in Appendix A.

The air in our atmosphere at the surface of the Earth naturally exerts a pressure on every surface due to the weight of air above the surface. This pressure is referred to as the barometric or atmospheric pressure. The actual atmospheric pressure varies slightly with atmospheric conditions and elevation, but standard barometric pressure at sea level is defined to be 1 atm, which is equivalent to $P_{atm} = 101{,}325$ Pa (101.325 kPa) or 14.696 psi.

Many of the pressure measurement devices (i.e., pressure gages) that we commonly use directly measure the difference in pressure relative to atmospheric pressure. For example, Figure 1-7 illustrates a manometer. A manometer is a tube (typically glass) that is bent to form a U and filled with a liquid (e.g., water or mercury). One of the ends of the manometer is connected to the location where you would like to measure pressure (e.g., a tank or a pipe) and the other end is left open to the atmosphere. The difference in the elevation of the liquid in the two legs of the tube (L in Figure 1-7) is proportional to the difference in the pressure of interest (P) and atmospheric pressure (P_{atm}). A force

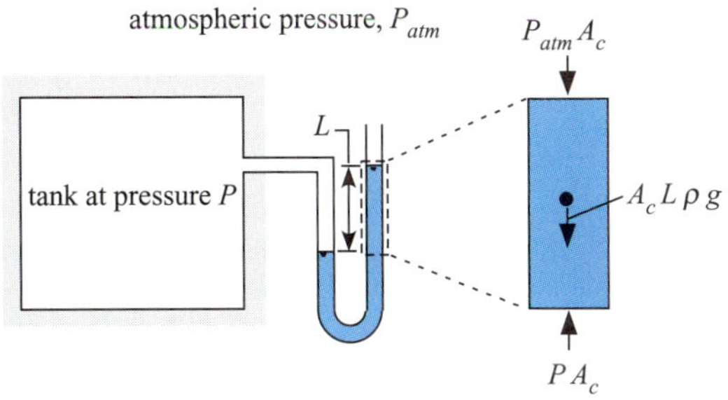

Figure 1-7: Manometer.

balance on the liquid column is shown in Figure 1-7 and provided below:

$$\underbrace{P\,A_c}_{\substack{\text{upward pressure force}\\ \text{on bottom of column}}} = \underbrace{A_c\,L\,\rho\,g}_{\text{weight of column}} + \underbrace{P_{atm}\,A_c}_{\substack{\text{downward pressure}\\ \text{force on top of column}}} \tag{1-21}$$

where A_c is the cross-sectional area of the tube, ρ is the density of the fluid in the tube, and g is the acceleration of gravity. Equation (1-21) can be rearranged:

$$P - P_{atm} = L\,\rho\,g \tag{1-22}$$

Both ρ and g are known and therefore measurement of L is sufficient to provide the pressure relative to the atmospheric pressure ($P - P_{atm}$). Manometers are in such common use that units for pressure include inches of water (inH2O) and inches of mercury (inHg); these units are included in Table A-4 and are recognized by EES.

Many other measurement devices also measure pressure relative to atmospheric pressure. Various types of elastic pressure transducers determine pressure by measuring the displacement caused when atmospheric pressure is applied to one side of a mechanical element (e.g., a membrane) while the pressure of interest is applied to the other. The result is clearly a direct measurement of the pressure relative to the atmospheric pressure ($P - P_{atm}$) rather than the pressure alone. The term *gage pressure* refers to the pressure measured with respect to atmospheric pressure:

$$P_{gage} = P - P_{atm} \tag{1-23}$$

The *absolute pressure* (P) can be obtained from a gage pressure measurement by adding the atmospheric pressure. Unless otherwise stated, all pressures used within this text will be absolute pressure. Gage pressures are sometimes reported by adding the letter g to the end of the unit and absolute pressures by adding the letter a. For example, 5 psig should be read as 5 pounds per square inch gage pressure and 19.7 psia should be read as 19.7 pounds per square inch absolute pressure; these two pressures are the same provided that atmospheric pressure is 14.7 psi.

There are specialized instruments that measure the absolute atmospheric pressure. For example, a barometer is an inverted glass tube that has been evacuated and placed in a pool of liquid, as shown in Figure 1-8. The height of the liquid in the tube relative to the pool of liquid (L) is proportional to the absolute pressure above the pool (i.e., the local atmospheric pressure). A force balance on the column of liquid is shown in Figure 1-8:

$$\underbrace{P_{atm}\,A_c}_{\substack{\text{upward pressure force}\\ \text{on bottom of column}}} = \underbrace{A_c\,L\,\rho\,g}_{\text{weight of column}} \tag{1-24}$$

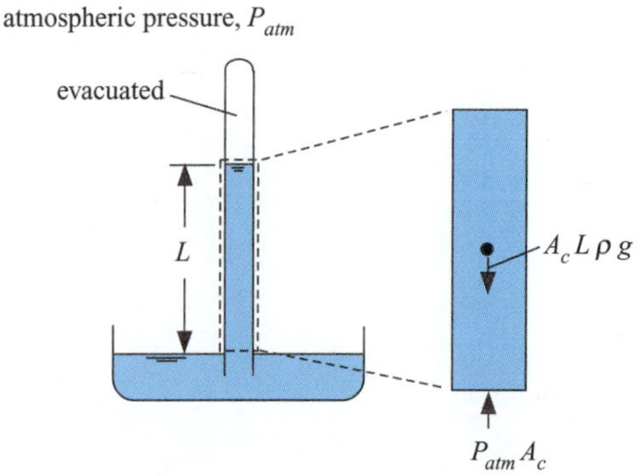

Figure 1-8: Barometer.

Equation (1-24) can be rearranged:

$$P_{atm} = L \rho g \tag{1-25}$$

Most laboratories include a barometer so that gage pressure measurements can be accurately converted to absolute pressure.

1.7.3 Temperature

The precise meaning of temperature is not easily stated. From a molecular viewpoint, the temperature of a substance is related to the average energy or velocity of the individual molecules of the substance. When high velocity (i.e., hot) molecules are in the presence of low velocity (i.e., cold) molecules, then collisions between the molecules tend to cause an energy transfer that we call heat from the hot molecules to the cold molecules.

Most of us are familiar with either the Fahrenheit or Celsius temperature scales. Depending on what country you live in, the weather forecasts are provided in these scales. However, these are not absolute temperature scales in the same way that gage pressure is not absolute pressure. Zero gage pressure (e.g., 0 psig) does not correspond to zero force per unit area but rather to atmospheric pressure. Zero degrees Fahrenheit (or Celsius) does not correspond to an absence of molecular motion (i.e., absolute zero).

The different temperature scales are of interest because they are in common use and their origin reflects the problems that are encountered when setting up a unit system. How might you set up a scale for temperature? At a minimum, it is necessary to select two fixed or reference temperatures. One point fixes the zero for the scale and the second allows the degree size to be specified. In the early 1700's, Anders Celsius in Sweden chose the freezing and boiling points of pure water as the fixed points for a temperature scale (although he apparently did not initially choose the zero point to be the freezing point of water, as we do today). Around the same time, Daniel Fahrenheit in Holland developed a scale in which the zero point was defined as the freezing point of a salt-water solution. There are different stories about how the second point was selected. One explanation is that he used normal human body temperature as the second point. Normal human body temperature is now accepted to be 98.6°F but, for some reason, he initially assigned this temperature to 96°F. An alternative story suggests that he used the body temperature of a cow (approximately 100°F) as the second fixed point (which is perhaps more understandable than using the body temperature of a human). Yet another story suggests that he assigned 212°F to be the boiling point of water in order to fix the scale. It is not clear why 212°F would be chosen for this point as opposed to a more round number.

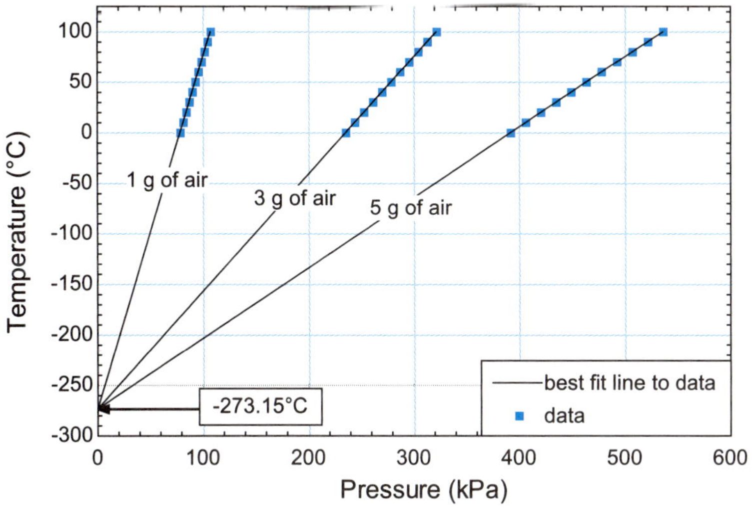

Figure 1-9: Temperature as a function of pressure in a 1 liter container containing various masses of air.

The Kelvin and Rankine scales are absolute temperature scales. They have the same degree size as the Celsius and Fahrenheit scales, respectively, but because they are absolute scales, the zero point corresponds to absolutely no molecular motion (i.e., absolute zero – an unattainable condition). The zero point for these scales was first identified from observations of the pressure of a gas in a fixed volume container, as reported by Boyle and Charles in the 1600's. Recall that pressure is the manifestation of molecular motion through the force exerted by collisions of molecules with a surface. Therefore, pressure is related to temperature. Researchers noticed that the pressure and temperature of a gas are linearly related to one another under some conditions (i.e., conditions where the ideal gas law applies). For example, Figure 1-9 illustrates measurements of temperature as a function of pressure for 3 g of air contained in a 1 liter rigid volume. Extrapolation of these data to zero pressure (corresponding to some theoretical condition where there is no molecular motion – i.e., absolute zero) leads to an ultimate lower limit on temperature of –273.15°C.

Figure 1-9 shows that this lower limit on the temperature applies regardless of the mass of air that is contained in the tank (provided, of course, that the ideal gas law continues to hold). Data taken in this manner for other gases also extrapolates to –273.15°C. All of this evidence suggests that the absolute temperature scale corresponding to the Celsius scale should be defined based on adding 273.15 to the value of the temperature in °C. The Kelvin temperature scale is therefore defined according to:

$$T\,[\mathrm{K}] = T\,[^{\circ}\mathrm{C}] + 273.15 \tag{1-26}$$

A similar experiment accomplished using temperature measurements taken on the Fahrenheit scale would suggest that -459.67°F is the lowest possible temperature. Therefore, the Rankine temperature scale is an absolute temperature defined according to:

$$T\,[\mathrm{R}] = T\,[^{\circ}\mathrm{F}] + 459.67 \tag{1-27}$$

Equations for converting between any of the temperature scales are summarized in Table A-3 in Appendix A. Most unit conversions require only a multiplicative correction that can be provided by the convert function in EES, as discussed in Example 1.6-2. However,

Eqs. (1-26) and (1-27) (and the other relationships listed in Table A-3) show that both multiplication and addition operations are required to convert between temperature scales. Therefore, the convert function in EES cannot be used to convert from one temperature scale to another temperature scale. Instead, the converttemp function in EES should be used for this purpose. The converttemp function accepts three arguments. The first argument indicates the temperature scale to convert from, the second argument indicates the temperature scale to convert to, and the third argument is the numerical value, variable, or expression to be converted. The temperature scale indicator can be F, C, K or R, representing the Fahrenheit, Celsius, Kelvin, and Rankine scales, respectively. For example, the following EES code converts a temperature of 70°F to the Celsius temperature scale.

```
T_F=70 [F]                          "temperature in F"
T_C=converttemp(F,C,T_F)            "temperature in C"
```

Note that EES will indicate that there are potential unit errors when you solve these equations unless the units for the variable T_F is set to F and the units for the variable T_C is set to C (using any of the techniques discussed in Example 1.6-2).

REFERENCES

Tribus, M., *Thermodynamics and Thermostatics: An Introduction to Energy, Information and States of Matter, with Engineering Applications*, D. Van Nostrand Company Inc., New York, (1961).

Problems

The problems included here have been selected from a larger set of problems that are available at the website associated with this book (www.cambridge.org/kleinandnellis).

A. Balances

1.A-2 A mixing tank in a chemical processing plant is shown in Figure 1.A-2.

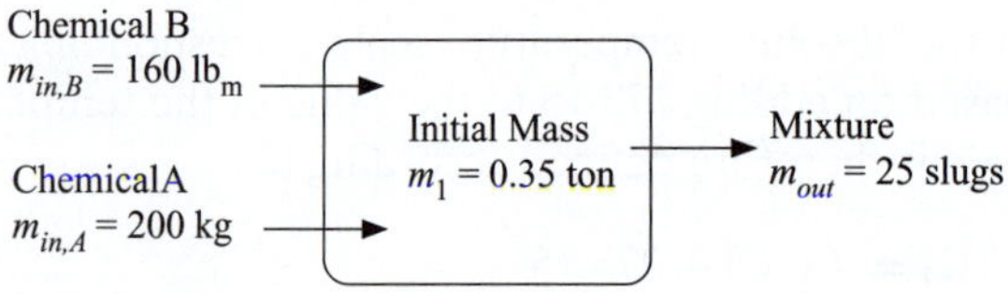

Figure 1.A-2: Mixing tank in a chemical processing plant.

Two chemicals enter the tank, mix, and leave. Initially, the tank contains $m_1 = 0.35$ tons. Over an hour of operation, $m_{in,A} = 200$ kg of chemical A enter and $m_{in,B} =$ 160 lb_m of chemical B enter while $m_{out} = 25$ slug of the mixture leaves.

a) Sketch the system that you will use to carry out a mass balance for this problem. Is your system open or closed?
b) What is the final mass in the tank (ton)?
c) Is the tank operating at steady state?

1.A-3 Figure 1.A-3 illustrates a company that manufactures gadgets.

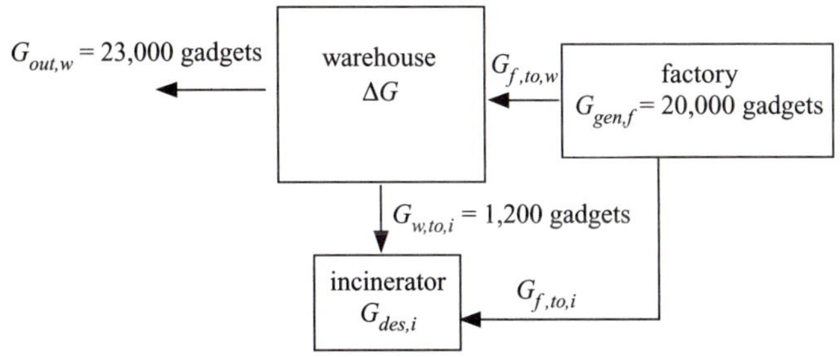

Figure 1.A-3: A company manufacturing gadgets.

Initially we will analyze the company on an incremental basis; the increment of time for the analysis will be one day. Raw materials enter the factory where $G_{gen,f} = 20{,}000$ gadgets are produced during the day. The functional gadgets are sent to the warehouse building ($G_{f,to,w}$) and the defective gadgets are sent to the incinerator to be destroyed ($G_{f,to,i}$). Approximately 5% of the gadgets produced in the factory are found to be defective. Neither the factory nor the incinerator can store gadgets (i.e., they operate always at steady state). Gadgets can be stored in the warehouse. The number of gadgets that are shipped from the warehouse during the day is $G_{out,w} = 23{,}000$ gadgets. Every day, $G_{w,to,i} = 1200$ gadgets that are old (beyond their expiration date) are pulled from the warehouse shelves and sent to the incinerator to be destroyed.

a) Determine the number of gadgets destroyed in the incinerator during the day, $G_{des,i}$.
b) Is the warehouse operating at steady state? If not then determine the storage of gadgets in the warehouse during the day, ΔG_w.

Let's next analyze the company on a rate basis. At a particular instant of time, the rate at which gadgets are shipped from the warehouse is $\dot{G}_{out,w} = 1500$ gadgets/hr and the rate at which gadgets are being pulled off the warehouse shelves and sent to the incinerator is $\dot{G}_{w,to,i} = 50$ gadgets/hr. Assume that the defect rate associated with gadgets produced in the factory remains at 5% (i.e., 5% of the gadgets produced by the factory are sent to the incinerator rather than the warehouse).

c) Determine the rate at which gadgets must be produced in the factory and destroyed in the incinerator in order for the warehouse to operate at steady state (i.e., in order to maintain a constant inventory of gadgets).

B. Introduction to EES

1.B-1 a) Solve the equation $a/x + b + c\,x^2 = 1$ using EES with $a = 1$, $b = 2$, and $c = 0.5$.
b) Make a plot showing how the solution varies as the value of c changes from 0.1 to 10. Label your axes.

1.B-3 A thermistor is an electrical resistor that is made of temperature-dependent materials. Properly calibrated, the thermistor can be used measure temperature. The relation between resistance (R, ohms) and temperature (T, K) for a thermistor is given by:

$$R = R_o \exp\left[\alpha\left(\frac{1}{T} - \frac{1}{T_o}\right)\right]$$

where R_o is the resistance in ohms at temperature T_o in K and α is a material constant. For a particular resistor, it is known that $R_o = 2.6$ ohm with $T_o = 298.15$ K (25°C). A calibration test indicates that $R_1 = 0.72$ ohm at $T_1 = 60$°C.

a) Determine the value of α.
b) Prepare a plot of R versus T (in °C) for temperatures between 0° and 100°C. Indicate the range over which this instrument will work best.

C. Dimensions and Units

1.C-1 The damage that a bullet does to a target is largely dictated by the kinetic energy of the bullet. A 0.22 caliber bullet is fired from a handgun with a muzzle velocity of approximately $\tilde{V} = 1060$ ft/s and has a mass of $m = 40$ grains. A 0.357 magnum bullet is fired with a muzzle velocity of approximately $\tilde{V} = 1450$ ft/s and has a mass of $m = 125$ grains. Grains is the typical unit that is used to report the mass of a bullet, there are 7000 grains in a lb_m. The kinetic energy of an object (EK) is given by:

$$EK = \frac{m\,\tilde{V}^2}{2}$$

a) Determine the kinetic energy of the 0.22 and 0.357 caliber bullets (in lb_f-ft) as they are fired.

1.C-3 Figure 1.C-3 illustrates a spring-loaded pressure relief valve.

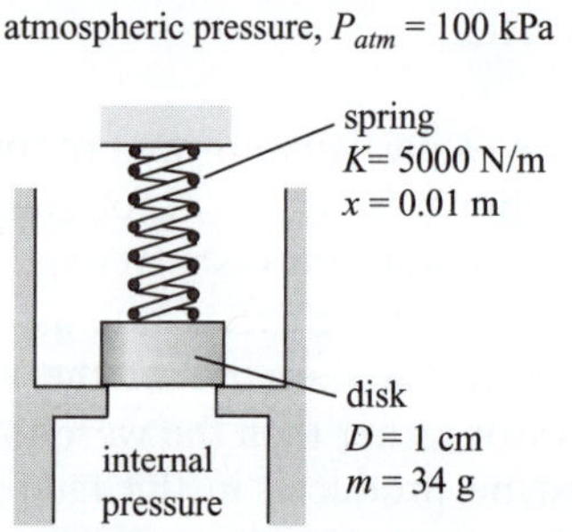

Figure 1.C-3: Spring-loaded valve.

When the valve is seated as shown, the spring is compressed by $x = 0.01$ m and pushes down on a disk having a $D = 1$ cm diameter. The disk has a mass of $m = 34$ grams. One side of the valve is exposed to atmospheric pressure at $P_{atm} = 100$ kPa and the other is exposed to an elevated internal pressure. The spring constant is $K = 5000$ N/m.

a) Determine the pressure at which the valve opens (in kPa and lb_f/in^2).
b) Generate a plot showing how the opening pressure and spring constant are related.

1.C-5 One of the main purposes of a seat belt is to ensure that the passenger stops with the car during a crash rather than flying freely, only to be stopped more quickly by a hard object. Assume that a vehicle traveling at $\tilde{V}_{ini} = 30$ mph comes to a halt in a distance of $L = 0.75$ ft during a crash. Assume that the vehicle and passenger experience a constant rate of deceleration during the crash.

a) The stopping distance of a passenger that is not wearing a seat belt is estimated to be 20% of the stopping distance of the vehicle. Calculate the force experienced by a $m = 180$ lb_m passenger (in lb_f and N) if he is not wearing a seat belt during the crash.
b) Calculate the force experienced by a $m = 180$ lb_m passenger (in lb_f and N) if he is wearing a seat belt that does not stretch during the crash.
c) Many seat belts are designed to stretch during a crash in order to increase the stopping distance of the passenger relative to that of the vehicle. Calculate the force experienced by the passenger (in lb_f and N) if his seat belt stretches by $L_{stretch} = 0.5$ ft during the crash.
d) Plot the force experienced by the passenger as a function of the stretch in the seat belt.

1.C-6 Some hybrid and electric car manufacturers have begun to advertise that their cars are "green" because they can incorporate solar photovoltaic panels on their roof and use the power generated by the panel to directly power the wheels. For example, a solar array can be installed on the roof of the Fisker Karma car (see www.fiskerautomotive.com). In this problem we will assess the value of a solar panel installed on the roof of a car. Assume that the panel is $L = 5$ ft long and $W = 4$ ft wide. On a very sunny day (depending on your location), the rate of solar energy per area hitting the roof of the car is $sf = 750$ W/m^2. Assume that the panel's efficiency relative to converting solar energy to electrical energy is $\eta_p =$ 10% (0.10). The cruising power required by the car is $\dot{W}_{car} = 20$ hp.

a) Estimate the rate of electrical power produced by the panel.
b) Calculate the fraction of the power required by the car that is produced by the solar panel.
c) If the car (and therefore the panel) sits in the sun for $time_{sit} = 6$ hours during a typical day then determine the total amount of electrical energy produced by the panel and stored in the car's battery.
d) If the car is driven for $time_{drive} = 30$ minutes during a typical day then determine the total amount of energy required by the car.
e) Calculate the fraction of the energy required by the car that is produced by the solar panel during a typical day.
f) Create a plot showing the fraction of energy required by the car that is produced by the solar panel, your answer from part (e), as a function of the panel efficiency; note that available photovoltaic panels operate at efficiencies less than 20%.

D. Pressure, Volume and Temperature

1.D-2 A manometer is a device that is commonly used to measure gage pressure and a barometer is used to measure absolute pressure. In your lab there is a mercury barometer that is used to measure ambient pressure and a water manometer that is used to measure the gage pressure in a tank, as shown in Figure 1.D-2. The temperature in the tank is also measured.

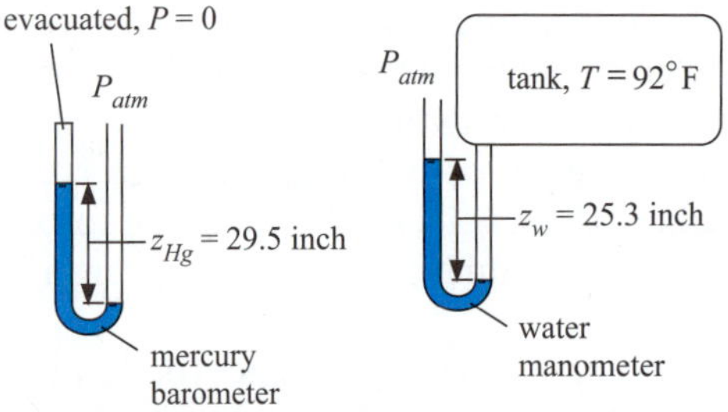

Figure 1.D-2: Mercury barometer and water manometer.

The mercury barometer consists of a tube that is open at one end and contains a column of mercury. The closed end of the tube is evacuated. Therefore, the height of the mercury column can be related to the absolute pressure that is applied to the open end using a force balance. The height of the mercury column is $z_{Hg} =$ 29.5 inch and the density of mercury is $\rho_{Hg} = 13{,}530$ kg/m^3. The water manometer is a U-shaped tube that is open at both ends. One end of the tube is exposed to the pressure in the tank and the other end is exposed to local atmospheric pressure. Therefore, the height of the column of water can be related to the pressure difference between the tank and ambient (i.e., the gage pressure). The height of the water column is $z_w = 25.3$ inch and the density of water is $\rho w =$ 996.6 kg/m^3. The temperature in the tank is 92°F.

a) Determine the gage pressure in the tank in Pa and psi.
b) Determine the absolute pressure in the tank in Pa and psi.
c) Determine the temperature in the tank in °C, K, and R.

1.D-4 You are working on an experimental fission reactor that is fueled by deuterium. The deuterium gas is very valuable and is delivered in $V = 1$ liter tanks, as shown in Figure 1.D-4. You would like to transfer as much of the deuterium that is initially in the tank to your experiment as possible.

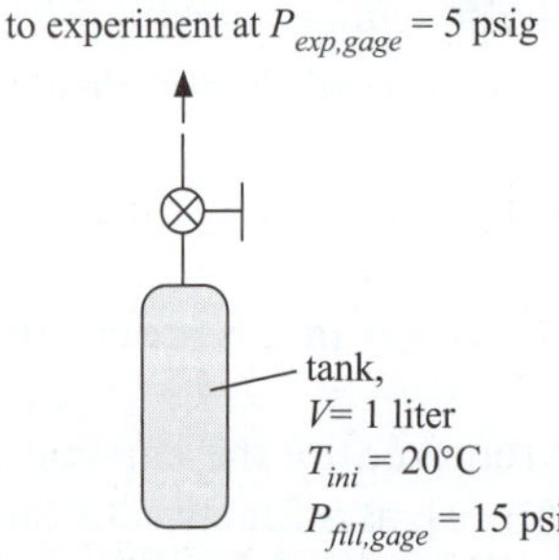

Figure 1.D-4: Deuterium tank connected to your experiment.

The initial temperature of the tank is $T_{ini} = 20°$C and the initial pressure is $P_{fill,gage} = 15$ psig (i.e., pounds per square inch *gage* pressure). The experiment is always maintained at pressure $P_{exp,gage} = 5$ psig (also a gage pressure). The atmospheric pressure is $P_{atm} = 1$ atm. You should model deuterium as an ideal gas for this problem; the ideal gas law relates pressure, temperature, and specific volume according to:

$$v = \frac{RT}{P}$$

where $R = 2064$ N-m/kg-K is the ideal gas constant for deuterium and T and P are the absolute temperature and pressure.

a) Determine the absolute pressure of the deuterium initially in the tank and the absolute pressure of the deuterium in the experiment (in Pa).

Initially, the valve connecting the tank to the experiment is opened and deuterium flows from the tank to the experiment until the pressure in the tank reaches P_{exp}. (Note that the pressure in the experiment does not change during this process). The temperature of the tank remains at T_{ini} during this process due to heat transfer with the surroundings.

b) Determine the mass of deuterium that passes from the tank to the experiment.
c) Determine the fraction of the mass of deuterium that is initially in the tank that passes to the experiment during this process.

In order to get an additional transfer of deuterium from the tank to the experiment you have decided to heat the tank to $T_{heat} = 50°\text{C}$ while the valve remains open. The tank is connected to the experiment during this process so that the pressure in the tank remains at P_{exp}.

d) Determine the mass of deuterium that passes from the tank to the experiment during the heating process.
e) Determine the total fraction of the mass of deuterium initially in the tank that is transferred to the experiment during the entire process.
f) Plot the total fraction of the mass of deuterium that is initially in the tank that is transferred to the experiment during the entire process as a function of the temperature of the tank at the conclusion of the heating process for $20°\text{C} < T_{heat} < 100°\text{C}$.

2 Thermodynamic Properties

2.1 Equilibrium and State Properties

Every thermodynamics problem starts by specifying the system. Once selected, it is necessary to determine the state of the system. Thermodynamics deals with *equilibrium* states. A system is in equilibrium if there are no unbalanced potentials or driving forces within the system. For example, a system is in thermal equilibrium if the temperature within the system is spatially uniform. A system is in mechanical equilibrium if the pressure within the system is spatially uniform. In Chapters 10 and 11 we will learn about phase equilibrium and in Chapter 14, chemical equilibrium is discussed. If a system is in equilibrium then it is internally balanced and it will experience no macroscopic changes except those caused by external effects. A system in equilibrium will not change its state if it is isolated from its surroundings.

The state of a substance that is in internal equilibrium can be specified completely provided that the values of a relatively small number of internal intensive properties are known. Examples of internal properties are mass, temperature, pressure, and volume. Velocity and elevation (in a gravitational field) are properties of a system; however they are external properties. External properties do not depend on the molecular structure of the substance composing the system whereas internal properties do. Extensive properties of a system, e.g., volume (V), internal energy (U), enthalpy (H), and entropy (S), are linearly-dependent on mass. Intensive properties, such as temperature (T) and pressure (P), do not depend on the mass of the system. A specific property is defined as the ratio of an extensive property to the mass of the system. Examples of specific properties include the specific volume, (v), specific internal energy (u), specific enthalpy (h) and specific entropy (s). Specific properties are intensive properties, i.e., they are independent of mass. Our concern in this chapter is with the internal intensive properties that are functionally related through their dependence on the molecular structure of the substance.

The First and Second Laws of Thermodynamics can be used to determine the number of internal intensive properties that are required in order to fix the state of a substance that is in equilibrium (F). The result is called the *phase rule* and the derivation of the phase rule is discussed in Section 11.6. The phase rule for non-reacting systems is given by:

$$F = C - \Pi + 2 \tag{2-1}$$

where C is the number of distinguishable chemical species (i.e., types of molecules) in the system and Π is the number of phases (i.e., solid, liquid, or vapor) that are present.

Consider the piston-cylinder apparatus shown in Figure 2-1. The apparatus contains water vapor that is in internal equilibrium (i.e., the pressure and temperature of the water vapor are spatially uniform).

Figure 2-1: Water vapor enclosed in a piston-cylinder device.

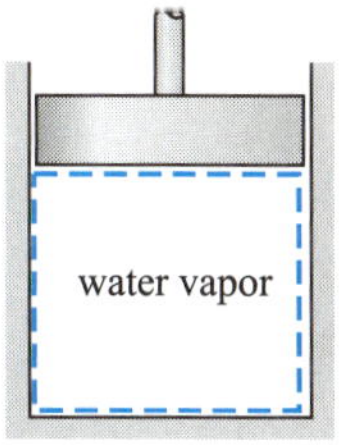

There is only one chemical species present in the system (i.e., water) and therefore $C = 1$. There is only one phase present (vapor) so $\Pi = 1$. Applying the phase rule, Eq. (2-1), to this system leads to:

$$F = C - \Pi + 2 = 1 - 1 + 2 = 2 \tag{2-2}$$

Therefore, only two internal intensive properties are required to completely specify the internal intensive state of this system. The state is fixed if, for example, we specify the temperature and pressure of the water. The word 'fixed' here means that we are no longer free to arbitrarily specify the values of any other internal intensive property. For example, if the temperature and pressure of the water are known, all other specific internal properties, such as the specific volume, specific internal energy, and specific entropy, have fixed values.

Suppose that the piston cylinder device contains a mixture of liquid water in equilibrium with water vapor, as shown in Figure 2-2. In this case, the number of phases is $\Pi = 2$ but the number of chemical species remains $C = 1$. Applying the phase rule to this situation:

$$F = C - \Pi + 2 = 1 - 2 + 2 = 1 \tag{2-3}$$

indicates that it is only necessary to specify a single internal intensive property in order to fix the intensive state of both the liquid and vapor phases in the apparatus. For example, if we specify that the temperature is 100°C, then every other internal intensive property is fixed. The pressure in the piston-cylinder device must be 101.325 kPa. We know this to be true because of careful experiments that have repeatedly been conducted in order to measure this property. The specific volumes of the liquid and vapor phases are also fixed as are the specific internal energy, specific enthalpy, and specific entropy of each phase.

One purpose of this chapter is to explain how properties can be obtained from existing data sources once the state of the substance has been fixed. The internal properties of many engineering substances of general interest have been measured and are recorded in tables, equations, or graphs. These databases can be used to retrieve the relationships between the internal properties for these substances. Under some conditions, the behavior of the substance is so simple that these databases are not necessary and very simple models (e.g., the ideal gas or incompressible substance models) can be applied with little loss of accuracy.

Figure 2-2: Water in liquid and vapor phases enclosed in a piston-cylinder device.

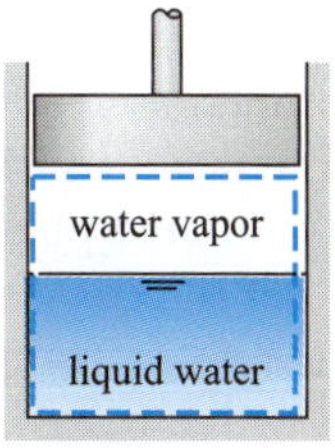

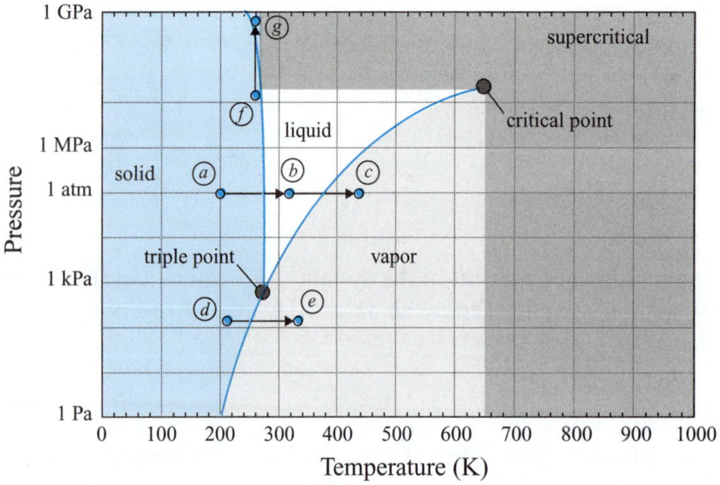

Figure 2-3: Pressure-temperature diagram for water.

Some of the properties that are tabulated in these databases (e.g., specific internal energy or specific entropy) are not directly measurable. If that is the case, how were these values determined and recorded? Internal properties are functionally related by the First and Second Laws of Thermodynamics. Therefore it is possible to infer the value of non-measurable properties from the values of other properties that are measurable. For example, we will see in Chapter 10 that the specific internal energy, specific enthalpy and specific entropy of a substance can be derived provided that the specific heat capacity and specific volume of the substance (both of which are measurable properties) are accurately known as a function of temperature and pressure.

2.2 General Behavior of Fluids

All pure substances exhibit qualitatively similar behavior and therefore it is only necessary to completely understand the behavior of one fluid in order to have a firm grasp of the general behavior of all fluids. Water is a fluid that we are familiar with, so we will examine its behavior in some detail in order to develop our understanding of fluids in general.

Figure 2-3 illustrates the phase of water in the parameter space of pressure and temperature. The phase rule dictates that two independent properties (e.g., pressure and temperature) are required to specify the state of a single phase (i.e., $\Pi = 1$), pure ($C = 1$) fluid. Therefore, single phase liquid water can exist over a range of both temperature and pressure; this area is labeled 'liquid' in Figure 2-3. The regions of pressure and temperature where single phase water vapor and single phase solid water (i.e., ice) exist are also labeled in Figure 2-3.

At a specified pressure (e.g., 1 atm or 101.325 kPa) and a sufficiently low temperature water exists as a solid, as indicated by point *a* in Figure 2-3. As the temperature of water is increased at constant pressure from its value at point *a*, the water remains a solid until it reaches its freezing point at one atmosphere (which lies on the line separating the solid and liquid phases) at which point it transitions from a solid to a liquid. Figure 2-3 shows that the freezing point of water at atmospheric pressure is 273.15 K (0°C). Eventually, the water completely melts and becomes a single phase liquid, as indicated by point *b*. Further heating at constant pressure causes the liquid temperature to rise until eventually the water reaches its boiling temperature and begins to *evaporate* (i.e., to turn from liquid to vapor) when it reaches the line separating the liquid and vapor phases. With additional heating, the water becomes a single phase vapor, as indicated by point *c*. The *normal boiling point* is defined as the boiling point at atmospheric pressure; the normal boiling point of water is 373.15 K (100°C).

At very low pressures, the solid phase of water will transition directly to vapor phase without passing through a liquid phase. This process is called *sublimation* and is shown in Figure 2-3 by the process of moving from point *d* to point *e*. A familiar substance that sublimes at atmospheric pressure is carbon dioxide, which is also called *dry ice*.

An interesting and unusual characteristic of water is that the saturation line delineating its solid phase from its liquid phase actually has a slightly negative slope (i.e., the freezing temperature of water decreases slightly with increasing pressure). This characteristic is a result of the fact that solid water (ice) is less dense than liquid water. This is the reason ice chunks float rather than sink. One consequence of this behavior is that it is possible to melt ice at constant temperature by increasing the pressure that is applied to it; this process is shown in Figure 2-3 by the process of moving from state *f* to state *g*. This characteristic of water is the reason that ice skates work. Your body weight is concentrated on the very small area associated with the metal skate blade, substantially increasing the pressure exerted on the ice below the blade. The increase in pressure causes the ice under the blade to melt, forming a thin film of liquid water that lubricates the blade. Relatively few substances exhibit this characteristic; some examples include bismuth, gallium, antimony, silicon and acetic acid.

Notice that the information presented in Figure 2-3 is consistent with the phase rule. When water exists in a single phase, both temperature and pressure may vary independently and so single phase water exists over regions in the pressure-temperature diagram. However, there are two phases present during a transition process (i.e., sublimation, melting, or evaporation). During evaporation water is a mixture of liquid and vapor, both phases existing at the same temperature and pressure. The phase rule requires that a single intensive property (e.g., either temperature or pressure, but not both) fixes the state of a two-phase system. The phase transitions in Figure 2-3 appear as lines because the temperature at which a phase transition occurs is uniquely defined once the pressure is given (or vice versa). When a mixture of two or more phases is present, the substance is said to be in a *saturated* condition. The term saturated applies to the coexistence of any phases, but it will most often be used in this textbook to refer to a liquid and vapor mixture.

It is possible for a pure fluid to exist in three phases simultaneously. The phase rule, Eq. (2-1), indicates that the number of degrees of freedom (F) for a pure fluid in a saturated state for which three phases (solid, liquid, and vapor) are present is zero. Zero degrees of freedom implies that there is only one such state and it is impossible to independently specify the value of any property. The single condition at which solid, liquid and vapor phases coexist is called the *triple point*. Figure 2-3 shows that the triple point temperature of water is 273.16 K and the triple point pressure of water is 611.7 Pa. The triple point state is often used as a reference condition because it is uniquely and precisely defined for any fluid.

One last feature of Figure 2-3 that should be examined is the *critical point*. The existence of separate liquid and vapor phases is observable if the fluid is placed in a transparent tube under appropriate conditions. The liquid phase has higher density (lower specific volume) and will therefore tend to pool at the bottom of the tube (assuming that the experiment is carried out in a gravitational field). The boundary between the liquid and vapor is called a *meniscus*; the meniscus can have a concave or convex shape, depending on the interaction of the fluid with the tube wall. However, at a sufficiently high

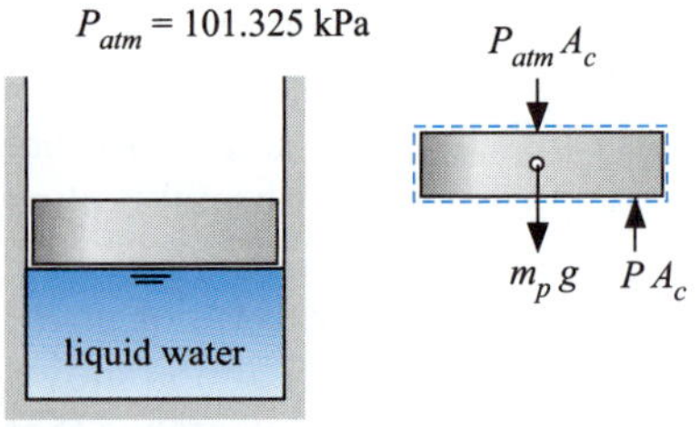

Figure 2-4: Liquid water in a piston-cylinder device.

temperature we find that liquid and vapor cannot coexist under any conditions. The highest temperature at which liquid and vapor can coexist is called the *critical temperature*. The specific volume of the liquid phase at the critical temperature is identical to the specific volume of the vapor phase; therefore, the two phases cannot be distinguished from one another. Similarly, when the pressure is at the *critical pressure* it is no longer possible to distinguish between the liquid and vapor phases. Fluid at a temperature above the critical temperature or a pressure above the critical pressure can only exist in one phase, and it is sometimes referred to as *supercritical*. The supercritical region is shown in gray in Figure 2-3.

The qualitative shape of a pressure-temperature diagram for any substance will be similar to the diagram for water shown in Figure 2-3, although the values of temperature and pressure corresponding to the different transitions and regions will be different. For example, the triple point pressure of carbon dioxide is a little over 5 atm. Therefore, solid carbon dioxide (dry ice) will sublimate rather than melt at atmospheric pressure.

The relationship that exists between the temperature, pressure, and specific volume of water can be understood by carrying out the following experiment. Liquid water at room temperature, $T_1 = 20°\text{C}$, is placed in a piston-cylinder apparatus as shown in Figure 2-4. The mass of water is $m = 1$ kg. The piston-cylinder apparatus is assumed to be "perfect"; that is, the piston moves without friction, it has negligible mass, and it does not allow any leakage. A force balance on the piston (see Figure 2-4) is:

$$P A_c = P_{atm} A_c + m_p g \tag{2-4}$$

where A_c is the cross-sectional area of the piston. If the mass of the piston (m_p) is negligible, then Eq. (2-4) indicates that the pressure in the piston-cylinder apparatus will be equal to the atmospheric pressure, $P = 101.325$ kPa, regardless of the volume occupied by the water. The temperature and pressure together are two independent intensive internal properties that fix the state of the single phase liquid water. The volume of the liquid can be measured, $V = 0.001$ m^3 (1 liter), therefore the specific volume associated with this state is $v_1 = 0.001$ m^3/kg.

The water is slowly heated, causing its temperature to rise. According to Eq. (2-4), the pressure remains constant during this process and we can continue to measure the volume of the water (by noting the position of the piston) as the temperature of the system changes. Since the mass of water is known, it is possible to calculate the specific volume of the water associated with each volume measurement. Figure 2-5 illustrates the measured temperature as a function of the measured specific volume for this process; Figure 2-5(a) provides this information on a linear scale for specific volume while Figure 2-5(b) provides the same information on a logarithmic scale.

Extremely careful measurements would show that the specific volume of the liquid water increases very slightly with temperature in the progression from state 1 to state *f*. At state *f*, which corresponds to a temperature of 100°C, the behavior exhibited by our experiment changes drastically. The specific volume, which up to this point was nearly constant, now begins to increase dramatically while the temperature remains constant.

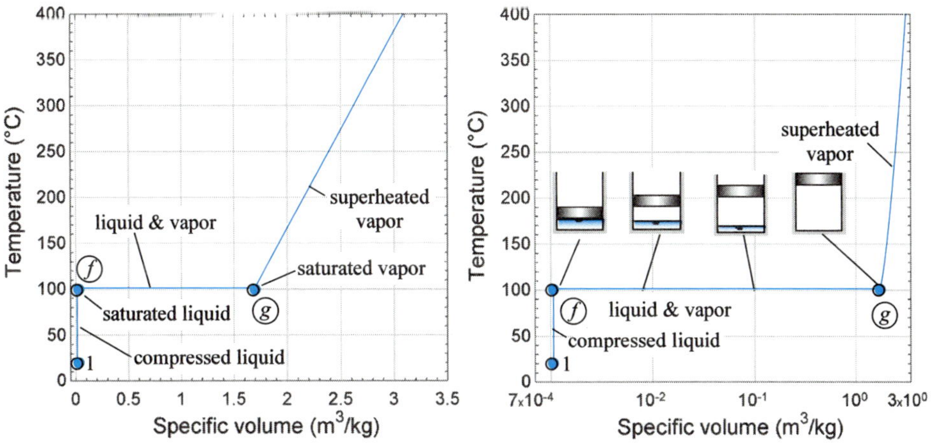

Figure 2-5: Temperature as a function of specific volume for water at 101.3 kPa using (a) a linear and (b) a logarithmic scale for specific volume.

Between states *f* and *g*, the water remains in a saturated state where liquid and vapor phases coexist, both at 100°C and 101.325 kPa. At state *f*, the water is a liquid that is just ready to start evaporating; this state is referred to as *saturated liquid*. The specific volume at state *f* is designated as v_f. The subscript *f* originates from the German word Flüssigkeit, meaning liquid. As the heating progresses, some of the water transitions from the liquid to the vapor phase; however the temperature remains at 100°C during this process. This temperature at which both liquid and vapor coexist is referred to as the *saturation temperature*. With continued heating, all of the liquid eventually becomes vapor; this state is referred to as *saturated vapor* and corresponds to state *g* where the specific volume is $v_g = 1.674$ m³/kg. The subscript *g* originates from the German word *Gas*, which translates as vapor. A rather large amount of thermal energy is required to accomplish this boiling process.

Another distinct change in the behavior of our experiment is observed at state *g*. At state *g*, all of the liquid has evaporated and only water vapor exists in the cylinder. Further heating causes both the specific volume and the temperature of the vapor to increase. Vapor that is heated to a temperature above its saturation temperature is referred to as *superheated vapor*. The liquid that originally existed below the saturation temperature (between state 1 and state *f*) is referred to as *compressed liquid*.

In principle, if we carefully measure and record the temperature as a function of specific volume at this pressure (101.325 kPa) then we never need to do this experiment again for water. The relation between specific volume and temperature at 101.325 kPa is fixed. Figure 2-5(b) shows that the specific volume of saturated water vapor at 100°C, state *g*, is more than 1000 times greater than the specific volume of saturated liquid, state *f*. This enormous change in specific volume would complicate the process of actually conducting this experiment; the volume of the piston cylinder device must increase from its initial value of 1 liter to a volume of 1674 liters at state *g*.

This experiment can be repeated at different pressures by placing a mass onto the piston. A force balance shows that the pressure exerted by the piston on the water is:

$$P = P + \frac{m_p g}{A_c} \tag{2-5}$$

where m_p is the mass of the piston and g is the acceleration of gravity. Figure 2-6 illustrates the results of repeating this experiment with a piston mass that is sufficiently large to increase the pressure in the piston-cylinder apparatus to 500 kPa and to 2 MPa.

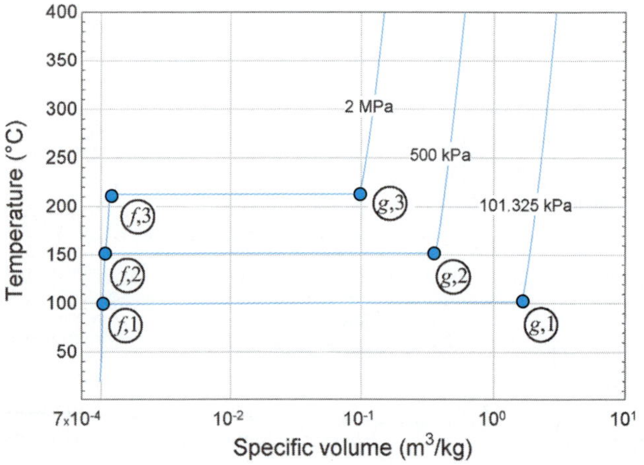

Figure 2-6: Temperature as a function of specific volume for water measured at several different pressures.

Upon heating, the initial behavior of the experiment is essentially the same as it was when the pressure was 101.325 kPa. However, when the pressure is held at 500 kPa, the water does not begin to evaporate until the temperature reaches 151.8° C; the saturation temperature of water at 500 kPa is 151.8° C. The specific volume of the saturated liquid water at 500 kPa, state *f*,2, is only slightly larger than it was at 101.325 kPa, state *f*,1. The specific volume of saturated vapor at 500 kPa, state *g*,2, is substantially less than it was at 101.325 kPa, state *g*,1. The specific volumes of saturated liquid and saturated vapor approach one another as the pressure increases. The temperature as a function of specific volume measured at $P = 2$ MPa is also shown in Figure 2-6 in order to demonstrate this behavior.

If we repeat this experiment for many different pressures then it is possible to generate the property plot that is referred to as a temperature-specific volume (or *T*-*v*) diagram; the *T*-*v* diagram for water is shown in Figure 2-7. Each line of constant pressure is referred to as an *isobar*. The locus of the states at which saturated liquid exists (e.g., states *f*,1 and *f*,2 in Figure 2-6) forms one side of the *vapor dome*. The locus of the states at which saturated vapor exists (e.g., states *g*,1 and *g*,2) forms the other side.

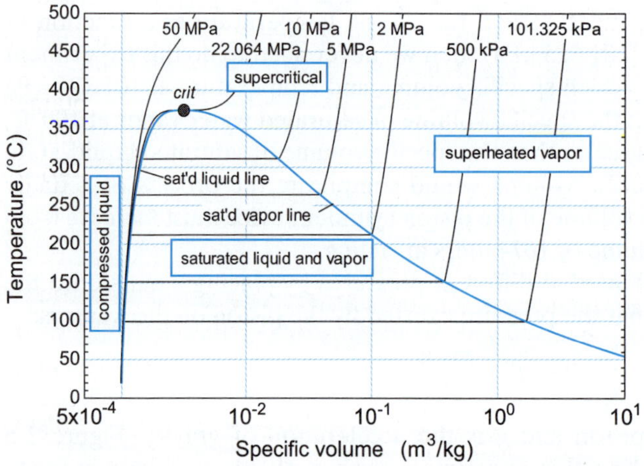

Figure 2-7: Temperature–specific volume (*T*-*v*) diagram for water.

The top of the vapor dome is terminated by the critical point, state *crit* in Figure 2-7. All states that lie within the vapor dome consist of both liquid and vapor phases. The left boundary of the vapor dome is the saturated liquid line. States on this line consist entirely of saturated liquid that is ready to evaporate. The right boundary of the vapor dome is the saturated vapor line. All states falling on this line consist entirely of water vapor that is ready to condense. Several isobars are displayed in Figure 2-7 in order to show their trajectory. Note that the isobars are horizontal when they pass through the vapor dome. An important implication of the behavior in the saturated region is that it is not possible to determine the relative amounts of liquid and vapor for a state that lies within the vapor dome by specifying only the temperature (or only the pressure). All states between saturated liquid and saturated vapor at a given pressure exhibit the same temperature.

States that lie to the right of the vapor dome (i.e., states with specific volume greater than the specific volume of saturated vapor at the same temperature) are referred to as *superheated vapor*. At the other extreme, states that lie to the left of the vapor dome (i.e., states with specific volume less than the specific volume of the saturated liquid at the same temperature) are referred to as *compressed liquid*. Notice that the specific volume of compressed liquid is smaller than that of saturated liquid at the same temperature, but only by a small amount. Liquids are nearly incompressible. Unless an enormous pressure is exerted on the liquid, its specific volume cannot be changed significantly (i.e., it cannot be compressed). Compressed liquid states are also called *subcooled liquid*.

The isobar that passes through state *crit* in Figure 2-7 is the critical pressure, which corresponds to $P_{crit} = 22.064$ MPa for water. The critical temperature for water is $T_{crit} = 374.0°$C and the critical specific volume for water is $v_{crit} = 0.003106$ m^3/kg. A fluid that is at a pressure or temperature above its critical pressure or temperature cannot exist in two phases.

Figure 2-8 illustrates, qualitatively, the three-dimensional surface that is formed when pressure is plotted as a function of temperature and specific volume. Notice that the *P*-*T* and *T*-*v* plots, shown in Figure 2-3 and Figure 2-7 for water, are projections of the *P*-*v*-*T* surface viewed from different directions.

2.3 Property Tables

Fluid property data of the type represented in Figure 2-7 have traditionally been made available in tables. The tables developed for water are commonly referred to as *steam tables*, but of course these tables provide information for liquid water as well as for steam. A detailed set of property tables for water is contained in Appendix B as Tables B-1 through B-4. Data tables for refrigerant R134a are provided in Appendix C. Section 2.4 describes how the property data for these substances and many others can be obtained using built-in property functions in EES. These computer functions are more convenient to use than property tables and therefore most of the examples in this book will rely on the property data obtained using EES.

2.3.1 Saturated Liquid and Vapor

Tables B-1 and B-2 in Appendix B provide property data for water that lie on the vapor dome (i.e., the saturated liquid and saturated vapor lines). Recall that the number of degrees of freedom for a pure substances existing in two phases is one; therefore, the specification of a single intensive variable fixes the intensive state of both phases. Table B-1 provides information at regular intervals of temperature whereas Table B-2 uses regular intervals of pressure. A small excerpt of Table B-1 is provided as Table 2-1. The

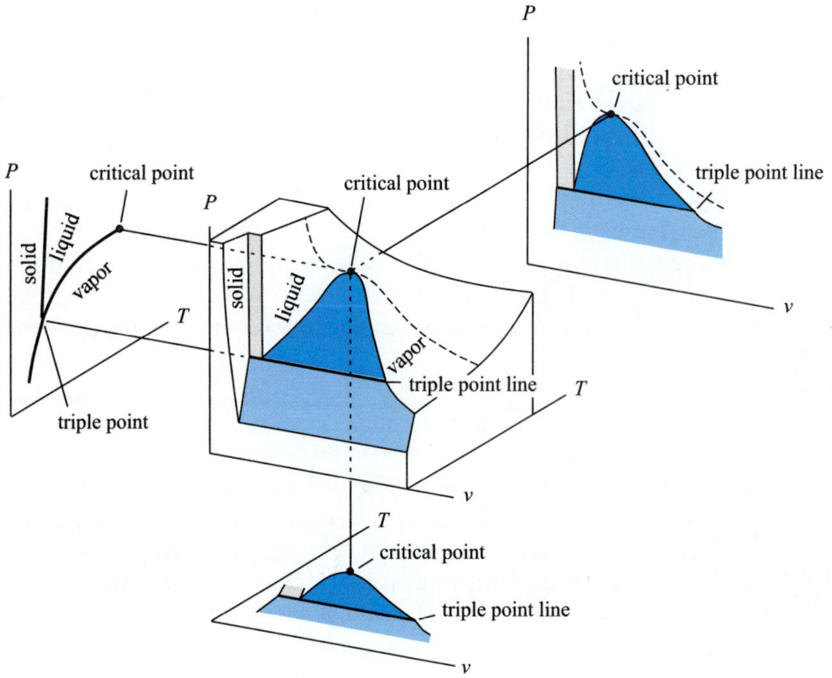

Figure 2-8: Pressure as a function of specific volume and temperature for a pure substance that contracts upon freezing, based on Myers (2007).

first column in Table 2-1 is the temperature of interest while the second column is the saturation pressure that corresponds to this temperature; notice in Figure 2-7 that there is a unique saturation pressure corresponding to each temperature. After pressure, the table provides four sets of two columns. Each set of columns corresponds to a different property: specific volume (v), specific internal energy (u), specific enthalpy (h), and specific entropy (s). Within each set, the first column corresponds to the value of the property for the saturated liquid (designated by the subscript f) and the second column is the value for the saturated vapor (designated by the subscript g). We have thus far only discussed specific volume; however, it will also be necessary to know the values of u, h, and s in order to apply the laws of thermodynamics. Table B-2 provides exactly the same information as Table B-1, except that the order of columns 1 and 2 is switched and the data are provided for equal increments of pressure. Note that the value of the specific volume of saturated liquid reported in these tables is multiplied by 1000 in order to minimize the space required by the column for v_f. Therefore, the specific volume of saturated liquid water at 8°C is 0.0010002 m^3/kg (and not 1.0002 m^3/kg).

The data in Tables B-1 and B-2 provide property information for states that are entirely saturated liquid (subscript f) and entirely saturated vapor (subscript g). However, a state that lies under the vapor dome will consist of both liquid and vapor phases that are in equilibrium. For example, Figure 2-9 illustrates a tank with volume V. The water contained in the tank is in a two-phase state at $T = 20°\text{C}$.

According to Table 2-1, the pressure in the tank must be 2.3392 kPa, which is the saturation pressure of water at 20°C. The liquid contained in the tank in Figure 2-9 is saturated liquid at 20°C and 2.3392 kPa; the properties of the liquid phase are $v_f =$ 0.0010018 m^3/kg, $u_f = 83.913$ kJ/kg, $h_f = 83.915$ kJ/kg and $s_f = 0.29649$ kJ/kg-K. The vapor contained in the tank is saturated vapor at 20°C and 2.3392 kPa; the properties of the vapor phase are $v_g = 57.762$ m^3/kg, $u_g = 2402.3$ kJ/kg, $h_g = 2537.4$ kJ/kg, and

Table 2-1: An excerpt of Table B-1, which provides saturation data for water at regular intervals of temperature.

Temp. T (°C)	Pressure P (kPa)	Specific volume (m^3/kg)		Specific internal energy (kJ/kg)		Specific enthalpy (kJ/kg)		Specific entropy (kJ/kg-K)		T (°C)
		$10^3\ v_f$	v_g	u_f	u_g	h_f	h_g	s_f	s_g	
0.01	0.6117	1.0002	206.00	0	2374.9	0.000	2500.9	0	9.1556	0.01
2	0.7060	1.0001	179.78	8.3911	2377.7	8.3918	2504.6	0.03061	9.1027	2
4	0.8135	1.0001	157.14	16.812	2380.4	16.813	2508.2	0.06110	9.0506	4
6	0.9353	1.0001	137.65	25.224	2383.2	25.225	2511.9	0.09134	8.9994	6
8	1.0729	1.0002	120.85	33.626	2385.9	33.627	2515.6	0.12133	8.9492	8
10	1.2281	1.0003	106.32	42.020	2388.7	42.022	2519.2	0.15109	8.8999	10
12	1.4028	1.0006	93.732	50.408	2391.4	50.410	2522.9	0.18061	8.8514	12
14	1.5989	1.0008	82.804	58.791	2394.1	58.793	2526.5	0.20990	8.8038	14
16	1.8187	1.0011	73.295	67.169	2396.9	67.170	2530.2	0.23898	8.7571	16
18	2.0646	1.0015	65.005	75.542	2399.6	75.544	2533.8	0.26784	8.7112	18
20	2.3392	1.0018	57.762	83.913	2402.3	83.915	2537.4	0.29649	8.6661	20
22	2.6452	1.0023	51.422	92.280	2405.1	92.283	2541.1	0.32493	8.6217	22

$s_g = 8.6661$ kJ/kg-K. The relative amount of each phase is expressed in terms of the *quality*. Quality, x, is defined as:

$$x = \frac{\text{mass of vapor}}{\text{total mass}} = \frac{m_g}{m} \tag{2-6}$$

It is useful to derive relationships between the average specific property associated with the two-phase mixture (i.e., the system shown in Figure 2-9), the quality, and the specific properties of the saturated liquid and saturated vapor that are present. For example, the total volume of the tank in Figure 2-9 is equal to the sum of the volume occupied by the liquid (V_f) and the volume occupied by the vapor (V_g).

$$V = V_f + V_g \tag{2-7}$$

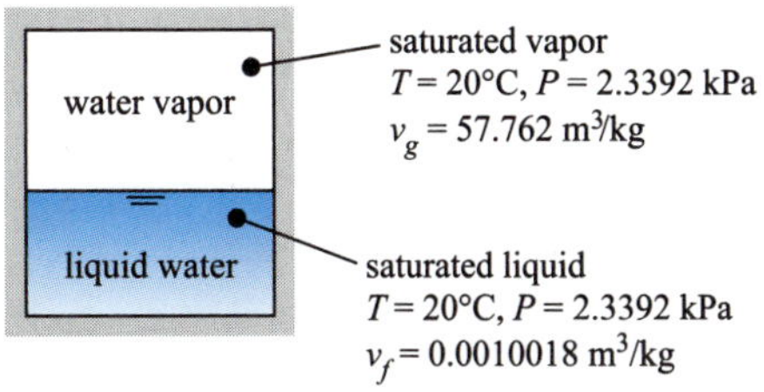

Figure 2-9: A tank that contains water in a two-phase state.

The volume of liquid is the product of the mass of liquid (m_f) and its specific volume (v_f):

$$V_f = m_f\, v_f \tag{2-8}$$

The volume of gas can similarly be expressed as:

$$V_g = m_g\, v_g \tag{2-9}$$

Equations (2-8) and (2-9) are substituted into Eq. (2-7):

$$V = m_f\, v_f + m_g\, v_g \tag{2-10}$$

Equation (2-10) is divided by the total mass of water:

$$\frac{V}{m} = \underbrace{\left(\frac{m_f}{m}\right)}_{(1-x)} v_f + \underbrace{\left(\frac{m_g}{m}\right)}_{x} v_g \tag{2-11}$$

The definition of quality, Eq. (2-6), is substituted into Eq. (2-11):

$$\frac{V}{m} = (1-x)\, v_f + x\, v_g \tag{2-12}$$

The quantity on the left side of Eq. (2-12) is the ratio of the total volume of the liquid and vapor to the total mass of the liquid and vapor; it is the mass-average specific volume of the two-phase mixture (v).

$$v = (1-x)\, v_f + x\, v_g \tag{2-13}$$

Rearranging Eq. (2-13) leads to:

$$\boxed{v = v_f + x(v_g - v_f)} \tag{2-14}$$

or

$$\boxed{x = \frac{(v - v_f)}{(v_g - v_f)}} \tag{2-15}$$

The average specific internal energy, specific enthalpy and specific entropy for states consisting of both liquid and vapor can be expressed in the same manner:

$$u = (1-x)\, u_f + x\, u_g \tag{2-16}$$

$$h = (1-x)\, h_f + x\, h_g \tag{2-17}$$

$$s = (1-x)\, s_f + x\, s_g \tag{2-18}$$

Equations (2-16) through (2-18) can be rearranged:

$$x = \frac{(u - u_f)}{(u_g - u_f)} \tag{2-19}$$

$$x = \frac{(h - h_f)}{(h_g - h_f)} \tag{2-20}$$

$$x = \frac{(s - s_f)}{(s_g - s_f)} \tag{2-21}$$

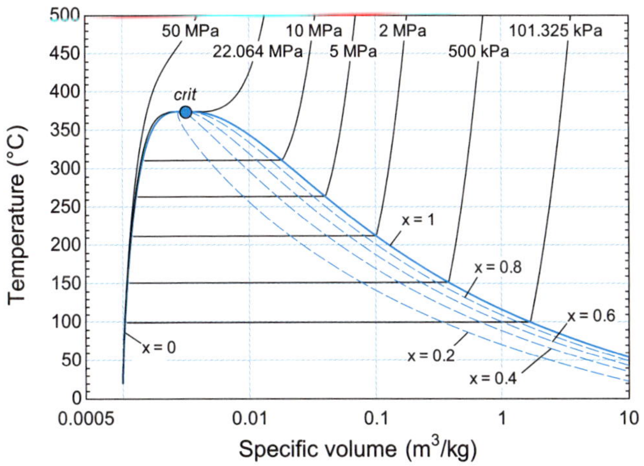

Figure 2-10: A *T-v* diagram for water showing lines of constant quality.

Quality is only relevant for states under the vapor dome and therefore its value must lie between zero and one. The quality is equal to zero for saturated liquid and equal to one for saturated vapor. Equation (2-15) shows that the quality associated with a state represents the fractional progression of the state across the vapor dome along an isobar. Figure 2-10 illustrates a *T-v* diagram for water that includes lines of constant quality in addition to isobars. Because the abscissa is logarithmic, the lines of constant quality appear to be compressed towards the right side of the vapor dome.

EXAMPLE 2.3-1: PRODUCTION OF A VACUUM BY CONDENSATION

In 1601, Giambattista della Porta described a method by which a *vacuum* (i.e., a pressure lower than the pressure of the atmosphere) can be produced through the condensation of steam. A small amount of liquid water at $T_1 = 20°\text{C}$ and $P_1 = 1$ atm is placed in a rigid container with total volume $V = 0.35$ liter. The water is heated until it boils. The boiling continues for some time in order to allow the water vapor generated by the boiling process to push all of the air out of the container. Then the container is sealed and weighed. The total mass of water in the container is $m_1 = 1.25$ g.

a) What is the volume of liquid water and the volume of water vapor in the container at the time that it is sealed?

The specific volume of the water in the container at the time that it is sealed is:

$$v_1 = \frac{V}{m_1} = \frac{0.35\,\text{liter}}{} \left| \frac{}{1.25\,\text{g}} \right\| \frac{1000\,\text{g}}{1\,\text{kg}} \left| \frac{0.001\,\text{m}^3}{1\,\text{liter}} \right. = 0.28\,\text{m}^3/\text{kg}$$

The pressure and specific volume together specify state 1. Whenever you solve a problem, it is useful to sketch a *T-v* diagram (or some other type of property diagram) and qualitatively locate each state that is involved in the problem on the diagram. This exercise will usually help you to better understand the problem.

Figure 1 illustrates a qualitative sketch of a *T-v* diagram. State 1 lies at the intersection of the isobar $P_1 = 101.325$ kPa and the line of constant specific volume (also called an *isochor*) $v_1 = 0.28$ m³/kg. On a *T-v* diagram, an isochor is a vertical line (and an *isotherm* is a horizontal line). According to Table B-1 in Appendix B, at 1 atm (101.3 kPa, corresponding to a saturation temperature of 100°C) the specific volume of saturated liquid is $v_{f,1} = 0.0010435$ m³/kg and the specific volume of saturated vapor is $v_{g,1} = 1.6720$ m^3/kg. Because $v_{f,1} < v_1 < v_{g,1}$, state 1 must lie within the vapor dome as shown in Figure 1.

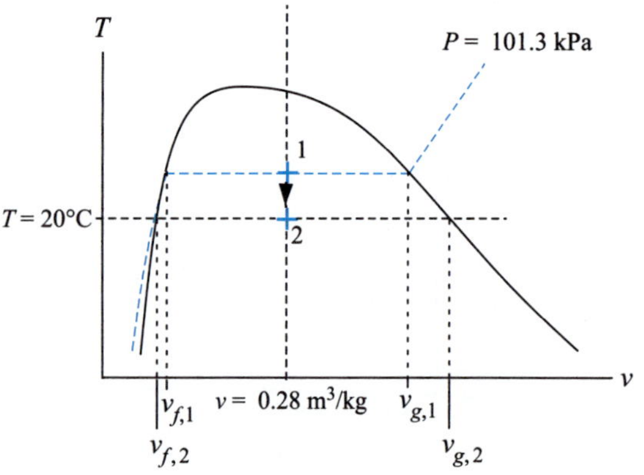

Figure 1: Qualitative sketch of the T-v diagram.

The quality of the water in the container at the time that it is sealed is:

$$x_1 = \frac{(v_1 - v_{f,1})}{(v_{g,1} - v_{f,1})} = \frac{(0.28 - 0.0010435)}{(1.6720 - 0.0010435)} = 0.1669$$

The mass of vapor in the container is obtained from Eq. (2-6):

$$m_{g,1} = x_1\, m_1 = \frac{0.1669}{}\left|\frac{1.25\text{ g}}{}\right. = 0.2087\text{ g}$$

and the mass of liquid in the container is:

$$m_{f,1} = (1 - x_1)\, m_1 = \frac{(1 - 0.1669)}{}\left|\frac{1.25\text{ g}}{}\right. = 1.041\text{ g}$$

The volume occupied by the vapor is the product of the mass of the vapor and its specific volume:

$$V_{g,1} = m_{g,1}\, v_{g,1} = \frac{0.2087\text{ g}}{}\left|\frac{1.6720\text{ m}^3}{\text{kg}}\right\|\frac{1\text{ kg}}{1000\text{ g}}\left|\frac{\text{liter}}{0.001\text{ m}^3}\right. = 0.349\text{ liter}$$

The volume of liquid is calculated according to:

$$V_{f,1} = m_{f,1}\, v_{f,1} = \frac{1.041\text{ g}}{}\left|\frac{0.0010435\text{ m}^3}{\text{kg}}\right\|\frac{1\text{ kg}}{1000\text{ g}}\left|\frac{\text{liter}}{0.001\text{ m}^3}\right. = 0.0011\text{ liter}$$

b) The sealed container is allowed to cool to $T_2 = 20°$C. What is the resulting pressure in the container? What is the quality of the water in the container? Assume that the container volume does not change.

In this process, both the volume and the mass of water in the container remain constant. Therefore, the specific volume at state 2 is equal to its value at state 1, $v_2 = v_1 = 0.28$ m^3/kg. State 2 is specified by the intersection of an isochor, $v_2 = 0.28$ m^3/kg, and an isotherm, $T_2 = 20°$C, as shown in Figure 1. The specific volume falls between the values of saturated liquid water at 20°C, $v_{f,2} = 0.0010018$ m^3/kg, and saturated water vapor at 20°C, $v_{g,2} = 57.762$ m^3/kg (both of these values are provided in Table B-1). Therefore, state 2 also lies inside the vapor dome (as shown in

Figure 1) and the pressure in the container must be $P_2 = 2.3392$ kPa (the saturation pressure at 20°C). The quality at state 2 is:

$$x_2 = \frac{(v_2 - v_{f,2})}{(v_{g,2} - v_{f,2})} = \frac{(0.28 - 0.0010018)}{(57.762 - 0.0010018)} = 0.00483$$

Note that the pressure in the container is very much lower than atmospheric and, unless the container is very sturdy, it will likely be crushed by the force exerted from the atmospheric pressure.

c) What are the volumes of vapor and liquid in the container at the final state?

The mass of vapor is calculated according to:

$$m_{g,2} = x_2\, m_2 = \frac{0.00483}{} \left| \frac{1.25\,\text{g}}{} \right. = 0.006038\ \text{g}$$

and the mass of liquid is:

$$m_{f,2} = (1 - x_2)\, m_2 = \frac{(1 - 0.00483)}{} \left| \frac{1.25\,\text{g}}{} \right. = 1.244\ \text{g}$$

The volume of vapor is:

$$V_{g,2} = m_{g,2}\, v_{g,2} = \frac{0.006038\,\text{g}}{} \left| \frac{57.762\,\text{m}^3}{\text{kg}} \right\| \frac{1\,\text{kg}}{1000\,\text{g}} \left| \frac{\text{liter}}{0.001\,\text{m}^3} \right. = 0.349\ \text{liter}$$

The volume of liquid is:

$$V_{f,2} = m_{f,2}\, v_{f,2} = \frac{1.244\,\text{g}}{} \left| \frac{0.0010018\,\text{m}^3}{\text{kg}} \right\| \frac{1\,\text{kg}}{1000\,\text{g}} \left| \frac{\text{liter}}{0.001\,\text{m}^3} \right. = 0.0012\ \text{liter}$$

Notice that the specific volume of saturated water vapor at low temperature is orders of magnitude larger than the specific volume of saturated liquid water at the same temperature. As a result, even though the water in the container has a very small quality (only 0.483% of the mass is vapor) the vapor takes up most of the volume (99.7% of the volume is vapor).

2.3.2 Superheated Vapor

The term *superheated vapor* is used to describe a state for which the specific volume is larger than the specific volume of saturated vapor at the same temperature. The superheated vapor region lies to the right of the saturation dome on a T-v diagram, as shown in Figure 2-7. Saturated vapor cannot exist for temperatures above the critical temperature; states that are above the critical point are referred to as supercritical. The properties of water in the superheated vapor and supercritical regions are provided in Table B-3 of Appendix B. These states are single phase ($\Pi = 1$) for a pure fluid ($C = 1$); therefore, according to phase rule in Eq. (2-1), there are two degrees of freedom ($F = 2$). The data in Table B-3 are presented as an array of information in the dimensions of pressure and temperature. An excerpt of Table B-3 is shown in Table 2-2.

The individual columns in Table 2-2 correspond to different temperatures and each set of four rows corresponds to a different pressure. The first row in each set is the specific volume, the second is the specific internal energy, the third is the specific enthalpy, and the fourth is the specific entropy. Therefore, it is possible to obtain the properties v, u, h, and s at a given pressure and temperature. For example, the specific volume of superheated water vapor at $T = 300$°C and $P = 40$ kPa is $v = 6.6067$ m^3/kg.

Table 2-2: An excerpt of Table B-3, which provides superheated vapor and supercritical properties for water over a matrix of temperature and pressure.

P (kPa)		Temperature, *T* (°C)									
		50	100	150	200	300	400	500	600	800	1000
10	*v* (m^3/kg)	14.867	17.196	19.513	21.826	26.446	31.063	35.68	40.296	49.527	58.758
	u (kJ/kg)	2443.3	2515.5	2587.9	2661.4	2812.3	2969.3	3132.9	3303.3	3665.4	4055.3
	h (kJ/kg)	2592.0	2687.5	2783.0	2879.6	3076.7	3280.0	3489.7	3706.3	4160.6	4642.8
	s (kJ/kg-K)	8.1741	8.4489	8.6893	8.9049	9.2827	9.6094	9.8998	10.163	10.631	11.043
20	*v* (m^3/kg)		8.5855	9.7486	10.907	13.220	15.530	17.839	20.147	24.763	29.379
	u (kJ/kg)		2514.5	2587.4	2661.0	2812.1	2969.2	3132.8	3303.3	3665.3	4055.2
	h (kJ/kg)		2686.2	2782.3	2879.2	3076.5	3279.8	3489.6	3706.2	4160.6	4642.8
	s (kJ/kg-K)		8.1263	8.3681	8.5843	8.9625	9.2893	9.5798	9.8432	10.311	10.723
40	*v* (m^3/kg)		4.2799	4.8662	5.448	6.6067	7.7628	8.9179	10.073	12.381	14.689
	u (kJ/kg)		2512.5	2586.3	2660.3	2811.7	2969.0	3132.7	3303.2	3665.3	4055.2
	h (kJ/kg)		2683.7	2780.9	2878.2	3076.0	3279.5	3489.4	3706.1	4160.5	4642.7
	s (kJ/kg-K)		7.8011	8.0456	8.2629	8.642.0	8.9691	9.2597	9.5231	9.9913	10.403
60	*v* (m^3/kg)		2.8445	3.2387	3.6283	4.4023	5.1739	5.9444	6.7144	8.2538	9.7927
	u (kJ/kg)		2510.5	2585.2	2659.6	2811.4	2968.8	3132.5	3303.0	3665.2	4055.1
	h (kJ/kg)		2681.1	2779.5	2877.3	3075.5	3279.2	3489.2	3705.9	4160.4	4642.7
	s (kJ/kg-K)		7.6084	7.8559	8.0743	8.4542	8.7816	9.0723	9.3359	9.8041	10.216

Interpolation

A problem that often arises when using tabulated property data is that the state of interest will not exactly coincide with the states that are reported in the table. In this case, it is necessary to use interpolation. Although higher order interpolation schemes are available, data are usually obtained from tables using linear interpolation. Linear interpolation uses information obtained from the property table for two states that bracket the state of interest. The general relation for accomplishing linear interpolation is:

$$\frac{Y - Y_1}{Y_2 - Y_1} = \frac{X - X_1}{X_2 - X_1} \tag{2-22}$$

where Y and X are two thermodynamic properties, such as specific volume and temperature, and the subscripts 1 and 2 indicate the two states for which these properties are known from the table.

For example, suppose you need the specific volume of superheated water vapor at $T = 115°\text{C}$ and $P = 20$ kPa. Examination of Table B-3 (or Table 2-2) indicates that this state is not exactly represented. The two states that bracket the state of interest lie in the row corresponding to $P = 20$ kPa: $T_1 = 100°\text{C}$, $v_1 = 8.5855$ m^3/kg and $T_2 = 150°\text{C}$, $v_2 = 9.7486$ m^3/kg. The specific volume at $T = 115°\text{C}$ is estimated by applying the linear interpolation relation, Eq. (2-22), where the variable Y is replaced with specific volume and X is replaced with temperature:

$$\frac{v - v_1}{v_2 - v_1} = \frac{T - T_1}{T_2 - T_1} \tag{2-23}$$

Equation (2-23) is solved for the unknown value of specific volume, v:

$$v = v_1 + \frac{(T - T_1)}{(T_2 - T_1)}(v_2 - v_1) \tag{2-24}$$

Substituting the values from the table into Eq. (2-24) results in:

$$v = 8.5855\,\frac{\text{m}^3}{\text{kg}} + \frac{(115°\text{C} - 100°\text{C})}{(150°\text{C} - 100°\text{C})}\left(9.7486\,\frac{\text{m}^3}{\text{kg}} - 8.5855\,\frac{\text{m}^3}{\text{kg}}\right) = 8.9344\,\frac{\text{m}^3}{\text{kg}} \tag{2-25}$$

The same process can be applied to estimate the values of specific internal energy, specific enthalpy and specific entropy at 115°C and 20 kPa. The interpolated values for these properties are: u = 2536.4 kJ/kg, h = 2716.1 kJ/kg, and s = 8.199 kJ/kg-K. More accurate estimates of these properties can be obtained using the thermodynamic property functions in EES, as discussed in Section 2.4: v = 8.9352 m^3/kg, u = 2536.3 kJ/kg, h = 2715.0 kJ/kg, and s = 8.2021 kJ/kg-K. The agreement between the interpolated and accepted values is very good in this case.

It is even more common that neither the temperature nor the pressure of interest is exactly represented in the tables. In this case, a double-interpolation process is required to provide estimated property values. The double-interpolation process requires three separate linear interpolations. To illustrate the process, suppose that we are interested in the specific volume at $T = 115°\text{C}$ and $P = 32$ kPa. Table B-3 (or Table 2-2) does not provide data for either 115°C or 32 kPa. The first step is to determine two pressures for which data are available that bracket the state of interest; in this case, these pressures are 20 kPa and 40 kPa. Next, the property values corresponding to 115°C at these two pressures are determined by linear interpolation at each pressure. This process has already been completed for 20 kPa. At 40 kPa, the interpolated value of specific volume at 115°C is obtained from:

$$\begin{aligned} v &= 4.2799\,\text{m}^3/\text{kg} + \frac{(115^\circ\text{C} - 100^\circ\text{C})}{(150^\circ\text{C} - 100^\circ\text{C})}(4.8662\,\text{m}^3/\text{kg} - 4.2799\,\text{m}^3/\text{kg}) \\ &= 4.4558\,\text{m}^3/\text{kg} \end{aligned} \tag{2-26}$$

The other interpolated properties at $T = 115^\circ\text{C}$ and $P = 40$ kPa are $u = 2534.6$ kJ/kg, $h = 2712.9$ kJ/kg, and $s = 7.8744$ kJ/kg-K. Finally, the interpolated values corresponding to 115°C at pressures of 20 kPa and 40 kPa are interpolated for a pressure of 32 kPa. Applying Eq. (2-22) for specific volume provides:

$$\frac{v - \overbrace{8.9344\,\text{m}^3/\text{kg}}^{v\text{ at }115^\circ\text{C and }20\,\text{kPa}}}{\underbrace{4.4558\,\text{m}^3/\text{kg}}_{v\text{ at }115^\circ\text{C and }40\,\text{kPa}} - \underbrace{8.9344\,\text{m}^3/\text{kg}}_{v\text{ at }115^\circ\text{C and }20\,\text{kPa}}} = \frac{32\,\text{kPa} - 20\,\text{kPa}}{40\,\text{kPa} - 20\,\text{kPa}} \tag{2-27}$$

or

$$\begin{aligned} v &= 8.9344\,\text{m}^3/\text{kg} + \frac{(32\,\text{kPa} - 20\,\text{kPa})}{(40\,\text{kPa} - 20\,\text{kPa})}(4.4558\,\text{m}^3/\text{kg} - 8.9344\,\text{m}^3/\text{kg}) \\ &= 6.2472\,\text{m}^3/\text{kg} \end{aligned} \tag{2-28}$$

Repeating the process for the other properties results in $u = 2535.3$ kJ/kg, $h = 2713.7$ kJ/kg, and $s = 8.0042$ kJ/kg-K. The accepted values at this state obtained using EES are: $v = 5.5763$ m^3/kg, $u = 2535.4$ kJ/kg, $h = 2713.8$ kJ/kg, and $s = 7.9826$ kJ/kg-K. Notice that the interpolated and accepted values of u, h, and s agree quite well. However, the interpolated and accepted values of specific volume differ by 12%. This discrepancy occurs because specific volume varies approximately according to the inverse of pressure and the table entries are too far apart to allow linear interpolation to provide an accurate answer. The specific volume estimated from interpolation could be improved if the interpolation were carried out using the inverse of pressure, as suggested by the ideal gas law. In this case, Eq. (2-22) is applied by replacing X with the inverse of pressure:

$$\frac{v - 8.9344\,\text{m}^3/\text{kg}}{4.4558\,\text{m}^3/\text{kg} - 8.9344\,\text{m}^3/\text{kg}} = \frac{1/32\,\text{kPa} - 1/20\,\text{kPa}}{1/40\,\text{kPa} - 1/20\,\text{kPa}} \tag{2-29}$$

Solving Eq. (2-29) results in $v = 5.5754$ m^3/kg, which is within 0.014% of the accepted value.

2.3.3 Compressed Liquid

The term *compressed liquid* is used to describe a state for which the specific volume is less than the specific volume of saturated liquid at the same temperature. The compressed liquid region lies to the left of the saturation dome on a T-v diagram, as shown in Figure 2-7. Compressed liquid property data for water are provided in Table B-4 in Appendix B. The compressed liquid table is arranged in a manner that is similar to the superheated vapor table, Table B-3 (or Table 2-2); data are provided over a matrix of temperatures and pressures.

Most liquids at a temperature much lower than their critical temperature are nearly incompressible; that is, the specific volume of the liquid at a specified temperature is essentially constant regardless of pressure. When plotted to scale on a temperature-specific volume plot, each isobar nearly collapses onto a single, nearly vertical line that coincides with the saturated liquid side of the vapor dome. This characteristic behavior is highlighted in Figure 2-11. Notice that even at relatively high temperatures, the various isobars tend to collapse onto the saturated liquid line of the vapor dome;

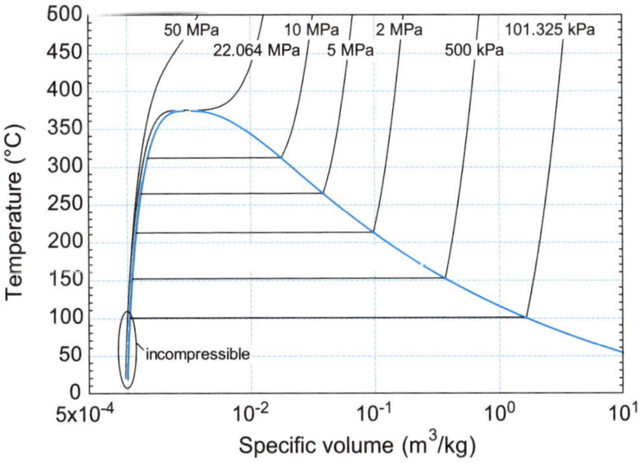

Figure 2-11: Temperature-specific volume plot for water showing the incompressible liquid behavior.

therefore, the specific volume of a compressed liquid depends slightly on temperature, but is almost completely independent of pressure. One useful result of this behavior is that the properties of a substance in the compressed liquid region can usually be estimated fairly accurately using the properties of saturated liquid evaluated at the same temperature.

2.4 EES Fluid Property Data

You do not have to solve many thermodynamics problems before the limitations associated with the use of the property tables become evident. Looking up property values in tables usually requires either single or double interpolation. The process is time-consuming and likely to introduce math errors. It is not easy or even practical to carry out the parametric studies that are required for optimization or design using property tables. Fortunately, computer programs are available that automate the process of determining property values. EES is the most convenient program available in this regard. This section will describe how EES can be used to provide the property information for many substances.

2.4.1 Thermodynamic Property Functions

EES provides a number of built-in thermodynamic and transport property functions. In this textbook we will focus on the thermodynamic property functions. The transport property functions that provide properties such as viscosity and thermal conductivity are useful for problems that you may encounter in a Fluid Dynamics or Heat Transfer class. The thermodynamic property functions that are of greatest interest are listed in Table 2-3, together with the thermodynamic property that is returned by the function and the possible units of the returned value.

It is essential that you pay close attention to the units of the variables that are used to access the built-in thermodynamic property functions in EES. The units that EES will expect in the call to the thermodynamic property functions can be specified with the Unit System menu command in the Options menu. The Unit System dialog, shown in Figure 2-12, allows either SI or English units to be selected (but it does not allow a combination of SI and English units to be used). Within each unit system, there are options for the temperature scale, the energy units and the pressure units. Specific property values can be defined on either a mass or molar basis. The molar

Table 2-3: Commonly used EES thermodynamic property functions and their units.

Function Name	Returns	SI Units	English Units
Enthalpy	specific enthalpy	J/kg, kJ/kg	Btu/lb_m
Entropy	specific entropy	J/kg-K, kJ/kg-K	Btu/lb_m-R
IntEnergy	specific internal energy	J/kg, kJ/kg	Btu/lb_m
Pressure	absolute pressure	Pa, kPa, bar, MPa	psia, atm
Quality	quality	none	none
Temperature	temperature	°C, K	°F, R
Volume	specific volume	m^3/kg	ft^3/lb_m
Density	density	kg/m^3	lb_m/ft^3
Phase$	phase, e.g, 'superheated'	none	none
MolarMass[1]	molecular weight	kg/kmol	lb_m/lbmol
T_sat[2]	saturation temperature	°C, K	°F, R
P_sat[3]	saturation pressure	Pa, kPa, bar, MPa	psia, atm
T_crit[1]	critical temperature	°C, K	°F, R
P_crit[1]	critical pressure	Pa, kPa, bar, MPa	psia, atm
v_crit[1]	critical specific volume	m^3/kg	ft^3/lb_m

1. The functions MolarMass, T_crit, P_crit, and v_crit require only the name of the fluid.
2. The function T_sat requires the name of the fluid and the pressure.
3. The function P_sat requires the name of the fluid and the temperature.

basis is particularly convenient for combustion and chemical equilibrium calculations, as discussed in Chapters 13 and 14.

The unit system that EES uses can alternatively be specified using the $UnitSystem directive, which is a command that is entered directly in the Equations window. For example, the following line:

```
$UnitSystem SI, K, Pa, J, Mass, Radian
```

entered at the top of the Equations window will specify that the SI unit system should be used with the Kelvin temperature scale and units of J and Pa for energy and pressure, respectively. It also specifies that specific property values must be entered and returned on a mass basis and angles must be provided to trigonometric functions in radians rather than in degrees. These specifications are consistent with the choices that would otherwise be made in the Unit System dialog shown in Figure 2-12. The entries in the $UnitSystem directive over-ride any entries made in the Unit System dialog. The advantage of using the $UnitSystem directive is that the units associated with the EES solution are visible and will not change if the EES code is copied and pasted to a different EES file that may have a different default set of unit specifications. Unit setting are saved with other information when the EES file is saved.

All property values that are returned by the EES built-in property functions will have the units that are specified with the Unit System dialog or the $UnitSystem directive. Further, any values provided to the built-in property functions will be interpreted as if

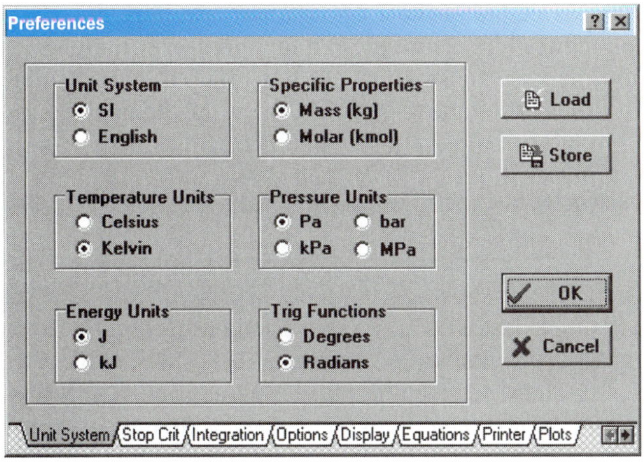

Figure 2-12: Unit System dialog.

they have these units. EES will not automatically convert the values of the properties that you supply to the built-in functions in order to achieve unit consistency (although EES will identify unit inconsistencies when the Check Units option is selected from the Calculate menu). This is an important point: EES will identify unit errors but you must fix them.

The thermodynamic property functions that are listed in Table 2-3 can be used to determine the properties of many different engineering fluids. A complete list of the fluids that are included in EES can be accessed by selecting Function Information from the Options menu and selecting the Fluid properties radio button, as shown in Figure 2-13. A list of built-in property functions is shown on the left. Choose a particular property function (e.g., the Volume function, which returns the specific volume) by clicking on it. Click the Function Info button above the list in order to obtain information about the selected property function. The list on the right is populated by the names of all of the substances that the selected property function can be applied to. Specific information about the source of the property information and its range of applicability are provided by clicking the Fluid Info button.

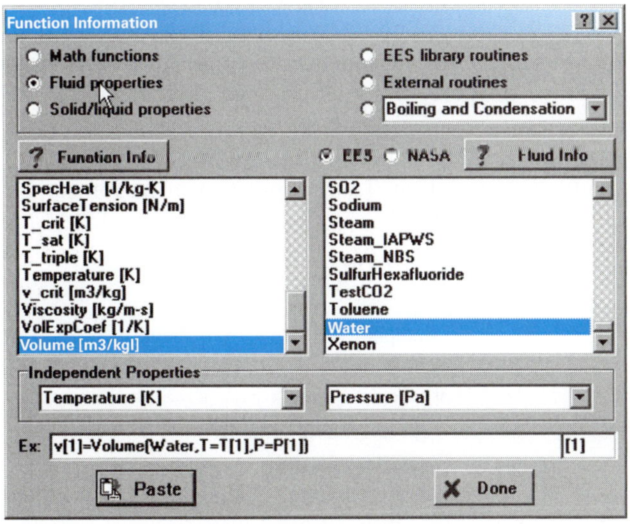

Figure 2-13: Function Information dialog showing fluid property information.

Almost all of the fluids that are available in the EES database are pure fluids (or mixtures of fixed composition that behave like a pure fluid) and therefore two intensive internal properties must be specified in order to fix the state, allowing the determination of any of other property. With this in mind, the general form of a thermodynamic property function call in EES is:

Value = Function Name (Fluid Name, Property 1 = Value 1, Property 2 = Value 2) (2-30)

The first argument required by all built-in property functions is the fluid name. The fluid names that are recognized by EES are shown in the list that appears on the right side of the Function Information dialog (Figure 2-13). Examination of this list will show what appears to be a number of duplicate fluids. For example, the list contains both 'N2' and 'Nitrogen', 'O2' and 'Oxygen', etc. These apparent duplicates actually correspond to very different models of the behavior of the same fluid. Whenever a chemical symbol notation is used to specify the fluid (e.g., 'N2' and 'O2'), the fluid is modeled as an ideal gas, as described in Section 2.5. Whenever the substance name is spelled out (e.g., 'Nitrogen' and 'Oxygen'), the fluid is modeled as a real fluid that has compressed liquid, saturated, superheated vapor and supercritical behavior depending on its state. Clearly then the fluids 'N2' and 'Nitrogen' correspond to very different models of the behavior of nitrogen. One exception to this rule is the fluid 'Air' which corresponds to dry air modeled as an ideal gas. The fluid 'Air_ha' corresponds to the real fluid mixture of the gases that together form air. The fluid 'AirH2O' is the notation for an air-water vapor mixture, which is discussed in Chapter 12.

There are some fluids that are truly duplicates. For example, the fluids 'Steam', 'Water', 'R718' and 'Steam_NBS' all use the property correlations for pure water published by Harr, Gallagher, and Kell (1984). The fluid 'Steam_IAPWS' implements a newer and somewhat more accurate, but also more computationally intensive, formulation for the properties of water that is presented by Wagner and Pruss (1993) and Wagner and Pruss (2002). The fluid 'H2O' corresponds to water vapor modeled as an ideal gas.

The next two arguments to the thermodynamic property function call in Eq. (2-30) provide the two intensive internal properties that must be specified in order to fix the state of the fluid. These property values can be supplied in any order and there is considerable flexibility relative to the properties that can be used for this purpose. Each property specification consists of a single letter (case-insensitive) that designates the property (corresponding to the parameters Property 1 and Property 2) followed by an equal sign and the value of that property (corresponding to Value 1 and Value 2). The value of the property can be specified using a numeric constant, variable, or an algebraic expression. The recognized single letter property designations and their meaning are summarized in Table 2-4.

The properties of water at 115°C and 32 kPa were determined in Section 2.3 by using data in Table B-3 to carry out a double interpolation. This property information can be obtained more conveniently using the following EES code:

```
$UnitSystem SI, K, Pa, J, Mass, Radian
T=converttemp(C,K,115 [C])              "temperature"
P=32 [kPa]*convert(kPa,Pa)              "pressure"
v=volume(Water,T=T, P=P)                "specific volume"
u=intenergy(Water,T=T, P=P)             "specific internal energy"
h=enthalpy(Water,T=T,P=P)               "specific enthalpy"
s=entropy(Water,T=T,P=P)                "specific entropy"
```

Table 2-4: Property designations used in the EES thermodynamic property functions.

Property Designator	Property
H or h	specific enthalpy
P or p	pressure
S or s	specific entropy
T or t	temperature
U or u	specific internal energy
V or v	specific volume
X or x	quality

which produces $v = 5.576$ m^3/kg, $u = 2.535\times10^6$ J/kg, $h = 2.713\times10^6$ J/kg, and $s = 7981$ J/kg-K; these values agree with the results of the interpolation. Notice that the functions volume, intenergy, enthalpy, and entropy all required that temperature be provided in units of K and pressure in units of Pa based on the settings in the $UnitSystem directive. The units of the variables v, u, h, and s returned by the thermodynamic functions are m^3/kg, J/kg, J/kg, and J/kg-K, respectively, consistent with the unit system settings. EES knows what the units of these variables should be, based on the unit system that has been set, and will check to ensure that the units entered for each variable are consistent.

EXAMPLE 2.4-1: THERMOSTATIC EXPANSION VALVE

A thermostatic expansion valve is a passive device that is used to control the flow of refrigerant to an evaporator (a heat exchanger that allows the refrigerant to accept a heat transfer from a refrigerated space), as shown in Figure 1.

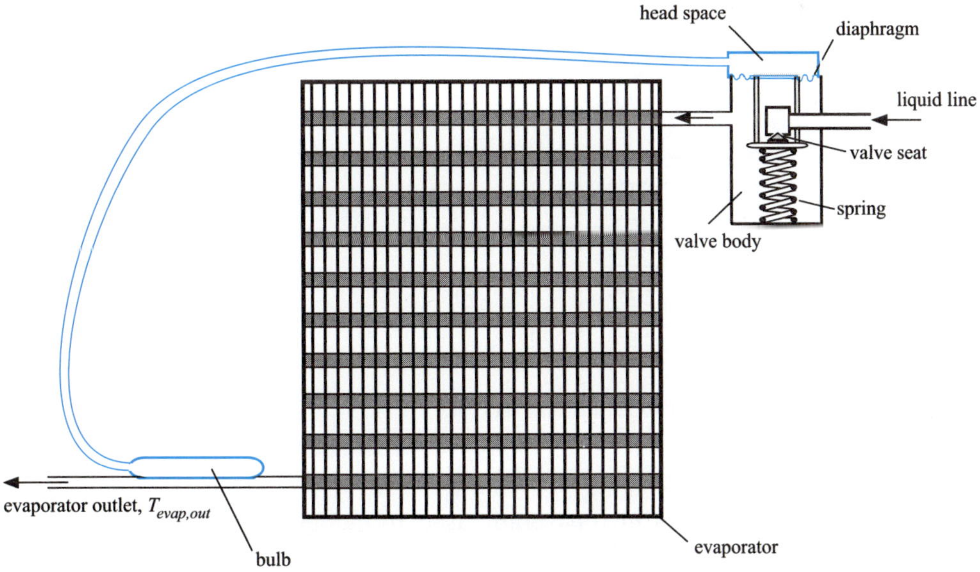

Figure 1: Thermostatic expansion valve controlling the flow of a refrigerant to an evaporator.

Saturated (or nearly saturated) refrigerant enters the thermostatic expansion valve through the liquid line. The flow rate of refrigerant is controlled by the position of the valve seat. The refrigerant subsequently flows through the evaporator at nearly constant pressure. The heat transfer causes the refrigerant to completely evaporate and so it leaves the evaporator as superheated vapor at temperature $T_{evap,out}$. The job of the thermostatic expansion valve is to regulate the refrigerant flow so that it leaves at a precisely controlled temperature even if the load on the evaporator (i.e., the heat transfer to the refrigerant) changes. If the load on the evaporator decreases then $T_{evap,out}$ will tend to decrease and the thermostatic valve should act to reduce the refrigerant flow rate. Alternatively, if the load increases then $T_{evap,out}$ will tend to increase and this should result in the thermostatic valve increasing the flow rate.

A bulb filled with refrigerant is attached to the evaporator outlet. The bulb is connected, through a tube, to the head space of the valve. The pressure in the bulb and the head space are the same; therefore, an increase in $T_{evap,out}$ tends to increase the pressure in the head space of the valve which causes the diaphragm in the thermostatic expansion valve to move downward and open the valve. If $T_{evap,out}$ decreases then the opposite happens; the head space pressure decreases which, along with the force exerted by the spring, causes the diaphragm to move up thereby closing the valve.

Figure 2 provides a simplified schematic of the thermostatic expansion valve and bulb. The bulb volume is V_{bulb} = 10 cm^3 and the head space volume is V_{head} = 2 cm^3, the volume of the connecting tube can be neglected. The temperature of the refrigerant contained in the head space is equal to the evaporator inlet temperature, T_{evap} = –5°C. Under nominal operating conditions, the temperature of the refrigerant leaving the evaporator (and therefore the temperature of the refrigerant contained in the bulb) is $T_{evap,out,nom}$ = –2°C and the bulb and head space pressure are both $P_{bulb,nom}$ = 700 kPa. The refrigerant is R22. (The naming code for refrigerants is discussed in Chapter 9.)

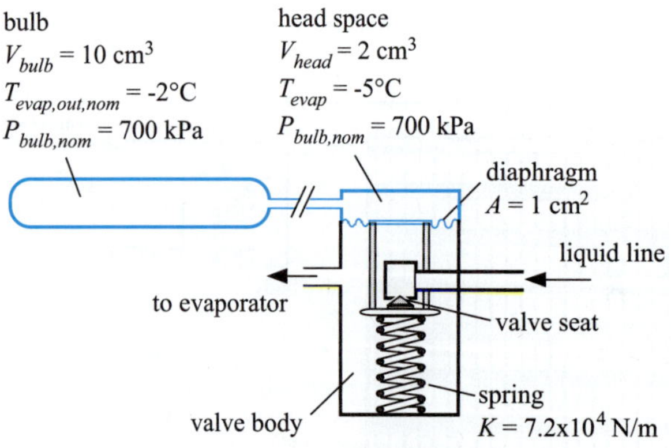

Figure 2: Schematic of thermostatic expansion valve and bulb.

a) What is the total mass of refrigerant used to charge the bulb and the head space?

The inputs are entered in EES and the unit system is specified using the $UnitSystem directive.

```
$UnitSystem SI, K, Pa, J, Mass, Radian
"Inputs"
Vol_bulb=10 [cm^3]*convert(cm^3,m^3)                "Volume of bulb"
Vol_head=2 [cm^3]*convert(cm^3,m^3)                 "Volume of head space"
T_evap=converttemp(C,K,-5 [C])                      "Temperature of refrigerant in head space"
T_evap_out_nom=converttemp(C,K,-2 [C])              "Nominal evaporator outlet temperature"
P_bulb_nom=700 [kPa]*convert(kPa,Pa)                "Nominal bulb pressure"
```

We will start by identifying the state of the refrigerant in the bulb and the state of the refrigerant in the head space. The pressure and temperature are known for both of these spaces. The saturation temperature of the refrigerant at the nominal bulb pressure (T_{sat}) is obtained using the T_sat function; note that the T_sat function requires the name of the fluid and the pressure of interest.

```
T_sat=T_sat(R22,P=P_bulb_nom)      "saturation temperature at bulb pressure"
T_sat_C=converttemp(K,C,T_sat)     "in C"
```

Solving provides $T_{sat} = 10.91^\circ$C. Because both T_{evap} and $T_{evap,out,nom}$ are both less than T_{sat}, we know that the state of the refrigerant in the bulb and the state of the refrigerant in the head space both lie in the compressed liquid region. The specific volume of the refrigerant in the bulb under nominal operating conditions ($v_{bulb,nom}$) is obtained using the thermodynamic function Volume by specifying the state with the pressure ($P_{bulb,nom}$) and temperature ($T_{evap,out,nom}$). The mass of refrigerant in the bulb is:

$$m_{bulb,nom} = \frac{V_{bulb}}{v_{bulb,nom}}$$

```
v_bulb_nom=volume(R22,T=T_evap_out_nom,P=P_bulb_nom)    "specific volume of refrigerant in bulb"
m_bulb_nom=Vol_bulb/v_bulb_nom                          "mass of refrigerant in bulb"
```

The specific volume of the refrigerant in the head space under nominal operating conditions ($v_{head,nom}$) is obtained using the Volume function by specifying the state with the pressure ($P_{bulb,nom}$) and temperature (T_{evap}). The mass of refrigerant in the head space is:

$$m_{head,nom} = \frac{V_{head}}{v_{head,nom}}$$

```
v_head_nom=volume(R22,T=T_evap,P=P_bulb_nom)    "specific volume of refrigerant in head space"
m_head_nom=Vol_head/v_head_nom                  "mass of refrigerant in head space"
```

The total mass of refrigerant used to charge the bulb and the head space is:

$$m = m_{bulb,nom} + m_{head,nom}$$

```
m=m_bulb_nom+m_head_nom    "total mass of refrigerant"
```

which leads to $m = 0.01549$ kg (15.49 g).

b) The evaporator outlet temperature rises from $T_{evap,out,nom} = -2°C$ to $T_{evap,out} = -1°C$. Determine the increase in bulb pressure that results from this change.

The additional input is entered in EES:

```
T_evap_out_C=-1 [C]                          "evaporator outlet temperature, in C"
T_evap_out=converttemp(C,K,T_evap_out_C)     "evaporator outlet temperature"
```

The specific volume of the refrigerant in the bulb (v_{bulb}) is specified by the new evaporator outlet temperature and bulb pressure (P_{bulb}, an unknown). The mass of refrigerant in the bulb is:

$$m_{bulb} = \frac{V_{bulb}}{v_{bulb}}$$

```
v_bulb=volume(R22,T=T_evap_out,P=P_bulb)    "specific volume of refrigerant in bulb"
m_bulb=Vol_bulb/v_bulb                      "mass of refrigerant in bulb"
```

The specific volume of the refrigerant in the head space (v_{head}) is specified by the evaporator temperature and bulb pressure. The mass of refrigerant in the head space is:

$$m_{head} = \frac{V_{head}}{v_{head}}$$

```
v_head=volume(R22,T=T_evap,P=P_bulb)    "specific volume of refrigerant in head space"
m_head=Vol_head/v_head                  "mass of refrigerant in head space"
```

The total mass of refrigerant in the head space and bulb cannot change relative to the mass determined in (a).

$$m = m_{bulb} + m_{head}$$

```
m=m_bulb+m_head    "mass balance"
```

These last five equations entered in EES are a set of 5 equations in 5 unknowns (P_{bulb}, m_{bulb}, m_{head}, v_{bulb}, and v_{head}), which EES can solve simultaneously. The result is $P_{bulb} = 1.447 \times 10^6$ Pa, a large increase from the nominal value $P_{bulb,nom} = 700$ kPa.

c) The spring constant is $K = 7.2 \times 10^4$ N/m and the area of the diaphragm that the bulb pressure acts on is $A = 1$ cm^2. Determine the change in the valve opening that is caused by the increase in evaporator outlet temperature (from its nominal value of $T_{evap,out,nom} = -2°C$ to its new value $T_{evap,out} = -1°C$).

The additional inputs are entered in EES:

```
A=1 [cm^2]*convert(cm^2,m^2)    "area of diaphragm"
K=7.2e4 [N/m]                   "spring constant"
```

The increase in pressure force pushing down on the diaphragm must be balanced by an increase in the spring force pushing up (the pressure in the valve body is unchanged since the evaporator temperature is constant). A force balance provides:

$$A\left(P_{bulb} - P_{bulb,nom}\right) = K\,\Delta x$$

where Δx is the change in the compression of the spring (which corresponds to the change in the position of the valve seat).

```
(P_bulb-P_bulb_nom)*A=K*DELTAx          "force balance"
DELTAx_mm=DELTAx*convert(m,mm)          "change in valve opening, in mm"
```

Solving leads to $\Delta x = 1.037$ mm.

d) Prepare a plot of the change in the valve position as a function of the evaporator outlet temperature.

Construct a Parametric table with columns for variables DELTAx and T_evap_out_C. Fill the T_evap_out_C column with values between -2°C and 1°C. Comment out the T_evap_out_C = -1 [C] equation. Solve the table and plot the results. Figure 3 illustrates Δx as a function of $T_{evap,out}$. Notice that an increase in the evaporator outlet temperature from its nominal value (-2°C) causes an increase in the valve opening that will result in an increase in the flow rate.

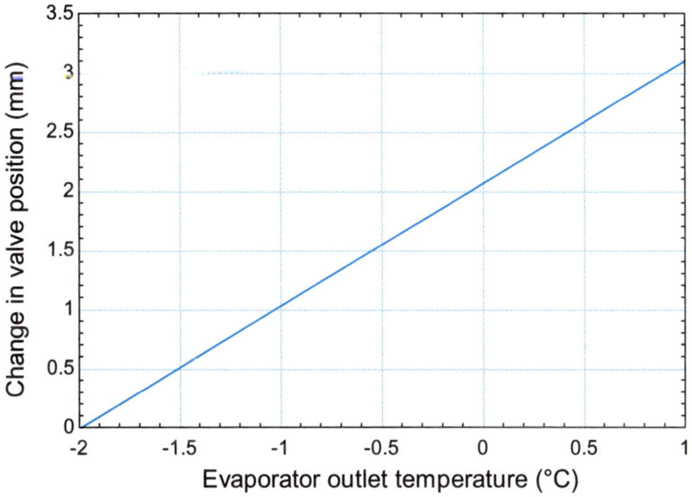

Figure 3: Change in valve position as a function of the evaporator outlet temperature.

2.4.2 Arrays and Property Plots

So far, we have only dealt with scalar variables in EES; that is, each variable had a single value. Arrays are groups of variables with the same name and the same units. It is often useful to use arrays in order to keep track of the various states that are involved in a thermodynamics problem. For example, the pressures at each state might be stored in an array P; the pressure for state 1 is the first element of P, the pressure for state 2 is the second element of P, etc.

Suppose we have a problem involving propane at three different states. All of the states are at the same pressure, $P_1 = P_2 = P_3 = 100$ kPa. The statements below assign these pressures to the array P. Notice that the number contained in square brackets after the variable name (e.g., P[1]) indicates the element of the array that is being assigned.

```
$UnitSystem SI, K, Pa, J, Mass, Radian
P[1]=100 [kPa]*convert(kPa,Pa)
P[2]=100 [kPa]*convert(kPa,Pa)
P[3]=100 [kPa]*convert(kPa,Pa)
```

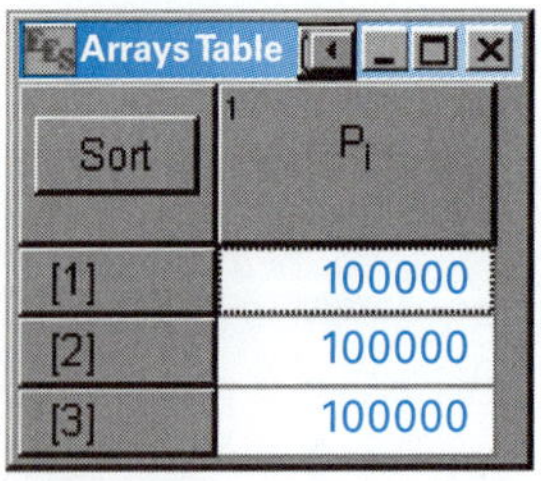

Figure 2-14: Arrays window.

If you solve the problem (select Solve from the Calculate menu or press F2) you will find that the Solution window is empty. By default, the arrays are displayed in a separate window called the Arrays window that can be accessed by selecting Arrays from the Windows menu (Figure 2-14).

Each element of array P is shown in the Arrays window. The units for all of the elements of P can be set at once by right-clicking on the column heading and entering the units in the dialog that appears (Figure 2-15).

The temperatures of the propane at the three states are $T_1 = 300$ K, $T_2 = 350$ K, and $T_3 = 375$ K. These values are entered into the array T:

```
T[1]=300 [K]
T[2]=350 [K]
T[3]=375 [K]
```

The elements of an array can be manipulated like any other variable in EES. For example, we can use the Volume function to obtain the specific volume of each state:

```
v[1]=Volume(Propane,T=T[1],P=P[1])
v[2]=Volume(Propane,T=T[2],P=P[2])
v[3]=Volume(Propane,T=T[3],P=P[3])
```

The phase associated with each state can be found using the Phase$ function:

```
Phase$[1]=Phase$(Propane,T=T[1],P=P[1])
Phase$[2]=Phase$(Propane,T=T[2],P=P[2])
Phase$[3]=Phase$(Propane,T=T[3],P=P[3])
```

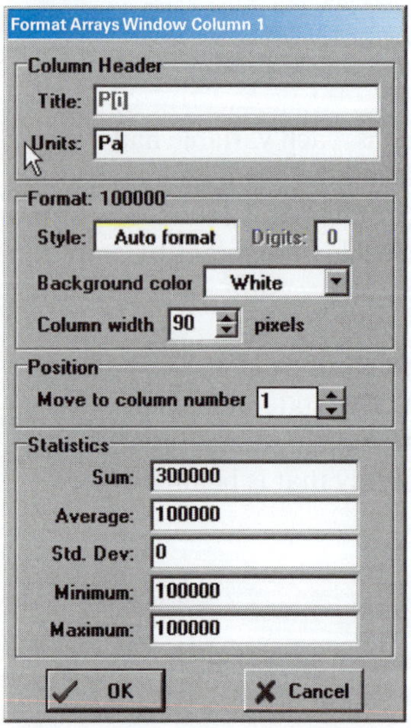

Figure 2-15: Enter units for array.

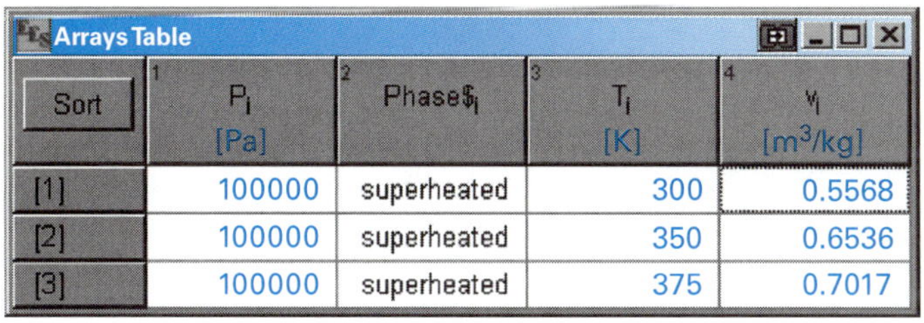

Arrays Table	1 P_i [Pa]	2 $Phase\$_i$	3 T_i [K]	4 v_i [m³/kg]
[1]	100000	superheated	300	0.5568
[2]	100000	superheated	350	0.6536
[3]	100000	superheated	375	0.7017

Figure 2-16: Arrays window.

The Phase$ function returns a string that describes the state (e.g., 'liquid' or 'superheated') and therefore the array used to store the output of the Phase$ function must contain string variables. Variable names that end (before the array index) with the $ character indicate that the variable will store a character string rather than a number.

After setting the units for the arrays T and v, the Arrays window should appear as shown in Figure 2-16. Figure 2-16 illustrates one of the advantages of using arrays to solve a thermodynamics problem: the information about each state is organized in tabular form. It is possible to quickly examine the properties associated with each state by examining each row of the Arrays table rather than having to gather this information from a list of variables in the Solution window.

In Section 2.2, we discussed the importance of property plots. Thus far we have dealt exclusively with T-v plots; however, we will encounter other types of property plots that are also useful. The process of generating property plots can be automated in EES. For example, we can use EES to generate a T-v plot for propane and locate states 1, 2 and 3 on this plot. Select Property Plot from the Plot menu and select Propane from the fluid list. Click the T-v radio button in order to create a temperature-specific volume plot. Up to six lines of constant pressure and six lines of constant specific entropy can optionally be placed on the plot. Since we know that a pressure of interest is 100 kPa, enter that isobar (Figure 2-17). Also, enter isobar pressures of 2.8 MPa, 1.1 MPa, and 350 kPa. Click on [×] in the list on the right to unselect all of the constant specific entropy lines.

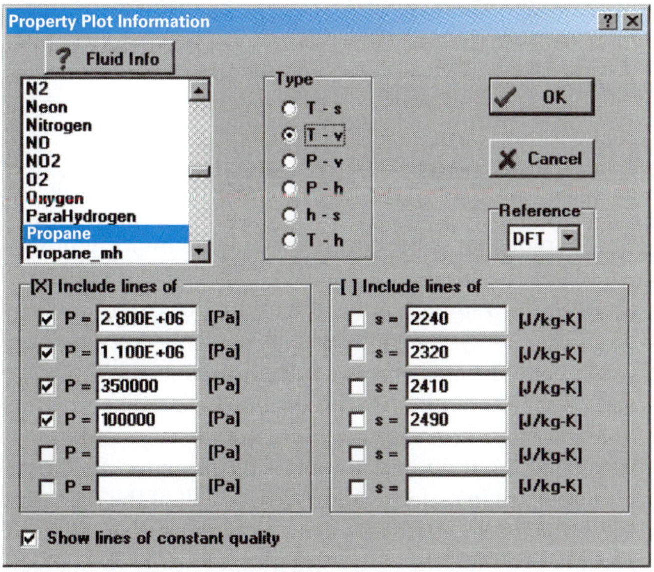

Figure 2-17: Property plot information dialog.

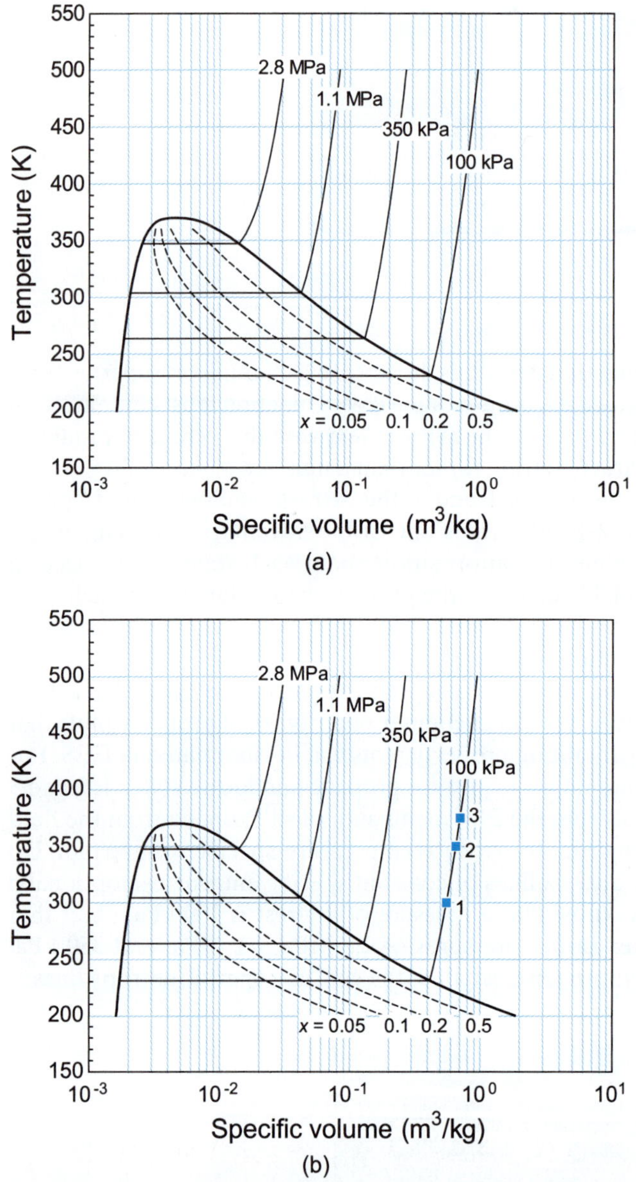

Figure 2-18: (a) *T-v* plot for propane and (b) with states overlaid.

Figure 2-18(a) illustrates the *T-v* plot that is produced (with some minor editing). Because the properties at each state are stored in arrays rather than as scalar variables, it is possible to overlay the states onto the *T-v* plot. Select Overlay Plot from the Plots menu. Data should be plotted from the Arrays table. Select v[i] as the X-Axis variable and T[i] as the Y-axis variable. Check the Show array indices selection, as shown in Figure 2-19, in order to identify each state with its number on the property plot. The result is shown in Figure 2-18(b).

It is possible to setup the plot so that the states are automatically updated each time the program is run. Double-click on the plot in order to access a list of all of the data that are being presented. The last item in the list should be the data from the Arrays table. If the Automatic update selection is checked then the states will automatically re-position

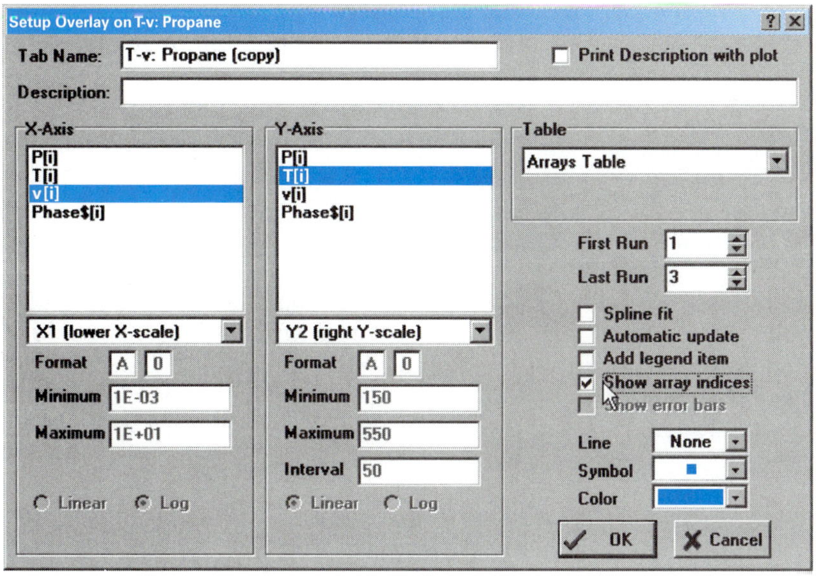

Figure 2-19: Overlay Plot dialog.

themselves in response to changes in the inputs. For example, select Automatic update and then change the assignment of the pressure for state (2) according to:

```
P[2]=350 [kPa]*convert(kPa,Pa)
```

Solve the equations and return to your property plot. You should find that the position of state 2 has jumped from the 100 kPa isobar to the 350 kPa isobar in response to the updated calculations. This automatic update capability is often useful when you are trying to understand how changes in the inputs affect the result of a thermodynamic analysis.

EXAMPLE 2.4-2: LIQUID OXYGEN TANK

The engine of a small, unmanned launch vehicle requires fuel (propane) and an oxidizer (oxygen). You have been asked to assess a method of storing oxygen and providing it to the engine. The oxygen will be stored in a Dewar (an insulated tank) in liquid form as shown in Figure 1; this approach requires that the oxygen be kept at very low temperature.

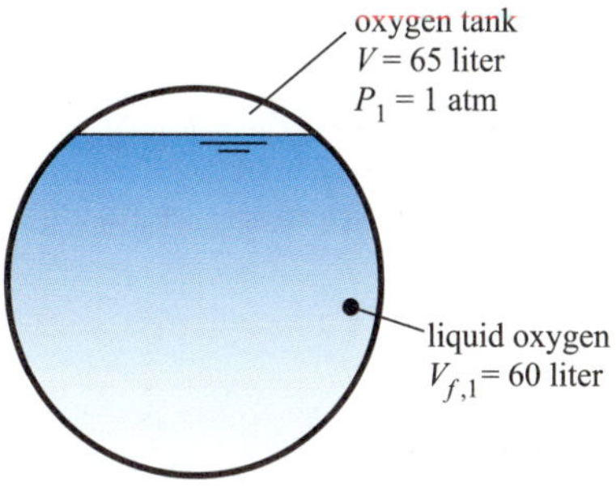

Figure 1: Liquid oxygen tank.

The internal volume of the tank is $v = 65$ liter and the pressure in the tank is initially $P_1 = 1$ atm. The oxygen in the tank is initially in a two-phase state. The volume of liquid oxygen in the tank is $V_{f,1} = 60$ liter.

a) Determine the temperature, quality and specific volume of the oxygen that is initially in the tank. What is the total mass of oxygen stored in the tank?

The inputs are entered in EES:

```
$UnitSystem SI, K, Pa, J, Mass, Radian
Vol=65 [liter]*convert(liter,m^3)          "tank volume"
P[1]=1 [atm]*convert(atm,Pa)               "initial tank pressure"
Vol_f[1]=60 [liter]*convert(liter,m^3)     "volume of liquid in tank"
```

The specific volume of the liquid ($v_{f,1}$) and vapor ($v_{g,1}$) are obtained using EES' built-in thermodynamic property functions. The state of the saturated liquid is specified by its pressure ($P = P_1$) and quality ($x = 0$). Likewise, the state of the saturated vapor is also specified by its pressure and quality ($x = 1$). The temperature of the liquid and vapor are the same and equal to the saturation temperature at the given pressure (T_{sat}).

```
v_f[1]=volume(Oxygen,x=0,P=P[1])    "specific volume of saturated liquid"
v_g[1]=volume(Oxygen,x=1,P=P[1])    "specific volume of saturated vapor"
T[1]=T_sat(Oxygen,P=P[1])           "temperature of oxygen in tank"
```

The volume of the vapor is:

$$V_{g,1} = V - V_{f,1}$$

The mass of liquid and vapor are:

$$m_{f,1} = \frac{V_{f,1}}{v_{f,1}}$$

$$m_{g,1} = \frac{V_{g,1}}{v_{g,1}}$$

```
Vol_g[1]=Vol-Vol_f[1]      "volume of vapor in tank"
m_f[1]=Vol_f[1]/v_f[1]     "mass of liquid in the tank"
m_g[1]=Vol_g[1]/v_g[1]     "mass of vapor in the tank"
```

The total mass of oxygen in the tank is:

$$m_1 = m_{g,1} + m_{f,1}$$

The quality is:

$$x_1 = \frac{m_{g,1}}{m_1}$$

The average specific volume of all of the oxygen initially in the tank is:

$$v_1 = \frac{V}{m_1}$$

```
m[1]=m_g[1]+m_f[1]    "total mass of oxygen in tank"
x[1]=m_g[1]/m[1]      "quality"
v[1]=Vol/M[1]         "specific volume"
```

Solving provides $T_1 = 90.2$ K, $m_1 = 68.5$ kg, $x_1 = 3.26 \times 10^{-4}$, and $v_1 = 0.00948$ m^3/kg.

b) The engine requires liquid oxygen at a pressure of $P_{engine} = 25$ atm for $t_{burn} = 120$ s. The obvious method for providing oxygen to the engine is to pull liquid from the bottom of the tank using a pump that discharges to the engine at P_{engine}, as shown in Figure 2.

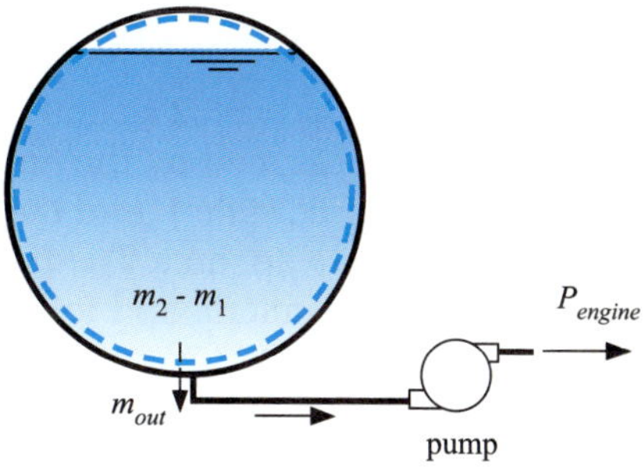

Figure 2: Liquid oxygen provided to the engine with a pump.

The emptying process ends when the liquid in the tank is completely removed. Assume that the tank pressure remains at $P = 1$ atm during this process. Determine the mass of oxygen provided to the engine. What fraction of the mass of oxygen that is initially contained in the tank is provided to the engine?

When the tank is empty of liquid, the quality has reached unity; $x_2 = 1$. The pressure is also given, $P_2 = 1$ atm. The state of the oxygen that remains in the tank at state 2 is specified by the quality and pressure. Therefore, the specific volume (v_2) and temperature (T_2) are obtained using EES' internal property routines.

```
P_engine_atm=25 [atm]                          "engine pressure, in atm"
P_engine=P_engine_atm*convert(atm,Pa)          "engine pressure"
t_burn = 120 [s]                               "burn time"
x[2]=1 [-]                                     "quality when tank is empty of liquid"
P[2]=P[1]                                      "pressure when tank is empty of liquid"
v[2]=volume(Oxygen,x=x[2],P=P[2])              "specific volume when tank is empty of liquid"
T[2]=temperature(Oxygen,x=x[2],P=P[2])         "temperature when tank is empty of liquid"
```

The mass of oxygen that remains in the tank is:

$$m_2 = \frac{V}{v_2}$$

Mass is a conserved quantity (it is neither produced nor destroyed). Therefore, for any system it is possible to write a mass balance:

$$In = Out + Stored$$

A mass balance on a system that is defined as the internal volume of the tank is shown in Figure 2 and provides:

$$0 = m_{out} + (m_2 - m_1)$$

The fraction of oxygen that is initially contained in the tank (m_1) that is provided to the engine (m_{out}) is:

$$f_{pump} = \frac{m_{out}}{m_1}$$

```
m[2]=Vol/v[2]              "mass of oxygen in the tank when it is empty of liquid"
0=m_out+m[2]-m[1]          "mass balance"
f_pump=m_out/m[1]          "fraction of oxygen in tank that is provided to the engine"
```

which leads to $m_{out} = 68.2$ kg and $f_{pump} = 0.9958$ (i.e., 99.58% of the oxygen initially in the tank is consumed by the engine). Figure 3 illustrates the state of the oxygen initially in the tank, state 1, and the state of the oxygen in the tank when the liquid is gone, labeled state 2p, on a T-v diagram for oxygen generated by selecting Property Plot from the Plots menu and following the procedure discussed in Section 2.4.2.

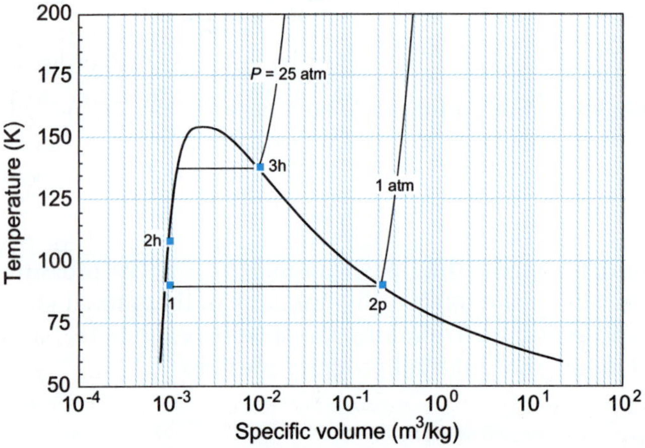

Figure 3: A T-v diagram for oxygen showing the states associated with the problem.

c) What is the volumetric flow rate of oxygen at the suction port of the pump? Assume that the mass flow rate of liquid oxygen is constant during the emptying process.

The mass flow rate is given by:

$$\dot{m} = \frac{m_{out}}{t_{burn}}$$

The oxygen entering the pump is always saturated liquid at pressure P_1 and therefore it has specific volume $v_{f,1}$. The volumetric flow rate at the pump inlet is:

$$\dot{V} = \dot{m}\, v_{f,1}$$

```
m_dot=m_out/t_burn                      "mass flow rate at pump suction port"
V_dot=m_dot*v_f[1]                      "volumetric flow rate at pump suction port"
V_dot_cfm=V_dot*convert(m^3/s,ft^3/min) "in cfm"
```

which results in $\dot{V} = 1.055$ cfm.

d) The pump shown in Figure 2 has some serious drawbacks. The temperature of the oxygen passing through the pump is very low (around 90.2 K according to Figure 3) and therefore the pump needs to be a specialized piece of cryogenic equipment that is expensive and potentially unreliable. An alternative method of providing oxygen to the engine is shown in Figure 4.

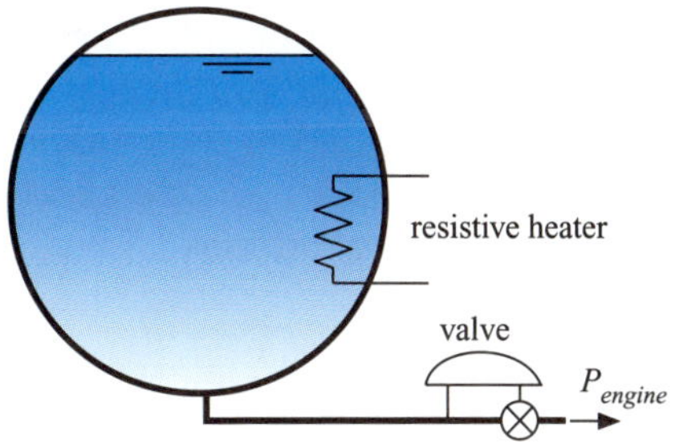

Figure 4: Liquid oxygen tank emptied with a heater.

A resistive heater is placed in the tank and activated until the pressure in the tank reaches $P_2 = P_{engine}$. At that time, a valve placed in the bottom of tank opens and allows liquid oxygen to exit at a rate that maintains the pressure in the tank at P_{engine}. This continues until the tank is emptied of liquid at state 3. Determine the mass of oxygen provided to the engine using the system shown in Figure 3. What fraction of the mass of oxygen initially in the tank is utilized by this system?

The lines of EES code corresponding to the analysis of the pumped system shown in Figure 2, are commented out (i.e., highlight these lines of code, right-click, and select Comment).

```
{x[2]=1 [-]                               "quality when tank is empty of liquid"
P[2]=P[1]                                 "pressure when tank is empty of liquid"
v[2]=volume(Oxygen,x=x[2],P=P[2])         "specific volume when tank is empty of liquid"
T[2]=temperature(Oxygen,x=x[2],P=P[2])    "temperature when tank is empty of liquid"
m[2]=Vol/v[2]                             "mass of oxygen in the tank when it is empty of liquid"
0=m_out+m[2]-m[1]                         "mass balance"
f_pump=m_out/m[1]                         "fraction of oxygen in tank that is provided to the engine"
m_dot=m_out/t_burn                        "mass flow rate at pump suction port"
V_dot=m_dot*v_f[1]                        "volumetric flow rate at pump suction port"
V_dot_cfm=V_dot*convert(m^3/s,ft^3/min)   "in cfm"}
```

State 2 is defined to be the state of the oxygen at the time that the valve opens. The pressure at state 2 is specified to be equal to the engine pressure. A mass balance on the tank for the period of time when it is heated from its initial condition, state 1, to the time when the valve opens, state 2, provides:

$$0 = m_2 - m_1$$

Therefore, the mass in the tank does not change. Since neither the mass nor the volume of oxygen in the tank changes, the specific volume at state 2 must be equal to its value at state 1. State 2 is specified by P_2 and v_2. The temperature of the oxygen at the time that the valve opens (T_2) is computed using EES' internal property function Temperature.

```
P[2]=P_engine                            "pressure when the valve opens"
0=m[2]-m[1]                              "mass balance on tank"
v[2]=v[1]                                "specific volume of oxygen does not change"
T[2]=temperature(Oxygen,v=v[2],P=P[2])   "temperature of oxygen when valve opens"
```

State 3 is defined as the state of the oxygen that remains in the tank at the time that the tank is empty of liquid. The pressure in the tank does not change once the valve opens; therefore, the pressure at state 3 is the same as the pressure at state 2. If the tank is just emptied of liquid then the quality is $x_3 = 1$. State 3 is specified by P_3 and x_3. The specific volume and temperature (v_3 and T_3) are obtained using EES' internal property routines.

```
P[3]=P[2]                                "pressure remains the same"
x[3]=1 [-]                               "tank contains only saturated vapor"
v[3]=volume(Oxygen,x=x[3],P=P[3])        "specific volume when tank is empty"
T[3]=temperature(Oxygen,x=x[3],P=P[3])   "temperature when tank is empty"
```

The mass of oxygen in the tank at state 3 is:

$$m_3 = \frac{V}{v_3}$$

A mass balance on the tank for the period of time between the opening of the valve and the emptying of the tank provides:

$$0 = m_{out} + m_3 - m_2$$

The fraction of the oxygen that is initially contained in the tank that is provided to the engine is:

$$f_{heater} = \frac{m_{out}}{m_1}$$

```
m[3]=Vol/v[3]              "mass when tank is empty"
0=m_out+m[3]-m[2]          "mass balance"
f_heater=m_out/m[1]        "fraction of oxygen initially in tank that is provided to the engine"
```

which results in $f_{heater} = 0.9015$ (90.15% of the oxygen is provided to the engine). The states of the oxygen in the tank at the time when the valve opens and when the tank is emptied are shown in Figure 3 as states 2h and 3h, respectively. The fraction of the oxygen that is supplied to the engine using this technique is reduced because the specific volume of the vapor that remains in the tank at state 3h is much lower than the specific volume that remains in the tank at state 2p.

e) Plot the fraction of the mass of oxygen that is initially contained in the tank that is provided to the engine using the heating method shown in Figure 4 as a function of the required engine pressure.

Create a Parametric table with columns for variables P_engine_atm and f_heater. Fill the P_engine_atm column with values between 1 atm and 50 atm. Comment out the P_engine_atm = 25 [atm] equation. Solve the table and plot the result. Figure 5 illustrates the fraction of the mass of oxygen that is initially contained in the tank that is provided to the engine (f_{heater}) as a function of the engine pressure (P_{engine}). Note that the method becomes less attractive as the engine pressure increases because the specific volume of the vapor decreases (i.e., its density increases), which results in a larger fraction of the oxygen remaining in the tank.

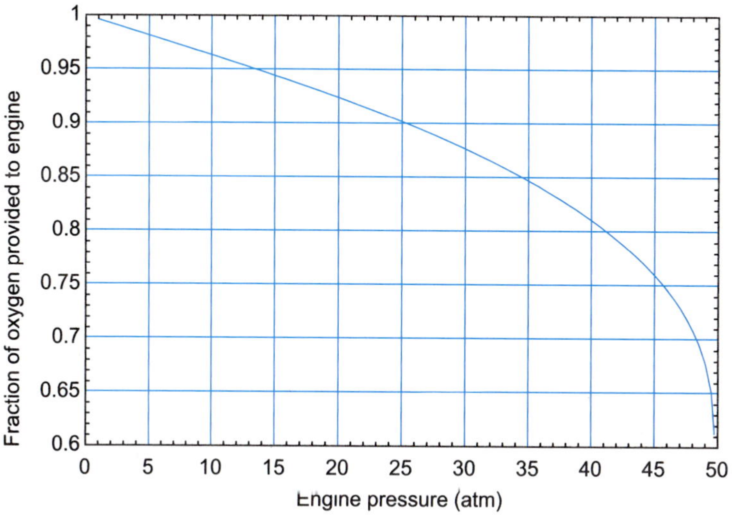

Figure 5: Fraction of oxygen provided to the engine as a function of engine pressure for the system where the liquid oxygen tank is emptied with a heater.

2.5 The Ideal Gas Model

A macroscopic description of a system does not require consideration of the elementary fluid particles (molecules) in the system. However, it is useful to consider these particles in order to understand the behavior of the fluid properties. The force between two molecules as a function of their center-to-center distance is shown qualitatively in Figure 2-20. When two molecules are very close to one another, they exhibit a strong repulsive force that results from overlapping electron orbitals. The magnitude of the repulsive force between the molecules diminishes as the distance between their centers increases and the intermolecular force becomes zero at a characteristic distance that is referred to as the *Lennard-Jones length potential*, σ. At larger values of the center-to-center distance there is an attractive force between the molecules. This attractive force increases with center-to-center distance until it reaches a maximum value that is referred to as the *Lennard-Jones energy potential*, ε. Further increase in the distance between the molecules causes the attractive force to decrease asymptotically towards zero. When

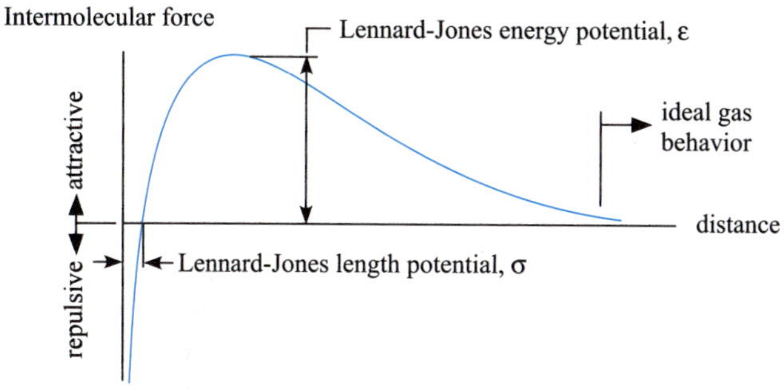

Figure 2-20: Force between two molecules as a function of the distance between their centers.

Table 2-5: Universal gas constant.

SI Unit System	English Unit System
8.314 kJ/kmol-K	1.987 Btu/lbmol-R
0.08314 bar-m^3/kmol-K	0.7302 atm-ft^3/lbmol-R
8314 J/kmol-K	1545 ft-lb$_f$/lbmol-R

the molecules are very far apart (relative to the Lennard-Jones length potential, σ), the substance is in a low density vapor phase and Figure 2-20 shows that the intermolecular forces are very weak, approaching zero. It is in this condition (i.e., a very low density gas) that the ideal gas law can be used to approximate the behavior of a fluid.

The ideal gas law relates the pressure, specific volume and temperature of a low density gas. The behavior of gases was the subject of many early studies by researchers such as Boyle, Charles, Avogadro, Dalton and others, and so the ideal gas law was established based on experimental observations well before thermodynamics became a recognized science. The ideal gas law can also be derived from statistical thermodynamics by assuming that there are no attractive or repulsive forces between molecules and that each molecule individually occupies zero volume. The assumptions used to derive the ideal gas law become increasingly appropriate as the distance between molecules (which is related to the specific volume of the gas) increases.

The ideal gas can be expressed as:

$$P\,V = n\,R_{univ}\,T \tag{2-31}$$

where n is the number of moles of gas, R_{univ} is the universal gas constant (which does not depend on the type of gas), and T is the absolute temperature (i.e., the temperature expressed in either the Kelvin or Rankine scale). The value of R_{univ} in several sets of units is provided in Table 2-5 and in Appendix A.

A mole is defined as a particular number of molecules, which is referred to as Avogadro's number; many people can recall Avogadro's number to four significant figures (6.023×10^{23}) from their high school physics or chemistry classes. However, there are actually different units for moles (e.g., kmol, gmol, lbmol). Each of these units for mole is associated with a different number of molecules. A mole is more generally defined as the amount of a substance that is equal to its molar mass (MW) in whatever unit system you are using. For example, consider a gas such as nitrogen. The nitrogen doublet molecule has a molar mass of 28. Therefore, a gram-mole (gmol) of nitrogen must have a mass of 28 g, a kilogram-mole (kmol or kgmol) of nitrogen is 28 kg, and a pound-mole (lbmol) of nitrogen is 28 lb$_m$. Clearly 28 lb$_m$ of nitrogen contains more molecules than 28 grams of nitrogen. The familiar number that most people remember, 6.023×10^{23}, is the number of molecules in a gram-mole.

It will generally be more convenient to work on a mass rather than a molar basis, at least until we study mixtures and chemical reactions in Chapters 11 and 13. The molar mass (also called the molecular weight) is the ratio of the number of moles to the mass of a substance:

$$MW = \frac{m}{n} \tag{2-32}$$

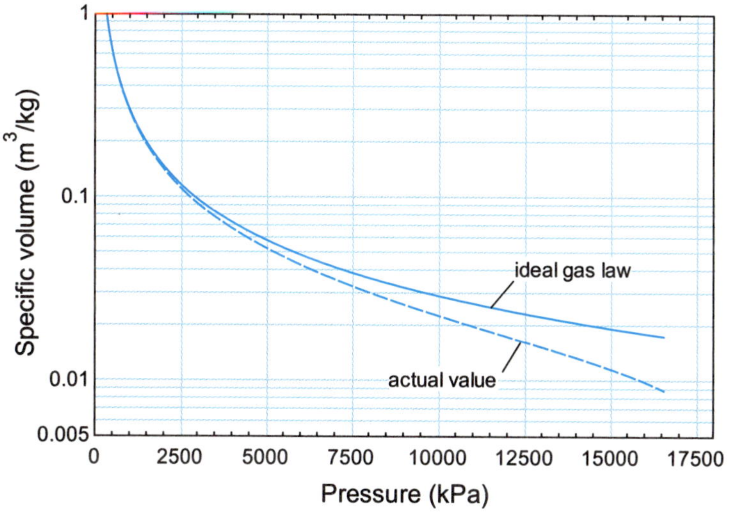

Figure 2-21: Specific volume of water vapor at 350°C as a function of pressure; also shown is the value predicted by the ideal gas law.

Substituting Eq. (2-32) into the ideal gas law, Eq. (2-31), leads to:

$$P\,V = m \underbrace{\left(\frac{R_{univ}}{MW}\right)}_{R} T \tag{2-33}$$

or

$$P\,V = m\,R\,T \tag{2-34}$$

where R is the ideal gas constant expressed on a mass basis. Examination of Eqs. (2-33) and (2-34) shows that R is related to the universal gas constant and the molar mass according to:

$$R = \frac{R_{univ}}{MW} \tag{2-35}$$

Notice that the ideal gas constant expressed on a mass basis depends on the molar mass of the specific gas that is being modeled. The molar mass and ideal gas constant for several gases are included in Table D-1 in Appendix D. The ideal gas law can be expressed in terms of specific volume by dividing Eq. (2-34) by the mass of the gas.

$$P\,v = R\,T \tag{2-36}$$

Figure 2-21 illustrates the specific volume predicted by the ideal gas law and the actual specific volume of water vapor at 350°C as a function of pressure. Notice that the accuracy of the ideal gas law increases as the specific volume increases, which corresponds to decreasing pressure. For pressures below 2500 kPa, the ideal gas law estimate is nearly perfect.

Figure 2-22 illustrates the deviation of the actual specific volume of water vapor from the value predicted by the ideal gas law as a function of pressure for various values of temperature. Notice that the ideal gas law becomes more accurate (i.e., the deviation becomes smaller) as the temperature increases and as the pressure is reduced.

Many common gases, such as air, helium, oxygen, and carbon dioxide, behave according to the ideal gas law at atmospheric pressure and temperature. Therefore, we will often

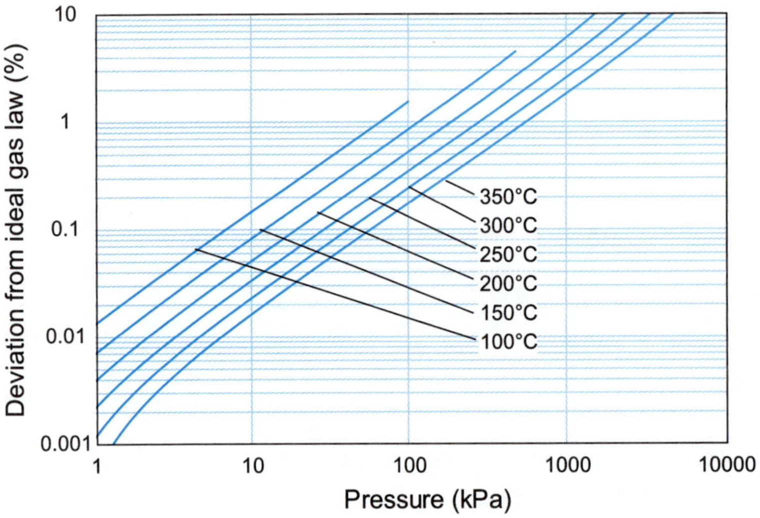

Figure 2-22: Deviation of the specific volume of water vapor from the ideal gas law as a function of pressure at various values of temperature.

use the ideal gas law to estimate the thermodynamic properties of these gases. However, as seen in Figure 2-22, the ideal gas law becomes less accurate with decreasing temperature or with increasing pressure. More advanced equations of state that provide improved estimates of the specific volume as a function of temperature and pressure, as well as general relations between other thermodynamic properties, are the subject of Chapter 10.

EXAMPLE 2.5-1: THERMALLY-DRIVEN COMPRESSOR

Micro-electro-mechanical systems (MEMS) are fabricated at the micro-scale using the same manufacturing technology that is used to make computer chips. A MEMS compressor would be useful for a variety of applications but it is not easy to fabricate the moving parts that are required by conventional compressor designs (e.g., a reciprocating piston) using MEMS technology. However, it is relatively easy to heat and cool components quickly in a MEMS device. With this in mind, a conceptual drawing of a thermally-driven MEMS compressor is shown in Figure 1.

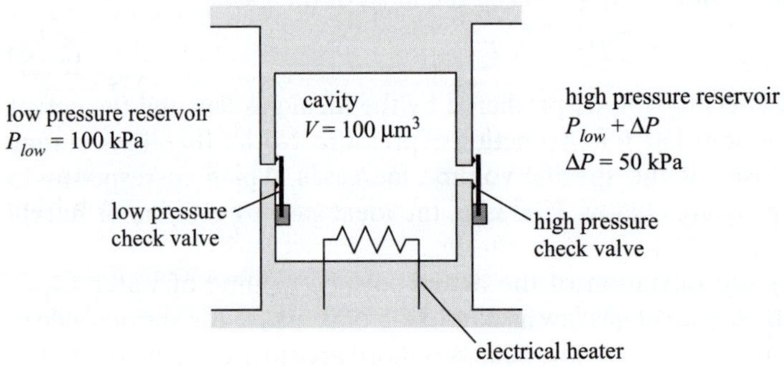

Figure 1: Concept for a thermally-driven MEMS compressor.

The compressor consists of a closed cavity with volume $V = 100\ \mu m^3$. The suction port of the compressor is connected to a reservoir of air at P_{low} = 100 kPa and the discharge port is connected to a reservoir of air at $P_{low} + \Delta P$, where $\Delta P = 50$ kPa. Two check valves separate the cavity volume from the reservoirs. The check valves are "flapper valves". If the pressure in the cavity exceeds the pressure in the high pressure reservoir then the high pressure check valve is pushed open; otherwise it is pressed against its seat by the pressure force. The low pressure check valve operates in the same way, allowing flow only in one direction and preventing the pressure in the cavity from ever going below the pressure in the low pressure reservoir. Figure 2 illustrates the operation of the compressor.

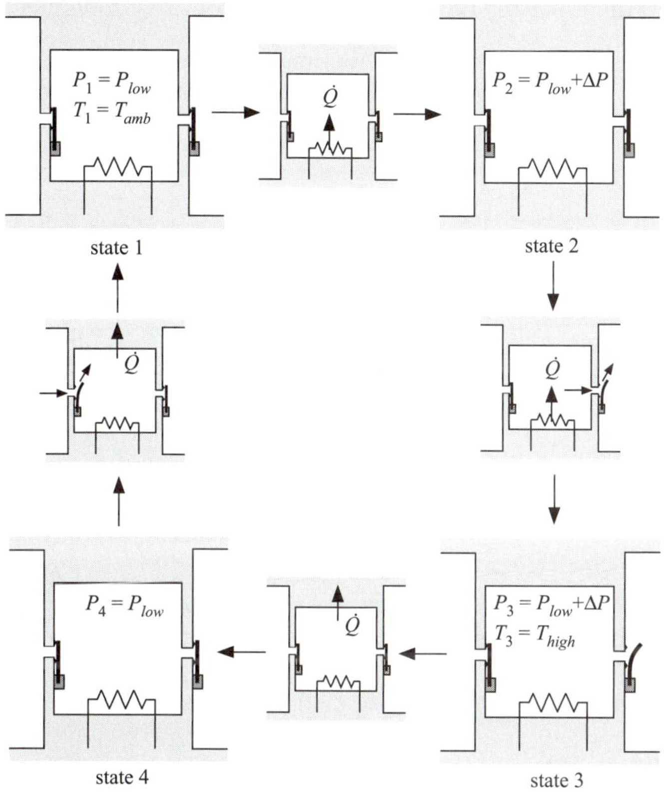

Figure 2: Operation of the compressor.

At state 1, the cavity is filled with air at $P_1 = P_{low}$ and ambient temperature, $T_1 = T_{amb} = 20^\circ$C. Both check valves are shut. An electrical heater placed in the cavity causes the air temperature, and therefore also its pressure, to rise. This process continues until the pressure reaches $P_2 = P_{low} + \Delta P$, at which time the high pressure check valve opens. This condition is referred to as state 2. The temperature of the air continues to rise until it reaches $T_3 = T_{high} = 700^\circ$C at state 3. The high pressure check valve is open during this process, therefore the air pressure in the cavity remains at $P_3 = P_2 = P_{low} + \Delta P$ and air is forced out of the cavity into the high pressure reservoir. Next, the heater is deactivated and therefore the temperature of the air begins to drop rapidly. The reduction in temperature causes the pressure to drop as well and therefore the high pressure check valve immediately shuts. This process continues until the pressure reaches $P_4 = P_{low}$, at which time the low pressure check valve opens. This condition is referred to as state 4. The temperature continues to drop until it reaches $T_1 = T_{amb} = 20^\circ$C. The low pressure check valve remains open during this process, therefore the air pressure remains at $P_1 = P_4 = P_{low}$ as air is drawn into the cavity from the low pressure reservoir.

Having returned to state 1, the compressor has executed a cycle (i.e., it ends at the same state at which it began) so it is ready to begin the process of activating and deactivating the heater again. Each time the cycle shown in Figure 2 executes, a small amount of air is drawn into the cavity from the low pressure reservoir during the process of going from state 4 to state 1, and transferred to the high pressure reservoir during the process of going from state 2 to state 3. Because the device is so small, it is estimated that the cycle can be repeated at a frequency $f = 10$ kHz. Air can be assumed to obey the ideal gas law at the conditions of this problem.

a) Determine the mass of gas removed from the low pressure reservoir and delivered to the high pressure reservoir during a single cycle. What is the average mass flow rate produced by the compressor?

The inputs are entered in EES:

```
$UnitSystem SI,Rad,Mass,Pa,K,J
T_amb=converttemp(C,K,20 [C])                "ambient temperature"
T_high=converttemp(C,K,700 [C])              "high temperature"
f=10e3 [Hz]                                  "frequency of operation"
P_low=100 [kPa]*convert(kPa,Pa)              "low pressure"
DELTAP_kPa=50 [kPa]                          "pressure rise produced by compressor, in kPa"
DELTAP=DELTAP_kPa*convert(kPa,Pa)            "pressure rise produced by compressor"
Vol=100 [micron^3]*convert(micron^3,m^3)     "volume of cavity"
```

The ideal gas constant for air is obtained from Appendix D, $R = 287.0$ J/kg-K.

```
R=287.0 [J/kg-K]    "ideal gas constant"
```

Note that the ideal gas constant for air could also be obtained using Eq. (2-35):

$$R = \frac{R_{univ}}{MW}$$

where the molar mass, MW, is obtained using the MolarMass function in EES and the universal gas constant is obtained using the built-in constant R# in EES. Many commonly encountered natural constants are programmed in EES; the # indicates that R# is a pre-defined constant. To examine a list of these constants select Constants from the Options menu.

```
R=287.0 [J/kg-K]              "ideal gas constant"
R_check=R#/MolarMass(Air)     "check"
```

Solving the EES code will reveal that the value of the variable R_check is 287.0 J/kg-K.

At state 1, the pressure and temperature are known, $P_1 = P_{low}$ and $T_1 = T_{amb}$. The pressure and temperature together are sufficient to specify the state of a single phase substance and therefore all other internal intensive properties can be determined. In this case, the ideal gas law is used to compute the specific volume:

$$v_1 = \frac{R\,T_1}{P_1}$$

The mass of air in the cavity at state 1 is:

$$m_1 = \frac{V}{v_1}$$

```
"state 1 – low temperature achieved"
P[1]=P_low                  "pressure"
T[1]=T_amb                  "temperature"
v[1]=R*T[1]/P[1]            "specific volume"
m[1]=Vol/v[1]               "mass"
```

Note that the specific volume could also be evaluated using the function **Volume** in EES with the substance 'Air' (i.e., air modeled as an ideal gas, as discussed in Section 2.4.1).

```
v_1_check=volume(Air,T=T[1],P=P[1])    "check using Volume function in EES"
```

Solving the problem will show that the variables v[1] and v_1_check are both equal to 0.8414 m^3/kg.

At state 2, the air has been warmed sufficiently such that the pressure in the cavity is equal to the pressure in the high pressure reservoir. The pressure at state 2 is therefore known: $P_2 = P_{low} + \Delta P$. During the warming process, the pressure in the cavity is between P_{low} and $P_{low} + \Delta P$; therefore, neither the high or the low pressure check valves in the cavity will be open (see Figure 1, both flaps are pushed shut by pressure differences). A mass balance on the cavity for the process of moving from state 1 to state 2 is:

$$m_2 = m_1$$

The specific volume of the air in the cavity at state 2 must be:

$$v_2 = \frac{V}{m_2}$$

The pressure and specific volume together fix state 2. The temperature at state 2 is computed using the ideal gas law:

$$T_2 = \frac{P_2\, v_2}{R}$$

```
"state 2 – high pressure valve ready to open"
P[2]=P_low+DELTAP           "pressure"
m[2]=m[1]                   "mass"
v[2]=Vol/m[2]               "specific volume"
T[2]=P[2]*v[2]/R            "temperature"
```

At state 3, the pressure remains at $P_3 = P_{low} + \Delta P$ because any increase in pressure will cause the high pressure check valve to open and allow gas to leak out. The temperature at state 3 is known, $T_3 = T_{high}$. The pressure and temperature together fix state 3. The specific volume at state 3 is calculated using the ideal gas law:

$$v_3 = \frac{R\, T_3}{P_3}$$

The mass of air in the cavity at state 3 is:

$$m_3 = \frac{V}{v_3}$$

```
"state 3 – high temperature achieved"
T[3]=T_high                    "temperature"
P[3]=P_low+DELTAP              "pressure"
v[3]=R*T[3]/P[3]               "specific volume"
m[3]=Vol/v[3]                  "mass"
```

A mass balance on the cavity for the process of going from state 2 to state 3 is shown in Figure 3.

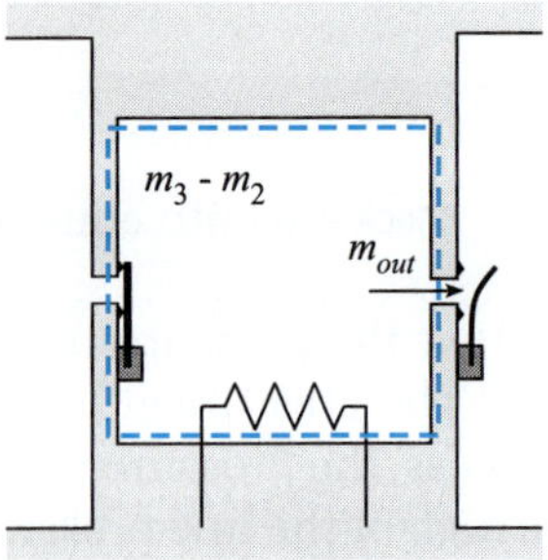

Figure 3: Mass balance on the cavity volume for the process associated with going from state 2 to state 3.

The mass balance is:

$$0 = m_{out} + m_3 - m_2$$

```
0=m_out+m[3]-m[2]    "mass balance from 2 to 3"
```

which results in $m_{out} = 6.52 \times 10^{-17}$ kg of air delivered to the high pressure reservoir during each cycle.

Once state 3 has been achieved, the heater is switched off so that the temperature and pressure in the cavity decrease. The pressure at state 4 is equal to $P_4 = P_{low}$ so that the low pressure check valve will open. Assuming no leakage through the check valves, the mass of air in the cavity at state 4 is $m_4 = m_3$. The specific volume of the air in the cavity at state 4 is:

$$v_4 = \frac{V}{m_4}$$

The pressure and specific volume together fix state 4. The temperature at state 4 is computed using the ideal gas law:

$$T_4 = \frac{P_4\, v_4}{R}$$

```
"state 4 – low pressure valve ready to open"
P[4]=P_low                     "pressure"
m[4]=m[3]                      "mass"
v[4]=Vol/m[4]                  "specific volume"
T[4]=P[4]*v[4]/R               "temperature"
```

To complete the cycle, the temperature in the cavity is allowed to return to $T_1 = T_{amb}$. The low pressure check valve remains open during this process so the pressure is $P_1 = P_{low}$. A mass balance on the system as it goes from state 4 to state 1 is:

$$m_{in} = m_1 - m_4$$

```
m_in=m[1]-m[4]    "mass balance from 4 to 1"
```

which results in $m_{in} = 6.52 \times 10^{-17}$ kg of air drawn into the cavity. Note that m_{in} has to be equal to m_{out} because the system undergoes a cycle; that is, the final state is equal to the initial state. A mass balance on the cavity for the duration of time associated with a complete cycle (i.e., the process of going from state 1 to 2 to 3 to 4 and back to state 1 again) is:

$$m_{in} = m_{out} + \underbrace{m_1 - m_1}_{storage}$$

The storage term must be zero (there can be no change in the mass contained in the system if the initial and final states are the same); therefore:

$$m_{in} = m_{out}$$

Any other result would indicate that we had made an error in our calculations. The time required to undergo one cycle is:

$$t_{cycle} = \frac{1}{f}$$

The average mass flow rate produced by the MEMS compressor is:

$$\dot{m}_{avg} = \frac{m_{out}}{t_{cycle}}$$

```
t_cycle=1/f                                             "cycle time"
m_dot_bar=m_out/t_cycle                                 "average mass flow rate"
m_dot_bar_microgphr=m_dot_bar*convert(kg/s,microgram/hr) "in microgram/hr"
```

which leads to $\dot{m}_{avg} = 2.345$ μg/hr. Note that EES will accept microgram/hr or μg/hr as the units for the variable m_dot_bar_microgphr. The μ character can be entered on most keyboards by holding the Alt key down while entering 230 on the numeric keypad.

b) Generate a head/flow curve for the compressor. That is, plot the pressure rise produced by the compressor as a function of the mass flow rate. Overlay on your plot the curve associated with different values of T_{high} in order to assess the importance of this temperature limit on the performance of the compressor.

A Parametric table is generated that includes the variables DELTAP_kPa and m_dot_avg_microgphr. The value of DELTAP_kPa is varied from 0 to 300 kPa and the value of DELTAP_kPa specified in the Equations window is commented out. The table is run and the results are plotted in Figure 4, which shows the pressure rise produced by the compressor as a function of the average mass flow rate. The process is repeated for different values of T_{high}, resulting in the different curves in Figure 4. Notice that a higher value of T_{high} leads to better performance.

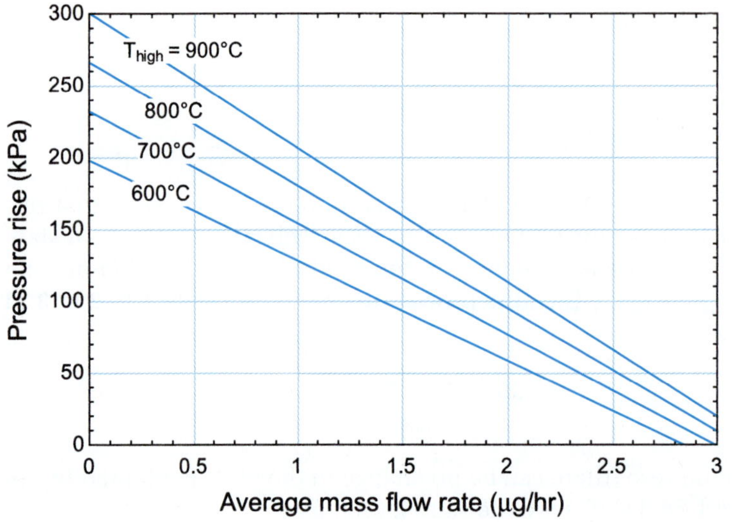

Figure 4: Pressure rise as a function of the average mass flow rate for various values of high temperature.

2.6 The Incompressible Substance Model

The specific volume of solids and liquids that are at a temperature well below the critical temperature are not strongly dependent on pressure. Such substances are referred to as being *incompressible*. The specific volume of an incompressible substance (or, equivalently, the density) is independent of pressure, although it may change with temperature:

$$\rho = 1/v = f(T) \tag{2-37}$$

Values of the density (and other properties) of solids and liquids are available from many sources. Table D-2 in Appendix D includes a summary of the properties of many solid and liquid substances that can be modeled as being nearly incompressible at room temperature and pressure. Another convenient source of information for incompressible substances is the EES Solid/Liquid property library functions. Information about these functions is accessed with the Function Information dialog window in the Options menu. Click the Solid/liquid properties button at the upper left of the window in order to view the properties and substances that are available (Figure 2-23).

There are many property functions available; at this point, the one that is of interest is the density function (rho_). The Function Info button provides information about each of the functions while the Property Info button provides information about the source of the property information. Click on a property function name in the list on the left (e.g., rho_) and then on a substance name in the list on the right (e.g., Aluminum). An example of the property function call is shown in the example box at the bottom of the window; you can paste this function call into the Equations window and edit it as necessary for your problem.

Note that the name of the substance is a string and therefore the substance can be represented by a string constant. A string constant must be enclosed in single quotes. For example, the code below determines the density of aluminum at 20°C:

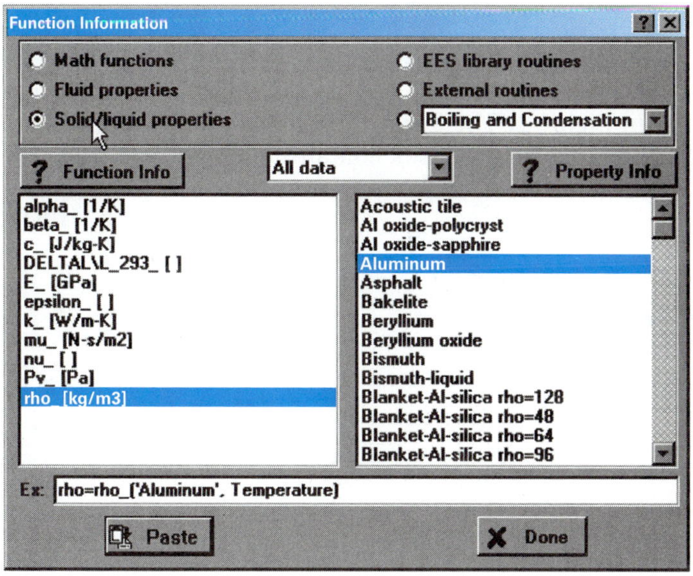

Figure 2-23: Function Information dialog showing the Solid/Liquid property functions.

```
$UnitSystem SI,Mass,Radian,J,K,Pa
T=converttemp(C,K,20[C])
rho=rho_('Aluminum',T)
```

which results in $\rho = 2703$ kg/m^3. The substance may also be represented by a string variable. A string variable is identified by a $ after the last character in its name. For example, the code below also determines the density of aluminum at 20°C.

```
$UnitSystem SI,Mass,Radian,J,K,Pa
T=converttemp(C,K,20[C])
S$='Aluminum'
rho=rho_(S$,T)
```

The use of a string variable is convenient if you would like to create an EES program that is flexible with regard to the thermodynamic substance being considered. Use a string variable to store the name of the substance so that changing the value of the string variable (e.g., from 'Aluminum' to 'Asphalt') will change the substance everywhere that it occurs in the program. There are many substances available in the Solid/Liquid property library. You can shorten the list by selecting the class of substance from the pull-down list in the center of the dialog window. By default it shows 'all data', but you can selectively examine only metals, liquid metals, building materials, etc. For consistency with the fluid property functions, EES will also accept the temperature input to these property functions with the T= notation, but this notation is optional for incompressible substances as temperature is the only input argument that is required or accepted for these functions. The property functions will return values and expect the temperature input in units that are consistent with the specified unit system. For example, in the code listed above the units are specified using the $UnitSystem directive; these units require that the temperature provided to the rho_ function be in K and the rho_ function will return density in kg/m^3.

EXAMPLE 2.6-1: FIRE EXTINGUISHING SYSTEM

A fire control system in a building consists of $V = 50$ gallon tanks that are partially filled with water, as shown in Figure 1. The fraction of the tank that is filled with water is $f = 0.8$, the remaining volume in the tank contains air.

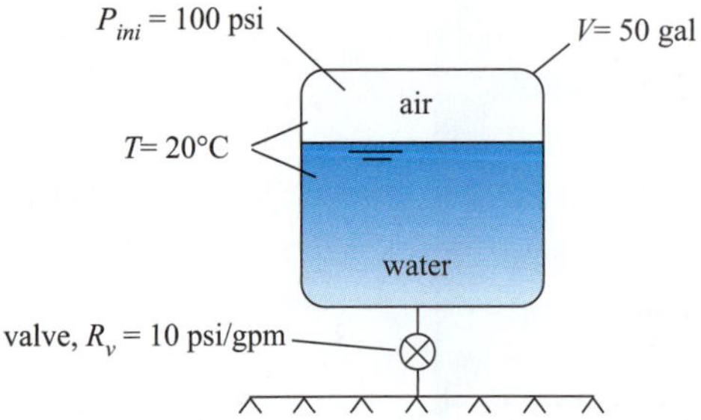

Figure 1: Fire control system.

The air in the tank is pressurized to $P_{ini} = 100$ psi. When a fire is detected, the tank valve is opened and the water is quickly forced out of the tank and provided to the sprinkler system by the air pressure. The volumetric flow through the valve ($\dot{V}_f$) depends on the instantaneous tank pressure (P) according to:

$$\dot{V}_f = \frac{(P - P_{atm})}{R_v} \tag{1}$$

where $R_v = 10$ psi/gpm is the valve resistance parameter and $P_{atm} = 1$ atm is the atmospheric pressure that exists downstream of the valve. The air and water in the tank are both maintained at $T = 20^{\circ}$C during the process. Model the water as an incompressible substance and the air as an ideal gas.

a) Use EES' Integral command to develop a numerical model that is capable of predicting the volume of water, volume of air, and the pressure in the tank as a function of time. Prepare a plot showing the volume of water and the pressure in the tank as a function of time. Appendix G provides an introduction to numerical integration and the EES Integral command.

The inputs are entered in EES.

```
$UnitSystem SI,Rad,Mass,Pa,K,J
V=50 [gal]*convert(gal,m^3)                       "tank volume"
T=converttemp(C,K,20 [C])                         "temperature"
f=0.8 [-]                                         "fraction of tank initially filled with water"
P_ini=100 [psi]*convert(psi,Pa)                   "initial pressure"
R_v=10 [psi/gpm]*convert(psi/gpm,Pa-s/m^3)        "valve resistance"
P_atm=1 [atm]*convert(atm,Pa)                     "atmospheric pressure"
```

The ideal gas constant for air is $R = 287.0$ J/kg-K and the density of liquid water at 20°C is $\rho = 998.2$ kg/m^3 according to Table D-2 in Appendix D.

```
R=287.0 [J/kg-K]       "ideal gas constant"
rho=998.2 [kg/m^3]     "density of water"
```

The initial volume of water in the tank is:

$$V_{f,ini} = f\, V$$

and the initial volume of air in the tank is:

$$V_{g,ini} = (1 - f)\ V$$

```
V_f_ini=f*V          "initial volume of water"
V_g_ini=(1-f)*V      "initial volume of air"
```

The state variables for this problem include the volume of water (V_f), the volume of air (V_g), and the tank pressure (P). Using the procedure outlined in Section G.5 of Appendix G, arbitrary values of the state variables (the dependent, or solution, variables) and the integration variable (time, t) are assumed in order to develop the state equations (i.e., the equations that determine the rate of change of the state variables).

```
"arbitrary values of state and integration variables"
P=P_ini                          "pressure"
V_f=V_f_ini                      "liquid volume"
V_g=V_g_ini                      "gas volume"
time=0 [s]                       "time"
```

A mass balance on the water in the tank is shown in Figure 2.

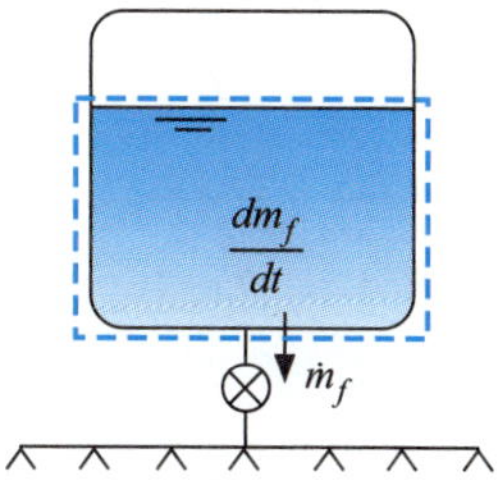

Figure 2: Mass balance on the water in the tank.

The mass balance provides:

$$0 = \dot{m}_f + \frac{dm_f}{dt} \tag{2}$$

The mass flow rate of water leaving the tank is given by:

$$\dot{m}_f = \dot{V}_f\,\rho_f \tag{3}$$

where $\dot{V}_f$ is the volumetric flow rate of water leaving the tank, given by Eq. (1). The rate of change of the mass of water in the tank is given by:

$$\frac{dm_f}{dt} = \frac{d}{dt}(V_f\,\rho_f)$$

The density of an incompressible substance is constant, therefore:

$$\frac{dm_f}{dt} = \rho_f\,\frac{dV_f}{dt} \tag{4}$$

Substituting Eqs. (3) and (4) into Eq. (2) provides:

$$0 = \dot{V}_f + \frac{dV_f}{dt} \tag{5}$$

Substituting Eq. (1) into Eq. (5) provides an expression for the time rate of change of the volume of fluid in the tank:

$$0 = \frac{(P - P_{atm})}{R_v} + \frac{dV_f}{dt}$$

```
0=(P-P_o)/R_v+dVfdt    "state equation for V_f"
```

When the code is solved, $\frac{dV_f}{dt} = -0.0005382$ m^3/s. The volume of liquid is decreasing.

The overall volume of the tank is constant:

$$V = V_g + V_f \tag{6}$$

Taking the derivative of Eq. (6) leads to:

$$\frac{dV_g}{dt} + \frac{dV_f}{dt} = 0$$

which provides another state equation, an expression for the time rate of change of the volume of gas in the tank.

```
dVfdt+dVgdt=0    "state equation for V_g"
```

When the code is solved, $\frac{dV_g}{dt} = 0.0005382$ m^3/s, indicating that the volume of air is increasing. Figure 3 shows a mass balance on the air.

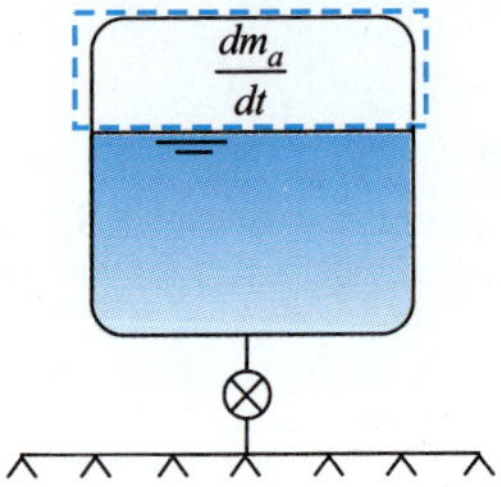

Figure 3: Mass balance on the air in the tank.

The mass balance provides:

$$0 = \frac{dm_a}{dt} \tag{7}$$

Substituting the ideal gas law into Eq. (7) shows that:

$$0 = \frac{d}{dt}\left(\frac{P\, V_g}{R\, T}\right)$$

The gas constant and temperature of the air are constant, therefore:

$$0 = V_g \frac{dP}{dt} + P \frac{dV_g}{dt}$$

which provides the final state equation, an expression for the time rate of change of the pressure in the tank.

```
V_g*dPdt+P*dVgdt=0    "state equation for P"
```

When the code is solved, $\frac{dP}{dt} = -9803$ Pa/s, indicating that the pressure of the air is decreasing.

Once the state equations have been implemented, remove the arbitrary values that were set for the state variables and the integration variable:

```
{"arbitrary values of state and integration variables"
P=P_ini                                  "pressure"
V_f=V_f_ini                              "liquid volume"
V_g=V_g_ini                              "gas volume"
time=0 [s]                               "time"}
```

and integrate the state equations according:

$$P = P_{ini} + \int_0^{t_{sim}} \frac{dP}{dt} dt$$

$$V_f = V_{f,ini} + \int_0^{t_{sim}} \frac{dV_f}{dt} dt$$

$$V_g = V_{g,ini} + \int_0^{t_{sim}} \frac{dV_g}{dt} dt$$

These integrals are evaluated numerically using the **Integral** command in EES, as described in Section G.5 of Appendix G.

```
t_sim=500 [s]                              "simulation time"
P=P_ini+Integral(dPdt,time,0,t_sim)        "pressure"
V_g=V_g_ini+Integral(dVgdt,time,0,t_sim)   "gas volume"
V_f=V_f_ini+Integral(dVfdt,time,0,t_sim)   "water volume"
P_psi=P*convert(Pa,psi)                    "pressure, in psi"
V_f_gal=V_f*convert(m^3,gal)               "water volume, in gal"
```

The results are stored in an Integral table using the **$IntegralTable** directive:

```
$IntegralTable time,P,V_g,V_f,P_psi,V_g_gal,V_f_gal,V_dot_out_gpm
```

and the numerical integration parameters are (optionally) specified using the **$IntegralAutoStep** directive, as detailed in Section G.5 of Appendix G.

```
$IntegralAutoStep Vary=1 Min=5 Max=2000 Reduce=1e-3 Increase=1e-5
```

The water volume and tank pressure as a function of time are shown in Figure 4.

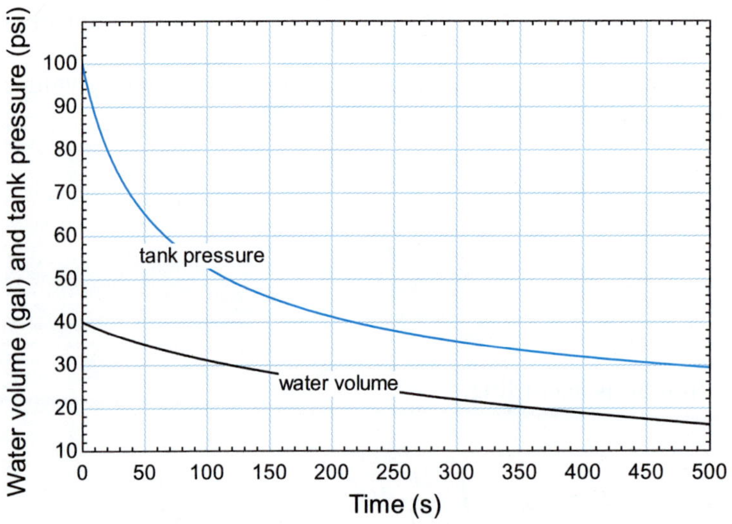

Figure 4: Water volume and tank pressure as a function of time.

b) Plot the volume of water expelled from the tank during $t_{sim} = 120$ seconds of operation as a function of the fraction of the tank that is initially filled with water, f.

The volume of expelled water is given by:

$$V_{expel} = V_{f,ini} - V_{f,t=120\,\mathrm{s}}$$

Note that our model is not valid when the amount of liquid in the tank reaches zero (in fact, the equations allow the amount of liquid in the tank to become negative, which makes no sense). Therefore, the volume of expelled water is calculated according to:

$$V_{expel} = V_{f,ini} - \mathrm{Max}(0, V_{f,t=120\,\mathrm{s}})$$

where the Max function in EES takes the maximum value of the arguments that are provided to it.

```
t_sim=120 [s]                                        "simulation time"
V_expel_gal=(V_f_ini-MAX(V_f,0 [m^3]))*convert(m^3,gal)   "volume of water expelled during t_sim,
                                                       in gal"
```

A Parametric table is created that includes the variables V_expel_gal and f. The volume of water expelled from the tank as a function of the fraction of the tank volume that is initially filled with water is shown in Figure 5. At low values of f, the tank is completely emptied and therefore V_{expel} increases with f. At high values of f, there is not enough gas to push the liquid out and therefore V_{expel} decreases with f. The optimal value of f is approximately 28% according to Figure 5.

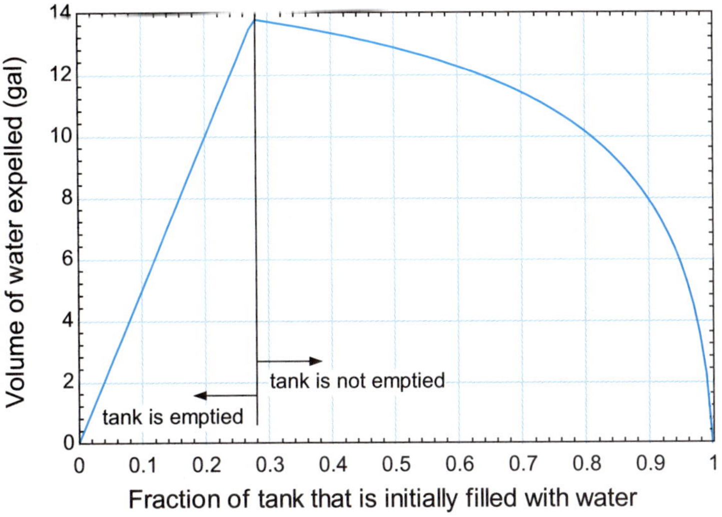

Figure 5: Volume of water expelled from the tank as a function of the fraction of the tank that is initially filled with water.

REFERENCES

Harr, L., J.S. Gallagher, and G.S. Kell, *NBS/NRC Steam Tables*, Hemisphere Publishing Co., New York (1984).

Myers, G.E., *Engineering Thermodynamics, 2nd Edition*, AMCHT Publications (2007).

Wagner, W. and A. Pruss, "The IAPWS Formulation 1995 for the Thermodynamic Properties of Ordinary Water Substance for General and Scientific Use," *J. Phys. Chem. Ref. Data*, Vol. 22, pp. 783–787 (1993) and *J. Phys. Chem. Ref. Data*, Vol. 31, pp. 387–535 (2002).

Problems

The problems included here have been selected from a larger set of problems that are available on the website associated with this book (www.cambridge.org/kleinandnellis).

A. Property Data from Tables

Use the property information from the tables in the appendix to do the problems in this section.

2.A-1 Pure water is held in a container. The temperature of the water is $T_1 = 520°\text{C}$ and the pressure is $P_1 = 800$ kPa.

a) Sketch a T-v diagram and locate the state of the water on the diagram. Your sketch should be qualitatively correct, but it does not have to be to scale.

b) Determine the specific volume of the water (m^3/kg) and the density of the water (kg/m^3).

c) If the mass of water in the container is $m = 7.2$ kg then what is the volume of the container?

2.A-2 Determine the boiling temperature (°F) of water:

a) at normal atmospheric pressure ($P_{atm} = 14.7$ psi, which corresponds to sea-level),

b) in Denver, where the pressure is $P = 24.58$ inch Hg, and

c) at the summit of Mount Everest, where the pressure is $P = 30$ kPa.

2.A-3 A rigid tank with volume $V = 8000\ \text{cm}^3$ is filled with water with quality $x_1 = 0.05$ and temperature $T_1 = 140°\text{C}$.

a) What is the specific volume (m^3/kg) and the pressure (kPa) of the water.
b) What is the total mass of water in the tank (kg)? What is the mass of liquid (kg) and the mass of vapor (kg) in the tank?
c) What are the volumes of liquid and vapor in the tank (m^3)?

The water in the tank is heated to $T_2 = 200°\text{C}$. The tank is rigid (i.e., its volume doesn't change) and leak tight.

a) What is the pressure (kPa) and quality of the water in the tank at state 2?
b) What is the mass of liquid (kg) in the tank at state 2?

2.A-7 Figure 2.A-7 illustrates a pressure cooker.

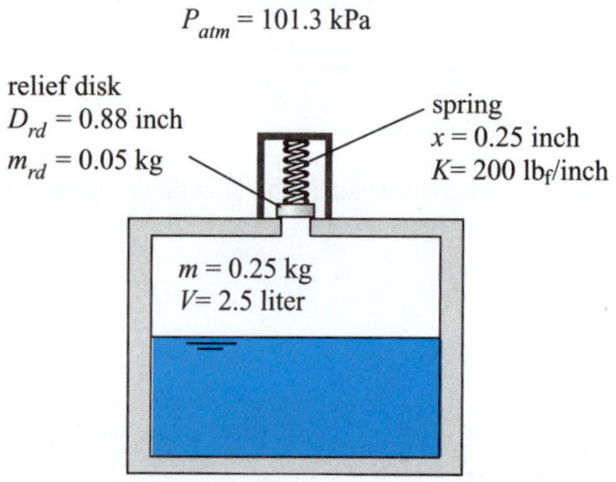

Figure 2.A-7: A pressure cooker.

The pressure cooker has an internal volume of $V = 2.5$ liter and contains $m = 0.25$ kg of pure water. The pressure relief valve consists of a spring loaded disk that is positioned over a hole in the top of the cooker. The disk has diameter $D_{rd} = 0.88$ inch and mass $m_{rd} = 0.05$ kg. The spring is compressed $x = 0.25$ inch and has a spring constant of $K = 200$ lb$_f$/inch. The atmospheric pressure is $P_{atm} = 101.3$ kPa.

a) Determine the internal pressure in the pressure cooker that is required to open the pressure relief valve.
b) The pressure relief valve allows vapor to escape in order to maintain the pressure that you calculated in (a). What is the temperature of the water remaining in the pressure cooker after the relief valve opens, assuming that some liquid remains?

You have been asked to examine the possibility that the relief valve fails to open. In this case, no water can escape and therefore the temperature and pressure of the contents of the pressure cooker will continue to rise until the device fails. Assume that the pressure cooker is rigid (i.e., the volume does not change).

c) What is the temperature in the pressure cooker when the pressure reaches $P_2 = 24$ MPa?

2.A-10 Refrigerant R134a is contained in a small recharge tank with volume $V = 2$ liter. Initially, the tank is filled with $m_1 = 2.0$ kg of R134a at $T_1 = 15°\text{C}$.

a) Sketch a T-v diagram and locate the state of the R134a on your sketch. The sketch can be approximate, but it should clearly show the intersection of the two property lines that define the state.
b) What is the pressure in the tank at state 1?
c) What is the quality of the R134a in the tank?

d) What is the mass of liquid R134a in the tank? What is the mass of vapor R134a in the tank?
e) What is the volume of liquid R134a in the tank? What is the volume of vapor R134a in the tank?

The recharge tank is equipped with a pressure relief valve that allows refrigerant to escape when the pressure within the tank reaches $P_2 = 9.0$ bar. The tank is accidentally transported in an un-conditioned truck where the temperature is $T_{truck} = 40°$C. As a result, the temperature of the R134a in the tank slowly starts to rise causing the pressure to rise. State 2 is defined to be the state of the refrigerant where the pressure relief valve just opens.
f) On your *T-v* diagram from (a) overlay state 2. Indicate on your diagram what two properties define state 2.
g) What is the temperature of the R134a at the time that it reaches state 2?

The refrigerant continues to increase in temperature until finally it reaches $T_3 = T_{truck}$. During this time, the pressure relief valve vents refrigerant in order to maintain the pressure in the tank always at $P_3 = P_2 = 9.0$ bar.
h) On your *T-v* diagram from (a) overlay state 3. Indicate on your diagram what two properties define state 3.
i) What mass of refrigerant is vented from the tank during the time that it goes from state 2 to state 3?

Eventually, the tank is unloaded from the truck and cooled to $T_4 = 20°$C.
j) On your *T-v* diagram from (a) overlay state 4. It should be clear on your diagram what two properties define state 4.
k) What is the volume of liquid refrigerant in the tank at state 4?

2.A-11 A rigid tank with volume $V_{tank} = 5.0$ gallon is maintained at $T_1 = 300°$C and initially contains $m_1 = 0.08$ kg of water, as shown in Figure 2.A-11. At some time, a valve is opened allowing $V_{in} = 0.5$ gallons of water at $T_{in} = 20°$C and $P_{in} = 20.0$ MPa to enter the tank. The valve is shut and eventually all of the water in the tank comes to $T_2 = 300°$C.

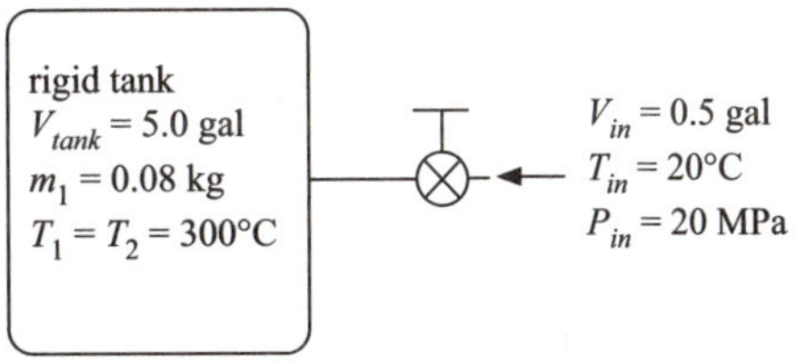

Figure 2.A-11: Rigid tank of water.

a) Locate state 1 and state *in*, the states of the water initially in the tank and the water added to the tank, respectively, on a sketch of a *T-v* diagram.
b) What is the initial pressure in the tank (MPa)?
c) What is the mass of water added to the tank (kg)?
d) Locate state 2, the final state of the water in the tank, on the *T-v* diagram from (a).
e) What is the final pressure in the tank (MPa)?

2.A-13 Figure 2.A13(a) illustrates a pressure cooker with the pressure relief valve removed. The pressure cooker has an internal volume of $V = 2$ liter. Because the relief valve is removed, the contents are initially at atmospheric pressure, $P_{atm} =$ 100 kPa. The pressure cooker contains water in a two-phase state. The bottom

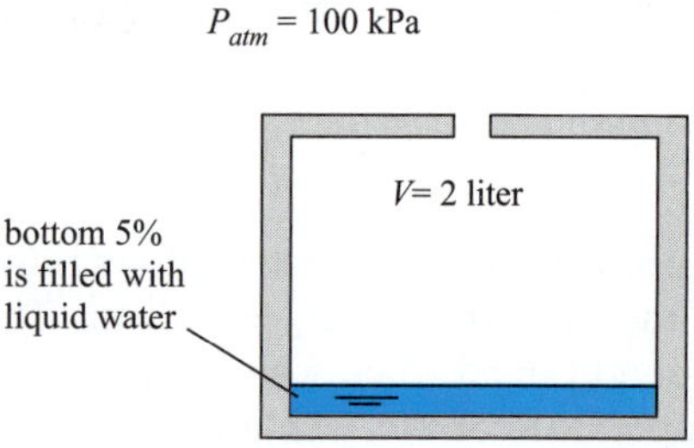

Figure 2.A-13(a): A pressure cooker with the pressure relief valve removed.

5% of the volume of the vessel is filled with liquid water while the remainder of the vessel contains water vapor.

a) Determine the initial temperature of the water in the cooker.
b) Determine the quality of the water initially in the cooker.
c) Locate the initial state of the water (state 1) on a sketch of a T-v diagram.

The pressure relief valve is installed on the pressure cooker, as shown in Figure 2.A-13(b).

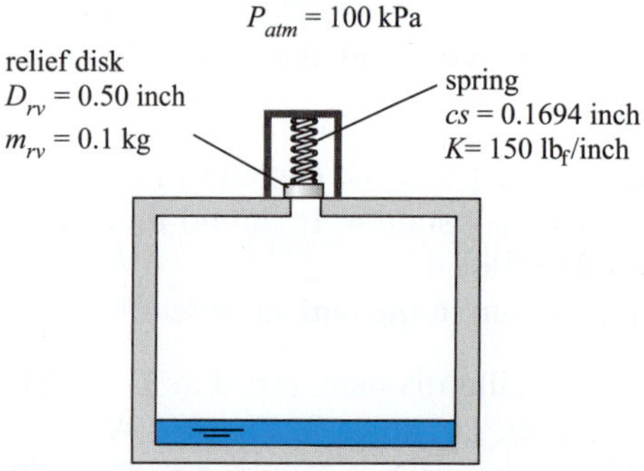

Figure 2.A-13(b): A pressure cooker with the pressure relief valve installed.

The pressure relief valve consists of a spring loaded disk that is positioned over a hole in the top of the cooker. The disk has diameter $D_{rv} = 0.5$ inch and mass $m_{rv} = 0.1$ kg. The spring is compressed $cs = 0.1694$ inch and has a spring constant of $K = 150$ lb_f/inch.

d) Determine the internal pressure that is required to open the pressure relief valve.
e) Heat is added to the pressure cooker with the relief valve in place. The pressure relief valve opens when the pressure reaches the value calculated in (d) and allows vapor to escape in order to maintain the pressure at this value. What are the temperature and quality of the water in the pressure cooker at the instant that the pressure relief valve opens at state 2? Add state 2 to your T-v sketch from (c).
f) What is the fraction of the volume of the pressure cooker that is filled with liquid at the instant that the pressure relief valve opens (at state 2).
g) Heat continues to be added to the pressure cooker until all of the liquid disappears at state 3. What is the mass of water that has passed through the pressure relief valve at this instant? Locate state 3 on your sketch from (c).
h) Heat continues to be added to the pressure cooker until the temperature reaches $T_4 = 400°$C. Locate state 4 on your sketch from (c). What is the mass of water that passes through the pressure relief valve between the time that the pressure cooker is at state 3 and at state 4?

i) The pressure cooker is removed from the source of heat and begins to cool. The pressure begins to drop and therefore the pressure relief valve closes. At what temperature does liquid water begin to form in the vessel? Locate this state 5 in your sketch from part (c).

j) The pressure cooker is cooled until the temperature reaches $T_6 = 20^\circ$C. Determine the pressure and quality of the water at state 6 and locate state 6 in your sketch from part (c).

2.A-14 Refrigerant R134a is contained at its critical point in a piston-cylinder device.

a) What is the temperature and the pressure of the R134a?

b) If the fluid is cooled slightly at constant volume, how many phases will be present in the cylinder? Provide a sketch to show your result.

c) Repeat part (b) assuming that pressure rather than the volume remains constant as the fluid is cooled.

2.A-15 At state 1, superheated water vapor is contained in a sealed glass vial at $T_1 = 200^\circ$C. You would like to know the pressure at this state, but have no means of measuring it directly. However, when the vial is slowly cooled to state 2, $T_2 = 120^\circ$C, you notice that droplets of liquid begin to form on the glass walls.

a) Use this information to determine the pressure at state 1.

b) Sketch the process on a temperature-volume diagram.

B. Property Data from EES

Use the property functions in EES to do the problems in this section.

2.B-1 A piston-cylinder device with volume $V_1 = 40$ cm^3 is filled with saturated vapor R134a at temperature $T_1 = -20^\circ$C, as shown in Figure 2.B-1(a).

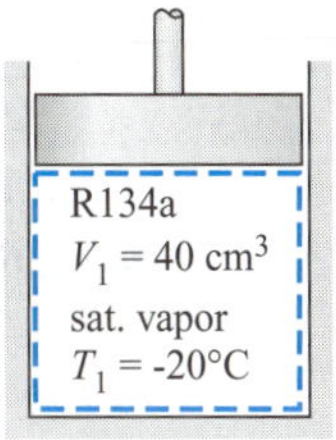

Figure 2.B-1(a): Piston-cylinder device at state 1.

a) Use EES' internal property routines to determine the pressure (kPa) of the R134a at state 1.

b) Use the tables in the appendix of your book to determine the pressure (kPa) of the R134a at state 1 and compare your answer with part (a).

c) What is the mass of R134a (kg) in the piston?

The piston is moved so that the volume decreases to $V_2 = 10$ cm^3, as shown Figure 2.B-1(b). At the conclusion of this process, the pressure in the cylinder is $P_2 = 950$ kPa.

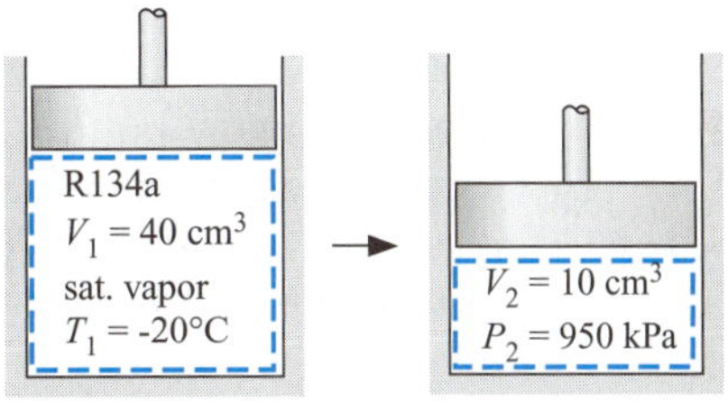

Figure 2.B-1(b): Piston-cylinder device at state 2.

d) What is the temperature of the R134a (°C) at state 2?

The piston is locked in place and the contents are cooled until the R134a reaches $T_3 = -20°C$.

e) What is the pressure of the R134a (kPa) at state 3?
f) Sketch the locations of states 1, 2, and 3 on a *T-v* diagram that includes the vapor dome. Make sure that you determine whether these states are sub-cooled liquid, two-phase, or superheated vapor and place them accordingly.
g) Using EES, prepare a plot showing the pressure at state 3 as a function of T_3 for $-40°C < T_3 < 100°C$; assume all other parameters do not change. Your plot should have pressure in kPa as the y-axis and temperature in °C as the x-axis. Explain the shape of your plot.

2.B-2 One problem facing people who build large-scale refrigeration systems that include a lot of piping and components is that it is not easy to estimate the total volume contained in the system. This is important because you must order refrigerant in a quantity that is sufficient to charge the system. If you order too much then you'll have to ship the excess back (if the company will take it back) and if you order too little then your project faces delays as you end up re-ordering. You have come up with a technique that you think will allow the volume enclosed in an installed refrigeration system to be measured. The idea is to attach a high pressure air bottle to the system, as shown in Figure 2.B-2.

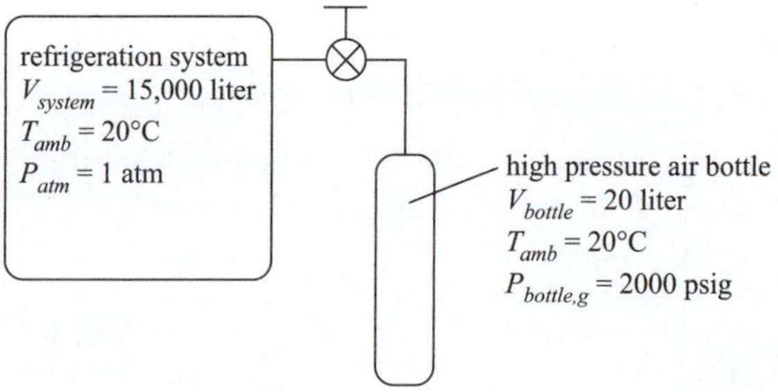

Figure 2.B-2: High pressure air bottle connected to the system.

The air in the system is initially at ambient temperature and atmospheric pressure, $T_{amb} = 20°C$ and $P_{atm} = 1$ atm, respectively. Assume for now that the system volume is $V_{system} = 15{,}000$ liter. The air in the bottle is initially at ambient temperature and a high pressure, $P_{bottle,g} = 2000$ psig (note that the gage mounted on the bottle measures the gage pressure of the air). The volume of the bottle is $V_{bottle} = 20$ liter. The valve connecting the bottle to the system is opened, allowing high pressure air from the bottle to flow into the system. Eventually the air in the bottle and the system come to the same final pressure, P_2. The final temperature of the air in the bottle and in the system is $T_2 = T_{amb}$. You would like to use the measurement of the final pressure, P_2, in order to determine the volume of the system. Use the substance 'Air_ha' in EES to model the air for this problem (i.e., do not model the air as an ideal gas).

a) Determine the final pressure (in psig, the gage pressure) that exists in the system and in the bottle.
b) Generate a graph showing the system volume (the quantity you are interested in) as a function of the final pressure (the quantity that you can measure). Does this technique work? Can you determine the system volume if you measure the pressure?

c) Your EES program uses system volume as an input and computes the final pressure as an output. However, you can comment out the system volume that you input and instead specify the measured output pressure (i.e., you can turn an input into an output). If you measure a final pressure of $P_2 = 4.5$ psig (again, a gage pressure) then what is the system volume?

d) During the commissioning process you want to charge your system (after first evacuating it to remove all of the air) with refrigerant R134a at pressure $P_{charge} = 450$ kPa and temperature $T_{charge} = T_{amb}$. Based on your calculation from (c), how much refrigerant (in lb_m) do you need to order?

You are in the process of specifying the pressure and temperature instrumentation that are required to carry out the procedure described above.

e) If you can measure the bottle pressure to within $\delta P_{bottle} = 25$ psi then what is the related uncertainty in your measurement of the system volume? One easy way to evaluate this is to change the specified value of P_{bottle} by the uncertainty δP_{bottle} and see how much the calculated system volume changes.

f) If you can measure the final pressure to within $\delta P_2 = 0.25$ psi then what is the related uncertainty in your measurement of the system volume?

g) If you can measure the ambient temperature to within $\delta T_{amb} = 2.5°\text{C}$ then what is the related uncertainty in your measurement of the system volume?

C. The Ideal Gas and Incompressible Fluid Models

2.C-1 A piston-cylinder device has volume $V_1 = 1.0\ \text{m}^3$ and contains air at $T_1 = 20°\text{C}$. The piston is frictionless and has no mass. The piston-cylinder device is in an environment with atmospheric pressure $P_{atm} = 1$ atm. Therefore, as long as the piston is free to move, the pressure of the air in the cylinder is also at $P = 1$ atm. Assume that the air in the tank can be modeled as an ideal gas.

a) What is the mass of air in the cylinder (kg)?

b) How many moles of air are in the cylinder (kmol)? How many air molecules (molecules) are in the cylinder?

The air is heated to $T_2 = 100°\text{C}$ with the piston free to move. No air leaks out of the piston during this process.

c) What is the volume of the cylinder at state 2 (m^3)?

The piston is locked in place so that the volume cannot change. The air is cooled to $T_3 = 20°\text{C}$; again, no air leaks out during this process.

d) What is the final pressure in the tank, P_3 (kPa)?

e) Sketch states 1, 2, and 3 on a T-v diagram for the air. Your diagram should clearly show the intersecting lines that define each state.

3 Energy and Energy Transport

3.1 Conservation of Energy Applied to a Closed System

Section 1.4 discusses the application of a balance of an arbitrary quantity applied to a system. The First Law of Thermodynamics, also referred to as conservation of energy, requires that (in the absence of nuclear reactions) energy, like mass, is conserved (i.e., it is neither generated nor destroyed). Therefore, an energy balance expressed on a rate basis for any system can be stated as:

$$\dot{E}_{in} = \dot{E}_{out} + \frac{dE}{dt} \tag{3-1}$$

where $\dot{E}_{in}$ is the rate of energy transfer into the system across its boundary, $\dot{E}_{out}$ is the rate of energy transfer out of the system across its boundary, and $\frac{dE}{dt}$ is the rate of change of energy contained in the system. Integrated over a specified time period, the energy balance can be expressed on an incremental basis as:

$$E_{in} = E_{out} + \Delta E \tag{3-2}$$

where E_{in} is the total amount of energy transferred into the system across its boundary, E_{out} is the total amount of energy transferred out of the system across its boundary, and ΔE is the change in the amount of energy contained in the system during the time period.

The remainder of this chapter discusses each of the terms in the energy balance. Energy can be stored in a system in external forms (as kinetic energy, KE, and potential energy, PE) and as internal energy (U). These forms of energy are discussed in Sections 3.2 and 3.3. Energy can be transferred across the boundary of a system as heat (Q) and work (W). Heat is discussed in Section 3.4 and work is discussed in Section 3.5. Mass that crosses the boundary of an open system will carry some energy with it. However, this chapter is concerned only with closed systems (i.e., systems in which mass does not cross the system boundary). Open systems are discussed in Chapter 4. With these substitutions, an energy balance on a closed system on a rate basis, Eq. (3-1), becomes:

$$\boxed{\dot{Q}_{in} + \dot{W}_{in} = \dot{Q}_{out} + \dot{W}_{out} + \frac{dKE}{dt} + \frac{dPE}{dt} + \frac{dU}{dt}} \tag{3-3}$$

where $\dot{Q}_{in}$ and $\dot{W}_{in}$ are the rates that heat and work, respectively, enter the system across its boundary and $\dot{Q}_{out}$ and $\dot{W}_{out}$ are the rates that heat and work, respectively, leave the system across its boundary. Integrating the energy balance on a rate basis, Eq. (3-3), over a specified time period leads to:

$$\boxed{Q_{in} + W_{in} = Q_{out} + W_{out} + \Delta KE + \Delta PE + \Delta U} \tag{3-4}$$

where Q_{in} and W_{in} are the amounts of heat and work, respectively, that enter the system across its boundary and Q_{out} and W_{out} are the amounts of heat and work, respectively, that leave the system across its boundary during the time period. The terms ΔKE, ΔPE, and ΔU refer to the storage of kinetic energy, potential energy, and internal energy within the system, respectively.

Equations (3-3) and (3-4) are general statements of the First Law of Thermodynamics for a closed system. These equations will be used as the starting point for many problems. Equations (3-3) and (3-4) can be expressed in an equivalent manner by assigning a sign convention to the heat and work terms. In this approach, heat is set to a positive value when energy flows *into* the system. A heat flow out of the system is then represented by a negative value. A common sign convention for the work term is to assign it a positive value when energy flows *out* of the system, i.e., when the system does work on its surroundings. Work done on the system is then represented with a negative value. Using these sign conventions, Eq. (3-3) can be written as

$$\boxed{\sum \dot{Q} - \sum \dot{W} = \frac{dKE}{dt} + \frac{dPE}{dt} + \frac{dU}{dt}} \tag{3-5}$$

and Eq. (3-4) can be written as:

$$\boxed{\sum Q - \sum W = \Delta KE + \Delta PE + \Delta U} \tag{3-6}$$

where the summations refer to the addition of every heat and work effect that is acting on the system. Regardless of the choice of sign convention, the best approach is to draw a sketch of the system that indicates all of the energy flows and the assumed direction of each flow using arrows. The energy balance can then be written so that it corresponds to the sketch. This practice will be illustrated in the examples provided throughout this textbook.

3.2 Forms of Energy

The equilibrium state of a system is described in terms of a small set of characteristics that are called properties. Mass, temperature, pressure, and volume are all properties that were addressed in Chapters 1 and 2. Energy is also a property. A system can possess energy in several forms, including kinetic energy (KE), potential energy (PE), and internal energy (U). The total energy of a system (E) is the sum of these various forms of energy:

$$E = U + KE + PE \tag{3-7}$$

3.2.1 Kinetic Energy

The term *kinetic energy* is used to describe the energy of a system associated with its motion. Motion and energy are directly related. For example, we understand electrical energy is associated with the movement of electrical charges. Electromagnetic energy is associated with the movement of photons. Matter consists of elementary particles and these particles each move with a velocity that contributes to their energy. However, the term kinetic energy, as used in macroscopic thermodynamics, refers to the energy associated with the bulk translational or rotational motion (i.e., the bulk velocity) of a system. The kinetic energy (KE) of a system due to its translational motion is given by:

$$KE = \frac{1}{2} m \tilde{V}^2 \tag{3-8}$$

where m is the mass of the system and $\tilde{V}$ is the bulk velocity of the system in a specified reference frame. Note that kinetic energy is an extensive property in that it linearly depends on mass. Also, note that kinetic energy has no natural zero value since velocity is a relative term. It is common to specify the velocity of a system relative to an observer who is stationary on the surface of Earth. However, this is an arbitrary choice of reference frame. The Earth itself completes one full rotation about its axis every 24 hours. Therefore, if the surface of the Earth were viewed from an appropriate vantage point it would appear to be moving at a rather high velocity. The Earth is moving around the sun, completing one revolution each year. It is not really possible to specify an absolute value of velocity and, as a result, it is not possible to specify an absolute value of the kinetic energy of a system. Fortunately, it is only necessary to determine the change (or rate of change) in the kinetic energy of a system and therefore the choice of reference frame does not matter.

3.2.2 Potential Energy

The term *potential energy* refers to the energy associated with a mass that is located at a specified position in a force field. There are many different types of force fields including magnetic, electrical, and gravitational. In this text, potential energy (PE) will primarily be used to refer to the energy of a mass (m) due to its elevation in a gravitational force field, defined by:

$$PE = m\,g\,z \tag{3-9}$$

where g is the local acceleration of gravity and z is the elevation of the mass in the direction opposing gravity. Potential energy, like kinetic energy, is an extensive property because it is linearly related to mass. Potential energy also has no natural zero value because elevation is a relative term. In order to specify a numerical value of potential energy, it is necessary to arbitrarily specify a reference elevation at which the potential energy is defined to be zero. The value of z is then measured from the reference position. Fortunately, we are only concerned with changes in the potential energy of a system and therefore the choice of a reference position does not matter.

3.2.3 Internal Energy

Kinetic and potential energy are external forms of energy because they are independent of the molecular structure of matter. There are many other types of energy, e.g., chemical, nuclear, magnetic, electrical, and thermal. These forms of energy differ from kinetic and potential energy in that they all depend in some way on the molecular structure of the substance that is being considered. We group all of these forms of energy together and refer to their collective energies as the *internal energy* of a system, U.

Internal energy is an extensive property because its value depends of the mass of the system. The specific internal energy, u, is the ratio of internal energy to mass:

$$u = \frac{U}{m} \tag{3-10}$$

Specific internal energy is an intensive, internal property of matter (like pressure, temperature, and specific volume). Therefore, the specific internal energy is a property that is fixed once the state of the system is known. The determination of specific internal energy for engineering substances is discussed in Section 3.3.

3.3 Specific Internal Energy

Internal energy, like kinetic and potential energy, has no natural zero value. Therefore, it is necessary to arbitrarily define the specific internal energy of a substance to be zero at

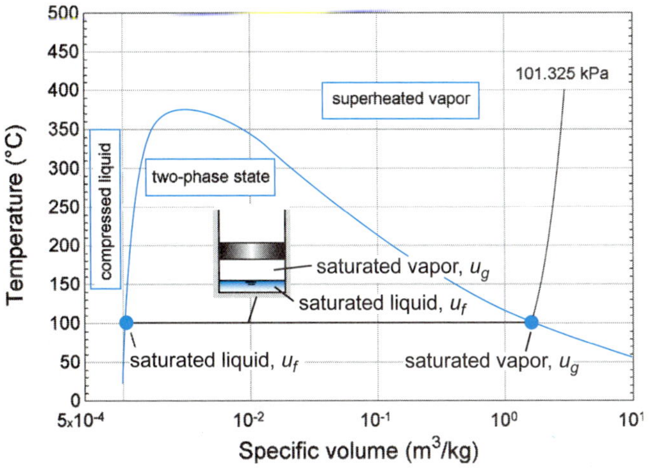

Figure 3-1: Temperature-specific volume diagram for water.

some state that is referred to as the *reference state*. The absolute value of specific internal energy always depends on the choice of the reference state. However, the difference in the specific internal energy of a substance at two specified states is unaffected by the choice of the reference state. In thermodynamic analyses that do not involve changes in molecular structure (i.e., non-reacting systems), we are only concerned with changes (or rates of change) in the specific internal energy of a substance rather than its absolute value and therefore the choice of the reference state does not matter.

The determination of specific internal energy for pure substances follows naturally from the discussion of properties in Chapter 2. The techniques used to determine the specific volume of a substance can be used with little modification to determine its specific internal energy.

3.3.1 Property Tables

Section 2.3 discusses the use of fluid property data that are available in tabular form, e.g., the tables for water that appear in Appendix B. Figure 3-1 illustrates a T-v diagram for water. Property information for the saturated region within the vapor dome is provided in Tables B-1 and B-2. When a state is located in the superheated vapor region (i.e., to the right of the vapor dome), property information is found in Table B-3 (for water). Table B-3 provides an array of property information over a range of pressure and temperature; specific internal energy is one of the properties that are provided at each temperature and pressure in the table. When a state is located in the compressed liquid region (i.e., to the left of the vapor dome), information is found in Table B-4 (for water). The properties of R134a are provided in a similar format in Appendix C.

A two-phase state is located underneath the vapor dome and consists of some liquid in equilibrium with some vapor, as shown in Figure 3-1. The liquid is saturated liquid and the vapor is saturated vapor. The properties of saturated liquid water and saturated vapor water are found in Tables B-1 (in equal increments of temperature) and B-2 (in equal increments of pressure). The specific internal energy of saturated liquid is referred to as u_f and the specific internal energy of saturated vapor is referred to as u_g. As discussed in Section 2.3.1, the relative amount of each phase is expressed in terms of quality:

$$x = \frac{\text{mass of vapor}}{\text{total mass}} = \frac{m_g}{m} \tag{3-11}$$

The mass-average specific internal energy, u, is the ratio of the total internal energy (of both phases) to the total mass (of both phases). The mass-average specific internal energy is analogous to the mass-average specific volume, v, discussed in Chapter 2. The relationship between u and x is:

$$u = u_f + x\,(u_g - u_f) \tag{3-12}$$

Equation (3-12) can be rearranged:

$$x = \frac{(u - u_f)}{(u_g - u_f)} \tag{3-13}$$

3.3.2 EES Fluid Property Data

Section 2.4 discussed the use of the built-in thermodynamic property functions in EES. The IntEnergy function provides the specific internal energy and it is accessed using the same protocol as the Volume, Pressure, or Temperature functions. For example, the EES code below provides the specific internal energy of the refrigerant R134a at $P = 250 \times 10^3$ Pa and $T = 325$ K:

```
$UnitSystem SI, K, Pa, J, Mass, Radian
u=IntEnergy(R134a,P=250000 [Pa],T=325 [K])
```

which results in $u = 272{,}118$ J/kg. Specific internal energy is an intensive internal property and therefore the state of a pure single phase substance can be specified using the specific internal energy and one other intensive internal property. For example, the specific volume of R134a at $P = 250 \times 10^3$ Pa and $u = 300 \times 10^3$ J/kg is obtained from:

```
$UnitSystem SI, K, Pa, J, Mass, Radian
v=volume(R134a,P = 250000 [Pa], u = 300000 [J/kg])
```

which leads to $v = 0.1136$ m^3/kg.

EXAMPLE 3.3-1:

EXAMPLE 3.3-1: HOT STEAM EQUILIBRATING WITH COLD LIQUID WATER

An accident scenario that is being considered in the design of a nuclear power plant involves a steam leak into a rigid insulated tank that is partially filled with liquid water. A simple model of the accident is shown in Figure 1.

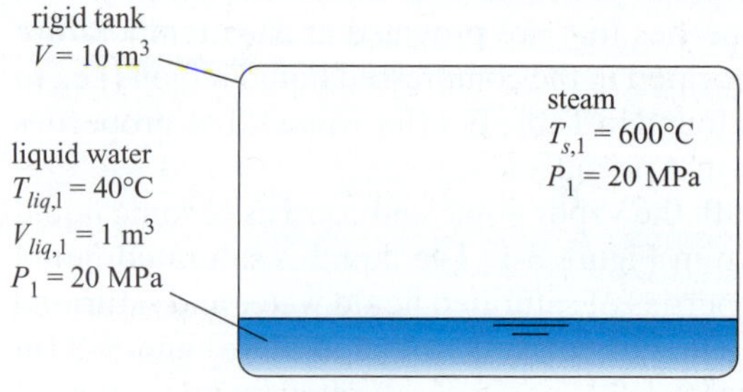

Figure 1: Steam in contact with liquid water.

The steam and liquid water are not initially in thermal equilibrium although they have come to the same pressure. The steam is at temperature $T_{s,1} = 600^\circ$C and pressure $P_1 = 20$ MPa. The liquid water is initially at $T_{liq,1} = 40^\circ$C and pressure $P_1 = 20$ MPa. The total volume of the tank is $V = 10$ m^3 and the volume of the liquid water initially in the tank is $V_{liq,1} = 1\ m^3$. Eventually, the contents of the tank reach a uniform temperature and pressure, T_2 and P_2. The tank is well-insulated and rigid.

a) Determine the final temperature and pressure of the water in the tank using the property tables in Appendix B.

The first step in solving this problem is to locate the initial state of the steam, state *s*,1, and the initial state of the liquid, state *liq*,1. This process is facilitated by sketching a *T*-*v* diagram. The temperature and pressure of the steam ($T_{s,1}$ and P_1) are provided. According to Table B-2, the saturation temperature at $P_1 = 20$ MPa is $T_{sat} = 365.75^\circ$C. Since $T_{s,1} > T_{sat}$, state *s*,1 must lie in the superheated vapor region, as shown in Figure 2.

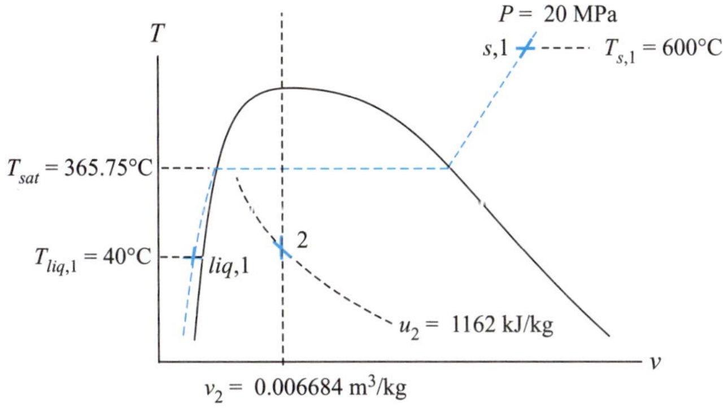

Figure 2: Qualitative $T - v$ diagram for locating states.

The properties for state *s*,1 are found in Table B-3; the specific volume is $v_{s,1} = 0.01817\ m^3/kg$ and the specific internal energy is $u_{s,1} = 3175.3$ kJ/kg. The temperature and pressure of the liquid water ($T_{liq,1}$ and P_1) are provided in the problem statement. Because $T_{liq,1} < T_{sat}$, state *liq*,1 must be in the compressed liquid region, as shown in Figure 2. The properties for state *liq*,1 are found in Table B-4; the specific volume is $v_{liq,1} = 0.9992 \times 10^{-3}\ m^3/kg$ and the specific internal energy is $u_{liq,1} = 165.17$ kJ/kg.

The mass of liquid water in the tank is:

$$m_{liq,1} = \frac{V_{liq,1}}{v_{liq,1}} = \frac{1\ m^3}{} \left| \frac{kg}{0.9992 \times 10^{-3}\ m^3} \right. = 1001\ kg$$

The volume of steam in the tank is:

$$V_{s,1} = V - V_{liq,1} = 10\,m^3 - 1\,m^3 = 9\,m^3$$

The mass of steam in the tank is:

$$m_{s,1} = \frac{V_{s,1}}{v_{s,1}} = \frac{9\,m^3}{} \left| \frac{kg}{0.01817\,m^3} \right. = 495.3\ kg$$

The total mass of water (steam plus liquid) contained in the tank is:

$$m_1 = m_{s,1} + m_{liq,1} = 495.3\,kg + 1000.7\ kg = 1496.1\,kg$$

A mass balance on the tank for the process of going from the initial state (state 1) to the final state that exists after equilibration has occurred (state 2), provides:

$$0 = m_2 - m_1$$

Therefore, the mass of water in the tank does not change:

$$m_2 = m_1 = 1496.1\ \text{kg}$$

The specific volume of the water at state 2 is:

$$v_2 = \frac{V}{m_2} = \frac{10\ \text{m}^3}{}\left|\frac{}{1496.1\ \text{kg}}\right. = 0.006684\ \text{m}^3/\text{kg}$$

The tank experiences no heat or work. Therefore, an energy balance on the tank for the process of going from state 1 to state 2 is:

$$0 = U_2 - U_1$$

or

$$0 = u_2\, m_2 - \left(u_{liq,1}\, m_{liq,1} + u_{s,1}\, m_{s,1}\right)$$

The specific internal energy associated with the final state is:

$$u_2 = \frac{\left(u_{liq,1}\, m_{liq,1} + u_{s,1}\, m_{s,1}\right)}{m_2}$$

$$= \frac{\left(165.17\frac{\text{kJ}}{\text{kg}}\, 1001\ \text{kg} + 3175.3\frac{\text{kJ}}{\text{kg}}\, 495.3\ \text{kg}\right)}{1496.1\ \text{kg}} = 1162\ \frac{\text{kJ}}{\text{kg}}$$

State 2 is specified by the specific volume and specific internal energy. It lies on the line of constant specific volume, $v_2 = 0.006684\ \text{m}^3/\text{kg}$ at the point where $u_2 = 1162\ \text{kJ/kg}$ (see Figure 2). It is not easy to determine the pressure and temperature associated with this state using tabular data. The process is necessarily iterative and tedious. The data in Appendix B are presented over regular intervals of pressure and temperature. Therefore, it is easiest to pick a candidate pressure and determine the specific internal energy at that pressure and $v_2 = 0.006684\ \text{m}^3/\text{kg}$. The discrepancy between the specific internal energy at that pressure and the required value of u_2 is used to select the next candidate pressure.

The final pressure is expected to be lower than the initial pressure. Therefore, our first candidate pressure might be $P_2 = 15$ MPa. According to Appendix B-2, the specific volume of saturated liquid at 15 MPa is $v_f = 0.0016572\ \text{m}^3/\text{kg}$ and the specific volume of saturated vapor at 15 MPa is $v_g = 0.010341\ \text{m}^3/\text{kg}$. Therefore, the quality associated with $v_2 = 0.006684\ \text{m}^3/\text{kg}$ is:

$$x = \frac{(v_2 - v_f)}{(v_g - v_f)} = \frac{(0.006684 - 0.0016572)}{(0.010341 - 0.0016572)} = 0.5789$$

The specific internal energy associated with this quality and the candidate pressure is:

$$u = u_f + x(u_g - u_f) = 1585.5\ \frac{\text{kJ}}{\text{kg}} + 0.5789(2455.7 - 1585.5)\frac{\text{kJ}}{\text{kg}} = 2089\frac{\text{kJ}}{\text{kg}}$$

The specific internal energy is much too large compared to the required value, $u_2 = 1162$ kJ/kg; the discrepancy between the calculated and required value of u is 927.4 kJ/kg. Therefore, the candidate pressure should be reduced. The results of this first iteration are recorded in the first row of Table 1. Guided by the results of the first iteration, a much lower candidate pressure is selected for iteration

Table 1: Iterations required to estimate P_2 and T_2

Iteration	P (kPa)	v_f (m^3/kg)	v_g (m^3/kg)	x	u (kJ/kg)	$u - u_2$ (kJ/kg)	T
1	15×10^3	0.0016572	0.010341	0.5789	2089	927.4	342.16°C
2	2×10^3	0.0011767	0.099587	0.05598	1001	−161.1	212.38°C
3	4×10^3	0.0012524	0.049779	0.1120	1253	90.52	250.35°C
4	3.25×10^3	0.0012258	0.061508	0.09058	1168	6.50	238.33°C

#2, $P_2 = 2000$ kPa. Repeating the calculations (see Table 1, row 2) leads to $u =$ 1001 kJ/kg, which is 161.1 kJ/kg less than the required value.

The two candidate pressures associated with iterations #1 and #2 bracket the actual pressure; the first pressure was too high and therefore led to $u > u_2$ while the second was too low resulting in $u < u_2$. Iteration #3 proceeds by linearly interpolating between iterations #1 and #2:

$$P = 15{\times}10^3\ \text{kPa} + \frac{(0 - 927.4)}{(-161.1 - 927.4)}(2{\times}10^3 - 15{\times}10^3)\ \text{kPa} = 3924\ \text{kPa}$$

Repeating the calculations for $P = 4000$ kPa (see row 3 of Table 1) results in $u =$ 1253 kJ/kg which is 90.52 kJ/kg in excess of the required value. One final iteration linearly interpolates between iterations #2 and #3:

$$P = 2{\times}10^3\ \text{kPa} + \frac{(0 - (-161.1))}{(90.52 - (-161.1))}(4{\times}10^3 - 2{\times}10^3)\ \text{kPa} = 3281\ \text{kPa}$$

Repeating the calculations for $P = 3250$ kPa (see row 4 of Table 1) results in $u =$ 1168 kJ/kg which is 6.50 kJ/kg in excess of the required value. We could continue iterating if a more accurate result were required.

The process of determining the properties associated with state 2 by hand using the property tables is tedious. One could argue that the thermodynamics portion of the problem was complete once Figure 2 had been sketched; we knew that state 2 was specified by u_2 and v_2. The iterations are only required because the property data were available in a tabular form that is not convenient for this problem. This process is not required if computerized property data are used.

b) Repeat the problem using EES.

The inputs are entered in EES:

```
$UnitSystem SI, K, Pa, J, Mass, Radian
T_s[1]=converttemp(C,K,600 [C])          "initial temperature of steam"
P[1]=20000 [kPa]*convert(kPa,Pa)         "initial pressure"
T_liq[1]=converttemp(C,K,40 [C])         "initial temperature of liquid"
Vol=10 [m^3]                             "volume of tank"
Vol_liq[1]=1 [m^3]                       "initial volume of liquid water"
```

States s,1 and liq,1 are determined by their temperature and pressure. The properties of the steam and liquid ($u_{s,1}$, $v_{s,1}$, $u_{liq,1}$, and $v_{liq,1}$) are obtained using EES' internal property routines.

```
u_s[1]=intenergy(water,T=T_s[1],P=P[1])      "specific internal energy of the steam"
v_s[1]=volume(water,T=T_s[1],P=P[1])         "specific volume of the steam"
u_liq[1]=intenergy(water,T=T_liq[1],P=P[1])  "specific internal energy of the liquid water"
v_liq[1]=volume(water,T=T_liq[1],P=P[1])     "specific volume of the liquid water"
```

The mass of liquid in the tank is:

$$m_{liq,1} = \frac{V_{liq,1}}{v_{liq,1}}$$

The volume of steam in the tank is:

$$V_{s,1} = V - V_{liq,1}$$

The mass of steam in the tank is:

$$m_{s,1} = \frac{V_{s,1}}{v_{s,1}}$$

The total mass of water contained in the tank is:

$$m_1 = m_{s,1} + m_{liq,1}$$

```
m_liq[1]=Vol_liq[1]/v_liq[1]   "initial mass of liquid water"
Vol_s[1]=Vol-Vol_liq[1]        "initial volume of steam"
m_s[1]=Vol_s[1]/v_s[1]         "initial mass of steam"
m[1]=m_s[1]+m_liq[1]           "total mass of water (liquid and steam)"
```

A mass balance on the tank for the process of going from state 1 to state 2 provides:

$$m_2 = m_1$$

The specific volume of the water at state 2 is:

$$v_2 = \frac{V}{m_2}$$

An energy balance on the tank for the process of going from state 1 to state 2 is:

$$0 = U_2 - U_1$$

or

$$0 = u_2\, m_2 - (u_{liq,1}\, m_{liq,1} + u_{s,1}\, m_{s,1})$$

```
m[2]=m[1]                                   "mass balance on tank"
v[2]=Vol/m[2]                               "final specific volume"
0=m[2]*u[2]-m_s[1]*u_s[1]-m_liq[1]*u_liq[1] "energy balance – leads to final specific internal energy"
```

State 2 is fixed by the specific volume and the specific internal energy. The other properties associated with the state (P_2, T_2, and x_2) can be obtained using EES' internal property routines.

```
P[2]=pressure(water,u=u[2],v=v[2])            "final pressure"
P_2_kPa=P[2]*convert(Pa,kPa)                  "in kPa"
T[2]=temperature(water,u=u[2],v=v[2])         "final temperature"
T_2_C=converttemp(K,C,T[2])                   "in C"
x[2]=quality(water,u=u[2],v=v[2])             "final quality"
```

which results in $P_2 = 3189$ kPa, $T_2 = 237.3°C$, and $x_2 = 0.0888$. Note that EES is solving the same set of non-linear equations that we solved iteratively by hand in part (a). Therefore EES must carry out a similar set of iterations. These iterations, like ours, must start from some initial guess values (analogous to our initial candidate pressure) and proceed to the final solution. In many problems, it will be necessary to adjust the guess values used by EES so that they are reasonable. (The default guess values were sufficient to provide the solution to this problem.) To access the guess values, select Variable Info from the Options menu (Figure 3).

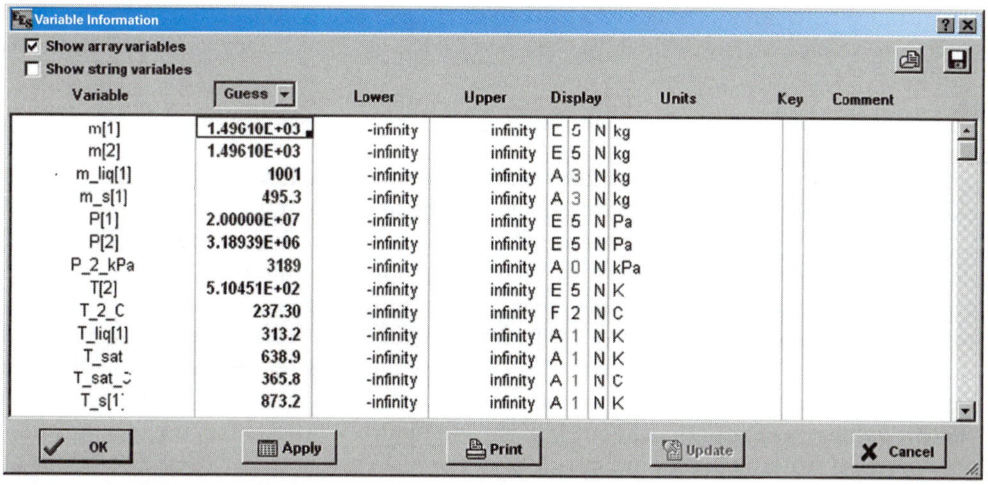

Figure 3: Variable Information window.

Each variable in the problem is listed in the Variable Information dialog. The user can adjust the guess value for each variable as well as the allowable lower and upper limits; these settings can sometimes help EES find the solution to a difficult problem.

3.3.3 Ideal Gas

According to the phase rule, Eq. (2-1), a pure single-phase substance has two degrees of freedom. Therefore, the specific internal energy of a pure single-phase substance is fixed if two intensive properties, such as temperature and specific volume, are known. The specific internal energy (u) of a substance can be expressed as a function of temperature (T) and specific volume (v):

$$u = f(T, v) \tag{3-14}$$

Taking the total derivative of Eq. (3-14) leads to:

$$du = \left(\frac{\partial u}{\partial T}\right)_v dT + \left(\frac{\partial u}{\partial v}\right)_T dv \tag{3-15}$$

The partial derivative of specific internal energy with respect to temperature at constant specific volume is defined as the *specific heat capacity at constant volume*:

$$\boxed{c_v = \left(\frac{\partial u}{\partial T}\right)_v} \tag{3-16}$$

The specific heat capacity at constant volume is a measurable thermodynamic property, like temperature or pressure. The specific heat capacity at constant volume is returned by the EES function cv, which is used like any of the other thermodynamic functions in EES. Substituting Eq. (3-16) into Eq. (3-15) leads to:

$$du = c_v\, dT + \left(\frac{\partial u}{\partial v}\right)_T dv \tag{3-17}$$

The ideal gas model is discussed in Section 2.5. The model is applicable in the limit that the distance between molecules of a gas becomes large and therefore the intermolecular forces are negligible. In this limit, the behavior of the gas can be represented by the ideal gas law:

$$P\,v = R\,T \tag{3-18}$$

where R is the gas constant. If the ideal gas law is valid then the partial derivative of specific internal energy with respect to specific volume at constant temperature, the second term in Eq. (3-17), must also be zero. This result is derived in Chapter 10. Figure 3-2 illustrates the specific internal energy of water vapor as a function of specific volume at a fixed temperature of 250°C. As the specific volume increases, the slope of the line, which corresponds to the quantity $(\frac{\partial u}{\partial v})_T$, approaches zero. Recall that the accuracy of the ideal gas law, Eq. (3-18), improves as specific volume increases (i.e., as the distance between molecules increases). Water vapor obeys the ideal gas law at 250°C when the specific volume is greater than about 1 m^3/kg and Figure 3-2 shows that the specific internal energy of water becomes independent of specific volume under these conditions as well.

Figure 3-3 illustrates $(\frac{\partial u}{\partial v})_T$ as a function of pressure for water vapor at various values of temperature. Notice the similarity between Figure 3-3 and Figure 2-22 in Section 2.5, which showed the deviation of the specific volume from the ideal gas law as a function of pressure for water vapor at various values of temperature. The ideal gas law is most appropriate at high temperature and low pressure and this is also the region where $(\frac{\partial u}{\partial v})_T$ approaches zero. Therefore, according to Eq. (3.17), the specific internal energy of an ideal gas is independent of specific volume and only a function of temperature.

$$u = f(T) \text{ for an ideal gas} \tag{3-19}$$

As discussed in Section 2.4.1, there are many fluids in EES that appear to be listed twice. For example, both 'N2' and 'Nitrogen' are included in EES. The chemical symbol notation ('N2') indicates that the fluid is modeled as an ideal gas whereas the full name ('Nitrogen') indicates that the fluid is modeled as a real fluid and therefore exhibits

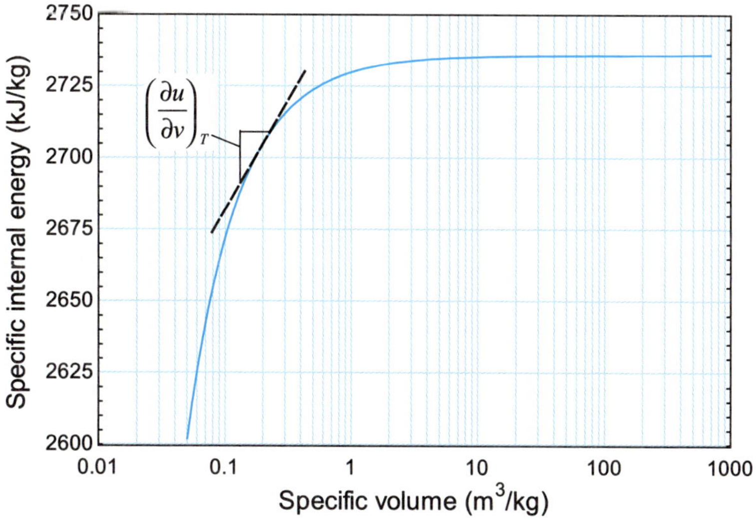

Figure 3-2: Specific internal energy as a function of specific volume for water vapor at 250°C.

compressed liquid, saturated, superheated vapor and supercritical behavior. It is computationally simpler to use the ideal gas model for problems where the substance can be adequately modeled as an ideal gas, although both models should work under these conditions. Equation (3-19) indicates that only temperature is required to determine the specific internal energy of an ideal gas. Therefore, when the IntEnergy function in EES is accessed for an ideal gas substance (e.g., 'N2'), it is not necessary (or even permitted) to specify both temperature and pressure or temperature and specific volume. The EES code below:

```
$UnitSystem SI, K, Pa, J, Mass, Radian
u=IntEnergy(N2,P=100000 [Pa],T=300[K])
```

will lead to the error shown in Figure 3-4.

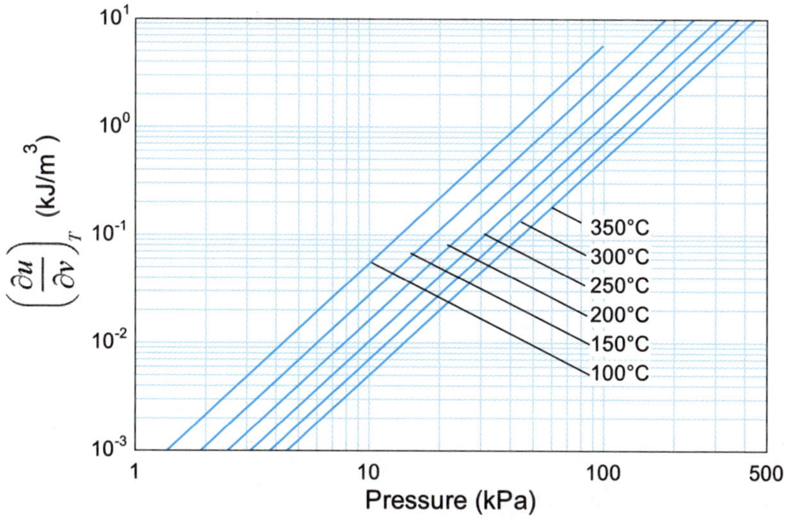

Figure 3-3: Partial derivative of specific internal energy with respect to specific volume at constant temperature as a function of pressure for water vapor at various values of temperature.

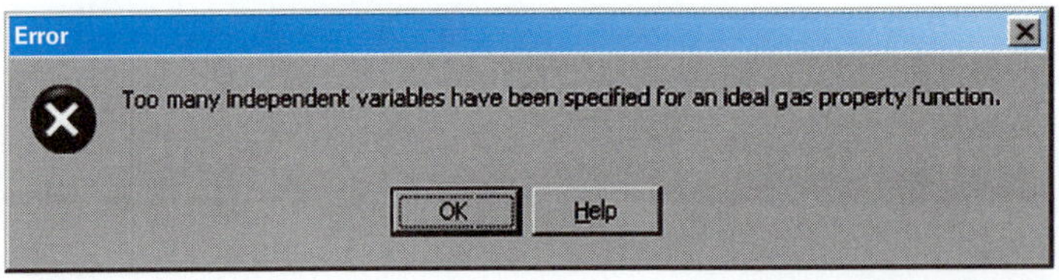

Figure 3-4: Error associated with accessing the IntEnergy function for an ideal gas with too many independent variables.

The internal energy of an ideal gas is only a function of temperature and therefore the correct call to the IntEnergy function for the substance 'N2' is:

```
$UnitSystem SI, K, Pa, J, Mass, Radian
u=IntEnergy(N2,T=300[K])
```

which provides $u = -87121$ J/kg. Note that the specific internal energy returned by EES is negative due to the choice of reference state used by EES for this substance.

Figure 3-3 and Figure 3-4 shows that $(\frac{\partial u}{dv})_T$ becomes zero under conditions where a substance behaves as an ideal gas; therefore, for an ideal gas Eq. (3-17) becomes:

$$du = \frac{du}{dT}\,dT = c_v\,dT \quad \text{for an ideal gas} \tag{3-20}$$

Equations (3-19) and (3-20) together show that the specific heat capacity at constant volume of an ideal gas must be a function of temperature only. Figure 3-5 illustrates c_v as a function of temperature for several gases in the ideal gas limit (i.e., in the limit that specific volume approaches infinity). The behavior of c_v with temperature in the ideal gas limit can be understood using statistical thermodynamics, as discussed in Chapter 15. In general, c_v is an increasing function of temperature, but it is essentially constant for monatomic gases such as helium or argon.

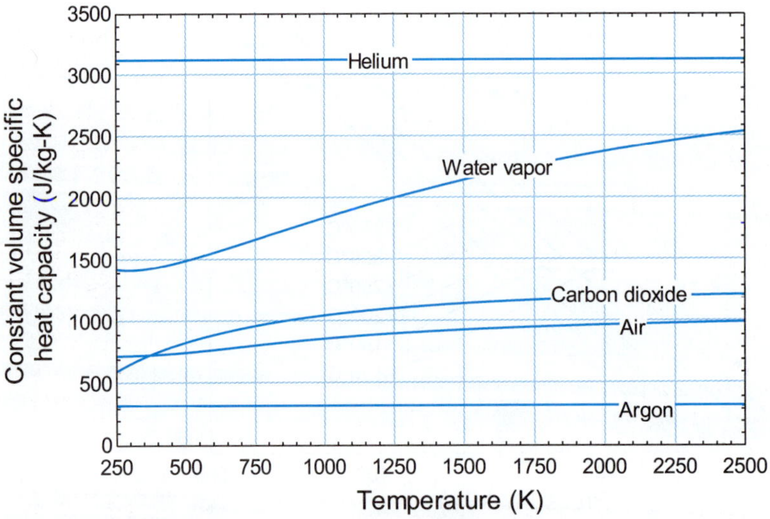

Figure 3-5: Specific heat capacity at constant volume as a function of temperature for several gases in the ideal gas limit (i.e., as $v \rightarrow \infty$).

The internal energy of an ideal gas is obtained by the integration of Eq. (3-20):

$$u - u_{ref} = \int_{T_{ref}}^{T} c_v(T)\, dT \quad \text{for an ideal gas} \tag{3-21}$$

where T_{ref} is the reference temperature and u_{ref} is the value of specific internal energy at the reference temperature. The values of T_{ref} and u_{ref} are arbitrary and it is often convenient to set u_{ref} to zero.

Appendix E provides the specific heat capacity at constant volume for air, in the limit that air behaves as an ideal gas, as a function of temperature. Appendix E also provides the specific internal energy for air in the ideal gas limit calculated using Eq. (3-21). The reference temperature used for Appendix E is $T_{ref} = 0$ K and the reference value of the specific internal energy, u_{ref}, is defined to be zero.

Figure 3-5 shows that it is often acceptable to assume that c_v is constant, particularly if the temperature change associated with the problem is small or if the substance is a monatomic gas. The values of c_v for several gases at room temperature are provided in Table D-1 of Appendix D. In the limit that c_v is assumed to be constant, Eq. (3-21) can be simplified to:

$$u - u_{ref} = c_v \left(T - T_{ref}\right) \quad \text{for an ideal gas with } c_v \text{ constant} \tag{3-22}$$

If u_{ref} is defined to be zero at $T_{ref} = 0$ K and c_v is constant then Eq. (3-22) reduces to:

$$u = c_v\, T \quad \text{for an ideal gas with } c_v \text{ constant, } T_{ref} = 0\,\text{K, and } u_{ref} = 0 \tag{3-23}$$

Equation (3-23) is a particularly simple model for the specific internal energy of an ideal gas.

It should be clear that there are several options available to compute the specific internal energy change of a gas. For example, suppose you need to compute the change in the specific internal energy of air at $P = 500$ kPa as it changes temperature from $T_1 = 500$ K to $T_2 = 1270$ K. The most accurate method of computing $\Delta u = u_2 - u_1$ is to use the substance 'Air_ha' in EES; this option accesses the complete set of property information for air, making no assumption about ideal gas behavior.

```
$UnitSystem SI, K, Pa, J, Mass, Radian
P=500 [kPa]*convert(kPa,Pa)
T_2=1270 [K]
T_1=500 [K]
DELTAu=IntEnergy(Air_ha,T=T_2,P=P)-IntEnergy(Air_ha,T=T_1,P=P)
```

Using substance 'Air_ha' provides $\Delta u = 636{,}765$ J/kg.

Alternatively, a slightly less accurate estimate of the change in specific internal energy can be obtained using the substance 'Air' in EES; this option assumes that air behaves according to the ideal gas law, but it does not assume that the specific heat capacity at constant volume is constant. That is, the specific internal energy is computed according to Eq. (3-21).

```
$UnitSystem SI, K, Pa, J, Mass, Radian
P=500 [kPa]*convert(kPa,Pa)
T_2=1270 [K]
T_1=500 [K]
DELTAu_IG_1=IntEnergy(Air,T=T_2)-IntEnergy(Air,T=T_1)
```

Using substance 'Air' leads to $\Delta u_{IG,1} = 636{,}175$ J/kg. This answer is only 0.1% in error relative to Δu calculated with the substance 'Air_ha' at these conditions. Notice that the value of the pressure played no role in the calculation of $\Delta u_{IG,1}$ because the specific internal energy of an ideal gas is only a function of temperature.

An equivalent estimate of the change in specific internal energy can be obtained using the ideal gas properties of air that are summarized in Appendix E. By interpolation from Table E-1, the value of specific internal energy at 1270 K is $u_2 = 996.0$ kJ/kg. The specific internal energy in Table E-1 at 500 K is $u_1 = 359.8$ kJ/kg. Therefore, the change in specific internal energy is:

$$\Delta u_{IG,2} = u_2 - u_1 = \frac{(996.0 - 359.8)\,\text{kJ}}{\text{kg}} \left\| \frac{1000\,\text{J}}{\text{kJ}} \right. = 636{,}200\,\text{J/kg} \qquad (3\text{-}24)$$

which is essentially the same as the value obtained using the substance 'Air' in EES. Any difference between $\Delta u_{IG,2}$ and $\Delta u_{IG,1}$ is due to the interpolation required to use Table E-1, as the underlying model used to obtain both quantities is the same.

Finally, it is possible to estimate the specific internal energy change by approximating c_v as being constant. The average temperature of the air is 885 K and, by interpolation in Appendix E, the specific heat capacity at constant volume at 885 K is $c_v = 0.8305$ kJ/kg-K. The change in specific internal energy obtained from Eq. (3-23) is:

$$\Delta u_{IG,3} = c_v(T_2 - T_1) = \frac{0.8305\,\text{kJ}}{\text{kg}} \left| \frac{(1270 - 500)\,\text{K}}{} \right\| \frac{1000\,\text{J}}{\text{kJ}} = 636{,}466\,\text{J/kg} \qquad (3\text{-}25)$$

which is 0.4% in error relative to Δu. For many problems, any of these methods will provide acceptable accuracy.

A word of caution. It is important that you pick a model of the substance involved in the problem and then stick with it for the duration of the problem. For example, it is not acceptable or appropriate to calculate u_2 using the substance 'Air_ha' and calculate u_1 using the substance 'Air'. Because the properties of the two substances may be defined using different reference states, this mistake can lead to major errors.

3.3.4 Incompressible Substances

No real substance is truly incompressible, but liquids and solids are often nearly incompressible since very large pressure changes are needed to produce a measurable change in their specific volume. The specific internal energy of an incompressible substance, like that of an ideal gas, is only a function of temperature.

$$u = f(T) \quad \text{for an incompressible substance} \qquad (3\text{-}26)$$

The specific heat capacity of an incompressible substance is defined as:

$$\boxed{c = \frac{du}{dT} \quad \text{for an incompressible substance}} \qquad (3\text{-}27)$$

Note that there is no need for a subscript v on the specific heat capacity of an incompressible substance because the specific heat at constant volume and the specific heat at constant pressure (defined in Chapter 4) are nearly identical. The specific heat capacity of an incompressible substance depends only on temperature. Section 2.6 discussed the Solid/Liquid property library that is available in EES for incompressible substances. The specific heat capacity of these substances can be obtained using the `c_` function. For example, the specific heat capacity of aluminum at 20°C can be obtained according to:

```
$UnitSystem SI, K, Pa, J, Mass, Radian
T=converttemp(C,K,20[C])
c=c_('Aluminum',T)
```

which results in $c = 895.8$ J/kg-K. Table D-2 in Appendix D provides the specific heat capacity for several common liquid and solid substances that can be modeled as being incompressible at room temperature.

By rearranging and integrating Eq. (3-27), the change in the specific internal energy associated with an incompressible substance is obtained according to:

$$u_2 - u_1 = \int_{T_1}^{T_2} c(T)\, dT \quad \text{for an incompressible substance} \tag{3-28}$$

We will often assume that the specific heat capacity is constant over the temperature range of interest, which allows Eq. (3-28) to be simplified.

$$u_2 - u_1 = c\,(T_2 - T_1) \quad \text{for an incompressible substance with constant } c \tag{3-29}$$

EXAMPLE 3.3-2: AIR IN A TANK

Hot air is stored in a spherical metal tank, as shown in Figure 1.

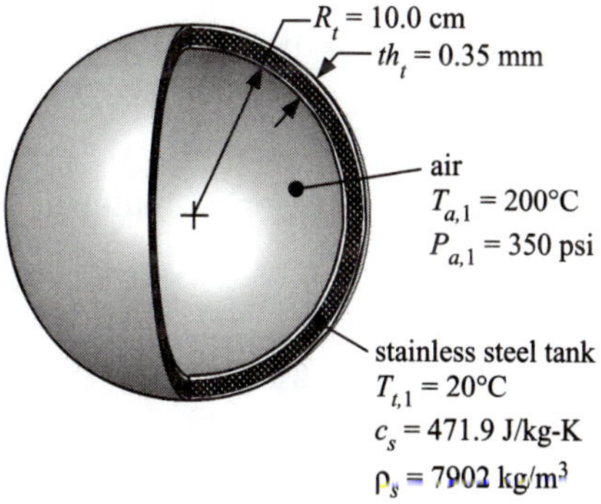

Figure 1: Hot air in a spherical metal tank.

The air in the tank is initially at $T_{a,1} = 200$°C and $P_{a,1} = 350$ psi and the tank material is initially at $T_{t,1} = 20$°C. The inner radius of the tank is $R_t = 10$ cm and the tank wall is $th_t = 0.35$ mm thick. The tank is made of stainless steel with specific heat capacity $c_s = 471.9$ J/kg-K and density $\rho_s = 7902$ kg/m^3. The outer surface of the tank is insulated. The air in the tank may be modeled as an ideal gas and the stainless steel tank wall may be modeled as an incompressible substance.

a) Eventually, the air thermally equilibrates with the tank material. Determine the final temperature of the air and the tank after they equilibrate. Determine the final pressure of the air in the tank.

The inputs are entered in EES:

```
$UnitSystem SI, K, Pa, J, Mass, Radian
c_s=471.9 [J/kg-K]                          "specific heat capacity of tank material"
rho_s=7902 [kg/m^3]                         "density of tank material"
R_t=10 [cm]*convert(cm,m)                   "inner radius of tank"
th_t=0.35 [mm]*convert(mm,m)                "thickness of tank wall"
T_t[1]=converttemp(C,K,20 [C])              "initial tank temperature"
T_a[1]=converttemp(C,K,200 [C])             "initial air temperature"
P_a[1]=350 [psi]*convert(psi,Pa)            "initial air pressure"
```

The initial state of the air is fixed by its temperature and pressure. The specific volume and specific internal energy of the air ($v_{a,1}$ and $u_{a,1}$) are obtained using the internal property functions in EES with the substance 'Air', which corresponds to air modeled as an ideal gas.

```
v_a[1]=volume(Air,T=T_a[1],P=P_a[1])        "initial specific volume of air"
u_a[1]=intenergy(Air,T=T_a[1])              "initial specific internal energy of air"
```

The volume of the air within the tank is:

$$V_a = \frac{4\pi}{3} R_t^3$$

and the mass of air in the tank is:

$$m_a = \frac{V_a}{v_{a,1}}$$

```
Vol_a=4*pi*(R_t)^3/3                        "volume of air"
m_a=Vol_a/v_a[1]                            "mass of air"
```

The volume of tank material is:

$$V_t = \frac{4\pi}{3}\left[(R_t + th_t)^3 - R_t^3\right]$$

and the mass of the tank material (stainless steel) is:

$$m_t = V_t\, \rho_s$$

```
Vol_t=4*pi*((R_t+th_t)^3-R_t^3)/3           "volume of tank material"
m_t=Vol_t*rho_s                             "mass of tank material"
```

The specific volume of the air in the tank does not change during the equilibration process:

$$v_{a,2} = v_{a,1} \tag{1}$$

An energy balance on the system that consists of both the tank material and the air can be written as:

$$0 = m_t\,(u_{t,2} - u_{t,1}) + m_a\,(u_{a,2} - u_{a,1})$$

Substituting the relationship for the change in the specific internal energy of an incompressible substance (stainless steel) with a constant specific heat capacity, Eq. (3-29), allows the energy balance to be written as:

$$0 = m_t\, c_s\,(T_{t,2} - T_{t,1}) + m_a\,(u_{a,2} - u_{a,1}) \tag{2}$$

The temperatures of the tank material and the air will be equal when a condition of thermal equilibrium has been reached.

$$T_{t,2} = T_{a,2} \tag{3}$$

Equations (1), (2), and (3) provide three equations with four unknowns: $v_{a,2}$, $T_{t,2}$, $T_{a,2}$, and $u_{a,2}$. We need one additional equation in order to solve the problem. The last relationship comes from the recognition that two properties are sufficient to fix the state of the air. Therefore, the properties $v_{a,2}$, $T_{a,2}$, and $u_{a,2}$ are not independent; they must be related according to the property behavior of air. Because the air is being modeled as an ideal gas, $u_{a,2}$ can be evaluated given $T_{a,2}$.

$$u_{a,2} = f(T_{a,2}) \tag{4}$$

Equation (4) provides the final equation.

```
v_a[2]=v_a[1]                                   "specific volume of air"
0=m_t*c_s*(T_t[2]-T_t[1])+m_a*(u_a[2]-u_a[1])   "energy balance"
T_a[2]=T_t[2]                                   "thermal equilibration"
u_a[2]=intenergy(Air,T=T_a[2])                  "state of air is given by T and v"
```

These four equations are non-linear and implicit. Therefore, EES will attempt to solve them iteratively, starting from an initial guess solution in the manner that is discussed in Example 3.3-1. Solving the EES code at this point may produce the error message shown in Figure 2.

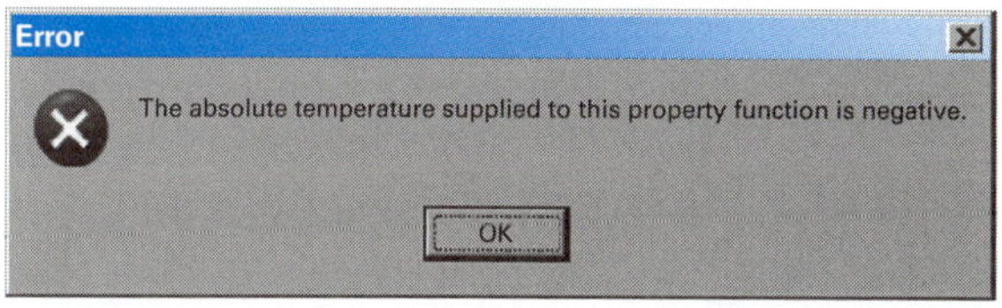

Figure 2: Error message.

The error message does not necessarily mean that there is an error in the equations that have been entered in EES. In this case, the guess values of the variables that are used to start the process of iteratively solving the equations are not very good. Select Variable Info from the Options menu in order to display the Variable Information dialog shown in Figure 3.

The variables involved in the problem are listed in the Variable Information dialog in alphabetical order. The second column lists the guess value associated with each variable. In the time required for this dialog to appear, EES has pre-calculated the value of many of the variables; these values are shown in bold type. The guess values of the variables that have not been explicitly set in the Equations window (e.g., P_a[2], T_2_C, T_a[2], and T_t[2]) are, by default, equal to 1, which is not a great guess value for many physical parameters. In this problem, it is not possible for the final temperature of the tank material or the air to be 1 K and it is not even possible to determine property values for air at this low temperature. Therefore, it is necessary to reset some of these guess values to more reasonable numbers. For example, a reasonable guess value for the variables T_a[2] and T_t[2] is 300 K. With this change, the problem should converge to a solution with no difficulty. Once a solution has been obtained, it is useful to select Update Guesses from the Calculate menu; this action will set each of the guess values according to the current solution.

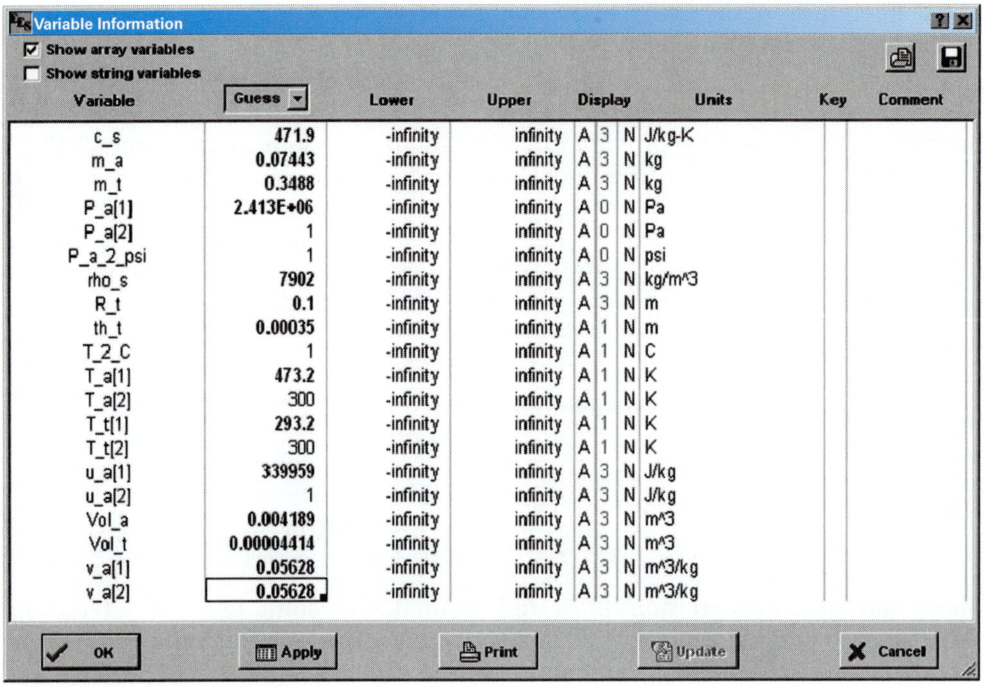

Variable Information

☑ Show array variables

☐ Show string variables

Variable	Guess	Lower	Upper	Display	Units	Key	Comment
c_s	471.9	-infinity	infinity	A 3 N	J/kg-K		
m_a	0.07443	-infinity	infinity	A 3 N	kg		
m_t	0.3488	-infinity	infinity	A 3 N	kg		
P_a[1]	2.413E+06	-infinity	infinity	A 0 N	Pa		
P_a[2]	1	-infinity	infinity	A 0 N	Pa		
P_a_2_psi	1	-infinity	infinity	A 0 N	psi		
rho_s	7902	-infinity	infinity	A 3 N	kg/m^3		
R_t	0.1	-infinity	infinity	A 3 N	m		
th_t	0.00035	-infinity	infinity	A 1 N	m		
T_2_C	1	-infinity	infinity	A 1 N	C		
T_a[1]	473.2	-infinity	infinity	A 1 N	K		
T_a[2]	300	-infinity	infinity	A 1 N	K		
T_t[1]	293.2	-infinity	infinity	A 1 N	K		
T_t[2]	300	-infinity	infinity	A 1 N	K		
u_a[1]	339959	-infinity	infinity	A 3 N	J/kg		
u_a[2]	1	-infinity	infinity	A 3 N	J/kg		
Vol_a	0.004189	-infinity	infinity	A 3 N	m^3		
Vol_t	0.00004414	-infinity	infinity	A 3 N	m^3		
v_a[1]	0.05628	-infinity	infinity	A 3 N	m^3/kg		
v_a[2]	0.05628	-infinity	infinity	A 3 N	m^3/kg		

OK Apply Print Update Cancel

Figure 3: Variable Information dialog.

The pressure of the air in the tank, $P_{a,2}$, is determined using the internal property routines in EES.

```
T_2_C=converttemp(K,C,T_a[2])              "final temperature, in C"
P_a[2]=pressure(Air,T=T_a[2],v=v_a[2])     "final pressure"
P_a_2_psi=P_a[2]*convert(Pa,psi)           "in psi"
```

Solving results in $T_2 = 64.6°\text{C}$ and $P_{a,2} = 249.8$ psi. It is interesting to notice that the heat capacity of the tank wall, often neglected in thermodynamics problems, is actually very important in real problems involving tanks filled with gas. The final temperature of the air is much closer to the initial temperature of the tank than it is to the initial temperature of the air, even for this very thin walled stainless steel tank.

3.4 Heat

Heat is defined as the energy that transfers across the boundary of a system as a result of a temperature difference between the surface of the system and its surroundings. Heat is a quantity of energy with units of J or Btu that is represented by the symbol Q. In this text, the sign associated with a heat transfer will usually be defined with a system schematic that includes an arrow indicating the direction of energy flow term of interest. A positive value of heat indicates that the heat transfer is positive in the direction shown while a negative value means that it is actually in the opposite direction. An alternative method that is equally valid is to use a uniform sign convention for all problems. The most common sign convention defines Q to have a positive value when energy transfers from the surroundings to the system and a negative value when energy transfers from the system to the surroundings. In any case, it is necessary to consider the direction of the heat flow when doing an energy balance. A system with a boundary that is not experiencing a heat transfer (i.e., a system for which $Q = 0$) is referred to as an *adiabatic* system.

Energy transfer processes, such as heat, are necessarily rate phenomena. Time is required for a finite amount of energy to transfer across the boundary of a system as heat. Heat is the time integral of the heat transfer rate, $\dot{Q}$:

$$Q = \int_{o}^{t_{process}} \dot{Q}\, dt \tag{3-30}$$

where $t_{process}$ is the duration of the process.

It is necessary to clearly define a system and specify its boundary before it is possible to consider heat because heat is defined as the transfer of energy across the boundary of a system. If you have not chosen the system, there is no boundary and the definition of heat is ambiguous.

3.4.1 Heat Transfer Mechanisms

There are two mechanisms that cause heat transfer, referred to as *conduction* and *radiation*. Heat is important to the study of thermodynamics, but a detailed discussion of heat transfer mechanisms and the rate equations that can be used to calculate the rate of heat transfer for a specific situation is beyond the scope of this textbook. Entire text books are dedicated to this subject; in fact, the authors of this textbook have also written a heat transfer textbook that may be of interest (Nellis and Klein, 2009). Only a brief description of heat transfer mechanisms is provided here.

Heat transfer due to conduction is the result of the interactions between micro-scale energy carriers within a material. The type of energy carriers depends upon the structure of the material. For example, in a gas or a liquid the energy carriers are individual molecules. The energy carriers in a solid may be electrons or phonons (i.e., vibrations in the structure of the solid). The transfer of energy by conduction is fundamentally related to the interactions of these energy carriers; more energetic (i.e., higher temperature) energy carriers interact with and transfer energy to less energetic (i.e., lower temperature) energy carriers. The result is a net flow of energy from hot to cold. Regardless of the type of energy carriers involved, conduction can be characterized by Fourier's Law provided that the length and time scales of the problem are large relative to the distance and time between energy carrier interactions (which is almost always the case). Fourier's Law relates the *heat flux* (i.e., the heat transfer rate per unit area) to the temperature gradient. Fourier's Law written in the x-direction provides:

$$\frac{\dot{Q}}{A} = -k\,\frac{\partial T}{\partial x} \tag{3-31}$$

where $\dot{Q}/A$ is the heat flux in the x-direction due to conduction and k is the thermal conductivity of the material. Fourier's Law actually provides the definition of thermal conductivity, which is a transport property of matter. Thermal conductivity varies widely depending on the type of material and its state. The property thermal conductivity has been extensively measured and values have been tabulated in various references, for example NIST (2005). The thermal conductivity of many substances can be obtained using EES with the Conductivity function (for engineering fluids) and the k_ function (for incompressible solids and liquids).

Radiation heat transfer is the process in which energy transfer occurs between surfaces due to the emission and absorption of electromagnetic waves. Energy transfer by radiation occurs without the benefit of any molecular interactions. Indeed, radiation energy exchange can occur over long distances through a complete vacuum. For example, the energy that our planet receives from the sun is a heat transfer that is the result of radiation exchange. Radiation heat transfer is complex when many surfaces at different temperatures are involved. However, in the limit that a single surface at temperature T interacts with surroundings at temperature T_{sur} the rate of radiation heat transfer to the surface can be calculated according to:

$$\frac{\dot{Q}}{A} = \sigma\, \varepsilon \left(T_{sur}^4 - T^4\right) \tag{3-32}$$

where A is the area of the surface, σ is the Stefan-Boltzmann constant (5.67×10^{-8} W/m^2-K^4), and ε is the emissivity of the surface. Emissivity is a parameter that ranges between near 0 (for highly reflective surfaces) to near 1 (for highly absorptive surfaces). Note that both T and T_{sur} must be expressed as absolute temperatures (e.g., in units of K rather than °C) in Eq. (3-32).

Depending on your point of view, convection may be classified as a third heat transfer mechanism. *Convection* is the process in which a solid surface experiences heat transfer with a moving fluid. Fundamentally, the heat transfer between the solid and the fluid is the result of conduction at the fluid surface. However, the motion of the fluid complicates the problem substantially and therefore it is necessary to either carry out precise experiments or solve a complex set of partial differential equations in order to determine the convection heat transfer rate. The rate equation for convection heat transfer to the surface is Newton's Law of Cooling, which can be expressed as:

$$\frac{\dot{Q}}{A} = h_{conv}\left(T_{\infty} - T\right) \tag{3-33}$$

where T is the surface temperature, T_{∞} is the temperature of the fluid (at a location far from the surface, where it is not affected by the surface) and h_{conv} is a convection heat transfer coefficient. Equation (3-33) should be viewed as the definition of the convection heat transfer coefficient, which is a complex function of the fluid properties, flow condition, and surface geometry.

EXAMPLE 3.4-1: RUPTURE OF A HELIUM DEWAR

A container that is used to store liquid cryogens such as helium is called a Dewar. Liquid helium at atmospheric pressure has a saturation temperature of 4.2 K and a relatively small latent heat of vaporization. Therefore, the insulation used to thermally isolate the liquid helium in a Dewar from the room temperature surroundings consists of an evacuated space that is filled with many radiation shields. Figure 1 illustrates a liquid helium Dewar with an internal volume of $V = 100$ liter. Initially,

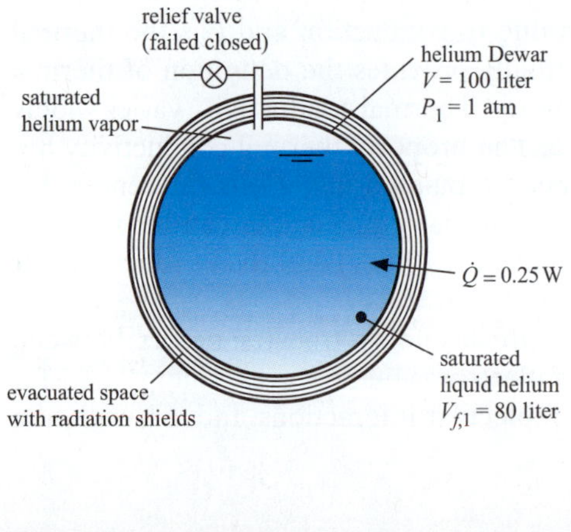

Figure 1: Liquid helium Dewar.

the volume of saturated liquid helium is $V_{f,1} = 80$ liter. The remainder of the Dewar contains saturated helium vapor. The rate of heat transfer to the helium through the insulation is $\dot{Q} = 0.25$ W.

During normal operation, the heat transfer causes the saturated liquid helium to evaporate and a pressure relief valve vents helium vapor to the atmosphere in order to maintain the pressure in the Dewar at $P_1 = 1$ atm. You are analyzing an accident scenario in which the pressure relief valve has failed in a closed state so that no helium vapor can be vented.

a) How long does it take for the pressure in the Dewar to reach the rupture pressure, $P_2 = 150$ psi?

The inputs are entered in EES:

```
$UnitSystem SI K Pa J mass Radian
Vol=100 [liter]*convert(liter,m^3)                "volume of Dewar"
Q_dot=0.25 [W]                                    "rate of heat transfer to helium"
Vol_f_1_liter=80 [liter]                          "volume of liquid initially in Dewar, in liter"
Vol_f[1]=Vol_f_1_liter*convert(liter,m^3)         "volume of liquid initially in Dewar"
P[1]=1 [atm]*convert(atm,Pa)                      "initial pressure"
P[2]=150 [psi]*convert(psi,Pa)                    "rupture pressure of Dewar"
```

The specific volume of the saturated liquid and vapor at state 1 ($v_{f,1}$ and $v_{g,1}$, respectively) are obtained from EES' internal property routines. Note that the substance 'Helium' rather than 'He' must be used for this problem because saturated liquid is present and therefore helium obviously cannot obey the ideal gas law under these conditions. The mass of liquid initially in the Dewar is:

$$m_{f,1} = \frac{V_{f,1}}{v_{f,1}}$$

The volume of vapor at state 1:

$$V_{g,1} = V_1 - V_{f,1}$$

and the mass of vapor initially in the Dewar is:

$$m_{g,1} = \frac{V_{g,1}}{v_{g,1}}$$

The total mass of helium at state 1 is:

$$m_1 = m_{g,1} + m_{f,1}$$

The initial quality is:

$$x_1 = \frac{m_{g,1}}{m_1}$$

```
v_f[1]=volume(Helium,x=0,P=P[1])         "specific volume of liquid at state 1"
v_g[1]=volume(Helium,x=1,P=P[1])         "specific volume of vapor at state 1"
m_f[1]=Vol_f[1]/v_f[1]                   "mass of liquid at state 1"
Vol_g[1]=Vol-Vol_f[1]                    "volume of vapor initially in Dewar"
m_g[1]=Vol_g[1]/v_g[1]                   "mass of vapor initially in Dewar"
m[1]=m_f[1]+m_g[1]                       "total mass of helium initially in Dewar"
x[1]=m_g[1]/m[1]                         "initial quality of helium"
```

The specific internal energy, temperature, and specific volume of the helium initially in the Dewar (u_1, T_1, and v_1) are obtained from EES' internal property routines.

```
u[1]=intenergy(Helium,P=P[1],x=x[1])          "specific internal energy at state 1"
T[1]=temperature(Helium,P=P[1],x=x[1])        "temperature at state 1"
v[1]=volume(Helium,P=P[1],x=x[1])             "specific volume at state 1"
```

A mass balance on the helium in the Dewar for the process associated with going from state 1 to state 2 (immediately before the Dewar ruptures) is:

$$m_2 = m_1$$

The specific volume at the time that the Dewar ruptures is:

$$v_2 = \frac{V}{m_2}$$

The state immediately before the Dewar ruptures is fixed by the specific volume and pressure. The specific internal energy and temperature (u_2 and T_2) are obtained from EES' internal property routines.

```
m[2]=m[1]                                     "mass balance on helium"
v[2]=Vol/m[2]                                 "specific volume at state 2"
u[2]=intenergy(Helium,P=P[2],v=v[2])          "specific internal energy at state 2"
T[2]=temperature(Helium,P=P[2],v=v[2])        "temperature at state 2"
```

An energy balance on the helium in the Dewar for the period of time required to go from state 1 to state 2 is shown in Figure 2.

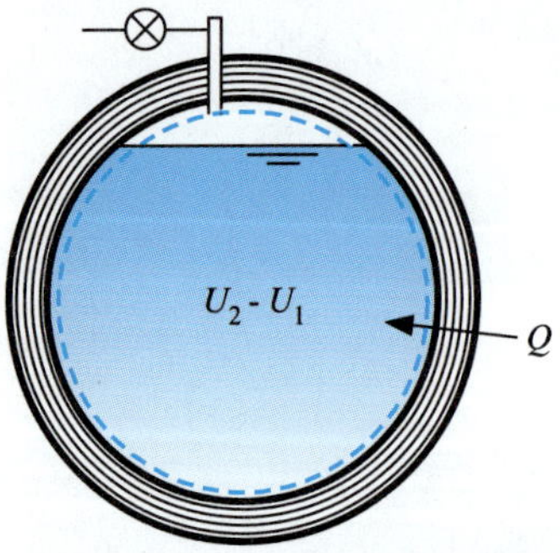

Figure 2: Energy balance.

The energy balance is based on simplifying Eq. (3-4):

$$Q = U_2 - U_1$$

which can be expressed in terms of specific internal energies according to:

$$Q = m_2\, u_2 - m_1\, u_1$$

```
Q=m[2]*u[2]-m[1]*u[1]          "heat transfer required to rupture Dewar"
```

The total heat transfer is the integral of the rate of heat transfer with respect to time:

$$Q = \int_0^{t_{rupture}} \dot{Q}\, dt \tag{1}$$

where $t_{rupture}$ is the time required for the Dewar to rupture. The rate of heat transfer is constant and it therefore can be removed from the integrand in Eq. (1):

$$Q = \dot{Q} \int_{0}^{t_{rupture}} dt = \dot{Q}\, t_{rupture}$$

```
Q=Q_dot*time_rupture                                "time until Dewar ruptures"
time_rupture_day=time_rupture*convert(s,day)        "in days"
```

which leads to $t_{rupture} = 5.30$ day.

b) Plot the time required for the Dewar to rupture as a function of the initial volume of saturated liquid.

A Parametric table is created with columns for the variables time_rupture_day and Vol_f_1_liter. The equation that specifies the initial volume of liquid is commented out. Solving the table provides results that are plotted in Figure 3, which illustrates the time required for the Dewar to rupture as a function of the initial volume of saturated liquid helium. The non-linear relation between the required time for rupture and the initial volume of liquid is a result of competing effects. Increasing the amount of liquid increases the mass and thus the thermal capacity of the helium. However, it also reduces the available volume for the gas.

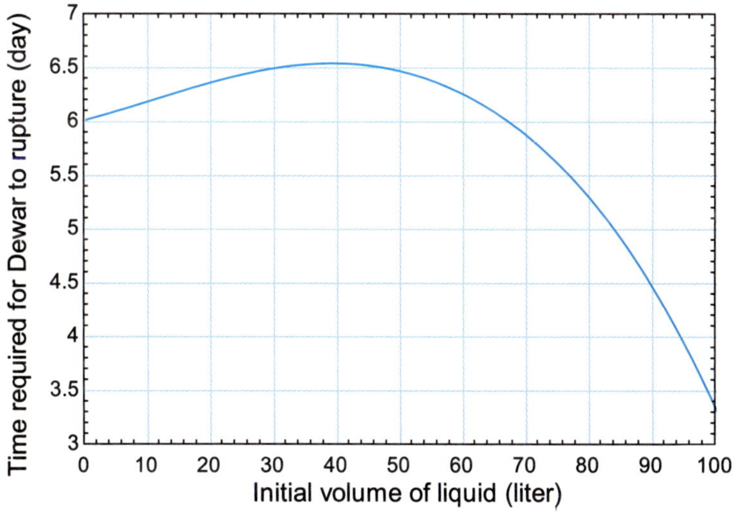

Figure 3: Time required for Dewar to rupture as a function of the initial volume of liquid helium.

3.4.2 The Caloric Theory

The idea of heat may appear intuitive, but the concepts underlying heat are actually very complex. This fact becomes evident if we examine the history of thermodynamics. The science of calorimetry (the measurement of heat) was well-established in the late 1700s, but researchers did not understand exactly what was being transferred to a substance that caused its temperature to increase. Various theories were postulated in order to explain the concept that today we call heat. The *caloric theory* proposed that an invisible substance called *caloric* is transferred to a system when it is heated (Keenan, 1958).

Experiments were conducted that seemed to show that caloric is conserved during any process. That is, the amount of caloric that is transferred to a system when it is heated is exactly equal to the amount of caloric that is transferred from the system when it is subsequently cooled to its original temperature. This is an obvious precursor to what we call today the First Law of Thermodynamics. However, caloric was also believed to have mass. This belief was the result of careful experiments that seemed to show that the mass of materials increased slightly when they were heated to very high temperatures. Eventually, it was found that the measured increase in mass was actually due to an oxidation process occurring at the surface of the materials. The caloric theory was later modified to include latent and sensible types of the caloric substance in order to explain how caloric can be transferred in some cases and not increase the temperature of a system (as for example, in the boiling of water at constant pressure). However, the caloric theory could not explain the apparent heating effect that results from friction. Therefore, in the middle of the 19th century, after about 75 years of existence, the caloric theory was abandoned and the new science of thermodynamics was born.

Note that the common use of the word heat, like the word energy, differs from its formal thermodynamic definition. This difference can occasionally be a source of confusion. Heat is formally defined as an energy transfer process. Heat is not a property of a system; it is defined only at the instant and location where energy crosses the boundary between the system and surroundings. Since heat is not a property, its value cannot be determined by knowing only the initial and final states of a system.

From a thermodynamics perspective, terms such as "heat content", "heat loss", or "heat transfer" are inconsistent with the modern definition of heat. Heat is not a property of a system and therefore a system cannot contain heat. Heat also cannot be lost from a system because it is not a property and therefore does not 'belong' to the system. The term "heat transfer" is pervasive. (In fact, this is the title of a textbook written by the authors.) However, the definition of heat indicates that it is an energy transfer process; therefore the word "transfer" appearing in the term "heat transfer" is redundant.

The confusion surrounding the word "heat" is understandable since the concept of heat is not intuitive. The terms that were originally proposed to explain observed phenomena have taken on new definitions as our understanding of the underlying concepts has increased. On one hand, these differences in terminology can just be dismissed as semantics. Do the terms that we use really matter once we understand the concept? On the other hand, the common use of the word *heat* reinforces the now discredited theory that heat is a substance that can be transferred between a system and its surroundings as if it were mass or energy. Change the word from *heat* to *caloric* and there is no apparent difference between our common use of the term heat today and the caloric theory that was abandoned in the mid-19th century.

3.5 Work

In mechanics, work is defined as the integral of a force over a displacement. In mathematical form, work (W) is a scalar quantity defined according to:

$$W = \int \vec{F} \cdot d\vec{x} \tag{3-34}$$

where $\vec{F}$ is a force vector and $d\vec{x}$ is the differential displacement vector. The vector dot product of $\vec{F}$ and $d\vec{x}$ is the product of their magnitude and the cosine of the angle between them (θ).

$$W = \int F \cos\theta \, dx \tag{3-35}$$

Although this definition of work is correct, a more general definition of work is needed for thermodynamics. Work, like heat, is defined as an energy transfer across the boundary of a system. Like heat, it is necessary to carefully keep track of the direction of work. In this text, the sign associated with a work term will usually be defined by drawing a system schematic that includes an arrow indicating the assumed direction of each energy flow term. A positive value of work indicates that the work is in the direction indicated by the arrow while a negative value means that it is actually in the opposite direction. An alternative method that is equally valid is to use a uniform sign convention for all problems. The most common sign convention sets work to be a positive quantity when it represents an energy flow *out* of the system, i.e., work done by the system on the surroundings.

Heat results from a temperature difference between the surface of the system and its surroundings. Work is the result of a difference in any other potential (e.g., pressure, voltage, or chemical potential) between the system and the surroundings. When we discuss the Second Law of Thermodynamics in Chapter 5 it will become clear that temperature is a unique property. Therefore, heat and work are fundamentally different energy transport processes even though they appear to be equivalent in the First Law of Thermodynamics.

Work can take on a number of forms (e.g., electrical, mechanical, or magnetic) since it can result from a variety of potential differences. The following general definition attempts to accommodate all forms of work in a general manner.

Work (and only work) is performed by a system on its surroundings during a process if the only effect external to the system could be the raising of a mass in a gravitational field.

The words *could be* in the above definition allow the placement of perfect energy converters, linkages, or other mechanisms in the surroundings in order to affect the process of raising the weight.

The concept of work in the context of this definition is best explained with a few simple examples. Consider the battery and electric motor shown in Figure 3-6(a). The battery and motor both operate perfectly; that is, the battery provides current and voltage (electrical energy) to the motor, which transforms the electrical energy into rotation and torque (mechanical energy) with negligible losses. The shaft of the motor is connected to a fan blade that spins in the atmosphere in order to move air. There are several possible system boundaries that could be used to analyze this problem. In Figure 3-6(a), the system is defined so that it includes the battery and the motor without any of the surrounding air. The shaft of the motor crosses the boundary between the system and its surroundings and there is an energy transfer associated with this interaction. Can this interaction be classified as work?

The definition of work that is provided above stipulates that the interaction is work if the only effect caused by the interaction could be the raising of a mass in a gravitational field. In Figure 3-6(b), the fan blade is replaced by a frictionless pulley mechanism that is designed to provide exactly the same torque that was associated with the fan blade in Figure 3-6(a). The fan blade is not part of the system and it can thereby be replaced with alternative energy converters, according to the definition. The pulley mechanism in Figure 3-6(b) raises the weight in a gravity field. No other effect is apparent as a result of this process. Therefore the energy transfer process associated with the motor shaft is classified as a work interaction.

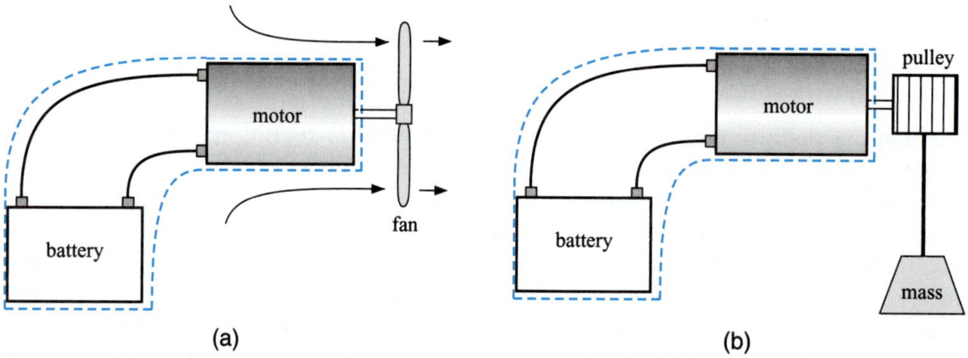

Figure 3-6: A system consisting of a battery and motor that (a) turns a fan blade and (b) turns a pulley mechanism that raises a weight in a gravitational field.

The system choice in a thermodynamics problem is always arbitrary. Suppose we now choose the system boundary so that the fan blades as well as the battery and motor are included in the system, as shown in Figure 3-7. The surroundings consist of air at atmospheric pressure and temperature. We observe a semi-organized air flow at the boundary between the system and its surroundings. This interaction would not be classified as work because it involves a transfer of mass (air) across the system boundary; this type of interaction is discussed in Chapter 4. The fact that the interaction is related to mass crossing a boundary does not mean that it would be impossible to do any work external to the system. We could, for example, erect a wind turbine at the boundary of the system and use the flow of air crossing the boundary to spin the turbine and thereby do work, raising a mass in a gravitational field. Experiments show, however, that the amount of work that could be obtained in this way would be significantly less than the energy flow that is associated with the moving air.

The battery-motor systems shown in Figure 3-6 and Figure 3-7 demonstrate two important points. The first is that, in theory, any form of work can be converted to any other form of work with 100% efficiency. In Figure 3-6(b), electrical power was converted to mechanical power in the form of a rotating shaft, which was then used to lift the mass. Although these conversion processes never exhibit 100% efficiency in practice, there is no thermodynamic limitation to the conversion efficiency from one form of work to another. The second point is that the type of interaction and its magnitude are dependent on the definition of the system. The system boundary is imaginary. Nothing in the process itself changed by including the fan blades in the system shown in Figure 3-7 and excluding them in the system shown in Figure 3-6(a), yet in one case we have a work interaction

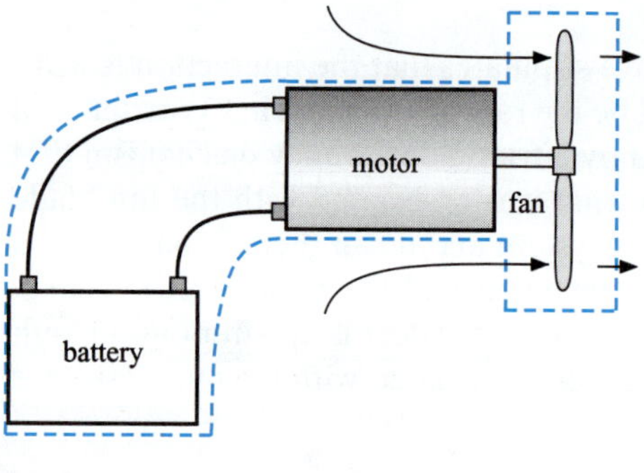

Figure 3-7: System consisting of a battery, motor and fan blade

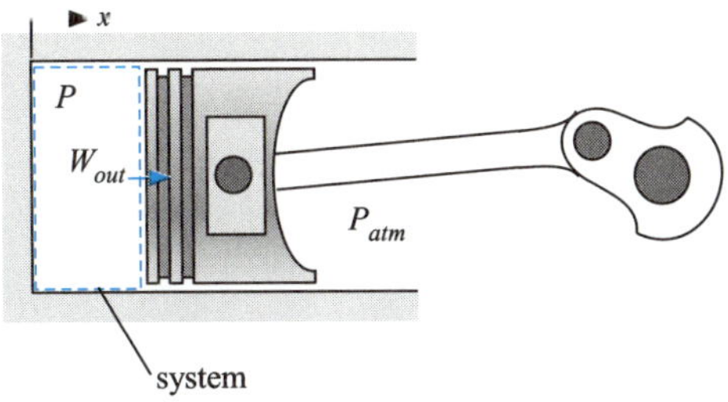

Figure 3-8: Expansion of a gas in a piston-cylinder device.

and in the other, we do not. Choosing the system is the first step in any thermodynamics problem and its specification is necessary in order to quantify heat and work.

Many of the problems in this textbook will involve mechanical work that is the result of compressing or expanding a fluid. This type of work is often referred to as *P-V work*. For example, consider the piston-cylinder device shown in Figure 3-8. A system is defined that includes the gas contained in the cylinder, as shown. The gas exerts a force on the piston face that is equal to the product of the gas pressure in the cylinder and the surface area of the piston. The process considered here is one in which the combined force exerted by atmospheric pressure acting on the back-side of the piston and the connecting rod is smaller than the force exerted by the gas. Therefore, the piston moves to the right, causing the gas to expand. We wish to calculate the work done by the gas on the piston during this process (i.e., the term W_{out} in Figure 3-8). The mechanical definition of work provided in Eq. (3-34) can be applied:

$$W_{out} = \int \vec{F} \cdot d\vec{x} = \int_{x_1}^{x_2} \underbrace{F \cos\theta}_{F_x} \, dx = \int_{x_1}^{x_2} F_x \, dx \tag{3-36}$$

where F_x is the force on the piston face (i.e., the system boundary that is experiencing the work) in the direction of piston motion (the x-direction) and dx is the differential change in the position of the piston. Equation (3-36) is written with an implied sign convention that work done by the system is a positive quantity. Note that because the force and the motion of the system boundary are in the same direction, the energy flow due to work calculated using Eq. (3-36) is positive. Work flows out of the system and into the surroundings (the direction of the work should also be intuitive – in this case, the gas is expanding so it is doing work on its surroundings). If the piston were moving to the left (i.e., the gas were being compressed), then dx would be in the negative x-direction while the force due to the pressure of the gas, F_x, would still be in the positive x-direction. Therefore, W_{out} calculated using Eq. (3-36) is negative for a compression process, indicating that work is being done on the gas.

The force in the x-direction is the product of the gas pressure at the piston face (P) and the piston cross-sectional area (A_c). Therefore, Eq. (3-36) can be written as:

$$W_{out} = \int_{x_1}^{x_2} P \underbrace{A_c \, dx}_{dV} \tag{3-37}$$

The product $A_c \, dx$ in Eq. (3-37) is equal to the differential change in the volume contained in the piston, dV.

$$\boxed{W_{out} = \int_{V_1}^{V_2} P \, dV} \tag{3-38}$$

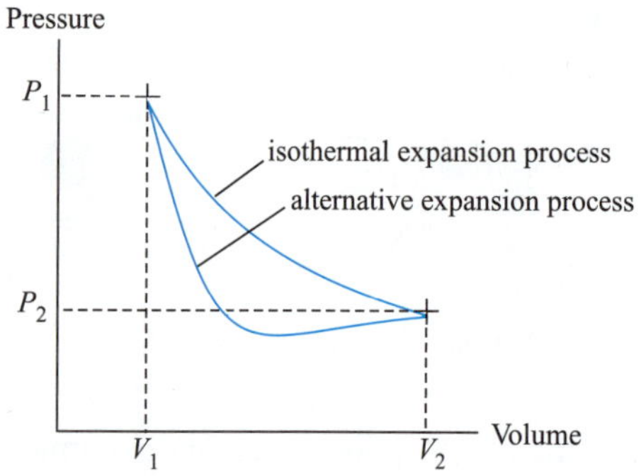

Figure 3-9: Pressure-volume diagram for an isothermal expansion process and an alternative expansion process.

Equation (3-38) relates the work out of a system to the pressure at the boundary of the system and the change in the volume of the system that is associated with the motion of the boundary. The work evaluated using Eq. (3-38) is defined as being out of the system. If the volume change of the system is negative (i.e., $dV < 0$) then the work evaluated using Eq. (3-38) will be negative, indicating that the work corresponds to an energy transfer into the system.

It is not possible to evaluate the integral in Eq. (3-38) even if we know the initial and final volumes (V_1 and V_2, respectively), unless we also know how the pressure at the piston face is related to the volume of the system during the process. For example, perhaps we know the mass of gas in the cylinder and also that the process is isothermal (i.e., the temperature of the gas remains constant during the expansion process). If the expansion process is sufficiently slow, then the gas within the piston remains always in an equilibrium state. For this process it is possible to compute two intensive variables at any piston position: temperature and specific volume. Since the equilibrium state of the gas is fixed by two intensive variables, as discussed in Section 2.1, the pressure can be determined for any piston volume. As the gas expands at constant temperature, it passes through a continuous set of equilibrium states that can be represented on a pressure-volume diagram, as shown in Figure 3-9. The work done by the gas on the piston can be computed using the integral provided by Eq. (3-38). Note that the work integral corresponds to the area under the curve that defines the pressure-volume trajectory in Figure 3-9.

The integration required to evaluate the work can be accomplished in different ways, depending on the available information. For example, the gas may obey the ideal gas law under the conditions associated with the expansion process. Substituting the ideal gas law into Eq. (3-38) leads to:

$$W = \int_{V_1}^{V_2} \frac{m\,R\,T}{V}\,dV \tag{3-39}$$

Recognizing that mass, the ideal gas constant, and temperature are all constant for this process allows an analytical solution to Eq. (3-39):

$$W = m\,R\,T \int_{V_1}^{V_2} \frac{dV}{V} = m\,R\,T \ln\left(\frac{V_2}{V_1}\right) \tag{3-40}$$

An energy balance would show that heat must be transferred to the system in order to maintain the gas at a constant temperature during this expansion process.

We could conduct a different expansion process that has a different pressure-volume trajectory than the isothermal process previously considered. One such alternative process is also shown in Figure 3-9. Note that the initial and final states of the gas in the piston for this alternative expansion process are exactly the same as they were for the isothermal process. However, the work done by the system is different for this process, as evident by the fact that the area under the pressure-volume curve is smaller. There are an infinite number of possible trajectories that start and end at the same states, each of these trajectories is characterized by a unique value of work. Knowing the end states of the process is not sufficient to determine the work; it is also necessary to know the trajectory.

EXAMPLE 3.5-1: COMPRESSION OF AMMONIA

Ammonia is a popular refrigerant for industrial refrigeration because it is cheap and leads to a very efficient system, as discussed in Chapter 9. During one of the processes required by the refrigeration cycle, ammonia vapor must be compressed using a reciprocating compressor. The compressor consists of a piston within a cylinder that is outfitted with intake and exhaust valves, as shown in Figure 1.

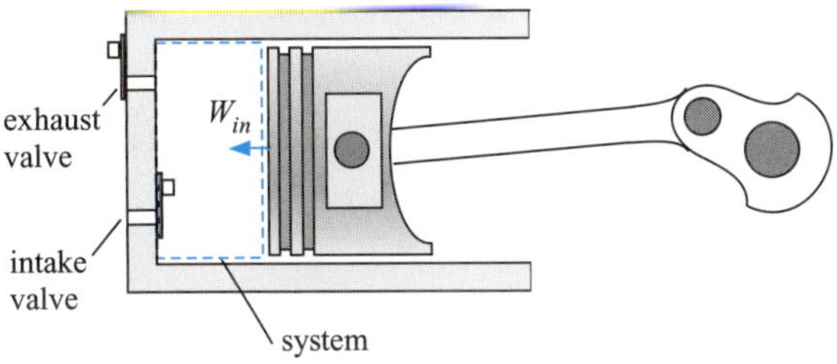

Figure 1: Reciprocating compressor used to compress ammonia in a refrigeration cycle.

The pressure-volume data listed in Table 1 have been experimentally measured during the compression stroke of the piston in the reciprocating compressor.

Table 1: Experimentally measured pressure as a function of piston volume.

Pressure (psi)	Volume (in^3)
65.1	80.0
80.5	67.2
93.2	60.1
110	52.5
134	44.8
161	37.6
190	32.5

a) Determine the work done on the ammonia by the piston (in units of both ft-lb$_f$ and Joules) during the compression process by numerically integrating the data in Table 1. Note that the EES Integral command can be used to do the integration.

The ammonia contained in the piston is selected as the system, as shown in Figure 1. The work done by the ammonia to the piston is the integral of pressure with respect to volume, as given by Eq. (3-38). However, the problem statement asks for the work done by the piston to the ammonia (W_{in} in Figure 1); therefore the negative of the integral in Eq. (3-38) must be used:

$$W_{in} = -\int_{V_1}^{V_2} P\, dV \tag{1}$$

The measured pressure and volume in the table can be numerically integrated using EES. The data are entered into a Lookup table. Select New Lookup Table from the Tables menu and specify that the table has 2 columns and 7 rows. Manually enter the data into each cell, as shown in Figure 2. Add a title and units to each column by right-clicking the column header and selecting Properties. Change the title of the Lookup table by right-clicking on the tab associated with the table.

Lookup Table

PV Data

	1 P [psi]	2 V [in³]
Row 1	65.1	80
Row 2	80.5	67.2
Row 3	93.2	60.1
Row 4	110	52.5
Row 5	134	44.8
Row 6	161	37.6
Row 7	190	32.5

Figure 2: Lookup table with measured pressure and volume data.

The use of the Integral command is discussed in Section G.5 of Appendix G. The first step in doing a numerical integration is to ensure that you can compute the integrand, P in Eq. (1), given any arbitrary value of the integration variable, V in Eq. (1). An arbitrary value of volume in specified:

```
$UnitSystem SI, Radian, Mass, Pa, J, K
V_in3=50 [in^3]                          "volume, in in^3"
V=V_in3*convert(in^3,m^3)                "volume"
```

The pressure at the specified volume is obtained from the data in the Lookup Table via interpolation. The Interpolate function is used for this purpose. The Interpolate function requires four arguments, as shown below:

Interpolate(TableName, ColName1, ColName2, ColName = Value)

where TableName is the name of the table to be used for the interpolation and ColName1 and ColName2 are the names of the two columns in the table that are to be used for the interpolation process. TableName, ColName1 and ColName2 must be either string constants or string variables. String constants are enclosed within single quotes and string variables are variable names that end with the $ character and store a string value. Either of the columns can be the independent variable and the other column will be the dependent variable. The last argument of the Interpolate command includes both the name of the column that contains the independent variable, ColName, as well as its value, Value. The value of the dependent variable at the specified value of the independent variable is returned by the function. The Interpolate function uses cubic interpolation; alternative interpolation methods are used by the functions Interpolate1 and Interpolate2. For this problem, the table name is 'PV Data' and the two columns are 'P' and 'V'. The independent variable is V and its value is V_in3. Note, that the Interpolate function accepts but does not require the quotes for the column names.

```
P_psi=Interpolate('PV Data','P','V','V'=V_in3)        "interpolate to get pressure, in psi"
P=P_psi*convert(psi,Pa)                               "pressure"
```

Solving results in $P = 117.1$ psi. Having established that the integrand can be computed, comment out the arbitrary value of the integration variable:

```
{V_in3=50 [in^3]          "volume, in in^3"}
```

and instead use the Integral command to vary V and carry out the numerical integration:

```
V_1=80 [in^3]*convert(in^3,m^3)          "initial volume"
V_2=32.5 [in^3]*convert(in^3,m^3)        "final volume"
W=-Integral(P,V,V_1,V_2)                 "work"
W_ftlbf=W*convert(J,ft-lbf)              "in ft-lbf"
```

which results in $W_{in} = 586$ J (432.2 ft-lb$_f$).

b) Fit the experimental data to an equation of the form $P\,V^n = C$ where C and n are both constants. This equation is referred to as a polytropic equation and the exponent n is the polytropic exponent

Add two columns to the Lookup table that contains the experimental data by right-clicking on the header of one column and selecting Insert Column to the Right. Right-click on the header of the first new column, select Alter Values and then select Enter Equation (Figure 3). Enter an equation in the equation box in order to fill in the column values using an equation. It is possible to refer to data contained in the same row of a different column in the equation by using the parameter #C where C is the column number. In column 3, convert the pressure (contained in column 1 and therefore referred to in the equation as #1) from psi to Pa using the convert function (see Figure 3). In column 4, convert the volume from in^3 to m^3. The Lookup Table should appear as shown in Figure 4. Prepare a plot of the pressure as a function of the volume (in SI units), as shown by the symbols in Figure 5.

Rearranging the polytropic equation provides an explicit equation for pressure:

$$P = C\,V^{\,n} \tag{2}$$

Select Curve Fit from the Plots menu and then click the Enter/edit equation button (Figure 6). Enter an equation that has the form of Eq. (2) with the unknown parameters, C and n, indicated by the coefficients a0 and a1, respectively. Click the Fit button, provide reasonable guess values for a0 and a1 (e.g., 1×10^3 and 1, respectively) and select OK. EES will determine the best fit values of a0 and a1 (i.e., a0 $= C = 241.034$ and a1 $= n = 1.13941$). Select Plot in order to overlay the best fit curve, as shown by the solid line in Figure 5.

c) Analytically integrate the best-fit polytropic equation identified in part (b) in order to calculate the work done on the ammonia. Compare your result with the result determined in part (a).

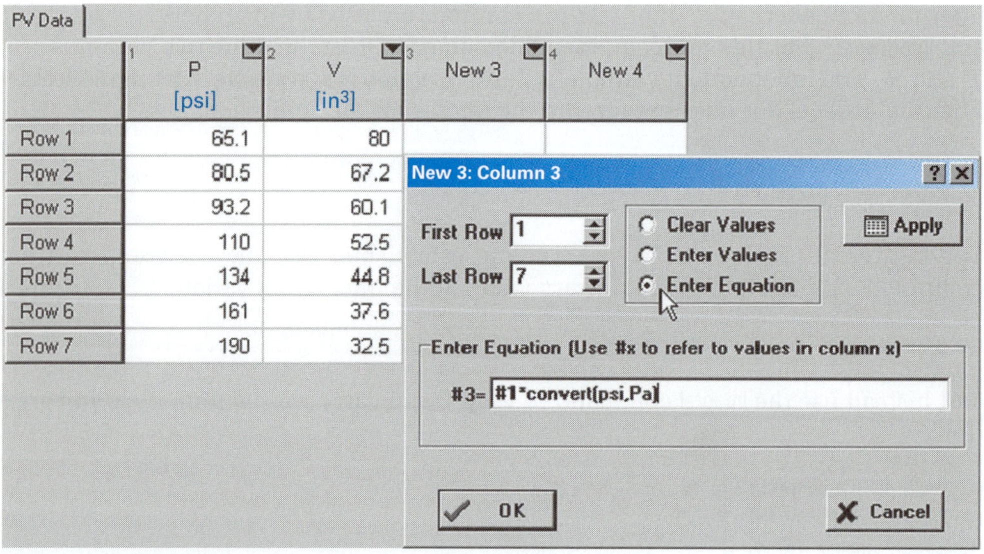

Figure 3: Alter values dialog showing the Enter Equation box.

The work done by the piston on the ammonia is obtained by substituting Eq. (2) into Eq. (1):

$$W_{in} = -\int_{V_1}^{V_2} C\,V^{-n}\,dV$$

Carrying out the integration provides:

$$W_{in} = -\frac{C}{(1-n)}\left(V_2^{1-n} - V_1^{1-n}\right)$$

Lookup Table

PV Data

	P [psi]	V [in3]	P_{SI} [Pa]	V_{SI} [m3]
Row 1	65.1	80	448849	0.001311
Row 2	80.5	67.2	555028	0.001101
Row 3	93.2	60.1	642591	0.0009849
Row 4	110	52.5	758423	0.0008603
Row 5	134	44.8	923898	0.0007341
Row 6	161	37.6	1.110E+06	0.0006162
Row 7	190	32.5	1.310E+06	0.0005326

Figure 4: Lookup table with pressure and volume converted to SI units.

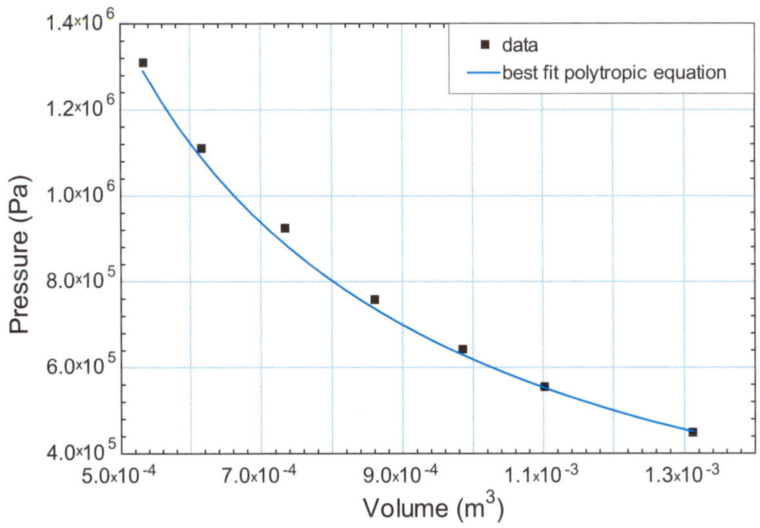

Figure 5: Pressure as a function of volume; both the data in Table 1 and the best fit polytropic equation are shown.

```
C=241.034 [Pa-(m^3)^1.13941]                "constant"
n=1.13941                                   "polytropic exponent"
W_2=-C*(V_2^(1-n)-V_1^(1-n))/(1-n)          "work calculated by integrating polytropic equation"
W_2_ftlbf=W_2*convert(J,ft-lbf)             "in ft-lbf"
```

which leads to $W_{in} = 583.6$ J (430.4 ft-lb$_f$). This result compares well with the value found in part (a); the difference is related to errors resulting from the numerical integration and interpolation as well as deviations between the best fit polytropic curve and the data.

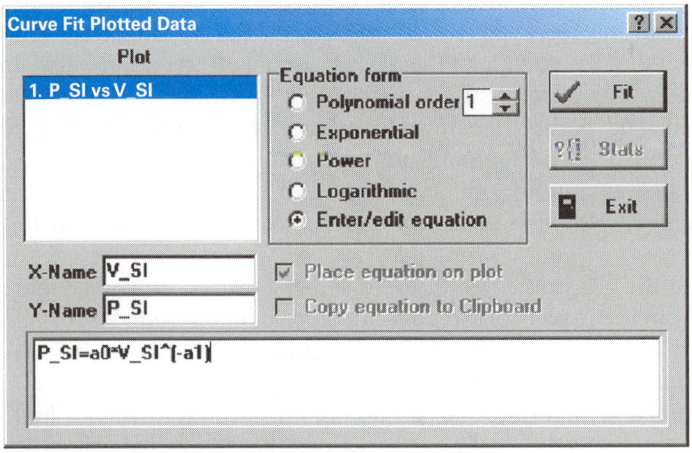

Figure 6: Curve Fit Plotted Data dialog.

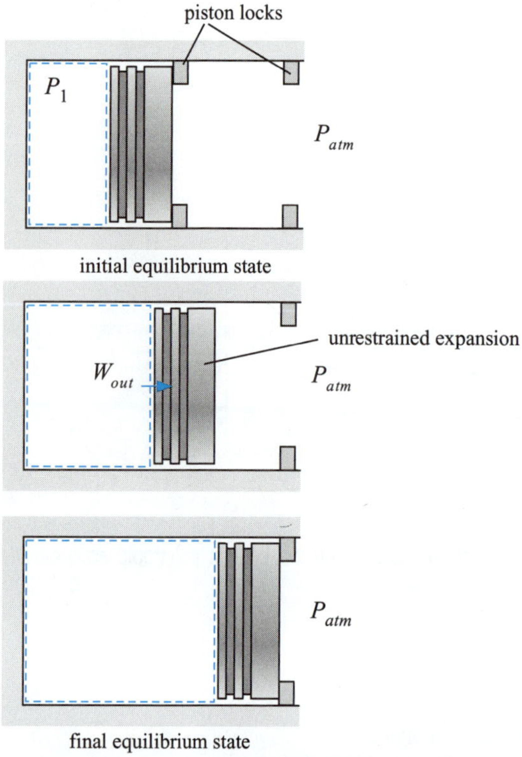

Figure 3-10: Gas undergoing a partially unrestrained expansion in a piston.

The slow expansion of a gas in a piston-cylinder apparatus is not typical of actual work production processes. More often, the gas undergoes a partially unrestrained expansion, as shown in Figure 3-10. The gas in the cylinder is initially at a pressure P_1 that is above atmospheric pressure; however, the piston cannot move because of the piston locks that are in place. The gas temperature is equal to the temperature of the surroundings, $T_1 = T_{amb}$. The enclosed gas is in a state of constrained equilibrium and all of the intensive properties at state 1 can be determined knowing the temperature and pressure. The piston locks are removed and the force exerted by the gas on the piston causes it to accelerate to the right, increasing the volume of the gas. Frictional effects occur between the piston and cylinder as the gas expands. Eventually, the piston hits the second set of locks with some velocity and perhaps chatters for a short time before coming to rest. Eventually, the gas returns to the temperature of the surroundings, $T_2 = T_{amb}$. The final state, state 2, is also an equilibrium state and all of the intensive properties of the gas can be determined. We wish to determine the work associated with this process.

Initially we will choose the gas in the piston to be the system, as shown in Figure 3-10. The work resulting from the expansion process can be calculated using Eq. (3-38). The pressure appearing in Eq. (3-38) corresponds to the pressure on the piston face, the boundary of the system that is experiencing the expansion. But here, we encounter a problem. We know the initial and final states but we do not know that pressure-volume trajectory that defines the process and therefore it is not possible to evaluate the work using Eq. (3-38).

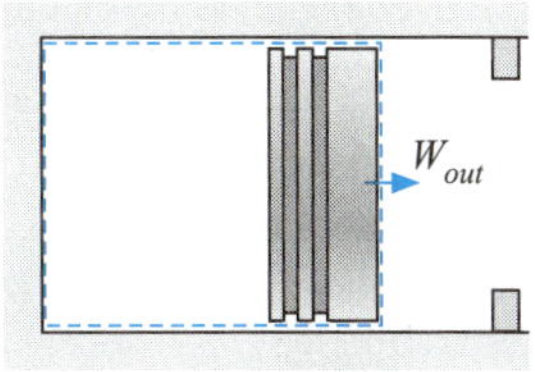

Figure 3-11: System that includes the gas, piston and cylinder.

If we choose the gas, piston, and the cylinder together to be the system, as shown in Figure 3-11, then it is possible to compute the work done by this alternative system. The system boundary is located on the outside surface of the piston and it moves with the piston. The work for this process is again given by Eq. (3-38); however, for this system the pressure on the boundary that is moving is atmospheric pressure, P_{atm}, which remains constant during the process. Therefore, the work can be computed according to:

$$W_{out} = \int_{V_1}^{V_2} P\,dV = \int_{V_1}^{V_2} P_{atm}\,dV = P_{atm}\,(V_2 - V_1) \tag{3-41}$$

We were unable to calculate the work when the system was chosen to be the gas within the system. However, the work can be determined from Eq. (3-41) when we choose the gas, piston and cylinder together to be the system. This example illustrates the importance of properly choosing a system. Both of the system choices shown in Figure 3-10 and Figure 3-11 are valid; however, the problem is easier to solve with the system choice in Figure 3-11. Note that the work done by the gas to the piston (W_{out} in Figure 3-10, which we are not able to calculate) would not be the same as the work done by the gas, piston and cylinder (W_{out} in Figure 3-11). The work depends on the choice of system because work is defined as energy crossing a system boundary.

The *P-V* work associated with expanding a fluid, described by Eq. (3-38), is the most common form of work that will be encountered in this text. However, there are many other forms of work that can also be considered and some of these are summarized in Table 3-1 with their associated rate equation. Note that the equations provided in Table 3-1 are consistent with work transferred *out* of the system.

In addition to moving a force through a linear distance and expanding a fluid, which have already been discussed, work can be transferred from a system in several other forms. When a thin film that is under tension (e.g., a soap bubble or balloon) is stretched then work is required to increase the surface area (A_s) against the action of the surface tension in the film (γ). Work is also associated with the rotation of a shaft through an angle (θ) under the influence of a torque (τ). Work can occur electrically by the motion of Coulomb forces (i.e., electrical charges). The electrical work occurs via the transfer of current (i) across a voltage potential (E). All of the work mechanisms in Table 3-1 share the characteristic that the interaction is equivalent to moving a force through a distance and therefore the action could be used to raise a mass in a gravitational field. There may be more than one form of work involved in a process. In this case, we assume that their effects are additive:

$$W = \sum W_i \tag{3-42}$$

and that the different work mechanisms are independent.

Table 3-1: Some forms of work and their rate equations.

Form of work	Incremental basis	Rate basis		Parameters and notes
Expansion or compression of a fluid (*P-V* work)	$W=\int P\,dV$	$\dot{W}=P\dfrac{dV}{dt}$	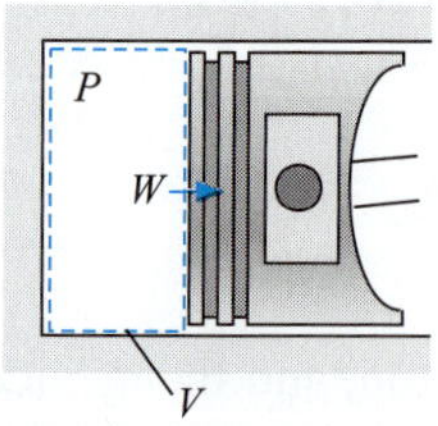	$P=$ pressure (N/m^2) $V=$ volume (m^3)
Stretching a thin film	$W=-\int \gamma\,dA_s$	$\dot{W}=-\gamma\dfrac{dA_s}{dt}$	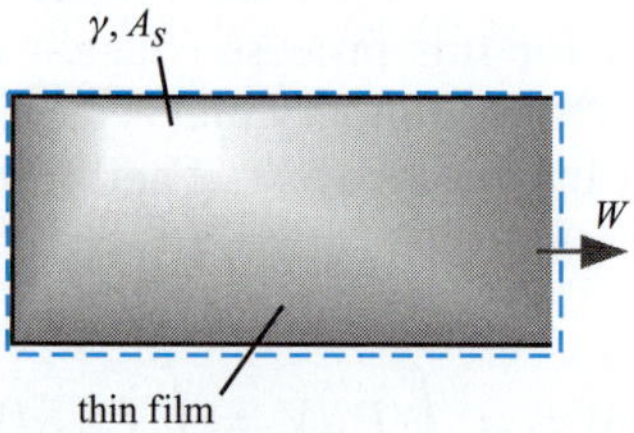	$\gamma=$ surface tension (N/m) $A_s=$ surface area (m^2)
Moving a force through a linear distance	$W=\int F\,dx$	$\dot{W}=F\tilde{V}$	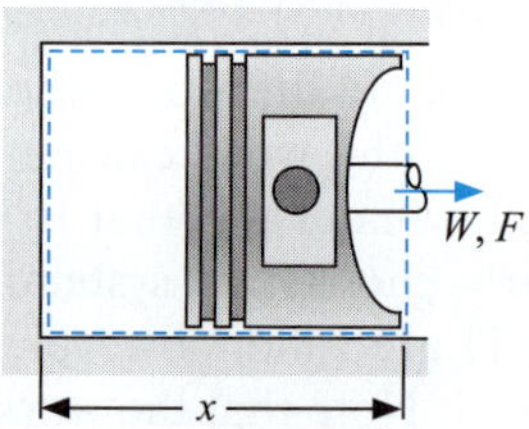	$F=$ force (N) $x=$ distance (m) $\tilde{V}=$ velocity (m/s)
Rotating a shaft against a torque	$W=\int \tau\,d\theta$	$\dot{W}=\tau\,\omega$	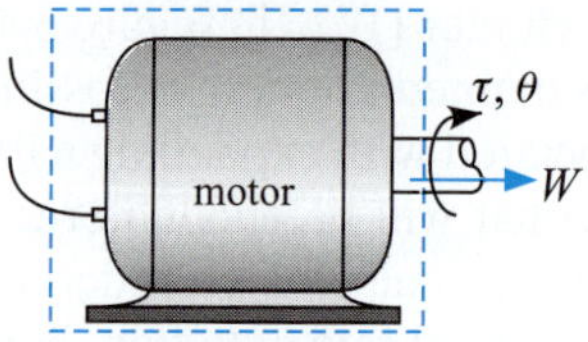	$\tau=$ torque (N-m) $\theta=$ angular position (rad) $\omega=$ angular velocity (rad/s)
Current flowing through a voltage difference	$W=\int E\,i\,dt$	$\dot{W}=E\,i$	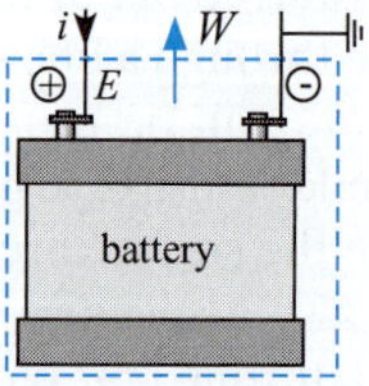	$E=$ voltage potential (Volt) $i=$ current (Ampere)

EXAMPLE 3.5-2: HELIUM BALLOON

A balloon is filled with helium. The initial diameter of the balloon is $D_{b,1} = 8$ cm. The initial temperature of the helium in the balloon is $T_1 = 20°$C. The balloon is surrounded by air at pressure $P_{atm} = 1$ atm. The balloon is a thin film with surface tension that depends on the diameter according to:

$$\gamma = K\, D_b^2 \tag{1}$$

where $K = 5.2 \times 10^4$ N/m^3. The helium within the balloon is heated until the balloon diameter is $D_{b,2} = 9$ cm.

a) Determine the initial pressure in the balloon and the mass of helium in the balloon.

The inputs are entered in EES:

```
$UnitSystem SI MASS RAD PA K J
"Inputs"
D_b_1=8 [cm]*convert(cm,m)          "initial diameter of balloon"
K=5.2e4 [N/m^3]                     "balloon constant"
T_1=converttemp(C,K,20 [C])         "initial temperature"
P_atm=1 [atm]*convert(atm,Pa)       "atmospheric pressure"
D_b_2=9 [cm]*convert(cm,m)          "final diameter"
```

A force balance on the balloon if it were cut along its mid-section is shown in Figure 1.

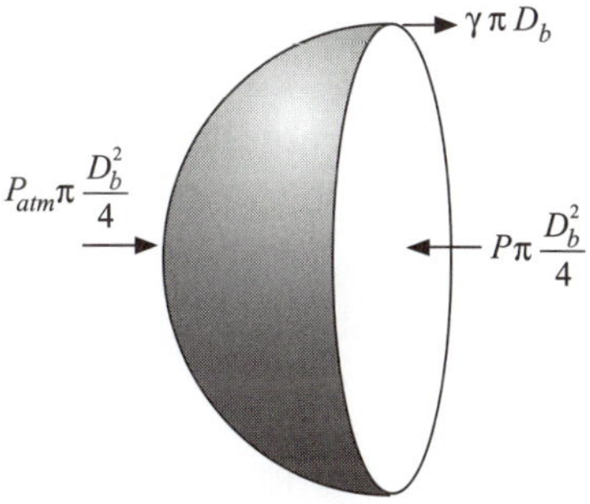

Figure 1: Force balance on balloon.

The force balance can be written as:

$$P_{atm}\,\pi\,\frac{D_b^2}{4} + \gamma\,\pi\,D_b = P\,\pi\,\frac{D_b^2}{4}$$

Solving for the balloon pressure, P, indicates that:

$$P = P_{atm} + \frac{\gamma\,4\,\pi\,D_b}{\pi\,D_b^2}$$

Substituting Eq. (1) for the surface tension and simplifying leads to:

$$P = P_{atm} + 4\,K\,D_b \tag{2}$$

Therefore, the initial pressure in the balloon is:

$$P_1 = P_{atm} + 4\,K\,D_{b,1}$$

```
P_1=P_atm+4*K*D_b_1          "initial pressure"
```

which evaluates to $P_1 = 117{,}965$ Pa. The pressure and temperature together fix the state of the helium; here we will model helium as an ideal gas (i.e., using the substance 'He' in EES rather than 'Helium'). The specific internal energy and specific volume of the helium at state 1 (u_1 and v_1) are determined using EES' internal property functions.

```
u_1=intenergy(He,T=T_1)          "specific internal energy"
v_1=volume(He,T=T_1,P=P_1)       "specific volume"
```

The initial volume of the balloon is:

$$V_1 = \frac{4}{3}\pi\left(\frac{D_{b,1}}{2}\right)^3$$

and the mass of helium in the balloon is:

$$m = \frac{V_1}{v_1}$$

```
Vol_1=4*pi*(D_b_1/2)^3/3          "initial volume"
m=Vol_1/v_1                       "mass of helium"
```

which results in $m = 5.194 \times 10^{-5}$ kg.

b) What is the work done by the helium to the balloon material during the heating process?

The final pressure in the balloon is given by Eq. (2) using the final diameter:

$$P_2 = P_{atm} + 4\,K\,D_{b,2}$$

The final volume of the balloon is:

$$V_2 = \frac{4}{3}\pi\left(\frac{D_{b,2}}{2}\right)^3$$

The specific volume of the helium within the balloon is:

$$v_2 = \frac{V_2}{m}$$

The final state of the helium is fixed by its pressure and specific volume. The final temperature and specific internal energy (T_2 and u_2) are obtained using the internal property routines in EES.

```
P_2=P_atm+4*K*D_b_2                  "final pressure"
Vol_2=4*pi*(D_b_2/2)^3/3             "final volume"
v_2=Vol_2/m                          "final specific volume"
T_2=temperature(He,v=v_2,P=P_2)      "final temperature"
T_2_C=converttemp(K,C,T_2)           "in C"
u_2=intenergy(He,v=v_2,P=P_2)        "specific internal energy"
```

The work done by the helium to the balloon material (i.e., the work out of the system that consists of the helium contained within the balloon, $W_{out,A}$ shown in Figure 2) is given by:

$$W_{out,A} = \int_{V_1}^{V_2} P\, dV \tag{3}$$

where P is the pressure of the helium within the balloon.

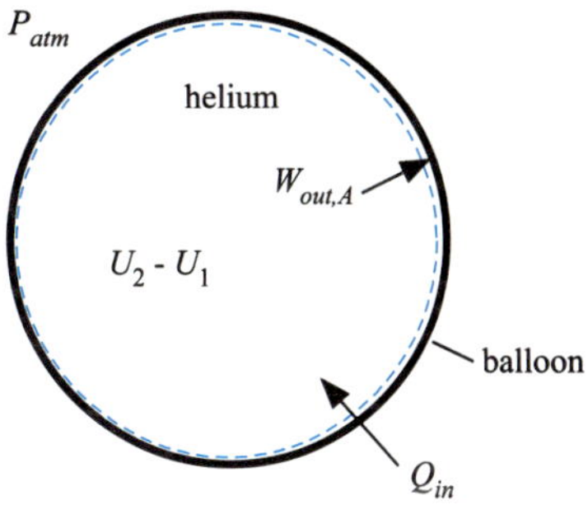

Figure 2: Energy balance on the system containing the helium within the balloon.

Substituting Eq. (2) into Eq. (3) leads to:

$$W_{out,A} = \int_{V_1}^{V_2} (P_{atm} + 4\, K\, D_b)\, dV$$

or

$$W_{out,A} = \int_{V_1}^{V_2} P_{atm}\, dV + \int_{V_1}^{V_2} 4\, K\, D_b\, dV \tag{4}$$

The diameter of the balloon is related to the volume of the system according to:

$$V = \frac{\pi}{6} D_b^3$$

Therefore, a differential change in volume is related to a differential change in diameter according to:

$$dV = \frac{\pi}{2} D_b^2\, dD_b \tag{5}$$

Substituting Eq. (5) into Eq. (4) allows the second integral to be expressed in terms of the balloon diameter:

$$W_{out,A} = \int_{V_1}^{V_2} P_{atm}\, dV + \int_{D_{b,1}}^{D_{b,2}} 2\,\pi\, K\, D_b^3\, dD_b$$

Carrying out the integration leads to:

$$W_{out,A} = P_{atm}\,(V_2 - V_1) + \frac{\pi}{2}\, K\,(D_{b,2}^4 - D_{b,1}^4)$$

```
W_out_A=P_atm*(Vol_2-Vol_1)+pi*K*(D_b_2^4-D_b_1^4)/2          "work done by helium on balloon"
```

which results in $W_{out,A} = 13.53$ J.

c) Determine the heat transfer to the balloon.

An energy balance on the helium is shown in Figure 2:

$$Q_{in} = W_{out,A} + m\,(u_2 - u_1)$$

```
Q_in=W_out_A+m*(u_2-u_1)          "heat transfer to balloon"
```

which results in $Q_{in} = 34.82$ J.

d) Determine the work done by the balloon material to the atmosphere.

The system shown in Figure 3 contains the balloon material and the helium.

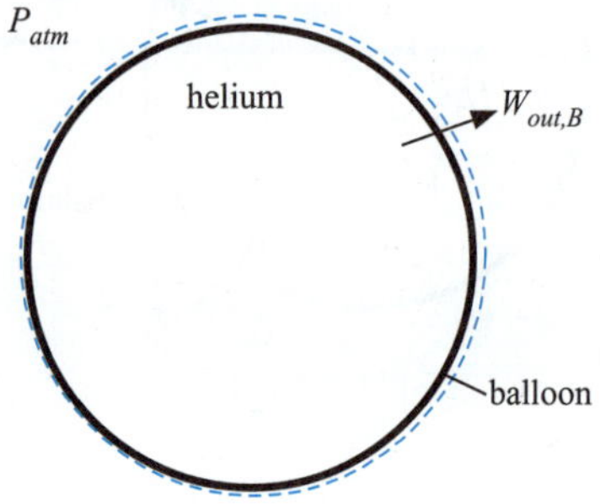

Figure 3: System containing the helium and the balloon.

The work done by the system shown in Figure 3 is given by:

$$W_{out,B} = \int_{V_1}^{V_2} P\, dV$$

where P is the pressure at the interface between the atmosphere and the balloon (i.e., P_{atm}). Carrying out the integration leads to:

$$W_{out,B} = P_{atm}\,(V_2 - V_1)$$

```
W_out_B=P_atm*(Vol_2-Vol_1)          "work done by balloon on atmosphere"
```

which results in $W_{out,B} = 11.51$ J.

e) Use the equation for the work associated with stretching a thin film, listed in Table 3-1, to compute the net work done by the balloon. Check your answers against those obtained in parts (b) and (d).

According to Table 3-1, the net work done by a thin film is given by:

$$W_{net} = -\int_{A_{s,1}}^{A_{s,2}} \gamma\, dA_s \tag{6}$$

where A_s is the surface area of the balloon. Substituting Eq. (1) into Eq. (6):

$$W_{net} = -\int_{A_{s,1}}^{A_{s,2}} K\, D_b^2\, dA_s \tag{7}$$

The surface area of the balloon is:

$$A_s = \pi\, D_b^2$$

Therefore, the differential change in surface area is related to a differential change in diameter according to:

$$dA_s = 2\,\pi\, D_b\, dD_b \tag{8}$$

Equation (8) is substituted into Eq. (7):

$$W_{net} = -\int_{D_{b,1}}^{D_{b,2}} 2\,\pi\, K\, D_b^3\, dD_b$$

Carrying out the integration leads to:

$$W_{net} = -\frac{\pi}{2} K\left(D_{b,2}^4 - D_{b,1}^4\right)$$

```
W_net=-pi*K*(D_b_2^4-D_b_1^4)/2          "net work done by the balloon"
```

which results in $W_{net} = -2.013$ J (i.e., net work is done on the balloon). The net work done by the balloon on its surroundings must be the difference between the work done by the balloon on the atmosphere, $W_{out,B}$ calculated in (d), and the work done by the helium on the balloon, $W_{out,A}$ calculated in (b):

$$W_{net} = W_{out,B} - W_{out,A}$$

```
W_net_check=W_out_B-W_out_A          "check using results from (c) and (d)"
```

which also results in −2.013 J.

3.6 What is Energy and How Can you Prove that it is Conserved?

The central idea that has been presented in this chapter appears to be simple: energy is a conserved quantity; it is neither generated nor destroyed. However, conservation of energy is actually a difficult concept to truly understand. What is energy? Why do we believe that it is conserved? Can we experimentally prove that energy is conserved?

Energy is defined in Section 3.2 to be a property of a system that may exist in different forms. Notice that the word *energy* was used in the definition of the different forms of energy (kinetic energy, potential energy, and internal energy); this is a somewhat unsatisfying way of explaining what energy really is.

Thermodynamics is defined as the science that studies energy transformations. It is therefore reasonable to look to thermodynamics textbooks for a definition of energy. Çengel and Boles (2002) acknowledge that "*it is difficult to give a precise definition for it* (energy)" and offer the definition that energy is "*the ability to cause changes*". Moran and Shapiro (2008) state that "*energy is a familiar notion and you already know a great deal about it*"; no formal definition is provided. Energy has often been defined as "*the ability to do work*". However, that definition seems inconsistent with the statement of the First Law of Thermodynamics that requires that energy must be conserved. Experience shows that the ability to do work is not a conserved quantity. Doolittle (1983) states that "*Energy is that which produces a thermodynamic change in a substance; it is only manifested by the results it produces; it cannot be seen and does not have substance.*" But, what is "*that*"? It has been suggested that energy is an invented function that is defined

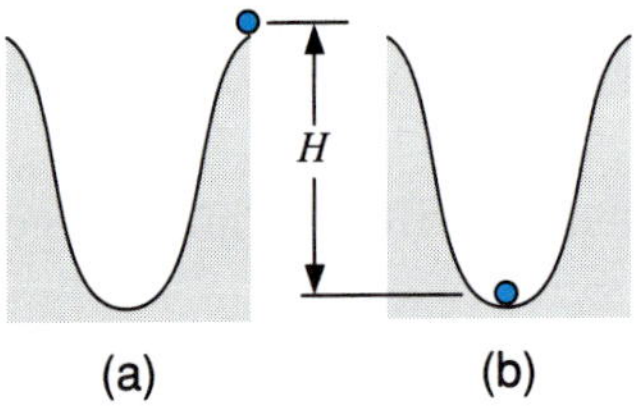

Figure 3-12: An experiment designed to demonstrate the First Law; (a) is the initial position of the ball and (b) is the final position.

so that it is conserved. Quoting from Lewis and Randall's classic thermodynamics book (1923), "*So as the science has progressed, it has been necessary to invent other forms of energy and indeed an unfriendly critic may claim, with some reason, that the law of conservation of energy is true because we make it true by assuming the existence of forms of energy for which there is no other justification than the desire to retain energy as a conservative quantity.*" This train of thought is disturbing because it implies that the First Law of Thermodynamics is a tautology; that is, the First Law is true because we defined it to be true and therefore the First Law is useless. We will see that this not the case.

It would seem that the validity of the First Law of Thermodynamics could be readily verified by some simple experiments. For example, the First Law of Thermodynamics for a closed system is given by Eq. (3-4), which is repeated below:

$$Q_{in} + W_{in} = Q_{out} + W_{out} + \Delta KE + \Delta PE + \Delta U \tag{3-43}$$

All that is necessary to experimentally validate the First Law is to measure each term in Eq. (3-43) and demonstrate that the values on the left side of the equation are equal to the values on the right side of the equation, to within the experimental uncertainty of the measurements.

Designing an experiment that will demonstrate the validity of Eq. (3-43) presents an interesting problem. How might you go about doing this? What measurements would you make and what are the uncertainties associated with these measurements? A simple experiment that is often suggested is one that involves only kinetic and potential energy terms. A ball that is initially at rest is positioned at the top of a hill, as shown in Figure 3-12(a). The ball is allowed to roll down the hill. As the ball descends, its kinetic energy increases and its potential energy decreases. The ball achieves a maximum velocity at the bottom of the hill and then slows as it travels up the other side of the hill. In your experiment, you could simultaneously measure the velocity and position of the ball and use these quantities to compute the change in the kinetic energy and the potential energy of the ball, relative to their initial values. It may be possible to show that the sum of the changes in the kinetic and potential energy of the ball do indeed sum to zero, to within experimental uncertainty, during the first few cycles. However, regardless of how carefully you design this experiment, the ball will eventually come to rest at the bottom of the hill, as shown in Figure 3-12(b). The change in the kinetic energy of the ball between its initial position, Figure 3-12(a), and its final position, Figure 3-12(b), is zero (the ball is at rest in both cases). However, the change in the potential energy of the ball is:

$$\Delta PE = -m\,g\,H \tag{3-44}$$

where m is the mass of the ball, g is the acceleration of gravity, and H is the height of the hill. You may understand that the potential energy that was originally available when the ball was placed at the top of the hill has been transferred out of the system in some way as heat. However, it would not be easy to measure the heat transfer from the ball to the surroundings in order to prove that Eq. (3-43) is valid. Therefore, to a skeptical observer, this experiment might appear to actually disprove the First Law.

Figure 3-13: Heating a fluid at constant volume.

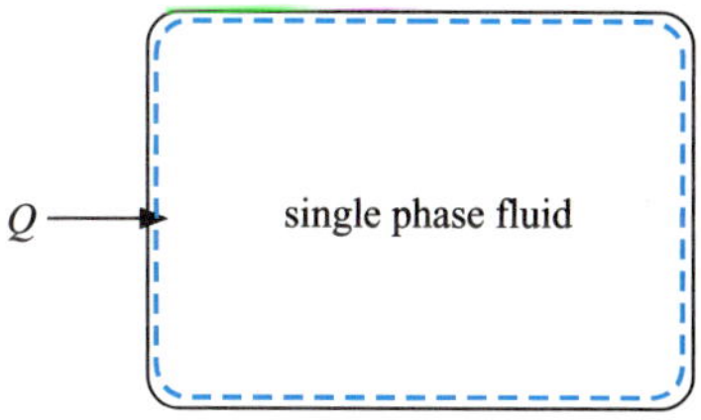

It is instructive to consider, in general, how one would measure each of the terms in Eq. (3-43). The kinetic and potential energy terms can be determined relatively easily by measuring the velocity and elevation of a system. There are many forms of work; perhaps the easiest to measure is electrical work, as described in Table 3-1. Measuring heat is not as easy. The science called calorimetry has been developed to measure heat. Calorimetry originated in the eighteenth century with the invention of the calorimeter. While there are many different designs for a calorimeter, these devices typically relate the measurement of heat to the increase in the temperature of a known mass of a reference substance. All that remains in Eq. (3-43) is to measure the change in internal energy. What instrument would you use to measure the change in the internal energy of a system? The fact is, you cannot directly measure the change in the internal energy of a substance in a manner that is independent from the measurement of heat or the measurement of work. As a consequence, you cannot validate (or invalidate) the First Law of Thermodynamics in this manner. So, what then is the basis of the First Law? Why do we believe it is valid?

Consider an experiment in which a single phase fluid is heated in a rigid container, as shown in Figure 3-13. The state of the fluid is fixed by two properties; the two properties that are easily measured in this experiment are temperature and specific volume. The specific volume remains constant during the experiment because neither the mass nor the volume of the fluid change. The methods of calorimetry can be used to measure the heat required to change the state of the fluid. Notice that the heat required to change the state of the container could confound this experiment. Therefore, we must design the container so that it has a negligible effect on the measurements. Alternatively, we could run separate experiments on the empty container and subtract the heat required to change the state of the empty container from any subsequent measurements.

We find (by repeated measurements) that the heat required to change the fluid from a specified initial state, state 1, to a different state, state 2, is a unique function of the initial and final states of the fluid. The states can be easily identified in this experiment in terms of the fluid temperature. We can mathematically express this observation as:

$$Q_{1\to2} = f(1,\ 2) \tag{3-45}$$

Suppose we now heat the system from state 2 to state 3. In this case, the measured heat depends only upon states 2 and 3:

$$Q_{2\to3} = f(2,\ 3) \tag{3-46}$$

We run a third set of experiments in which the fluid is heated directly from state 1 to state 3 and find that the measured heat depends only on these states:

$$Q_{1\to3} = f(1,\ 3) \tag{3-47}$$

We would also find that, to within experimental uncertainty:

$$Q_{1\to3} = Q_{1\to2} + Q_{2\to3} \tag{3-48}$$

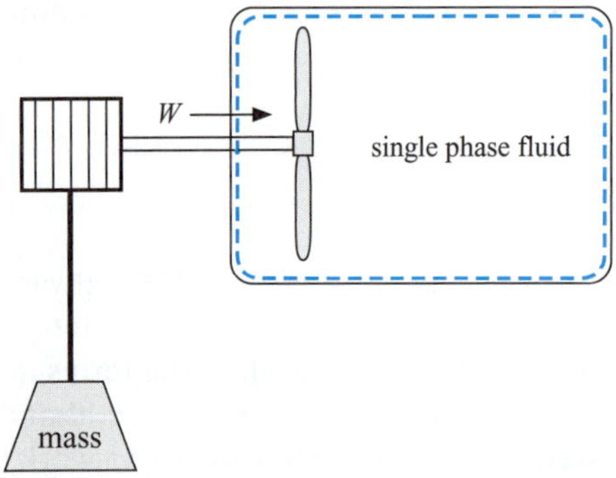

Figure 3-14: Using work to change the state of a fluid at constant volume.

Substituting Eqs. (3-45), (3-46), and (3-47) into Eq. (3-48) leads to:

$$f(1,\ 3) = f(1,\ 2) + f(2,\ 3) \tag{3-49}$$

We do not know anything about function f, other than what is observed in Eq. (3-49). However, the left side of Eq. (3-49) depends on states 1 and 3 and the right side depends on states 1, 2, and 3. The only possible way that this result can be generally true is if the function f has a form that allows state 2 to be canceled from the right side. Mathematically, this requirement can be expressed as:

$$f(i,\ j) = U(j) - U(i) \tag{3-50}$$

where $U(i)$ is another function that we know nothing about other than it depends only on state i.

Equation (3-50) is of profound significance. The function $f(i,\ j)$ depended on the conditions of the fluid at two different states and therefore it cannot be a property of the fluid. However, the function $U(i)$ depends only on the properties that describe state i and therefore U must itself be a property.

One of the major advancements in the development of the science of thermodynamics was the realization that work can bring about the same change in a system as heat. For example, we could construct the alternative experiment shown in Figure 3-14 that uses a paddle wheel driven by a descending weight to change the state of the fluid. The work is easily measured in this experiment directly from its definition, Eq. (3-34), where the force is the product of the mass and gravitational acceleration of the weight.

Again, careful repeated experiments would show that the work required to change the state of the fluid from state 1 to state 2 is given by the same function f that is measured for the experiment involving heat, shown in Figure 3-13:

$$W_{1\to 2} = f(1,\ 2) = U(2) - U(1) \tag{3-51}$$

Since the change in the fluid state that is observed by turning the paddle wheel (i.e., the increase in its temperature at fixed volume) is exactly the same as the change that is observed by the heating process, the function $f(1,\ 2)$ must be the same regardless of whether heat or work is applied. In addition, heat and work must both transfer the same thing to the fluid since they bring about identical changes.

Superposition of these experiments (which do not involve kinetic or potential energy terms or any outflow of work or heat) allows the generalization that:

$$Q_{in} + W_{in} = \Delta U \tag{3-52}$$

Further experiments can include kinetic and internal energy terms. Generalization of these experiments results in the statement of the First Law for a closed system given by Eq. (3-4). The function $U(i)$ referred to in these experiments is the internal energy of the fluid. We see that the internal energy has a unique value at a given state of the fluid and therefore it fulfills the requirement of being a property. Internal energy does not need to have any further significance.

So, here's how you might go about validating the First Law of Thermodynamics, even if you cannot measure internal energy or its change, ΔU. Fix the initial and final states of the system and measure all of the terms in Eq. (3-4) except ΔU. All of the other terms are, in theory, measurable and thus $Q_{in} - Q_{out} + W_{in} - W_{out} - \Delta KE - \Delta PE$ should always be a constant, within experimental uncertainty, regardless of how the change of state is brought about. Perhaps the easiest option is allow the experiment to operate in a cycle so that the initial and final states are identical. In this case, we know that $\Delta U = 0$ and we do not need to measure it. Confirmation of the First Law for a closed system that operates in a cycle results from showing that $Q_{in} - Q_{out} + W_{in} - W_{out} - \Delta KE - \Delta PE = 0$ regardless of the other details of the cycle.

Let's close this section with one last comment about the processes that can transfer energy across a boundary. These have been neatly classified as heat and work. In Chapter 4 we will see that energy is also transferred with a mass flow across the boundary. In some cases, however, the classification of an energy transfer is not clear. For example, we have classified the energy transfer that results from a flow of electrical current across a system boundary as work. Work is defined as an energy transfer that is independent of mass transfer. However, electrical current flow is a result of small charged particles (electrons) that are moving across the system boundary. These particles do have some (very small) mass and therefore perhaps we should classify electrical energy as an energy transfer associated with a mass flow rather than as a work interaction. We classify the energy transfer process that results from visible light (e.g., radiation from an incandescent or fluorescent light bulb) crossing a system boundary as heat. Heat, like work, is defined as energy transfer that is independent of mass transfer. However, we understand light to be electromagnetic radiation that consists of photons and photons (like electrons) have a (very small) equivalent mass. Therefore, perhaps radiation heat transfer should also be considered to be an energy transfer associated with a flow of mass rather than a heat transfer interaction. In any case, the First Law places all of the energy transformations experienced by a system as being equivalent and therefore what difference does it make how an energy transfer is classified? The answer to this question is the subject of Chapter 5.

REFERENCES

Çengel, Y.A., and Boles, M.A., *Thermodynamics – An Engineering Approach*, 4th Edition, McGraw-Hill, ISBN 0–07-112177-3, (2002).

Doolittle, J.S., *Thermodynamics for Engineers*, John Wiley & Sons, ISBN 9780471058052, (1983).

Keenan, J.H., "Adventure in Science", *Mechanical Engineering*, May, (1958).

Lewis, G.N. and Randall, M., *Thermodynamics and the Free Energy of Chemical Substances*, McGraw-Hill, (1923).

NIST Standard Reference Database Number 69, http://webbook.nist.gov/chemistry, June 2005 Release, (2005).

Moran, M.J. and Shapiro, H.N., *Fundamentals of Thermodynamics*, 6th Edition, John Wiley & Sons, ISBN 978–0471-78735-8, (2008).

Nellis, G.F. and S.A. Klein, Heat Transfer, Cambridge University Press, New York, (2009).

Problems

The problems included here have been selected from a much larger set of problems that are available on the website associated with this book (www.cambridge.org/kleinandnellis).

A. Heat and Work

3.A-1 Cryogenic liquids (e.g., liquid helium or liquid neon) are sometimes used to keep instruments at cryogenic (i.e., very cold) temperatures for space science missions. As the liquid boils off due to heat transfer it is vented to space. When all of the cryogenic liquid is gone, the temperature of the instrument increases and the mission is over. Flight operations engineers need to be able to check the amount of liquid that is left in the tank from the ground while the spacecraft is in orbit. In microgravity, the mixture of liquid and vapor in the tank is not stratified by gravity in the same way that it is on earth. Therefore, traditional liquid level measurement techniques do not work. One alternative technique that has been used by NASA is referred to as mass gauging. In order to accomplish mass gauging, a heater is activated for a short time and the temperature rise of the neon and the tank material is measured. The magnitude of the temperature rise can be used to calculate the mass of liquid that is left in the tank. Consider the spherical, aluminum cryogenic tank shown in Figure 3.A-1.

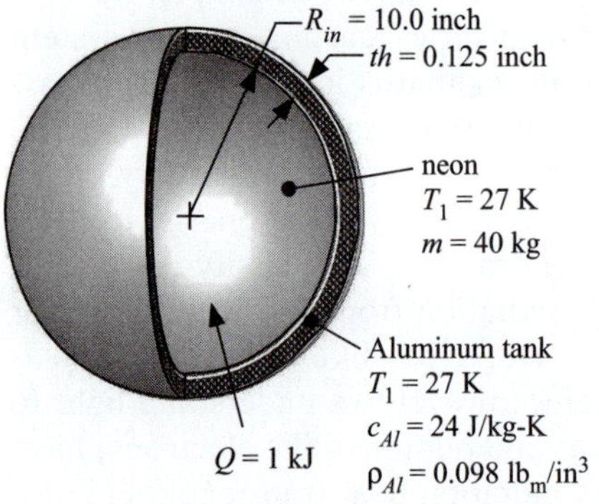

Figure 3.A-1: A spherical cryogenic tank.

The tank has an inner radius of R_{in} = 10 inch and a wall thickness of th = 0.125 inch. The density and specific heat capacity of aluminum (at the cryogenic temperatures associated with this problem) are $\rho_{Al} = 0.098$ lb$_m$/in^3 and c_{Al} = 24 J/kg-K, respectively. The tank contains a mixture of saturated liquid and saturated vapor neon at $T_1 = 27$ K. The mass of neon in the tank is $m = 40$ kg. For these calculations you may assume that no mass leaves the tank during the short time that it takes to complete the mass gauging process. A heater in the tank is activated and provides $Q = 1$ kJ of heat in order to accomplish the mass gauging. You may assume that the temperature of the neon is uniform and that the tank and the neon are at the same temperature. There is no work done on or by the tank or its contents during this process.

a) What is the mass of the aluminum?
b) What is the quality of the neon initially in the tank? What is the pressure of the neon initially in the tank?
c) What is the temperature of the neon and aluminum at the conclusion of the mass gauging process (T_2)? What is the increase in temperature detected by the operators?
d) What are the quality and pressure of the neon in the tank at the conclusion of the mass gauging process?
e) Generate a calibration curve that an operator could use for the mass gauging process. That is, generate a plot showing the mass of neon in the tank (m) as a

function of the temperature rise ($\Delta T = T_2 - T_1$) for neon mass ranging from 1 kg to 80 kg.

f) If your temperature sensors are capable of resolving a temperature difference of approximately 1 mK (that is, the uncertainty in your measurement of the temperature rise is $\delta\Delta T = 0.001$ K) then estimate the resolution of the mass gauging process (that is, how well can you measure the mass of neon in the tank?). Use your engineering judgment to answer this question; you may want to refer to the calibration curve generated in part (e).

3.A-2 Figure 3.A-2 is the pressure-volume diagram for a thermodynamic cycle that is executed by $m = 18$ lb$_m$ of nitrogen gas. Assume that nitrogen behaves according to the ideal gas law.

a) Determine the temperature at each of the states shown in Figure 3.A-2.

b) Determine the work done by the nitrogen gas for each of the processes shown in Figure 3.A-2 (i.e., process 1 to 2, 2 to 3, 3 to 4, and 4 to 5). What is the net work done by the nitrogen during cycle?

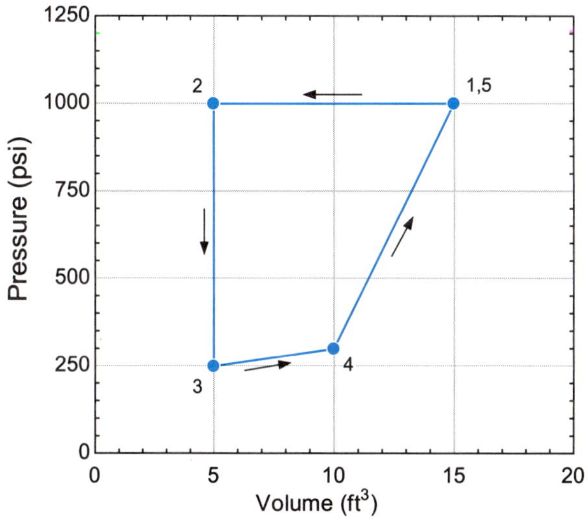

Figure 3.A-2: Pressure-volume diagram for a thermodynamic cycle.

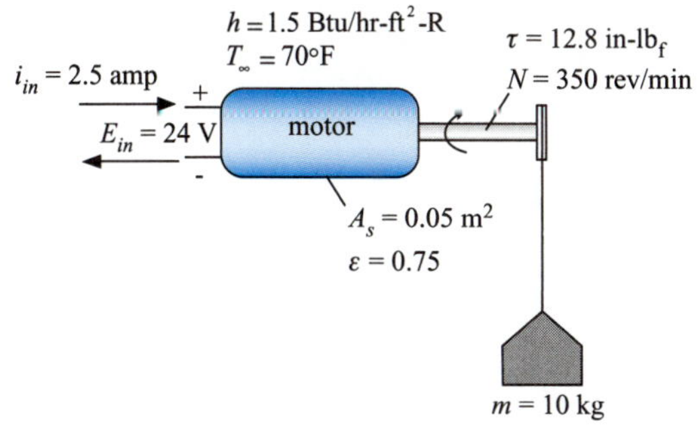

Figure 3.A-4: Motor being used to lift a mass.

3.A-4 Figure 3.A-4 illustrates a small motor that is being used to lift an $m = 10$ kg mass.

The motor is operating at steady state. The surface area of the motor is $A_s = 0.05$ m^2 and the motor is surrounded by air at $T_\infty = 70°$F. The surface of the motor has an emissivity of $\varepsilon = 0.75$. The heat transfer coefficient between the motor and the air is approximately $h = 1.5$ Btu/hr-ft^2-R. The motor radiates to surroundings that are also at T_∞. The motor is provided electrical input power with a voltage of $E_{in} = 24$ V and a current $i_{in} = 2.5$ amp. The motor shaft is rotating at $N = 350$ rev/min and the torque on the shaft is $\tau = 12.8$ inch-lb$_f$.

a) Carry out an energy balance on a system that consists of the motor. What is the rate at which the energy in the system is changing? What is the rate of heat transfer from the motor?
b) What is the surface temperature of the motor?
c) Assume that all of the mechanical power carried by the shaft is being used to increase the potential energy of the mass. What is the velocity at which the mass is rising?

3.A-6 Water is contained in a piston-cylinder device, as shown in Figure 3.A-6. You may neglect friction between the piston and the wall and assume that the piston does not leak.

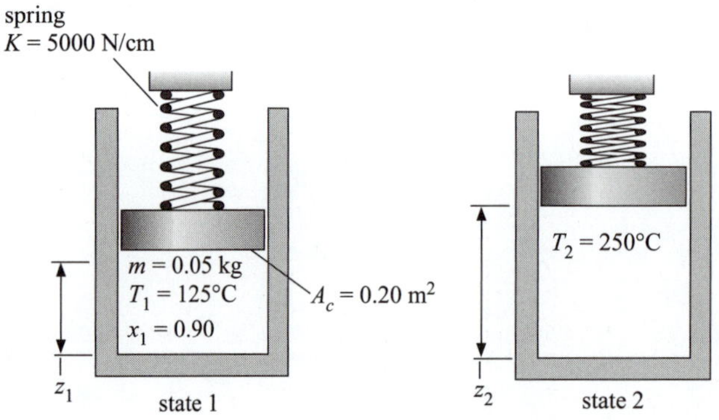

Figure 3.A-6: Water in a spring-loaded piston-cylinder device at state 1 and state 2.

The mass of the water is $m = 0.05$ kg and the area of the piston face is $A_c = 0.2$ m^2. Initially, the water is at $T_1 = 125°$C with a quality of $x_1 = 0.90$. The water is heated and the piston begins to rise; as this occurs, the spring is compressed and so the pressure in the cylinder begins to rise. The pressure rise is proportional to the amount that the spring is compressed, according to:

$$P_2 = P_1 + K\frac{(z_2 - z_1)}{A_c}$$

where $K = 5000$ N/cm is the spring constant and z_2 and z_1 are the final and initial positions of the piston, respectively, as shown in Figure 3.A-2. The heating stops when the water temperature reaches $T_2 = 250°$C.

a) What is the initial pressure in the piston? What is the initial position of the piston, z_1?
b) What is the pressure in the piston at the end of the heating process?
c) Using EES, prepare a T-v diagram for water using the Property Plot option from the Plots menu and overlay states 1 and 2 on this plot.

B. Closed System Energy Balances

3.B-1 A well-insulated rigid container with an internal volume of $V = 0.01\ \text{m}^3$ holds $m = 2$ kg of R134a, as shown in Figure 3.B-1.

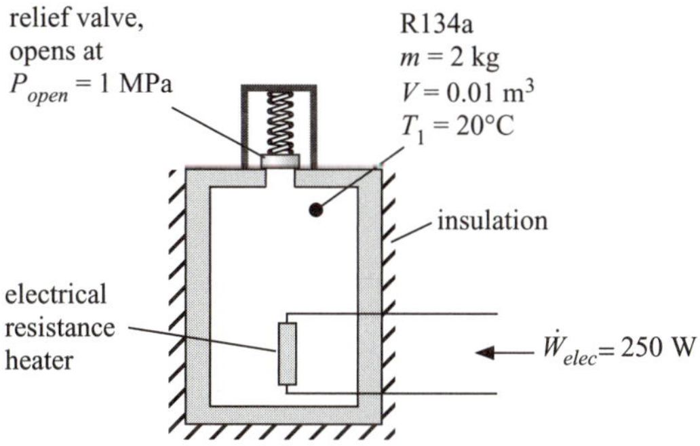

Figure 3.B-1: Container holding R134a.

The container is initially at room temperature ($T_1 = 20°\text{C}$). The container is fitted with an electrical resistance heater as shown in the figure. The heater draws $\dot{W}_{elec} = 250$ W when it is switched on. The pressure relief valve opens when the internal pressure in the container reaches $P_{open} = 1$ MPa. Assume that the mass of the container is negligible.

a) Determine the initial pressure in the container.
b) Determine the time that the electrical heater can operate before the pressure relief valve opens.
c) Determine the temperature of the R134a at the time that the pressure relief just opens.

You have carried out some experiments and measured the time required to open the relief valve for the container analyzed in Figure 3.B-1. You have found that the measured time required is much longer than what you predicted in part (b). It is suspected that the discrepancy is related to the energy required to change the temperature of the container walls. The container is made of AISI 304 stainless steel and has a total mass of $m_{cyl} = 7.25$ kg. The specific heat of this material is $c_{cyl} = 478.2$ J/kg-K. (Note that this property is also available in the EES Solid/Liquid Property library.) Assume that the container wall temperature is uniform and that the container is in thermal equilibrium with the R134a (i.e., they are at the same temperature).

d) Determine the time required for the pressure relief valve to open if the container wall is included in the analysis.

3.B-4 A piston-cylinder device is used to fill a balloon with helium, as shown in Figure 3.B-4. The piston diameter is $D_p = 3$ mm and the piston position is initially $z_1 = 5$ cm from the bottom of the cylinder. At this point, the balloon diameter is $D_{b,1} = 3.5$ mm and the pressure and temperature of the helium in the piston and the balloon are $P_1 = 103$ kPa and $T_1 = 22°\text{C}$, respectively. The balloon stretches as it is filled and so the pressure in the balloon increases. The relationship between the pressure within the balloon and the volume of the balloon is given by:

$$P = P_{atm} + K_b V_b$$

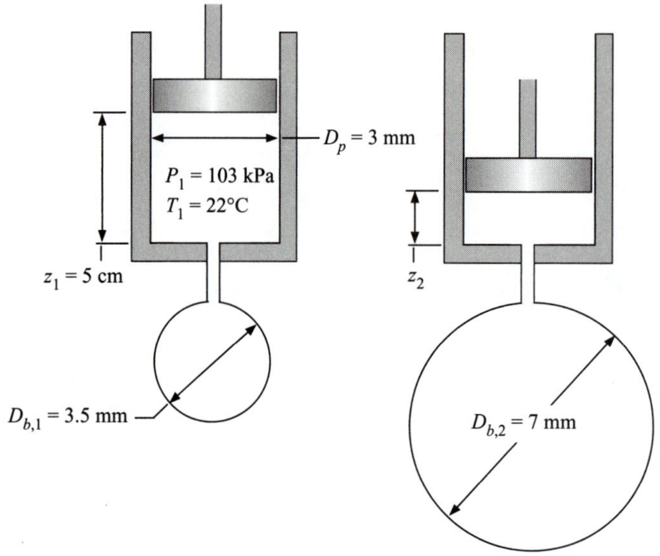

Figure 3.B-4: Piston-cylinder device used to fill a balloon.

where P_{atm} is atmospheric pressure (1 atm) and K_b is a constant. The final diameter of the balloon is $D_{b,2} = 7$ mm. The temperature of the helium in the piston is maintained at $T_{in} = T_1$ during the process (i.e., all of the helium leaving the piston and entering the balloon is at T_{in}). The final temperature of the helium in the balloon is $T_2 = 30°$C. Model helium as an ideal gas with $R = 2076.9$ J/kg-K. Assume that the piston and balloon are both leak-tight. You may ignore the volume of the tube connecting the balloon to the piston. The pressure in the piston and the pressure in the balloon are always the same.

a) What is the work done by the helium on the balloon during the filling process?
b) What is the mass of helium that was added to the balloon in order to inflate it?
c) What is the final position of the piston (z_2 in Figure 3.B-4)?
d) What is the heat transfer from the balloon to the surroundings?
e) Plot the diameter of the inflated balloon as a function of the final position of the piston.

3.B-6 At the end of a manufacturing process, you are left with small pieces of scrap metal that are relatively hot, $T_H = 700$ K, and have mass $m_b = 2.0$ kg. Rather than just throw the metal pieces away, you have come up with an idea for a simple machine that you hope can produce useful work. The machine is a piston-cylinder device that is filled with air and it goes through four processes in order to lift a mass (i.e., produce useful work). The processes take the air in the piston from state 1 to state 2 to state 3 to state 4 and then back to state 1. The cumulative effect of the four processes is referred to as a cycle. Model the metal pieces as incompressible with a constant specific heat capacity, $c_b = 900$ J/kg-K. Model the air as an ideal gas with $R = 287.1$ N-m/kg-K and $c_v = 717.6$ J/kg-K.

The piston-cylinder device is initially filled with air at atmospheric pressure, $P_1 = P_{atm} = 1$ atm and ambient temperature $T_1 = T_{amb} = 300$ K. The piston area is $A_c = 0.05$ m^2 and the distance between the piston and the bottom of the cylinder is initially $z_1 = 0.5$ m. The piston is massless, frictionless, and leak-tight. A mass, $m_p = 100$ kg, is slowly placed on the piston so that the piston moves downward. During this process, the piston is transfers heat to the ambient atmosphere and therefore the air in the piston is compressed isothermally (i.e., the temperature of the air in the piston during this process is always $T = T_{amb}$). This process of going from state 1 to state 2 is shown in Figure 3.B-6(a).

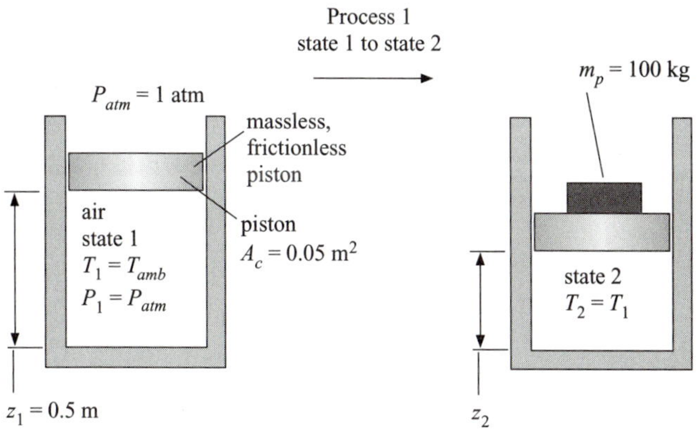

Figure 3.B-6(a): Process of going from state 1 to state 2.

a) What is the final pressure of the air, P_2? What is the final position of the piston, z_2?
b) Determine the work done **by** the air, W_{12}; note that this quantity may be negative if work is transferred to the air.
c) Determine the heat transfer **to** the air, Q_{12}; note that this quantity may be negative if heat is transferred from the air.

Next, one of the pieces of hot scrap metal is brought into thermal contact with the piston; this causes a heat transfer from the metal to the air in the piston. The air temperature rises while the metal temperature drops and this continues until the air and metal come to the same temperature, T_3. The increase in the air temperature causes the piston to rise, lifting the weight. Assume that there is no heat transfer to the ambient atmosphere during this process (i.e., the only heat transfer is from the metal piece to the air in the cylinder). This process of going from state 2 to state 3 is shown in Figure 3.B-6(b).

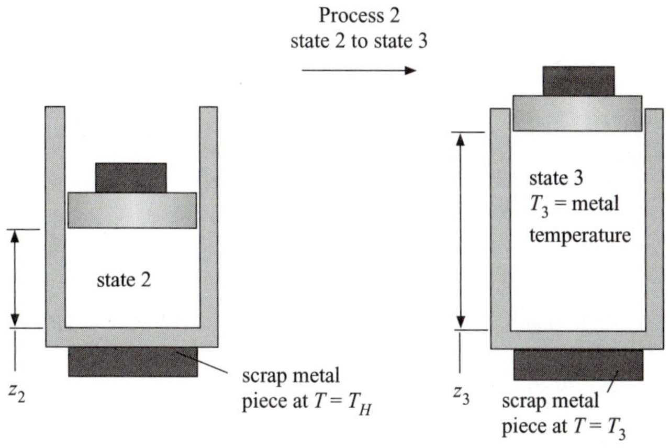

Figure 3.B-6(b): Process of going from state 2 to state 3.

d) What is the final temperature of the air and the scrap metal piece, T_3?
e) What is the work done **by** the air, W_{23}?

f) What is the heat transfer from the scrap metal piece **to** the air, Q_{23}?
g) What is the final position of the piston, z_3?

Next, the piston is locked in place and the mass is removed from the piston. The scrap metal piece is taken out of thermal contact with the piston. With the piston locked in place, the air in the piston is allowed to cool by transferring heat to the ambient atmosphere until the pressure in the piston reaches atmospheric pressure, $P_4 = P_{atm}$. This process of going from state 3 to state 4 is shown in Figure 3.B-6(c).

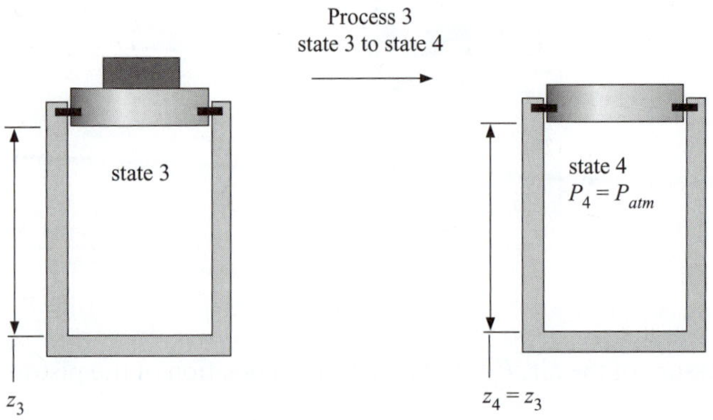

Figure 3.B-6(c): Process of going from state 3 to state 4.

h) What is the final temperature of the air, T_4?
i) What is the work done **by** the air, W_{34}?
j) What is the heat transfer **to** the air, Q_{34}?

Finally, the piston is unlocked and allowed to float freely (note that the mass has been removed) and the air in the piston is allowed to cool by transferring heat to the ambient environment until it returns to state 1, $P_1 = P_{atm}$ and $T_1 = T_{amb}$. This process of going from state 4 to state 1 is shown in Figure 3.B-6(d).

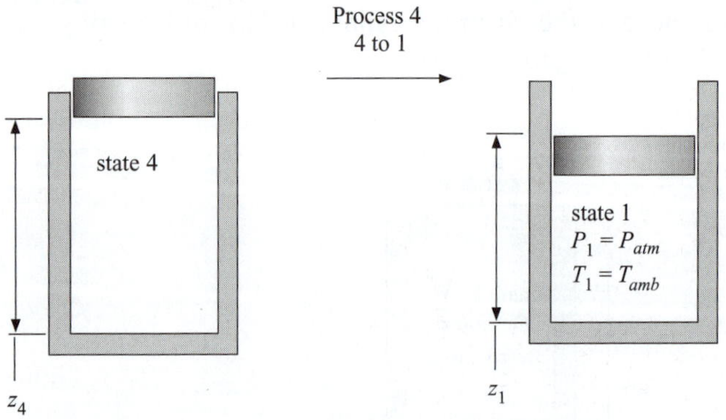

Figure 3.B-6(d): Process of going from state 4 to state 1.

k) What is the work done **by** the air, W_{41}?
l) What is the heat transfer **to** the air, Q_{41}?

Note that your machine has gone through a cycle – the final state of the air is equal to its initial state and so it is ready to go through the cycle again.

m) Sketch states 1 through 4 on a T-v diagram for air (or use EES to draw one for you).

n) What is the net work done by your machine (i.e., what is the sum of the work terms done by the air that you calculated for each process)?
o) If you can run your system at a frequency of $f = 5$ Hz (i.e., the piston goes back and forth 5 × per second) then what is the average power produced by the machine?
p) One way to double-check your answer is to calculate the net heat transfer **to** the air (the sum of the heat transfers to the air that you calculated for each process) and compare it with the net work done **by** the air (the sum of the work done by the air that you calculated for each process). These two quantities ought to be the same – are they? Why should these quantities be equal?
q) The efficiency of your machine is defined as the ratio of what you get out, the net work calculated in part (n), to what you put in, the heat transfer from the metal calculated in part (f). What is the efficiency of your machine?
r) Prepare a plot showing the efficiency of your machine as a function of the initial temperature of the scrap metal, T_H.

3.B-8 Water is held in a piston/cylinder apparatus. The pressure in the cylinder is initially $P_1 = 1.5$ bar and the temperature is $T_1 = 150°$C. The volume of water contained in the cylinder is $V_1 = 1$ m^3.
a) On a T-v sketch, locate state 1. This qualitative sketch does not need to be accurate, but it should clearly show the two properties that define the state.
b) Determine the mass of water in the cylinder (kg).

The piston is pushed in so that the volume of the water is reduced until liquid water droplets just start to form in the cylinder at state 2. During this process, the contents of the cylinder are maintained at a constant temperature ($T_2 = 150°$C) by heat transfer with the surroundings.
c) On the T-v sketch from (a), locate state 2. This qualitative sketch does not need to be accurate, but it should clearly show the two properties that define the state.
d) What is the volume in the cylinder (m^3) at the instant that liquid water begins to form?

The piston is pushed in further, reducing the volume to $V_3 = 0.05$ m^3 at state 3. During this process, the contents of the cylinder are maintained at a constant temperature ($T_3 = 150°$C) by heat transfer with the surroundings.
e) On the T-v sketch from (a), locate state 3. This qualitative sketch does not need to be accurate, but it should clearly show the two properties that define the state.
f) What is the pressure in the cylinder at state 3?
g) What is the volume of liquid in the cylinder at state 3?
h) Determine the heat transfer required to maintain the contents of the cylinder at 150°C as the piston is pushed in from state 2 to state 3 (kJ). Be sure to indicate clearly whether the heat transfer that you calculate is into or out of the water.

3.B-9 Consider the piston-cylinder device shown in Figure 3.B-9. The diameter of the piston is $D_p = 0.10$ m. Initially, the piston is resting on a set of stops and the distance from the bottom of the cylinder is $z = 0.1$ m. The entire apparatus is at $T_1 = 25°$C and is in thermal equilibrium with its surroundings. The cylinder is evacuated except for a spherical capsule having an inner diameter of $D_c = 2.5$ cm containing carbon dioxide at T_1 and a known high pressure, P_1. The piston has a mass of $m_p = 75$ kg. The capsule then ruptures and the carbon dioxide is rapidly released into the cylinder.

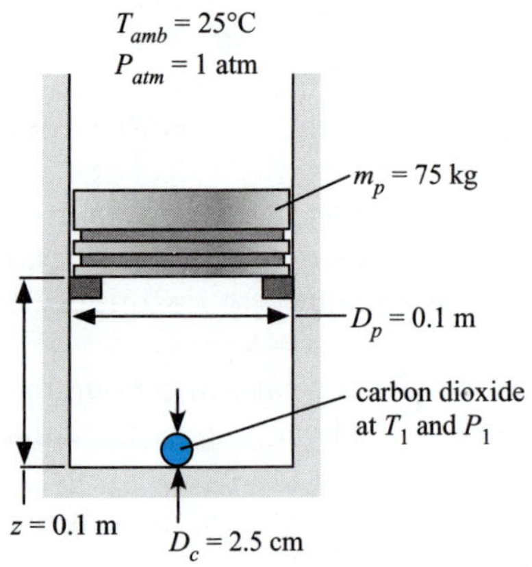

Figure 3.B-9: Piston-cylinder apparatus with capsule containing CO_2.

a) Prepare a plot indicating the work done by the piston-cylinder apparatus during this process as a function of the initial pressure of the carbon dioxide in the capsule for pressures ranging 1 MPa $< P_1 <$ 10 MPa. Clearly indicate your system and state any assumptions you employ. Note that carbon dioxide does not obey the ideal gas law under the conditions that it exists in the capsule.

3.B-12 Immediately after a high pressure tank of air is filled, the air in the tank is hot, $T_{a,1} = 200°$C, and the tank material itself is at room temperature, $T_{t,1} = 20°$C. The valve on the tank is shut and the volume of air in the tank is $V_a = 5$ liter. The initial pressure of the air is $P_1 = 350$ psi. The tank material has a mass of $m_t = 0.45$ kg and a specific heat capacity of $c_t = 471.9$ J/kg-K. Model the air as an ideal gas, $R = 287$ J/kg-K, with constant specific heat capacity at constant specific volume, $c_v = 726$ J/kg-K. There is a transfer of heat between the tank material and the air that continues until they come to the same temperature, T_2. Assume that there is no heat transfer from the tank to the surroundings during this process.

a) Determine the final temperature of the air and the tank and the final pressure of the air.
b) What is the heat transfer to the tank that occurs during this process?

Eventually, the air and the tank both return to room temperature, $T_3 = 20°$C, due to heat transfer with the surroundings.

c) Determine the final pressure of the air.
d) What is the heat transfer to the surroundings that occurs during this process?

3.B-13 An emergency flotation device is made by attaching a high pressure canister of air to a balloon via a valve. Initially, the balloon is deflated (i.e., its volume is zero) and the canister is pressurized to $P_{c,1} = 6000$ psi. The cylindrical canister has inner radius $R_c = 2.5$ cm and length $L_c = 20$ cm. In order to activate the flotation device, the valve is opened allowing air to flow into the balloon causing it to inflate. The internal pressure in the balloon is higher than atmospheric pressure due to the tension in the balloon material. The internal pressure is given by:

$$P_b = P_{atm} + K_b\, V_b$$

where $P_{atm} = 100$ kPa is the atmospheric pressure, V_b is the balloon volume, and $K_b = 1 \times 10^6$ N/m^5. The inflation process is complete when the pressure within the canister and the pressure within the balloon are the same. Assume that the air within the canister and the balloon is maintained at $T_{atm} = 15°C$ by heat transfer with the surroundings. Model air as an ideal gas.

a) Determine the final radius of the balloon.
b) What is the work done by the air on the balloon? What is the heat transfer from the surroundings to the air?
c) Determine the buoyancy force associated with the flotation device once it is activated. Assume that the density of water is $\rho_w = 1000$ kg/m^3.
d) Plot the buoyancy force produced by the flotation device as a function of the initial pressure in the canister.

3.B-17 An elevator is shown in Figure 3.B-17. The air in the $V = 30$ m^3 tank is initially at $T_1 = 25°C$ and $P_1 = 100$ kPa. The piston has mass $m_p = 230$ kg and diameter $D_p = 0.75$ m. A casting with mass $m_c = 1000$ kg is slid onto the platform at its lower level (level 1). Then saturated steam at $T_s = 150°C$ is provided to raise the temperature of the air in the tank by heat exchange. The heating continues until the platform reaches its upper level (level 2) that is $z = 6$ m above the lower level, at which point the platform hits a stop and heating is stopped. Here, the casting is slid off of the platform. Then, cooling water is provided to lower the air temperature until the platform returns to the initial level where the piston rests on a stop. Cooling continues until the air in the tank is returned to T_1. Please answer the following questions and state any assumptions that you employ.

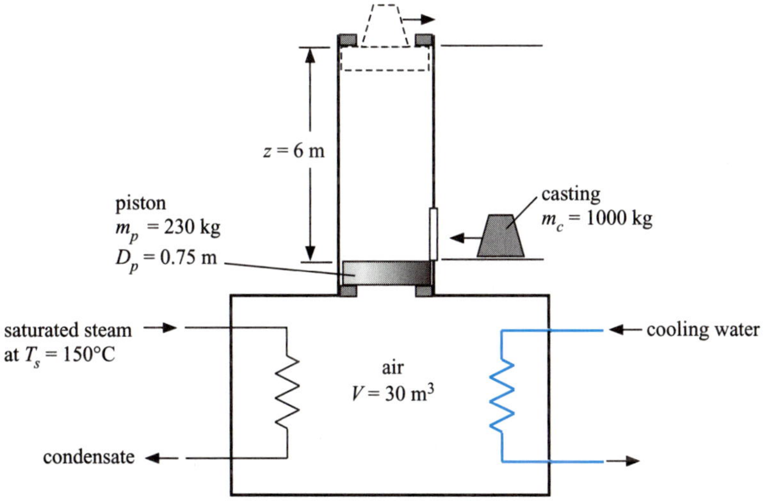

Figure 3.B-17: Heat powered elevator.

a) Determine the temperature and pressure of the air in the tank when the platform *just* reaches level 2.
b) Determine the work done by the air in raising the platform and casting from level 1 to level 2.
c) What is the heat transfer to the air from the steam while raising the platform and casting from level 1 to level 2?
d) What is the heat transfer between the air and the cooling water during the process in which the platform returns from level 2 to level 1?

e) What is the overall efficiency of the elevator for the completion of one cycle, lifting the casting to level 2 and returning to level 1?
f) What is the maximum casting mass that can be lifted by this elevator?

C: Advanced Problems

3.C-3 The purpose of this problem is to analyze the dynamic behavior of a vertical piston-cylinder device containing a gas. The gas is placed in the cylinder at an initial pressure that is above atmospheric and the piston is locked in place. In a particular case, the cylinder contains carbon dioxide gas initially at $T_{ini} = 25°C$ and $P_{ini} = 1.5$ bar. Assume that the carbon dioxide obeys the ideal gas law (i.e., if you are using EES, then the fluid name should be 'CO2' rather than 'CarbonDioxide'). The internal radius of the cylinder is $R = 0.05$ m. The piston, which has a mass of $m_p = 20$ kg, is initially locked at a position $H_{ini} = 0.15$ m above the bottom of the cylinder. The piston and cylinder are made of metal and initially this metal is also at T_{ini}. When the piston locks are released, the piston moves in an oscillatory manner that is damped by frictional effects. Prepare an analysis of this experiment. Use your analysis to calculate and plot the position of the piston above the bottom of the cylinder for a simulation time of $t_{sim} = 2$ s for the following cases:

a) No friction and no heat exchange between gas and metal. Determine the frequency of the oscillations.
b) Include frictional effects (without heat transfer) by setting the frictional force resisting piston motion to a value of $F_f = 10$ [N-s/m] $\tilde{V}_p$ where $\tilde{V}_p$ is the instantaneous piston velocity. Indicate how the amplitude and frequency are affected by frictional effects based on examination of your plot.
c) Include heat transfer between the gas and cylinder walls with a convection heat transfer coefficient of $h_{conv} = 100$ W/m^2-K. Ignore friction for this simulation. You may assume that the wall temperature remains constant at T_{ini} due to its large thermal mass. Indicate how the amplitude and frequency are affected by heat transfer based on examination of your plot.

3.C-10 The experiment shown in Figure 3.C-10 attempts to determine the constant relating mechanical and thermal energy transfer. Air, initially at $T_1 = 75°F$ and $P_1 = 15$ psi, is contained in a well-insulated cylinder that has diameter $D = 2$ ft and length $L = 4$ ft. The cylinder is fitted with a fan blade (that occupies a negligibly small volume) connected by a shaft to a spool of wire. The fan blade rotates as an $m_w = 50$ lb$_m$ weight descends a total vertical distance of $H = 30$ ft in $t_e = 3.5$ seconds as it unwinds the wire and spins the spool. The air within the cylinder experiences convection heat transfer with the internal cylinder walls, which remain at $T_1 = 75°F$ during this experiment. The convection coefficient between the air and the cylinder walls is estimated to be $h_{conv} = 23$ W/m^2-K.

a) What is the final temperature of the air in cylinder that you would expect if heat transfer between the air and cylinder walls did not occur?
b) What is the accepted value of the constant relating ft-lb$_f$ of work to Btu of thermal energy?
c) Prepare a plot of the temperature of the air in the cylinder as a function of time for the duration of the experiment; include heat transfer with the cylinder walls in your analysis.
d) If you did not know that heat transfer was occurring between the air and the internal cylinder walls, what value would you have reported for the constant relating ft-lb$_f$ of work to Btu of thermal energy based on this experiment?

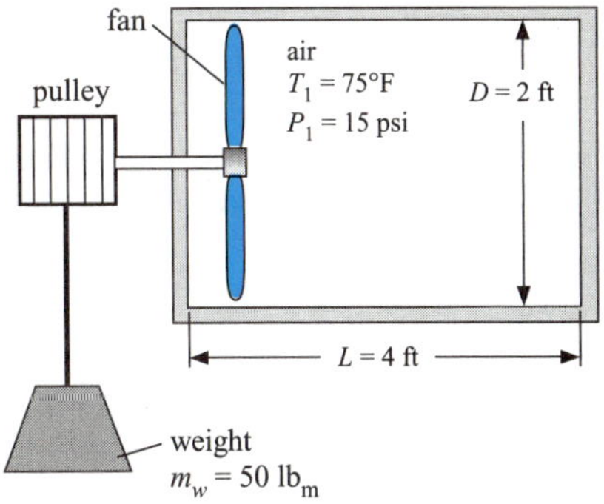

Figure 3.C-10: Experiment to relate mechanical and thermal energy transfer.

3.C-12 A spherical balloon containing air is located with its top surface at the air-water interface in a pool of water at $T_w = 25°$C. The diameter of the balloon at this location is $D_{b,1} = 0.35$ m. The balloon is made of an elastic material that expands or contracts such that the difference in pressure between the air in the balloon and its surroundings is directly proportional to the surface area of the balloon. The mass of the elastic material is $m_b = 0.015$ kg. The atmospheric pressure at the surface of the pool is $P_{atm} = 101.3$ kPa and the pressure of the air in the balloon at this location is $P_1 = 104.8$ kPa. The balloon is now slowly lowered a distance of $H = 10$ m into the pool of water. You may assume ideal gas behavior for air. Water may be assumed to be incompressible with constant density. The entire process may be assumed to be isothermal at T_w. State any other assumptions you employ.

a) Prepare a plot of the diameter of the balloon and the air pressure in the balloon as a function of depth between 0 (the surface) and 10 m.
b) The balloon is buoyant in water. Calculate the work required to move the balloon from the surface to a depth of 10 m.
c) The volume of the air changes as it descends into the water. Calculate the work done on the air in the balloon during this process.
d) The balloon material contracts as the balloon descends into the pool. What is the work done on the balloon material during this process?
e) Is the process adiabatic? If not, estimate the heat transfer to the balloon.

3.C-14 One type of household humidifier operates by expelling water droplets into the air. The droplet radii are assumed to be normally distributed with a mean radius of $\bar{r} = 1\ \mu$m and a standard deviation of $\sigma = 0.1\ \mu$m. These small droplets then evaporate into $T = 25°$C air to provide humidification. If the droplets follow a normal distribution, the probability density function of a droplet having radius less than or equal to r is given by:

$$f = \frac{1}{\sigma\sqrt{2\pi}} \exp\left[-\frac{(r-\bar{r})^2}{2\sigma^2}\right]$$

The integral $\int_{r_{min}}^{r_{max}} f\, dr = 1$; the limits of integration can be approximated as $\bar{r} \pm 4\sigma$.

a) What are the total mass, volume, and surface area of a sample of $N = 1 \times 10^{14}$ droplets?
b) Assuming that water is supplied to the humidifier at temperature T and atmospheric pressure, estimate the mechanical work required to produce these droplets.

c) What would the energy requirement be if the same mass of water were directly evaporated in the humidifier by adding heat (as many types of humidifiers do)?
d) Comment on the energy use and the advantages and disadvantages of these two different designs. If it is relevant, assume that the humidifier is in use in a home on a winter day when the outdoor temperature is –5°C.

4 General Application of the First Law

4.1 General Statement of the First Law

The concept of energy was introduced in Chapter 3 as a system property that consists of internal and external contributions. The energy balances considered in Chapter 3 are applied to closed systems, that is, systems in which there is no mass flow into or out of the system. In this chapter, a more general statement of the First Law is developed that can be applied to both open and closed systems. In order to derive the general statement of the First Law of Thermodynamics, it is necessary to relate a balance carried out on a closed system that encloses a fixed amount of mass (i.e., a *control mass*) to a more general balance on an open system that encloses a specified volume (i.e., a *control volume*). This process is illustrated first for a mass balance and then for an energy balance.

Consider a tank that has an inlet port and an outlet port, as shown in Figure 4-1. An open system is defined that includes the internal volume of the tank. The system is open because mass is flowing across its boundary at two locations. Mass enters the tank at one location with mass flow rate $\dot{m}_{in}$, temperature T_{in}, pressure P_{in}, velocity $\tilde{V}_{in}$, and elevation z_{in}. Mass leaves the tank at another location with mass flow rate $\dot{m}_{out}$, temperature T_{out}, pressure P_{out}, velocity $\tilde{V}_{out}$, and elevation z_{out}. The tank experiences no other interactions with the surroundings.

Figure 4-2(a) illustrates a control mass containing a fixed amount of mass at time t and Figure 4-2(b) shows the same control mass at time $t + \Delta t$. Notice that the shape of the control mass changes as the boundary moves in order to follow the fixed mass. At time t, the control mass shown in Figure 4-2(a) includes all of the mass contained in the control volume (the internal volume of the tank), $m_{cv,t}$, as well as some mass that has not yet entered the tank through the inlet port, Δm_{in}. At time $t + \Delta t$, the control mass shown in Figure 4-2(b) includes all of the mass contained in the control volume, $m_{cv,t+\Delta t}$, as well as some mass that has left through the outlet port, Δm_{out}.

The control mass is constant; therefore, a mass balance on the closed system requires that:

$$\underbrace{m_{cv,t} + \Delta m_{in}}_{\text{mass in control mass at time } t} = \underbrace{m_{cv,t+\Delta t} + \Delta m_{out}}_{\text{mass in control mass at time } t+\Delta t} \tag{4-1}$$

Equation (4-1) is rearranged:

$$\Delta m_{in} = \Delta m_{out} + m_{cv,t+\Delta t} - m_{cv,t} \tag{4-2}$$

and divided through by the time interval, Δt:

$$\frac{\Delta m_{in}}{\Delta t} = \frac{\Delta m_{out}}{\Delta t} + \frac{m_{cv,t+\Delta t} - m_{cv,t}}{\Delta t} \tag{4-3}$$

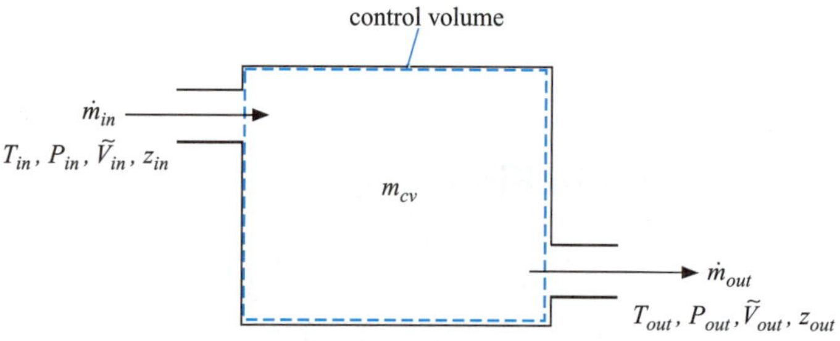

Figure 4-1: A tank that includes an inlet port and an outlet port.

In the limit that Δt approaches zero, the first two terms in Eq. (4-3) correspond to the mass flow rate into and out of the tank, respectively, and the third term is the rate of change of the mass contained within the control volume shown in Figure 4-1.

$$\underbrace{\lim_{\Delta t\to 0}\left(\frac{\Delta m_{in}}{\Delta t}\right)}_{\dot{m}_{in}} = \underbrace{\lim_{\Delta t\to 0}\left(\frac{\Delta m_{out}}{\Delta t}\right)}_{\dot{m}_{out}} + \underbrace{\lim_{\Delta t\to 0}\left(\frac{m_{cv,t+\Delta t}-m_{cv,t}}{\Delta t}\right)}_{\frac{dm_{cv}}{dt}} \tag{4-4}$$

This process has led to the familiar mass balance on an open system that was introduced in Section 1.4:

$$\dot{m}_{in} = \dot{m}_{out} + \frac{dm_{cv}}{dt} \tag{4-5}$$

The same process is used to derive the general form of an energy balance that is valid for an open or closed system. Figure 4-3(a) and (b) illustrate the same fixed mass system that is shown in Figure 4-2; i.e., Figure 4-3(a) shows the fixed mass at time t and Figure 4-3(b) shows the fixed mass at time $t + \Delta t$.

In Chapter 3, energy balances for closed systems are discussed. An energy balance for the control mass shown in Figure 4-3(a) and (b) is:

$$Q_{in} + W_{in} = Q_{out} + W_{out} + \Delta E \tag{4-6}$$

The tank experiences no interactions with its surroundings other than the mass flows and the associated changes in volume which result in work terms. Therefore, there is no heat transfer and Eq. (4-6) can be simplified:

$$W_{in} = W_{out} + \Delta E \tag{4-7}$$

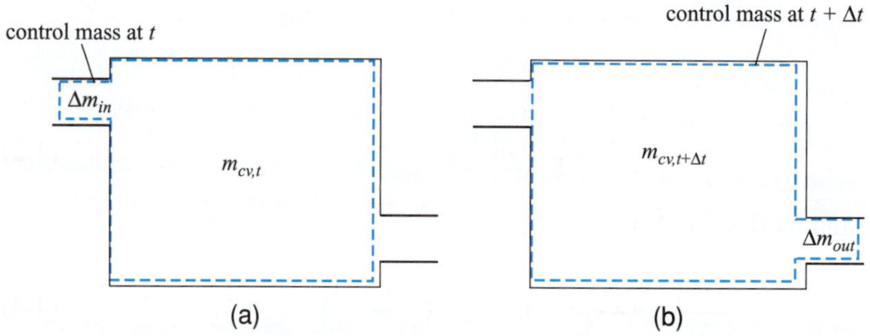

Figure 4-2: Closed system enclosing a fixed amount of mass (a) at time t, and (b) at time $t + \Delta t$.

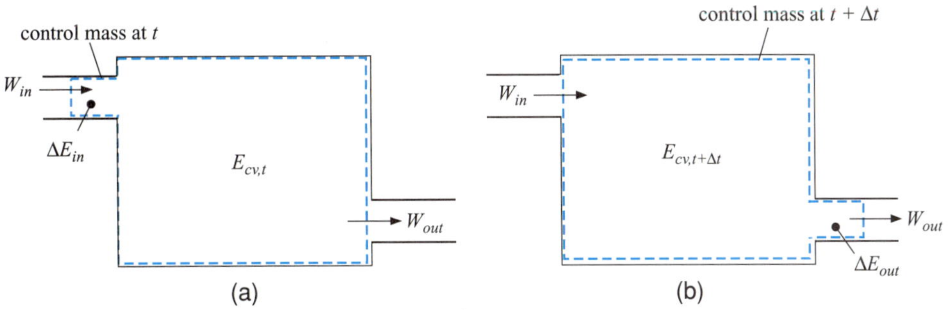

Figure 4-3: (a) Closed system enclosing a fixed amount of mass at time t, (b) the same fixed mass at time $t + \Delta t$.

The change in the energy of the mass is given by:

$$\Delta E = \underbrace{(E_{cv,t+\Delta t} + \Delta E_{out})}_{\text{energy of mass at } t+\Delta t} - \underbrace{(E_{cv,t} + \Delta E_{in})}_{\text{energy of mass at } t} \tag{4-8}$$

where $E_{cv,t+\Delta t}$ and $E_{cv,t}$ are the total energies of the control volume at times $t + \Delta t$ and t, respectively, and ΔE_{in} and ΔE_{out} are the energies associated with the mass that enters and leaves the control volume during Δt, respectively. Note that the boundary of the control mass is being compressed at the inlet port. Therefore, there is compression work (i.e., the work done on the system) at the inlet port:

$$W_{in} = P_{in} \underbrace{v_{in}\, \Delta m_{in}}_{\text{decrease in volume}} \tag{4-9}$$

where v_{in} is the specific volume of the mass entering the tank. Similarly, the boundary of the control mass is being expanded at the outlet port. The expansion work (i.e., the work done by the system) at the outlet port is:

$$W_{out} = P_{out} \underbrace{v_{out}\, \Delta m_{out}}_{\text{increase in volume}} \tag{4-10}$$

where v_{out} is the specific volume associated with the mass leaving the tank. Substituting Eqs. (4-8) through (4-10) into Eq. (4-7) results in:

$$P_{in}\, v_{in}\, \Delta m_{in} = P_{out}\, v_{out}\, \Delta m_{out} + (E_{cv,t+\Delta t} + \Delta E_{out}) - (E_{cv,t} + \Delta E_{in}) \tag{4-11}$$

Equation (4-11) is rearranged:

$$\underbrace{P_{in}\, v_{in}\, \Delta m_{in} + \Delta E_{in}}_{\text{energy entering at inlet port}} = \underbrace{P_{out}\, v_{out}\, \Delta m_{out} + \Delta E_{out}}_{\text{energy leaving at outlet port}} + \underbrace{(E_{cv,t+\Delta t} - E_{cv,t})}_{\text{change in energy in tank volume}} \tag{4-12}$$

Notice that, from the standpoint of the open system shown in Figure 4-1, the energy entering at the inlet port includes not only the internal and external energy forms associated with the mass that is entering (ΔE_{in}), but also a compression work that is associated with pushing the mass into the system ($P_{in}\, v_{in}\, \Delta m_{in}$). Similarly, the amount of energy leaving at the exit port includes some expansion work.

The internal and external energy associated with the mass that is entering is given by:

$$\Delta E_{in} = \Delta m_{in} \left(u_{in} + \frac{1}{2} \tilde{V}_{in}^2 + g\, z_{in} \right) \tag{4-13}$$

Similarly, the internal and external energy associated with the mass that is leaving is:

$$\Delta E_{out} = \Delta m_{out} \left(u_{out} + \frac{1}{2} \tilde{V}_{out}^2 + g\, z_{out} \right) \tag{4-14}$$

Substituting Eqs. (4-13) and (4-14) into Eq. (4-12) provides:

$$\Delta m_{in} \left(\underbrace{u_{in} + P_{in}\, v_{in}}_{\text{specific enthalpy, } h_{in}} + \frac{1}{2} \tilde{V}_{in}^2 + g\, z_{in} \right) = \Delta m_{out} \left(\underbrace{u_{out} + P_{out}\, v_{out}}_{\text{specific enthalpy, } h_{out}} + \frac{1}{2} \tilde{V}_{out}^2 + g\, z_{out} \right) + \left(E_{cv,t+\Delta t} - E_{cv,t} \right) \tag{4-15}$$

The group of properties $u+Pv$ appears often in thermodynamic analyses of open systems. Since specific internal energy, pressure, and specific volume are themselves internal properties, the group $u+Pv$ is also an internal property, referred to as *specific enthalpy*, h:

$$\boxed{h = u + P\, v} \tag{4-16}$$

Equation (4-15) can be rewritten in terms of specific enthalpy:

$$\Delta m_{in} \left(h_{in} + \frac{1}{2} \tilde{V}_{in}^2 + g\, z_{in} \right) = \Delta m_{out} \left(h_{out} + \frac{1}{2} \tilde{V}_{out}^2 + g\, z_{out} \right) + \left(E_{cv,t+\Delta t} - E_{cv,t} \right) \tag{4-17}$$

Dividing Eq. (4-17) by Δt:

$$\frac{\Delta m_{in}}{\Delta t} \left(h_{in} + \frac{1}{2} \tilde{V}_{in}^2 + g\, z_{in} \right) = \frac{\Delta m_{out}}{\Delta t} \left(h_{out} + \frac{1}{2} \tilde{V}_{out}^2 + g\, z_{out} \right) + \frac{\left(E_{cv,t+\Delta t} - E_{cv,t} \right)}{\Delta t} \tag{4-18}$$

and taking the limit as Δt approaches zero results in a rate expression for energy conservation applied to the open system shown in Figure 4-1:

$$\dot{m}_{in} \left(h_{in} + \frac{1}{2} \tilde{V}_{in}^2 + g\, z_{in} \right) = \dot{m}_{out} \left(h_{out} + \frac{1}{2} \tilde{V}_{out}^2 + g\, z_{out} \right) + \frac{dE_{cv}}{dt} \tag{4-19}$$

The last term in Eq. (4-19) is the rate at which the energy contained in the open system changes and includes the rate of change of the internal, kinetic, and potential energy of the system:

$$\dot{m}_{in} \left(h_{in} + \frac{1}{2} \tilde{V}_{in}^2 + g\, z_{in} \right) = \dot{m}_{out} \left(h_{out} + \frac{1}{2} \tilde{V}_{out}^2 + g\, z_{out} \right) + \frac{dU}{dt} + \frac{dKE}{dt} + \frac{dPE}{dt} \tag{4-20}$$

A more general energy balance on an open system includes the possibility of heat and work interactions with the surroundings, beyond the compression and expansion work terms that were considered in the derivation of Eq. (4-19). Figure 4-4 illustrates a general energy balance on an open system, which can be represented as:

$$\boxed{\dot{m}_{in} \left(h_{in} + \frac{1}{2} \tilde{V}_{in}^2 + g\, z_{in} \right) + \dot{Q}_{in} + \dot{W}_{in} = \dot{m}_{out} \left(h_{out} + \frac{1}{2} \tilde{V}_{out}^2 + g\, z_{out} \right) + \dot{Q}_{out} + \dot{W}_{out} + \frac{dU}{dt} + \frac{dKE}{dt} + \frac{dPE}{dt}} \tag{4-21}$$

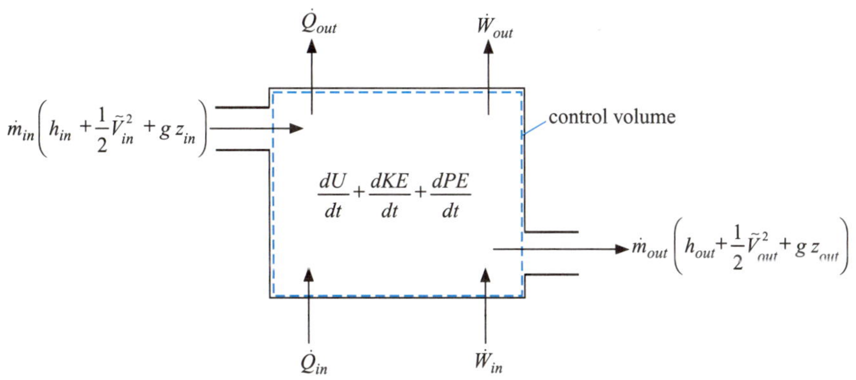

Figure 4-4: General energy balance on an open system.

In general there may be multiple locations on a control volume where mass, heat, or work is entering; in this case, they must all be summed. Equation (4-22) includes all of the possible energy contributions and employs a sign convention for heat and work terms. Heat is considered to be a positive quantity when it results in an energy transfer *into* the system. Work is considered to be a positive quantity when it results in an energy transfer *out* of the system.

$$\sum_{i=1}^{\#\text{ inlets}} \dot{m}_{in,i}\left(h_{in,i}+\frac{1}{2}\tilde{V}_{in,i}^2+g\,z_{in,i}\right)+\sum_{i=1}^{\#\text{ heat terms}}\dot{Q}_i = \sum_{i=1}^{\#\text{ outlets}} \dot{m}_{out,i}\left(h_{out,i}+\frac{1}{2}\tilde{V}_{out,i}^2+g\,z_{out,i}\right)+\sum_{i=1}^{\#\text{ work terms}}\dot{W}_i+\frac{dU}{dt}+\frac{dKE}{dt}+\frac{dPE}{dt} \tag{4-22}$$

The general energy balance is a very important equation. It is the most important equation in this textbook and perhaps the most important equation in all of physical science. The general energy balance provides the starting point for every thermodynamics problem that we will do.

4.2 Specific Enthalpy

The determination of specific enthalpy for pure substances follows naturally from the discussion regarding specific volume in Sections 2.3 and 2.4 and specific internal energy in Section 3.3. The techniques introduced in these sections can be used with little modification to determine specific enthalpy.

4.2.1 Property Tables

Fluid property data are often available in tabular form, e.g., the tables for water that appear in Appendix B. The appropriate table to use depends on the phase of the fluid. When a state is located in the superheated vapor region, information is found in Table B-3 (for water) which provides an array of properties, including specific enthalpy, over a range of pressure and temperature. When a state is located in the compressed liquid region, information is found in Table B-4; again, specific enthalpy is one of the properties listed for an array of temperatures and pressures.

The properties of saturated liquid and saturated vapor are found in Tables B-1 (in equal increments of temperature) and B-2 (in equal increments of pressure). The specific enthalpy of saturated liquid is referred to as h_f and the specific enthalpy of saturated vapor is referred to as h_g. The mass-average specific enthalpy of a two-phase state, h, is the ratio of the total enthalpy (of both phases) to the total mass (of both phases). The relationships between h and quality, x, are:

$$h = h_f + x\,(h_g - h_f) \tag{4-23}$$

and:

$$x = \frac{(h - h_f)}{(h_g - h_f)} \tag{4-24}$$

Tables for R134a that include specific enthalpy are provided in Appendix C.

4.2.2 EES Fluid Property Data

Section 2.4 discusses the use of the built-in thermodynamic property functions in EES. The Enthalpy function provides the specific enthalpy and it is accessed using the same protocol as the Volume, Pressure, IntEnergy or Temperature functions that have been described previously. For example, the EES code below provides the specific enthalpy of carbon dioxide at $P = 5\times10^6$ Pa and $T = 350$ K.

```
$UnitSystem SI K Pa J Mass
h=Enthalpy(CarbonDioxide, P=5e6 [Pa],T=350 [K])
```

Solving results in $h = 8033$ J/kg. Enthalpy, like internal energy, has a value that is relative to a selected reference state. The reference state for each fluid used by EES is indicated in the on-line help. The choice of reference state may result in negative values of specific enthalpy.

Specific enthalpy is an intensive, internal property. Therefore, the state of a substance can be fixed using specific enthalpy and one other intensive internal property. For example, the specific volume of carbon dioxide at $P = 5\text{x}10^6$ Pa and $h = 8033$ J/kg is obtained from:

```
$UnitSystem SI K Pa J Mass
v=volume(CarbonDioxide,P=5e6,h=8033 [J/kg])
```

which results in $v = 0.01116$ m^3/kg.

4.2.3 Ideal Gas

According to the phase rule in Eq. (2-1), a pure single-phase substance has two degrees of freedom. Therefore, the specific enthalpy of a pure single-phase substance is fixed by the specification of two intensive properties, such as temperature and pressure:

$$h = f\,(T, P) \tag{4-25}$$

The total derivative of Eq. (4-25) is:

$$dh = \left(\frac{\partial h}{\partial T}\right)_P dT + \left(\frac{\partial h}{\partial P}\right)_T dP \tag{4-26}$$

The partial derivative of specific enthalpy with respect to temperature at constant pressure is defined as the *specific heat capacity at constant pressure*:

$$\boxed{c_P = \left(\frac{\partial h}{\partial T}\right)_P} \tag{4-27}$$

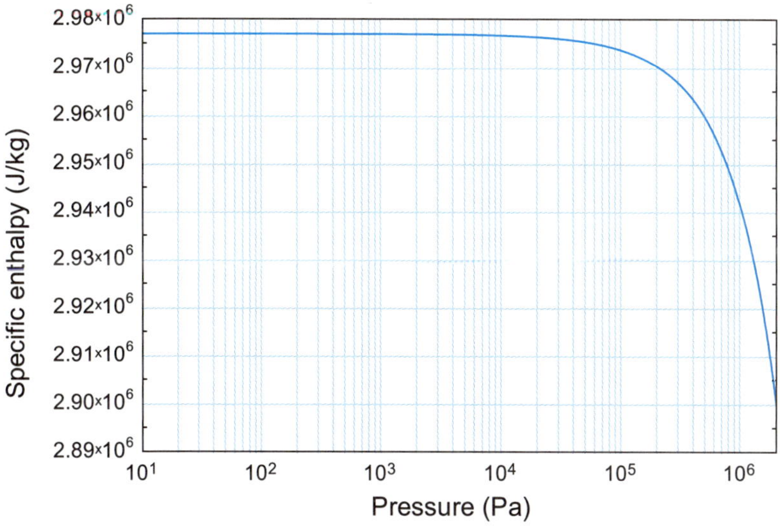

Figure 4-5: Specific enthalpy as a function of pressure for water vapor at 250° C.

The specific heat capacity at constant pressure is returned by the EES function CP, which is used like any of the other thermodynamic functions in EES. Substituting Eq. (4-27) into Eq. (4-26) leads to:

$$dh = c_P\, dT + \left(\frac{\partial h}{\partial P}\right)_T dP \tag{4-28}$$

The ideal gas model, discussed in Section 2.4, applies when the distance between molecules of a gas becomes large and therefore the intermolecular forces are negligible. In this limit, the behavior of the gas is represented by the ideal gas law:

$$P v = R T \tag{4-29}$$

where R is the gas constant. Under conditions where the ideal gas law is valid, the partial derivative of specific enthalpy with respect to pressure at constant temperature must be zero; this is shown formally in Chapter 10. Figure 4-5 illustrates the specific enthalpy as a function of pressure for water vapor at 250°C. Notice that, as the pressure decreases, the slope of the line in Figure 4-5, $\left(\frac{\partial h}{\partial P}\right)_T$, approaches zero. Recall that the accuracy of the ideal gas law also improves as pressure decreases. Water vapor approximates ideal gas behavior at 250°C for values of pressure that are less than about 1×10^4 Pa. Figure 4-5 shows that in this region, specific enthalpy becomes independent of pressure and is therefore only a function of temperature:

$$h = f\,(T) \text{ for an ideal gas} \tag{4-30}$$

As discussed in Sections 2.4.1 and 3.3.3, fluids in EES that are specified by their chemical symbol notation (e.g., 'N2') indicate that an ideal gas model is used to determine the properties whereas the full name (e.g., 'Nitrogen') indicates that the fluid is modeled as a real fluid (with compressed liquid, saturated, superheated vapor and supercritical behavior). Equation (4-30) indicates that only temperature is required to determine the specific enthalpy of an ideal gas. Therefore, if the Enthalpy function in EES is accessed for an ideal gas fluid (e.g., 'N2'), it is only necessary (or permitted) to specify the temperature (or combinations from which temperature can be determined, such as specific volume and pressure).

Since specific enthalpy is not a function of pressure for an ideal gas, Eq. (4-28) can be simplified to:

$$dh = c_P\, dT \text{ for an ideal gas} \tag{4-31}$$

Equations (4-30) and (4-31) show that the specific heat capacity at constant pressure for a substance that obeys the ideal gas law depends only on temperature. The specific enthalpy of an ideal gas is obtained by the integration of Eq. (4-31):

$$\boxed{h - h_{ref} = \int_{T_{ref}}^{T} c_P(T)\, dT \quad \text{for an ideal gas}} \tag{4-32}$$

where $c_P(T)$ is the constant pressure specific heat capacity as a function of temperature, T_{ref} is the reference temperature and h_{ref} is the value of specific enthalpy at the reference temperature. The values of T_{ref} and h_{ref} are arbitrary; however, it is often convenient to set h_{ref} to zero.

Appendix E provides the ideal gas specific heat capacity at constant pressure for air as a function of temperature as well as the specific enthalpy calculated using Eq. (4-32). The reference temperature used for Appendix E is $T_{ref} = 0$ K at which point u_{ref} is defined to be zero and therefore h_{ref} must also be zero:

$$h_{ref} = u_{ref} + \underbrace{P\,v}_{R\,T_{ref}} = \underbrace{u_{ref}}_{=0} + R\underbrace{T_{ref}}_{=0} = 0 \tag{4-33}$$

The specific heat capacity at constant volume and the specific heat capacity at constant pressure are related for an ideal gas. The definition of specific enthalpy is:

$$h = u + P\,v \tag{4-34}$$

The ideal gas law can be substituted into Eq. (4-34) to provide:

$$h = u + R\,T \tag{4-35}$$

Taking the derivative of Eq. (4-35) with respect to temperature results in:

$$\underbrace{\frac{dh}{dT}}_{c_P} = \underbrace{\frac{du}{dT}}_{c_v} + R \tag{4-36}$$

Substituting Eq. (4-31) and Eq. (3-20) into Eq. (4-36) leads to:

$$c_P = c_v + R \quad \text{for an ideal gas} \tag{4-37}$$

It is sometimes acceptable to assume that c_P is constant, particularly if the temperature change associated with the problem is small. The specific heat capacities at constant pressure for several gases at room temperature are provided in Table D-1 of Appendix D. In the limit that c_P is constant, Eq. (4-32) can be simplified to:

$$h - h_{ref} = c_P\left(T - T_{ref}\right) \quad \text{for an ideal gas with } c_P \text{ constant} \tag{4-38}$$

If the reference temperature is $T_{ref} = 0$ K and h_{ref} is defined to be zero, then Eq. (4-38) further simplifies to:

$$h = c_P\, T \quad \text{for an ideal gas with } c_P \text{ constant, } T_{ref} = 0 \text{ K and } h_{ref} = 0 \tag{4-39}$$

which is a particularly simple model for the specific enthalpy. As with specific internal energy, it is important to pick a model of the substance and then stick with it for the duration of the problem.

4.2.4 Incompressible Substance

Specific enthalpy is defined as:

$$h = u + P\,v \tag{4-40}$$

The change in the specific enthalpy of an incompressible substance is given by:

$$h_2 - h_1 = (u_2 + P_2\,v_2) - (u_1 + P_1\,v_1) = u_2 - u_1 + P_2\,v_2 - P_1\,v_1 \tag{4-41}$$

Recognizing that the specific volume of an incompressible substance is constant requires that:

$$h_2 - h_1 = u_2 - u_1 + v\,(P_2 - P_1) \quad \text{for an incompressible substance} \tag{4-42}$$

Substituting Eq. (3-28) into Eq. (4-42) leads to:

$$\boxed{h_2 - h_1 = \int_{T_1}^{T_2} c\,(T)\,dT + v\,(P_2 - P_1) \quad \text{for an incompressible substance}} \tag{4-43}$$

where $c(T)$ is the specific heat capacity as a function of temperature. In many cases, we will assume that the specific heat capacity is constant which allows Eq. (4-43) to be simplified to:

$$h_2 - h_1 = c\,(T_2 - T_1) + v\,(P_2 - P_1) \quad \text{for an incompressible substance with constant } c \tag{4-44}$$

The second term in Eq. (4-44) is often small relative to the first term. For example, the specific heat capacity and specific volume of liquid water at $T_1 = 25^\circ$C and $P_1 = 100$ kPa are $c = 4.181$ kJ/kg-K and $v = 0.00101$ m^3/kg, respectively. If the state of the water is changed to $T_2 = 75^\circ$C and $P_2 = 1000$ kPa then the first term in Eq. (4-44) is 209.1 kJ/kg while the second term is only 0.911 kJ/kg. For this reason, the second term is often neglected when computing the change in enthalpy associated with an incompressible substance that is undergoing only modest changes in pressure.

4.3 Methodology for Solving Thermodynamics Problems

There is a general methodology for solving thermodynamics problems that can be summarized by the following steps.

1. *Carefully review the problem statement and the information that is known*
   This step is obvious, but nevertheless, it is very important. It is not likely that you will be able to solve a problem if you do not clearly understand the problem description and the specific information that is provided. It is usually helpful to sketch the problem.

2. *Choose the system*
   All thermodynamic system analyses require the careful identification of a system before energy transfers such as heat and work can be defined. In some cases, the system choice may be obvious or even specified in the problem statement. However, typically there are several possible choices for a system. A carefully chosen system will often simplify the problem. Some experience is helpful in choosing the system, as will be demonstrated in the example problems throughout this textbook. The system choice should be clearly indicated in the sketch that was drawn in step 1.

3. *Apply a mass balance on the chosen system*
The mass balance in Equation (4-5) can be generalized for any number of inlet and outlet ports so that it can be applied to any system:

$$\sum_{i=i}^{\#\,inlets} \dot{m}_{in,i} = \sum_{i=1}^{\#\,outlets} \dot{m}_{out,i} + \frac{dm}{dt} \tag{4-45}$$

Note that the mass balance cannot be applied until the system boundary is identified. Equation (4-45) can often be simplified depending on the choice of the system and the details of the process. For example, if no mass crosses the system boundary (i.e., the system is closed), then the mass flow rates across the system boundary are zero and Eq. (4-45) shows that the time rate of change of the system mass is zero, indicating that the system mass is constant.

4. *Apply an energy balance on the chosen system*
The general energy balance, Eq.(4-22), is repeated below:

$$\begin{aligned}&\sum_{i=1}^{\#inlets} \dot{m}_{in,i}\left(h_{in,i} + \frac{1}{2}\tilde{V}_{in,i}^2 + g\,z_{in,i}\right) + \sum_{i=1}^{\#\text{ heat terms}} \dot{Q}_i \\ &= \sum_{i=1}^{\#\text{ outlets}} \dot{m}_{out,i}\left(h_{out,i} + \frac{1}{2}\tilde{V}_{out,i}^2 + g\,z_{out,i}\right) + \sum_{i=1}^{\#\text{ work terms}} \dot{W}_i + \frac{dU}{dt} + \frac{dKE}{dt} + \frac{dPE}{dt}\end{aligned} \tag{4-46}$$

Equation (4-46) can be used as the starting point for every thermodynamics problem that requires an energy balance. Note that an energy balance also cannot be written until the system is clearly identified. Once the system choice is clear and the process is described, Eq. (4-46) can be simplified by removing terms that are obviously zero. For example, applying Eq. (4-46) to a closed system results in:

$$\dot{Q}_{in} + \dot{W}_{in} = \dot{Q}_{out} + \dot{W}_{out} + \frac{dU}{dt} + \frac{dKE}{dt} + \frac{dPE}{dt} \tag{4-47}$$

where the summations for the heat and power terms have been equivalently written in terms of the corresponding energy flows into and out of the system. Note that Eq. (4-47) is identical to the energy balance for a closed system that was provided in Eq. (3-3). Other simplifications are also possible. For example, the heat transfer terms must be zero for a system that is adiabatic. The kinetic and potential energy terms appearing in Eq. (4-47) are often neglected as being small in comparison with the other terms in the energy balance.

Many processes of engineering interest operate at steady-state conditions. *Steady-state* is the term used to indicate that the system properties do not change with time. Energy is a system property and therefore if the system is at steady-state then the storage terms in Eq. (4-46) are zero. The resulting energy balance is then an algebraic, rather than a differential equation, and usually easier to solve. The possibilities and permutations are endless and it is pointless to attempt to remember the special cases of Eq. (4-46) that apply to each possibility. Rather, learn to work with the general energy balance in Eq. (4-46) and adjust it to your specific problem.

5. *Solve the resulting set of equations*
The above steps will identify a set of equations corresponding to mass and energy balances. These equations are usually accompanied by additional equations that provide relations between property values. Depending on the system and process,

the result may be a coupled set of algebraic or differential equations that must be solved simultaneously. The solution to these equations can be the most challenging aspect of the problem. The EES program can be particularly helpful with this step.

EXAMPLE 4.3-1: PORTABLE COOLING SYSTEM

The military is interested in providing portable cooling to dismounted soldiers who are deployed in hot and humid environments. One proposed technology utilizes canisters with volume $V = 2$ liter that are charged with refrigerant R134a at $T_{charge} = 20^{\circ}$C and quality $x_{charge} = 0.05$. Heat is transferred from the soldier to the canister. A relief valve on the canister allows only saturated vapor to escape when the pressure in the canister reaches a value corresponding to a temperature of $T_{rv} = 30^{\circ}$C. When the canister becomes completely empty of liquid refrigerant, it must be discarded.

a) Determine the cooling density of the system. The cooling density is defined as the ratio of the amount of energy that can be absorbed before the canister must be discarded to the initial system mass. Compare your answer to the cooling density of an ice pack. The latent heat of fusion of ice (i.e., the amount of energy required to melt ice per unit mass) is $\Delta h_{fus} = 333.6$ J/g.

The inputs are entered in EES:

```
$UnitSystem SI K Pa J Mass Radian
Vol=2 [liter]*convert(liter,m^3)            "volume"
x_charge=0.05 [-]                           "initial quality"
T_charge=converttemp(C,K,20 [C])            "initial temperature"
T_rv=converttemp(C,K,30 [C])                "temperature at which relief valve opens"
F$='R134a'                                  "refrigerant"
```

The open system used for this analysis consists of the volume contained within the canister, as shown in Figure 1.

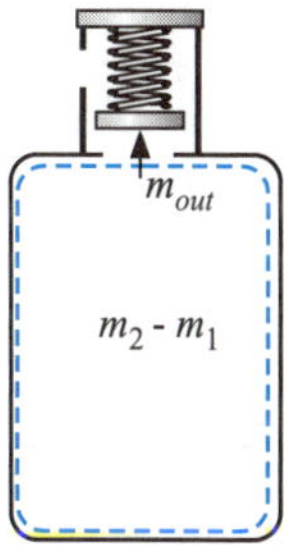

Figure 1: Mass balance on the open system defined as the volume within the canister.

The initial state of the refrigerant in the system is fixed by its quality, $x_1 = x_{charge}$, and temperature, $T_1 = T_{charge}$. The pressure, specific volume, and specific internal energy (P_1, v_1, and u_1) are obtained using EES' internal property functions.

```
"state 1"
x[1]=x_charge                               "quality"
T[1]=T_charge                               "temperature"
P[1]=pressure(F$,x=x[1],T=T[1])             "pressure"
v[1]=volume(F$,x=x[1],T=T[1])               "specific volume"
u[1]=intenergy(F$,x=x[1],T=T[1])            "specific internal energy"
```

The final state of the refrigerant in the system is fixed by its temperature, $T_2 = T_{rv}$, and quality, $x_2 = 1$. The pressure, specific volume, and specific internal energy (P_2, v_2, and u_2) are obtained.

```
"state 2"
T[2]=T_rv                              "temperature"
x[2]=1 [-]                             "tank is empty of liquid"
P[2]=pressure(F$,T=T[2],x=x[2])        "temperature"
u[2]=intenergy(F$,T=T[2],x=x[2])       "specific internal energy"
v[2]=volume(F$,T=T[2],x=x[2])          "specific volume"
```

The initial mass of refrigerant in the system is:

$$m_1 = \frac{V}{v_1}$$

The final mass of refrigerant in the system is:

$$m_2 = \frac{V}{v_2}$$

A mass balance on the system is shown in Figure 1 and can be written as:

$$0 = m_{out} + m_2 - m_1$$

where m_{out} is the mass of refrigerant that leaves the canister.

```
m[1]=Vol/v[1]             "initial mass"
m[2]=Vol/v[2]             "final mass"
0=m_out+m[2]-m[1]         "mass balance"
```

Figure 2 illustrates an energy balance on the system.

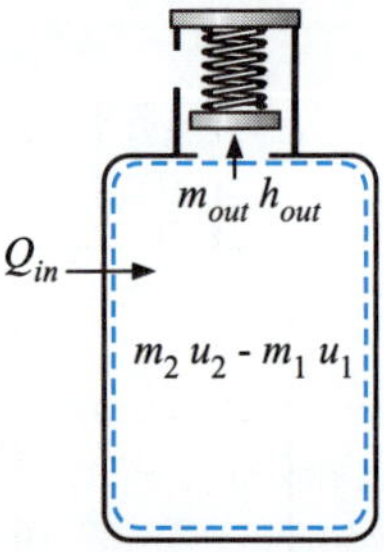

Figure 2: Energy balance.

The refrigerant leaving the system through the relief valve is saturated vapor at temperature T_{rv}; therefore, the state of this refrigerant is fixed by its temperature and its quality, which is 1. The specific enthalpy of the refrigerant leaving the system (h_{out}) is obtained:

```
h_out=enthalpy(F$,T=T_rv,x=1 [-])    "specific enthalpy of saturated vapor leaving"
```

The energy balance suggested by Figure 2 is:

$$Q_{in} = m_{out}\, h_{out} + m_2\, u_2 - m_1\, u_1$$

Note that the kinetic and potential energy contributions of the refrigerant are neglected for this problem.

```
Q_in=m_out*h_out+m[2]*u[2]-m[1]*u[1]    "energy balance"
```

The cooling density of the system (neglecting the mass of the canister) is:

$$CD = \frac{Q_{in}}{m_1}$$

```
CD=Q_in/m[1]                              "cooling density"
CD_Whrpkg=CD*convert(J/kg,W-hr/kg)        "in W-hr/kg"
```

which results in a value of $CD = 49.4$ W-hr/kg. For comparison, the cooling density associated with ice (CD_{ice}) is equal to its latent heat of fusion:

```
"Ice"
Dh_fus_ice=333.6 [J/g]*convert(J/g,J/kg)               "latent heat of fusion"
CD_ice_Whrpkg=Dh_fus_ice*convert(J/kg,W-hr/kg)         "energy density in W-hr/kg"
```

Solving shows that $CD_{ice} = 92.7$ W-hr/kg.

4.4 Thermodynamic Analyses of Steady-State Applications

Many processes and equipment operate under conditions for which the properties of the system do not change with time; this condition is referred to as steady-state. In a steady-state process, the mass of the system is constant. Therefore, the general mass balance provided in Eq. (4-45) reduces to:

$$\sum_{i=i}^{\#\,inlets} \dot{m}_{in,i} = \sum_{i=1}^{\#\,outlets} \dot{m}_{out,i} \tag{4-48}$$

The energy of the system is also constant and therefore the general energy balance reduces to:

$$\begin{aligned}\sum_{i=1}^{\#\,inlets} \dot{m}_{in,i}\left(h_{in,i} + \frac{1}{2}\tilde{V}_{in,i}^2 + g\,z_{in,i}\right) + \sum_{i=1}^{\#\,\text{heat terms}} \dot{Q}_i \\ = \sum_{i=1}^{\#\,\text{outlets}} \dot{m}_{out,i}\left(h_{out,i} + \frac{1}{2}\tilde{V}_{out,i}^2 + g\,z_{out,i}\right) + \sum_{i=1}^{\#\,\text{work terms}} \dot{W}_i\end{aligned} \tag{4-49}$$

It is often possible to simplify Eq. (4-49) further based on the details of the equipment or process under consideration. For example, many types of equipment that operate at steady conditions only have one mass flow inlet and one outlet, making the summations for these terms unnecessary. Similarly, there is often only one heat or power term, eliminating the need for the summations. In this section, a number of common energy conversion devices that typically are assumed to operate at steady-state are described and the simplifications that are appropriate in their analyses are discussed.

4.4.1 Turbines

The purpose of a *turbine* is to expand a high pressure fluid stream through a set of rotating blades that are attached to a shaft in order to produce power. Turbines are commonly used in power plants to produce mechanical power that may be used directly or provided to a generator that produces electrical power. Turbines are a key component in power generation cycles, as discussed in Chapter 8. A schematic of a turbine is shown in Figure 4-6(a). The fluid, which is often steam or combustion gases, enters the turbine

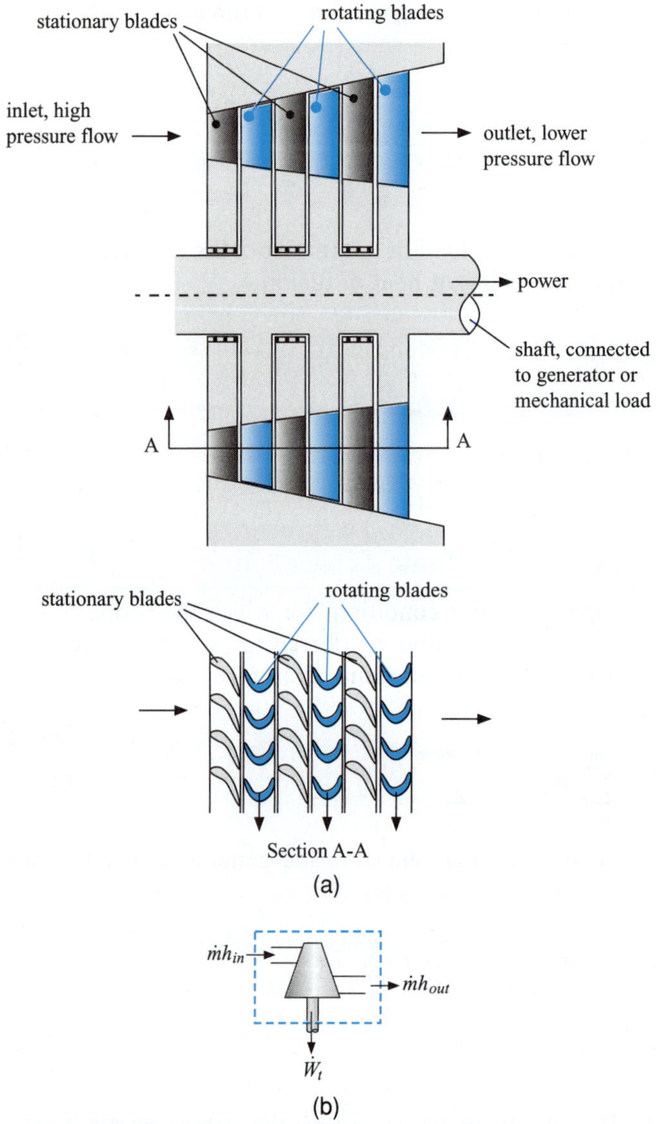

Figure 4-6: (a) Schematic of an axial turbine and (b) the symbol for the turbine component.

at high pressure and temperature. The fluid is directed by stationary blades so that it impinges on the rotating turbine blades. During this process, momentum is transferred from the fluid causing a torque on the rotating shaft. Stationary blades are used to redirect the fluid between each set of rotating blades. The pressure of the fluid decreases as it flows from the inlet to the outlet of the turbine. The specific enthalpy of the fluid also decreases as it flows through the device since energy is transferred from the fluid to the rotating shaft. Figure 4-6(b) illustrates the typical symbol that is used to represent the turbine component in a cycle schematic.

Turbines having more than one mass flow inlet or outlet are possible, but in the simplest case, there is only one inlet and one outlet. In this case, the mass balance becomes:

$$\dot{m}_{in} = \dot{m}_{out} = \dot{m} \tag{4-50}$$

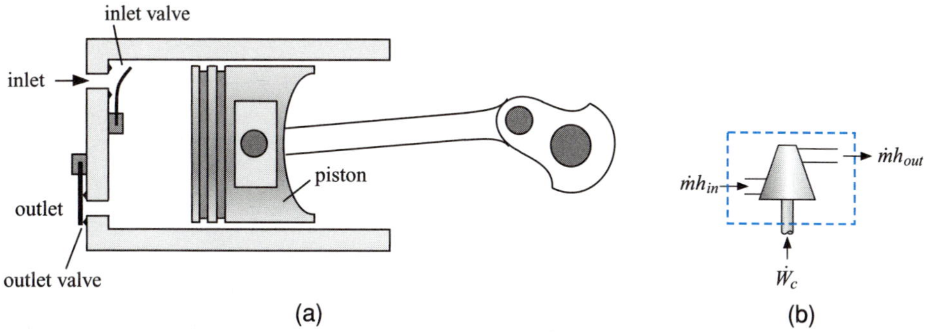

Figure 4-7: (a) Schematic of a reciprocating compressor and (b) the symbol for the compressor component.

where $\dot{m}$ is the constant mass flow rate in the device. The kinetic and potential energy terms are rarely significant for the flow entering or leaving a turbine (with the exception of turbines used for jet engine cycles, discussed in Section 8.3.3). Turbines are typically assumed to be adiabatic because the rate of heat transfer to or from a turbine is usually small relative to the power produced. With these assumptions, the energy balance on the turbine shown in Figure 4-6(b) provides:

$$\dot{m}\, h_{in} = \dot{m}\, h_{out} + \dot{W}_t \tag{4-51}$$

where $\dot{W}_t$ is the mechanical power produced by the turbine. Turbine efficiency is defined in Section 6.6.1 based on the ratio of the actual power produced by a turbine to the maximum possible power that could be produced. The maximum possible power must be determined using the Second Law of Thermodynamics.

4.4.2 Compressors

Compressors use a mechanical power input to increase the pressure of a compressible fluid (i.e., to reduce its specific volume). You are probably familiar with compressed air, which is used for purposes such as operating machine tools, spraying paint, and filling automobile tires. Compressors are also a major component in refrigeration cycles, as described in Chapter 9.

There are many different types of compressors, e.g., reciprocating, rotary vane, scroll, screw, and centrifugal compressors. Most compressors are classified as *positive displacement* compressors because they displace a fixed volume of fluid during each cycle. A centrifugal compressor is classified as a *dynamic* compressor because the increase in pressure results by increasing the velocity of the fluid and then a diffuser (which is discussed in Section 4.4.5) is used to convert the increased velocity to an increase in pressure. Although each compressor type operates differently, they all accomplish the objective of increasing the pressure of a fluid.

A schematic of a reciprocating compressor is shown in Figure 4-7(a). The fluid enters the compressor cylinder through an inlet valve as the piston is moved to the right in order to increase the volume in the cylinder. The inlet valve closes when the piston reaches the end of its stroke. The fluid pressure increases as the piston is then moved to the left in order to reduce the volume of gas in the cylinder. The outlet valve eventually opens so that the piston pushes the high pressure fluid out. A net work input must be provided by the piston to the fluid during each cycle in order to accomplish this compression process. The fluid temperature will increase during the compression process by an amount that depends on the heat that is transferred from the fluid to the surroundings. Typically, the duration of the cycle is sufficiently short that there is not time for any substantial heat

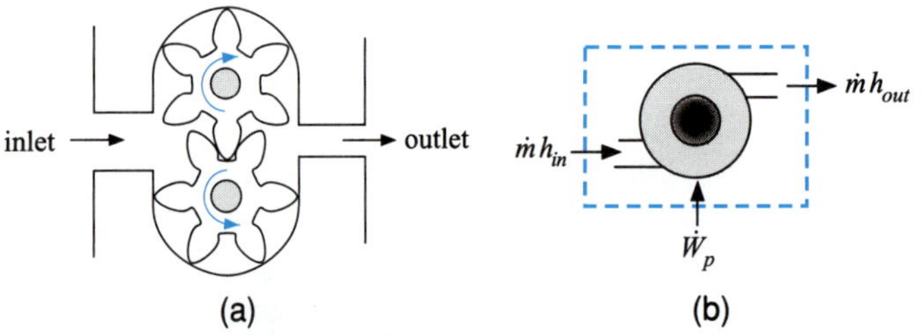

Figure 4-8: (a) Schematic of a gear pump and (b) the symbol for the pump component.

transfer from the fluid and therefore the compressor is often modeled as being adiabatic. Figure 4-7(b) illustrates the typical symbol used to represent the compressor component in a cycle schematic.

Compressors normally operate at steady-state (or cyclic steady-state) conditions. Although multiple inlets and exits are possible, most compressors have only one inlet and one outlet and therefore the steady-state mass balance is described by Eq. (4-50). Because kinetic and potential energy terms are usually negligible and heat transfer is often small, the steady-state energy balance on a compressor shown in Figure 4-7(b) provides:

$$\dot{W}_c + \dot{m}\,h_{in} = \dot{m}\,h_{out} \tag{4-52}$$

where $\dot{W}_c$ is the mechanical power provided to the compressor. Compressor efficiency is defined in Section 6.6.2 as the ratio of the minimum possible power required to run the compressor to the actual power required by the compressor. The minimum power required must be determined using the Second Law of Thermodynamics.

4.4.3 Pumps

Pumps are similar to compressors in that they use a power input in order to raise the pressure of a fluid. However, pumps are designed to operate with incompressible fluids, whereas compressors operate with compressible fluids. As a result, the design of a pump is generally quite different than that of a compressor. Indeed, many compressors would fail if an incompressible fluid were drawn into the suction port.

There are many different types of pumps which, like compressors, are classified as positive displacement or dynamic (i.e., centrifugal). An example of a positive displacement pump is the gear pump shown in Figure 4-8(a). The fluid enters at the left and is captured by the pockets formed between the rotating gear teeth. The rotational motion of the gears causes the fluid to be pushed through the pump and it exits at the right at an elevated pressure. Figure 4-8(b) illustrates the typical symbol used to represent the pump component in a cycle schematic.

With one inlet and one outlet, the steady-state mass balance for a pump is given by Eq. (4-50). The steady state energy balance (assuming negligible kinetic and potential energy terms at the inlet and outlet and no heat transfer to or from the pump) is shown in Figure 4-8(b) and given by:

$$\dot{W}_p + \dot{m}\,h_{in} = \dot{m}\,h_{out} \tag{4-53}$$

where $\dot{W}_p$ is the shaft power required to operate the pump. The pump efficiency is defined in Section 6.6.3 based on the Second Law of Thermodynamics.

4.4.4 Nozzles

The purpose of a *nozzle* is to convert a high pressure fluid into a low pressure fluid with a higher velocity. A familiar example of a nozzle is the spray controller that is often attached to the end of a garden hose. Nozzles are required in many applications and devices. For example, a turbine will internally use nozzles to convert the high pressure fluid at the inlet into a high velocity flow that impinges on the rotating blades attached to the shaft. The stationary blades shown in Figure 4-6(a) act as nozzles. Nozzles are also used in the propulsion systems, as discussed in Section 8.3.3.

A detailed analysis of nozzles is provided in Chapter 16. Nozzles are classified as *subsonic* and *supersonic*. Supersonic nozzles have a converging-diverging geometry that allows the fluid velocity to exceed the speed of sound. The velocity of the fluid in a subsonic nozzle never exceeds the speed of sound and therefore the area for flow always decreases as the velocity increases (i.e., it has a converging shape).

Nozzles typically have one inlet and one outlet and operate under steady-state conditions. The mass balance is therefore given by Eq. (4-50). The general steady-state energy balance, Eq. (4-49), can often be simplified for application to a nozzle. The potential energy terms of the inlet and outlet flows are not usually significant. Nozzles neither require nor produce power ($\dot{W}_{in} = \dot{W}_{out} = 0$) and are very nearly adiabatic. Note that the kinetic energy terms associated with the flows, particularly the outlet flow, cannot be neglected as the purpose of the nozzle is to increase the kinetic energy of the fluid. With these simplifications, an energy balance on the nozzle becomes:

$$h_{in} + \frac{1}{2}\tilde{V}_{in}^2 = h_{out} + \frac{1}{2}\tilde{V}_{out}^2 \tag{4-54}$$

Nozzle efficiency is defined in Section 6.6.4 based on the ratio of the actual kinetic energy of the fluid leaving the nozzle to the maximum possible kinetic energy. The maximum possible kinetic energy is determined using the Second Law of Thermodynamics.

4.4.5 Diffusers

A *diffuser* is basically a nozzle that is operating in reverse. The purpose of a diffuser is to transform a high velocity fluid into a fluid with lower velocity but higher pressure. Diffusers are required by a number of applications and components. One use for a diffuser was noted in the description of the centrifugal compressor in Section 4.4.2. A high velocity fluid results from the action of the impeller. This fluid subsequently flows through a diffuser in order to increase its pressure at the expense of reduced velocity.

Diffusers are commonly used in heating, ventilating and air conditioning systems in order to reduce the velocity of the air entering a room from conditioning ducts so as to reduce noise and increase occupant comfort. Diffusers are also used in the jet aircraft engines discussed in Section 8.3.3. The air entering the aircraft engine has a large kinetic energy due to the high speed of the aircraft itself. This high velocity air may pass through a diffuser in order to increase the pressure of the air before it enters the compressor that is required by the gas turbine cycle. The increase in pressure obtained from the diffuser reduces the power required by the compressor and therefore improves the efficiency of the engine.

Diffuser efficiency is defined in Section 6.6.5 based on the ratio of the actual pressure rise produced by the diffuser to the maximum possible pressure rise. The maximum possible pressure rise must be determined using the Second Law of Thermodynamics.

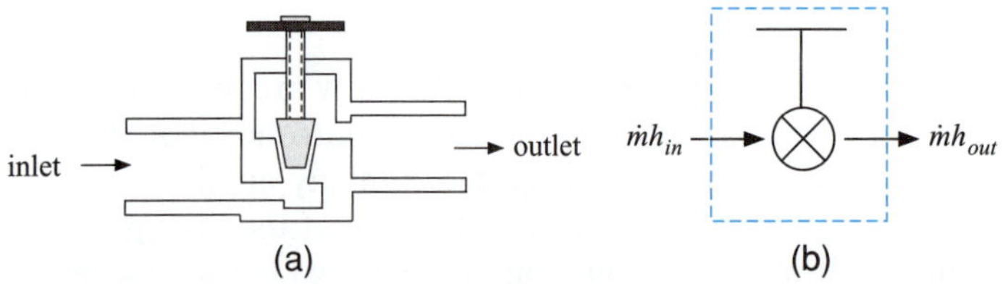

Figure 4-9: (a) Schematic of a controlled throttle and (b) the symbol for the throttle component.

4.4.6 Throttles

The purpose of a *throttle* (also called a *valve* or a *restrictor*) is to reduce the pressure of a flowing fluid, often in order to control the flow rate. There are many different throttle designs but they all impose a restriction that impedes flow and thereby reduces the fluid pressure. For example, pressure may be reduced as mass flows through a porous material such as a wad of cotton or a bed of closely packed spheres. Pressure is also reduced when mass flows through a tube of small diameter, often referred to as a *capillary tube*. The porous material and capillary tube throttle designs are classified as uncontrolled throttles since the pressure difference at specified conditions can not be actively adjusted in order to control the flow. Figure 4-9(a) shows a schematic of a controlled throttle; the position of the valve seat can be adjusted in order to control the flow area and therefore the flow rate. Figure 4-9(b) illustrates the typical symbol used to represent the throttle component in a cycle schematic.

The most common purpose of a throttle is to control a mass flow rate. You use a controllable throttle whenever you open the faucet valve in your bathroom sink or step on the accelerator in your car. However, there are other applications for throttles. Chapter 9 discusses the use of a throttle in vapor compression refrigeration cycles. The saturation pressure and saturation temperature of a two-phase fluid are related; therefore, a refrigeration cycle reduces the pressure of a two-phase fluid with a throttle in order to reduce its temperature and therefore provide refrigeration.

Throttles normally operate at steady-state with one inlet and one exit. The mass balance for this situation is therefore given by Eq. (4-50). The energy balance provided by Eq. (4-49) can be simplified substantially for application to a throttle. Throttles neither require nor produce power and they can usually be assumed to operate adiabatically. The kinetic and potential energy of the entering and exiting fluid streams are normally small enough to neglect. With these simplifications, an energy balance on a throttle becomes:

$$h_{in} = h_{out} \tag{4-55}$$

Equation (4-55) indicates that the specific enthalpy of a fluid does not change as it passes through a throttle. That is, throttles are *isenthalpic*.

4.4.7 Heat Exchangers

Heat exchangers are used to transfer thermal energy (heat) from one fluid to another. Heat exchangers are perhaps the most common device employed in energy systems. Several types of heat exchangers are shown in Figure 4-10(a) through (c). The heat exchanger design in Figure 4-10(a) is commonly used to exchange energy between two liquid streams, for example oil and cooling water in the oil cooler for an engine. The radiator in a car is a cross-flow heat exchanger that transfers energy from the glycol-water solution in the engine cooling system to the air in the environment. The fluids move

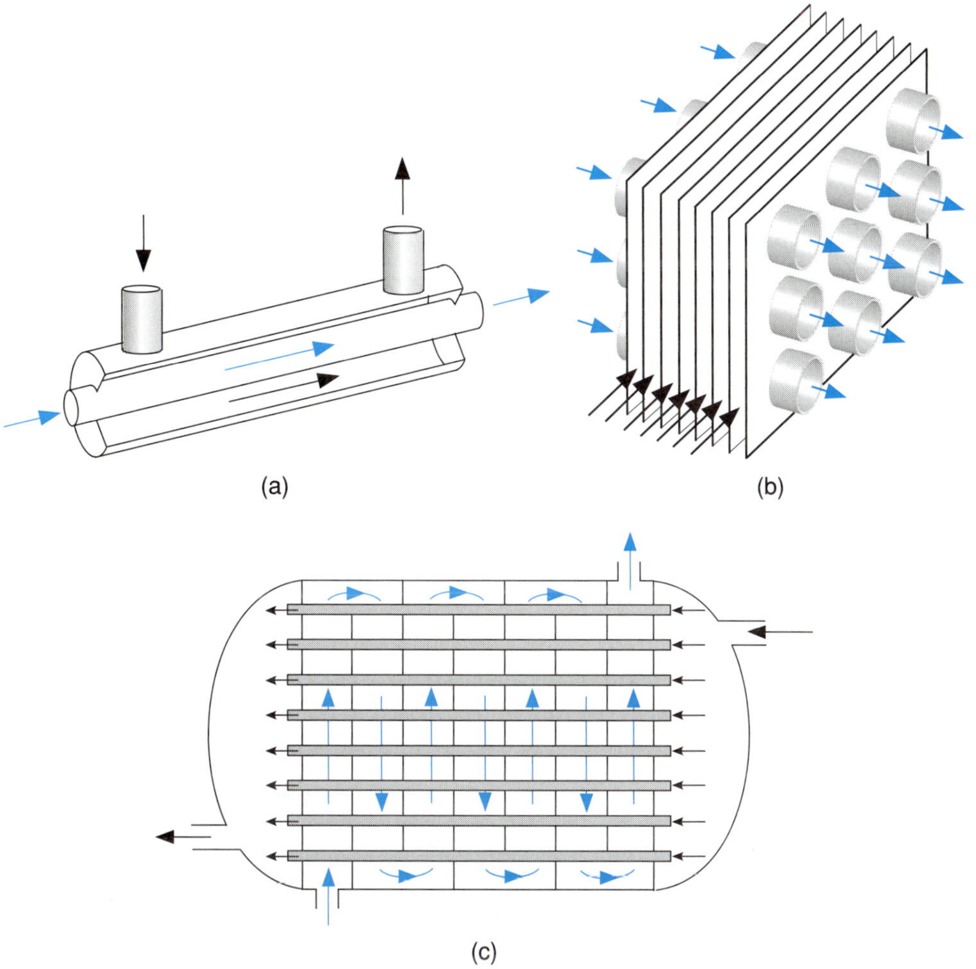

Figure 4-10: Schematic of (a) a concentric tube heat exchanger, (b) a cross-flow heat exchanger, and (c) a shell-and-tube heat exchanger. From Nellis and Klein (2009).

perpendicular to one another in a cross-flow heat exchanger, as shown in Figure 4-10(b). Refrigeration cycles employ at least two heat exchangers, as explained in Chapter 9. One of the heat exchangers, the *evaporator*, transfers energy from the fluid being cooled to the evaporating refrigerant (typically an organic fluid such as R134a contained in a hermetically-sealed circuit). The second heat exchanger, the *condenser*, transfers energy from the condensing refrigerant to the surroundings. The shell-and-tube heat exchanger shown in Figure 4-10(c) could be used for a condenser in a large refrigeration system.

Regardless of the details of the heat exchanger geometry and configuration, there are two streams, typically referred to as the hot stream and the cold stream, that flow through the device while transferring heat from one to another without mixing. Figure 4-11(a) illustrates a mass balance on each side of the heat exchanger.

$$\dot{m}_{H,in} = \dot{m}_{H,out} = \dot{m}_H \tag{4-56}$$

$$\dot{m}_{C,in} = \dot{m}_{C,out} = \dot{m}_C \tag{4-57}$$

Heat exchangers do not involve a power input and the kinetic and potential energy terms associated with the flows are ordinarily negligible. Thermal losses from the jacket

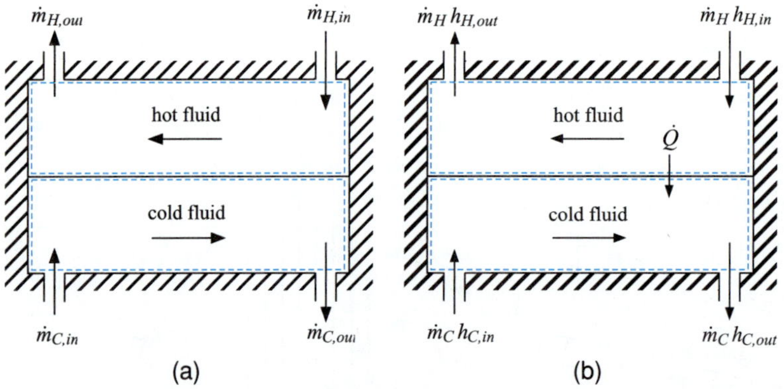

Figure 4-11: (a) Mass balances and (b) energy balances on each side of a heat exchanger.

of the heat exchanger are also usually neglected. Energy balances on systems that encompass only the hot fluid and only the cold fluid, as shown in Figure 4-11(b), lead to:

$$\dot{Q} = \dot{m}_H \left(h_{H,in} - h_{H,out} \right) \tag{4-58}$$

$$\dot{Q} = \dot{m}_C \left(h_{C,out} - h_{C,in} \right) \tag{4-59}$$

where $\dot{Q}$ is the rate of heat transfer from the hot fluid to the cold fluid. The effectiveness of a heat exchanger is defined as the ratio of the actual to the maximum possible rate of heat transfer between the two streams, as discussed in Section 6.6.6.

EXAMPLE 4.4-1

EXAMPLE 4.4-1: DE-SUPERHEATER IN AN AMMONIA REFRIGERATION SYSTEM

In a refrigeration system, it is necessary to recompress the saturated vapor that is produced as thermal energy is transferred to the evaporating refrigerant in the evaporator heat exchanger. The compression process increases the pressure of the refrigerant, which allows it to condense at a temperature that is above the ambient temperature. This type of refrigeration cycle is discussed more completely in Chapter 9. Ammonia is often used as the working fluid in large industrial refrigeration systems because it is an efficient refrigerant that is also inexpensive in large quantities.

In a particular system, ammonia must be compressed from state 1 at $T_1 = -32^\circ\text{C}$ and $P_1 = 0.9$ bar to a final pressure $P_6 = 1.70$ MPa (where the saturation temperature is sufficiently high, 43.3°C, that the ammonia can be condensed by heat transfer to the outdoor air). Rather than compress the ammonia in a single compressor, it is more efficient to use two separate compressors that are separated by a de-superheater, as shown in Figure 1. The refrigeration system operates at steady-state and the kinetic and potential energies of the fluid streams entering each component are negligible. In addition, all of the components operate adiabatically.

Ammonia enters compressor #1 at state 1 with mass flow rate $\dot{m}_1 = 2.5$ kg/min. The ammonia leaves compressor #1 at $T_2 = 75^\circ\text{C}$ and $P_2 = 750$ kPa and subsequently enters the de-superheater. The de-superheater is a large, insulated mixing tank. The purpose of the de-superheater is to cool the superheated vapor leaving compressor #1 so that ammonia enters compressor #2 at state 5 as saturated vapor. It requires less compressor power to compress saturated vapor than to compress the superheated vapor leaving compressor #1 at state 2. The de-superheater mixes the ammonia at

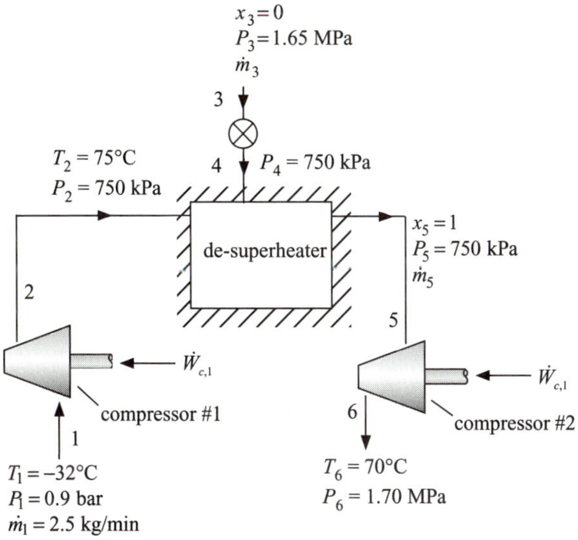

Figure 1: Two-stage compression process using a de-superheater.

state 2 with a stream of saturated liquid ammonia that is extracted from elsewhere in the system at state 3. The pressure at state 3 is $P_3 = 1.65$ MPa (slightly less than the discharge pressure of compressor #2). The ammonia at state 3 is passed through a throttle valve and leaves at state 4 with pressure $P_4 = 750$ kPa. If the de-superheater is properly designed (and the throttle is properly controlled), then the ammonia leaves the de-superheater at state 5 as saturated vapor with pressure $P_5 = 750$ kPa. The ammonia leaving the de-superheater flows to compressor #2 where it is compressed to $P_6 = 1.70$ MPa. The temperature of the ammonia leaving compressor #2 is $T_6 = 70°$C.

a) Determine the power required by compressor #1, $\dot{W}_{c,1}$.

The inputs are entered in EES and converted into standard SI units.

```
$UnitSystem SI K Pa J Mass Radian
"Inputs"
T[1]=converttemp(C,K,-32 [C])                   "inlet temperature to compressor 1"
P[1]=0.90 [bar]*convert(bar,Pa)                  "inlet pressure to compressor 1"
m_dot[1]=2.5 [kg/min]*convert(kg/min,kg/s)       "mass flow rate of ammonia entering compressor 1"
P[2]=750 [kPa]*convert(kPa,Pa)                   "exit pressure of compressor 1"
T[2]=converttemp(C,K,75 [C])                     "exit temperature of compressor 1"
P[3]=1.65 [MPa]*convert(MPa,Pa)                  "pressure entering valve"
x[3]=0.0 [-]                                     "saturated liquid enters valve"
P[4]=P[2]                                        "pressure leaving valve"
P[5]=P[2]                                        "pressure leaving de-superheater"
x[5]=1.0 [-]                                     "saturated vapor leaves valve"
P[6]=1.7 [MPa]*convert(MPa,Pa)                   "exit pressure of compressor 2"
T[6]=converttemp(C,K,70 [C])                     "exit temperature of compressor 2"
```

States 1 and 2 are both fixed by their given pressure and temperature. The specific enthalpy and specific volume at each state are obtained using EES' internal property routines (h_1, v_1, h_2, and v_2). The phase (i.e., superheated, saturated, or liquid) can also be obtained using the Phase$ property function, which shows that states 1 and 2 are both superheated.

```
h[1]=enthalpy(Ammonia,T=T[1],P=P[1])          "specific enthalpy of ammonia entering compressor 1"
v[1]=volume(Ammonia,T=T[1],P=P[1])            "specific volume of ammonia entering compressor 1"
Phase$[1]=Phase$(Ammonia,T=T[1],P=P[1])       "phase at state 1"
h[2]=enthalpy(Ammonia,T=T[2],P=P[2])          "specific enthalpy of ammonia leaving compressor 1"
v[2]=volume(Ammonia,T=T[2],P=P[2])            "specific volume of ammonia leaving compressor 1"
Phase$[2]=Phase$(Ammonia,T=T[2],P=P[2])       "phase at state 2"
```

A mass balance on compressor #1 requires that:

$$\dot{m}_1 = \dot{m}_2$$

```
m_dot[1]=m_dot[2]                             "mass balance on compressor 1"
```

An energy balance on compressor #1 is:

$$\dot{W}_{c,1} + \dot{m}_1\, h_1 = \dot{m}_2\, h_2$$

```
W_dot_c_1+m_dot[1]*h[1]=m_dot[2]*h[2]         "energy balance on compressor 1"
```

so that $\dot{W}_{c,1} = 8.67$ kW.

b) What is the quality of the ammonia leaving the valve at state 4, x_4?

State 3 is specified by the given pressure and quality. Therefore, the specific enthalpy, temperature, and specific volume of the ammonia at state 3 can be obtained (h_3, T_3, and v_3).

```
h[3]=enthalpy(Ammonia,x=x[3],P=P[3])          "specific enthalpy of ammonia entering valve"
v[3]=volume(Ammonia,x=x[3],P=P[3])            "specific volume of ammonia entering valve"
T[3]=temperature(Ammonia,x=x[3],P=P[3])       "temperature of ammonia entering valve"
```

A mass balance on the throttle valve provides:

$$\dot{m}_3 = \dot{m}_4 \tag{1}$$

An energy balance on the throttle valve leads to:

$$\dot{m}_3\, h_3 = \dot{m}_4\, h_4 \tag{2}$$

Substituting Eq. (1) into Eq. (2) results in:

$$h_3 = h_4$$

```
h[4]=h[3]                                     "valve is isenthalpic"
```

State 4 is specified by the specific enthalpy and pressure. Therefore, the quality, temperature, and specific volume at state 4 can be obtained (x_4, T_4, and v_4) along with the phase of this state.

```
x[4]=quality(Ammonia,h=h[4],P=P[4])           "quality of ammonia leaving valve"
T[4]=temperature(Ammonia,h=h[4],P=P[4])       "temperature of ammonia leaving valve"
v[4]=volume(Ammonia,h=h[4],P=P[4])            "specific volume of ammonia leaving valve"
Phase$[4]=Phase$(Ammonia,h=h[4],P=P[4])       "phase at state 4"
```

Solving these equations shows that $x_4 = 0.1058$.

c) What is the mass flow rate of ammonia leaving the de-superheater, $\dot{m}_5$?

State 5 is fixed by the given pressure and quality. Therefore, the specific enthalpy, temperature, and specific volume at state 5 can be obtained (h_5, T_5, and v_5) along with the phase of this state.

```
h[5]=enthalpy(Ammonia,x=x[5],P=P[5])        "specific enthalpy of ammonia leaving de-superheater"
v[5]=volume(Ammonia,x=x[5],P=P[5])          "specific volume of ammonia leaving de-superheater"
T[5]=temperature(Ammonia,x=x[5],P=P[5])     "temperature of ammonia leaving de-superheater"
Phase$[5]=Phase$(Ammonia,x=x[5],P=P[5])     "phase at state 5"
```

A steady-state mass balance on the de-superheater is:

$$\dot{m}_2 + \dot{m}_4 = \dot{m}_5$$

An energy balance on the de-superheater, assuming it to be adiabatic, leads to:

$$\dot{m}_2\, h_2 + \dot{m}_4\, h_4 = \dot{m}_5\, h_5$$

```
m_dot[2]+m_dot[4]=m_dot[5]                   "mass balance on de-superheater"
m_dot[2]*h[2]+m_dot[4]*h[4]=m_dot[5]*h[5]    "energy balance on de-superheater"
m_dot_5_kgpm=m_dot[5]*convert(kg/s,kg/min)   "mass flow rate leaving de-superheater, in kg/min"
```

Solving results in $\dot{m}_5 = 2.859$ kg/min.

d) Determine the power required by compressor #2.

State 6 is fixed by its pressure and temperature. The specific enthalpy and specific volume of state 6 are computed (h_6 and v_6). The phase is also determined in order to verify that the ammonia is superheated.

```
h[6]=enthalpy(Ammonia,P=P[6],T=T[6])      "specific enthalpy of ammonia leaving compressor #2"
v[6]=volume(Ammonia,P=P[6],T=T[6])        "specific volume of ammonia leaving compressor #2"
Phase$[6]=Phase$(Ammonia,P=P[6],T=T[6])   "phase at state 6"
```

A mass balance on compressor #2 is:

$$\dot{m}_5 = \dot{m}_6$$

An energy balance on compressor #2 is:

$$\dot{m}_5\, h_5 + \dot{W}_{c,2} = \dot{m}_6\, h_6$$

```
m_dot[5]=m_dot[6]                         "mass balance on compressor #2"
m_dot[5]*h[5]+W_dot_c_2=m_dot[6]*h[6]     "energy balance on compressor #2"
```

which solves to $\dot{W}_{c,2} = 4.77$ kW.

e) Refrigeration processes are commonly represented on pressure-specific enthalpy (*P-h*) plots. Use EES to generate a *P-h* diagram. Label the states on the diagram and draw lines to show processes that link the states. The processes should include compressor #1, compressor #2, the valve, and the de-superheater.

Select the Property Plot menu command in the Plots menu. Choose ammonia as the fluid and select P-h as the plot type. We will include lines of constant temperature for 241 K (−32°C) and 348 K (75°C) as these are the two temperature levels provided in the problem statement. Uncheck the box that provides lines of constant specific entropy. The Property Plot dialog should appear as shown in Figure 2.

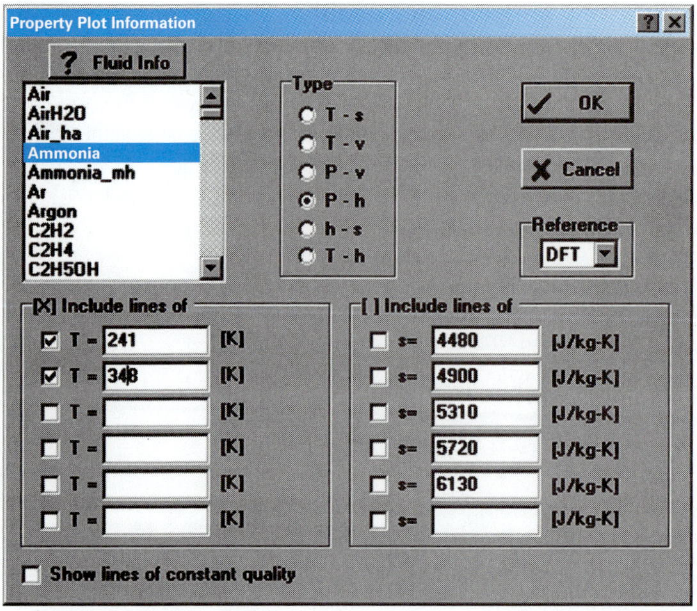

Figure 2: Property plot dialog.

The property plot will be created when you select OK. Next select Overlay Plot from the Plots menu. There is only one table, the Arrays table, and that will be selected for plotting. Choose specific enthalpy (h[i]) for the X-axis and pressure (P[i]) for the Y-axis. Choose a symbol and color for the points, but do not connect the points with a line; the correct states will be connected manually in order to represent the processes. The state points will be identified with the state number if the "Show array indices" check box is enabled. The completed dialog will appear as shown in Figure 3.

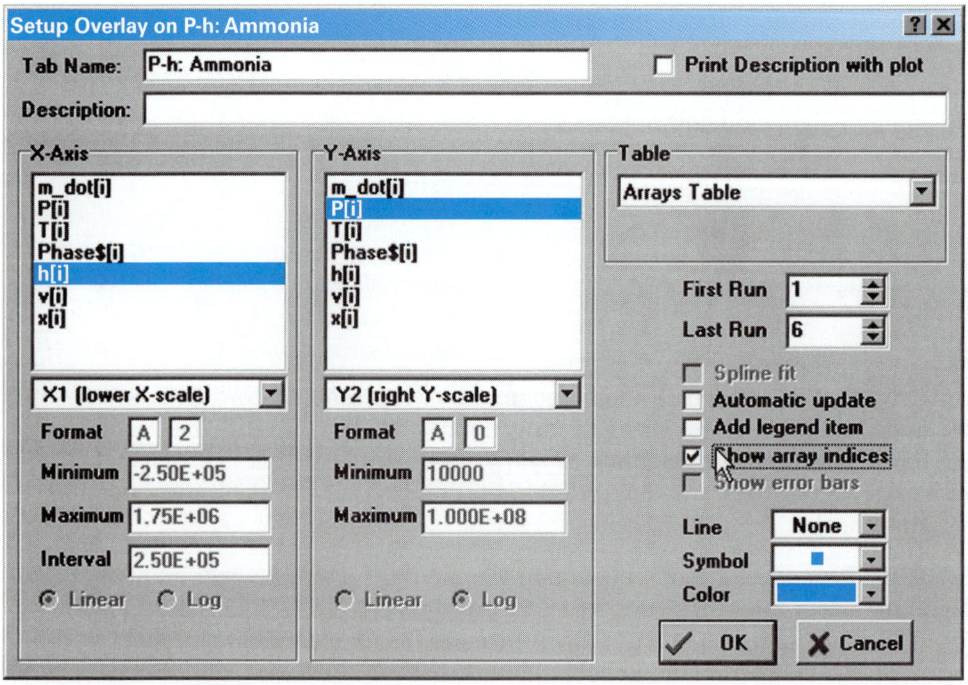

Figure 3: Overlay Plot dialog.

Finally, use the line tool in the Plot tool palette to connect points 1 and 2 (compressor #1), 3 and 4 (the throttle valve), 4, 5 and 2 (the de-superheater), and 5 and 6 (compressor #2). The completed plot is shown in Figure 4.

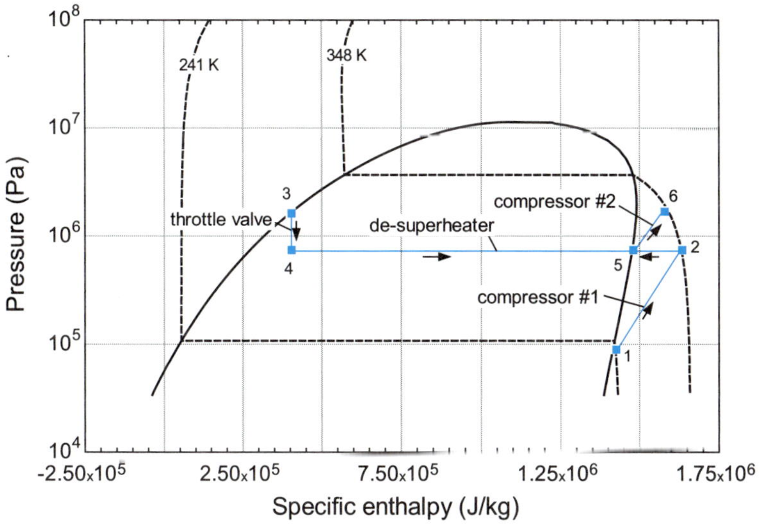

Figure 4: Pressure-specific enthalpy plot showing the states and processes.

4.5 Analysis of Open Unsteady Systems

Section 4.4 described a number of components that typically operate at steady-state. In this section, open systems undergoing unsteady processes will be considered. The methodology for identifying the governing equations for these processes is exactly the same as for steady-state problems. However, the governing equations for an unsteady process will involve storage terms and therefore they will often involve one or more differential equations that must be solved either analytically or numerically.

In general, the process and the properties of the fluid are sufficiently complex that a numerical solution will be required. The numerical solution of one or more differential equations is discussed in Appendix G. The first step in any such solution is the identification of one or more state equations; these equations describe the time rate of change of the state variables in terms of their values and the values of the integration parameter (typically time). In a thermodynamics problem, the state equations will typically involve mass and energy balances and therefore the state variables will naturally be the mass and internal energy of the system. The numerical solution of an unsteady problem will therefore begin by developing the equations that provide the time rates of change of the mass and internal energy of the system, given its mass and internal energy at a specific time. When this is done, it is relatively simple to integrate the state equations through time using any of the numerical methods discussed in Appendix G. The most convenient and powerful method for implementing the numerical solution is to use the Integral command in EES, which is discussed in Section G.5. There are some situations where numerical integration is not required because the differential equations that govern the problem are sufficiently simple that they can be solved analytically.

The problems presented in this section are only a small subset of the possible open, unsteady processes that you may encounter. However, these problems illustrate the solution steps and methodology.

EXAMPLE 4.5-1: HYDROGEN STORAGE TANK FOR A VEHICLE

There has been increasing interest in the use of hydrogen as a fuel for fuel cell powered vehicles. Hydrogen is an attractive fuel from a global warming perspective because it reacts to form water and it does not produce any carbon dioxide. Unfortunately, hydrogen (unbound to other substances) does not occur naturally on earth's surface and therefore it must be generated using other energy sources.

One concern associated with hydrogen-fueled vehicles is the compromise between storage tank size and vehicle range. One proposed fuel tank design uses $N_{tank} = 3$ heavy-walled steel fuel cylinders. The tanks are cylindrical with a height of $H_{tank} = 65$ cm and an internal diameter of $D_{tank} = 28$ cm. Additional tanks could be added in order to increase vehicle range. However, the tanks are expensive, heavy, and reduce the cargo capacity of the vehicle. Filling stations will provide hydrogen from a large fuel reservoir at $P_{supply} = 300$ bar and ambient temperature, $T_{supply} = T_{amb} = 25°\text{C}$.

A vehicle enters a filling station with its fuel tanks at ambient temperature, $T_{ini} = T_{amb}$, and an initial pressure $P_{ini} = 60$ bar. The tanks are connected to the fuel reservoir and the fill valve is opened for $t_{fill} = 3$ minutes, at which time the valve is closed. The mass flow rate through the fill valve has been found to scale with the square root of the pressure difference between the supply and fuel tank according to:

$$\dot{m} = C_{valve}\sqrt{(P_{supply} - P)} \tag{1}$$

where P is the instantaneous pressure in the tank and $\dot{m}$ is the instantaneous mass flow rate. Measurements on the valve set used for the fuel tanks indicate that the valve coefficient is $C_{valve} = 2.68\times10^{-6}$ kg/s-Pa$^{0.5}$. The heat transfer rate to the hydrogen in the tanks from the tank walls is given by:

$$\dot{Q} = h_{conv}\, A_s\, (T_{wall} - T) \tag{2}$$

where $h_{conv} = 40$ W/m^2-K is the convective heat transfer coefficient between the hydrogen in the fuel tanks and the cylinder walls and A_s is the internal surface area of the tank walls. Because of the large thermal mass of the steel wall, you may assume that the temperature of the tank walls (T_{wall}) remains at T_{amb} during the entire filling process.

a) Plot the pressure and temperature of the hydrogen in the tanks as a function of time during the filling process.

The inputs are entered in EES:

```
$UnitSystem SI MASS RAD PA K J
"Inputs"
N_tank=3 [-]                                "number of tanks"
H_tank = 65 [cm]*convert(cm,m)              "height of tanks"
D_tank=28 [cm]*convert(cm,m)                "inner diameter of tanks"
P_supply = 300 [bar]*convert(bar,Pa)        "pressure of supply hydrogen"
T_amb=converttemp(C,K,25 [C])               "temperature of ambient"
T_supply=T_amb                              "temperature of supply hydrogen"
P_ini=60 [bar]*convert(bar,Pa)              "initial tank pressure"
C_valve=2.68e-6 [kg/s-Pa^0.5]               "valve coefficient"
```

```
T_ini=T_amb                          "initial temperature of the hydrogen in the tank"
T_wall=T_amb                         "temperature of tank wall"
h_conv=40 [W/m^2-K]                  "convective heat transfer coefficient"
t_fill_min=3 [min]                   "filling time, in min"
t_fill = t_fill_min*convert(min,s)   "filling time"
```

We will choose the system to be the internal volume of the tanks, as shown in Figure 1.

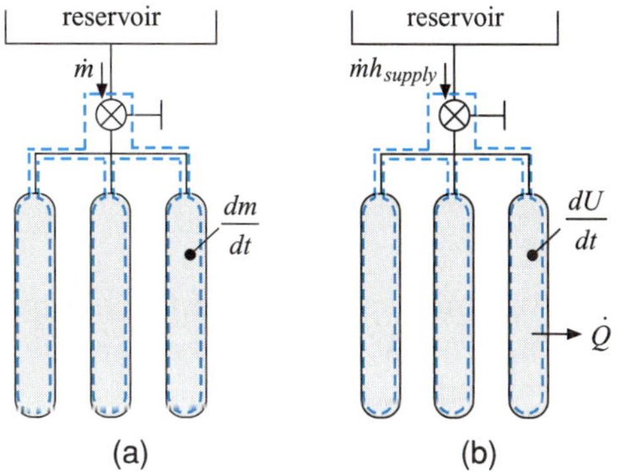

Figure 1: (a) Mass balance and (b) energy balance.

The connecting tubing and valve are assumed to have negligible volume relative to the tanks. The tank material is not included in the system; therefore the heat transfer rate between the hydrogen and the tank walls ($\dot{Q}$) must be considered in the energy balance, as shown in Figure 1(b).

The volume of a single tank is determined according to:

$$V_{tank} = \pi \frac{H_{tank}\, D_{tank}^2}{4}$$

```
V_tank=pi*D_tank^2*H_tank/4   "get tank volume"
```

The initial state of the hydrogen in the tank is fixed by the temperature and pressure. Therefore, the specific volume and specific internal energy (v_{ini} and u_{ini}) can be computed. The initial mass and internal energy are computed according to:

$$m_{ini} = \frac{N_{tank}\, V_{tank}}{v_{ini}}$$

$$U_{ini} = u_{ini}\, m_{ini}$$

```
"Initial conditions"
v_ini=volume(Hydrogen,T=T_ini,P=P_ini)      "initial specific volume of hydrogen"
u_ini=intenergy(Hydrogen,T=T_ini,P=P_ini)   "initial specific internal energy of hydrogen"
m_ini=V_tank*N_tank/v_ini                   "initial mass of hydrogen"
Utot_ini=u_ini*m_ini                        "initial internal energy"
```

The natural state variables that result from the governing mass and energy balances are the system mass and internal energy, m and U. We will initially set arbitrary values for these state variables and the integration variable, time, and then develop the equations that are required to compute the time rate of change of m and U. Once this is done (i.e., once the state equations have been formulated), the problem will be solved by integrating the time rate of change of the state variables forward through time from the beginning to the end of the filling process.

```
"arbitrary state variables and time"
Utot=Utot_ini                          "total energy of the system"
m=m_ini                                "mass"
time=0 [s]                             "time"
```

The specific volume and specific internal energy correspond to the specified values of m and U:

$$v = \frac{N_{tank}\, V_{tank}}{m}$$

$$u = \frac{U}{m}$$

The instantaneous state of the hydrogen in the tank is fixed by its specific volume and specific internal energy. The temperature and pressure (T and P) can be determined:

```
u=Utot/m                               "specific internal energy"
v=V_tank*N_tank/m                      "specific volume"
P=pressure(Hydrogen, v=v,u=u)          "pressure in tank"
P_bar=P*convert(Pa,bar)                "in bar"
T=temperature(Hydrogen,v=v,u=u)        "temperature in tank"
T_C=converttemp(K,C,T)                 "in C"
```

The mass flow rate into the tank is determined from Eq. (1). A mass balance on the system is shown in Figure 1(a):

$$\dot{m} = \frac{dm}{dt}$$

```
m_dot=C_valve*sqrt(P_supply-P)         "mass flow rate into tank"
m_dot=dmdt                             "mass balance on tank"
```

At this point, the time rate of change of one of the state variables, m, has been determined.

The specific enthalpy of the hydrogen entering the valve from the reservoir is constant because the conditions in the reservoir do not change; this assumption is consistent with a supply tank that is much larger than the fuel tank. The specific enthalpy of the entering hydrogen (h_{supply}) is fixed by the temperature and pressure in the reservoir.

```
h_supply=enthalpy(Hydrogen,T=T_supply,P=P_supply)   "specific enthalpy of hydrogen entering the valve"
```

The internal surface area of the N_{tank} cylindrical tanks is computed:

$$A_s = N_{tank}\left(2\,\pi\,\frac{D_{tank}^2}{4} + \pi\, D_{tank}\, H_{tank}\right)$$

The rate of heat transfer from the hydrogen to the tank wall is computed using Eq. (2).

```
A_s=N_tank*(pi*D_tank*H_tank+2*pi*D_tank^2/4)    "get tank surface area"
Q_dot=h_conv*A_s*(T-T_wall)                      "rate of heat transfer to the tank wall"
```

An energy balance on the system is shown in Figure 1(b):

$$\dot{m}\,h_{supply} = \dot{Q} + \frac{dU}{dt}$$

```
m_dot*h_supply=Q_dot+dUtotdt    "energy balance on tank"
```

At this point, we have determined the time rate of change of the remaining state variable, U. The assumed values for mass, energy, and time are commented out:

```
"arbitrary state variables and time"
{Utot=Utot_ini                 "total energy of the system"
m=m_ini                        "mass"
time=0 [s]                     "time"}
```

and the state equations are integrated through time using the **Integral** command in EES. (See Section G.5 of Appendix G for more information on the use of the **Integral** command.) Initially, an integration step size of 1 second (the 5th argument in the **Integral** command) is chosen. The effect of the integration step size on the result should be evaluated; note that the step size can be automatically adjusted by EES in order to control the accuracy of the integration by removing the 5th argument of the **Integral** command or by setting its value to zero. An Integral table is created using the **$IntegralTable** directive in order to record the temperature and pressure in the tank as a function of time.

```
m=m_ini+Integral(dmdt,time,0,t_fill,1 [s])             "integrate dmdt"
Utot=Utot_ini+Integral(dUtotdt,time,0,t_fill,1 [s])    "integral dUtotdt"
$IntegralTable time,T_C,P_bar
```

Figure 2 illustrates the pressure and temperature of the hydrogen in the tank as a function of time during the filling process.

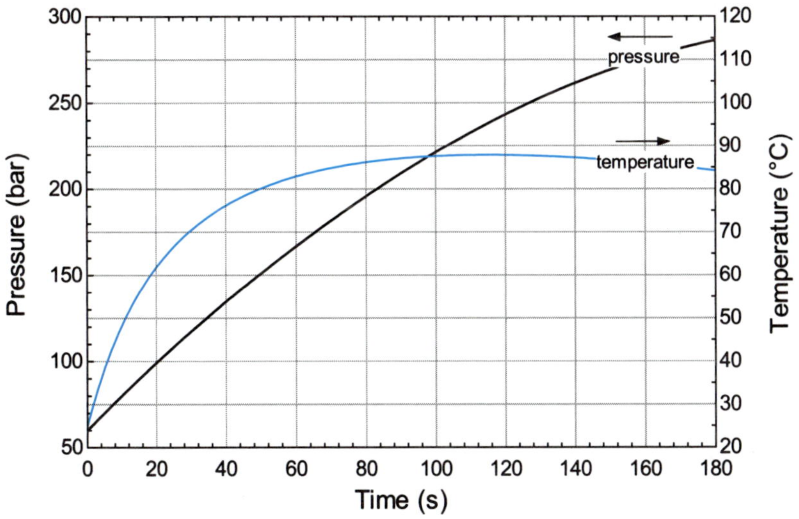

Figure 2: Pressure and temperature of the hydrogen in the tank as a function of time.

b) Plot the mass of hydrogen that is added to the tank during the filling process as a function of the fill time.

The mass added to the tank is given by:

$$m_{fuel} = m - m_{ini}$$

```
m_fuel=m-m_ini    "fuel added"
```

The mass of fuel that is added is included in the Integral table:

```
$IntegralTable time,T_C,P_bar, m_fuel
```

Figure 3 illustrates the mass of fuel added to the tank as a function of fill time.

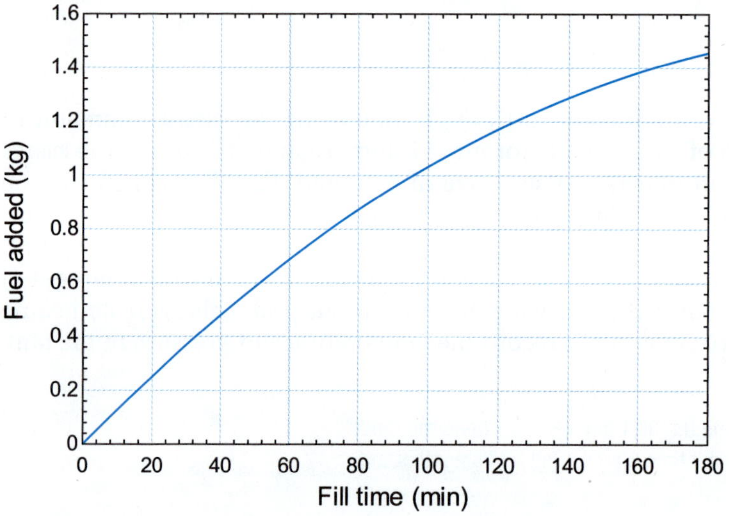

Figure 3: Mass of fuel added to tanks as a function of fill time.

EXAMPLE 4.5-2: EMPTYING AN ADIABATIC TANK FILLED WITH IDEAL GAS

A tank is filled with air at $P_1 = 100$ psi and $T_1 = 20°$C. The air can be modeled as an ideal gas with constant $c_v = 717.6$ J/kg-K. The tank is opened and air escapes until the pressure in the tank reaches $P_2 = 14.7$ psi. Assume that the air within the tank undergoes an adiabatic process.

a) Determine the final temperature of the air in the tank.

The inputs are entered in EES and converted to standard SI units.

```
$UnitSystem SI MASS RAD PA K J
"Inputs"
T_1=converttemp(C,K,20 [C])         "initial temperature"
P_1=100 [psi]*convert(psi,Pa)       "initial pressure"
c_v=717.6 [J/kg-K]                  "specific heat capacity at constant volume"
P_2_psi=14.7 [psi]                  "final pressure, in psi"
P_2=P_2_psi*convert(psi,Pa)         "final pressure"
```

The ideal gas constant is determined according to:

$$R = \frac{R_{univ}}{MW}$$

where MW is the molar mass of air, obtained using the MolarMass function. The specific heat capacity at constant pressure is given by:

$$c_P = c_v + R$$

```
R=R#/MolarMass(Air)     "ideal gas constant"
c_P=c_v+R               "specific heat capacity at constant pressure"
```

A mass balance on the open system defined as the internal volume of the tank is shown in Figure 1(a) and provides:

$$0 = \dot{m}_{out} + \frac{dm}{dt} \tag{1}$$

where $\dot{m}_{out}$ is the mass flow rate leaving the tank. An energy balance on the same system is shown in Figure 1(b) and provides:

$$0 = \dot{m}_{out}\, h + \frac{dU}{dt} \tag{2}$$

where h is specific enthalpy of the air leaving the tank; we assume here that the air leaving the tank has the same state as the air in the tank at any point in time, an assumption that implies that there are no temperature or pressure gradients within the tank.

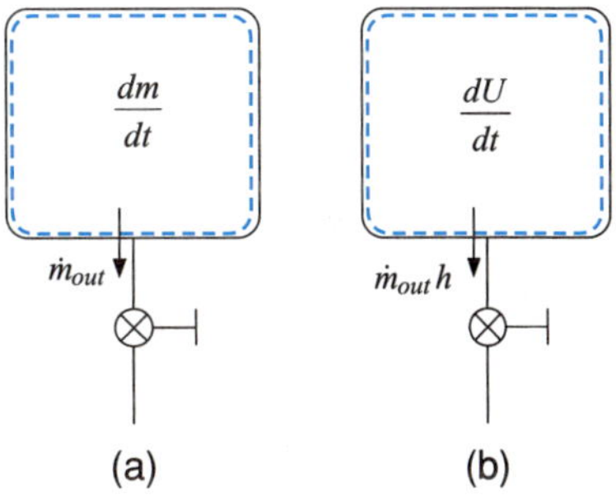

Figure 1: (a) Mass balance and (b) energy balance on the open system consisting of the internal volume of the tank.

The internal energy of the air in the tank is the product of the mass and the specific internal energy:

$$U = m u$$

The time rate of change of U is:

$$\frac{dU}{dt} = u\frac{dm}{dt} + m\frac{du}{dt} \tag{3}$$

Equations (1) and (3) are substituted into Eq. (2) to yield:

$$0 = \underbrace{-\frac{dm}{dt}}_{\dot{m}_{out}} h + \underbrace{u\frac{dm}{dt} + m\frac{du}{dt}}_{\frac{dU}{dt}}$$

which can be rearranged:

$$\frac{dm}{dt}(h-u) = m\frac{du}{dt} \tag{4}$$

The definition of specific enthalpy is:

$$h = u + P\,v \tag{5}$$

Substituting Eq. (5) into Eq. (4) results in:

$$\frac{dm}{dt}P\,v = m\frac{du}{dt} \tag{6}$$

The ideal gas law is:

$$P\,v = R\,T \tag{7}$$

According to Eq. (3-23), the specific internal energy of an ideal gas with constant c_v can be expressed as:

$$u = c_v\,T$$

Therefore, the time rate of change of the specific internal energy is:

$$\frac{du}{dt} = c_v\frac{dT}{dt} \tag{8}$$

Substituting Eqs. (7) and (8) into Eq. (6) leads to:

$$\frac{dm}{dt}\underbrace{R\,T}_{P\,v} = \underbrace{m\,c_v\frac{dT}{dt}}_{\frac{du}{dt}} \tag{9}$$

Equation (9) can be separated:

$$R\frac{dm}{m} = c_v\frac{dT}{T}$$

and integrated:

$$R\int_{m_1}^{m_2}\frac{dm}{m} = c_v\int_{T_1}^{T_2}\frac{dT}{T}$$

which results in:

$$R\ln\left(\frac{m_2}{m_1}\right) = c_v\ln\left(\frac{T_2}{T_1}\right) \tag{10}$$

The ideal gas law is substituted into Eq. (10) for the mass:

$$R \ln\left(\underbrace{\frac{P_2\, V}{R\, T_2}}_{m_2}\ \underbrace{\frac{R\, T_1}{P_1\, V}}_{1/m_1}\right) = c_v \ln\left(\frac{T_2}{T_1}\right)$$

and simplified:

$$R \ln\left(\frac{P_2\, T_1}{P_1\, T_2}\right) = c_v \ln\left(\frac{T_2}{T_1}\right) \tag{11}$$

Equation (11) is solved for the final temperature:

$$R \ln\left(\frac{P_2}{P_1}\right) - R \ln\left(\frac{T_2}{T_1}\right) = c_v \ln\left(\frac{T_2}{T_1}\right)$$

$$R \ln\left(\frac{P_2}{P_1}\right) = \underbrace{(c_v + R)}_{c_P} \ln\left(\frac{T_2}{T_1}\right)$$

$$c_P \ln\left(\frac{T_2}{T_1}\right) = R \ln\left(\frac{P_2}{P_1}\right)$$

$$T_2 = T_1 \left(\frac{P_2}{P_1}\right)^{R/c_P}$$

```
T_2=T_1*(P_2/P_1)^(R/c_P)       "final temperature"
T_2_C=converttemp(K,C,T_2)      "in C"
```

After solving, $T_2 = 169.5$ K (-103.6°C).

b) Plot the temperature in the tank as a function of the final pressure, P_2.

A Parametric table is created that includes P_2 and T_2. The temperature of the air in the tank as a function of pressure is shown in Figure 2.

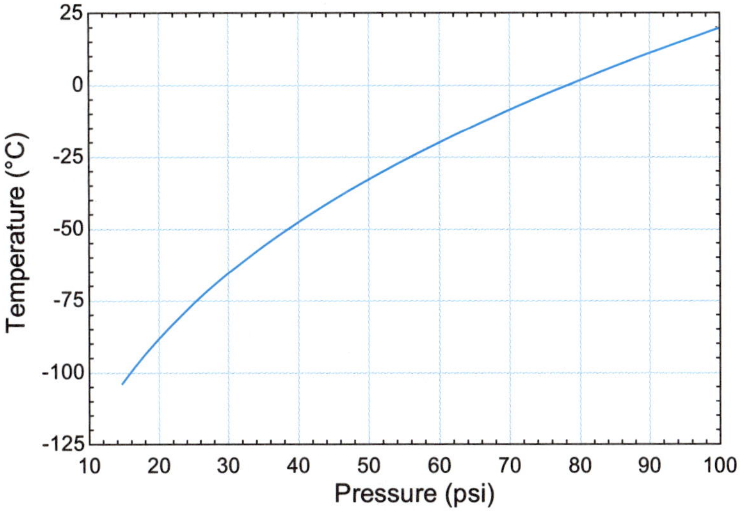

Figure 2: Temperature as a function of pressure.

EXAMPLE 4.5-3: EMPTYING A BUTANE TANK

A camp stove uses butane fuel (n-butane) that is supplied from a canister that has a volume of $V = 0.54$ liter. Initially, the canister contains $m_1 = 0.23$ kg of butane at $T_1 = 25°$C. At these conditions, both liquid and vapor are present in the tank. The butane is provided to the burner of the stove at a constant mass flow rate $\dot{m}_{out} =$ 0.035 g/s through a valve located at the top of the tank. If the tank is positioned normally, only butane vapor will exit through the valve. However, if the canister is inverted then only liquid butane will exit through the valve. The mass flow rate is the same in either case, since it is controlled by the burner.

a) Calculate and plot the temperature and pressure of the butane in the canister as a function of time for these two operating conditions during a 20 minute operating period. Assume that the process is adiabatic during this time.

This tank emptying problem differs from the one discussed in EXAMPLE 4.5-2 because the butane can not be approximated as an ideal gas. The known information is entered into EES and converted to standard SI units.

```
$UnitSystem SI K Pa J Mass Radian
"Inputs"
Vol=0.54 [liter]*convert(liter,m^3)              "tank volume"
m_1=0.23 [kg]                                    "initial mass of butane"
T_1=converttemp(C,K,25 [C])                      "initial temperature"
m_dot_out=0.035 [g/s]*convert(g/s,kg/s)          "mass flow rate"
```

The initial specific volume of the butane is given by:

$$v_1 = \frac{V}{m_1}$$

The initial specific volume and temperature fix the initial state of the butane. The pressure and specific internal energy of the butane (P_1 and u_1) are determined as well as its phase. The total internal energy of the butane initially in the tank is:

$$U_1 = u_1\, m_1$$

```
v_1=Vol/m_1                                      "initial specific volume"
P_1=pressure('n-Butane',T=T_1,v=v_1)             "initial pressure"
u_1=intenergy('n-Butane',T=T_1,v=v_1)            "initial specific internal energy"
Phase_1$=Phase$('n-Butane',T=T_1,v=v_1)          "initial phase"
U_tot_1=u_1*m_1                                  "initial internal energy"
```

A mass balance on the system is:

$$0 = \dot{m}_{out} + \frac{dm}{dt} \tag{1}$$

Equation (1) is integrated:

$$0 = \int_0^t \dot{m}_{out}\, dt + \int_0^t \frac{dm}{dt}\, dt \tag{2}$$

Because the mass flow rate is constant, Eq. (2) can be integrated analytically:

$$0 = \dot{m}_{out} \int_0^t dt + \int_0^m dm$$

$$0 = \dot{m}_{out}\, t + m - m_1 \tag{3}$$

where m is the mass in the tank at time t. An energy balance on the system is:

$$0 = \dot{m}_{out}\, h_{out} + \frac{dU}{dt} \tag{4}$$

Because the specific enthalpy of the exiting butane (h_{out}) varies as the temperature and pressure in the tank changes, it is not possible to integrate Eq. (4) analytically. Therefore, a numerical solution will be obtained. The state variable suggested by Eq. (4) is the total internal energy, U. Arbitrary values of the state variable (U) and time are assumed so that we can enter and test the equations that calculate the time derivative of internal energy:

```
"arbitrary state variable"
U_tot=U_tot_1                      "Internal energy"
time=0 [s]                         "time"
```

The mass in the tank is computed using Eq. (3).

```
0=m_dot_out*time+m-m_1             "mass"
```

The specific volume and specific internal energy of the butane in the tank (v and u) are computed:

$$v = \frac{V}{m}$$

$$u = \frac{U}{m}$$

```
v=Vol/m                            "specific volume"
u=U_tot/m                          "specific internal energy"
```

The state of the butane is fixed by the specific volume and specific internal energy. The temperature and pressure (T and P) are computed:

```
T=temperature('n-Butane',v=v,u=u)  "temperature"
T_C=converttemp(K,C,T)             "in C"
P=pressure('n-Butane',v=v,u=u)     "pressure"
P_kPa=P*convert(Pa,kPa)            "in kPa"
```

The specific enthalpy of the butane exiting the tank (h_{out}) is computed assuming that the tank is positioned so that the valve is at the top and therefore only saturated vapor exits:

```
h_out=enthalpy('n-Butane',P=P,x=1) "specific enthalpy leaving, assuming only saturated vapor exits"
```

The rate of change of the internal energy of the system is computed using the energy balance, Eq. (4).

```
0=m_dot_out*h_out+dUdt     "energy balance"
```

The arbitrary value of the state variable (U) and time are commented out:

```
{U_tot=U_tot_1     "Internal energy"
time=0 [s]         "time"}
```

The Integral command is used to numerically integrate the state equation through time and an Integral table is created in order to record the temperature and pressure in the tank as a function of time.

```
U_tot=U_tot_1+Integral(dUdt,time,0, 20 [min]*convert(min,s))     "integrate state equation"
time_min=time*convert(s,min)
$IntegralTable time, time_min, T_C,P_kPa
```

Figure 1 illustrates the temperature of the butane in the tank as a function of time and Figure 2 illustrates the pressure as a function of time. In order to simulate emptying the tank by removing saturated liquid (i.e., positioning the valve at the bottom) it is only necessary to recompute the value of the specific enthalpy leaving using the instantaneous pressure and a quality of zero.

```
{h_out=enthalpy('n-Butane',P=P,x=1)}     "saturated vapor leaves"
h_out=enthalpy('n-Butane',P=P,x=0)       "saturated liquid leaves"
```

Both methods of emptying the tank are shown in Figures 1 and 2.

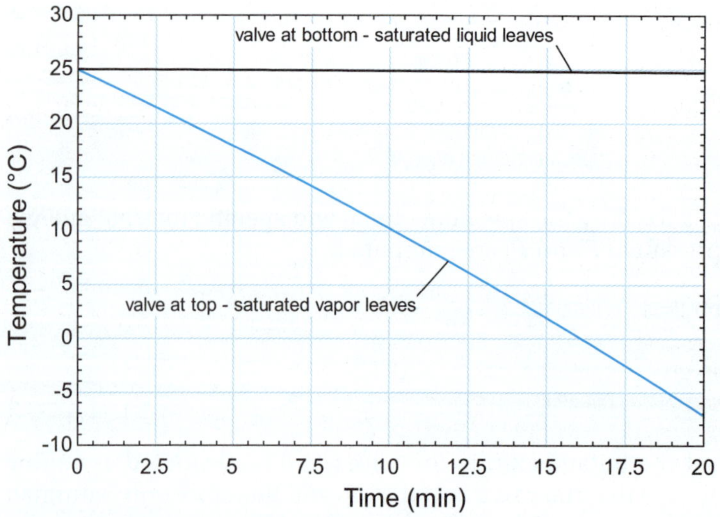

Figure 1: Temperature of butane as a function of time for both methods of emptying the tank.

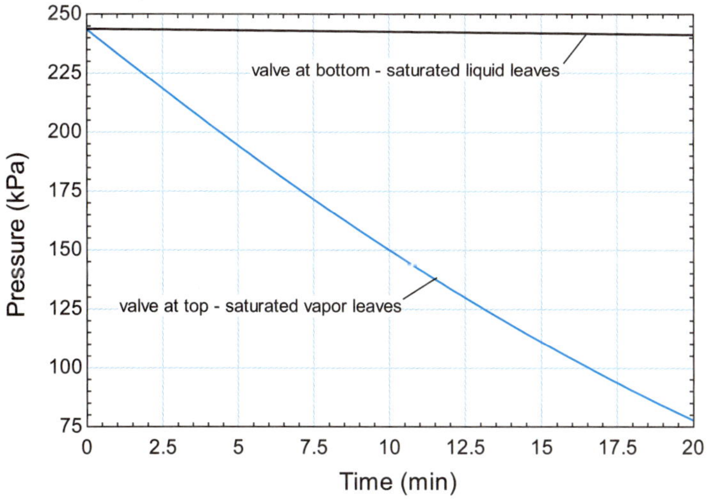

Figure 2: Pressure of butane as a function of time for both methods of emptying the tank.

REFERENCE

Nellis, G.F. and S.A. Klein, *Heat Transfer*, Cambridge University Press, New York (2009).

Problems

The problems included here have been selected from a much larger set of problems that are available on the website associated with this book (www.cambridge.org/kleinandnellis).

A: Thermodynamic Analyses of Steady-State Applications

4.A-1 Air conditioning systems in large building are required to bring in 20 cubic feet per minute (cfm) of outdoor air per occupant, according to ASHRAE Standard 62. In a particular case, the Heating, Ventilating and Air Conditioning (HVAC) system for a building is being designed for a capacity of $N_p = 200$ persons. As shown in Figure 4.A-1, outdoor air at $T_1 = 92^\circ$F is brought into the building through a $D_1 = 2.5$ ft diameter circular duct and mixed with chilled air at $T_2 = 55^\circ$F that is flowing through a $D_2 = 4$ ft diameter duct at velocity $\tilde{V}_2 = 600$ ft/min. The pressure of all of the air streams is $P = 1$ atm.

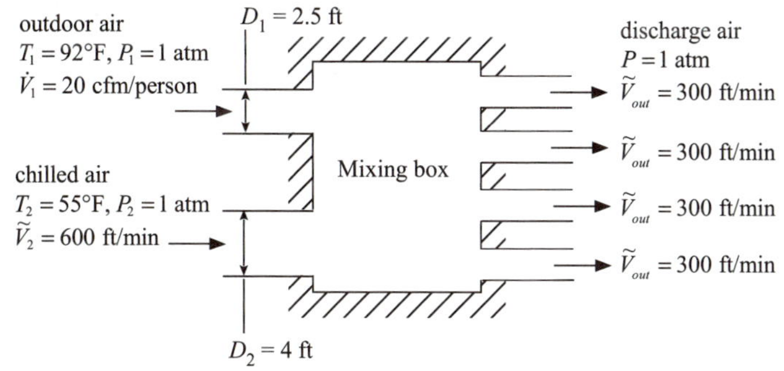

Figure 4.A-1: Mixing box.

The mixture is discharged with velocity $\tilde{V}_{out} = 300$ ft/min through 4 identical exit ducts in order to reduce duct noise. Assume that the mixing process is adiabatic and steady-state and neglect the humidity in the air. Do not neglect the kinetic energy of the streams.

a) What is the temperature of the discharge, mixed air?
b) Determine the necessary diameter of the four discharge ducts.

4.A-2 Figure 4.A-2 illustrates an air-breathing gas turbine power plant. Air at $T_1 = 20°$C and $P_1 = 1$ atm is drawn into the compressor with a volumetric flow rate of $\dot{V}_1 = 42{,}000$ ft^3/min. The compressor increases the pressure and temperature of the air. The temperature of the air leaving the compressor is $T_2 = 260°$C. The compressor requires shaft work at a rate $\dot{W}_c$. The air passes from the compressor to a combustor where a heat addition at rate $\dot{Q}_c$ elevates the temperature of the air to T_3. The air passes from the combustor to a gasifier turbine where the pressure and temperature are reduced. The output shaft power from the gasifier turbine ($\dot{W}_{gt}$) is used to drive the compressor. Finally, the air passes from the gasifier turbine to the power turbine where its temperature and pressure are again reduced. The power turbine exhausts to atmospheric pressure and the temperature of the air leaving the power turbine is $T_5 = 550°$C. The power output produced by the power turbine is $\dot{W}_{pt} = 9{,}500$ kW. Neglect the potential and kinetic energy terms and assume that the compressor and turbines are adiabatic. Also assume that the system is operating at steady state.

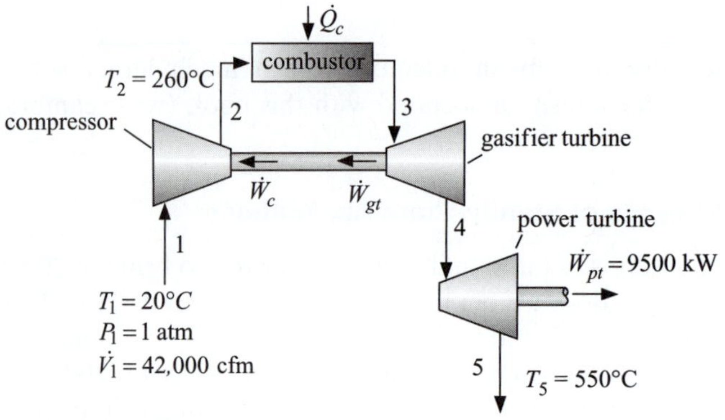

Figure 4.A-2: Gas turbine power plant.

a) What is the shaft power required by the compressor?
b) What is the temperature of the air entering the power turbine, T_4?
c) What is the temperature of the air entering the gasifier turbine, T_3?
d) What is the rate of heat transfer to the combustor?
e) What is the efficiency of the gas turbine power plant? The efficiency is defined as the ratio of the work produced by the power turbine to the heat transfer required by the combustor.

4.A-6 Figure 4.A-6 illustrates a vapor separator that is part of a refrigeration system. Saturated liquid R134a leaves the condenser at state 1 with mass flow rate $\dot{m}_{cond} = 0.05$ kg/s and pressure $P_{cond} = 890$ kPa. This refrigerant enters the vapor separator (a large, insulated tank) through a valve. The pressure in the tank is $P_{tank} = 500$ kPa. There is no pressure drop due to flow through the tank (i.e., the pressure at states 2, 3, and 5 are all the same). The flow at state 2 is two-phase. The liquid portion of this flow falls to the bottom of the tank due to gravity while the vapor

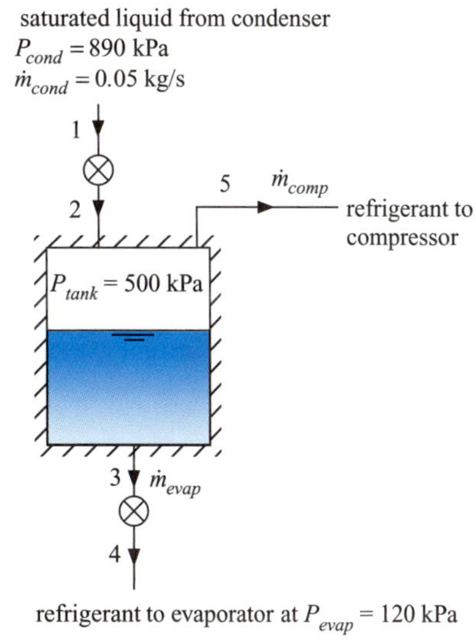

Figure 4.A-6: Vapor separator.

portion remains at the top. Saturated liquid is drawn from the bottom of the tank (at state 3) and expands through a valve to state 4 where it enters the evaporator at P_{evap} = 120 kPa. The saturated vapor is drawn off of the top of the tank at state 5 and fed to a compressor. Assume that the vapor separator is adiabatic, rigid, and operating at steady state.

a) Determine the quality of the refrigerant at state 2.
b) Determine the mass flow rate of refrigerant that goes to the evaporator ($\dot{m}_{evap}$) and the mass flow rate of refrigerant that goes to the compressor ($\dot{m}_{comp}$).
c) Use EES to generate a temperature-specific volume diagram. Overlay the state points indicated in Figure 4.A-6 onto this diagram. Number the states.

4.A-8 At the factory where you work there is a diesel generator. The exhaust gas leaving the generator is quite hot, $T_{ex,1} = 660°$F, and it has a large volumetric flow rate, $\dot{V}_{ex,1} = 2{,}640$ cfm. You are evaluating the feasibility of utilizing this hot flow of exhaust gas to heat water using a heat exchanger, as shown in Figure 4.A-8.

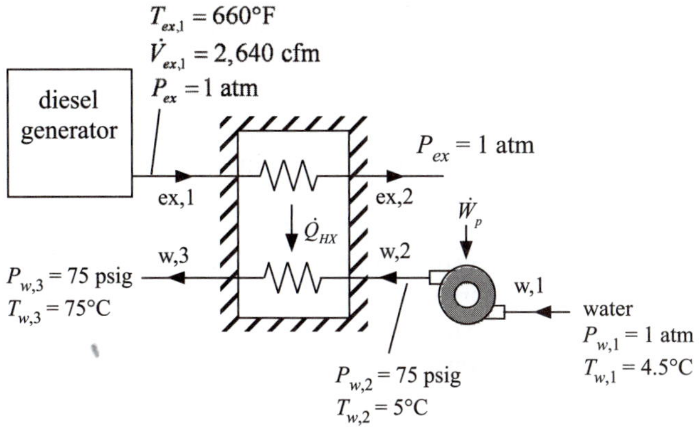

Figure 4.A-8: Diesel generator interfaced with heat exchanger to provide hot water.

The heat exchanger will take in water at $T_{w,2} = 5°\text{C}$ and heat it to $T_{w,3} = 75°\text{C}$ by transferring heat from the exhaust gas to the water. Model the exhaust gas as an ideal gas with $R = 287.1$ J/kg-K and a constant $c_P = 1005$ J/kg-K. Assume that the pressure of the exhaust gas entering and exiting the heat exchanger is $P_{ex} =$ 1 atm. Model the water as an incompressible substance with $\rho_w = 1000$ kg/m^3 and constant $c_w = 4200$ J/kg-K. The pressure of the water entering and exiting the heat exchanger is $P_{w,2} = P_{w,3} = 75$ psig. Initially, assume that the approach temperature difference for the heat exchanger is $\Delta T_{HX} = 50$ K and that the pinch point occurs at the cold end (i.e., the end where the water enters and exhaust gas leaves). This implies that:

$$T_{ex,2} = T_{w,2} + \Delta T_{HX}$$

a) Determine the rate of heat transfer in the heat exchanger ($\dot{Q}_{HX}$) and the volumetric flow rate of water that can be heated (in gpm).
b) The pressure of the water entering the pump that provides water to the heat exchanger is $P_{w,1} = 1$ atm. The temperature of the water entering the pump is $T_{w,1} = 4.5°\text{C}$. Determine the power consumed by the pump, $\dot{W}_p$.
c) You would like to examine the economic feasibility of installing the energy recovery system (which consists of the pump and the heat exchanger). Currently, water is heated by burning natural gas. The cost of natural gas is $gc =$ 0.92 \$/therm. (A therm is 1×10^5 Btu; it is a built-in unit in EES). The electricity required to run the pump costs $ec = 0.12$ \$/kW-hr. The plant operates 360 days/year and 16 hr/day. Determine the savings per year that is associated with using the energy recovery system.
d) You would like to select the best heat exchanger for the application. A large heat exchanger will provide a small approach temperature difference, ΔT_{HX}, and therefore lead to large savings per year. However, the capital cost associated with a large heat exchanger is high. On the other hand, a small heat exchanger will give a large ΔT_{HX} which reduces the savings per year but has a smaller capital cost. There must be a trade-off between these extremes. The cost of the heat exchanger depends on the approach temperature difference according to the formula:

$$Cost_{HX} = \frac{a}{\Delta T_{HX}}$$

where $a = 5 \times 10^5$ \$-K. Determine the net savings associated with purchasing and operating the system for $N_{year} = 3$ years. Neglect the time value of money in your analysis.
e) Plot the net savings associated with purchasing and operating the system system for $N_{year} = 3$ years as a function of ΔT_{HX}. You should see that there is an economically optimal heat exchanger for this application.

4.A-11 In the evaporator of an air conditioning system, dry air at $\dot{m}_a = 0.75$ kg/s, $T_{a,in} =$ 40°C and $P_{a,in} = 1.04$ bar passes over finned tubes through which refrigerant R134a flows. The air exits the evaporator at $T_{a,out} = 25°\text{C}$ and $P_{a,out} = 1.01$ bar. The refrigerant enters the tubes with a quality of $x_{r,in} = 0.2$ at $T_r = 12°\text{C}$ and exits as saturated vapor at the same pressure. The jacket of the evaporator is well-insulated.

a) Determine the volumetric flow rate of the air.
b) Determine the mass flow rate of the refrigerant.

c) Determine the rate of energy transfer from the air to the refrigerant.
d) Before entering the heat exchanger, the refrigerant existed as saturated liquid leaving a condenser. This saturated liquid was throttled through an expansion valve to the evaporator pressure. Determine the pressure and temperature of the saturated liquid refrigerant that existed upstream of the valve.

B: Thermodynamic Analyses of Open Unsteady Systems

4.B-1 A scuba tank having an internal volume of $V = 3.5$ liter initially contains air at $P_1 = 100$ kPa and $T_1 = 22°$C. The tank is connected to a supply line through a valve, as shown in Figure 4.B-1. Air is available in the supply line at $P_s = 600$ kPa and $T_s = 22°$C. The valve is opened and air quickly fills the tank until the pressure in the tank reaches the supply line pressure, at which point the valve is closed. A thermometer indicates that the air temperature in the scuba tank just after the valve is closed is $T_2 = 47°$C. Assume that air behaves in accordance with the ideal gas law.

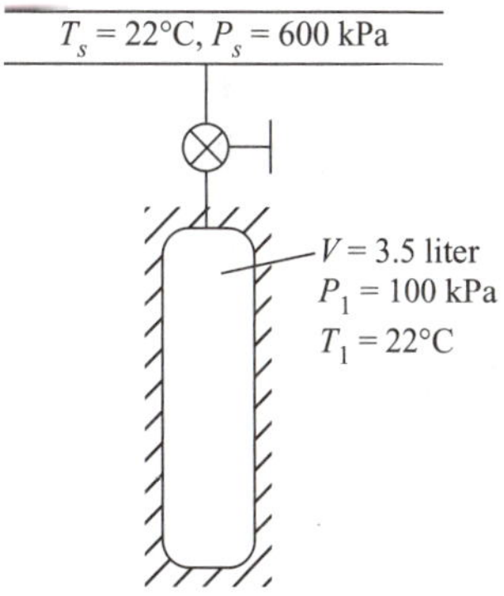

Figure 4.B-1: Scuba tank being filled from a supply line.

a) Determine the mass of air that enters the tank.
b) Determine the heat transfer from the air that occurs during the filling process.
c) After the valve is closed, the tank eventually thermally equilibrates with the ambient and reaches $T_3 = 22°$C. What is the pressure in the tank at this time?

4.B-2 A rigid tank with internal volume $V = 0.3$ m^3 is filled with saturated liquid water at $T = 200°$C. A valve at the bottom of the tank is opened, and saturated liquid only is withdrawn from the tank. Heat is transferred to the water in the tank so that the temperature of the water remains constant during this process. The process is complete when half of the total mass of water in the tank is withdrawn.
a) Determine the quality at the final state.
b) Determine the amount of heat that must be transferred to the water in the tank.

4.B-3 Oxygen is used as a propellant on a missile. You are investigating a cryogenic technique for storing oxygen on the missile and providing that oxygen at high pressure to the combustor. A $V_{tank} = 25$ liter cryogenic tank (called a Dewar) is filled with oxygen at atmospheric pressure, $P_{fill} = 1$ atm. The filling is stopped when the tank holds $V_{f,1} = 15$ liter of liquid oxygen; the remainder of the tank contains saturated vapor, as shown in Figure 4.B-3.

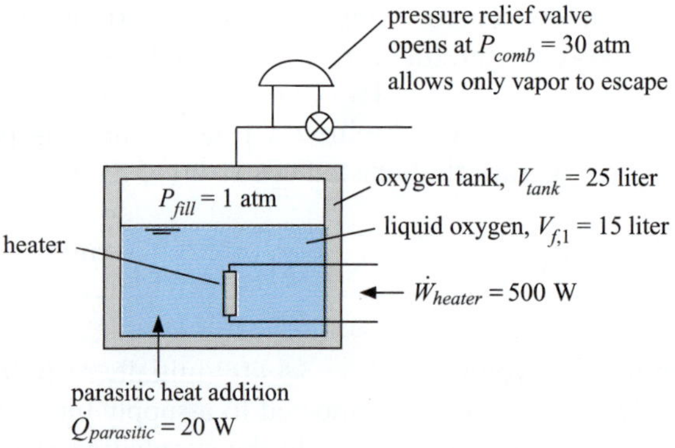

Figure 4.B-3: Oxygen tank.

a) What is the mass of liquid in the tank (kg) after it is filled, $m_{f,1}$?
b) What is the quality of the oxygen in the tank after it is filled, x_1?
c) What are the temperature and specific volume of the oxygen in the tank after it is filled?

Rather than pump the liquid in the tank to the high pressure required by the combustor ($P_{comb} = 30$ atm), you are designing a system that will self-pressurize in order to avoid the need for an expensive and potentially unreliable cryogenic pump. The tank is closed after it is filled. A pressure relief valve opens once the pressure in the tank reaches P_{comb}. The tank is cold, and therefore it experiences a heat transfer rate (sometimes referred to as a parasitic heat leak) of $\dot{Q}_{parasitic} =$ 20 W from the surrounding missile components. You have also installed an electrical heater (a resistor which converts electrical power to heat) that provides $\dot{W}_{heater} = 500$ W to the tank. The heater is activated until the pressure relief valve opens at state 2.

d) What is the temperature in the tank when the relief valve opens, T_2?
e) What is the mass of liquid in the tank at the time that the relief valve opens?
f) What is the total amount of energy (the sum of the heater power and parasitic heat transfer) that is required to self-pressurize the tank (i.e., to go from state 1 to state 2)? How long does this process take (min)?

Once the tank is self-pressurized, the heater is deactivated and the system is ready to launch. However, it is possible that the missile must remain on standby in the launch bay for a substantial period of time before it is finally launched. Even though the heater is deactivated, the oxygen in the tank continues to experience the parasitic heat leak, $\dot{Q}_{parasitic}$, and therefore the liquid in the tank continues to boil away. The pressure relief valve maintains the pressure in the tank at P_{comb} by allowing only saturated oxygen vapor to escape from the tank. At state 3, the missile can no longer be launched because too much of the liquid has boiled away; this situation occurs when the quality of the oxygen in the tank reaches $x_3 = 0.5$.

g) How much mass leaves the tank between states 2 and 3?
h) What is the total amount of heat transfer required to go from state 2 to state 3?
i) How long can the missile sit in the launch bay without being fired before the oxygen tank will need to be refilled (hr)?
j) Generate a T-v diagram using EES and overlay your states onto this diagram.

4.B-4 A spring-loaded accumulator is connected to a steam line, as shown in Figure 4.B-4. The spring constant is $K = 4.1\text{x}10^4$ N/m. The spring is connected to a piston with cross-sectional area $A_p = 0.025$ m^2. The spring is positioned so that it is uncompressed when the volume in the accumulator is zero. The pressure on the top surface of the piston is $P_{atm} = 101$ kPa and the mass of the piston is negligible. Initially, the valve is closed and the position of the piston is $z_1 = 0.05$ m. The temperature of the water that is initially in the accumulator is $T_1 = 135°$C. The steam line contains saturated vapor at $P_s = 60$ psi. The valve is opened and saturated vapor is allowed to enter until the pressure within the accumulator reaches $P_2 = P_s$. The final temperature of the contents of the accumulator is $T_2 = 150°$C.

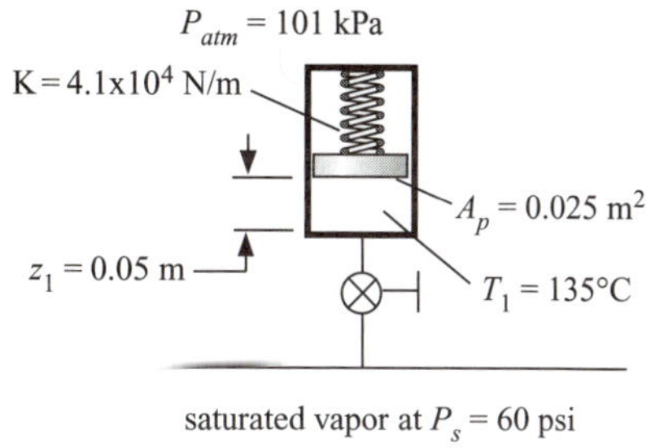

Figure 4.B-4: Spring-loaded accumulator.

a) Determine the heat transfer from the accumulator during the process.
b) Use EES to construct a *T-v* diagram and overlay your states onto this diagram.

4.B-5 Air at $T_1 = 20°$C and $P_1 = 100$ kPa enters a compressor with a mass flow rate of $\dot{m} = 0.025$ kg/s through a circular inlet pipe having an inner diameter of $D_1 = 1$ cm. The compressor operates at steady state. The mechanical power input to the compressor is $\dot{W}_c = 3.5$ kW. Air exits the compressor at $T_2 = 50°$C and $P_2 = 650$ kPa. The diameter of the exit pipe is large and therefore the velocity of the air leaving the compressor is small and its kinetic energy negligible. However, the kinetic energy of the air entering the compressor is not negligible. The outlet of the compressor is connected to a rigid storage tank having a volume of $V_{tank} = 1.5$ m^3, as shown in Figure 4.B-5. The tank initially contains air at $P_{ini} = 100$ kPa. The pressure of the air within the tank rises as it is filled, but heat transfer between the tank and the surroundings keeps the temperature of the air in the tank always at $T_{tank} = 25°$C. This compressor is not adiabatic.

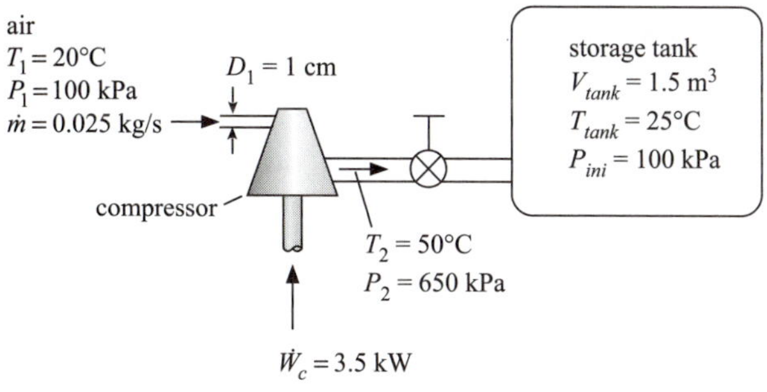

Figure 4.B-5: Compressor connected to a tank.

Assume that the air obeys the ideal law with $R = 287$ J/kg-K. Assume that the specific heat capacities of air are constant and equal to $c_v = 717$ J/kg-K and $c_P = 1005$ J/kg-K. State and justify any other assumptions that you employ.

a) Determine the velocity of the air entering the compressor (m/s).
b) Determine the rate of heat transfer to the compressor (W).
c) Determine the mass of air in the tank after $t = 200$ sec of operation.
d) Determine the pressure of the air in the tank after 200 sec of operation.
e) Determine the total heat transfer from the tank to the room during 200 sec of operation.

4.B-7 A rigid tank contains both air and a block of copper, as shown in Figure 4.B-7. The internal volume of the tank is $V = 1\ \text{m}^3$. The initial pressure of the air in the tank is $P_1 = 5000$ Pa and the initial temperature of both the air and the copper block is $T_1 = 200$ K.

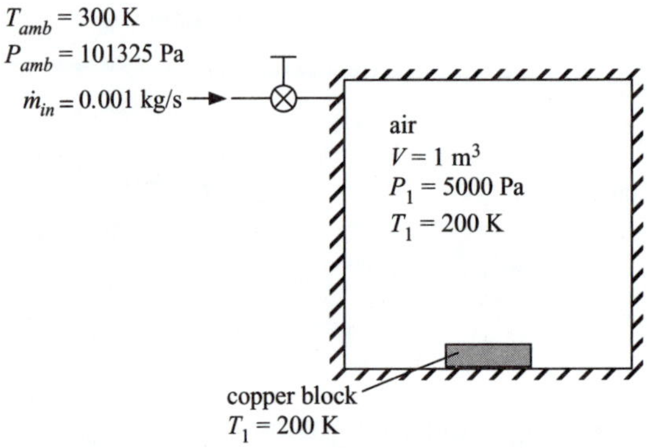

Figure 4.B-7: A rigid tank containing air and a block of copper.

At some time the tank valve starts to leak. Air from ambient conditions, $T_{amb} =$ 300 K and $P_{amb} = 101325$ Pa, flows through the valve and enters the tank. The valve leaks at a constant mass flow rate $\dot{m}_{in} = 0.001$ kg/s for $time = 15$ minutes. After 15 minutes, the leak stops and the temperature of both the air in the tank and the block of copper is $T_2 = 280$ K. You may assume that the contents of the tank are perfectly insulated from the outside environment and the tank walls (i.e., it is adiabatic). Model air as an ideal gas with $R = 287$ J/kg-K and constant specific heat capacities $c_v = 717$ J/kg-K and $c_P = 1004$ J/kg-K. Model copper as an incompressible substance with constant specific heat capacity $c_b = 370$ J/kg-K.

a) What is the mass of air that is initially in the tank?
b) What is the pressure in the tank after 15 minutes?
c) What is the heat transfer from the air to the copper block during the leak process?
d) What is the mass of the copper block?
e) What is the temperature of the air leaving the valve and entering the tank? Justify your answer.

4.B-8 A letter from a former student reads:

> Dear Professor: I'm embarrassed to admit it, but 8 years away from doing any calculations remotely complicated has left me depleted of my integral calculus skills. I remember doing a similar but more complicated calculation in your thermodynamics class at some point, but I can't, for the life of me, remember how to set it up or solve it and I can't find it in the notes I saved for reference. I don't want to waste your time but if you could help a helpless engineer out when you get a chance, I'd be grateful.

> I've got a leaking medical gas (oxygen) system that I'm trying to figure out the leakage rate of. The system has a total internal volume of 3.9 ft^3. The atmospheric pressure is 12.5 psia. The initial system pressure is 62.5 psia and the final system pressure (10 minutes later) is 46.5 psia. I'm assuming that the average temperature of the oxygen in the tank stayed nearly constant at 70°F on the test day, but I did not take any temperature measurements. I'm trying to figure out what the oxygen leakage rate (in scfm) is when the system pressure is 62.5 psia. I know that non-compressible flow rate through a fixed orifice equation usually takes the form of $\dot{V} = C_D\sqrt{\Delta P}$ but I think that's throwing me off here. I was hoping to even brush up on my analysis skills by working this out in the EES environment, but I can't even get an integrand form figured out. Pathetic, I know. Do you think that it matters if oxygen temperature does not remain at 70°F? Can you help me get started?

Let's try to help him out.

a) Determine the discharge coefficient (C_D) and oxygen leakage rate in scfm (standard cubic feet per minute) at 62.5 psia assuming that the tank contents are isothermal.
b) Repeat your calculations assuming that the tank contents are adiabatic.

4.B-9 Liquids are often transferred from a tank by pressurization with a gas. Consider the situation shown in Figure 4.B-9. Tanks A and B both contain refrigerant R134a. The volumes of tanks A and B are $V_A = 0.10$ m^3 and $V_B = 0.075$ m^3, respectively. The fluid in tank A is initially saturated vapor at $T_A = 30$°C. The fluid in tank B is initially at $T_{B,1} = -40$°C with a quality of $x_{B,1} = 0.01$.

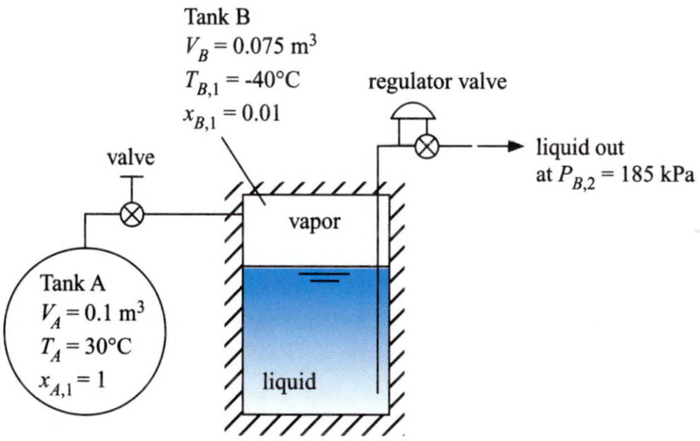

Figure 4.B-9: Liquid transfer equipment.

The valve between tanks A and B is opened, allowing refrigerant to flow from tank A to tank B. When the pressure in tank B exceeds $P_{B,2} = 185$ kPa, the regulator valve opens allowing liquid to flow from tank B to its destination. The fluid in tank A remains at T_A during the entire process by heat transfer to/from the fluid, as needed. Tank B is well-insulated. This process continues until the pressure in tank A has dropped to $P_{A,2} = 185$ kPa. Using engineering principles and judgment, answer the following questions. Assume that the heat capacity of the tanks is negligible. State any other assumptions you employ.

a) What is the initial mass and pressure of the refrigerant in tanks A and B?
b) What is the mass of refrigerant that flows from tank A to tank B?
c) What is the heat transfer required to maintain the contents of tank A at T_A?
d) What is the mass of liquid refrigerant that is transferred from tank B through the regulator valve?

4.B-10 Liquid nitrogen is stored in a spherical Dewar tank that has an inner diameter of $D = 0.7$ m. The tank is initially filled so that 50% of the volume of the Dewar is liquid. The tank is equipped with a pressure regulator that is set to $P_r = 250$ kPa. Vapor escapes from the tank through the regulator valve as needed to maintain this pressure in the tank. The liquid nitrogen is stored for $time$ = 1 week (168 hours). At the end of this period, 42% of the volume in the Dewar is liquid. Determine the rate of heat transfer to the nitrogen.

4.B-11 You want to build a potato gun. You've spent some time researching this subject and found that potato guns fall into two main categories: those that use compressed air and those that use lighter fluid. Being a super-geek you decide that yours should use both methods. A schematic of your design is shown in Figure 4.B-11(a).

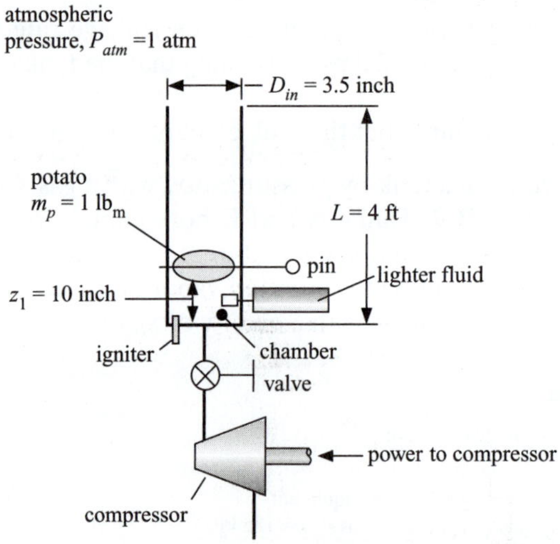

Figure 4.B-11(a): The super-potato-gun.

The pipe that you've located has an inner diameter $D_{in} = 3.5$ inch and length $L =$ 4 ft. The potato is jammed into the barrel so that it sits a distance $z_1 = 10$ inch from the bottom of the pipe, creating a trapped volume (called the chamber) that will be filled with pressurized air. The potato is held in place by a pin that is eventually removed in order to initiate the launch process. The potato has mass $m_p = 1$ lb$_m$. You have attached a source of lighter fluid and an igniter to the chamber and also located a small compressor in order to charge the chamber with compressed air. At state 1, the chamber volume is initially filled with ambient air at $P_{atm} = 100$ kPa and $T_{amb} = 20°$C. The compressor is activated and the charging process takes the chamber from state 1 to state 2 as shown in Figure 4.B-11(b).

The compressor provides a volumetric flow rate at its suction port (i.e., the inlet port) $\dot{V}_{suction} = 2.5$ liter/min. The compressor draws air from the environment at P_{atm} and T_{amb} into the suction port and compresses it to $P_{discharge} = 950$ kPa. The temperature of the air leaving the compressor is $T_{discharge} = 85°$C. The air leaving the compressor passes through a valve and enters the chamber. The valve is shut when the chamber reaches a pressure $P_2 = P_{charge} = 750$ kPa. The temperature of the air in the chamber at this time is $T_2 = T_{charge} = 102°$C. The compressor operates at steady state and is adiabatic. Model air as an ideal gas but do not assume that it has a constant specific heat capacity (i.e., use the substance 'Air' in EES).

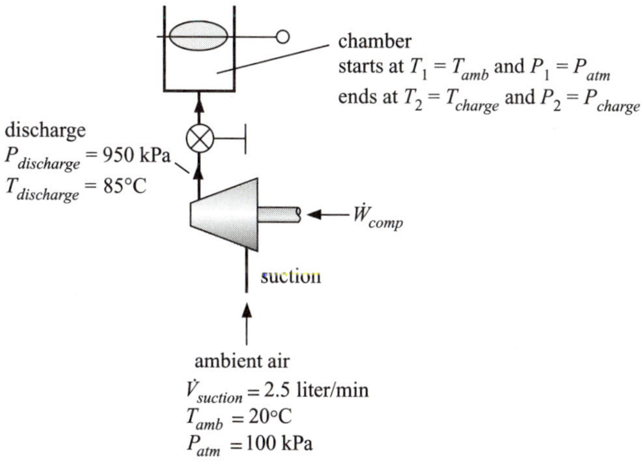

Figure 4.B-11(b): The charging process.

a) What is the mass flow rate provided by the compressor?
b) What is the volumetric flow rate of air leaving the compressor (in liter/min)?
c) What is the power that is required by the compressor?
d) How long must the compressor operate in order to charge the chamber?
e) What is the amount of heat transferred from the chamber to the surroundings during the charging process?

The next phase of the process is combustion, as shown in Figure 4.B-11(c). The combustion process takes the chamber from state 2 to state 3.

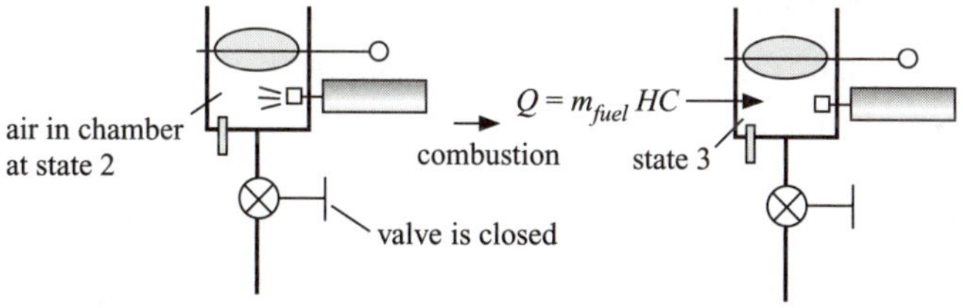

Figure 4.B-11(c): The combustion process.

Lighter fluid is injected into the chamber air and the air/fuel mixture is ignited. The mass of lighter fluid that is injected is $m_{fuel} = 0.05$ g. Model the combustion process as a heat transfer to the air in the chamber; the heat transfer per mass of fuel is referred to as the heat of combustion of the fuel. For lighter fluid, the heat of combustion is $HC = 46$ MJ/kg. Assume that the combustion process is so fast that no heat is transferred to the surroundings. You can model the combustion products as pure air and neglect the small amount of mass associated with the fuel.

f) What is the temperature and pressure of the air in the chamber at the conclusion of the combustion process?

The next phase of the process is launch, as shown in Figure 4.B-11(d). The pin is pulled and the potato is pushed from its initial position to the end of the barrel. The launch process takes the chamber from state 3 to state 4 (where the potato is just about to exit the gun).

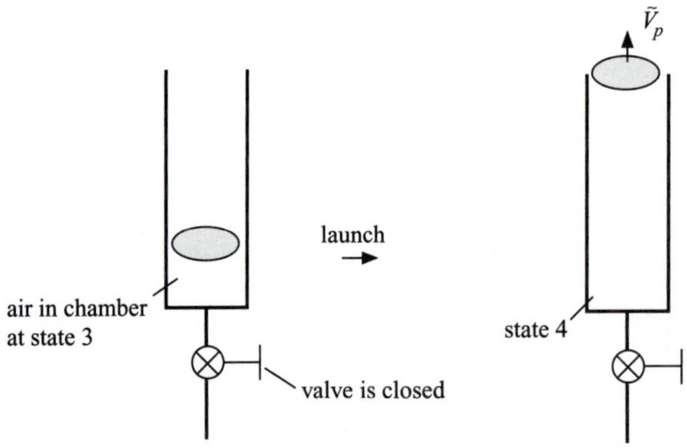

Figure 4.B-11(d): The launch process.

The launch process occurs so quickly that the heat transferred to the surroundings can be neglected. Further, assume that the air undergoes a polytropic expansion process during the launch process where the pressure and volume of the air in the chamber are related according to:

$$P\,V^n = C$$

where n =1.4 is the polytropic exponent and C is a constant.

g) Determine the work done by the air in the chamber on the potato during the launch process.
h) Determine the temperature and pressure of the air in the chamber at the conclusion of the launch process.
i) Determine the velocity that the potato has as it exits the barrel of the gun ($\tilde{V}_p$).
j) Estimate the height that the potato will reach after it is launched. Neglect air resistance for this calculation.
k) Plot the height that the potato will reach as a function of the initial position of the pin (z_1). You should see that there is an optimal position to place the potato. Explain why this is so.
l) Overlay on your plot from (k) the height as a function of pin position that you would get if you did not use lighter fluid (i.e., a compressed air potato gun).
m) Overlay on your plot from (k) the height as a function of pin position that you would get if you did not use compressed air (i.e., a lighter-fluid potato gun).

4.B-16 An engineering firm has designed an elevator that operates between two floors and is powered by compressed air from a supply line, as shown in Figure 4.B-16 The line temperature and pressure are maintained constant at $T_s = 25°\text{C}$ and $P_s = 2.8$ bar. The cross-sectional area of the platform is $A_c = 4.5\ \text{m}^2$. When the elevator is on the ground floor, the platform rests on the bottom of the cylinder (as shown). The mass of the elevator platform (without the load) is such that it is lifted when the pressure in the cylinder is $\Delta P_p = 4.2$ kPa greater than the atmospheric pressure ($P_{atm} = 100$ kPa). During normal operation, an $m = 5{,}000$ kg load is placed on the platform when the elevator is at ground level. The valve is opened and the elevator platform slowly rises the entire distance ($z = 3$ m) between floors. The load is removed and a valve is opened that expels the air from the cylinder, allowing the elevator platform to return to the ground floor. Assume that the device is well insulated, the platform is frictionless, and the air behaves as an ideal gas.

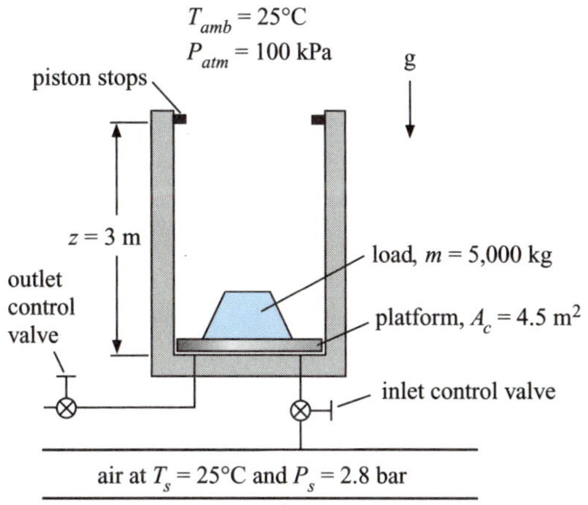

Figure 4.B-16: Elevator system.

a) What is the mass of the elevator platform?
b) Determine the total work done by the air in the cylinder during the process of raising the platform and the load.
c) Assume that the time required for the elevator to rise is short and therefore heat transfer from the air during the process is negligible. What is the temperature of the air in the cylinder when the elevator reaches elevation z (just before touching the stops)?
d) The inlet control valve is left open after the elevator platform reaches the stops at elevation z. Assuming that the process is adiabatic, what will the temperature of the air in the cylinder be when the cylinder air pressure equals the line pressure (2.8 bar)?

C: Advanced Problems

4.C-1 The specific heat ratio, $k = c_P/c_v$, can be easily determined for gases that obey the ideal gas law using the following simple experiment. A large bottle fitted with a cork stopper is filled with the gas of interest to a pressure that is slightly above atmospheric pressure. The gas is allowed to attain room temperature and the pressure in the bottle and atmospheric pressure are both carefully measured. Then the cork is removed and quickly replaced. After sufficient time has elapsed for the gas to again attain room temperature, the new pressure in the bottle is measured.

a) Show how the value of k can be determined from this experiment. Assume that c_P and c_v are both constant and state any other assumptions that you employ. (Note that solution to this problem does not require the use of the Second Law of Thermodynamics.)

4.C-2 An understanding of the dynamic behavior of room temperature and pressure is important in the design of HVAC (Heating, Ventilating, and Air-Conditioning) equipment. Consider the following case. A $W^2 = 20$ m × 20 m operating room in a hospital with a ceiling height of $H = 2.5$ m is initially at $T_1 = 25°C$ and $P_1 = 100$ kPa. Just before occupancy, a supply fan is turned on so that air at $P_{in} = 110$ kPa and $T_{in} = 15°$C is blown into the room at a volumetric rate of $\dot{V} = 2.78$ m^3/sec. Air within the room escapes to the adjoining hallway, which is maintained at $P_{atm} = 100$ kPa. (Note that during an operation, it is important to pressurize the operating room relative to the surroundings so that bioaerosols tend to flow out of the room, reducing the possibility of infection to the patient.) Experiments have shown that the mass flow rate of air from the operating room to the hallway through cracks around the doors and other paths ($\dot{m}$) can be described by a simple orifice equation of the form:

$$\dot{m} = K \, \Delta P^{0.65}$$

where K is a constant found experimentally to be 0.0157 kg/s-Pa$^{0.65}$ and ΔP is the pressure difference between the room and the hallway. The lights and equipment in the room act as a $\dot{Q} = 3.5$ kW thermal energy source. The heat transfer coefficient between the air in the room and the walls is estimated to be $h_{conv} = 15$ W/m^2-K. You may assume the air in the room to be fully mixed at any time and that the air obeys the ideal gas law. State and justify any other assumptions that you employ.

a) Plot the temperature and pressure of the room air as a function of time over a 5 minute period and explain the behavior of these variables.
b) Determine the equilibrium temperature and pressure in the operating room.

4.C-7 The USDA (in Kozempel et. al., *Journal of Food Protection*, Vol. 63, No. 4, 2000) has invented a new method of surface pasteurization called "flash" or "VSV" pasteurization that uses cycles of vacuum–steam–vacuum to reduce bacteria on the surface of prepared meats. A small scale experiment has been designed to test this process. The purpose of this problem is to provide an engineering analysis of this experiment. The cylindrical chamber has a diameter of $D = 240$ mm and a height of $H = 200$ mm and it is made of stainless steel. The chamber is equipped with two ball valves that connect it to a vacuum line and to a steam line. The ball valves are electronically actuated and have a rapid response time when triggered. A single hot dog ($D_h = 2.5$ cm in diameter and $L = 13$ cm length) is placed on a screen located in the center of the test chamber. At the start of the experiment, the chamber contains steam at $T_1 = 160°$C and $P_1 = 1$ atm. The steam is evacuated for a period of $t_e = 0.2$ sec by opening the valve that connects the chamber to the vacuum line. The mass flow rate from the chamber ($\dot{m}$) can be represented by the equation:

$$\dot{m} = C_d \, A_o \, P \sqrt{\frac{k \, MW}{R_{univ} \, T} \left[\frac{2}{k+1}\right]^{\frac{(k+1)}{2(k-1)}}}$$

where $C_d = 0.8$ is the discharge coefficient, A_o is the orifice area (the orifice diameter is $D_o = 1.25$ cm), P and T are the pressure and temperature in the chamber, $k = 1.33$ is the specific heat ratio, R_{univ} is the universal gas constant, and MW is the molar mass of steam.

At the conclusion of the evacuation process, the valve to the vacuum line is closed and the valve that connects the chamber to the steam line is opened. Saturated steam at $T_s = 140°$C is injected into the chamber in order to pasteurize the hot dog surfaces. The steam injection process is stopped when the pressure in the chamber is equal to the pressure in the steam line. The exposure times of the hot dog to the steam are deliberately short to prevent any cooking or changes in the surface properties of the product. Consequently, you may assume that heat transfer between the steam and the hot dog or the chamber surfaces is negligible during the test period.

a) Plot the pressure and temperature of the steam in the chamber as a function of time for the evacuation process. What are the pressure and temperature at the end of the process?
b) Estimate the temperature of the steam in the chamber at the end of the steam injection process (when the pressure in the chamber has reached the pressure of the saturated steam in the line).

4.C-9 A schematic of the primary coolant loops in a pressurized water nuclear reactor is shown in Figure 4.C-9. The coolant is normally maintained at a pressure greater than the saturation pressure corresponding to the maximum coolant temperature in the reactor so that the coolant in the primary loop is always in a subcooled state. However, the near incompressibility of the coolant in the subcooled state introduces a problem. Small changes in volume resulting from a change in the average temperature in the primary coolant loop can result in enormous oscillatory pressure changes. If the average loop temperature increases, the resulting pressure increase can burst the piping, whereas flashing, pump cavitation, and possible burnout of the reactor fuel can result from decreases in the average loop temperature. To avoid these problems, the primary coolant loop is equipped with a vapor pressurizer. The vapor pressurizer is essentially a small boiler in which liquid, the same liquid as used in the primary coolant loop, can be maintained at a controlled temperature by electrical heaters. The pressure in the vapor pressurizer is the same as the pressure in the primary coolant loop. The pressurizer normally contains about 50% liquid by volume. During a positive pressure surge, the vapor above the liquid is compressed and condensation occurs, limiting the pressure rise in the primary loop. During a negative pressure surge, some coolant in the pressurizer flashes to vapor and the electrical heaters are actuated; this mitigates the pressure reduction in the primary loop.

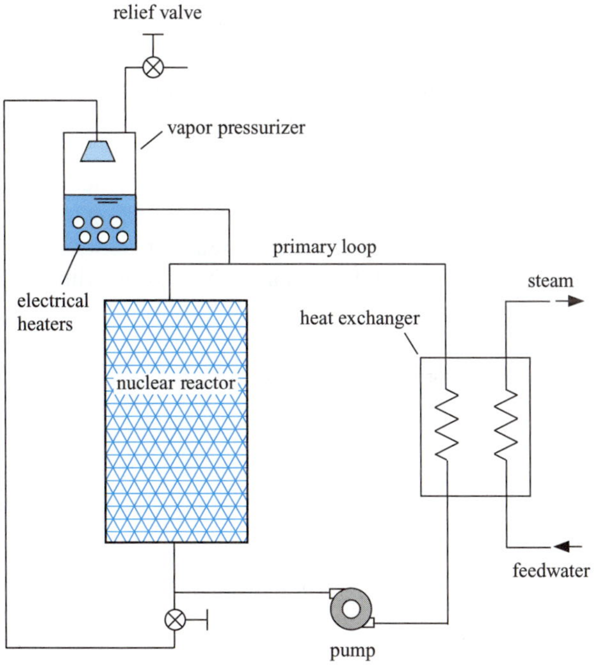

Figure 4.C-9: Schematic of primary loop in a nuclear reactor with a vapor pressurizer.

In a particular operating condition, the primary coolant contains water at $P_1 =$ 155 bar. The volume of the primary loop is $V_{pl} = 225\ m^3$ and the pressurizer volume is $V_p = 24\ m^3$. The pressurizer liquid volume is 60% of its total volume. Assume operation is steady. The average temperature of the primary coolant is $T_1 = 315°C$. Now suppose something happens in the reactor or heat exchanger such that the average temperature in the primary coolant loop quickly changes by ΔT, where ΔT can be either positive or negative. The changes occur so quickly, that the process is adiabatic.

a) Calculate and plot the electrical energy required and volume fraction of liquid in the vapor pressurizer for $-10\ K < \Delta T < 0\ K$ if the primary loop pressure is to remain constant.
b) Calculate and plot the primary loop pressure and volume fraction of the pressurizer for $0\ K < \Delta T < 10\ K$.
c) A relief valve is attached to the top of the vapor pressurizer in order to protect against pressure surges that are beyond the capacity of the vapor pressurizer. What is your recommendation for the pressure setting at which the relief valve should open?

4.C-10 A small rotary vacuum pump is used to evacuate some laboratory equipment that has an internal volume of $V = 150$ liters. The equipment initially contains air at $P_1 = 1$ atm and $T_1 = 25°C$. The pump produces a suction volumetric flow rate that is constant and equal to $\dot{V}_{in} = 30$ liters/min. The evacuation process would be an adiabatic process except for the heat input of $\dot{Q} = 20$ W from the motor.

a) Prepare a plot of the temperature and pressure of the air remaining inside of the equipment as a function of time for a period of 10 minutes. Explain the behavior of the plots.
b) Determine the lowest pressure that can be attained using this vacuum pump.

4.C-15 A perfectly-insulated cylinder is fitted with a freely-floating piston made of a non-conducting material. To the left of the piston is $m_h = 0.35$ kg of helium that is initially at $T_{h,1} = 25°C$ and $P_1 = 4$ bar. To the right of the piston is air, initially at $T_{a,1} = 25°C$ and $P_1 = 4$ bar. The total volume of the helium and air together is $V = 1.25\ m^3$. The valve is opened allowing air from the pipeline to enter the right side of the cylinder through the valve; this forces the piston to slowly move to the left increasing the pressure on the helium. The valve is closed when the pressure in the cylinder is $P_2 = 8$ bar. The air in the pipeline is maintained at $T_s = 25°C$ and $P_s = 20$ bar. Note that helium and air both behave according to the ideal gas law at these conditions and have constant specific heat capacities. Assume that there is no heat interaction between the gases and the piston-cylinder surfaces. State any other assumptions you employ.

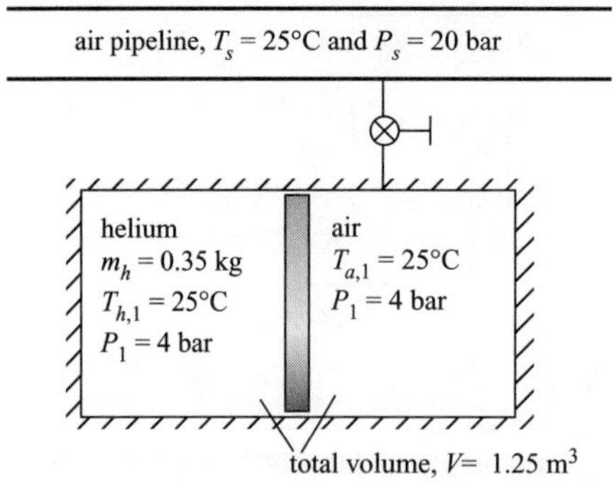

Figure 4.C-15: Air pipeline connected to piston-cylinder device with free-floating piston.

a) What is the volume of air in the cylinder before the valve is opened?
b) What is the temperature of the helium and its volume at the final state?
c) What is the work done on the helium during this process?
d) What is the temperature of the air and its volume at the final state?

4.C-16 A natural gas metering system consists of a chamber having a volume of $V =$ 4.1 liters. This chamber initially contains natural gas (assumed to pure methane) at $P_1 = 34$ bar and the ambient temperature of $T_{amb} = 25°$C. The chamber is connected to a natural gas supply pipeline that carries natural gas at $T_s =$ 34°C and $P_s = 57$ bar. The supply line has an internal diameter of $D_s = 2$ cm. When the ball valve between the supply line and the chamber is opened, natural gas flows into the chamber through an orifice that has a $D_o = 4$ mm diameter. The mass flow rate entering the chamber is given by $\dot{m}_{in} = K\sqrt{P_s - P}$ where the orifice constant has been determined to have a value of $K = 0.0117$ kg/s-bar$^{0.5}$. Simultaneously, natural gas flows out of the chamber with mass flow rate $\dot{m}_{out} = K\sqrt{P - P_d}$ through a $D_d = 2$ cm pipe containing a second orifice having the same diameter and orifice constant (D_o and K). The downstream pressure is maintained at $P_d = 34$ bar. The chamber is made of metal that remains at the ambient temperature for the $t_m = 10$ sec duration of this metering operation. The heat transfer coefficient between the flowing natural gas and the metal is estimated to be $h_{conv} = 26$ W/m^2-K. The total surface area of the chamber is A_s 1.3 m^2.

a) Plot the temperature and pressure of the natural gas in the chamber as a function of time. Note any unusual or non-intuitive behavior in these plots and provide a detailed explanation.

5 The Second Law of Thermodynamics

The Second Law of Thermodynamics is probably the most famous law in all of science with implications that extend beyond the thermal energy systems that are the focus of this textbook. The Second Law dictates the direction in which processes will occur. For thermodynamic systems, the Second Law assigns a quality to energy; it explains why heat transfer always occurs from hot to cold and it places an upper limit on the efficiency of energy conversion processes. More generally, the Second Law explains why we age, why things break, and why time travel cannot happen. An elegant explanation of the observations that we attribute to the Second Law is provided by statistical thermodynamics, as described by Boltzmann in the latter part of the 19th century. Starting with a model of the energy storage and interactions associated with elementary particles of matter (molecules), statistical thermodynamics dictates that a process will occur in the direction that is most probable. Statistical thermodynamics is discussed more completely in Chapter 15. However, it is not necessary to adopt a molecular viewpoint in order to explain the Second Law. The first formal statement of the Second Law, which is attributed to Carnot in 1824, is not based upon statistical thermodynamics at all but rather on macroscopic observations about the efficiency with which heat can be converted into work.

There are many ways to introduce the Second Law of Thermodynamics to engineers. Some textbooks choose to introduce the concepts and implications in the same order that they were discovered, starting with Carnot's theorem. Others prefer a more general approach in which the thermodynamic property *entropy* is introduced immediately. All of the concepts and implications associated with the Second Law can be explained by the generation of entropy. There is no right or wrong way to present the Second Law since all of its statements are equivalent and self-consistent. In this chapter, the Second Law will be introduced by generalizing the observation that heat flows spontaneously from a system of higher temperature to a system of lower temperature, a statement that is usually attributed to Rudolf Clausius in 1850. Clausius later introduced the concept of entropy, which is the focus of Chapter 6.

5.1 The Second Law of Thermodynamics

One of the major developments that eventually led to the First Law of Thermodynamics is the recognition that heat and work are both energy transfer processes. The equivalence of heat and work, from a First Law standpoint, is evident by noting that the same change in the state of a system can be accomplished using either heat or work. A relatively simple experiment that illustrates this point is shown in Figure 5-1. Figure 5-1(a) shows a heating process for a fluid that is maintained at constant volume. The temperature and pressure of the fluid increase as the heating process progresses. An energy balance for this process results in:

$$Q_{in} = m\,(u_2 - u_1) \tag{5-1}$$

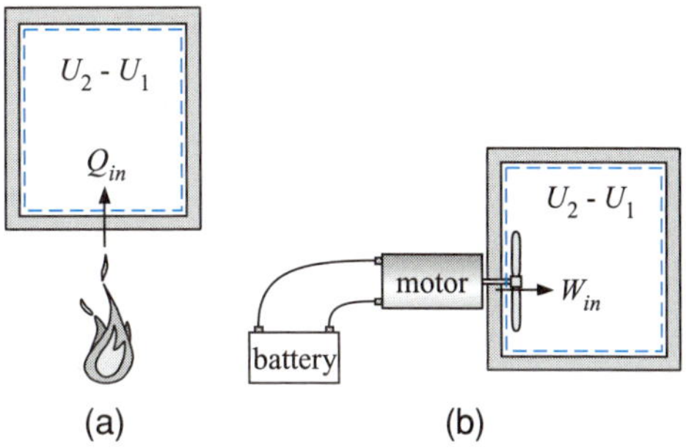

Figure 5-1: Changing the state of a fluid (a) with heat and (b) with work.

where m is the mass of fluid. Knowledge of the specific volume (which does not change) and specific internal energy at the end of the process fixes state 2 and therefore the final temperature and pressure of the fluid can be determined. Figure 5-1(b) shows a process in which work is transferred to the same system using a paddle wheel that is turned by a motor. Again, the temperature and pressure of the fluid are observed to increase as the process proceeds. An energy balance for this process results in:

$$W_{in} = m\,(u_2 - u_1) \tag{5-2}$$

Provided that the amount of heat and work are the same for these two processes, the final state of the fluid will be identical. Equations (5-1) and (5-2) show that the final specific internal energy is the same and the specific volume of the fluid does not change. From a First Law standpoint, heat and work are equivalent.

Joule conducted experiments similar to the one shown in Figure 5-1 during the 1840's. His experiments demonstrated an equivalence of heat and work that is remarkably close to the currently accepted value of 778 ft-lb_f = 1 Btu (British Thermal Unit); a Btu is defined as the amount of thermal energy required to increase the temperature of 1 lb_m of water by 1°F. These experiments challenged the "caloric theory" that had previously been accepted, which postulated that heat (caloric) was an invisible substance that transferred from one system to another as a result of a temperature difference. How could "caloric" have been transferred to the fluid through the paddle wheel when the paddle wheel was at the same temperature as the fluid?

With the understanding that heat and work are both forms of energy transfer, it is reasonable to ask why we continue to distinguish between them. Heat (Q) is defined as an energy transfer across a system boundary as a result of a temperature difference. Work (W) is defined as an energy transfer across a system boundary as a result of a difference in any property other than temperature. Heat and work are treated equally by the First Law. Why bother to make the distinction in the first place?

The answer to this question is the essence of the Second Law of Thermodynamics. Heat and work are actually fundamentally different and this difference becomes evident if we try to run the experiments shown in Figure 5-1 in reverse. The thermal energy provided to the system during the process shown in Figure 5-1(a) can be completely recovered when the process is reversed. That is, if the system changes from state 2 to state 1 then an equal and opposite heat transfer will occur from the fluid to the surroundings. However, we cannot recover all of the work that was provided to the system during the process shown in Figure 5-1(b) if the process is reversed. In fact, the process shown in Figure 5-1(b) cannot be reversed (it is "irreversible") even though the First Law of Thermodynamics does not disallow the reverse process. The Second Law of Thermodynamics formally deals with irreversible processes and thereby dictates the direction in which processes can occur.

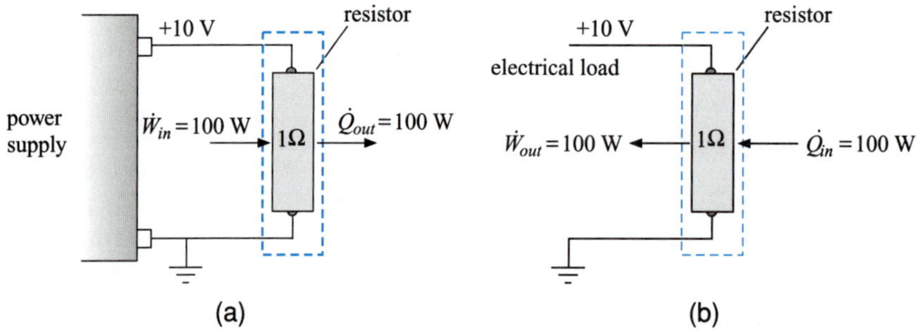

Figure 5-2: Electrical resistor (a) converting electrical work into heat and (b) converting heat into electrical work.

There are many processes that are allowed by the First Law of Thermodynamics but cannot actually happen in nature. For example, consider the $R = 1\ \Omega$ electrical resistor shown in Figure 5-2(a). The resistor is connected to a power supply and an electrical voltage of $E = 10$ V is applied across its leads. The current that flows through the resistor is:

$$i = \frac{E}{R} = \frac{10\,\text{V}}{}\bigg|\frac{}{1\,\Omega} = 10\,\text{A} \tag{5-3}$$

The rate of electrical work provided to the resistor is:

$$\dot{W}_{in} = E\,i = \frac{10\,\text{V}}{}\bigg|\frac{10\,\text{A}}{} = 100\,\text{W} \tag{5-4}$$

At steady state, the electrical power provided to the resistor must be transferred from the resistor as heat. The energy balance on the resistor is shown in Figure 5-2(a) provides:

$$\dot{W}_{in} = \dot{Q}_{out} = 100\,\text{W} \tag{5-5}$$

This heat transfer can only occur if the resistor temperature increases relative to the surroundings. The resistor will become hot to the touch and may actually be destroyed depending on the power rating of the device. The process shown in Figure 5-2(a) in which electrical work is converted into heat is a familiar one to most of us. The experiment is easy to set up and you can verify that indeed, the resistor becomes hot as the electrical power is converted to heat transfer.

The reverse of the process shown in Figure 5-2(a) is shown in Figure 5-2(b). A source of heat (e.g., a blow torch or a hair dryer) is used to provide $\dot{Q}_{in} = 100$ W of heat transfer to the resistor. In Figure 5-2(b), this heat transfer is converted to electrical power at a rate that is defined by an energy balance:

$$\dot{Q}_{in} = \dot{W}_{out} = 100\,\text{W} \tag{5-6}$$

Any attempt to set up the experiment shown in Figure 5-2(b) will be disappointing as it is not possible to get 100 W of electrical power out of the resistor by applying heat. In fact, there will be no voltage or current generated at all. Instead, you will get a hot resistor.

Why is it that the process shown in Figure 5-2(a) can occur but the one shown in Figure 5-2(b) cannot? Neither process violates the First Law of Thermodynamics. However, the process shown in Figure 5-2(a) is *irreversible* – it cannot be run in reverse and therefore the process shown in Figure 5-2(b) is not possible. How do we know this? Common sense and physical intuition are helpful since we've seen one process and not the other. More formally, the Second Law of Thermodynamics dictates the directionality of processes.

5.1.1 Second Law Statements

Our observation that the process shown in Figure 5-2(b) does not occur is, in fact, an example of the Kelvin-Planck statement of the Second Law that was developed in the latter part of the 19th century.

> *It is impossible to build an engine that will operate in a cycle (i.e., continuously) that will provide no effect except the raising of a weight and the cooling of a single thermal reservoir.*

There are other statements of the Second Law. The earliest Second Law statement is attributed to Carnot, who indicated in his 1824 Ph.D. thesis that:

> *Heat cannot be converted completely and continuously into work.*

The Second Law statement attributed to Clausius in 1850 is:

> *Heat cannot pass spontaneously from a body of lower temperature to a body of higher temperature.*

These (and other) statements of the Second Law are equivalent in that none of them can be independently proven. However, if we accept any one of these statements to be true then it is possible to prove any of the others. In this chapter, we will accept the Clausius statement to be the Second Law of Thermodynamics. Heat transfer is directly measurable using calorimetric techniques and temperatures are also measurable. Therefore, the Clausius statement of the Second Law can be directly tested. In addition, the Clausius statement is intuitive and understandable to most people even without the benefit of any formal thermodynamics training. In Chapter 6, we will generalize the Second Law and make it more directly useful for engineering problems by expressing the Second Law in terms of the property entropy.

In the remainder of this section, we will show that both the Kelvin-Planck and the Carnot statements of the Second Law can be proven from the Clausius statement. In order to do this, it is necessary to clarify two of the terms that are found in these alternative Second Law statements: *continuous operation* and a *thermal reservoir.*

5.1.2 Continuous Operation

Both the Carnot and Kelvin-Planck statements of the Second Law refer to continuous operation. The Kelvin-Planck statement involves the word *cycle*. A cyclic process is one in which the state of the system is periodically restored to its initial condition. All energy conversion processes that are designed to operate continuously must operate in a cycle and it is the cyclic or continuous performance of the process that is governed by the Carnot and Kelvin-Planck statements of the Second Law.

Consider the work-producing process shown in Figure 5-3. A gas that obeys the ideal gas law is contained in a piston-cylinder device. The gas is slowly heated, which results in a small increase in its temperature. However, as the temperature rises, its pressure

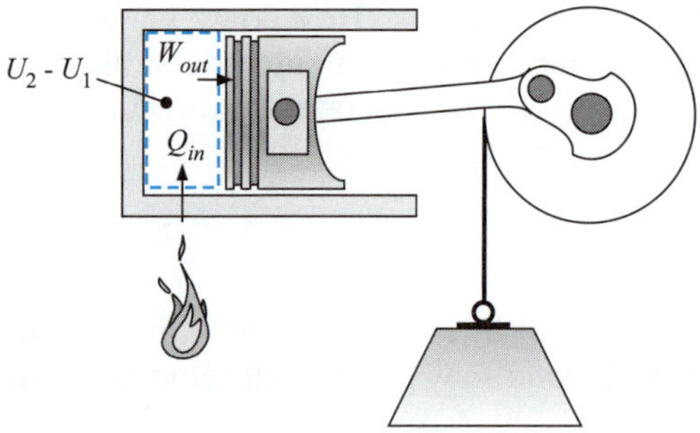

Figure 5-3: A non-continuous process that converts heat to work with 100% efficiency.

also rises and the increased pressure acts (through some mechanical linkage) to raise a weight, thereby doing work. The work done by the gas is an energy transfer out of the system that results in a reduction in its temperature. The heat and work processes progress in this way so that the temperature of the gas remains constant during the process (i.e., $T_2 = T_1$).

An energy balance on the closed system that encompasses the ideal gas in the cylinder is shown in Figure 5-3 and leads to:

$$Q_{in} = W_{out} + U_2 - U_1 \tag{5-7}$$

The total internal energy is expressed as the product of the mass and the specific internal energy:

$$Q_{in} = W_{out} + m(u_2 - u_1) \tag{5-8}$$

The specific internal energy of an ideal gas depends only on its temperature. Since the temperature does not change during this process, the change in the specific internal energy of the gas is zero and Eq. (5-8) becomes:

$$Q_{in} = W_{out} \tag{5-9}$$

The net effect of this process is that the heat provided to the system is converted with 100% efficiency into work. It would appear that this process violates the Carnot statement of the Second Law, which indicates that heat cannot be continuously and completely converted into work. In fact this process does not violate the Second Law; it is a physically possible process that could be approximately accomplished with the correct set of components. It is possible to convert heat into work with 100% efficiency. It is not possible, however, for this process to operate continuously. You will notice that the system does not undergo a cycle; the initial and final states of the ideal gas in the cylinder are not the same. Sooner or later, the enclosed gas will occupy such a large volume that it can no longer be contained in the cylinder and the process can no longer continue. In order for a process to operate continuously, the state of the system must periodically be restored to its initial condition. The Second Law statements of Kelvin-Planck and Carnot apply to processes that operate continuously.

5.1.3 Thermal Reservoir

The Kelvin-Planck statement of the Second Law refers to a thermal reservoir. A thermal reservoir is defined as a system that can provide any amount of heat without its

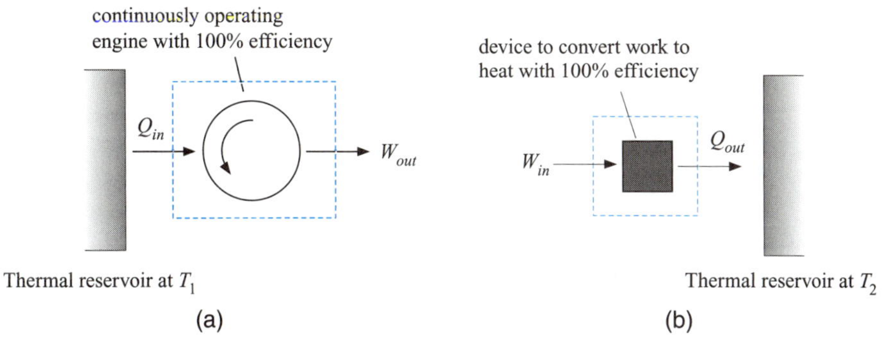

Figure 5-4: (a) Process in which heat from a thermal reservoir at temperature T_1 is converted into work with 100% efficiency by a continuously operating engine. (b) The work is subsequently converted to heat and delivered to a thermal reservoir at $T_2 > T_1$. The process shown in (a) violates the Second Law of Thermodynamics.

temperature changing. A thermal reservoir is a theoretical construct, but it is one that can be closely approached in practice. For example, the environment can often be considered to be a thermal reservoir. The temperature of the environment is unaffected by the energy transfer processes from our energy systems because the environment is massive compared to the magnitude of these energy transfer processes.

5.1.4 Equivalence of the Second Law Statements

The Carnot statement of the Second Law precludes the possibility of heat being continuously converted into work with 100% efficiency. The Clausius statement prohibits spontaneous heat flow from a system at low temperature to a system at higher temperature. The following thought experiment shows that these statements are consistent. Consider the process shown in Figure 5-4(a) in which a heat transfer, Q_{in}, is obtained from a thermal reservoir at temperature T_1. The heat is transferred to an engine that operates continuously (i.e., in a cycle) and produces work, W_{out}, with 100% efficiency. The *efficiency* (η) of a system is defined in general as the ratio of the desired output (what you want) to the required input (what you pay for). The efficiency of an engine is therefore the ratio of the work provided (which is the desired output) to the heat required (which is the required input).

$$\eta = \frac{W_{out}}{Q_{in}} \tag{5-10}$$

The work produced in the process shown in Figure 5-4(a) can be converted to other forms of work or to heat with 100% efficiency. We know from experience that work in one form (e.g., mechanical) can be converted to another form of work (e.g., electrical); thermodynamics places no upper limit on the efficiency of these processes. We also know that work can be converted with 100% efficiency into heat; the resistor studied in Figure 5-2(a) is an example of this conversion process. Figure 5-4(b) illustrates the conversion of the work produced by the engine to heat that is then transferred to a second thermal reservoir at temperature T_2. There is no restriction on the temperature of this reservoir. By properly insulating the device shown in Figure 5-4(b), it is possible to reach extremely high temperatures. Therefore, T_2 could be higher than T_1. In this case, the net effect of the processes shown in Figure 5-4 is the transfer of heat from the lower temperature thermal reservoir (at T_1) to the higher temperature thermal reservoir (at T_2) with no

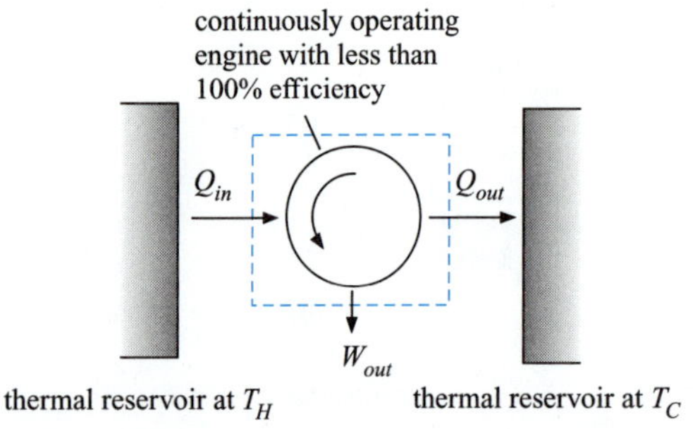

Figure 5-5: Continuous conversion of heat into work with an efficiency that is less than 100%.

other effects. This process is a violation of the Clausius statement of the Second Law. We know that the process of converting work to heat with 100% efficiency, shown in Figure 5-4(b), is possible and it does not violate the Second Law. As a result, the Second Law violation must occur when heat is continuously converted to work with 100% efficiency in Figure 5-4(a). The Carnot statement of the Second Law prohibits the continuous and complete conversion of heat into work and we see that such a process also results in a violation of the Clausius statement of the Second Law.

The Second Law does not prohibit the continuous conversion of heat into work. However, the Second Law does place an upper bound on the efficiency of this process. It is not possible to convert all of the heat that is provided to an engine into work in a continuous or cyclic process. Consequently, part of the thermal energy that is provided must ultimately be rejected as heat to another thermal reservoir. Figure 5-5 shows an engine that receives heat, Q_{in}, from a thermal reservoir at temperature T_H. The engine continuously produces some work, W_{out}, but according to the Carnot statement of the Second Law, the efficiency of the engine must be less than unity:

$$\eta = \frac{W_{out}}{Q_{in}} < 1 \tag{5-11}$$

An energy balance on the system is shown in Figure 5-5:

$$Q_{in} = \underbrace{W_{out}}_{<Q_{in}} + Q_{out} \tag{5-12}$$

Equation (5-12) requires that some heat, Q_{out}, be rejected from the engine.

Where does the rejected heat go? The short answer to this question is: somewhere else. However, the Clausius statement of the Second Law requires that the heat must flow to a system of lower temperature, which is represented in Figure 5-5 as a thermal reservoir at temperature T_C, where $T_C < T_H$. The Kelvin-Planck statement of the Second Law can be interpreted as requiring a minimum of two thermal reservoirs at different temperatures in order to accomplish any cyclic process in which heat is converted into work. We will see in Section 5.3 that the Second Law provides a very specific upper limit on the efficiency of the engine shown in Figure 5-5.

5.2 Reversible and Irreversible Processes

The First Law of Thermodynamics deals with the quantity of energy and treats all energy as being equivalent. The Second Law of Thermodynamics assigns a quality to energy. For example, work represents a higher quality energy transfer than heat. Work can be converted into heat with 100% efficiency. However, the converse is not true.

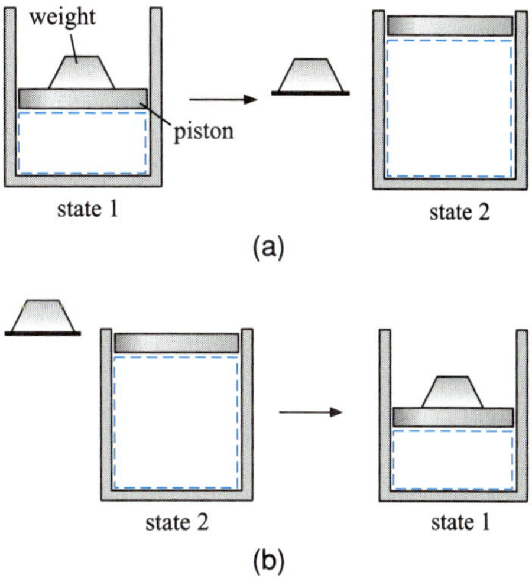

Figure 5-6: Piston-cylinder apparatus undergoing (a) an unconstrained expansion and (b) an unconstrained compression.

Therefore, whenever work is converted into heat, the quality of the energy is reduced. A thermodynamically ideal process never uses work to accomplish a task that could have been accomplished using heat. Therefore the quality of the energy is not degraded and such a process is *reversible*.

A thermodynamically reversible process is defined as a process which, once having taken place, can be reversed in order to restore the system and the surroundings to their original states without violating any of the laws of thermodynamics. A reversible process is a theoretical construct – it cannot actually occur and it is seldom even approached in reality. Thermodynamics is not directly concerned with rate mechanisms; however, it is evident that a process would have to progress at an infinitely slow rate in order to be reversible. Processes that occur at infinitely slow rates are not of practical interest. The primary purpose for studying reversible processes is that they provide an absolute upper bound on the efficiency of energy conversion processes.

It is not easy to recognize whether a process is reversible or not. For example, consider the piston cylinder apparatus in Figure 5-6. A system is defined that consists of the gas that is contained in the cylinder. Initially, the system is in equilibrium at state 1 where the gas is at the temperature of the surroundings and at a pressure that supports both the piston and a weight that is placed on the piston. The weight is suddenly removed and the resulting force imbalance on the piston causes it to accelerate upwards and rise to a new position. After some time, the system reaches equilibrium state 2 where the gas is again at the temperature of the surroundings and a lower pressure that just supports the piston, as shown in Figure 5-6(a). In Figure 5-6(b), the weight is placed back onto the piston causing it to move downwards. Eventually, the system returns to state 1 where the gas is at the temperature of the surroundings and the pressure is at the original pressure that is required to support the piston and the weight. Are the processes shown in Figure 5-6 reversible?

In order for the processes to be reversible, both the system and the surroundings must return to their original states. Clearly, the system has returned to its original condition; it starts and ends at state 1. It is less obvious, but the surroundings does not return to its original state. In order to accomplish the second process, returning from state 2 to state 1, the weight had to be lifted from its initial height, corresponding to the position of the piston at state 1, up to the height corresponding to the position of the piston at state 2, as evident by observing the position of the weight in Figure 5-6(b) as compared to its position in Figure 5-6(a). Work is required to lift the weight and that work had

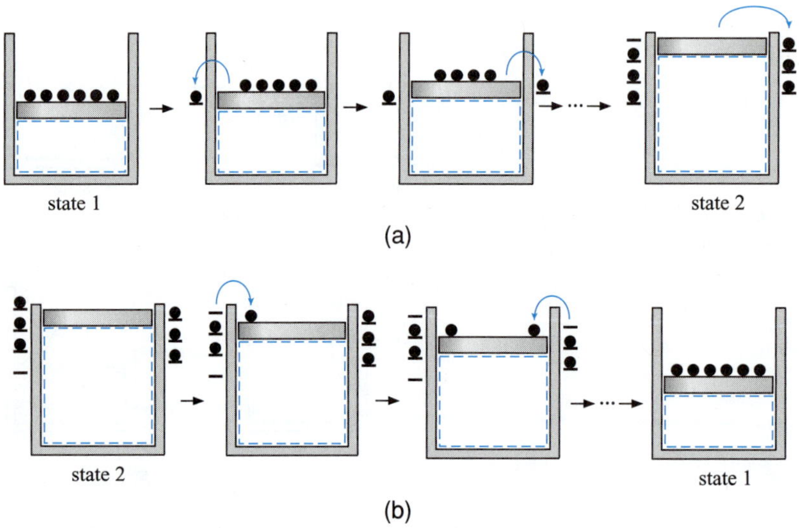

Figure 5-7: Piston-cylinder apparatus undergoing (a) a slow and incremental expansion and (b) a slow and incremental compression.

to be provided by the surroundings. The surroundings therefore do not end at exactly the same condition that they started at and the processes shown in Figure 5-6 are not reversible.

How could we accomplish the process in a reversible manner? Suppose that the weight resting on the piston in state 1 is replaced with millions of grains of sand that collectively have the same total mass as the weight, as shown in Figure 5-7(a) (note that only six of the millions of grains are shown).

A grain of sand is removed from the piston and stored in the surroundings at the current elevation of the piston. The reduction of the mass on the piston causes the piston and all of the remaining grains of sand to rise slightly to a new equilibrium state. A second grain of sand is removed from the pile and carefully stored in the surroundings at the new elevation of the piston. This process is repeated, one grain at a time, until all of the sand has been removed and the system has achieved state 2, as shown in Figure 5-7(a). To reverse the process, the very first grain of sand that was removed is lifted and placed on the piston at state 2 causing it to descend very slightly so that it reaches the elevation at which the last grain of sand that was removed is stored. This grain is placed on the piston causing it to move down to the elevation of the next grain of sand. We can continue to move the sand grains one-by-one from their stored positions in the surroundings onto the piston. Each grain of sand that we place on the piston causes it to descend slightly allowing the next grain of sand to be placed on the piston without having to change its elevation. Finally, all of the grains of sand have been replaced and the original state of the system has been restored, as shown in Figure 5-7(b). This process is very nearly reversible because the surroundings must only provide sufficient work to lift one grain of sand from the height of the piston at state 1 to its height at state 2. However, the reversible process would clearly take a long time to complete. This is a characteristic of all reversible processes; they occur at an infinitely slow rate.

It is usually easier to identify an irreversible process than a reversible process. An irreversible process is one in which there is an unconstrained equalization (i.e., equilibration) of the difference in any potential. For example, any process that exhibits one or more of the phenomena listed in Table 5-1 is irreversible. All irreversible processes result in a degradation of the quality of an energy transfer; these processes produce less

Table 5-1: Phenomena that identify an irreversible process.

Phenomena	Unconstrained potential difference
Friction	Force
Unrestrained expansion	Pressure
Heat transfer across a finite temperature difference	Temperature
Current flow across finite voltage	Voltage
Mixing	Chemical potential

work (or require more work) than a reversible process that achieves the same change in state.

Friction is the direct conversion of mechanical work into heat. This process is clearly irreversible since it is not possible to convert the heat back to an equivalent amount of work. Doing so would violate the Kelvin-Planck statement of the Second Law. Friction obviously and intuitively reduces our ability to do work (or it results in additional work being required to accomplish a task, depending on the application) and thereby lowers the quality of energy. Friction can be decreased by reducing the velocity difference between the surfaces that are in contact. If the velocity difference between the surface approaches zero then there will be no friction and the process will become nearly reversible; however, the process will also take forever to occur.

Current flow across a finite voltage results in effects that are similar to friction. In this case, however, it is electrical work rather then mechanical work that is converted into heat. The process is irreversible because the heat cannot be entirely converted back to electrical work.

An example of partially unrestrained expansion and compression are the processes depicted in Figure 5-6(a) and (b), respectively. A force imbalance acts on a mass and causes it to accelerate; this is the situation that occurs when the weight is removed from the piston. The piston rapidly rises and then likely oscillates for some time until friction finally brings it to rest at its new equilibrium position. The quality of the energy transfer from the gas is reduced due to the partially unrestrained expansion process and the ability of the system to produce work is diminished. The only way to avoid this degradation in the quality of energy is to reduce the force imbalance on the piston until it approaches zero, as shown in Figure 5-7. However, the rate at which the process occurs will then also approach zero. The amount of work done by the system on the surroundings during the irreversible expansion in Figure 5-6(a) will be smaller than the amount of work obtained during the nearly reversible expansion in Figure 5-7(a). This subject is examined more completely in Example 5.2-1. A familiar example of an unrestrained expansion is the flow of a fluid through a pipe or duct due to a pressure gradient. The pressure gradient represents the force imbalance and the fluid is the mass being acted on.

A process in which heat transfers from a system of higher temperature to a system of lower temperature is necessarily an irreversible process since the energy transfer cannot be reversed without violating the Clausius statement of the Second Law. The quality of the energy transfer is reduced whenever heat is transferred from higher to lower temperature. This point will be become evident in the following section where the maximum efficiency with which heat can be converted into work is shown to be a function only of the temperatures of the heat source and heat sink. Reducing the temperature of the heat source reduces the maximum conversion efficiency.

The characteristics of reversible and irreversible processes can be summarized with the following observations:

1. An irreversible process occurs whenever there is an unconstrained (or partially unconstrained) difference in a potential (pressure, temperature, force, etc.).
2. An irreversible process proceeds at a finite rate whereas a reversible process proceeds at an infinitely slow rate.
3. An irreversible process proceeds in a definite direction. Reversible processes do not have an obvious or preferred direction.
4. During an irreversible process, a system may start in an equilibrium state and, as a consequence of removal of a constraint, proceed to another equilibrium state. However, the system is not in a true equilibrium state during the process. A reversible process passes though a continuous set of equilibrium states.
5. An irreversible process cannot be reversed without causing changes in the state of the system or the surroundings. Work producing capability (i.e., the quality of energy) is reduced in an irreversible process whereas it is conserved in a reversible process.

EXAMPLE 5.2-1: REVERSIBLE AND IRREVERSIBLE WORK

In this example, we will calculate the work associated with the reversible and irreversible expansion processes that are shown in Figure 5-7(a) and Figure 5-6(a), respectively, under a specific set of conditions. The pressure outside of the cylinder is constant at $P_{atm} = 100$ kPa. The gas in the cylinder can be assumed to obey the ideal gas law. The initial pressure of the gas in the cylinder is $P_1 = 450$ kPa and the initial volume is $V_1 = 0.45$ m^3. The cross-sectional area of the piston is $A_c = 0.18$ m^2.

a) In the reversible expansion shown in Figure 5-7(a), the weight on the piston is very slowly reduced (by removing grains of sand, one at a time) until the volume of gas doubles, $V_2 = 2V_1$. The process occurs so slowly that the gas temperature is constant during the process. Determine the work done by the gas on the piston during the reversible expansion process.

The inputs are entered in EES:

```
$UnitSystem SI K Pa J Mass Radian
$TabStops 0.5 3.5 in

P_atm=100 [kPa]*convert(kPa,Pa)          "pressure in surroundings"
A_c=0.18 [m^2]                           "cross-sectional area of piston"
P_1=450 [kPa]*convert(kPa,Pa)            "initial pressure"
V_1=0.45 [m^3]                           "initial volume"
V_2=2*V_1                                "final volume"
```

The gas in the cylinder obeys the ideal gas law, therefore:

$$PV = mRT \tag{1}$$

where m is the mass of gas, R is the gas constant, and T is the temperature. The mass, gas constant, and temperature are constant for the process and therefore Eq. (1) shows that the product of pressure and volume must be constant:

$$P_1 V_1 = mRT \tag{2}$$

The work done by the gas is calculated according to:

$$W_{rev} = \int_{V_1}^{V_2} P \, dV \tag{3}$$

Substituting Eq. (1) into Eq. (3) leads to:

$$W_{rev} = \int_{V_1}^{V_2} \frac{m R T}{V} dV$$

The mass and temperature are constant and therefore can be removed from the integrand:

$$W_{rev} = m R T \int_{V_1}^{V_2} \frac{dV}{V}$$

The integration can be carried out analytically:

$$W_{rev} = m R T \ln\left(\frac{V_2}{V_1}\right) \tag{4}$$

Substituting Eq. (2) into Eq. (4) provides:

$$W_{rev} = P_1 V_1 \ln\left(\frac{V_2}{V_1}\right)$$

```
W_rev=P_1*V_1*ln(V_2/V_1)          "work accomplished during reversible process"
```

which results in $W_{rev} = 140.36$ kJ.

b) Determine the pressure in the cylinder at state 2, after all of the mass placed on the piston (i.e., all of the sand) has been removed and the gas is supporting the mass of the piston alone. What is the mass of the piston alone?

The pressure at state 2 can be calculated by applying the ideal gas law at state 2:

$$P_2 V_2 = m R T \tag{5}$$

Substituting Eq. (2) into Eq. (5) provides:

$$P_2 V_2 = P_1 V_1$$

```
P_2*V_2=P_1*V_1                    "final pressure"
```

which results in $P_2 = 225.0$ kPa. The mass of the piston is obtained using a force balance on the piston at state 2:

$$P_{atm} A_c + m_p g = P_2 A_c$$

```
P_atm*A_c+m_p*g#=P_2*A_c           "mass of piston"
```

The force balance requires that $m_p = 2{,}294$ kg.

c) In the irreversible expansion shown in Figure 5-6(a), the weight on the piston is quickly removed and the piston is allowed to achieve a new position. The final pressure and temperature of the gas at state 2 are the same as for the reversible process. Determine the work from the gas to the piston during the irreversible expansion process.

The work done by the gas during this irreversible process, W_{irrev}, is obtained by doing an energy balance on the piston, as shown in Figure 1.

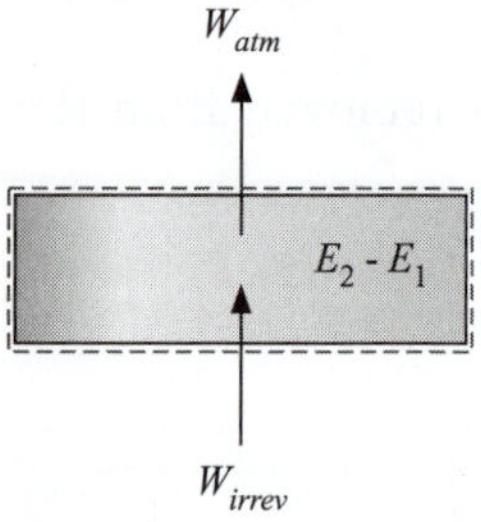

Figure 1: Energy balance on piston.

An energy balance on the piston leads to:

$$W_{irrev} = W_{atm} + E_2 - E_1 \tag{6}$$

where W_{atm} is the work done by the piston on the surroundings and $E_2 - E_1$ is the change in the energy of the piston. The work done by the piston on the surroundings is:

$$W_{atm} = P_{atm}\,(V_2 - V_1) \tag{7}$$

The change in energy of the piston is related to the change in its potential energy:

$$E_2 - E_1 = m_p\, g\,(z_2 - z_1)$$

where z_1 and z_2 are the initial and final height of the piston, respectively. The height of the piston is the ratio of the volume to the cross-sectional area:

$$E_2 - E_1 = \frac{m_p\, g}{A_c}\,(V_2 - V_1) \tag{8}$$

Substituting Eqs. (7) and (8) into Eq. (6) leads to:

$$W_{irrev} = P_{atm}(V_2 - V_1) + \frac{m_p\, g}{A_c}\,(V_2 - V_1)$$

```
W_irrev=P_atm*(V_2-V_1)+m_p*g#*(V_2-V_1)/A_c    "work accomplished during irreversible process"
```

which results in $W_{irrev} = 101.3$ kJ. The irreversible expansion has led to a smaller amount of work than the reversible expansion. There is no force imbalance in the reversible expansion process; the force imposed by the gas on the piston is always equal to the net force exerted by the combined effects of the surrounding pressure, the piston, and the weight on the piston. This expansion process is reversible and reversible work-producing processes always result in the maximum possible work. The irreversible expansion process is characterized by a force imbalance resulting from the sudden reduction in the weight that is placed on the piston. The irreversible process results in less work, but it occurs much more rapidly.

5.3 Maximum Thermal Efficiency of Heat Engines and Heat Pumps

The Carnot statement of the Second Law discussed in Section 5.1.1 is:

Heat cannot be converted completely and continuously into work.

Continuous conversion of heat into work with 100% efficiency is not possible. However, this statement does not place an upper bound on the efficiency. Can we achieve a 90% conversion of heat into work? How about 99%?

This question was addressed by Carnot in his Ph.D. thesis published in 1824. He considered the heat engine shown in Figure 5-8(a) that receives heat Q_H from a source at temperature T_H and rejects heat Q_C to a sink at temperature T_C. Carnot's reasoning has led to the conclusion that the maximum possible efficiency of the heat engine shown in Figure 5-8(a) is:

$$\boxed{\eta_{max} = \frac{W_{out}}{Q_H} = 1 - \frac{T_C}{T_H}} \tag{5-13a}$$

Equation (5-13a) is a cornerstone of classical thermodynamics and we will use it repeatedly throughout the remainder of the textbook; it is often referred to as the Carnot efficiency of a heat engine. Let's review some of the facts that Carnot may have used to arrive at this result.

1. Equation (5-13a) only applies to a process that is operating *continuously*. Any process in which heat is continuously converted into work must operate in a cycle. Otherwise the state of the working substance will eventually reach an impractical condition for which further work production is not possible.
2. Any continuous process in which heat is converted into work must involve a minimum of two heat transfers at two different temperatures. One heat transfer is from a heat source and the other is to a heat sink. If there is no heat sink, then all of the heat provided from the source is converted into work, which is in disagreement with the experimental data that were available during Carnot's time and all additional data taken subsequently.
3. The processes that are used to convert heat into work at maximum efficiency must necessarily be reversible processes because irreversible processes degrade the quality of energy and therefore reduce the capability to produce work.
4. Carnot recognized that heat (referred to at that time as caloric) flows spontaneously from a system of higher temperature to a system of lower temperature, but the reverse process cannot occur.

Note that the heat engine shown in Figure 5-8(a) could be operated in reverse, as shown in Figure 5-8(b). In this case, work (W_{in}) is provided to the engine and heat from the low temperature thermal reservoir (Q_C) is transferred to the engine. In cyclic operation, the energy represented by the work input and heat from the low temperature reservoir are rejected as heat (Q_H) to the high temperature reservoir. This reverse operation provides

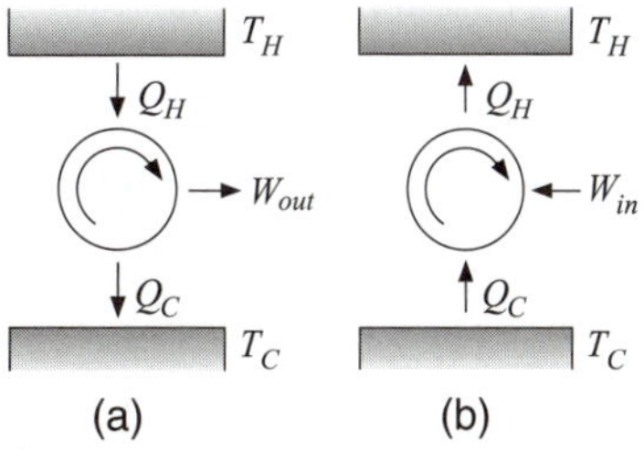

Figure 5-8: (a) Heat engine and (b) Heat pump or refrigerator.

a heat pump or refrigerator. In essence, the input work is used to pump heat from low temperature to high temperature, opposite of the direction it would normally flow. The performance of a heat pump or refrigerator is quantified in terms of its *Coefficient of Performance* (*COP*) which, like thermal efficiency, is defined as the ratio of the desired product to the required input. If the desired product is refrigeration at temperature T_C, the coefficient of performance is defined as the ratio of the cooling provided to the net work required. The maximum possible value of the *COP* for a refrigerator is:

$$COP_R = \frac{Q_C}{W_{in}} = \frac{T_C}{T_H - T_C} \tag{5-13b}$$

If the desired product is heat at temperature T_H, the coefficient of performance is defined as the ratio of the heating provided to the net work required. The maximum possible value of the *COP* for a heat pump is:

$$COP_H = \frac{Q_H}{W_{in}} = \frac{T_H}{T_H - T_C} \tag{5-13c}$$

This subject is presented in more detail in Section 9.1.

Steam engines were operating during Carnot's time and experimental data from these engines showed that the greater the temperature difference between the supply steam (i.e., the heat source) and the cooling water (i.e., the heat sink), the higher the efficiency of the engine. Therefore, Carnot surely knew that temperature is an important factor in determining the maximum efficiency of a heat engine. What was less clear is whether temperature is the only important factor in determining the maximum possible efficiency. Are other parameters also important? For example, does the pressure or the type of working fluid used in the engine also affect the maximum efficiency?

Carnot answered this question using the following simple reasoning. Consider the two reversible heat engines, engine *A* and engine *B*, shown in Figure 5-9. Both engines operate between the same thermal reservoirs at T_H and T_C, where T_H is greater than T_C. The thermal efficiencies of the two engines are defined as

$$\eta_A = \frac{W_A}{Q_{H,A}} \tag{5-14}$$

$$\eta_B = \frac{W_B}{Q_{H,B}} \tag{5-15}$$

where subscript *A* refers to engine *A* and subscript *B* refers to engine *B*. If the temperature of the heat source and heat sink are the only factors that affect the maximum possible efficiency of a heat engine, then the efficiencies of engines *A* and *B* must be the same: $\eta_A = \eta_B$.

The processes occurring within the engines are necessarily cyclic in order to allow continuous operation. Also the processes are reversible so that they result in the maximum possible work or require the minimum possible work. One implication of an engine operating in a thermodynamically reversible manner is that it literally could be operated in reverse. Assume that engine *B* is operating in reverse, as shown in Figure 5-10, in which case it acts as a heat pump or refrigerator.

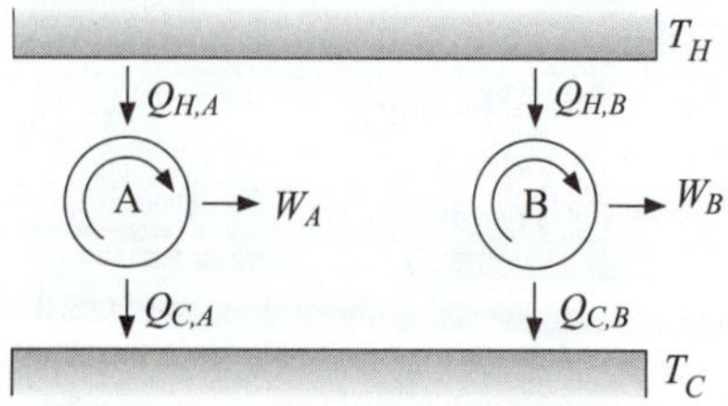

Figure 5-9: Two reversible engines operating between the same thermal reservoirs.

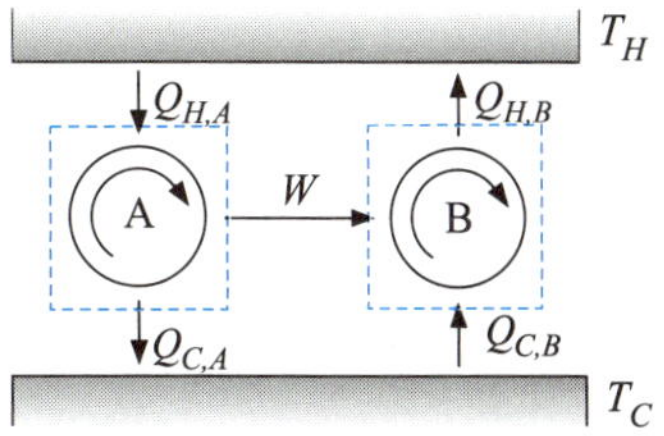

Figure 5-10: Two reversible engines operating between the same thermal reservoirs, engine B is operating in reverse.

The work produced by engine A is used by engine B in order to move thermal energy from the lower temperature thermal reservoir to the higher temperature thermal reservoir; therefore:

$$W_A = W_B = W \tag{5-16}$$

Equations (5-14) and (5-15) are solved for $Q_{H,A}$ and $Q_{H,B}$, respectively.

$$Q_{H,A} = \frac{W_A}{\eta_A} \tag{5-17}$$

$$Q_{H,B} = \frac{W_B}{\eta_B} \tag{5-18}$$

If the efficiencies of engine A and engine B are equal, then Eqs. (5-16) through (5-18) require that:

$$Q_{H,A} = Q_{H,B} \tag{5-19}$$

An energy balance on the two engines (see Figure 5-10) requires that:

$$Q_{H,A} = W + Q_{C,A} \tag{5-20}$$

$$W + Q_{C,B} = Q_{H,B} \tag{5-21}$$

Note that there are no energy storage terms in Eqs. (5-20) and (5-21) because the engines operate in a cyclic manner. Substituting Eq. (5-19) into Eq. (5-20) leads to the conclusion that:

$$Q_{C,B} = Q_{C,A} \tag{5-22}$$

Equations (5-19) and (5-22) suggest that the net effect of the combined operation of the two engines is... nothing. That is, the thermal energy supplied to engine A from the high temperature thermal reservoir is exactly equal to the thermal energy provided to a high temperature thermal reservoir from engine B. The same is true for the low temperature reservoir.

Now, suppose that there is some factor other than temperature that affects the maximum possible efficiency and causes engine A to operate with a higher efficiency than engine B, $\eta_A > \eta_B$. Perhaps engine A is operating at a higher pressure or engine A uses ammonia as the working fluid while engine B uses water. The work produced by engine A is still entirely used to drive engine B in Figure 5-10; therefore:

$$W_A = W_B = W \tag{5-23}$$

According to Eqs. (5-17) and (5-18), if $\eta_A > \eta_B$ then it must be true that:

$$Q_{H,A} < Q_{H,B} \tag{5-24}$$

The energy balances on the engines, Eqs. (5-20) and (5-21), require that:

$$Q_{C,A} = Q_{H,A} - W \tag{5-25}$$

$$Q_{C,B} = Q_{H,B} - W \tag{5-26}$$

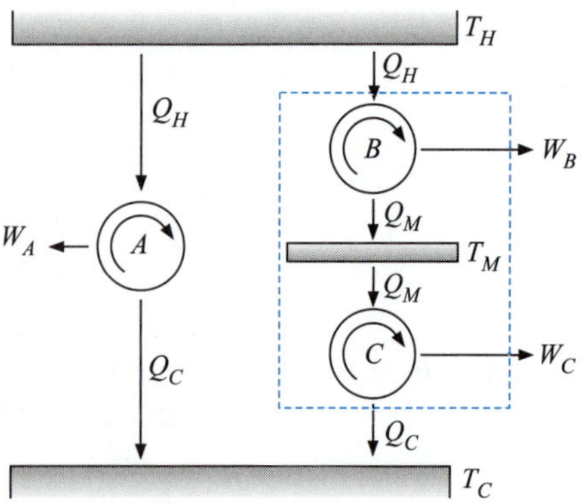

Figure 5-11: Reversible heat engines.

Substituting Eq. (5-24) into Eqs. (5-25) and (5-26) requires that:

$$Q_{C,A} < Q_{C,B} \tag{5-27}$$

and therefore more energy is provided to the thermal reservoir at T_C by engine B than was removed by engine A. Now, the net effect of the combined operation of both engines is a continuous transfer of thermal energy in amount $Q_{C,B} - Q_{C,A}$ from the low temperature thermal reservoir to the high temperature thermal reservoir. No other effects are evident since the work produced by engine A is entirely consumed by engine B. The transfer of thermal energy from the low temperature to the high temperature thermal reservoir, with no other effect, violates the Clausius statement of the Second Law. Of course, Carnot's research predates the Clausius statement by about 25 years, but he must have intuitively understood this principle.

The fundamental conclusion that results from this reasoning is that *the maximum thermal efficiency of a (reversible) heat engine operating in a cycle depends only on the source and sink temperatures and it is independent of all other factors.*

5.4 Thermodynamic Temperature Scale

Carnot's reasoning, described in Section 5.3, shows that the maximum efficiency of a continuously operating heat engine is a function only of the source and sink temperatures, T_H and T_C, respectively. However, this conclusion does not lead to the result given in Eq. (5-13) and repeated below:

$$\eta_{max} = \frac{W_{out}}{Q_H} = 1 - \frac{T_C}{T_H} \tag{5-13a}$$

This section describes how Carnot likely arrived at the maximum efficiency relationship that has led to the thermodynamic temperature scale that we use today.

Figure 5-11 shows three reversible heat engines that are operating in a continuous manner. The energy source for engine A is heat Q_H provided from a thermal reservoir at temperature T_H. Engine A produces work W_A with the maximum possible efficiency and rejects heat Q_C to a thermal reservoir at temperature T_C. Engine B receives the same amount of heat Q_H from a thermal reservoir at temperature T_H. Engine B produces work W_B and rejects heat Q_M to a thermal reservoir at temperature T_M, where T_M is intermediate between T_H and T_C ($T_C < T_M < T_H$). The heat rejected by engine B at T_M is subsequently supplied to engine C which produces work W_C and rejects heat Q_C to the low temperature thermal reservoir at T_C. Note that T_M, the temperature of the intermediate thermal reservoir, is arbitrary and there is no net energy transfer to the intermediate thermal reservoir.

Engines B and C could be combined (as shown by the dashed line in Figure 5-11) and considered to be a single engine operating under conditions that are identical to those experienced by engine A. Because all three of the engines are reversible, the combined efficiency of engines B and C must be the same as engine A. The same amount of heat is supplied to engines A and B. Therefore, the combined work produced by engines B and C must be the same as the work produced by engine A and the heat transfer to the low temperature thermal reservoir is the same for engines A and C.

The efficiency of engine A is:

$$\eta_A = \frac{W_A}{Q_H} \tag{5-28}$$

An energy balance on engine A provides:

$$W_A = Q_H - Q_C \tag{5-29}$$

Substituting Eq. (5-29) into Eq. (5-28) leads to:

$$\eta_A = \frac{Q_H - Q_C}{Q_H} = 1 - \frac{Q_C}{Q_H} \tag{5-30}$$

We know from Section 5.3 that the maximum efficiency of a heat engine depends only on the temperatures of the source and sink with which it communicates. For reversible engine A, the efficiency must therefore depend only on T_H and T_C

$$\eta_A = 1 - \frac{Q_C}{Q_H} = f(T_H, T_C) \tag{5-31}$$

where $f(T_H, T_C)$ is some unknown function of T_H and T_C. Engine B operates between thermal reservoirs at temperatures T_H and T_M and therefore its efficiency is a function only of these temperatures:

$$\eta_B = \frac{W_B}{Q_H} = \frac{Q_H - Q_M}{Q_H} = 1 - \frac{Q_M}{Q_H} = f(T_H, T_M) \tag{5-32}$$

Engine C operates between thermal reservoirs at temperatures T_M and T_C and therefore its efficiency is a function only of these temperatures:

$$\eta_C = \frac{W_C}{Q_M} = \frac{Q_M - Q_C}{Q_M} = 1 - \frac{Q_C}{Q_M} = f(T_M, T_C) \tag{5-33}$$

To simplify the algebra, we will define another function, f', such that:

$$f' = 1 - f \tag{5-34}$$

If f is only a function of T_H and T_C then the function f' must also be only a function of T_H and T_C. Substituting the definition of f' into Eqs. (5-31), (5-32), and (5-33) leads to:

$$f'(T_H, T_C) = \frac{Q_C}{Q_H} \tag{5-35}$$

$$f'(T_H, T_M) = \frac{Q_M}{Q_H} \tag{5-36}$$

$$f'(T_M, T_C) = \frac{Q_C}{Q_M} \tag{5-37}$$

We know that the following algebraic identity must be true.

$$\frac{Q_C}{Q_H} = \frac{Q_M}{Q_H}\frac{Q_C}{Q_M} \tag{5-38}$$

Substituting Eqs. (5-35) through (5-37) into Eq. (5-38) requires that:

$$f'(T_H, T_C) = f'(T_H, T_M)\; f'(T_M, T_C) \tag{5-39}$$

The left side of Eq. (5-39) only involves temperatures T_H and T_C whereas the right side involves temperatures T_H, T_M, and T_C. The only way that Eq. (5-39) can be true for any value of T_M is if the function f' has a form that causes the effect of T_M on the right side to cancel out, as in Eq. (5-40):

$$f'(T_i, T_j) = \frac{g(T_j)}{g(T_i)} \tag{5-40}$$

where the function $g(T)$ is another unknown function that depends on only one temperature. The quantity f' is not a state property since it depends to two temperatures. However, $g(T)$ depends only on one temperature and so it is a state property.

The function $g(T)$ defines a *thermodynamic temperature scale*. There are an infinite number of possible forms for the thermodynamic temperature scale. For example, the function $g(T)$ could be set to $\sqrt{T}$ or to $\exp(T)$. However, the simplest possible form is:

$$g(T) = T \tag{5-41}$$

The efficiency of a continuously operating, reversible heat engine can be obtained by substituting Eqs. (5-40) and (5-34) into Eq. (5-31):

$$\eta_{max} = f(T_H, T_C) = 1 - f'(T_H, T_C) = 1 - \frac{g(T_C)}{g(T_H)} \tag{5-42}$$

With the definition for the thermodynamic temperature scale given by Eq. (5-41), Eq. (5-42) becomes:

$$\eta_{max} = 1 - \frac{T_C}{T_H} \tag{5-43}$$

which is Eq. (5-13). One major benefit of choosing $g(T) = T$ is that the thermodynamic temperature scale is equivalent to the absolute (or ideal gas) temperature scale discussed in Section 1.7.3. The implications of this choice of thermodynamic temperature scale and an alternative choice are explored in Example 5.4-1.

EXAMPLE 5.4-1

EXAMPLE 5.4-1: THERMODYNAMIC TEMPERATURE SCALES

The thermodynamic temperature scale has been (arbitrarily) chosen so that $g(T) = T$, Eq. (5-41).

a) Show that this choice results in a temperature scale where each degree of temperature difference between the hot and cold thermal reservoirs contributes equally to the work produced by a reversible heat engine.

Consider a reversible heat engine operating between two thermal reservoirs that differ in temperature by N degrees K, as shown in Figure 1.

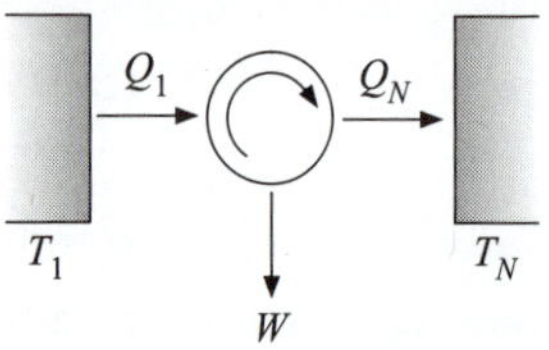

Figure 1: Reversible heat engine operating between temperatures T_1 and T_N

The same total work and the same net efficiency could be obtained by using $N - 1$ reversible heat engines operating in series, as shown in Figure 2. Each of these engines operates between thermal reservoirs that differ in temperature by 1 K.

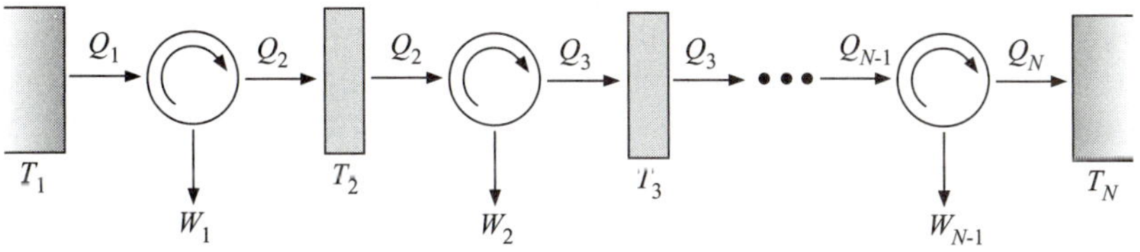

Figure 2: Series of reversible heat engines operating between temperature T_1 and T_N, the temperature of each subsequent temperature reservoir is reduced by 1 K.

The work produced by the first engine in the series, W_1, can be determined using the definition of the efficiency of a heat engine:

$$W_1 = Q_1 \eta_1 \tag{1}$$

The efficiency of engine 1 is the maximum possible efficiency because it is a reversible engine. The thermodynamic temperature scale defined by Eq. (5-41) allows us to write the maximum possible efficiency according to Eq. (5-43):

$$\eta_1 = 1 - \frac{T_2}{T_1} = \frac{T_1 - T_2}{T_1}$$

Note that the temperature difference between reservoirs 1 and 2 (or any two adjacent reservoirs) is $T_1 - T_2 = 1$ K; therefore:

$$\eta_1 = \frac{1\text{K}}{T_1} \tag{2}$$

Substituting Eq. (2) into Eq. (1) leads to:

$$W_1 = \frac{Q_1}{T_1} 1\text{K}$$

A similar analysis for any of the $N-1$ engines in Figure 2 will show that:

$$W_i = \frac{Q_i}{T_i} 1\text{K} \tag{3}$$

Substituting Eq. (5-40) into Eq. (5-35) leads to:

$$f'(T_H, T_C) = \frac{Q_C}{Q_H} = \frac{g(T_C)}{g(T_H)} \tag{4}$$

Using the thermodynamic temperature scale definition, Eq. (5-41), Eq. (4) becomes:

$$\frac{Q_C}{Q_H} = \frac{T_C}{T_H}$$

or, for engine 1 in Figure 2:

$$\frac{Q_2}{Q_1} = \frac{T_2}{T_1} \tag{5}$$

Rearranging Eq. (5) results in:

$$\frac{Q_1}{T_1} = \frac{Q_2}{T_2}$$

A similar analysis for any of the engines in Figure 2 will provide:

$$\frac{Q_i}{T_i} = \text{constant} \tag{6}$$

Substituting Eq. (6) into Eq. (3) proves that the work output from each of the engines in Figure 2 must be the same. The thermodynamic temperature scale that we use leads to an equal amount of work produced by a reversible heat engine operating with a 1 K temperature difference, regardless of the absolute temperatures involved. However, Eq. (2) shows that the efficiency of each engine is different.

b) Define an alternative thermodynamic temperature scale, $g(T) = S$, in which each degree of temperature difference between the hot and cold reservoirs results in equal thermal efficiency for a reversible heat engine.

The thermodynamic temperature scale given in Eq. (5-41) results in the same work output for each degree difference. An alternative temperature scale (t) can be defined such that the efficiency of each of the engines shown in Figure 3 is identical:

$$\eta_1 = \eta_2 = \cdots = \eta_i \tag{7}$$

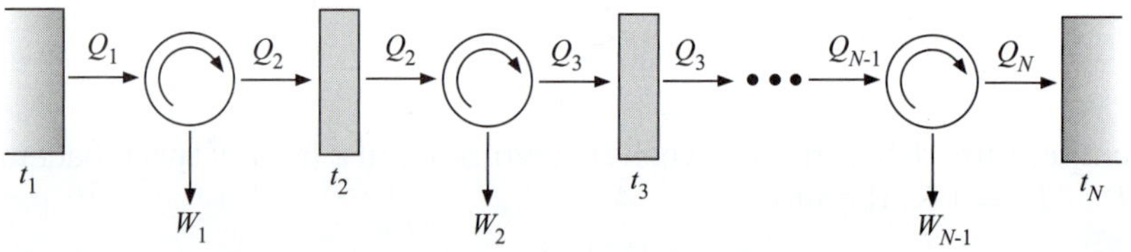

Figure 3: Series of reversible heat engines operating between temperature t_1 and t_N, the temperature of each subsequent temperature reservoir decreases by $t_i - t_{i+1} = 1^\circ\text{S}$ where t is a new temperature scale that is defined so that the efficiency of each engine is identical.

We need to identify the nature of the function $g(t)$ that results in Eq. (7). In Figure 3, the temperatures of successive thermal reservoirs differ by 1 degree; however, the scale is not the thermodynamic temperature scale (i.e., Kelvin) as assumed in Figure 2, but rather in some new thermodynamic scale (referred to as $^\circ$S). The new temperature scale will be represented with symbol t rather than T in order to avoid confusion.

The efficiency of any of the heat engines can be expressed in terms of the function $g(t)$ using Eq. (5-42):

$$\eta_i = 1 - \frac{g(t_{i+1})}{g(t_i)} \tag{8}$$

The temperature of reservoir $i+1$ is 1°S less than the temperature of reservoir i:

$$t_{i+1} = t_i - 1^\circ\text{S} \tag{9}$$

Substituting Eq. (9) into Eq. (8) requires that:

$$\eta_i = 1 - \frac{g\,(t_i - 1^\circ\text{S})}{g\,(t_i)} \tag{10}$$

Rearranging Eq. (10) leads to:

$$\frac{g\,(t_i - 1^\circ\text{S})}{g\,(t_i)} = 1 - \eta_i \tag{11}$$

Our objective is to identify a function $g(t)$ such that the right side of Eq. (11) is constant for all of the engines in Figure 3, regardless of the value of t_i. One function that satisfies this requirement is the exponential:

$$g(t_i) = \exp(b\,t_i) \tag{12}$$

where b is a constant. Substituting Eq. (12) into Eq. (11) requires that:

$$\frac{\exp\left[b\left(t_i - 1^\circ\text{S}\right)\right]}{\exp\left[b\,t_i\right]} = 1 - \eta_i \tag{13}$$

Rearranging Eq. (13) leads to:

$$\frac{\exp\left(b\,t_i\right)\exp\left(-b\right)}{\exp\left(b\,t_i\right)} = 1 - \eta_i \tag{14}$$

The temperature, t_i, can be cancelled out of Eq. (14), so that:

$$\exp(-b) = 1 - \eta_i \tag{15}$$

Equation (15) shows that the choice of a temperature scale defined by Eq. (12) results in each degree (i.e., each 1°S) providing an equal efficiency for a reversible heat engine.

c) Determine the temperature values in the alternative temperature scale defined in part (b) that correspond to 0 K, 1 K, and infinity.

The relationship between temperatures on the K and $^\circ$S temperature scales can be obtained by setting the definition of the S scale in Eq. (12) equal to the corresponding temperature in K, Eq. (5-41):

$$\underbrace{g(t_i)}_{\exp(bt_i)} = \underbrace{g(T_i)}_{T_i}$$

which leads to:

$$\exp(b\,t_i) = T_i \tag{16}$$

Solving Eq. (16) for t_i results in:

$$t_i = \frac{\ln(T_i)}{b} \tag{17}$$

Solving Eq. (17) for $T = 0$ K, 1 K, and infinity results in temperatures in the S scale of $t = -\infty^\circ$S, 0°S and ∞°S, respectively. The alternative S temperature scale does not have an absolute zero representing the lowest possible temperature in the same way that the Kelvin temperature scale does. Viewed from this perspective, absolute zero temperature (in K) occurs only because of the arbitrary choice of the temperature scale associated with Eq. (5-41). In some ways, the S temperature scale makes more sense than the Kelvin scale, but we will discontinue using it here.

5.5 The Carnot Cycle

The maximum efficiency with which thermal energy at temperature T_H can be continuously converted into work while rejecting thermal energy to a heat sink at T_C is given by Eq. (5-43); this efficiency is often called the Carnot efficiency. Efficiency is an important consideration for any heat engine, but the rate of work production (i.e., the power output) is perhaps an even more important consideration for most applications.

Table 5-2: Processes in the Carnot Cycle.

Process	Description
1-2	Reversible, isothermal expansion at temperature T_H
2-3	Reversible, adiabatic expansion from T_H to T_C
3-4	Reversible, isothermal compression at temperature T_C
4-1	Reversible, adiabatic compression from T_C to T_H

Heat engines that operate at the maximum (Carnot) efficiency must also have a power output that is zero because all of the processes involved must be reversible and reversible processes can only occur at infinitely slow rates. This issue is examined in more detail in Section 8.6 where the relationship between power and efficiency for a heat engine is studied and used to identify the conditions that lead to maximum power (rather than maximum efficiency).

At this point, it is of interest to describe how a cycle could theoretically achieve the Carnot efficiency. Any heat engine that achieves the Carnot efficiency given by Eq. (5-43) must satisfy the following requirements.

1. All of the processes involved in the conversion of heat to work must be reversible. Irreversible processes will reduce the amount of work that is produced and will therefore not allow a cycle to operate at maximum efficiency.
2. The Carnot efficiency limit applies to a heat engine that operates continuously. In order for an engine to operate continuously, the state of the working fluid must periodically be restored to its initial condition; this requirement implies that the heat engine must operate in a cyclic manner.
3. Heat is transferred to the engine from a temperature reservoir at T_H and from the engine to a temperature reservoir at T_C. The heat transfer processes must occur without a temperature gradient (and therefore require infinite time) because heat transfer that occurs through a temperature difference is necessarily an irreversible process.

A heat engine cycle that can (in theory) satisfy these requirements is the Carnot cycle, which consists of the four processes listed in Table 5-2. The Carnot engine consists of a fluid contained in a piston-cylinder device. The fluid is initially at state 1 with temperature T_H, high pressure, and small volume. The engine moves through the processes listed in Table 5-2 in sequence, as shown in Figure 5-12. During process 1-2, heat is transferred from a thermal reservoir at temperature T_H. As the heating process occurs, the gas expands causing the piston to move. The heat transfer from the high temperature reservoir is $Q_{in,1\text{-}2}$ and the work accomplished during the process is $W_{out,1\text{-}2}$. The fluid remains at temperature T_H during this heat transfer process. At state 2, the thermal communication between the fluid and the thermal reservoir at T_H is removed. During process 2-3, the force exerted by the mechanism is very slowly reduced allowing the fluid to expand. This process is analogous to the reversible expansion that is shown in Figure 5-7(a). During this adiabatic process, additional work is obtained from the fluid, $W_{out,2\text{-}3}$. At the conclusion of the expansion process, the temperature of the fluid has been reduced to T_C. The piston is now at its most withdrawn position and it is necessary to return the fluid to its original state in order for the heat engine to operate cyclically; this process is accomplished in two steps. In process 3-4, the fluid is allowed to transfer heat in amount $Q_{out,3\text{-}4}$ to a thermal reservoir at temperature T_C. The temperature of the fluid remains unchanged during this heat transfer process and the volume of the fluid is reduced, requiring work input $W_{in,3\text{-}4}$. Finally, the fluid is thermally isolated from the low

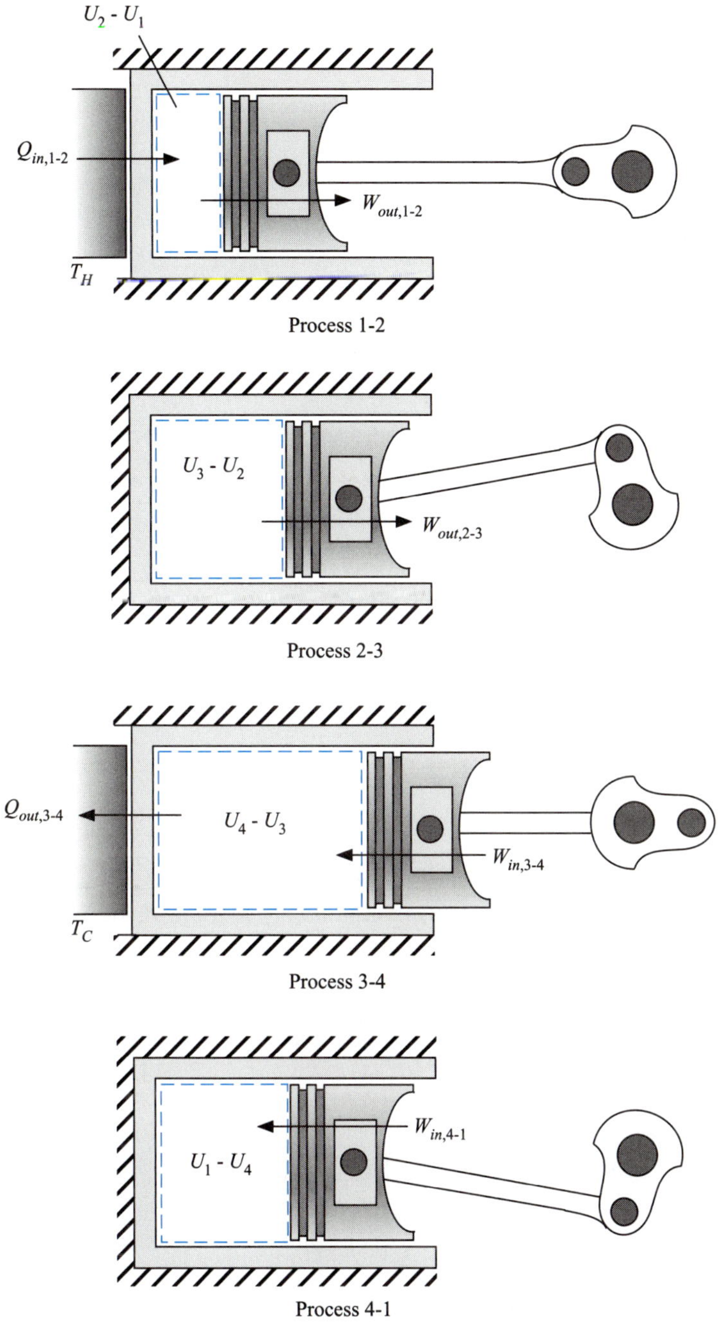

Figure 5-12: Fluid in a piston-cylinder device undergoing the processes in a Carnot cycle.

temperature reservoir and returned to its initial state by an adiabatic compression that is accomplished by slowly increasing the force on the mechanism; this process is analogous to the reversible compression that is shown in Figure 5-7(b). The compression continues until the temperature reaches T_H. The work done on the fluid during the compression process is $W_{in,4\text{-}1}$.

The net work done by the fluid during one cycle is:

$$W_{net} = W_{out,1-2} + W_{out,2-3} - W_{in,3-4} - W_{in,4-1} \tag{5-44}$$

The work terms in Eq. (5-44) have the sign convention identified in Figure 5-12. The efficiency of the engine is the ratio of the net work to the heat transfer provided to the engine from the high temperature reservoir:

$$\eta = \frac{W_{net}}{Q_{in,1-2}} \tag{5-45}$$

Because the cycle satisfies all of the requirements of a reversible heat engine, the efficiency of the engine must be the maximum efficiency appearing in Eq. (5-43). The calculations required to analyze a Carnot cycle are relatively easy to solve analytically if the working fluid obeys the ideal gas law.

Process 1-2: Constant temperature expansion at T_H.

An energy balance on the system for the constant temperature expansion process, process 1-2, is:

$$Q_{in,1-2} = W_{out,1-2} + U_2 - U_1 \tag{5-46}$$

Process 1-2 occurs at constant temperature. Since the specific internal energy of an ideal gas depends only on temperature, the change in the internal energy is zero:

$$U_2 - U_1 = m(u_2 - u_1) = 0 \tag{5-47}$$

The work done by the gas is given by Eq. (3-38):

$$W_{out,1-2} = \int_{V_1}^{V_2} P\,dV \tag{5-48}$$

The ideal gas law relates pressure, volume and temperature according to:

$$P = \frac{m\,R\,T}{V} \tag{5-49}$$

Substituting Eq. (5-49) into Eq. (5-48) leads to:

$$W_{out,1-2} = \int_{V_1}^{V_2} \frac{m\,R\,T_H}{V}\,dV \tag{5-50}$$

Carrying out the integration in Eq. (5-50) results in:

$$W_{out,1-2} = m\,R\,T_H \ln\left(\frac{V_2}{V_1}\right) \tag{5-51}$$

Substituting Eqs. (5-47) and (5-51) into Eq. (5-46) shows that:

$$Q_{in,1-2} = W_{out,1-2} = m\,R\,T_H \ln\left(\frac{V_2}{V_1}\right) \tag{5-52}$$

Process 2-3: Adiabatic expansion

An energy balance on the system for the adiabatic expansion process, process 2-3, provides:

$$0 = W_{out,2-3} + U_3 - U_2 \tag{5-53}$$

The temperature of the gas changes during the adiabatic expansion from T_H (at state 2) to T_C (at state 3). The change in the internal energy is given by:

$$U_3 - U_2 = m(u_3 - u_2) \tag{5-54}$$

Substituting Eq. (3-21) into Eq. (5-54) provides:

$$U_3 - U_2 = m \int_{T_H}^{T_C} c_v \, dT \tag{5-55}$$

where c_v is the specific heat capacity at constant volume as a function of temperature. Substituting Eq. (5-55) into Eq. (5-53) requires that:

$$W_{out,2-3} = m \int_{T_C}^{T_H} c_v \, dT \tag{5-56}$$

The volumes of the gas at states 2 and 3 are related to the temperatures at these states and this relationship will be needed in order to analytically determine the cycle efficiency and is derived from an energy balance on a differential basis. At any point during the expansion process, the differential energy balance provides:

$$0 = \delta W_{out} + dU \tag{5-57}$$

The differential work done by the gas is:

$$\delta W_{out} = P \, dV \tag{5-58}$$

The differential change in internal energy is:

$$dU = m \, c_v \, dT \tag{5-59}$$

Substituting Eqs. (5-58) and (5-59) into Eq. (5-57) leads to:

$$0 = P \, dV + m \, c_v \, dT \tag{5-60}$$

The pressure can be expressed in terms of temperature and volume using the ideal gas law. Substituting, Eq. (5-49) into Eq. (5-60) provides an equation relating the change in volume to the change in temperature:

$$0 = \frac{m \, R \, T}{V} dV + m \, c_v \, dT \tag{5-61}$$

Equation (5-61) can be rearranged:

$$\frac{dV}{V} = -\frac{c_v}{R} \frac{dT}{T} \tag{5-62}$$

and integrated:

$$\int_{V_2}^{V_3} \frac{dV}{V} = -\frac{1}{R} \int_{T_H}^{T_C} \frac{c_v}{T} dT \tag{5-63}$$

which leads to:

$$\ln\left(\frac{V_3}{V_2}\right) = \frac{1}{R} \int_{T_C}^{T_H} \frac{c_v}{T} dT \tag{5-64}$$

Process 3-4: Constant temperature compression at T_C

The analysis for process 3-4 is essentially the same as for process 1-2 except that the heat removal occurs at T_C. The heat and work for this process are:

$$W_{in,3-4} = Q_{out,3-4} = -\int_{V_3}^{V_4} P dV = -\int_{V_3}^{V_4} \frac{m\,R\,T_C}{V}\,dV = -m\,R\,T_C\,\ln\left(\frac{V_4}{V_3}\right) \quad (5\text{-}65)$$

Process 4-1: Adiabatic compression

In the final process, the temperature of the gas changes from T_C (at state 4) to T_H (at state 1) during the adiabatic compression. The work done on the gas during this process is:

$$W_{in,4-1} = U_1 - U_4 = m\int_{T_C}^{T_H} c_v\,dT \quad (5\text{-}66)$$

Note that the work done on the gas during process 4-1 is exactly equal to the work done by the gas during process 2-3, $W_{out,2\text{-}3}$ from Eq. (5-56). A relationship between the volumes at states 4 and 1 and the temperatures at these states can be developed from integration of a differential process, just as was done for process 3-4:

$$-\ln\frac{V_1}{V_4} = \frac{1}{R}\int_{T_C}^{T_H} \frac{c_v\,dT}{T} \quad (5\text{-}67)$$

Cycle efficiency

The net work for the cycle is obtained by substituting Eqs. (5-51), (5-56), (5-65), and (5-66) into Eq. (5-44),

$$W_{net} = m\,R\,T_H\,\ln\left(\frac{V_2}{V_1}\right) + \xcancel{m\int_{T_C}^{T_H} c_v\,dT} + m\,R\,T_C\,\ln\left(\frac{V_4}{V_3}\right) - \xcancel{m\int_{T_C}^{T_H} c_v\,dT} \quad (5\text{-}68)$$

which can be simplified to:

$$W_{net} = m\,R\,T_H\ln\left(\frac{V_2}{V_1}\right) + m\,R\,T_C\ln\left(\frac{V_4}{V_3}\right) \quad (5\text{-}69)$$

The cycle efficiency is obtained by substituting Eqs. (5-69) and (5-52) into Eq. (5-45):

$$\eta = \frac{m\,R\,T_H\ln\left(\frac{V_2}{V_1}\right) + m\,R\,T_C\ln\left(\frac{V_4}{V_3}\right)}{m\,R\,T_H\ln\left(\frac{V_2}{V_1}\right)} \quad (5\text{-}70)$$

Equation (5-70) can be rearranged:

$$\eta = 1 + \frac{T_C\ln\left(\frac{V_4}{V_3}\right)}{T_H\ln\left(\frac{V_2}{V_1}\right)} = 1 - \frac{T_C\ln\left(\frac{V_3}{V_4}\right)}{T_H\ln\left(\frac{V_2}{V_1}\right)} \quad (5\text{-}71)$$

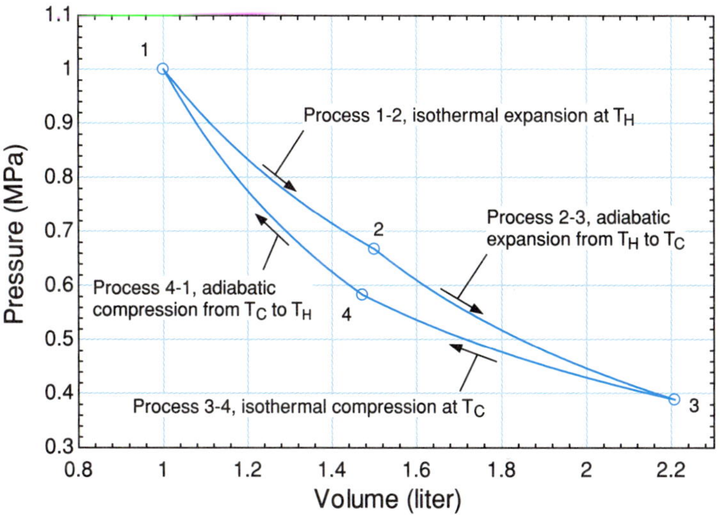

Figure 5-13: Pressure-volume diagram for air undergoing a Carnot cycle between temperatures $T_H = 350$ K and $T_C = 300$ K with $P_1 = 1$ MPa, $V_1 = 1.0$ liter, and $V_2/V_1 = 1.5$.

Equation (5-71) can be further simplified using the relationships between volume and temperature that were derived in Eqs. (5-64) and (5-67). The integrals on the right side of both of these equations are identical, therefore:

$$\ln\left(\frac{V_3}{V_2}\right) = -\ln\left(\frac{V_1}{V_4}\right) = \ln\left(\frac{V_4}{V_1}\right) \tag{5-72}$$

Taking the antilog of both sides and rearranging leads to:

$$\left(\frac{V_2}{V_1}\right) = \left(\frac{V_3}{V_4}\right) \tag{5-73}$$

Substituting Eq. (5-73) into Eq. (5-71) results in the final equation for efficiency:

$$\eta = 1 - \frac{T_C}{T_H} \tag{5-74}$$

The efficiency in Eq. (5-74) is the maximum efficiency that can be achieved under these conditions because all of the processes associated with the Carnot cycle are reversible. The result agrees with Eq. (5-43). A Carnot cycle analysis for fluids that do not behave according to the ideal gas law would be more complicated, but it will also result in the efficiency given by Eq. (5-74).

The four processes associated with the Carnot cycle are shown on a pressure-volume diagram in Figure 5-13. Figure 5-13 is drawn to scale assuming that air is the working fluid. The maximum pressure and minimum volume (i.e., the conditions at state 1) are $P_1 = 1$ MPa and $V_1 = 1.0$ liter, respectively. The source and sink temperatures are $T_H =$ 350 K and $T_C = 300$ K, respectively. The ratio of the volumes is $V_2/V_1 = 1.5$. Recall that work out of a system is the integral of pressure with respect to volume. Therefore, the net work associated with a single cycle is represented by the area enclosed by the P-V diagram. In Chapter 6, we will discuss the property entropy in the context of the Carnot cycle shown in Figure 5-13.

Problems

The problems included here have been selected from a larger set of problems that are available on the website associated with this book (www.cambridge.org/kleinandnellis).

A: Maximum Efficiency

5.A-1 A well-insulated cylindrical tank of water with $D = 3$ m diameter is filled to a depth of $H = 6$ m. It is desired to lower the temperature of the water in the tank from $T_1 = 20°\text{C}$ to $T_2 = 2°\text{C}$ in two hours using a heat pump. Heat is to be rejected to the atmospheric at $T_{amb} = 38°\text{C}$. What is the minimum average power that must be supplied to achieve this cooling. State all assumptions that you employ.

5.A-2 After graduation, you take a job with the Acme energy services company. Your first job is to purchase high efficiency heat pumps. The heat pumps employ mechanical power and a heat engine to provide heat at $T_H = 100°\text{C}$. A low temperature heat sink at $T_C = 20°\text{C}$ is also available. A salesman shows you two models. Performance data for each model operating at steady state are provided in Figure 5.A-2.

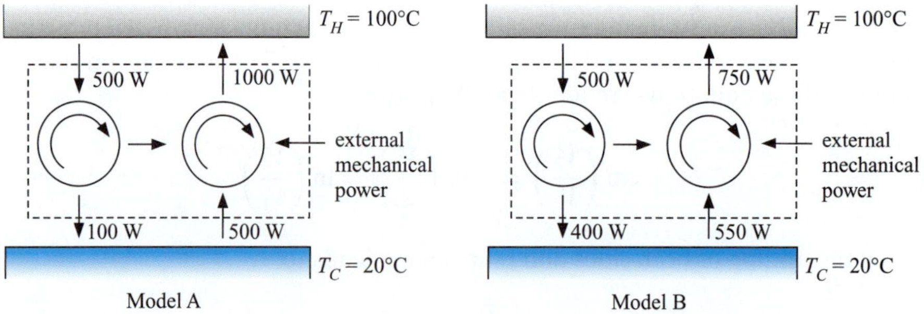

Figure 5.A-2: High efficiency heat pumps; Model A is on left and Model B is on the right.

a) What is the mechanical power that is needed to operate Model A?
b) What is the mechanical power needed to operate Model B?
c) What is the Coefficient of Performance (*COP*) for heating for Model A? The *COP* for heating is defined as the ratio of the net heat provided to T_H to the mechanical power required.
d) What is the *COP* for heating for Model B?
e) What is the maximum possible heating *COP* that can be obtained under these conditions?
f) Which heat pump system would you recommend that your company purchase?

5.A-3 A Carnot refrigeration cycle uses water contained in a piston-cylinder device. Process 1-2 is a reversible isothermal compression in which the water is transformed from saturated vapor to saturated liquid at a pressure of $P_1 = 250$ kPa. Process 2-3 is a reversible adiabatic expansion that proceeds until the water reaches $T_3 = 7°\text{C}$. The quality at state 3 is $x_3 = 0.169$. Process 3-4 is a reversible isothermal expansion that proceeds until the quality reaches $x_4 = 0.784$. Process 4-1 is a reversible adiabatic compression. The mass of water in the cylinder is $m = 1$ kg and the period of the cycle is $time = 20$ s.

a) Complete the table below.

State	Pressure (kPa)	Temperature (°C)	Specific internal energy (J/kg)	Quality	Specific volume (m^3/kg)
1	250			0	
2	250			1	
3		7		0.169	
4		7		0.784	

b) Determine the Coefficient of Performance (*COP*) for refrigeration for this cycle.
c) What is the refrigeration effect, in kW?
d) What is the power input required to operate this cycle, in hp?
e) Overlay the state points for the cycle on a *P-v* diagram for water.

5.A-4 The refrigeration system shown in Figure 5.A-4 transfers energy from cylinder A to cylinder B. Cylinder A contains $m_A = 1.75$ kg of R-134a, which is initially saturated vapor at a pressure of $P_A = 280$ kPa. Cylinder B contains $m_B = 0.5$ kg of water at $T_B = 45°$C and an initial quality of $x_{B,1} = 0.25$. The pressures in cylinders A and B remain constant during the process in which the R134a changes from saturated vapor to saturated liquid.

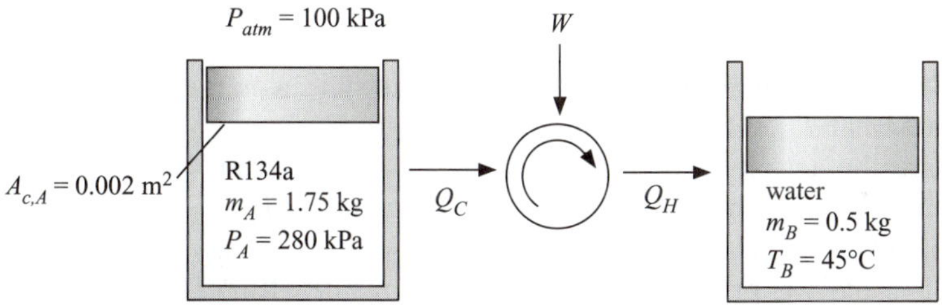

Figure 5.A-4: Refrigeration cycle.

a) The refrigeration cycle operates with a coefficient of performance, $COP = 1.5$. The *COP* is defined as the ratio of the refrigeration provided to the work required. Determine the required work input.
b) The cross-sectional area of the piston for cylinder A is $A_{c,A} = 0.002$ m². Determine the required mass of the piston. The atmospheric pressure is $P_{atm} = 100$ kPa.
c) Determine the work done on or by the water during this process.
d) What is the lower limit on the work input required for this process, assuming that all irreversible processes could be replaced with reversible processes?

5.A-6 A refrigeration system for freezing freshly-baked bread is under design. The system must be able to continuously cool $N = 4{,}250$ loaves of bread per 8-hour working day from $T_1 = 25°$C (the ambient temperature) to $T_2 = -10°$C. Tests show that the average bread loaf has a mass of $m = 450$ g and a specific heat capacity of $c = 2.93$ kJ/kg-K. In addition, when the bread is cooled below $T_f = 0°$C, the water in the bread freezes releasing an additional $\Delta H_{fs} = 109.3$ kJ per loaf. The proposed design circulates air past cooling coils that cools the air to $T_{air} = -20^\circ$C. The cold air is then circulated past the bread and returned to the cooling coils.

a) As the engineer in charge of the project, determine the required steady rate of cooling capacity for this process
b) To maintain consistent product, the change in temperature of the air blown by the bread must be less than $\Delta T_{air} = 8^\circ$C. Determine the minimum required volumetric flow rate of air (assume that the air is at atmospheric pressure, P_{atm} = 100 kPa).
c) A refrigeration contractor has indicated that he can supply refrigeration equipment for the bread cooling project that has a $COP = 2.1$. Note that COP is defined as the ratio of the refrigeration provided to the power required. Determine the electrical service that will be needed for the motor used to operate the cycle.
d) Compare the COP with the maximum possible COP for this application.

5.A-7 A heat pump is to be used to heat a house in the winter and then reversed to cool the house in the summer. The interior house temperature is to be maintained at $T_{in} = 70^\circ$F during both summer and winter. Heat transfer through the roof, walls, and floor is estimated to occur at a rate of $UA = 870$ Btu/hr per degree temperature difference (in $^\circ$F) between inside and outside.

a) Determine the required cooling capacity to maintain the interior temperature on a summer day in which the outdoor temperature is $T_{out} = 95^\circ$F
b) Determine the minimum electrical power required to provide the cooling capacity determined in part (a).
c) Now consider a heating application with the heat pump operating with the same *power* determined in part (b). For this heating application, determine the lowest outdoor temperature for which the interior temperature could be maintained at T_{in}.
d) Compare the actual costs of heating for a 24 hour period using a gas furnace and a heat pump when the outdoor temperature is $T_{out} = 20^\circ$F. The gas furnace operates at $\eta_f = 92\%$ efficiency based on the energy content of the gas, which is usually expressed in therms. (1 therm=10^5 Btu.) The heat pump has a coefficient of performance that is 35% of the Carnot limit. Obtain the unit cost of gas and electricity from your local utility.

B: Advanced Problems

5.B-2 You have probably seen the 'drinking bird' toy (http://en.wikipedia.org/wiki/Drinking_bird). If examined closely, we could see that this toy is simply two hollow glass bulbs that are separated by a tube and mounted on a swivel joint. The lower bulb is partially filled with a volatile liquid such as isopentane.

a) Explain how the drinking bird works.
b) The drinking bird can be viewed as a machine that converts thermal energy into mechanical energy. Do you agree with this characterization? If so, determine the maximum efficiency of the bird. Please indicate the assumptions employed in your assessment.
c) Is the efficiency an important performance indicator for this device? Please provide a justification for your answer.
d) During its operation, the bird raises the working fluid from the lower bulb to the upper bulb. It is possible for the bird to produce more work than required

to raise the working fluid? Do you see any practical potential for useful power production with devices of this nature?

5.B-4 Dr. Alan Williams describes a constant volume power cycle that he claims can produce mechanical power with an efficiency that approaches 100%. The analysis and below was excerpted from the website prepared by Dr. Williams:

from: http://www.globalwarmingsolutions.co.uk/energy_cycles_at_constant_volume.htm

Consider the energy cycle depicted in the pressure-volume diagram shown in Figure 5.B-4 that consists of two processes at constant volume and two processes at constant pressure.

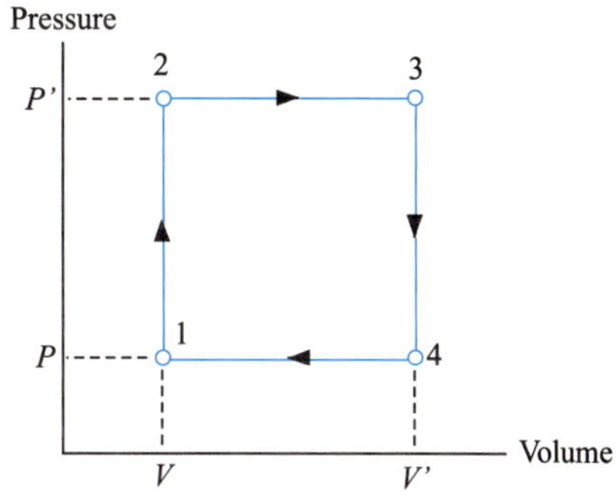

Figure 5.B-4: Pressure-volume diagram.

At state 1 the system consists of a fixed mass of gas at volume V, pressure P, and temperature T_1. The gas then undergoes the following four processes:

State 1 to 2: The gas is heated at constant volume to temperature T_2 raising its pressure to P'. The heat input required per unit mass for this process is $c_v\,(T_2 - T_1)$.

State 2 to 3: The gas is heated at constant pressure to temperature T_3 and volume V'. The heat input required per unit mass for this process is $c_P\,(T_3 - T_2)$.

State 3 to 4: Work is done by the gas at constant volume so that its pressure falls to P and its temperature falls to T_4. There is no heat input for this process. An energy balance is therefore: $Q = \Delta U + W = 0$ and the work done by the gas during this process per unit mass is $c_v\,(T_3 - T_4)$

State 4 to 1: Heat is transferred out of the gas to an energy sink during a constant pressure process so that the volume returns to V and the temperature to T_1.

Efficiency = (useful work achieved) / (total heat input) = $[c_v\,(T_3 - T_4)]\,/\,[c_v\,(T_2 - T_1) + c_P\,(T_3 - T_2)]$

Now consider the energy cycle portrayed approximating towards a constant volume heat engine. As $V' \rightarrow V$, $T_4 \rightarrow T_1$, $T_3 \rightarrow T_2$ and $(T_3 - T_2) \rightarrow 0$, therefore, efficiency $\rightarrow [c_v\,(T_2 - T_1)]\,/\,[c_v\,(T_2 - T_1)] \rightarrow 100\%$.

Therefore – if an energy cycle can be devised with the entire heat input and extraction of work at constant volume it will have a maximum theoretical efficiency of 100%.

Please review the analysis above and then answer the following questions.

a) Calculate the efficiency of the cycle with 1 kg of air as the working fluid for the following conditions: $T_1 = 300$ K, $P = 100$ kPa, $P' = 2P$, $v_3 = 1.5\, v_1$
b) What is the Carnot efficiency for the conditions of part (a)? Compare the efficiencies.
c) Do you agree with the analysis presented by Dr. Williams? If your answer is no, indicate where you see a flaw in his logic or a violation of a physical law – other than the Second Law of Thermodynamics. Dr. Williams does not believe that the Second Law is applicable to this process.

6 Entropy

The Second Law of Thermodynamics was introduced in Chapter 5 based on the observation that heat flows spontaneously from a system of higher temperature to a system of lower temperature. This statement of the Second Law was used to prove that heat cannot be continuously converted into power with 100% efficiency and to identify an upper limit for this conversion efficiency. These observations demonstrate that energy has quality as well as quantity. In this chapter, the property entropy is introduced in order to quantify the quality of energy. A statement of the Second Law based on entropy is shown to be entirely consistent with, but more general than, the statements presented in Chapter 5. Entropy is not conserved in any real process. Instead, entropy must always be generated by a real process. An entropy balance, similar in concept and form to mass and energy balances, is used to analyze thermodynamics problems using the Second Law.

6.1 Entropy, a Property of Matter

Figure 6-1 illustrates a continuously operating engine that receives heat Q_H from a high temperature reservoir at T_H and produces work W_{out} while rejecting heat Q_C to a lower temperature reservoir at T_C. The Carnot cycle presented and analyzed in Section 5.5 can, in theory, be implemented so that the heat engine in Figure 6-1 achieves the maximum possible efficiency.

$$\eta_{max} = \frac{W_{out}}{Q_H} = 1 - \frac{T_C}{T_H} \tag{6-1}$$

Equation (6-1) is only applicable if all of the processes that occur in the engine, including the heat transfer processes, are reversible. The Carnot cycle consists of four reversible processes for a compressible fluid system of fixed mass. Process 1-2 is an isothermal expansion of the working fluid at T_H in which the system receives thermal energy from a thermal reservoir that is also at T_H. Process 2-3 is an adiabatic and reversible expansion in which the working fluid temperature is reduced from T_H to T_C. Process 3-4 is an isothermal compression in which the working fluid remains at T_C by rejecting heat to the thermal reservoir at T_C. Process 4-1 is an adiabatic and reversible compression that

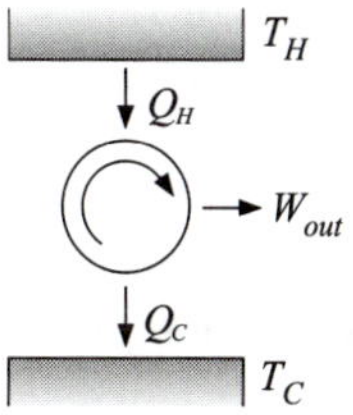

Figure 6-1: A heat engine.

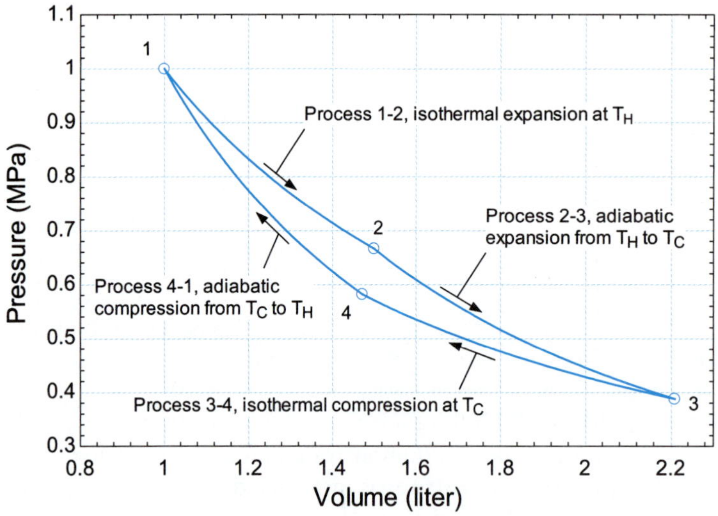

Figure 6-2: Pressure-volume diagram for air undergoing a Carnot cycle between temperatures $T_H = 350$ K and $T_C = 300$ K with $P_1 = 1$ MPa, $V_1 = 1.0$ liter, and $V_2/V_1 = 1.5$.

causes the working fluid temperature to increase from T_C to T_H. These processes are shown on the pressure-volume diagram in Figure 6-2, which is drawn to scale assuming that air is the working fluid. The maximum pressure and minimum volume (i.e., the conditions at state 1) are $P_1 = 1$ MPa and $V_1 = 1.0$ liter, respectively. The source and sink temperatures are $T_H = 350$ K and $T_C = 300$ K, respectively, and $V_2/V_1 = 1.5$.

The line joining states 1 and 2 in Figure 6-2 represents an isotherm at T_H. The line joining states 3 and 4 represents an isotherm at T_C. These isotherms do not appear to be parallel; therefore, it is interesting to speculate whether extensions of these lines could possibly cross. This possibility is hypothesized in Figure 6-3, where the extensions of lines 1-2 and 3-4 are shown to intersect at state 5. Can this happen? If so, what would the temperature at state 5 be? Can the temperature at state 5 be both T_H and T_C? Can a state have two (or more) temperatures? A pure gas in an equilibrium state at fixed pressure and specific volume has never been observed to exhibit more than one temperature. Therefore, two isotherms can never cross on a pressure-volume plot and the extensions of lines 1-2 and 3-4 will never intersect.

The lines connecting states 2 and 3 and states 4 and 1 in Figure 6-2 each represent reversible adiabatic processes. These lines also do not appear to be parallel. Is it possible that extensions of these reversible adiabatic lines can cross? This situation is shown in Figure 6-4, where the extensions of lines 2-3 and 4-1 appear to intersect at state 6. Based on the previous discussion, it seems likely that this intersection cannot occur. But, why not? What physical principle is violated if the reversible adiabatic process lines cross?

Suppose that the extensions of lines 2-3 and 4-1 do intersect at state 6, as shown in Figure 6-4. It would then be possible to construct a cycle that starts at point 1 and proceeds through states 2, 3, 6, 4 and then back to 1. Such a cycle must violate the Second Law. Process 1-2 is a reversible isothermal process in which heat Q_H is provided to the system at temperature T_H. Process 2-3-6 is a reversible adiabatic expansion process. Process 6-4-1 is a reversible adiabatic compression process. These three processes complete the cycle. The net work per cycle, W_{net}, is the area enclosed within the *P-V* diagram. The net work produced by the cycle proceeding through states 1-2-3-6-4-1 is greater than zero because the area enclosed by the cycle diagram in Figure 6-4 is greater than zero.

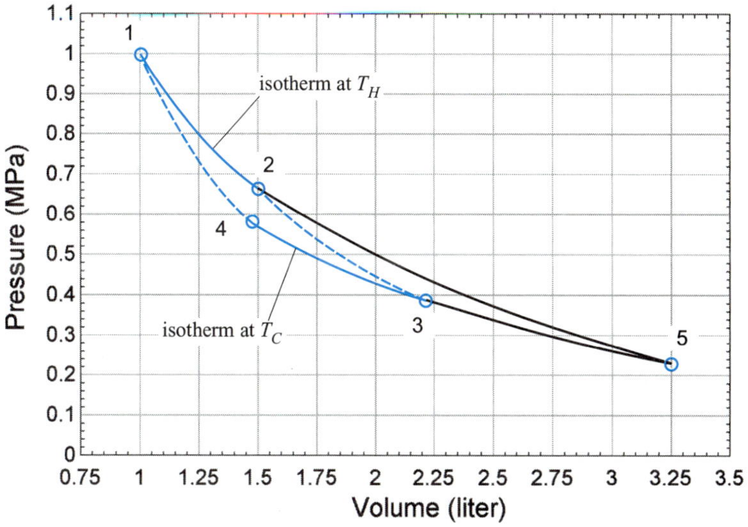

Figure 6-3: Pressure-volume plot showing the Carnot cycle processes. The lines corresponding to the constant temperature processes (i.e., the isotherms) are extended and assumed to cross at state 5.

However, the only heat transfer to the cycle is Q_H, transferred during process 1-2. An energy balance for the entire cycle then requires that $W_{net} = Q_H$ and the efficiency of the cycle must be 100% (i.e., all of the heat is converted to work). This cycle directly violates the Second Law of Thermodynamics, which limits the efficiency to the value indicated in Eq. (6-1). Consequently, extensions of lines 2-3 and 4-1, which represent reversible adiabatic processes, can never intersect.

At this point, you must be wondering where this discussion is going. Why do we care about lines on a pressure-volume diagram that can or cannot intersect? Lines that cannot

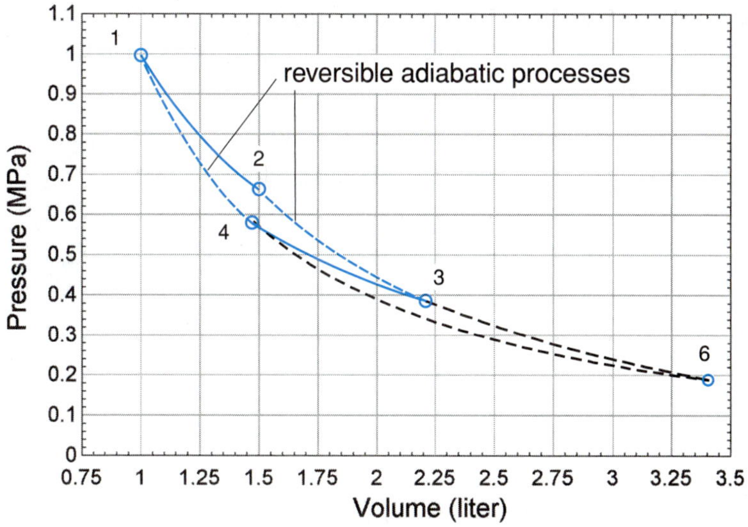

Figure 6-4: Pressure-volume plot showing the Carnot cycle processes. The lines corresponding to the reversible adiabatic processes are extended and assumed to cross at state 6.

intersect on a pressure-volume, or any other type of property diagram, represent lines of some constant property. Two different horizontal lines in Figure 6-2 represent two different pressures and they will never intersect. Two vertical lines in Figure 6-2 represent two different specific volumes and they also will never intersect. Lines representing two different temperatures, for example, extensions of lines 1-2 and 3-4, also can never intersect. Now we find that the lines representing reversible adiabatic processes, for example lines 2-3 and 4-1, cannot intersect. Therefore, these lines must also represent a constant value of some property; this is the property that we call *entropy*.

Entropy as a property was first proposed by Rudolf Clausius in 1865. Quoting from Clausius (1865), as noted by Bejan (1998):

> *"Each state of matter can be assigned a value of this new property such that all states designated by the same value of entropy can be reached by a reversible adiabatic process."*
>
> *"I have felt it more suitable to take the names of important scientific quantities from the ancient languages in order that they may appear unchanged in all contemporary languages. Hence I propose that we call S, the Entropy of the body after the Greek word meaning transformation."*
>
> *"I have intentionally formed the word entropy to be as similar as possible to the word energy since the two quantities that are given these names are so closely related in the physical significance that a certain likeness in these names seems appropriate."*

We have established the existence of a property of matter called entropy. In order to be useful, it is necessary to set up a scale for entropy so that it can be assigned numerical values at different states. What has changed between the states that lie on line 2-3 (one value of entropy) and the states that lie on line 4-1 (a different value of entropy)? What stays the same for states that lie on either of these lines? Could it be the amount of heat that has been transferred to the fluid? Consider moving from line 4-1 to line 2-3 in Figure 6-2 by traveling from state 1 to state 2 along line 1-2. The heat transferred to the fluid during this process is Q_H. Alternatively, it is possible to move from line 4-1 to line 2-3 by traveling from state 4 to state 3 along line 4-3 in Figure 6-2. The heat transferred to the fluid during this process is Q_C, which is different from Q_H. Therefore, the heat transferred to the fluid must not be the answer.

Whatever the correct scale for entropy is, it must lead to the same change of the property if we move along the line joining states 1 and 2 and along the line joining states 4 and 3. Recall that the efficiency of the Carnot cycle is given by Eq. (6-1):

$$\eta_{max} = \frac{W_{out}}{Q_H} = 1 - \frac{T_C}{T_H} \tag{6-2}$$

An energy balance on the Carnot cycle provides:

$$W_{out} = Q_H - Q_C \tag{6-3}$$

Substituting Eq. (6-3) into Eq. (6-2) leads to:

$$\frac{Q_H - Q_C}{Q_H} = 1 - \frac{T_C}{T_H} \tag{6-4}$$

which can be simplified:

$$\frac{Q_C}{Q_H} = \frac{T_C}{T_H} \tag{6-5}$$

Equation (6-5) is rearranged:

$$\frac{Q_C}{T_C} = \frac{Q_H}{T_H} \tag{6-6}$$

Examination of Eq. (6-6) shows that the ratio of the heat transferred to the absolute temperature at which the heat is transferred is the same for the processes represented by line 1-2 (Q_H/T_H) and line 4-3 (Q_C/T_C) in Figure 6-2. Both of these processes are reversible. This observation leads to the definition of entropy on a differential basis:

$$dS = \left.\frac{\delta Q}{T}\right|_{rev} \tag{6-7}$$

where S is the entropy of a system. The differential change in the property entropy is defined as the ratio of the differential amount of heat transferred to the system in a reversible process (δQ) to the temperature at which the heat is transferred. Note that the definition in Eq. (6-7) is consistent with the observation that entropy is a property that remains constant in a reversible adiabatic process. Equation (6-7) is a differential equation and therefore a boundary condition (i.e., a reference state) is needed to complete the specification of entropy. For systems in which chemical reactions do not occur, we are free to specify a zero value for entropy at an arbitrary reference state. The choice of the reference state does not matter since the reference value of entropy always cancels out of a properly formulated analysis. This situation is similar to the reference value of internal energy that is discussed in Section 3.3. Note that entropy is an extensive property. It has units of J/K in the SI system and Btu/R in the English system. We will often make use of the intensive property specific entropy (s), defined as:

$$s = \frac{S}{m} \tag{6-8}$$

where m is the mass of the system.

6.2 Fundamental Property Relations

Entropy is an internal property of matter, which means that its value depends of the molecular structure of the matter composing the system. Internal properties are interrelated. The phase rule in Eq. (2-1) formalizes this statement by indicating the number of degrees of freedom, i.e., the number of intensive internal properties that are required to fix the state of the system. A pure, single phase system, for example, has two degrees of freedom. Specifying any two internal, intensive properties determines the state of the system and therefore fixes the values of all other internal, intensive properties. If the temperature and pressure of a single phase pure substance are specified, then the specific entropy can be determined. This section shows how the value of entropy at a specified state is related to other property values.

Figure 6-5 shows a piston-cylinder device that contains a pure fluid. The piston can be moved in order to compress or expand the fluid. Consider a very small period of time dt during which a differential amount of heat, δQ_{in}, is transferred to the fluid and a differential amount of work, δW_{out}, is done by the fluid. Kinetic and potential energy effects are negligible for this process. An energy balance for the differential process is:

$$\delta Q_{in} = \delta W_{out} + dU \tag{6-9}$$

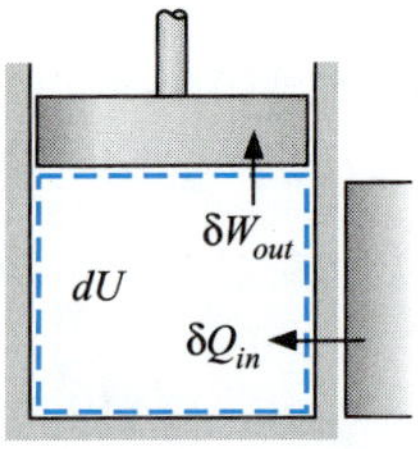

Figure 6-5: Fluid in a piston-cylinder device.

We will assume that the process undergone by the fluid is reversible. In order to be reversible, the force exerted by the fluid on the piston must be balanced to within a differential amount by the forces exerted on the outside surface of the piston. Also, the heat must be from a source that is at the same temperature as the fluid. Said differently, the fluid must always be in internal equilibrium; i.e., it has uniform properties with no internal gradients in pressure or temperature. Clearly, these are impractical constraints and the attempt to carry out such a process would require an infinite amount of time.

The differential work done by the expanding fluid during the process is:

$$\delta W_{out} = P\,dV \tag{6-10}$$

Because the process is reversible, the differential heat to the fluid is related to the change in the entropy of the fluid according to Eq. (6-7).

$$\delta Q_{in} = T\,dS \tag{6-11}$$

Substituting Eqs. (6-10) and (6-11) into Eq. (6-9) results in:

$$dU = T\,dS - P\,dV \tag{6-12}$$

It is often more convenient to write Eq. (6-12) in terms of specific properties by dividing through by the system mass.

$$\boxed{du = T\,ds - P\,dv} \tag{6-13}$$

Equation (6-13) is referred to as a *fundamental property relation*. Note that the fundamental property relation does not involve heat or work; it only involves the properties of the fluid. Properties are, by definition, independent of the history of the system and therefore do not depend on the trajectory of a process.

Equation (6-13) can be expressed in an alternative form that is often more convenient. Specific enthalpy, h, is defined in Eq. (4-16) as:

$$h = u + P\,v \tag{6-14}$$

Taking the total derivative of Eq. (6-14) leads to:

$$dh = du + P\,dv + v\,dP \tag{6-15}$$

Substituting Eq. (6-13) into Eq. (6-15) provides an alternative fundamental property relation.

$$\boxed{dh = T\,ds + v\,dP} \tag{6-16}$$

Equations (6-13) and (6-16), together with other similar property relations, are the starting point for determining the values of properties such as specific entropy that cannot be directly measured, as discussed in Chapter 10.

6.3 Specific Entropy

The determination of specific entropy for pure substances follows naturally from the discussions in Section 2.2, Section 3.3, and Section 4.2 regarding specific volume, specific internal energy, and specific enthalpy, respectively. The techniques presented in these sections can be used with little modification to determine specific entropy.

6.3.1 Property Tables

Fluid property data are often provided in tabular form, organized according to the phase of the fluid, e.g., saturated or superheated. Property data for saturated water are provided in Tables B-1 and B-2. Property information for superheated water is found in Table B-3, which provides an array of property information over a range of pressure and temperature; specific entropy is one of the properties provided at each temperature and pressure. Similar property tables for R134a appear in Appendix C.

The properties of saturated liquid water and saturated vapor water are found in Tables B-1 (in equal increments of temperature) and B-2 (in equal increments of pressure). The specific entropy of saturated liquid is referred to as s_f and the specific entropy of saturated vapor is referred to as s_g. The mass-average specific entropy of a two-phase state, s, is the ratio of the total entropy (of both phases) to the total mass (of both phases). The relationships between s and quality, x, are:

$$s = s_f + x\,(s_g - s_f) \tag{6-17}$$

and

$$x = \frac{(s - s_f)}{(s_g - s_f)} \tag{6-18}$$

6.3.2 EES Fluid Property Data

The built-in thermodynamic property functions in EES are introduced in Section 2.4. The Entropy function provides the specific entropy and is called using the same protocol as the Volume, Pressure, IntEnergy, Enthalpy or Temperature functions. For example, the EES code below provides the specific entropy of carbon dioxide at $P = 5 \times 10^6$ Pa and $T = 350$ K:

```
$UnitSystem SI, Mass, J, Pa, K
s=Entropy(CarbonDioxide, P=5e6 [Pa],T=350 [K])
```

which results in $s = -674.7$ J/kg-K. Note that specific entropy can have negative values depending on the arbitrary choice of the reference state at which specific entropy is assigned to be zero. The state of a substance can be specified using specific entropy as one of the two required properties. For example, the specific volume of carbon dioxide at $P = 5 \times 10^6$ Pa and $s = -600$ J/kg is obtained from:

```
v=volume(CarbonDioxide,P=5e6,s=-600 [J/kg-K])
```

which results in $v = 0.01241$ m^3/kg.

EXAMPLE 6.3-1: ENTROPY CHANGE DURING A PHASE CHANGE

When a pure fluid, such as water, is boiled at constant pressure its temperature remains constant.

a) Using this fact, apply a fundamental property relation in order to calculate the change in the specific entropy of water as it changes state from saturated liquid to saturated vapor at atmospheric pressure. Compare the result with the values obtained from Table B-2 in Appendix B and from the property functions in EES.

Either Eq. (6-13) or Eq. (6-16) can be used for this problem. However, Eq. (6-16) is more convenient because the pressure and temperature remain constant during the phase change:

$$dh = T\,ds + v\,\underbrace{dP}_{=0} \tag{1}$$

The change in pressure is zero for the phase change. Therefore Eq. (1) can be simplified to:

$$dh = T\,ds \tag{2}$$

Integrating Eq. (2) for a process that starts in a saturated liquid state (designated with subscript f) and ends in a saturated vapor state (subscript g) at the same pressure results in:

$$\int_{h_f}^{h_g} dh = \int_{s_f}^{s_g} T\,ds$$

Because the temperature is constant, it can be pulled out of the integrand:

$$\int_{h_f}^{h_g} dh = T\int_{s_f}^{s_g} ds$$

Carrying out the integration leads to:

$$h_g - h_f = T\left(s_g - s_f\right) \tag{3}$$

Equation (3) is rearranged:

$$s_g - s_f = \frac{(h_g - h_f)}{T}$$

At atmospheric pressure, the temperature of the water remains constant at 373.16 K (100°C) during the boiling process. Using specific enthalpy data in Appendix B from either Table B-1 at 100°C or Table B-2 at 101.325 kPa results in:

$$(s_g - s_f) = \frac{(h_g - h_f)}{T} = \frac{2256.4\,\text{kJ/kg}}{373.15\,\text{K}} = 6.0470\,\text{kJ/kg - K}$$

Note that the same result is obtained by subtracting the values of s_g and s_f that are provided in the table. Alternatively, the following lines in EES can be used to determine the change in the specific entropy of water as it changes phase:

```
$UnitSystem SI K Pa J
S$='Steam'
T=convertTemp(C,K,100 [C])
DELTAs_EES=(entropy(S$,T=T,x=1)-entropy(S$,T=T,x=0))
```

which results in $s_g - s_f = 6.048$ kJ/kg-K. Note that if the fluid name in the string variable S$ is changed from 'Steam' to 'Steam_IAPWS' then EES will return exactly the same result that is obtained from the tables in Appendix B. The fact that the result obtained using the fundamental property relation is in agreement with the data in Appendix B and provided by EES should not be a surprise. The fundamental property relation is used to determine the specific entropy values that are provided by these databases.

6.3.3 Entropy Relations for Ideal Gases

Equation (6-13) can be rearranged to determine the differential specific entropy change of a pure fluid:

$$ds = \frac{du}{T} + \frac{P}{T}dv \tag{6-19}$$

If the fluid obeys the ideal gas law then integration of Eq. (6-19) is relatively simple. The specific internal energy of an ideal gas is only a function of temperature, as noted in Section 3.3.3, and therefore the differential change in the specific internal energy of an ideal gas is given by:

$$du = c_v\, dT \tag{6-20}$$

where c_v is the specific heat capacity at constant volume, which is a measurable property. Substituting Eq. (6-20) and the ideal gas law, $P\,v = R\,T$, into Eq. (6-19) results in:

$$ds = \frac{c_v}{T}dT + \frac{R}{v}dv \tag{6-21}$$

Integration of Eq. (6-21) between two states provides:

$$\int_{s_1}^{s_2} ds = \int_{T_1}^{T_2} \frac{c_v}{T}dT + R\int_{v_1}^{v_2} \frac{dv}{v} \tag{6-22}$$

which leads to:

$$\boxed{s_2 - s_1 = \int_{T_1}^{T_2} \frac{c_v}{T}dT + R\ln\left(\frac{v_2}{v_1}\right) \quad \text{for an ideal gas}} \tag{6-23}$$

In general, the specific heat capacity at constant volume is a function of temperature. However, it is often appropriate to approximate c_v as being constant. This approximation is valid when c_v is a weak function of temperature and/or the difference between T_1 and T_2 is relatively small. In this case, Eq. (6-23) can be simplified to:

$$\boxed{s_2 - s_1 = c_v\ln\left(\frac{T_2}{T_1}\right) + R\ln\left(\frac{v_2}{v_1}\right) \quad \text{for an ideal gas with constant } c_v} \tag{6-24}$$

The same algebra can be applied to the fundamental relation written in terms of specific enthalpy, Eq. (6-16), in order to obtain specific entropy as a function of temperature and pressure. The specific enthalpy of an ideal gas is only a function of temperature, as discussed in Section 4.2.3; therefore:

$$dh = c_P\, dT \tag{6-25}$$

where c_P is the specific heat capacity at constant pressure. For an ideal gas, c_P and c_v are related by:

$$c_P - c_v = R \tag{6-26}$$

as shown in Eq. (4-37). Substituting Eq. (6-25) and the ideal gas law into Eq. (6-16) provides a relation for the differential change in specific entropy in terms of differential changes in temperature and pressure for an ideal gas:

$$ds = \frac{c_P}{T}dT - \frac{R}{P}dP \tag{6-27}$$

Integration of Eq. (6-27) from state 1 to state 2 leads to:

$$\boxed{s_2 - s_1 = \int_{T_1}^{T_2} \frac{c_P}{T}dT - R\ln\left(\frac{P_2}{P_1}\right) \quad \text{for an ideal gas}} \tag{6-28}$$

For small changes in temperature it may be appropriate to neglect the temperature dependence of the constant pressure specific heat capacity. In this case, Eq. (6-28) can be simplified to:

$$\boxed{s_2 - s_1 = c_P \ln\left(\frac{T_2}{T_1}\right) - R\ln\left(\frac{P_2}{P_1}\right) \quad \text{for an ideal gas with constant } c_P} \tag{6-29}$$

The specific heat capacities at constant volume and constant pressure are provided in Appendix D for several gases at 25° C. Note that there is no reason to use the approximate equations (6-24) or (6-29) when solving a problem in EES because the Entropy function in EES includes the temperature dependence of the specific heat capacity, as indicated in Eqs. (6-23) or (6-28), for substances that are modeled as an ideal gas. However, these equations are convenient when an approximate result is needed and EES is not available.

Substituting the ideal gas law into Eq. (6-29) and simplifying leads to a relationship between specific entropy, specific volume, and pressure:

$$\boxed{s_2 - s_1 = c_v \ln\left(\frac{P_2}{P_1}\right) + c_P \ln\left(\frac{v_2}{v_1}\right) \quad \text{for an ideal gas with constant } c_v \text{ and } c_P} \tag{6-30}$$

In many problems we will be interested in isentropic processes (i.e., processes in which the specific entropy remains constant). Equations (6-24), (6-29), and (6-30) can be used to develop relationships between the temperature, pressure, and specific volume of an ideal gas with constant specific heat capacities undergoing an isentropic process. For example, setting the entropy change to zero in Eq. (6-29) provides a relation between pressure and temperature:

$$c_P \ln\left(\frac{T_2}{T_1}\right) = R\ln\left(\frac{P_2}{P_1}\right) \tag{6-31}$$

Equation (6-31) can be rearranged:

$$\ln\left(\frac{T_2}{T_1}\right) = \ln\left[\left(\frac{P_2}{P_1}\right)^{\frac{R}{c_P}}\right] \tag{6-32}$$

Taking the anti-log of both sides of Eq. (6-32) allows the relation between pressure and temperature for an isentropic process involving an ideal gas with constant specific heat capacities to be expressed as:

$$\boxed{\frac{T_2}{T_1} = \left(\frac{P_2}{P_1}\right)^{\frac{R}{c_P}} \quad \begin{array}{l}\text{for an ideal gas with constant } c_P\\ \text{undergoing an isentropic process}\end{array}} \tag{6-33}$$

Equation (6-33) is often written in terms of the ratio of the specific heat capacities:

$$k = \frac{c_P}{c_v} \tag{6-34}$$

Substituting Eq. (6-34) into Eq. (6-33) provides:

$$\boxed{\frac{T_2}{T_1} = \left(\frac{P_2}{P_1}\right)^{\frac{k-1}{k}} \quad \text{for an ideal gas with constant } k \text{ undergoing an isentropic process}} \tag{6-35}$$

Similar relationships can be found from Eqs. (6-24) and (6-30):

$$\boxed{\frac{T_2}{T_1} = \left(\frac{v_2}{v_1}\right)^{\frac{-R}{c_v}} = \left(\frac{v_2}{v_1}\right)^{1-k} \quad \text{for an ideal gas with constant } k \text{ undergoing an isentropic process}} \tag{6-36}$$

$$\boxed{\frac{P_2}{P_1} = \left(\frac{v_2}{v_1}\right)^{\frac{-c_P}{c_v}} = \left(\frac{v_2}{v_1}\right)^{-k} \quad \text{for an ideal gas with constant } k \text{ undergoing an isentropic process}} \tag{6-37}$$

EXAMPLE 6.3-2: SPECIFIC ENTROPY CHANGE FOR NITROGEN

The constant pressure specific heat of nitrogen gas at low pressure is given by:

$$c_P = a + b\,\theta^{-1.5} + c\,\theta^{-2} + d\,\theta^{-3} \tag{1}$$

where $a = 1{,}394.3$ J/kg-K, $b = -18{,}302$ J/kg-K, $c = 38{,}292$ J/kg-K, and $d = -29{,}286$ J/kg-K. The dimensionless temperature, θ, in Eq. (1) is defined according to:

$$\theta = \frac{T}{T_{ref}} \tag{2}$$

where $T_{ref} = 100$ K. This specific heat relation is applicable for temperatures between 300 K and 3500 K according to Van Wylen and Sonntag (1986).

a) Use Eqs. (1) and (2) to calculate the change in specific entropy of nitrogen for a constant specific volume process in which the temperature is changed from $T_1 = 30°\text{C}$ to $T_2 = 150°\text{C}$.

The inputs are entered in EES:

```
$UnitSystem SI K Pa J
"Coefficients of cP function"
a=1394.3 [J/kg-K]
b=-18302 [J/kg-K]
c=38292 [J/kg-K]
d=-29286 [J/kg-K]
T_ref=100 [K]                      "reference temperature"
T_1=converttemp(C,K,30[C])         "temperature at state 1"
T_2=converttemp(C,K,150[C])        "temperature at state 2"
```

Either Eq. (6-23) or Eq. (6-28) can be used to determine the specific entropy change of the nitrogen. Equation (6-23) is convenient because the second term on the right side is zero for a constant specific volume process, allowing it to be simplified to:

$$s_2 - s_1 = \int_{T_1}^{T_2} \frac{c_v}{T}\, dT \tag{3}$$

Equation (3) requires c_v whereas Eq. (1) provides c_P. These two specific heat capacities are related by Eq. (4-37) for an ideal gas:

$$c_P - c_v = R \tag{4}$$

where R is the gas constant.

```
R=R#/MolarMass(N2)          "gas constant"
```

Substituting Eqs. (4) and (2) into Eq. (1) provides a relation for c_v:

$$c_v = (a - R) + \left(b\, T_{ref}^{1.5}\right) T^{-1.5} + \left(c\, T_{ref}^{2}\right) T^{-2} + \left(d\, T_{ref}^{3}\right) T^{-3} \tag{5}$$

Equation (5) is substituted into Eq. (3):

$$s_2 - s_1 = \int_{T_1}^{T_2} \left[(a - R)\, T^{-1} + \left(b\, T_{ref}^{1.5}\right) T^{-2.5} + \left(c\, T_{ref}^{2}\right) T^{-3} + \left(d\, T_{ref}^{3}\right) T^{-4}\right] dT$$

and integrated analytically:

$$s_2 - s_1 = (a - R)\ln\left(\frac{T_2}{T_1}\right) - \frac{\left(b\, T_{ref}^{1.5}\right)}{1.5}\left(T_2^{-1.5} - T_1^{-1.5}\right) - \frac{\left(c\, T_{ref}^{2}\right)}{2}\left(T_2^{-2} - T_1^{-2}\right) - \frac{\left(d\, T_{ref}^{3}\right)}{3}\left(T_2^{-3} - T_1^{-3}\right) \tag{6}$$

```
DELTAs=(a-R)*ln(T_2/T_1)-(b*T_ref^1.5)*(T_2^(-1.5)-T_1^(-1.5))/1.5&
  -(c*T_ref^2)*(T_2^(-2)-T_1^(-2))/2-(d*T_ref^3)*(T_2^(-3)-T_1^(-3))/3      "change in specific entropy"
```

The change in specific entropy calculated using Eq. (6) is 248.6 J/kg-K. This result can be compared to the property routine programmed in EES for nitrogen modeled as an ideal gas. The Entropy function in EES is called using the specified temperatures and a constant, arbitrary value of the specific volume.

```
DELTAs_EES=entropy(N2,T=T_2,v=1 [m^3/kg])-entropy(N2,T=T_1,v=1 [m^3/kg])
                                              "change in specific entropy, from EES"
```

which provides $s_2 - s_1 = 248.0$ J/kg-K. This good agreement should not be surprising since internally, EES is doing the same calculation but with a slightly different relation for c_P.

b) Calculate the specific entropy change for the process described in part (a) by assuming that the specific heat capacity is constant.

Equation (5) is used to evaluate the specific heat capacity at constant volume at the average temperature, $c_{v,avg}$:

```
T_avg=(T_1+T_2)/2                                          "average temperature"
c_v_avg=(a-R)+(b*T_ref^1.5)*T_avg^(-1.5)+(c*T_ref^2)*T_avg^(-2)+(d*T_ref^3)*T_avg^(-3)
                          "constant volume specific heat capacity at average temperature"
```

which results in $c_{v,avg} = 744.9$ J/kg-K. Equation (6-24) is used to compute the entropy change (recognizing that the specific volume does not change):

$$s_2 - s_1 = c_{v,avg} \ln\left(\frac{T_2}{T_1}\right)$$

```
DELTAs_app=c_v_avg*ln(T_2/T_1)          "change in specific entropy with constant c_v"
```

which provides $s_2 - s_1 = 248.4$ J/kg-K. Using a constant value of c_v evaluated at the average temperature usually produces an accurate result when the difference in temperatures is small, as it is in this case.

6.3.4 Entropy Relations for Incompressible Substances

The specific volume of an incompressible substance is constant. Therefore, for an incompressible substance, the fundamental property relation given by Eq. (6-13) becomes:

$$du = T\,ds \tag{6-38}$$

In Section 3.3.4, it was shown that the specific internal energy of an incompressible substance is only a function of temperature and that the differential change in the specific internal energy of an incompressible substance is given by:

$$du = c\,dT \tag{6-39}$$

where c is the specific heat capacity. Substituting Eq. (6-39) into Eq. (6-38) allows it to be expressed as:

$$ds = \frac{c}{T}dT \tag{6-40}$$

Integrating Eq. (6-40) leads to:

$$\int_{s_1}^{s_2} ds = \int_{T_1}^{T_2} \frac{c}{T}dT \tag{6-41}$$

or

$$\boxed{s_2 - s_1 = \int_{T_1}^{T_2} \frac{c}{T}dT \quad \text{for an incompressible substance}} \tag{6-42}$$

Equation (6-42) implies that an incompressible substance undergoing an isentropic process will not experience a temperature change. In many cases, it is appropriate to assume that the specific heat capacity is constant, leading to:

$$\boxed{s_2 - s_1 = c\ln\left(\frac{T_2}{T_1}\right) \quad \text{for an incompressible substance with constant } c} \tag{6-43}$$

6.4 A General Statement of the Second Law of Thermodynamics

The Clausius statement of the Second Law of Thermodynamics, presented in Chapter 5, is:

> *Heat cannot pass spontaneously from a body of lower temperature to a body of higher temperature.*

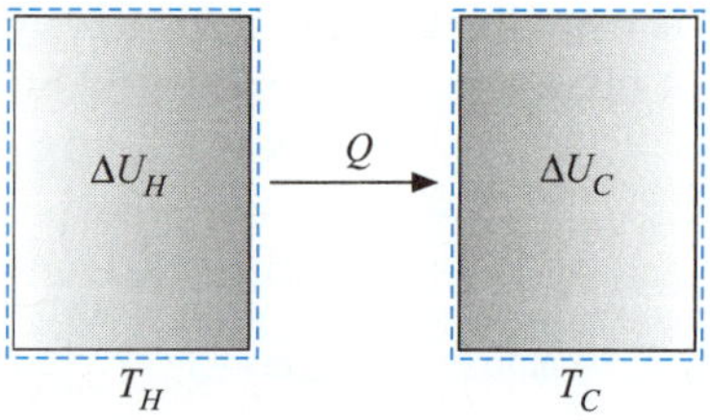

Figure 6-6: Two thermal reservoirs undergoing an unconstrained heat transfer process.

If this statement is accepted, then the other Second Law statements discussed in Chapter 5 can be proven. The Clausius statement was favored from those listed in Chapter 5 because it is intuitive. However, the Second Law is not limited to processes involving heat. A more general statement of the Second Law can be formulated in terms of the property entropy defined in Section 6.1.

$$\boxed{\Delta S_{total} = \Delta S_{system} + \Delta S_{surr} \geq 0} \tag{6-44}$$

Equation (6-44) indicates that the total amount of entropy (i.e., the entropy of the system and its surroundings) must always increase. Surely, there is no other law in the physical sciences that can be stated so simply or applied so generally.

Let's first demonstrate that Eq. (6-44) is consistent with the Clausius statement of the Second Law of Thermodynamics by considering two thermal reservoirs that are at different temperatures, T_H and T_C where $T_H > T_C$, as shown in Figure 6-6. Recall that a thermal reservoir is a system that remains at a constant temperature even as heat is transferred to it or from it.

The two thermal reservoirs are initially in a state of constrained equilibrium as a result of perfect insulation that prevents heat transfer between them. However, we will disturb this equilibrium condition by removing the insulation between the thermal reservoirs for a period of time that is sufficient for heat Q to transfer from the higher temperature reservoir at T_H to the lower temperature reservoir at T_C. Based on the discussion in Section 5.2, this heat transfer process must be irreversible.

An energy balance on the higher temperature reservoir (i.e., the system indicated by the dashed lines on the left side of Figure 6-6) is:

$$0 = Q + \Delta U_H \tag{6-45}$$

where ΔU_H is the change in the internal energy of the high temperature reservoir. The change in entropy of the high temperature reservoir can be determined by integrating Eq. (6-7):

$$\Delta S = \int \left. \frac{\delta Q}{T} \right|_{rev} \tag{6-46}$$

where δQ is the differential amount of heat transfer to the system during a reversible process and T is the temperature at which heat crosses the system boundary. The temperature of the thermal reservoir and the system boundary shown in Figure 6-6 both remain at temperature T_H during the heat transfer process and that simplifies the integration in Eq. (6-46). This simplification is the primary reason for using thermal reservoirs in this example. In the process shown in Figure 6-6, the heat transfer is from the system and it occurs at constant temperature, T_H; therefore, Eq. (6-46) becomes:

$$\Delta S_H = -\frac{Q}{T_H} \tag{6-47}$$

There is another aspect of Eq. (6-46) that requires explanation. The definition of the entropy change in Eq. (6-46) is only applicable for a reversible process, as indicated by the *rev* subscript. However, we concluded that the process occurring in Figure 6-6 is irreversible, so why can Eq. (6-46) be applied in this case? Clearly, the overall process is irreversible because heat is transferred from a higher to a lower temperature. However, each of the systems defined in Figure 6-6 are internally in equilibrium. Temperature is assumed to be uniform everywhere within the high temperature reservoir. Temperature is also uniform everywhere within the low temperature reservoir. Therefore there is no irreversibility due to heat flow across a temperature difference occurring within either of these systems and Eq. (6-46) can be applied to both the high temperature and the low temperature thermal reservoirs. The irreversibility must occur between the two thermal reservoirs.

An energy balance on the low temperature thermal reservoir provides:

$$Q = \Delta U_C \tag{6-48}$$

The change in entropy of the low temperature reservoir is calculated according to:

$$\Delta S_C = \int \left.\frac{\delta Q}{T}\right|_{rev} = \frac{Q}{T_C} \tag{6-49}$$

The total energy change resulting from the process shown in Figure 6-6 is the sum of the energy change in each of the two thermal reservoirs:

$$\Delta U_{total} = \Delta U_H + \Delta U_C \tag{6-50}$$

Substituting Eqs. (6-45) and (6-48) into Eq. (6-50) provides:

$$\Delta U_{total} = -Q + Q = 0 \tag{6-51}$$

The total change in energy for this process is zero. Energy is a conserved property and this process does not violate the First Law of Thermodynamics.

The total entropy change resulting from this process is the sum of the entropy changes in each of the two thermal reservoirs:

$$\Delta S_{total} = \Delta S_H + \Delta S_C \tag{6-52}$$

Substituting Eqs. (6-47) and (6-49) into Eq. (6-52) leads to:

$$\Delta S_{total} = -\frac{Q}{T_H} + \frac{Q}{T_C} \tag{6-53}$$

Equation (6-53) can be rearranged:

$$\Delta S_{total} = \frac{Q\,(T_H - T_C)}{T_H\,T_C} \tag{6-54}$$

Note that the value of ΔS_{total} must be greater than zero provided that $Q > 0$ (i.e., the heat transfer is from T_H to T_C, as shown in Figure 6-6). Both temperatures in the denominator of Eq. (6-54) must be greater than zero on an absolute temperature scale (K or R) and the temperature difference in the numerator of Eq. (6-54) is greater than zero. This result is in agreement with the general statement of the Second Law provided in Eq. (6-44) and shows that this statement is consistent with the Clausius statement of the Second Law. The total change in entropy for any process will always be greater than or equal to zero. Entropy is not a conserved property.

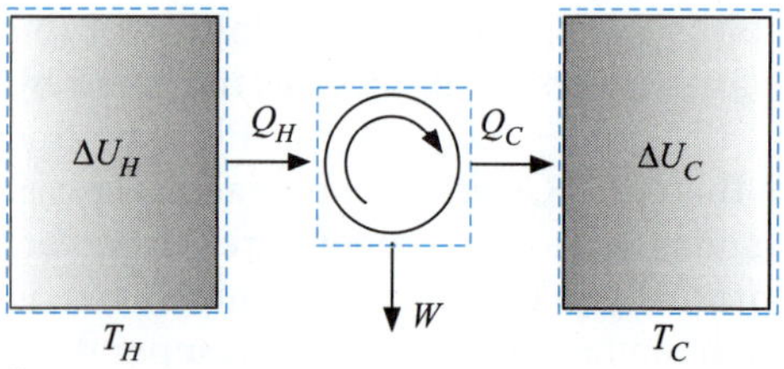

Figure 6-7: Two thermal reservoirs separated by a heat engine.

The total change in entropy will be zero only in the limiting case that the entire process is reversible. This result is also is evident in Eq. (6-54). If the temperatures of the two reservoirs approach one another then the heat transfer process becomes more reversible and the numerator of Eq. (6-54) approaches zero. All real processes are to some extent irreversible and therefore the total entropy change will be greater than zero for any real process. The magnitude of the total entropy change is related to the degree of irreversibility associated with the process.

Consider the theoretical, reversible process shown in Figure 6-7, which uses the same two temperature reservoirs shown in Figure 6-6. This time, the thermal interaction is constrained and used to produce work by inserting a cyclic, reversible heat engine between them. An amount of heat, Q_H, is provided to the heat engine from the reservoir at temperature T_H. The engine is operating at its maximum possible efficiency in order to produce work W_{out}. Heat is rejected to the low temperature thermal reservoir at temperature T_C in the amount Q_C. Based on the discussion in Section 5.3, the efficiency of the heat engine is:

$$\eta_{max} = \frac{W_{out}}{Q_H} = 1 - \frac{T_C}{T_H} \tag{6-55}$$

An energy balance on the heat engine (the system shown in the center of Figure 6-7) provides:

$$Q_H = W_{out} + Q_C \tag{6-56}$$

Substituting Eq. (6-56) into Eq. (6-55) and rearranging results in:

$$\frac{Q_H}{T_H} = \frac{Q_C}{T_C} \tag{6-57}$$

The total energy change resulting from this process is:

$$\Delta U_{total} = \Delta U_H + \Delta U_C + \Delta U_{engine} + \Delta U_{surr} \tag{6-58}$$

An energy balance on the high and low temperature reservoirs (the left and right control volumes in Figure 6-7) leads to:

$$0 = Q_H + \Delta U_H \tag{6-59}$$

$$Q_C = \Delta U_C \tag{6-60}$$

An energy balance on the engine for an integral number of cycles is:

$$\Delta U_{engine} = 0 \tag{6-61}$$

An energy balance on the surroundings (everything not included in the dashed lines in Figure 6-7) requires that:

$$W_{out} = \Delta U_{surr} \tag{6-62}$$

Equations (6-59) through (6-62) are substituted into Eq. (6-58) to yield:

$$\Delta U_{total} = -Q_H + Q_C + W_{out} \tag{6-63}$$

Comparing Eq. (6 63) with Eq. (6-56) leads to the conclusion:

$$\Delta U_{total} = 0 \tag{6-64}$$

As expected, the total energy change of the system and its surroundings for this process is zero. Energy is a conserved quantity.

The total entropy change resulting from this process is:

$$\Delta S_{total} = \Delta S_H + \Delta S_C + \Delta S_{engine} + \Delta S_{surr} \tag{6-65}$$

The entropy change for each thermal reservoir is determined by integrating Eq. (6-7), recognizing that the process occurring within each thermal reservoir is internally reversible.

$$\Delta S_H = -\frac{Q_H}{T_H} \tag{6-66}$$

$$\Delta S_C = \frac{Q_C}{T_C} \tag{6-67}$$

The change in entropy of the engine is zero because it is operated for an integral number of cycles. All of the properties of the engine are the same at the end of the process as they were at the start. Entropy is one of these properties.

$$\Delta S_{engine} = 0 \tag{6-68}$$

The entropy change of the surroundings is also zero. There is no heat transfer with the surroundings. Note that work is not associated with an entropy flow whereas heat is always accompanied by an entropy flow. This observation is the major reason for distinguishing heat and work as different types of energy transfers.

$$\Delta S_{surr} = 0 \tag{6-69}$$

Substituting Eqs. (6-66) through (6-69) into Eq. (6-65) requires that:

$$\Delta S_{total} = -\frac{Q_H}{T_H} + \frac{Q_C}{T_C} \tag{6-70}$$

Comparing Eq. (6-70) with Eq. (6-57) leads to:

$$\Delta S_{total} = 0 \tag{6-71}$$

The total entropy change is zero for this entirely reversible process. This result is consistent with the Second Law statement in Eq. (6-44). The word *total* appearing as a subscript in Eq. (6-44) is sometimes replaced with the word *universe* since total refers to the combination of the system and its surroundings, which is everything. If we take this definition literally, we are in a situation in which the energy of the universe is constant but the entropy of the universe is increasing. This realization is sobering because increased entropy results in reduced energy quality and thus reduced capability of doing work. The Second Law statement, $\Delta S_{universe} \geq 0$, is so powerful that it has been used by some theologians in an attempt to prove the existence of God; others have used it to attempt to disprove the existence of God. The fact that anyone might try to use the laws of thermodynamics for such a purpose is a testimony to the power of thermodynamics; surely no other physical science has been used in this manner. However, the First and Second Laws are nothing more than ingenious methods to mathematically describe observable phenomena.

EXAMPLE 6.4-1: ENTROPY GENERATED BY HEATING WATER

A perfectly insulated rigid tank contains m = 4 kg of water at P_1 = 100 kPa, as shown in Figure 1. Initially, three-quarters of the mass is in the liquid phase (i.e., the quality is $x_1 = 0.25$). An electric heater in the tank is activated and operated until all of the liquid in the tank is vaporized. Neglect the heat capacity of the tank material and the electrical heater.

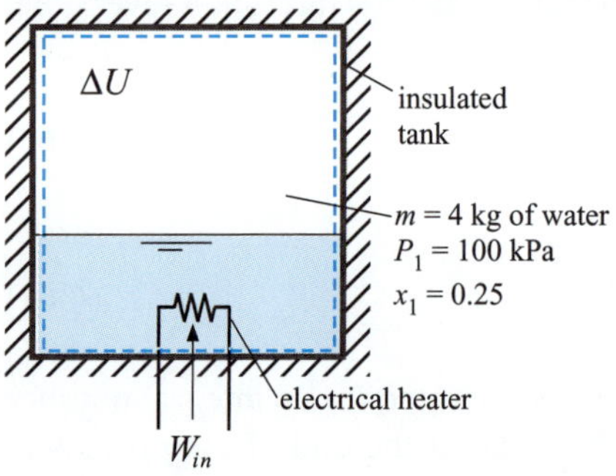

Figure 1: Rigid, insulated tank containing water and an electric heater.

a) Determine the final temperature and pressure of the water.

The known information is entered into EES.

```
$UnitSystem SI Radian Mass J K Pa
"Inputs"
m=4 [kg]                                "total mass of water in the tank"
P[1]=100 [kPa]*convert(kPa,Pa)          "initial pressure in the tank"
x[1]=0.25                               "initial quality"
x[2]=1.0                                "final quality"
```

State 1 is fixed by the given pressure and quality. The remaining thermodynamic properties (v_1, u_1, s_1, and T_1) are obtained:

```
v[1]=volume(Water,x=x[1],P=P[1])          "specific volume at state 1"
u[1]=intenergy(Water,x=x[1],P=P[1])       "specific internal energy at state 1"
s[1]=entropy(Water,x=x[1],P=P[1])         "specific entropy at state 1"
T[1]=temperature(Water,x=x[1],P=P[1])     "temperature at state 1"
```

Because neither the mass nor the volume of the water changes during the heating process, the specific volume must be constant:

$$v_2 = v_1$$

State 2 is fixed by the specific volume and quality ($x_2 = 1$). The remaining thermodynamic properties (P_2, u_2, s_2, and T_1) are obtained:

```
v[2]=v[1]                                  "specific volume at state 2"
u[2]=intenergy(Water,x=x[2],v=v[2])        "specific internal energy at state 2"
s[2]=entropy(Water,x=x[2],v=v[2])          "specific entropy at state 2"
T[2]=temperature(Water,x=x[2],v=v[2])      "temperature at state 2"
P[2]=pressure(Water,x=x[2],v=v[2])         "pressure at state 2"
```

which results in $T_2 = 420.1$ K and $P_2 = 438$ kPa.

b) Determine the electrical work required by the process.

The system used for the analysis includes the water in the tank and the electrical heater, as shown in Figure 1. The tank wall is assumed to have negligible heat capacity and, consequently, there is no heat transfer to the tank material. An energy balance on the system is:

$$W_{in} = m(u_2 - u_1)$$

```
W_in=m*(u[2]-u[1])          "electrical work"
```

which leads to $W_{in} = 6.47$ MJ.

c) Determine the total change in entropy associated with this process.

The total change in entropy is the change in entropy of the system (the water and electrical heater) plus the change in entropy of the surroundings.

$$\Delta S_{total} = \Delta S_{system} + \Delta S_{surr} \tag{1}$$

The change in entropy of the electrical heater and the tank material are both zero since their respective heat capacities are assumed to be zero. The system interacts with the surroundings only as a result of the electrical work. There is no entropy flow associated with work and therefore the entropy change of the surroundings is zero, allowing Eq. (1) to be simplified:

$$\Delta S_{total} = \Delta S_{system} = m(s_2 - s_1)$$

```
DELTAS_total=m*(s[2]-s[1])          "total entropy change associated with process"
```

Solving shows that $\Delta S_{total} = 16.2$ kJ/K. This is an irreversible process which is consistent with the increase in total entropy.

d) Plot the state points for the water on a temperature-specific entropy diagram.

A temperature-specific entropy (T-s) diagram for water can be produced in EES by selecting the Property Plot menu item in the Plots menu. It is helpful to show the lines of constant pressure corresponding to the initial (100 kPa) and final (437.88 kPa) states of the water. A constant specific volume line corresponding to the specific volume of the water (0.4247 m^3/kg) is also helpful. Check the boxes for lines of constant pressure and volume so that the Property Plot dialog appears as shown in Figure 2(a).

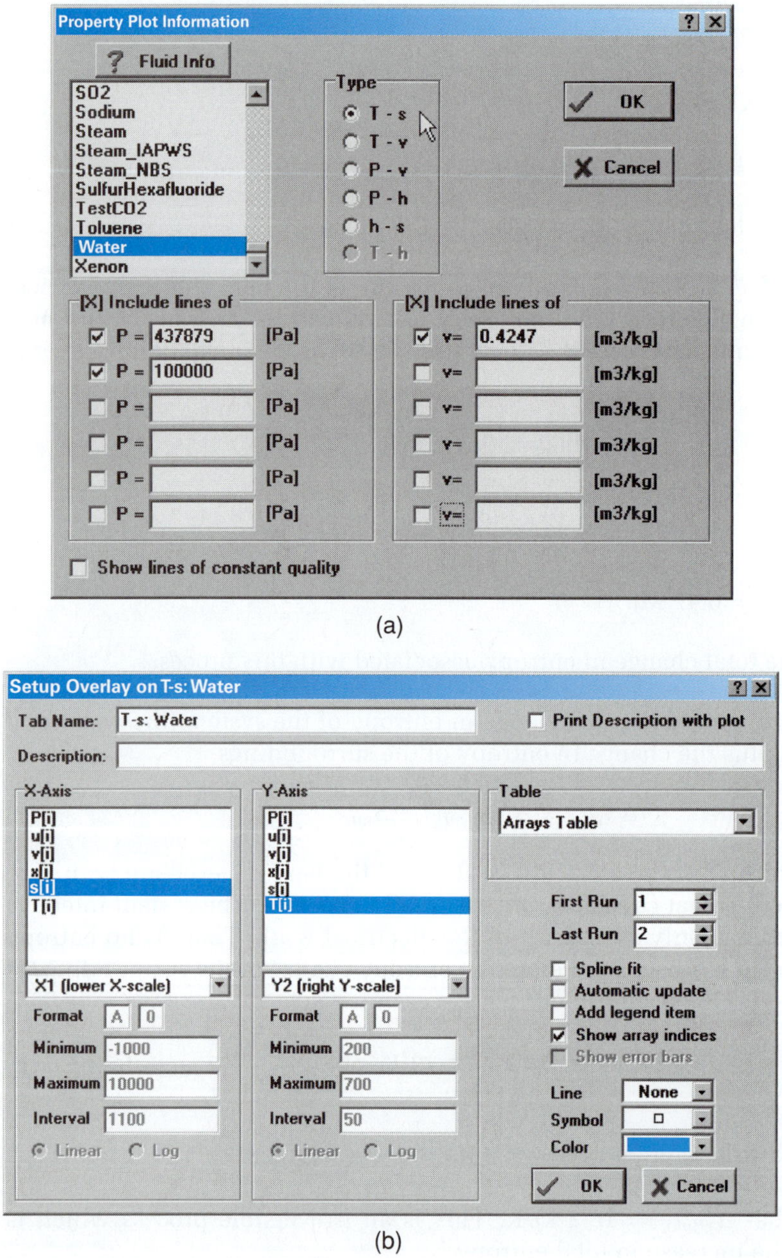

Figure 2: (a) Property Plot dialog and (b) Overlay Plot dialog.

Use the Overlay Plot menu item to overlay the state point information onto the property plot. Check the Show array indices option, as shown in Figure 2(b), in order to identify the state points on the plot. The resulting *T-s* diagram is shown in Figure 3.

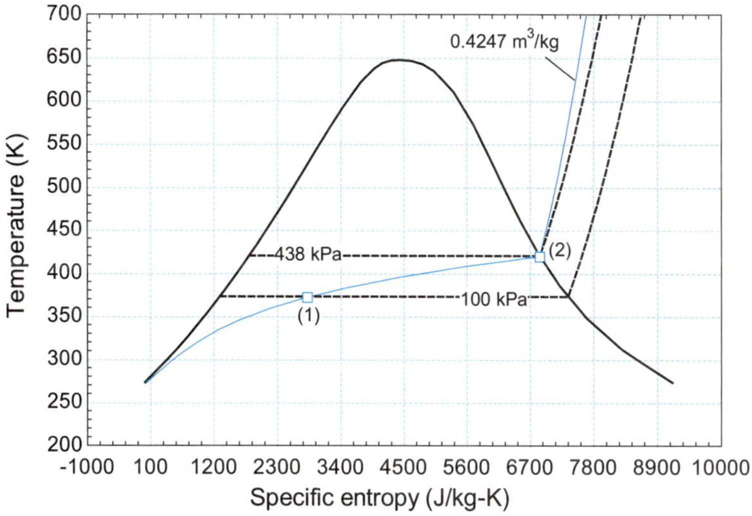

Figure 3: Temperature–specific entropy plot for the water heating process.

6.5 The Entropy Balance

Entropy is a property of matter, just like mass and energy. However, unlike mass and energy, entropy is not conserved. The Second Law requires that the total change in entropy (i.e., the change in the entropy of a system and its surroundings) be greater than zero in any real process. In this section, an entropy balance is introduced as a mechanism for conveniently including Second Law considerations in a thermodynamics problem.

6.5.1 Entropy Generation

Entropy is not a conserved quantity. It may be helpful to point out that most thermodynamic properties are also not conserved. Volume, for example, is not a conserved quantity. Consider a high pressure air tank that is slowly vented to the atmosphere. The volume of air leaving the tank (at the point where it enters the atmosphere) will be many times the volume of the tank itself. Temperature, pressure, and enthalpy are also not conserved. Only mass and energy are conserved quantities.

In order to solve problems involving entropy, it will be necessary to keep track of the entropy changes, generation, and transfers that occur during a process. The most convenient way to do this accounting is with an entropy balance, analogous to the mass and energy balances that have been used to solve problems in Chapters 1 through 4. The general balance equation was presented in Section 1.4:

$$In + Generated = Out + Destroyed + Stored \tag{6-72}$$

Entropy is not conserved. Within any system, entropy can be generated as a result of the irreversible processes that are occurring within the system. However, entropy can never be destroyed. Therefore, an entropy balance on a system written for a finite period of time is:

$$S_{in} + S_{gen} = S_{out} + \Delta S \tag{6-73}$$

where S_{in} and S_{out} are the entropy transfers into and out of the system across its boundary, respectively, S_{gen} is the entropy generated within the system as a consequence of irreversible processes occurring within the system, and ΔS is the change in the entropy of the system. An entropy balance written on a rate basis is:

$$\dot{S}_{in} + \dot{S}_{gen} = \dot{S}_{out} + \frac{dS}{dt} \tag{6-74}$$

where $\dot{S}_{in}$ and $\dot{S}_{out}$ are the rates at which entropy is transferred into and out of the system across its boundary, respectively, and $\dot{S}_{gen}$ is the rate at which entropy is generated within the system.

Any of the irreversible phenomena listed in Table 5-1 will contribute to the entropy generation term in Eqs. (6-73) or (6-74) if they occur within the system. The storage and rate of storage terms in Eqs. (6-73) and (6-74), respectively, represent the change and rate of change of the entropy of the system. The remaining terms in Eqs. (6-73) and Eq. (6-74) are related to the transfer of entropy across a boundary. Entropy can cross a system boundary in only two ways: with mass transfer across the boundary ($S_{m,in}$ and $S_{m,out}$) and with heat ($S_{Q,in}$ and $S_{Q,out}$). Therefore, Eqs. (6-73) and (6-74) can be written, respectively, as:

$$S_{m,in} + S_{Q,in} + S_{gen} = S_{m,out} + S_{Q,out} + \Delta S \tag{6-75}$$

$$\dot{S}_{m,in} + \dot{S}_{Q,in} + \dot{S}_{gen} = \dot{S}_{m,out} + \dot{S}_{Q,out} + \frac{dS}{dt} \tag{6-76}$$

Notice that there is no entropy associated with work because work carries no entropy; this is the primary distinction between heat and work and the reason for distinguishing these two forms of energy transfer.

Entropy is a property of matter. When mass enters or leaves an open system, entropy enters or leaves with it. The amount of entropy transfer associated with mass transfer is the product of the amount of mass crossing the boundary (m) and the specific entropy (s) of the mass at the point where it crosses the boundary. Therefore the amount of entropy entering the system due to mass transfer through a single inlet is:

$$S_{m,in} = m_{in}\, s_{in} \tag{6-77}$$

where m_{in} is the amount of mass entering the system and s_{in} is the specific entropy of that mass at the location it crosses the boundary. The rate of entropy transfer entering a system due to a mass flow rate through a single inlet is:

$$\dot{S}_{m,in} = \dot{m}_{in}\, s_{in} \tag{6-78}$$

where $\dot{m}_{in}$ is the mass flow rate entering the system. Similar equations can be written for the entropy transfer out of the system due to mass leaving the system. Note that the velocity or elevation of the mass entering the system (i.e., its kinetic and potential energy) have no effect on the entropy that is transferred with the mass.

According to Eq. (6-7), the amount of entropy transferred across a system boundary is the ratio of the amount of heat transferred across the boundary to the absolute temperature at the boundary. Therefore, the amount of entropy transferred into the system having a constant temperature boundary due to a heat transfer to the system (Q_{in}) is:

$$S_{Q,in} = \frac{Q_{in}}{T_b} \tag{6-79}$$

where T_b is the temperature at the boundary where Q_{in} is transferred. The rate of entropy transfer into the system is:

$$\dot{S}_{Q,in} = \frac{\dot{Q}_{in}}{T_b} \tag{6-80}$$

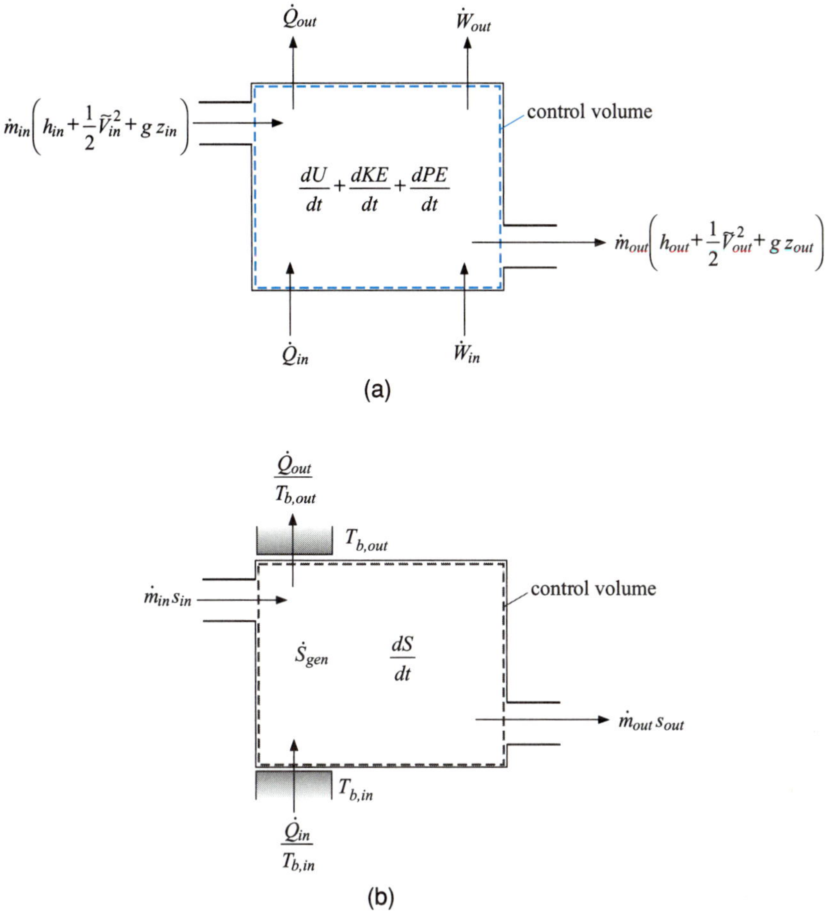

Figure 6-8: (a) General energy balance and (b) general entropy balance on an open system with a single inlet and outlet.

where $\dot{Q}_{in}$ is the rate of heat transfer into the system. Similar equations can be written for the entropy transfer out of the system due to heat transfer. Careful attention must be paid to the system definition and its boundary in order to properly apply Eqs. (6-77) through (6-80).

Figure 6-8(a) illustrates a general energy balance written on a rate basis for an open system with a single inlet and outlet. The energy balance for this system is:

$$\dot{m}_{in}\left(h_{in}+\frac{1}{2}\tilde{V}_{in}^2+g\,z_{in}\right)+\dot{Q}_{in}+\dot{W}_{in}=\dot{m}_{out}\left(h_{out}+\frac{1}{2}\tilde{V}_{out}^2+g\,z_{out}\right)$$
$$+\dot{Q}_{out}+\dot{W}_{out}+\frac{dU}{dt}+\frac{dKE}{dt}+\frac{dPE}{dt} \tag{6-81}$$

The entropy balance on the same open system is shown in Figure 6-8(b) and it leads to:

$$\boxed{\dot{m}_{in}\,s_{in}+\frac{\dot{Q}_{in}}{T_{b,in}}+\dot{S}_{gen}=\dot{m}_{out}\,s_{out}+\frac{\dot{Q}_{out}}{T_{b,out}}+\frac{dS}{dt}} \tag{6-82}$$

where $T_{b,in}$ and $T_{b,out}$ are the temperatures at the boundaries where heat is transferred into and out of the system, respectively. $T_{b,in}$ and $T_{b,out}$ are assumed to be constant.

An entropy balance written on an increment basis for the system shown in Figure 6-8 is:

$$\boxed{m_{in}\, s_{in} + \frac{Q_{in}}{T_{b,in}} + S_{gen} = m_{out}\, s_{out} + \frac{Q_{out}}{T_{b,out}} + \Delta S} \tag{6-83}$$

In general there may be multiple locations on a control volume where mass and heat are crossing the boundary; in this case, these transfer terms must be summed. Using the convention that heat transfer rate $\dot{Q}_i$ is a positive quantity when it represents an energy transfer to the system, Eq. (6-82) becomes:

$$\boxed{\sum_{i=1}^{\#inlets} \dot{m}_{in,i}\, s_{in,i} + \sum_{i=1}^{\#heat\ terms} \frac{\dot{Q}_i}{T_{b,i}} + \dot{S}_{gen} = \sum_{i=1}^{\#outlets} \dot{m}_{out,i}\, s_{out,i} + \frac{dS}{dt}} \tag{6-84}$$

6.5.2 Solution Methodology

An additional step can now be added to the general methodology for solving thermodynamics problems that is presented in Section 4.3 in order to include Second Law considerations. A summary of the revised methodology is as follows:

1. Carefully review the problem statement and the information that is known
2. Choose the system
3. Apply a mass balance on the system, Eq. (4-5)
4. Apply an energy balance on the system, Eq. (4-22)
5. Apply an entropy balance on the system, Eq. (6-84)

6.5.3 Choice of System Boundary

The entropy generation term in an entropy balance is a result of irreversible processes occurring within the system boundaries. In many cases, the purpose of the entropy balance is to determine the rate of entropy generation; this term in the entropy balance cannot be determined in any other independent manner. Alternatively, if it is known that there are no irreversible processes occurring within the system (i.e., the process occurring within the system can be approximated as being reversible) then the rate of entropy generation can be set to zero and the entropy balance can be used to solve for some other quantity. The choice of the system boundary used to solve the problem is important. Although an entropy balance can be applied to any system, careful selection of the system is required in order for it to be useful. In almost all cases, the best choice for a system boundary is one that either (1) encompasses all of the irreversibilities or (2) excludes all of the irreversibilities associated with the process. The first choice will allow the calculation of the total entropy generation associated with the process. The second choice allows the entropy generation term to be set to zero and therefore the entropy balance can be used to solve for some other quantity of interest.

System Encloses all Irreversible Processes

Application of the Second Law to irreversible processes allows the calculation of the entropy generation rate. Depending on the system choice, entropy can be generated in the system and in the surroundings. The total entropy generation is the sum of the entropy generation in the system and the surroundings. It is almost always the total entropy generation associated with a particular process that is of interest. The total entropy generation is directly obtained from Eq. (6-84) only if the system boundary is chosen to enclose all irreversible processes that occur as a result of the process being considered.

The total entropy generation associated with a process is of interest because entropy is related to the quality of energy and increased entropy corresponds to the degradation of this quality. Reduced quality of energy is related to a reduction in the capability to do work. We may interpret the effects that lead to entropy generation as an unexploited equilibration that could otherwise have been used to produce work. For example, a thermal equilibration could have been used to drive a heat engine or a pressure equilibration could have been used to energize a turbine. The loss of work producing capability (i.e., the degradation in the quality of energy) is directly related to the total entropy generation according to:

$$\boxed{\text{Amount of lost work potential} = T_0\, S_{gen}} \tag{6-85}$$

where T_0 is the temperature of the surroundings, sometimes referred to as the *dead state temperature*. The basis of Eq. (6-85) is related to the property *exergy* that is discussed in Chapter 7; however, Eq. (6-85) can be used without formally considering exergy. The product $T_0\, S_{gen}$ is often called the *"lost work"*. However, the work itself (a form of energy transfer) is not lost; rather it is the capability to do work that is lost when entropy is generated.

There is also a direct relationship between time and entropy generation that can be summarized with the following facts:

1. The total entropy generation is zero for processes that are completely reversible. Processes that are reversible occur at an infinitely slow rate (i.e., they do not occur at all). A zero entropy generation rate can only occur for infinitely slow processes.
2. Irreversible processes occur at a finite rate. Everything else being equal, the greater the rate of the process, the greater the rate of entropy generation. The design of real systems must then provide a compromise between reducing entropy generation (leading to higher efficiency) and increasing entropy generation (leading to higher power output). These design compromises for power cycles are the subject of Section 8.6.
3. According to the Second Law, entropy generation can never be negative. Negative entropy generation would allow a process to run backwards in time and thereby permit travel back in time. The Second Law effectively states that time travel is not possible.

Example 6.5-1 illustrates how it is possible to select different systems in order to separately consider different irreversible processes and concludes by selecting a system that includes all of the irreversible processes.

EXAMPLE 6.5-1: AIR HEATING SYSTEM

Air enters a duct with a volumetric flow rate of $\dot{V}_{in} = 0.27$ m³/s. The air is heated from $T_{in} = 15°C$ to $T_{out} = 35°C$ at atmospheric pressure using an electrical heater that draws $i = 15$ amps through an $R = 30$ ohm electrical resistor, as shown in Figure 1. The surface temperature of the heater has been measured and found to be $T_{htr} = 46°C$. The air outside of the duct is at $T_{sur} = 15°C$.

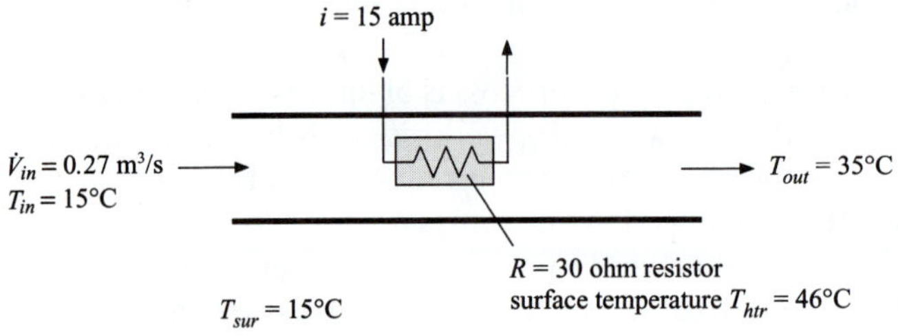

Figure 1: Air heating system.

a) Determine the heat transfer from the duct to the surroundings.

The inputs are entered in EES:

```
$UnitSystem SI K Pa J Mass Radian
V_dot_in=0.27 [m^3/s]                  "volumetric flow rate of air entering duct"
T_in=converttemp(C,K,15 [C])           "inlet temperature"
T_out=converttemp(C,K,35 [C])          "outlet temperature"
i=15 [amp]                             "heater current"
R=30 [ohm]                             "heater resistance"
T_htr=converttemp(C,K,46 [C])          "heater surface temperature"
T_sur=converttemp(C,K,15 [C])          "temperature of the surroundings"
```

The state of the air entering the duct is fixed by its temperature and pressure. The specific volume, specific enthalpy, and specific entropy of the entering air (v_{in}, h_{in}, and s_{in}) are computed:

```
v_in=volume(Air,T=T_in,P=Po#)          "specific volume of air at the inlet"
h_in=enthalpy(Air,T=T_in)              "specific enthalpy"
s_in=entropy(Air,T=T_in,P=Po#)         "specific entropy"
```

The mass flow rate of air entering the duct is:

$$\dot{m}_{in} = \frac{\dot{V}_{in}}{v_{in}}$$

The electrical power provided to the heater is:

$$\dot{W}_{htr} = i^2 R$$

```
m_dot=V_dot_in/v_in          "mass flow rate"
W_dot_htr=i^2*R              "electrical power"
```

The state of the air leaving the duct is fixed by its temperature and pressure. The specific enthalpy and specific entropy of the leaving air (h_{out} and s_{out}) are computed:

```
h_out=enthalpy(Air,T=T_out)            "specific enthalpy of air at the outlet"
s_out=entropy(Air,T=T_out,P=Po#)       "specific entropy"
```

An energy balance on the entire duct, shown in Figure 2, is:

$$\dot{m}\,h_{in} + \dot{W}_{htr} = \dot{m}\,h_{out} + \dot{Q}_{duct}$$

where $\dot{Q}_{duct}$ is the rate of heat transfer from the duct to the surroundings.

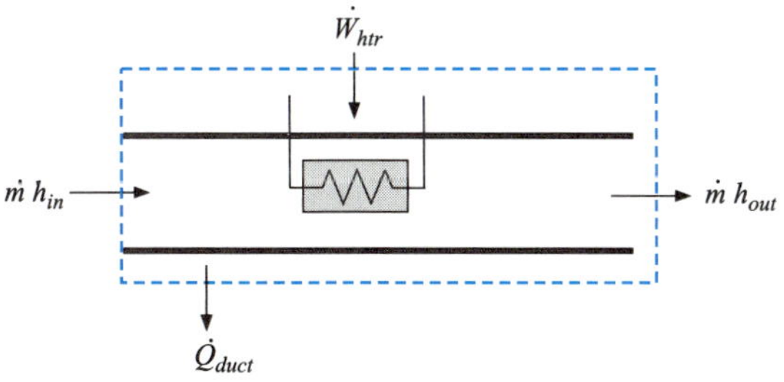

Figure 2: Energy balance on entire duct.

```
m_dot*h_in+W_dot_htr=m_dot*h_out+Q_dot_duct          "energy balance"
```

Solving results in $\dot{Q}_{duct} = 103.8$ W.

b) Calculate the rate of entropy generation in the heater alone.

An energy balance on the heater alone, the closed system shown in Figure 3(a), provides:

$$\dot{W}_{htr} = \dot{Q}_{htr}$$

where $\dot{Q}_{htr}$ is the rate of heat transfer from the heater to the air in the duct. Since the process occurs at steady-state, the surface temperature of the heater and all other properties of the heater are constant so an entropy balance on the heater alone, shown in Figure 3(b), is:

$$\dot{S}_{gen,htr} = \frac{\dot{Q}_{htr}}{T_{htr}}$$

```
W_dot_htr=Q_dot_htr                    "energy balance on heater"
S_dot_gen_htr=Q_dot_htr/T_htr          "entropy balance on heater"
```

which results in $\dot{S}_{gen,htr} = 21.15$ W/K.

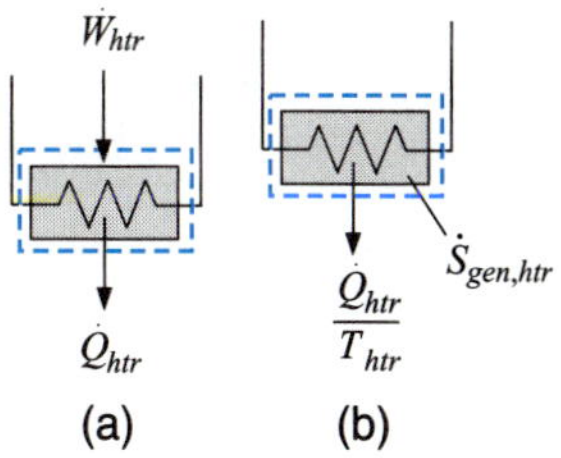

Figure 3: (a) Energy balance and (b) entropy balance on the heater.

c) Determine the entropy generation external to the heater.

An entropy balance on the duct, excluding the heater, is shown in Figure 4. Note that the boundary of the system through which $\dot{Q}_{htr}$ passes is placed at T_{htr} and the boundary of the system through which $\dot{Q}_{duct}$ passes is placed at T_{sur}. There are then two separate entropy transfers associated with these two heat transfer processes. An entropy balance on this system is:

$$\dot{m}\,s_{in} + \frac{\dot{Q}_{htr}}{T_{htr}} + \dot{S}_{gen,duct} = \frac{\dot{Q}_{duct}}{T_{sur}} + \dot{m}\,s_{out}$$

```
m_dot*s_in+Q_dot_htr/T_htr+S_dot_gen_duct=m_dot*s_out+Q_dot_duct/T_sur
"entropy balance on duct, excluding heater"
```

which results in $\dot{S}_{gen,duct} = 1.526$ W/K.

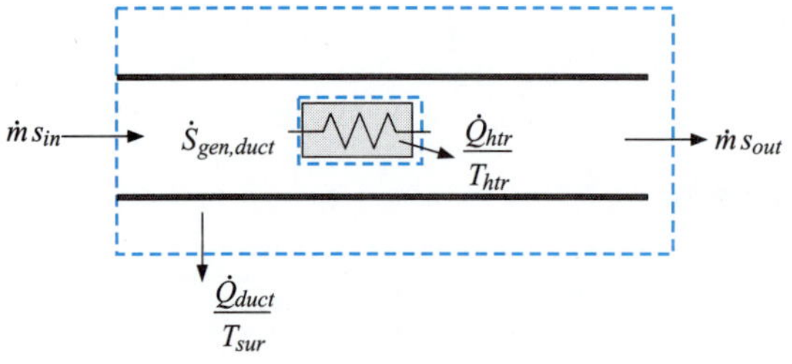

Figure 4: Entropy balance on duct, excluding the heater.

d) Calculate the total rate of entropy generation for this process.

The total rate of entropy generation is the sum of the entropy generation rates determined using the systems shown in Figure 3(b) and Figure 4.

$$\dot{S}_{gen,total} = \dot{S}_{gen,duct} + \dot{S}_{gen,htr}$$

```
S_dot_gen_total=S_dot_gen_duct+S_dot_gen_htr          "total rate of entropy generation"
```

which provides $\dot{S}_{gen,total} = 22.68$ W/K. Alternatively, the total rate of entropy generation can be calculated more directly with an entropy balance on a system that includes all of the irreversibilities. Figure 5 illustrates an entropy balance on a system that includes both the heater and the duct, with its boundaries extending to T_{sur}. An entropy balance on this system is:

$$\dot{m}\, s_{in} + \dot{S}_{gen,total} = \frac{\dot{Q}_{duct}}{T_{sur}} + \dot{m}\, s_{out}$$

```
m_dot*s_in+S_dot_gen_total=m_dot*s_out+Q_dot_duct/T_sur    "entropy balance on duct and heater"
```

which also results in $\dot{S}_{gen,total} = 22.68$ W/K.

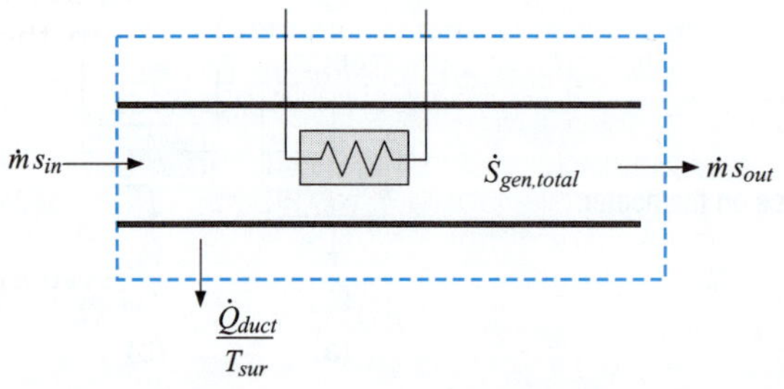

Figure 5: Entropy balance on duct and heater.

System Excludes all Irreversible Processes

Occasionally, it is useful to select system boundaries so that all of the irreversible processes occur outside of the system. Such a system is considered to be *internally reversible*. The tank emptying process in Example 6.5-2 is an example of a situation where this system choice is useful.

EXAMPLE 6.5-2: EMPTYING AN ADIABATIC TANK WITH IDEAL GAS (REVISITED)

In this problem, Example 4.5-2 is revisited and solved using an entropy balance. A tank is filled with air at $P_1 = 100$ psi and $T_1 = 20^\circ$C. Assume that the air can be modeled as an ideal gas with constant $c_v = 717.6$ J/kg-K. The tank is opened and air exits until the pressure in the tank reaches $P_2 = 14.7$ psi. Assume that the air within the tank undergoes an adiabatic process.

a) Determine the final temperature of the air in the tank.

In Example 4.5-2, the emptying process was analyzed using mass and energy balances on the air in the tank. By assuming that there are no pressure or temperature gradients in the air that is contained in the tank, it was possible to derive an expression for the final temperature:

$$T_2 = T_1 \left(\frac{P_2}{P_1}\right)^{\frac{R}{c_P}} \tag{1}$$

The same result can be obtained much more easily using the Second Law of Thermodynamics. The system illustrated in Figure 1 is a closed system that includes the mass of air that remains in the tank at the conclusion of the emptying process.

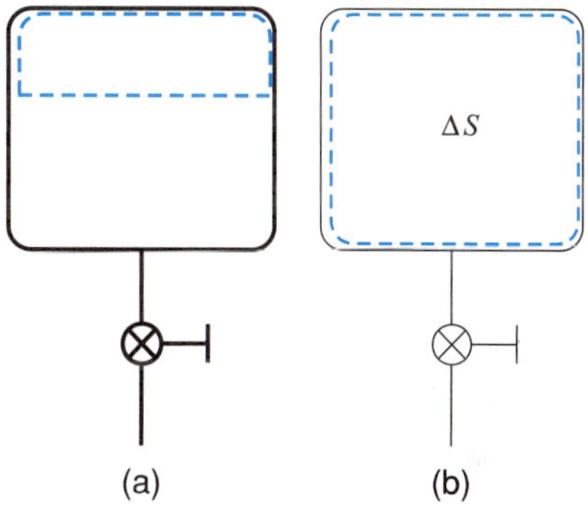

Figure 1: Entropy balance on the mass of air that remains in the tank. The closed system is shown (a) at the beginning of the emptying process and (b) at the conclusion of the process.

The system shown in Figure 1 is internally reversible; there are no temperature or pressure gradients within the air that remains in the tank. Therefore, the entropy generated within this system approaches zero. There is no heat transfer to the system and any work associated with the expansion of the air carries no entropy. Therefore, an entropy balance on the system is:

$$0 = \Delta S \tag{2}$$

Equation (2) indicates that the entropy of the air that remains within the system does not change. The air undergoes an isentropic (i.e., constant entropy) process:

$$0 = m(s_2 - s_1)$$

In Section 6.3.3, the relationship between pressure and temperature for an ideal gas with constant specific heat capacities undergoing an isentropic process, Eq. (6-33), was derived:

$$T_2 = T_1 \left(\frac{P_2}{P_1}\right)^{\frac{R}{c_P}} \tag{3}$$

Equation (3) agrees with Eq. (1), the result obtained in Example 4.5-2 using only a First Law analysis assuming an adiabatic process, negligible kinetic and potential energy effects and no property gradients within the tank. These assumptions together eliminate the possibility of irreversible processes occurring within the mass of air that remains in the tank. We could have immediately deduced that the air that remains in the tank during the venting process undergoes a process that is nearly isentropic using an appropriately defined entropy balance. This approach greatly simplifies the analysis.

6.6 Efficiencies of Thermodynamic Devices

Most energy conversion systems are assembled from a relatively small number of basic components: turbines, compressors, pumps, nozzles, diffusers, throttles, and heat exchangers. The basic function and form of these devices are discussed in Section 4.4 together with the typical simplifications and assumptions that are applied in order to analyze their performance. The efficiency of these components is defined as the ratio of actual performance to some limiting or "best-possible" performance. Combined with the First Law, the Second Law of Thermodynamics provides a method that can be used to estimate the ideal performance of a component, which is typically the performance that the device would exhibit if it operated reversibly. None of these components are expected to operate in a reversible manner; however, this limiting behavior provides a natural basis for defining the thermodynamic efficiencies that are used to describe the actual performance of these devices.

6.6.1 Turbine Efficiency

The purpose of a turbine is to convert the energy in a high pressure fluid stream into mechanical power, as described in Section 4.4.1. Turbine analyses will typically assume adiabatic, steady-state operation with a single fluid inlet and outlet. The kinetic and potential energy changes of the fluid as it passes through the turbine are usually negligible. The performance of a turbine can be specified by a *turbine isentropic efficiency*, η_t, defined as:

$$\boxed{\eta_t = \frac{\dot{W}_t}{\dot{W}_{s,t}}} \tag{6-86}$$

where $\dot{W}_t$ is the power produced by the turbine and $\dot{W}_{s,t}$ is the power produced by a reversible turbine that is operating under the same conditions as the actual turbine. The phrase, "operating under the same conditions" implies that the reversible turbine experiences the same mass flow rate, inlet state, and outlet pressure, as shown in Figure 6-9.

An energy balance on the actual turbine is shown in Figure 6-10(a) and provides:

$$\dot{m}\, h_{in} = \dot{m}\, h_{out} + \dot{W}_t \tag{6-87}$$

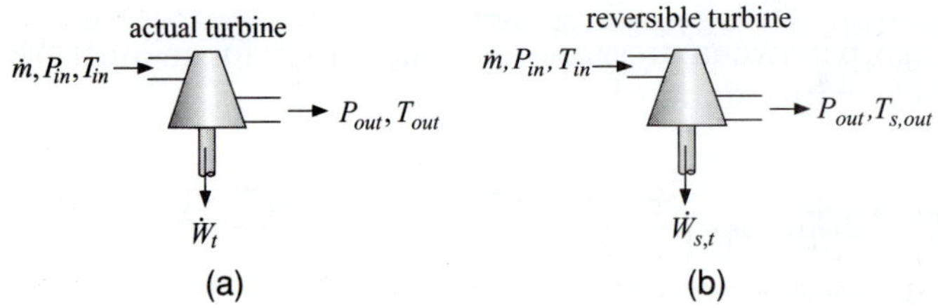

Figure 6-9: (a) An actual turbine and (b) a reversible turbine operating under the same conditions.

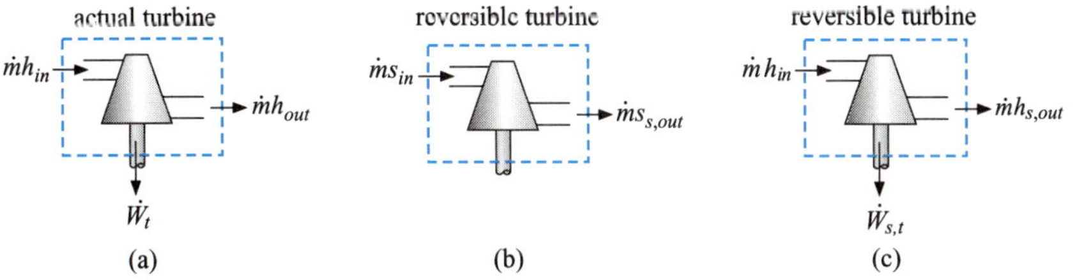

Figure 6-10: (a) An energy balance on the actual turbine and (b) an entropy balance and (c) an energy balance on the reversible turbine that is used to bound the performance.

where $\dot{W}_t$ is the power produced by the turbine, $\dot{m}$ is the mass flow rate passing through the turbine, and h_{in} and h_{out} are the specific enthalpies of the fluid at the turbine inlet and outlet, respectively. In the usual case, the turbine inlet state is known and therefore h_{in} can be determined. The pressure that is applied at the turbine outlet is known, but knowing only the pressure is not sufficient to determine h_{out}. Consequently Eq. (6-87) cannot be solved to determine the turbine power without additional information.

The additional information that is usually used to determine the turbine power is the turbine efficiency. Figure 6-10(b) shows an entropy balance on the reversible turbine shown in Figure 6-9(b). The reversible turbine is assumed to operate adiabatically with no friction or other irreversible phenomena and therefore there is no entropy generation. Since the turbine operates at steady state, the entropy balance reduces to:

$$s_{in} = s_{s,out} \tag{6-88}$$

where $s_{s,out}$ is the specific entropy of the fluid leaving the reversible turbine. The outlet state of the reversible turbine is fixed by the outlet pressure and specific entropy and therefore the specific enthalpy of the fluid leaving the reversible turbine, $h_{s,out}$, can be determined. Figure 6-10(c) illustrates an energy balance on the reversible turbine:

$$\dot{m}\, h_{in} = \dot{m}\, h_{s,out} + \dot{W}_{s,t} \tag{6-89}$$

Equation (6-89) allows the power produced by the reversible turbine to be computed. This is the maximum possible power and application of the turbine efficiency, Eq. (6-86) is sufficient to determine the actual power. The energy balance in Eq. (6-87) can then used to determine the outlet specific enthalpy of the actual turbine. The specific enthalpy and pressure fix the outlet state of the actual turbine.

EXAMPLE 6.6-1: TURBINE ISENTROPIC EFFICIENCY

Figure 1 illustrates a steam turbine with an isentropic efficiency $\eta_t = 0.84$. The inlet pressure is $P_1 = 4$ MPa and the inlet temperature is $T_1 = 650°$C. The exit pressure is $P_2 = 10$ kPa. The mass flow rate passing through the turbine is $\dot{m} = 100$ kg/s.

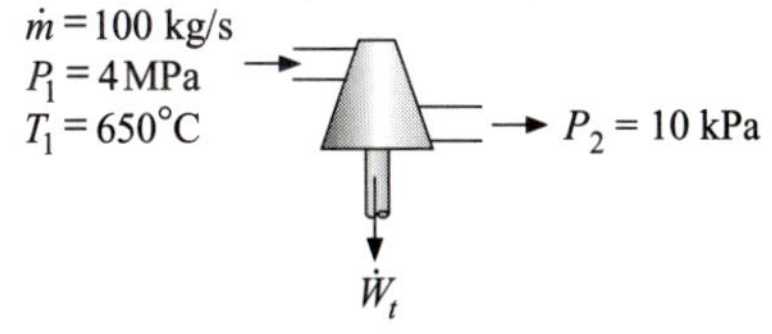

Figure 1: Turbine.

a) Determine the power produced by the turbine and the rate of entropy generation associated with the turbine.

The inputs are entered into EES:

```
$UnitSystem SI K Pa J mass radian
F$='Water'                              "fluid"
P[1]=4 [MPa]*convert(MPa,Pa)            "inlet pressure"
T[1]=converttemp(C,K,650 [C])           "inlet temperature"
P[2]=10 [kPa]*convert(kPa,Pa)           "exit pressure"
eta_t=0.84 [-]                          "isentropic efficiency"
m_dot=100 [kg/s]                        "mass flow rate"
```

In order to carry out a turbine analysis using the isentropic efficiency it is necessary to analyze the two turbines shown in Figure 6-9; the hypothetical, reversible turbine and the actual turbine. The inlet conditions for both turbines are the same. The inlet specific enthalpy and specific entropy (h_1 and s_1) are determined from the temperature and pressure.

```
h[1]=enthalpy(F$,T=T[1],P=P[1])         "inlet specific enthalpy"
s[1]=entropy(F$,T=T[1],P=P[1])          "inlet specific entropy"
```

An entropy balance on the reversible turbine reduces to:

$$s_{s,2} = s_1$$

The specific entropy and pressure ($s_{s,2}$ and P_2) fix the state of the steam leaving the reversible turbine. The temperature and specific enthalpy ($T_{s,2}$ and $h_{s,2}$) are obtained.

```
s_s[2]=s[1]                              "entropy balance on reversible turbine"
h_s[2]=enthalpy(F$,P=P[2],s=s_s[2])      "specific enthalpy leaving reversible turbine"
T_s[2]=temperature(F$,P=P[2],s=s_s[2])   "temperature leaving reversible turbine"
```

The energy balance for the steady-state reversible and adiabatic turbine provides:

$$\dot{m}\,h_1 = \dot{m}\,h_{s,2} + \dot{W}_{s,t}$$

```
h[1]*m_dot=h_s[2]*m_dot+W_dot_s_t        "energy balance on reversible turbine"
```

The power produced by the reversible turbine is $\dot{W}_{s,t} = 141.4$ MW. This is the maximum possible performance of any turbine operating under these conditions. The power produced by the actual turbine is computed using the turbine efficiency:

$$\dot{W}_t = \eta_t\,\dot{W}_{s,t}$$

```
W_dot_t=W_dot_s_t*eta_t                  "definition of isentropic efficiency"
```

which results in $\dot{W}_t = 118.7$ MW. An energy balance on the actual turbine, which is assumed to operate adiabatically, is:

$$\dot{m}\, h_1 = \dot{m}\, h_2 + \dot{W}_t$$

```
h[1]*m_dot=h[2]*m_dot+W_dot_t          "energy balance on actual turbine"
```

The pressure and specific enthalpy (P_2 and h_2) fix the state of the steam leaving the actual turbine. The temperature and specific entropy (T_2 and s_2) are determined.

```
T[2]=temperature(F$,h=h[2],P=P[2])          "exit temperature"
s[2]=entropy(F$,h=h[2],P=P[2])              "exit specific entropy"
```

An entropy balance on the actual turbine provides:

$$\dot{m}\, s_1 + \dot{S}_{gen} = \dot{m}\, s_2$$

```
m_dot*s[1]+S_dot_gen=m_dot*s[2]          "entropy balance on actual turbine"
```

which results in $\dot{S}_{gen} = 70.8$ kW/K. An *h-s* diagram is shown in Figure 2; this plot is generated by preparing a property plot for water and overlaying onto it the states associated with both the actual and reversible turbine.

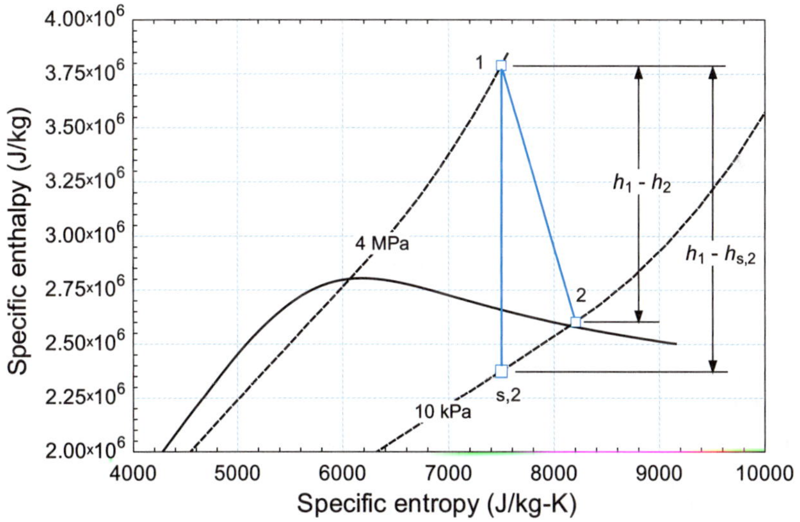

Figure 2: An *h-s* diagram showing the states for the actual and reversible turbine.

Notice that the specific entropy leaving the actual turbine (state 2) is increased relative to the inlet specific entropy (state 1) due to the entropy generation that occurs within the turbine. As a result, state 2 lies at a higher specific enthalpy than state s,2 with both states at the same pressure. The specific enthalpy change (which is proportional to the power produced) associated with the actual turbine is less than the specific enthalpy change associated with the reversible turbine.

EXAMPLE 6.6-2: TURBINE POLYTROPIC EFFICIENCY

In Figure 1, the single turbine analyzed in Example 6.6-1 has been replaced by N turbines, each with the same isentropic efficiency, $\eta_t = 0.84$. The pressure ratio (i.e., the ratio of the inlet pressure to the outlet pressure) associated with each stage (i.e., each turbine) is the same.

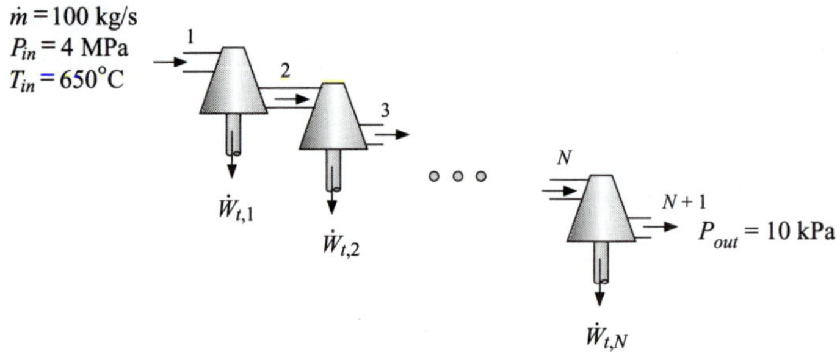

Figure 1: Multiple turbines with the same isentropic efficiency.

a) Determine the total power produced by $N = 2$ turbines.

The calculations presented in Example 6.6-1 for a single turbine must be repeated N times in order to solve this problem. In order to facilitate this process, the EES code from Example 6.6-1 is converted to a procedure. A procedure in EES is placed at the top of the Equations window using the following format:

```
Procedure Name(Input 1, Input 2, ..., Input N: Output 1, Output 2, ..., Output M)
    Code
    Output 1 = ...
    Output 2 = ...
    etc.
End
```

The procedure is called from the Equations window using the following format:

```
Call Name(Input 1, Input 2, ..., Input N: Output 1, Output 2, ..., Output M)
```

The procedure requires one or more inputs (Input 1 through Input N) and returns one or more outputs (Output 1 through Output M). The input list is separated from the output list by a colon. The code that makes up the procedure is placed between the Procedure and End statements. Writing code in a procedure is somewhat different than solving equations in the Equations Window. The procedure code is similar to a typical formal programming language in that assignment statements (rather than equalities) are used. Anything to the right of the equal sign in an assignment statement must already be known. The assignment sets the value of the variable on the left of the equal sign to the value of the expression on the right of the equal sign. Systems of implicit equations cannot be solved in a procedure (although Modules and Subprograms are available in EES for this purpose). Also, the workspace associated with a procedure only has access to variables that are explicitly passed to it or to variables that are assigned within the procedure. A procedure cannot access other variables that reside in the Equations window unless they are specifically indicated in a $Common directive. Finally, it is possible to use many of the logical constructs that are associated with a formal programming language (e.g., if-then-else, repeat-until, goto, and return statements) in a procedure. These logic statements cannot be applied in the Equations window.

For this example, we will write a procedure named Turbine that requires as inputs the operating conditions (T_in, P_in, P_out, and m_dot) as well as a string variable identifying the fluid (F$), and the turbine efficiency (eta_t). The procedure calculates and returns the outlet temperature and specific entropy of the fluid (T_out and s_out), the specific entropy of the fluid leaving the reversible turbine (s_s_out), and the power produced by the turbine (W_dot_t). Therefore, the format of the procedure is:

```
Procedure Turbine(T_in,P_in,P_out,m_dot, F$, eta_t: T_out, s_out, s_s_out, W_dot_t)
end
```

The original set of EES statements used to solve the turbine problem in Example 6.6-1 is shown below:

```
h[1]=enthalpy(F$,T=T[1],P=P[1])              "inlet specific enthalpy"
s[1]=entropy(F$,T=T[1],P=P[1])               "inlet specific entropy"
s_s[2]=s[1]                                  "entropy balance on reversible turbine"
h_s[2]=enthalpy(F$,P=P[2],s=s_s[2])          "specific enthalpy leaving reversible turbine"
T_s[2]=temperature(F$,P=P[2],s=s_s[2])       "temperature leaving reversible turbine"
h[1]*m_dot=h_s[2]*m_dot+W_dot_s_t            "energy balance on reversible turbine"
W_dot_t=W_dot_s_t*eta_t                      "definition of isentropic efficiency"
h[1]*m_dot=h[2]*m_dot+W_dot_t                "energy balance on actual turbine"
T[2]=temperature(F$,h=h[2],P=P[2])           "exit temperature"
s[2]=entropy(F$,h=h[2],P=P[2])               "exit specific entropy"
m_dot*s[1]+S_dot_gen=m_dot*s[2]              "entropy balance on actual turbine"
```

The code is copied and pasted into the procedure block with appropriate modifications:

```
Procedure Turbine(T_in,P_in,P_out,m_dot, F$, eta_t: T_out, s_out, s_s_out, W_dot_t)
   h_in=enthalpy(F$,T=T_in,P=P_in)              "inlet specific enthalpy"
   s_in=entropy(F$,T=T_in,P=P_in)               "inlet specific entropy"
   s_s_out=s_in                                 "entropy balance on reversible turbine"
   h_s_out=enthalpy(F$,P=P_out,s=s_s_out)       "specific enthalpy leaving reversible turbine"
   W_dot_t_s=m_dot*(h_in-h_s_out)               "energy balance on reversible turbine"
   W_dot_t=W_dot_t_s*eta_t                      "definition of isentropic efficiency"
   h_out=h_in-W_dot_t/m_dot                     "energy balance on actual turbine"
   T_out=temperature(F$,h=h_out,P=P_out)        "exit temperature"
   s_out=entropy(F$,h=h_out,P=P_out)            "exit specific entropy"
   S_dot_gen=m_dot*(s_out-s_in)                 "entropy balance on actual turbine"
end
```

Notice that there are a few changes between the code as it appeared in the Equations window in Example 6.6-1 and the code that has been moved into the procedure block. First, the names of the variables are changed in order to match the names of the inputs and outputs. Second, those statements that were implicit are rearranged so that they are explicit. For example, the statement:

```
h[1]*m_dot=h_s[2]*m_dot+W_dot_s_t            "energy balance on reversible turbine"
```

in the Equations window of Example 6.6-1 is explicitly solved for variable W_dot_s_t in the procedure:

```
W_dot_t_s=m_dot*(h_in-h_s_out)          "energy balance on reversible turbine"
```

The Call Turbine statement can now be used to accomplish the turbine analysis that was carried out in Example 6.6-1:

```
F$='Water'                              "fluid"
P_in=4 [MPa]*convert(MPa,Pa)            "inlet pressure"
T_in=converttemp(C,K,650 [C])           "inlet temperature"
P_out=10 [kPa]*convert(kPa,Pa)          "exit pressure"
eta_t=0.84 [-]                          "isentropic efficiency of each stage"
m_dot=100 [kg/s]                        "mass flow rate"
Call Turbine(T_in,P_in,P_out,m_dot, F$, eta_t: T_out, s_out, s_s_out, W_dot_t)
```

which leads to $\dot{W}_t = 118.7$ MW, the same answer that was obtained in Example 6.6-1.

The procedure can now be used to quickly analyze the series of N turbines shown in Figure 3. The overall pressure ratio for the system is:

$$PR = \frac{P_{in}}{P_{out}} \tag{1}$$

The pressure ratio for each stage is given by:

$$PR_{stage} = PR^{\frac{1}{N}} \tag{2}$$

The inlet pressure for the first stage turbine is:

$$P_1 = P_{in} \tag{3}$$

and the exit pressure for each stage is set according to:

$$P_{i+1} = \frac{P_i}{PR_{stage}} \text{ for } i = 1..N \tag{4}$$

Equation (4) can be easily implemented with a duplicate clause that duplicates the statements contained between the duplicate and end keywords while setting the index variable i to each integer value between 1 and N.

```
N=2 [-]                      "number of compression stages"
PR=P_in/P_out                "overall pressure ratio"
PR_stage=PR^(1/N)            "stage pressure ratio"
P[1]=P_in                    "inlet pressure"
duplicate i=1,               "exit pressure"
   P[i+1]=P[i]/PR_stage
end
```

The inlet temperature for the first stage is given, $T_1 = T_{in}$. The inlet state for the first stage is fixed by the temperature and pressure (T_1 and P_1). The inlet specific entropy and specific enthalpy (s_1 and h_1) are determined. The Turbine procedure is used to analyze each turbine in the series. The specific enthalpies at the exit of each reversible and actual turbine stage (h_{i+1} and $h_{s,i+1}$) are found from the specific entropy returned by the procedure and the exit pressure. These values are not directly needed, but will eventually be used to generate an *h-s* diagram.

```
T[1]=T_in                                "inlet temperature for 1st stage turbine"
s[1]=entropy(F$,T=T[1],P=P[1])           "inlet specific entropy for 1st stage turbine"
h[1]=enthalpy(F$,T=T[1],P=P[1])          "inlet specific enthalpy for 1st stage turbine"
duplicate i=1,N
    Call Turbine(T[i],P[i],P[i+1],m_dot, F$, eta_t: T[i+1], s[i+1], s_s[i+1], W_dot_t[i])
    h[i+1]=enthalpy(F$,s=s[i+1],P=P[i+1])          "specific enthalpy leaving actual turbine"
    h_s[i+1]=enthalpy(F$,s=s_s[i+1],P=P[i+1])      "specific enthalpy leaving reversible turbine"
end
```

The total power produced by the series of turbines is:

$$\dot{W}_{total} = \sum_{i=1}^{N} \dot{W}_{t,i} \tag{5}$$

```
W_dot_t_total=sum(W_dot_t[i],i=1,N)      "total turbine power"
```

which leads to $\dot{W}_{total} = 123.3$ MW for $N = 2$ turbines that each have an isentropic efficiency of 0.84. This is surprising since the power produced by a single turbine with the same overall pressure ratio operating under identical conditions and having the same isentropic efficiency was found 118.7 MW in Example 6.6-1. The apparent performance of the two turbines operating in series is higher than the performance of a single turbine. The apparent isentropic efficiency of the two turbine stages operating together can be computed according to:

$$\eta_{total} = \frac{\dot{W}_{total}}{\dot{W}_{s,total}} \tag{6}$$

where $\dot{W}_{s,total}$ is the power produced by a reversible turbine operating between the pressures P_{in} and P_{out}:

$$\dot{W}_{s,total} = \dot{m}\left(h_1 - h_{s,out}\right) \tag{7}$$

where $h_{s,out}$ is the specific enthalpy evaluated at P_{out} and s_1.

```
h_s_out=enthalpy(F$,s=s[1],P=P_out)
                    "specific enthalpy leaving a rev. turbine replacing the series of turbines"
W_dot_s_total=m_dot*(h[1]-h_s_out)"isentropic turbine power"
eta_total=W_dot_total/W_dot_s_total"isentropic efficiency of all of the stages when considered together"
```

These calculations indicate that $\eta_{total} = 0.873$ for two turbine stages, each with isentropic efficiency $\eta_t = 0.84$.

Why should the power produced by a single turbine with $\eta_t = 0.84$ (118.7 MW) be different than the power produced by two turbines in series (123.3 MW), when each of the two turbines has $\eta_t = 0.84$? This effect can be understood by examining Figure 2, which shows an *h-s* diagram for the system shown in Figure 1 with $N = 2$. Note that Figure 2 is generated by overlaying the states in the Arrays table onto an *h-s* diagram for water generated by selecting Property Plot from the Plots menu.

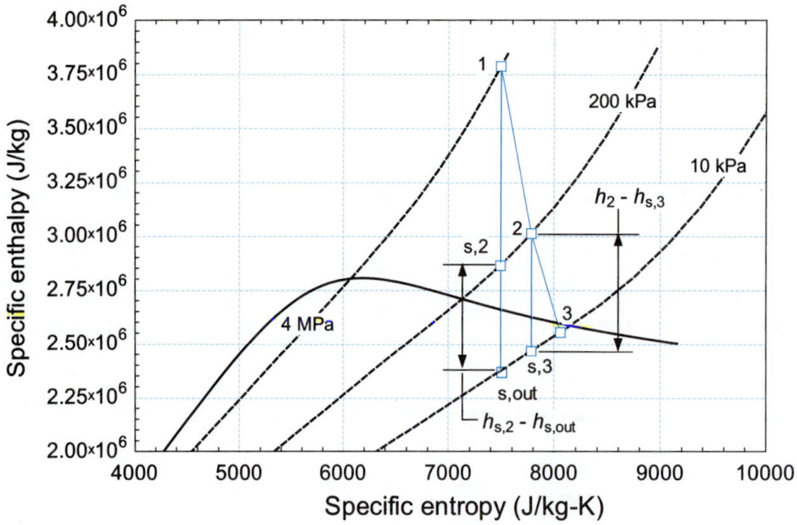

Figure 2: An *h-s* diagram for the 2-stage system showing the states for the actual and reversible turbines.

Figure 2 shows that the isobars diverge as specific entropy increases. That is, an isentropic expansion from $P_{in} = 4$ MPa to $P_{out} = 10$ kPa will produce a larger change in specific enthalpy at higher specific entropy than the same isentropic expansion will produce at lower specific entropy. One consequence of this behavior is that the isentropic specific enthalpy change associated with the second stage turbine is larger when the inlet is state 2 than when the inlet state is state s,2. That is, $h_2 - h_{s,3}$ is larger than $h_{s,2} - h_{s,out}$ in Figure 2. The entropy that is generated in the first stage turbine causes the starting point for the expansion from P_2 to P_3 to move to the right for the second stage turbine. Therefore, the isentropic specific enthalpy change from P_2 to P_3 is larger in the second stage turbine than it is in the single stage turbine. The power produced by turbine 2 in the two-stage configuration shown in Figure 1 is $\eta_t\,(h_2 - h_{s,3})$ whereas the power produced by the expansion from P_2 to P_3 in the single turbine considered in Example 6.6-1 is $\eta_t\,(h_{s,2} - h_{s,out})$.

b) The fact that a series of turbines, each with the same isentropic efficiency can produce results that, taken together, are characterized by a different isentropic efficiency is not desirable. This observation motivates the definition of the *polytropic efficiency*, $\eta_{poly,t}$, as the isentropic efficiency in the limit that the pressure ratio across the turbine approaches unity (i.e., the pressure change across the turbine approaches zero). The advantage of using the polytropic efficiency is that the performance of the turbine is independent of the pressure ratio. That is, the polytropic efficiency of two or more turbines operating in series will be the same as a single turbine with the same polytropic efficiency operating between the same inlet and exit pressure. Plot the isentropic efficiency of a turbine with polytropic efficiency $\eta_{poly,t} = 0.84$ as a function of the exit pressure, assuming the inlet conditions shown in Figure 1.

In order to simulate a turbine given its polytropic efficiency it is necessary to divide the expansion process into many small steps and use the polytropic efficiency as the isentropic efficiency for each of the expansion steps. The polytropic efficiency is defined as the isentropic efficiency as the pressure difference approaches zero. The EES code developed in part (a) can be used for this purpose by setting N to a large number.

The input conditions are set in EES:

```
F$='Water'                              "fluid"
P_in=4 [MPa]*convert(MPa,Pa)            "inlet pressure"
T_in=converttemp(C,K,650 [C])           "inlet temperature"
P_out_kPa=10 [kPa]                      "exit pressure, in kPa"
P_out=P_out_kPa*convert(kPa,Pa)         "exit pressure"
eta_poly_t=0.84 [-]                     "polytropic efficiency is the isentropic efficiency of each expansion"
m_dot=100 [kg/s]                        "mass flow rate"
N=20 [-]                                "number of stages"
```

Note that the polytropic efficiency is equal to the isentropic efficiency of each small expansion process when N, the number of stages, is set to a large number. The pressure ratio and pressures associated with each small expansion process are specified according to Eqs. (1) through (4).

```
PR=P_in/P_out                           "overall pressure ratio"
PR_stage=PR^(1/N)                       "stage pressure ratio"
P[1]=P_in                               "inlet pressure"
duplicate i=1,N
    P[i+1]=P[i]/PR_stage                "exit pressure"
end
```

The simulation of each expansion process is accomplished using the Turbine procedure, which is provided with the polytropic efficiency.

```
T[1]=T_in                               "inlet temperature for 1st stage turbine"
s[1]=entropy(F$,T=T[1],P=P[1])          "inlet specific entropy for 1st stage turbine"
h[1]=enthalpy(F$,T=T[1],P=P[1])         "inlet specific enthalpy for 1st stage turbine"
duplicate i=1,N
    Call Turbine(T[i],P[i],P[i+1],m_dot, F$, eta_poly_t: T[i+1], s[i+1], s_s[i+1], W_dot_t[i])
    h[i+1]=enthalpy(F$,s=s[i+1],P=P[i+1])         "specific enthalpy leaving actual turbine"
    h_s[i+1]=enthalpy(F$,s=s_s[i+1],P=P[i+1])     "specific enthalpy leaving reversible turbine"
end
```

The total power and isentropic efficiency are computed using Eqs. (5) through (7).

```
W_dot_total=sum(W_dot_t[i],i=1,N)       "total turbine power"
h_s_out=enthalpy(F$,s=s[1],P=P_out)     "specific enthalpy leaving a reversible turbine"
W_dot_s_total=m_dot*(h[1]-h_s_out)      "isentropic turbine power"
eta_t=W_dot_total/W_dot_s_total         "isentropic efficiency"
```

Figure 3 illustrates the overall isentropic efficiency as a function of the number of expansion processes used to simulate the turbine, N, with an exit pressure of $P_{out} =$ 100 kPa. It is not possible to use an infinitesimally small pressure drop across each expansion as this would require an infinite value of N. However, Figure 3 shows that $N = 100$ is sufficient to represent the behavior of an infinite number of turbine stages.

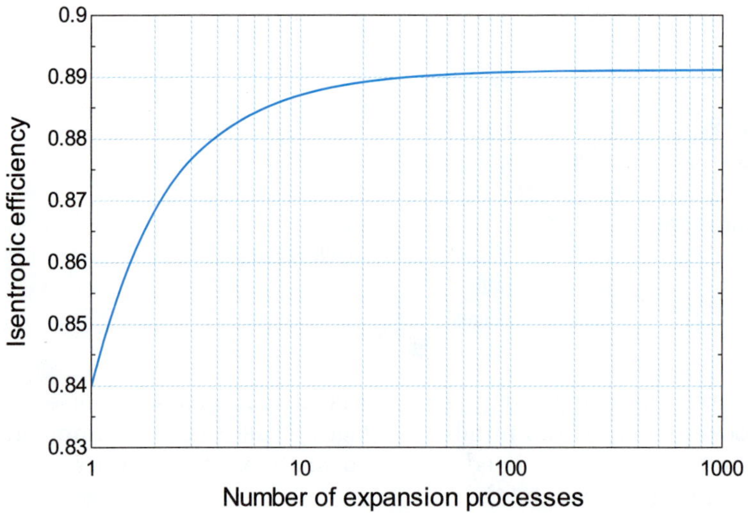

Figure 3: Isentropic efficiency as a function of *N*, the number of stages used to simulate the turbine for a polytropic efficiency of 0.84.

Figure 4 illustrates the isentropic efficiency as a function of the outlet pressure obtained using $N = 100$ for a turbine having a polytropic efficiency of 0.84. Notice that as P_{out} approaches P_{in} (4000 kPa), the isentropic efficiency approaches the polytropic efficiency. However, as the pressure difference across the turbine increases, the isentropic and polytropic efficiencies diverge.

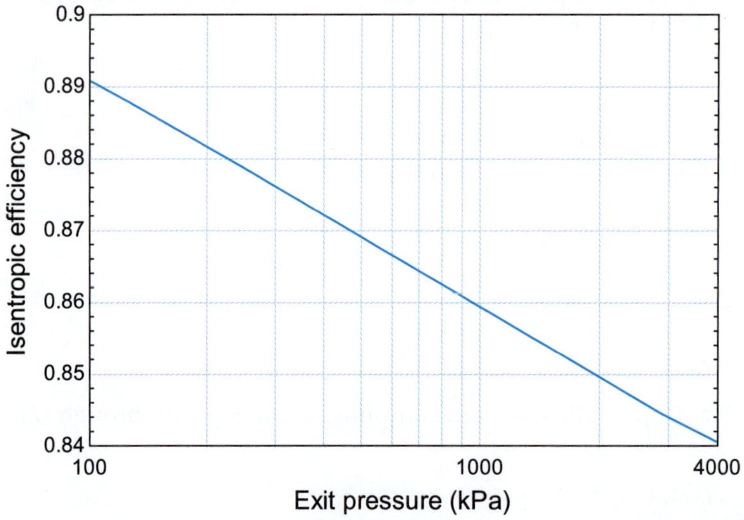

Figure 4: Isentropic efficiency as a function of the outlet pressure for a polytropic efficiency of 0.84.

Clearly it is more convenient to use the isentropic efficiency for turbine analyses and in most cases this is sufficient, particularly if there is only one turbine stage and the pressure ratio is not changing. However, the polytropic efficiency is a more meaningful measure of the true performance of the turbine as it is not confounded by the effect of pressure ratio. The concept of a polytropic efficiency can be applied to compressors as well as turbines.

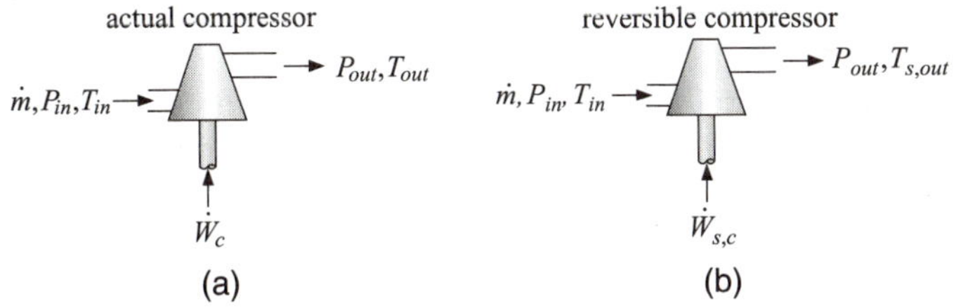

Figure 6-11: (a) An actual compressor and (b) a reversible compressor operating under the same conditions.

6.6.2 Compressor Efficiency

A compressor uses a mechanical power input to increase the pressure of a gas, as described in Section 4.4.2. Compressor performance is improved if there is significant heat transfer from the fluid during the compression process. However, compressor operation is normally assumed to be adiabatic since the time that the fluid spends in the compressor is too short to allow significant heat transfer in most cases. The kinetic and potential energy changes of the fluid as it passes through the compressor are typically small and usually neglected. A compressor normally operates at steady state. The performance of a compressor is specified by a *compressor isentropic efficiency*, η_c, defined as:

$$\boxed{\eta_c = \frac{\dot{W}_{s,c}}{\dot{W}_c}} \tag{6-90}$$

where $\dot{W}_c$ is the power required by the compressor and $\dot{W}_{s,c}$ is the power required by a reversible compressor operating under the same conditions as the actual compressor; the same conditions means that the mass flow rate, inlet state, and outlet pressure are the same, as shown in Figure 6-11. Notice that the definition of compressor efficiency, Eq. (6-90), is the inverse of the definition of the turbine efficiency, Eq. (6-86), because the reversible compressor will require less power than the actual compressor. The reversible compressor establishes the minimum amount of power that is required to accomplish the adiabatic compression process.

An energy balance on the actual compressor is shown in Figure 6-12(a) and can be written as:

$$\dot{m}\, h_{in} + \dot{W}_c = \dot{m}\, h_{out} \tag{6-91}$$

where $\dot{W}_c$ is the power required by the compressor, $\dot{m}$ is the mass flow rate processed by the compressor, and h_{in} and h_{out} are the specific enthalpies of the fluid at the compressor inlet and outlet, respectively. In the usual case, the compressor inlet state is known, therefore h_{in} can be determined. The pressure rise required by the compressor is known, but knowledge of the outlet pressure is not sufficient to determine h_{out}. Consequently

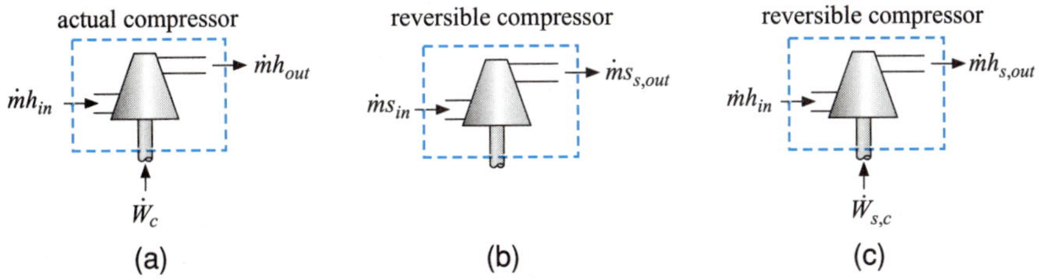

Figure 6-12: (a) An energy balance on the actual compressor and (b) an entropy balance and (c) an energy balance on the reversible compressor that is used to bound the performance.

Eq. (6-91) cannot be solved to determine the compressor power without additional information.

The additional information that is usually used to determine the compressor power is the compressor efficiency. Figure 6-12(b) shows an entropy balance on the reversible compressor:

$$s_{in} = s_{s,out} \tag{6-92}$$

where $s_{s,out}$ is the specific entropy of the fluid leaving the reversible compressor. The outlet state of the reversible compressor is fixed by the outlet pressure and specific entropy and therefore the specific enthalpy of the fluid leaving the reversible compressor, $h_{s,out}$, can be determined. Figure 6-12(c) illustrates an energy balance on the reversible compressor:

$$\dot{m}\, h_{in} + \dot{W}_{s,c} = \dot{m}\, h_{s,out} \tag{6-93}$$

Equation (6-93) allows the calculation of the power required by a reversible compressor, which is the minimum possible power that must be provided under these conditions. Application of the compressor efficiency, Eq. (6-90), to the isentropic power calculated using Eq. (6-93) determines the actual power. The energy balance in Eq. (6-91) is then used to determine the outlet specific enthalpy and therefore the outlet state of the actual compressor.

EXAMPLE 6.6-3: INTERCOOLED COMPRESSION

Figure 1 illustrates an air compressor with an aftercooler that is used to provide compressed air to a factory. The compressor has an isentropic efficiency $\eta_c = 0.74$. Air at atmospheric pressure and ambient temperature, $P_{atm} = 100$ kPa and $T_{amb} = 20°$C, is drawn into the suction port of the compressor. The discharge pressure is $P_{out} = 1$ MPa. The mass flow rate provided by the compressor is $\dot{m} = 0.5$ kg/s. The aftercooler is a perfect heat exchanger that rejects heat to the ambient surroundings. Therefore, there is no pressure loss in the aftercooler and the exit temperature of the air is T_{amb}.

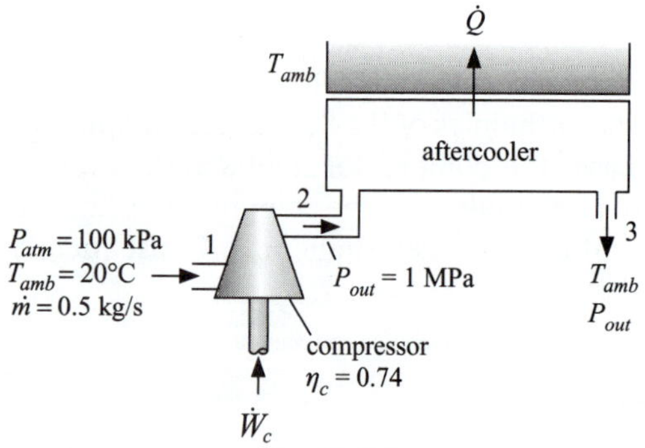

Figure 1: Compressor with aftercooler.

a) Determine the power required by the compressor.

The inputs are entered in EES:

```
$UnitSystem SI K Pa J mass radian
F$='Air'                                   "fluid"
P_atm=100 [kPa]*convert(kPa,Pa)            "atmospheric pressure"
T_amb=converttemp(C,K,20[C])               "ambient temperature"
P_out=1 [MPa]*convert(MPa,Pa)              "discharge pressure"
eta_c=0.74 [-]                             "isentropic efficiency"
m_dot=0.5 [kg/s]                           "mass flow rate"
```

In order to carry out a compressor analysis using the isentropic efficiency, it is necessary to analyze the two compressors shown in Figure 6-11: the hypothetical, reversible compressor and the actual compressor. The inlet conditions for both compressors are the same and both compressors are adiabatic. The inlet specific enthalpy and specific entropy (h_1 and s_1) are determined from the temperature ($T_1 = T_{amb}$) and pressure ($P_1 = P_{atm}$).

```
T[1]=T_amb                                 "inlet temperature"
P[1]=P_atm                                 "inlet pressure"
h[1]=enthalpy(F$,T=T[1])                   "inlet specific enthalpy"
s[1]=entropy(F$,T=T[1],P=P[1])             "inlet specific entropy"
```

An entropy balance on the reversible compressor simplifies to:

$$s_{s,2} = s_1$$

The specific entropy and pressure ($s_{s,2}$ and $P_2 = P_{out}$) fix the state of the air leaving the reversible compressor. The temperature and specific enthalpy ($T_{s,2}$ and $h_{s,2}$) are obtained.

```
s_s[2]=s[1]                                "entropy balance on reversible compressor"
P[2]=1 [MPa]*convert(MPa,Pa)               "compressor exit pressure"
h_s[2]=enthalpy(F$,P=P[2],s=s_s[2])        "specific enthalpy leaving reversible compressor"
T_s[2]=temperature(F$,P=P[2],s=s_s[2])     "temperature leaving reversible compressor"
```

An energy balance on the steady-state, reversible, adiabatic compressor provides:

$$\dot{m}\,h_1 + \dot{W}_{s,c} = \dot{m}\,h_{s,2}$$

```
h[1]*m_dot+W_dot_s_c=h_s[2]*m_dot          "energy balance on reversible compressor"
```

which results in $\dot{W}_{s,c} = 136.8$ kW. The power required by the actual compressor is computed using the compressor efficiency:

$$\dot{W}_c = \frac{\dot{W}_{s,c}}{\eta_c}$$

```
W_dot_c=W_dot_s_c/eta_c                    "definition of isentropic efficiency"
```

The actual compressor power is $\dot{W}_c = 184.9$ kW.

b) Determine the rate of entropy generation in the compressor.

An energy balance on the actual compressor is:

$$\dot{m}\,h_1 + \dot{W}_c = \dot{m}\,h_2$$

```
h[1]*m_dot+W_dot_c=h[2]*m_dot          "energy balance on actual compressor"
```

The pressure and specific enthalpy (P_2 and h_2) fix the state of the air leaving the actual compressor. The temperature and specific entropy (T_2 and s_2) are determined.

```
T[2]=temperature(F$,h=h[2])            "exit temperature"
s[2]=entropy(F$,h=h[2],P=P[2])         "exit specific entropy"
```

An entropy balance on the actual compressor leads to:

$$\dot{m}\, s_1 + \dot{S}_{gen,c} = \dot{m}\, s_2$$

```
m_dot*s[1]+S_dot_gen_c=m_dot*s[2]          "entropy balance on actual compressor"
```

which results in $\dot{S}_{gen,c} = 79.3$ W/K.

c) Determine the rate of heat transfer from the aftercooler.

The state of the air leaving the aftercooler is fixed by its pressure and temperature ($P_3 = P_2$ and $T_3 = T_{amb}$). The specific entropy and specific enthalpy (s_3 and h_3) are obtained.

```
T[3]=T_amb                             "aftercooler exit temperature"
P[3]=P[2]                              "aftercooler exit pressure"
s[3]=entropy(Air,P=P[3],T=T[3])        "specific entropy leaving aftercooler"
h[3]=enthalpy(Air,T=T[3])              "specific enthalpy leaving aftercooler"
```

An energy balance on the aftercooler is shown in Figure 2(a):

$$\dot{m}\, h_2 = \dot{Q} + \dot{m}\, h_3$$

```
Q_dot=m_dot*(h[2]-h[3])          "energy balance on aftercooler"
```

The rate of heat transfer from the aftercooler is $\dot{Q} = 184.9$ kW.

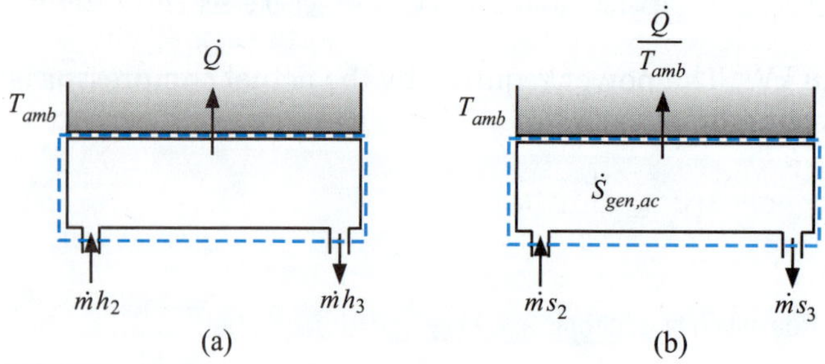

Figure 2: (a) Energy and (b) entropy balance on the aftercooler.

d) Determine the rate of entropy generation in the aftercooler.

Figure 2(b) illustrates an entropy balance on the aftercooler; note that the system boundary is chosen so that the temperature is T_{amb} at the point where heat is rejected in order to capture all of the entropy generation associated with this device. The entropy balance on the aftercooler is:

$$\dot{m}\,s_2 + \dot{S}_{gen,ac} = \frac{\dot{Q}}{T_{amb}} + \dot{m}\,s_3$$

```
m_dot*s[2]+S_dot_gen_ac=Q_dot/T_amb+m_dot*s[3]          "entropy balance on aftercooler"
```

which leads to $\dot{S}_{gen,ac} = 221.1$ W/K. A *T-s* diagram is shown in Figure 3; this plot is generated by preparing a property plot for air and overlaying the states associated with both the actual and reversible compressor and the aftercooler outlet state.

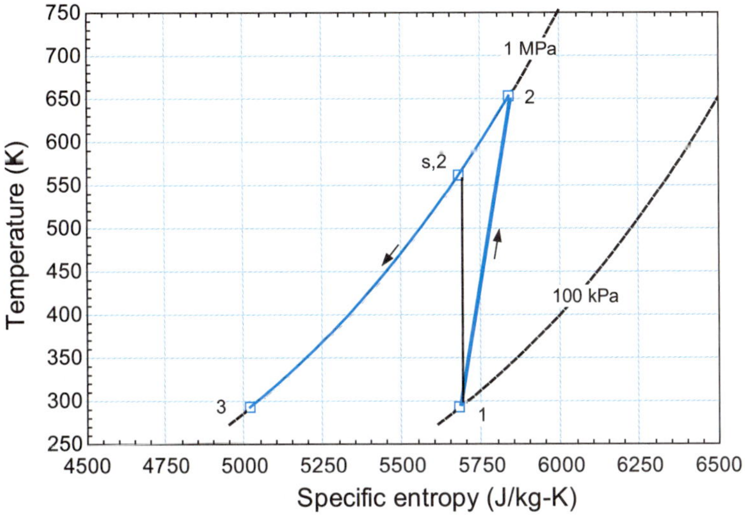

Figure 3: A *T-s* diagram showing the states for the actual and reversible compressor.

Notice that the specific entropy leaving the actual compressor (state 2) is increased relative to the inlet specific entropy (state 1) due to the entropy generation associated within the compressor. As a result, state 2 lies at a higher temperature (and therefore higher specific enthalpy) than state s,2 with both states at the same pressure. The change in the specific enthalpy (which is proportional to the power required) associated with the actual compressor is larger than the change in specific enthalpy associated with the reversible compressor.

e) Write a procedure that can accomplish the compressor analysis carried out in parts (a) and (b). The procedure should require as inputs the operating conditions (T_{in}, P_{in}, P_{out}, $\dot{m}$), the compressor efficiency (η_c), and a string variable that specifies the fluid. The procedure should be capable of dealing with substances that are modeled as an ideal gas (e.g., the substance 'Air' in EES) and as a real fluid (e.g., the substance 'Air_ha' in EES).

The development of procedures in EES is discussed in Example 6.6-1. The EES code developed in parts (a) and (b) is shown below:

```
T[1]=T_amb                                  "inlet temperature"
P[1]=P_amb                                  "ambient pressure"
h[1]=enthalpy(F$,T=T[1])                    "inlet specific enthalpy"
s[1]=entropy(F$,T=T[1],P=P[1])              "inlet specific entropy"
s_s[2]=s[1]                                 "entropy balance on reversible compressor"
P[2]=1 [MPa]*convert(MPa,Pa)                "compressor exit pressure"
h_s[2]=enthalpy(F$,P=P[2],s=s_s[2])         "specific enthalpy leaving reversible compressor"
T_s[2]=temperature(F$,P=P[2],s=s_s[2])      "temperature leaving reversible compressor"
h[1]*m_dot+W_dot_s_c=h_s[2]*m_dot           "energy balance on reversible compressor"
W_dot_c=W_dot_s_c/eta_c                     "definition of isentropic efficiency"
h[1]*m_dot+W_dot_c=h[2]*m_dot               "energy balance on actual compressor"
T[2]=temperature(F$,h=h[2])                 "exit temperature"
s[2]=entropy(F$,h=h[2],P=P[2])              "exit specific entropy"
m_dot*s[1]+S_dot_gen_c=m_dot*s[2]           "entropy balance on actual compressor"
```

This code is used as the basis of a procedure that is named Compressor:

```
Procedure Compressor(T_in,P_in,P_out,m_dot, F$, eta_c:T_out, s_out, s_s_out, W_dot_c)
    h_in=enthalpy(F$,T=T_in)                    "inlet specific enthalpy"
    s_in=entropy(F$,T=T_in,P=P_in)              "inlet specific entropy"
    s_s_out=s_in                                "entropy balance on reversible compressor"
    h_s_out=enthalpy(F$,P=P_out,s=s_s_out)      "specific enthalpy leaving reversible compressor"
    W_dot_s_c=m_dot*(h_s_out-h_in)              "energy balance on reversible compressor"
    W_dot_c=W_dot_s_c/eta_c                     "definition of isentropic efficiency"
    h_out=h_in+W_dot_c/m_dot                    "energy balance on actual compressor"
    T_out=temperature(F$,h=h_out)               "exit temperature"
    s_out=entropy(F$,h=h_out,P=P_out)           "exit specific entropy"
    S_dot_gen=m_dot*(s_out-s_in)                "entropy balance on actual turbine"
end
```

Note that the only major modification of the original EES code is related to the fact that EES procedures use assignments rather than equalities. Also, subscripted variables are not needed in the procedure, so the notation for variables ending with array indices [1] and [2] have been changed to _in and _out, respectively. Therefore, the original statement in the EES window:

```
h[1]*m_dot+W_dot_s_c=h_s[2]*m_dot           "energy balance on reversible compressor"
```

had to be rearranged and solved for the unknown variable W_dot_s_c, in the procedure:

```
W_dot_s_c=m_dot*(h_s_out-h_in)              "energy balance on reversible compressor"
```

The analysis accomplished in parts (a) and (b) can be carried out using the procedure Compressor.

```
F$='Air'                                        "fluid"
P_amb=100 [kPa]*convert(kPa,Pa)                 "ambient pressure"
T_amb=converttemp(C,K,20[C])                    "ambient temperature"
P_out=1 [MPa]*convert(MPa,Pa)                   "compressor exit pressure"
eta_c=0.74 [-]                                  "isentropic efficiency"
m_dot=0.5 [kg/s]                                "mass flow rate"

Call Compressor(T_amb,P_amb,P_out,m_dot, F$, eta_c: T_out, s_out, s_s_out, W_dot_c)
```

which results in $\dot{W}_c = 184.9$ kW, the same answer obtained in part (a). However, the procedure that has been developed cannot be used to analyze a compressor working with a substance that is not modeled as an ideal gas. For example, if the string variable F$ is changed so that the fluid Air_ha is used rather than the fluid Air (i.e., we are not making the simplification that the air can be modeled as an ideal gas):

```
F$='Air_ha'          "fluid"
```

then the procedure will generate an error message indicating that specific enthalpy cannot be determined knowing just the temperature, as shown in Figure 4.

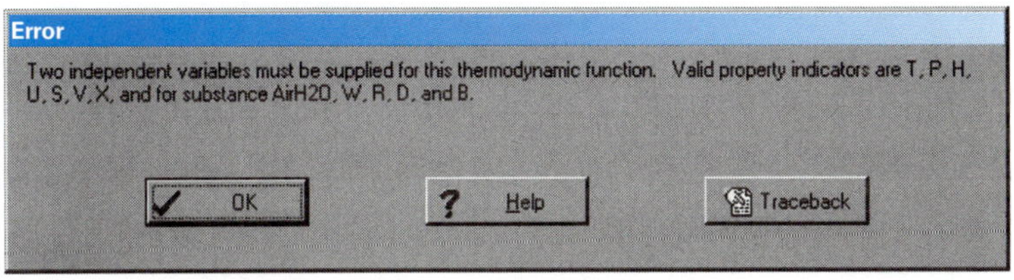

Figure 4: Error message.

The error message refers to the line:

```
h_in=enthalpy(F$,T=T_in)          "inlet specific enthalpy"
```

in the procedure Compressor. The call to the function Enthalpy only requires temperature when the substance is modeled as an ideal gas but an additional property is required to fix the state for a real fluid. A similar error is associated with the line:

```
T_out=temperature(F$,h=h_out)          "exit temperature"
```

An additional property (beyond specific enthalpy) is required to determine the temperature if the substance is not an ideal gas.

One of the advantages of procedures is that they can utilize logic statements such as If-Then-Else type decision structures. To make this procedure robust to the type of substance that is being analyzed, it is necessary to check whether the substance provided to the procedure in the string variable F$ is an ideal gas using the IsIdealGas function. The IsIdealGas function takes a single argument, the string containing the fluid of interest, and returns true if the fluid is an ideal gas and false otherwise.

The two lines that led to error messages are replaced in the procedure below:

```
Procedure Compressor(T_in,P_in,P_out,m_dot, F$, eta_c: T_out, s_out, s_s_out, W_dot_c)
    {h_in=enthalpy(F$,T=T_in)}                    "inlet specific enthalpy"
    If (IsIdealGas(F$)) then
        h_in=enthalpy(F$,T=T_in)                  "inlet specific enthalpy"
    else
        h_in=enthalpy(F$,T=T_in,P=P_in)           "inlet specific enthalpy"
    endif
    s_in=entropy(F$,T=T_in,P=P_in)                "inlet specific entropy"
    s_s_out=s_in                                  "entropy balance on reversible compressor"
    h_s_out=enthalpy(F$,P=P_out,s=s_s_out)        "specific enthalpy leaving reversible compressor"
    W_dot_s_c=m_dot*(h_s_out-h_in)                "energy balance on reversible compressor"
    W_dot_c=W_dot_s_c/eta_c                       "definition of isentropic efficiency"
    h_out=h_in+W_dot_c/m_dot                      "energy balance on actual compressor"
    {T_out=temperature(F$,h=h_out)}               "exit temperature"
    if(IsIdealGas(F$)) then
        T_out=temperature(F$,h=h_out)             "exit temperature"
    else
        T_out=temperature(F$,h=h_out,P=P_out)     "exit temperature"
    endif
    s_out=entropy(F$,h=h_out,P=P_out)             "exit specific entropy"
    S_dot_gen=m_dot*(s_out-s_in)                  "entropy balance on actual turbine"
end
```

The modified procedure can be run using the substance 'Air_ha'. Solving shows that $\dot{W}_c = 185.1$ kW, which is within 0.01% of the value obtained using 'Air'. The ideal gas model is clearly appropriate for these conditions.

f) The results of parts (b) and (d) showed that the rate of entropy generation in the aftercooler is nearly three times larger than the rate of entropy generation in the compressor. Entropy generation in the aftercooler is associated with the transfer of heat across a large temperature difference. The air leaving the compressor is quite hot and the heat transfer is to the ambient temperature, which is much colder. The rate of entropy generation can be reduced by intercooling, as explained below; an intercooled multi-stage compression process is shown in Figure 5.

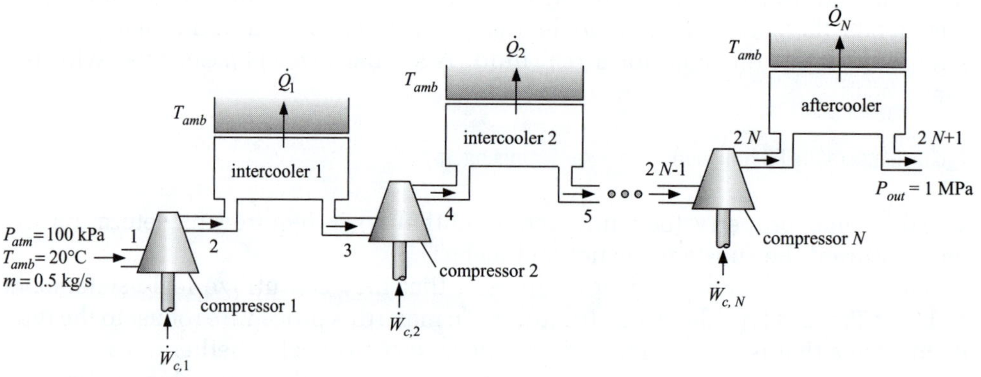

Figure 5: Intercooled compression process.

The overall pressure ratio is divided into several (N) stages. Between each stage, the air is cooled to near ambient temperature in an intercooler. Because the pressure ratio for each stage is smaller than the overall pressure ratio, the temperature of the air leaving each stage is much lower than the exit temperature of a single stage compression process and therefore the rate of entropy generation due to heat transfer is also reduced. We have seen in Eq. (6-85) that entropy generation is accompanied by "lost work". In this case, the lost work corresponds to increased compression power and therefore less power is required to accomplish the intercooled compression process shown in Figure 5 than the non-intercooled compression process shown in Figure 1.

Use the procedure developed in part (e) to analyze the intercooled compression process shown in Figure 5. Assume that the pressure ratio across each compressor is the same, that the isentropic efficiency of each compressor is $\eta_c = 0.74$, and that the intercoolers are perfect (i.e., they have no pressure loss and the exit temperature from each intercooler is T_{amb}). Plot the power required as a function of the number of stages of intercooling, N.

The overall pressure ratio for the system is:

$$PR = \frac{P_{in}}{P_{out}}$$

The pressure ratio for each stage can be obtained according to:

$$PR_{stage} = PR^{\frac{1}{N}}$$

The inlet state for compressor i is numbered $2i$-1 in Figure 5. The inlet pressure for each compressor is therefore:

$$P_{2\,i-1} = P_{atm}\, PR_{stage}^{(i-1)} \text{ for } i = 1..N$$

and the inlet temperature is:

$$T_{2\,i-1} = T_{amb} \text{ for } i = 1..N$$

The specific entropy at the inlet to each compressor ($s_{2i\text{-}1}$) is obtained from the temperature and pressure. The exit pressure for each stage is set according to:

$$P_{2\,i} = P_{atm}\, PR_{stage}^{i} \text{ for } i = 1..N$$

```
N=2 [-]                                                "number of intercooled compression stages"
PR=P_out/P_atm                                         "total pressure ratio"
PR_stage=PR^(1/N)                                      "stage pressure ratio"
duplicate i=1,N
   P[2*i-1]=P_atm*PR_stage^(i-1)                       "inlet pressure"
   T[2*i-1]=T_amb                                      "inlet temperature"
   s[2*i-1]=entropy(F$,T=T[2*i-1],P=P[2*i-1])          "inlet specific entropy"
   P[2*i]=P_atm*PR_stage^i                             "exit pressure"
end
```

The Compressor procedure is used to analyze each of the compressors.

```
duplicate i=1,N
   Call Compressor(T[2*i-1],P[2*i-1],P[2*i],m_dot, F$, eta_c: T[2*i], s[2*i], s_s[2*i], W_dot_c[i])
end
```

The pressure and temperature at the exit of the final heat exchanger, the aftercooler in Figure 5, are specified ($T_{2N+1} = T_{amb}$ and $P_{2N+1} = P_{out}$). The specific entropy (s_{2N+1}) is determined from the temperature and pressure.

```
"exit of aftercooler"
T[2*N+1]=T_amb
P[2*N+1]=P_out
s[2*N+1]=entropy(F$,T=T[2*N+1],P=P[2*N+1])
```

The total compressor power required is:

$$\dot{W}_{c,total} = \sum_{i=1}^{N} \dot{W}_{c,i}$$

```
W_dot_c_total=sum(W_dot_c[i],i=1,N)          "total compression work"
```

For two-stages of compression, $N = 2$, the total compression power required is $\dot{W}_{c,total} = 154.8$ kW, a significant reduction relative to the power required by a non-intercooled system (184.9 kW). The *T-s* diagram for the intercooled system with $N = 2$ is shown in Figure 6.

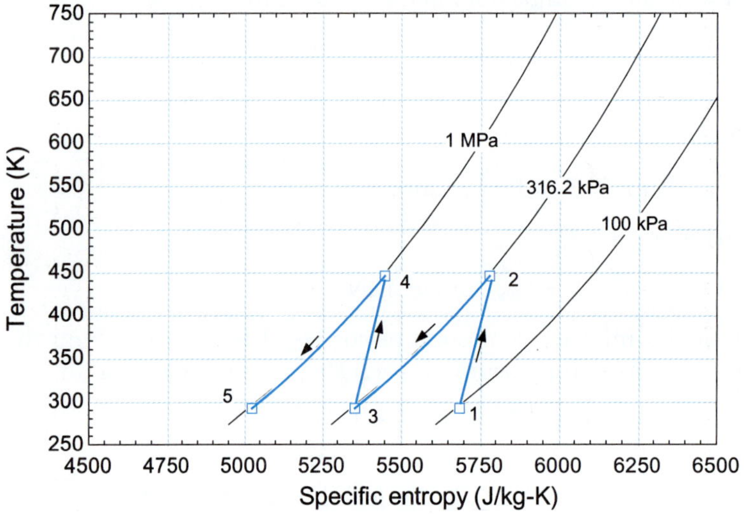

Figure 6: *T-s* diagram for a two-stage intercooled process.

Notice that the temperature difference between the hot air leaving the compressor and the ambient temperature that is used for heat rejection has been reduced substantially in Figure 6 as compared to the non-intercooled cycle shown in Figure 3. The smaller temperature difference reduces the rate of entropy generation and therefore the amount of power required. The benefit associated with compressing the air at, on average, a lower value of specific entropy can be viewed in another way. As noted in Example 6.6-2, the specific enthalpy difference between two values of pressure tends to increase with increasing specific entropy. The compression from 316.2 kPa to 1 MPa occurs at lower specific entropy in Figure 6 than it does in Figure 3 and therefore the power is reduced. Figure 7 illustrates the total compression power required as a function of the number of intercooled compression stages. As the number of stages becomes large, the compression process approaches an isothermal process.

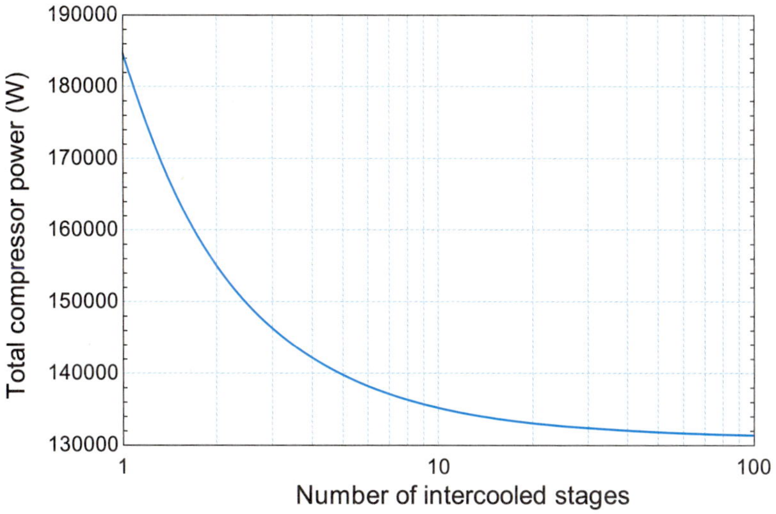

Figure 7: Total compression power as a function of the number of intercooled compression stages.

6.6.3 Pump Efficiency

The functionality of a pump is similar to that of a compressor. The pump uses a mechanical power input to increase the pressure of a fluid. However, pumps operate on nearly incompressible liquids whereas compressors operate on compressible gases. Although the design requirements for pumps are very different than for compressors, the energy and entropy balances are identical and all of the relations presented in Section 6.6.2 for compressors also apply to pumps. Indeed, if a pump is operating with a real fluid then the analysis of a pump is essentially identical to the analysis of a compressor. However, pumps often operate on fluids that can be modeled as being incompressible. In this case, the analysis can be simplified.

Figure 6-13(a) illustrates an entropy balance on a reversible, adiabatic pump operating at steady conditions:

$$\dot{m}\, s_{in} = \dot{m}\, s_{s,out} \tag{6-94}$$

Equation (6-94) can be rearranged to provide:

$$s_{s,out} - s_{in} = 0 \tag{6-95}$$

The specific entropy of an incompressible substance is considered in Section 6.3.4. Substituting Eq. (6-42) for the change in the specific entropy of an incompressible substance into Eq. (6-95) leads to:

$$s_{s,out} - s_{in} = \int_{T_{in}}^{T_{s,out}} \frac{c}{T}\, dT = 0 \tag{6-96}$$

Equation (6-96) can only be satisfied if:

$$T_{s,out} = T_{in} \tag{6-97}$$

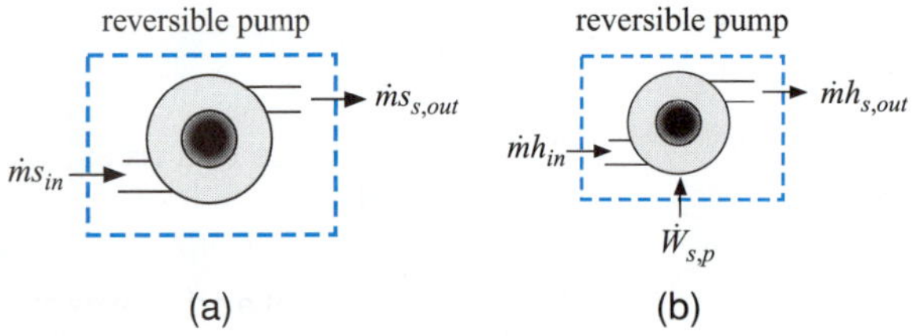

Figure 6-13: (a) Entropy balance on a reversible pump and (b) energy balance on a reversible pump.

Therefore, the temperature of the fluid remains constant for a reversible pump operating on an incompressible fluid.

Figure 6-13(b) illustrates an energy balance on a reversible adiabatic pump:

$$\dot{m}\, h_{in} + \dot{W}_{s,p} = \dot{m}\, h_{s,out} \tag{6-98}$$

or

$$\dot{W}_{s,p} = \dot{m}\,(h_{s,out} - h_{in}) \tag{6-99}$$

Substituting Eq. (4-43) for the change in the specific enthalpy of an incompressible substance into Eq. (6-99) leads to:

$$\dot{W}_{s,p} = \dot{m}\left[\int_{T_{in}}^{T_{s,out}} c\,dT + v\,(P_{out} - P_{in})\right] \tag{6-100}$$

Substituting Eq. (6-97) into Eq. (6-100) provides:

$$\dot{W}_{s,p} = \underbrace{\dot{m}\,v}_{\dot{V}}\,(P_{out} - P_{in}) \tag{6-101}$$

Equation (6-101) indicates that the minimum power required to pump an incompressible fluid is the product of the volumetric flow rate and the pressure rise. This relationship can be used to accomplish quick estimates of the pump power requirement for a variety of applications. The actual pump power required is greater than Eq. (6-101) by an amount that depends on the pump efficiency. The pump efficiency is defined in the same way as the compressor efficiency; it is the ratio of the power required by a reversible pump to the power required by the actual pump.

$$\boxed{\eta_p = \frac{\dot{W}_{s,p}}{\dot{W}_p}} \tag{6-102}$$

The actual power required by a pump is found by substituting Eq. (6-101) into Eq. (6-102):

$$\dot{W}_p = \frac{\dot{m}\,v\,(P_{out} - P_{in})}{\eta_p} \tag{6-103}$$

EXAMPLE 6.6-4: SOLAR POWERED LIVESTOCK PUMP

Figure 1 illustrates a pump that is used to transfer water from a source to a livestock watering tank. Because the watering tank is located remotely, it is convenient to power the system using a photovoltaic (PV) solar panel. The solar collector panel converts solar energy into electrical power with efficiency $\eta_{PV} = 0.08$ and stores the electrical energy in a battery. At the location of interest, the 24-hour annual average solar flux on a flat collector is $SF = 225\ \text{W/m}^2$. The pump operates continuously using electrical energy drawn from the battery.

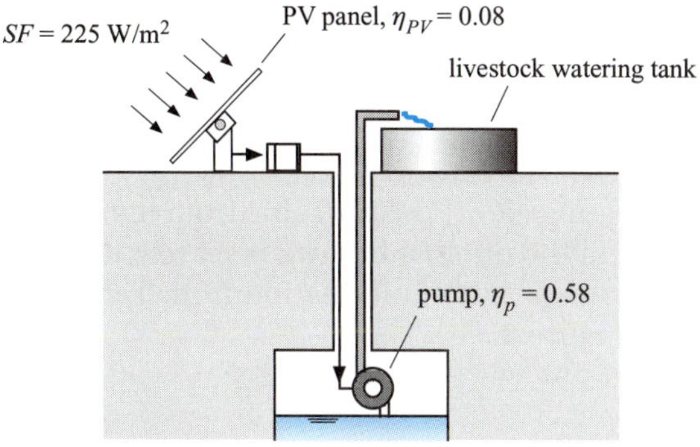

Figure 1: PV-powered pump serving a livestock watering tank.

The manufacturer provides information about the pump performance in Table 1. The pressure rise that can be produced (head) in units of ft of water is listed for various values of the flow rate. This information is referred to as the pump curve. The isentropic efficiency of the pump is approximately constant and equal to $\eta_p = 0.58$.

Table 1: Data from the manufacturer's pump curve.

Head (ft of water)	Flow (gpm)
60 ft	0 gpm
60 ft	2 gpm
57 ft	4 gpm
52 ft	6 gpm
44 ft	8 gpm
33 ft	10 gpm
18 ft	12 gpm
0 ft	14 gpm

The source of water used to fill the tank is located 20 ft below the livestock tank. Therefore, the pump must provide a pressure rise that is equal to 20 ft of water in order to overcome the hydrostatic pressure due to the weight of the water in the pipe. In addition, there is a pressure drop associated with the friction between the flowing water and the pipe surface that is proportional to the volumetric flow rate to the second power. The total pressure drop between the inlet and outlet of the pipe can be expressed as:

$$\Delta P_{pipe} = 20\left[\text{ftH2O}\right] + 0.085\left[\frac{\text{ftH2O}}{\text{gpm}^2}\right]\dot{V}^2 \tag{1}$$

where ΔP_{pipe} is the pressure drop and $\dot{V}$ is the volumetric flow rate of water. Assume that water is an incompressible substance.

a) Determine the flow rate of water that the pump will deliver to the livestock tank. If each animal requires nominally $\dot{V}_{animal} = 20$ gal/day then determine the number of animals that can be serviced by the system.

The inputs are entered in EES.

```
$UnitSystem SI MASS RAD PA K J
"Inputs"
V_dot_animal=20 [gal/day]*convert(gal/day,m^3/s)     "required flow of water per animal"
SF=225 [W/m^2]                                      "average solar flux"
eta_p=0.58 [-]                                      "pump efficiency"
eta_PV=0.08 [-]                                     "efficiency of photovoltaic array"
```

A Lookup table is generated in EES by selecting New Lookup Table from the Tables menu. The number of columns is set to 2 and the number of rows is set to 8 in order to correspond to the data in Table 1. The pump curve information from Table 1 is entered in the Lookup table, as shown in Figure 2.

Lookup Table

Pump curve

	Head [ftH2O]	Flow [gpm]
Row 1	60	0
Row 2	60	2
Row 3	57	4
Row 4	52	6
Row 5	44	8
Row 6	33	10
Row 7	18	12
Row 8	0	14

Figure 2: Lookup table containing the pump curve data.

By right-clicking on each heading and selecting Properties from the pop up menu it is possible to add a title and units to each column in the Lookup table. The pump curve data are plotted in Figure 3 by selecting New Plot Window from the Plots menu. The source of data in the New Plot Setup dialog is specified to be the Lookup table.

In order to model the pump, it is useful to have an equation that represents the pump curve data. Select Curve Fit from the Plots menu and fit a third order polynomial to the pump curve data; select Fit and then Plot to overlay the curve fit on the plot (Figure 3). Select Copy equation to Clipboard and then paste the equation into the Equations Window:

```
Head=60.0606061 + 0.398809524*Flow - 0.255681818*Flow^2 - 0.00568181818*Flow^3
"best fit pump curve"
```

Change the variable names in the EES equation above in order to reflect the non-SI units that are used in the pump curve. Include units for the constants in the pump curve formula:

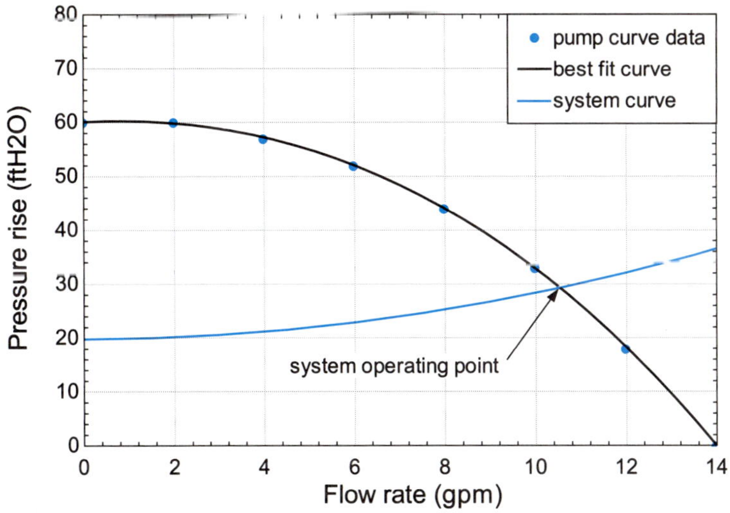

Figure 3: Pump curve data with best fit curve and system resistance curve. The intersection is the operating point.

```
DP_pump_ftH2O=60.0606061 [ftH2O] + 0.398809524 [ftH2O/gpm]*V_dot_gpm &
-0.255681818 [ftH2O/gpm^2]*V_dot_gpm^2 - 0.00568181818 [ftH2O/gpm^3]*V_dot_gpm^3
                    "best fit pump curve"
```

The best fit curve can be used to estimate the pump performance at an arbitrary flow rate. For example, with a flow rate of $\dot{V} = 2.1$ gpm,

```
V_dot_gpm=2.1 [gpm]          "arbitrary flow"
```

the pump will provide approximately $\Delta P_{pump} = 59.7$ ftH2O. Equation (1) is used to determine the pressure drop across the pipe:

```
DP_pipe_ftH2O=20 [ftH2O]+0.085 [ftH2O/gpm^2]*V_dot_gpm^2     "System resistance curve"
```

which leads to $\Delta P_{pipe} = 20.4$ ftH2O. For the flow rate that was set above, $\dot{V} = 2.1$ gpm, the pipe requires $\Delta P_{pipe} = 20.4$ ftH2O and the pump is capable of providing $\Delta P_{pump} = 59.7$ ftH2O. The pump pressure rise is larger than is required by the pipe and therefore the flow rate will increase. The increased flow rate will cause the pressure provided by the pump to decrease and the pressure drop across the pipe to increase. The flow rate where these quantities coincide is the operating point for the system. The system resistance curve is overlaid onto the pump curve in Figure 3 by creating a Parametric table that includes the variables DP_pipe_ftH2O and V_dot_gpm. According to Figure 3, the flow to the tank will be about $\dot{V} = 10.5$ gpm because this is the point where the pump curve intersects with the system resistance curve. A more precise estimate can be obtained by commenting out the specified value of the flow rate and setting the pressure rise produced by the pump to be equal to the pressure drop across the pump:

```
{V_dot_gpm=2.1 [gpm]}                  "arbitrary flow"
DP_pipe_ftH2O=DP_pump_ftH2O            "find operating point"
```

which results in $\dot{V} = 10.51$ gpm and $\Delta P_{pump} = 29.4$ ftH2O; note that it is necessary to specify a reasonable guess value for the variable V_dot_gpm in order to arrive at this result. The flow rate is converted to SI units and used to determine the number of animals than can be supported by the system:

```
V_dot=V_dot_gpm*convert(gpm,m^3/s)          "flow rate"
N_animal=V_dot/V_dot_animal                 "number of animals"
```

which shows, when solved, that $N_{animal} = 757$ animals.

b) Determine the area of the solar panel required by the system.

The power required by the pump is computed using Eq. (6-103):

$$\dot{W}_p = \frac{\dot{V}\,\Delta P_{pump}}{\eta_p}$$

Assuming that the electrical conversion and storage system is 100% efficient, the motor power must be obtained from the solar panel:

$$SF\,A\,\eta_{PV} = \dot{W}_p$$

where A is the area of the solar panel.

```
DP_pump=DP_pump_ftH2O*convert(ftH2O,Pa)     "pressure rise"
W_dot_pump=V_dot*DP_pump/eta_p              "power required by pump"
SF*A*eta_PV=W_dot_pump                      "collector area"
```

Solving these equations provides $A = 5.6\ \text{m}^2$.

6.6.4 Nozzle Efficiency

A nozzle uses a reduction in the pressure of a fluid to cause an increase in its velocity, as described in Section 4.4.4. Nozzles are typically assumed to be adiabatic since the fluid passes through the nozzle very quickly. The change in the potential energy of the fluid is typically neglected and a nozzle is assumed to operate at steady state. The performance of a nozzle is specified by the *nozzle isentropic efficiency*, η_n, defined as:

$$\boxed{\eta_n = \frac{\tilde{V}^2_{out}}{\tilde{V}^2_{s,out}}} \qquad (6\text{-}104)$$

where $\tilde{V}_{out}$ is the velocity of the fluid leaving the nozzle (and therefore $\tilde{V}^2_{out}$ is proportional to the kinetic energy of the exiting flow) and $\tilde{V}_{s,out}$ is the velocity leaving a reversible nozzle that is operating under the same conditions as the actual nozzle. In this case, the 'same conditions' implies that the reversible nozzle experiences the same mass flow rate, inlet state, inlet velocity, and outlet pressure, as shown in Figure 6-14.

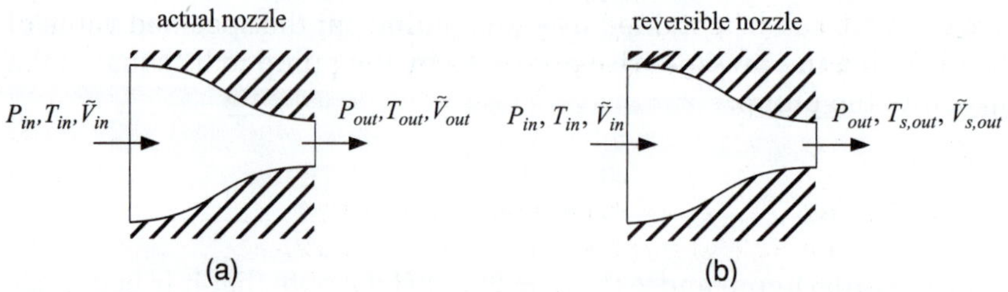

Figure 6-14: (a) An actual nozzle and (b) a reversible nozzle that are operating under the same conditions.

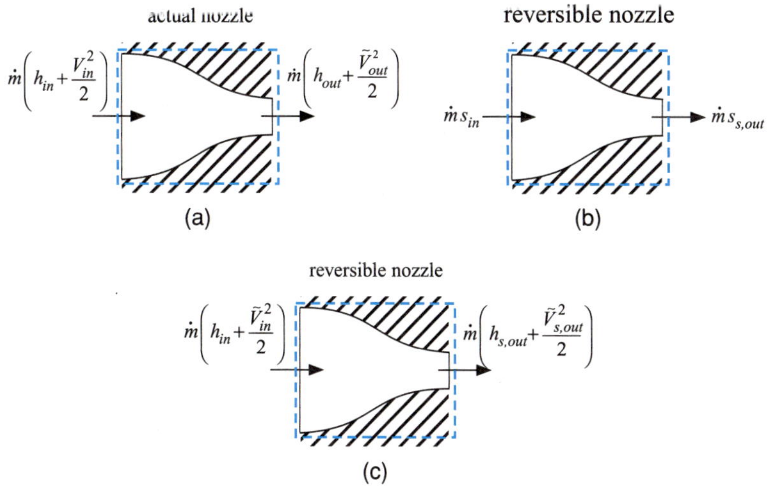

Figure 6-15: (a) An energy balance on the actual nozzle and (b) an entropy balance and (c) an energy balance on the reversible nozzle that is used to bound the performance.

An energy balance on the actual adiabatic nozzle is shown in Figure 6-15(a) and simplifies to:

$$h_{in} + \frac{\tilde{V}_{in}^2}{2} = h_{out} + \frac{\tilde{V}_{out}^2}{2} \tag{6-105}$$

where $\tilde{V}_{in}$ is the inlet velocity and h_{in} and h_{out} are the specific enthalpies of the fluid at the nozzle inlet and outlet, respectively. In the usual case, the nozzle inlet velocity and state are known so h_{in} can be determined. The outlet pressure is also known, but knowing only the pressure is not sufficient to determine h_{out}. Consequently Eq. (6-105) cannot be solved to determine the nozzle exit velocity without additional information.

The additional information that is usually used to determine the nozzle exit velocity is the nozzle efficiency. Figure 6-15(b) shows an entropy balance on the reversible nozzle:

$$s_{in} = s_{s,out} \tag{6-106}$$

where $s_{s,out}$ is the specific entropy of the fluid leaving the reversible nozzle. The outlet state of the reversible nozzle is fixed by the outlet pressure and specific entropy and therefore the specific enthalpy of the fluid leaving the reversible nozzle, $h_{s,out}$, can be determined. Figure 6-15(c) illustrates an energy balance on the reversible adiabatic nozzle:

$$h_{in} + \frac{\tilde{V}_{in}^2}{2} = h_{s,out} + \frac{\tilde{V}_{s,out}^2}{2} \tag{6-107}$$

Equation (6-107) allows the calculation of the velocity leaving the reversible nozzle, which is the maximum possible velocity that can be achieved under these conditions. The nozzle efficiency defined by Eq. (6-104) can then be used to determine the actual velocity. The energy balance in Eq. (6-105) is then used to determine the outlet state of the actual nozzle. It is not uncommon for well-designed nozzles to operate with efficiencies greater than 90%. The fluid in a nozzle flows from high pressure to low pressure, which is the natural direction for fluid flow. Nozzles do not experience losses related to separation in the same way that diffusers do and therefore the efficiency of a well-designed nozzle will typically be higher than the efficiency of a diffuser. Nozzles can be used to accelerate the flow to supersonic speeds, as discussed in Chapter 16.

EXAMPLE 6.6-5: JET-POWERED WAGON

You have decided to build an air-powered vehicle by attaching a high pressure air bottle to your little red wagon, as shown in Figure 1. The air from the bottle is exhausted through a nozzle in order to produce a high velocity flow that will provide thrust and cause your wagon to accelerate forwards.

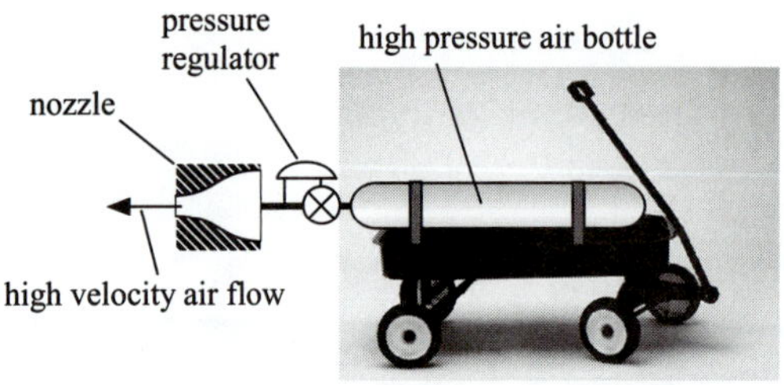

Figure 1: Jet-powered wagon.

Figure 2 illustrates a schematic of the propulsion system.

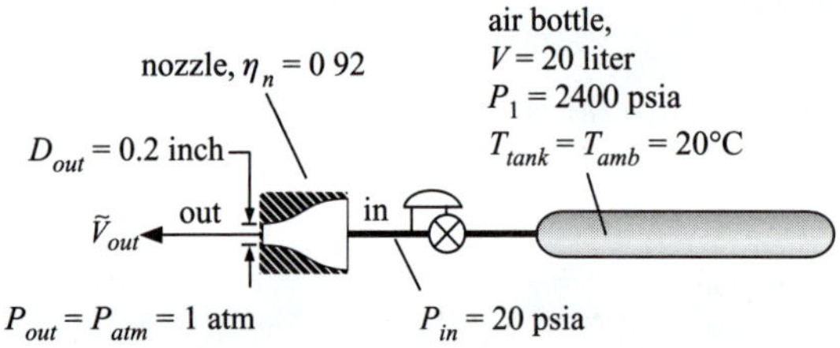

Figure 2: Propulsion system.

The air bottle has volume $V = 20$ liter and is initially charged to a pressure $P_1 = 2400$ psia. Heat transfer from the ambient keeps the air in the tank always at $T_{tank} = T_{amb} = 20°C$. The air leaves the tank and passes through a pressure regulator that reduces the pressure of the air from the instantaneous tank pressure to a fixed pressure at the inlet to the nozzle, $P_{in} = 20$ psia. The propulsion system stops working when the bottle pressure reaches $P_2 = P_{in}$. The nozzle has an efficiency $\eta_n = 0.92$. The nozzle exit pressure is atmospheric, $P_{out} = P_{atm} = 1$ atm. The outlet diameter of the nozzle is $D_{out} = 0.2$ inch. Model the air as an ideal gas ($R = 287.1$ N-m/kg-K) with a constant $c_P = 1004$ J/kg-K.

a) Determine the velocity of the air leaving the nozzle, $\tilde{V}_{out}$, and the mass flow rate of air passing through the nozzle, $\dot{m}$. Neglect the kinetic energy of the air entering the nozzle.

   The inputs are entered in EES:

```
$UnitSystem SI K Pa J Mass Radian
Vol=20 [liter]*convert(liter,m^3)            "volume of air bottle"
T_amb=converttemp(C,K,20[C])                 "ambient temperature"
P_atm=1 [atm]*convert(atm,Pa)                "atmospheric pressure"
T_tank=T_amb                                 "temperature of air in tank"
P_out=P_atm                                  "pressure of air leaving nozzle"
P[1]=2400 [psi]*convert(psi,Pa)              "initial charge pressure"
R=287.0[J/kg-K]                              "gas constant for air"
c_P=1004 [J/kg-K]                            "constant pressure specific heat capacity of air"
```

```
P_in_psi=20 [psi]                          "nozzle inlet pressure, in psi"
P_in=P_in_psi*convert(psi,Pa)              "nozzle inlet pressure"
D_out_inch=0.2 [inch]                      "nozzle outlet diameter, in inch"
D_out=D_out_inch*convert(inch,m)           "nozzle outlet diameter"
```

The specific heat capacity of air at constant volume is calculated according to:

$$c_v = c_P - R$$

```
c_v=c_P-R          "constant volume specific heat capacity"
```

The specific enthalpy of the gas entering the regulator is:

$$h_{tank} = c_P\, T_{tank} \qquad (1)$$

Equation (1) is appropriate because the air is being modeled as an ideal gas with a constant specific heat capacity. Equation (1) assumes a reference state at which specific enthalpy is zero to be 0 K. An energy balance on the regulator (which is a valve) is:

$$h_{tank} = h_{in} \qquad (2)$$

The specific enthalpy of the gas entering the nozzle is:

$$h_{in} = c_P\, T_{in} \qquad (3)$$

Substituting Eqs. (1) and (3) into Eq. (2) shows that:

$$T_{tank} = T_{in}$$

Therefore, the temperature of the air entering the nozzle is equal to the tank temperature, regardless of the instantaneous tank pressure.

```
h_tank=c_P*T_tank      "specific enthalpy of air leaving the tank and entering the regulator"
h_in=h_tank            "energy balance on regulator"
h_in=c_P*T_in          "nozzle inlet temperature"
```

In order to utilize the nozzle efficiency, it is necessary to analyze both the reversible nozzle that provides the limiting performance, shown in Figure 6-14(a), as well as the actual nozzle shown in Figure 6-14(b). An entropy balance on the reversible adiabatic nozzle provides:

$$s_{in} = s_{s,out}$$

Section 6.3.3 provides equations that relate the properties of an ideal gas with constant specific heat capacities undergoing an isentropic process. Equation (6-33) relates the pressures and temperatures at two states with the same specific entropy:

$$T_{s,out} = T_{in} \left(\frac{P_{out}}{P_{in}} \right)^{\frac{R}{c_P}}$$

The temperature and pressure specify the state of the air leaving the reversible nozzle. The specific enthalpy ($h_{s,out}$) is computed from:

$$h_{s,out} = c_P\, T_{s,out}$$

```
T_s_out=T_in*(P_out/P_in)^(R/c_P)     "reversible nozzle outlet temperature"
h_s_out=c_P*T_s_out                   "reversible nozzle outlet specific enthalpy"
```

An energy balance on the reversible nozzle provides:

$$h_{in} = h_{s,out} + \frac{\tilde{V}^2_{s,out}}{2}$$

The velocity leaving the actual nozzle is obtained from the definition of nozzle efficiency:

$$\tilde{V}^2_{out} = \eta_n \tilde{V}^2_{s,out}$$

```
h_in=h_s_out+0.5*Vel_s_out^2          "energy balance on reversible nozzle"
Vel_out^2=eta_n*Vel_s_out^2           "nozzle efficiency"
```

which leads to $\tilde{V}_{out}$ = 213.7 m/s. An energy balance on the actual nozzle provides:

$$h_{in} = h_{out} + \frac{\tilde{V}^2_{out}}{2}$$

The outlet specific enthalpy and pressure specify the state of the air leaving the actual nozzle. The outlet temperature is obtained from:

$$h_{out} = c_P \, T_{out}$$

and the outlet specific volume is computed according to:

$$v_{out} = \frac{R\, T_{out}}{P_{out}}$$

The mass flow rate leaving the nozzle is given by:

$$\dot{m} = \frac{\pi \, D^2_{out}}{4 \, v_{out}} \tilde{V}_{out}$$

```
h_in=h_out+0.5*Vel_out^2                   "energy balance on actual nozzle"
h_out=c_P*T_out                            "outlet temperature"
v_out=R*T_out/P_out                        "specific volume at outlet"
m_dot=(pi*D_out^2/4)*Vel_out/v_out         "mass flow rate leaving nozzle"
```

which results in $\dot{m} = 0.00566$ kg/s.

b) Determine the thrust force pushing your wagon forward. If we neglect the (small) velocity of the wagon, the thrust force is given by the product of the nozzle exit velocity and the mass flow rate.

The thrust force that is propelling the wagon is:

$$F_{thrust} = \dot{m} \, \tilde{V}_{out}$$

```
F_thrust=m_dot*vel_out                     "thrust force"
```

which provides $F_{thrust} = 1.208$ N.

c) Determine the time that the propulsion system will operate before the tank pressure reaches the nozzle inlet pressure.

The initial state of the air in the tank (state 1) is fixed by the temperature and pressure, $T_1 = T_{tank}$ and P_1. The specific volume and specific internal energy at state 1 are determined from:

$$v_1 = \frac{R\,T_1}{P_1}$$

$$u_1 = c_v\,T_1$$

The mass of air initially in the tank is:

$$m_1 = \frac{V}{v_1}$$

```
T[1]=T_tank            "inital tank temperature"
v[1]=R*T[1]/P[1]       "specific volume"
u[1]=c_V*T[1]          "specific internal energy"
m[1]=Vol/v[1]          "mass"
```

The process is complete when the tank pressure reaches the nozzle inlet pressure, $P_2 = P_{in}$. The tank temperature does not change during the process, $T_2 = T_{tank}$. State 2 is fixed by its temperature and pressure. The specific volume and specific internal energy are calculated from:

$$v_2 = \frac{R\,T_2}{P_2}$$

$$u_2 = c_v\,T_2$$

The mass of air in the tank at the conclusion of the process is:

$$m_2 = \frac{V}{v_2}$$

```
P[2]=P_in              "final tank pressure"
T[2]=T_tank            "final tank temperature"
v[2]=R*T[2]/P[2]       "specific volume"
u[2]=c_V*T[2]          "internal energy"
m[2]=Vol/v[2]          "mass"
```

A mass balance on the tank is shown in Figure 3 and expressed in Eq. (4):

$$0 = m_{out} + m_2 - m_1 \tag{4}$$

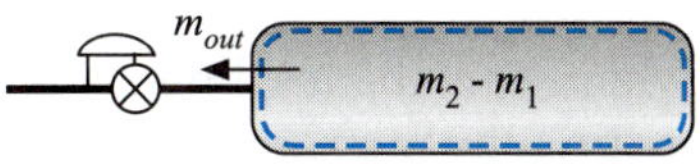

Figure 3: Mass balance on tank.

The time that the wagon will operate is given by:

$$m_{out} = \dot{m}\,t$$

```
0=m_out+m[2]-m[1]      "mass balance on tank"
m_out=m_dot*time       "operating time"
```

Solving indicates that $t = 689.8$ s or about 11.5 minutes.

d) The combined mass of the wagon and the rider is about $m_{wagon} = 200$ lb$_m$. If the wagon experiences no resistance of any type (i.e., no friction or air resistance) then determine the velocity that it will obtain at the time calculated in part (c).

In the absence of resistance, the wagon will accelerate steadily according to:

$$m_{wagon} \frac{d\tilde{V}_{wagon}}{dt} = F_{thrust} \tag{5}$$

where $\tilde{V}_{wagon}$ is the velocity of the wagon. Integrating Eq. (5) leads to a final velocity of:

$$\tilde{V}_{wagon,max} = \frac{F_{thrust}}{m_{wagon}} t$$

```
m_wagon=200 [lbm]*convert(lbm,kg)                    "mass of wagon"
vel_wagon_max=F_thrust*time/m_wagon                  "maximum velocity"
vel_wagon_max_mph=vel_wagon_max*convert(m/s,mph)     "in mph"
```

which leads to $\tilde{V}_{wagon,max} = 9.19$ m/s (20.55 mph).

e) List the phenomena that cause entropy generation for this process.

Entropy is generated by flow through the pressure regulator (a valve), heat transfer to the air leaving the nozzle as it thermally equilibrates with ambient, and the deceleration of the flow irreversibly to zero velocity with the ambient. There is also heat transfer to the tank; however, the tank and the ambient are assumed to be at the same temperature and therefore this process does not generate entropy.

f) Determine the total entropy generated by the process of emptying the tank.

An energy balance on the entire system is shown in Figure 4. Note that the boundary of the system is placed at a location where the air leaving the nozzle has thermally equilibrated with the ambient and decelerated to near zero velocity in order to capture all of the irreversibility associated with the process. The energy balance for the time in which the tank provides air to the nozzle is:

$$Q = m_{out}\, h_{atm} + m_2\, u_2 - m_1\, u_1$$

where Q is the total heat transfer from the ambient (to the tank and to the cold gas exiting the nozzle) and h_{atm} is the specific enthalpy of the air after it has thermally equilibrated with the ambient:

$$h_{atm} = c_P\, T_{amb}$$

```
h_atm=c_P*T_amb                       "specific enthalpy of air after it has equilibrated with ambient"
Q=m_out*h_atm+m[2]*u[2]-m[1]*u[1]     "energy balance"
```

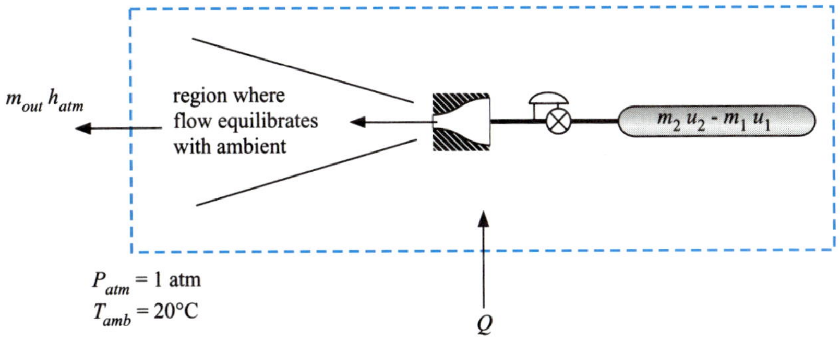

Figure 4: Energy balance on the entire system.

An entropy balance on the entire system is shown in Figure 5; note that the boundary of the system is selected so that the heat transfer enters the system at T_{amb} in order to capture all of the irreversibilities that are associated with this process.

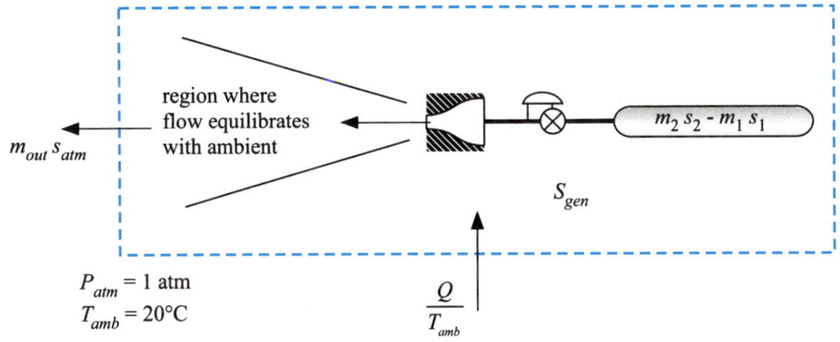

Figure 5: Entropy balance on the entire system.

The entropy balance is:

$$\frac{Q}{T_{amb}} + S_{gen} = m_{out}\, s_{atm} + m_2\, s_2 - m_1\, s_1 \tag{6}$$

Substituting Eq. (4) into Eq. (6) leads to:

$$\frac{Q}{T_{amb}} + S_{gen} = (m_1 - m_2)\, s_{atm} + m_2\, s_2 - m_1\, s_1$$

Rearranging:

$$\frac{Q}{T_{amb}} + S_{gen} = m_1\,(s_{atm} - s_1) + m_2\,(s_2 - s_{atm}) \tag{7}$$

The specific entropy differences that appear in Eq. (7) can be expressed in terms of pressure and temperature using Eq. (6-29):

$$\frac{Q}{T_{amb}} + S_{gen} = m_1 \left[c_P \ln\left(\frac{T_{amb}}{T_1}\right) - R \ln\left(\frac{P_{atm}}{P_1}\right)\right] + m_2 \left[c_P \ln\left(\frac{T_2}{T_{amb}}\right) - R \ln\left(\frac{P_2}{P_{atm}}\right)\right]$$

```
Q/T_amb+S_gen=m[1]*(c_P*ln(T_amb/T[1])-R*ln(P_atm/P[1]))+&
m[2]*(c_P*ln(T[2]/T_amb)-R*ln(P[2]/P_atm))                    "entropy balance"
```

Solving results in $S_{gen} = 4630$ J/K.

g) Plot the maximum velocity achieved by the wagon as a function of the nozzle inlet pressure that is set by the pressure regulator. You should see that an optimal nozzle inlet pressure exists. Explain why this optimum occurs. What practical considerations will actually limit the nozzle exit velocity?

Figure 6 illustrates the maximum velocity achieved by the wagon as a function of the nozzle inlet pressure. When the nozzle inlet pressure is too low, the exit velocity is low and therefore the thrust force is small. When the nozzle inlet pressure is too high, the exit velocity is high but the bottle pressure is reduced to the nozzle inlet pressure quickly and therefore there is little time for the wagon to accelerate. The optimal nozzle inlet pressure balances these two effects. It is not likely that the nozzle will function well when the exit velocity reaches the speed of sound, which will limit the practical range of nozzle inlet pressures.

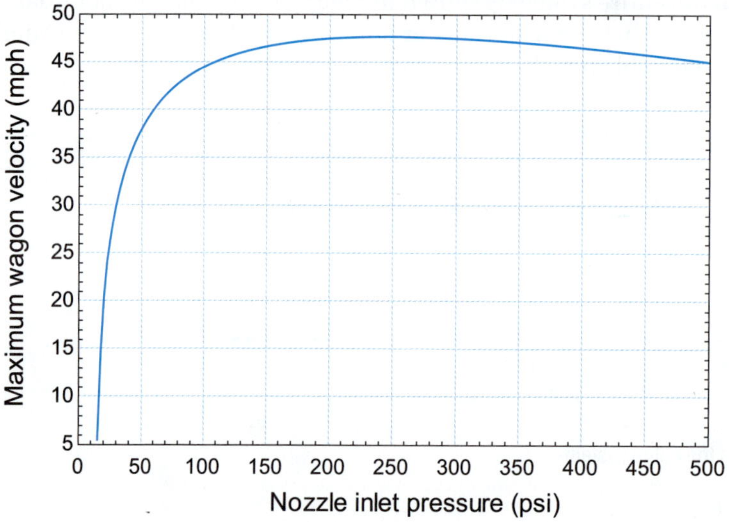

Figure 6: Maximum wagon velocity as a function of the nozzle inlet pressure.

6.6.5 Diffuser Efficiency

A diffuser uses a reduction in the velocity of a fluid to cause an increase in its pressure, as described in Section 4.4.5. Diffusers are typically assumed to be adiabatic since the fluid passes through the device very quickly. The change in the potential energy of the fluid is neglected and a diffuser is assumed to operate at steady state. The performance of a diffuser can be specified by the *diffuser isentropic efficiency*, η_d, defined as:

$$\boxed{\eta_d = \frac{P_{out} - P_{in}}{P_{s,out} - P_{in}}} \tag{6-108}$$

where P_{in} and P_{out} are the inlet and exit pressures associated with the diffuser, respectively, and $P_{s,out}$ is the pressure at the exit of a reversible diffuser that operates under the same conditions as the actual diffuser. The 'same conditions' implies that the reversible diffuser experiences the same mass flow rate, inlet state, inlet velocity, and the same outlet velocity, as shown in Figure 6-16.

An energy balance on the actual adiabatic diffuser is shown in Figure 6-17(a):

$$h_{in} + \frac{\tilde{V}_{in}^2}{2} = h_{out} + \frac{\tilde{V}_{out}^2}{2} \tag{6-109}$$

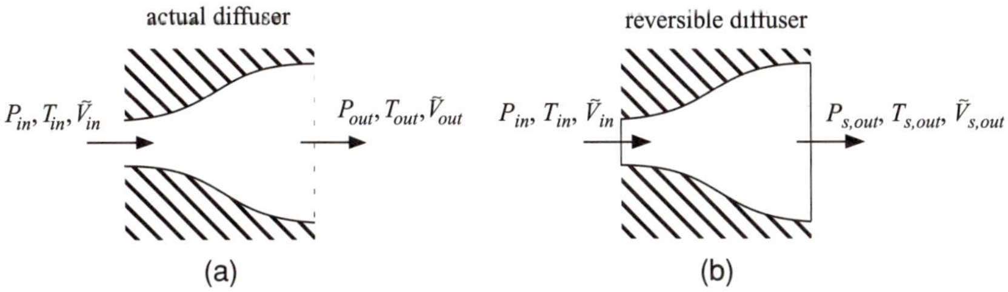

Figure 6-16: (a) An actual diffuser and (b) a reversible diffuser that is operating under the same conditions.

where $\tilde{V}_{in}$ is the inlet velocity and h_{in} and h_{out} are the specific enthalpies of the fluid at the diffuser inlet and outlet, respectively. In the usual case, the diffuser inlet velocity and state are known so $\tilde{V}_{in}$ and h_{in} can be determined. The outlet velocity, $\tilde{V}_{out}$, is also known and therefore Eq. (6-109) can be used to determine the outlet specific enthalpy. However, knowing only the outlet specific enthalpy is not sufficient to determine P_{out}. Consequently Eq. (6-109) cannot be solved to determine the diffuser exit pressure without additional information.

The additional information that can be used to determine the diffuser exit pressure may be the diffuser efficiency. Figure 6-17(b) shows an entropy balance on the reversible diffuser:

$$s_{in} = s_{s,out} \tag{6-110}$$

Figure 6-17(c) illustrates an energy balance on the reversible diffuser:

$$h_{in} + \frac{\tilde{V}_{in}^2}{2} = h_{s,out} + \frac{\tilde{V}_{out}^2}{2} \tag{6-111}$$

Equation (6-110) can be solved for $s_{s,out}$, the specific entropy leaving the reversible diffuser, and Eq. (6-111) can be solved for $h_{s,out}$, the specific enthalpy leaving the reversible diffuser. The outlet state of the reversible diffuser is fixed by the specific entropy and specific enthalpy and therefore the pressure of the fluid leaving the reversible diffuser, $P_{s,out}$, can be determined. The pressure $P_{s,out}$ is the maximum possible pressure that can be achieved by a diffuser under these conditions. Application of the diffuser efficiency,

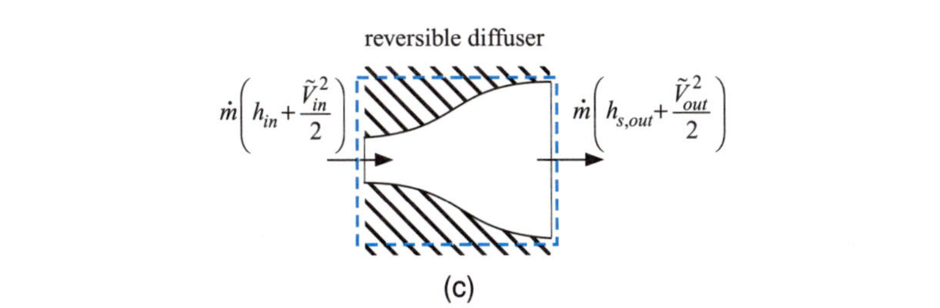

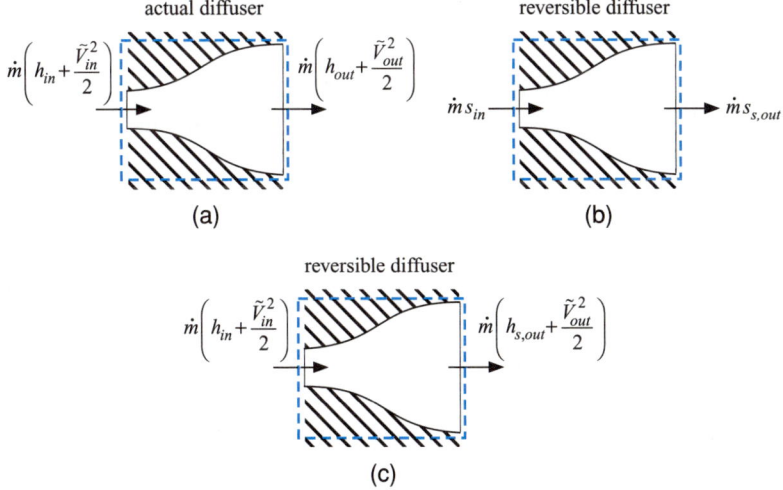

Figure 6-17: (a) An energy balance on the actual diffuser and (b) an entropy balance and (c) an energy balance on the reversible diffuser that is used to bound the performance.

Eq. (6-108), determines the actual outlet pressure. The energy balance in Eq. (6-109) is then used to determine the outlet specific enthalpy of the actual nozzle. The outlet specific enthalpy and pressure fix the outlet state.

It is difficult to build a high efficiency diffuser because the fluid in a diffuser must flow from low pressure to high pressure, which is not a natural direction for fluid flow. As a result, diffusers must be carefully designed in order to avoid losses related to the fluid separating from the diffuser wall. In particular, it is not possible to achieve large changes in the cross-sectional area over short distances as these changes result in large, adverse pressure gradients that promote separation. Therefore, the outlet kinetic energy of a diffuser is often not negligible.

The performance of a diffuser is more commonly expressed in terms of the *coefficient of pressure recovery*, K_P, than the isentropic efficiency. The coefficient of pressure recovery is defined as the ratio of the pressure rise produced by the diffuser to the kinetic energy of the fluid entering the diffuser:

$$K_P = \frac{P_{out} - P_{in}}{\left(\dfrac{\rho_{in} \tilde{V}_{in}^2}{2}\right)} \tag{6-112}$$

Unlike the isentropic diffuser efficiency defined in Eq. (6-108), the coefficient of pressure recovery will not necessarily reach unity even if the diffuser is reversible.

EXAMPLE 6.6-6: DIFFUSER IN A GAS TURBINE ENGINE

A diffuser is placed at the inlet to a gas turbine engine in order to provide some ram-air compression ahead of the first compressor stage, as shown in Figure 1.

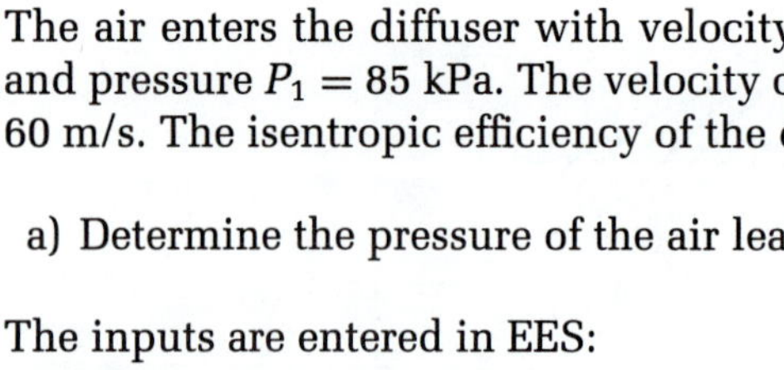

Figure 1: Diffuser.

The air enters the diffuser with velocity $\tilde{V}_1 = 300$ mph, temperature $T_1 = -5°C$, and pressure $P_1 = 85$ kPa. The velocity of the air is reduced in the diffuser to $\tilde{V}_2 = 60$ m/s. The isentropic efficiency of the diffuser is $\eta_d = 0.6$.

a) Determine the pressure of the air leaving the diffuser.

The inputs are entered in EES:

```
$UnitSystem SI K Pa J Mass Radian
Vel[1]=300 [mph]*convert(mph,m/s)        "inlet velocity"
T[1]=converttemp(C,K,-5 [C])             "inlet temperature"
P[1]=85 [kPa]*convert(kPa,Pa)            "inlet pressure"
Vel[2]=60 [m/s]                          "exit velocity"
eta_d=0.6 [-]                            "diffuser isentropic efficiency"
```

State 1 is fixed by the temperature and pressure. The specific enthalpy, specific entropy, and specific volume (h_1, s_1, and v_1) are determined.

```
h[1]=enthalpy(Air,T=T[1])                "inlet specific enthalpy"
s[1]=entropy(Air,T=T[1],P=P[1])          "inlet specific entropy"
v[1]=volume(Air,T=T[1],P=P[1])           "inlet specific volume"
```

An entropy balance on the reversible diffuser, shown in Figure 6-17(b), is:

$$s_{s,2} = s_1$$

An energy balance on the reversible adiabatic diffuser, shown in Figure 6-17(c), is:

$$h_1 + \frac{\tilde{V}_1^2}{2} = h_{s,2} + \frac{\tilde{V}_2^2}{2} \tag{1}$$

The state of the air leaving the diffuser is fixed by $h_{s,2}$ and $s_{s,2}$. The pressure at the outlet of the reversible diffuser, $P_{s,2}$, is determined. The actual outlet pressure is obtained from the definition of the diffuser efficiency:

$$P_2 = P_1 + \eta_d \left(P_{s,2} - P_1\right)$$

```
s_s[2]=s[1]                                "entropy balance on reversible diffuser"
h[1]+Vel[1]^2/2=h_s[2]+Vel[2]^2/2          "energy balance on reversible diffuser"
P_s[2]=pressure(Air,s=s_s[2],h=h_s[2])     "pressure at the exit of a reversible diffuser"
P[2]=P[1]+eta_d*(P_s[2]-P[1])              "pressure at the exit of the actual diffuser"
```

which results in $P_2 = 89.9$ kPa.

b) What is the rate of entropy generation in the diffuser per unit of mass flow rate?

An energy balance on the actual diffuser is shown in Figure 6-17(a):

$$h_1 + \frac{\tilde{V}_1^2}{2} = h_2 + \frac{\tilde{V}_2^2}{2} \tag{2}$$

The pressure and specific enthalpy at the exit of the diffuser fix the state. The specific entropy and specific volume (s_2 and v_2) are determined. An entropy balance on the actual diffuser is:

$$\dot{m} s_1 + \dot{S}_{gen} = \dot{m} s_2$$

Therefore, the entropy generation per mass flow rate is:

$$\frac{\dot{S}_{gen}}{\dot{m}} = s_2 - s_1$$

```
h[1]+Vel[1]^2/2=h[2]+Vel[2]^2/2          "energy balance on the actual diffuser"
s[2]=entropy(Air,P=P[2],h=h[2])          "specific entropy leaving actual diffuser"
v[2]=volume(Air,P=P[2],h=h[2])           "specific volume leaving actual diffuser"
S_dot_gen\m_dot=s[2]-s[1]                "entropy balance on the actual diffuser"
```

which results in $\dot{S}_{gen}/\dot{m} = 10.3$ J/kg-K. Figure 2 illustrates an *h-s* diagram that shows the states at the inlet and outlet of the actual and reversible diffuser. Note that the exit specific enthalpy is the same for both the reversible and actual diffuser, as evident by comparing Eqs. (1) and (2). However, because the specific entropy at the exit of the actual diffuser is larger, the pressure is lower. Figure 2 shows that the actual diffuser achieved an exit pressure of 89.9 kPa compared with the value of 93.2 kPa that could be obtained with a reversible diffuser.

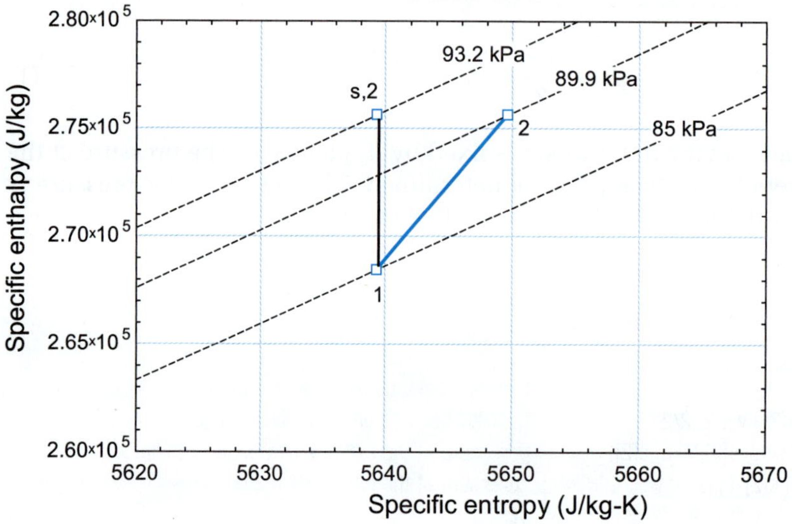

Figure 2: An *h-s* diagram for air showing the states for the actual and reversible diffuser.

c) Determine the area ratio for the diffuser. The area ratio is defined as the ratio of the diffuser exit area to the diffuser inlet area.

The mass flow rate at the inlet and exit of the diffuser must be the same. The mass flow rate can be expressed in terms of the velocity, area, and specific volume:

$$\dot{m} = \frac{\tilde{V}_1 A_1}{v_1} = \frac{\tilde{V}_2 A_2}{v_2} \tag{3}$$

Rearranging Eq. (3) provides the area ratio:

$$AR = \frac{A_2}{A_1} = \frac{v_2 \tilde{V}_1}{v_1 \tilde{V}_2}$$

```
AR=v[2]*Vel[1]/(v[1]*Vel[2])          "area ratio, ratio of exit-to-inlet areas"
```

$AR = 2.17$.

d) Determine the coefficient of pressure recovery for the diffuser.

The coefficient of pressure recovery is determined from Eq. (6-112):

$$K_P = \frac{P_2 - P_1}{\left(\dfrac{\tilde{V}_1^2}{2\, v_1}\right)}$$

```
K_P=(P[2]-P[1])/(Vel[1]^2/(2*v[1]))          "coefficient of pressure recovery"
```

which provides $K_P = 0.496$.

e) Plot the coefficient of pressure recovery as a function of the area ratio for several values of diffuser isentropic efficiency.

Figure 3 illustrates K_P as a function of AR for several values of η_d. Notice that when the area ratio becomes very large, the coefficient of pressure recovery approaches the isentropic diffuser efficiency. In this limit, the exit kinetic energy of the fluid becomes negligibly small.

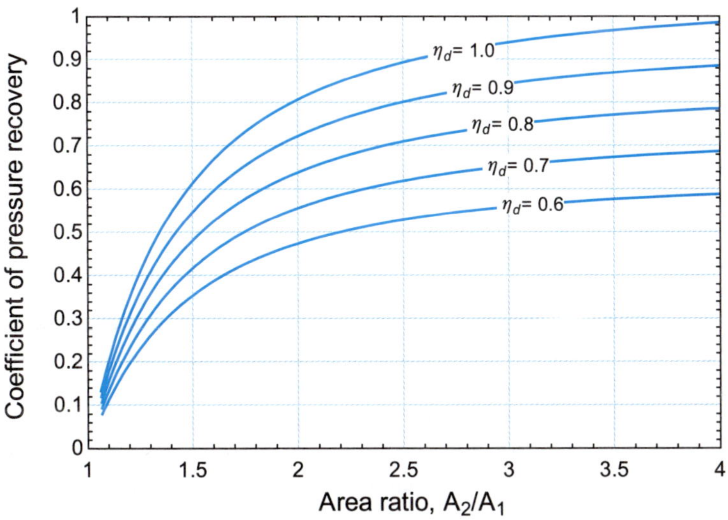

Figure 3: Coefficient of pressure recovery as a function of the area ratio for various values of isentropic diffuser efficiency.

6.6.6 Heat Exchanger Effectiveness

Heat exchangers are perhaps the most common component in energy systems. Their purpose is to exchange thermal energy between two (or possibly more) fluid streams without allowing the streams to mix. There are many different configurations of heat exchangers that are classified based on flow pattern (e.g., parallel flow or cross flow), as discussed in Section 4.4.7.

The efficiencies of the other components that have been considered in this section are defined by comparison with a reversible device that accomplishes the same function. This definition is not appropriate for heat exchangers because heat exchangers are inherently irreversible devices. A temperature difference between the two streams will occur somewhere within the heat exchanger, regardless of how "good" the heat exchanger is. Heat transfer through a temperature difference is always accompanied by

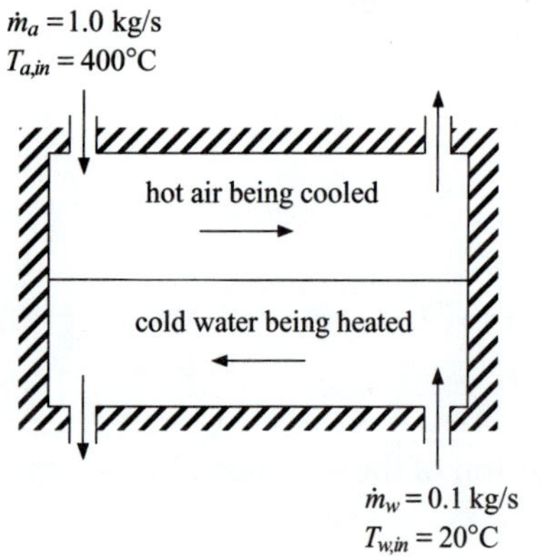

Figure 6-18: Counterflow heat exchanger using a flow of hot air to produce steam.

entropy generation. Even a very large and well-designed heat exchanger will operate irreversibly.

It is still possible to identify a limiting performance for a heat exchanger based on the Second Law of Thermodynamics. This limiting performance is used to define the heat exchanger effectiveness (ε_{hx}):

$$\boxed{\varepsilon_{hx} = \frac{\dot{Q}}{\dot{Q}_{max}}} \tag{6-113}$$

where $\dot{Q}_{max}$ is the maximum possible rate of heat transfer between the streams. We will examine the maximum possible heat transfer rate in the context of the counterflow heat exchanger shown in Figure 6-18.

The heat exchanger shown in Figure 6-18 uses a flow of hot air to heat water and produce steam. The air has mass flow rate $\dot{m}_a = 1.0$ kg/s and inlet temperature $T_{a,in} = 400°$C. The water is at atmospheric pressure with mass flow rate $\dot{m}_w = 0.1$ kg/s and inlet temperature $T_{w,in} = 20°$C. Under these conditions, the temperature distribution within the heat exchanger is shown in Figure 6-19.

The temperature of the air decreases as the air passes through the heat exchanger from left to right and transfers energy to the water. The temperature of the liquid water initially increases as it passes through the heat exchanger from right to left and receives heat from the air. When the water reaches its saturation temperature (100°C at 1 atm)

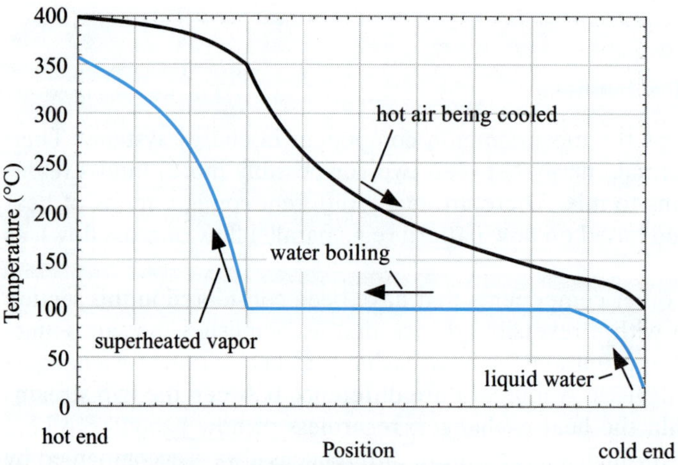

Figure 6-19: Temperature distribution in the heat exchanger shown in Figure 6-18.

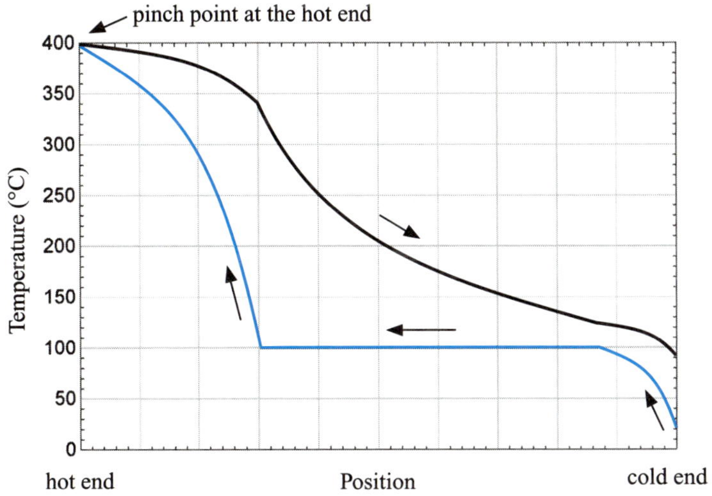

Figure 6-20: Temperature distribution in the heat exchanger shown in Figure 6-18 as the size of the heat exchanger approaches infinity.

it begins to boil and the temperature of the water remains constant (assuming that the pressure change is small). Eventually, the water is converted entirely to steam and the temperature of the superheated vapor continues to rise until it exits the heat exchanger at about 360°C.

What happens to the temperature distribution as the performance of the heat exchanger improves (e.g., a larger and more effective heat exchanger is installed)? Is there a limit to the rate of heat transfer that can occur as the performance of the heat exchanger becomes perfect (e.g., the heat exchanger becomes infinitely large)? As the heat exchanger performance is improved, the rate of heat transfer will increase. As a result, the temperature of the air leaving the device (at the right side) will decrease and the temperature of the water leaving the device (at the left side) will increase. However, this trend cannot continue indefinitely. Eventually, the temperatures of the air and water will coincide at some location within the heat exchanger. Figure 6-20 illustrates the temperature distribution in the heat exchanger that occurs if the heat exchanger size approaches infinity. In this case, the temperature difference between the two streams at the end of the heat exchanger where the hot air enters approaches zero.

The location within the heat exchanger where the temperature difference between the fluids approaches zero is referred to as the *pinch point* and the temperature difference at this location is referred to as the *pinch point temperature difference* or the *approach temperature difference*. Often, the approach or pinch point temperature difference for a heat exchanger is used to specify the performance of a heat exchanger. When the pinch point temperature difference approaches zero, the maximum possible heat transfer rate, $\dot{Q}_{max}$, has been achieved. Notice that the infinitely large heat exchanger that has the temperature distribution shown in Figure 6-20 is characterized by a finite rate of entropy generation because there is heat transfer through a temperature difference occurring within the device.

The pinch point in a heat exchanger does not have to occur at either of the two ends. For example, Figure 6-21 illustrates the temperature distribution within the heat exchanger shown in Figure 6-18 as the size approaches infinity and the pressure of the water is increased from 1 atm to 5 atm. Notice that the temperature at which the water boils has increased to approximately 150°C and, as a result, the pinch point has moved from the left edge to a position within the heat exchanger.

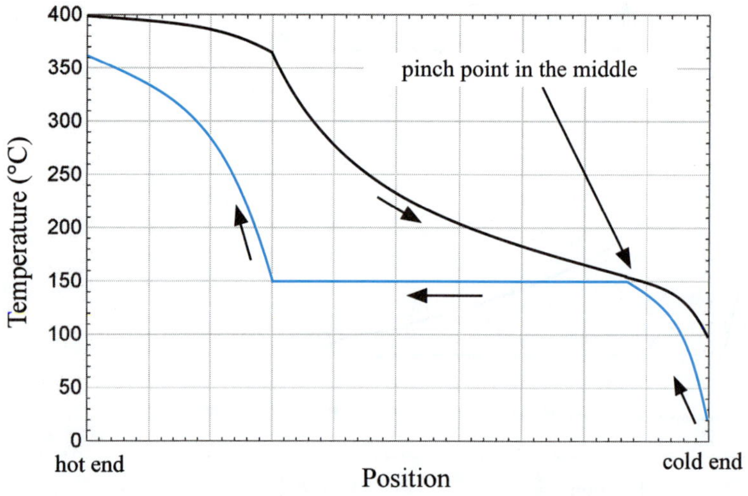

Figure 6-21: Temperature distribution in the heat exchanger shown in Figure 6-18 as the size of the heat exchanger approaches infinity and the pressure of the water is increased to 5 atm.

A thorough analysis of heat exchangers requires a background in heat transfer and it is therefore beyond the scope of this book. A detailed treatment of heat exchangers is provided by Nellis and Klein (2009).

EXAMPLE 6.6-7: ARGON REFRIGERATION CYCLE

The simple refrigeration cycle shown in Figure 1 has been designed to provide a refrigeration load at $T_{load} = 150$ K. Pressurized argon gas at $P_{high} = 6.5$ MPa and $T_{in} = 20°$C enters a counterflow heat exchanger (referred to as a recuperator) at state 1 with mass flow rate $\dot{m} = 0.01$ kg/s. The argon is cooled in the heat exchanger by the transfer of energy to the low pressure argon returning from the load. The pre-cooled argon gas exits the heat exchanger at state 2 and expands through a valve to $P_{low} = 100$ kPa at state 3. Cooling is provided as the gas flows through the load heat exchanger where it is heated to $T_{load} = 150$ K at state 4. The low pressure gas then enters the cold side of the heat exchanger where it is heated by the high pressure argon and exits at state 5.

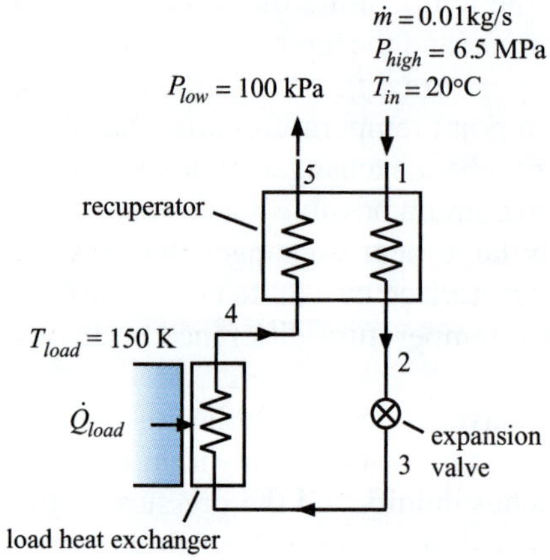

Figure 1: Schematic of the argon refrigeration cycle.

The pinch point for the recuperator occurs at the hot end (i.e, at the location where the high pressure argon enters at state 1 and the low pressure argon leaves at state 5). The approach temperature difference for the recuperator is $\Delta T = 10$ K. Assume that there is no pressure drop in either of the heat exchangers.

a) Determine the refrigeration provided by the cycle, $\dot{Q}_{load}$.

The inputs are entered in EES:

```
$UnitSystem SI K Pa J Mass Radian
T_load=150 [K]                          "refrigeration load temperature"
P_high=6.5 [MPa]*convert(MPa,Pa)        "high pressure"
T_in=converttemp(C,K,20[C])             "inlet temperature"
P_low=100 [kPa]*convert(kPa,Pa)         "low pressure"
m_dot=0.01 [kg/s]                       "mass flow rate"
DT=10 [K]                               "approach temperature difference"
```

State 1 is fixed by the temperature and pressure, $T_1 = T_{in}$ and $P_1 = P_{high}$. The specific enthalpy and specific entropy (h_1 and s_1) are determined.

```
P[1]=P_high                             "pressure at state 1"
T[1]=T_in                               "temperature at state 1"
h[1]=enthalpy(Argon,P=P[1],T=T[1])      "specific enthalpy at state 1"
s[1]=entropy(Argon,P=P[1],T=T[1])       "specific entropy at state 1"
```

The temperature of state 5 is calculated using the approach temperature difference:

$$T_5 = T_1 - \Delta T$$

State 5 is fixed by the temperature and pressure, $P_5 = P_{low}$. The specific enthalpy and specific entropy (h_5 and s_5) are determined.

```
T[5]=T[1]-DT                            "temperature at state 5"
P[5]=P_low                              "pressure at state 5"
h[5]=enthalpy(Argon,P=P[5],T=T[5])      "specific enthalpy at state 5"
s[5]=entropy(Argon,P=P[5],T=T[5])       "specific entropy at state 5"
```

State 4 is fixed by the temperature and pressure, $T_4 = T_{load}$ and $P_4 = P_{low}$. The specific enthalpy and specific entropy (h_4 and s_4) are determined.

```
T[4]=T_load                             "temperature at state 4"
P[4]=P_low                              "pressure at state 4"
h[4]=enthalpy(Argon,P=P[4],T=T[4])      "specific enthalpy at state 4"
s[4]=entropy(Argon,P=P[4],T=T[4])       "specific entropy at state 4"
```

An energy balance on the cold side of the heat exchanger is used to determine the heat transfer rate in the heat exchanger:

$$\dot{Q} = \dot{m}\left(h_5 - h_4\right)$$

Losses through the jacket of the heat exchanger are assumed to be negligible, so the heat transfer rate is the same for the hot and cold fluids. An energy balance on the hot side of the heat exchanger provides the specific enthalpy at state 2:

$$\dot{Q} = \dot{m}\left(h_1 - h_2\right)$$

State 2 is fixed by the specific enthalpy and pressure, $P_2 = P_{high}$. The temperature and specific entropy (T_2 and s_2) are determined.

```
Q_dot=m_dot*(h[5]-h[4])                      "energy balance on cold side of HX"
Q_dot=m_dot*(h[1]-h[2])                      "energy balance on hot side of HX"
P[2]=P_high                                  "pressure at state 2"
T[2]=temperature(Argon,P=P[2],h=h[2])        "temperature at state 2"
s[2]=entropy(Argon,P=P[2],h=h[2])            "specific entropy at state 2"
```

An energy balance on the valve provides:

$$h_3 = h_2$$

State 3 is fixed by the specific enthalpy and pressure, $P_3 = P_{low}$. The temperature and specific entropy (T_3 and s_3) are determined.

```
h[3]=h[2]                                    "energy balance on the valve"
P[3]=P_low                                   "pressure at state 3"
T[3]=temperature(Argon,P=P[3],h=h[3])        "temperature at state 3"
s[3]=entropy(Argon,P=P[3],h=h[3])            "specific entropy at state 3"
```

An energy balance on the load heat exchanger is:

$$\dot{Q}_{load} = \dot{m}\left(h_4 - h_3\right)$$

```
Q_dot_load=m_dot*(h[4]-h[3])         "refrigeration load"
```

which leads to $\dot{Q}_{load} = 69.86$ W.

b) Determine the effectiveness of the recuperator.

The effectiveness is defined in Eq. (6-113) as the actual rate of heat transfer in the heat exchanger to the maximum possible rate of heat transfer. The maximum possible rate of heat transfer results if the temperatures of the two streams coincide at the pinchpoint which, for the recuperator, occurs at the hot end. Therefore, the maximum possible rate of heat transfer will result if the cold fluid is heated to $T_{5,max} = T_{in}$ (i.e., if $\Delta T = 0$ K). The maximum possible specific enthalpy of the argon leaving the recuperator at state 5 ($h_{5,max}$) is fixed by the temperature ($T_{5,max}$) and pressure (P_{low}).

$$\dot{Q}_{max} = \dot{m}\left(\underbrace{h_{P=P_{low},T=T_{in}}}_{h_{5,max}} - h_4\right)$$

The effectiveness is computed according to:

$$\varepsilon_{hx} = \frac{\dot{Q}}{\dot{Q}_{max}}$$

```
Q_dot_max=m_dot*(enthalpy(Argon,P=P_low,T=T_in)-h[4])   "maximum possible rate of heat transfer"
eff_hx=Q_dot/Q_dot_max                                   "heat exchanger effectiveness"
```

which results in $\varepsilon_{hx} = 0.930$.

c) Plot the refrigeration load as a function of the effectiveness of the recuperator. What is the minimum recuperator effectiveness required in order to produce any refrigeration?

Figure 2 illustrates the refrigeration load as a function of the heat exchanger effectiveness. The minimum required heat exchanger effectiveness is approximately $\varepsilon_{hx} = 0.837$.

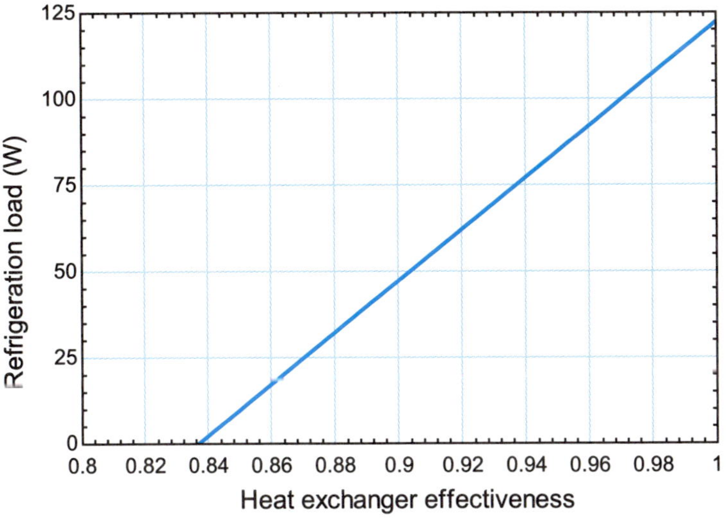

Figure 2: Refrigeration load as a function of the recuperator effectiveness.

A more precise answer can be obtained by updating the guess values and setting the variable Q_dot_load to zero. Comment out the specified value of DT and solve the problem in order to find the minimum required effectiveness, $\varepsilon_{hx} = 0.8372$.

d) What is the rate of entropy generation in the recuperator, the rate of entropy generation in the valve, and the rate of entropy generation in the load heat exchanger?

An entropy balance on the recuperator is:

$$\dot{m}\,s_1 + \dot{m}\,s_4 + \dot{S}_{gen,hx} = \dot{m}\,s_2 + \dot{m}\,s_5$$

An entropy balance on the valve is:

$$\dot{m}\,s_2 + \dot{S}_{gen,v} = \dot{m}\,s_3$$

An entropy balance on the load heat exchanger is:

$$\dot{m}\,s_3 + \frac{\dot{Q}_{load}}{T_{load}} + \dot{S}_{gen,load} = \dot{m}\,s_4$$

```
m_dot*s[1]+m_dot*s[4]+S_dot_gen_hx=m_dot*s[2]+m_dot*s[5]
                                   "entropy balance on heat exchanger"
m_dot*s[2]+S_dot_gen_v=m_dot*s[3]  "entropy balance on valve"
m_dot*s[3]+Q_dot_load/T_load+S_dot_gen_load=m_dot*s[4]
                                   "entropy balance on load heat exchanger"
```

which results in $\dot{S}_{gen,hx} = 0.39$ W/K, $\dot{S}_{gen,v} = 7.96$ W/K, and $\dot{S}_{gen,load} = 0.022$ W/K.

Heat Exchangers with Constant Specific Heat Capacity
In many situations, it is appropriate to assume that the constant pressure specific heat capacities of the two fluids passing through the heat exchanger are constant. Under this limiting condition, the solution to the heat exchanger exchanger equations can be obtained analytically or numerically. These solutions have been developed for a variety of heat exchanger configurations and are presented without derivation in this section. The derivation of these solutions can be found in heat transfer texts such as Nellis and Klein (2009).

If the specific heat capacities of the fluids are constant, then the pinch point can only occur at one of the two ends of the heat exchanger; which end depends on the relative magnitude of the capacitance rates of the two fluids. The *capacitance rate* ($\dot{C}$) is defined as the product of the mass flow rate and the specific heat capacity. The capacitance rates of the hot and cold fluids are defined as:

$$\dot{C}_H = \dot{m}_H \, c_{P,H} \tag{6-114}$$

$$\dot{C}_C = \dot{m}_C \, c_{P,C} \tag{6-115}$$

An energy balance on either side of a counterflow heat exchanger is shown in Figure 6-22 and can be written as:

$$\dot{Q} = \dot{m}_H \underbrace{(h_{H,in} - h_{H,out})}_{\dot{C}_H(T_{H,in}-T_{H,out})} = \dot{m}_C \underbrace{(h_{C,out} - h_{C,in})}_{\dot{C}_C(T_{C,out}-T_{C,in})} \tag{6-116}$$

Substituting Eqs. (6-114) and (6-115) into Eq. (6-116) and using the fact that the specific heat capacity is constant leads to:

$$\dot{Q} = \dot{C}_H \,(T_{H,in} - T_{H,out}) = \dot{C}_C \,(T_{C,out} - T_{C,in}) \tag{6-117}$$

Equation (6-117) indicates that the fluid with the smaller capacitance rate will experience the larger change in temperature. Therefore, the pinch point will occur at the cold end of the heat exchanger (i.e., the end where the cold fluid enters) if the capacitance rate of the hot fluid is lower than the capacitance rate of the cold fluid, as shown in Figure 6-23(a) for a counterflow heat exchanger. In this case, the hot fluid is cooled to the cold fluid inlet temperature as the heat exchanger size is increased to infinity and the maximum possible heat transfer rate is:

$$\dot{Q}_{max} = \dot{C}_H \,(T_{H,in} - T_{C,in}) \tag{6-118}$$

If the capacitance rate of the cold fluid is less than that of the hot fluid then the pinch point will occur at the hot end of the heat exchanger, as shown in Figure 6-23(b) for a counterflow heat exchanger. In this case, the cold fluid is heated to the hot fluid inlet

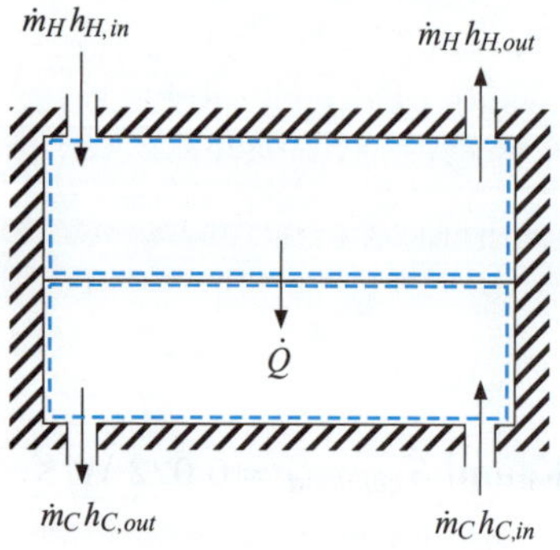

Figure 6-22: Energy balance on a heat exchanger.

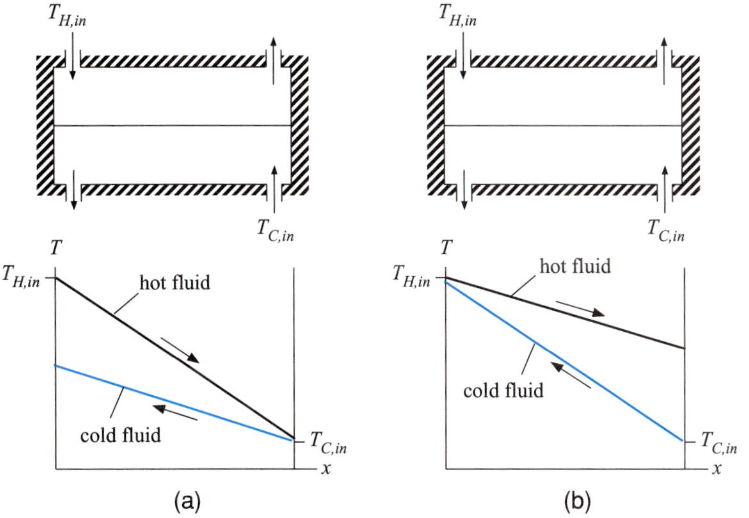

Figure 6-23: Counterflow exchanger operating with constant specific heat capacities where (a) $\dot{C}_H < \dot{C}_C$ so the pinch point occurs at the cold end, and (b) $\dot{C}_C < \dot{C}_H$ so the pinch point occurs at the hot end.

temperature as the heat exchanger size is increased to infinity and the maximum possible heat transfer rate is:

$$\dot{Q}_{max} = \dot{C}_C \left(T_{H,in} - T_{C,in}\right) \tag{6-119}$$

In general, the maximum possible rate of heat transfer is the minimum of the values predicted by Eqs. (6-118) and (6-119):

$$\dot{Q}_{max} = \dot{C}_{min} \left(T_{H,in} - T_{C,in}\right) \tag{6-120}$$

where $\dot{C}_{min}$ is the minimum of the two capacitance rates.

The solution to a heat exchanger problem with constant capacity rates is typically provided in effectiveness-NTU format. The effectiveness, defined in Eq. (6-113), is related to the number of transfer units, NTU, and the capacity ratio, C_R. The number of transfer units is the dimensionless size of the heat exchanger, defined as:

$$NTU = \frac{UA}{\dot{C}_{min}} \tag{6-121}$$

where UA is the conductance of the heat exchanger. The conductance of the heat exchanger has units W/K and can be interpreted as a measure of the physical size of the heat exchanger. The determination of UA for a particular heat exchanger geometry and operating condition is typically the result of a detailed heat transfer analysis. The capacity ratio is defined as:

$$C_R = \frac{\dot{C}_{min}}{\dot{C}_{max}} \tag{6-122}$$

where $\dot{C}_{max}$ is the maximum capacitance rate.

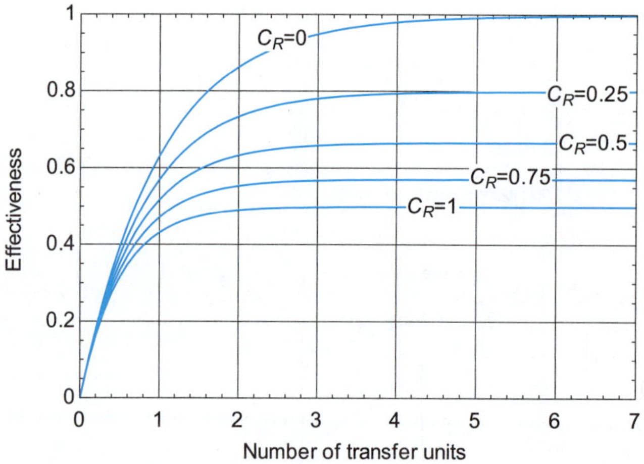

Figure 6-24: Effectiveness of a parallel flow heat exchanger as a function of number of transfer units for various values of the capacity ratio.

Each heat exchanger configuration has its own, specific effectiveness-NTU solution. For example, the effectiveness-NTU solution that characterizes a parallel flow heat exchanger (where both streams enter on the same side and flow in the same direction) is:

$$\varepsilon_{hx} = \frac{1 - \exp\left[-NTU\,(1 + C_R)\right]}{1 + C_R} \tag{6-123}$$

Equation (6-123) can be arranged to provide the NTU as a function of effectiveness.

$$NTU = \frac{\ln\left[1 - \varepsilon_{hx}\,(1 + C_R)\right]}{1 + C_R} \tag{6-124}$$

Figure 6-24 illustrates the effectiveness of a parallel flow heat exchanger as a function of the number of transfer units for various values of the capacity ratio. Notice that the effectiveness increases with the number of transfer units but at high NTU the effectiveness approaches some fixed value that depends on the capacity ratio. This behavior is typical of most heat exchanger configurations.

The ε-NTU solutions for a few different heat exchanger configurations are summarized in Table 6-1 in the form ε_{hx} as a function of NTU and C_R and in Table 6-2 in the form NTU as a function of ε_{hx} and C_R. EES functions are available that implement the solutions listed in Table 6-1 and Table 6-2 as well as other solutions that are less easy to express analytically. These functions are implemented in the two forms that are most useful: $\varepsilon_{hx}\,(NTU, C_R)$ and $NTU\,(\varepsilon_{hx}, C_R)$. They can be accessed from the Function Information window by selecting Heat Exchangers and then either NTU-> Effectiveness or Effectiveness -> NTU, respectively. The first form assumes that UA and the capacitance rates of the two streams are known (and therefore NTU and C_R can be directly computed) and this information is used to determine the effectiveness, which can be used to provide the outlet temperatures. The second form assumes that the outlet temperatures and the capacitance rates of the two streams are known (and therefore ε_{hx} and C_R can be directly computed) and this information is used to determine the NTU which can be used to provide the UA that is required.

Table 6-1: Effectiveness-NTU relations for various heat exchanger configurations in the form effectiveness as a function of number of transfer units and capacity ratio.

Flow Arrangement		$\varepsilon_{hx}\,(NTU, C_R)$
One fluid (or any configuration with $C_R = 0$)		$\varepsilon_{hx} = 1 - \exp(-NTU)$
Counter-flow		$\varepsilon_{hx} = \begin{cases} \dfrac{1 - \exp[-NTU(1 - C_R)]}{1 - C_R \exp[-NTU(1 - C_R)]} & \text{for } C_R < 1 \\ \dfrac{NTU}{1 + NTU} & \text{for } C_R = 1 \end{cases}$
Parallel-flow		$\varepsilon_{hx} = \dfrac{1 - \exp[-NTU(1 + C_R)]}{1 + C_R}$
Cross-flow	both fluids unmixed	$\varepsilon_{hx} = 1 - \exp\left[\dfrac{NTU^{0.22}}{C_R}\left\{\exp\left(-C_R\,NTU^{0.78}\right) - 1\right\}\right]$
	both fluids mixed	$\varepsilon_{hx} = \left[\dfrac{1}{1 - \exp(-NTU)} + \dfrac{C_R}{1 - \exp(-C_R\,NTU)} - \dfrac{1}{NTU}\right]^{-1}$
	$\dot{C}_{max}$ mixed & $\dot{C}_{min}$ unmixed	$\varepsilon_{hx} = \dfrac{1 - \exp[C_R\{\exp(-NTU) - 1\}]}{C_R}$
	$\dot{C}_{min}$ mixed & $\dot{C}_{max}$ unmixed	$\varepsilon_{hx} = 1 - \exp\left[-\dfrac{1 - \exp(-C_R\,NTU)}{C_R}\right]$

Table 6-2: Effectiveness-NTU relations for various heat exchanger configurations in the form of number transfer units as a function of effectiveness and capacity ratio.

Flow Arrangement		$NTU\,(\varepsilon, C_R)$
One fluid (or any configuration with $C_R = 0$)		$NTU = -\ln(1 - \varepsilon_{hx})$
Counter-flow		$NTU = \begin{cases} \dfrac{\ln\left[\dfrac{1 - \varepsilon_{hx} C_R}{1 - \varepsilon_{hx}}\right]}{1 - C_R} & \text{for } C_R < 1 \\ \dfrac{\varepsilon_{hx}}{1 - \varepsilon_{hx}} & \text{for } C_R = 1 \end{cases}$
Parallel-flow		$NTU = \dfrac{\ln[1 - \varepsilon_{hx}(1 + C_R)]}{1 + C_R}$
Cross-flow	$\dot{C}_{max}$ mixed & $\dot{C}_{min}$ unmixed	$NTU = -\ln\left[1 + \dfrac{\ln(1 - \varepsilon_{hx} C_R)}{C_R}\right]$
	$\dot{C}_{min}$ mixed & $\dot{C}_{max}$ unmixed	$NTU = -\dfrac{\ln[C_R \ln(1 - \varepsilon_{hx}) + 1]}{C_R}$

EXAMPLE 6.6-8: ENERGY RECOVERY HEAT EXCHANGER

Figure 1(a) illustrates the heating system for a small building.

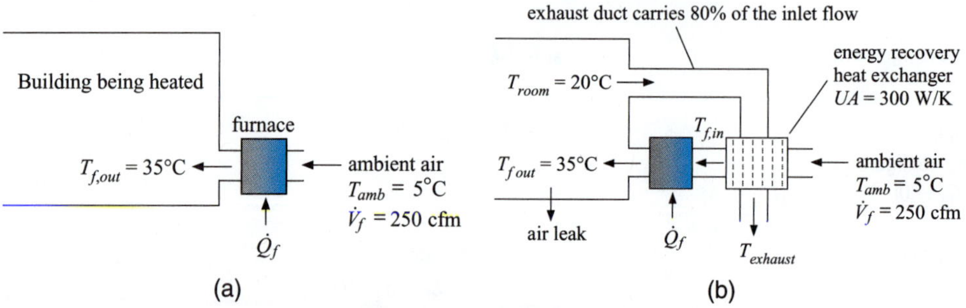

Figure 1: (a) Heating system without energy recovery, (b) heating system with energy recovery.

Outdoor air is drawn into the building at temperature $T_{amb} = 5°\text{C}$ with a volumetric flow rate $\dot{V}_f = 250$ cfm. The air is heated in a furnace and leaves at $T_{f,out} = 35°\text{C}$. Model air as an ideal gas with $R = 287.1$ N-m/kg-K and $c_P = 1005$ J/kg-K. Assume that the pressure of the air is constant and equal to $P_{atm} = 100$ kPa.

a) Determine the rate of heat transfer to the air in the furnace, $\dot{Q}_f$.

The inputs are entered in EES:

```
$UnitSystem SI K Pa J mass radian
T_amb=converttemp(C,K,5 [C])                 "outdoor air temperature"
V_dot_f=250 [cfm]*convert(cfm,m^3/s)         "volumetric flow rate of outdoor air"
T_f_out=converttemp(C,K,35 [C])              "furnace exit air"
c_P=1005 [J/kg-K]                            "specific heat capacity of air"
R=287.1 [N-m/kg-K]                           "gas constant for air"
P_atm=100 [kPa]*convert(kPa,Pa)              "pressure"
```

The specific volume of the air entering the furnace is:

$$v_{out} = \frac{R\,T_{amb}}{P_{atm}}$$

The mass flow rate of air entering the furnace is:

$$\dot{m}_f = \frac{\dot{V}_f}{v_{amb}}$$

An energy balance on the furnace is:

$$\dot{Q}_f = \dot{m}_f\, c_P\, \left(T_{f,out} - T_{amb}\right)$$

```
v_amb=R*T_amb/P_atm                  "specific volume of outdoor air"
m_dot_f=V_dot_f/v_amb                "mass flow rate of outdoor air"
Q_dot_f=m_dot_f*c_P*(T_f_out-T_amb)  "furnace heat transfer rate"
```

which leads to $\dot{Q}_f = 4455$ W.

b) The cost of the natural gas used to provide the heat transfer in the furnace is $ngc = 1.55$ \$/therm and the operating time per year is $time = 130$ day/year. Determine the yearly cost of heating the building assuming that the average ambient temperature is 5°C.

Assuming the rate of heat transfer is constant when the furnace is operating, the total amount of energy consumed in the furnace per year is:

$$Q = \dot{Q}_f \, time$$

and the cost of this energy is:

$$Cost_{year} = Q_f \, ngc$$

```
time=130 [day/year]*convert(day/year,s/year)        "operating time per year"
ngc=1.55 [$/therm]*convert($/therm,$/J)             "natural gas cost"
Q_f=Q_dot_f*time                                    "total energy used per year"
Cost_year=Q_f*ngc                                   "heating cost per year"
```

which results in $Cost_{year} = 735.1$ \$/year.

c) Figure 1(b) illustrates the same building with an energy recovery heat exchanger installed. The energy recovery heat exchanger is a cross-flow heat exchanger with both fluids unmixed. The term unmixed means that the heat exchanger is designed with barriers to flow in the direction perpendicular to the flow. The total conductance of the heat exchanger is $UA = 300$ W/K. The heated air that is provided to the building either leaks from the building or is captured in an exhaust duct. The fraction of the air that leaks from the building is $f_{leak} = 0.2$. The remaining air flows through the exhaust duct where it is directed to the energy recovery heat exchanger with inlet temperature $T_{room} = 20°$C. Heat transfer from the exhaust air to the outdoor air preheats the incoming air before it enters the furnace. The energy consumption of the furnace is therefore reduced. Determine the rate of heat transfer in the energy recovery heat exchanger.

The additional inputs are entered in EES:

```
f_leak=0.2 [-]                          "fraction of air leaking from building"
T_room=converttemp(C,K,20[C])           "room temperature"
UA=300 [W/K]                            "conductance of recovery HX"
```

The mass flow rate of air in the exhaust duct is determined:

$$\dot{m}_{exhaust} = \left(1 - f_{leak}\right) \dot{m}_f$$

The capacitance rates of the hot and cold fluids are determined:

$$\dot{C}_H = \dot{m}_{exhaust} \, c_P$$

$$\dot{C}_C = \dot{m}_f \, c_P$$

The maximum and minimum capacitance rates ($\dot{C}_{max}$ and $\dot{C}_{min}$, respectively) are determined using the MAX and MIN functions in EES.

```
m_dot_exhaust=(1-f_leak)*m_dot_f            "mass flow rate of exhaust flow"
C_dot_H=m_dot_f*c_P                         "capacitance rate of hot flow"
C_dot_C=m_dot_exhaust*c_P                   "capacitance rate of cold flow"
C_dot_max=MAX(C_dot_H,C_dot_C)              "maximum capacitance rate"
C_dot_min=MIN(C_dot_H,C_dot_C)              "minimum capacitance rate"
```

The number of transfer units is determined:

$$NTU = \frac{UA}{\dot{C}_{min}}$$

The capacity ratio is determined:

$$C_R = \frac{\dot{C}_{min}}{\dot{C}_{max}}$$

```
NTU=UA/C_dot_min                    "number of transfer units"
C_R=C_dot_min/C_dot_max             "capacitance rate"
```

The solution for a cross-flow heat exchanger with both fluids unmixed is obtained from Table 6-1:

$$\varepsilon_{hx} = 1 - \exp\left[\frac{NTU^{0.22}}{C_R}\left\{\exp\left(-C_R\, NTU^{0.78}\right) - 1\right\}\right]$$

```
eff_hx=1-exp(NTU^0.22*(exp(-C_R*NTU^0.78)-1)/C_R)        "effectiveness of heat exchanger"
```

which results in $\varepsilon_{hx} = 0.71$. The same solution for the effectiveness is provided in an EES function that can be accessed by selecting Heat Exchangers from the drop down menu in the Function Information window. Select the option NTU -> Effectiveness and then scroll to the correct configuration, as shown in Figure 2. Select Paste in order to put the function call in the Equations Window.

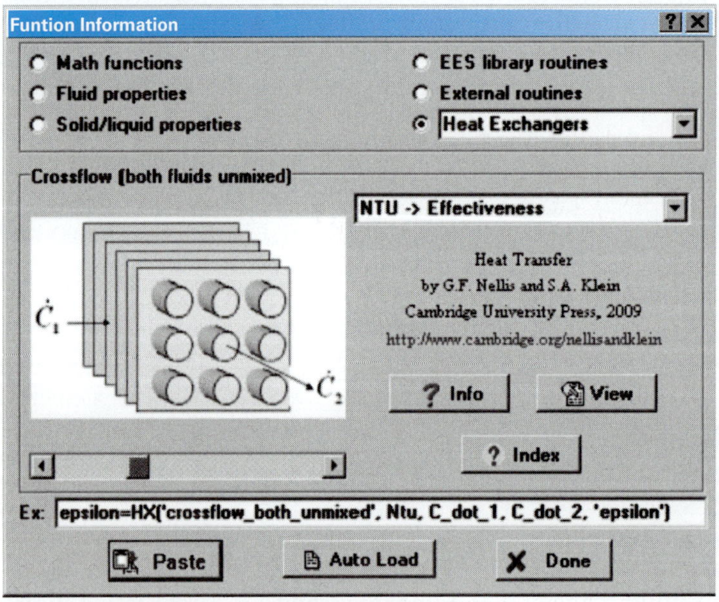

Figure 2: Heat exchanger function information.

```
eff_hx_EES=HX('crossflow_both_unmixed', NTU, C_dot_H, C_dot_C, 'epsilon')        "from EES"
```

Solving shows that $\varepsilon_{hx} = 0.71$, as expected. The maximum possible rate of heat transfer in the heat exchanger is computed using Eq. (6-120):

$$\dot{Q}_{max} = \dot{C}_{min}\left(T_{room} - T_{amb}\right)$$

The actual rate of heat transfer is:

$$\dot{Q}_{hx} = \varepsilon_{hx}\, \dot{Q}_{max}$$

```
Q_dot_max=C_dot_min*(T_room-T_amb)   "maximum possible heat transfer"
Q_dot_hx=eff_hx*Q_dot_max            "actual heat transfer in recovery HX"
```

which leads to $\dot{Q}_{hx} = 1265$ W.

d) Determine the yearly cost of heating the building with the energy recovery heat exchanger installed.

The operating cost calculated in part (b) is reduced by the value of the heat that is recovered in the heat exchanger.

$$Cost_{r,year} = Cost_{year} - (\dot{Q}_{hx}\, time)\, ngc$$

```
Operating_Cost_r_year=Cost_year-Q_dot_hx*time*ngc   "operating cost per year"
```

which shows that $Cost_{r,year} = 526.3$ \$/year. The operating cost has been reduced substantially relative to 735.1 \$/year, the answer that is determined in (b) for the situation without the energy recovery heat exchanger.

e) The cost of the heat exchanger on a per unit conductance basis is $uac = 2.50$ \$-K/W. Determine the total cost of operating and owning the system over a $N_{year} = 5$ year period.

The heat exchanger cost is:

$$HX_{cost} = uac\, UA$$

The total operating cost for 5 years is:

$$Cost_r = Cost_{r,year}\, N_{year}$$

The total cost of operating the system and purchasing the equipment is:

$$TotalCost = Cost_r + HX_{cost}$$

```
uac=2.5 [$-K/W]                                "cost of heat exchanger per unit of conductance"
N_year=5 [year]                                "number of years"
HX_cost=uac*UA                                 "cost of recovery heat exchanger"
Operating_Cost_r=N_year*Operating_Cost_r_year  "total operating cost"
Total_Cost=Operating_Cost_r+HX_cost            "total cost"
```

which results in $TotalCost = \$3382$.

f) Plot the total cost of operating the system and purchasing the equipment as a function of UA, the heat exchanger conductance. Determine the optimal conductance of the heat exchanger for this application.

Figure 3 illustrates the total cost of purchasing and operating the system as a function of the heat exchanger conductance. Also shown in Figure 3 is the total operating cost for 5 years and the cost of the heat exchanger as a function of the heat exchanger conductance. Notice that the operating cost decreases with UA but at a decreasing rate, eventually asymptoting to a constant value as the heat exchanger reaches the limit of its effectiveness. The heat exchanger cost increases linearly with UA. There is a clear optimal value of conductance, around $UA = 150$ W/K, where the reduction in operating cost is balanced by an equal increase in heat exchanger cost.

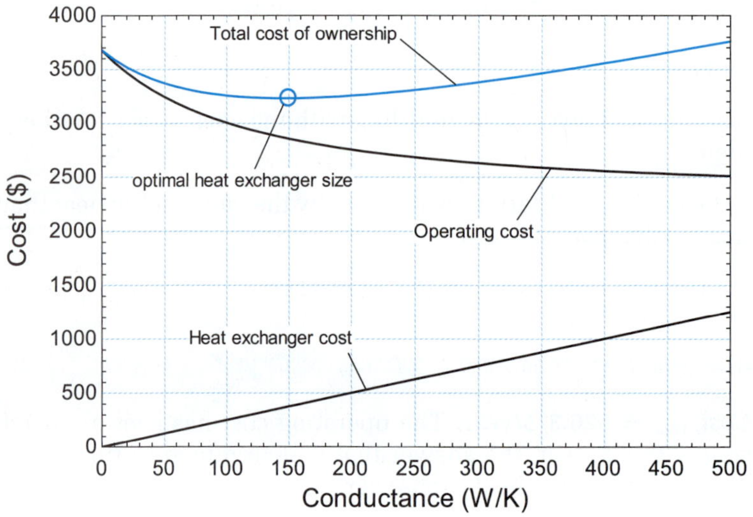

Figure 3: Operating cost for 5 years, heat exchanger cost, and total cost of ownership as a function of conductance.

A more precise identification of the optimal UA can be obtained using the internal optimization algorithms that are programmed in EES. Comment out the specified value of UA:

```
{UA=300 [W/K]}          "conductance of recovery HX"
```

Select Min/Max from the Calculate menu in order to obtain the dialog shown in Figure 4. The objective is to minimize the total cost of ownership. Therefore, select the Minimize radio button and select the variable Total_Cost from the list of variables. The independent variable that can be varied in order to accomplish the minimization is the conductance. Therefore, select the variable UA from the list of independent variables. You will need to select the Bounds button and indicate an initial guess for UA (300 W/K) as well as reasonable upper and lower bounds (1000 W/K and 1 W/K, respectively). Select OK and the optimizer will identify an optimal value of $UA = 145.8$ W/K resulting in a total cost of ownership of $3233.

g) Determine the rate of entropy generation in the heat exchanger. Plot the rate of entropy generation as a function of conductance.

An energy balance on the hot air passing through the heat exchanger is used to determine the temperature of the air that is being exhausted from the building:

$$\dot{Q}_{hx} = \dot{C}_H \left(T_{room} - T_{exhaust}\right)$$

An energy balance on the cold air is used to determine the temperature of the air entering the furnace:

$$\dot{Q}_{hx} = \dot{C}_C \left(T_{f,in} - T_{amb}\right)$$

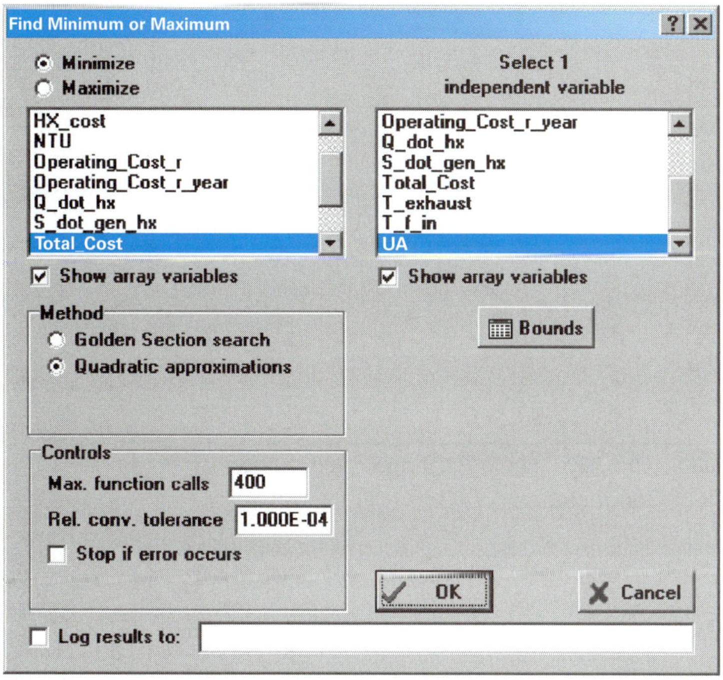

Figure 4: Min/Max dialog.

An entropy balance on the heat exchanger (which is assumed to be externally adiabatic) provides:

$$\dot{m}_f \, s_{amb} + \dot{m}_{exhaust} \, s_{room} + \dot{S}_{gen,hx} = \dot{m}_f \, s_{f,in} + \dot{m}_{exhaust} \, s_{exhaust} \tag{1}$$

Rearranging Eq. (1) provides:

$$\dot{S}_{gen,hx} = \dot{m}_f \left(s_{f,in} - s_{amb}\right) + \dot{m}_{exhaust} \left(s_{exhaust} - s_{room}\right) \tag{2}$$

Substituting the equation for the entropy change of an ideal gas with constant heat capacity, Eq. (6-29), into Eq. (2) and recognizing that the pressure of the air is constant leads to:

$$\dot{S}_{gen,hx} = \dot{m}_f \, c_P \, \ln\left(\frac{T_{f,in}}{T_{amb}}\right) + \dot{m}_{exhaust} \, c_P \, \ln\left(\frac{T_{exhaust}}{T_{room}}\right)$$

```
Q_dot_hx=C_dot_H*(T_room_T_exhaust)
Q_dot_hx=C_dot_C*(T_f_in_T_amb)
S_dot_gen_hx=m_dot_f*c_P*ln(T_f_in/T_amb)+m_dot_exhaust*c_P*ln(T_exhaust/T_room)
"rate of entropy generation"
```

which results in $\dot{S}_{gen,hx} = 2.075$ W/K. Figure 5 illustrates the rate of entropy generation in the heat exchanger as a function of conductance. Notice that even as the heat exchanger becomes very large (and therefore has the maximum possible performance) it is not reversible. Rather, the entropy generation rate approaches a constant value that is unavoidable due to the inherent temperature differences that occur at every location in the heat exchanger except at the pinchpoint. The entropy generation approaches zero as the conductance approaches zero. In this limit, there is no heat transferred in the heat exchanger and therefore no entropy generated. This is a typical result: the reversible process is not optimal as it accomplishes nothing.

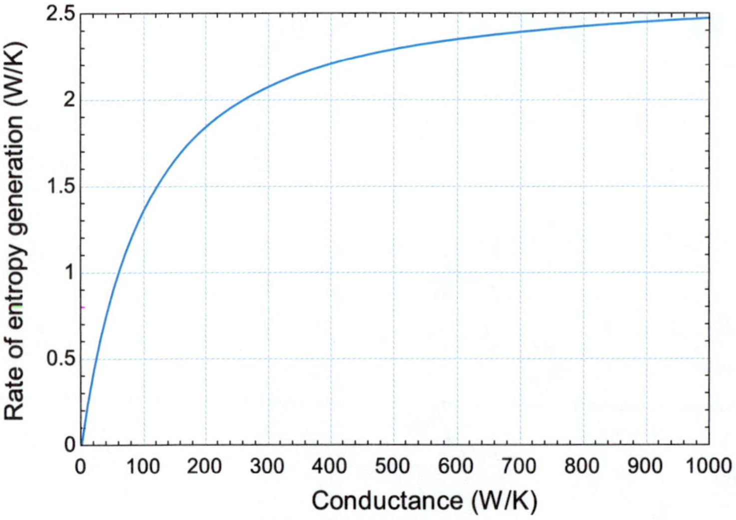

Figure 5: Rate of entropy generation in the heat exchanger as a function of the conductance.

REFERENCES

Bejan, A., "Research into the origins of engineering thermodynamics," *Int. Communications in Heat and Mass Transfer*, Vol. 15, No. 5, pp. 571–580 (1998).

Nellis, G. F. and S.A. Klein, *Heat Transfer*, Cambridge University Press (2009).

Clausius, R., "Über die Wärmeleitung gasförmiger Körper", *Annalen der Physik*, 125: 353–400 (1865).

Van Wylen, G. J., and R. E. Sonntag, *Fundamentals of Thermodynamics*, 3rd edition, John Wiley & Sons (1986).

Problems

The problems included here have been selected from a much larger set of problems that are available on the website associated with this book (www.cambridge.org/kleinandnellis).

A. Entropy Balances

6.A-1 The two places that thermal energy is transferred from a car are through the engine block (to the coolant) and through the exhaust pipe. A company is marketing a device that scavenges heat from your car in order to produce some auxiliary power. The device is shown schematically in Figure 6.A-1.

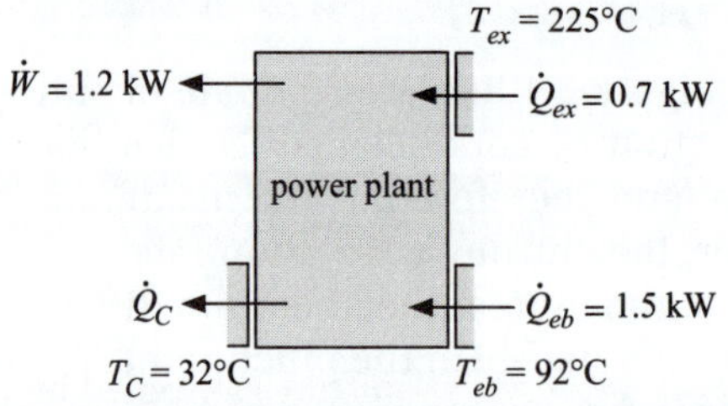

Figure 6.A-1: Device for scavenging heat from an engine to produce auxiliary power.

You have been hired to assess the feasibility of this device for some potential investors. The company claims that the device receives $\dot{Q}_{eb} = 1.5$ kW from the engine block and the engine block temperature is $T_{eb} = 92°C$. The device also receives $\dot{Q}_{ex} = 0.7$ kW from the exhaust pipe and the exhaust pipe temperature is $T_{ex} = 225°C$. The device rejects heat to surrounding air and is supposed to continue to function as advertised even in climates where the air temperature reaches $T_C = 32°C$. The company claims that they have measured a power output equal to $\dot{W} = 1.2$ kW under these conditions. The device operates at steady state.

a) Assess the company's claim using the Second Law of Thermodynamics – would you suggest that the investors invest money or not?
b) What is the maximum rate at which the engine can produce power under these conditions?

6.A-4 One method for storing hydrogen involves cooling it so that it liquefies. Figure 6.A-4(a) illustrates a single-stage refrigerator that is used for this purpose. A refrigerator that operates at very low temperature is referred to as a cryocooler.

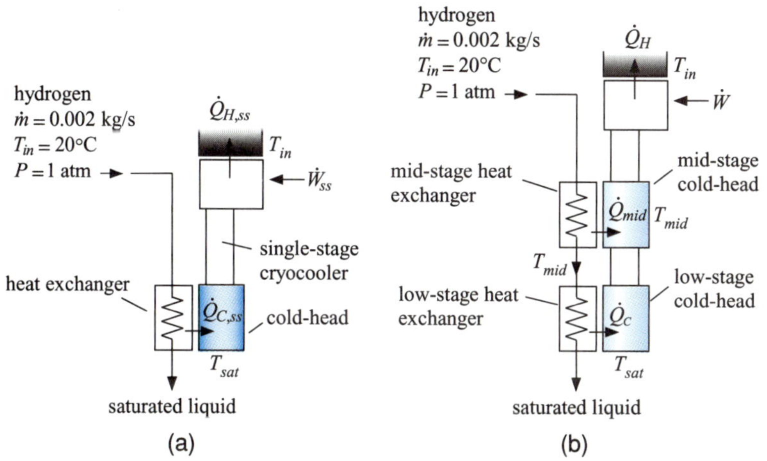

Figure 6.A-4: (a) Single-stage refrigerator and (b) multi-stage refrigerator used to liquefy hydrogen.

A stream of gaseous hydrogen at $\dot{m} = 0.002$ kg/s, $T_{in} = 20°C$, and $P = 1$ atm is fed to the device. The gaseous hydrogen is liquefied by flowing through a heat exchanger that interfaces with the cold-head of the cryocooler. The cold head is held at the saturation temperature of the hydrogen, T_{sat}, and hydrogen exits the heat exchanger as saturated liquid. Assume that the cryocooler operates at steady state, is internally reversible, and rejects heat at temperature T_{in}. Assume that there is no pressure loss associated with the flow of hydrogen.

a) Determine the rate of heat transfer to the cold head at T_{sat}, $\dot{Q}_{C,ss}$.
b) Determine the rate at which heat is rejected from the cryocooler to T_{in}, $\dot{Q}_{H,ss}$.
c) Determine the rate of power consumed by the cryocooler, $\dot{W}_{ss}$. What is the coefficient of performance of the cryocooler?

Figure 6.A-4(b) illustrates an alternative, two-stage cryocooler used for liquefaction. A two-stage cryocooler can provide refrigeration at two separate temperatures. The hydrogen is cooled to an intermediate temperature, $T_{mid} = 100$ K, by flowing through the mid-stage heat exchanger that interfaces with the mid-stage cold head of the cryocooler. The mid-stage cold head is maintained at T_{mid}. The hydrogen leaving the mid-stage heat exchanger is further cooled by flowing through the low-stage heat exchanger that interfaces with the low-stage cold-head of the cryocooler. The low-stage cold head is held at the saturation temperature of the hydrogen, T_{sat}, and hydrogen exits the low-stage heat exchanger as saturated liquid. Assume that the two-stage cryocooler operates at steady state, is internally reversible, and rejects heat at temperature T_{in}. Assume that there is no pressure loss associated with the flow of hydrogen.

d) Determine the rate of heat transfer to the mid-stage cold head at T_{mid}, $\dot{Q}_{mid}$.
e) Determine the rate of heat transfer to the low-stage cold head at T_{sat}, $\dot{Q}_C$.
f) Determine the rate at which heat is rejected from the two-stage cryocooler to T_{in}, $\dot{Q}_H$.
g) Determine the power consumed by the two-stage cryocooler, $\dot{W}$.
h) Plot the power consumed by the two-stage cryocooler, $\dot{W}$, as a function of the mid-stage temperature, T_{mid}.
i) Your plot from (h) should show that there is an optimal mid-stage temperature. Explain why this optimal value exists.
j) Use the optimization capability in EES to precisely determine the optimal value of T_{mid}.

6.A-5 You are designing the power plant shown in Figure 6.A-5.

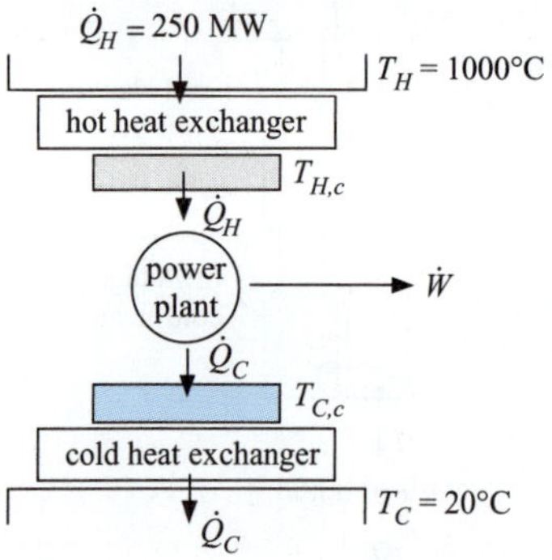

Figure 6.A-5: Power plant.

The power plant receives heat at a rate of $\dot{Q}_H = 250$ MW from a heat source at $T_H = 1000°C$ and rejects heat at a rate $\dot{Q}_C$ to an environment at $T_C = 20°C$. The power plant requires two heat exchangers. The hot heat exchanger is used to transfer $\dot{Q}_H$ from the thermal reservoir at T_H into the cycle and the cold heat exchanger is used to transfer $\dot{Q}_C$ out of the cycle to the thermal reservoir at T_C. The size of a heat exchanger is indicated by its total conductance, UA. A heat exchanger with a large value of UA will be physically very large but will allow heat transfer into or out of the system with a small temperature drop. Therefore, a large hot heat exchanger will allow $\dot{Q}_H$ to enter the cycle at a temperature, $T_{H,c}$, that is close to T_H.

$$\dot{Q}_H = (T_H - T_{H,c})\, UA$$

A large cold heat exchanger will allow $\dot{Q}_C$ to leave the cycle with a small temperature drop – therefore the cold temperature that is actually seen by the power plant ($T_{C,c}$ in Figure 6.A-5) will be very close to T_C.

$$\dot{Q}_C = (T_{C,c} - T_C)\, UA$$

Assume that the hot and cold heat exchangers have the same conductance, $UA = 2 \times 10^6$ W/K. Also, assume that the power cycle has an efficiency that is 50% of the efficiency that a reversible power plant would exhibit if it were operating between $T_{H,c}$ and $T_{C,c}$ (this is sometimes called the Second Law efficiency of the power plant). The system operates at steady state.

a) Determine the efficiency of the power plant and the rate of power production, $\dot{W}$.
b) Determine the rate of entropy generation in the hot heat exchanger, the rate of entropy generation in the cold heat exchanger, and the rate of entropy generation in the power plant itself.
c) Plot the efficiency of the power plant as a function of UA (for 750×10^3 W/K $< UA < 1 \times 10^7$ W/K). Provide a brief explanation for the shape of your plot.

You would like to optimize your power plant design by correctly sizing your heat exchangers. The result from (c) should have shown that the efficiency of you plant increases as UA increases. However, heat exchangers with very large conductance are expensive. You estimate that the cost of the heat exchanger per unit of conductance is UAc = 10$-K/W. The capital cost of the mechanical components in the power cycle is fixed and equal to $EngineCost$ = $35 × 10^6. You can sell the electrical power that is produced at a rate of ec = 0.065$/kW-hr.

d) Plot the revenue (i.e., the money you receive by selling power) over a $t = 5$ year period (assume the plant operates continuously) as a function of UA. Ignore the time value of money. Explain the shape of your plot.
e) Overlay on your plot from (d) the capital cost (i.e., the initial investment required to buy the power plant and the heat exchangers) as a function of UA. Explain the shape of your plot.
f) Overlay on your plot from (d) the profit for $t = 5$ years of operation (i.e., the revenue less the capital cost). You should see that there is a value of UA that maximizes your profit – explain why this is the case.
g) What is the economically optimal value of UA obtained using the Min/Max function in EES? What is the corresponding value of profit?

6.A-6 Figure 6.A-6 illustrates a rigid, insulated tank that is divided into two parts by an interior partition.

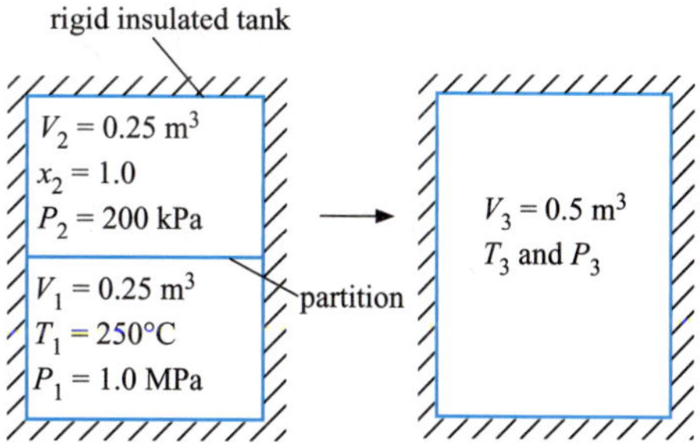

Figure 6.A-6: Rigid tank divided into two parts by a partition.

The volume below the partition is $V_1 = 0.25$ m^3 and contains water at $T_1 = 250°$C and $P_1 = 1.0$ MPa. The volume above the partition is $V_2 = 0.25$ m^3 and contains saturated water vapor at $P_2 = 200$ kPa. At some time, the partition is ruptured allowing the water below the partition to mix with the water above the partition. Eventually, all of the water in the tank comes to new equilibrium at a uniform temperature, T_3, and pressure, P_3.

a) What is the final temperature and pressure in the tank, T_3 and P_3?
b) Determine S_{gen}, the entropy generated by this process.

c) Use EES to prepare a T-v and a T-s diagram for water and overlay your states on these diagrams.

6.A-8 In order to measure the loss associated with eddy currents in a motor stator designed to operate at low temperature you have developed the experiment shown in Figure 6.A-8. The stator core is placed in a $V = 1.5$ liter tank of two-phase nitrogen that is placed on a sensitive scale. The tank is equipped with a relief valve that maintains the pressure at $P_{rv} = 2$ atm by allowing only saturated vapor to escape to ambient at $P_{amb} = 1$ atm. You have separately measured the mass of the tank and stator alone; therefore, you know that at the start of the experiment the mass of nitrogen in the tank is $m_1 = 119$ g. The initial pressure in the tank is $P_1 = P_{rv}$. You energize the core with an oscillating voltage and allow it to operate for $t = 50$ min. During this time, the mass of nitrogen decreases by $m_{out} = 50$ g due to the boil off induced by the eddy current losses in the stator as well as heat transfer to the tank from surroundings at $T_{amb} = 20°$C. You have separately measured the rate of heat transfer to the tank to be $\dot{Q}_{amb} = 0.92$ W. The eddy current losses manifest themselves as an electrical power to the core, $\dot{W}_e$, that is subsequently transferred as heat to the nitrogen.

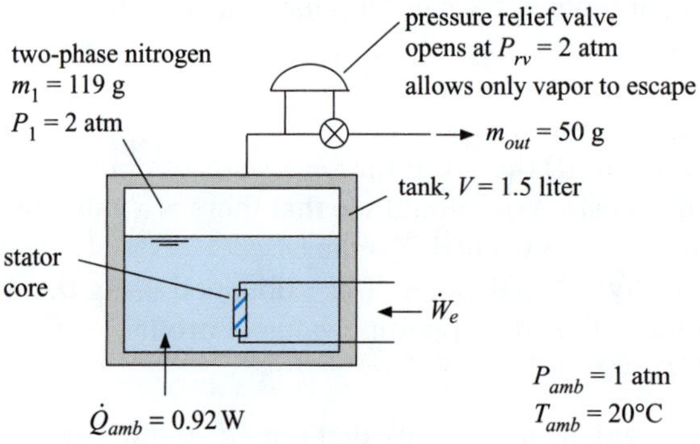

Figure 6.A-8: Experiment to measure eddy current losses in a motor stator.

a) What is the rate of eddy current loss suggested by the experimental results (i.e., what is $\dot{W}_e$)?
b) Determine the entropy generated (J/K) due to the conversion of the electrical work to heat transfer in the stator. Clearly draw the system that you use to accomplish this calculation.
c) Determine the entropy generated (J/K) due to the heat transfer from ambient through the tank wall to the liquid nitrogen. Clearly draw the system that you use to accomplish this calculation.
d) Determine the entropy generated (J/K) due to the flow through the valve. Clearly draw the system that you use to accomplish this calculation.
e) Determine the entropy generated (J/K) due to the re-equilibration of the cold nitrogen leaving the valve with the atmosphere. Clearly draw the system that you use to accomplish this calculation.
f) Draw a system that encompasses all of the sources of entropy generation that were separately considered in parts (b) through (e). Carry out an entropy balance on this system and show that the total entropy generation is the sum of what you calculated in parts (b) through (e).

6.A-12 Figure 6.A-12 illustrates a "vortex tube".

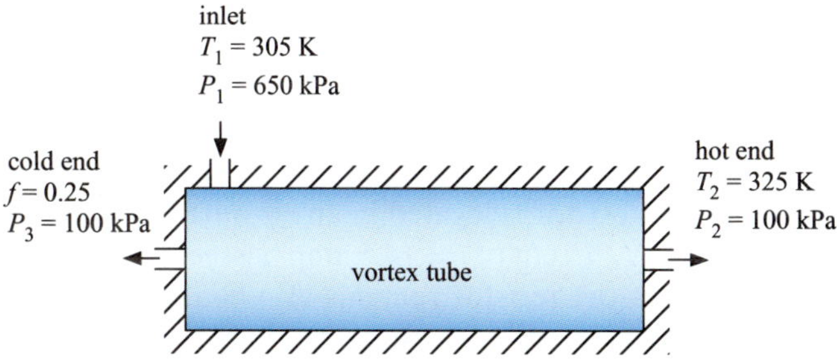

Figure 6.A-12: A vortex tube.

According to the company that makes the vortex tube, the device takes in high pressure air at pressure $P_1 = 650$ kPa and temperature $T_1 = 305$ K and splits it into two streams of air that both leave at lower pressure, $P_2 = P_3 = 100$ kPa. One of the streams exits through the cold end of the device at a low temperature, T_3, and the other exits through the warm end at a high temperature, $T_2 = 325$ K. The fraction of the entering mass that leaves through the cold end is $f = 0.25$. The vortex tube operates at steady-state, it is adiabatic and there is no work done on or by the device. Model air as an ideal gas with $R = 287$ J/kg-K and constant $c_P = 1004$ J/kg-K.

a) What is the temperature of the air leaving through the cold end, T_3?
b) Is this device possible? Justify your answer.

6.A-16 A piston cylinder device is shown in Figure 6.A-16.

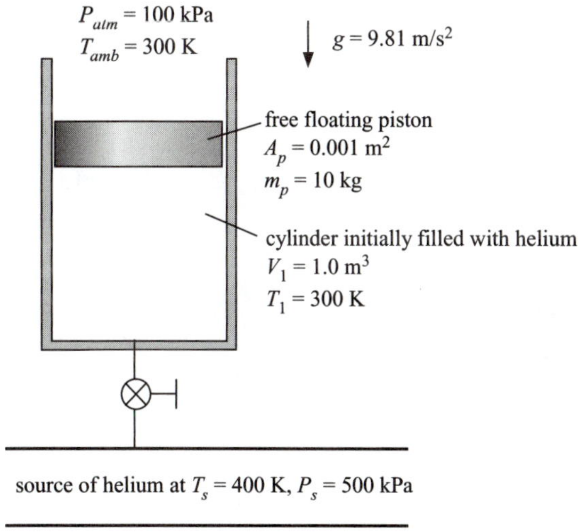

Figure 6.A-16: Piston cylinder device.

The piston is free floating with area $A_p = 0.001$ m^2 and mass $m_p = 10$ kg. The acceleration of gravity is $g = 9.81$ m/s^2. The ambient pressure is $P_{atm} = 100$ kPa and the ambient temperature is $T_{amb} = 300$ K. The cylinder volume is initially $V_1 = 1.0$ m^3. The cylinder is initially filled with helium at $T_1 = T_{amb}$. The cylinder is connected to a source of high pressure helium at $P_s = 500$ kPa and $T_s = 400$ K through a valve. The valve is opened, allowing $m_{in} = 0.5$ kg of helium to enter the cylinder. Then the valve is closed. Eventually the temperature of the helium in the cylinder returns to $T_2 = T_{amb}$ due to heat transfer with the environment. Model helium as an ideal gas with $R = 2077$ J/kg-K and constant $c_P = 5200$ J/kg-K.

a) What is the heat transfer to the surroundings?
b) What is the total entropy generated by the process?

6.A-17 An $R_e = 30$-ohm electrical resistor is located in an air duct through which air is flowing with a steady rate. A current of $i = 14.5$ amp passes through the resistor. The surface temperature of the resistor remains at a constant temperature of $T_s = 43^\circ$C. Air from the surroundings enters the duct at $T_{in} = 15^\circ$C with a volumetric flow rate of $\dot{V}_{in} = 0.285$ m^3/s. Air exits the duct at $T_{out} = 32^\circ$C. The air is at atmospheric pressure.

a) Determine the rate of heat transfer from the duct to the surroundings.
b) Determine the First-Law efficiency of this heating process.
c) Choosing the resistor to be the system, determine the rate of entropy generation in this system.
d) For a control volume that includes the resistor and the air duct, determine the rate of entropy generation.
e) If your answers for parts (c) and (d) differ, provide an explanation for the difference.

6.A-23 You have the air-powered water rocket toy shown in Figure 6.A-23(a). A simplified schematic of the toy is shown in Figure 6.A-23(b).

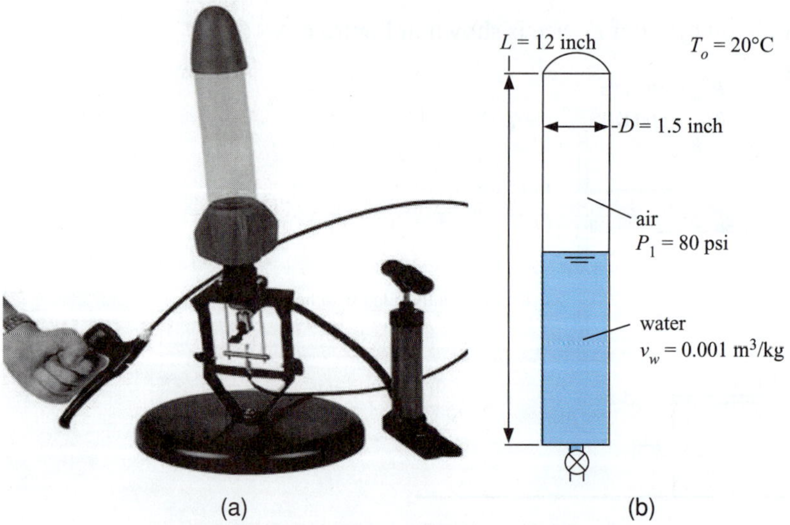

Figure 6.A-23: (a) A picture and (b) a schematic of an air-powered water rocket.

The rocket operates using water and air. The rocket itself is a hollow plastic tube with an inner diameter $D = 1.5$ inch and a length $L = 12$ inch. In a typical launch, the rocket is filled part way with liquid water and then connected to a hand-powered air pump that pressurizes the air and water to about $P_1 = 80$ psi. The fraction of the rocket that is filled with water is $f = 0.5$. The surroundings are at $T_o = 20^\circ$C and the air and water are initially at this temperature. When the trigger is released, the valve is opened and the pressurized water shoots out of the nozzle and propels the rocket upwards. This process continues until all of the water has been pushed out, at which time the rocket is spent and falls back to the ground. The total time associated with the flight is *time* = 2 s. Model air as an ideal gas with $R = 287.1$ J/kg-K and constant $c_v = 717.4$ J/kg-K. The specific volume of liquid water is constant and equal $v_w = 0.001$ m^3/kg.

a) Assume that the air in the rocket undergoes a reversible and adiabatic expansion as the water is pushed out. Determine the work done by the air onto the water (J).

The thrust produced by the water varies with time and requires the consideration of the details of the nozzle and the instantaneous pressure in the rocket. However, a very simple and approximate model assumes that the work done by the air on the water, W, goes entirely to increasing the kinetic energy of the water and pushing the atmosphere out of the way. With this model, the average velocity of the water leaving the rocket can be approximately determined according to:

$$W = m_w \frac{\tilde{V}^2}{2} + P_{atm}\, V_w$$

where m_w and V_w are the mass and volume, respectively, of water initially in the rocket, $\tilde{V}$ is the velocity of the water leaving the rocket, and P_{atm} is atmospheric pressure.

b) Use the equation above to estimate the velocity of the water leaving the rocket.
c) If we ignore the velocity of the rocket itself, then the thrust force produced by the rocket is the product of the velocity and mass flow rate of the water; estimate the thrust force produced by the rocket (N).
d) Plot the thrust as a function of f, the fraction of the rocket volume initially filled with water. You should see that an optimal value of f exists. Explain why there is an optimal value for f.

An alternative model for the rocket assumes that the air in the rocket is not adiabatic, but rather isothermal. Rather than having no heat transfer during the launch, the isothermal model assumes that the heat transfer is sufficient that the air in the rocket is always at T_o. The adiabatic and isothermal models will bound the actual performance of the rocket.

e) Repeat your calculations using an isothermal model of the launch. Overlay the thrust force as a function of f for the isothermal model onto the plot generated in (d).

6.A-27 An industrial plant has a compressed air stream at $P_1 = 2.5$ atm and $T_1 = 35^\circ$C that is presently vented to the atmosphere. Due to increased fuel costs, the management is very interested in a device that is claimed to separate the air stream into two equal mass flow rates: a hot stream at $T_2 = 230^\circ$C and a cold stream at $T_3 = -160^\circ$C, both at atmospheric pressure. These two streams could be used to satisfy some of the heating and cooling requirements in other parts of the plant. The device is also claimed to be self-sustaining, requiring no additional heat or work. Is such a device possible? Explain why or why not. If the device is possible, suggest a process (or series of processes) which would produce the same effect as this device.

6.A-29 A company advertises a heat-driven refrigerator, shown in Figure 6.A-29.

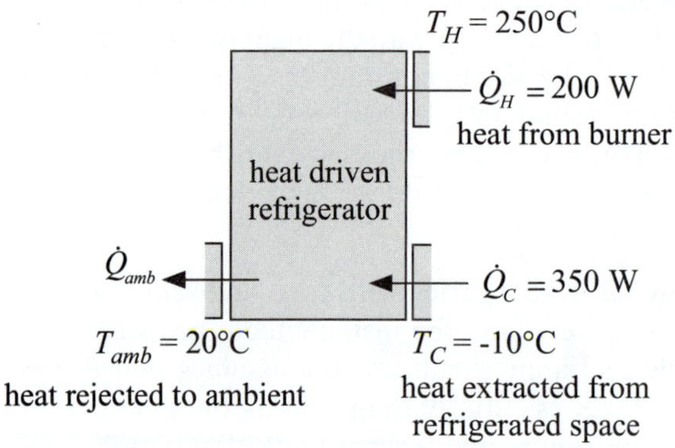

Figure 6.A-29: Heat-driven refrigerator.

The refrigerator is driven by a heat transfer from a burner at $T_H = 250°\text{C}$ at a rate of $\dot{Q}_H = 200$ W. The refrigerator removes $\dot{Q}_C = 350$ W from a refrigerated space at $T_C = -10°\text{C}$ and rejects heat to the surroundings at $T_{amb} = 20°\text{C}$. The refrigerator operates at steady state and requires no work transfer.

a) Is the device possible? Does it violate the Second Law of Thermodynamics?
b) A reversible, heat-driven refrigerator obtains $\dot{Q}_H = 200$ W from a thermal reservoir at $T_H = 250°\text{C}$ and provides refrigeration at $T_C = -10°\text{C}$ with heat rejection to ambient at $T_{amb} = 20°\text{C}$. What is the rate of refrigeration provided by this reversible device?
c) What is the Coefficient of Performance (COP) of the reversible device that you examined in (b)? The Coefficient of Performance for this system is defined as the ratio of what you want ($\dot{Q}_C$) to what you pay for ($\dot{Q}_H$) and the COP of the reversible device will be the maximum possible COP for this type of device operating with these temperatures.
d) Plot the Coefficient of Performance of the reversible device as a function of T_H.

6.A-31 A piston-cylinder device is shown in Figure 6.A-31.

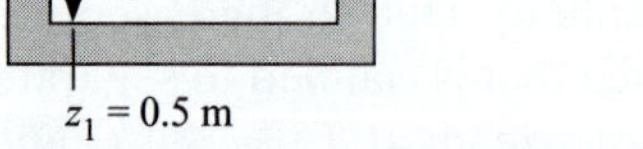

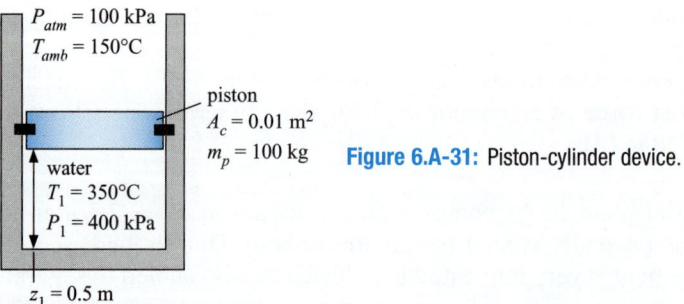

Figure 6.A-31: Piston-cylinder device.

The initial position of the piston is $z_1 = 0.5$ m and the piston cross-sectional area is $A_c = 0.01\ \text{m}^2$. The mass of the piston is $m_p = 100$ kg. The cylinder contains water that is initially at $T_1 = 350°\text{C}$ and $P_1 = 400$ kPa. The surroundings are at $P_{atm} = 100$ kPa and $T_{amb} = 150°\text{C}$. The piston is initially held in place by a pin to prevent it from moving due to the internal pressure. At some time, the pin is removed and the piston quickly moves upward under the action of the internal pressure. The piston motion continues for some time until eventually the oscillations are damped out and the piston obtains a new equilibrium position at state 2 where it is in mechanical equilibrium with the surroundings (i.e., a force balance on the piston can be used to provide the internal pressure at state 2). There is no heat transfer between the contents of the piston and the surroundings during the time required by the equilibration. Note: this is an irreversible mechanical equilibration process. You do not know, nor is there any way to determine, the pressure of the water acting on the lower surface of the cylinder during this process. However, you do know the pressure of the atmosphere acting on the upper surface of the piston during the process. Your system selection should be informed by these facts.

a) Determine the position of the piston at state 2, z_2.
b) Determine the temperature of the water at state 2, T_2.
c) What is the entropy generated by the process of moving from state 1 to state 2, $S_{gen,1\text{-}2}$?
d) What is the work transfer from the water to the piston during the process of going from state 1 to state 2, $W_{out,1\text{-}2}$?

After some time has passed, heat transfer between the water to the surroundings causes the water to come to a final temperature that is equal to the temperature of the surroundings, T_{amb}. This is an irreversible thermal equilibration process that must result in entropy generation because heat is being transferred through a temperature gradient. The piston is allowed to move freely during this process.

e) Determine the position of the piston at state 3, z_3.
f) Determine the heat transferred from the water to the surroundings during this process, $Q_{out,2\text{-}3}$.
g) Determine the entropy generated by the process of moving from state 2 to state 3, $S_{gen,2\text{-}3}$.
h) Using EES, generate a temperature-entropy diagram that shows states 1, 2, and 3.
i) Plot the entropy generated by the process of moving from state 1 to state 2 as a function of P_1 for $100 < P_1 < 500$ kPa. You should see that there is an optimal pressure at which the entropy generated by this process is minimized. Explain why this is the case.
j) What initial pressure and temperature should you use if you want to minimize the total entropy generated by the equilibration processes (i.e., you want to minimize $S_{gen} = S_{gen,12} + S_{gen,23}$). Why?

6.A-33 Figure 6.A-33 illustrates a cryogenic refrigerator that is used to maintain a detector at a very low temperature. The refrigerator utilizes the reverse-Brayton cycle. The reverse-Brayton refrigeration cycle utilizes a gas working fluid so that it can continue to operate even at very low temperatures where more conventional refrigerants will freeze. The device shown in Figure 6.A-33 utilizes neon. Model neon as an ideal gas with $R = 412$ J/kg-K and $c_P = 1030$ J/kg-K. Neon enters the compressor at state 1 with pressure $P_{low} = 152$ kPa and mass flow rate $\dot{m} = 0.0006$ kg/s. The compressor is reversible and adiabatic. The neon leaves the compressor at state 2 with pressure $P_{high} = 532$ kPa. The neon leaving the compressor is cooled in an aftercooler that rejects heat at rate $\dot{Q}_H$ to the ambient temperature at $T_H = 20°$C. The aftercooler approach temperature difference is $\Delta T_{ac} = 10$ K; this means that the neon leaving the aftercooler at state 3 has been cooled to within ΔT_{ac} of T_H (i.e., $T_3 = T_H + \Delta T_{ac}$). Neon then flows through the recuperative heat exchanger where it is cooled by heat transfer with the cold neon returning from the cold end of the device. Cold neon leaves the recuperator at state 4 and enters the turbine. Assume that the turbine is reversible and adiabatic. The neon

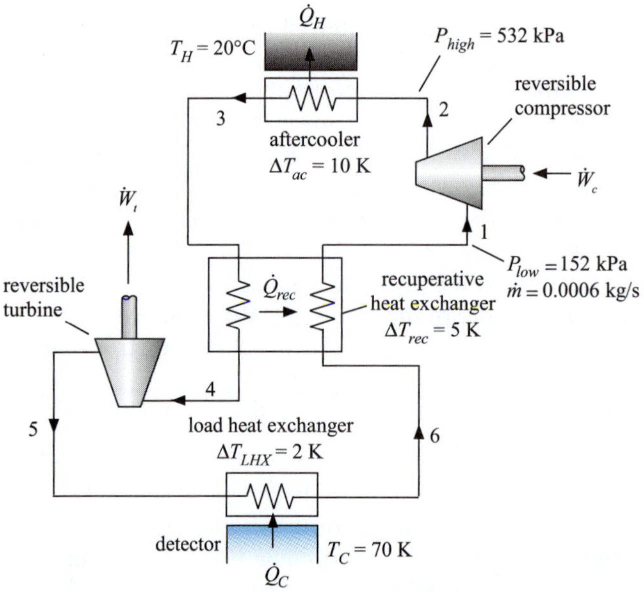

Figure 6.A-33: Reverse-Brayton refrigeration cycle.

exits the turbine at state 5 with pressure P_{low}. The neon leaving the turbine is quite cold (this is the coldest point in the cycle). This neon enters the load heat exchanger where it is warmed by a heat transfer from the detector which is at temperature $T_C = 70$ K. The heat transfer rate from the detector to the neon is $\dot{Q}_C$ and the approach temperature difference for the load heat exchanger is $\Delta T_{LHX} = 2$ K. That is, the neon passing through the load heat exchanger is warmed to within ΔT_{LHX} of T_C ($T_6 = T_C - \Delta T_{LHX}$). Finally, the neon leaving the load heat exchanger at state 6 passes through the recuperator where it is warmed by heat transfer with the high pressure neon flowing in the opposite direction. The recuperator has an approach temperature difference of $\Delta T_{rec} =$ 5 K. Therefore, the neon leaving the recuperator at state 1 has been warmed to within ΔT_{rec} of T_3 ($T_1 = T_3 - \Delta T_{rec}$). Assume that there is no loss of pressure associated with flow through any of the heat exchangers.

a) Determine the temperature and pressure at each of the six numbered states in Figure 6.A-33.
b) Determine the power required by the compressor ($\dot{W}_c$), the power produced by the turbine ($\dot{W}_t$), the rate of heat transfer from the neon in the aftercooler ($\dot{Q}_H$), and the rate of heat transfer to the neon in the load heat exchanger ($\dot{Q}_C$).
c) Check your solution by drawing a system that encompasses the entire cryocooler. Verify that energy balances for this overall system.
d) Assuming that the turbine power is used to help drive the compressor, determine the Coefficient of Performance (COP) of the cryocooler.
e) What is the maximum possible COP for a refrigerator operating between T_C and T_H.
f) Determine the total rate of entropy generation in the cryocooler.
g) Plot the refrigeration power ($\dot{Q}_C$) as a function of the load temperature (T_C). You should see that your refrigerator can provide less refrigeration as the temperature is reduced.

h) The detector, however, will require more refrigeration as its temperature is reduced since the rate of heat transfer is related to the temperature difference between the detector and its surroundings. Assume that the refrigeration required by the detector is given by:

$$\dot{Q}_{C,req} = 2\ [\mathrm{W}] + 0.1\ [\mathrm{W/K}]\,(T_H - T_C)$$

Overlay the required refrigeration as a function of T_C onto your plot of available refrigeration from part (g). Based on your plot – what is the lowest possible temperature that the system can achieve?

6.A-39 A perfectly-insulated tank contains $m_1 = 25$ kg of refrigerant R134a that is initially at $P_{tank} = 300$ kPa with a quality (vapor mass fraction) of $x_1 = 0.80$. The tank pressure is maintained constant by nitrogen gas acting against a flexible bladder, as shown in Figure 6.A-39. The valve is opened between the tank and the supply line that carries R134a at $T_s = 120°$C and $P_s = 1$ MPa, allowing R134a to enter the tank. The tank pressure remains at P_{tank} during this process. The valve is closed when all of the R134a in the tank has been vaporized.

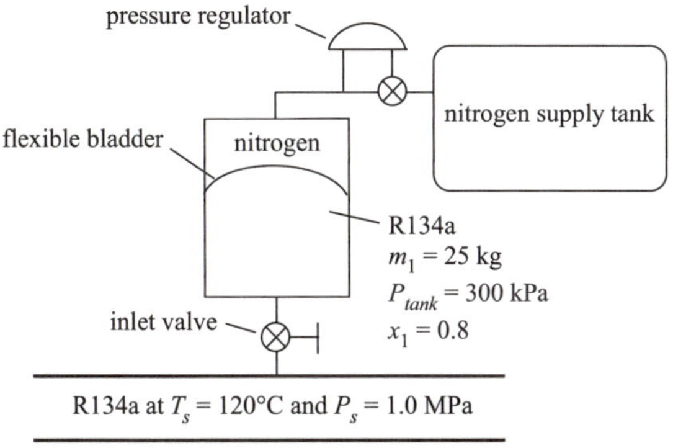

Figure 6.A-39: Tank holding pressurized R134a and nitrogen.

a) What is the mass of R134a that enters the tank?
b) Calculate the entropy produced in this process.

6.A-40 A tank of fixed volume is to be evacuated. The tank has a volume of $V = 1.4$ m^3 and it initially contains air at $T_{amb} = 25°$C and $P_{atm} = 1$ atm. Evacuation is carried out slowly using a small vacuum pump which, like all real devices, does not operate in a reversible manner. Because the process occurs slowly, the air that remains in the tank stays at T_{amb} as a result of heat transfer through the tank walls.

a) Calculate the total heat transfer to the contents of the tank during the time required to reduce the pressure in the tank to $P_2 = 0.1$ atm.
b) Determine the minimum possible work required to accomplish the process in part (a).
c) Is it possible to completely evacuate the tank? If you believe that this is not possible, indicate what physical law is violated if this were to happen. Otherwise, determine the minimum work that would be needed for complete evacuation.

6.A-46 An insulated piston-cylinder apparatus is fitted with an electrical heating element. The cylinder initially contains $m_{s,1} = 4.0$ kg of ice and $m_{f,1} = 1.0$ kg of liquid water, both at $T_m = 0°$C. A thermocouple on the surface of the heating element registers at steady temperature of $T_{htr} = 168°$C. After $t = 400$ seconds of operation there are $m_{s,2} = 2.6$ kg of ice remaining in the cylinder and $m_{f,2} = 2.4$ kg of water, both at T_m. The cylinder contents remain at constant pressure during this process. The temperature and pressure of the surroundings are $T_{amb} = 25°C$ and $P_{atm} = 101.3$ kPa, respectively.

a) Determine the steady power dissipation in the heating element.
b) Determine the change in entropy of the heating element during the process.
c) Determine the change in entropy of the cylinder contents (i.e., the ice and liquid water) during the process.
d) Determine the total entropy generation during the process.

6.A-49 Water can be cooled by pulling a vacuum over liquid water in a Dewar flask using the equipment that is shown in Figure 6.A-49. In this case, a $V = 5$ liter Dewar flask initially contains $V_{f,1} = 2.5$ liters of saturated liquid water at $T_1 = 25°C$ with the remainder of the volume occupied by saturated water vapor. The vacuum pump is turned on causing water vapor to exit the top of the flask. The temperature of the water remaining in the flask decreases as the process proceeds and at some time later, the water temperature is $T_2 = 2°C$. Measurements indicate that the volume of liquid water remaining in the flask at this time is 92% of the original liquid volume. The surroundings are at $T_{amb} = 25°C$ and $P_{atm} =$ 101.3 kPa.

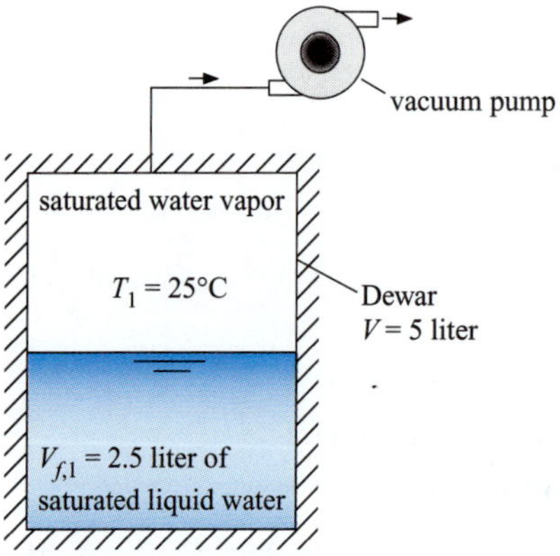

Figure 6.A-49: Pulling a vacuum over water in a Dewar flask.

a) Determine the mass and quality of water in the flask at the start of the experiment.
b) Determine the mass and quality of water in the flask at the end the experiment.
c) The Dewar flask is very well insulated so heat transfer during this experiment can be considered to be negligible. However, the flask is made of a material that has some thermal capacitance. Estimate the effective mass – specific-heat product of the flask.
d) Determine the total entropy generation resulting from this process.

B. Isentropic Efficiencies and Heat Exchangers

6.B-1 Figure 6.B-1 illustrates a compressor that is providing air to a tank. The tank is leaking air through a valve; the air that is leaking re-enters the atmosphere and eventually returns to atmospheric conditions.

The compressor is adiabatic and operating at steady state. The compressor efficiency is $\eta_c = 0.82$. The compressor draws in atmospheric air at $T_{amb} = 20°C$ and $P_{atm} = 100$ kPa and compresses it to the tank pressure, $P_{tank} = 600$ kPa. The

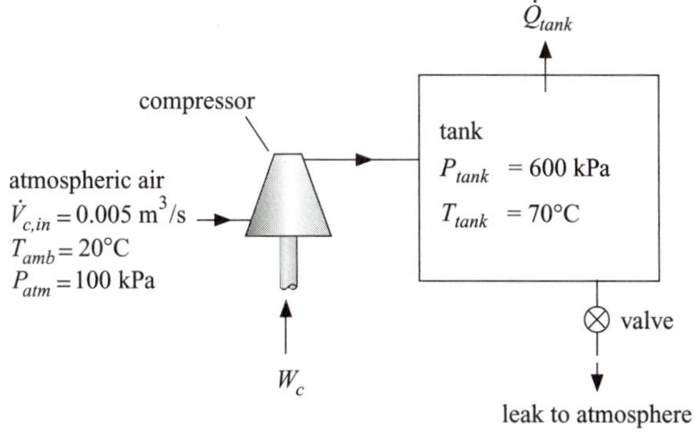

Figure 6.B-1: Compressor providing air to a tank that is leaking through a valve.

volumetric flow rate of air at the compressor inlet is $\dot{V}_{c,in} = 0.005$ m^3/s. The temperature of the air in the tank is $T_{tank} = 70°$ C. Both the pressure and temperature of the air in the tank are constant with time (i.e., the tank is at steady state). Model air as an ideal gas with constant specific heat capacities; $R = 287$ J/kg-K, $c_P = 1005$ J/kg-K, and $c_v = 718$ J/kg-K.

a) What is the mass flow rate provided by the compressor (kg/s)?
b) What is the power required by the compressor (kW)?
c) What is the temperature of the air leaving the compressor?
d) What is the rate of entropy generation in the compressor?
e) Determine the rate of heat transfer from the tank.
f) List and describe as many of the specific sources of entropy generation as you can for this problem.
g) Calculate the total rate that entropy is generated by the system, the sum of all of the sources of entropy generation from (f).

6.B-2 Figure 6.B-2 illustrates a heat exchanger in which hot air is used to generate steam. Air enters the heat exchanger at $T_{a,in} = 320°$ C, $P_a = 100$ kPa, and $\dot{m}_a = 0.5$ kg/s. Model air as an ideal gas with constant specific heat capacity ($R_a = 0.287$ kJ/kg-K, $c_{P,a} = 1.01$ kJ/kg-K). Water enters the heat exchanger at $T_{w,in} = 20°$ C and $P_w = 100$ kPa. The mass flow rate of the water is $\dot{m}_w = 0.025$ kg/s. The pinch point associated with the heat exchanger is on the hot side (i.e., where the air enters and water leaves). The approach temperature difference is $\Delta T = 160$ K. Neglect pressure drop on the air and the water sides of the heat exchanger.

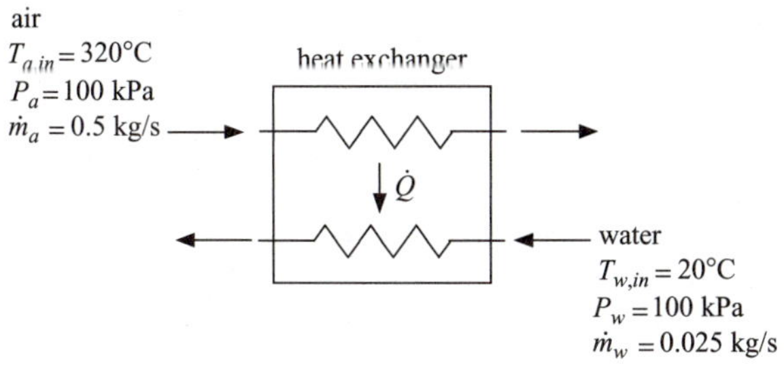

Figure 6.B-2 Heat exchanger.

a) Determine the rate of heat transfer from the air to the water in the heat exchanger.
b) Determine the total rate entropy generation in the heat exchanger.

c) Determine the effectiveness of the heat exchanger.
d) If the approach temperature difference of the heat exchanger were reduced to $\Delta T = 0$ K (i.e., the heat exchanger became "perfect") then would the entropy generation in the heat exchanger be reduced to zero? Justify your answer.

6.B-5 You have access to heat that is the byproduct of an industrial operation. The heat transfer rate is $\dot{Q}_H = 5$ kW and the temperature of the heat source is $T_H = 500°$C. Currently, the heat is transferred directly to ambient temperature at $T_o = 20°$C, as shown in Figure 6.B-5(a).

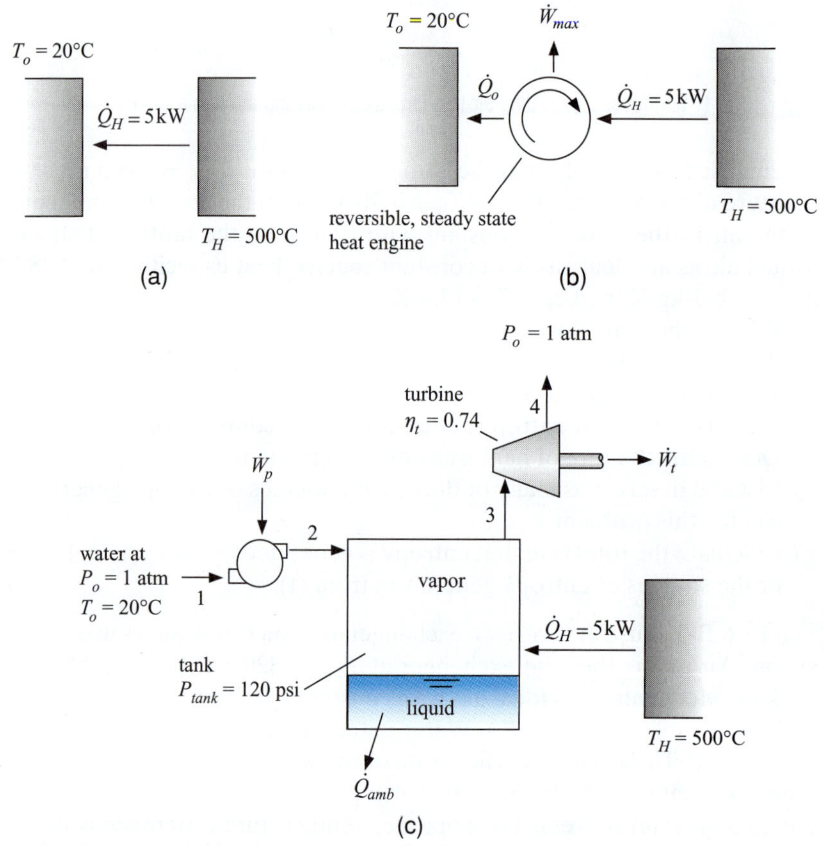

Figure 6.B-5: (a) Heat transferred directly to ambient, (b) heat source being used by a steady-state, reversible heat engine, and (c) heat source being used by the heat engine that you've designed.

a) Determine the rate at which entropy is generated by rejecting the heat to ambient (W/K).

Because you have taken thermodynamics you know that where entropy generation occurs there is a lost potential to produce work.

b) Evaluate the lost potential for power production by determining the power that would be produced (W) if the same heat source were used in a reversible, steady state heat engine, as shown in Figure 6.B-5(b).

You have designed the energy scavenging system shown in Figure 6.B-5(c) in order to produce some power using the heat source. The system consists of a tank of water that contains a two-phase liquid/vapor mixture. The tank is heated by the heat source. The saturated vapor that is produced is fed to a turbine with efficiency $\eta_t = 0.74$. The turbine exhausts to atmospheric pressure, $P_o = 1$ atm.

In order to maintain the tank at a steady operating condition, it is fed with liquid that is provided by a reversible pump. The pump takes in liquid at atmospheric conditions, T_o and P_o, and increases its pressure it to the tank pressure, $P_{tank} =$ 120 psi. The flow rate of the pump is controlled so that the tank pressure never changes and the level of the liquid water in the tank never varies. The tank is not perfectly insulated. There is a heat transfer from the tank to the surroundings given by:

$$\dot{Q}_{amb} = UA\,(T_{tank} - T_o)$$

where $UA = 8.2$ W/K and T_{tank} is the temperature of the contents of the tank.

c) Determine the net power produced by your system (W); this is the turbine output power less the power required by the pump.
d) Determine the total entropy generation rate associated with operating the system (W/K).

The "lost power" associated with the system in Figure 6.B-5(c) is equal to the maximum power that could have been obtained, calculated in (b), less the net power that was obtained, calculated in (c).

e) Determine the lost power associated with the system (W).
f) If you've done the problem right, then the product of the entropy generation rate, from (d), and the ambient temperature should equal the lost power, from (e). Show that this is so. This is the most direct and intuitive meaning of entropy generation – it is the lost power divided by the ambient temperature.
g) Determine the total efficiency of the system.
h) Plot the efficiency of the system as a function of the tank pressure. You should see that an optimal pressure exists. Clearly explain why this is so.

6.B-6 Supercritical carbon dioxide power cycles are being considered for the next generation of nuclear power plants. These cycles operate with carbon dioxide in the vicinity of the critical point. There are some potential advantages associated with this cycle including high efficiency and small equipment. However, there are several engineering challenges that must be overcome as well. The leakage of carbon dioxide through the seals in the turbomachinery required by the cycle must be well-understood in order to design the system. Your company has received a contract from the Dept. of Energy to build a test facility into which sub-scale seal designs can be tested. You need to design an experiment that is capable of applying a specific inlet condition and exit pressure to a seal and measure the leakage flow rate through various seal geometries. This problem addresses the design of the equipment required for this facility. Your preliminary system design is shown in Figure 6.B-6. The seal (labeled test facility in the figure) acts like a valve – it is a restriction to the flow and the purpose of the system is to characterize this restriction. The nominal design calls for an inlet pressure and temperature (at state 1) of $P_{in} = 9$ MPa and $T_{in} = 310$ K and an exit pressure (at state 2) of $P_{out} = 2$ MPa. In order to design the system you have estimated the flow through the test facility to be $\dot{m}_{test} = 0.12$ kg/s. Eventually, this is the quantity to be measured.

The carbon dioxide leaving the test facility at state 2 lies under the vapor dome and therefore is a mixture of liquid and vapor. The liquid cannot be allowed to enter the compressors that energize the system. Therefore, the carbon dioxide at state 2 is captured in a large insulated tank equipped with an electrical heater. The vapor from the top of the tank at state 3 passes through another electrical heater in order to elevate its temperature by $\Delta T_{htr} = 10$ K and to therefore be absolutely sure that no liquid enters the compressor. The low pressure compressor has a suction volumetric flow rate (i.e., the volumetric flow rate at state 4) of $\dot{V}_{LPc} = 10.2$ ft^3/min and an efficiency of $\eta_{LPc} = 0.72$. The carbon dioxide leaving the low pressure compressor at state 5 is intercooled by rejecting heat to the atmosphere at $T_o = 20°$C. The intercooler has an approach temperature difference of $\Delta T_{ic} = 5$ K; that is, the temperature of the carbon dioxide leaving the intercooler has been cooled to within ΔT_{ic} of T_o ($T_6 = T_o + \Delta T_{ic}$). The carbon dioxide leaving the intercooler is compressed again in the high pressure compressor. The high pressure compressor has a suction volumetric flow rate (i.e., the volumetric flow rate at state 6) of $\dot{V}_{HPc} = 4.5$ ft^3/min and an efficiency of $\eta_{HPc} = 0.76$. The carbon dioxide leaving the high pressure compressor at state 7 is divided into two streams. The mass flow that is not required by the test facility passes through a throttle valve and returns to the tank at state 8. The remaining mass flow rate passes through a cooler that rejects heat to T_o in order to reduce its temperature to the inlet temperature required by the test facility. Assume that there is no loss of pressure due to flow through the tank, heater, intercooler, or cooler (i.e., $P_2 = P_8 = P_3 = P_4$, $P_5 = P_6$, and $P_7 = P_1$). Also assume that the system operates at steady state and each component is adiabatic except the intercooler and cooler.

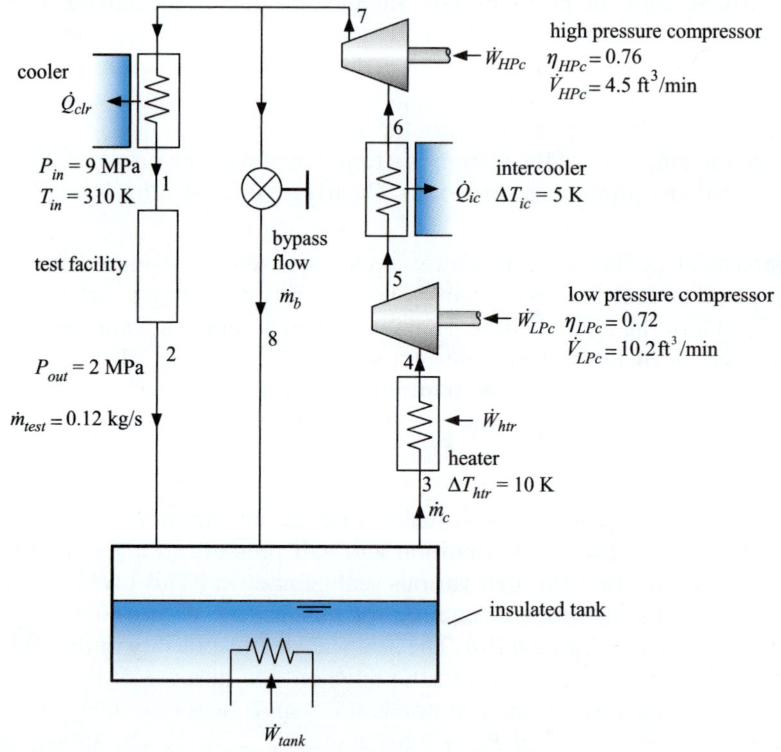

Figure 6.B-6: System to measure leakage of carbon dioxide through sub-scale turbomachinery seals.

a) Determine the properties at states 1, 2, 3, and 4.
b) Determine the mass flow rate that is passing through the compressors, $\dot{m}_c$.
c) Determine the specific enthalpy, pressure, specific entropy, temperature, and specific volume at state 6. Hint, the specific volume at state 6 is related to the mass flow rate and suction volumetric flow rate of the high pressure compressor.

d) You should now have the inlet conditions and exit pressure for both compressors. Use this information together with the compressor efficiency in order to determine states 5 and 7. What is the compressor power required by the low pressure and high pressure compressors?
e) Determine the properties at state 8.
f) Prepare a temperature-entropy plot that shows each of the states.
g) Determine the heater power required by the tank ($\dot{W}_{tank}$) and the heater power required by the heater immediately before the low pressure compressor ($\dot{W}_{htr}$).
h) Determine cooling required in the intercooler ($\dot{Q}_{ic}$) and the cooler ($\dot{Q}_{clr}$).
i) Check your solution by carrying out an overall energy balance on the system.
j) Determine the total rate of entropy generation in the system.
k) Plot the bypass mass flow rate as a function of the exit pressure from the test facility, P_{out} (assume that the mass flow rate through the test facility does not change). Based on this graph, what is the lowest pressure that your system can achieve?
l) Plot the total electrical power required (compressors and heaters) as a function of the test facility exit pressure. If your laboratory electrical service can only provide 60 kW of electrical power then what is the maximum value of the test facility exit pressure that you can achieve?

6.B-7 You own a swimming pool that must be heated to $T_{pool} = 80°$F using the system shown in Figure 6.B-7.

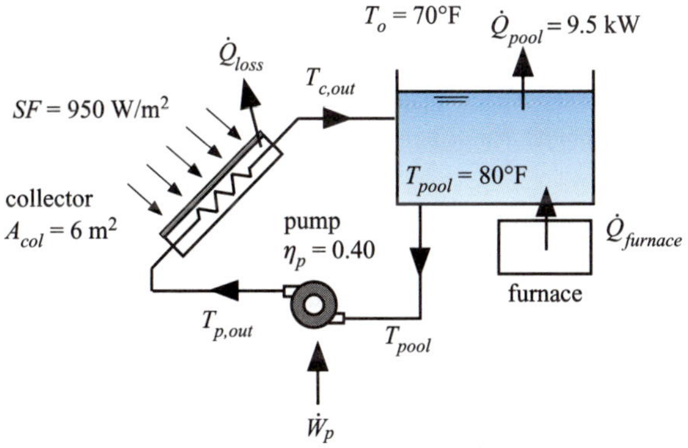

Figure 6.B-7: Swimming pool heating system.

The system uses both a conventional natural gas-fired boiler as well as a low temperature, unglazed solar collector. On this day, the outdoor air temperature is $T_o = 70°$F and the solar flux is $SF = 950$ W/m^2. The total heat loss from the pool under these conditions is $\dot{Q}_{pool} = 9.5$ kW. The collector has a total surface area $A_{col} = 6$ m^2. A pump pulls water out of the pool and pumps it through the solar collector where it is heated to $T_{c,out}$ and returned to the pool. The collector absorbs all of the incident solar radiation. The rate at which the collector transfers heat to the surroundings is given by:

$$\dot{Q}_{loss} = UA\left(T_{c,out} - T_o\right)$$

where $UA = 120$ W/K is the loss coefficient for the collector. The pressure drop associated with flow through the collector has been characterized by the manufacturer and is given by:

$$\Delta P_{col} = 5 \times 10^{12} \left[\frac{\text{Pa} - \text{s}^2}{\text{m}^6}\right] \dot{V}^2$$

where $\dot{V}$ is the volumetric flow rate of water through the collector (in m^3/s) and ΔP_{col} is the pressure drop across the collector (in Pa). The pump is currently operating at $N = 1000$ rev/min. The dead-head pressure rise produced by the pump (i.e., the pressure rise with no flow) depends on the rotational speed and is given by:

$$\Delta P_{dh,pump} = 150 \left[\frac{\text{Pa} - \text{min}}{\text{rev}}\right] N$$

where $\Delta P_{dh,pump}$ is the dead head pressure rise (in Pa) and N is the rotational speed (in rev/min). The open-circuit volumetric flow rate produced by the pump (i.e., the flow rate produced with no pressure rise) also depends on rotational speed and is given by:

$$\dot{V}_{oc,pump} = 5 \times 10^{-7} \left[\frac{\text{m}^3\text{–min}}{\text{s} - \text{rev}}\right] N$$

where $\dot{V}_{oc,pump}$ is the open circuit flow rate (in m^3/s) and N is the rotational speed (in rev/min). Assume that the pump curve is linear between $\Delta P_{dh,pump}$ at zero flow and $\dot{V}_{oc,pump}$ at zero pressure rise. That is, the pressure rise produced by the pump (ΔP_{pump}) as a function of flow rate ($\dot{V}$) is given by:

$$\Delta P_{pump} = \Delta P_{dh,pump} \left(1 - \frac{\dot{V}}{\dot{V}_{oc,pump}}\right)$$

The pump efficiency is relatively constant and equal to $\eta_p = 0.40$. The cost of the electricity required to run the pump is $ec = 0.12$ \$/kW-hr. The portion of the total pool heating load that is not met by the solar collector is met using a conventional natural gas-fired furnace ($\dot{Q}_{furnace}$). The cost of natural gas is $ngc = 1.25$ \$/therm. Model the pool water as an incompressible substance with $v_w = 0.001$ m^3/kg and $c_w = 4200$ J/kg-K.

a) Determine the volumetric flow rate of water pumped through the collector ($\dot{V}$).
b) Determine the pump power required ($\dot{W}_p$).
c) Determine the temperature of the water leaving the pump ($T_{p,out}$).
d) Determine the temperature of the water leaving the solar collector ($T_{c,out}$).
e) What is the rate at which energy is provided to the pool from the solar collector system? What is the cost (\$/hr) associated with providing this energy (i.e., the cost of running the pump)?
f) What is the remaining rate of heat transfer required by the pool that is provided by the furnace ($\dot{Q}_{furnace}$)? What is the cost (\$/hr) associated with providing this energy (i.e., the cost of burning natural gas)?
g) What is the total cost (\$/hr) associated with running the pool heating system.
h) You have purchased a variable speed drive for the pump so that you can adjust the rotational speed, N, in order to optimally run the system. Plot the total cost associated with running the system as a function of the rotational speed

of the pump. You should see that an optimal pump exists – clearly explain why this is so.

6.B-8 Many detectors must be maintained at cryogenic temperatures in order to operate. One method of keeping space-borne detectors cold is to launch them immersed in a tank of liquid cryogen (e.g., liquid nitrogen, liquid neon, etc.) that slowly boils off due to unavoidable but hopefully small, parasitic heat transfer from the surroundings. As long as the cryogenic tank (referred to as a Dewar) contains some liquid, the detector can operate. Eventually, the tank runs out of liquid and the mission must be terminated. Figure 6.B-8(a) illustrates a Dewar of liquid oxygen that is vented to space. Model space as being at pressure $P_{out} = 0.1$ atm.

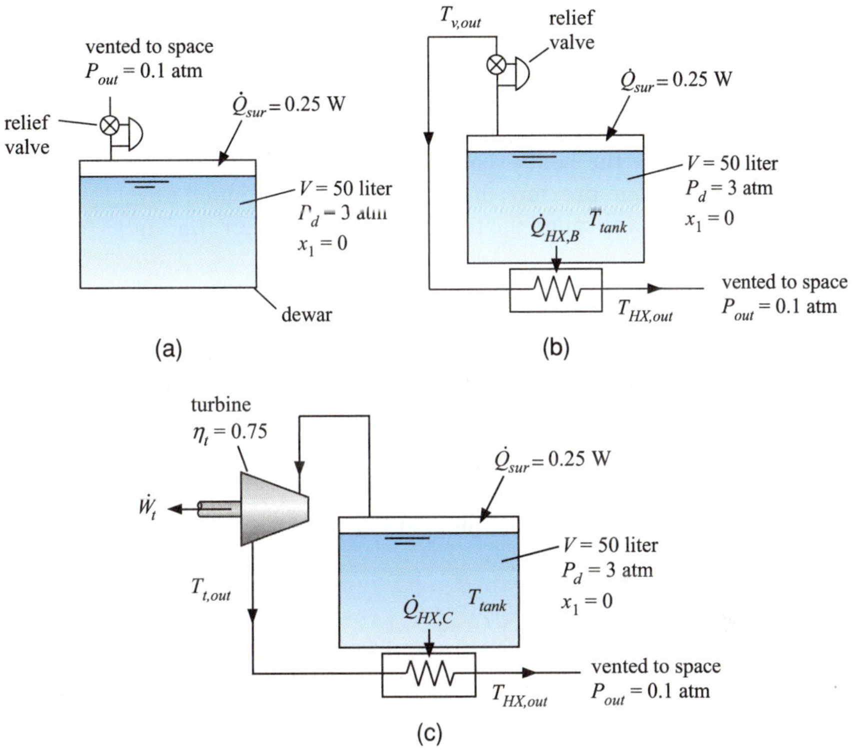

Figure 6.B-8: (a) Dewar vented to space, (b) Dewar with heat exchanger, and (c) Dewar with turbine and heat exchanger.

The volume of the Dewar is $V = 50$ liter and the Dewar is initially filled with saturated liquid oxygen at $P_d = 3$ atm. A relief valve maintains the pressure within the Dewar at P_d by venting saturated vapor to space. The mission is terminated when all of the liquid is gone and the Dewar contains saturated vapor. The Dewar experiences a parasitic heat transfer rate from the surroundings of $\dot{Q}_{sur} = 0.25$ W.

a) Determine the mission duration (days) for the vented Dewar shown in Figure 6.B-8(a).

An engineer has proposed the alternative system shown in Figure 6.B-8(b). He noticed that the temperature of the oxygen leaving the Dewar drops somewhat as it passes through the valve. Therefore, rather than vent the oxygen directly to space it could be passed through a heat exchanger that is interfaced with the tank in order to extract some heat from the tank as it warmed back up to near the tank temperature. The approach temperature difference for the heat exchanger is $\Delta T_{HX} = 0.5$ K. That is, the temperature of the vapor leaving the heat exchanger is $T_{HX,out} = T_{tank} - \Delta T_{HX}$ where T_{tank} is the temperature of the contents of the tank. Assume that the pressure downstream of the valve (at state *v,out*) and downstream of the heat exchanger (at state *HX,out*) are the same and are equal to the pressure of space, P_{out}.

b) Determine the temperature of the oxygen leaving the valve, $T_{v,out}$.
c) Determine the heat transfer per unit mass of oxygen extracted from the tank by the heat exchanger ($\dot{Q}_{HX,B}/\dot{m}$).
d) What is the effectiveness of the heat exchanger?
e) Determine the mission duration (in units of day) for the system shown in Figure 6.B-8(b).

The engineer is disappointed with the results of his first idea and therefore has proposed the system shown in Figure 6.B-8(c). The relief valve is replaced by a turbine with efficiency $\eta_t = 0.75$ in order to provide a larger temperature reduction prior to entering the heat exchanger and therefore provide more cooling effect to the Dewar. The pressure in the Dewar is maintained at P_d by appropriately metering the flow through the turbine. All of the other conditions for the problem remain the same.

f) Determine the temperature of the oxygen leaving the turbine, $T_{t,out}$.
g) Determine the rate of heat transfer to the heat exchanger per mass of oxygen passing through the heat exchanger.
h) Determine the mission duration (day) for the system shown in Figure 6.B-8(c).
i) Plot the mission duration for the system shown in Figure 6.B-8(c) as a function of the turbine efficiency.

6.B-10 Figure 6.B-10 illustrates a proposed design for a very small refrigeration system that can be used to provide cooling at low temperatures with no moving parts in the cold portion of the system.

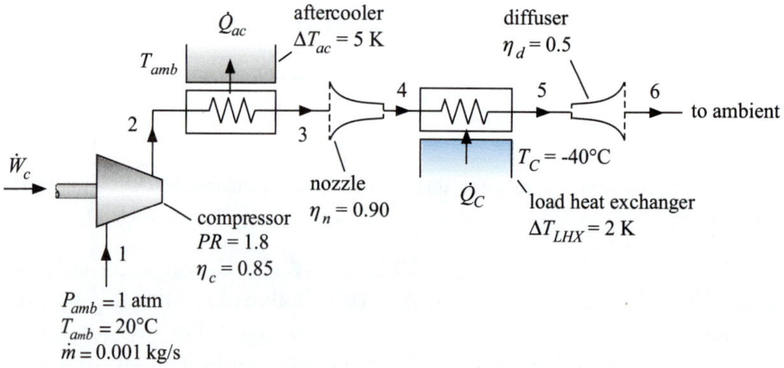

Figure 6.B-10: Proposed cooling system.

Ambient air is drawn into the compressor at $P_{amb} = 1$ atm and $T_{amb} = 20°C$ with mass flow rate $\dot{m} = 0.001$ kg/s. The compressor has a pressure ratio $PR = 1.8$ (that is, the ratio of the discharge pressure to the suction pressure is PR). The efficiency of the compressor is $\eta_c = 0.85$. The hot air leaving the compressor passes through an aftercooler that has an approach temperature difference of $\Delta T_{ac} =$ 5 K. Assume that there is no pressure loss in the aftercooler. The air passes through a nozzle, causing its velocity to increase and its temperature to drop.

The nozzle efficiency is $\eta_n = 0.90$. The cold, high velocity air passes through the load heat exchanger that accepts heat from the refrigeration load at $T_C = -40°$C. The load heat exchanger has an approach temperature difference of $\Delta T_{LHX} = 2$ K. Assume that neither the velocity nor the pressure of the air passing through the load heat exchanger change. Finally, the high velocity air passes through a diffuser with efficiency $\eta_d = 0.5$. The air leaving the diffuser is exhausted to the ambient. Neglect kinetic energy everywhere in the cycle except at states 4 and 5. Model air as an ideal gas but do not assume that it has a constant specific heat capacity.

a) Determine the power required by the compressor and the temperature of the air leaving the compressor.
b) Determine the rate of heat transfer from the aftercooler.

It is next necessary to analyze the nozzle, load heat exchanger, and diffuser. This process is made more complicated by the fact that the pressure at state 4 is not known and the velocity at state 5 is not known. Therefore, it is not really possible to directly and sequentially analyze the nozzle, load heat exchanger, and then finally the diffuser. The best way to proceed is to guess a value of P_4 and analyze the nozzle – this guess will eventually have to be removed in order to complete the problem. Set the guess for $P_4 = 40000$ Pa and proceed with the problem.

c) Using $P_4 = 40000$ Pa, determine the exit velocity and temperature of the nozzle.
d) Determine the rate of heat transfer from T_C provided in the load heat exchanger.
e) Determine the pressure leaving the diffuser. Note that the actual pressure leaving the diffuser must be atmospheric pressure. However, it won't be for your assumed value of P_4 from part (c). We will force P_6 to be equal to P_{amb} in part (f).
f) The pressure that you calculated in part (e) was not equal to atmospheric pressure and yet the diffuser exhausts to the atmosphere. You need to adjust the assumed value of P_4 from part (c) until $P_6 = P_{amb}$. You can do this manually but EES makes the process much easier. Update your guess values (select Update Guesses from the Calculate menu) and then comment out (i.e., remove) your guessed value for P_4. Add the specification that $P_6 = P_{amb}$ and solve the problem. What is the value of $\dot{Q}_C$ and the temperature of the air leaving the diffuser?
g) What is *COP* of the refrigerator? Compare this to the *COP* of a reversible refrigerator providing refrigeration at T_C and rejecting heat to T_{amb}.
h) Use EES to draw a *T-s* diagram of the cycle. Label each of the points and include isobars for each of the pressures in the cycle.
i) Plot the *COP* and refrigeration power ($\dot{Q}_C$) of the cycle as a function of the compressor pressure ratio. You should see that there is a pressure ratio that optimizes the *COP*.
j) What is the highest *COP* that the cycle shown in Figure 6.B-10 can have? That is, if each of the components were performing at their maximum possible level, what would the *COP* be? Note that the cycle would still not be reversible because the heat exchangers cannot be reversible devices.

6.B-13 Figure 6.B-13 illustrates the cooling system for an engine. Antifreeze is circulated through the engine block at a rate of $\dot{V}_{af} = 30$ gpm where it receives heat at a rate $\dot{Q}_{eb}$. Model antifreeze as an incompressible substance with $\rho_{af} = 1039$ kg/m^3 and

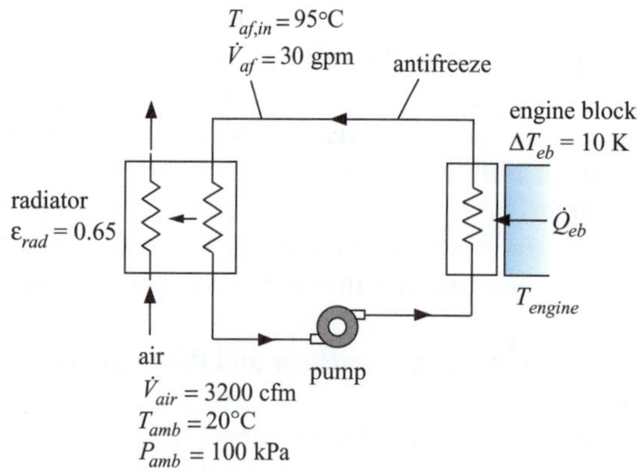

Figure 6.B-13: Cooling system for an engine.

$c_{af} = 3740$ J/kg-K. The antifreeze leaves the engine block and enters the radiator at $T_{af,in} = 95^{\circ}$C. The antifreeze is cooled in the radiator by a flow of ambient air that enters at $P_{amb} = 100$ kPa and $T_{amb} = 20^{\circ}$C with an inlet volumetric flow rate $\dot{V}_{air,in} = 3200$ cfm (cubic feet per minute). Model air as an ideal gas with $R_{air} = 287.1$ N-m/kg-K and $c_{P,air} = 1005$ J/kg-K. The effectiveness of the radiator is $\varepsilon_{rad} = 0.65$ and the pinch point is at the warm end (i.e., at the end where the coolant enters). The approach temperature difference for the engine block is $\Delta T_{eb} = 10$ K. Neglect the effect of the pump on the state of the antifreeze (i.e., assume that the state of the antifreeze entering and leaving the pump is essentially the same).

a) Determine the rate of heat from the antifreeze to the air in the radiator.
b) Determine the temperature of the air and the antifreeze leaving the engine (in °C).
c) What is the approach temperature difference associated with the radiator?
d) Determine the engine temperature (T_{engine} in °C) and the effectiveness of the engine block.
e) Plot the engine temperature and the temperature of the coolant entering the radiator as a function of the rate of heat transfer from the engine block (on the same plot). If the maximum temperature that the coolant can reach before the engine overheats is 100°C then what is the maximum rate of heat transfer from the engine that can be accommodated?
f) Cars tend to overheat more readily in high altitudes where the atmospheric pressure is lower than its value at sea level. Plot $\dot{Q}_{eb}$ as a function of P_{amb} for $T_{af,in} = 95^{\circ}$C for P_{amb} between 80 kPa and 100 kPa to show how the cooling capacity of the system is affected.

6.B-14 You have access to a flow of saturated steam (i.e., saturated vapor) that is continuously vented from an open feedwater heater in a coal-fired heating plant. The mass flow rate of the steam is $\dot{m}_{stm} = 0.35$ kg/s and the steam is at $P_{stm} = 50$ psi. The environment is at pressure $P_{amb} = 1$ atm and $T_{amb} = 20^{\circ}$C. You are interested in trying to utilize the steam to provide some power. Initially, you would like to estimate the maximum possible rate at which power could be produced using this resource in this environment. The maximum possible rate of power that could be produced occurs if the steam is provided to a set of reversible equipment that exchanges heat with the atmosphere. The steam must be brought to the temperature and pressure of the environment (called the dead state) as it exhausts from the equipment. This hypothetical system is shown in Figure 6.B-14(a).

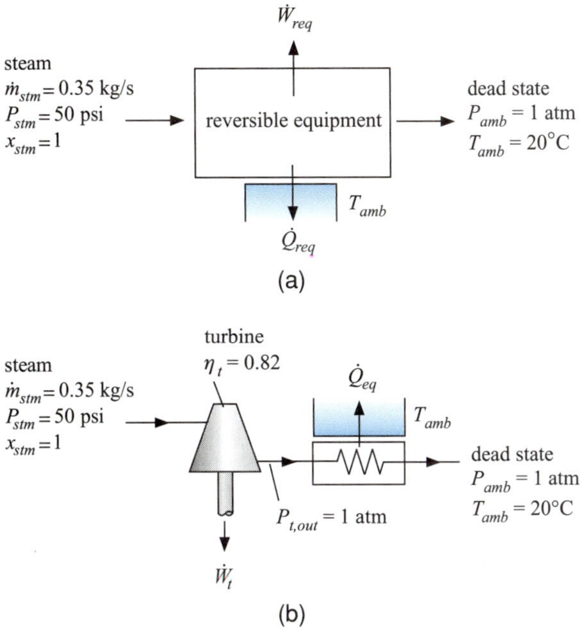

Figure 6.B-14: Steam being (a) used by a reversible set of equipment to produce the maximum possible amount of power and (b) used with a real turbine to produce some power.

a) Analyze the system shown in Figure 6.B-14(a) in order to determine the maximum possible rate at which power can be produced using the steam.

In order to use the steam to provide some power, you have decided to feed it to a turbine, which exhausts to the atmosphere, as shown in Figure 6.B-14(b). The efficiency of the turbine is $\eta_t = 0.82$. Note that the steam leaving the turbine is at atmospheric pressure but not atmospheric temperature. Therefore, the steam will eventually re-equilibrate thermally with the atmosphere and finally come to the dead state temperature.

b) Analyze the system shown in Figure 6.B-14(b) and determine the power produced by the turbine.
c) Your answer from (b) should have been much less than your answer from (a). The difference is the lost power related to entropy generation. Determine the lost power by subtracting your answer from (b) from your answer from (a).
d) The rate of lost power is related to the entropy generation in the system. Determine the total rate of entropy generation for the system shown in Figure 6.B-14(b) as well as the rate of entropy generation that can be attributed to the irreversible turbine and the rate of entropy generation that can be attributed to the re-equilibration process.
e) If you've done the problem correctly, the product of the total rate of entropy generation and the dead state temperature should be equal to the value of the lost power computed in (c). Verify that this is true.
f) Based on your calculations in (d), what modification to your actual cycle would make the most sense in order to capture more of the power producing potential of the steam?

6.B-15 Steam enters a nozzle at $T_1 = 520°\text{C}$ and $P_1 = 100$ bar with a velocity $\tilde{V}_1 = 13$ m/s. It leaves the nozzle at $P_2 = 20$ bar and the speed of sound at this condition. Steam does not obey the ideal gas law at these conditions.

a) Determine the temperature of the steam exiting the nozzle
b) Determine the nozzle efficiency
c) Determine the pressure, temperature and velocity at the nozzle exit that would result in a nozzle efficiency of 100%.

6.B-20 A portable power system consists of an un-insulated evacuated tank made of steel with a volume of $V = 0.75\ \text{m}^3$. The tank is attached to the exhaust of an air turbine-driven grinding wheel. The plan is to drive the air turbine with the pressure difference between the air at atmospheric pressure and $T_{amb} = 20°\ \text{C}$ and the (lower) pressure in the tank. As air enters the tank, the pressure in the tank will rise and the process will eventually stop when the pressure in the tank is atmospheric pressure. The turbine operates adiabatically with an isentropic efficiency of $\eta_t = 0.7$. The air in the tank is always at T_{amb} due to heat transfer with the tank walls. The mass flow rate of air through the turbine varies with the pressure in the tank according to

$$\dot{m} = 0.01[\text{kg/s}]\sqrt{\frac{(P_{atm} - P_{tank})}{P_{atm}}}$$

where P_{atm} is atmospheric pressure and P_{tank} is the pressure in the tank.

a) How many seconds of operation will be provided by this system?
b) Determine the total mechanical work that can be obtained from the system during the operation period.
c) Calculate the total entropy generated during the operation period.

C. Advanced Problems

6.C-3 Compressed air at $T_s = 30°\text{C}$ and $P_s = 500$ kPa is available in a factory that is not air conditioned. One of the engineers is tired of working in an uncomfortably hot environment and he has suggested that an air-conditioning effect can be achieved using this compressed air supply. His plan is illustrated in Figure 6.C-3. A receiver tank with volume $V_{tank} = 30\ \text{m}^3$ is placed on the roof of the building. The tank is initially filled with ambient air at $T_{amb} = 30°\text{C}$ and $P_{atm} = 100$ kPa. During the initial filling process (state 1 to 2), valves B and C are closed and valve A is opened so that air from the compressed air line rushes into the receiver tank. The filling process continues until the pressure in the tank reaches P_s. The filling process occurs very quickly and therefore can be approximated as being adiabatic.

a) Determine the temperature of the air in the tank at the conclusion of the filling process and the mass of air that enters the tank during the process.

During the constant pressure cooling process (state 2 to state 3), valve A remains open while the air in the tank cools to $T_3 = 35°$ due to heat transfer with the surroundings at T_{amb}. The pressure remains constant during this process.

b) Determine the mass of air that enters the tank and the heat transfer from the tank during the constant pressure cooling process.

At the conclusion of the cooling process, the air in the receiver tank is still warmer than the building. In order to avoid having this warm air enter the building, valve A is closed, valve C is opened and air is vented from the tank. The first venting process (state 3 to 4) continues until the temperature of air remaining in the tank

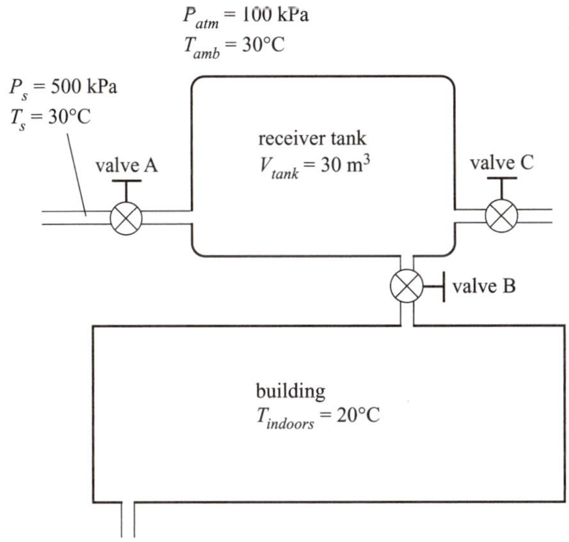

Figure 6.C-3: Proposed air-conditioning system using compressed air.

reaches $T_{indoors} = 20°$C. The venting process occurs very quickly and therefore it is appropriate to neglect heat transfer during this process.

c) Determine the mass of air in the tank at the conclusion of the first venting process.

Finally, valve C is closed and valve B is opened so that air from the receiver tank rushes into the building. The second venting process (state 4 to 5) continues until the pressure in the receiver tank reaches P_{atm}. The venting process is rapid and therefore the tank can be modeled as being adiabatic. The air passing through the building is warmed and leaves the building at $T_{indoors}$, providing cooling to the building.

f) Determine the cooling provided to the building during the second venting process.

e) The compressor installed in the building has an isentropic efficiency of $\eta_c = 0.65$. Determine the work required by the compressor in order to accomplish this cycle and the coefficient of performance (defined as the ratio of cooling provided to work required) associated with this cycle.

6.C-10 A long horizontal hollow stainless steel cylinder has inner and outer radii of $R_{in} = 1$ cm and $R_{out} = 4$ cm, respectively. Saturated steam at $P_s = 15$ bar flows steadily through the cylinder. The heat transfer coefficient between the steam and the inner radius of the cylinder is estimated to be $h_{conv,s} = 11{,}000$ W/m^2-K. Heat transfer from the outer surface of the cylinder occurs by convection to air at $T_a = 25°$C. The convection coefficient is between the air and the outside surface of the cylinder is $h_{conv,a} = 55$ W/m^2-K. Assume the properties of the stainless steel are constant. State and justify any other assumptions that you employ.

a) What is the steady-state rate of heat flow to the air per meter of cylinder?

b) What is the steady-state rate of entropy generation per meter of cylinder in this process?

c) Plot the steady state temperature distribution of the cylinder as a function of radius.

d) The steam flow is stopped and the cylinder eventually returns to a uniform temperature of T_a. What is the resulting change in energy and entropy of the cylinder per meter for this process?

6.C-12 We have three objects in our possession: A, B, and C. The mass, heat capacity and initial temperature of each object are listed in Table 6.C-12.

Table 6.C-12: Specific heat capacity, mass, and initial temperature of objects A, B, and C.

Object	Specific heat capacity (kJ/kg-K)	Mass (kg)	Initial temperature (°C)
A	0.46 [kJ/kg-K] + 0.052 [kJ/kg-K^2] T	25	200
B	2.40 [kJ/kg-K] + 0.120 [kJ/kg-K^2] T	8	50
C	0.84 [kJ/kg-K] + 0.091 [kJ/kg-K^2] T	20	75

The objective of this problem is to determine the minimum temperature that can be achieved in any one of the three objects by a sequence of heat or work interactions between them without a net change in the energy of the environment. Each object can be considered to be 'lumped' (i.e., isothermal at any point in time) but the heat capacities are dependent on temperature as indicated in the table. Your solution should indicate the minimum temperature, which of the three objects achieves this temperature, and a sequence of processes which, in theory, would produce this change.

6.C-15 Ocean Thermal Energy Conversion (OTEC) plants attempt to use the temperature difference that naturally occurs between deep water and surface water in order to generate power. In a particular case, water at $T_{C,in} = 4.5°C$ can be pumped to the OTEC equipment from the depths to provide a thermal sink while water at $T_{H,in} = 26.7°C$ can be pumped from the surface. A plant schematic is shown in Figure 6.C-15.

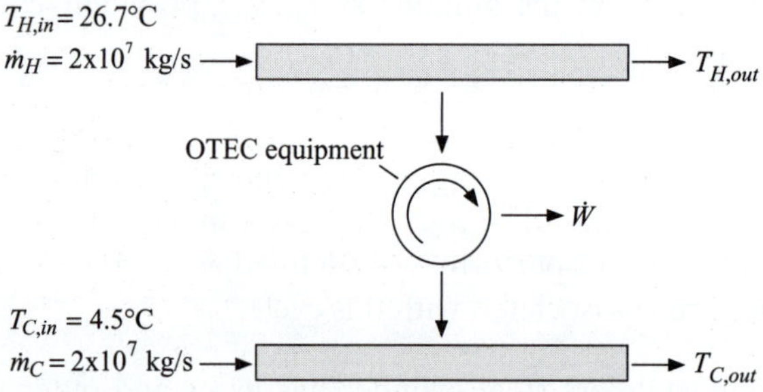

Figure 6.C-15: Ocean thermal energy conversion plant.

a) What is the maximum power that can be produced if the mass flow rates of both the hot and cold flows are $\dot{m}_H = \dot{m}_C = 2.0 \times 10^7$ kg/s?
b) What are the outlet temperatures of both streams for the conditions of part (a)?
c) What is the thermal efficiency of the power production process in part (a)?
d) If the total pumping capacity that is available for both streams is $\dot{m}_{total} = 4 \times 10^7$ kg/s then what would you recommend for the hot and cold flows in order to maximize the power production?

6.C-21 Most electric utilities in the U.S. provide base load electrical power with generators that are driven by Rankine steam cycles. Peak electrical loads are typically met with gas turbine-driven generators. Gas turbine systems offer the advantage of being able to start quickly, but they typically operate at lower efficiency than Rankine cycles. It has been proposed to use off-peak electrical power produced by the more efficient Rankine steam cycle to compress air and store it in an underground cavern. When peak electricity is needed, the air can be withdrawn and heated to $T_{t,in} = 800^\circ C$ by combusting natural gas. The hot, high pressure air can then be passed through turbines in order to produce power. The turbine exhausts to the environment. In a particular case, the cavern has a volume of $V = 0.365 \times 10^6$ m^3. The cavern initially contains air at $T_1 = 25^\circ C$ and $P_1 = 14$ bar. During operation, the air pressure in the cavern is reduced to $P_2 = 11$ bar in two hours while the air temperature remains at T_1 as a result of heat transfer with the cavern walls. Since the air-fuel ratio is large for gas turbine systems, assume that the gas passing through the turbine is pure air that can be modeled as an ideal gas with $R = 287$ J/kg-K and $c_P = 1030$ J/kg-K. Neglect the mass of the fuel.

a) Estimate the maximum total amount of electrical energy that can be generated during the two hour period using the compressed air from the cavern to drive the gas turbine.
b) Determine the thermal efficiency of the turbine for this 2 hour operation period.

7 Exergy

Energy is sometimes defined as the 'capability to do work'; this definition differs from the thermodynamic definition of energy that is conserved according to the First Law of Thermodynamics (i.e., energy is a property of matter that is conserved). Although energy is conserved, we know that the 'capability to do work' is not conserved. Therefore, energy must have both quantity and quality. The Second Law states that the quality of energy, i.e., the capability to do work, is reduced in all real processes. The property entropy introduced in Chapter 6 is related to the quality of energy. The energy and entropy together determine both the quantity and quality of energy. However, the relationship between the 'capability to do work' and the property entropy is indirect. In this chapter, the thermodynamic concept of *exergy* is introduced. Exergy provides a direct relationship between the thermodynamic state of a system and its capability to do useful work. The introduction of exergy does not involve any new thermodynamic concepts, but rather is a re-packaging of concepts that have already been presented. The concept of exergy also allows the definition of a *Second Law efficiency* of a system or process that is often more meaningful than an efficiency based only on energy transfers.

7.1 Definition of Exergy and Second Law Efficiency

The difference between the commonly held meaning of energy and its thermodynamic definition can lead to confusion. According to the First Law of Thermodynamics, energy is conserved in all (non-nuclear) processes, yet it is common to refer to *energy shortages* or an *energy crisis*. How can we have a shortage of a quantity that is conserved? Of course we don't actually have a shortage of energy. Rather, we are running out of various resources that have the 'capability to do work' (e.g., oil). The 'capability to do work' is not conserved and it is always reduced by any real process.

The thermodynamic definition of energy is complex. We saw in Chapter 5 that energy has both quantity and quality and the definition of entropy in Chapter 6 attempts to provide a scale for the quality of energy. We know that the quality of energy transferred as work is higher than the quality of the same amount of energy transferred as heat. The difference in quality is evident when we recognize that work can be continuously converted with 100% efficiency into heat, but the reverse process is not possible.

Since heat and work reflect different energy qualities, it does not make sense to directly compare them. However, this is exactly what we do in the usual definition of the efficiency of energy conversion processes. For example, a power plant that produces electrical power at a rate of $\dot{W}_{out}$ from a heat transfer $\dot{Q}_H$ operates with an efficiency that is defined according to:

$$\eta = \frac{\dot{W}_{out}}{\dot{Q}_H} \tag{7-1}$$

The efficiency of a conventional coal-fired power plant may be about 35%; this low value may lead you to conclude that the plant is woefully inefficient. However, we know that the maximum possible efficiency of this steady-state energy conversion process is limited by the Second Law of Thermodynamics. If, for example, the heat is provided to the power plant from a thermal reservoir at temperature T_H and heat is rejected from the power plant to a thermal reservoir at temperature T_C, then the maximum possible efficiency of the power plant is:

$$\eta_{max} = 1 - \frac{T_C}{T_H} \tag{7-2}$$

as shown in Chapter 5. Equation (7-2) shows that the efficiency of the power plant can never approach 100% for any finite values of T_H and T_C. The thermal efficiency defined in Eq. (7-1) is not entirely meaningful because it does not approach 100%, even in the limit of a thermodynamically perfect process. The concepts introduced in this chapter provide a way to define a more meaningful, thermodynamic efficiency that is referred to as the *Second Law efficiency*. A general technique for evaluating the Second Law efficiency of energy conversion processes is presented and a method of quantifying the work producing capability that is lost due to irreversible processes is developed.

The thermodynamic concept *exergy* (also called *availability* or *available energy*) is defined as the capability to do useful work using thermodynamically perfect (i.e., reversible) processes. The capability of a system to do useful work is related to its energy and its entropy; exergy quantifies this work production capability into a single thermodynamic property. Exergy is not a conserved quantity; some exergy is destroyed in all real processes. Exergy destruction is related to entropy generation. Exergy must be destroyed in order to allow processes to proceed at a finite rate. All else being the same, the faster a process proceeds, the greater will be the rate of exergy destruction.

The concept of exergy is useful in several ways. Exergy provides a rational basis for the valuation of fuels and resources and it allows us to rank the processes occurring within a system according to how they contribute to the loss of work producing capability. Exergy also allows the calculation of meaningful process efficiencies. The Second Law efficiency of a process or system, η_2, can be defined as the ratio of the exergy of the desired output of the system to the exergy that is supplied to the system:

$$\eta_2 = \frac{\text{exergy of desired output}}{\text{exergy supplied}} \tag{7-3}$$

Because exergy is defined as the capability to produce work, the Second Law efficiency can approach unity for a thermodynamically perfect device and it therefore provides a true indication of the efficiency of the system.

7.2 Exergy of Heat

The Second Law of Thermodynamics prohibits heat from being continuously converted into work with 100% efficiency, although the conversion from heat to work at a lower efficiency may be possible. The maximum useful work that can be produced from a specified quantity of heat is defined as the *exergy of the heat*. We can quantify the relationship between exergy and heat by considering the system shown in Figure 7-1. A thermal reservoir at temperature T_H provides heat at rate $\dot{Q}_H$ to an engine that converts some of the heat into power, $\dot{W}_{out}$.

Maximum power is produced if the engine is reversible and therefore operating at its maximum efficiency. The reasoning provided in Chapter 5 demonstrated that the maximum efficiency is given by Eq. (7-2). However, we need to carefully consider the temperature of the thermal sink, T_C. According to Eq. (7-2), lower values of T_C will result in higher efficiency and greater work production. If T_C could be reduced to 0 K,

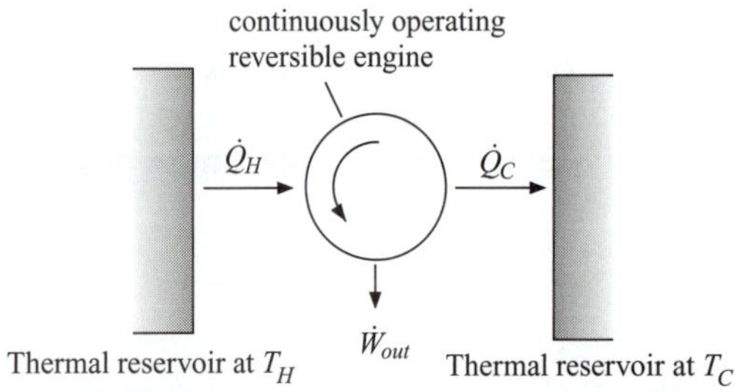

Figure 7-1: Thermal reservoir providing heat at rate $\dot{Q}_H$ at temperature T_H to a reversible, continuously operating engine.

then the maximum efficiency would reach 100%; but where would one find a convenient thermal sink at 0 K?

The lowest practical temperature that heat can be rejected to is the temperature of the environment, T_0. The temperature of the environment is referred to as the *dead state temperature*. The term 'dead state' is related to the fact that the work producing capacity of heat obtained from the environment is zero. If $T_H = T_0$ and $T_C = T_0$ then Eq. (7-2) shows that the maximum possible efficiency of any process that uses heat obtained from the environment must be 0%. Heat delivered at T_0 is effectively 'dead' as it has no ability to produce useful work.

The maximum power that can be produced using a heat transfer rate $\dot{Q}_H$ delivered at temperature T_H with heat rejection to the dead state is:

$$\dot{Q}_H\left(1-\frac{T_0}{T_H}\right) \tag{7-4}$$

The maximum possible power is, by definition, the exergy of the heat transfer rate. In general, the rate of exergy transfer resulting from a heat transfer rate $\dot{Q}$ that crosses a boundary at temperature T is given by:

$$\boxed{\dot{X}_Q = \dot{Q}\left(1-\frac{T_0}{T}\right)} \tag{7-5}$$

Equation (7-5) defines the flow rate of exergy, $\dot{X}_Q$ that is associated with a rate of heat transfer. Note that the environment is effectively a thermal reservoir at temperature T_0 so the exergy of a heat flow to or from the environment is zero.

A more detailed analysis is required if the temperature at the system boundary where heat crosses is not constant. For example, consider fluid flowing through a pipe as shown in Figure 7-2(a). Fluid enters the pipe at T_H and is cooled to T_C by heat transfer to the environment at T_0. The temperature and local heat flux ($\dot{Q}''$, the rate of heat transfer per unit area) both vary with position when the system boundary is selected to coincide with the surface of the pipe, as shown in Figure 7-2(a).

In order to evaluate the exergy of the heat transfer passing through the system boundary shown in Figure 7-2(a), it is necessary to integrate the exergy per unit area over the entire area of the system,

$$\dot{X}_Q = \int_A \dot{Q}''\left(1-\frac{T_0}{T}\right) dA \tag{7-6}$$

where $\dot{Q}''$ is the local heat flux at the boundary, T is the local temperature at the boundary, and A is the surface area of the system.

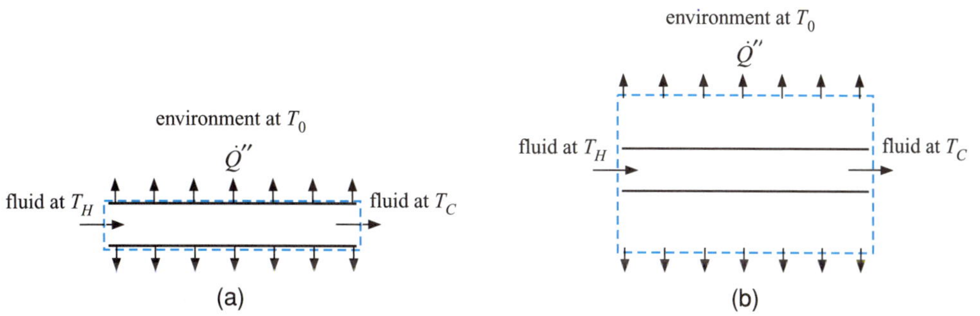

Figure 7-2: Heat transfer from fluid flowing through a pipe analyzed using a system that has its boundaries (a) at the surface of the pipe and (b) far from the surface of the pipe.

Note that the integral in Eq. (7-6) can be simplified by changing the choice of the system boundary. For example, Figure 7-2(b) shows an alternative system in which the boundary is selected so that it is far from the surface of the pipe and therefore at the dead-state temperature at the locations where heat transfer occurs. The evaluation of the exergy transfer associated with the heat transfer through the boundary of the system in Figure 7-2(b) using Eq. (7-6) will result in zero because heat transfer at T_0 has no exergy. The heat transfer from the pipe surface had some exergy at its source, but all of the exergy was destroyed once it reached the environment due to the irreversible heat transfer processes that occurred within the system.

EXAMPLE 7.2-1: SECOND LAW EFFICIENCY

A power plant receives two heat inputs, as shown in Figure 1.

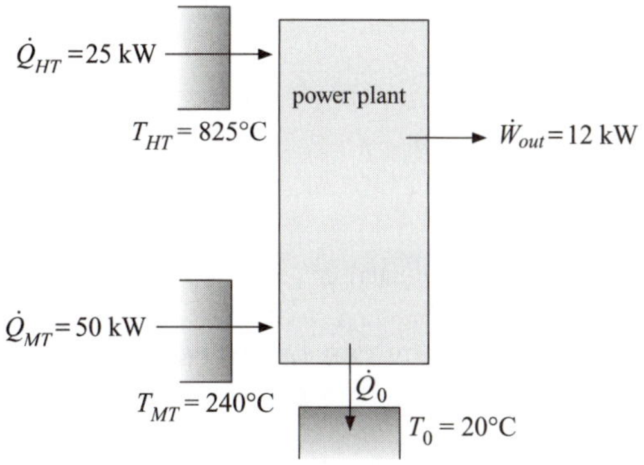

Figure 1: Power plant with two heat inputs.

The plant receives $\dot{Q}_{HT} = 25$ kW at temperature $T_{HT} = 825°\text{C}$. The plant also receives $\dot{Q}_{MT} = 50$ kW at temperature $T_{MT} = 240°\text{C}$. The plant rejects heat to the environment at $T_0 = 20°\text{C}$ and produces power $\dot{W}_{out} = 12$ kW.

a) Determine the Second Law efficiency of the power plant.

The inputs are entered in EES:

```
$UnitSystem SI K Pa J mass radian
Q_dot_HT=25 [kW]*convert(kW,W)          "rate of heat transfer from the high temperature reservoir"
Q_dot_MT=50 [kW]*convert(kW,W)          "rate of heat transfer from the mid-temperature reservoir"
T_HT=converttemp(C,K,825 [C])           "temperature of high temperature reservoir"
T_MT=converttemp(C,K,240 [C])           "temperature of mid-temperature reservoir"
T_0=converttemp(C,K,20 [C])             "dead state temperature"
W_dot_out=12 [kW]*convert(kW,W)         "power produced"
```

The rate at which exergy enters the plant with the heat transfer from the reservoir at T_{HT} is:

$$\dot{X}_{Q_{HT}} = \dot{Q}_{HT}\left(1 - \frac{T_0}{T_{HT}}\right)$$

The rate at which exergy enters the plant with the heat transfer from the reservoir T_{MT} is:

$$\dot{X}_{Q_{MT}} = \dot{Q}_{MT}\left(1 - \frac{T_0}{T_{MT}}\right)$$

```
X_dot_Q_HT=Q_dot_HT*(1-T_0/T_HT)        "rate of exergy associated Q_dot_HT from T_HT"
X_dot_Q_MT=Q_dot_MT*(1-T_0/T_MT)        "rate of exergy associated Q_dot_MT from T_MT"
```

The Second Law efficiency of the system is given by Eq. (7-3). The desired exergy output is the net power output of the plant, $\dot{W}_{out}$. The rate of exergy entering the plant is the sum of the rates of exergy associated with the heat transfers at T_{HT} and T_{MT}. There is no exergy leaving the plant with the heat flow to the environment because the exergy associated with the heat transfer transferred from the plant to the surroundings at T_0 is zero. Therefore, the Second Law efficiency of the system is:

$$\eta_2 = \frac{\dot{W}}{\dot{X}_{Q_{HT}} + \dot{X}_{Q_{MT}}}$$

```
eta_2=W_dot_out/(X_dot_Q_HT+X_dot_Q_MT)          "second law efficiency"
```

which provides $\eta_2 = 30.2\%$. The Second Law efficiency can be thought of as the ratio of the actual power produced to the maximum possible power that could be produced.

It is possible to solve this problem without using the concept of exergy. We can determine the maximum power, which is the power that would be produced by a reversible power plant operating under these conditions and use this value to compute a Second Law efficiency. The answer determined in this manner will be identical to the result obtained using an exergy analysis. Figure 2(a) illustrates an energy balance on a reversible power plant that is provided with the same heat inputs at the same temperatures as the system shown in Figure 1. The energy balance is:

$$\dot{Q}_{HT} + \dot{Q}_{MT} = \dot{Q}_{0,rev} + \dot{W}_{out,rev}$$

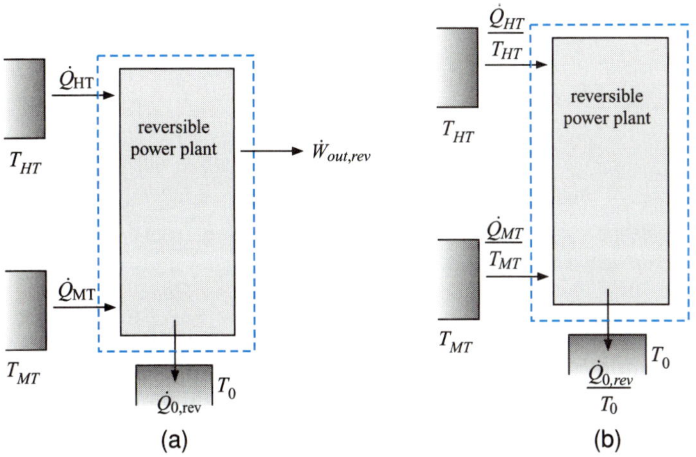

Figure 2: (a) Energy and (b) entropy balance on a reversible power plant with two heat inputs.

Figure 2(b) illustrates an entropy balance on a reversible power plant, which can be expressed as:

$$\frac{\dot{Q}_{HT}}{T_{HT}} + \frac{\dot{Q}_{MT}}{T_{MT}} = \frac{\dot{Q}_{0,rev}}{T_0}$$

The Second Law efficiency is the ratio of the actual power produced, $\dot{W}_{out}$ in Figure 1, to the maximum possible power produced, $\dot{W}_{out,rev}$ in Figure 2.

$$\eta_2 = \frac{\dot{W}_{out}}{\dot{W}_{out,rev}}$$

```
Q_dot_HT/T_HT+Q_dot_MT/T_MT=Q_dot_0_rev/T_0      "entropy balance on reversible system"
Q_dot_HT+Q_dot_MT=W_dot_out_rev+Q_dot_0_rev      "energy balance on reversible system"
eta_2_check=W_dot_out/W_dot_out_rev              "second law efficiency check"
```

which also results in $\eta_2 = 30.2\%$. Notice that it was not necessary to use the concept of exergy to calculate the Second Law efficiency. Exergy provides a convenient method of combining the First and Second Laws into a concept that is physically meaningful. However, exergy does not provide information that could not already be obtained using the First and Second Laws.

7.3 Exergy of a Flow Stream

The energy carried by a flowing stream can be converted, in part, into power. Consider, for example, a fluid with mass flow rate $\dot{m}$ at temperature T_1, pressure P_1, velocity $\tilde{V}_1$ and elevation z_1. The maximum possible rate at which useful power can be produced using this flow stream is the exergy rate associated with the flow stream, $\dot{X}_f$. To determine this exergy rate, the flow must be processed by a steady-state system that has the highest possible efficiency (i.e., it involves only reversible processes) while interacting with the environment at the dead state, as shown in Figure 7-3. The dead state corresponds to the temperature, T_0, and pressure, P_0, of the environment. It is not necessary to specify the arrangement of the equipment in the system. However, we can imagine that the system must involve reversible heat engines in order to utilize the temperature potential

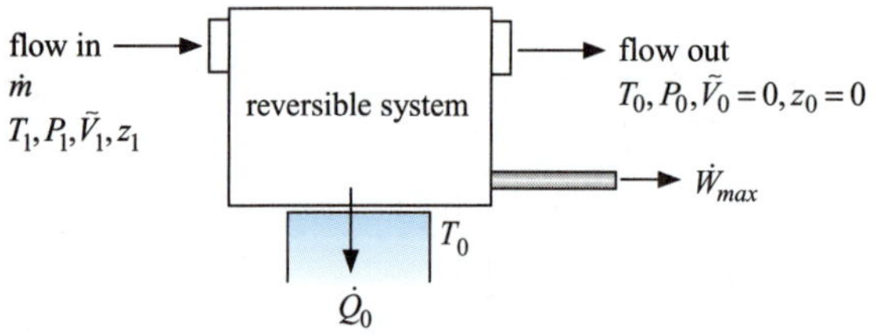

Figure 7-3: Flow stream at state 1 entering a steady-state, reversible system that interacts with the environment at the dead state and will therefore produce the maximum possible power.

between the fluid and the dead state as well as some type of mechanism that can utilize the pressure potential between the fluid and the dead state.

If the system is producing the maximum possible power, then the fluid must exit with its potentials exhausted. For example, the temperature of the exiting fluid must be the dead state temperature, T_0. If the fluid exits at any other temperature (either higher or lower) then there is a temperature difference between the fluid and the surroundings. A heat engine could utilize this temperature difference in order to produce additional power. The fluid must also exit at the dead state pressure, P_0, so that it does not exert a force relative to the surroundings that could be used to produce additional power. The kinetic and potential energies of the fluid must also be zero; therefore, the fluid exits the system at zero velocity ($\tilde{V}_0 = 0$) and elevation ($z_0 = 0$) relative to the surroundings. Of course, the velocity cannot be truly zero or the fluid would not exit the system. However, if the exit port of the system is very large then the velocity will be small and the kinetic energy of the exiting flow will be negligible. An energy balance on the system, shown in Figure 7-4(a), provides:

$$\dot{m}\left(h_1 + \frac{\tilde{V}_1^2}{2} + g\,z_1\right) = \dot{Q}_0 + \dot{W}_{max} + \dot{m}\,h_0 \tag{7-7}$$

where h_1 and h_0 are the specific enthalpies of the fluid at the inlet and dead states, respectively. The mass flow rate and the properties of the entering fluid are known. The dead state is also known and therefore the specific enthalpy of the fluid at the dead state can be determined. However, Eq. (7-7) has two unknowns, the maximum power $\dot{W}_{max}$ and the heat transfer rate from the system to the environment, $\dot{Q}_0$. An entropy balance on the system, shown in Figure 7-4(b), provides:

$$\dot{m}\,(s_1 - s_0) = \frac{\dot{Q}_0}{T_0} \tag{7-8}$$

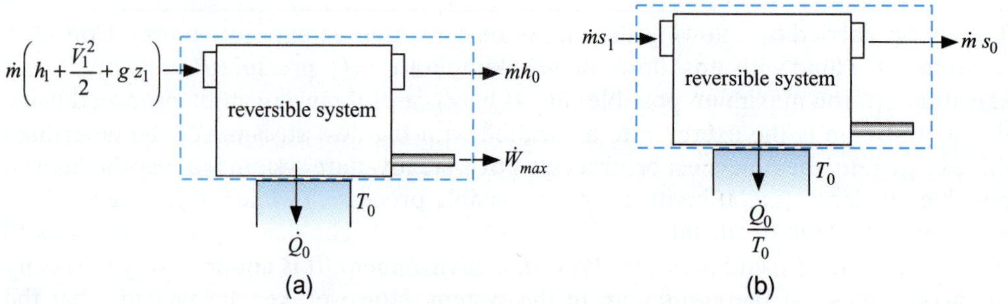

Figure 7-4: (a) Energy balance and (b) entropy balance on the system shown in Figure 7-3.

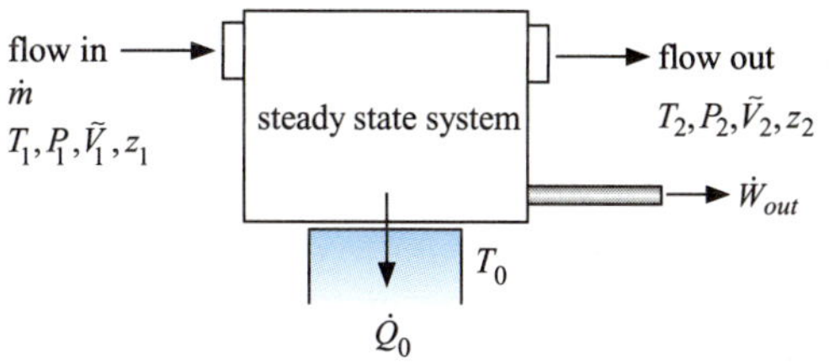

Figure 7-5: System that is using a flow stream to produce power.

where s_1 is the specific entropy of the fluid entering at temperature T_1 and pressure P_1 and s_0 is the specific entropy of the fluid at the temperature and pressure of the dead state.

The heat transfer rate can be found by solving Eq. (7-8):

$$\dot{Q}_0 = \dot{m}\, T_0 (s_1 - s_0) \tag{7-9}$$

Substituting Eq. (7-9) into Eq. (7-7) leads to:

$$\dot{m}\left(h_1 + \frac{\tilde{V}_1^2}{2} + g\, z_1\right) = \dot{m}\, T_0\,(s_1 - s_0) + \dot{W}_{max} + \dot{m}\, h_0 \tag{7-10}$$

Solving Eq. (7-10) for the maximum power that can be produced by the flow, $\dot{W}_{max}$, is equivalent to determining the exergy rate associated with the flow at state 1, $\dot{X}_{f,1}$:

$$\dot{X}_{f,1} = \dot{W}_{max} = \dot{m}\left[(h_1 - h_0) - T_0\,(s_1 - s_0) + \frac{\tilde{V}_1^2}{2} + g\, z_1\right] \tag{7-11}$$

It is usually more convenient to express the exergy rate of a flow stream on a per unit mass basis. The specific flow exergy, $x_{f,1}$, is obtained by dividing Eq. (7-11) by the mass flow rate $\dot{m}$.

$$x_{f,1} = (h_1 - h_0) - T_0(s_1 - s_0) + \frac{\tilde{V}_1^2}{2} + g\, z_1 \tag{7-12}$$

The specific flow exergy, $x_{f,1}$, is the maximum amount of useful work per unit mass that can be done by a system that is provided with a flow stream at state 1 and interacts with an environment at the dead state temperature and pressure, T_0 and P_0, respectively. The values of h_0 and s_0 that appear in Eq. (7-12) are the specific enthalpy and specific entropy that the fluid has at the dead state temperature and pressure. Once the dead state has been specified, the values of T_0, s_0, and h_0 in Eq. (7-12) are all determined. Therefore, Eq. (7-12) shows that $x_{f,1}$ is only a function of the values of other thermodynamic properties of the fluid at state 1. Therefore $x_{f,1}$ is, itself, a thermodynamic property of the fluid that can be determined based on its state.

Consider the system shown in Figure 7-5. Fluid enters the system at state 1 and exits at state 2. The system operates at steady state and produces power, $\dot{W}_{out}$. The maximum power that could be produced by this system is the product of the mass flow rate and the difference in the specific flow exergies between the inlet and exit states. The Second Law efficiency of the process, η_2, is given by:

$$\eta_2 = \frac{\dot{W}_{out}}{\dot{m}(x_{f,1} - x_{f,2})} \tag{7-13}$$

Equation (7-13) is consistent with the definition of the Second Law efficiency in Eq. (7-3).

EXAMPLE 7.3-1: HEATING SYSTEM

On a winter day, outdoor air at $T_0 = 20°\text{F}$ and $P_0 = 1$ atm is drawn through a heat exchanger at a steady flow rate of $\dot{V} = 370$ cfm. The air is heated to $T_1 = 110°\text{F}$ at constant pressure, $P_1 = P_0$, and discharged to a building in order to provide heating. The building temperature is maintained at $T_b = 75°\text{F}$. The energy transferred to the air, $\dot{Q}$, is provided by combustion of a fuel. We will represent the heat obtained from the combustion process as heat transfer from a thermal reservoir at a constant temperature of $T_H = 1900°\text{F}$, as shown in Figure 1.

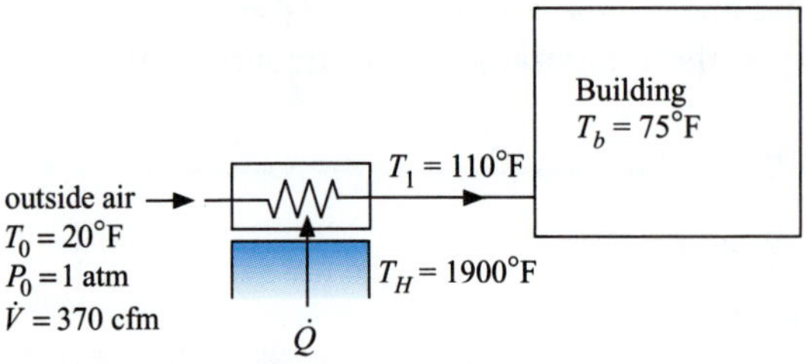

Figure 1: Building heating process using a thermal reservoir at 1900°F.

a) Determine the rate at which heat is provided from the thermal reservoir.

The inputs are entered in EES:

```
$UnitSystem SI K Pa J
T_0=convertTemp(F,K,20 [F])                 "dead state temperature"
T_1=convertTemp(F,K,110 [F])                "exit temperature of the air"
P_0=1 [atm]*convert(atm,Pa)                 "dead state pressure"
V_dot=370 [cfm]*convert(cfm,m^3/s)          "volumetric flow rate of entering air"
T_b=converttemp(F,K,75 [F])                 "temperature of building"
T_H=convertTemp(F,K,1900 [F])               "temperature of the thermal reservoir"
```

The dead state, state 0, is fixed by the temperature and pressure. The properties at state 0 (h_0, s_0, and v_0) are determined. The mass flow rate entering the heat exchanger is:

$$\dot{m} = \frac{\dot{V}}{v_0}$$

```
h_0=enthalpy(Air,T=T_0)              "specific enthalpy of the inlet air"
s_0=entropy(Air,T=T_0,P=P_0)         "specific entropy of the inlet air"
v_0=volume(Air,T=T_0,P=P_0)          "specific volume of the inlet air"
m_dot=V_dot/v_0                      "mass flow rate of entering air"
```

State 1 is fixed by the temperature and pressure. The properties at state 1 (h_1 and s_1) are determined. An energy balance on the heat exchanger in Figure 1 is:

$$\dot{m}(h_0 - h_1) + \dot{Q} = 0$$

```
h_1=enthalpy(Air,T=T_1)              "specific enthalpy of the air exiting the heat exchanger"
s_1=entropy(Air,T=T_1,P=P_0)         "specific entropy of the exit air"
m_dot*(h_0-h_1)+Q_dot=0              "energy balance on the heat exchanger"
```

The heat provided to the air from the thermal reservoir is $\dot{Q} = 11{,}617$ W.

b) What is the rate at which exergy is provided to the system from the thermal reservoir?

The rate at which exergy is provided to the system is given by Eq. (7-5):

$$\dot{X}_Q = \dot{Q}\left(1 - \frac{T_0}{T_H}\right)$$

where T_0 is the dead state temperature.

```
X_dot_Q=Q_dot*(1-T_0/T_H)          "rate of exergy provided by thermal reservoir"
```

The rate at which exergy is provided is $\dot{X}_Q = 9{,}255$ W; 79.6% of the heat from the thermal reservoir is exergy because the temperature of the reservoir is high relative to the dead state temperature.

c) What is the increase in the rate of exergy carried by the air stream?

The specific exergy of a flow stream is given in Eq. (7-12) where the dead state condition is the state of the outdoor air. Therefore, the specific exergy of the flow entering the heat exchanger at the dead state must be zero. The specific exergy of the air leaving the heat exchanger at state 1 is:

$$x_{f,1} = h_1 - h_0 - T_0(s_1 - s_0)$$

```
x_f_1=h_1-h_0-T_0*(s_1-s_0)          "specific exergy of air exiting the heat exchanger"
```

The increase in the rate of exergy carried by the air is:

$$\Delta\dot{X}_{f,air} = \dot{m}\,x_{f,1}$$

```
DELTAX_dot_f_air=m_dot*x_f_1          "rate of exergy increase of the air"
```

which results in $\Delta\dot{X}_{f,air} = 965.1$ W.

d) Determine the Second Law efficiency for this process.

The Second Law efficiency is defined in Eq. (7-3) as the ratio of the exergy of the desired output to the exergy provided to the system. In this case, the desired output is the increase in the rate at which exergy is carried by the air and the exergy provided to the system is associated with the heat transfer from the thermal reservoir.

$$\eta_2 = \frac{\Delta\dot{X}_{f,air}}{\dot{X}_Q}$$

```
eta_2=DELTAX_dot_f_air/X_dot_Q          "Second Law efficiency"
```

The Second Law efficiency of the process is $\eta_2 = 0.104$. This process is highly irreversible because of the large temperature difference that exists between the thermal reservoir and the air being heated. Therefore, the process is characterized by a high rate of exergy destruction which results in the low Second Law efficiency.

e) The fact that the Second Law efficiency is so low indicates that the space heat could be provided to the building much more efficiently. One possibility is to use a heat pump driven by a heat engine, as shown in Figure 2.

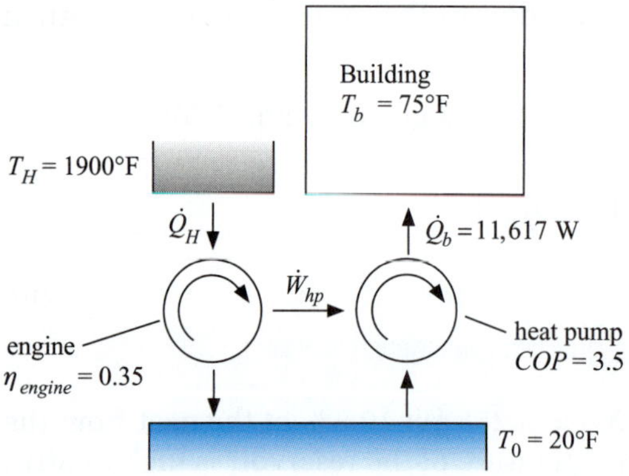

Figure 2: Space heating system employing heat engine and heat pump.

Heat at rate $\dot{Q}_H$ from the high temperature reservoir at T_H is used to drive a heat engine. The First Law efficiency of the engine is $\eta_{engine} = 0.35$. The power produced by the engine is used to run a heat pump that moves heat from the environment at T_0 to the building at temperature $T_b = 75^\circ$F. The rate of heat transfer to the building is $\dot{Q}_b = 11{,}617$ W. Note that heat is provided to the building at the same temperature and rate as in Figure 1. The First Law efficiency of the heat pump is specified in terms of its Coefficient of Performance (*COP*), which is defined as the ratio of the heat provided, $\dot{Q}_b$, to the power required, $\dot{W}_{hp}$. The Coefficient of Performance of the heat pump is $COP = 3.5$. Determine the First Law and the Second Law efficiencies of the heating system shown in Figure 2.

The additional inputs are entered in EES:

```
eta_engine=0.35 [-]        "efficiency of engine"
COP_hp=3.5 [-]             "coefficient of performance of the heat pump"
Q_dot_b=11617 [W]          "rate of heat transfer required by building"
```

The power required by the heat pump is determined using the definition of the heat pump *COP*:

$$\dot{W}_{hp} = \frac{\dot{Q}_b}{COP}$$

The rate of heat transfer required by the engine is determined using the definition of the engine efficiency:

$$\dot{Q}_H = \frac{\dot{W}_{hp}}{\eta_{engine}}$$

```
W_dot_hp=Q_dot_b/COP_hp           "power required by the heat pump"
Q_dot_H=W_dot_hp/eta_engine       "rate of heat transfer required by the engine"
```

The First Law efficiency of the system is the ratio of the heat provided to the building to the heat obtained from the thermal reservoir at T_H.

$$\eta_1 = \frac{\dot{Q}_b}{\dot{Q}_H}$$

```
eta_1=Q_dot_b/Q_dot_H        "First Law Efficiency"
```

which results in $\eta_1 = 1.225$. This is a very good efficiency for a space heating application (it is analogous to having a furnace with an efficiency of 122.5%) that can be achieved with existing equipment. Because the heat pump moves a portion of the energy needed to heat the building from outdoors, the amount of fuel energy that must be provided is less than the heat provided to the house. An economic analysis is needed to determine whether the increased first cost of the system is justified by the reduction in the fuel expense.

The Second Law efficiency is defined as the ratio of the exergy associated with the heat provided to the building:

$$\dot{X}_{Q_b} = \dot{Q}_b \left(1 - \frac{T_0}{T_b}\right)$$

to the exergy associated with the heat obtained from the high temperature reservoir:

$$\dot{X}_{Q_H} = \dot{Q}_H \left(1 - \frac{T_0}{T_H}\right)$$

The Second Law efficiency is therefore:

$$\eta_2 = \frac{\dot{X}_{Q_b}}{\dot{X}_{Q_H}}$$

```
X_dot_Q_b=Q_dot_b*(1-T_0/T_b)        "rate of exergy provided to building"
X_dot_Q_H=Q_dot_H*(1-T_0/T_H)        "rate of exergy provided by thermal reservoir"
eta_2_alt=X_dot_Q_b/X_dot_Q_H        "Second Law efficiency of alternative system"
```

which is $\eta_2 = 0.158$. This result is 50% higher than the Second Law efficiency of the original heating system shown in Figure 1.

7.4 Exergy of a System

Figure 7-6 shows a non-reacting system of fixed mass that is at a uniform temperature T_1 and pressure P_1. Although the temperature of the system differs from the temperature of the surroundings, T_0, heat transfer between the system and the surroundings cannot occur due to the perfect insulation. The pressure in the system also differs from the pressure of the surroundings, P_0. However, the piston cannot move because it is locked in place. The system is in a state of constrained equilibrium.

This system is capable of doing some work because it has properties (T_1 and P_1) that differ from those of the surroundings. The exergy of the system (X_s) is defined as the maximum amount of useful work that this system can do by interacting with the dead state. In order to determine the exergy of the system, it is necessary to imagine a process in which reversible energy conversion equipment is utilized in order to exploit

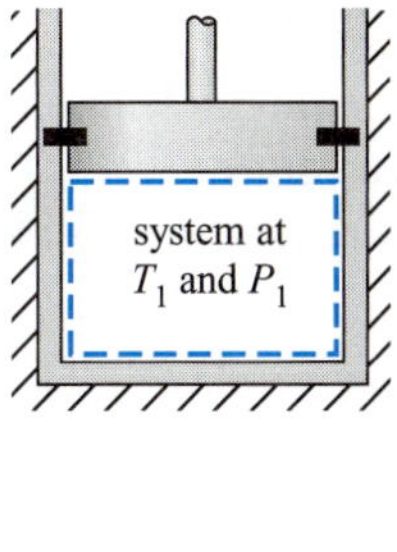

Figure 7-6: System at temperature T_1 and pressure P_1 in a state of constrained equilibrium.

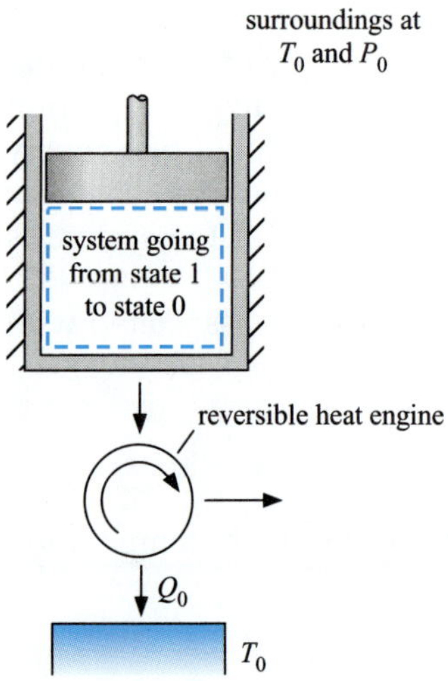

Figure 7-7: Extracting the maximum amount of useful work from the system shown in Figure 7-6.

the differences in temperature and pressure that exist between the system and the surroundings. For example, we can replace the piston locks with a frictionless linkage that always matches the force exerted by the system on the piston. Any movement of the piston is used to lift a weight and therefore accomplishes some work. The process will continue until the pressure of the gas in the system is the same as the dead state pressure. The temperature difference between the system and the surrounding can also be exploited to produce work by allowing heat to transfer through a reversible heat engine that rejects thermal energy to the surroundings. At the conclusion of these processes, the temperature and pressure of the system must be equal to the dead state temperature and pressure and therefore no further work can be done. Such a system is shown in Figure 7-7.

Figure 7-8(a) illustrates an entropy balance on the system shown in Figure 7-7 and the reversible heat engine. Notice that there is no entropy generation within the reversible system that is required in order to extract the maximum useful work from the system. The entropy balance is:

$$0 = \frac{Q_0}{T_0} + m(s_0 - s_1) \tag{7-14}$$

where Q_0 is the heat rejected from the heat engine to the environment, s_1 is the specific entropy of the system at its initial state, s_0 is the specific entropy of the system at the dead state (i.e., the conclusion of the process), and m is the mass of the system. Figure 7-8(b) illustrates an energy balance on the system:

$$0 = Q_0 + W_p + W_{engine} + W_{sur} + m(u_0 - u_1) \tag{7-15}$$

where W_p is the work done by the mechanism connected to the piston rod, W_{engine} is the work done by the heat engine, W_{sur} is the work done on the surroundings, u_1 is the initial specific internal energy of the system, and u_0 is the internal energy of the system at the dead state.

Equation (7-14) is solved for the heat transfer to the environment:

$$Q_0 = -m\, T_0(s_0 - s_1) \tag{7-16}$$

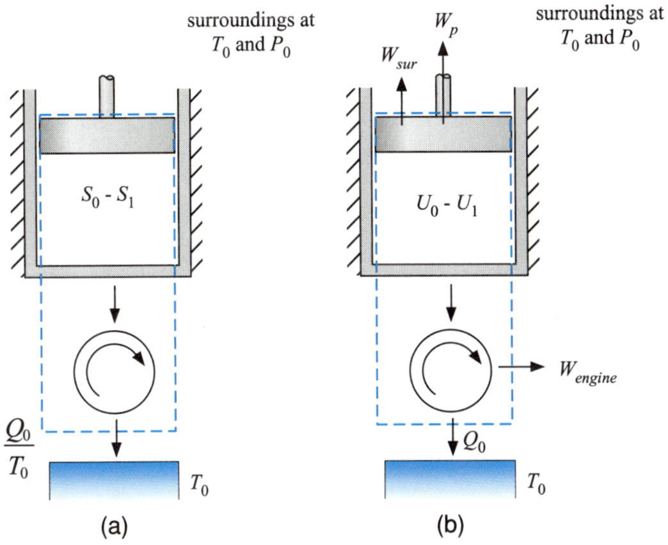

Figure 7-8: (a) Entropy balance and (b) energy balance on a reversible system that is used to extract the maximum useful work from the system shown in Figure 7-6.

Equation (7-16) is substituted into Eq. (7-15):

$$0 = -m\,T_0(s_0 - s_1) + \underbrace{W_p + W_{engine}}_{W_{max}} + W_{sur} + m(u_0 - u_1) \tag{7-17}$$

The useful work accomplished by the process, W_{max}, is the sum of the work done by the mechanism connected to the piston rod and the work produced by the heat engine. Equation (7-17) is solved for W_{max}:

$$W_{max} = m(u_1 - u_0) - m\,T_0(s_1 - s_0) - W_{sur} \tag{7-18}$$

The work done by the system on the surroundings is, by definition, not useful. This work is calculated according to:

$$W_{sur} = \int_{V_1}^{V_0} P_0\,dV \tag{7-19}$$

where V_1 is the initial volume of the system and V_0 is the volume that the system has when it is at the dead state temperature and pressure. The dead state pressure is constant and therefore Eq. (7-19) can be written as:

$$W_{sur} = P_0(V_0 - V_1) \tag{7-20}$$

The volume of the system is expressed as the product of the mass and the specific volume of the system:

$$W_{sur} = m\,P_0(v_0 - v_1) \tag{7-21}$$

Substituting Eq. (7-21) into Eq. (7-18) provides:

$$W_{max} = m(u_1 - u_0) - m\,T_0(s_1 - s_0) + m\,P_0(v_1 - v_0) \tag{7-22}$$

Equation (7-22) is the maximum useful work that can be obtained from a system (i.e., the exergy of the system) at state 1:

$$X_{s,1} = W_{max} = m\left[(u_1 - u_0) - T_0\,(s_1 - s_0) + P_0\,(v_1 - v_0)\right] \tag{7-23}$$

The exergy of a system is expressed on a unit mass basis by dividing Eq. (7-23) by the mass of the system:

$$x_{s,1} = \frac{X_{s,1}}{m} = (u_1 - u_0) - T_0(s_1 - s_0) + P_0(v_1 - v_0) \tag{7-24}$$

The specific exergy, $x_{s,1}$, is the maximum useful work per unit mass that a system at state 1 can accomplish. The dead state is specified; therefore, the quantities u_0, s_0, T_0, and P_0 in Eq. (7-24) are all constants and the specific exergy of the mass at state 1 is a function of other thermodynamic properties at state 1. The specific exergy is therefore also a thermodynamic property of the system. Equation (7-24) was derived assuming that the kinetic energy and potential energy of the system in its initial state are negligible. If these external energy forms are also considered then the specific exergy of the system is given by:

$$\boxed{x_{s,1} = (u_1 - u_0) - T_0(s_1 - s_0) + P_0(v_1 - v_0) + \frac{\tilde{V}_1^2}{2} + g\, z_1} \tag{7-25}$$

EXAMPLE 7.4-1: COMPRESSED AIR POWER SYSTEM

The volumetric energy density of lithium-ion batteries is approximately 250 W-hr/liter. This energy density limits the deployment of many power-consuming devices that might otherwise be useful for dismounted soldiers and emergency personnel. Therefore, there is substantial research into alternative power systems that can achieve higher volumetric energy density. One system that has been proposed is shown in Figure 1.

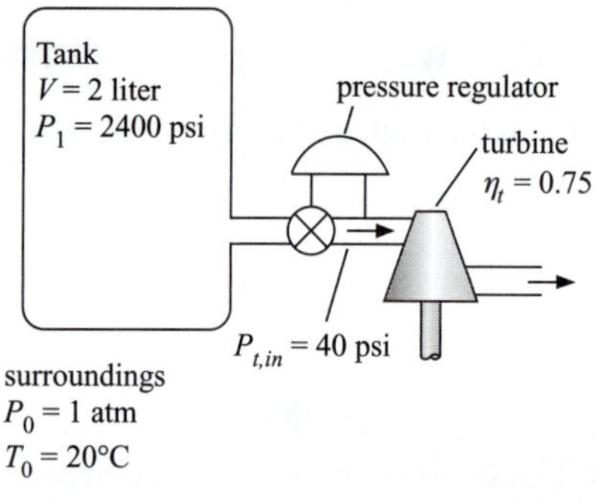

Figure 1: Proposed compressed air power system.

Air is stored in a $V = 2$ liter compressed air tank that is initially charged to $P_1 = 2400$ psi. The air is drawn from the tank through a pressure regulator and then it enters a turbine. The regulator maintains the inlet pressure to the turbine at $P_{t,in} = 40$ psi regardless of the tank pressure. Because the process occurs very slowly, the air in the tank is always maintained at the temperature of the surroundings, $T_0 = 20^\circ$C. The turbine exhausts to ambient pressure, $P_0 = 1$ atm. The isentropic efficiency of the turbine is $\eta_t = 0.75$. The process ends when the tank pressure reaches the turbine inlet pressure. Although high pressures are involved, assume air obeys the ideal gas law.

a) Determine the volumetric energy density of the power system. The volumetric energy density is defined as the ratio of the total work produced to the volume of the consumable energy source (in this case, the volume of the tank).

The inputs are entered in EES:

```
$UnitSystem SI K Pa J Mass Radian
P_1=2400 [psi]*convert(psi,Pa)              "initial tank pressure"
P_t_in_psi=40 [psi]                         "turbine inlet pressure, in psi"
P_t_in=P_t_in_psi*convert(psi,Pa)           "turbine inlet pressure"
P_0=1 [atm]*convert(atm,Pa)                 "environment pressure"
T_0=converttemp(C,K,20 [C])                 "environment temperature"
Vol=2 [liter]*convert(liter,m^3)            "volume of tank"
eta_t=0.75 [-]                              "isentropic turbine efficiency"
```

State 1, the state of the air initially in the tank, is fixed by the temperature, $T_1 = T_0$, and the pressure. The properties (v_1, u_1, and s_1) are calculated. The mass of air initially in the tank is:

$$m_1 = \frac{V}{v_1}$$

```
"state 1"
T_1=T_0                                     "initial temperature"
v_1=volume(Air,P=P_1,T=T_1)                 "specific volume"
u_1=intenergy(Air,T=T_1)                    "specific internal energy"
s_1=entropy(Air,P=P_1,T=T_1)                "specific entropy"
m_1=Vol/v_1                                 "initial mass of air in the tank"
```

State 2, the state of the air in the tank at the end of the process, is specified by the temperature, $T_2 = T_0$, and pressure, $P_2 = P_{t,in}$. The properties (v_2, u_2, and s_2) are calculated. The final mass of air in the tank is:

$$m_2 = \frac{V}{v_2}$$

A mass balance on the tank provides the mass of air that leaves the tank and passes through the turbine (m_{out}).

$$0 = m_{out} + m_2 - m_1$$

```
"state 2"
T_2=T_0                                     "final temperature"
P_2=P_t_in                                  "final pressure"
v_2=volume(Air,P=P_2,T=T_2)                 "specific volume"
u_2=intenergy(Air,T=T_2)                    "specific internal energy"
s_2=entropy(Air,P=P_2,T=T_2)                "specific entropy"
m_2=Vol/v_2                                 "initial mass of air in the tank"
0=m_out+m_2-m_1                             "mass balance"
```

An energy balance on the regulator provides:

$$h_{v,in} = h_{t,in}$$

where $h_{v,in}$ is the specific enthalpy of the air leaving the tank and $h_{t,in}$ is the specific enthalpy of the air entering the turbine. The air leaving the tank is always at temperature T_0. Because the specific enthalpy of an ideal gas is only a function of temperature, the specific enthalpy of the air entering the turbine is constant and equal to its value at temperature T_0. The state of the air entering the turbine is fixed by the specific enthalpy and pressure. The specific entropy of the air entering the turbine ($s_{t,in}$) is determined at these conditions. The exit pressure of the turbine is atmospheric, $P_{t,out} = P_0$. An entropy balance on a reversible turbine provides:

$$s_{t,in} = s_{s,t,out}$$

where $s_{s,t,out}$ is the specific entropy of air leaving the reversible turbine. The state of the air leaving a reversible turbine is fixed by the specific entropy and pressure. The specific enthalpy at the exit of the reversible turbine ($h_{s,t,out}$) is determined. An energy balance on the reversible turbine leads to:

$$h_{t,in}\, m_{out} = W_{s,t} + h_{s,t,out}\, m_{out}$$

where $W_{s,t}$ is the total amount of work produced by the reversible turbine. The work produced by the actual turbine is:

$$W_t = W_{s,t}\, \eta_t$$

```
"turbine inlet"
h_t_in=enthalpy(Air,T=T_0)                      "inlet specific enthalpy"
s_t_in=entropy(Air,h=h_t_in,P=P_t_in)           "inlet specific entropy"
P_t_out=P_0                                     "exit pressure"
s_s_t_out=s_t_in                                "entropy balance on reversible turbine"
h_s_t_out=enthalpy(Air,P=P_t_out,s=s_s_t_out)   "specific enthalpy of air leaving reversible turbine"
h_t_in*m_out=W_s_t+h_s_t_out*m_out              "energy balance on reversible turbine"
W_t=eta_t*W_s_t                                 "actual turbine work"
```

The volumetric energy density of the device is:

$$VED = \frac{W_t}{V}$$

```
VED=W_t/Vol                                     "volumetric energy density"
VED_WhrpL=VED*convert(J/m^3,W-hr/liter)         "in W-hr/liter"
```

which leads to VED = 2.95 W-hr/liter. This is nearly two orders of magnitude smaller than batteries and therefore it is unlikely that this system will be practical.

b) Plot the volumetric energy density as a function of the turbine inlet pressure.

Figure 2 illustrates the volumetric energy density as a function of the turbine inlet pressure. Notice that there exists an optimal turbine inlet pressure that maximizes the volumetric energy density. This behavior occurs because a large turbine inlet pressure results in a lot of high pressure gas (and therefore exergy) left in the tank. However, a small turbine inlet pressure results in a large amount of exergy destruction in the valve. The optimal turbine inlet pressure balances these effects.

c) Determine the total exergy associated with the high pressure air that is initially in the tank and, based on this, the maximum possible volumetric energy density that could be obtained by a system using this air.

The properties at the dead state (v_0, u_0, and s_0) are obtained. The total exergy of the air that is initially in the tank is obtained by multiplying the specific exergy, Eq. (7-24), with the initial mass of air in the tank:

$$X_{s,1} = m_1[(u_1 - u_0) - T_0(s_1 - s_0) + P_0(v_1 - v_0)]$$

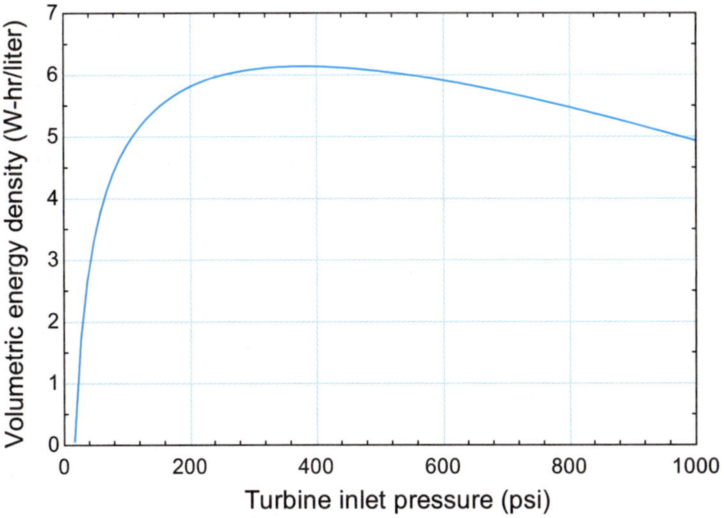

Figure 2: Volumetric energy density as a function of turbine inlet pressure.

The maximum possible volumetric energy density is:

$$VED_{max} = \frac{X_{s,1}}{V}$$

```
"dead state"
v_0=volume(Air_ha,P=P_0,T=T_0)                  "specific volume at the dead state"
u_0=intenergy(Air_ha,P=P_0,T=T_0)               "specific internal energy at the dead state"
s_0=entropy(Air_ha,P=P_0,T=T_0)                 "specific entropy at the dead state"
X_s_1=m_1*(u_1-u_0-T_0*(s_1-s_0)+P_0*(v_1-v_0)) "exergy of air in the tank"
VED_max=X_s_1/Vol                               "maximum possible volumetric energy density"
VED_max_WhrpL=VED_max*convert(J/m^3,W-hr/liter) "in W-hr/liter"
```

which results in $VED_{max} = 37.5$ W-hr/liter. This energy density is still only 16% of what can be achieved using batteries.

7.5 Exergy Balance

Exergy is defined as the capability to do useful work and it corresponds to the quantity that most people think of when they use the word 'energy' in everyday language. Exergy is not conserved; it is literally consumed or destroyed in order to allow processes to proceed at finite rates. All else being the same, the faster a process occurs, the greater the rate of exergy destruction. Exergy can never be generated, it can only be destroyed. It is useful to calculate the amount or the rate of *exergy destruction* in order to understand or to optimize thermodynamic systems. The best way to determine the amount of exergy destruction is with an exergy balance.

The first step in setting up an exergy balance is to define the system. Once the system boundaries have been identified, the general balance equation introduced in Section 1.4:

$$In + Generated = Out + Destroyed + Stored \tag{7-26}$$

is written for exergy. Note that exergy can never be generated and therefore the exergy balance on a rate basis can be written as:

$$\dot{X}_{in} = \dot{X}_{out} + \dot{X}_{des} + \frac{dX_s}{dt} \tag{7-27}$$

where $\dot{X}_{in}$ and $\dot{X}_{out}$ are the rates at which exergy enters and leaves the system, $\dot{X}_{des}$ is the rate of exergy destruction (which is also called the *irreversibility rate*) and the final term is the rate at which exergy is stored in the system. Exergy can cross the boundary of a system with heat, work, or mass. For simplicity, in this section all heat transfer is defined as being positive for energy transfer into the system and all work terms are defined to be positive for energy transfer out of the system.

The exergy transfer associated with the total heat transfer into the system is defined in Section 7.2:

$$\dot{X}_{Q_{in}} = \int_A \dot{Q}''_{in} \left(1 - \frac{T_0}{T}\right) dA \tag{7-28}$$

Heat is a surface phenomenon. It is necessary to consider the entire surface area of the system, A, in the determination of the exergy flow associated with heat. The heat flux entering the system ($\dot{Q}''_{in}$) and the local boundary temperature (T) in Eq. (7-28) may vary with position, which is why Eq. (7-28) is expressed as an integral. In practice, the system boundary is usually selected so that the boundary temperature at the location where heat enters the system is uniform, which simplifies the evaluation of Eq. (7-28). For a single heat input at a fixed boundary temperature, T_b, Eq. (7-28) can be simplified to:

$$\dot{X}_{Q_{in}} = \dot{Q}_{in} \left(1 - \frac{T_0}{T_b}\right) \tag{7-29}$$

Note that if the boundary is selected so that its temperature is the same as the dead state temperature, T_0, then there is no exergy flow associated with heat transfer across the boundary.

Exergy is defined as the ability to do *useful* work. This definition stipulates that work done to expand a system against the force exerted by the environment is not useful. Therefore, it is necessary to subtract the work done by the system on the environment from the total work done by the system in order to determine the exergy flow associated with work. The rate at which work is done on the surroundings due to a volume change of the system is:

$$\dot{W}_{sur} = P_0 \frac{dV}{dt} \tag{7-30}$$

where P_0 is the dead state pressure and $\frac{dV}{dt}$ is the rate of change of the system volume. If the volume of the system is constant, as it must be in a steady-state application, the exergy flow associated with power is identical to the net power. However, if the volume of the system is changing, then the exergy flow rate associated with power (defined as being out of the system) must be defined as:

$$\dot{X}_{W_{out}} = \dot{W}_{out} - P_0 \frac{dV}{dt} \tag{7-31}$$

The specific exergy associated with a flowing stream crossing a boundary at an arbitrary state is shown in Section 7.3 to be:

$$x_f = (h - h_0) - T_0(s - s_0) + \frac{\tilde{V}^2}{2} + g\,z \tag{7-32}$$

The rate at which exergy flows into or out of an open system with mass crossing the boundary is therefore the product of the mass flow rate, $\dot{m}$, and the specific flow exergy that it has at the point where it crosses the boundary, x_f.

The specific exergy associated with a system at an arbitrary state is defined in Section 7.4.

$$x_s = (u - u_0) - T_0(s - s_0) + P_0(v - v_0) + \frac{\tilde{V}^2}{2} + g\,z \tag{7-33}$$

The rate of exergy storage is the rate of change of the product of the mass of the system and its specific exergy. With these exergy terms defined, the general exergy balance can be written as:

$$\sum_{i=1}^{\#\,\text{inlets}} \dot{m}_{in,i}\, x_{f,in,i} + \sum_{i=1}^{\#\,\text{heat terms}} \dot{X}_{Q_{in},i} = \sum_{i=1}^{\#\,\text{exits}} \dot{m}_{out,i}\, x_{f,out,i} + \sum_{i=1}^{\#\,\text{work terms}} \dot{X}_{W_{out},i} + \dot{X}_{des} + \frac{dX_s}{dt} \tag{7-34}$$

The exergy balance equation resembles the energy and entropy balances in Eqs. (4-22) and (6-84), respectively, and it is used in the same manner. In most problems, many of the terms in Eq. (7-34) will be zero, which simplifies the balance equation. For example, if the general exergy balance is applied to a system that is operating at steady-state conditions then the exergy of the system X_s is constant and its time derivative is zero. Closed systems cannot have any exergy flows associated with mass flow. The exergy transfer associated with heat transfer will, of course, be zero in an adiabatic process. The exergy transfer associated with heat transfer will also be zero in a non-adiabatic process in which heat transfer occurs only through a boundary that is at the dead state temperature.

The exergy balance is usually applied in order to determine the rate of exergy destruction, $\dot{X}_{des}$, which is also called the *irreversibility rate*. The rate of exergy destruction can be thought of as the rate at which the capability to generate useful work is being destroyed due to the irreversible processes occurring within the system. Expressed in this manner, irreversibility is an intuitive concept that has physical meaning. However, it is worth keeping in mind that exergy destruction is unavoidable as all real processes are irreversible to some extent and exergy destruction must occur in order for the process to proceed at a finite rate. It is often useful to determine the rates of exergy destruction that can be attributed to each of the components that make up a system in order to understand the most important impediments to high efficiency. Such an analysis might clearly point out where additional equipment investment could have the largest effect on performance.

EXAMPLE 7.5-1: EXERGY ANALYSIS OF A COMMERCIAL LAUNDRY FACILITY

A commercial laundry requires $\dot{m}_w = 2{,}500$ kg/hr of water at $T_2 = 65°\text{C}$ in addition to $\dot{W}_e = 32$ kW of electricity in order to operate washers and dryers. Currently, electricity is purchased from the local utility at a rate of $C_{elec} = 0.15$ \$/kW-hr. Hot water is produced on site in a steady-flow process using a boiler, as shown in Figure 1. Natural gas is combusted in the boiler, producing combustion products. Assume that the specific exergy of the natural gas is approximately equal to its lower heating value, $LHV = 5\times10^4$ J/kg. The heat transfer resulting from combustion of the natural gas is equal to the product of the lower heating value and the mass flow rate of fuel. The cost of natural gas is $C_{fuel} = 1.3$ \$/therm. The boiler efficiency is $\eta_b = 0.88$ based on the lower heating value of the fuel. The boiler takes in water at pressure $P_3 = 1250$ psi and produces steam at temperature $T_4 = 450°\text{C}$. The pressure drop associated with the flow of water through the boiler is $\Delta P_b = 430$ kPa. The steam enters a heat exchanger where it transfers heat to the supply water entering at $P_1 = 3$ atm and $T_1 = 10°\text{C}$. The condensed steam leaves the heat exchanger as saturated liquid ($x_5 = 0$). The pressure drop on the steam side of the heat exchanger is $\Delta P_{s,hx} = 350$ kPa and the pressure drop on the water side of the heat exchanger is $\Delta P_{w,hx} = 50$ kPa. The pump that circulates water between the boiler and the heat exchanger has an isentropic efficiency of $\eta_p = 0.45$. The dead state associated with the environment is $T_0 = 10$°C and $P_0 = 1$ atm.

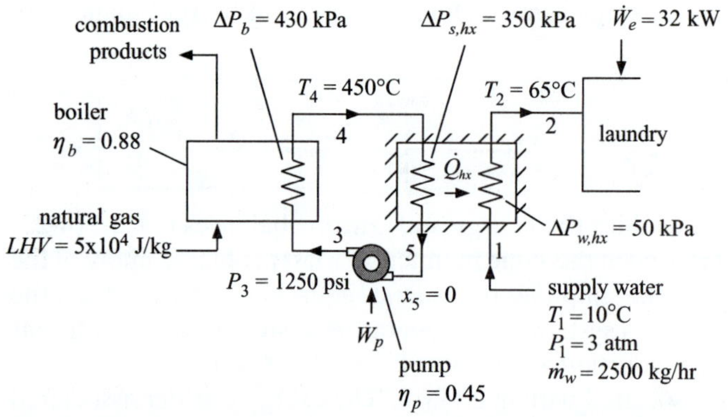

Figure 1: Boiler providing hot water to a commercial laundry.

a) Determine the Second Law efficiency of the on-site system for producing hot water.

The function x_f is developed in order to conveniently calculate the specific flow exergy of the water at the various states shown in Figure 1 with respect to the dead state. The function is placed at the top of the EES code and it takes as input parameters the specific enthalpy and specific entropy of the fluid stream. Note that the function references only variables that are passed to it as input arguments. Also, the equations used in the function are assignments rather than equalities; that is, the right hand side of each equation refers only to previously assigned variables. The specific flow exergy is computed using Eq. (7-32).

```
$UnitSystem SI K Pa J Mass Radian
function x_f(h,s)
    T_0=convertTemp(C,K,10 [C])          "dead state temperature is 10 [C]"
    P_0=1 [atm]*convert(atm,Pa)          "dead state pressure"
    h_0=enthalpy(Steam,T=T_0,P=P_0)      "dead state specific enthalpy"
    s_0=entropy(Steam,T=T_0,P=P_0)       "dead state specific entropy"
    x_f=h-h_0-T_0*(s-s_0)                "specific flow exergy"
end
```

The function x_f can be used like any mathematical function. For example, to obtain the flow exergy at a state where $h = 1 \times 10^5$ J/kg and $s = 1000$ J/kg it is only necessary to type:

```
h_1=1e5 [J/kg]                "specific enthalpy"
s_1=1e3 [J/kg-K]              "specific entropy"
x_f_1=x_f(h_1,s_1)            "specific flow exergy"
```

Notice that the addition of the function has led to an additional tab in the Solutions window; the tab labeled 'Main' contains the variables in the Equations window and the tab labeled 'x_f' contains the variables that are used in the function x_f.

The units for the variables that are used in the main EES program can be set by right-clicking each variable in the Main tab of the Solutions window in the usual manner. The units for the variables in the function x_f can be set in the 'x_f' tab so that EES will check the unit consistency of the input arguments and the equations used to implement the function. The code demonstrating the use of the function is removed and the inputs are entered in EES.

```
"Inputs"
T[1]=converttemp(C,K,10 [C])                      "temperature of supply water"
P[1]=3 [atm]*convert(atm,Pa)                      "pressure of supply water"
DP_w_hx=50 [kPa]*convert(kPa,Pa)                  "pressure drop on water side of the heat exchanger"
T[2]=converttemp(C,K,65 [C])                      "temperature of water provided to laundry"
m_dot_w=2500 [kg/hr]*convert(kg/hr,kg/s)          "mass flow rate of water provided to laundry"
eta_p=0.45 [-]                                    "efficiency of the pump"
eta_b=0.88 [-]                                    "efficiency of the boiler"
LHV=5e4 [kJ/kg]*convert(kJ/kg,J/kg)               "lower heating value of the fuel"
W_dot_e=32 [kW]*convert(kW,W)                     "electrical power consumption"
P[3]=1250 [psi]*convert(psi,Pa)                   "pressure of water provided to boiler"
T[4]=converttemp(C,K,450 [C])                     "temperature of water leaving boiler"
x[5]=0 [-]                                        "quality of water leaving heat exchanger"
DP_s_hx=350 [kPa]*convert(kPa,Pa)                 "pressure drop on steam side of heat exchanger"
DP_b=430 [kPa]*convert(kPa,Pa)                    "pressure drop associated with boiler"
C_elec=0.15 [$/kW-hr]*convert($/kW-hr,$/J)        "electricity cost"
C_fuel=1.3 [$/therm]*convert($/therm,$/J)         "fuel cost"
```

State 1 is fixed by the pressure and temperature of the supply water. The specific enthalpy, specific entropy, and specific flow exergy (h_1, s_1, and $x_{f,1}$) are determined.

```
"state 1"
h[1]=enthalpy(Water,T=T[1],P=P[1])                "specific enthalpy of supply water"
s[1]=entropy(Water,T=T[1],P=P[1])                 "specific entropy of supply water"
x_f[1]=x_f(h[1],s[1])                             "specific flow exergy of supply water"
```

The pressure at state 2 is reduced by the pressure drop on the water side of the heat exchanger:

$$P_2 = P_1 - \Delta P_{w,hx}$$

State 2 is fixed by the pressure and temperature. The specific enthalpy, specific entropy, and specific flow exergy (h_2, s_2, and $x_{f,2}$) are determined:

```
"state 2"
P[2]=P[1]-DP_w_hx                                 "pressure of water provided to laundry"
h[2]=enthalpy(Water,T=T[2],P=P[2])                "specific enthalpy of water provided to laundry"
s[2]=entropy(Water,T=T[2],P=P[2])                 "specific entropy of water provided to laundry"
x_f[2]=x_f(h[2],s[2])                             "specific flow exergy of water provided to laundry"
```

An energy balance on the water side of the heat exchanger provides:

$$\dot{Q}_{hx} = \dot{m}_w (h_2 - h_1)$$

```
Q_dot_HX=m_dot_w*(h[2]-h[1])                      "energy balance on water side of HX"
```

The pressure at state 4 is the pump exit pressure less the pressure drop across the boiler.

$$P_4 = P_3 - \Delta P_b$$

State 4 is fixed by the temperature and pressure. The specific enthalpy, specific entropy, and specific flow exergy (h_4, s_4, and $x_{f,4}$) are determined.

```
"state 4"
P[4]=P[3]-DP_b                        "pressure of steam leaving boiler"
h[4]=enthalpy(Water,T=T[4],P=P[4])    "specific enthalpy of steam leaving boiler"
s[4]=entropy(Water,T=T[4],P=P[4])     "specific entropy of steam leaving boiler"
x_f[4]=x_f(h[4],s[4])                 "specific flow exergy of steam leaving boiler"
```

The pressure at state 5 is the pressure leaving the boiler less the pressure drop on the steam side of the heat exchanger.

$$P_5 = P_4 - \Delta P_{s,hx}$$

State 5 is fixed by the pressure and quality, which allows the temperature, specific enthalpy, specific entropy, and specific flow exergy (x_5, h_5, s_5, and $x_{f,5}$) to be determined.

```
"state 5"
P[5]=P[4]-DP_s_hx                          "pressure"
T[5]=temperature(Water,x=x[5],P=P[5])      "temperature"
h[5]=enthalpy(Water,x=x[5],P=P[5])         "specific enthalpy"
s[5]=entropy(Water,x=x[5],P=P[5])          "specific entropy"
x_f[5]=x_f(h[5],s[5])                      "specific flow exergy"
```

An energy balance on the steam side of the heat exchanger provides the mass flow rate of steam:

$$\dot{m}_s\, h_4 = \dot{Q}_{hx} + \dot{m}_s\, h_5$$

```
m_dot_s*h[4]=Q_dot_HX+m_dot_s*h[5]   "energy balance on steam side of HX"
```

An entropy balance on the reversible pump provides:

$$s_{s,3} = s_5$$

where $s_{s,3}$ is the specific entropy of the water leaving the reversible pump. The state of the water leaving the reversible pump is fixed by the specific entropy and pressure. The specific enthalpy ($h_{s,3}$) is determined. An energy balance on the reversible pump is:

$$\dot{m}_s\, h_5 + \dot{W}_{s,p} = \dot{m}_s\, h_{s,3}$$

where $\dot{W}_{s,p}$ is the power consumed by the reversible pump. The power consumed by the actual pump is given by:

$$\dot{W}_p = \frac{\dot{W}_{s,p}}{\eta_p}$$

An energy balance on the actual pump provides the specific enthalpy at state 3.

$$\dot{m}_s\, h_5 + \dot{W}_p = \dot{m}_s\, h_3$$

```
s_s[3]=s[5]                              "specific entropy balance on reversible pump"
h_s[3]=enthalpy(Water,P=P[3],s=s_s[3])   "specific enthalpy of water leaving reversible pump"
m_dot_s*h[5]+W_dot_s_p=m_dot_s*h_s[3]    "energy balance on reversible pump"
W_dot_p=W_dot_s_p/eta_p                  "actual pump power"
m_dot_s*h[5]+W_dot_p=m_dot_s*h[3]        "energy balance on actual pump"
```

State 3 is fixed by the pressure and specific enthalpy. The specific entropy, temperature, and specific flow exergy (s_3, T_3, and $x_{f,3}$) are determined.

```
"state 3"
s[3]=entropy(Water,P=P[3],h=h[3])        "specific entropy of water leaving pump"
T[3]=temperature(Water,P=P[3],h=h[3])    "temperature of water leaving pump"
x_f[3]=x_f(h[3],s[3])                    "specific flow exergy of water leaving pump"
```

An energy balance on the steam passing through the boiler is:

$$\dot{m}_s\, h_3 + \dot{Q}_b = \dot{m}_s\, h_4$$

The heat transfer from the fuel is higher than the heat transfer to the water due to the boiler inefficiency:

$$\dot{Q}_f = \frac{\dot{Q}_b}{\eta_b}$$

The mass flow rate of fuel can be determined:

$$\dot{m}_f\, LHV = \dot{Q}_f$$

```
m_dot_s*h[3]+Q_dot_b=m_dot_s*h[4]        "energy balance on steam side of boiler"
Q_dot_f=Q_dot_b/eta_b                    "boiler efficiency"
m_dot_f*LHV=Q_dot_f                      "fuel flow rate"
```

The Second Law efficiency of the system is the ratio of the increase in the exergy of the water provided to the laundry to the exergy provided to the system; the exergy provided to the system includes the exergy associated with the fuel and the pump power.

$$\eta_2 = \frac{\dot{m}_w (x_{f,2} - x_{f,1})}{\dot{m}_f\, LHV + \dot{W}_p}$$

```
eta_2=m_dot_w*(x_f[2]-x_f[1])/(m_dot_f*LHV+W_dot_p)   "Second Law efficiency"
```

which leads to $\eta_2 = 0.07563$.

b) Determine the exergy destruction within the boiler, pump, and heat exchanger.

The rate at which exergy is supplied to the boiler by the fuel is $\dot{m}_f\, LHV$. An exergy balance on the boiler is therefore:

$$\dot{m}_f\, LHV + \dot{m}_s\, x_{f,3} = \dot{m}_s\, x_{f,4} + \dot{X}_{des,b}$$

Note that any heat loss from the boiler (which must be present since its efficiency is less than 100%) will ultimately be transferred to the atmosphere and it therefore has no exergy. An exergy balance on the pump is:

$$\dot{m}_s\, x_{f,5} + \dot{W}_p = \dot{m}_s\, x_{f,3} + \dot{X}_{des,p}$$

An exergy balance on the heat exchanger is:

$$\dot{m}_s\, x_{f,4} + \dot{m}_w\, x_{f,1} = \dot{m}_s\, x_{f,5} + \dot{m}_w\, x_{f,2} + \dot{X}_{des,hx}$$

```
m_dot_f*LHV+m_dot_s*x_f[3]=m_dot_s*x_f[4]+X_dot_des_b          "Exergy balance on boiler"
m_dot_s*x_f[5]+W_dot_p=m_dot_s*x_f[3]+X_dot_des_p              "Exergy balance on pump"
m_dot_s*x_f[4]+m_dot_w*x_f[1]=m_dot_s*x_f[5]+m_dot_w*x_f[2]+X_dot_des_hx
                                                               "Exergy balance on heat exchanger"
```

Solving these equations results in $\dot{X}_{des,b} = 98.93$ kW, $\dot{X}_{des,p} = 0.053$ kW, and $\dot{X}_{des,hx} = 68.8$ kW. Clearly most of the power producing capability of the fuel is destroyed in the boiler and the heat exchanger. Further, the rate of exergy destruction is many times greater than the rate at which the laundry consumes electrical power. Therefore, in theory, the fuel used to produce the hot water has sufficient exergy content to meet both the heating and power needs of the laundry if only it could be used more efficiently.

c) What is the operating cost associated with the laundry?

The operating cost is the combined cost of the fuel and the electrical power:

$$Cost = \dot{m}_f\, LHV\, C_{fuel} + (\dot{W}_e + \dot{W}_p)C_{elec}$$

```
Cost=m_dot_f*LHV*C_fuel+(W_dot_e+W_dot_p)*C_elec     "operating cost"
Cost_dph=Cost*convert($/s,$/hr)                      "in $/hr"
```

which results in $Cost = 12.87$ \$/hr.

d) The owners of the laundry have been approached by a company that proposes the installation of the electrical generation equipment shown in Figure 2. In the proposed system, the steam leaving the existing boiler at state 4 passes through a turbine that is connected to a generator in order to produce electrical power. The turbine exhaust pressure is $P_5 = 65$ psia. The steam leaving the turbine enters the heat exchanger in order to provide the hot water required by the laundry. The condensate leaving the heat exchanger is saturated liquid, $x_6 = 0$. The isentropic efficiency of the turbine is $\eta_t = 0.72$ and the efficiency of the generator is $\eta_g = 0.92$. Additional electricity, if needed, is purchased from the utility. Determine the rate at which the generator produces electricity and the Second Law efficiency for the system shown in Figure 2.

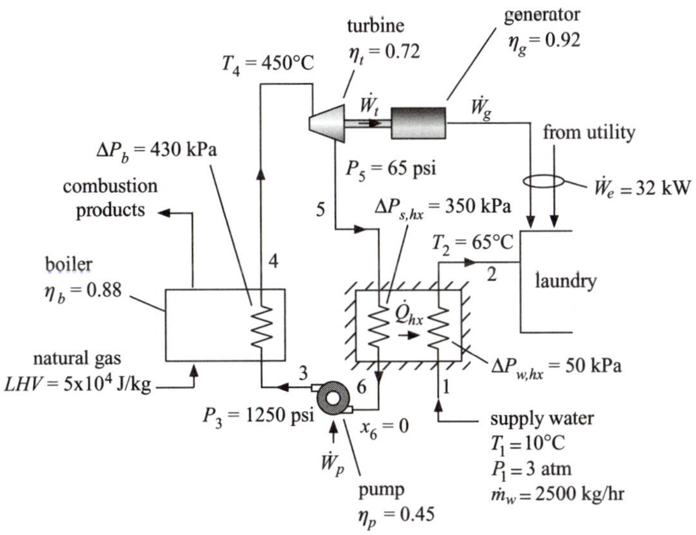

Figure 2: Proposed system providing hot water and electricity to a commercial laundry.

A new EES program is used for this analysis. The specific flow exergy function x_f developed in part (a) is placed at the top of the Equations window.

```
$UnitSystem SI K Pa J Mass Radian
function x_f(h,s)
    T_0=convertTemp(C,K,10 [C])             "dead state temperature is 10 [C]"
    P_0=1 [atm]*convert(atm,Pa)             "dead state pressure"
    h_0=enthalpy(Steam,T=T_0,P=P_0)         "dead state specific enthalpy"
    s_0=entropy(Steam,T=T_0,P=P_0)          "dead state specific entropy"
    x_f=h-h_0-T_0*(s-s_0)                   "specific flow exergy"
end
```

The inputs are entered:

```
T[1]=converttemp(C,K,10 [C])                  "temperature of supply water"
P[1]=3 [atm]*convert(atm,Pa)                  "pressure of supply water"
DP_w_hx=50 [kPa]*convert(kPa,Pa)              "pressure drop on water side of the heat exchanger"
T[2]=converttemp(C,K,65 [C])                  "temperature of water provided to laundry"
m_dot_w=2500 [kg/hr]*convert(kg/hr,kg/s)      "mass flow rate of water provided to laundry"
eta_p=0.45 [-]                                "efficiency of the pump"
eta_t=0.72 [-]                                "turbine efficiency"
eta_g=0.92 [-]                                "generator efficiency"
eta_b=0.88 [-]                                "efficiency of the boiler"
LHV=5e4 [kJ/kg]*convert(kJ/kg,J/kg)           "lower heating value of the fuel"
W_dot_e=32 [kW]*convert(kW,W)                 "electrical power consumption"
P[3]=1250 [psi]*convert(psi,Pa)               "pressure of water provided to boiler"
P[5]=65 [psi]*convert(psi,Pa)                 "turbine discharge pressure"
T[4]=converttemp(C,K,450 [C])                 "temperature of water leaving boiler"
x[6]=0 [-]                                    "quality of water leaving heat exchanger"
DP_s_hx=350 [kPa]*convert(kPa,Pa)             "pressure drop on steam side of heat exchanger"
DP_b=430 [kPa]*convert(kPa,Pa)                "pressure drop associated with boiler"
C_elec=0.15 [$/kW-hr]*convert($/kW-hr,$/J)    "electricity cost"
C_fuel=1.3 [$/therm]*convert($/therm,$/J)     "fuel cost"
```

The analysis of the water side of the heat exchanger remains unchanged:

```
"state 1"
h[1]=enthalpy(Water,T=T[1],P=P[1])          "specific enthalpy of supply water"
s[1]=entropy(Water,T=T[1],P=P[1])           "specific entropy of supply water"
x_f[1]=x_f(h[1],s[1])                       "specific flow exergy of supply water"

"state 2"
P[2]=P[1]-DP_w_hx                           "pressure of water provided to laundry"
h[2]=enthalpy(Water,T=T[2],P=P[2])          "specific enthalpy of water provided to laundry"
s[2]=entropy(Water,T=T[2],P=P[2])           "specific entropy of water provided to laundry"
x_f[2]=x_f(h[2],s[2])                       "specific flow exergy of water provided to laundry"
Q_dot_HX=m_dot_w*(h[2]-h[1])                "energy balance on water side of HX"
```

State 4 is determined as before:

```
"state 4"
P[4]=P[3]-DP_b                              "pressure of steam leaving boiler"
h[4]=enthalpy(Water,T=T[4],P=P[4])          "specific enthalpy of steam leaving boiler"
s[4]=entropy(Water,T=T[4],P=P[4])           "specific entropy of steam leaving boiler"
x_f[4]=x_f(h[4],s[4])                       "specific flow exergy of steam leaving boiler"
```

An entropy balance on a reversible turbine is:

$$s_{s,5} = s_4$$

where $s_{s,5}$ is the specific entropy of the fluid leaving a reversible turbine. The state of the fluid leaving a reversible turbine is fixed by the specific entropy and pressure. The specific enthalpy ($h_{s,5}$) is determined. An energy balance on the reversible turbine is:

$$h_4 = \frac{\dot{W}_{s,t}}{\dot{m}_s} + h_{s,5}$$

The actual power produced by the turbine per unit of mass flow rate is given by:

$$\frac{\dot{W}_t}{\dot{m}_s} = \eta_t \frac{\dot{W}_{s,t}}{\dot{m}_s}$$

An energy balance on the actual turbine provides the specific enthalpy at state 5:

$$h_4 = \frac{\dot{W}_t}{\dot{m}_s} + h_5$$

```
s_s[5]=s[4]                                 "entropy balance on reversible turbine"
h_s[5]=enthalpy(Water,s=s_s[5],P=P[5])      "specific enthalpy leaving reversible turbine"
h[4]=W_dot_s_t\m_dot+h_s[5]                 "energy balance on reversible turbine"
W_dot_t\m_dot=eta_t*W_dot_s_t\m_dot         "actual turbine power per mass flow rate"
h[4]=W_dot_t\m_dot+h[5]                     "energy balance on actual turbine"
```

State 5 is fixed by the specific enthalpy and pressure. The temperature, specific entropy, and specific flow exergy (T_5, s_5, and $x_{f,5}$) are determined.

```
"state 5"
T[5]=temperature(Water,h=h[5],P=P[5])       "temperature"
s[5]=entropy(Water,h=h[5],P=P[5])           "specific entropy"
x_f[5]=x_f(h[5],s[5])                       "specific flow exergy"
```

The pressure at state 6 is the pressure leaving the turbine less the pressure drop on the steam side of the heat exchanger.

$$P_6 = P_5 - \Delta P_{s,hx}$$

State 5 is fixed by the pressure and quality. The temperature, specific enthalpy, specific entropy, and flow exergy (x_6, h_6, s_6, and $x_{f,6}$) are determined.

```
"state 6"
P[6]=P[5]-DP_s_hx                           "pressure"
T[6]=temperature(Water,x=x[6],P=P[6])       "temperature"
s[6]=entropy(Water,x=x[6],P=P[6])           "specific entropy"
h[6]=enthalpy(Water,x=x[6],P=P[6])          "specific enthalpy"
x_f[6]=x_f(h[6],s[6])                       "specific flow exergy"
```

An energy balance on the steam side of the heat exchanger provides the mass flow rate of steam.

$$\dot{m}_s\, h_5 = \dot{Q}_{hx} + \dot{m}_s\, h_6$$

The pump is analyzed as before.

```
s_s[3]=s[6]                                 "entropy balance on reversible pump"
h_s[3]=enthalpy(Water,P=P[3],s=s_s[3])      "specific enthalpy of water leaving reversible pump"
m_dot_s*h[6]+W_dot_s_p=m_dot_s*h_s[3]       "energy balance on reversible pump"
W_dot_p=W_dot_s_p/eta_p                     "actual pump work"
m_dot_s*h[6]+W_dot_p=m_dot_s*h[3]           "energy balance on actual pump"
```

State 3 is specified as before.

```
"state 3"
s[3]=entropy(Water,P=P[3],h=h[3])           "specific entropy of water leaving pump"
T[3]=temperature(Water,P=P[3],h=h[3])       "temperature of water leaving pump"
x_f[3]=x_f(h[3],s[3])                       "specific flow exergy of water leaving pump"
```

The boiler is analyzed as before.

```
m_dot_s*h[3]+Q_dot_b=m_dot_s*h[4]           "energy balance on steam side of boiler"
Q_dot_f=Q_dot_b/eta_b                       "boiler efficiency"
m_dot_f*LHV=Q_dot_f                         "fuel flow rate"
```

The power produced by the turbine is:

$$\dot{W}_t = \frac{\dot{W}_t}{\dot{m}_s}\,\dot{m}_s$$

and the electrical power produced by the generator is:

$$\dot{W}_g = \dot{W}_t \, \eta_g$$

```
W_dot_t=W_dot_t\m_dot*m_dot_s    "power produced by turbine"
W_dot_g=W_dot_t*eta_g            "electrical power produced by generator"
```

which leads to $\dot{W}_g = 29.3$ kW. The Second Law efficiency of the system is the ratio of the sum of the exergy associated with the electrical energy produced and the exergy increase of the water to the sum of the exergy associated with the fuel and the exergy associated with the pump power:

$$\eta_2 = \frac{\dot{m}_w (x_{f,2} - x_{f,1}) + \dot{W}_g}{\dot{m}_f \, LHV + \dot{W}_p}$$

```
eta_2=(m_dot_w*(x_f[2]-x_f[1])+W_dot_g)/(m_dot_f*LHV+W_dot_p)    "Second Law efficiency"
```

which results in $\eta_2 = 0.198$. This Second Law efficiency is much higher than the value calculated for the original system.

e) Determine the rate of exergy destruction in the boiler, pump, heat exchanger, and turbine.

The exergy balances on the boiler, pump, and heat exchanger are unchanged.

```
m_dot_f*LHV+m_dot_s*x_f[3]=m_dot_s*x_f[4]+X_dot_des_b    "Exergy balance on boiler"
m_dot_s*x_f[6]+W_dot_p=m_dot_s*x_f[3]+X_dot_des_p        "Exergy balance on pump"
m_dot_s*x_f[5]+m_dot_w*x_f[1]=m_dot_s*x_f[6]+m_dot_w*x_f[2]+X_dot_des_hx
                                                          "Exergy balance on heat exchanger"
```

The exergy balance on the turbine is:

$$\dot{m}_s \, x_{f,4} = \dot{W}_t + \dot{m}_s \, x_{f,5} + \dot{X}_{des,t}$$

```
m_dot_s*x_f[4]=W_dot_t+m_dot_s*x_f[5]+X_dot_des_t    "Exergy balance on turbine"
```

Solving will show that $\dot{X}_{des,b} = 125.1$ kW, $\dot{X}_{des,p} = 0.551$ kW, $\dot{X}_{des,hx} = 38.0$ kW, and $\dot{X}_{des,t} = 8.27$ kW.

f) Determine the operating cost associated with the system shown in Figure 2.

The operating cost is:

$$Cost = \dot{m}_f \, LHV \, C_{fuel} + (\dot{W}_e + \dot{W}_p - \dot{W}_g) C_{elec}$$

```
Cost=m_dot_f*LHV*C_fuel+(W_dot_e+W_dot_p-W_dot_g)*C_elec    "operating cost"
Cost_dph=Cost*convert($/s,$/hr)                             "in $/hr"
```

which leads to $Cost = 10.19$\$/hr, a 20% reduction from the system shown in Figure 1.

7.6 Relation Between Exergy Destruction and Entropy Generation

This extended section of the book can be found on the website (www.cambridge.org/kleinandnellis). It shows how the exergy balance derived in Section 7.5 can also be obtained more directly by algebraically rearranging the energy and entropy balances previously discussed in Chapters 4 and 6, respectively. The result of this section is the formal proof that exergy destruction is equivalent to the product of entropy generation and the dead state temperature, i.e:

$$\boxed{\dot{X}_{des} = T_0\,\dot{S}_{gen}} \tag{7-49}$$

Problems

The problems included here have been selected from a larger set of problems that are available on the website associated with this book (www.cambridge.org/kleinandnellis).

A. Exergy and Exergy Balances

7.A-1 A tank with volume V has been completely evacuated. What is the exergy of the tank if the surroundings consist of air at temperature T_0 and pressure P_0?

7.A-2 A cold fluid cannot be stored for long periods because thermal gains inevitably occur, even in a Dewar (a vacuum-insulated container). An alternative is to store a high pressure gas (e.g., air) and then release it as needed to generate the cold source. In a particular application, air at $P_1 = 100$ atm is stored at $T_1 = 25°\text{C}$ in a $V = 15$ liter tank. Note that air does not obey the ideal gas law at this high pressure so use fluid 'Air_ha' in EES, which provides real gas properties for air.

a) What is the exergy of the air in the tank?
b) What is the maximum possible cooling (in J) that can be provided at $T_c = 0°\text{C}$?
c) Indicate how this cooling might be accomplished.

7.A-3 At a particular instant of time, a square metal bar has an axial temperature distribution given by:

$$T(x) = 50(1 + 8x^2)$$

where x is the distance (in meters) measured from one end and T is the local temperature (in °C). Due to its high thermal conductivity, the temperature in the bar may be assumed to be uniform at any cross-section. The cross-section of the bar has width $W = 2.5$ cm and the length of the bar is $L = 0.3$ m. The density and specific heat of the metal are $\rho = 2700$ kg/m^3 and $c = 0.90$ J/kg-K, respectively.

a) Is the average bar temperature rising or falling at this instant of time? (Assume that the bar can only transfer energy at its end points; i.e., the sides are insulated.)
b) Calculate the change in internal energy if the bar is cooled to a uniform temperature of $T_f = 20°\text{C}$.
c) Calculate the change in entropy of the bar for the process in part (b).
d) What is the change in exergy of the bar for the process in part (b) given a large heat sink at 20°C?
e) What is the maximum thermal efficiency at which work could be produced for the conditions in part (d)?

7.A-4 On a day in which the outdoor temperature is $T_0 = 10°F$, an energy input of $\dot{Q} = 75{,}000$ Btu/hr is needed to maintain the indoor temperature of a house at a comfortable $T_{in} = 70°F$.

a) Electric resistance heaters are employed to supply the heating load. The surface temperature of the heating elements under steady operating conditions is measured to be $T_s = 560°F$. Determine the rate of exergy destruction for this process.

b) The electric resistance heaters are replaced with a heat source at $T_H = 560°C$ that provides $\dot{Q} = 75{,}000$ Btu/hr in order to meet the heating load. Determine the rate of exergy destruction for this process.

c) Determine the minimum possible electrical energy required to provide the heating load to the house. Indicate how this electrical energy would be used to supply the heating load.

7.A-5 Spent steam is exhausted at a rate of $\dot{m}_s = 2{,}350$ kg/hr from an industrial process at $T_{s,in} = 110°C$ and $P_s = 1$ atm and is currently condensed and cooled to $T_{s,out} = 40°C$ at constant pressure by heat transfer to cooling water in a heat exchanger. The cooling water enters the heat exchanger at $T_{cw,in} = 12°C$ and $P_{cw} = 1$ atm and exits at $T_{cw,out} = 18°C$ and the same pressure. An engineer in the plant has recognized that this heat exchange process is wasteful and he has proposed an alternative process in which power is generated. He claims that in this alternative process, the temperature and pressures of both streams remain exactly the same as they were in the original process.

a) What is the rate at which exergy is destroyed in the existing process?

b) Do you believe that this alternative process is possible? If so, determine the maximum possible rate that power could be produced.

7.A-7 An elevator system consists of a well-insulated tank having a volume of $V = 0.28$ m^3 containing two phase water at $P_1 = 250$ kPa. At the start of the lifting process, the liquid in the tank occupies 50% of the tank volume, with the remainder being vapor. At the top of the tank is a short pipe that connects the tank to a piston-cylinder device, as shown in Figure 7.A-7.

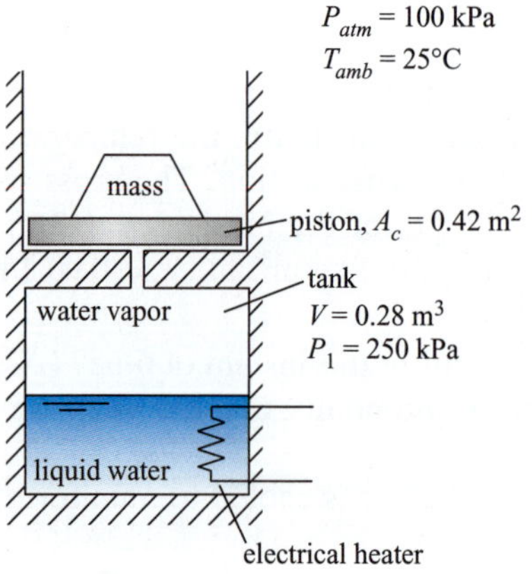

Figure 7.A-7: Elevator system.

There is negligible pressure loss when vapor flows through the pipe and the volume of the pipe is negligible. The cross-sectional area of the piston is $A_c = 0.42$ m^2. At the start of this process, the piston is floating very near (but not touching) the bottom of the cylinder. Then, the electric heater in the tank is switched on. The piston is observed to rise $z = 2.5$ m in the cylinder, at which time the heater is switched off. Thermal losses may be assumed to be negligible during the lifting process.

a) Determine the combined mass of the piston and the mass that is on the piston.
b) Determine the quality of the water in the tank before the heating element is engaged.
c) Determine the quality of the water at the completion of the lifting process.
d) Determine the work done by the steam as a result of expansion during the lifting process.
e) Determine the total electrical energy provided to the water.
f) Determine the temperature of the steam in the piston-cylinder device at the completion of the process.
g) Determine the total exergy destroyed during the lifting process.
h) Determine the Second-Law efficiency of the lifting process.

7.A-8 Before the widespread use of mechanical refrigeration, cooling was provided by ice, which was delivered by icemen and stored in an icebox. In a current application, blocks of ice at $T_{ice} = 32^\circ$F with a total mass of $m_{ice} = 500$ lb$_m$ are placed in a large food-storage icebox that is used to keep food at $T_{food} = 45^\circ$F on a day in which the outdoor temperature is $T_{amb} = 78^\circ$F. The rate of heat loss through the walls of the ice box is $\dot{Q} = 2900$ Btu/hr.

a) Estimate the amount of time the ice it will take this ice to melt.
b) Compare the rate of cooling provided in ton units to the mass of ice in tons.
c) Define and calculate an efficiency for this process.
d) Determine the change in exergy of the ice in this process.
e) Calculate the Second-Law efficiency for this process.

7.A-10 Water at $T_{out} = 60^\circ$C, $P_{out} = 1$ bar and a volumetric flow rate of $\dot{V} = 0.25$ liters/sec is needed for use in a commercial laundry operation. The cold water is supplied to the heating system at $T_{in} = 10^\circ$C and $P_{in} = 1$ bar. Energy losses from the water tanks and associated piping to the surroundings (at $T_{amb} = T_{in}$) have been measured to be $\dot{Q}_{amb} = 3.2$ kW during steady operation. Two methods have been proposed to heat the water. In Method 1, steam at $P_s = 1.5$ bar with $x_s = 90\%$ quality is directly mixed with the cold water. In Method 2, electricity is used to heat the water with a resistance heater.

a) Determine the required mass flow rate of steam for Method 1.
b) What is the rate of exergy destruction for Method 1?
c) Determine the electrical power required for Method 2.
d) What is the rate of exergy destruction for Method 2?
e) Calculate the Second-Law efficiencies for Methods 1 and 2.

7.A-13 An electric water heater having a $V = 300$ liter capacity employs a $\dot{W}_{htr} = 4.5$ kW electric resistance heater to raise the temperature of the water from $T_1 = 25$ to $T_2 = 55^\circ$C. The water in the tank can be assumed to be well-mixed and thus at a uniform temperature at any instant of time. The outer surface of the resistor heater remains at an average temperature of $T_{htr} = 80^\circ$C during the entire process. The outer surface of the tank is insulated, but imperfectly so. During this process, the average heat transfer rate through the tank walls to the $T_{amb} = 25^\circ$C, $P_{atm} = 101.3$ kPa surroundings is $\dot{Q}_{amb} = 250$ W. The thermal capacity of the tank itself is negligible relative to thermal capacity of the water.

a) How much time is required to heat the water?

b) What is the efficiency of this water heating process based on the First Law?
c) What is the efficiency of this water heating process based on the Second Law?
d) Explain why the First and Second law efficiencies are not the same for this process.

7.A-18 A counterflow heat exchanger is used in a cryogenic refrigeration system to heat exchange two neon gas streams having equal mass flow rates. The hot stream enters at $T_{H,in} = 300$ K, $P_H = 400$ bar and exits at $T_{H,out} = 100$ K. The cold stream enters at $T_{C,in} = 40$ K, $P_C = 1$ bar. Heat losses through the jacket of the heat exchanger and pressure drop in each stream may be neglected. The environment is at $T_{amb} = 300$ K and $P_{atm} = 1$ bar.
a) At what temperature does the cold stream exit the heat exchanger?
b) What is the lowest temperature that the hot stream could reach assuming that the heat exchanger was perfect (i.e., it had an infinite surface area for heat transfer)?
c) What is the heat exchanger effectiveness for this heat exchanger?
d) What is the Second Law efficiency for this process?

B. Advanced Problems

7.B-1 Two streams enter a counterflow liquid-to-liquid heat exchanger. The hot stream enters at $T_{H,in} = 140°$C and $\dot{m}_H = 4.2$ kg/sec. The hot stream is a 40% ethylene glycol-water solution that has a specific heat of $c_H = 3.76$ kJ/kg-K and a density of $\rho_H = 1017$ kg/m^3. The pressure loss for this stream as it passes through the heat exchanger is $\Delta P_H = 18.5$ kPa. The cold stream is pure water that enters at $T_{C,in} = 5°$C and $\dot{m}_C = 8.5$ kg/sec with a pressure loss of $\Delta P_C = 12.2$ kPa. The exit pressures of the hot and cold streams are both $P_{atm} = 1$ atm. The heat exchanger effectiveness is $\varepsilon_{hx} = 0.90$.
a) Calculate the outlet temperatures and the rate of exergy destruction.
b) Calculate the Second Law efficiency for this process. How is the Second Law efficiency related to the heat exchanger effectiveness?
c) If the heat exchanger effectiveness were 1 (indicative of a perfect heat exchanger) would the Second Law efficiency also be 1? (Why or why not?)
d) What would the outlet temperatures and heat exchanger effectiveness be for a device that has a Second Law efficiency of 1.0? Is it possible to approach this limit? Is so, how would you construct a device to accomplish this performance?

7.B-3 The buildings on a large university campus are heated with steam that is centrally produced. The steam is produced at $P_s = 600$ psia, $T_s = 500°$F in a $\eta_b = 90\%$ efficient coal-fired boiler (based on a heating value of $HV = 14{,}100$ Btu/lb$_m$ for coal). The actual flow sheet of the campus heating system is quite complicated, but for the purpose of this problem, it can be represented as shown in Figure 7.B-3(a). The steam is throttled to $P_{out} = 175$ psia and then piped to the campus buildings. Condensate at $T_{cond} = 70°$F, $P_{cond} = 14.7$ psia is returned to the plant, pumped to P_s using pump with efficiency $\eta_p = 0.42$ and then sent back to the boiler to complete a cycle.

An alternative to the configuration shown in Figure 7.B-3(a) is to utilize a steam turbine, rather than a throttle, to reduce the pressure of the steam from P_s to P_{out}, as shown in Figure 7.B-3(b). The turbine has an estimated isentropic efficiency of $\eta_t = 0.70$ and it drives a $\eta_g = 90\%$ efficient electric generator. Both systems provide $Q = 2.3 \times 10^{12}$ Btu of heat to campus buildings during an average heating

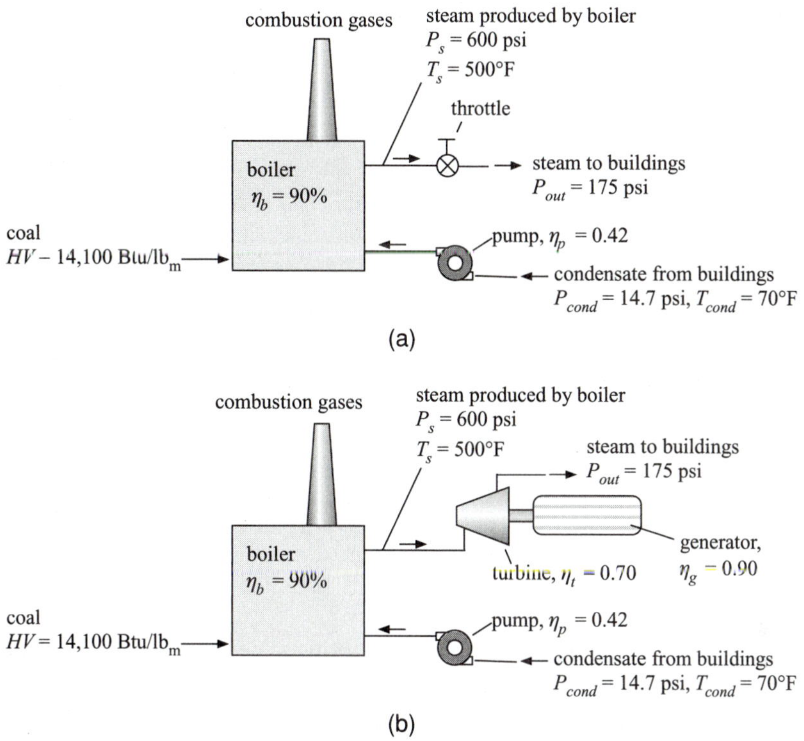

Figure 7.B-3: Schematic of (a) existing system and (b) proposed system.

season. Electricity is purchased to operate the pump. The purpose of this problem is to compare the two systems shown in Figure 7.B-3. Specifically, your solution should answer the following questions.

a) How many tons of coal are used by the systems in Figures 7.B-3(a) and (b) during the heating season?
b) How much electrical energy can be produced by the system in Figure 7.B-3(b) during the heating season?
c) What is the annual cost to heat campus buildings with the system in Figure 7.B-3(a) if coal costs cc = \$45/ton? Also determine the cost associated with the system in Figure 7.B-3(b) assuming that the electricity that is generated is valued at ec = \$0.10/kWhr.
d) What are the major sources of exergy destruction for the systems in Figures 7.B-3(a) and (b)?
e) What are the Second-Law efficiencies of the systems in Figures 7.B-3(a) and (b)? Assume that the exergy of coal is equal to its heating value and that the dead state is $T_0 = 70^\circ$F and $P_0 = 1$ atm.

7.B-4 An electric heating system is used for heating a house. When in operation, air enters the heating system at $\dot{V} = 250$ cfm, $T_{in} = 60^\circ$F, and $P_{in} = 1$ atm through the return ducts and exits at $T_{out} = 110^\circ$F and essentially the same pressure. The electrical heaters draw $\dot{W}_{htr} = 4.2$ kW and the outdoor temperature is $T_{amb} = 5^\circ$F.

a) Estimate the rate of exergy destruction and the First and Second-Law efficiencies for this electrical heating process during steady-state operation at these conditions.

b) An electric driven heat pump having an overall *COP* that is 45% of the Carnot *COP* could be used to supply space heat to the building. Estimate the First and Second-Law efficiencies for this heat pump heating process during steady-state operation at the conditions for part (a).
c) Suppose the house were heated with natural gas instead of electricity. The lower heating value of the natural gas is $LHV = 21{,}500$ Btu/lb$_m$ and the furnace efficiency based on the lower heating value is $\eta_f = 82\%$. The exergy of natural gas is nearly equal to its lower heating value. Estimate the First and Second-Law efficiencies for the heating with natural gas for the same conditions as in part (a).

The annual heating load for the building can be estimated as the product of the building loss coefficient and the annual heating degree days. The annual heating degree-days (using a 65°F base) for the building location is 7,800°F-days. The loss coefficient for the building is 9000 Btu/°F-day.

d) Compare the annual costs of heating with electrical resistance, the heat pump, and natural gas. Obtain electricity and gas costs from the local utility.

Each mole of natural gas that is combusted produces one mole of carbon dioxide. The carbon dioxide is of concern because it is believed to contribute to global warming.

e) Calculate the amount of carbon dioxide produced per year if natural gas is used. Would more or less carbon dioxide be produced if the house were heated with electrical resistance heaters?
f) Space heating is responsible for about 13% of the total energy consumed in the U.S each year. What is your overall recommendation of this technology, based on your answers to the previous questions?

8 Power Cycles

Systems that are designed to continuously transform heat into power are referred to as power cycles. Many different types of power cycles are in common use and these cycles can be classified in several ways. *Closed power cycles* use a working fluid that is recirculated within the equipment whereas *open power cycles* continuously draw a working fluid (usually air) through the equipment and discharge it to the environment. *Reciprocating* cycles operate in a cyclic manner so that the thermodynamic state of the working fluid at any location varies with time whereas the state of the working fluid at a particular location is constant in *steady flow cycles*. *Internal combustion cycles* obtain the energy that is to be converted into power from a combustion process that occurs within the equipment. *Externally heated* equipment can process thermal energy from any available source; these systems can use solar or nuclear energy in addition to the energy released in an external combustion process. Each type of power cycle offers some specific operating advantage, e.g., high efficiency, high power per unit weight, rapid startup, or low capital cost. The goal of any power cycle is to maximize the power output and efficiency for a specific task, given economic and other constraints. The chapter will review several common types of power cycles and demonstrate how the thermodynamic principles developed in earlier chapters can be used to analyze and optimize these cycles.

8.1 The Carnot Cycle

We have already encountered a power cycle in Chapter 5 during our study of the Second Law of Thermodynamics. The Carnot cycle was presented as a method of continuously converting heat from a constant temperature source into work with the highest possible efficiency. However, there are implementation issues that prevent the Carnot cycle from being practical and it is not a power cycle that is used. Nevertheless, the Carnot cycle provides a good starting point for our discussion of power cycles.

The Carnot cycle, as presented in Section 5.5, is a closed, reciprocating, externally heated power cycle that receives heat from a high temperature reservoir at T_H and rejects heat to a low temperature reservoir at T_C. The Carnot cycle consists of the four processes summarized in Table 8-1. The Carnot cycle is shown on a pressure-volume diagram in Figure 5-13. An alternative way to represent the cycle is with a temperature-specific entropy diagram, as shown in Figure 8-1. The *T-s* diagram provides a useful method for analyzing all of the power cycles that are presented in this chapter.

Heat is provided to the cycle (reversibly) from a constant temperature source at T_H in process 1-2. An entropy balance on the system for this process is:

$$\frac{Q_{1-2}}{T_H} = m(s_2 - s_1) \tag{8-1}$$

Table 8-1: Processes in the Carnot Cycle

Process	Description
1-2	Reversible, isothermal expansion at temperature T_H
2-3	Reversible, adiabatic expansion from T_H to T_C
3-4	Reversible, isothermal compression at temperature T_C
4-1	Reversible, adiabatic compression from T_C to T_H

Equation (8-1) can be rearranged to be provide:

$$\frac{Q_{1-2}}{m} = T_H(s_2 - s_1) \tag{8-2}$$

The heat transfer to the cycle per unit mass of working fluid during process 1-2, Q_{1-2}/m, is therefore the area under the line 1-2 in Figure 8-1. Processes 2-3 and 4-1 are reversible and adiabatic and therefore the entropy of the working fluid does not change during these processes. Process 2-3 is a constant temperature, reversible heat rejection process. An entropy balance on the system for process 3-4 provides:

$$\frac{Q_{3-4}}{m} = s_3 - s_4 \tag{8-3}$$

The heat transfer from the cycle per unit mass of working fluid during process 3-4, Q_{3-4}/m, is the area under line 3-4 in Figure 8-1. An energy balance on the system for one complete cycle determines the net work provided per cycle, W_{net}:

$$W_{net} = Q_{1-2} - Q_{3-4} \tag{8-4}$$

Therefore, the net work produced per mass of fluid, W_{net}/m, is the area enclosed within the temperature-specific entropy diagram for the cycle shown in Figure 8-1. This result is generally true for any internally reversible cycle and is one of the reasons why it is useful to represent a cycle on a T-s diagram.

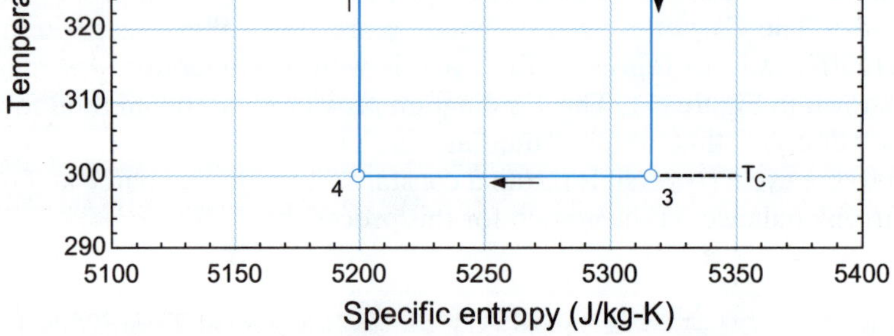

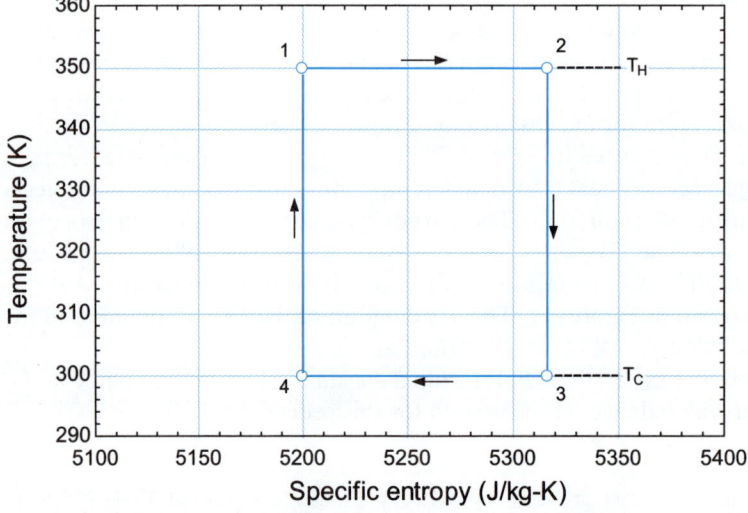

Figure 8-1: Temperature-specific entropy diagram for air undergoing a Carnot cycle between temperatures $T_H = 350$ K and $T_C = 300$ K with $P_1 = 1$ MPa, $V_1 = 1.0$ liter, and $V_2/V_1 = 1.5$.

The efficiency of the Carnot cycle is:

$$\eta = \frac{W_{net}}{Q_{1-2}} \tag{8-5}$$

Substituting Eq. (8-4) into Eq. (8-5) leads to:

$$\eta = \frac{Q_{1-2} - Q_{3-4}}{Q_{1-2}} = 1 - \frac{Q_{3-4}}{Q_{1-2}} \tag{8-6}$$

Substituting Eqs. (8-1) and (8-3) into Eq. (8-6) allows the efficiency to be expressed as:

$$\eta = 1 - \frac{T_C\, m(s_3 - s_4)}{T_H\, m(s_2 - s_1)} \tag{8-7}$$

Recognizing from Figure 8-1 that $(s_3 - s_4) = (s_2 - s_1)$ allows Eq. (8-7) to be reduced to:

$$\eta = 1 - \frac{T_C}{T_H} \tag{8-8}$$

Equation (8-8) is the same result that was obtained in Eq. (5-74) in Section 5.5, by considering each process in detail for an ideal gas. Equation (8-8) provides the maximum efficiency that can be achieved by a reversible heat engine operating between two fixed temperatures, T_H and T_C. The efficiency increases with increasing heat source temperature, T_H, and with decreasing heat sink temperature, T_C. This general trend is observed for all power cycles.

The Carnot cycle achieves the highest possible efficiency of any power cycle operating between two fixed temperatures. There are, however, a number of considerations that cause the Carnot cycle to be impractical. The primary problem with the Carnot cycle is related to the fact that two of the processes, the isothermal expansion and compression processes, involve the transfer of heat and work simultaneously. Most equipment separate processes involving heat from those involving work. Consider the components that we have studied. Turbines and compressors involve work but are typically nearly adiabatic. Heat exchangers involve heat transfer but do not involve work. The reason that these processes are separated in practical components is related to the large amount of time that is required for a fluid to transfer heat compared to the short amount of time that is required for a fluid to transfer work. For example, the time that a fluid particle spends in a turbine is much less than the time that it spends in a typical heat exchanger (for an equivalent energy transfer rate). By design, the heat exchanger will consist of a long, typically circuitous flow path that places the fluid in intimate contact with the wall for a long time in order to accomplish the heat transfer. In contrast, large power production equipment operates in a near adiabatic manner because the time required for the fluid to move through the equipment is too short to allow heat transfer to occur to any significant extent.

An easy way to implement the constant temperature heat transfer processes that are required in the Carnot cycle is to accomplish these processes under two-phase conditions. The heat addition and heat rejection processes (1-2 and 3-4) can be accomplished rather easily and without any work transfer for a pure two-phase fluid because the temperature remains constant if the pressure is held constant. A Carnot cycle that uses water as the working fluid and operates entirely under the vapor dome is shown on a temperature–specific entropy diagram in Figure 8-2. The cycle accepts heat from a high temperature reservoir at $T_H = 500$ K and rejects heat to a low temperature reservoir at $T_C = 325$ K.

The Carnot cycle shown in Figure 8-2 is not practical because process 2-3 is a work-producing expansion process that involves both liquid and vapor phases and

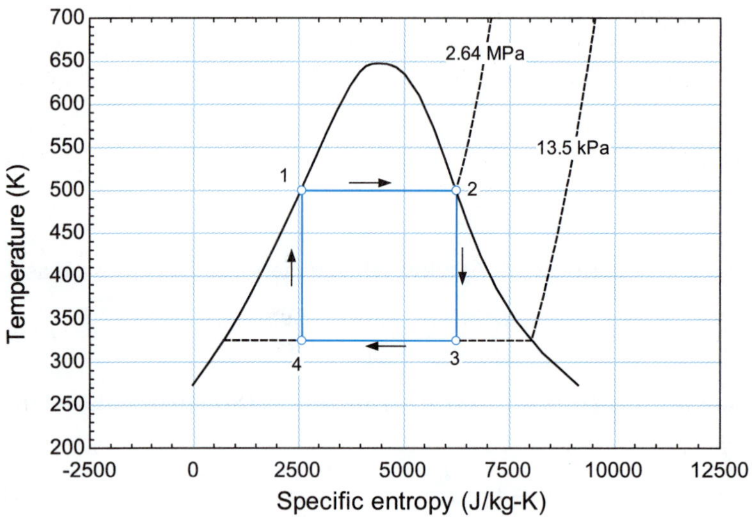

Figure 8-2: Carnot cycle using water as the working fluid under two-phase conditions.

process 4-1 is a compression process that also involves both liquid and vapor. It is difficult to design equipment that can efficiently handle significant amounts of both liquid and vapor phases during compression or expansion processes. The Rankine cycle presented in the following section provides one method of dealing with these implementation problems while retaining many of the characteristics of the Carnot cycle shown in Figure 8-2.

8.2 The Rankine Cycle

The Rankine cycle is the power cycle that is most often used in utility-scale electrical generation. It is an externally heated, steady flow system. The source of external heat is often provided by the combustion of coal or natural gas, but it can also be provided by nuclear, geothermal, or solar energy.

8.2.1 The Ideal Rankine Cycle

The basic Rankine cycle can be viewed as a modification of the Carnot cycle. The two-phase compression and expansion processes required by the Carnot cycle shown in Figure 8-2 (processes 4-1 and 2-3) are impractical. Therefore, the Rankine cycle moves the compression and expansion processes out of the vapor dome. The need to compress a mixture of liquid and vapor in process 4-1 is eliminated by rejecting additional heat from the working fluid until state 4 is saturated liquid. The need to expand a mixture of liquid and vapor in process 2-3 is nearly eliminated by adding additional heat to the working fluid until state 3 approaches a saturated vapor state. Figure 8-3 illustrates an ideal Rankine cycle using water as the working fluid that operates between the same pressures as the Carnot cycle shown in Figure 8-2.

The Rankine cycle is a closed, steady flow, externally heated cycle in which the working fluid changes state as it circulates through various pieces of equipment. Figure 8-4 illustrates a schematic of a simple Rankine cycle that is receiving heat from a high temperature reservoir at T_H and rejecting heat to a low temperature reservoir at T_C. The liquid leaving the condenser is compressed to a high pressure in the pump. The fluid passes through the boiler/superheater, heat exchangers where heat is added in order to

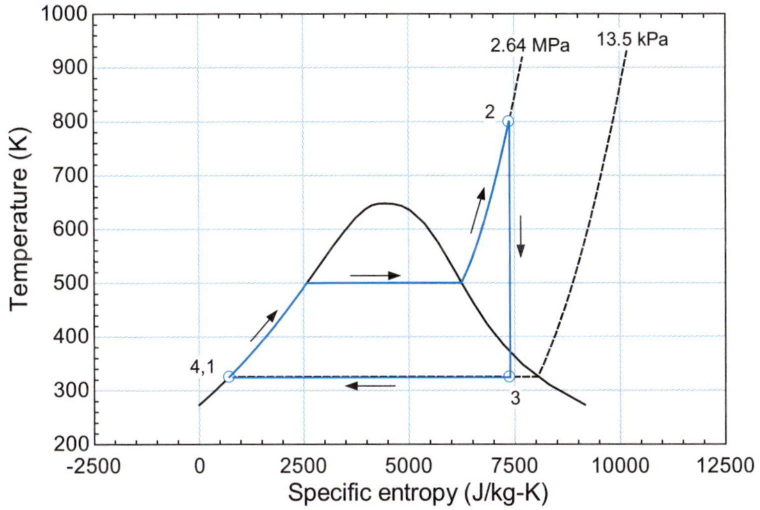

Figure 8-3: Ideal Rankine cycle using water as the working fluid operating between the same pressures as the Carnot cycle shown in Figure 8-2; notice that the compression and expansion processes are moved out of the vapor dome.

accomplish evaporation (boiling) and subsequent heating so that the fluid leaves as a superheated vapor. The high pressure superheated vapor is expanded in a turbine in order to produce power. The vapor is condensed by rejecting heat in the condenser (another heat exchanger) so that it enters the pump as liquid and starts this continuous flow cycle again.

The ideal Rankine cycle refers to the simple Rankine cycle configuration shown in Figure 8-4 operating with "perfect" components (i.e., components that individually have the highest possible performance). In order to completely specify the ideal Rankine cycle, it is only necessary to select the working fluid, the temperature of the heat source (T_H), the temperature of the heat rejection (T_C), and the pressure in the boiler (P_b).

```
$UnitSystem SI K Pa J mass radian
T_H=800 [K]                          "high temperature reservoir"
T_C=325 [K]                          "low temperature reservoir"
P_b=2.64 [MPa]*convert(MPa,Pa)       "boiler pressure"
F$='Water'                           "working fluid"
```

Note that the working fluid is specified by the string variable, F$, so that it is easy to change the working fluid in the model.

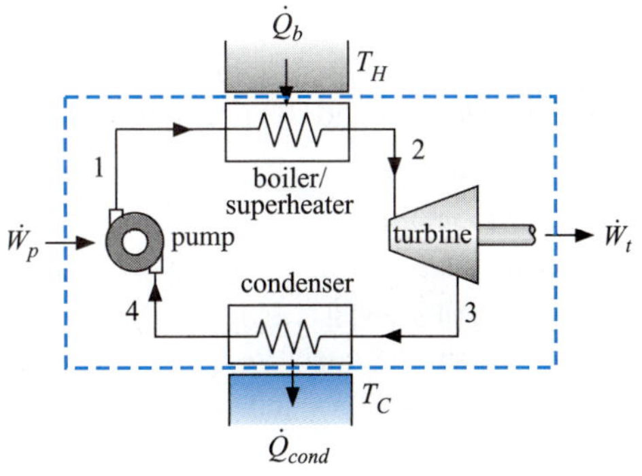

Figure 8-4: Simple Rankine cycle.

The boiler/superheater in an ideal Rankine cycle is a set of perfect heat exchangers. The term "perfect" here implies that there is no loss of pressure associated with the flow and that the temperature of the fluid leaving the boiler will be equal to the temperature of the hot reservoir. (i.e., the approach temperature difference is zero). State 2, the state of the fluid leaving the boiler, is fixed by the temperature, $T_2 = T_H$, and pressure, $P_2 = P_b$. The remaining properties (h_2 and s_2) are obtained.

```
"State 2, Boiler exit/Turbine inlet"
P[2]=P_b                            "pressure"
T[2]=T_H                            "temperature"
h[2]=enthalpy(F$,P=P[2],T=T[2])     "specific enthalpy"
s[2]=entropy(F$,P=P[2],T=T[2])      "specific entropy"
```

The condenser in the ideal Rankine cycle is also a perfect heat exchanger. Therefore, the saturated liquid leaving the condenser does so at the temperature of the cold reservoir (i.e., the approach temperature difference is zero). State 4, the state of the fluid leaving the condenser, is specified by the quality and temperature, $x_4 = 0$ and $T_4 = T_C$. The remaining properties (h_4, s_4, and P_4) are determined.

```
"State 4, Condenser exit/Pump inlet"
x[4]=0 [-]                          "quality"
T[4]=T_C                            "temperature"
s[4]=entropy(F$,x=x[4],T=T[4])      "specific entropy"
h[4]=enthalpy(F$,x=x[4],T=T[4])     "specific enthalpy"
P[4]=pressure(F$,x=x[4],T=T[4])     "pressure"
```

The pump in the ideal Rankine cycle has an isentropic efficiency of unity; it is a reversible and adiabatic device. Therefore, an entropy balance on the pump provides:

$$s_1 = s_4 \tag{8-9}$$

State 1 is fixed by the specific entropy and pressure, $P_1 = P_2$. The remaining properties at state 1 are determined (h_1 and T_1).

```
"State 1, Pump exit/Boiler inlet"
s[1]=s[4]                               "specific entropy"
P[1]=P[2]                               "pressure"
h[1]=enthalpy(F$,s=s[1],P=P[1])         "specific enthalpy"
T[1]=temperature(F$,s=s[1],P=P[1])      "temperature"
```

The turbine in an ideal Rankine cycle has an isentropic efficiency of unity; it is a reversible device. Therefore, an entropy balance on the turbine provides:

$$s_3 = s_2 \tag{8-10}$$

State 3 is fixed by the specific entropy and pressure, $P_3 = P_4$. The remaining properties at state 3 (h_3, T_3, and x_3) are determined.

```
"State 3, Turbine exit/Condenser inlet"
s[3]=s[2]                               "specific entropy"
P[3]=P[4]                               "pressure"
h[3]=enthalpy(F$,s=s[3],P=P[3])         "specific enthalpy"
T[3]=temperature(F$,s=s[3],P=P[3])      "temperature"
x[3]=quality(F$,s=s[3],P=P[3])          "quality"
```

With each of the states in the cycle determined, it is convenient to analyze the components on a per unit mass flow rate basis. This method is typical of cycle analysis; the states associated with the interfaces between different pieces of equipment can be specified independent of the overall size of the system. The mass flow rate can then be selected to achieve a desired output. An energy balance on the pump provides:

$$\dot{m}\,h_4 + \dot{W}_p = \dot{m}\,h_1 \tag{8-11}$$

Equation (8-11) can be rearranged:

$$\frac{\dot{W}_p}{\dot{m}} = h_1 - h_4 \tag{8-12}$$

Energy balances on the boiler, turbine, and condenser are:

$$\frac{\dot{Q}_b}{\dot{m}} = h_2 - h_1 \tag{8-13}$$

$$\frac{\dot{W}_t}{\dot{m}} = h_2 - h_3 \tag{8-14}$$

$$\frac{\dot{Q}_{cond}}{\dot{m}} = h_3 - h_4 \tag{8-15}$$

```
"Component energy balances"
W_dot_p\m_dot=h[1]-h[4]          "pump"
Q_dot_b\m_dot=h[2]-h[1]          "boiler"
W_dot_t\m_dot=h[2]-h[3]          "turbine"
Q_dot_cond\m_dot=h[3]-h[4]       "condenser"
```

Note that the backslash (\) character used in the variables W_dot_p\m_dot, Q_dot_b\m_dot, etc. is part of the EES variable name with no other significance as far as EES is concerned.

It is a good idea to check your work on a cycle problem by performing an overall energy balance on the system. An overall energy balance for a system consisting of all of the components is shown by the dashed line in Figure 8-4:

$$\frac{\dot{W}_p}{\dot{m}} + \frac{\dot{Q}_b}{\dot{m}} = \frac{\dot{W}_t}{\dot{m}} + \frac{\dot{Q}_{cond}}{\dot{m}} \tag{8-16}$$

Each of the terms in Eq. (8-16) has been previously determined by considering the components individually. Therefore, Eq. (8-16) provides a check on our solution:

$$check_1 = \frac{\dot{W}_t}{\dot{m}} + \frac{\dot{Q}_{cond}}{\dot{m}} - \frac{\dot{W}_p}{\dot{m}} - \frac{\dot{Q}_b}{\dot{m}} \tag{8-17}$$

```
check_1=W_dot_t\m_dot+Q_dot_cond\m_dot-W_dot_p\m_dot-Q_dot_b\m_dot
                              "check on solution using an overall energy balance"
```

Solving these equations indicates that $check_1 = 0$ J/kg and therefore our solution satisfies an overall energy balance on the system.

The net power provided by the Rankine cycle is the turbine power less the pump power:

$$\frac{\dot{W}_{net}}{\dot{m}} = \frac{\dot{W}_t}{\dot{m}} - \frac{\dot{W}_p}{\dot{m}} \tag{8-18}$$

```
W_dot_net\m_dot=W_dot_t\m_dot-W_dot_p\m_dot    "net power per mass"
```

The mass flow rate can be selected in order to achieve a desired net power output. For example, if the power plant must provide $\dot{W}_{net} = 50$ MW then the mass flow rate must be selected according to:

$$\dot{W}_{net} = \left(\frac{\dot{W}_{net}}{\dot{m}}\right)\dot{m} \tag{8-19}$$

```
W_dot_net=50 [MW]*convert(MW,W)        "net power output"
W_dot_net=W_dot_net\m_dot*m_dot        "mass flow rate"
```

which indicates that $\dot{m} = 43.81$ kg/s. The rates of energy transfer associated with each component can be determined according to:

$$\dot{W}_p = \left(\frac{\dot{W}_p}{\dot{m}}\right)\dot{m} \tag{8-20}$$

$$\dot{Q}_b = \left(\frac{\dot{Q}_b}{\dot{m}}\right)\dot{m} \tag{8-21}$$

$$\dot{W}_t = \left(\frac{\dot{W}_t}{\dot{m}}\right)\dot{m} \tag{8-22}$$

$$\dot{Q}_{cond} = \left(\frac{\dot{Q}_{cond}}{\dot{m}}\right)\dot{m} \tag{8-23}$$

```
W_dot_p=W_dot_p\m_dot*m_dot            "pump power consumption"
Q_dot_b=Q_dot_b\m_dot*m_dot            "rate of boiler heat input"
W_dot_t=W_dot_t\m_dot*m_dot            "turbine power output"
Q_dot_cond=Q_dot_cond\m_dot*m_dot      "rate of condenser heat rejection"
```

which leads to $\dot{W}_p = 0.12$ MW, $\dot{Q}_b = 144.6$ MW, $\dot{W}_t = 50.12$ MW, and $\dot{Q}_{cond} = 94.6$ MW. Notice that the pump power consumed by the cycle is a small fraction of the turbine power produced by the cycle. This is an attractive characteristic of the Rankine cycle; it has a small *back work ratio* (*bwr*). The back work ratio is defined as the power consumed by the cycle to the gross (rather than net) power produced by the cycle. For a Rankine cycle, the back work ratio is the ratio of the pump power to the turbine power:

$$bwr = \frac{\dot{W}_p}{\dot{W}_t} \tag{8-24}$$

```
bwr=W_dot_p/W_dot_t                    "back work ratio"
```

which results in $bwr = 0.0023$; the pump power required to run the cycle is 0.23% of the turbine power produced. Obviously, a small back work ratio is a good thing. If the back work ratio reaches unity then the efficiency must be zero since the net power output, Eq. (8-18), will be zero.

The efficiency of any power cycle is the ratio of the net work provided by the cycle to the heat input to the cycle (i.e., the ratio of what you want to what you must pay for). For the Rankine cycle, the efficiency is:

$$\eta_{Rankine} = \frac{\dfrac{\dot{W}_{net}}{\dot{m}}}{\dfrac{\dot{Q}_b}{\dot{m}}} \tag{8-25}$$

```
eta_Rankine=W_dot_net\m_dot/Q_dot_b\m_dot     "efficiency"
```

The performance of a power cycle is sometimes characterized by the *heat rate*. The heat rate is defined as the inverse of the efficiency:

$$\text{heat rate} = \frac{\dot{Q}_b}{\dot{W}_{net}} \left(\text{typically reported in units } \frac{\text{Btu}}{\text{kW–hr}}\right) \tag{8-26}$$

The heat rate is the rate of heat transfer required per unit of power produced and therefore, even though the quantity is dimensionless, it is usually provided in units of Btu/kW-hr because heat transfer rates are often given in Btu/hr and power in kW.

```
heat_rate_BtupkWhr=Q_dot_b/W_dot_net*convert(-,Btu/kW-hr)     "heat rate, in Btu/kW-hr"
```

Running the EES code developed in this section with the given inputs (water at T_H = 800 K, T_C = 325 K, P_b = 2.64 MPa) leads to the *T-s* diagram shown in Figure 8-3. The cycle has an efficiency $\eta_{Rankine} = 0.346$ (34.6%) and a heat rate 9869 Btu/kW-hr. The efficiency of the Rankine cycle is less than the efficiency of the Carnot cycle that is discussed in Section 5.5:

$$\eta_{max} = 1 - \frac{T_C}{T_H} \tag{8-27}$$

```
eta_max=1-T_C/T_H     "maximum possible efficiency"
```

which provides $\eta_{max} = 0.594$ (59.4%) for these conditions.

The Rankine cycle modeled in this section receives heat at a rate of $\dot{Q}_b = 144.6$ MW to the boiler from a thermal reservoir at $T_H = 800$ K and rejects heat from the condenser to a sink at $T_C = 325$ K. If the Rankine cycle were replaced by a reversible power cycle that operates with the Carnot efficiency given by Eq. (8-27), then it would produce the maximum possible power that could be obtained using this resource:

$$\dot{W}_{max} = \dot{Q}_b \, \eta_{max} \tag{8-28}$$

```
W_dot_max=Q_dot_b*eta_max     "maximum possible rate of power"
```

which leads to $\dot{W}_{max} = 85.9$ MW. The Rankine cycle actually produced power at a lower rate, $\dot{W}_{net} = 50$ MW. The difference between these quantities is *lost power production capacity*:

$$\dot{W}_{lost} = \dot{W}_{max} - \dot{W}_{net} \tag{8-29}$$

```
W_dot_lost=W_dot_max-W_dot_net     "rate at which power producing capability is lost"
```

which leads to $\dot{W}_{lost} = 35.9$ MW. The loss of power producing capability must be related to entropy generation within the cycle (or exergy destruction, according to Chapter 7).

An entropy balance on a closed system that consists of the entire cycle, as shown by the dashed line in Figure 8-4, is:

$$\frac{\dot{Q}_b}{T_H} + \dot{S}_{gen} = \frac{\dot{Q}_{cond}}{T_C} \tag{8-30}$$

```
Q_dot_b/T_H+S_dot_gen=Q_dot_cond/T_C    "overall entropy balance"
```

which leads to $\dot{S}_{gen} = 110.3$ kW/K. A second check on our solution can be obtained by confirming that the product of the rate of entropy generation, $\dot{S}_{gen}$, and the heat rejection temperature, T_C (i.e., the dead state temperature) is equal to the lost power production capacity, $\dot{W}_{lost}$.

$$check_2 = T_C\,\dot{S}_{gen} - \dot{W}_{lost} \tag{8-31}$$

```
check_2=T_C*S_dot_gen-W_dot_lost    "check on solution with lost work"
```

Solving the equation shows that $check_2 = 0$ W (or a number close to zero, to within numerical precision). The two checks on our solution, $check_1$ from Eq. (8-17) and $check_2$ from Eq. (8-31), are considered best-practice when carrying out a cycle analysis and provide a powerful method for catching calculation errors.

Where is entropy being generated within the ideal Rankine cycle? All of the components in the ideal Rankine cycle are modeled as being "perfect". The pump and turbine are both modeled as reversible devices and therefore no entropy is generated within these components. The condenser is a perfect heat exchanger that is transferring heat from fluid that is always at 325 K (see Figure 8-3) to a cold reservoir that is also at 325 K. Therefore, the condenser is also reversible. An entropy balance on the condenser will confirm this result:

$$\dot{m}\,s_3 + \dot{S}_{gen,cond} = \dot{m}\,s_4 + \frac{\dot{Q}_{cond}}{T_C} \tag{8-32}$$

```
m_dot*s[3]+S_dot_gen_cond=m_dot*s[4]+Q_dot_cond/T_C    "entropy balance on condenser"
```

which results in $\dot{S}_{gen,cond} = 0$ W/K. Note that $\dot{S}_{gen,cond}$ may not be exactly zero depending on the precision of the property data and solving tolerances in EES, but it will certainly be very small compared to the total rate of entropy generation in the cycle, $\dot{S}_{gen} = 110.3$ kW/K.

The boiler is also assumed to be a perfect heat exchanger. However, as discussed in Section 6.6.6, a perfect heat exchanger will rarely be a reversible device. The condenser in this cycle is only reversible because there is no temperature difference between the condensing steam and the cold reservoir. The boiler successfully heats the fluid to the hot reservoir temperature with no pressure drop; it is therefore doing the best that it can do. Nevertheless, the boiler is not reversible. There is an unavoidable heat transfer through a temperature difference that occurs within the boiler. The temperature difference becomes evident when you examine Figure 8-3 closely. Water enters the boiler at $T_1 = 325$ K and experiences a heat transfer from $T_H = 800$ K. The temperature difference is reduced as the fluid is warmed in the boiler; however, there is clearly a substantial entropy generation rate associated with the operation of the boiler. An entropy balance on the boiler provides:

$$\dot{m}\,s_1 + \frac{\dot{Q}_b}{T_H} + \dot{S}_{gen,b} = \dot{m}\,s_2 \tag{8-33}$$

```
m_dot*s[1]+Q_dot_b/T_H+S_dot_gen_b=m_dot*s[2]    "entropy balance on boiler"
```

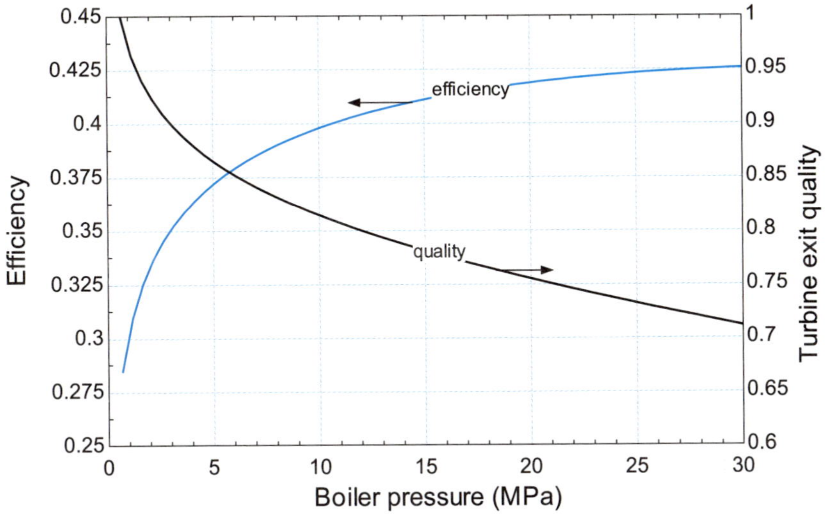

Figure 8-5: Efficiency of the ideal Rankine cycle as a function of boiler pressure. The cycle is operating with water at $T_H = 800$ K and $T_C = 325$ K. Also shown is the quality of the fluid leaving the turbine as a function of boiler pressure.

which results in $\dot{S}_{gen,b} = 110.3$ kW/K. This entropy generation is equal to the total rate of entropy generation in the cycle calculated in Eq. (8-30), indicating that the boiler is the only source of entropy generation for this ideal Rankine cycle. Many of the modifications to the Rankine cycle that are discussed subsequently are directed at reducing the entropy generation that is otherwise inherent in the transfer of heat from a high temperature source to the low temperature fluid entering the boiler. The simple model that has been developed in this section for the ideal Rankine cycle can be used to understand the effect of the boiler pressure, heat source temperature, and heat sink temperature on the cycle performance.

Effect of Boiler Pressure

Figure 8-5 illustrates the efficiency of the ideal Rankine cycle shown in Figure 8-3 as a function of the boiler pressure. The efficiency increases with boiler pressure because the saturation temperature of the water increases and therefore heat is being added to water in the boiler at a higher average temperature. The result is a reduction in the rate of entropy generation and an increase in the efficiency of the cycle.

Figure 8-6 illustrates the states associated with three different ideal Rankine cycles (A, B, and C) overlaid onto a *T-s* diagram. These cycles each use water with $T_H = 800$ K and $T_C = 325$ K. However, the boiler pressure is varied; cycle A has $P_b = 2.64$ MPa (the same as in Figure 8-3) while cycles B and C use elevated boiler pressures, $P_b =$ 10 MPa and 20 MPa, respectively. The elevation of the average temperature of the water in the boiler is apparent in Figure 8-6 and tends to reduce the entropy generation and increase the efficiency. However, there is a practical limitation on the boiler pressure for a cycle that uses water as the working fluid. As the boiler pressure increases, the state of the fluid leaving the turbine, state 3, moves further into the vapor dome. Figure 8-5 shows the quality of the fluid leaving the turbine, x_3, as a function of boiler pressure. The presence of liquid in the turbine is undesirable. A turbine converts the energy of a fluid at high pressure and temperature into mechanical power. Nozzles within the turbine convert the high pressure into high velocity, as discussed in Section 4.4.4.

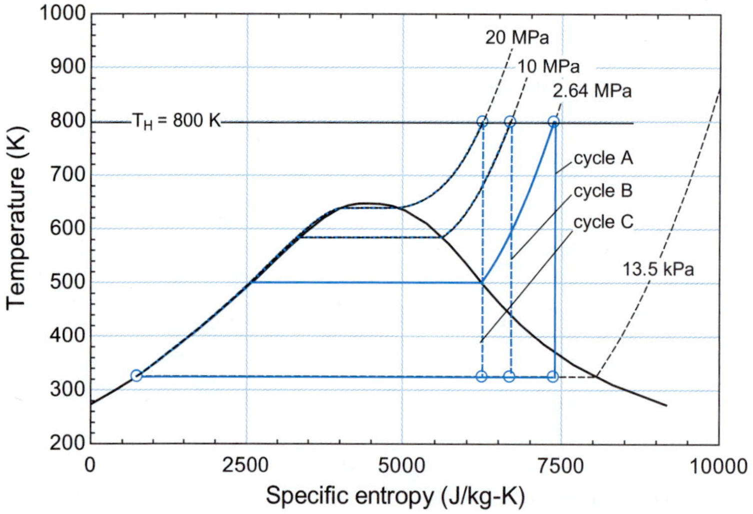

Figure 8-6: Three ideal Rankine cycles on a *T-s* diagram, each with a different boiler pressure. The cycle is operating with water at $T_H = 800$ K and $T_C = 325$ K. Cycle A is at $P_b = 2.64$ (i.e., the same condition as Figure 8-3), cycle B is at $P_b = 10$ MPa, and cycle C is at $P_b = 20$ MPa.

The high velocity fluid leaving the nozzles impinges on the turbine blades and thereby transfers the momentum of the fluid to the rotating blades, causing a torque. The blades are connected to the turbine shaft that transfers the mechanical power out of the turbine (e.g., to a generator where it is converted into electrical power). The turbine blades spin at a high rotational velocity. If there are liquid droplets in the fluid, as there will be if the turbine operates under the vapor dome, then they will tend to cause the turbine blades to erode. Liquid droplets form when the quality is less than one, but the droplet size depends on the quality. Operation of the turbine is not recommended when the outlet state has a quality lower than about 0.90. According to Figure 8-5, this constraint will limit the boiler pressure in the ideal Rankine cycle to about 3 MPa.

There are several ways to ensure that the quality at the turbine exit is sufficiently high to avoid turbine blade erosion. One alternative is to use a different working fluid. The shape of the vapor dome in temperature-specific entropy coordinates differs significantly for different fluids. Figure 8-7, for example, shows a *T-s* diagram for an ideal Rankine cycle using toluene as the working fluid with $T_C = 325$ K, $T_H = 500$ K, and $P_b = 1.171$ MPa. Notice that the boiler pressure has been set to the saturation pressure of toluene at T_H. However, due to the concave nature of the right side of the vapor dome, the turbine outlet state lies in the superheat region. This feature of toluene completely eliminates concerns about turbine blade erosion. Toluene is a non-toxic organic fluid that is used in some geothermal power cycle systems.

Although the problem of operating in the two-phase region can be eliminated by using an appropriate fluid, water is the working fluid of choice in most Rankine cycles. Water is non-flammable, non-toxic and it can be heated to higher temperatures than almost any other organic compound without dissociating. Also, you can't beat the cost. The Rankine cycle with reheat, discussed in Section 8.2.3, provides one method of operating a Rankine cycle using water at high boiler pressure while avoiding operation of the turbine in a two-phase condition. Even with the reheat modification, there are other limitations on the boiler pressure. For example, the critical pressure of water is about 22 MPa and the boiler pressure cannot exceed this limit. Doing so would result in a "supercritical" cycle, which is possible, but it would need to be designed in a different

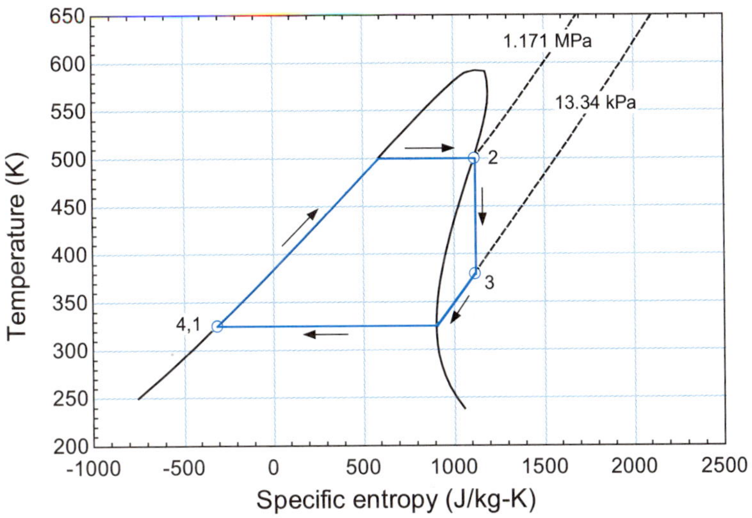

Figure 8-7: Ideal Rankine cycle using toluene as the working fluid with $T_H = 500$ K, $T_C = 325$ K, and $P_b = 1.171$ MPa.

manner than the Rankine cycle. As the boiler pressure increases, the specific enthalpy change of vaporization decreases and eventually becomes zero at the critical point. Increasing the boiler pressure therefore reduces the specific enthalpy change of the fluid as it flows through the boiler, $h_2 - h_1$. As a result, the working fluid must be circulated at a higher rate in order to maintain a fixed power output. However, the major limitation on the boiler pressure is economics. The boiler is a huge heat exchanger and the higher the pressure that it must withstand, the more it will cost to build. The boiler pressure in large modern Rankine cycle systems is typically between 9 and 13 MPa.

Effect of Heat Source Temperature

Equation (8-27) indicates that the efficiency of a reversible power cycle will be increased by increasing the heat source temperature. The efficiency of the ideal Rankine cycle will also increase with the heat source temperature. Figure 8-8 illustrates the efficiency as a function of T_H for the ideal Rankine cycle using water as a working fluid with $P_b =$ 2.64 MPa and $T_C = 325$ K. There is a practical limit to the heat source temperature that is dictated by the turbine metallurgy. The turbine blades are continuously exposed to high temperature steam at state 2 in Figure 8-4. The maximum temperature that they can continuously withstand without significant reduction in their time to failure is in the range of 500°C to 600°C (930°F to 1110°F or 773 K to 873 K) which corresponds to the creep limit of stainless steel. Improvements in metallurgy have resulted in a gradual but continuous increase in the maximum allowable operating temperature over the years with a corresponding improvement in efficiency of the Rankine cycle. The maximum operating temperature for modern steam turbines is approximately 840 K (1050°F).

Effect of Heat Sink Temperature

Equation (8-27) also indicates that the efficiency of a reversible power cycle will be increased by decreasing the heat sink temperature. This is true for the Rankine cycle as well. Figure 8-9 illustrates the efficiency of the ideal Rankine cycle using water with $P_b = 2.64$ MPa and $T_H = 800$ K as a function of T_C.

Clearly, it is advantageous to operate at the lowest possible sink temperature. Reducing the sink temperature will lead to a reduction in the condenser pressure (i.e., the

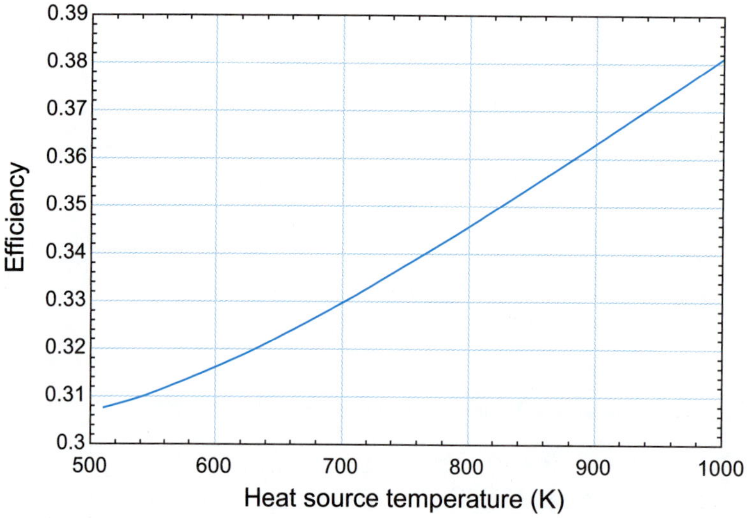

Figure 8-8: Efficiency of the ideal Rankine cycle as a function of T_H. The cycle operates with water at $P_b = 2.64$ MPa and $T_C = 325$ K.

pressure of the fluid leaving the turbine at state 3 and entering the pump at state 4). The condenser pressure as a function of the heat sink temperature is also shown in Figure 8-9. Utilities will try to use the lowest temperature heat sink that is available. For example, many large power plants are often located near a lake or river that provides a heat sink temperature that is usually lower than the air temperature, particularly during the summer when the peak loads on the utility occur. Power plants will generally operate more efficiently in the winter due to the lower available heat sink temperature. If a lake or river is not available, the sink temperature can also be lowered by evaporation of water in a cooling tower, as discussed in Section 12.5. Because the cycle is sensitive to the heat sink temperature, the design of the condenser can have a substantial impact on the cycle efficiency. The condenser in a Rankine cycle consists of many small diameter

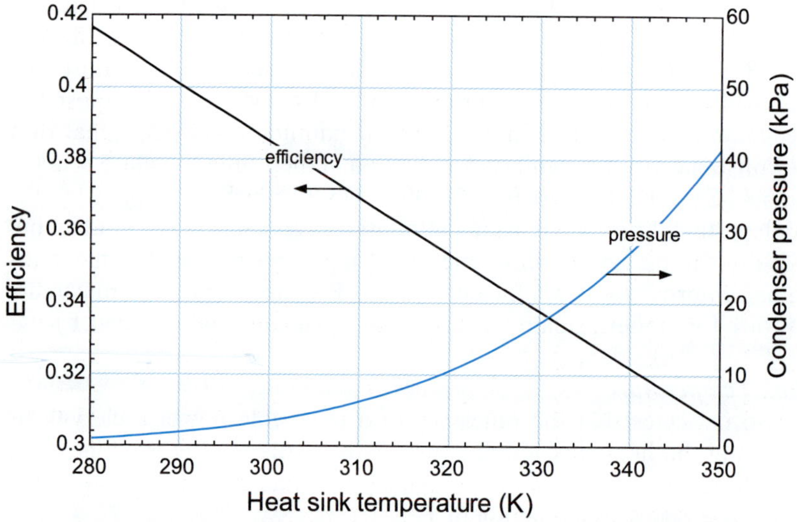

Figure 8-9: Efficiency of the ideal Rankine cycle as a function of T_C. The cycles operates with water at $P_b = 2.64$ MPa and $T_H = 800$ K. Also shown is the condenser pressure as a function of T_C.

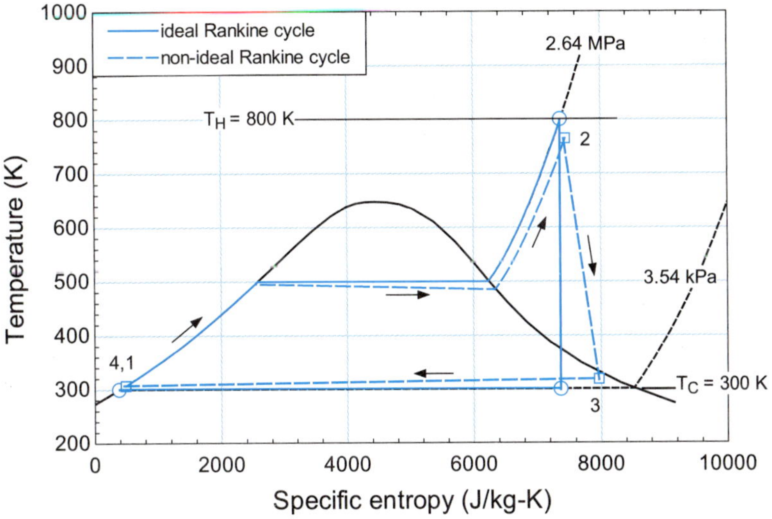

Figure 8-10: Temperature-specific entropy diagram for an ideal Rankine cycle using water with $T_H = 800$ K, $T_C = 300$ K, $P_b = 2.64$ MPa. The non-ideal Rankine cycle operating at the same conditions is overlaid onto the *T-s* diagram ($\eta_t = 0.85$, $\eta_p = 0.65$, $\Delta T_{cond} = 7$ K, $\Delta T_b = 35$ K, $\Delta P_{cond} = 4.5$ kPa, $\Delta P_b = 792$ kPa).

tubes that provide a large surface area for heat transfer between the condensing steam and the cooling water. A larger surface area will reduce the temperature difference that is required between the steam and cooling sink and increase the cycle efficiency. One disadvantage of lowering the sink temperature is that it tends to decrease the quality of the steam at the turbine outlet which may lead to increased turbine blade erosion.

8.2.2 The Non-Ideal Rankine Cycle

The components that are used to implement a Rankine cycle will not be perfect, as was assumed in the ideal Rankine cycle analysis presented in Section 8.2.1. The turbine and pump will have isentropic efficiencies (η_t and η_p, respectively) that are less than unity. Flow through boiler and condenser will be accompanied by pressure drops (ΔP_b and ΔP_{cond}, respectively). Finally, the approach temperature differences characterizing the boiler and condenser (ΔT_b and ΔT_{cond}) will not be zero. Rather, the fluid passing through the boiler will not be heated entirely to T_H and the fluid passing through the condenser will not be cooled entirely to T_C.

Figure 8-10 illustrates ideal and non-ideal Rankine cycles overlaid onto a temperature-specific entropy diagram and drawn to scale. Both cycles have the same operating conditions; $T_H = 800$ K, $T_C = 300$ K, and $P_b = 2.64$ MPa with water as the working fluid. The impact of the non-ideal boiler performance is evident in the pressure loss ($P_2 < 2.64$ MPa) and approach temperature difference ($T_2 < T_H$). The same attributes can be identified for the condenser. The non-ideal turbine causes the specific entropy of the fluid to rise as it passes through the turbine ($s_3 > s_2$). One beneficial effect of the reduced turbine efficiency is an increase in the quality of the fluid at the turbine exit, which reduces the possibility of blade erosion. A specific entropy increase also occurs in the pump, although it cannot be detected on the *T-s* diagram. All of these effects will act to reduce the efficiency of the cycle.

The analysis of the non-ideal Rankine cycle begins with the specification of the operating conditions; these include the working fluid, hot and cold reservoir temperatures

(T_H and T_C), and boiler inlet pressure (P_b):

```
$UnitSystem SI K Pa J Mass Radian
"Cycle operating conditions"
T_H=800 [K]                          "high temperature reservoir"
T_C=325 [K]                          "low temperature reservoir"
P_b_MPa=2.64 [MPa]                   "boiler inlet pressure, in MPa"
P_b=P_b_MPa*convert(MPa,Pa)          "boiler inlet pressure"
F$='Water'                           "working fluid"
```

The performance parameters that characterize the non-ideal components (η_t, η_p, ΔT_{cond}, ΔP_{cond}, ΔT_b and ΔP_b) must be specified.

```
"Performance parameters"
eta_t=0.85 [-]                       "turbine efficiency"
eta_p=0.65 [-]                       "pump efficiency"
DT_cond=7 [K]                        "condenser approach temperature difference"
DT_b=35 [K]                          "boiler approach temperature difference"
DP_cond=4.5 [kPa]*convert(kPa,Pa)    "condenser pressure drop"
DP_b=792 [kPa]*convert(kPa,Pa)       "boiler pressure drop"
```

The pressure of the fluid leaving the boiler at state 2 is the boiler inlet pressure less the boiler pressure drop:

$$P_2 = P_b - \Delta P_b \tag{8-34}$$

The temperature at state 2 is the hot reservoir temperature less the boiler approach temperature difference:

$$T_2 = T_H - \Delta T_b \tag{8-35}$$

The state of the fluid leaving the boiler, state 2, is fixed by the pressure and temperature. The specific enthalpy and specific entropy at state 2 (h_2 and s_2) are computed.

```
"State 2, Boiler exit/Turbine inlet"
P[2]=P_b-DP_b                        "pressure"
T[2]=T_H-DT_b                        "temperature"
h[2]=enthalpy(F$,P=P[2],T=T[2])      "specific enthalpy"
s[2]=entropy(F$,P=P[2],T=T[2])       "specific entropy"
```

The fluid leaving the condenser at state 4 is assumed to be saturated liquid ($x_4 = 0$). The temperature of the fluid at state 4 is the sum of the cold reservoir temperature and the condenser approach temperature difference:

$$T_4 = T_C + \Delta T_{cond} \tag{8-36}$$

The state of the fluid leaving the condenser is fixed by the quality and temperature. The remaining properties (s_4, h_4, P_4, and v_4) are determined.

```
"State 4, Condenser exit/Pump inlet"
x[4]=0 [-]                           "quality"
T[4]=T_C+DT_cond                     "temperature"
s[4]=entropy(F$,x=x[4],T=T[4])       "specific entropy"
h[4]=enthalpy(F$,x=x[4],T=T[4])      "specific enthalpy"
P[4]=pressure(F$,x=x[4],T=T[4])      "pressure"
v[4]=volume(F$,x=x[4],T=T[4])        "specific volume"
```

The pressure of the fluid leaving the pump is specified, $P_1 = P_b$. However, the fluid passing through the non-ideal pump does not undergo an isentropic process. An entropy balance on a reversible pump operating under the same conditions provides:

$$s_{s,1} = s_4 \tag{8-37}$$

The specific entropy and pressure specify the state of the fluid leaving the reversible pump and allow the determination of the specific enthalpy leaving the reversible pump ($h_{s,1}$). An energy balance on the reversible pump provides:

$$\frac{\dot{W}_{s,p}}{\dot{m}} = h_{s,1} - h_4 \tag{8-38}$$

The actual pump power is larger than the reversible pump power by an amount that is related to the pump efficiency:

$$\frac{\dot{W}_p}{\dot{m}} = \frac{\dot{W}_{s,p}}{\dot{m}} \frac{1}{\eta_p} \tag{8-39}$$

An energy balance on the actual adiabatic pump is:

$$h_1 = h_4 + \frac{\dot{W}_p}{\dot{m}} \tag{8-40}$$

The specific enthalpy and pressure fix the state of the fluid leaving the pump, state 4. The remaining properties (s_1 and T_1) are computed.

```
"State 1, Pump exit/Boiler inlet"
P[1]=P_b                                  "boiler inlet pressure"
s_s[1]=s[4]                               "specific entropy leaving reversible pump"
h_s[1]=enthalpy(F$,s=s_s[1],P=P[1])       "specific enthalpy leaving reversible pump"
W_dot_s_p\m_dot=h_s[1]-h[4]               "work per mass for reversible pump"
W_dot_p\m_dot=W_dot_s_p\m_dot/eta_p       "work per mass for actual pump"
h[1]=h[4]+W_dot_p\m_dot                   "specific enthalpy leaving actual pump"
s[1]=entropy(F$,h=h[1],P=P[1])            "specific entropy leaving actual pump"
T[1]=temperature(F$,h=h[1],P=P[1])        "temperature leaving actual pump"
```

Solving these equations results in $\dot{W}_p/\dot{m} = 4097$ J/kg. Note that the liquid processed by the pump can often be modeled as an incompressible substance. Section 6.6.3 discusses the incompressible substance model applied to a pump:

$$\frac{\dot{W}_{p,IC}}{\dot{m}} = \frac{v_4 (P_1 - P_4)}{\eta_p} \tag{8-41}$$

```
W_dot_p_IC\m_dot=v[4]*(P[1]-P[4])/eta_p   "work per mass for pump using incompressible model"
```

which leads to $\dot{W}_{p,IC}/\dot{m} = 4099$ J/kg. The incompressible model leads to a pump power that is within 0.05% of the actual answer in this case.

The pressure at state 3, the entrance to the condenser and exit of the turbine, is specified according to:

$$P_3 = P_4 + \Delta P_{cond} \tag{8-42}$$

A turbine analysis is required to completely specify state 3. An entropy balance on a reversible turbine operating under the same conditions provides:

$$s_{s,3} = s_2 \tag{8-43}$$

The specific entropy and pressure fix the state of the fluid leaving the reversible turbine and allow the determination of the specific enthalpy leaving the reversible turbine ($h_{s,3}$). An energy balance on the reversible turbine is:

$$\frac{\dot{W}_{s,t}}{\dot{m}} = h_2 - h_{s,3} \tag{8-44}$$

The actual turbine power is less than the reversible turbine power by an amount that is related to the turbine efficiency:

$$\frac{\dot{W}_t}{\dot{m}} = \frac{\dot{W}_{s,t}}{\dot{m}}\eta_t \tag{8-45}$$

An energy balance on the actual turbine is:

$$h_3 = h_2 + \frac{\dot{W}_t}{\dot{m}} \tag{8-46}$$

The specific enthalpy and pressure fix the state of the fluid leaving the turbine, state 3. The remaining properties (s_3, T_3, and x_3) are computed.

```
"State 3, Turbine exit/Condenser inlet"
P[3]=P[4]+DP_cond                       "turbine exit pressure"
s_s[3]=s[2]                             "specific entropy leaving reversible turbine"
h_s[3]=enthalpy(F$,s=s_s[3],P=P[3])     "specific enthalpy leaving reversible turbine"
W_dot_s_t\m_dot=h[2]-h_s[3]             "work per mass for reversible turbine"
W_dot_t\m_dot=W_dot_s_t\m_dot*eta_t     "work per mass for actual turbine"
h[3]=h[2]-W_dot_t\m_dot                 "specific enthalpy leaving actual turbine"
s[3]=entropy(F$,h=h[3],P=P[3])          "specific entropy leaving actual turbine"
T[3]=temperature(F$,h=h[3],P=P[3])      "temperature leaving actual turbine"
x[3]=quality(F$,h=h[3],P=P[3])          "quality leaving actual turbine"
```

Energy balances on the boiler and condenser provide:

$$\frac{\dot{Q}_b}{\dot{m}} = h_2 - h_1 \tag{8-47}$$

$$\frac{\dot{Q}_{cond}}{\dot{m}} = h_3 - h_4 \tag{8-48}$$

```
"Component energy balances"
Q_dot_b\m_dot=h[2]-h[1]                 "boiler"
Q_dot_cond\m_dot=h[3]-h[4]              "condenser"
```

We should check our model using an overall energy balance applied to the system shown in Figure 8-4.

$$check_1 = \frac{\dot{W}_t}{\dot{m}} + \frac{\dot{Q}_{cond}}{\dot{m}} - \frac{\dot{W}_p}{\dot{m}} - \frac{\dot{Q}_b}{\dot{m}} \tag{8-49}$$

```
check_1=W_dot_t\m_dot+Q_dot_cond\m_dot-W_dot_p\m_dot-Q_dot_b\m_dot
                        "check on solution with overall energy balance"
```

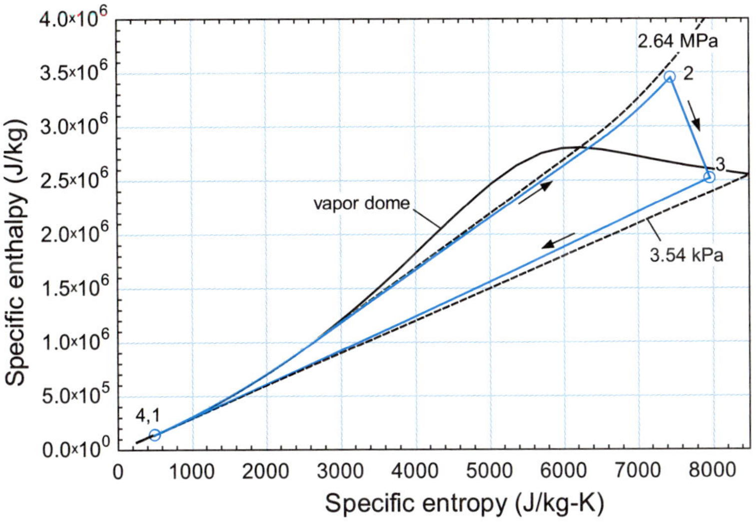

Figure 8-11: The non-ideal Rankine cycle shown in Figure 8-10 overlaid onto an *h-s* diagram.

Solving shows that $check_1 = 0$ J/kg. The net work per unit mass, overall efficiency, back work ratio, and heat rate of the non-ideal Rankine cycle are computed in the same way as for the ideal Rankine cycle, using Eqs. (8-18), (8-25), (8-24) and (8-26).

```
W_dot_net\m_dot=W_dot_t\m_dot-W_dot_p\m_dot          "net power per mass"
eta_Rankine=W_dot_net\m_dot/Q_dot_b\m_dot             "efficiency"
bwr=W_dot_p\m_dot/W_dot_t\m_dot                       "back work ratio"
heat_rate_BtupkWhr=Q_dot_b\m_dot/W_dot_net\m_dot*convert(-,Btu/kW-hr)
                                                      "heat rate, in Btu/kW-hr"
```

After solving, $\eta_{Rankine} = 0.257$ (25.7%) with a heat rate of 13,266 Btu/kW-hr. We can compare this performance to that of an ideal Rankine cycle operating under the same conditions by setting the performance of each component to be "perfect".

```
"Performance parameters for ideal cycle"
eta_t=1 [-]                          "turbine efficiency"
eta_p=1 [-]                          "pump efficiency"
DT_cond=0 [K]                        "condenser approach temperature difference"
DT_b=0 [K]                           "boiler approach temperature difference"
DP_cond=0 [kPa]*convert(kPa,Pa)      "condenser pressure drop"
DP_b=0 [kPa]*convert(kPa,Pa)         "boiler pressure drop"
```

The performance of the ideal Rankine cycle operating under these conditions is $\eta_{Rankine} = 0.346$ (34.6%) with a corresponding heat rate of 9869 Btu/kW-hr, which agree with the results found in Section 8.2.1 for the ideal Rankine cycle.

The small back work ratio associated with an ideal Rankine cycle was noted in Section 8.2.1; the back work ratio for the non-ideal Rankine cycle (obtained by re-setting each of the performance parameters to their original values) is also small, $bwr = 0.00495$. The reason for this characteristic has to do with the density that the working fluid has when it is expanded in a Rankine cycle as compared to its density when it is compressed. Figure 8-11 illustrates the non-ideal Rankine cycle shown in Figure 8-10 on specific enthalpy-specific entropy coordinates.

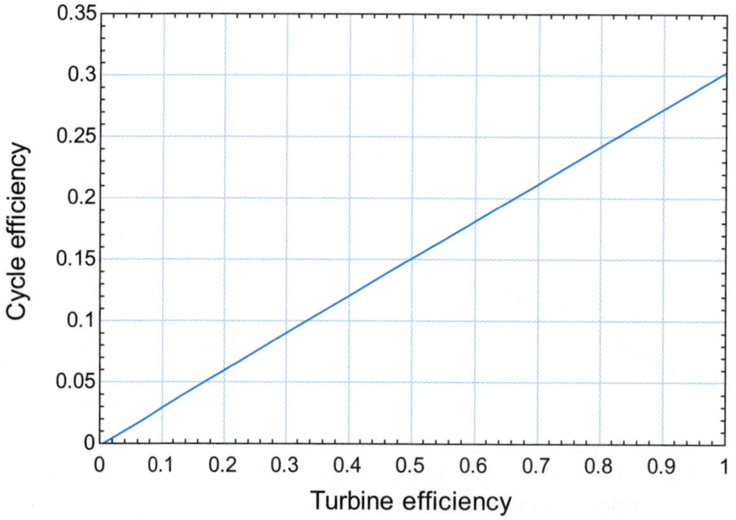

Figure 8-12: Efficiency of the non-ideal Rankine cycle as a function of turbine efficiency. The cycle is operating with water at $T_H = 800$ K, $T_C = 325$ K, $P_b = 2.64$ MPa, $\eta_p = 0.65$, $\Delta T_{cond} = 7$ K, $\Delta T_b = 35$ K, $\Delta P_{cond} = 4.5$ kPa, $\Delta P_b = 792$ kPa.

Notice that the isobars in Figure 8-11 (3.54 kPa and 2.64 MPa) tend to converge as they reach the saturated liquid side of the vapor dome and extend into the compressed liquid region. This behavior occurs because the specific volume of liquid water is extremely small (i.e., its density is large). Conversely, the isobars diverge as they reach the saturated vapor side of the vapor dome and extend into the superheated vapor region. This behavior occurs because the specific volume of vapor is large. To see why the specific volume causes this behavior, recall the fundamental property relationship derived in Section 6.2:

$$dh = T\,ds + v\,dP \tag{8-50}$$

Integrating Eq. (8-50) along a line of constant entropy provides the isentropic specific enthalpy change between any two isobars (e.g., the boiler pressure and the condenser pressure):

$$\Delta h_s = \int_{P_{cond}}^{P_b} v\,dP \tag{8-51}$$

In the compressed liquid region, evaluation of Eq. (8-51) results in a very small value because the specific volume is small; this behavior is seen by the closely spaced isobars in Figure 8-11. The small change in specific enthalpy indicates that the work per unit mass required to compress liquid water is also small. In the superheated vapor region, evaluation of Eq. (8-51) results in a much larger value because the specific volume of vapor is large, as seen by the highly separated isobars. As a result, the work per unit mass obtained by the expansion of water vapor is large.

The small back work ratio that is associated with the Rankine cycle allows the cycle to continue to produce power even as the efficiency of the turbine is reduced. Figure 8-12 illustrates the cycle efficiency as a function of the turbine efficiency for the conditions that are used to generate Figure 8-10 and Figure 8-11. Notice that the cycle efficiency remains positive, albeit small, even if the turbine efficiency is as low as $\eta_t = 1\%$. The power produced by the turbine need only be larger than that required by the pump in order for the cycle to produce net power. This characteristic is likely the reason that the Rankine cycle was used to produce power before high efficiency turbines were available.

8.2.3 Modifications to the Rankine Cycle

Rankine cycles are used extensively by utilities to generate electrical power, primarily to meet baseload demand (i.e., to meet the electricity demand that is present continuously throughout the year). Therefore, a large Rankine cycle may produce 100's of MW of electrical power continuously for many years. If a Rankine power plant runs continuously at 500 MW for one year and the electricity that it produces is purchased at an average price of \$0.10/kW-hr then this electricity is worth approximately \$438 million dollars. Even a small improvement in the efficiency of the plant will lead to millions of dollars in additional revenue or reduced fuel costs. This section investigates some of the cycle modifications that are used to increase the efficiency of the Rankine cycle.

Reheat

The effect of the boiler pressure on the performance of a Rankine cycle is investigated in Section 8.2.1. It was shown that increasing the boiler pressure tends to improve the performance of the Rankine cycle. However, increasing the boiler pressure also moves the turbine exit state further into the vapor dome and therefore increases the possibility that liquid droplets will erode the turbine blades. This trend is shown in Figure 8-6. The reheat modification addresses this limitation. A schematic of the Rankine cycle with reheat is shown in Figure 8-13(a). The ideal reheat cycle (i.e., the reheat cycle with perfect components) is shown in Figure 8-13(b) on a temperature-specific entropy diagram. The reheat cycle allows the use of a high boiler pressure by terminating the expansion at state 3 at an intermediate pressure. The intermediate pressure is sufficiently high that the expansion process remains in or near the superheated vapor region, as shown in Figure 8-13(b). The fluid leaving the high pressure turbine is subsequently heated at nearly constant pressure in a reheater. The fluid is then expanded to the condenser pressure in a low pressure turbine. The exit of the low pressure turbine remains at a sufficiently high quality so that blade erosion is prevented. There is also some increase in efficiency in the reheat cycle because the heat transfer from the high temperature reservoir to the fluid occurs with the fluid at a higher average temperature.

In this section, a simple model of the ideal Rankine cycle with reheat is developed and used to understand the impact of the reheat pressure. The operating conditions associated with the ideal Rankine cycle with reheat are specified by the working fluid, hot and cold temperatures (T_H and T_C), boiler pressure (P_b) and reheat pressure (P_{rh}).

```
$UnitSystem SI K Pa J Mass Radian
T_H=800 [K]                         "temperature of the high temperature reservoir"
T_C=325 [K]                         "temperature of the low temperature reservoir"
P_b=10e6 [Pa]                       "boiler pressure"
P_b_MPa=P_b*convert(Pa,MPa)         "in MPa"
P_rh=750e3 [Pa]                     "reheat pressure"
P_rh_MPa=P_rh*convert(Pa,MPa)       "in MPa"
F$='Water'                          "working fluid"
```

The ideal cycle utilizes a reversible pump, reversible turbines (high pressure and low pressure), and perfect heat exchangers (boiler, condenser and reheater). The state of the fluid leaving the boiler (state 2) is fixed by the pressure, $P_2 = P_b$, and temperature, $T_2 = T_H$. The specific enthalpy and specific entropy (h_2 and s_2) are computed. The state of the

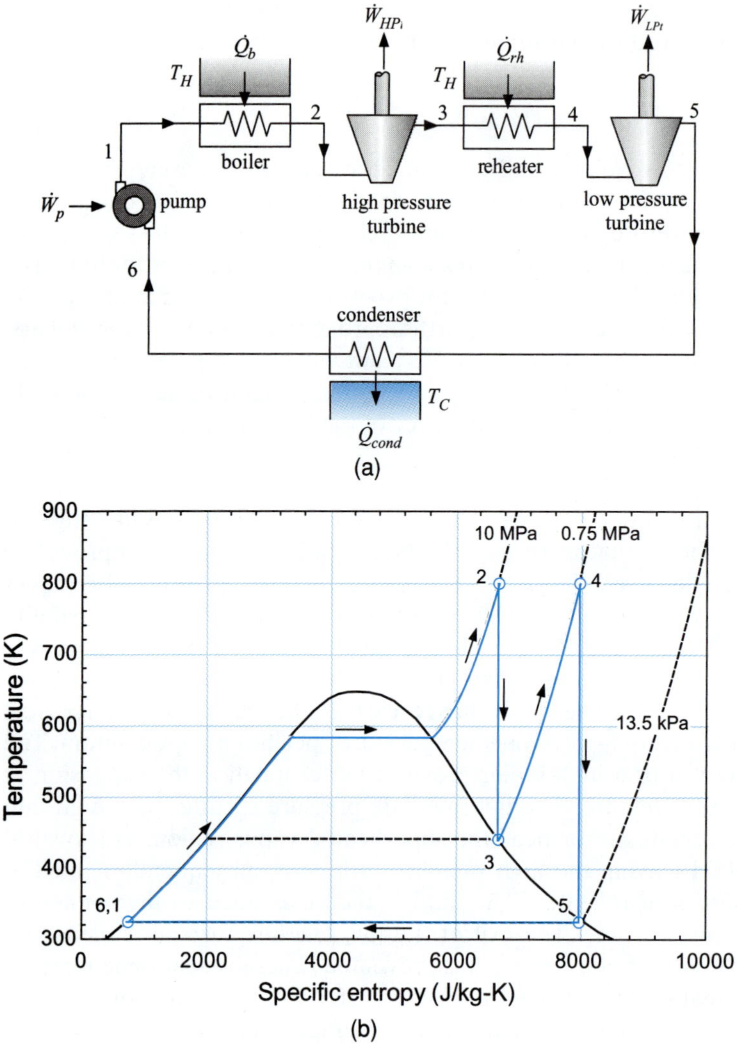

Figure 8-13: (a) Schematic of the Rankine cycle with reheat and (b) the ideal Rankine cycle with reheat on a *T-s* diagram; the cycle is operating with water at $T_H = 800$ K, $T_C = 325$ K, $P_b = 10$ MPa, and $P_{rh} = 0.75$ MPa.

fluid leaving the condenser (state 6) is fixed by the temperature, $T_6 = T_C$, and quality, $x_6 = 0$. The specific entropy, specific enthalpy, and pressure (s_6, h_6, and P_6) are computed.

```
"State 2, Boiler exit/High pressure turbine inlet"
P[2]=P_b                                   "pressure"
T[2]=T_H                                   "temperature"
h[2]=enthalpy(F$,P=P[2],T=T[2])            "specific enthalpy"
s[2]=entropy(F$,P=P[2],T=T[2])             "specific entropy"

"State 6, Condenser exit/Pump inlet"
x[6]=0 [-]                                 "quality"
T[6]=T_C                                   "temperature"
s[6]=entropy(F$,x=x[6],T=T[6])             "specific entropy"
h[6]=enthalpy(F$,x=x[6],T=T[6])            "specific enthalpy"
P[6]=pressure(F$,x=x[6],T=T[6])            "pressure"
```

The ideal cycle utilizes a reversible pump, therefore $s_1 = s_6$. State 1 is fixed by the specific entropy and pressure, $P_1 = P_b$. The specific enthalpy and temperature (h_1 and T_1) are determined.

```
"State 1, Pump exit/Boiler inlet"
s[1]=s[6]                                  "specific entropy"
P[1]=P_b                                   "pressure"
h[1]=enthalpy(F$,s=s[1],P=P[1])            "specific enthalpy"
T[1]=temperature(F$,s=s[1],P=P[1])         "temperature"
```

The specific entropy does not change as the fluid passes through the reversible high pressure turbine, therefore $s_3 = s_2$. State 3 is fixed by the specific entropy and pressure, $P_3 = P_{rh}$. The specific enthalpy, temperature, and quality (h_3, T_3, and x_3) are computed.

```
"State 3, High pressure turbine exit/Condenser inlet"
s[3]=s[2]                                  "specific entropy"
P[3]=P_rh                                  "pressure"
h[3]=enthalpy(F$,s=s[3],P=P[3])            "specific enthalpy"
T[3]=temperature(F$,s=s[3],P=P[3])         "temperature"
x[3]=quality(F$,s=s[3],P=P[3])             "quality"
```

The reheater exit state, state 4, is fixed by the temperature, $T_4 = T_H$, and pressure, $P_4 = P_{rh}$. The specific enthalpy and specific entropy (h_4 and s_4) are computed.

```
"State 4, Reheater exit/Low pressure turbine inlet"
T[4]=T_H                                   "temperature"
P[4]=P_rh                                  "pressure"
h[4]=enthalpy(F$,T=T[4],P=P[4])            "specific enthalpy"
s[4]=entropy(F$,T=T[4],P=P[4])             "specific entropy"
```

The specific entropy does not change as the fluid passes through the reversible low pressure turbine, therefore $s_5 = s_4$. State 5 is fixed by the specific entropy and pressure, $P_5 = P_6$. The specific enthalpy, temperature, and quality (h_5, T_5, and x_5) are computed.

```
"State 5, Low pressure turbine exit/Condenser inlet"
P[5]=P[6]                                  "pressure"
s[5]=s[4]                                  "specific entropy"
h[5]=enthalpy(F$,P=P[5],s=s[5])            "specific enthalpy"
T[5]=temperature(F$,P=P[5],s=s[5])         "temperature"
x[5]=quality(F$,P=P[5],s=s[5])             "quality"
```

The computed states are shown in Figure 8-13(b). Energy balances for each component provide:

$$\frac{\dot{W}_p}{\dot{m}} = h_1 - h_6 \tag{8-52}$$

$$\frac{\dot{Q}_b}{\dot{m}} = h_2 - h_1 \tag{8-53}$$

$$\frac{\dot{W}_{HPt}}{\dot{m}} = h_2 - h_3 \tag{8-54}$$

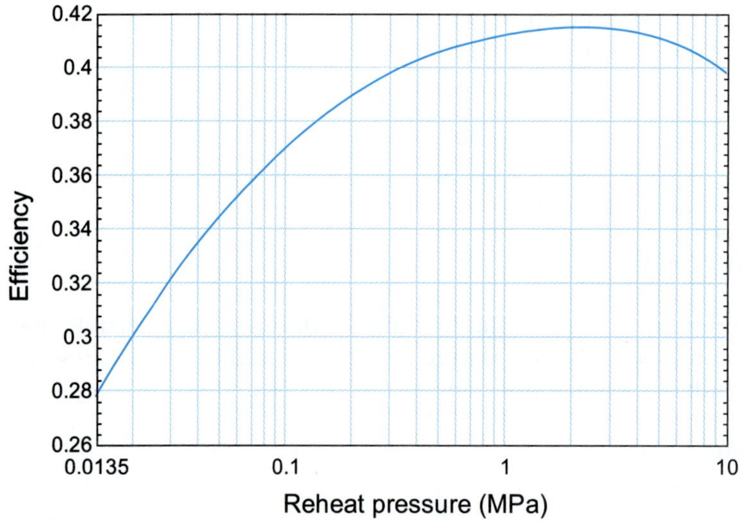

Figure 8-14: Efficiency of the ideal Rankine cycle with reheat as a function of the reheat pressure. The cycle is operating with water with $T_H = 800$ K, $T_C = 325$ K, and $P_b = 10$ MPa.

$$\frac{\dot{Q}_{rh}}{\dot{m}} = h_4 - h_3 \tag{8-55}$$

$$\frac{\dot{W}_{LPt}}{\dot{m}} = h_4 - h_5 \tag{8-56}$$

$$\frac{\dot{Q}_{cond}}{\dot{m}} = h_5 - h_6 \tag{8-57}$$

```
"Component energy balances"
W_dot_p\m_dot=h[1]-h[6]          "pump"
Q_dot_b\m_dot=h[2]-h[1]          "boiler"
W_dot_HPt\m_dot=h[2]-h[3]        "high pressure turbine"
Q_dot_rh\m_dot=h[4]-h[3]         "reheater"
W_dot_LPt\m_dot=h[4]-h[5]        "low pressure turbine"
Q_dot_cond\m_dot=h[5]-h[6]       "condenser"
```

The efficiency of the Rankine cycle with reheat is:

$$\eta_{Rankine} = \frac{\dfrac{\dot{W}_{HPt}}{\dot{m}} + \dfrac{\dot{W}_{LPt}}{\dot{m}} - \dfrac{\dot{W}_p}{\dot{m}}}{\dfrac{\dot{Q}_b}{\dot{m}} + \dfrac{\dot{Q}_{rh}}{\dot{m}}} \tag{8-58}$$

```
eta_Rankine=(W_dot_HPt\m_dot+W_dot_LPt\m_dot-W_dot_p\m_dot)/&
  (Q_dot_b\m_dot+Q_dot_rh\m_dot)                              "efficiency"
```

which results in $\eta_{Rankine} = 0.4098$ (40.98%). The reheat pressure can be adjusted in order to optimize the performance of the cycle. Figure 8-14 illustrates the efficiency of the ideal Rankine cycle with reheat as a function of the reheat pressure. The optimal reheat pressure under these conditions is about $P_{rh} = 2$ MPa.

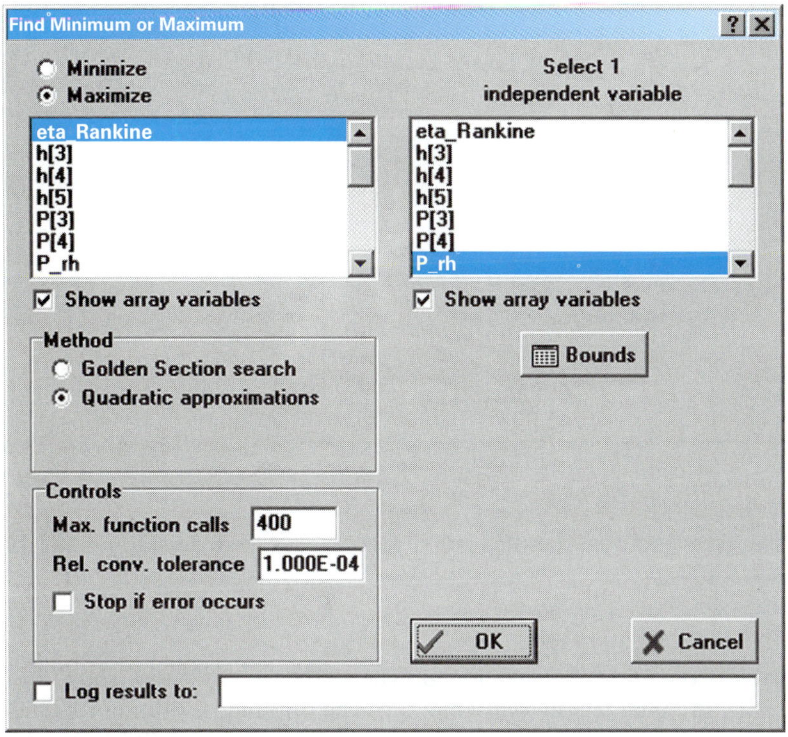

Figure 8-15: Find Minimum or Maximum dialog.

The optimal reheat pressure can be obtained using EES' internal optimization algorithms. Comment out the specified value of the reheat pressure:

```
{P_rh=750e3}   "reheat pressure"
```

Select Min/Max from the Calculate menu. The Find Minimum or Maximum dialog is shown in Figure 8-15. Select Maximize and the variable eta_Rankine. The independent variable is P_rh. Select the Bounds button and specify a reasonable guess and upper and lower bounds. The optimizer will identify an optimal reheat pressure, $P_{rh} = 2.21$ MPa, and the associated maximum efficiency, $\eta_{Rankine} = 0.4149$ (41.49%). The values of all of the variables obtained in the optimization process are provided in the Solutions Window.

It is possible to parametrically change a variable and carry out a series of optimizations, one for each value of the variable. This process is useful in order to identify the sensitivity of the optimal operating condition or solution to various system parameters. For example, it is of interest to examine how the optimal reheat pressure and efficiency vary with the boiler pressure. Create a Parametric table that includes the variables P_b, P_rh, eta_Rankine, and x_3. The value of the boiler pressure in the table is set so that it ranges from 1 MPa to 25 MPa. The reheat pressure is optimized for each value of the boiler pressure by commenting out both the specified boiler pressure and the specified reheat pressure in the Equations window:

```
{P_b=10e6 [Pa]}   "boiler pressure"
{P_rh=750e3}      "reheat pressure"
```

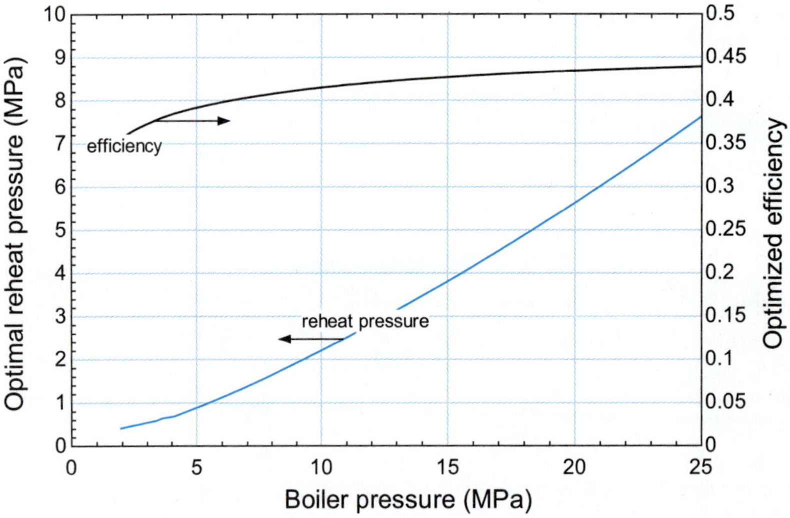

Figure 8-16: Optimal reheat pressure and the associated cycle efficiency as a function of the boiler pressure (for water at $T_C = 325$ K and $T_H = 800$ K).

Select Min/Max Table from the Calculate menu. The value of eta_Rankine should be maximized by adjusting the independent variable P_rh. The optimized solution for each value of boiler pressure is recorded in each row of the table. Figure 8-16 illustrates the optimal reheat pressure and the associated efficiency as a function of the boiler pressure.

Regeneration

The analysis of the ideal Rankine cycle presented in Section 8.2.1 showed that all of the entropy generation in the cycle occurred in the boiler. Specifically, the heat transfer from the high temperature reservoir to the cold fluid leaving the pump and entering the boiler occurs through a large temperature gradient and it is therefore quite irreversible. In addition, the cold fluid entering the boiler at state 1 can cause thermal stress to the boiler. The regenerative modification to the Rankine cycle reduces this entropy generation by using a lower temperature heat source to accomplish the initial heating of the cold fluid leaving the pump before it enters the boiler. The fluid leaving the pump is preheated using fluid from within the cycle that is extracted from the turbine at an intermediate pressure.

Figure 8-17(a) provides a schematic of the Rankine cycle with regeneration and the ideal cycle is shown on a temperature-specific entropy diagram in Figure 8-17(b). The hot fluid leaving the high pressure turbine is extracted at an intermediate pressure, as was done in the reheat modification. The fluid leaving the high pressure turbine at state 3 is at an intermediate temperature (between T_C and T_H). Therefore, heat transfer from the fluid at state 3 to the cold fluid leaving the pump at state 1 will generate less entropy than in the conventional Rankine cycle where the heat transfer is obtained directly from the high temperature reservoir at temperature T_H.

The fluid leaving the high pressure turbine at state 3 is split into two streams. Some fraction of the fluid stream (f) is diverted to the heat exchanger that is used for the preheating process, referred to as a feedwater heater. The remaining fluid continues to the low pressure turbine and the condenser. The closed feedwater heater shown in Figure 8-17(a) keeps the two streams separate, allowing them to transfer heat across a

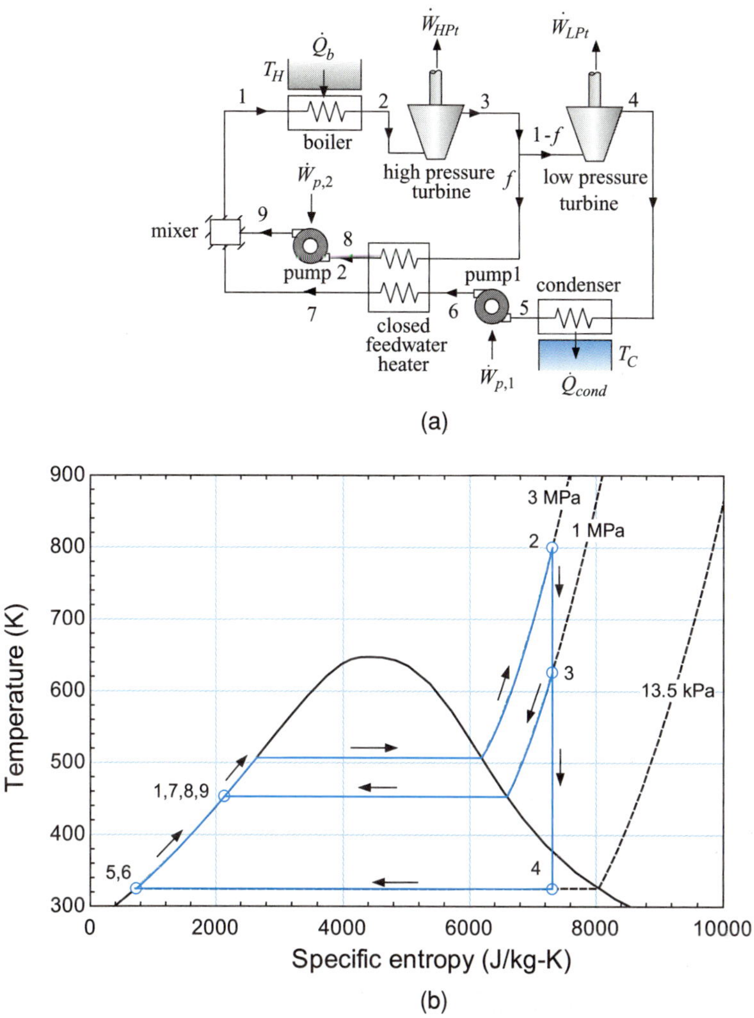

Figure 8-17: (a) Schematic of a Rankine cycle with regeneration accomplished by a closed feedwater heater, and (b) the ideal cycle shown on a temperature-specific entropy diagram.

pressure boundary. A closed feedwater heater is a conventional heat exchanger of the type discussed in Sections 4.4.7 and 6.6.6. The fraction of flow that is diverted from the main flow and fed to the closed feedwater heater (f) is typically controlled so that the fluid exits at state 8 as saturated liquid. This information, coupled with the approach temperature difference or effectiveness of the closed feedwater heater, can be used to fix the states of each of the flows entering and leaving the device. An energy balance on the closed feedwater heater provides:

$$f\, h_3 + (1 - f)\, h_6 = f\, h_8 + (1 - f)\, h_7 \qquad (8\text{-}59)$$

Equation (8-59) is used to determine the value of f. The extracted liquid leaving the closed feedwater heater is then pumped up to the boiler pressure at state 9 and mixed with the main flow stream, as shown in Figure 8-17(a). The temperature of the fluid leaving the closed feedwater heater at state 7 depends on the approach temperature difference for the heat exchanger. Figure 8-17(b) shows the ideal cycle on a temperature-specific entropy diagram. In Figure 8-17(b) the approach temperature difference of the

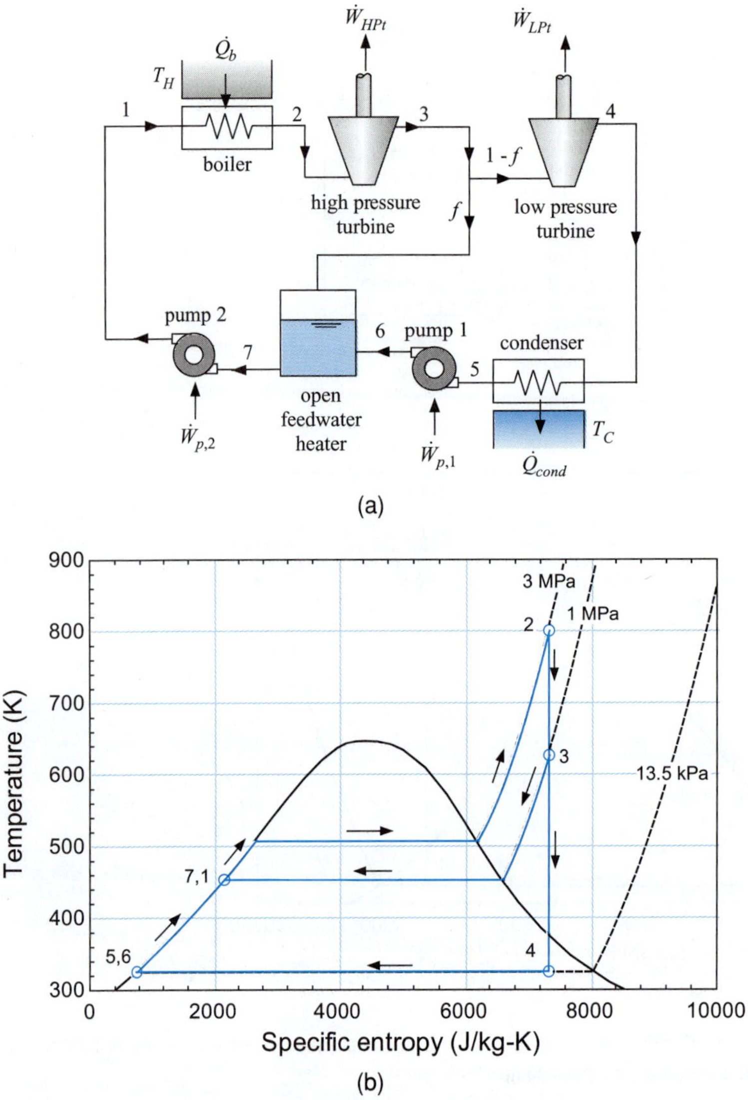

Figure 8-18: (a) Schematic of a Rankine cycle with regeneration accomplished with an open feedwater heater and (b) the ideal cycle shown on a temperature-specific entropy diagram.

feedwater heater is zero and therefore the main flow is heated to the temperature of the condensate entering pump 2, i.e., $T_7 = T_8$.

The regenerative modification can also be accomplished using open feedwater heaters, as shown in Figure 8-18(a). The ideal cycle is shown on a temperature-specific entropy diagram in Figure 8-18(b). The hot fluid leaving the high pressure turbine is extracted at an intermediate pressure and split into two streams. Some fraction of the fluid stream (f) is diverted to the open feedwater heater, which is essentially a mixing tank. The remaining flow is sent to the low pressure turbine and condenser. The main flow leaving the condenser is pumped to the extraction pressure and fed to the open feedwater heater where it mixes with the vapor extracted from the turbine. The fraction of flow diverted through the open feedwater heater is typically controlled so that the fluid exits the open feedwater heater at state 7 as saturated liquid. The feedwater heater is typically assumed to be adiabatic. An energy balance on the open feedwater heater determines the fraction of the stream (f) that must be extracted.

$$f\,h_3 + (1 - f)\,h_6 = h_7 \tag{8-60}$$

The liquid leaving the open feedwater heater is then pumped to the boiler pressure at state 1 by pump 1.

Figure 8-17(a) and Figure 8-18(a) both show a power plant with a single feedwater heater. In practice, large power plants that use water as the working fluid may use five or more feedwater heaters. The performance advantage of using feedwater heaters is not obvious. Using one or more feedwater heaters increases the temperature of the fluid entering the boiler (state 1) which, all else being the same, results in less input energy required to bring the fluid to state 2. This effect also reduces thermal stresses resulting from adding cold fluid to the hot metal parts of the boiler. However, by extracting a portion of the working fluid between the turbines, less fluid enters the downstream turbine(s) and therefore less power is produced. Thermodynamic models are particularily useful to quantify the benefits of various cycle modifications. Example 8.2-1 illustrates the analysis of a regenerative cycle that utilizes both reheat and regeneration. The optimum extraction pressures for this type of cycle are determined through a thermodynamic analysis.

EXAMPLE 8.2-1: SOLAR TROUGH POWER PLANT

A solar parabolic trough power plant is a Rankine cycle that uses solar energy as its heat input. Incident solar energy is focused by parabolic mirrors onto a receiver pipe that carries a heat transfer fluid. Figure 1 illustrates a large field of parabolic trough receivers that are necessary to capture a large amount of solar energy.

Figure 1: Parabolic solar trough collectors used to collect solar energy. The pipe located at the focal point of the receiver carries a heat transfer fluid.

The heat transfer fluid is heated as it flows through the field and then returns to the power plant. A heat exchanger is used to transfer heat from the fluid circulating through the collectors to the working fluid used by the power plant in order to provide the thermal energy that drives the power cycle. The Rankine cycle shown in Figure 2 is used to convert the thermal energy into electrical power. The cycle employs both reheat and regeneration with both open and closed feedwater heaters.

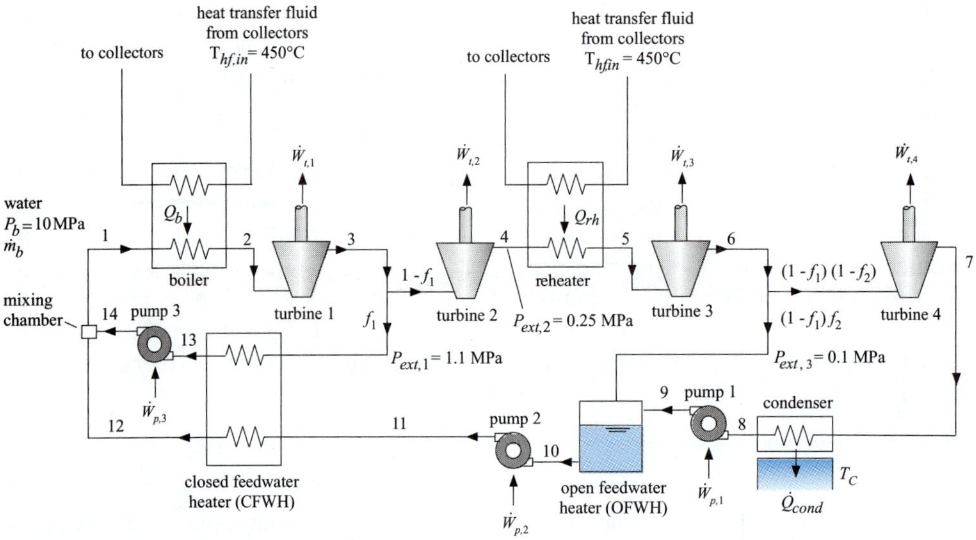

Figure 2: Schematic of the Rankine cycle power plant with reheat and regeneration using both an open and a closed feedwater heater.

The Rankine cycle uses water as the working fluid. The boiler pressure is P_b = 10 MPa. The heat transfer fluid returning from the solar collector field is at $T_{hf,in}$ = 450°C. The plant rejects heat to a temperature reservoir at T_C = 30°C. We will optimize the extraction pressures as part of this problem. However, initially assume that the fluid is extracted at state 3 from turbine 1 at $P_{ext,1}$ = 1.1 MPa. The fraction of the flow fed to the closed feedwater heater is f_1. The fluid is subsequently extracted at state 4 from turbine 2 at $P_{ext,2}$ = 0.25 MPa and reheated. The fluid is finally extracted at state 6 from turbine 3 at $P_{ext,3}$ = 0.1 MPa. The fraction of the flow extracted at state 6 that is fed to the open feedwater heater is f_2. The remainder of the flow passes through turbine 4 to the condenser. The efficiencies of the four turbines are $\eta_{t,1}$ = 0.87, $\eta_{t,2}$ = 0.90, $\eta_{t,3}$ = 0.92, and $\eta_{t,4}$ = 0.93. The flow leaving the condenser is pumped to $P_{ext,3}$ with pump 1 having efficiency $\eta_{p,1}$ = 0.65. Saturated liquid is pulled from the open feedwater heater at state 10 and pumped to the boiler pressure with pump 2 having efficiency $\eta_{p,2}$ = 0.67. The flow through the closed feedwater heater is controlled so that the extracted fluid leaving at state 13 is saturated liquid. The approach temperature difference associated with the closed feedwater heater is ΔT_{CFHW} = 2 K; therefore, the flow leaving the closed feedwater heater at state 12 is at $T_{12} = T_{13} - \Delta T_{CFHW}$. The extracted fluid leaving the closed feedwater heater at state 13 is pumped to the boiler pressure at state 14 with pump 3 having efficiency $\eta_{p,3}$ = 0.69. The approach temperature differences associated with the boiler and the reheater are ΔT_b = 15 K and ΔT_{rh} = 10 K, respectively. The pinch points for both of these heat exchangers occur at their warm end. Therefore, water leaves the boiler at $T_2 = T_{hf,in} - \Delta T_b$ and it leaves the reheater at

$T_5 = T_{hf,in} - \Delta T_{rh}$. The approach temperature difference associated with the condenser is $\Delta T_{cond} = 5$ K. Neglect the pressure loss that occurs through all of the heat exchangers.

a) Develop two procedures to facilitate the analysis. One procedure should be capable of analyzing any of the four turbines in the system and the other should be applicable to any of the three pumps in the system.

There are four turbines and three pumps in the system. Each pump and turbine is analyzed in the same manner. In order to facilitate the analysis of the cycle, the turbine and pump calculations are placed in procedures that can be called from the main section of the EES program. The development of procedures is discussed in Example 6.6-2. The Turbine procedure is placed at the top of the Equations window and carries out a generic turbine analysis that can be applied to any of the four turbines in the system. The procedure takes as inputs the specific enthalpy and pressure at the turbine inlet (h_{in} and P_{in}), the outlet pressure (P_{out}), the fluid name (in the string variable F$), and the turbine efficiency (η_t). Note that specific enthalpy and pressure are used to fix the inlet state (rather than temperature and pressure, as was done in Example 6.6-2) because the inlet to the turbine may fall under the vapor dome as the extraction pressures are optimized; it is not possible to fix a state under the vapor dome knowing only temperature and pressure. The outputs returned by the procedure include the outlet specific enthalpy (h_{out}), temperature (T_{out}), and specific entropy (s_{out}).

```
Procedure Turbine(h_in,P_in,P_out, F$, eta_t: h_out, T_out, s_out)
```

The inlet specific entropy (s_{in}) is computed from the specific enthalpy and pressure. The specific entropy leaving a reversible turbine ($s_{s,out}$) is equal to the inlet specific entropy. The specific enthalpy leaving a reversible turbine ($h_{s,out}$) is specified by $s_{s,out}$ and P_{out}.

```
    s_in=entropy(F$,h=h_in,P=P_in)              "inlet entropy"
    s_s_out=s_in                                "specific entropy balance on reversible turbine"
    h_s_out=enthalpy(F$,P=P_out,s=s_s_out)      "specific enthalpy leaving reversible turbine"
```

The work per unit mass associated with a reversible turbine is obtained from an energy balance:

$$\frac{\dot{W}_{s,t}}{\dot{m}} = h_{in} - h_{s,out}$$

The work per unit mass associated with the actual turbine is:

$$\frac{\dot{W}_t}{\dot{m}} = \eta_t \frac{\dot{W}_{s,t}}{\dot{m}}$$

The specific enthalpy leaving the actual turbine is obtained from an energy balance:

$$h_{out} = h_{in} - \frac{\dot{W}_t}{\dot{m}}$$

The outlet state is specified by h_{out} and P_{out}. The specific entropy and temperature (s_{out} and T_{out}) are computed.

```
    W_dot_t_s\m_dot=h_in-h_s_out              "energy balance on reversible turbine"
    W_dot_t\m_dot=W_dot_t_s\m_dot*eta_t       "definition of isentropic efficiency"
    h_out=h_in-W_dot_t\m_dot                  "energy balance on actual turbine"
    T_out=temperature(F$,h=h_out,P=P_out)     "exit temperature"
    s_out=entropy(F$,h=h_out,P=P_out)         "exit specific entropy"
end
```

The Pump procedure is developed in a similar way. It takes as inputs the inlet specific enthalpy and pressure (h_{in} and P_{in}), the outlet pressure (P_{out}), the working fluid (in the string variable F$), and the pump efficiency (η_p). The outputs are the outlet specific enthalpy, temperature, and specific entropy (h_{out}, T_{out}, and s_{out}).

```
Procedure Pump(h_in,P_in,P_out, F$, eta_p: h_out, T_out, s_out)
```

The inlet specific entropy (s_{in}) is computed from the specific enthalpy and pressure. The specific entropy leaving a reversible pump ($s_{s,out}$) is equal to the inlet specific entropy. The specific enthalpy leaving a reversible pump ($h_{s,out}$) is specified by $s_{s,out}$ and P_{out}.

```
    s_in=entropy(F$,h=h_in,P=P_in)             "inlet specific entropy"
    s_s_out=s_in                               "entropy balance on reversible pump"
    h_s_out=enthalpy(F$,P=P_out,s=s_s_out)     "specific enthalpy leaving reversible pump"
```

The work per unit mass associated with a reversible pump is obtained from an energy balance:

$$\frac{\dot{W}_{s,p}}{\dot{m}} = h_{s,out} - h_{in}$$

The work per unit mass associated with the actual pump is:

$$\frac{\dot{W}_p}{\dot{m}} = \frac{1}{\eta_p}\frac{\dot{W}_{s,p}}{\dot{m}}$$

The specific enthalpy leaving the actual pump is obtained from an energy balance:

$$h_{out} = h_{in} + \frac{\dot{W}_p}{\dot{m}}$$

The outlet state is specified by h_{out} and P_{out}. The specific entropy and temperature (s_{out} and T_{out}) are computed.

```
    W_dot_s_p\m_dot=h_s_out-h_in              "energy balance on reversible pump"
    W_dot_p\m_dot=W_dot_s_p\m_dot/eta_p       "definition of isentropic efficiency"
    h_out=h_in+W_dot_p\m_dot                  "energy balance on actual compressor"
    T_out=temperature(F$,h=h_out,P=P_out)     "exit temperature"
    s_out=entropy(F$,h=h_out,P=P_out)         "exit specific entropy"
end
```

Notice that the equations in the procedures are explicit (i.e., all of the variables on the right side of the equations are known). The units for the variables in the procedures can be specified either in the Solutions window or in the Variable

Information dialog by selecting the appropriate tab ('Turbine' or 'Pump').

b) Determine the cycle efficiency.

The procedures, operating conditions and known cycle information are entered in EES:

```
$UnitSystem SI K Pa J Mass Radian
Procedure Turbine(h_in,P_in,P_out, F$, eta_t: h_out, T_out, s_out)
   s_in=entropy(F$,h=h_in,P=P_in)                 "inlet specific entropy"
   s_s_out=s_in                                   "entropy balance on reversible turbine"
   h_s_out=enthalpy(F$,P=P_out,s=s_s_out)         "specific enthalpy leaving reversible turbine"
   W_dot_t_s\m_dot=h_in-h_s_out                   "energy balance on reversible turbine"
   W_dot_t\m_dot=W_dot_t_s\m_dot*eta_t            "definition of isentropic efficiency"
   h_out=h_in-W_dot_t\m_dot                       "energy balance on actual turbine"
   T_out=temperature(F$,h=h_out,P=P_out)          "exit temperature"
   s_out=entropy(F$,h=h_out,P=P_out)              "exit specific entropy"
end

Procedure Pump(h_in,P_in,P_out,F$, eta_p: h_out, T_out, s_out)
   s_in=entropy(F$,h=h_in,P=P_in)                 "inlet specific entropy"
   s_s_out=s_in                                   "entropy balance on reversible pump"
   h_s_out=enthalpy(F$,P=P_out,s=s_s_out)         "specific enthalpy leaving reversible pump"
   W_dot_s_p\m_dot=h_s_out-h_in                   "energy balance on reversible pump"
   W_dot_p\m_dot=W_dot_s_p\m_dot/eta_p            "definition of isentropic efficiency"
   h_out=h_in+W_dot_p\m_dot                       "energy balance on actual compressor"
   T_out=temperature(F$,h=h_out,P=P_out)          "exit temperature"
   s_out=entropy(F$,h=h_out,P=P_out)              "exit specific entropy"
end

"operating conditions"
F$='Water'                                        "working fluid"
P_b=10 [MPa]*convert(MPa,Pa)                      "boiler pressure"
T_hf_in=converttemp(C,K,450 [C])                  "heat transfer fluid inlet temperature"
T_C=converttemp(C,K,30 [C])                       "cold reservoir temperature"
P_ext_1=1.1 [MPa]*convert(MPa,Pa)                 "first extraction pressure"
P_ext_2=0.25 [MPa]*convert(MPa,Pa)                "second extraction pressure"
P_ext_3=0.1 [MPa]*convert(MPa,Pa)                 "third extraction pressure"

"performance parameters"
eta_T_1=0.87 [-]                                  "first turbine efficiency"
eta_T_2=0.90 [-]                                  "second turbine efficiency"
eta_T_3=0.92 [-]                                  "third turbine efficiency"
eta_T_4=0.93 [-]                                  "fourth turbine efficiency"
eta_p_1=0.65 [-]                                  "first pump efficiency"
eta_p_2=0.67 [-]                                  "second pump efficiency"
eta_p_3=0.69 [-]                                  "third pump efficiency"
DT_b=15 [K]                                       "boiler approach temperature difference"
DT_rh=10 [K]                                      "reheater approach temperature difference"
DT_cond=5 [K]                                     "condenser approach temperature difference"
DT_CFHW=2 [K]                                     "closed feedwater heater approach temp. difference"
```

The cycle analysis proceeds by specifying as many states as possible given the input conditions. The state of the fluid leaving the boiler, state 2, is fixed by the pressure, $P_2 = P_b$, and temperature:

$$T_2 = T_{hf,in} - \Delta T_b$$

The specific enthalpy and specific entropy (h_2 and s_2) are determined.

```
"state 2"
P[2]=P_b                              "pressure"
T[2]=T_hf_in-DT_b                     "temperature"
h[2]=enthalpy(F$,T=T[2],P=P[2])       "specific enthalpy"
s[2]=entropy(F$,T=T[2],P=P[2])        "specific entropy"
```

The state of the fluid leaving the reheater, state 5, is fixed by the pressure, $P_5 = P_{ext,2}$, and temperature:

$$T_5 = T_{hf,in} - \Delta T_{rh}$$

The specific enthalpy and specific entropy (h_5 and s_5) are determined.

```
"state 5"
P[5]=P_ext_2                          "pressure"
T[5]=T_hf_in-DT_rh                    "temperature"
h[5]=enthalpy(F$,T=T[5],P=P[5])       "specific enthalpy"
s[5]=entropy(F$,T=T[5],P=P[5])        "specific entropy"
```

The state of the fluid leaving the condenser, state 8, is fixed by the quality, $x_8 = 0$, and temperature:

$$T_8 = T_C + \Delta T_{cond}$$

The pressure, specific enthalpy, and specific entropy (P_8, h_8 and s_8) are determined.

```
"state 8"
T[8]=T_C+DT_cond                      "temperature"
x[8]=0 [-]                            "quality"
s[8]=entropy(F$,T=T[8],x=x[8])        "specific entropy"
h[8]=enthalpy(F$,T=T[8],x=x[8])       "specific enthalpy"
P[8]=pressure(F$,T=T[8],x=x[8])       "pressure"
```

The state of the fluid leaving the open feedwater heater, state 10, is fixed by the quality, $x_{10} = 0$, and pressure, $P_{10} = P_{ext,3.}$ The temperature, specific enthalpy, and specific entropy (T_{10}, h_{10} and s_{10}) are determined.

```
"state 10"
x[10]=0 [-]                               "quality"
P[10]=P_ext_3                             "pressure"
s[10]=entropy(F$,P=P[10],x=x[10])         "specific entropy"
h[10]=enthalpy(F$,P=P[10],x=x[10])        "specific enthalpy"
T[10]=temperature(F$,P=P[10],x=x[10])     "temperature"
```

The state of the extracted fluid leaving the closed feedwater heater, state 13, is fixed by the quality, $x_{13} = 0$, and pressure, $P_{13} = P_{ext,1.}$ The temperature, specific enthalpy, and specific entropy (T_{13}, h_{13} and s_{13}) are determined.

```
"state 13"
x[13]=0 [-]                                       "quality"
P[13]=P_ext_1                                     "pressure"
s[13]=entropy(F$,P=P[13],x=x[13])                 "specific entropy"
h[13]=enthalpy(F$,P=P[13],x=x[13])                "specific enthalpy"
T[13]=temperature(F$,P=P[13],x=x[13])             "temperature"
```

The state of the main fluid leaving the closed feedwater heater, state 12, is fixed by the pressure, $P_{12} = P_b$, and temperature:

$$T_{12} = T_{13} - \Delta T_{CFHW}$$

The specific enthalpy and specific entropy (h_{12} and s_{12}) are determined.

```
"state 12"
T[12]=T[13]-DT_CFHW                               "temperature"
P[12]=P_b                                         "pressure"
h[12]=enthalpy(F$,T=T[12],P=P[12])                "specific enthalpy"
s[12]=entropy(F$,T=T[12],P=P[12])                 "specific entropy"
```

The analysis proceeds by working through each of the four turbines using the Turbine procedure. The exit pressure for turbine 1 is given, $P_3 = P_{ext,1}$. The exit specific enthalpy, temperature, and specific entropy for turbine 1 (h_3, T_3, and s_3) are returned from the Turbine procedure.

```
"State 3"
P[3]=P_ext_1                                                "pressure"
Call Turbine(h[2],P[2],P[3], F$, eta_t_1: h[3], T[3], s[3])  "turbine procedure determines exit state"
```

The exit pressure for turbine 2 is given, $P_4 = P_{ext,2}$. The exit specific enthalpy, temperature, and specific entropy for turbine 2 (h_4, T_4, and s_4) are returned from the Turbine procedure.

```
"State 4"
P[4]=P_ext_2                                                "pressure"
Call Turbine(h[3],P[3],P[4], F$, eta_t_2: h[4], T[4], s[4])  "turbine procedure determines exit state"
```

The exit pressure for turbine 3 is given, $P_6 = P_{ext,3}$. The exit specific enthalpy, temperature, and specific entropy for turbine 3 (h_6, T_6, and s_6) are returned from the Turbine procedure.

```
"State 6"
P[6]=P_ext_3                                                "pressure"
Call Turbine(h[5],P[5],P[6], F$, eta_t_3: h[6], T[6], s[6])  "turbine procedure determines exit state"
```

The exit pressure for turbine 4 is equal to the condenser pressure, $P_7 = P_8$. The exit specific enthalpy, temperature, and specific entropy for turbine 4 (h_7, T_7, and s_7) are returned from the Turbine procedure.

```
"State 7"
P[7]=P[8]                                                   "pressure"
Call Turbine(h[6],P[6],P[7], F$, eta_t_4: h[7], T[7], s[7])  "turbine procedure determines exit state"
```

The analysis continues by working through each of the three pumps using the Pump procedure. The exit pressure for pump 1 is given, $P_9 = P_{ext,3}$. The exit specific enthalpy, temperature, and specific entropy for pump 1 (h_9, T_9, and s_9) are returned from the Pump procedure.

```
"State 9"
P[9]=P_ext_3                                              "pressure"
Call Pump(h[8],P[8],P[9], F$, eta_p_1: h[9], T[9], s[9])  "pump procedure determine exit state"
```

The exit pressure for pump 2 is given, $P_{11} = P_b$. The exit specific enthalpy, temperature, and specific entropy for pump 2 (h_{11}, T_{11}, and s_{11}) are returned from the Pump procedure.

```
"State 11"
P[11]=P_b                                                     "pressure"
Call Pump(h[10],P[10],P[11], F$, eta_p_2: h[11], T[11], s[11])  "pump procedure determine exit state"
```

The exit pressure for pump 3 is given, $P_{14} = P_b$. The exit specific enthalpy, temperature, and specific entropy for pump 3 (h_{14}, T_{14}, and s_{14}) are obtained from the Pump procedure.

```
"State 14"
P[14]=P_b                                                     "pressure"
Call Pump(h[13],P[13],P[14], F$, eta_p_3: h[14], T[14], s[14])  "pump procedure determine exit state"
```

The second extraction fraction, f_2, can be determined from an energy balance on the open feedwater heater:

$$(1 - f_1) f_2 \dot{m}_b h_6 + (1 - f_1)(1 - f_2) \dot{m}_b h_9 = (1 - f_1) \dot{m}_b h_{10} \tag{1}$$

where $\dot{m}_b$ is the mass flow rate passing through the boiler (the maximum mass flow rate in the cycle). Equation (1) can be simplified by dividing through by the common factor, $(1 - f_1) \dot{m}_b$:

$$f_2 h_6 + (1 - f_2) h_9 = h_{10}$$

```
f_2*h[6]+(1-f_2)*h[9]=h[10]     "second extraction fraction"
```

which leads to $f_2 = 0.0914$ (9.14%). The first extraction fraction, f_1, can be determined from an energy balance on the closed feedwater heater:

$$f_1 \dot{m}_b h_3 + (1 - f_1) \dot{m}_b h_{11} = f_1 \dot{m}_b h_{13} + (1 - f_1) \dot{m}_b h_{12} \tag{2}$$

Equation (2) can be simplified by dividing through by the common factor, $\dot{m}_b$:

$$f_1 h_3 + (1 - f_1) h_{11} = f_1 h_{13} + (1 - f_1) h_{12}$$

```
f_1*h[3]+(1-f_1)*h[11]=f_1*h[13]+(1-f_1)*h[12]     "first extraction fraction"
```

which results in $f_1 = 0.1482$ (14.82%). An energy balance on the mixing chamber downstream of the closed feedwater heater leads to:

$$f_1 \dot{m}_b h_{14} + (1 - f_1) \dot{m}_b h_{12} = \dot{m}_b h_1 \tag{3}$$

Equation (3) can be simplified by dividing through by the common factor, $\dot{m}_b$:

$$f_1 h_{14} + (1 - f_1) h_{12} = h_1 \tag{4}$$

The state of the fluid leaving the mixing chamber and entering the boiler, state 1, is fixed by the specific enthalpy, h_1 from Eq. (4), and pressure, $P_1 = P_b$. The temperature and specific entropy (T_1 and s_1) are determined.

```
"State 1"
P[1]=P_b                                "pressure"
f_1*h[14]+(1-f_1)*h[12]=h[1]            "energy balance on mixer"
T[1]=temperature(F$,h=h[1],P=P[1])      "temperature"
s[1]=entropy(F$,h=h[1],P=P[1])          "entropy"
```

An energy balance on the water-side of the boiler is:

$$\dot{m}_b\, h_1 + \dot{Q}_b = \dot{m}_b\, h_2$$

which is rearranged to provide the boiler heat transfer per unit mass of boiler flow:

$$\frac{\dot{Q}_b}{\dot{m}_b} = h_2 - h_1$$

An energy balance is written for turbine 1:

$$\frac{\dot{W}_{t,1}}{\dot{m}_b} = h_2 - h_3$$

```
"Energy balances  per unit mass of boiler flow"
Q_dot_b\m_dot=h[2]-h[1]                 "boiler"
W_dot_t_1\m_dot=h[2]-h[3]               "turbine 1"
```

An energy balance on turbine 2 is:

$$(1 - f_1)\, \dot{m}_b\, h_3 = \dot{W}_{t,2} + (1 - f_1)\, \dot{m}_b\, h_4$$

which is rearranged to provide the work from turbine 2 per unit mass of boiler flow:

$$\frac{\dot{W}_{t,2}}{\dot{m}_b} = (1 - f_1)\, (h_3 - h_4)$$

The energy balance on the water-side of the reheater provides:

$$\frac{\dot{Q}_{rh}}{\dot{m}_b} = (1 - f_1)\, (h_5 - h_4)$$

An energy balance on turbine 3 is:

$$\frac{\dot{W}_{t,3}}{\dot{m}_b} = (1 - f_1)\, (h_5 - h_6)$$

```
W_dot_t_2\m_dot=(1-f_1)*(h[3]-h[4])     "turbine 2"
Q_dot_rh\m_dot=(1-f_1)*(h[5]-h[4])      "reheater"
W_dot_t_3\m_dot=(1-f_1)*(h[5]-h[6])     "turbine 3"
```

An energy balance on turbine 4 provides:

$$(1 - f_1)\,(1 - f_2)\, \dot{m}_b\, h_6 = \dot{W}_{t,4} + (1 - f_1)\,(1 - f_2)\, \dot{m}_b\, h_7$$

which is rearranged to provide the work from turbine 4 per unit mass of boiler flow:

$$\frac{\dot{W}_{t,4}}{\dot{m}_b} = (1 - f_1)(1 - f_2)(h_6 - h_7)$$

An energy balance on the condenser is:

$$\frac{\dot{Q}_{cond}}{\dot{m}_b} = (1 - f_1)(1 - f_2)(h_7 - h_8)$$

An energy balance on pump 1 is:

$$\frac{\dot{W}_{p,1}}{\dot{m}_b} = (1 - f_1)(1 - f_2)(h_9 - h_8)$$

```
W_dot_t_4\m_dot=(1-f_1)*(1-f_2)*(h[6]-h[7])      "turbine 4"
Q_dot_cond\m_dot=(1-f_1)*(1-f_2)*(h[7]-h[8])     "condenser"
W_dot_p_1\m_dot=(1-f_1)*(1-f_2)*(h[9]-h[8])      "pump 1"
```

Energy balances are written for pumps 2 and 3:

$$\frac{\dot{W}_{p,2}}{\dot{m}_b} = (1 - f_1)(h_{11} - h_{10})$$

$$\frac{\dot{W}_{p,3}}{\dot{m}_b} = f_1 (h_{14} - h_{13})$$

```
W_dot_p_2\m_dot=(1-f_1)*(h[11]-h[10])     "pump 2"
W_dot_p_3\m_dot=f_1*(h[14]-h[13])         "pump 3"
```

The solution is checked by carrying out an overall energy balance on the cycle:

$$check = \frac{\dot{Q}_b}{\dot{m}_b} + \frac{\dot{Q}_{rh}}{\dot{m}_b} + \frac{\dot{W}_{p,1}}{\dot{m}_b} + \frac{\dot{W}_{p,2}}{\dot{m}_b} + \frac{\dot{W}_{p,3}}{\dot{m}_b} - \frac{\dot{W}_{t,1}}{\dot{m}_b} - \frac{\dot{W}_{t,2}}{\dot{m}_b} - \frac{\dot{W}_{t,3}}{\dot{m}_b} - \frac{\dot{W}_{t,4}}{\dot{m}_b} - \frac{\dot{Q}_{cond}}{\dot{m}_b}$$

```
check=Q_dot_b\m_dot+Q_dot_rh\m_dot+W_dot_p_1\m_dot+W_dot_p_2\m_dot&
  +W_dot_p_3\m_dot-W_dot_t_1\m_dot-W_dot_t_2\m_dot-W_dot_t_3\m_dot&
  -W_dot_t_4\m_dot-Q_dot_cond\m_dot                "overall energy balance"
```

which leads to $check = 0$ J/kg (to within numerical precision). The efficiency of the cycle is:

$$\eta_{plant} = \frac{\frac{\dot{W}_{t,1}}{\dot{m}_b} + \frac{\dot{W}_{t,2}}{\dot{m}_b} + \frac{\dot{W}_{t,3}}{\dot{m}_b} + \frac{\dot{W}_{t,4}}{\dot{m}_b} - \frac{\dot{W}_{p,1}}{\dot{m}_b} - \frac{\dot{W}_{p,2}}{\dot{m}_b} - \frac{\dot{W}_{p,3}}{\dot{m}_b}}{\frac{\dot{Q}_b}{\dot{m}_b} + \frac{\dot{Q}_{rh}}{\dot{m}_b}}$$

```
eta_plant=(W_dot_t_1\m_dot+W_dot_t_2\m_dot+W_dot_t_3\m_dot&
  +W_dot_t_4\m_dot-W_dot_p_1\m_dot-W_dot_p_2\m_dot-W_dot_p_3\m_dot)&
  /(Q_dot_b\m_dot+Q_dot_rh\m_dot)                  "efficiency"
```

which results in $\eta_{plant} = 0.3984$ (39.84%).

c) Optimize the cycle efficiency by varying the three extraction pressures: $P_{ext,1}$, $P_{ext,2}$, and $P_{ext,3}$.

It is necessary to constrain the optimization so that $P_8 < P_{ext,3} < P_{ext,2} < P_{ext,1} < P_b$. The total pressure range between the condenser pressure, P_8, and the boiler pressure, P_b, can be broken into different ranges corresponding to the different extraction pressures. There are several methods for accomplishing these constraints. Here, the fraction of the total pressure range that lies between $P_{ext,1}$ and P_b is defined as y_1. Therefore, $P_{ext,1}$ is calculated from y_1 according to:

$$P_{ext,1} = P_b - y_1 \left(P_b - P_8\right)$$

Notice that the variable y_1 must lie between 0 and 1. If $y_1 = 0$ then the first extraction pressure is equal to the boiler pressure and if $y_1 = 1$ then it is equal to the condenser pressure. The initial specification of $P_{ext,1}$ is commented out and the first extraction pressure is set using y_1.

```
{P_ext_1=1.1 [MPa]*convert(MPa,Pa)   "first extraction pressure"}
y_1=0.5 [-]                           "fraction of total pressure range between P_ext_1 and P_b"
P_ext_1=P_b-y_1*(P_b-P[8])            "first extraction pressure, calculated with y_1"
```

With $P_{ext,1}$ assigned, the remaining pressure range spans from the condensing pressure to $P_{ext,1}$ and this range must be broken into ranges corresponding to $P_{ext,2}$ and $P_{ext,3}$. The fraction of the remaining pressure range that lies between $P_{ext,2}$ and $P_{ext,1}$ is defined as y_2. Therefore, $P_{ext,2}$ is calculated from y_2 according to:

$$P_{ext,2} = P_{ext,1} - y_2 \left(P_{ext,1} - P_8\right)$$

Notice that the variable y_2 must also lie between 0 and 1; this is important as it allows us to easily constrain the independent variables used in the optimization so that the extraction pressures are in the correct order. If $y_2 = 0$ then the second extraction pressure is equal to the first extraction pressure and if $y_2 = 1$ then it is equal to the condenser pressure. The initial specification of $P_{ext,2}$ is commented out and the second extraction pressure is set using y_2.

```
{P_ext_2=0.25 [MPa]*convert(MPa,Pa)   "second extraction pressure"}
y_2=0.5 [-]                            "fraction of remaining pressure range btwn P_ext_2 & P_ext_1"
P_ext_2=P_ext_1-y_2*(P_ext_1-P[8])     "second extraction pressure, calculated with y_2"
```

With $P_{ext,1}$ and $P_{ext,2}$ both assigned, the remaining pressure range spans from the condensing pressure to $P_{ext,2}$ and the final extraction pressure, $P_{ext,3}$, must lie within this range. The fraction of the remaining pressure range that lies between $P_{ext,3}$ and $P_{ext,2}$ is defined as y_3. Therefore, $P_{ext,3}$ is calculated from y_3 according to:

$$P_{ext,3} = P_{ext,2} - y_3 \left(P_{ext,2} - P_8\right)$$

Again, the variable y_3 must also lie between 0 and 1. The initial specification of $P_{ext,3}$ is commented out and the third extraction pressure is set using y_3.

```
{P_ext_3=0.1 [MPa]*convert(MPa,Pa)    "third extraction pressure"}
y_3=0.5 [-]                            "fraction of remaining pressure range btwn P_ext_3 & P_ext_2"
P_ext_3=P_ext_2-y_3*(P_ext_2-P[8])     "third extraction pressure, calculated with fP3"
```

In order to accomplish the optimization, the values of y_1, y_2, and y_3 are commented out and the Min/Max option is selected from the Calculate menu. Maximize the efficiency of the power plant by varying the independent variables y_1, y_2, and y_3. Set the bounds for each variable to be between 0 and 1 with reasonable guess values. The optimized efficiency using the Direct Search method is η_{plant} = 0.4245 (42.45%) which results when $y_1 = 0.5705$ ($P_{ext,1} = 4.298$ MPa), $y_2 = 0.1886$ ($P_{ext,2} = 3.488$ MPa), and $y_3 = 0.8763$ ($P_{ext,3} = 0.4363$ MPa). (Slightly different values will be obtained with other optimization algorithms.) These values are entered in the Equations window and used for the remainder of this problem.

```
y_1=0.5705 [-]                          "fraction of total pressure range between P_ext_1 and P_b"
P_ext_1=P_b-y_1*(P_b-P[8])              "first extraction pressure, calculated with y_1"
y_2=0.1886 [-]                          "fraction of remaining pressure range btwn P_ext_2 & P_ext_1"
P_ext_2=P_ext_1-y_2*(P_ext_1-P[8])      "second extraction pressure, calculated with y_2"
y_3=0.8763 [-]                          "fraction of remaining pressure range btwn P_ext_3 & P_ext_2"
P_ext_3=P_ext_2-y_3*(P_ext_2-P[8])      "third extraction pressure, calculated with y_3"
```

d) The solar trough field has a total collector area of $A_{collector} = 40{,}000\ \text{m}^2$ and is operating on a day when the solar flux is $\dot{Q}''_{solar} = 1000\ \text{W/m}^2$. Assume that all of the radiation is absorbed by the collector pipe. Determine the total rate of solar energy incident on the solar trough field.

The total rate of incident solar radiation is:

$$\dot{Q}_{solar} = A_{collector}\,\dot{Q}''_{solar}$$

```
A_collector=4000 [m^2]                  "total collector area"
Q``_dot_solar=1000 [W/m^2]              "solar flux"
Q_dot_solar=A_collector*Q``_dot_solar   "solar heat transfer"
```

Solving indicates that $\dot{Q}_{solar} = 4.0$ MW.

e) The collector pipe that carries the heat transfer fluid is separated from the ambient air by an evacuated space. The outer surface of the space is made out of glass so that the solar radiation can penetrate and strike the collector pipe. This design minimizes heat transfer from the heat transfer fluid to the atmosphere. Even so, there is a substantial amount of heat lost from the collector pipe. The rate of heat transfer from the oil in the collector pipe to the atmosphere is given by:

$$\dot{Q}_{loss} = LC\,A_{collector}\left(T_{hf,in} - T_{amb}\right) \tag{5}$$

where $LC = 0.35\ \text{W/m}^2\text{-K}$ is a loss coefficient that characterizes the trough design and $T_{amb} = 20°\text{C}$ is the ambient air temperature. The net heat transfer to the heat transfer fluid, and therefore to the power cycle shown in Figure 2, is the solar heat transfer, calculated in part (d), less the thermal loss, calculated using Eq. (5). Determine the net rate of heat transfer to the heat transfer fluid and the efficiency of the solar field, defined as the ratio of the energy delivered to the power plant to the incident solar energy.

The heat transfer delivered to the power plant is:

$$\dot{Q}_{field} = \dot{Q}_{solar} - \dot{Q}_{loss}$$

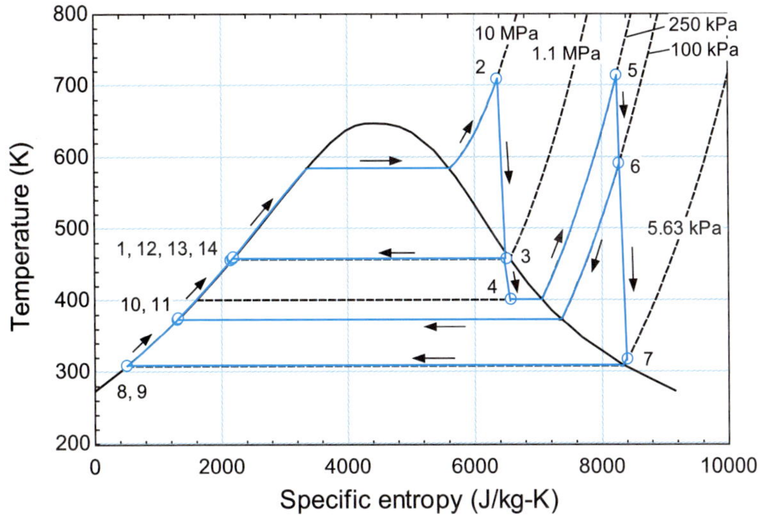

Figure 3: Temperature-specific entropy diagram showing the cycle states and processes.

The efficiency of the solar field is defined as the ratio of the energy delivered to the powerplant to the incident solar energy. The field efficiency is:

$$\eta_{field} = \frac{\dot{Q}_{field}}{\dot{Q}_{solar}}$$

```
LC=0.35 [W/m^2-K]                          "loss coefficient"
T_amb=converttemp(C,K,20[C])               "ambient temperature"
Q_dot_loss=LC*A_collector*(T_hf_in-T_amb)  "losses"
Q_dot_field=Q_dot_solar-Q_dot_loss         "heat transfer from field to plant"
eta_field=Q_dot_field/Q_dot_solar          "field efficiency"
```

Solving provides $\dot{Q}_{field} = 3.398$ MW and $\eta_{field} = 0.8495$ (84.95%).

f) Determine the total efficiency of the solar power system with the optimized extraction pressures. The total efficiency is defined as the ratio of the net power produced by the power plant to the solar heat transfer incident on the field.

The total efficiency is:

$$\eta_{total} = \frac{\dot{W}_{net}}{\dot{Q}_{solar}} = \underbrace{\frac{\dot{W}_{net}}{(\dot{Q}_b + \dot{Q}_{rh})}}_{\eta_{plant}} \underbrace{\frac{(\dot{Q}_b + \dot{Q}_{rh})}{\dot{Q}_{solar}}}_{\eta_{field}} = \eta_{plant}\,\eta_{field}$$

```
eta_total=eta_plant*eta_field    "total efficiency"
```

which results in $\eta_{total} = 0.3606$ (36.06%).

g) Plot the total efficiency as a function of the temperature to which the heat transfer fluid is heated, $T_{hf,in}$. You should see that there is an optimal value of $T_{hf,in}$; explain why this optimal value exists.

Figure 4 illustrates the total efficiency of the system as a function of the heat transfer fluid inlet temperature, $T_{hf,in}$. The optimal value of $T_{hf,in}$ balances the field efficiency against plant efficiency. The field efficiency decreases with increasing $T_{hf,in}$ because the losses increase. The plant efficiency increases with increasing $T_{hf,in}$.

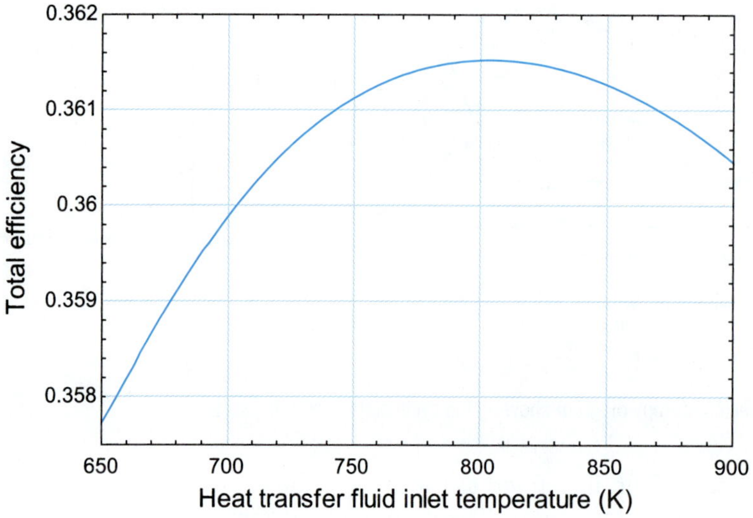

Figure 4: Total system efficiency as a function of the heat transfer fluid inlet temperature.

h) Determine the mass flow rate of water passing through the boiler in the power plant ($\dot{m}_b$) and the net power produced by the plant.

The heat transfer from the field must be transferred to the working fluid in the heater and re-heater. Therefore:

$$\left(\frac{\dot{Q}_h}{\dot{m}} + \frac{\dot{Q}_{rh}}{\dot{m}}\right)\dot{m} = \dot{Q}_{field}$$

The net power produced is:

$$\dot{W}_{net} = \eta_{plant}\,\dot{Q}_{field}$$

```
(Q_dot_b\m_dot+Q_dot_rh\m_dot)*m_dot=Q_dot_field     "mass flow rate"
W_dot_net=Q_dot_field*eta_plant                     "net power produced"
```

which provides $\dot{m} = 1.427$ kg/s and $\dot{W}_{net} = 1.442$ MW.

8.3 The Gas Turbine Cycle

Like the Rankine cycle, the gas turbine cycle is commonly used to produce electrical power. The gas turbine cycle is also used for mechanical power and propulsion. In its most common form, it is a steady flow, open, internal combustion cycle. Compared to the Rankine cycle, the gas turbine cycle offers the advantages of quick startup and shutdown. Gas turbine power plants are often started remotely when there is a need for electrical power above the base load level. The gas turbine cycle is characterized by very large power to weight ratio and this is a primary advantage of these cycles. There are a few reasons that the gas turbine cycle provides a high power to weight ratio. The turbomachinery used for compression and expansion spin continuously and can achieve very high velocity (near the speed of sound). The velocity of the fluid passing through these components scales with the velocity of the blades and therefore a relatively small compressor and turbine can process very large flow rates. Also, air-breathing gas turbine engines do not require any heat exchangers, which are otherwise the most massive components in a power cycle. The air is heated by mixing it with fuel that is combusted and the combustion products leaving the turbine are exhausted to the atmosphere. The high power to weight ratio makes jet engines and turbo-prop engines, which are variations of the gas turbine cycle, the engines of choice for modern aircraft. Compared to the reciprocating spark ignition engine (the Otto cyle) and compression ignition engine (the Diesel cycle), which will be discussed in following sections, the gas turbine cycle operates at higher air-fuel ratios which leads to lower combustion temperatures and cleaner exhaust (i.e., less carbon monoxide, unburned hydrocarbons, and oxides of nitrogen). The gas turbine cycle is however, typically less fuel efficient than the Otto and Diesel cycles. This behavior is particularly true at part-load conditions and therefore gas turbine engines are not used in automobiles and trucks.

8.3.1 The Basic Gas Turbine Cycle

The basic air-breathing (i.e., open) gas turbine cycle is shown schematically in Figure 8-19(a) and on a temperature-specific entropy diagram in Figure 8-19(b). The basic cycle is often called the Brayton cycle, named for the American inventor George Brayton who developed the power cycle in the 1870's. The original Brayton cycle used reciprocating compression and expansion processes rather than the rotary machines that are used in modern gas turbine cycles.

Outdoor air at state 1 is drawn into a compressor and compressed to a higher pressure at state 2. The ratio of the pressure at state 2 to the pressure at state 1 is called the *pressure ratio* and it is an important design parameter for this cycle. The compressed air at state 2 enters a combustion chamber where it is mixed with fuel and combusted. The combustion process is assumed to occur at nearly constant pressure. A detailed discussion of combustion is provided in Chapter 13. In this section, we will assume that the energy provided by combustion of the fuel can be expressed in terms of the *heat of combustion* of the fuel (HC) which is the energy released per unit mass of fuel. In this chapter, we will further assume that the properties of the combustion products are the same as the properties of pure air. The ratio of the mass flow rate of air to the the mass flow rate of fuel is referred to as the *air-fuel ratio* (AF). High air-fuel ratios are used in the combustion process (i.e., the combustion is lean) in order to control the temperature of the combustion products entering the turbine at state 3 and avoid damage to the blades. Higher turbine inlet temperatures lead to higher cycle efficiency. However the fatigue strength of the turbine blades that are continuously exposed to the high temperature exhaust products dictates a maximum temperature that is approximately 1200°C. Special alloys and/or enhanced turbine blade cooling methods may allow higher temperatures and a significant amount of research is devoted to increasing the maximum turbine inlet temperature in a gas turbine power cycle.

The high pressure and high temperature combustion gas entering the turbine at state 3 expands in a *gasifier* turbine that produces shaft power sufficient to drive the compressor. If the gas turbine power cycle is operating correctly, it is not necessary to expand the gas to atmospheric pressure in the gasifier turbine; that is, $P_4 > P_1$, as shown in Figure 8-19(b). Therefore, the gas can subsequently be expanded in a *power* turbine that produces some net useful power. Depending on the application, the output power may be used to provide mechanical power (e.g., to a helicopter rotor) or electrical power using a generator. In some cases the gasifier and power turbines are collocated on the

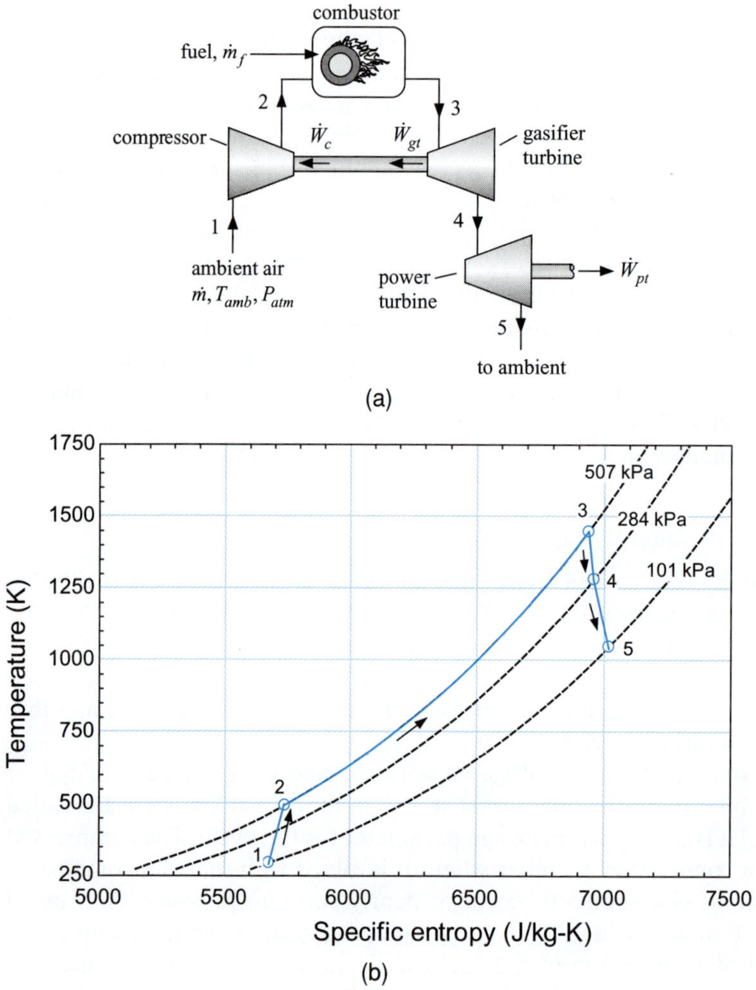

Figure 8-19: (a) Schematic of the basic gas turbine cycle and (b) a *T-s* diagram of the basic gas turbine cycle.

same shaft whereas in other cases it is useful to separate these functions, allowing the gasifier turbine to operate at its optimal speed while the speed of the power turbine varies in response to the mechanical load.

In this section, a simple model of the basic gas turbine cycle shown in Figure 8-19(a) is developed in order to illustrate the proper method of analysis as well as to explore some of the features and behaviors of the cycle. In order to specify the gas turbine cycle, it is necessary to set the ambient temperature and atmospheric pressure (T_{amb} and P_{atm}), the pressure ratio (PR) and the air-fuel ratio (AF). We will assume that the fuel is methane with a heat of combustion, $HC = 50$ MJ/kg. The performance parameters include the efficiency of the compressor (η_c) as well as the efficiencies of the gasifier and power turbines (η_{gt} and η_{pt}, respectively).

```
$UnitSystem SI K Pa J Mass Radian
"Input conditions"
T_amb=converttemp(C,K,20[C])          "ambient temperature"
P_atm=1 [atm]*convert(atm,Pa)         "atmospheric pressure"
AF=45 [-]                             "air-fuel ratio"
PR=5 [-]                              "pressure ratio"
HC=50e6 [J/kg]                        "heat of combustion of fuel"
```

```
"Performance parameters"
eta_gt=0.88 [-]                  "gasifier turbine efficiency"
eta_pt=0.82 [-]                  "power turbine efficiency"
eta_c=0.85 [-]                   "compressor efficiency"
```

State 1 is fixed by the ambient temperature and atmospheric pressure, $T_1 = T_{amb}$ and $P_1 = P_{atm}$. The specific entropy and specific enthalpy (s_1 and h_1) are determined.

```
"State 1"
T[1]=T_amb                            "temperature"
P[1]=P_atm                            "pressure"
s[1]=entropy(Air,T=T[1],P=P[1])       "specific entropy"
h[1]=enthalpy(Air,T=T[1])             "specific enthalpy"
```

The pressure leaving the compressor is specified by the pressure ratio:

$$P_2 = PR\, P_1 \tag{8-61}$$

The inlet conditions, efficiency, and exit pressure for the compressor are known. Therefore, the compressor analysis is straightforward. An entropy balance on a reversible compressor operating under the same conditions is:

$$s_{s,2} = s_1 \tag{8-62}$$

The state of the air leaving a reversible compressor is fixed by the specific entropy and pressure. The specific enthalpy leaving a reversible compressor ($h_{s,2}$) is determined. An energy balance on a reversible compressor is:

$$\frac{\dot{W}_{s,c}}{\dot{m}} = h_{s,2} - h_1 \tag{8-63}$$

where $\dot{m}$ is the mass flow rate of air drawn into the engine through the compressor (i.e., the inlet mass flow rate). The power per unit of inlet mass flow rate required by the actual compressor is given by:

$$\frac{\dot{W}_c}{\dot{m}} = \frac{1}{\eta_c}\frac{\dot{W}_{s,c}}{\dot{m}} \tag{8-64}$$

The specific enthalpy leaving the actual adiabatic compressor is obtained from an energy balance:

$$h_2 = h_1 + \frac{\dot{W}_c}{\dot{m}} \tag{8-65}$$

The state of the air leaving the actual compressor is fixed by the specific enthalpy and pressure. The specific entropy and temperature (s_2 and T_2) are obtained.

```
"State 2"
P[2]=PR*P[1]                                 "pressure"
s_s[2]=s[1]                                  "entropy balance on reversible compressor"
h_s[2]=enthalpy(Air,s=s_s[2],P=P[2])         "specific enthalpy leaving reversible compressor"
W_dot_s_c\m_dot=h_s[2]-h[1]                  "work per mass required by reversible compressor"
W_dot_c\m_dot=W_dot_s_c\m_dot/eta_c          "work per mass required by actual compressor"
h[2]=h[1]+W_dot_c\m_dot                      "specific enthalpy leaving actual compressor"
s[2]=entropy(Air,h=h[2],P=P[2])              "specific entropy leaving actual compressor"
T[2]=temperature(Air,h=h[2])                 "temperature leaving actual compressor"
```

A steady-state energy balance on the combustor can be written as:

$$\dot{m}\, h_2 + \dot{m}_f\, HC = (\dot{m} + \dot{m}_f)\, h_3 \tag{8-66}$$

where $\dot{m}_f$ is the mass flow rate of fuel provided to the combustor; note that the quantity $\dot{m}_f\, HC$ in Eq. (8-66) represents the rate of heat release by the combustion process. Equation (8-66) is divided through by the mass flow rate of inlet air:

$$h_2 + \underbrace{\frac{\dot{m}_f}{\dot{m}}}_{1/AF} HC = \left(1 + \frac{\dot{m}_f}{\dot{m}}\right) h_3 \tag{8-67}$$

The definition of the air-fuel ratio is substituted into Eq. (8-67):

$$h_2 + \frac{HC}{AF} = \left(1 + \frac{1}{AF}\right) h_3 \tag{8-68}$$

Equation (8-68) provides the specific enthalpy of the combustion products leaving the combustor at state 3. Assuming that the combustion products can be modeled using the properties of pure air, state 3 is specified by the specific enthalpy and pressure, $P_3 = P_2$. The temperature and specific entropy (T_3 and s_3) are determined.

```
"State 3"
h[2]+HC/AF=h[3]*(1+1/AF)          "combustor energy balance"
P[3]=P[2]                         "no pressure loss"
T[3]=temperature(Air,h=h[3])      "temperature"
s[3]=entropy(Air,h=h[3],P=P[3])   "specific entropy"
```

The gasifier turbine is analyzed next. Notice that, unlike most turbine analyses, the exit pressure (P_4) is not known. However, it is clear that the power produced by the gasifier turbine must be equal to the power required by the compressor. The gasifier turbine exit pressure must be selected so that this is true. Here, we will initially guess a value for the exit pressure and solve for the turbine output power per unit of inlet air mass flow rate. Then, we will allow EES to iteratively adjust the exit pressure until the turbine and compressor powers are equal. This approach to solving the equations has some advantages. The primary advantage is that it allows equations to be entered sequentially and solved individually so that it is less likely that a small error will lead to a large amount of frustration. The initial guess for the gasifier turbine exit pressure is taken to be the average of the compressor inlet and exit pressures (P_1 and P_2).

```
"State 4"
P[4]=(P[1]+P[2])/2               "guess for pressure leaving gasifier turbine"
```

Note that the equation specifying P_4 can be highlighted in order to emphasize the fact that it serves only as the starting point for the calculations and that this equation should eventually be removed in order to obtain the correct solution. Because P_4 is known (or at least a value has been assumed), the gasifier turbine analysis can proceed in the usual way. An entropy balance on a reversible gasifier turbine operating under the same conditions leads to:

$$s_{s,4} = s_3 \tag{8-69}$$

The state of the air leaving a reversible turbine is fixed by the specific entropy and pressure. The specific enthalpy leaving a reversible turbine ($h_{s,4}$) is determined. The power output provided by a reversible gasifier turbine is obtained from an energy balance:

$$\dot{W}_{s,gt} = (h_3 - h_{s,4})(\dot{m} + \dot{m}_f) \tag{8-70}$$

Dividing Eq. (8-70) through by $\dot{m}$ and substituting the definition of the air-fuel ratio leads to:

$$\frac{\dot{W}_{s,gt}}{\dot{m}} = (h_3 - h_{s,4})\left(1 + \frac{1}{AF}\right) \tag{8-71}$$

The power per unit of inlet air mass flow rate provided by the actual gasifier turbine is given by:

$$\frac{\dot{W}_{gt}}{\dot{m}} = \eta_{gt}\frac{\dot{W}_{s,gt}}{\dot{m}} \tag{8-72}$$

The specific enthalpy leaving the actual gasifier turbine is obtained from an energy balance:

$$h_4 = h_3 - \frac{\dfrac{\dot{W}_{gt}}{\dot{m}}}{\left(1 + \dfrac{1}{AF}\right)} \tag{8-73}$$

The state of the air leaving the actual gasifier turbine is fixed by the specific enthalpy and pressure. The specific entropy and temperature (s_4 and T_4) are obtained.

```
s_s[4]=s[3]                                  "entropy balance on reversible gasifier turbine"
h_s[4]=enthalpy(Air,s=s_s[4],P=P[4])         "specific enthalpy leaving reversible gasifier turbine"
W_dot_s_gt\m_dot=(h[3]-h_s[4])*(1+1/AF)      "work per mass associated with reversible gasifier turbine"
W_dot_gt\m_dot=eta_gt*W_dot_s_gt\m_dot       "work per mass associated with actual gasifier turbine"
h[4]=h[3]-W_dot_gt\m_dot/(1+1/AF)            "specific enthalpy leaving actual gasifier turbine"
s[4]=entropy(Air,h=h[4],P=P[4])              "specific entropy leaving actual gasifier turbine"
T[4]=temperature(Air,h=h[4])                 "temperature leaving actual gasifier turbine"
```

At this point, the guess values are updated in order to facilitate solving the set of nonlinear, implicit equations that will result when the assumed value of P_4 is commented out and the requirement that:

$$\frac{\dot{W}_{gt}}{\dot{m}} = \frac{\dot{W}_c}{\dot{m}} \tag{8-74}$$

is enforced. Select Update Guesses from the Calculate menu; this will set the guess value for each of the variables to the value obtained by the most recent solution. The assumed value of P_4 is removed:

```
{P[4]=(P[1]+P[2])/2}          "guess for pressure leaving gasifier turbine"
```

and Eq. (8-74) is added:

```
W_dot_gt\m_dot=W_dot_c\m_dot
                                "power from gasifier turbine must equal power required by compressor"
```

By solving the equations, EES will adjust P_4 until Eq. (8-74) is satisfied; the result is $P_4 =$ 284 kPa. The remaining pressure drop between P_4 and atmospheric pressure can be used to drive the power turbine. The pressure at state 5 is equal to the atmospheric pressure, $P_5 = P_{atm}$. The analysis of the power turbine proceeds as usual. An entropy balance on a reversible power turbine operating under the same conditions is:

$$s_{s,5} = s_4 \tag{8-75}$$

The state of the air leaving a reversible turbine is fixed by the specific entropy and pressure. The specific enthalpy leaving a reversible power turbine ($h_{s,5}$) is determined. An energy balance on a reversible power turbine is:

$$\frac{\dot{W}_{s,pt}}{\dot{m}} = (h_4 - h_{s,5})\left(1 + \frac{1}{AF}\right) \tag{8-76}$$

The power per unit of inlet air mass flow rate delivered by the actual power turbine is:

$$\frac{\dot{W}_{pt}}{\dot{m}} = \eta_{pt}\frac{\dot{W}_{s,pt}}{\dot{m}} \tag{8-77}$$

The specific enthalpy leaving the actual power turbine is obtained from an energy balance:

$$h_5 = h_4 - \frac{\dfrac{\dot{W}_{pt}}{\dot{m}}}{\left(1 + \dfrac{1}{AF}\right)} \tag{8-78}$$

The state of the air leaving the actual power turbine is fixed by the specific enthalpy and pressure. The specific entropy and temperature (s_5 and T_5) are obtained.

```
"State 5"
P[5]=P_atm                                  "exit pressure"
s_s[5]=s[4]                                 "entropy balance on reversible power turbine"
h_s[5]=enthalpy(Air,s=s_s[5],P=P[5])        "specific enthalpy leaving reversible power turbine"
W_dot_s_pt\m_dot=(h[4]-h_s[5])*(1+1/AF)     "work per mass associated with reversible power turbine"
W_dot_pt\m_dot=eta_pt*W_dot_s_pt\m_dot      "work per mass associated with actual power turbine"
h[5]=h[4]-W_dot_pt\m_dot/(1+1/AF)           "specific enthalpy leaving actual power turbine"
s[5]=entropy(Air,h=h[5],P=P[5])             "specific entropy leaving actual power turbine"
T[5]=temperature(Air,h=h[5])                "temperature leaving actual power turbine"
```

The temperature-specific entropy diagram resulting from the analysis is shown in Figure 8-19(b). The solution can be checked by carrying out an overall energy balance on the system shown in Figure 8-19(a).

$$\dot{m}\,h_1 + \dot{m}_{fuel}\,HC = \dot{W}_{pt} + (\dot{m} + \dot{m}_{fuel})\,h_5 \tag{8-79}$$

Equation (8-79) is rearranged to provide:

$$check_1 = h_1 + \frac{HC}{AF} - \frac{\dot{W}_{pt}}{\dot{m}} - \left(1 + \frac{1}{AF}\right)h_5 \tag{8-80}$$

```
check_1=h[1]+HC/AF-W_dot_pt\m_dot-h[5]*(1+1/AF)    "overall energy balance"
```

which provides $check_1 = 0$ J/kg (to within numerical precision). The efficiency of the gas turbine engine is defined as the ratio of the net power output to the thermal energy provided by combustion of the fuel:

$$\eta = \frac{\dot{W}_{pt}}{\dot{m}_{fuel}\, HC} \tag{8-81}$$

Equation (8-81) is rearranged:

$$\eta = \frac{\dfrac{\dot{W}_{pt}}{\dot{m}}}{\dfrac{HC}{AF}} \tag{8-82}$$

```
eta=W_dot_pt\m_dot/(HC/AF)    "efficiency"
```

which provides $\eta = 0.2528$ (25.28%) for the given conditions. The back work ratio for the gas turbine engine is the ratio of the power required by the compressor to the power produced by the two turbines:

$$bwr = \frac{\dfrac{\dot{W}_c}{\dot{m}}}{\dfrac{\dot{W}_{gt}}{\dot{m}} + \dfrac{\dot{W}_{pt}}{\dot{m}}} \tag{8-83}$$

```
bwr=W_dot_c\m_dot/(W_dot_pt\m_dot+W_dot_gt\m_dot)    "back work ratio"
```

which results in $bwr = 0.419$. This back work ratio is much higher than the back work ratio associated with a Rankine cycle, which makes the gas turbine cycle performance much more sensitive to the efficiency of its components.

Effect of Air-Fuel Ratio

The air-fuel ratio controls the amount of fuel and therefore the temperature rise that occurs in the combustor. Figure 8-20 illustrates the combustor exit turbine inlet

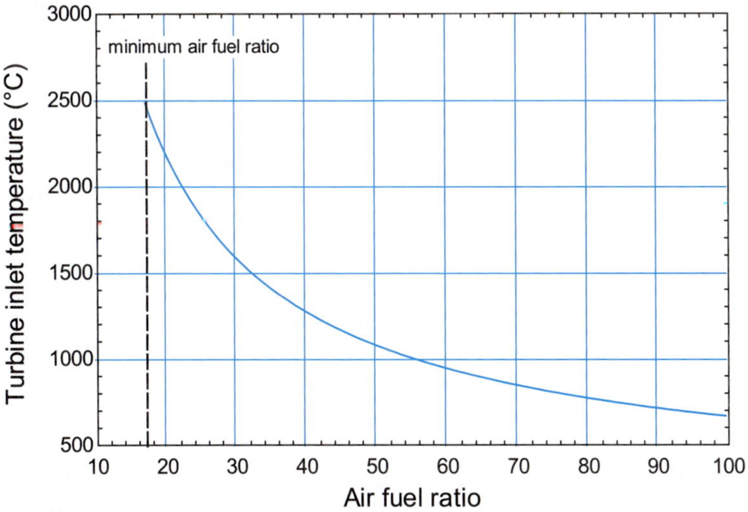

Figure 8-20: Combustor exit/turbine inlet temperature as a function of the air-fuel ratio ($T_{amb} = 20°$C, $P_{atm} = 1$ atm, $PR = 5$, methane fuel, $\eta_{gt} = 0.88$, $\eta_{pt} = 0.82$, $\eta_c = 0.85$).

temperature (T_3) as a function of the air-fuel ratio. As the air-fuel ratio increases, the amount of fuel per unit mass decreases, which reduces the combustor exit temperature.

There is a minimum air-fuel ratio below which there is insufficient oxygen present to completely combust the fuel. This limit will be discussed in more detail in Chapter 13; however, for methane the minimum air-fuel ratio is approximately 17.2 which leads to a maximum possible turbine inlet temperature of $T_3 = 2500°$C for the conditions used to develop the model. This temperature is far higher than what the turbine blades can withstand and therefore the air-fuel ratio is maintained at a much higher value in order to control T_3. In the EES model, we will set the turbine inlet temperature to $T_3 = 1200°$C and let EES calculate the associated air-fuel ratio:

```
{AF=45 [-]}                         "air-fuel ratio"
T_t_in_C=converttemp(K,C,T[3])      "turbine inlet temperature, C"
T_t_in_C=1200 [C]                   "turbine inlet temperature"
```

which provides $AF = 43.7$.

Effect of Pressure Ratio and Turbine Inlet Temperature

The effect of the pressure ratio and the turbine inlet temperature are examined by creating plots that show the cycle performance as a function of pressure ratio for various values of the turbine inlet temperature. The equations that set the values for the variables PR and T_t_in are commented out.

```
{PR=5 [-]}              "pressure ratio"
{T_t_in_C=1200 [C]}     "turbine inlet temperature"
```

A Parametric table is created that includes the variables PR, T_t_in, W_dot_pt\m_dot, eta, bwr and AF. The columns for bwr and AF are not needed, but are provided for interest. Fill in the PR column with values ranging from 2 to 100. Fill the T_t_in column with 900°C (select Alter Values and then specify that the first and last values are both 900°C). Solving the table should provide the calculated values of power turbine work per unit mass of inlet air, efficiency, back work ratio, and air-fuel ratio. We could plot these results and then change the maximum temperature to 1000°C and solve the table again, continuing this process until each temperature of interest has been examined. An alternative approach is to create separate tables for the different turbine inlet temperatures of interest: 900°C, 1000°C, 1100°C, 1200°C, 1300°C, and 1400°C. Right click on the Parametric table tab at the upper left of the table in order to display the dialog window shown in Figure 8-21(a). Change the title of the table to '900 C' and click OK.

Right click on the tab again and click the Duplicate button. A new Parametric table will appear with the title '900 C copy'. Right click on the tab and change the title to '1000 C'. Then fill in the T_t_in column in this new table with 1000°C for all rows. Repeat this process to create six Parametric tables, one for each of the following turbine inlet temperatures: 900°C, 1000°C, 1100°C, 1200°C, 1300°C, and 1400°C. The six parametric tables should appear as shown in Figure 8-21(b). All six parametric tables can be solved with a single Solve Table command. Select Solve Table from the Solve menu. Click in the pull-down menu to the right of the word Table and select all Parametric tables, as shown in Figure 8-21(c). Click the OK button and all of the Parametric tables will be solved. Plots of the efficiency and the work provided by the power turbine per unit mass of inlet air (the net specific power) as a function of pressure ratio can be prepared by

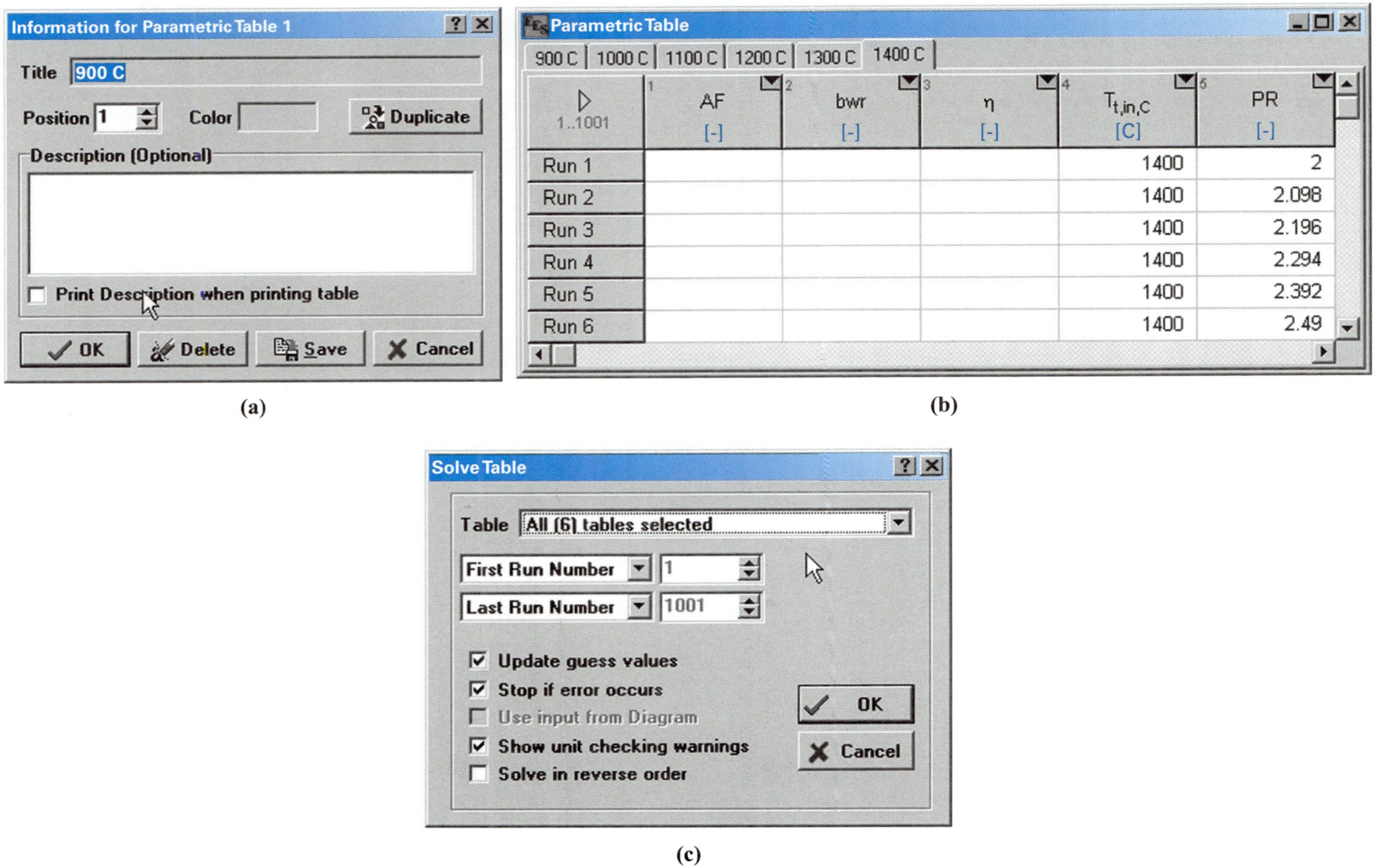

Figure 8-21: (a) Parametric table information dialog, (b) six parametric tables – one for each turbine inlet temperature, and (c) Solve Table dialog.

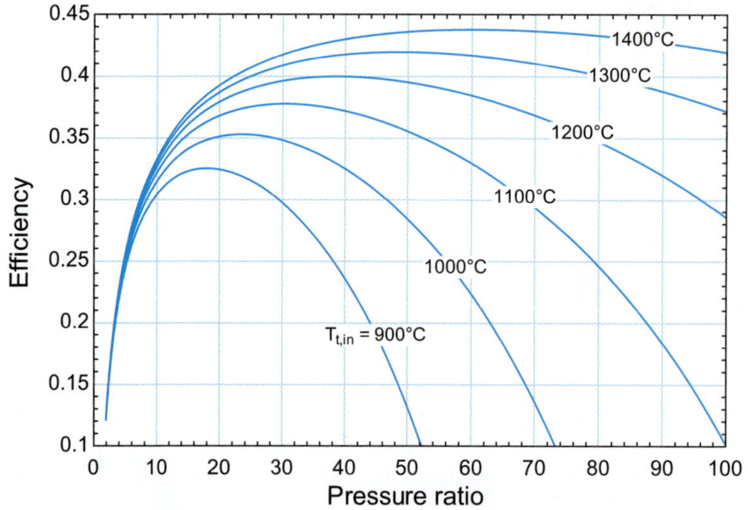

Figure 8-22: Efficiency as a function of pressure ratio for various values of turbine inlet temperature ($T_{amb} = 20^\circ$ C, $P_{atm} = 1$ atm, methane fuel, $\eta_{gt} = 0.88$, $\eta_{pt} = 0.82$, $\eta_c = 0.85$).

overlaying the data from the different parametric tables, as shown in Figure 8-22 and Figure 8-23.

Note that both the efficiency and the net specific work per cycle exhibit maximum values at a pressure ratio that depends on the turbine inlet temperature. The optimum pressure ratio for efficiency does not coincide with the optimum pressure ratio for net specific work; typically, gas turbine engines are operated at maximum power output. Both the power and efficiency increase dramatically with turbine inlet temperature, a trend that explains the intense effort that is devoted to developing turbine blade materials capable of withstanding higher temperatures.

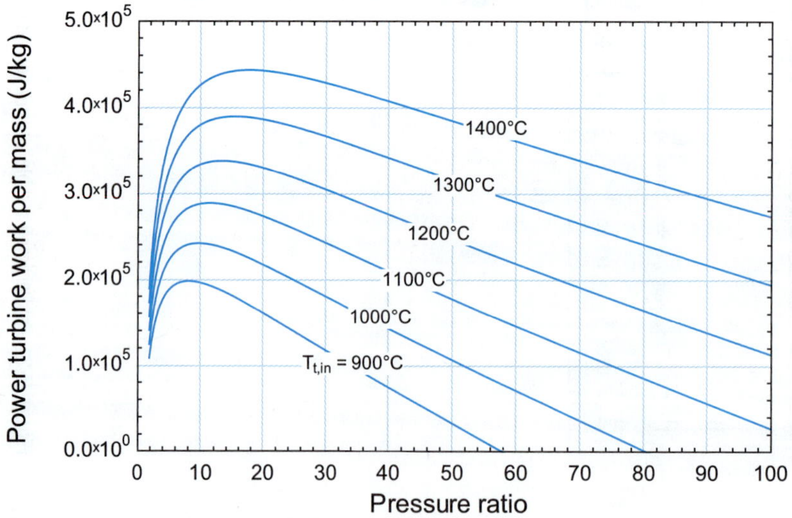

Figure 8-23: Power turbine work per unit mass of inlet flow as a function of pressure ratio for various values of turbine inlet temperature ($T_{amb} = 20^\circ$ C, $P_{atm} = 1$ atm, methane fuel, $\eta_{gt} = 0.88$, $\eta_{pt} = 0.82$, $\eta_c = 0.85$).

Effect of Compressor and Turbine Efficiencies

Decreasing the compressor and turbine efficiencies obviously tends to reduce the cycle efficiency and the power turbine work per unit mass. However, the impact is much more dramatic for the gas turbine engine than for the Rankine cycle. This behavior is due to the higher back work ratio associated with the gas turbine engine as compared to the Rankine cycle, as discussed in Section 8.2.1. The reason that the back work ratio is so large for the gas turbine cycle is that the cycle requires the compression of a gas rather than the pumping of a liquid. Recall the fundamental property relationship derived in Section 6.2:

$$dh = T\,ds + v\,dP \tag{8-84}$$

Integrating Eq. (8-84) along a line of constant entropy indicates that isentropic specific enthalpy change between any two isobars (e.g., the ambient pressure and the compressor discharge pressure) is:

$$\Delta h_s = \int_{P_1}^{P_2} v\,dP \tag{8-85}$$

In the compressed liquid region occupied by a pump in a Rankine cycle, evaluation of Eq. (8-85) results in a very small value because the specific volume is small; therefore the pumping power required in a Rankine cycle is small. However, in the superheated vapor region occupied by a compressor in a gas turbine cycle, evaluation of Eq. (8-85) results in a much larger value because the specific volume of gas is large; this leads to a much higher back work ratio for a gas turbine engine compared to the Rankine cycle. The large back work ratio associated with a gas turbine engine makes the cycle very sensitive to the performance of the compressor and turbine. The net power output is the (relatively) small difference between two large numbers, the total turbine output power and the compressor input power.

Figure 8-24 compares the cycle efficiency and the back work ratio for the gas turbine and the Rankine cycles as a function of the efficiency of the mechanical devices (i.e., the compressor, pump, and turbines are all assumed to have the same efficiency). Both cycles are simulated with the same turbine inlet temperature ($T_H = 800$ K for the Rankine cycle and $T_{t,in} = 800$ K for the gas turbine cycle). Notice that the back work ratio for the Rankine cycle is very small and therefore the efficiency of the cycle remains positive even if the efficiency of the turbine and pump become small. However, the large back work ratio for the gas turbine cycle makes the cycle much more sensitive to the efficiency of the compressor and the turbines. For the conditions used in Figure 8-24, the gas turbine cycle is not capable of producing power (i.e., the back work ratio approaches unity and the efficiency approaches zero) if the efficiency of the compressor and turbines is reduced below approximately 83%.

8.3.2 Modifications to the Gas Turbine Cycle

The modifications to the gas turbine cycle are similar to those discussed in Section 8.2.3 for the Rankine cycle.

Reheat and Intercooling

The reheat and intercooling modifications can be understood by examination of the basic gas turbine cycle on temperature-entropy coordinates, as shown in Figure 8-19(b). The

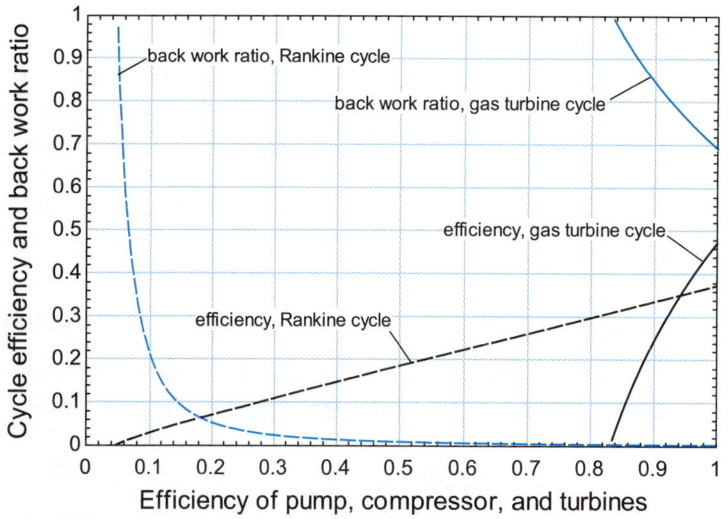

Figure 8-24: Cycle efficiency and back work ratio as a function of the efficiency of the mechanical devices (compressors, pumps, and turbines) for the Rankine and gas turbine cycles. Both cycles utilize a hot temperature of $T_H = T_{t,in} = 800$ K.

divergence of the isobars on the temperature-specific entropy diagram is related to the increase in the specific volume of the working fluid with increased temperature. It is this divergence that allows the gas turbine cycle to produce positive power; the expansion from P_{high} to P_{low} at high temperature produces more power than the compression from P_{low} to P_{high} requires at low temperature (provided that the efficiency of the turbines and compressors are sufficiently high). The reheat modification leads to higher turbine output power by pushing the low pressure portion of the expansion process into a region of higher specific volume. A schematic of the gas turbine cycle with the reheat modification is shown in Figure 8-25(a). Note that the cycle shown in Figure 8-25(a) has all of the compressors and turbines collocated on a single shaft as opposed to the separate gasifier and power turbines shown in Figure 8-19(a). The cycle with reheat is shown on a T-s diagram in Figure 8-25(b). The expansion process is terminated at an intermediate pressure and the working fluid is reheated in a second combustion process before completing the expansion process. The power output from the low pressure turbine is higher with the reheat process (i.e., from state 7 to state 8) than it would be without reheating (i.e., from state 6 to state 8') because of the divergence of the isobars.

Because the compressor power is a large fraction of the turbine power, as evident by the large back work ratio, it is also useful to utilize intercooling in a gas turbine cycle. Intercooling between compression stages is analogous to reheating between turbine stages. The intercooling modification is also shown in Figure 8-25. The air is not compressed from P_{low} to P_{high} in a single process. Instead, the compression process is terminated at an intermediate pressure and the air is subsequently cooled in an intercooler by heat transfer to the atmosphere. The compression process is then completed at a lower temperature, where the air has a smaller specific volume. The power required by the high pressure compressor is lower with the intercooling process (i.e., from state 3 to 4) than it would be without the intercooling (i.e., from state 2 to 4').

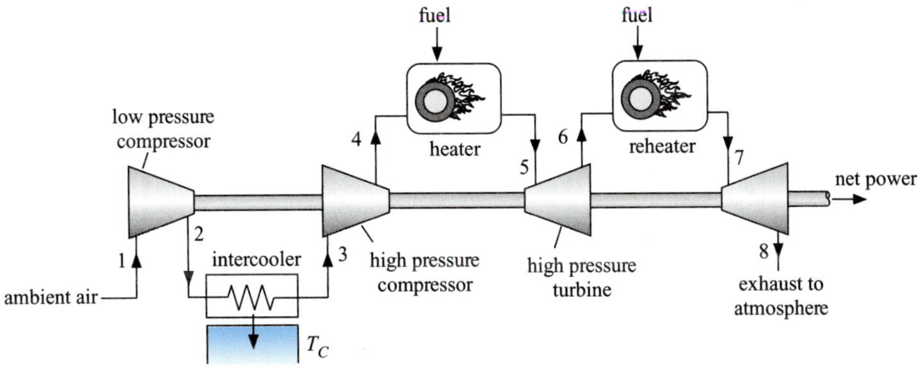

(a)

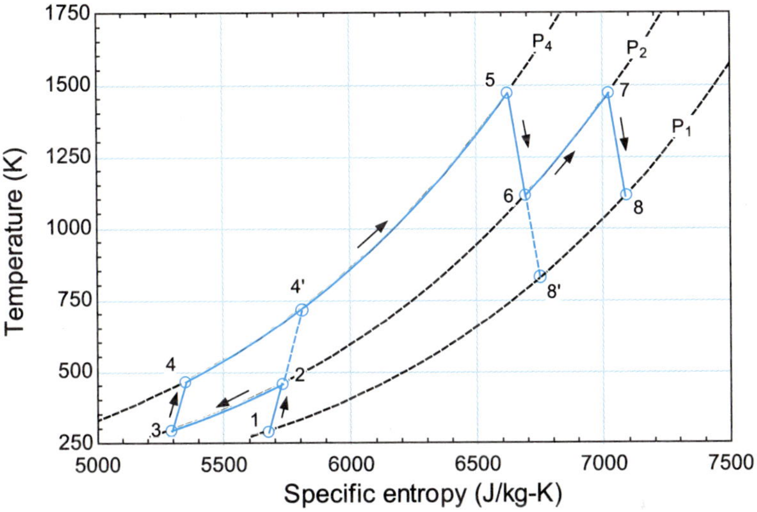

(b)

Figure 8-25: (a) Schematic of a gas turbine with reheat and intercooling modifications and (b) the cycle shown on a temperature-entropy diagram.

EXAMPLE 8.3-1: OPTIMAL INTERCOOLING PRESSURE

Figure 1 illustrates a two-stage, intercooled compressor that takes in gas at $T_1 = T_{amb}$ and $P_1 = P_{atm}$. The mass flow rate provided by the compressor is $\dot{m}$ and the outlet pressure is $P_4 = P_{out}$. The intercooler is perfect; therefore, the gas leaves the intercooler at $T_3 = T_{amb}$ and there is no pressure loss. Both compressor stages have efficiency η_c. Model the gas as an ideal gas with constant specific heat capacity at constant pressure.

a) What is the intercooling pressure, $P_{int} = P_2 = P_3$, that minimizes the total power consumed by the compressor?

An energy balance on a reversible compressor operating under the same conditions that are experienced by the first stage compressor is:

$$\dot{W}_{s,c,1} = \dot{m}\left(h_{s,2} - h_1\right)$$

EXAMPLE 8.3-1

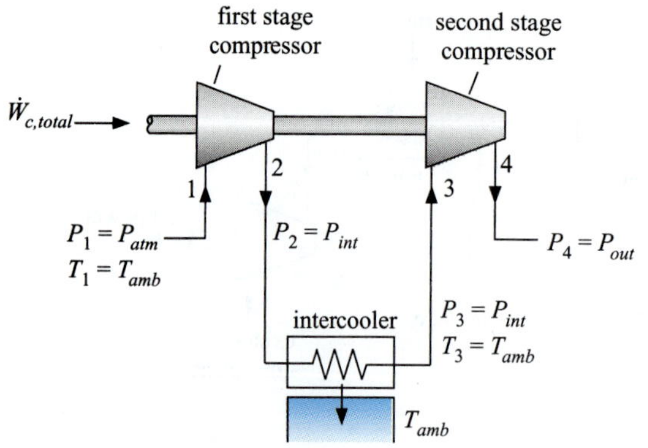

Figure 1: Two-stage, intercooled compressor.

The gas is modeled as having a constant c_P. Therefore, the specific enthalpy difference can be written as:

$$\dot{W}_{s,c,1} = \dot{m}\, c_P \left(T_{s,2} - T_1\right) \tag{1}$$

The specific entropy of the gas passing through the reversible first stage compressor does not change:

$$s_{s,2} = s_1$$

Using the isentropic relation for an ideal gas in Eq. (6-32) leads to:

$$T_{s,2} = T_1 \left(\frac{P_2}{P_1}\right)^{\frac{R}{c_P}} \tag{2}$$

Equation (2) is substituted into Eq. (1):

$$\dot{W}_{s,c,1} = \dot{m}\, c_P\, T_1 \left[\left(\frac{P_2}{P_1}\right)^{\frac{R}{c_P}} - 1\right] \tag{3}$$

Substituting $T_1 = T_{amb}$, $P_1 = P_{atm}$, and $P_2 = P_{int}$ into Eq. (3) results in:

$$\dot{W}_{s,c,1} = \dot{m}\, c_P\, T_{amb} \left[\left(\frac{P_{int}}{P_{atm}}\right)^{\frac{R}{c_P}} - 1\right] \tag{4}$$

The actual power required by the first stage compressor is obtained by dividing Eq. (4) by the compressor efficiency:

$$\dot{W}_{c,1} = \frac{\dot{m}\, c_P\, T_{amb}}{\eta_c} \left[\left(\frac{P_{int}}{P_{atm}}\right)^{\frac{R}{c_P}} - 1\right]$$

A similar analysis is carried out for the second stage compressor, which is assumed to have the same compressor efficiency:

$$\dot{W}_{c,2} = \frac{\dot{m}\, c_P\, T_{amb}}{\eta_c} \left[\left(\frac{P_{out}}{P_{int}}\right)^{\frac{R}{c_P}} - 1\right]$$

The total compressor power required is the sum of the power required by the two stages:

$$\dot{W}_{c,total} = \dot{W}_{c,1} + \dot{W}_{c,2} = \frac{\dot{m}\, c_P\, T_{amb}}{\eta_c} \left[\left(\frac{P_{int}}{P_{atm}}\right)^{\frac{R}{c_P}} - 1\right] + \frac{\dot{m}\, c_P\, T_{amb}}{\eta_c} \left[\left(\frac{P_{out}}{P_{int}}\right)^{\frac{R}{c_P}} - 1\right]$$

which can be rearranged:

$$\dot{W}_{c,total} = \frac{\dot{m}\, c_P\, T_{amb}}{\eta_c}\left[\left(\frac{P_{int}}{P_{atm}}\right)^{\frac{R}{c_P}} - 1 + \left(\frac{P_{out}}{P_{int}}\right)^{\frac{R}{c_P}} - 1\right]$$

The optimal intercooling pressure is obtained by computing the derivative of the total compressor power with respect to P_{int} and setting it equal to zero.

$$\frac{d\dot{W}_{c,total}}{dP_{int}} = \frac{d}{dP_{int}}\left\{\frac{\dot{m}\, c_P\, T_{amb}}{\eta_c}\left[\left(\frac{P_{int}}{P_{atm}}\right)^{\frac{R}{c_P}} - 1 + \left(\frac{P_{out}}{P_{int}}\right)^{\frac{R}{c_P}} - 1\right]\right\} = 0 \qquad (5)$$

Taking the derivative in Eq. (5) leads to:

$$\frac{\dot{m}\, c_P\, T_{amb}}{\eta_c}\left[\frac{R}{c_P}\frac{P_{int}^{\left(\frac{R}{c_P}-1\right)}}{P_{atm}^{\left(\frac{R}{c_P}\right)}} - \frac{R}{c_P}P_{out}^{\left(\frac{R}{c_P}\right)}P_{int}^{\left(-\frac{R}{c_P}-1\right)}\right] = 0 \qquad (6)$$

Equation (6) can be simplified:

$$\left[\frac{P_{int}^{\left(\frac{R}{c_P}\right)}}{P_{atm}^{\left(\frac{R}{c_P}\right)}P_{int}} - \frac{P_{out}^{\left(\frac{R}{c_P}\right)}}{P_{int}^{\left(\frac{R}{c_P}\right)}P_{int}}\right] = 0 \qquad (7)$$

Equation (7) can be rearranged to:

$$\frac{P_{int}^{\left(\frac{R}{c_P}\right)}}{P_{atm}^{\left(\frac{R}{c_P}\right)}} = \frac{P_{out}^{\left(\frac{R}{c_P}\right)}}{P_{int}^{\left(\frac{R}{c_P}\right)}}$$

which simplifies to:

$$\frac{P_{int}}{P_{atm}} = \frac{P_{out}}{P_{int}} \qquad (8)$$

Equation (8) suggests that the optimal intercooler pressure should be set so that the pressure ratio across each compression stage is identical. Equation (8) can also be written as:

$$P_{int}^2 = P_{atm}\, P_{out}$$

or

$$P_{int} = \sqrt{P_{atm}\, P_{out}} \qquad (9)$$

Equation (9) is divided through by the inlet pressure:

$$\frac{P_{int}}{P_{atm}} = \frac{\sqrt{P_{atm}\, P_{out}}}{P_{atm}}$$

which is the same as:

$$\frac{P_{int}}{P_{atm}} = \sqrt{\frac{P_{out}}{P_{atm}}} \qquad (10)$$

Equation (10) indicates that the pressure ratio across the first stage of the compressor should be set to the square root of the total pressure ratio. The pressure ratio across the second stage:

$$\frac{P_{out}}{P_{int}} = \frac{P_{out}}{\sqrt{P_{atm}\, P_{out}}} = \sqrt{\frac{P_{out}}{P_{atm}}}$$

must also be the square root of the total pressure ratio. Equation (10) would not be exactly correct if the compressors had different efficiencies, if the working fluid did not behave as an ideal gas with constant c_P, or if the temperature at state 3 were different than at state 1. However, the total power is relatively insensitive to the interstage pressure in the neighborhood of the optimum value.

Recuperation

The recuperation modification can be understood by examining the basic gas turbine cycle shown in Figure 8-19(b). Notice that air exits the power turbine at state 5 at $T_5 =$ 1046 K and is exhausted to the atmosphere. At the same time, air exits the compressor at state 2 at $T_2 = 492.7$ K and is heated in the combustor to $T_3 = 1447$ K. Clearly a large fraction of the energy that is obtained from combustion in the basic gas turbine cycle could be obtained by heat exchanging the air leaving the compressor with the air leaving the power turbine. This is the recuperative modification, shown schematically in Figure 8-26(a) and on a T-s diagram in Figure 8-26(b).

In a recuperated gas turbine cycle a heat exchanger, referred to as a recuperator or regenerator, is used to accomplish the heat transfer. The air leaving the compressor

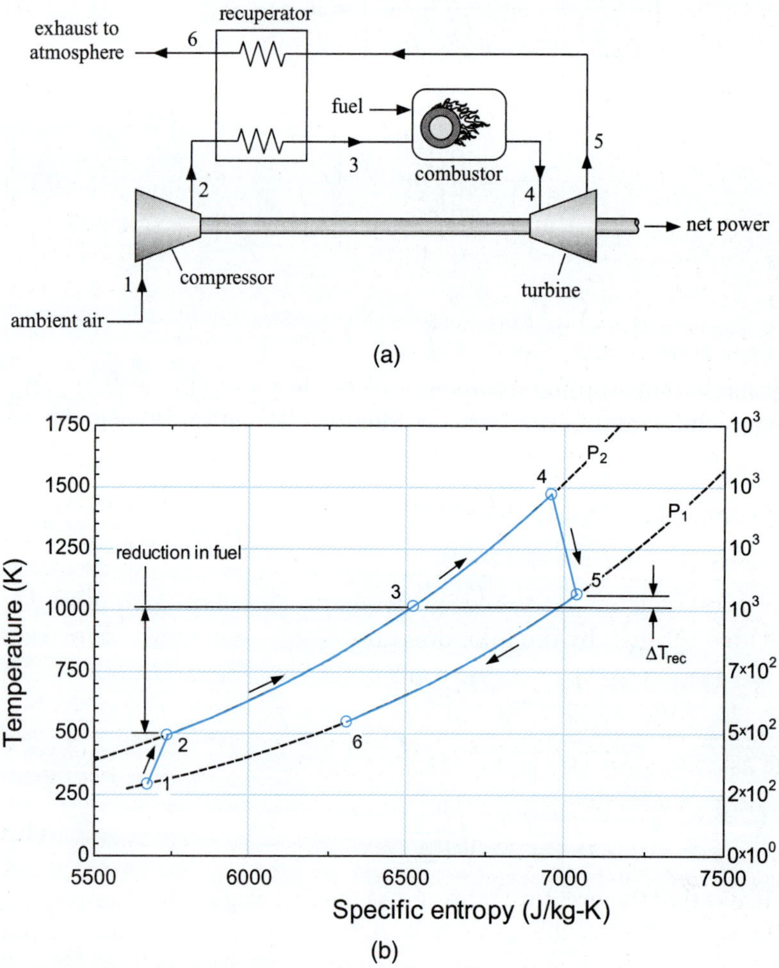

Figure 8-26: (a) Schematic of gas turbine cycle with recuperation and (b) the cycle on a temperature-specific entropy diagram.

at state 2 is heated in the recuperator to a temperature T_3 that is somewhat less than the turbine exit temperature, T_5, as discussed in Section 6.6.6. The pinch point in the recuperator will typically occur at the hot end. Therefore, the temperature of the high pressure air leaving the recuperator at state 3 is related to the turbine exit temperature according to:

$$T_3 = T_5 - \Delta T_{rec} \tag{8-86}$$

where ΔT_{rec} is the recuperator approach temperature difference. Figure 8-26(b) shows that the energy required by the combustor is substantially reduced relative to the unrecuperated gas turbine engine, which results in an increase in the efficiency of the cycle. In practice, the size of the recuperative heat exchanger required by the recuperated gas turbine cycle often discourages its use. Recall that one of the key advantages of the gas turbine cycle is its high power to weight ratio; a large heat exchanger will tend to reduce this advantage. Certainly recuperated gas turbine engines are not commonly used in aerospace applications.

The intercooling and reheat modifications tend to increase the advantage of recuperation by reducing the compressor outlet temperature and increasing the turbine outlet temperature, states 4 and 8, respectively, in Figure 8-25(b). The increased turbine outlet to compressor outlet temperature difference enables a higher fraction of the energy that would otherwise be required from the fuel to be provided from recuperation.

EXAMPLE 8.3-2: GAS TURBINE ENGINE FOR SHIP PROPULSION

Figure 1 is a schematic of a gas turbine engine power plant that is being considered for use on a small military ship that requires a power output, $\dot{W}_{pt} = 320$ hp.

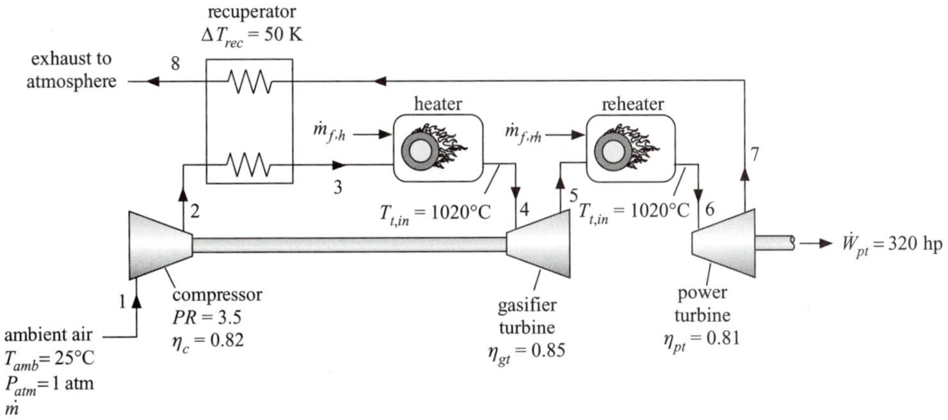

Figure 1: Gas turbine engine with reheat and recuperation.

The engine draws in ambient air at $T_{amb} = 25°C$ and $P_{atm} = 1$ atm through the compressor. The compressor pressure ratio is $PR = 3.5$ and the efficiency of the compressor is $\eta_c = 0.82$. The air leaving the compressor at state 2 is preheated in a recuperator. The pinch point for the recuperator occurs at its hot end and the approach temperature difference is $\Delta T_{rec} = 50$ K. The air enters the first combustor (referred to as the heater) at state 3. The air-fuel ratio for the heater is adjusted so that the air exiting the heater at state 4 is at temperature $T_{t,in} = 1020°C$. The air leaving the heater passes through the gasifier turbine which drives the compressor. The efficiency of the gasifier turbine is $\eta_{gt} = 0.85$. The air leaving the gasifier turbine at state 5 is fed to a second combustor (referred to as the reheater). The air-fuel ratio for the reheater is adjusted so that the air exiting the reheater at state 6 is at temperature $T_{t,in}$ = 1020°C. The air passes through the power turbine, which produces the mechanical power required by the ship. The efficiency of the power turbine is η_{pt} = 0.81. The hot air leaving the power turbine at state 7 is used to preheat the air exiting the compressor in the recuperator. The air exhausts to atmosphere at state 8. The gas turbine engine burns diesel fuel with a heat of combustion, HC = 48 MJ/kg. Neglect pressure losses in the heat exchangers. Model the working fluid in the cycle as ideal gas air.

a) Determine the efficiency of the engine and the rate at which it consumes fuel.

The inputs are entered in EES:

```
$UnitSystem SI K Pa J Mass Radian
"Input conditions"
T_amb=converttemp(C,K,25 [C])           "ambient temperature"
P_atm=1 [atm]*convert(atm,Pa)           "atmospheric pressure"
T_t_in=converttemp(C,K,1020 [C])        "turbine inlet temperature"
PR=3.5 [-]                              "compressor pressure ratio"
HC=48e6 [J/kg]                          "heat of combustion of fuel"
W_dot_pt=320 [hp]*convert(hp,W)         "power delivered to ship propulsion system"

"Performance parameters"
eta_c=0.82 [-]                          "compressor efficiency"
eta_gt=0.85 [-]                         "gasifier turbine efficiency"
eta_pt=0.81 [-]                         "power turbine efficiency"
DT_rec=50 [K]                           "recuperator approach temperature difference"
```

State 1 is specified by the pressure, $P_1 = P_{atm}$, and temperature, $T_1 = T_{amb}$. The specific enthalpy and specific entropy (h_1 and s_1) are obtained.

```
"State 1"
T[1]=T_amb                              "temperature"
P[1]=P_atm                              "pressure"
s[1]=entropy(Air,T=T[1],P=P[1])         "specific entropy"
h[1]=enthalpy(Air,T=T[1])               "specific enthalpy"
```

The pressure leaving the compressor is specified by the pressure ratio:

$$P_2 = PR\, P_1$$

An entropy balance on the reversible compressor requires that $s_{s,2} = s_1$. The state of the air leaving the reversible compressor is fixed by the specific entropy and pressure. The specific enthalpy leaving the reversible compressor ($h_{s,2}$) is determined. An energy balance on the reversible compressor is:

$$\frac{\dot{W}_{s,c}}{\dot{m}} = h_{s,2} - h_1$$

where $\dot{m}$ is the mass flow rate of air drawn into the engine through the compressor (the inlet mass flow rate). The power per unit of inlet mass required by the actual compressor is given by:

$$\frac{\dot{W}_c}{\dot{m}} = \frac{1}{\eta_c}\frac{\dot{W}_{s,c}}{\dot{m}}$$

The specific enthalpy leaving the actual compressor is obtained from an energy balance:

$$h_2 = h_1 + \frac{\dot{W}_c}{\dot{m}}$$

The state of the air leaving the actual compressor is fixed by the specific enthalpy and pressure. The specific entropy and temperature (s_2 and T_2) are obtained.

```
"State 2"
P[2]=PR*P[1]                              "pressure"
s_s[2]=s[1]                               "entropy balance on reversible compressor"
h_s[2]=enthalpy(Air,s=s_s[2],P=P[2])      "specific enthalpy leaving reversible compressor"
W_dot_s_c\m_dot=h_s[2]-h[1]               "work per mass required by reversible compressor"
W_dot_c\m_dot=W_dot_s_c\m_dot/eta_c       "work per mass required by actual compressor"
h[2]=h[1]+W_dot_c\m_dot                   "specific enthalpy leaving actual compressor"
s[2]=entropy(Air,h=h[2],P=P[2])           "specific entropy leaving actual compressor"
T[2]=temperature(Air,h=h[2])              "temperature leaving actual compressor"
```

State 4 is fixed by the pressure, $P_4 = P_2$, and temperature, $T_4 = T_{t,in}$. The specific enthalpy and specific entropy (h_4 and s_4) are determined.

```
"State 4"
T[4]=T_t_in                               "temperature"
P[4]=P[2]                                 "pressure"
h[4]=enthalpy(Air,T=T[4])                 "specific enthalpy"
s[4]=entropy(Air,T=T[4],P=P[4])           "specific entropy"
```

At this point in the analysis it is necessary to consider the gasifier turbine. However, we do not know the air-fuel ratio for the heater (AF_h) or the pressure at the exit of the gasifier turbine (P_5). The value of AF_h must eventually be set using an energy balance on the heater in order to achieve $T_4 = T_{t,in}$ and the value of P_5 must eventually be set so that the power produced by the gasifier turbine matches the power required by the compressor. Rather than enter all of the equations required to complete the problem at once and then try to solve, we will assume values of AF_h and P_5 in order to proceed in a sequential and explicit manner.

```
"State 5"
P[5]=(P[2]+P_atm)/2                       "guess for gasifier exit pressure"
AF_h=75 [-]                               "guess for the heater air-fuel ratio"
```

With these values set, the analysis of the gasifier turbine is straightforward. An entropy balance on the reversible gasifier turbine shows that $s_{s,5} = s_4$. The state of the air leaving the reversible gasifier turbine is fixed by the specific entropy and pressure. The specific enthalpy leaving the reversible gasifier turbine ($h_{s,5}$) is determined. An energy balance on the reversible gasifier turbine is:

$$\dot{W}_{s,gt} = (h_4 - h_{s,5})(\dot{m} + \dot{m}_{f,h}) \quad (1)$$

where $\dot{m}_{f,h}$ is the mass flow rate of fuel provided to the heater. Dividing Eq. (1) through by $\dot{m}$ and substituting the definition of the heater air-fuel ratio provides the isentropic turbine power per unit of inlet mass:

$$\frac{\dot{W}_{s,gt}}{\dot{m}} = (h_4 - h_{s,5})\left(1 + \frac{1}{AF_h}\right)$$

The power per unit of inlet mass delivered by the actual gasifier turbine is given by:

$$\frac{\dot{W}_{gt}}{\dot{m}} = \eta_{gt} \frac{\dot{W}_{s,gt}}{\dot{m}}$$

The specific enthalpy leaving the actual gasifier turbine is obtained from an energy balance:

$$h_5 = h_4 - \frac{\dfrac{\dot{W}_{gt}}{\dot{m}}}{\left(1 + \dfrac{1}{AF_h}\right)}$$

The state of the air leaving the actual gasifier turbine is fixed by the specific enthalpy and pressure. The specific entropy and temperature (s_5 and T_5) are obtained.

```
s_s[5]=s[4]                                        "entropy balance on reversible gasifier turbine"
h_s[5]=enthalpy(Air,s=s_s[5],P=P[5])               "specific enthalpy leaving reversible gasifier turbine"
W_dot_s_gt\m_dot=(h[4]-h_s[5])*(1+1/AF_h)          "work per mass associated with reversible gasifier turbine"
W_dot_gt\m_dot=eta_gt*W_dot_s_gt\m_dot             "work per mass associated with actual gasifier turbine"
h[5]=h[4]-W_dot_gt\m_dot/(1+1/AF_h)                "specific enthalpy leaving actual gasifier turbine"
s[5]=entropy(Air,h=h[5],P=P[5])                    "specific entropy leaving actual gasifier turbine"
T[5]=temperature(Air,h=h[5])                       "temperature leaving actual gasifier turbine"
```

The gasifier exit pressure must be adjusted so that the power consumed by the compressor equals the power produced by the gasifier turbine:

$$\dot{W}_c = \dot{W}_{gt} \tag{2}$$

Equation (2) is divided through by the mass flow rate of the inlet air:

$$\frac{\dot{W}_c}{\dot{m}} = \frac{\dot{W}_{gt}}{\dot{m}} \tag{3}$$

Select Update Guesses from the Calculate menu in order to set the guess value for each variable to the most recent solution. Comment out the assumed value of P_5 entered earlier and substitute Eq. (3) in order to complete the equation set.

```
{P[5]=(P[2]+P_atm)/2}               "guess for gasifier exit pressure"
W_dot_gt\m_dot=W_dot_c\m_dot        "gasifier turbine must drive the compressor"
```

State 6 is fixed by the pressure, $P_6 = P_5$, and temperature, $T_6 = T_{t,in}$. The specific enthalpy and specific entropy (h_6 and s_6) are determined.

```
"State 6"
T[6]=T_t_in                           "temperature"
P[6]=P[5]                             "pressure"
h[6]=enthalpy(Air,T=T[6])             "specific enthalpy"
s[6]=entropy(Air,T=T[6],P=P[6])       "specific entropy"
```

An energy balance on the reheater is:

$$(\dot{m} + \dot{m}_{f,h})\, h_5 + \dot{m}_{f,rh}\, HC = (\dot{m} + \dot{m}_{f,h} + \dot{m}_{f,rh})\, h_6 \tag{4}$$

where $\dot{m}_{f,rh}$ is the mass flow rate of fuel provided to the reheater. Dividing Eq. (4) through by the mass flow rate of inlet air and substituting the definitions of the air-fuel ratios leads to:

$$\left(1+\frac{1}{AF_h}\right)h_5+\frac{HC}{AF_{rh}}=\left(1+\frac{1}{AF_h}+\frac{1}{AF_{rh}}\right)h_6 \tag{5}$$

Equation (5) provides the air-fuel ratio required in the reheater that will result in the appropriate exit temperature.

```
h[5]*(1+1/AF_h)+HC/AF_rh=h[6]*(1+1/AF_h+1/AF_rh)    "energy balance on the reheater"
```

The pressure at the exit of the power turbine is atmospheric pressure, $P_7 = P_{atm}$. An entropy balance on the reversible power turbine is $s_{s,7} = s_6$. The state of the air leaving the reversible power turbine is fixed by the specific entropy and pressure. The specific enthalpy leaving the reversible power turbine ($h_{s,7}$) is determined. An energy balance on the reversible power turbine is:

$$\frac{\dot{W}_{s,pt}}{\dot{m}}=\left(h_6-h_{s,6}\right)\left(1+\frac{1}{AF_h}+\frac{1}{AF_{rh}}\right)$$

The power per unit of inlet mass delivered by the actual power turbine is given by:

$$\frac{\dot{W}_{pt}}{\dot{m}}=\eta_{pt}\frac{\dot{W}_{s,pt}}{\dot{m}}$$

The specific enthalpy leaving the actual power turbine is obtained from an energy balance:

$$h_7=h_6-\frac{\dfrac{\dot{W}_{pt}}{\dot{m}}}{\left(1+\dfrac{1}{AF_h}+\dfrac{1}{AF_{rh}}\right)}$$

The state of the air leaving the actual power turbine is fixed by the specific enthalpy and pressure. The specific entropy and temperature (s_7 and T_7) are obtained.

```
"State 7"
P[7]=P_amb                                     "power turbine exit pressure"
s_s[7]=s[6]                                    "entropy balance on reversible power turbine"
h_s[7]=enthalpy(Air,s=s_s[7],P=P[7])           "specific enthalpy leaving reversible power turbine"
W_dot_s_pt\m_dot=(h[6]-h_s[7])*(1+1/AF_h+1/AF_rh)
                                               "work per mass for rev. power turbine"
W_dot_pt\m_dot=eta_pt*W_dot_s_pt\m_dot         "work per mass for actual power turbine"
h[7]=h[6]-W_dot_pt\m_dot/(1+1/AF_h+1/AF_rh)    "specific enthalpy leaving actual power turbine"
s[7]=entropy(Air,h=h[7],P=P[7])                "specific entropy leaving actual power turbine"
T[7]=temperature(Air,h=h[7])                   "temperature leaving actual power turbine"
```

State 3 is fixed by the pressure, $P_3 = P_2$, and temperature. The temperature is obtained from the hot inlet temperature for the recuperator (the air leaving the power turbine at state 7) and the recuperator approach temperature difference:

$$T_3 = T_7 - \Delta T_{rec}$$

The specific enthalpy and specific entropy (h_3 and s_3) are determined.

```
"State 3"
T[3]=T[7]-DT_rec                    "temperature"
P[3]=P[2]                           "pressure"
h[3]=enthalpy(Air,T=T[3])           "specific enthalpy"
s[3]=entropy(Air,T=T[3],P=P[3])     "specific entropy"
```

An energy balance on the heater determines the air-fuel ratio in the heater, AF_h, which was set to an assumed value earlier.

$$h_3 + \frac{HC}{AF_h} = \left(1 + \frac{1}{AF_h}\right) h_4 \tag{6}$$

The guess values are updated, the assumed value of AF_h is commented out and Eq. (6) is entered.

```
{AF_h=75 [-]}                       "guess for the heater air-fuel ratio"
h[3]+HC/AF_h=h[4]*(1+1/AF_h)        "energy balance on the heater"
```

The value of AF_h determined from Eq. (6) is 180.7.

An energy balance on the recuperator is

$$h_2 + \left(1 + \frac{1}{AF_h} + \frac{1}{AF_{rh}}\right) h_7 = h_3 + \left(1 + \frac{1}{AF_h} + \frac{1}{AF_{rh}}\right) h_8 \tag{7}$$

Equation (7) provides the enthalpy leaving the recuperator at state 8. State 8 is specified by the enthalpy and pressure, $P_8 = P_7$. The temperature and specific entropy (T_8 and s_8) are determined.

```
"State 8"
h[2]+(1+1/AF_h+1/AF_rh)*h[7]=h[3]+(1+1/AF_h+1/AF_rh)*h[8]   "energy balance on recuperator"
P[8]=P[7]                                                   "pressure"
T[8]=temperature(Air,h=h[8])                                "temperature"
s[8]=entropy(Air,h=h[8],P=P[8])                             "specific entropy"
```

Figure 2 illustrates a temperature-specific entropy diagram for the cycle with the state points overlaid.

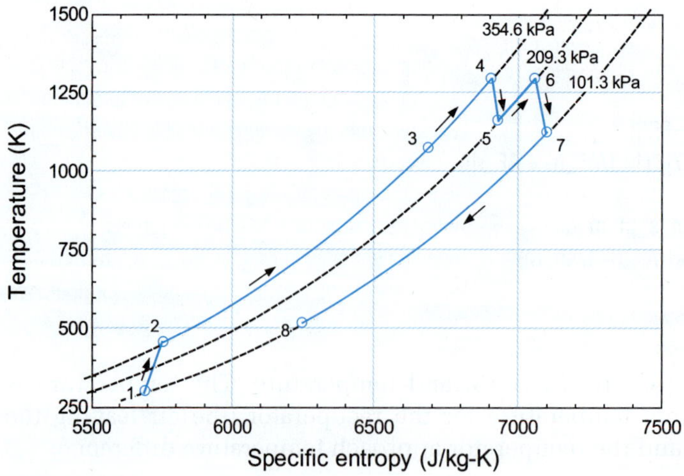

Figure 2: Temperature-entropy diagram for the cycle.

An overall energy balance is used to check the solution.

$$\dot{m}\,h_1 + \dot{m}_{f,h}\,HC + \dot{m}_{f,h}\,HC = \dot{W}_{pt} + \left(\dot{m} + \dot{m}_{f,h} + \dot{m}_{f,h}\right)\,h_8 \tag{8}$$

Equation (8) is divided through by the mass flow rate of inlet air and rearranged to provide a check on the solution:

$$check_1 = h_1 + \frac{HC}{AF_h} + \frac{HC}{AF_{rh}} - \frac{\dot{W}_{pt}}{\dot{m}} - \left(1 + \frac{1}{AF_h} + \frac{1}{AF_{rh}}\right)h_8$$

```
check_1=h[1]+HC/AF_h+HC/AF_rh-W_dot_pt\m_dot-(1+1/AF_h+1/AF_rh)*h[8]
                                                    "energy balance check"
```

Solving leads to $check_1 = 0$ J/kg. The efficiency of the cycle is computed according to:

$$\eta = \frac{\dot{W}_{pt}}{\dot{m}_{f,h}\,HC + \dot{m}_{f,rh}\,HC} \tag{9}$$

Equation (9) is divided through by the mass flow rate of inlet air:

$$\eta = \frac{\dfrac{\dot{W}_{pt}}{\dot{m}}}{\dfrac{HC}{AF_h} + \dfrac{HC}{AF_{rh}}}$$

```
eta=W_dot_gt\m_dot/(HC/AF_h+HC/AF_rh)     "efficiency"
```

which leads to $\eta = 0.3677$ (36.77%). The mass flow rate of air is determined from the required power output:

$$\dot{W}_{pt} = \frac{\dot{W}_{pt}}{\dot{m}}\dot{m}$$

The total mass flow rate of fuel is:

$$\dot{m}_f = \frac{\dot{m}}{AF_h} + \frac{\dot{m}}{AF_{rh}}$$

```
W_dot_pt=W_dot_pt\m_dot*m_dot                "mass flow rate of air"
m_dot_f=m_dot/AF_h+m_dot/AF_rh               "mass flow rate of fuel"
m_dot_f_kgphr=m_dot_f*convert(kg/s,kg/hr)    "in kg/hr"
```

Solving leads to $\dot{m}_f = 37.91$ kg/hr.

b) Determine the mass of fuel that the ship must have onboard in order to carry out missions with duration $time = 10$ day.

Assuming continuous operation, the mass of fuel required is:

$$m_f = \dot{m}_f\ time$$

```
time=10 [day]*convert(day,s)    "mission duration"
m_f=m_dot_f*time                "mass of fuel"
```

which results in $m_f = 9099$ kg of fuel.

c) Determine the effectiveness of the recuperative heat exchanger.

The heat transfer within the recuperator per mass of inlet air is obtained with an energy balance on the cold-side of the heat exchanger:

$$\frac{\dot{Q}_{rec}}{\dot{m}} = h_3 - h_2$$

The maximum possible value of h_3 occurs if the approach temperature difference approaches zero and therefore $T_3 = T_7$. The value of $h_{3,max}$ is determined and used to compute the maximum possible rate of heat transfer in the recuperator per mass of inlet air:

$$\frac{\dot{Q}_{rec,max}}{\dot{m}} = h_{3,max} - h_2$$

The recuperator effectiveness is:

$$\varepsilon_{rec} = \frac{\dfrac{\dot{Q}_{rec}}{\dot{m}}}{\dfrac{\dot{Q}_{rec,max}}{\dot{m}}}$$

```
Q_dot_rec\m_dot=h[3]-h[2]                      "recuperator heat transfer per unit mass"
h_3_max=enthalpy(Air,T=T[7])                   "maximum enthalpy leaving cold side"
Q_dot_rec_max\m_dot=h_3_max-h[2]               "maximum recuperator heat transfer per unit mass"
eff_rec=Q_dot_rec\m_dot/Q_dot_rec_max\m_dot    "effectiveness of recuperator"
```

which results in $\varepsilon_{rec} = 0.9209$ (92.09%).

d) Determine the mass of the recuperator required by the cycle assuming that it is proportional to the recuperator conductance with a proportionality constant of 0.05 kg-K/W.

The mass of the recuperator must be balanced against the mass of fuel that is saved in order to arrive at the optimal system design. The mass of recuperator is proportional to the total conductance of the recuperator. The total conductance of the recuperator can be estimated using the relations provided in Section 6.6.6 for a counterflow heat exchanger with constant capacitance rates. The specific heat capacity of air at constant pressure is assumed to be constant and equal to its value at the average temperature in the recuperator. The mass flow rate at state 7 is greater than the mass flow rate at state 2 due to the addition of the fuel. Therefore, the capacitance ratio for the recuperator is:

$$C_R = \frac{\dot{m}\, c_P}{\left(\dot{m} + \dot{m}_f\right) c_P} = \frac{1}{1 + \dfrac{1}{AF_h} + \dfrac{1}{AF_{rh}}} \tag{10}$$

The required number of transfer units is obtained using the appropriate relation in Table 6-2:

$$NTU_{rec} = \frac{\ln\left[\dfrac{(1 - \varepsilon_{rec}\, C_R)}{(1 - \varepsilon_{rec})}\right]}{(1 - C_R)} \tag{11}$$

The total conductance is:

$$UA_{rec} = NTU_{rec}\, \dot{m}\, c_P \tag{12}$$

where c_P is the average value of the constant pressure specific heat capacity in the recuperator. The mass of the recuperator is proportional to the total conductance.

The constant of proportionality that was provided in the problem statement is based on the specific design of the heat exchanger.

$$m_{rec} = 0.05 \left[\frac{\text{kg–K}}{\text{W}}\right] UA_{rec} \tag{13}$$

Equations (10) through (13) are used to determine the recuperator mass.

```
C_r=1/(1+1/AF_h+1/AF_rh)                        "capacity ratio"
NTU_rec=ln((1-eff_rec*C_r)/(1-eff_rec))/(1-C_r) "number of transfer units"
T_avg=(T[2]+T[3])/2                             "average temperature"
UA_rec=NTU_rec*m_dot*cP(Air,T=T_avg)            "total conductance"
m_rec=0.05 [kg-K/W]*UA_rec                      "mass of recuperator"
```

Solving provides $m_{rec} = 714.2$ kg.

e) Determine the total mass of the system (the sum of the fuel and recuperator mass). Plot the total mass as a function of the recuperator approach temperature difference. You should see that there is a specific value of the approach temperature difference that minimizes the total mass of the system; explain why this minimum occurs.

The total mass of the system is calculated according to:

$$m_{total} = m_{rec} + m_f$$

```
m_total=m_rec+m_f    "total mass"
```

which results in $m_{total} = 9813$ kg. Figure 3 shows how the total mass depends on the recuperator approach temperature difference, ΔT_{rec}. Also shown in Figure 3 are the recuperator mass and the fuel mass. Notice that a small value of ΔT_{rec} requires a massive recuperator but saves fuel due to the increased system efficiency. At the other extreme, a large value of ΔT_{rec} can be obtained with a small recuperator but the fuel mass increases due to the reduced system efficiency. An optimally designed recuperator balances these two effects.

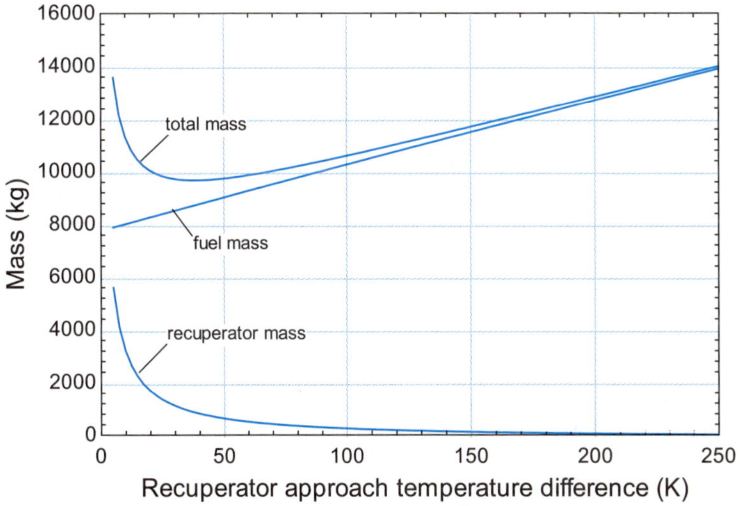

Figure 3: System mass as a function of the recuperator approach temperature difference. Also shown are the recuperator and fuel masses.

8.3.3 The Gas Turbine Engines for Propulsion

The gas turbine engine has the advantage of a high power to weight ratio which has made it the engine of choice for aircraft propulsion. There are a number of variants of the gas turbine engine cycle that are used for propulsion, including the turbojet, turbofan, and turboprop engines. Intercooling and regeneration are not usually employed in these systems due to the additional weight and space required for the associated heat exchangers.

Turbojet Engine

The turbojet engine is a simple modification to the basic air-breathing gas turbine engine that is discussed in Section 8.3.1. A simplified cross-section of the turbojet engine is shown in Figure 8-27(a). The compressor and turbine are attached to a single shaft and separated by a combustor. The turbojet cycle is shown schematically in Figure 8-27(b) and on a temperature-entropy diagram in Figure 8-27(c).

Air is drawn into the engine at relatively high velocity due to the speed of the aircraft. The high velocity air enters the diffuser at state 1. The diffuser reduces the velocity in order to provide a pressure rise at state 2, in advance of the compressor. The compressor increases the pressure and temperature of the air at state 3, which is mixed with fuel and burned in the combustor. The hot high pressure air leaving the combustor at state 4 is expanded through the turbine, which provides the power needed by the compressor. The pressure of the air leaving the turbine at state 5 is higher than atmospheric. In the basic gas turbine engine, shown in Figure 8-19(a), the air at leaving the gasifier turbine is directed to a power turbine that produces useful mechanical power. In the turbojet engine, the hot, high pressure air at state 5 enters a nozzle, which produces a high velocity flow of air leaving the turbojet at state 6.

The velocity of the air leaving the engine at state 6 is higher than the velocity of the air entering at state 1; therefore, a thrust force is produced. The net thrust produced by a turbojet engine is the difference between the momentum of the air leaving the engine and the momentum of the air entering the engine:

$$F_{thrust} = (\dot{m} + \dot{m}_f)\,\tilde{V}_6 - \dot{m}\,\tilde{V}_1 \tag{8-87}$$

where $\tilde{V}_1$ and $\tilde{V}_6$ are the velocities entering and exiting the engine, respectively, and $\dot{m}$ and $\dot{m}_f$ are the mass flow rates of entering air and fuel, respectively. Equation (8-87) can be expressed in terms of the air fuel ratio:

$$F_{thrust} = \dot{m}\left[\left(1 + \frac{1}{AF}\right)\tilde{V}_6 - \tilde{V}_1\right] \tag{8-88}$$

Several figures of merit are used to characterize the performance of a turbojet engine. One important figure is the *specific fuel consumption* (*SFC*), defined as the ratio of the fuel consumption to the thrust produced:

$$SFC = \frac{\dot{m}_f}{F_{thrust}} \tag{8-89}$$

The *thermal efficiency* of the engine is defined as the ratio of the increase in the kinetic energy of the flow (which is analogous to the power provided by the engine) to the thermal energy delivered to the engine from combusting fuel:

$$\eta_{th} = \frac{(\dot{m} + \dot{m}_f)\,\tilde{V}_6^2 - \dot{m}\,\tilde{V}_1^2}{2\,\dot{m}_f\,HC} \tag{8-90}$$

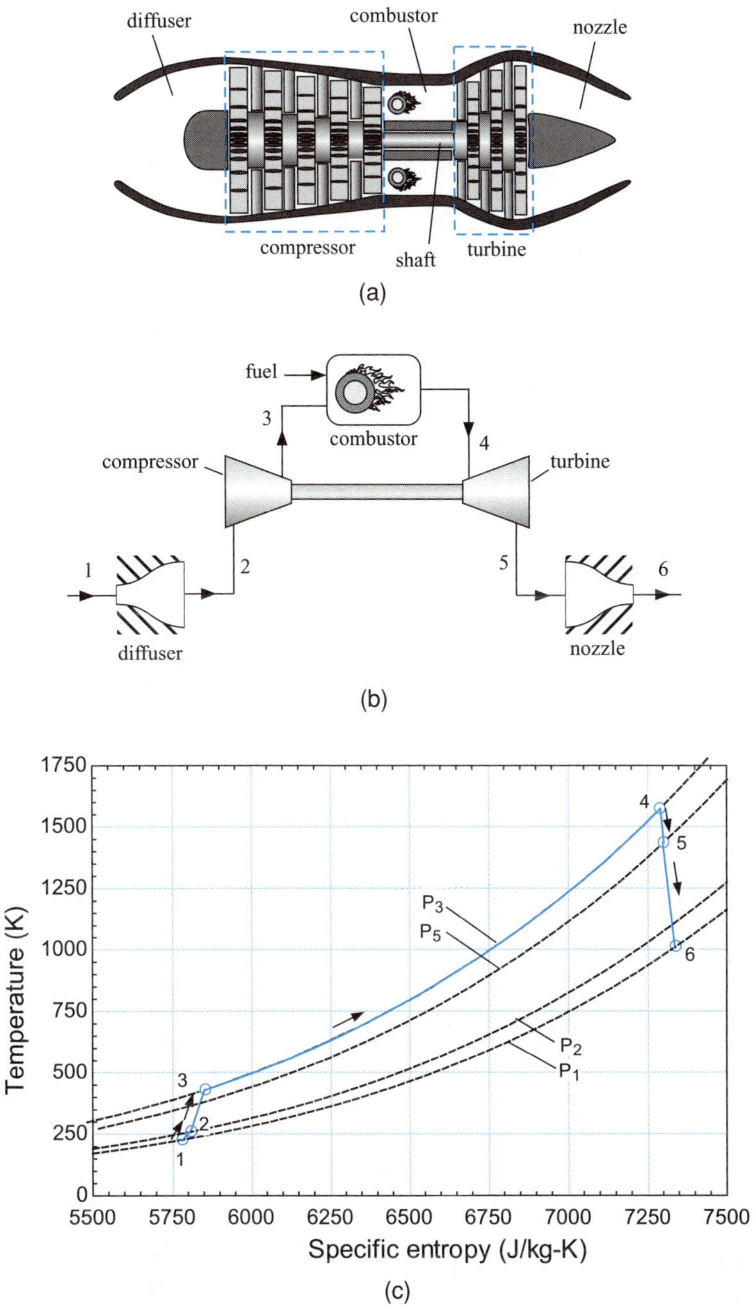

Figure 8-27: (a) Simplified cross-section, (b) schematic, and (c) *T-s* diagram for the turbojet engine.

The *propulsive efficiency* is the ratio of the power delivered to the aircraft (i.e., the product of thrust and velocity) to the increase in the kinetic energy of the flow:

$$\eta_p = \frac{2\,\tilde{V}_1\,F_{thrust}}{(\dot{m}+\dot{m}_f)\,\tilde{V}_6^2 - \dot{m}\,\tilde{V}_1^2} \tag{8-91}$$

The *overall efficiency* of the engine is the ratio of the power delivered to the aircraft to the thermal energy delivered to the engine; the overall efficiency is the product of the thermal and propulsive efficiencies.

$$\eta_o = \eta_{th}\,\eta_p = \frac{\tilde{V}_1\,F_{thrust}}{\dot{m}_f\,HC} \tag{8-92}$$

A turbojet engine can also use an afterburner in order to temporarily increase thrust at the expense of efficiency. The afterburner mixes the gas leaving the turbine with fuel and burns the mixture in order to elevate its temperature prior to expansion through the nozzle. The increase in temperature causes an increase in the velocity leaving the nozzle and therefore an increase in thrust.

EXAMPLE 8.3-3: TURBOJET ENGINE

A turbojet engine using the cycle shown in Figure 8-27(b) is being considered for use on an aircraft that is traveling at Mach 0.85 (i.e., $M = 0.85$, 85% of the speed of sound) at an elevation of 32,000 ft where $P_{atm} = 0.287$ atm and $T_{amb} = 227$ K. The diffuser coefficient of pressure recovery is $K_p = 0.65$, the compressor efficiency is $\eta_c = 0.89$, the turbine efficiency is $\eta_t = 0.91$, and the nozzle efficiency is $\eta_n = 0.93$. Assume that the working fluid has the same thermodynamic properties as air at all locations and that the air behaves as an ideal gas. The kinetic energy of the working fluid at all locations in the cycle is negligible, except at the inlet of the diffuser and the exit of the nozzle. The pressure ratio associated with the engine is $PR = 5$ and the turbine inlet temperature is $T_{t,in} = 1300°$C. The fuel for the engine has a heat of combustion of $HC = 48$ MJ/kg. The thrust required under these conditions is $F_{thrust} = 15$ kN. Neglect pressure loss in the combustor.

a) Analyze the turbojet engine and determine the required inlet air flow rate, air-fuel ratio, jet velocity, and specific fuel consumption. Determine the thermal, propulsive, and overall efficiency of the engine.

Procedures are used to analyze the turbine and compressor; these procedures are similar to those that were used in Example 8.2-1 to analyze the turbine and pump in a Rankine cycle.

```
$UnitSystem SI K Pa J Mass Radian
Procedure Turbine(T_in,P_in,P_out, F$, eta_t: h_out, T_out, s_out)
   s_in=entropy(F$,T=T_in,P=P_in)                    "inlet specific entropy"
   h_in=enthalpy(F$,T=T_in)                          "inlet specific enthalpy"
   s_s_out=s_in                                      "entropy balance on reversible turbine"
   h_s_out=enthalpy(F$,P=P_out,s=s_s_out)            "specific enthalpy leaving reversible turbine"
   W_dot_t_s\m_dot=h_in-h_s_out                      "energy balance on reversible turbine"
   W_dot_t\m_dot=W_dot_t_s\m_dot*eta_t               "definition of isentropic efficiency"
   h_out=h_in-W_dot_t\m_dot                          "energy balance on actual turbine"
   T_out=temperature(F$,h=h_out)                     "exit temperature"
   s_out=entropy(F$,h=h_out,P=P_out)                 "exit specific entropy"
end

Procedure Compressor(T_in,P_in,P_out, F$, eta_c: h_out, T_out, s_out)
   s_in=entropy(F$,T=T_in,P=P_in)                    "inlet specific entropy"
   h_in=enthalpy(F$,T=T_in)                          "inlet specific enthalpy"
```

```
  s_s_out=s_in                                   "entropy balance on reversible compressor"
  h_s_out=enthalpy(F$,P=P_out, s=s_s_out)        "specific enthalpy leaving reversible compressor"
  W_dot_s_c\m_dot=h_s_out-h_in                   "energy balance on reversible compressor"
  W_dot_c\m_dot=W_dot_s_c\m_dot/eta_c            "definition of isentropic efficiency"
  h_out=h_in+W_dot_c\m_dot                       "energy balance on actual compressor"
  T_out=temperature(F$,h=h_out)                  "exit temperature"
  s_out=entropy(F$,h=h_out,P=P_out)              "exit specific entropy"
end
```

The inputs are entered in EES:

```
"Inputs"
M=0.85 [-]                                       "Mach number of flight"
T_amb=227 [K]                                    "ambient temperature"
P_atm=0.287 [atm]*convert(atm,Pa)                "atmospheric pressure"
F$='Air'                                         "working fluid"
K_p=0.65 [-]                                     "coefficient of pressure recovery"
eta_c=0.89 [-]                                   "compressor efficiency"
eta_t=0.91 [-]                                   "turbine efficiency"
eta_n=0.93 [-]                                   "nozzle efficiency"
HC=48 [MJ/kg]*convert(MJ/kg,J/kg)                "heat of combustion of fuel"
PR=5 [-]                                         "pressure ratio"
T_t_in=converttemp(C,K,1300 [C])                 "turbine inlet temperature"
F_thrust=15 [kN]*convert(kN,N)                   "required thrust"
```

State 1 is fixed by the temperature and pressure, $T_1 = T_{amb}$ and $P_1 = P_{amb}$. The specific enthalpy, specific entropy, and specific volume (h_1, s_1, and v_1) are determined. The speed of sound at these conditions (c_1) is obtained using the SoundSpeed function in EES. The velocity of the air entering the engine is obtained from the definition of the Mach number, M:

$$\tilde{V}_1 = M\, c_1$$

```
"State 1"
T[1]=T_amb                                       "temperature"
P[1]=P_atm                                       "pressure"
h[1]=enthalpy(F$,T=T[1])                         "specific enthalpy"
s[1]=entropy(F$,T=T[1],P=P[1])                   "specific entropy"
v[1]=volume(F$,T=T[1],P=P[1])                    "specific volume"
c[1]=SoundSpeed(F$,T=T[1])                       "speed of sound"
Vel[1]=M*c[1]                                    "velocity"
```

The coefficient of pressure recovery for the diffuser is used to determine the pressure at the diffuser exit:

$$P_2 = P_1 + K_p \frac{\tilde{V}_1^2}{2\, v_1}$$

An energy balance on the adiabatic diffuser provides the specific enthalpy at state 2:

$$h_1 + \frac{\tilde{V}_1^2}{2} = h_2$$

The specific enthalpy and pressure fix state 2. The temperature and specific entropy (T_2 and s_2) are determined.

```
"State 2"
P[2]=P[1]+K_p*Vel[1]^2/(2*v[1])    "pressure"
h[1]+Vel[1]^2/2=h[2]               "specific enthalpy"
T[2]=Temperature(F$,h=h[2])        "temperature"
s[2]=entropy(F$,h=h[2],P=P[2])     "specific entropy"
```

The pressure at the exit of compressor is given by:

$$P_3 = PR\,P_2$$

The Compressor procedure is used to analyze the compressor and obtain the properties at state 3. An energy balance on the compressor is:

$$\frac{\dot{W}_c}{\dot{m}} = h_3 - h_2$$

```
"State 3"
P[3]=P[2]*PR                                              "pressure"
Call Compressor(T[2],P[2],P[3], F$, eta_c: h[3], T[3], s[3])  "compressor analysis"
W_dot_c\m_dot=h[3]-h[2]                                   "energy balance on compressor"
```

The pressure and temperature at state 4 are specified, $P_4 = P_3$ and $T_4 = T_{t,in}$. The specific enthalpy and specific entropy (h_4 and s_4) are obtained. An energy balance on the combustor provides the air fuel ratio:

$$h_3 + \frac{HC}{AF} = h_4\left(1 + \frac{1}{AF}\right)$$

```
"State 4"
P[4]=P[3]                          "pressure"
T[4]=T_t_in                        "temperature"
h[4]=enthalpy(F$,T=T[4])           "specific enthalpy"
s[4]=entropy(F$,T=T[4],P=P[4])     "specific entropy"
h[3]+HC/AF=h[4]*(1+1/AF)           "energy balance on combustor"
```

which results in $AF = 35.77$. The inlet conditions for the turbine are known, but the exit pressure must be adjusted until the turbine power output matches the required compressor power. A guess value for the turbine exit pressure is provided with an equation. This equation will be later removed and replaced by an equation that sets the turbine power output equal to the compressor power. The Turbine procedure is used to analyze the turbine. An energy balance on the turbine is:

$$\frac{\dot{W}_t}{\dot{m}} = \left(1 + \frac{1}{AF}\right)(h_4 - h_5)$$

```
"State 5"
P[5]=(P[4]+P[1])/2                                        "guess for turbine exit pressure"
Call Turbine(T[4],P[4],P[5], F$, eta_t: h[5], T[5], s[5])  "turbine analysis"
W_dot_t\m_dot=(1+1/AF)*(h[4]-h[5])                        "energy balance on turbine"
```

The guess values are updated by selecting Update Guesses from the Calculate menu. The assumed value of P_5 is commented out and the power produced by the turbine is set equal to the power consumed by the compressor:

$$\frac{\dot{W}_t}{\dot{m}} = \frac{\dot{W}_c}{\dot{m}}$$

```
{P[5]=(P[4]+P[1])/2}                  "guess for turbine exit pressure"
W_dot_t\m_dot=W_dot_c\m_dot           "compressor and turbine power must balance"
```

The pressure leaving the nozzle is the atmospheric pressure, $P_6 = P_{atm}$. An entropy balance on a reversible nozzle requires that $s_{s,6} = s_5$. The specific entropy and pressure fix the state of the air leaving the reversible nozzle and allow the calculation of $h_{s,6}$. An energy balance on a reversible nozzle provides the velocity leaving the reversible nozzle.

$$h_5 = h_{s,6} + \frac{\tilde{V}_{s,6}^2}{2}$$

The nozzle efficiency is used to compute the velocity leaving the actual nozzle.

$$\tilde{V}_6^2 = \eta_n \, \tilde{V}_{s,6}^2$$

An energy balance on the actual nozzle (which is assumed to be adiabatic) provides the specific enthalpy at state 6.

$$h_5 = h_6 = \frac{\tilde{V}_6^2}{2}$$

State 6 is fixed by the specific enthalpy and pressure. The temperature and specific entropy (T_6 and s_6) are determined.

```
"State 6"
P[6]=P_atm                              "pressure"
s_s[6]=s[5]                             "entropy balance on reversible nozzle"
h_s[6]=enthalpy(F$,P=P[6],s=s_s[6])     "specific enthalpy leaving reversible nozzle"
h[5]=h_s[6]+Vel_s[6]^2/2                "energy balance on reversible nozzle"
Vel[6]^2=eta_n*Vel_s[6]^2               "nozzle efficiency"
h[5]=h[6]+Vel[6]^2/2                    "energy balance on actual nozzle"
T[6]=temperature(F$,h=h[6])             "temperature"
s[6]=entropy(F$,h=h[6],P=P[6])          "specific entropy"
```

Solving leads to $\tilde{V}_6 = 973.4$ m/s; this supersonic velocity will require a converging/diverging nozzle, as discussed in Chapter 16. The mass flow rate is selected using Eq. (8-88) in order to provide the required thrust:

$$F_{thrust} = \dot{m}\left[\left(1 + \frac{1}{AF}\right)\tilde{V}_6 - \tilde{V}_1\right]$$

The mass flow rate of fuel is obtained using the air fuel ratio.

$$\dot{m}_f = \frac{\dot{m}}{AF}$$

```
"Performance of engine"
F_thrust=m_dot*(1+1/AF)*Vel[6]-m_dot*Vel[1]     "mass flow rate of air"
m_dot_f=m_dot/AF                                "mass flow rate of fuel"
```

Solving provides $\dot{m}$ = 20.17 kg/s. The specific fuel consumption is calculated according to:

$$SFC = \frac{\dot{m}_f}{F_{thrust}}$$

```
SFC=m_dot_f/F_thrust                          "specific fuel consumption"
SFC_English=SFC*convert(kg/s-N,lbm/hr-lbf)    "in English units"
```

which results in SFC = 1.327 $lb_m/hr\text{-}lb_f$. The thermal efficiency of the engine is calculated according to:

$$\eta_{th} = \frac{(\dot{m} + \dot{m}_f)\ \tilde{V}_6^2 - \dot{m}\,\tilde{V}_1^2}{2\,\dot{m}_f\,HC}$$

The propulsive efficiency is calculated according to:

$$\eta_p = \frac{2\,\tilde{V}_1\,F_{thrust}}{(\dot{m} + \dot{m}_f)\ \tilde{V}_6^2 - \dot{m}\,\tilde{V}_1^2}$$

The overall efficiency of the engine is the product of the thermal and propulsive efficiencies.

$$\eta_o = \eta_{th}\,\eta_p$$

```
eta_th=((m_dot+m_dot_f)*Vel[6]^2-m_dot*Vel[1]^2)/(2*m_dot_f*HC)    "thermal efficiency"
eta_p=2*Vel[1]*F_thrust/((m_dot+m_dot_f)*Vel[6]^2-m_dot*Vel[1]^2)  "propulsive efficiency"
eta_o=eta_th*eta_p                                                 "overall efficiency"
```

After solving, $\eta_{th} = 0.3383$ (33.8%), $\eta_p = 0.421$ (42.1%), and $\eta_o = 0.142$ (14.2%).

Turbofan Engine

In the turbojet engine, all of the air passes through the combustor and it therefore experiences a high pressure ratio and a large temperature change. The result is a very high velocity flow leaving the nozzle. In the turbojet engine considered in Example 8.3-3, the velocity of the air leaving the nozzle is $\tilde{V}_6 = 973.4$ m/s (several times the inlet velocity, $\tilde{V}_1 = 256.8$ m/s, and greater than the speed of sound). The high velocity of the air leaving the nozzle gives the turbojet engine a relatively low propulsive efficiency and therefore a low overall efficiency. This result is evident in Example 8.3-3; the turbojet engine exhibited a propulsive efficiency that is quite low (approximately 42%) leading to a low overall efficiency (approximately 14%). Substituting Eq. (8-88) for thrust into Eq. (8-91) for the propulsive efficiency leads to:

$$\eta_p = \frac{2\left[\left(1 + \frac{1}{AF}\right)\tilde{V}_6\,\tilde{V}_1 - \tilde{V}_1^2\right]}{\left[\left(1 + \frac{1}{AF}\right)\tilde{V}_6^2 - \tilde{V}_1^2\right]} \tag{8-93}$$

Examination of Eq. (8-93) suggests that the propulsive efficiency will be largest for an engine in which the nozzle exit velocity ($\tilde{V}_6$) is very close to the aircraft velocity ($\tilde{V}_1$); of course, in this limit the thrust produced per unit mass of air will be small, according to Eq. (8-88). As the nozzle exit velocity increases, the propulsive efficiency will decrease. The efficiency of the turbofan engine can be higher than the efficiency of the turbojet engine because it has a lower nozzle exit velocity. However, the turbofan engine is a larger and more complex engine that requires a higher mass flow rate to produce a

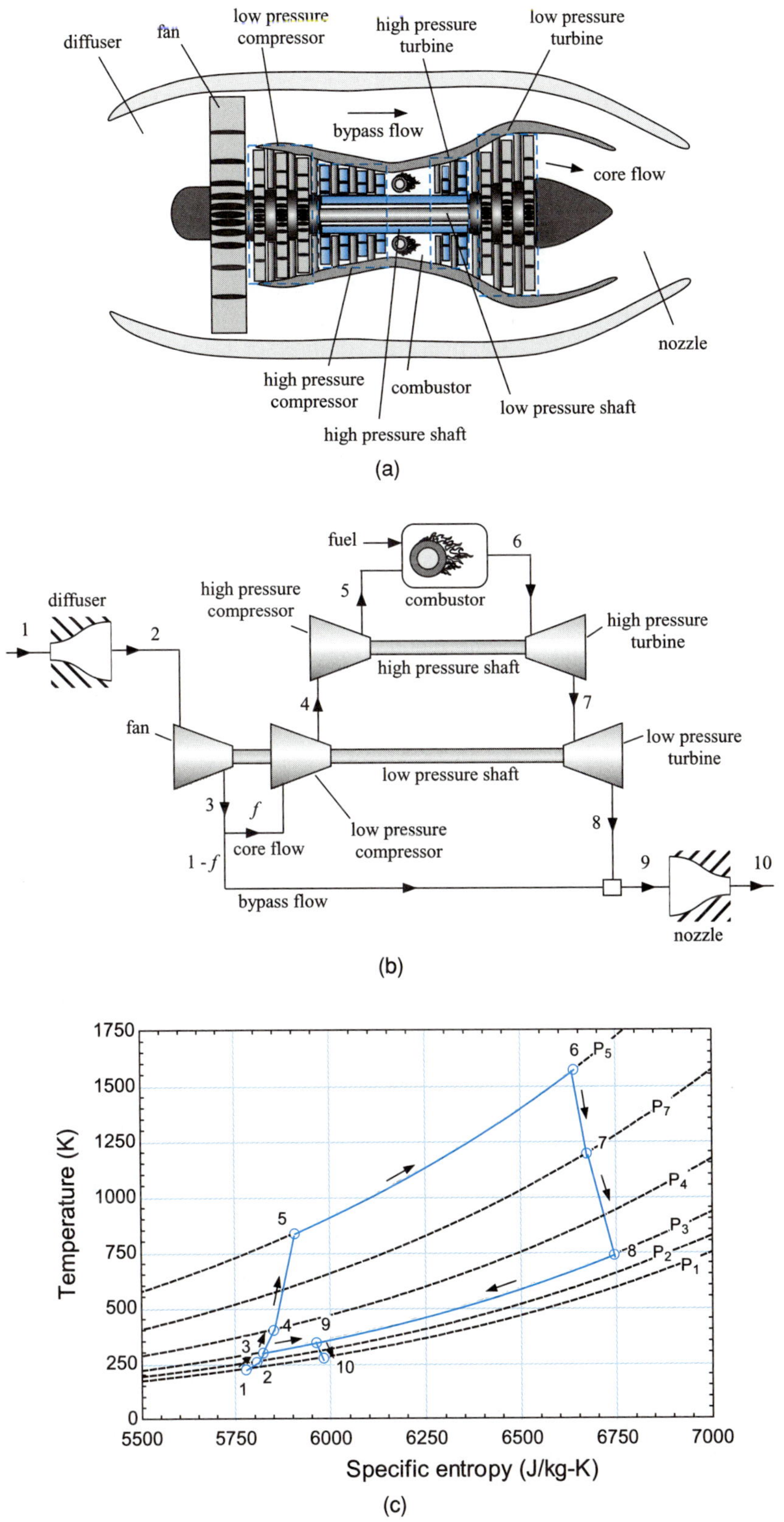

Figure 8-28: (a) Simplified cross-section, (b) schematic, and (c) *T-s* diagram for a two-shaft turbofan engine.

given amount of thrust. The turbofan engine has become of the engine of choice in large commercial aircraft because it is more fuel efficient and quieter than the turbojet engine.

A simplified cross-section of a two-shaft turbofan engine is shown in Figure 8-28(a). A schematic of the cycle is shown in Figure 8-28(b) and the cycle is shown on a temperature-entropy diagram in Figure 8-28(c). Air is drawn into the engine at state 1 and passes through a diffuser before it enters a fan at state 2. The fan is attached to the low pressure shaft and it has a relatively low pressure ratio. However, the fan is capable of processing a large amount of air. The air leaving the fan at state 3 is divided into two streams, sometimes referred to as the bypass flow and the core flow. The *bypass ratio* of the engine is the ratio of the bypass to the core flow rates.

The core flow passes through the low pressure compressor (also attached to the low pressure shaft) and the high pressure compressor (attached to the high pressure shaft) before entering the combustor at state 5. The hot high pressure air leaving the combustor at state 6 is expanded through the high pressure turbine (attached to the high pressure shaft) and the low pressure turbine (attached to the low pressure shaft). The high pressure turbine provides the power needed by the high pressure compressor and the low pressure turbine provides the power needed by the low pressure compressor and the fan. The core flow leaving the low pressure turbine at state 8 is mixed with the bypass flow at state 3 and the combined flow passes through the nozzle. The velocity of the air leaving the nozzle is higher than the inlet velocity, but not nearly as high as the nozzle exit velocity associated with a turbojet engine. However, the presence of the fan provides a much larger mass flow rate. Therefore it is possible to obtain a high thrust while maintaining a large propulsive efficiency.

EXAMPLE 8.3-4: TURBOFAN ENGINE

Example 8.3-3 considered a turbojet engine for application on an aircraft that is traveling at Mach 0.85 (i.e., $M = 0.85$, 85% of the speed of sound) at an elevation of 32,000 ft where $P_{atm} = 0.287$ atm and $T_{amb} = 227$ K. In this problem, we will consider the turbofan engine shown in Figure 8-28 for the same application. The diffuser has a coefficient of pressure recovery $K_p = 0.65$ and the nozzle has an efficiency of $\eta_n = 0.93$. The fan has a pressure ratio of $PR_{fan} = 1.6$ and an efficiency of $\eta_{fan} = 0.88$. The efficiency of the low and high pressure compressors are $\eta_{LPc} = 0.89$ and $\eta_{HPc} = 0.90$, respectively. The pressure ratios associated with the low and high pressure compressors are $PR_{LPc} = 2.5$ and $PR_{HPc} = 12$, respectively. The efficiencies of the low and high pressure turbines are $\eta_{LPt} = 0.91$ and $\eta_{HPt} = 0.92$, respectively. The kinetic energy of the working fluid is negligible at all locations in the cycle except at the inlet of the diffuser and the exit of the nozzle. The high pressure turbine inlet temperature is $T_{t,in} = 1300^\circ$C. The fuel for the engine has a heat of combustion of $HC = 48$ MJ/kg. The thrust required by the engine is $F_{thrust} = 15$ kN. Model the air as an ideal gas and model the combustion products as air. Neglect pressure loss in the combustor.

a) Analyze the turbofan engine and determine the required inlet air flow rate, bypass flow ratio, jet velocity, and specific fuel consumption. Determine the thermal, propulsive, and overall efficiency of the engine. Compare these values to the turbojet engine examined in Example 8.3-3.

The Compressor and Turbine procedures from Example 8.3-3 are used in this problem.

```
$UnitSystem SI K Pa J Mass Radian
Procedure Turbine(T_in,P_in,P_out, F$,eta_t: h_out, T_out, s_out)
  s_in=entropy(F$,T=T_in,P=P_in)                  "inlet specific entropy"
  h_in=enthalpy(F$,T=T_in)                        "inlet specific enthalpy"
  s_s_out=s_in                                    "entropy balance on reversible turbine"
  h_s_out=enthalpy(F$,P=P_out,s=s_s_out)          "specific enthalpy leaving reversible turbine"
  W_dot_t_s\m_dot=h_in-h_s_out                    "energy balance on reversible turbine"
  W_dot_t\m_dot=W_dot_t_s\m_dot*eta_t             "definition of isentropic efficiency"
  h_out=h_in-W_dot_t\m_dot                        "energy balance on actual turbine"
  T_out=temperature(F$,h=h_out)                   "exit temperature"
  s_out=entropy(F$,h=h_out,P=P_out)               "exit specific entropy"
end

Procedure Compressor(T_in,P_in,P_out, F$,eta_c: h_out, T_out, s_out)
  s_in=entropy(F$,T=T_in,P=P_in)                  "inlet specific entropy"
  h_in=enthalpy(F$,T=T_in)                        "inlet specific enthalpy"
  s_s_out=s_in                                    "entropy balance on reversible compressor"
  h_s_out=enthalpy(F$,P=P_out,s=s_s_out)          "specific enthalpy leaving reversible compressor"
  W_dot_s_c\m_dot=h_s_out-h_in                    "energy balance on reversible compressor"
  W_dot_c\m_dot=W_dot_s_c\m_dot/eta_c             "definition of isentropic efficiency"
  h_out=h_in+W_dot_c\m_dot                        "energy balance on actual compressor"
  T_out=temperature(F$,h=h_out)                   "exit temperature"
  s_out=entropy(F$,h=h_out,P=P_out)               "exit specific entropy"
end
```

The inputs are entered in EES:

```
"Inputs"
M=0.85 [-]                                "Mach number of flight"
T_amb=227 [K]                             "ambient temperature"
P_atm=0.287 [atm]*convert(atm,Pa)         "atmospheric pressure"
F$='Air'                                  "working fluid"
K_p=0.45 [-]                              "coefficient of pressure recovery"
eta_fan=0.88 [-]                          "fan efficiency"
eta_LPc=0.89 [-]                          "low pressure compressor efficiency"
eta_LPt=0.91 [-]                          "low pressure turbine efficiency"
eta_HPc=0.90 [-]                          "high pressure compressor efficiency"
eta_HPt=0.92 [-]                          "high pressure turbine efficiency"
eta_n=0.93 [-]                            "nozzle efficiency"
HC=48 [MJ/kg]*convert(MJ/kg,J/kg)         "heat of combustion of fuel"
PR_fan=1.6 [-]                            "fan pressure ratio"
PR_LPc=2.5 [-]                            "low pressure compressor pressure ratio"
PR_HPc=12 [-]                             "high pressure compressor pressure ratio"
T_t_in=converttemp(C,K,1300 [C])          "turbine inlet temperature"
F_thrust=15 [kN]*convert(kN,N)            "required thrust"
```

State 1 is fixed by the temperature and pressure, $T_1 = T_{amb}$ and $P_1 = P_{atm}$. The specific enthalpy, specific entropy, and specific volume (h_1, s_1, and v_1) are determined. The speed of sound at these conditions (c_1) is obtained using the SoundSpeed function in EES. The velocity of the air entering the engine is obtained from:

$$\tilde{V}_1 = M\, c_1$$

```
"State 1"
T[1]=T_amb                              "temperature"
P[1]=P_amb                              "pressure"
h[1]=enthalpy(F$,T=T[1])                "enthalpy"
s[1]=entropy(F$,T=T[1],P=P[1])          "entropy"
v[1]=volume(F$,T=T[1],P=P[1])           "specific volume"
c[1]=SoundSpeed(F$,T=T[1])              "speed of sound"
Vel[1]=M*c[1]                           "velocity"
```

The coefficient of pressure recovery for the diffuser is used to determine the pressure at the diffuser exit:

$$P_2 = P_1 + K_p \frac{\tilde{V}_1^2}{2\, v_1}$$

An energy balance on the diffuser provides the specific enthalpy at state 2:

$$h_1 + \frac{\tilde{V}_1^2}{2} = h_2$$

The specific enthalpy and pressure fix state 2. The temperature and specific entropy (T_2 and s_2) are determined.

```
"State 2"
P[2]=P[1]+K_p*Vel[1]^2/(2*v[1])     "pressure"
h[1]+Vel[1]^2/2=h[2]                "specific enthalpy"
T[2]=Temperature(F$,h=h[2])         "temperature"
s[2]=entropy(F$,h=h[2],P=P[2])      "specific entropy"
```

The pressure at the exit of fan is given by:

$$P_3 = P\,R_{fan}\, P_2$$

The Compressor procedure is used to analyze the fan and obtain the properties at state 3. An energy balance on the adiabatic fan is:

$$\frac{\dot{W}_{fan}}{\dot{m}} = h_3 - h_2$$

where $\dot{m}$ is the mass flow rate of inlet air.

```
"State 3"
P[3]=P[2]*PR_fan                                                "pressure"
Call Compressor(T[2],P[2],P[3], F$, eta_fan: h[3], T[3], s[3])  "fan analysis"
W_dot_fan\m_dot=h[3]-h[2]                                       "energy balance on fan"
```

The fraction of the flow that passes through the core (f) is not known. Eventually, the fraction must be selected so that the power obtained from the low pressure turbine is sufficient to drive the low pressure compressor and the fan. For now, in order to proceed with the problem solution in a sequential manner, a guess value for f is entered. The guess can be highlighted in order make it obvious that it must eventually be removed in order for the solution to be complete.

```
f=0.2 [-]    "guess for core fraction"
```

The pressure at the exit of the low pressure compressor is given by:

$$P_4 = PR_{LP\,c}\,P_3$$

The Compressor procedure is used to analyze the low pressure compressor and obtain the properties at state 4. An energy balance on the low pressure compressor is:

$$\frac{\dot{W}_{LP\,c}}{\dot{m}} = f\,(h_4 - h_3)$$

```
"State 4"
P[4]=PR_LPc*P[3]                                                     "pressure"
Call Compressor(T[3],P[3],P[4], F$, eta_LPc: h[4], T[4], s[4])   "low pressure compressor analysis"
W_dot_LPc\m_dot=f*(h[4]-h[3])                                        "energy balance on fan"
```

The pressure at the exit of the high pressure compressor is given by:

$$P_5 = PR_{HP\,c}\,P_4$$

The Compressor procedure is used to analyze the high pressure compressor and obtain the properties at state 5. An energy balance determines the power per unit mass for the high pressure compressor:

$$\frac{\dot{W}_{HP\,c}}{\dot{m}} = f\,(h_5 - h_4)$$

```
"State 5"
P[5]=PR_HPc*P[4]                                                     "pressure"
Call Compressor(T[4],P[4],P[5], F$, eta_HPc: h[5], T[5], s[5])   "high pressure compressor analysis"
W_dot_HPc\m_dot=f*(h[5]-h[4])                                        "energy balance on fan"
```

The pressure and temperature at state 6 are specified, $P_6 = P_5$ and $T_6 = T_{t,in}$. The specific enthalpy and specific entropy (h_6 and s_6) are obtained. An energy balance on the combustor provides the air fuel ratio:

$$f\,\dot{m}\,h_5 + \dot{m}_f\,HC = \left(f\,\dot{m} + \dot{m}_f\right)\,h_6 \tag{1}$$

where $\dot{m}_f$ is the mass flow rate of fuel. Equation (1) is divided through by $\dot{m}$ to yield:

$$f\,h_5 + \frac{HC}{AF} = \left(f + \frac{1}{AF}\right)\,h_6$$

```
"State 6"
T[6]=T_t_in                          "temperature"
P[6]=P[5]                            "pressure"
h[6]=enthalpy(F$,T=T[6])             "specific enthalpy"
s[6]=entropy(F$,T=T[6],P=P[6])       "specific entropy"
f*h[5]+HC/AF=h[6]*(f+1/AF)           "energy balance on combustor"
```

which results in $AF = 269.3$; note that AF is the ratio of the total air flow through the turbofan to the fuel flow, which is why it is such a large number. The air-fuel ratio characterizing the ratio of the air that passes through the combustor to the fuel will be much smaller. The inlet conditions for the high pressure turbine are known but the exit pressure must be adjusted until the high pressure turbine power output matches the required high pressure compressor power. The exit pressure is initially guessed; this guess will later be removed and the exit pressure will be adjusted by EES in order to set the high pressure turbine power equal to the high pressure compressor power. The Turbine procedure is used to analyze the high pressure turbine. An energy balance on the high pressure turbine is:

$$\frac{\dot{W}_{HPt}}{\dot{m}} = \left(f + \frac{1}{AF}\right)(h_6 - h_7)$$

```
"State 7"
P[7]=P[4]                                                "guess for high pressure turbine exit pressure"
Call Turbine(T[6],P[6],P[7], F$, eta_HPt: h[7], T[7], s[7])  "turbine analysis"
W_dot_HPt\m_dot=(f+1/AF)*(h[6]-h[7])                     "energy balance on turbine"
```

The guess values are updated by selecting Update Guesses from the Calculate menu. The assumed value of P_7 is commented out and the power produced by the high pressure turbine is set equal to the power consumed by the high pressure compressor:

$$\frac{\dot{W}_{HPt}}{\dot{m}} = \frac{\dot{W}_{HPc}}{\dot{m}}$$

```
{P[7]=P[4]}                            "guess for high pressure turbine exit pressure"
W_dot_HPt\m_dot=W_dot_HPc\m_dot        "high pressure compressor and turbine power must balance"
```

The pressure leaving the low pressure turbine must match the pressure leaving the fan, $P_8 = P_3$, so that the core and bypass flows can be mixed prior to passing through the nozzle. The Turbine procedure is used to analyze the low pressure turbine. An energy balance on the low pressure turbine is:

$$\frac{\dot{W}_{LPt}}{\dot{m}} = \left(f + \frac{1}{AF}\right)(h_7 - h_8)$$

```
"State 8"
P[8]=P[3]                                                "low pressure turbine exit pressure"
Call Turbine(T[7],P[7],P[8], F$, eta_LPt: h[8], T[8], s[8])  "turbine analysis"
W_dot_LPt\m_dot=(f+1/AF)*(h[7]-h[8])                     "energy balance on turbine"
```

The guess values are updated and the assumed value of the core flow fraction, f, is commented out. EES will solve for the value of f that sets the power from the low pressure turbine equal to the power to drive the low pressure compressor and the fan:

$$\frac{\dot{W}_{LPt}}{\dot{m}} = \frac{\dot{W}_{LPc}}{\dot{m}} + \frac{\dot{W}_{fan}}{\dot{m}}$$

```
{f=0.2 [-]}                                              "guess for core fraction"
W_dot_LPt\m_dot=W_dot_LPc\m_dot+W_dot_fan\m_dot
                              "low pressure compressor, fan, and turbine power must balance"
```

Solving provides $f = 0.100$ (10.0%). The bypass ratio of the engine is calculated from:

$$BPR = \frac{(1-f)}{f}$$

```
BPR=(1-f)/f                    "bypass ratio"
```

which results in $BPR = 8.96$. The pressure leaving the mixing junction is $P_9 = P_8$. An energy balance on the mixing junction leads to:

$$f\,h_8 + (1-f)\,h_3 = h_9$$

State 9 is fixed by the specific enthalpy and pressure. The temperature and specific entropy (T_9 and s_9) are determined.

```
"State 9"
P[9]=P[8]                            "pressure"
f*h[8]+(1-f)*h[3]=h[9]               "specific enthalpy"
T[9]=temperature(F$,h=h[9])          "temperature"
s[9]=entropy(F$,h=h[9],P=P[9])       "specific entropy"
```

The pressure leaving the nozzle is the ambient pressure, $P_{10} = P_{atm}$. An entropy balance on a reversible nozzle leads to $s_{s,10} = s_9$. The specific entropy and pressure fix the state of the air leaving the reversible nozzle and allow the calculation of $h_{s,10}$. An energy balance on a reversible nozzle provides the velocity leaving the reversible nozzle.

$$h_9 = h_{s,10} + \frac{\tilde{V}_{s,10}^2}{2}$$

The nozzle efficiency is used to compute the velocity leaving the actual nozzle.

$$\tilde{V}_{10}^2 = \eta_n\,\tilde{V}_{s,10}^2$$

An energy balance on the actual nozzle provides the specific enthalpy at state 10.

$$h_9 = h_{10} + \frac{\tilde{V}_{10}^2}{2}$$

State 10 is fixed by the specific enthalpy and pressure. The temperature and specific entropy (T_{10} and s_{10}) are determined.

```
"State 10"
P[10]=P_atm                                 "pressure"
s_s[10]=s[9]                                "entropy balance on reversible nozzle"
h_s[10]=enthalpy(F$,P=P[10],s=s_s[10])      "specific enthalpy leaving reversible nozzle"
h[9]=h_s[10]+Vel_s[10]^2/2                  "energy balance on reversible nozzle"
Vel[10]^2=eta_n*Vel_s[10]^2                 "nozzle efficiency"
h[9]=h[10]+Vel[10]^2/2                      "energy balance on actual nozzle"
T[10]=temperature(F$,h=h[10])               "temperature"
s[10]=entropy(F$,h=h[10],P=P[10])           "specific entropy"
```

Solving provides $\tilde{V}_{10} = 373.2$ m/s. The mass flow rate is selected in order to provide the required thrust:

$$F_{thrust} = \dot{m}\left[\left(1 + \frac{1}{AF}\right)\tilde{V}_{10} - \tilde{V}_1\right]$$

The mass flow rate of fuel is obtained using the air fuel ratio.

$$\dot{m}_f = \frac{\dot{m}}{AF}$$

```
"Performance of engine"
F_thrust=m_dot*(1+1/AF)*Vel[10]-m_dot*Vel[1]     "mass flow rate of air"
m_dot_f=m_dot/AF                                  "mass flow rate of fuel"
```

which results in $\dot{m} = 128.1$ kg/s. The specific fuel consumption is calculated according to:

$$SFC = \frac{\dot{m}_f}{F_{thrust}}$$

```
SFC=m_dot_f/F_thrust                              "specific fuel consumption"
SFC_English=SFC*convert(kg/s-N,lbm/hr-lbf)        "in English units"
```

which provides $SFC = 0.562$ $lb_m/hr\text{-}lb_f$. The thermal efficiency of the engine is calculated according to:

$$\eta_{th} = \frac{(\dot{m} + \dot{m}_f)\,\tilde{V}_{10}^2 - \dot{m}\,\tilde{V}_1^2}{2\,\dot{m}_f\,HC}$$

The propulsive efficiency is calculated according to:

$$\eta_p = \frac{2\,\tilde{V}_1\,F_{thrust}}{(\dot{m} + \dot{m}_f)\,\tilde{V}_{10}^2 - \dot{m}\,\tilde{V}_1^2}$$

The overall efficiency of the engine is the product of the thermal and propulsive efficiencies.

$$\eta_o = \eta_{th}\,\eta_p$$

```
eta_th=((m_dot+m_dot_f)*Vel[10]^2-m_dot*Vel[1]^2)/(2*m_dot_f*HC)   "thermal efficiency"
eta_p=2*Vel[1]*F_thrust/((m_dot+m_dot_f)*Vel[10]^2-m_dot*Vel[1]^2)  "propulsive efficiency"
eta_o=eta_th*eta_p                                                  "overall efficiency"
```

Solving provides $\eta_{th} = 0.411$ (41.1%), $\eta_p = 0.817$ (81.7%), and $\eta_o = 0.336$ (33.6%). Table 1 summarizes the calculated results for the turbojet engine from Example 8.3-3 and the turbofan engine considered in this example. Both engines deliver the

Table 1: Comparison of the turbojet engine and turbofan engine.

Parameter	Turbojet (Example 8.3-3)	Turbofan (Example 8.3-4)
Thrust force	15 kN	15 kN
Nozzle exit velocity	973.4 m/s	373.2 m/s
Mass flow rate of air	20.17 kg/s	128.1 kg/s
Specific fuel consumption	1.327 $lb_m/hr\text{-}lb_f$	0.562 $lb_m/hr\text{-}lb_f$
Propulsive efficiency	42.1%	81.7%
Thermal efficiency	33.8%	41.1%
Overall efficiency	14.2%	33.6%

same thrust. However, the turbofan engine uses a much larger mass flow rate and smaller nozzle exit velocity. The smaller jet velocity makes the turbofan engine much more efficient due primarily to the higher propulsive efficiency. This difference is evident in the overall efficiency as well as the specific fuel consumption. The turbofan engine is also quieter.

Turboprop Engine

The turboprop engine is similar to the turbofan engine. A simplified cross-section of the turboprop engine is shown in Figure 8-29. The core air flow is used to spin a propeller which moves a much larger flow of air; a gear box is employed to allow the propeller to spin at a lower speed than the compressor and turbine. Turboprop engines are usually used on small aircraft and are more efficient than the jet engines at lower flight speeds.

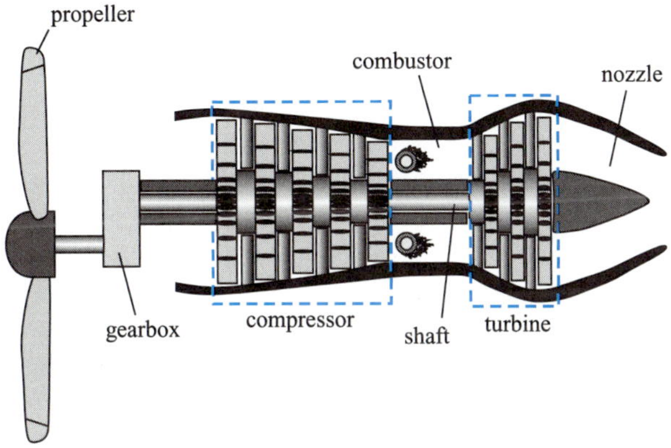

Figure 8-29: Simplified schematic of a turboprop engine.

8.3.4 The Combined Cycle and Cogeneration

The temperature of the combustion products leaving the turbine in a gas turbine engine is quite high. For example, the temperature of the products leaving the power turbine shown in Figure 8-19(b) is $T_5 = 1046$ K (773°C). This high temperature gas stream can be used to preheat the air leaving the compressor before it enters the compressor using the recuperated cycle illustrated in Figure 8-26. The high temperature gas stream can also be used as the heat source for another cycle or process. The temperature of the combustion products leaving the turbine is appropriate for use as the heat input to a Rankine cycle. The steam turbines in a Rankine cycle cannot tolerate inlet temperatures that are as high as those experienced by the gas turbines in a Brayton cycle. The high pressures encountered in the Rankine cycle lead to large pressure forces on the turbine blades and water is a corrosive fluid. Therefore, steam turbines must be made from relatively strong, corrosion-resistant materials that do not survive at very high temperatures. The maximum upper limit for a steam turbine inlet temperature is in the 600°C to 700°C range. Gas turbines can tolerate much larger temperatures because the pressures (and therefore pressure forces) are much lower and combustion gas is not as corrosive as steam. Therefore, more exotic high temperature materials can be used, allowing gas turbine temperatures in excess of 1100°C.

Figure 8-30 illustrates a combined cycle in which the combustion products leaving the gas turbine are used as the heat input for a boiler within a Rankine cycle. The efficiency of such a cycle (defined as the ratio of the net power from both cycles to the thermal energy supplied with the fuel) can exceed 50%. The combustion products leaving the gas

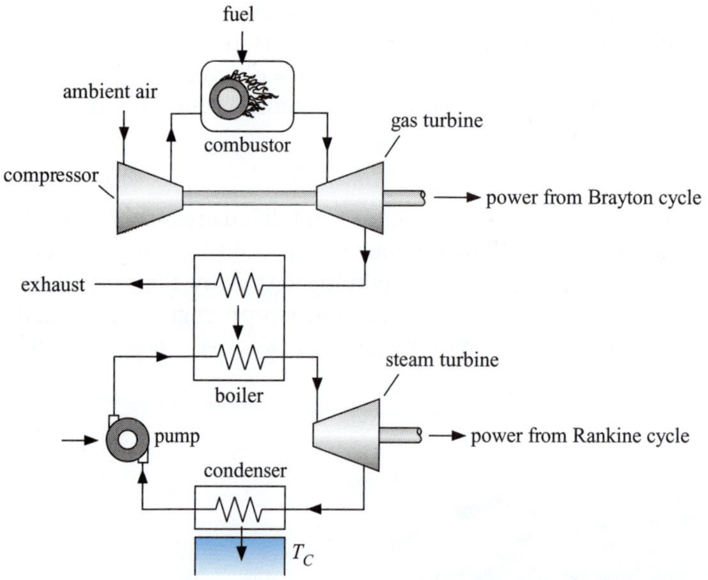

Figure 8-30: Combined cycle.

turbine can also be used to generate steam for heating (or even cooling, with a steam-driven refrigeration compressor). This type of multiple-use system is called a cogeneration system.

8.4 Reciprocating Internal Combustion Engines

Reciprocating internal combustion engines are used almost exclusively as the power source for ground transportation vehicles. These open cycle engines offer the advantages of high power to weight ratio (relative to the Rankine cycle), high efficiency over a range of power outputs, and an enviable record of reliability that has been achieved by more than a century of continuous engineering improvements. Reciprocating internal combustion engines can be categorized as either spark-ignition or compression-ignition engines.

The processes occurring in a reciprocating internal combustion engine are quite complicated and not amenable to a simple thermodynamic analysis. A detailed analysis must take into consideration the timing of the ignition process and subsequent combustion processes, the effect of intake and exhaust valves, mixing and heat transfer within the cylinder and many other factors. The analyses presented in this section do not include such detail and consequently provide only an approximate description of the processes. However, these simple models are sufficient to identify some important design parameters and their effects.

8.4.1 The Spark-Ignition Reciprocating Internal Combustion Engine

The spark-ignition reciprocating internal combustion engine is referred to as the *Otto cycle*, in honor of the German engineer Nicolaus Otto who first demonstrated this type of engine in the latter half of the 19th century. There are two variations of the spark-ignition engine: the four-stroke engine and the two-stroke engine.

Spark-Ignition, Four-Stroke Engine Cycle

The processes occurring in the four-stroke engine cycle are shown in Figure 8-31. These processes are also shown qualitatively on a pressure-volume diagram in Figure 8-32. The piston reciprocates, moving between two positions that are referred to as *top dead center* (where the cylinder volume is at its minimum) and *bottom dead center* (where the cylinder volume is at its maximum). The ratio of the volume in the cylinder at bottom dead center to the volume at top dead center is referred to as the *compression ratio*, *CR*.

$$CR = \frac{V_{BDC}}{V_{TDC}} \tag{8-94}$$

The crank angle refers the position of the crankshaft, which is related to the instantaneous volume in the cylinder. A crank angle of 0° corresponds to top dead center and 180° is bottom dead center. The four-stroke engine requires two complete revolutions of the crankshaft in order to complete a cycle, which corresponds to the crank angle going from 0° to 720°. The qualitative variation of the pressure and temperature of the gas in the cylinder as a function of the crank angle are shown in Figure 8-33.

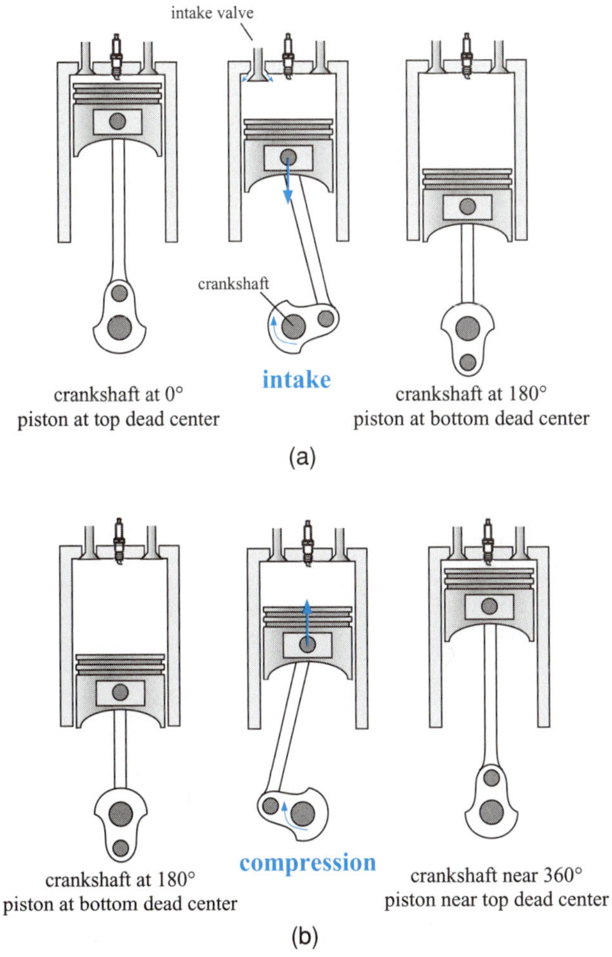

Figure 8-31: Processes occurring in the four-stroke spark-ignition engine: (a) intake, (b) compression, (c) combustion, (d) expansion, and (e) exhaust. An animation of this cycle can be obtained by selecting EES Example Problems from the Examples menu in EES. Select Animation and then Animation of a 4-stroke internal combustion engine.

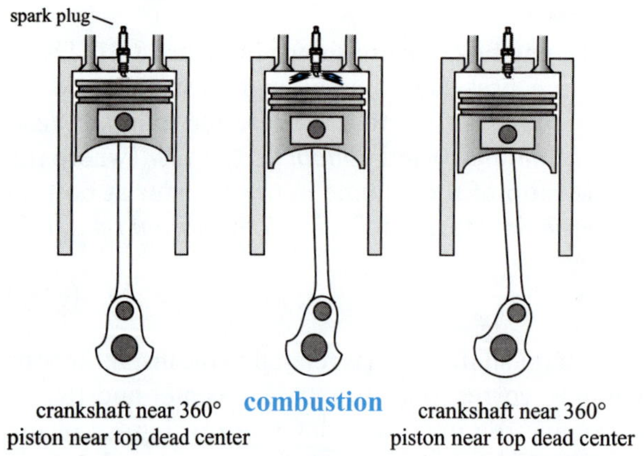
spark plug
crankshaft near 360°
piston near top dead center
combustion
crankshaft near 360°
piston near top dead center

(c)

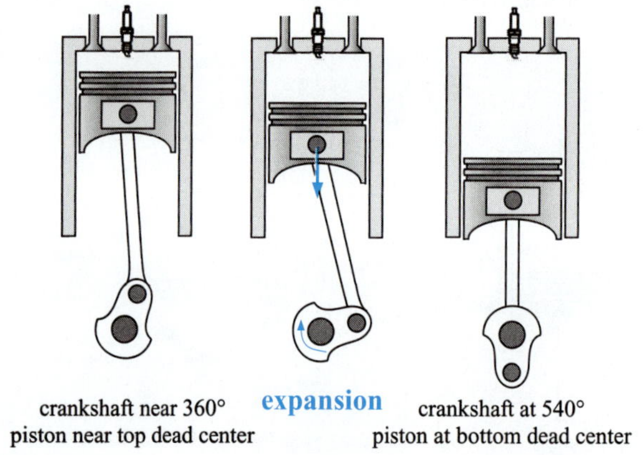
crankshaft near 360°
piston near top dead center
expansion
crankshaft at 540°
piston at bottom dead center

(d)

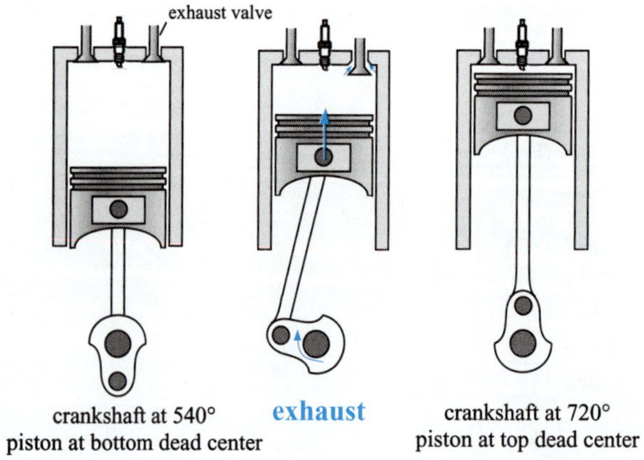
exhaust valve
crankshaft at 540°
piston at bottom dead center
exhaust
crankshaft at 720°
piston at top dead center

(e)

Figure 8-31: (continued)

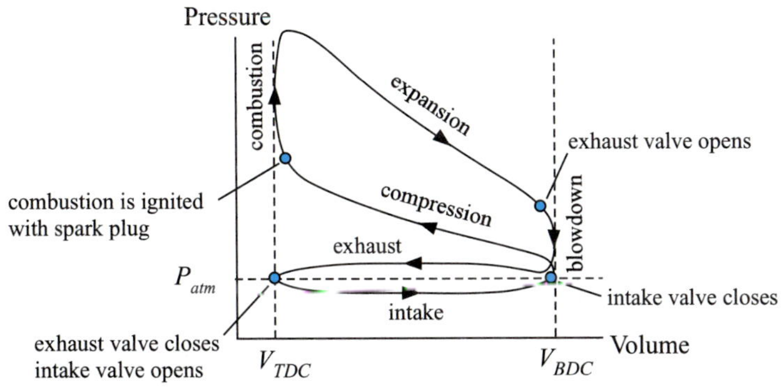

Figure 8-32: Processes occurring in the four-stroke spark-ignition engine shown qualitatively on a pressure-volume diagram.

During the intake stroke, the crankshaft rotates from 0° to 180° causing the piston to move from its top dead center to its bottom dead center position (increasing the cylinder volume from V_{TDC} to V_{BDC}). This process occurs with the intake valve open so that a gaseous mixture of air and fuel is drawn into the cylinder. The pressure in the cylinder may be slightly below ambient during the intake stroke due to pressure drop across the intake valve and intake manifold. The intake valve closes when the piston reaches bottom dead center, which concludes the intake stroke.

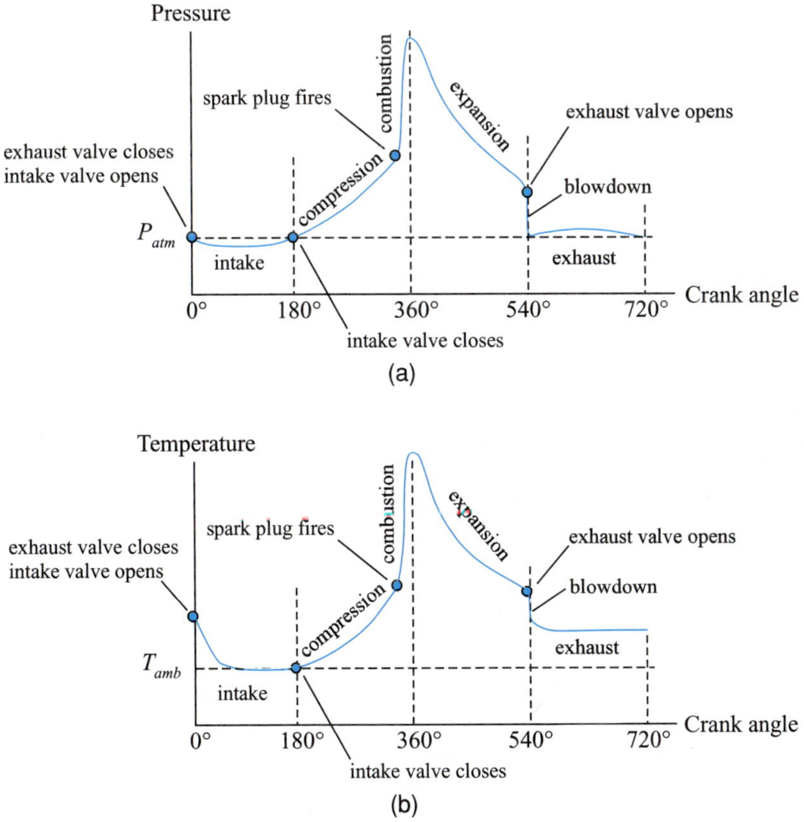

Figure 8-33: (a) Pressure and (b) temperature of the gas in the cylinder shown qualitatively as a function of crank angle.

During the compression process, the crankshaft rotates from 180° to (approximately) 360° causing the piston to move from its bottom dead center to near its top dead center position. The volume in the cylinder decreases causing the pressure and the temperature of the gas to increase. Work must be done on the air/fuel mixture in order to accomplish the compression process. When the piston approaches its top dead center position, the compression process is complete.

The combustion process is initiated with the spark plug when the piston is near top dead center and the crankshaft is near 360°. The timing of the ignition may deviate somewhat from top dead center depending on the engine design and load. During the combustion process, the hydrocarbon fuel reacts with the oxygen in the air, releasing chemical energy which significantly increases the temperature and pressure of the gas in the cylinder. The combustion process occurs quickly (relative to the time associated with the piston motion). Therefore, the pressure and temperature rise nearly instantaneously in Figure 8-33.

During the expansion process, the crankshaft rotates from 360° to 540° causing the piston to move from its top dead center to its bottom dead center position. The combustion products in the cylinder do work on the piston during the expansion process. Because the pressure in the cylinder is higher during the expansion process than it was during the compression process, more work is produced during the expansion process than is required by the compression process. Therefore, some of the energy produced during the expansion process is stored in the flywheel and used during the subsequent compression process while the remainder of the energy is provided to the transmission system.

At the beginning of the exhaust stroke, the exhaust valve is opened causing the cylinder contents to rapidly blow down to atmospheric pressure. During the exhaust stroke, the crankshaft rotates from 540° to 720° causing the piston to move from its bottom dead center to its top dead center position. This process occurs with the exhaust valve open so that most of the combustion products remaining in the cylinder are pushed out. The pressure in the cylinder may be slightly above ambient during the exhaust stroke due to pressure drop across the exhaust valve and exhaust system. The exhaust valve closes when the piston reaches top dead center, which concludes the exhaust stroke and also concludes the four-stroke engine process. Notice that four strokes (i.e., two complete rotations of the crankshaft corresponding to four linear motions of the piston) are required per cycle but only one of these strokes produces power.

Simple Model of Spark-Ignition, Four-Stroke Engine

The processes occurring in a reciprocating spark-ignition engine are complicated. It is possible to study these types of engines in more detail after we study combustion in Chapter 13. A simple model of the spark-ignition four-stroke engine can be used to provide some useful insight into their behavior. However, the simplifications associated with the model developed in this section prevent it from being quantitatively accurate.

We will assume that both the fuel/air mixture and the combustion products can be treated as pure air. The compression and expansion processes are modeled as being adiabatic and reversible. The effect of the residual combustion products that remain at the conclusion of the exhaust process is ignored; i.e., it is assumed that the compression process begins with ambient air. The combustion process is assumed to occur instantaneously and at top dead center. The combustion process is modeled using the heat of combustion of the fuel. Friction is neglected. Example 8.4-1 presents a more realistic model in which the compression and expansion processes are modeled using a polytropic exponent; this approach includes the possibility of some heat transfer from the gas to the cylinder walls. The effect of the residual combustion products that remain at the conclusion of the exhaust stroke is also considered in Example 8.4-1.

The model will consider an engine with $N_{cyl} = 4$ cylinders. Each cylinder has a diameter $bore = 4$ inch and a stroke, defined as the travel of the piston between top dead center and bottom dead center, of $stroke = 3.5$ inch. The compression ratio of the engine is $CR = 8.3$ and the crankshaft is rotating at $N = 3600$ rev/min. The air-fuel ratio is $AF = 16$ and the heat of combustion of the fuel is $HC = 44$ MJ/kg. The ambient air conditions are $T_{amb} = 32°$C and $P_{atm} = 100$ kPa. These inputs are entered in EES:

```
$UnitSystem SI K Pa J Mass Radian
bore=4.00 [inch]*convert(inch,m)            "bore"
stroke=3.5 [inch]*convert(inch,m)           "stroke"
CR=8.3 [-]                                  "compression ratio"
N_cyl=4 [-]                                 "number of cylinders"
N=3600 [1/min]*convert(1/min,1/s)           "engine speed"
AF=16 [-]                                   "air-fuel ratio"
HC=44 [MJ/kg]*convert(MJ/kg,J/kg)           "heat of combustion"
T_amb=converttemp(C,K,32 [C])               "outdoor air temperature"
P_atm=100 [kPa]*convert(kPa,Pa)             "atmospheric air pressure"
```

The displacement of a cylinder is defined as the volume that is swept out as the cylinder moves between its top dead center and its bottom dead center positions:

$$V_{dis,cyl} = \pi \frac{bore^2}{4} stroke \tag{8-95}$$

The compression ratio is defined according to Eq. (8-94), which is written in terms of the displacement and the clearance volume (V_{cl}, the volume in the cylinder at top dead center):

$$CR = \frac{V_{dis,cyl} + V_{cl}}{V_{cl}} \tag{8-96}$$

The clearance volume is determined by rearranging Eq. (8-96):

$$V_{cl} = \frac{V_{dis,cyl}}{(CR - 1)} \tag{8-97}$$

The volumes at bottom dead center and top dead center are determined:

$$V_{BDC} = V_{cl} + V_{dis,cyl} \tag{8-98}$$

$$V_{TDC} = V_{cl} \tag{8-99}$$

```
Vol_dis_cyl=pi*bore^2*stroke/4              "displacement of each cylinder"
Vol_cl=Vol_dis_cyl/(CR-1)                   "clearance volume"
Vol_BDC=Vol_dis_cyl+Vol_cl                  "bottom dead center volume"
Vol_TDC=Vol_cl                              "top dead center volume"
```

State 1 is defined as the state that exists immediately after the intake stroke is complete and before the compression process begins. The effect of the high temperature residual gas that remains in the cylinder at the beginning of the intake process is ignored; therefore, the air/fuel mixture at state 1 is assumed to be at ambient conditions, $T_1 = T_{amb}$ and $P_1 = P_{atm}$. State 1 is fixed by T_1 and P_1; the remaining properties (s_1, u_1, and v_1) are determined assuming that the properties of the air/fuel mixture are the same as those of pure air. The volume at state 1 is equal to the bottom dead center volume, $V_1 = V_{BDC}$, and the mass of air/fuel mixture at state 1 is computed according to:

$$m_1 = \frac{V_1}{v_1} \tag{8-100}$$

```
"State 1"
T[1]=T_amb                          "temperature"
P[1]=P_atm                          "pressure"
s[1]=entropy(Air,T=T[1],P=P[1])     "specific entropy"
u[1]=intenergy(Air,T=T[1])          "specific internal energy"
v[1]=volume(Air,T=T[1],P=P[1])      "specific volume"
Vol[1]=Vol_BDC                      "volume"
m[1]=Vol[1]/v[1]                    "mass"
```

State 2 is defined as the state that exists immediately after the compression process is completed. The mass of air/fuel mixture in the cylinder does not change, $m_2 = m_1$, and the volume at state 2 is equal to the top dead center volume, $V_2 = V_{TDC}$. The specific volume at state 2 is given by:

$$v_2 = \frac{V_2}{m_2} \tag{8-101}$$

The compression process is assumed to be adiabatic and reversible, therefore:

$$s_2 = s_1 \tag{8-102}$$

State 2 is fixed by v_2 and s_2; the remaining properties (T_2, P_2, and u_2) are determined.

```
"State 2"
m[2]=m[1]                              "mass balance"
Vol[2]=Vol_TDC                         "volume"
v[2]=Vol[2]/m[2]                       "specific volume"
s[2]=s[1]                              "entropy balance"
u[2]=intenergy(Air,v=v[2],s=s[2])      "specific internal energy"
T[2]=temperature(Air,v=v[2],s=s[2])    "temperature"
P[2]=pressure(Air,v=v[2],s=s[2])       "pressure"
```

Since the compression process is assumed to be adiabatic, the work done on the gas in the cylinder is obtained from an energy balance:

$$W_{comp} = m_2\, u_2 - m_1\, u_1 \tag{8-103}$$

```
W_comp=m[2]*u[2]m[1]*u[1]              "compression work"
```

The mass of fuel in the cylinder is related to the mass of fuel/air mixture that is drawn into the cylinder during the intake stroke:

$$m_{in} = \frac{(V_{BDC} - V_{TDC})}{v_{amb}} \tag{8-104}$$

where v_{amb} is the specific volume of the ambient air entering the cylinder. The inlet mass is composed of fuel and air:

$$m_{in} = m_a + m_f \tag{8-105}$$

Substituting the definition of the air-fuel ratio:

$$AF = \frac{m_a}{m_f} \tag{8-106}$$

into Eq. (8-105) leads to:

$$m_{in} = m_f\,(AF + 1) \tag{8-107}$$

Therefore, the mass of fuel contained in the cylinder is:

$$m_f = \frac{m_{in}}{(AF + 1)} \tag{8-108}$$

```
m_in=(Vol_BDC-Vol_TDC)/volume(Air,T=T_amb,P=P_atm)   "mass of incoming air"
m_f=m_in/(AF+1)                                      "mass of fuel"
```

State 3 is defined as the state that exists at the conclusion of the combustion process. The mass of gas in the cylinder does not change during the combustion process, $m_3 = m_2$. The combustion process is assumed to occur instantaneously at top dead center. Therefore, the volume at the conclusion of the combustion process is $V_3 = V_{TDC}$. The specific volume is given by:

$$v_3 = \frac{V_3}{m_3} \tag{8-109}$$

An energy balance for the combustion process is:

$$m_f\,HC = m_3\,u_3 - m_2\,u_2 \tag{8-110}$$

State 3 is fixed by the specific volume and the specific internal energy. The remaining properties (T_3, P_3, and s_3) are determined.

```
"State 3"
m[3]=m[2]                            "mass balance"
Vol[3]=Vol_TDC                       "volume"
v[3]=Vol[3]/m[3]                     "specific volume"
m_f*HC=m[3]*u[3]-m[2]*u[2]           "energy balance"
T[3]=temperature(Air,u=u[3])         "temperature"
P[3]=pressure(Air,u=u[3],v=v[3])     "pressure"
s[3]=entropy(Air,u=u[3],v=v[3])      "specific entropy"
```

State 4 is defined as the state that exists immediately after the expansion process. The mass of gas in the cylinder again does not change during the expansion process, $m_4 = m_3$, and the volume at state 4 is equal to the bottom dead center volume, $V_4 = V_{BDC}$. Therefore, the specific volume is given by:

$$v_4 = \frac{V_4}{m_4} \tag{8-111}$$

The expansion process is assumed to be adiabatic and reversible, therefore:

$$s_4 = s_3 \tag{8-112}$$

State 4 is fixed by v_4 and s_4; the remaining properties (T_4, P_4, and u_4) are determined.

```
"State 4"
m[4]=m[3]                              "mass balance"
Vol[4]=Vol_BDC                         "volume"
v[4]=Vol[4]/m[4]                       "specific volume"
s[4]=s[3]                              "entropy balance"
u[4]=intenergy(Air,v=v[4],s=s[4])      "specific internal energy"
T[4]=temperature(Air,v=v[4],s=s[4])    "temperature"
P[4]=pressure(Air,v=v[4],s=s[4])       "pressure"
```

The expansion process is assumed to be adiabatic. Therefore, the work done by the gas on the piston during the expansion process is given by:

$$0 = W_{exp} + m_4\, u_4 - m_3\, u_3 \tag{8-113}$$

```
0=W_exp+m[4]*u[4]m[3]*u[3]             "expansion work"
```

The net work accomplished during a complete cycle for one cylinder is:

$$W_{net} = W_{exp} - W_{comp} \tag{8-114}$$

The efficiency of the engine is the ratio of the net work to the heat input from combustion of the fuel for one cycle:

$$\eta = \frac{W_{net}}{m_f\, HC} \tag{8-115}$$

```
W_net=W_exp-W_comp                     "net work per power stroke"
eta=W_net/(m_f*HC)                     "efficiency"
```

which leads to $\eta = 0.495$ (49.5%). Note that this result is much larger than the actual efficiency of a four-stroke spark-ignition engine due to the idealizations that were used to develop the model. The period associated with a single rotation of the crankshaft is:

$$period = \frac{1}{N} \tag{8-116}$$

Therefore, the average power produced by the engine is:

$$\dot{W}_{net} = \frac{N_{cyl}\, W_{net}}{2\, period} \tag{8-117}$$

```
period=1/N                               "period of one rotation"
W_dot_net=W_net*N_cyl/(2*period)         "average power produced"
W_dot_net_hp=W_dot_net*convert(W,hp)     "in hp"
```

which provides $\dot{W}_{net} = 126.5$ kW (169.6 hp). Note that the factor of 2 in Eq. (8-117) is due to the fact that the time between each power stroke is twice the period of the rotation of the crankshaft because of the intake and exhaust strokes that occur during every other rotation in a 4-stroke engine. The back work ratio for reciprocating internal combustion engines is the ratio of the work required to compress the fuel-air mixture to the work done during the expansion process.

$$bwr = \frac{W_{comp}}{W_{exp}} \tag{8-118}$$

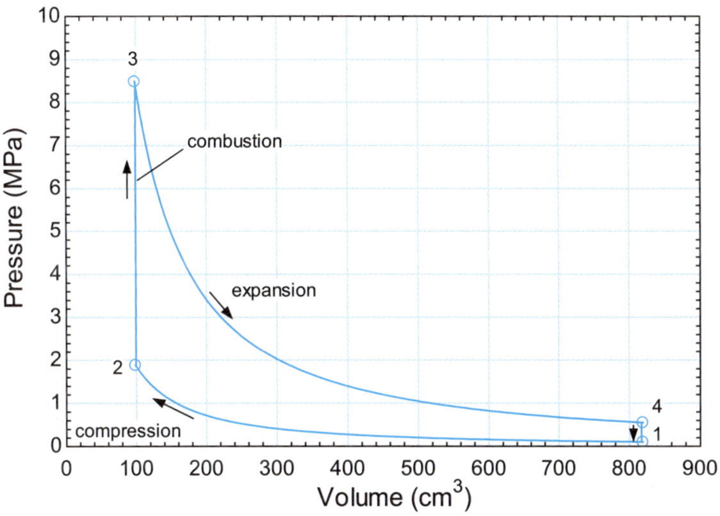

Figure 8-34: Pressure-volume diagram predicted by the simple model.

The mean effective pressure for the engine is defined as the ratio of the net work produced per cylinder to the displacement of each cylinder:

$$MEP = \frac{W_{net}}{V_{dis,cyl}} \tag{8-119}$$

```
bwr=W_comp/W_exp          "back work ratio"
MEP=W_net/Vol_dis_cyl     "mean effective pressure"
```

which results in $bwr = 0.204$ and $MEP = 1.46$ MPa. The back work ratio for the engine is much larger than that of the Rankine cycle, but considerably smaller than the back work ratio associated with the gas turbine cycle. The back work ratio is large because a gas must be compressed in the compression process. However, the temperatures occurring during the compression process are much lower than those occurring during the expansion process and therefore the difference in the specific volumes associated with the two processes is very large. Reciprocating internal combustion engines are able to operate at much higher temperatures than gas turbine engines because the cylinder walls are not continuously exposed to the high temperatures that occur during the cycle in the same way that the turbine blades are.

Figure 8-34 illustrates the idealized cycle on a pressure-volume diagram and Figure 8-35 illustrates the cycle on a temperature-specific entropy diagram. Figure 8-36 shows the power and efficiency of the engine as a function of the compression ratio. Notice that both the power and the efficiency of the spark ignition engine tend to increase with increasing compression ratio. The compression ratio employed in modern automobile engines is typically in the range from 7 to 9. There are several reasons that higher compression ratios are not used. One reason relates to the pre-ignition characteristics of the fuel-air mixture, which is discussed in the subsequent section.

Octane Number of Gasoline

In the spark ignition engine, the fuel and air are premixed in the proportion needed for combustion before the compression process begins. Compressing the mixture increases its temperature and higher temperatures result from higher compression ratios.

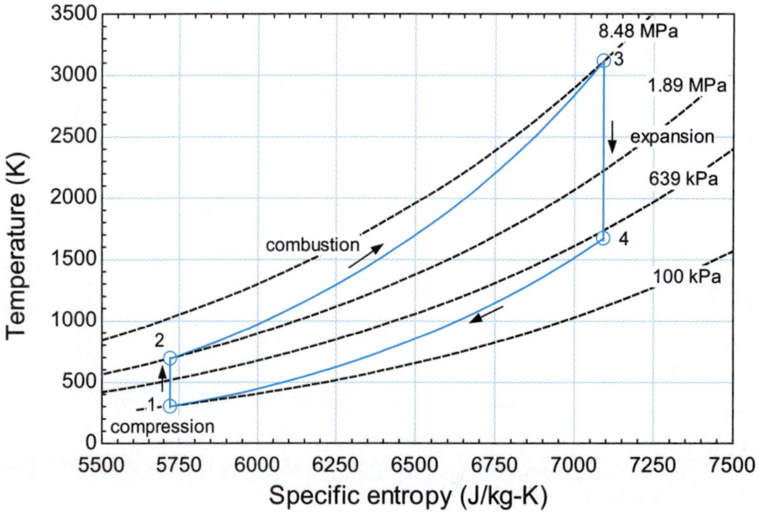

Figure 8-35: Temperature-specific entropy diagram predicted by simple model.

Figure 8-37 illustrates the temperature at the conclusion of the compression process, T_2, as a function of the compression ratio for the conditions considered in this section. The temperatures that occur as a result of compression can exceed the auto-ignition temperature of the fuel-air mixture, resulting in *pre-ignition*, i.e., combustion in advance of the planned ignition that is initiated with the spark plug. Depending on its extent, pre-ignition can result in a range of effects from minor performance degradation (i.e., engine knock) to engine damage. Pre-ignition is exacerbated by high engine temperatures that tend to occur when the engine is under heavy load.

Gasoline is an organic fluid mixture consisting of many different hydrocarbons. The tendency of a mixture of gasoline and air to pre-ignite can be controlled by altering the characteristics of the fuel mixture. The auto-ignition tendency of a fuel is quantified in terms of its *octane number*, which is based on the auto-ignition characteristics of n-heptane (C_7H_{16}) and iso-octane (C_8H_{18}). The chemical n-heptane auto-ignites at 215°C

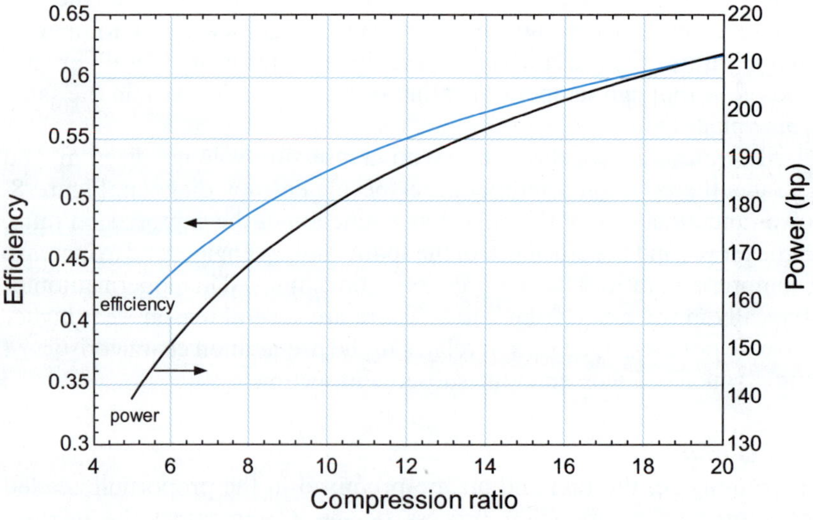

Figure 8-36: Power and efficiency as a function of compression ratio.

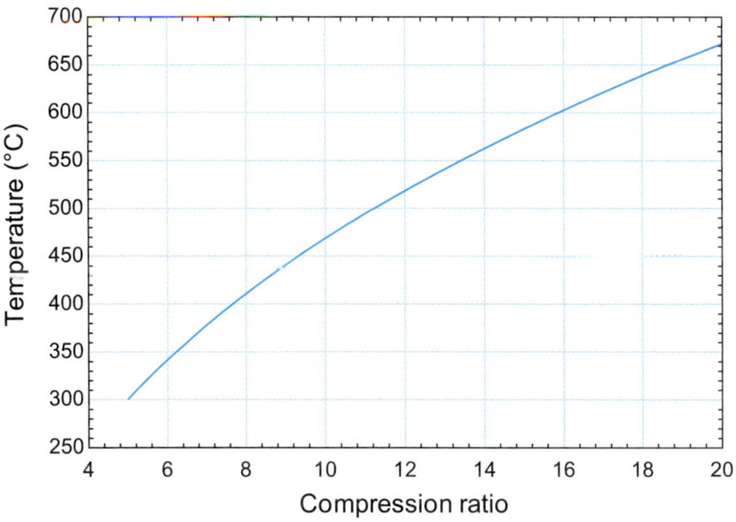

Figure 8-37: Temperature at the conclusion of the compression process as a function of the compression ratio.

and is assigned an octane number of 0 whereas iso-octane auto-ignites at a much higher temperature, 447°C, and it is defined to have an octane number of 100. The octane number of a gasoline is the volume percentage of iso-octane in a mixture of n-heptane and iso-octane that has the same auto-ignition temperature as the gasoline mixture of interest. Therefore, gasoline with an octane number of 86 has the same auto-ignition temperature as a mixture that is 86% iso-octane and 14% n-heptane, which will be closer to 447°C than 215°C (although the relationship between auto-ignition temperature and concentration of iso-octane is not linear). Two test methods, called the Research and the Motor tests, are used to measure the octane number for a gasoline mixture. These tests are conducted at different conditions and tend to provide different results. The octane rating that is advertised at filling stations in the U.S. is the average of these two tests, indicated by (R+M)/2. Gasolines with higher octane numbers have greater resistance to pre-ignition. Higher octane gasoline generally costs more because the resistance to pre-ignition requires more careful refining or additives. However, there are no benefits associated with a higher octane number other than resistance to pre-ignition. You will not see better gas mileage if you use a higher octane fuel in an engine that is designed to operate with lower octane fuel.

EXAMPLE 8.4-1: POLYTROPIC MODEL WITH RESIDUAL COMBUSTION GAS

Section 8.4.1 presented a simple model of a spark-ignition four-stroke engine in which the effect of the residual combustion products were not considered and the compression and expansion processes were assumed to be adiabatic. The model predicts cycle efficiencies that are much higher than those observed in practice. One reason for this discrepancy is that the compression and expansion processes are not adiabatic. In an actual engine, the cylinder wall is cooled by heat exchange with an engine coolant that is at a lower temperature than the gas in the cylinder, particularly during the expansion stroke. The gas in the cylinder will transfer heat to the cylinder wall; therefore, neither the compression nor the expansion process is isentropic. A more accurate representation of these processes is provided by assuming that the compression and expansion processes are *polytropic*. A polytropic process is described by Eq. (1):

$$PV^n = \text{constant} \tag{1}$$

where P is the pressure, V is the cylinder volume, and n is a constant referred to as the *polytropic index* or *polytropic exponent*. If the process is isentropic and the working fluid is an ideal gas with constant specific heat capacities, then the polytropic index is equal to k, the ratio of the specific heat capacity at constant pressure to the specific heat capacity at constant volume:

$$k = \frac{c_P}{c_v} \tag{2}$$

Equation (1) can be derived by algebraically rearranging the equations that relate pressure and specific volume for an isentropic process for an ideal gas with constant specific heat capacities, Eq. (6-37), which is repeated below:

$$\frac{P_2}{P_1} = \left(\frac{v_2}{v_1}\right)^{-k} \tag{3}$$

For a fixed mass of gas, the ratio of specific volumes in Eq. (3) is equivalent to the ratio of volumes:

$$\frac{P_2}{P_1} = \left(\frac{V_2}{V_1}\right)^{-k} \tag{4}$$

Equation (4) can be rearranged:

$$P_2 V_2^k = P_1 V_1^k \quad \text{or} \quad PV^k = \text{constant}$$

which is equivalent to Eq. (1) with $n = k$. A polytropic index of $n = 1$ corresponds to an isothermal process for an ideal gas. This result can be shown from the ideal gas law applied to an isothermal process:

$$PV = mRT \quad \text{or} \quad PV = \text{constant}$$

A polytropic index of $n = 0$ corresponds to an isobaric (constant pressure) process:

$$PV^0 = \text{constant or } P = \text{constant}$$

By adjusting the polytropic index, it is possible to simulate a process that falls between the limiting behaviors of adiabatic and isothermal processes; a polytropic index in the range $1 < n < k$ will correspond to a process that is not completely adiabatic but does not have sufficient heat transfer to maintain the gas at an isothermal condition. The polytropic index may be selected by fitting pressure-volume data to Eq. (1).

The work associated with a polytropic process can be determined by substituting Eq. (1) into the expression for P-V work:

$$W_{out} = \int_{V_1}^{V_2} P\, dV$$

Since $PV^n = \text{constant}$, the integral can be written as:

$$W_{out} = \text{constant} \int_{V_1}^{V_2} V^{-n}\, dV$$

Carrying out the integration:

$$W_{out} = \text{constant} \left[\frac{V^{(1-n)}}{(1-n)} \right]_{V_1}^{V_2}$$

Applying the limits:

$$W_{out} = \frac{\text{constant}}{(1-n)} \left[V_2^{(1-n)} - V_1^{(1-n)} \right] \tag{5}$$

The constant in Eq. (5) is given by:

$$\text{constant} = P_2\, V_2^n = P_1\, V_1^n \tag{6}$$

Substituting Eq. (6) into Eq. (5) leads to:

$$W_{out} = \frac{\left[P_2\, V_2^n\, V_2^{(1-n)} - P_1\, V_1^n\, V_1^{(1-n)} \right]}{(1-n)}$$

which can be simplified to:

$$W_{out} = \frac{(P_2\, V_2 - P_1\, V_1)}{(1-n)} \tag{7}$$

a) Model the four-stroke spark ignition engine considered in Section 8.4.1 using a polytropic model for the compression and expansion processes. The polytropic indices for the compression and expansion processes are $n_{comp} = 1.35$ and $n_{exp} = 1.40$, respectively. Do not neglect the impact of the residual gas that remains in the cylinder at the conclusion of the exhaust process. Compare the results of this model with the simple model from Section 8.4.1.

The same engine is considered. The number of cylinders is $N_{cyl} = 4$ cylinders, each cylinder has *bore* = 4 inch and *stroke* = 3.5 inch. The compression ratio of the engine is $CR = 8.3$ and the engine speed is $N = 3600$ rev/min. The air-fuel ratio is $AF = 16$ and the heat of combustion of the fuel is $HC = 44$ MJ/kg. The ambient air conditions are $T_{amb} = 32°\text{C}$ and $P_{atm} = 100$ kPa. The inputs are entered in EES:

```
$UnitSystem SI K Pa J Mass Radian
bore=4.00 [inch]*convert(inch,m)            "bore"
stroke=3.5 [inch]*convert(inch,m)           "stroke"
CR=8.3 [-]                                  "compression ratio"
N_cyl=4 [-]                                 "number of cylinders"
N=3600 [1/min]*convert(1/min,1/s)           "engine speed"
AF=16 [-]                                   "air-fuel ratio"
HC=44 [MJ/kg]*convert(MJ/kg,J/kg)           "heat of combustion"
T_amb=converttemp(C,K,32 [C])               "outdoor air temperature"
P_atm=100 [kPa]*convert(kPa,Pa)             "outdoor air pressure"
n_comp=1.35 [-]                             "polytropic exponent for compression"
n_exp=1.40 [-]                              "polytropic exponent for expansion"
```

The volumes in the cylinder at top dead center and bottom dead center are calculated using Eqs. (8-95) through (8-99).

```
Vol_dis_cyl=pi*bore^2*stroke/4      "displacement of each cylinder"
Vol_cl=Vol_dis_cyl/(CR-1)           "clearance volume"
Vol_BDC=Vol_dis_cyl+Vol_cl          "bottom dead center volume"
Vol_TDC=Vol_cl                      "top dead center volume"
```

State 1 is defined as the condition of the gas in the cylinder at the time that the exhaust stroke has concluded. The temperature of the residual gas that remains in the cylinder at this time is not known; therefore, the value of T_1 is initially guessed.

```
"State 1"
T[1]=400 [K]                        "guess for initial temperature"
```

The pressure in the cylinder at the conclusion of the exhaust stroke is atmospheric pressure, $P_1 = P_{atm}$. State 1 is fixed by the temperature and pressure. The remaining properties (s_1, u_1, and v_1) are determined. The volume in the cylinder at state 1 is the top dead center volume, $V_1 = V_{TDC}$. Therefore, the mass of residual gas remaining in the cylinder is:

$$m_1 = \frac{V_{TDC}}{v_1}$$

```
P[1]=P_atm                          "pressure"
s[1]=entropy(Air,T=T[1],P=P[1])     "specific entropy"
u[1]=intenergy(Air,T=T[1])          "specific internal energy"
v[1]=volume(Air,T=T[1],P=P[1])      "specific volume"
Vol[1]=Vol_TDC                      "volume"
m[1]=Vol[1]/v[1]                    "mass"
```

State 2 is defined as the state that exists when the intake stroke is concluded. The volume at state 2 is the bottom dead center volume, $V_2 = V_{BDC}$. The pressure is atmospheric, $P_2 = P_{atm}$. During the intake stroke, fuel/air mixture at ambient conditions is pulled through the open intake valve and mixed with the residual gas in the cylinder. The specific enthalpy of the gas that enters the cylinder during this process (h_{in}) is determined from the ambient temperature.

```
"State 2"
Vol[2]=Vol_BDC                      "volume"
P[2]=P_atm                          "pressure"
h_in=enthalpy(Air,T=T_amb)          "specific enthalpy of entering air"
```

A mass balance on the cylinder during the intake stroke is:

$$m_{in} = m_2 - m_1 \tag{8}$$

An energy balance on the cylinder is:

$$m_{in}\, h_{in} = W_{intake} + m_2\, u_2 - m_1\, u_1 \tag{9}$$

If the pressure losses associated with flow through the intake manifold and intake valve are neglected, then the intake stroke occurs at constant pressure; therefore, the work done by the gas in the cylinder during the intake stroke is:

$$W_{intake} = P_{atm}\left(V_{BDC} - V_{TDC}\right) \tag{10}$$

Substituting Eq. (10) into Eq. (9) leads to:

$$m_{in}\, h_{in} = P_{atm}\left(V_{BDC} - V_{TDC}\right) + m_2\, u_2 - m_1\, u_1 \tag{11}$$

Equations (8) and (11) provide two equations in three unknowns (m_{in}, m_2, and u_2). The mass in the cylinder at the conclusion of the intake stroke is:

$$m_2 = \frac{V_{BDC}}{v_2} \tag{12}$$

which provides another equation but also introduces another unknown (v_2). The final equation is provided from the fact that the properties P_2, u_2, and v_2 are related; for example, u_2 can be expressed as a function of v_2 and P_2. These equations are entered in EES and solved.

```
m_in=m[2]-m[1]                                          "mass balance"
m_in*h_in=P_atm*(Vol[2]-Vol[1])+m[2]*u[2]-m[1]*u[1]     "energy balance"
m[2]=Vol[2]/v[2]                                        "mass at the conclusion of the process"
u[2]=intenergy(Air,P=P[2],v=v[2])                       "property routine"
```

State 2 is fixed by the pressure and specific volume. The temperature and specific entropy (T_2 and s_2) are determined.

```
T[2]=temperature(Air,P=P[2],v=v[2])                     "temperature"
s[2]=entropy(Air,P=P[2],v=v[2])                         "specific entropy"
```

State 3 is defined as the state that exists at the conclusion of the compression process. The mass in the cylinder is unchanged, $m_3 = m_2$, and the volume is the top dead center volume, $V_3 = V_{TDC}$. Therefore, the specific volume is:

$$v_3 = \frac{V_{TDC}}{m_3}$$

The polytropic relationship, Eq. (1), is used to determine the pressure at state 3.

$$P_3\, V_3^{\,n_{comp}} = P_2\, V_2^{\,n_{comp}}$$

State 3 is fixed by the temperature and specific volume. The specific entropy and specific internal energy (s_3 and u_3) are determined.

```
"State 3"
m[3]=m[2]                                               "mass balance"
Vol[3]=Vol_TDC                                          "volume"
v[3]=Vol[3]/m[3]                                        "specific volume"
P[3]*Vol[3]^(n_comp)=P[2]*Vol[2]^(n_comp)               "polytropic relationship"
T[3]=temperature(Air,P=P[3],v=v[3])                     "temperature"
s[3]=entropy(Air,P=P[3],v=v[3])                         "specific entropy"
u[3]=intenergy(Air,P=P[3],v=v[3])                       "specific internal energy"
```

The work done on the gas in the cylinder during the compression process is calculated using Eq. (7). Note that Eq. (7) provides the work done by a system for a polytropic process; therefore, a negative sign is required to describe the work input.

$$W_{comp} = -\frac{(P_3 V_3 - P_2 V_2)}{(1 - n_{comp})}$$

The heat transfer from the gas during the compression process is obtained with an energy balance:

$$W_{comp} = Q_{comp} + m_3 u_3 - m_2 u_2$$

```
W_comp=-(P[3]*Vol[3]-P[2]*Vol[2])/(1-n_comp)    "work"
W_comp=Q_comp+m[3]*u[3]-m[2]*u[2]               "energy balance"
```

State 4 is defined as the state that exists immediately after the combustion process. The mass of fuel in the air/fuel mixture is computed using Eq. (8-108):

$$m_f = \frac{m_{in}}{(AF + 1)}$$

The mass of gas in the cylinder does not change during combustion, $m_4 = m_3$. The combustion process is assumed to occur instantaneously; therefore, the volume does not change, $V_4 = V_{TDC}$. The specific volume at state 4 is:

$$v_4 = \frac{V_{TDC}}{m_4}$$

An energy balance on the combustion process is:

$$m_f\, HC = m_4 u_4 - m_3 u_3$$

State 4 is fixed by v_4 and u_4. The remaining properties (T_4, P_4, and s_4) are obtained.

```
"State 4"
m_f=m_in/(AF+1)                          "mass of fuel"
m[4]=m[3]                                "mass balance"
Vol[4]=Vol_TDC                           "volume"
v[4]=Vol[4]/m[4]                         "specific volume"
m_f*HC=m[4]*u[4]-m[3]*u[3]               "energy balance"
T[4]=temperature(Air,u=u[4])             "temperature"
P[4]=pressure(Air,u=u[4],v=v[4])         "pressure"
s[4]=entropy(Air,u=u[4],v=v[4])          "specific entropy"
```

State 5 is defined as the state that exists at the conclusion of the expansion process. The mass in the cylinder is unchanged, $m_5 = m_4$, and the volume is the bottom dead center volume, $V_5 = V_{BDC}$. Therefore, the specific volume is:

$$v_5 = \frac{V_{BDC}}{m_5}$$

The polytropic relationship, Eq. (1), is used to determine the pressure at state 5.

$$P_5 V_5^{n_{exp}} = P_4 V_4^{n_{exp}}$$

State 5 is fixed by the temperature and specific volume. The specific entropy and specific internal energy (s_5 and u_5) are determined.

```
"State 5"
m[5]=m[4]                                  "mass balance"
Vol[5]=Vol_BDC                             "volume"
v[5]=Vol[5]/m[5]                           "specific volume"
P[5]*Vol[5]^(n_exp)=P[4]*Vol[4]^(n_exp)    "polytropic relationship"
T[5]=temperature(Air,P=P[5],v=v[5])        "temperature"
s[5]=entropy(Air,P=P[5],v=v[5])            "specific entropy"
u[5]=intenergy(Air,P=P[5],v=v[5])          "internal energy"
```

The work done by the gas in the cylinder during the expansion process is calculated using Eq. (7):

$$W_{exp} = \frac{(P_5\,V_5 - P_4\,V_4)}{(1 - n_{exp})}$$

The heat transfer from the gas during the expansion process is obtained with an energy balance:

$$0 = W_{exp} + Q_{exp} + m_5\,u_5 - m_4\,u_4$$

```
W_exp=(P[5]*Vol[5]-P[4]*Vol[4])/(1-n_exp)    "work"
0=W_exp+Q_exp+m[5]*u[5]-m[4]*u[4]            "energy balance"
```

State 6 is defined as the state that exists after the blowdown process is complete. The blowdown process occurs when the exhaust valve opens and the contents of the cylinder quickly equilibrate to atmospheric pressure. Therefore, $P_6 = P_{atm}$ and $V_6 = V_{BDC}$. The gas that remains in the cylinder at the conclusion of the blowdown process is assumed to have undergone an adiabatic and reversible expansion. Therefore, an entropy balance on this gas provides:

$$s_6 = s_5$$

State 6 is fixed by the specific entropy and pressure. The remaining properties (T_6, u_6, and v_6) are obtained. The mass of gas that remains in the cylinder at the conclusion of the blowdown process is:

$$m_6 = \frac{V_6}{v_6}$$

```
"State 6"
Vol[6]=Vol_BDC                           "volume"
P[6]=P_atm                               "pressure"
s[6]=s[5]                                "entropy balance on gas remaining in cylinder"
T[6]=temperature(Air,s=s[6],P=P[6])      "temperature"
u[6]=intenergy(Air,s=s[6],P=P[6])        "specific internal energy"
v[6]=volume(Air,s=s[6],P=P[6])           "specific volume"
m[6]=Vol[6]/v[6]                         "mass"
```

The exhaust stroke brings the cycle back to state 1. A mass balance on the exhaust stroke is:

$$0 = m_{out} + m_6 - m_5$$

```
0=m_out+m[6]-m[5]     "mass balance"
```

We initially guessed the temperature of the residual gas that remains in the cylinder at state 1 and we have finally calculated the temperature of the gas that remains at state 6. We need to let EES adjust the value of T_1 until it matches T_6. The guess values for the problem are updated by selecting Update Guesses from the Calculate menu. The specified value for T_1 is commented out:

```
{T[1]=400 [K]}          "guess for initial temperature"
```

and the final temperature is set equal to the initial temperature:

```
T[6]=T[1]               "initial gas temperature must match final gas temperature"
```

which results in $T_1 = 1002$ K. The net work provided by one cylinder during a single cycle is:

$$W_{net} = W_{exp} - W_{comp}$$

The efficiency of the engine is the ratio of the net work produced to the heat input from combustion of the fuel:

$$\eta = \frac{W_{net}}{m_f\, HC}$$

```
W_net=W_exp-W_comp    "net work per power stroke"
eta=W_net/(m_f*HC)    "efficiency"
```

which leads to $\eta = 0.428$ (42.8%). This value is much less than the efficiency predicted by the adiabatic compression and expansion model (49.5%) but still somewhat higher than is observed for typical engines. Other factors, such as friction, have not been considered in this model. The period associated with a single rotation of the crankshaft is:

$$period = \frac{1}{N}$$

Therefore, the average power produced by the engine is:

$$\dot{W}_{net} = \frac{N_{cyl}\, W_{net}}{2\, period}$$

```
period=1/N                              "period of one rotation"
W_dot_net=W_net*N_cyl/(2*period)        "net power produced"
W_dot_net_hp=W_dot_net*convert(W,hp)    "in hp"
```

which leads to $\dot{W}_{net} = 108.8$ kW (146 hp).

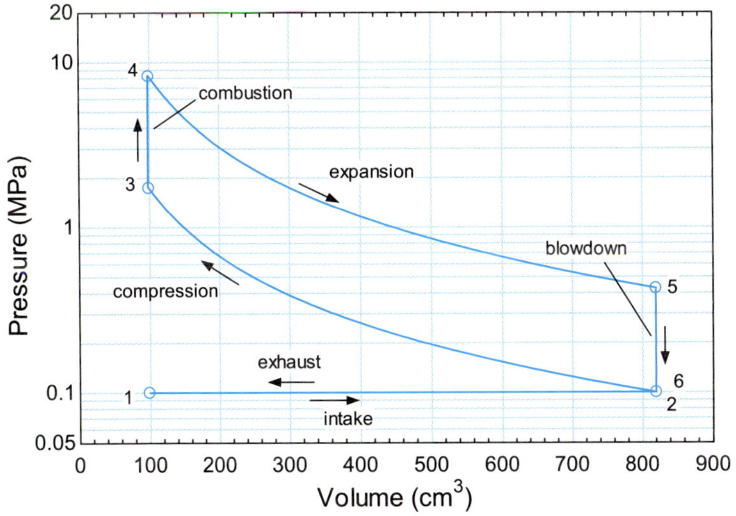

Figure 1: Pressure-volume diagram.

Figure 1 illustrates the cycle on a pressure-volume diagram and Figure 2 illustrates the cycle on a temperature-specific entropy diagram. Notice that the compression and expansion processes are no longer isentropic in Figure 2 because heat transfer from the gas in the cylinder causes the specific entropy to decrease during both of these processes.

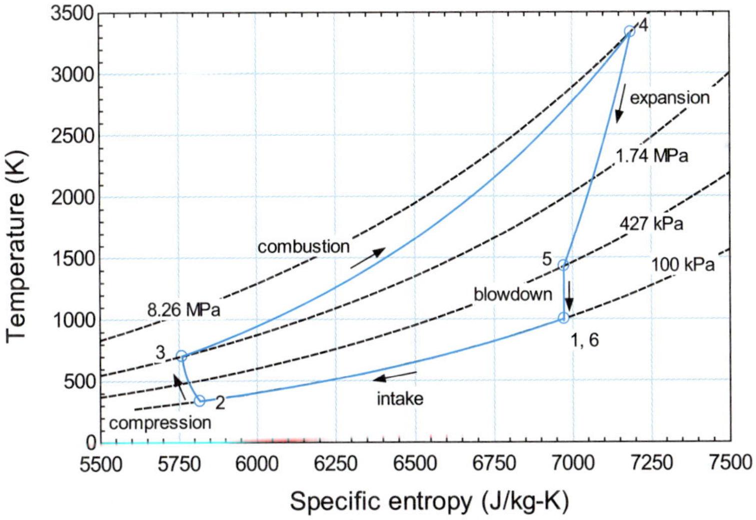

Figure 2: Temperature-specific entropy diagram.

Figure 3 illustrates the efficiency of the engine as a function of compression ratio predicted using the model developed in this example as well as the efficiency predicted by the adiabatic model presented in Section 8.4.1.

b) Determine the fraction of the energy content of the fuel that leaves as heat transfer to the cylinder wall, the fraction of the energy contained in the exhaust gas, and fraction that becomes work done on the piston.

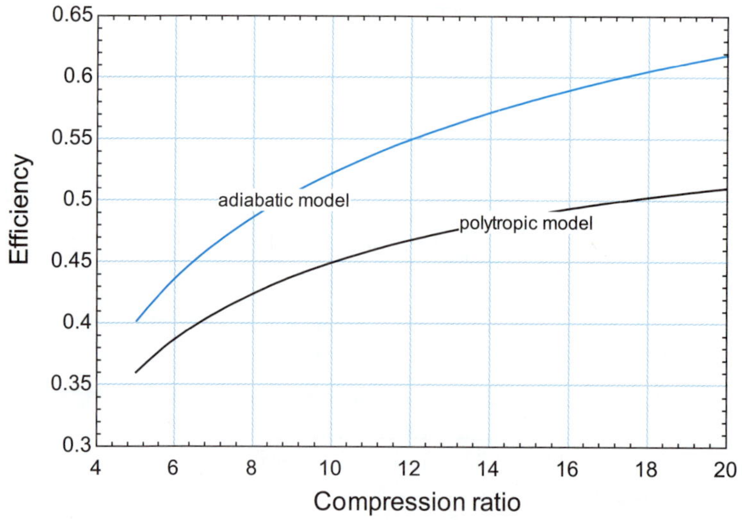

Figure 3: Engine efficiency as a function of the compression ratio for the polytropic and adiabatic models.

The fraction of the fuel energy content that leaves as net work done on the piston is equivalent to the efficiency of the engine:

$$f_{work} = \eta = \frac{W_{net}}{m_f\, HC}$$

The fraction of the fuel energy content that leaves as heat transfer to the cylinder walls is:

$$f_{heat} = \frac{(Q_{comp} + Q_{exp})}{m_f\, HC}$$

The remaining fuel energy content leaves with the high temperature gas pushed out of the cylinder during the exhaust stroke:

$$f_{exhaust} = 1 - f_{work} - f_{heat}$$

```
f_work=eta                           "fraction of fuel energy leaving engine as work"
f_heat=(Q_comp+Q_exp)/(m_f*HC)       "fraction of fuel energy leaving engine as cylinder heat transfer"
f_exhaust=1-f_heat-f_work            "fraction of fuel energy leaving as high temp. exhaust"
```

which results in $f_{work} = 0.428$ (42.8%), $f_{heat} = 0.2091$ (20.9%), and $f_{exhaust} = 0.363$ (36.3%). A more common distribution of energy for internal combustion engines is that the fuel energy content is approximately split equally between work, heat, and exhaust.

Spark-Ignition, Two-Stroke Internal Combustion Engine

The two-stroke internal combustion engine combines the compression and power strokes and the intake and exhaust strokes in a manner that allows the cycle to be completed with one up-down reciprocating motion of the piston. The processes occurring in a two-stroke engine are shown in Figure 8-38. The pressure in the cylinder and the pressure in the crankcase (the space below the cylinder) are shown qualitatively in Figure 8-39.

Figure 8-38(a) shows the combustion process. The process begins with the cylinder containing a compressed mixture of fuel and air and the piston near its top dead center position (i.e., the crank angle is near 0°). The pressure in the cylinder is high because the compression process has just been completed. The crankcase (the space below the cylinder) is filled with fuel/air mixture at near atmospheric pressure. The combustion process is initiated in the cylinder by firing the spark plug. The combustion process occurs quickly so that the piston moves very little during this process. The temperature and pressure in the cylinder increase dramatically during the combustion process.

Figure 8-38(b) shows the expansion process. The high pressure combustion products push the piston towards bottom dead center, causing the pressure in the cylinder to decrease. Work is done on the piston by the combustion products during the expansion

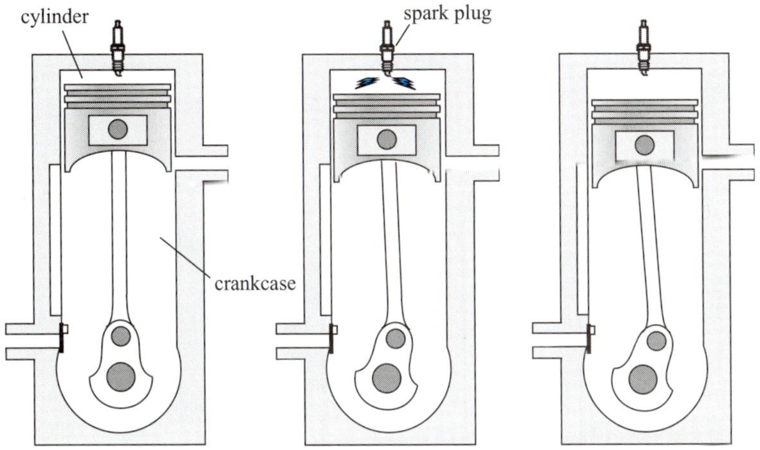

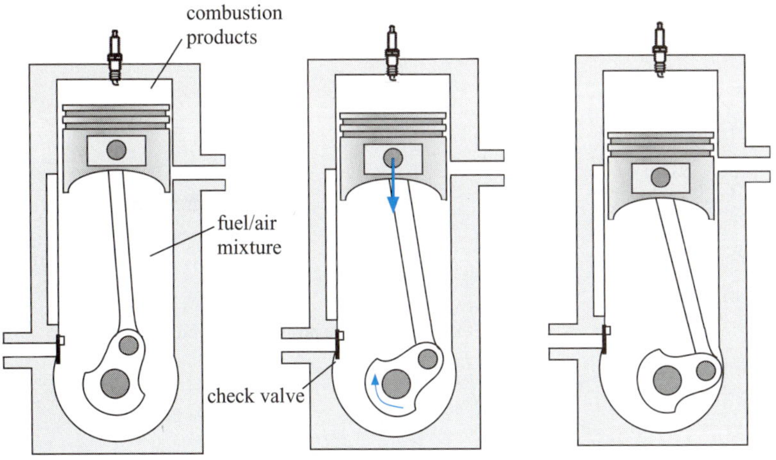

Figure 8-38: Processes occurring in the two-stroke spark-ignition engine: (a) combustion, (b) expansion, (c) intake and exhaust, and (d) compression. An animation of this cycle can be obtained by selecting EES Example Problems from the Examples menu in EES. Select Animation and then Animation of a 2-stroke internal combustion engine.

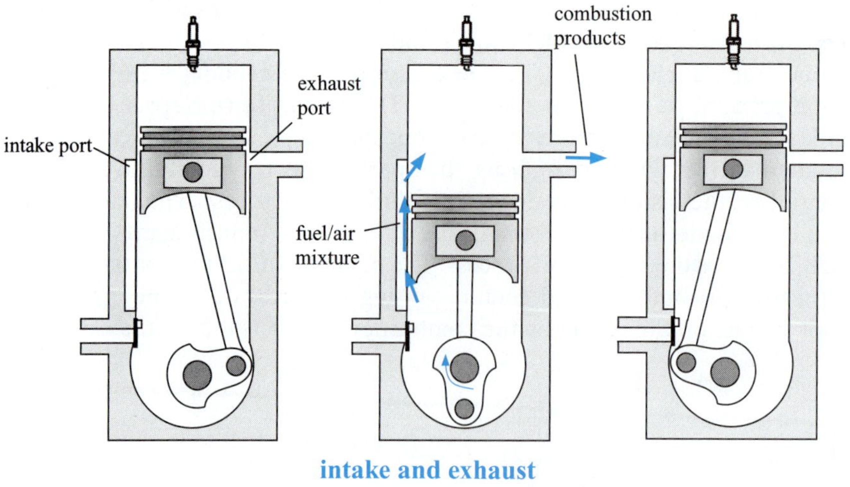

(c)

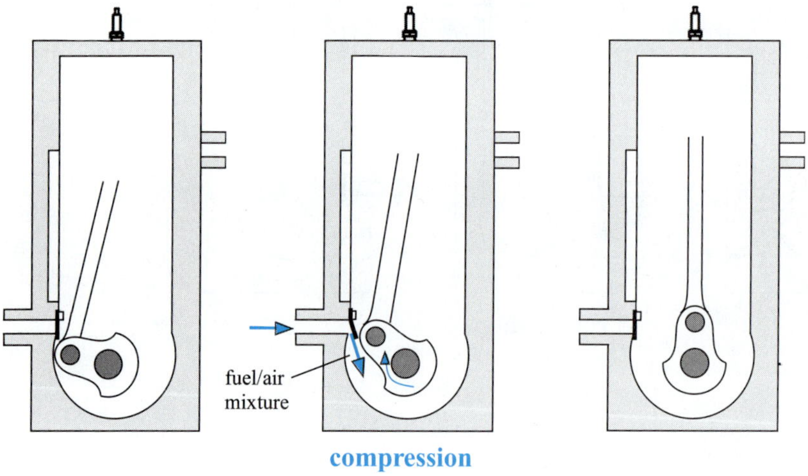

(d)

Figure 8-38: (continued)

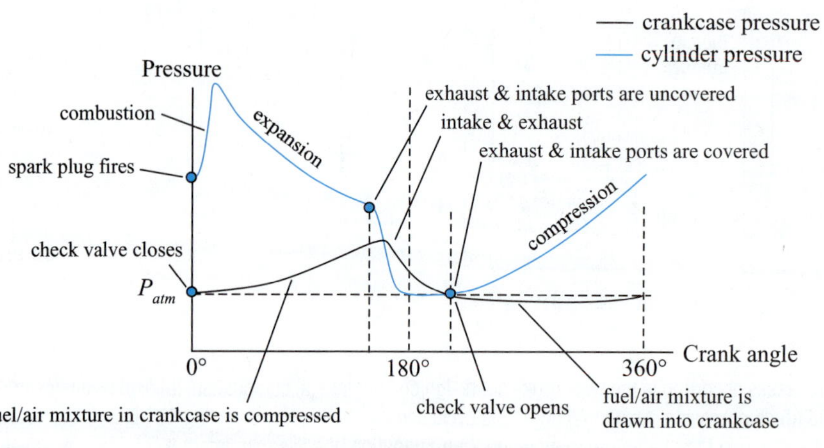

Figure 8-39: Qualitative variation of the cylinder and crankcase pressure as a function of crankangle.

process. A check valve prevents the fuel/air mixture contained in the crankcase from being pushed out; therefore, the pressure in the crankcase increases during the expansion process because the crankcase volume is being reduced.

Figure 8-38(c) shows the intake/exhaust process. At some point before bottom dead center, the intake and exhaust ports on the cylinder wall are uncovered. The cylinder pressure is greater than atmospheric pressure. Therefore, the combustion products in the cylinder are expelled through the exhaust port. Similarly, the pressure in the crankcase is higher than the cylinder pressure. Therefore, the air/fuel mixture in the crankcase rushes into the cylinder. At the conclusion of this process, the intake and exhaust ports become covered again by the piston skirt as the piston begins moving upwards.

Figure 8-38(d) shows the compression process. The air/fuel mixture in the cylinder is compressed as the piston moves towards top dead center. At the same time, additional air/fuel mixture is drawn through the check valve due to the motion of the piston which increases the volume in the crankcase. At the conclusion of the compression process, the piston is near top dead center and the cycle is repeated.

The major advantage of the two-stroke engine is its mechanical simplicity, which leads to low cost and high power-to-weight ratio. The simplicity results from the elimination of the cylinder valves, although rotary valves are used in some two-stroke engines to optimize the timing of the inlet and exhaust processes. The two-stroke engine completes a power stroke every revolution of the crank shaft, which contributes to its high power-to-weight ratio compared to the four-stroke engine. Two-stroke engines are commonly used for hand-held applications such as chain saws and weed-whackers as well as in some lawn mowers. A disadvantage of the two-stroke cycle is that the incoming fuel-air mixture and outgoing exhaust gases inevitably mix to some extent during the combined intake and exhaust process shown in Figure 8-38(c). This mixing results in relatively high emissions of unburned hydrocarbons. The two-stroke engine cycle used to be common in small motorcycles, but their use for this application has faded as the emissions requirements for all vehicles have become more stringent.

8.4.2 The Compression-Ignition Reciprocating Internal Combustion Engine

The compression process in a reciprocating engine results in an increase in the temperature of the gas contained in the cylinder. In the spark-ignition engine, this gas is a mixture of fuel and air and the increased temperature can cause the mixture to prematurely ignite. In the compression-ignition engine, the ignition of the fuel in the cylinder is intentionally caused by the high temperatures that occur during the compression process. The compression-ignition engine is referred to as the *Diesel cycle*, named after Rudolf Diesel, the German engineer who first developed a working compression-ignition engine.

The compression-ignition engine is similar to the four-stroke spark ignition engine. The processes that occur in a compression-ignition engine are shown in Figure 8-40 and on a pressure-volume diagram in Figure 8-41. During the intake stroke, the piston moves towards bottom dead center with the intake valve open in order to draw air into the cylinder, as shown in Figure 8-40(a). In the spark-ignition engine, a fuel-air mixture is drawn into the cylinder. However, in the compression-ignition engine, pure air passes through the intake valve. During the compression process, the intake valve closes and the piston moves towards top dead center, as shown in Figure 8-40(b). When the piston is near top dead center, fuel is injected into the cylinder, as shown in Figure 8-40(c). The compressed air is at a temperature above the auto-ignition temperature of the fuel, which causes the fuel to combust. The high temperature combustion products increase the pressure in the cylinder forcing the piston down and thereby producing work during the expansion process, shown in Figure 8-40(d). The exhaust valve opens as the piston

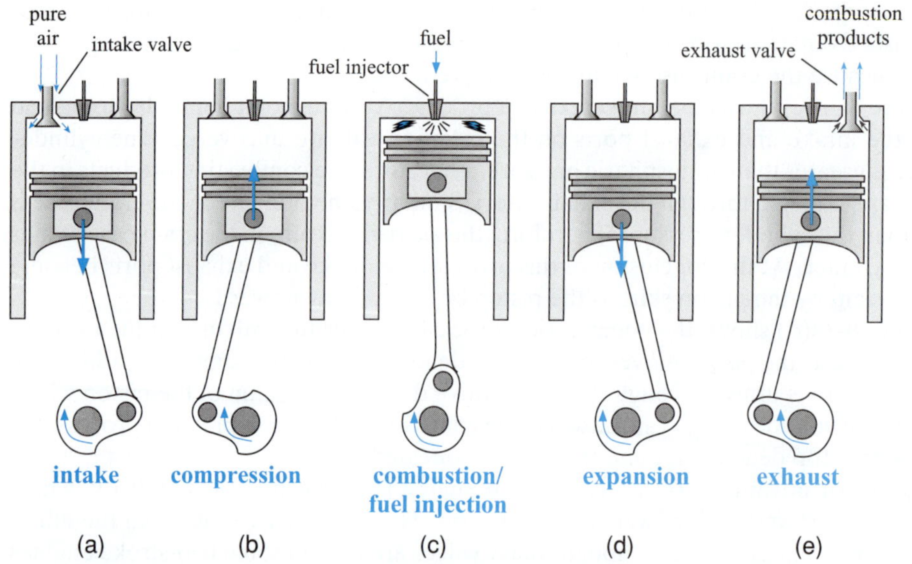

Figure 8-40: Processes occurring in the four-stroke compression-ignition engine: (a) intake, (b) compression, (c) fuel injection and combustion, (d) expansion, and (e) exhaust. An animation of this cycle can be obtained by selecting EES Example Problems from the Examples menu in EES. Select Animation and then Animation of a Diesel internal combustion engine.

moves back towards top dead center in order to push the combustion products out of the cylinder, as shown in Figure 8-40(e).

The timing of the fuel injection is crucial in the compression-ignition engine. If fuel is injected too quickly, then it will cool the air in the cylinder to a temperature below the auto-ignition temperature and combustion will not occur (i.e., the fuel will quench the combustion process). If the fuel is injected too slowly, then the power stroke will be completed with insufficient fuel. In addition, the fuel must mix with the air in the cylinder after injection in order for combustion to occur.

The fuel injection process is complicated and cannot be accurately represented with a simple model. However, as a first approximation, the fuel injection and combustion processes can be treated as a constant pressure heat release as shown approximately in Figure 8-41. This representation takes into account the fact that combustion during the fuel injection process is slower than in the spark-ignition process, due to the controlled rate at which fuel must be injected. The piston motion during combustion was neglected

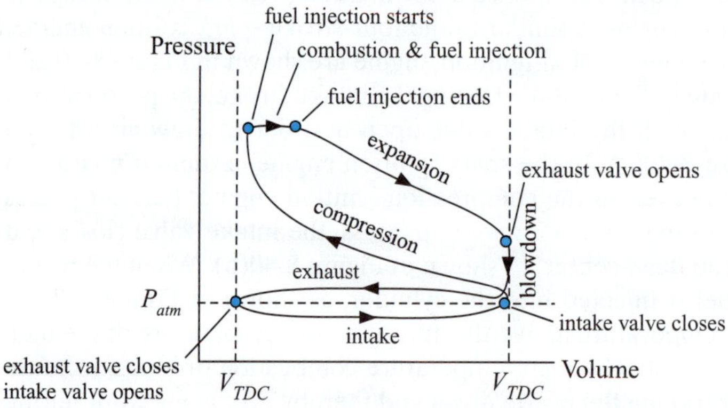

Figure 8-41: Processes occurring in the four-stroke compression-ignition engine shown qualitatively on a pressure-volume diagram.

in the spark-ignition models presented in Section 8.4.1, allowing the combustion process to be modeled as a constant volume heat release. In the compression-ignition engine, a much larger piston motion occurs during combustion as a result of the time required for the fuel injection to occur. The motion of the piston away from top dead center tends to reduce the pressure while the heat release associated with the combustion tends to increase the pressure. These two effects approximately cancel, leading to an approximately constant pressure in the cylinder during the fuel-injection/combustion process. The ratio of the cylinder volume at the time that the fuel-injection/combustion process is completed to the cylinder volume at top dead center is called the *cutoff ratio*. The cutoff ratio is typically in the range of 1.5 to 3.0. After the fuel-injection/combustion process is completed, the gas in the cylinder continues to expand thereby doing work as the temperature and pressure in the cylinder decrease.

The compression-ignition engine has advantages and disadvantages compared to the spark-ignition engine. A spark plug and the associated electronics are not needed to initiate the combustion process, which makes the engine cycle simpler and more reliable. (A heater, referred to as a glow coil, may be provided in some engine designs to assist in starting in cold weather conditions.) Since air, and not a fuel-air mixture, is compressed, there are no pre-ignition concerns in the compression-ignition engine. Therefore, high-octane fuel is not required (or desired) in a compression-ignition engine. The equivalent octane number for diesel fuel is around 20. Combustion cannot begin until the fuel is injected at the conclusion of the compression process. Therefore, the compression-ignition engine is able to operate with a higher compression ratio than the spark-ignition engine, increasing power and efficiency. The high compression ratios, however, result in increased forces on the engine parts and thus the engine must be made more robust, which in general increases its weight. Compression-ignition engines tend to have lower power to weight ratios than spark-ignition engines, but they can provide better fuel economy. Diesel engines have long been the engine of choice for trucks. Small Diesel engines are increasingly being used in automobiles because of their high fuel efficiency.

EXAMPLE 8.4-2: TURBOCHARGED DIESEL ENGINE

Figure 1 illustrates a large Diesel engine that is used to power an emergency generator. The Diesel engine is turbocharged in order to increase its power output. Air is drawn into the turbocharger compressor at $P_{atm} = 1$ atm and $T_{amb} = 20°$C. The compressor has a pressure ratio of $PR = 3.2$ and an efficiency of $\eta_c = 0.76$. The air leaving the compressor passes through an aftercooler that is cooled by a flow of ambient air. The approach temperature difference for the aftercooler is $\Delta T_{ac} = 10$ K. The air leaving the aftercooler is collected in the intake manifold and fed to the Diesel engine. The engine has $N_{cyl} = 4$ cylinders, each with $bore = 103$ mm and $stroke = 132$ mm. The compression ratio of the engine is $CR = 17.5$ and the engine speed is $N = 50$ rev/s. The fuel has a heat of combustion $HC = 45$ MJ/kg and the air-fuel ratio is $AF = 17$. The gas leaving the Diesel engine passes through the exhaust manifold. Assume that the pressure in the exhaust manifold is equal to the pressure in the inlet manifold and that the temperature of the air in the exhaust manifold is the average of the temperature at the conclusion of the expansion process and the temperature at the end of the intake stroke. The air in the exhaust manifold is fed to the turbine. The efficiency of the turbine is $\eta_t = 0.78$. Any gas flow that is not required by the turbine to drive the compressor is bypassed using the gate valve.

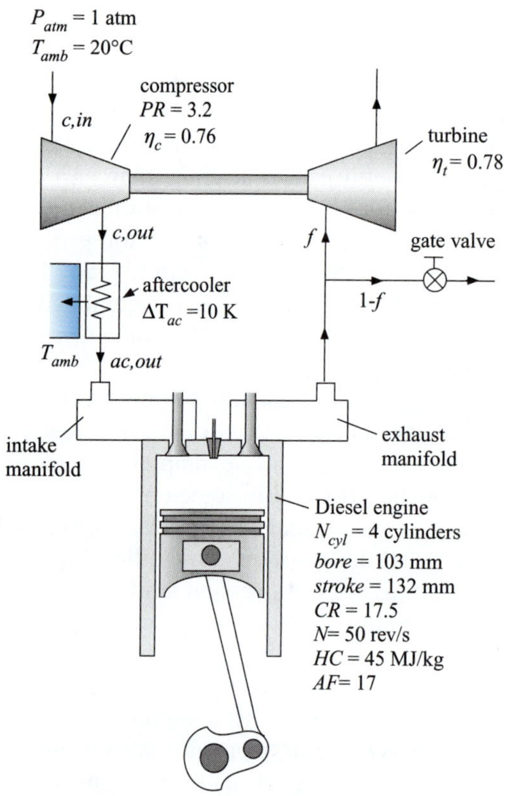

Figure 1: Turbocharged diesel engine.

a) Analyze the cycle and determine the cutoff ratio, efficiency, and net power output. What is the fraction of the air flow that passes through the turbine (f) rather than the gate valve? Assume adiabatic and reversible compression and expansion processes and a constant pressure fuel injection and combustion process. Neglect the impact of residual combustion gas.

The inputs are entered in EES:

```
$UnitSystem SI K Pa J Mass Radian
"Diesel engine"
N_cyl=4 [-]                                "number of cylinders"
bore=103 [mm]*convert(mm,m)                "bore"
stroke=132 [mm]*convert(mm,m)              "stroke"
CR=17.5 [-]                                "compression ratio"
N=50 [1/s]                                 "speed of engine"
HC=45 [MJ/kg]*convert(MJ/kg,J/kg)          "heat of combustion"
AF=17 [-]                                  "air fuel ratio"

"turbocharger/Aftercooler"
PR=3.2 [-]                                 "compressor pressure ratio"
eta_c=0.76 [-]                             "compressor efficiency"
eta_t=0.78 [-]                             "turbine efficiency"
DT_ac=10 [K]                               "aftercooler approach temperature difference"

T_amb=converttemp(C,K,20 [C])              "ambient temperature"
P_atm=1 [atm]*convert(atm,Pa)              "atmospheric pressure"
```

The compressor inlet state is fixed by the pressure, $P_{c,in} = P_{atm}$, and temperature, $T_{c,in} = T_{amb}$. The specific entropy and specific enthalpy ($s_{c,in}$ and $h_{c,in}$) are determined.

```
"compressor inlet"
P_c_in=P_atm                              "pressure"
T_c_in=T_amb                              "temperature"
s_c_in=entropy(Air,P=P_c_in,T=T_c_in)     "specific entropy"
h_c_in=enthalpy(Air,T=T_c_in)             "specific enthalpy"
```

The compressor outlet pressure is specified by the pressure ratio, $P_{c,out} = PR\, P_{c,in}$. The specific entropy leaving a reversible compressor is given by:

$$s_{s,c,out} = s_{c,in}$$

The specific enthalpy leaving a reversible compressor ($h_{s,c,out}$) is fixed by the specific entropy and pressure. The work per unit mass required by a reversible compressor is:

$$\frac{\dot{W}_{s,c}}{\dot{m}_c} = h_{s,c,out} - h_{c,in}$$

where $\dot{m}_c$ is the mass flow rate passing through the compressor. The actual work per unit mass required by the compressor is:

$$\frac{\dot{W}_c}{\dot{m}_c} = \frac{1}{\eta_c}\frac{\dot{W}_{s,c}}{\dot{m}_c}$$

An energy balance on the actual compressor is:

$$h_{c,out} = h_{c,in} + \frac{\dot{W}_c}{\dot{m}_c}$$

The state of the air leaving the compressor is fixed by the specific enthalpy and pressure. The temperature and specific entropy ($T_{c,out}$ and $s_{c,out}$) are determined.

```
"compressor outlet"
P_c_out=P_c_in*PR                                  "pressure"
s_s_c_out=s_c_in                                   "specific entropy leaving rev. compressor"
h_s_c_out=enthalpy(Air,P=P_c_out,s=s_s_c_out)      "specific enthalpy leaving rev. compressor"
W_dot_s_c\m_dot_c=h_s_c_out-h_c_in                 "work per mass for rev. compressor"
W_dot_c\m_dot_c=W_dot_s_c\m_dot_c/eta_c            "actual work per mass for compressor"
h_c_out=h_c_in+W_dot_c\m_dot_c                     "specific enthalpy leaving actual compressor"
T_c_out=temperature(Air,h=h_c_out)                 "temperature leaving actual compressor "
s_c_out=entropy(Air,h=h_c_out,P=P_c_out)           "specific entropy leaving actual compressor "
```

There is assumed to be no pressure loss in the aftercooler, $P_{ac,out} = P_{c,out}$. The temperature of the air leaving the aftercooler is specified by the approach temperature difference:

$$T_{ac,out} = T_{amb} + \Delta T_{ac}$$

The state of the air leaving the aftercooler is fixed by its temperature and pressure. The specific enthalpy and specific entropy ($h_{ac,out}$ and $s_{ac,out}$) are determined.

```
"aftercooler outlet"
P_ac_out=P_c_out                                "pressure"
T_ac_out=T_amb+DT_ac                            "temperature"
h_ac_out=enthalpy(Air,T=T_ac_out)               "specific enthalpy"
s_ac_out=entropy(Air,T=T_ac_out,P=P_ac_out)     "specific entropy"
```

Equations (8-95) through (8-99) are used to determine the volume in the cylinder at top dead center and bottom dead center.

```
Vol_dis_cyl=pi*bore^2*stroke/4                  "displacement of each cylinder"
Vol_cl=Vol_dis_cyl/(CR-1)                       "clearance volume"
Vol_BDC=Vol_dis_cyl+Vol_cl                      "bottom dead center volume"
Vol_TDC=Vol_cl                                  "top dead center volume"
```

State 1 is defined as the condition of the air immediately after the inlet stroke. If the effect of the residual combustion products that remain in the cylinder after the exhaust stroke is neglected then the temperature and pressure of the gas at state 1 are equal to the temperature and pressure of the gas leaving the aftercooler, $T_1 = T_{ac,out}$ and $P_1 = P_{ac,out}$. State 1 is fixed by the temperature and pressure. The remaining properties (s_1, u_1, and v_1) are determined. The volume at state 1 is the bottom dead center volume, $V_1 = V_{BDC}$. Therefore, the mass of air in the cylinder is:

$$m_1 = \frac{V_1}{v_1}$$

The mass of air that is pulled into the cylinder during the intake stroke is:

$$m_{in} = \frac{(V_{BDC} - V_{TDC})}{v_1}$$

```
"State 1"
T[1]=T_ac_out                                   "temperature"
P[1]=P_ac_out                                   "pressure"
s[1]=entropy(Air,T=T[1],P=P[1])                 "specific entropy"
u[1]=intenergy(Air,T=T[1])                      "specific internal energy"
v[1]=volume(Air,T=T[1],P=P[1])                  "specific volume"
Vol[1]=Vol_BDC                                  "volume"
m[1]=Vol[1]/v[1]                                "mass"
m_in=(Vol_BDC-Vol_TDC)/v[1]                     "mass of air pulled into the cylinder"
```

State 2 is defined as the condition of the air immediately after the compression process. The mass of the air does not change, $m_2 = m_1$, and the volume reaches top dead center, $V_2 = V_{TDC}$. Therefore, the specific volume is given by:

$$v_2 = \frac{V_2}{m_2}$$

The compression process is assumed to be reversible and adiabatic; therefore $s_2 = s_1$. State 2 is fixed by the specific volume and specific entropy. The remaining properties (u_2, T_2, and P_2) are determined. An energy balance on the compression process is:

$$W_{comp} = m_2\, u_2 - m_1\, u_1$$

```
"State 2"
m[2]=m[1]                                   "mass balance"
Vol[2]=Vol_TDC                              "volume"
v[2]=Vol[2]/m[2]                            "specific volume"
s[2]=s[1]                                   "entropy balance"
u[2]=intenergy(Air,v=v[2],s=s[2])           "specific internal energy"
T[2]=temperature(Air,v=v[2],s=s[2])         "temperature"
P[2]=pressure(Air,v=v[2],s=s[2])            "pressure"
W_comp=m[2]*u[2]-m[1]*u[1]                  "compression work"
```

State 3 is defined as the condition of the combustion products immediately after the fuel injection and combustion process. The mass of fuel that is injected into the cylinder is given by:

$$m_f = \frac{m_{in}}{AF}$$

```
m_f=m_in/AF                                 "mass of fuel"
```

This model will assume that the combustion process occurs isobarically due to piston motion during the heat release. Therefore, the pressure at state 3 is $P_3 = P_2$. The work done by the gas during the combustion process is:

$$W_{comb} = P_2\,(V_3 - V_2) \tag{1}$$

An energy balance on the combustion process is:

$$m_f\, HC = W_{comb} + m_3\, u_3 - m_2\, u_2 \tag{2}$$

The mass of combustion products is given by:

$$m_3 = m_2 + m_f \tag{3}$$

The specific volume of the combustion products is given by:

$$v_3 = \frac{V_3}{m_3} \tag{4}$$

Equations (1) through (4) are four equations in five unknowns (W_{comb}, V_3, u_3, m_3, and v_3). The remaining equation that is required comes from the property relationship that relates u_3, v_3, and P_3; that is, u_3 is a function of v_3 and P_3. State 3 is fixed by the specific volume and pressure. The remaining properties (s_3 and T_3) are determined. The cutoff ratio for the engine is defined as:

$$\text{cutoff ratio} = \frac{V_3}{V_2}$$

```
"State 3"
P[3]=P[2]                                   "constant pressure model"
W_comb=P[2]*(Vol[3]-Vol[2])                 "work out during combustion"
m_f*HC=W_comb+m[3]*u[3]-m[2]*u[2]           "energy balance"
m[3]=m[2]+m_f                               "mass"
v[3]=Vol[3]/m[3]                            "specific volume"
u[3]=intenergy(Air,v=v[3],P=P[3])           "specific internal energy"
s[3]=entropy(Air,v=v[3],P=P[3])             "specific entropy"
T[3]=temperature(Air,v=v[3],P=P[3])         "temperature"
CutoffRatio=Vol[3]/Vol[2]                   "cutoff ratio"
```

which leads to a cutoff ratio of 3.27. State 4 is defined as the condition of the air immediately after the expansion process. The mass of the air does not change, $m_4 = m_3$, and the volume reaches bottom dead center, $V_4 = V_{BDC}$. Therefore, the specific volume is given by:

$$v_4 = \frac{V_4}{m_4}$$

The expansion process is assumed to be reversible and adiabatic; therefore $s_4 = s_3$. State 4 is fixed by the specific volume and specific entropy. The remaining properties (u_4, T_4, and P_4) are determined. An energy balance on the expansion process is:

$$0 = W_{exp} + m_4\, u_4 - m_3\, u_3$$

```
"State 4"
m[4]=m[3]                                   "mass balance"
Vol[4]=Vol_BDC                              "volume"
v[4]=Vol[4]/m[4]                            "specific volume"
s[4]=s[3]                                   "entropy balance"
u[4]=intenergy(Air,v=v[4],s=s[4])           "specific internal energy"
T[4]=temperature(Air,v=v[4],s=s[4])         "temperature"
P[4]=pressure(Air,v=v[4],s=s[4])            "pressure"
0=W_exp+m[4]*u[4]-m[3]*u[3]                 "expansion work"
```

Figure 2 shows the pressure-volume diagram predicted by the model.

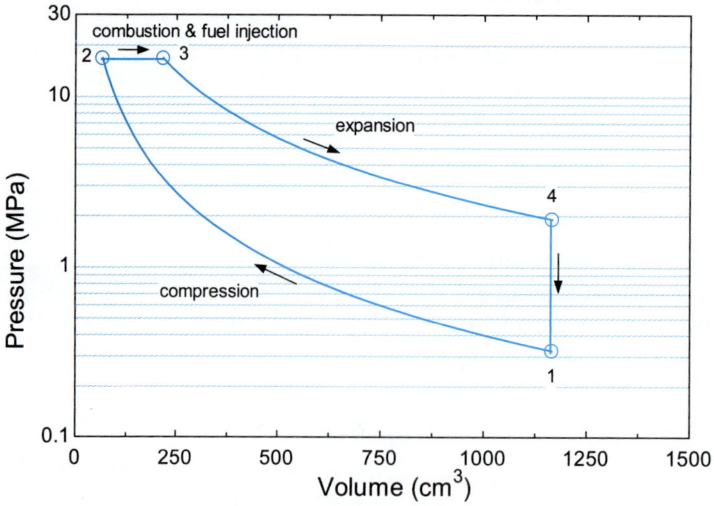

Figure 2: Pressure-volume diagram for Diesel cycle.

The turbine inlet pressure is assumed to be equal to the compressor outlet pressure, $P_{t,in} = P_{c,out}$. The turbine inlet temperature is taken to be the average of the temperatures in the cylinder at the completion of the expansion stroke and the beginning of the compression stroke:

$$T_{t,in} = \frac{(T_4 + T_1)}{2}$$

The turbine inlet state is fixed by the temperature and pressure. The remaining properties ($s_{t,in}$ and $h_{t,in}$) are determined.

```
"Turbine inlet"
P_t_in=P_c_out                          "pressure"
T_t_in=(T[4]+T[1])/2                    "temperature"
h_t_in=enthalpy(Air,T=T_t_in)           "specific enthalpy"
s_t_in=entropy(Air,T=T_t_in,P=P_t_in)   "specific entropy"
```

The turbine exit pressure is atmospheric, $P_{t,out} = P_{atm}$. The specific entropy leaving a reversible turbine is given by:

$$s_{s,t,out} = s_{t,in}$$

The specific enthalpy leaving a reversible turbine ($h_{s,t,out}$) is specified by the specific entropy and pressure. The work per unit mass provided by a reversible turbine is:

$$\frac{\dot{W}_{s,t}}{\dot{m}_t} = h_{t,in} - h_{s,t,out}$$

where $\dot{m}_t$ is the mass flow rate passing through the turbine. The actual work per unit mass provided by the turbine is:

$$\frac{\dot{W}_t}{\dot{m}_t} = \eta_t \frac{\dot{W}_{s,t}}{\dot{m}_t}$$

An energy balance on the turbine is:

$$h_{t,out} = h_{t,in} - \frac{\dot{W}_t}{\dot{m}_t}$$

The state of the air leaving the turbine is fixed by the specific enthalpy and pressure. The temperature and specific entropy ($T_{t,out}$ and $s_{t,out}$) are determined.

```
"Turbine exit"
P_t_out=P_atm                                  "pressure"
s_s_t_out=s_t_in                               "specific entropy leaving rev. turbine"
h_s_t_out=enthalpy(Air,P=P_t_out,s=s_s_t_out)  "specific enthalpy leaving rev. turbine"
W_dot_s_t\m_dot_t=h_t_in-h_s_t_out             "work per mass for rev. turbine"
W_dot_t\m_dot_t=eta_t*W_dot_s_t\m_dot_t        "actual work per mass for turbine"
h_t_out=h_t_in-W_dot_t\m_dot_t                 "specific enthalpy"
T_t_out=temperature(Air,h=h_t_out)             "temperature"
s_t_out=entropy(Air,h=h_t_out,P=P_t_out)       "specific entropy"
```

The fraction of the air passing through the turbine is determined by an energy balance on the compressor and turbine:

$$\dot{W}_t = \dot{W}_c \tag{5}$$

Equation (5) can be rearranged by dividing through by $\dot{m}_c$:

$$\frac{\dot{W}_t}{\dot{m}_t}\underbrace{\frac{\dot{m}_t}{\dot{m}_c}}_{f} = \frac{\dot{W}_c}{\dot{m}_c}$$

```
f*W_dot_t\m_dot_t=W_dot_c\m_dot_c      "fraction of flow passing through turbine"
```

After solving, $f = 0.678$ (67.8% of the flow passes through the turbine, the remaining 32.2% bypasses the turbine through the gate valve).

The net work produced per cylinder per cycle is:

$$W_{net} = W_{comb} + W_{exp} - W_{comp}$$

The efficiency of the engine is:

$$\eta = \frac{W_{net}}{m_f\,HC}$$

```
W_net=W_exp+W_comb-W_comp      "net work per power stroke"
eta=W_net/(m_f*HC)             "efficiency"
```

which leads to $\eta = 0.4997$ (49.97%). The average rate that power is produced by the Diesel engine is:

$$\dot{W}_{net} = \frac{W_{net}\,N_{cyl}}{2\,period}$$

where *period* is the time required for one crankshaft revolution:

$$period = \frac{1}{N}$$

```
period=1/N                             "period of one rotation"
W_dot_net=W_net*N_cyl/(2*period)       "net power produced"
W_dot_net_hp=W_dot_net*convert(W,hp)   "in hp"
```

which results in $\dot{W}_{net} = 542.2$ kW (727.1 hp).

b) Plot the fraction of the flow through the turbine, *f*, as a function of the turbocharger compressor pressure ratio, *PR*. Based on your plot, what is the maximum pressure ratio that can be used for the turbocharger?

Figure 3 illustrates *f* as a function of *PR*. It is not possible to have *f* greater than unity; therefore, the maximum pressure ratio is approximately $PR = 13$.

c) Plot the power and efficiency of the engine as a function of the compressor pressure ratio. Based on this plot, what is the benefit associated with turbocharging an engine?

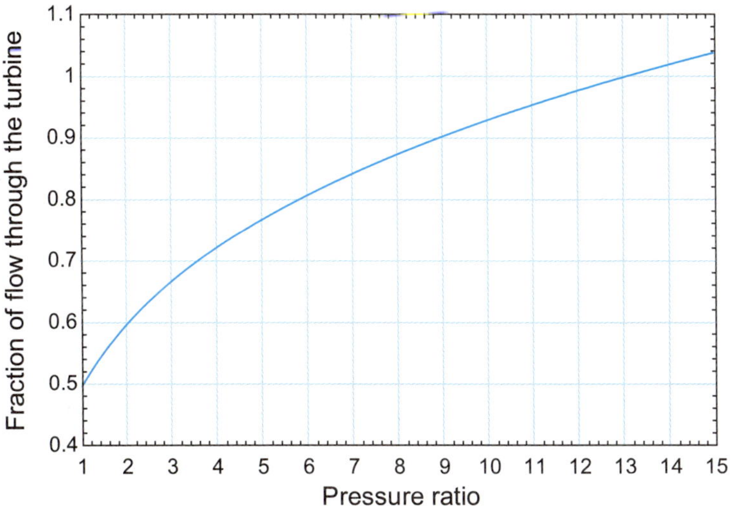

Figure 3: Fraction of the flow that passes through the turbine as a function of the compressor pressure ratio.

Figure 4 illustrates the power and efficiency of the engine as a function of the compressor pressure ratio. The model predicts that the power increases linearly due to the increased density of the air drawn into the cylinder but the efficiency does not change at all. Turbocharging allows a fixed engine size to produce more power but does not have a strong effect on the efficiency of the engine.

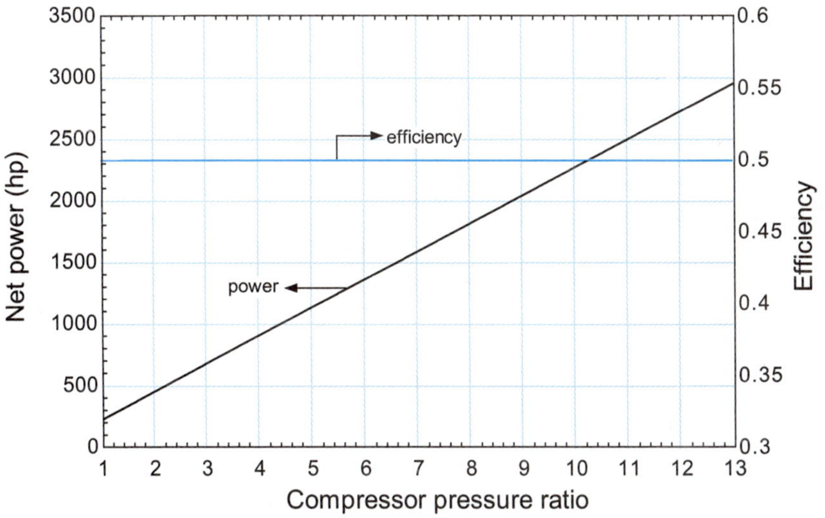

Figure 4: Power and efficiency as a function of the compressor pressure ratio.

8.5 The Stirling Engine

The Rankine cycle discussed in Section 8.2 is the most common design for large, external combustion power plants that provide baseload electrical power. The gas turbine engine cycle discussed in Section 8.3 is used for peaking power plants and for aircraft propulsion. The reciprocating combustion engines discussed in Section 8.4 provide power for most of the ground transportation in the world. This section discusses a much less commonly

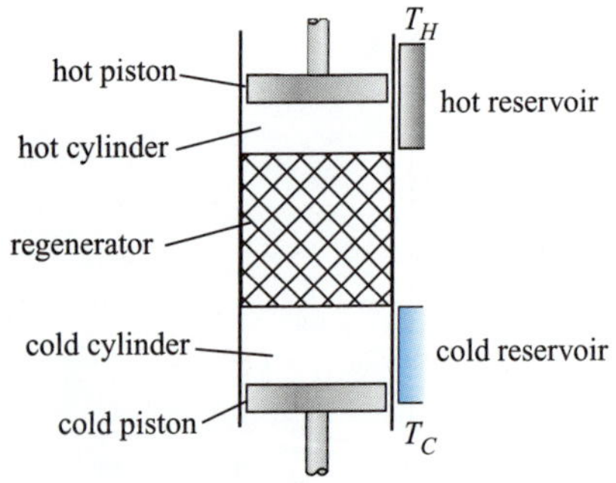

Figure 8-42: The Stirling engine.

used external combustion engine cycle, the *Stirling engine*. The Stirling engine was first suggested by Robert Stirling in 1816. Despite nearly two centuries of development, the engine has not found wide application.

8.5.1 The Stirling Engine Cycle

Figure 8-42 illustrates a schematic of the Stirling engine. The device consists of two opposed pistons, the hot piston and cold piston, which move back and forth within the hot and cold cylinders, respectively. The hot cylinder is thermally interfaced through a heat exchanger with a thermal reservoir at T_H that provides the high temperature heat source that energizes the cycle. The cold cylinder is thermally interfaced with a thermal reservoir at T_C that provides the low temperature sink for heat rejection. The Stirling engine is an externally heated engine (like the Rankine cycle) and therefore it is flexible with respect to the source of heat. Stirling engines can be energized by an external combustion process (e.g., the combustion of biofuels) or from other sources of heat (e.g., solar or nuclear energy).

The hot and cold cylinders are separated by a regenerator. Physically, the regenerator is a porous matrix of solid material that has a large capacity to store thermal energy. The fluid flows through the flow channels in the regenerator, transferring heat to and from the solid. Regenerators are generally simple devices. A container that is filled with lead shot or metal screens can be used as a regenerator in a Stirling cycle. Thermodynamically, the regenerator acts as a thermal storage medium. Heat is transferred from the gas to the solid during one part of the cycle and then from the solid back to the gas during a later part of the cycle. During normal operation there is a temperature gradient in the regenerator matrix so that the solid material adjacent to the hot cylinder is at (or near) the hot reservoir temperature and the solid material adjacent to the cold cylinder is near the cold reservoir temperature. In a well-designed regenerator, the heat capacity of the solid material (the matrix) will be much larger than the heat capacity of the fluid. Therefore, the temperature of the matrix will change very little as it absorbs and releases heat during each cycle. The regenerator behavior is similar to the heat exchanger used in the recuperated gas turbine cycle, shown in Figure 8-26(a). The recuperative heat exchanger operates continuously by transferring energy from gas flowing in one portion of the steady flow cycle to gas flowing in a different portion of the cycle. The regenerator shown in Figure 8-42 operates in a transient fashion by transferring energy from the gas at one time in the cycle to the same gas but at a different time in the cycle.

Figure 8-43 illustrates the processes that together make up a complete cycle of an ideal Stirling engine. State 1 is defined as the state of the working fluid immediately before the compression process, shown in Figure 8-43(a). The hot cylinder volume is zero and the cold cylinder volume is at its maximum value. In the limit that the regenerator has no

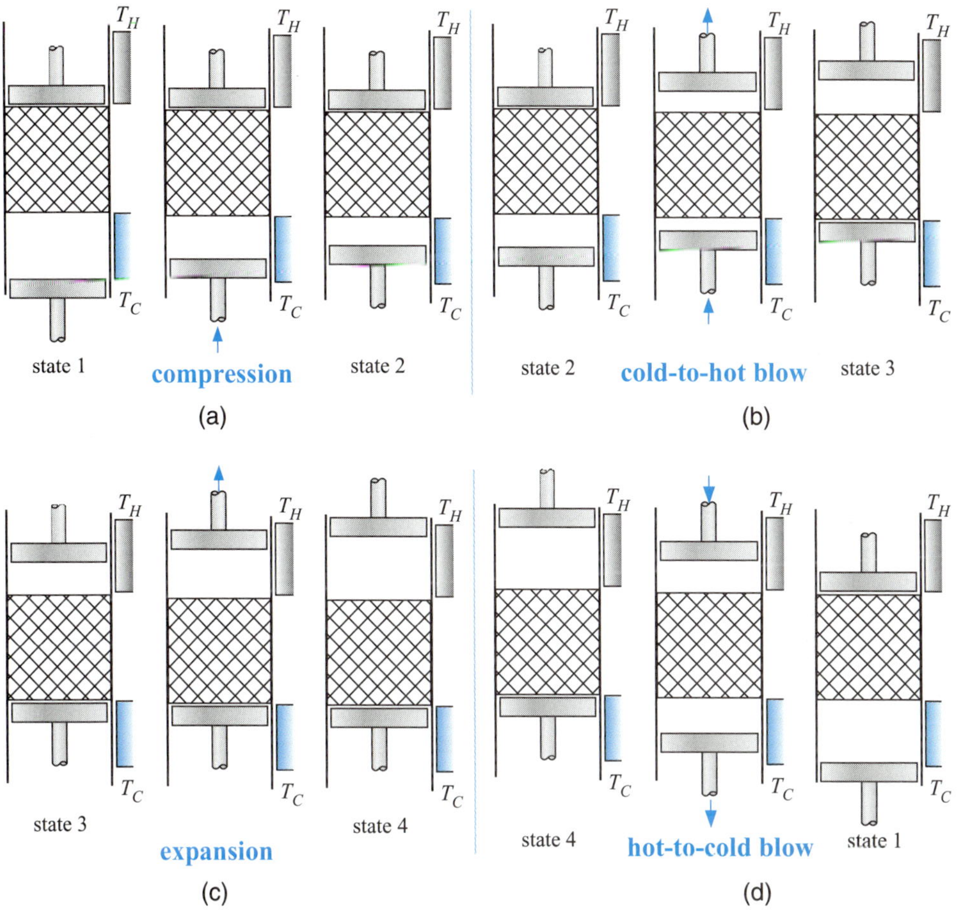

Figure 8-43: Processes that make up the ideal Stirling cycle: (a) compression, (b) cold-to-hot blow, (c) expansion, and (d) hot-to-cold blow.

void volume, the working fluid entirely resides in the cold cylinder at state 1. The fluid is assumed to be isothermal and at the cold reservoir temperature. During the compression process, the cold cylinder volume is decreased while the fluid is maintained at a constant temperature, T_C, due to heat transfer to the cold reservoir. The pressure of the fluid increases during this process.

State 2 is defined as the state of the working fluid immediately after the compression process. During the cold-to-hot blow process, the cold cylinder volume is decreased while the hot cylinder volume is increased by an equal amount. The cylinders move together, as shown in Figure 8-43(b). The total volume available for the fluid in the system is ideally constant during this process. At the conclusion of the cold-to-hot blow process, the cold cylinder volume is zero and therefore all of the fluid has been pushed through the regenerator and into the hot cylinder. The fluid is warmed from T_C to T_H as it flows through the regenerator by heat transfer from the solid matrix; ideally, the fluid enters the hot cylinder at T_H. The pressure of the fluid rises during this process.

State 3 is defined as the state of the fluid immediately after the cold-to-hot blow process. All of the fluid is in the hot cylinder at T_H. During the expansion process shown in Figure 8-43(c), the hot cylinder moves to its maximum volume and the gas is maintained at constant temperature, T_H, due to heat transfer from the hot reservoir. The pressure of the fluid decreases during this process.

State 4 is defined as the state of the working fluid immediately after the expansion process. During the hot-to-cold blow process, the hot cylinder volume is decreased while the cold cylinder volume is increased by an equal amount as shown in Figure 8-43(d). At the conclusion of the hot-to-cold blow process, all of the fluid has been pushed through the regenerator and into the cold cylinder. The fluid is cooled from T_H to T_C as it flows through the regenerator by heat transfer to the solid matrix; the regenerator stores this energy to be released during the subsequent cold-to-hot blow cycle. At the conclusion of the hot-to-cold blow cycle the system has been returned to state 1 and it is ready to begin another cycle.

Notice that none of the processes described in the ideal Stirling cycle are associated with any entropy generation. There is never heat transfer through a temperature gradient or flow through a pressure gradient for the ideal Stirling cycle. Indeed, it is this characteristic of the ideal Stirling cycle that has drawn so much attention; in theory, it can obtain the maximum possible efficiency for converting heat into work. However, the ideal Stirling cycle is also not practical. The compression and expansion processes combine heat and work during a single process. These processes would need to occur very slowly in order to approach a reversible limit. Notice that none of the components or processes in the other power cycles (i.e., those that are actually used, the Rankine cycle, gas turbine engine, and 4-stroke engine) have this characteristic. For example, in the Rankine cycle, work is done with some components (the pump and turbine) and heat transfer with different components (the condenser and boiler). There is a good reason for this separation: work and heat occur with very different time scales. Heat takes a substantial amount of time to transfer and requires components with large amounts of surface area.

The ideal Stirling cycle is never realized in practice. Instead, the hot and cold cylinders are nearly adiabatic and equipped with hot and cold heat exchangers that interface the working fluid thermally with the appropriate reservoirs. During the compression process, the cold piston moves in and compresses the working fluid nearly adiabatically, causing its temperature to rise above T_C. When the gas is subsequently pushed through the cold heat exchanger during the cold-to-hot blow process, it is cooled to T_C before it enters the regenerator. A similar process occurs in the hot cylinder and hot heat exchanger during the expansion and hot-to-cold blow process. Clearly, the temperature elevation and subsequent heat transfer results in irreversible processes. When these and other practical considerations are taken into account, the efficiency advantage associated with the Stirling cycle becomes less clear.

8.5.2 Simple Model of the Ideal Stirling Engine Cycle

This extended section can be found on the website www.cambridge.org/kleinandnellis. In this section, a simple model of the ideal Stirling cycle is developed and used to investigate its behavior. Even the idealized version of the Stirling cycle that is discussed in Section 8.5.1 is much more complex than it would appear at first glance. Most thermodynamics books present the Stirling cycle as two isothermal processes (corresponding to compression and expansion) separated by two constant specific volume processes (corresponding to the cold-to-hot and hot-to-cold blow processes). The analysis contained in this section shows that no fluid particle within the Stirling cycle actually undergoes this ideal cycle and, in fact, no two fluid particles in the system undergo the same cycle.

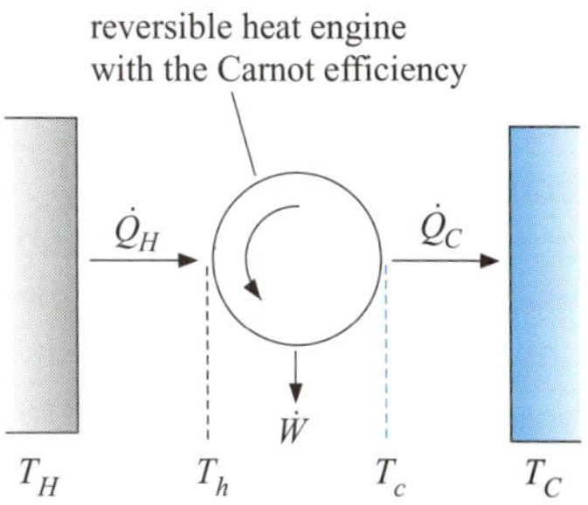

Figure 8-50: Reversible heat engine operating between an isothermal heat source at T_H and rejecting heat to an isothermal heat sink at T_C. The temperature of the working fluid in the engine receiving heat is T_h and the temperature of the working fluid rejecting heat is T_c.

8.6 Tradeoffs Between Power and Efficiency

The previous sections of this chapter have described several thermodynamic cycles that convert heat into power. Thermodynamic analyses have been developed in order to determine the thermal efficiency of each cycle as a function of operating conditions. However, the analyses did not specifically model the details of the heat transfer processes which, in fact, will have a significant effect on the cycle performance and the selection of an optimum cycle design. Thermodynamics dictates that efficiency will be maximized when temperature differences associated with the heat transfer processes are minimized. However, heat transfer rates increase with increasing temperature differences for given heat exchange equipment and the power output depends on the heat transfer rates. These competing processes result in a natural tradeoff between power and efficiency in the design of a power system.

8.6.1 The Heat Transfer Limited Carnot Cycle

The maximum efficiency at which heat can be converted into power depends only on the temperatures of the thermal energy source and sink. This maximum efficiency (the Carnot efficiency) is attained by a thermodynamic power cycle operating between an isothermal source and an isothermal sink in which all processes are thermodynamically reversible, as described in Section 8.1. The Carnot efficiency underscores the thermodynamic difference between heat and work. They are both forms of energy transport, but they are not equivalent because heat cannot be converted to work with 100% efficiency. This understanding is one of the foundations of classical thermodynamics.

The derivation of the Carnot efficiency in Section 8.1 does not involve any consideration of time or of the heat transfer mechanisms associated with the cycle. Time must pass in order for heat transfer to occur. In any practical cycle, we must be at least as concerned with the rates of the processes as we are with the overall efficiency. The Carnot efficiency is not a realistic design goal because it does not consider these heat transfer rates.

Consider the heat engine shown in Figure 8-50. The heat engine is receiving heat at rate $\dot{Q}_H$ from a thermal reservoir at temperature T_H. The temperature of the working fluid in the engine that is receiving the heat transfer is constant at T_h. The heat engine is rejecting heat at rate $\dot{Q}_C$ to a thermal reservoir at T_C. The temperature of the working fluid that is rejecting the heat is constant at T_c. To achieve the Carnot efficiency, heat transfer between the thermal reservoirs and the working fluid in the engine must occur in a thermodynamically reversible manner. The quality of thermal energy is associated with its temperature. When the temperature decreases, the quality of the thermal energy is degraded and the maximum efficiency is reduced (i.e., entropy is generated by heat transfer through a temperature difference). To avoid this degradation, the heat transfer must occur with no temperature difference. That is, the temperature of the hot reservoir (T_H) must be equal to the temperature of the working fluid receiving the heat (T_h) and the temperature of the working fluid rejecting the heat (T_c) must be equal to the temperature of the cold reservoir (T_C). However, heat transfer theory dictates that the rate of heat transfer is dependent upon the temperature difference between the thermal reservoir and the engine, as well as on the surface area available for heat transfer. For a heat engine with finite heat exchange surface area, the requirement that $T_H = T_h$ and $T_c = T_C$ necessarily results in zero heat transfer rates. The rate of heat transfer to the engine, $\dot{Q}_H$, and the rate of heat rejected by the engine, $\dot{Q}_C$, must both be zero. The power developed by the engine, $\dot{W}$, which is the difference between $\dot{Q}_H$ and $\dot{Q}_C$, must then also be zero. Thus, an engine operating at the Carnot efficiency achieves the highest possible efficiency but produces absolutely zero power! With this understanding, the Carnot efficiency can hardly be considered a realistic goal for design.

It is possible to explicitly consider the heat transfer mechanisms occurring in the Carnot cycle. The rate of heat transfer can be represented, approximately, by:

$$\dot{Q}_H = UA_H\,(T_H - T_h) \tag{8-155}$$

where UA_H is the total conductance characterizing the heat transfer equipment that interfaces the engine with the high temperature reservoir. Note that the fluid temperature, T_h, is the high temperature that is actually experienced by the Carnot cycle. In a realistic analysis of the engine it is clear that both $\dot{Q}_H$ and UA_H are finite; therefore, T_h will be less than T_H. Similarly, the rate of heat rejection from the engine can be expressed, approximately, by:

$$\dot{Q}_C = UA_C\,(T_c - T_C) \tag{8-156}$$

where UA_C is the total conductance characterizing the heat transfer equipment that interfaces the engine with the low temperature reservoir. The temperature T_c is the low temperature actually experienced by the Carnot cycle and T_c must be greater than T_C. An energy balance on the engine operating at steady-state requires that:

$$\dot{W} = \dot{Q}_H - \dot{Q}_C \tag{8-157}$$

Equations (8-155) and (8-156) lead to thermodynamic irreversibilities associated with heat transfer at finite rates. If no other dissipative (or thermodynamically irreversible) processes occur within the engine (e.g., friction) then the energy conversion process will occur at the Carnot efficiency; however, the engine efficiency must be computed using the temperatures T_h and T_c rather than temperatures T_H and T_C. A cycle operating in this manner is referred to as a *heat transfer-limited Carnot cycle* and has efficiency:

$$\eta = \frac{\dot{W}}{\dot{Q}_H} = 1 - \frac{T_c}{T_h} \tag{8-158}$$

For a specific engine operating between set thermal reservoir temperatures, the parameters UA_H, UA_C, T_H and T_C are specified. Equations (8-155) through (8-158) are then a set of five equations (note that Eq. (8-158) is actually two equations) with six unknown variables: $\dot{Q}_H$, $\dot{Q}_L$, $\dot{W}$, T_h, T_c, and η. With some algebra, any one of these six unknown quantities can be expressed as a function of any one of the other quantities. For example, the power produced by the engine, $\dot{W}$, can be expressed only as a function of the engine efficiency. Some algebra is required to eliminate the other four parameters ($\dot{Q}_H$, $\dot{Q}_L$, T_h, and T_c) from the equation set. The algebra is facilitated by the symbolic software Maple. An introduction to Maple is provided in Appendix H.

Each of the five equations presented in Eqs. (8-155) through (8-158) are entered in Maple as the variables Eq1 through Eq5:

```
> restart;
>Eq1: = W_dot = Q_dot_H – Q_dot_C;
```

$$Eq1 := W\_dot = Q\_dot\_H - Q\_dot\_C$$

```
>Eq2: = Q_dot_H = UA_H(T_H – T_h);
```

$$Eq2 := Q\_dot\_H = UA\_H(T\_H - T\_h)$$

```
>Eq4: = eta = W_dot/Q_dot_H;
```

$$Eq4 := \eta = \frac{W\_dot}{Q\_dot\_H}$$

```
>Eq5: = eta = 1 – T_c/T_h;
```

$$Eq5 := \eta = 1 - \frac{T\_c}{T\_h}$$

The eliminate command in Maple allows one or more variables to be eliminated from a system of equations. For our problem, we would like to eliminate the parameters $\dot{Q}_H$, $\dot{Q}_C$, T_h, and T_c from the five equations that have been entered in order to obtain a single relationship between $\dot{W}$ and η (as well as the other, fixed parameters, UA_H, UA_C, T_H and T_C). The eliminate command returns two outputs. The first is a list of equations for the eliminated variables in terms of the remaining variables and the second is the equation(s) that remain once the variables have been eliminated.

```
> List: = eliminate({Eq1,Eq2,Eq3,Eq4,Eq5},{Q_dot_H,Q_dot_C,T_h,T_c});
```

$$List := \left[\left\{T\_c = -\frac{-W\_dot\,\eta + UA\_H\,T\_H\,\eta^2 + W\_dot - UA\_H\,T\_H\eta}{UA\_H\,\eta}, T\_h = \frac{-W\_dot + UA\_HT\_H\,\eta}{UA\_H\eta}, Q\_dot\_C = -\frac{W\_dot(\eta-1)}{\eta}, Q\_dot\_H = \frac{W\_dot}{\eta}\right\}, \{-W\_dot\,\eta\,UA\_H + W\_dot\,UA\_H - UA\_C\,\eta W\_dot + UA\_C\eta^2\,UA\_H\,T\_H + UA\_C\,W\_dot - UA\_C\,UA\_H\,T\_H\,\eta + UA\_C\,T\_C\,\eta\,UA\_H\}\right]$$

Note that the first output, which is contained within the curly braces, consists of equations for T_c, T_h, $\dot{Q}_H$, $\dot{Q}_C$ each expressed in terms of $\dot{W}$ and η. This output of the eliminate command is referred to as List[1]. The second output, referred to as List[2], is the single equation that remains after T_c, T_h, $\dot{Q}_H$, and $\dot{Q}_C$ are eliminated. We can obtain an explicit expression for $\dot{W}$ as a function of η by using the solve command. Solve the second element of the variable List for the variable W_dot.

```
> solve(List[2],W_dot);
```

$$\left\{W\_dot = \frac{UA\_C\,\eta\,UA\_H(-T\_H + T\_C + \eta\,T\_H)}{UA\_H\,\eta - UA\_C - UA\_H + UA\_C\,\eta}\right\}$$

Therefore, the power produced by a heat transfer limited Carnot engine is expressed in terms of the engine efficiency, the fixed reservoir temperatures, and the conductances of the heat exchangers linking the engine to these thermal reservoirs according to:

$$\dot{W} = \frac{UA_C\,UA_H\,\eta\,[(1-\eta)\,T_H - T_C]}{[UA_H\,(1-\eta) + UA_C\,(1-\eta)]} \tag{8-159}$$

This result could be derived with some tedious algebra, but the process is much easier using Maple. An EES program is used to examine the behavior of the engine. Arbitrary

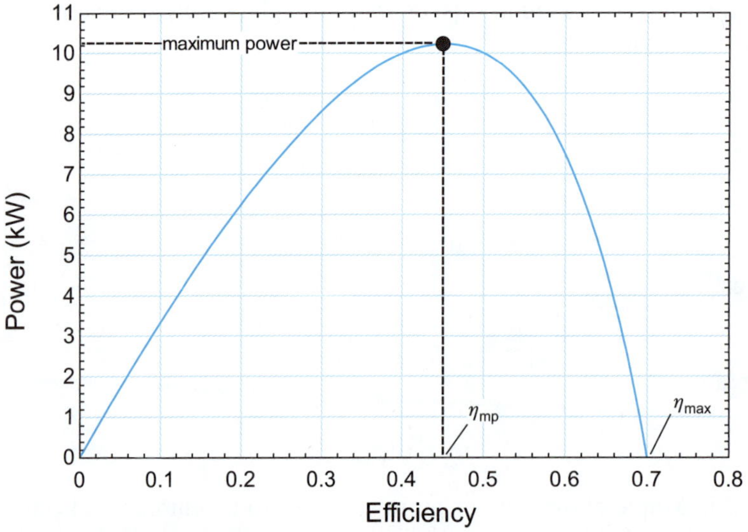

Figure 8-51: Power as a function of the efficiency for a heat transfer limited Carnot cycle with $T_H = 1000$ K, $T_C = 300$ K, $UA_H = UA_C = 100$ W/K.

values of the fixed parameters are specified ($T_H = 1000$ K, $T_C = 300$ K, $UA_C = UA_H = 100$ W/K).

```
$UnitSystem SI Mass Radian J K Pa
T_H=1000 [K]              "hot reservoir temperature"
T_C=300 [K]               "cold reservoir temperature"
UA_C=100 [W/K]            "conductance of cold heat exchanger"
UA_H=100 [W/K]            "conductance of hot heat exchanger"
```

The Maple solution for the variable W_dot is copied from Maple and pasted into EES with no modification:

```
W_dot = UA_C*eta*UA_H*(-T_H+T_C+eta*T_H)/(UA_H*eta-UA_C-UA_H+UA_C*eta)    "power"
W_dot_kW=W_dot*convert(W,kW)                                              "in kW"
```

Figure 8-51 illustrates the power as a function of efficiency. The Carnot efficiency for the selected reservoir temperatures is:

$$\eta_{max} = 1 - \frac{T_C}{T_H} \tag{8-160}$$

```
eta_max=1-T_C/T_H         "Carnot efficiency"
```

which evaluates to $\eta_{max} = 0.70$. Notice that the efficiency of the engine in Figure 8-51 can assume any value between zero and the Carnot efficiency (0.70). The plot of power as a function of efficiency will have the general shape shown in Figure 8-51 for any set of values of UA_H, UA_C, T_H and T_C.

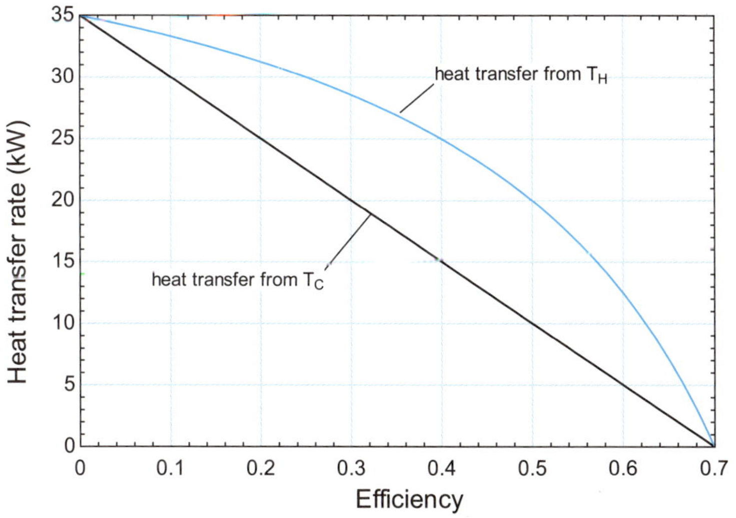

Figure 8-52: Heat transfer rate from hot and cold reservoirs as a function of efficiency with $T_H = 1000$ K, $T_C =$ 300 K, and $UA_H = UA_C = 100$ W/K.

When the efficiency is zero, the power is obviously zero. The power is also zero when the efficiency is at the Carnot limit. To see why this is so, copy and paste each of the equations for T_c, T_h, $\dot{Q}_H$, and $\dot{Q}_C$ from Maple into EES in order to examine how these quantities vary with efficiency.

```
T_c_e = -(-W_dot*eta+UA_H*T_H*eta^2+W_dot-UA_H*T_H*eta)/UA_H/eta
                                        "temperature of cold side working fluid"
T_h_e = (-W_dot+UA_H*T_H*eta)/UA_H/eta  "temperature of hot side working fluid"
Q_dot_C = -W_dot*(eta-1)/eta            "heat rejection rate"
Q_dot_H = W_dot/eta                     "heat transfer from heat source"
```

Figure 8-52 illustrates $\dot{Q}_H$ and $\dot{Q}_C$ and Figure 8-53 illustrates T_h and T_c as a function of η. Notice that when η is low, the heat transfer rates are very high and, as a result, there is a large temperature drop between T_H and T_h and between T_c and T_C. The efficiency reaches zero at the point where $T_h = T_c$ so that there is no temperature difference available to actually drive the heat engine. The power produced reaches zero in this limit. At the other extreme, when η is high, the heat transfer rates are low and there is very little temperature drop across the heat exchangers. The efficiency reaches the Carnot efficiency, η_{max}, at the point where $\dot{Q}_H = \dot{Q}_C = 0$, which corresponds to $T_h = T_H$ and $T_c = T_C$. Therefore, the power produced again reaches zero when $\eta = \eta_{max}$.

One interesting feature of Figure 8-51 is the point of maximum power that occurs at some efficiency between $\eta = 0$ and the Carnot efficiency. The efficiency at the maximum power point is identified as η_{mp} in Figure 8-51. It is possible to obtain an analytical solution for the maximum power and the efficiency at maximum power. The maximum power in Figure 8-51 occurs at the point where:

$$\frac{d\dot{W}}{d\eta} = 0 \tag{8-161}$$

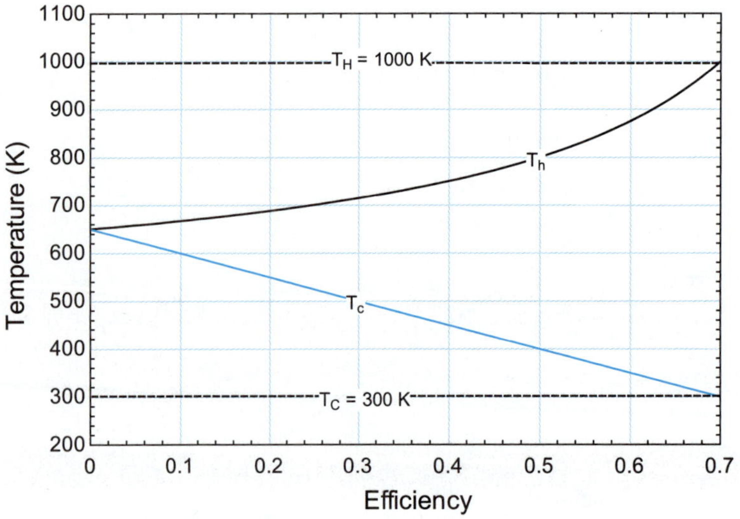

Figure 8-53: Temperature of the working fluid receiving heat (T_h) and rejecting heat (T_c) as a function of efficiency with $T_H = 1000$ K, $T_C = 300$ K, $UA_H = UA_C = 100$ W/K.

Maple is used to take the derivative required by Eq. (8-161) using the diff command.

```
> diff(solve(List[2],W_dot),eta);
```

$$\left\{0 = \frac{UA\_C\,UA\_H(-T\_H + T\_C + \eta\,T\_H)}{UA\_H\eta - UA\_C - UA\_H + UA\_C\eta} + \frac{UA\_C\eta\,UA\_H\,T\_H}{UA\_H\eta - UA\_C - UA\_H + UA\_C\eta} - \frac{UA - C\eta\,UA\_H(-T\_H + T\_C + \eta\,T\_H)\,(UA\_H + UA\_C)}{UA\_H\eta - UA\_C - UA\_H + UA\_C\eta)^2}\right\}$$

The expression is solved for the efficiency at maximum power.

```
solve(diff(solve(List[2],W_dot),eta),eta);
```

$$\left\{\eta = \frac{T\_H + \sqrt{T\_H\,T\_C}}{T\_H}\right\},\quad \left\{\eta = -\frac{-T\_H + \sqrt{T\_H\,T\_C}}{T\_H}\right\}$$

Maple has identified two solutions to Eq. (8-161):

$$\eta_{mp} = 1 \pm \sqrt{\frac{T_C}{T_H}} \tag{8-162}$$

The solution with the positive square root is not physically possible because it results in an efficiency that is greater than 1. As a result, the efficiency at which power is maximized, η_{mp}, is:

$$\eta_{mp} = 1 - \sqrt{\frac{T_C}{T_H}} \tag{8-163}$$

This is a remarkable result, first published by El-Wakil (1960) and again by Curzon and Ahlborn (1975). Equation (8-163) indicates that when the engine is operated at maximum power, the efficiency is a function only of the source and sink temperatures and is independent of the details of the heat exchangers (UA_H and UA_C). Substituting

Eq. (8-163) into Eq. (8-159) leads to an expression for the maximum possible power production from the engine.

```
subs(eta = -(-T_H+(T_H*T_C)^(1/2))/T_H,W_dot =UA_C*eta*UA_H*(-T_H+T_C+eta*T_H)
/(UA_H*eta-UA_C-UA_H+UA_C*eta));
```

$$W\_dot = -UA\_C(-T\_H + \sqrt{T\_H\,T\_C})UA\_H(T\_C - \sqrt{T\_H\,T\_C})/\left(T\_H\left(-\frac{UA\_H(-T\_H+\sqrt{T\_H\,T\_C})}{T\_H} - UA\_C - UA\_H - \frac{UA\_C(-T\_H+\sqrt{T\_H\,T\_C})}{T\_H}\right)\right)$$

```
> simplify(%);
```

$$W\_dot = -\frac{UA\_C(-T\_H+\sqrt{T\_H\,T\_C})\,UA\_H(-T\_C+\sqrt{T\_H\,T\_C})}{\sqrt{T\_H\,T\_C}\,(UA\_H+UA\_C)}$$

The result from Maple can be simplified to show that the maximum power that can be produced by the engine is:

$$\dot{W}_{max} = \frac{UA_H\,UA_C}{(UA_H + UA_C)}\left(\sqrt{T_H} - \sqrt{T_C}\right)^2 \tag{8-164}$$

Given a fixed power plant and fixed thermal reservoirs, an operator would naturally try to optimize the power production and not the efficiency. In this case, the efficiency of the plant would be expected to close to the result provided by Eq. (8-163) rather than Eq. (8-160).

8.6.2 Carnot Cycle using Fluid Streams as the Heat Source and Heat Sink

This extended section can be found online at www.cambridge.org/kleinandnellis. The Carnot cycle analysis in Section 8.6.1 assumes that heat is transferred to and from constant temperature thermal reservoirs for simplicity. However, practical power cycles do not utilize thermal reservoirs for either the energy source or sink. In general, the energy source can more accurately be regarded as a fluid stream that enters a high temperature heat exchanger and the energy sink as a fluid stream that enters a low temperature heat exchanger. This section extends the concepts developed in Section 8.6.1 to this situation and shows that for a fixed plant and fixed fluid inlet temperatures there is again a plant efficiency that is less than the Carnot efficiency which maximizes the power produced.

8.6.3 Internal Irreversibilities

This extended section can be found online at www.cambridge.org/kleinandnellis. The maximum power and the efficiency at maximum power examined in Sections 8.6.1 and 8.6.2 directly account for thermodynamic irreversibilities arising from temperature differences between the cycle and the external heat source and sink. However, in the development of these expressions, the power cycle itself has been assumed to operate reversibly. In this section the analysis is extended to include irreversibilities that are internal to the engine itself.

8.6.4 Application to other Cycles

The heat transfer limited Carnot cycle has an operating point that results in maximum power and the efficiency of cycle at maximum power is given by Eq. (8-163). The efficiency at maximum power is independent of the conductance of the heat exchangers and the capacitance rates of the fluid streams; it depends only on the temperatures of the external streams. Note that the maximum power point results in significant entropy generation or, equivalently, exergy destruction. Exergy destruction is necessary in order for any process to proceed at a finite rate. Equation (8-163) was shown to be applicable for an internally reversible Carnot cycle. However the general behavior remains similar, even when the Carnot cycle is irreversible. What makes these results of significant interest is that Eq. (8-163) and the other results from Sections 8.6.1 through 8.6.3 apply, at least approximately, to any of the power cycles that are studied in Sections 8.2 through 8.5.

Example 8.6-1 can be found online at www.cambridge.org/kleinandnellis. The example examines a Rankine cycle and shows that its behavior is consistent with the concepts discussed in this section.

REFERENCES

Curzon, F. L., and Ahlborn, B., "Efficiency of a Carnot Engine at Maximum Power Output," *Am. J. Phys.*, Vol. 43, pp. 22–24 (1975).

El-Wakil, M.M., *Nuclear Power Engineering*, McGraw-Hill, New York, (1962).

Ibrahim, O. M., Klein, S.A., and Mitchell J.W. "Optimum Heat Power Cycles For Specified Boundary Conditions", *ASME Journal of Power*, Vol. 113, No. 4, pp. 514–521 (1991).

Nellis, G. and Klein, S., *Heat Transfer*, Cambridge University Press, ISBN 978–0-521–88107-4 (2009).

Problems

The problems included here have been selected from a larger set of problems that are available from the website associated with this book (www.cambridge.org/kleinandnellis).

A. The Rankine cycle

8.A-1 The purpose of this problem is to determine the optimum pressure at which to reheat steam in the two-stage non-ideal Rankine cycle with reheat, shown in Figure 8-13(a). The cycle is operating with water. The hot reservoir temperature is $T_H = 825$ K and the cold reservoir is $T_C = 310$ K. The boiler pressure is $P_b = 6.11$ MPa. The pressure drop in the heat exchangers can be ignored. However, the condenser has an approach temperature difference of $\Delta T_{cond} = 5$ K, the boiler has an approach temperature difference of $\Delta T_b = 25$ K, and the reheater has an approach temperature difference of $\Delta T_{rh} = 20$ K. The efficiency of the pump is $\eta_p = 0.5$. The efficiencies of the high and low pressure turbines are $\eta_{HPt} = \eta_{LPt} = 0.9$.

a) Initially, assume that the reheat pressure is $P_{rh} = 1$ MPa. Determine the thermodynamic states at all points in the cycle and the efficiency of the power plant.
b) Plot the cycle efficiency as a function of the reheat pressure in order to identify the optimum reheat pressure.

8.A-3 Geothermal-based electrical power production is one alternative energy source that is attracting some attention. Cool water is pumped into the ground (several kilometers deep) where it is heated by geothermal energy and then returned to the surface. The hot water returning from the ground is used as the heat source to power some type of power plant. The down-side of geothermal energy is that heat is provided at low temperature (relative to burning coal or other, more conventional sources of energy). Even in the western U.S., where the geothermal resource is the best, the temperature of the ground does not exceed 250°C. In this problem you will try to identify an optimal working fluid for use in a low-temperature Rankine cycle that utilizes geothermal heat. In order to quickly change working fluids, define a string variable, F$, to be the name of the fluid; for example, if you want to use water then assign F$='Water'. Using this technique, each time you want to evaluate a property, use F$ to specify the substance. For example: h[1]=Enthalpy(F$,T=T[1],P=P[1]). Figure 8-4 illustrates a simple Rankine cycle. Assume that the components are ideal (i.e., reversible turbine and compressor and zero approach temperature difference for the boiler and condenser). Further, assume that the boiler pressure is selected so that the quality of the fluid leaving the turbine is sufficiently high so as to prevent damage, $x_3 = 0.9$.

a) Develop a model of the ideal Rankine cycle using water assuming $T_C = 30°C$ and $T_H = 200°C$. What is the efficiency of the cycle?
b) Plot the efficiency of the cycle as a function of T_H for the range of temperatures of interest for geothermal systems, $150°C < T_H < 250°C$. Overlay on your plot the efficiency for other potential working fluids (e.g., toluene, propylene, and ammonia).

8.A-6 In a geothermal system, water is pumped into the ground (several kilometers deep) where it is heated by geothermal energy and then returned to the surface to provide energy for a power plant. In 2005, the U.S. had 2851 MW of installed geothermal power production; most of this capacity is in California and Nevada. This problem examines the use of geothermal energy for power production with the Rankine cycle shown in Figure 8.A-6. The cycle employs both reheat and regeneration with two open feedwater heaters.

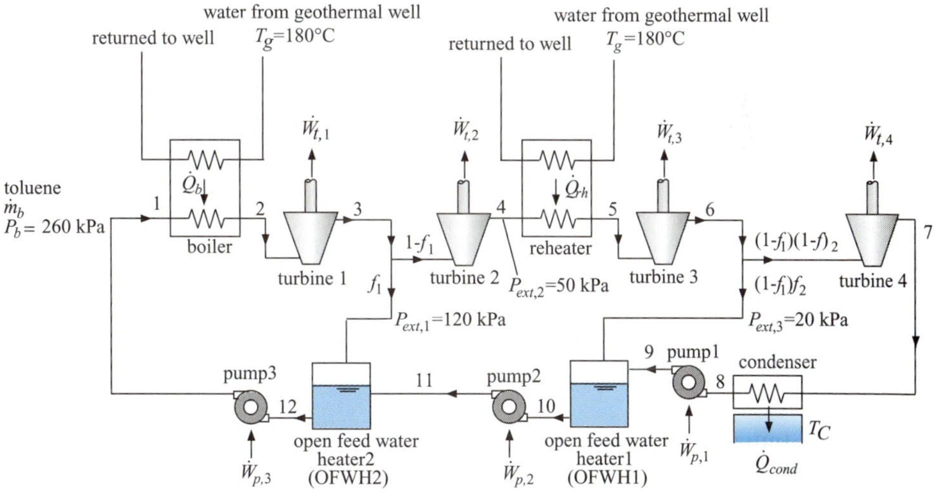

Figure 8.A-6. Schematic of the power plant cycle consisting of a Rankine cycle with reheat and regeneration using two open feedwater heaters.

The Rankine cycle uses toluene as the working fluid due to the low temperature of the heat source. The boiler pressure is $P_b = 260$ kPa. The water from the geothermal well is at $T_g = 180°C$. The plant rejects heat to a temperature reservoir at $T_C = 30°C$. The fluid is extracted at state 3 from turbine 1 at $P_{ext,1} = 120$ kPa and a fraction of the flow f_1 is fed to the open feedwater heater 1. The fluid is subsequently extracted at state 4 from turbine 2 at $P_{ext,2} = 50$ kPa and reheated.

The fluid is finally extracted at state 6 from turbine 3 at $P_{ext,3} = 20$ kPa and the fraction of the flow f_2 is fed to open feedwater heater 2. The remainder of the flow passes through turbine 4 to the condenser. The efficiencies of the four turbines are $\eta_{t,1} = 0.85$, $\eta_{t,2} = 0.86$, $\eta_{t,3} = 0.88$, and $\eta_{t,4} = 0.89$. The flow leaving the condenser is pumped to $P_{ext,3}$ with pump 1 having efficiency $\eta_{p,1} = 0.65$. Saturated liquid is pulled from open feedwater heater 1 and pumped to the first extraction pressure, $P_{ext,1}$, with pump 2 having efficiency $\eta_{p,2} = 0.67$. Saturated liquid is pulled from open feedwater heater 2 and pumped to the boiler pressure with pump 3 having efficiency $\eta_{p,3} = 0.69$. The approach temperature differences associated with the boiler and the reheater are $\Delta T_b = 15$ K and $\Delta T_{rh} = 10$ K, respectively. The pinch points for both of these heat exchangers occur at their warm end. Therefore, water leaves the boiler at $T_2 = T_{hf,in} - \Delta T_b$ and it leaves the reheater at $T_5 = T_{hf,in} - \Delta T_{rh}$. The approach temperature difference associated with the condenser is $\Delta T_{cond} = 5$ K. Neglect pressure loss for all of the heat exchangers.

a) Develop two procedures to facilitate the analysis. One procedure should be capable of analyzing any of the four turbines in the system and the other should be applicable to any of the three pumps in the system.
b) Determine each of the states associated with the cycle. Print out an Arrays table that includes the entropy, enthalpy, temperature, and pressure for each state. Plot your states on a *T-s* diagram for Toluene. Label each of your states.
c) Determine the efficiency of the cycle.
d) The total rate at which heat that can be extracted from the geothermal source is $\dot{Q}_g = 2.5$ MW. (This is the sum of the heat transfer to the boiler and reheater.) Determine the mass flow rate of toluene passing through the boiler and the net power produced by this power plant.
e) Determine the value of the electricity produced by the plant over a $time = 10$ year period. Assume that you can sell the electricity to the power company at a rate of $ec = 0.055$ \$/kW-hr and neglect the time value of money.
f) Determine the effectiveness of the boiler and the condenser.

The surface area required for the boiler and condenser are directly related to their effectiveneness. Your company has thoroughly tested heat exchangers operating with toluene and developed the following correlation for the surface area (A) as a function of the effectiveness (ε) and toluene mass flow rate ($\dot{m}$):

$$A = K_{HX}\dot{m}\left[-\ln\left(1-\varepsilon\right)\right]$$

where $K_{HX} = 170$ m^2-s/kg is an empirical constant that is appropriate for both the boiler and condenser. The cost of these heat exchangers scales linearly with their surface area according to:

$$Cost = C_{HX}\,A$$

where $C_{HX} = 50$ \$/m^2 is the cost coefficient that is also appropriate for both the boiler and condenser.

g) Determine the size (surface area) and cost of the boiler and the condenser.
h) You have estimated that the capital cost of the balance of the plant (the turbines, pumps, reheater, etc.) is $Cost_{mech} = 500\text{x}10^3$ \$. What is the net profit that you make over 10 years by building and operating the plant? Neglect the time value of money.
i) Plot the profit as a function of the approach temperature difference for the boiler for 2 K $< \Delta T_b <$ 30 K. You should see that there is an optimal value of the boiler approach temperature difference. Explain why this is true.

j) Use the Min/Max capability in EES to determine the optimal (from an economic standpoint) value of the condenser and boiler approach temperature differences. Note that you will need to provide reasonable bounds for these values (e.g., 2 K to 30 K).

8.A-9 A hot gas stream is available at $\dot{m}_{gas} = 10$ kg/s and $T_{gas,in} = 350°$C at ambient pressure. This gas has constant pressure specific heat capacity of $c_{P,gas} = 1.1$ kJ/kg-K. It has been proposed to use this gas to provide power with a simple Rankine cycle (Figure 8-4) that uses toluene as the working fluid. Assume that the turbine and pump efficiencies are 100%. Preliminary estimates indicate that the rate of heat transfer in the boiler can be computed according to:

$$\dot{Q}_B = \varepsilon_B \, \dot{m}_{gas} \, c_{P,gas} (T_{gas,in} - T_2)$$

where $\varepsilon_B = 0.45$ is the effectiveness of the boiler and T_2 is the saturation temperature at the boiler exit. Cooling water is available at $\dot{m}_w = 20$ kg/s at $T_{w,in} = 10°$C and $P_{atm} = 1$ atm. The condenser heat transfer rate is:

$$\dot{Q}_C = \varepsilon_C \, \dot{m}_w \, c_w \, (T_4 - T_{w,in})$$

where $\varepsilon_C = 0.70$ is the effectiveness of the condenser and T_4 is the saturation temperature at the condenser exit.

a) Prepare a plot of power versus efficiency for this Rankine cycle by varying the boiler pressure between 100 kPa and 4 MPa. Determine the maximum power that could be obtained by this plant and the corresponding efficiency from the plot.
b) Plot the optimum cycle on a *T*-*s* diagram for toluene.

8.A-11 Figure 8.A-11 illustrates a basic Rankine cycle using water as the working fluid.

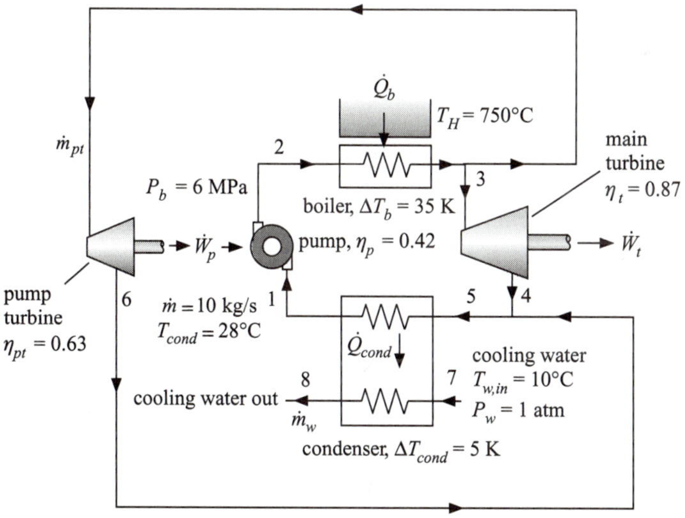

Figure 8.A-11. Rankine cycle with steam driven pump.

Saturated liquid at $T_{cond} = 28°$C leaves the condenser and enters the pump at state 1. The mass flow rate of water passing through the pump is $\dot{m} = 10$ kg/s and the pump efficiency is $\eta_p = 0.42$. The pressure of the water leaving the pump is $P_b = 6$ MPa. The high pressure liquid leaving the pump at state 2 enters the boiler. The boiler provides heat from a thermal reservoir at $T_H = 750°$C to the fluid and the boiler approach temperature difference is $\Delta T_b = 35$ K. Neglect any pressure loss in the boiler. The high pressure steam leaving the boiler at state 3 is split into two streams. A small fraction of the total mass flow rate is used to drive the pump turbine that is used to provide power to the pump. The mass flow rate that drives the pump turbine is $\dot{m}_{pt}$. The pump turbine efficiency is $\eta_{pt} = 0.63$. The remainder of the flow passes through the main power turbine which has efficiency $\eta_t = 0.87$. The two streams (the stream leaving the pump turbine at state 6 and the main turbine at state 4) are mixed upstream of the condenser so that water at state 5 enters the condenser. The condenser has an approach temperature difference of $\Delta T_{cond} = 5$ K. Neglect any pressure loss in the condenser or in the mixing process. Cooling water is used to provide the heat rejection for the power plant. The cooling water enters the condenser with temperature $T_{w,in} = 10°$C and pressure $P_w = 1$ atm.

a) Determine the power required by the pump, $\dot{W}_p$.
b) Determine the rate of heat transfer provided by the boiler, $\dot{Q}_b$.
c) Determine the mass flow rate required by the pump turbine, $\dot{m}_{pt}$.
d) Determine the rate of power produced by the main turbine, $\dot{W}_t$.
e) Determine the rate of heat transfer transferred in the condenser, $\dot{Q}_{cond}$.
f) Determine the mass flow rate of cooling water passing through the condenser, $\dot{m}_w$.
g) Determine the effectiveness of the condenser.
h) Check your answers by carrying out a total energy balance on the system.
i) Determine the efficiency of the Rankine power cycle.

8.A-12 A solar trough power plant is a Rankine cycle that uses solar energy as its heat input. Solar energy is focused by parabolic trough receivers onto a pipe that carries a heat transfer fluid, as shown in Example 8.2-1, Figure 1. The heat transfer fluid is heated as it flows through the field and then returns to the power plant. The fluid transfers heat to the working fluid of the power plant in order to provide the thermal energy that drives the power cycle, which is shown in Figure 8.A-12.

The heat transfer fluid leaves the field and enters the power plant at $T_{htf,in} = 288°$C. Some of the fluid enters the heater where it heats the working fluid for the cycle. Because the working temperature for the cycle is so low, water is not a very efficient working fluid. Instead, toluene is used in the cycle. The toluene leaves the heater at:

$$T_1 = T_{htf,in} - \Delta T_H$$

where $\Delta T_H = 20$ K is the heater approach temperature difference. The toluene enters the high pressure turbine at $P_1 = P_{high} = 1034$ kPa and is expanded to $P_2 = P_{reheat} = 250$ kPa. The efficiency of the high pressure turbine is $\eta_{HPt} = 0.81$. The remainder of the heat transfer fluid enters the re-heater where it re-heats the toluene leaving high pressure turbine to:

$$T_3 = T_{htf,in} - \Delta T_{RH}$$

where $\Delta T_{RH} = 20$ K is the re-heater approach temperature difference. The toluene leaving the re-heaters passes through the low pressure turbine which has an efficiency of $\eta_{LPt} = 0.78$. The condensing pressure is set so that the toluene leaving the condenser is saturated liquid at:

$$T_6 = T_{amb} + \Delta T_c$$

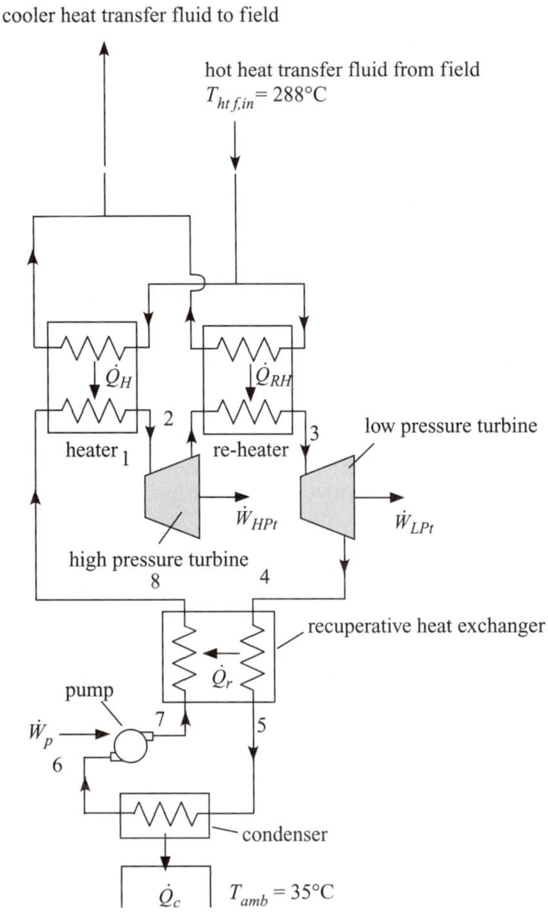

Figure 8.A-12. Rankine power cycle.

where $T_{amb} = 35^\circ$C is the ambient temperature and $\Delta T_c = 15$ K is the condenser approach temperature difference. The efficiency of the pump is $\eta_p = 0.6$. The toluene leaving the low pressure turbine is used to pre-heat the toluene before it reaches the heater using a recuperative heat exchanger. The temperature of the toluene leaving the recuperative heat exchanger and entering the condenser has temperature:

$$T_5 = T_7 + \Delta T_r$$

where $\Delta T_r = 20$ K is the recuperator approach temperature difference. Neglect the pressure drop through the heater, re-heater, recuperative heat exchanger, and condenser.

a) Set the pressures at each of the states shown in Figure 8.A-12.
b) Analyze the low pressure turbine. Determine the power per unit of mass flow rate of toluene obtained from the low pressure turbine ($\dot{W}_{LPt}/\dot{m}$) and the rate of entropy generation in the turbine per unit of mass flow rate ($\dot{S}_{gen,LPt}/\dot{m}$).
c) Analyze the pump. Determine the power per unit of mass flow rate of toluene required by the pump ($\dot{W}_p/\dot{m}$) and the rate of entropy generation in the pump per unit of mass flow rate ($\dot{S}_{gen,p}/\dot{m}$).

d) Analyze the recuperative heat exchanger. Determine the rate of heat transfer from the low pressure stream to the high pressure stream per unit of mass flow rate of toluene ($\dot{Q}_r/\dot{m}$) and the rate of entropy generation in the recuperator per unit of mass flow rate ($\dot{S}_{gen,r}/\dot{m}$).
e) Analyze the high pressure turbine. Determine the power per unit of mass flow rate of toluene obtained from the high pressure turbine ($\dot{W}_{HPt}/\dot{m}$) and the rate of entropy generation in the turbine per unit of mass flow rate ($\dot{S}_{gen,HPt}/\dot{m}$).
f) Determine the rate of heat transfer in the heater, re-heater, and condenser per unit of mass flow rate of toluene ($\dot{Q}_H/\dot{m}$, $\dot{Q}_{RH}/\dot{m}$, and $\dot{Q}_c/\dot{m}$, respectively).
g) Check your solution by drawing a system boundary that encompasses the cycle and showing that energy balances for the system.
h) Determine the thermal efficiency of the power plant.
i) Prepare a *T-s* plot for the cycle. Overlay your states (labeled) onto a *T-s* diagram for toluene. Make sure each process illustrated on the *T-s* diagram makes sense to you.
j) Plot the plant efficiency as a function of the reheat pressure, P_{reheat}. You should see that an optimal reheat pressure exists; use the Min/Max feature in EES to identify the optimal value of P_{reheat} and set the reheat pressure to its optimal value for the remainder of the problem.
k) Plot the plant efficiency as a function of the heat transfer fluid inlet temperature, $T_{htf,in}$ (i.e., the temperature of the heat transfer fluid returning from the solar field), for $250^\circ\text{C} < T_{htf,in} < 440^\circ\text{C}$.

B: Gas Turbine Cycles

8.B-1 Space-borne power systems often utilize radioisotope thermoelectric generators (RTGs) to produce power. RTGs release heat due to the radioactive decay of an isotope like plutonium-238. The half-life of plutonium-238 is about 90 years; therefore, the RTG produces heat at approximately a constant rate during the duration of the mission. The heat from an RTG is typically converted to electricity using a thermoelectric power production system because it has no moving parts and is therefore very reliable. However, the efficiency of such systems is very low (3-7%). Therefore, you are examining the alternative power production system shown in Figure 8.B-1.

The RTGs are used to provide heat to a reheated and recuperated gas turbine engine. The gas turbine engine is a closed cycle system as opposed to the typical air-breathing gas turbine engines used on earth. The working fluid is helium and the system rejects heat to space using a radiator that is at $T_{rad} = 290$ K. Helium enters the compressor at state 1 with pressure $P_{low} = 2$ atm and temperature $T_1 = T_{rad} + \Delta T_{rad}$ where $\Delta T_{rad} = 10$ K is the approach temperature difference associated with the heat exchanger that interfaces the radiator with the helium. The compressor pressure ratio is $PR = 5$ and its efficiency is $\eta_c = 0.72$. The helium leaves the compressor at state 2 and is preheated in a recuperative heat exchanger to state 3. The helium at state 3 is heated by an RTG and leaves the heater at $T_4 = T_H = 1000^\circ$C. The high pressure turbine has efficiency $\eta_{HPt} = 0.70$. The helium leaving the high pressure turbine at state 5 enters a reheater where it is reheated by another RTG to $T_6 = T_H = 1000^\circ$C. The low pressure turbine has efficiency $\eta_{LPt} = 0.68$. The pressure ratio across the two turbines is the same.

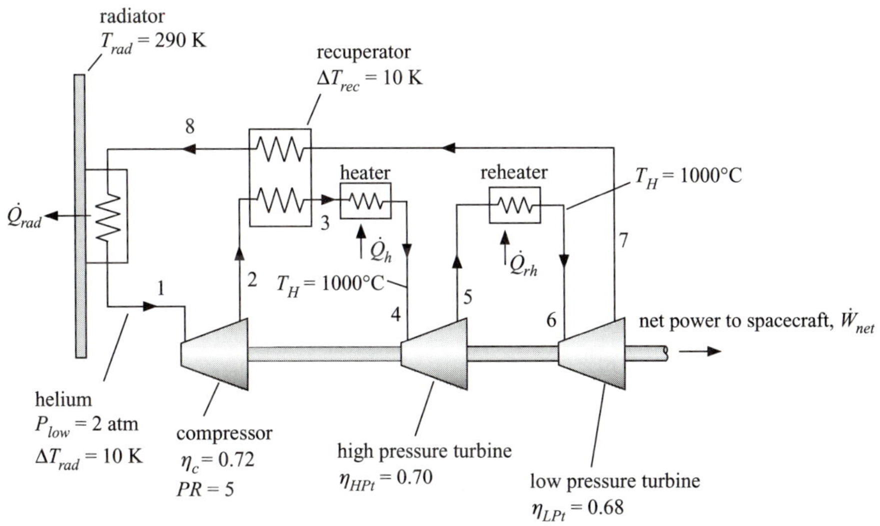

Figure 8.B-1: Gas turbine engine power by RTGs.

The helium is then cooled in the recuperative heat exchanger to state 8 and finally cooled by the radiator back to state 1. The pinch point for the recuperative heat exchanger occurs at the hot end and the approach temperature difference is $\Delta T_{rec} = 10$ K. You may neglect pressure drop in all of the heat exchangers. Model helium as an ideal gas (i.e., using the substance 'He' in EES).

a) Determine each of the state points in the cycle. Print out an Arrays table that contains the temperature, pressure, entropy, and enthalpy of each state point.
b) Use EES to generate a *T-s* diagram for helium. Overlay on this plot each of state points and label them.
c) Determine the efficiency of the cycle.
d) The spacecraft requires power at a rate of $\dot{W}_{net} = 250$ W. Determine the mass flow rate of helium required ($\dot{m}$) and the size of the RTG system required (the rate that heat is added to the cycle).
e) Determine the effectiveness of the recuperator (ε_{rec}).

The mass of the recuperator can be calculated according to:

$$m_{rec} = \frac{K_{HX}\,\dot{m}\,\varepsilon_{rec}}{(1-\varepsilon_{rec})}$$

where $K_{HX} = 200$ s is an empirical constant for the particular heat exchanger design being considered.

f) Determine the mass of the recuperator required.

The area of the radiator required can be calculated according to:

$$\dot{Q}_{rad} = \sigma\, A_{rad}\, T_{rad}^4$$

where $\sigma = 5.67\times10^{-8}$ W/m^2-K^4 is the Stefan-Boltzmann constant and $\dot{Q}_{rad}$ is the rate of heat transfer from the helium to the radiator (and therefore from the radiator to space). The mass of the radiator panel is calculated from:

$$m_{rad} = K_{rad}\, A_{rad}$$

where $K_{rad} = 1.85$ kg/m^2 is the mass of the panel per unit area.

g) Determine the area and mass of the radiator panel required.

The radiator and recuperator are the most massive parts of the system. The mass of the remainder of the system is relatively fixed and equal to $m_{misc} = 1.8$ kg.

h) Determine the total mass of the system.

System mass is the most important parameter for a space-borne power system. There are three free parameters that you, as a system designer, can vary in order to minimize the system mass: the recuperator performance (ΔT_{rec}), the radiator temperature (T_{rad}), and the pressure ratio (PR).

i) Plot the system mass as a function of ΔT_{rec} for 10 K $< \Delta T_{rec} <$ 150 K (with T_{rad} and PR set to their nominal values). You should see that an optimal value of ΔT_{rec} exists. Explain why this is true. You may want to generate additional plots to support your explanation.
j) Plot the system mass as a function of T_{rad} for 200 K $< T_{rad} <$ 350 K (with ΔT_{rad} and PR set to their nominal values). You should see that an optimal value of T_{rad} exists. Explain why this is true.
k) Use EES' multidimensional optimization capability to determine the optimal values of PR, T_{rad}, and ΔT_{rec} and the associated minimum possible system mass.

8.B-2 Figure 8.B-2 shows a gas turbine system with two-stage compression and inter-cooling and two-stage expansion with reheat. A recuperative heat exchanger is also provided to recover energy in the turbine exhaust. Air enters the compressor at $T_1 = 25°$C and atmospheric pressure, $P_1 = 1$ atm. Air exits the water-cooled intercooler at $T_3 = 45°$C. The fuel is methane, which has a heat of combustion $HC = 50{,}000$ kJ/kg. The turbine inlet temperature for both the high and low pressure turbines is $T_6 = T_8 = 1100°$C. The recuperator effectiveness is $\varepsilon = 0.40$ and the recuperator pinch point is at the warm end. The overall pressure ratio of the two compressors is $P_4/P_1 = 10$. Pressure losses in the heat exchange equipment are negligible.

Compressor and turbine performance is often expressed in terms of the isentropic efficiency. However, the isentropic efficiency varies with pressure ratio. A better representation of the performance is provided by the *polytropic efficiency*. The manufacturer has indicated that the polytropic efficiency of the compressors and turbines in the gas turbine engine are $\eta_{c,poly} = \eta_{t,poly} = 0.76$. The manufacturer indicates that the isentropic efficiency of the compressors, $\eta_{c,isen}$, can be related to the polytropic efficiency, $\eta_{c,poly}$, according to:

$$\eta_{c,isen} = \frac{(P_{out}/P_{in})^e - 1}{(P_{out}/P_{in})^{e/\eta_{c,poly}} - 1}$$

where P_{in} and P_{out} are the inlet and outlet pressures of the compressor and e is a property of the air that is defined in terms of the constant pressure and constant volume specific heat capacities:

$$e = 1 - \frac{c_v}{c_p} = \frac{k-1}{k}$$

where k is the ratio c_p/c_v. The isentropic efficiency of the turbines, $\eta_{t,isen}$ is related to the polytropic efficiency, $\eta_{t,poly}$, according to:

$$\eta_{t,isen} = \frac{1 - (P_{out}/P_{in})^{e\,\eta_{t,poly}}}{1 - (P_{out}/P_{in})^{e}}$$

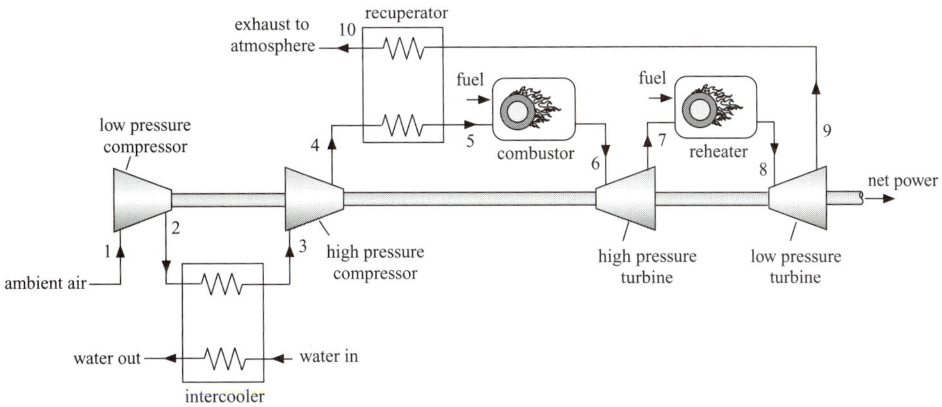

Figure 8.B-2: Gas turbine system with staged compression and expansion and regeneration.

Model air as an ideal gas and assume that the combustion products can be modeled as being air.

a) Assume that the intercooling and reheating pressures are both equal to $\sqrt{P_1\,P_4}$. Determine the net work per unit mass of inlet air, the combustor and reheater air fuel ratios, and the cycle efficiency. Print out the Arrays table that contains the properties (at least h, s, P, and T) at each state and develop a T-s diagram that shows the states.

b) Determine the intercooling and reheat pressures (i.e., P_2 and P_7) that maximize the net work. How well do these pressures compare with the ideal value determined assuming constant specific heat capacity, $\sqrt{P_1\,P_4}$?

8.B-3 A combined cycle power plant is shown in Figure 8.B-3. The turbine inlet temperature that can be tolerated for a gas turbine is much higher than for a steam turbine. Therefore, the Brayton cycle can operate at very high temperature. However, this leads to very hot gas leaving the gas turbine. Rather than waste this high temperature gas, it is first sent through a boiler in order to transfer heat to a Rankine cycle (which has higher efficiency than a Brayton cycle, but must operate at lower temperature). Finally, the low temperature gas leaving the boiler is used to produce steam for heating purposes. This cycle is an example of a combined heating and power (CHP) system.

Model the Brayton cycle assuming that the working fluid is air (i.e., use the substance 'Air' in EES). The mass flow rate of air is $\dot{m}_a = 73$ kg/s. Air enters the compressor at state 1 with $P_{atm} = 1$ atm and $T_{amb} = 20°$C. The compressor has a pressure ratio $PR = 7.5$ and an efficiency $\eta_c = 0.85$. Air enters the combustor and is heated to a temperature $T_H = 1250°$C. The turbine has an efficiency of $\eta_{t,1} = 0.87$. The air leaving the turbine enters the boiler at state 4 where it transfers heat to the water in the Rankine cycle. The ratio of the mass flow rate of water in the Rankine cycle to air in the Brayton cycle is $\dot{m}_w/\dot{m}_a = 0.1$. The boiler has an approach temperature difference of $\Delta T_b = 15$ K; therefore, the water leaving the boiler achieves temperature $T_9 = T_4 - \Delta T_b$. (Note that T_5 is not equal to T_8). The

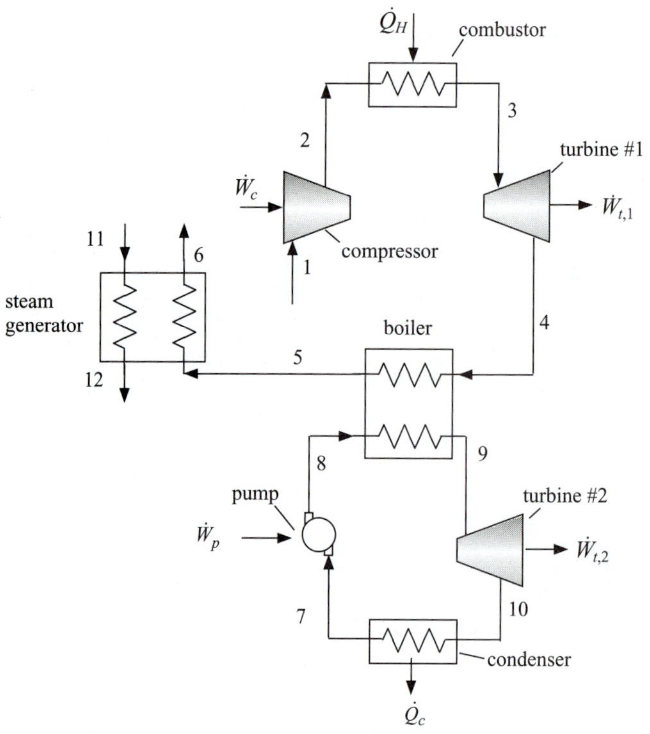

Figure 8.B-3: Combined cycle power plant.

air enters the steam generator at state 5. The air is used to heat water from the inlet state of $P_{11} = P_s = 5$ atm, $T_{11} = T_{w,in} = 20°$C to saturated vapor at $P_{12} = P_{11}$. The temperature of the air leaving the steam generator is $T_6 = T_{a,out} = 180°$C. In the Rankine cycle, the water enters the pump at state 7 as saturated liquid. The pump efficiency is $\eta_p = 0.65$ and the exit pressure of the pump is $P_{boiler} = 8.0$ MPa. The water enters the heat exchanger at state 8 where it is converted to steam by the heat transfer from the air. The water enters the steam turbine at state 9. The steam turbine has an efficiency of $\eta_{t,2} = 0.84$. The water enters the condenser at state 10. The condenser transfers heat from the water to a cooling water at $T_{amb} = 20°$C. The condenser approach temperature is $\Delta T_c = 5$ K; therefore, the water leaves the condenser at $T_c + \Delta T_c$. Neglect the pressure drop in all heat exchangers. Hint: analyze the compressor, combustor, and turbine #1. Then analyze the pump and boiler. Finally, analyze the steam generator.

a) Determine all of the state points for both cycles. You should have an Arrays table with at least the pressure, temperature, entropy and enthalpy of states 1 through 11.
b) Determine the mass flow rate of steam produced in the steam generator.
c) Check your solution by carrying out an overall energy balance on the entire cycle. (There are several ways to do this calculation.)
d) Determine the efficiency of the Brayton cycle, the efficiency of the Rankine cycle, and the efficiency of the combined cycle (relative to producing power do not include the value of the steam that is produced).
e) What is the net power produced by the cycle?

f) If natural gas is used to energize the cycle and the cost of natural gas is $NGc =$ 8 \$/million Btu, then determine the yearly fuel cost required to run the plant.

g) Determine the cost associated with producing the same amount of power using a conventional natural gas fired plant with efficiency $\eta_{conv} = 0.34$ and the same amount of steam using a natural gas fired boiler. What is the savings per year associated with using the combined cycle?

8.B-5 Many electric utilities rely on gas turbine generators to help supply their peak electrical demand during the summer months. Gas turbines operate less efficiently than the Rankine power cycle and this is one reason that electrical rates are higher during the summer than they are during the winter in many locations. If a utility is unable to supply their peak demand, they must purchase electricity from other utilities at a relatively high price. One way to ensure that the peak demand is met is to install extra gas turbine cycle generators. Another way, which may be more economical, is to cool the compressor inlet air using an ice storage unit, as shown in Figure 8.B-5. Cooling the inlet increases both the capacity and efficiency of the gas turbine cycle. For the specific case under consideration, the ice is produced by an electrically-driven vapor compression refrigeration cycle with $COP = 2.8$. The vapor compression cycle operates during off peak times and consumes electricity that has a utility cost of $ec = 0.04$ \$/kW-hr. The effective cost of the fuel (including a consideration for combustion efficiency) that is used to run the gas turbine is $fc = 5.5$ \$/GJ.

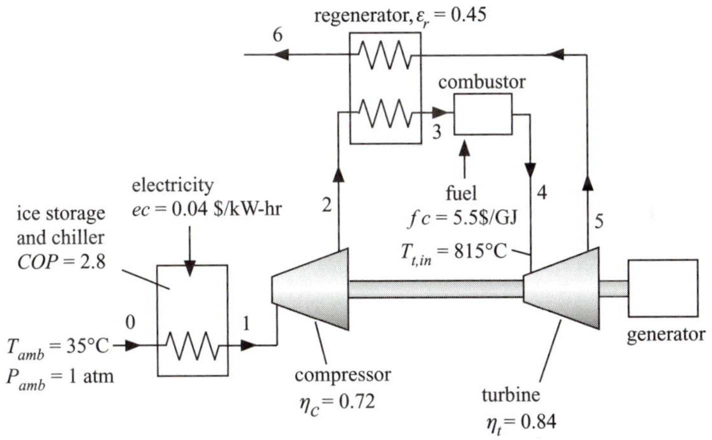

Figure 8.B-5: Gas turbine power system with inlet air cooled by ice storage system.

Assume the isentropic efficiencies of the compressor and turbine to be $\eta_c = 0.72$ and $\eta_t = 0.84$, respectively. The temperature at state 4 is $T_{t,in} = 815°C$. Pressure losses in the combustor and in each heat exchange operation are approximately $PL = 2\%$ of the inlet pressure to the component. The regenerator effectiveness is $\varepsilon_r = 0.45$ and the pinch point is at the hot end. With ice storage, the inlet air can be cooled to $T_{is} = 5°C$. Consider at peak day in which the ambient air is at $T_{amb} = 35°C$, $P_{amb} = 1$ atm. Assume the working fluid to be pure air. You may assume that the mass flow rate throughout the cycle is the same (i.e., neglect the small change in the mass flow rate associated with the addition of the fuel). State any other assumptions you employ.

a) Determine the pressure ratio that maximizes the efficiency of the gas turbine cycle assuming that there is no ice storage unit.
b) Determine the pressure ratio that maximizes the efficiency of the gas turbine cycle assuming that there is an ice storage unit that is sufficient to cool state 1 to 5°C. Compare the optimized efficiency with the result in part (a).
c) Estimate the utility cost (in $/kW-hr) of the electricity produced by the gas turbine system without and with the ice storage unit.
d) What is your assessment of the ice storage concept based on these calculations?

8.B-7 Supercritical carbon dioxide gas turbine cycles have been receiving a lot of attention as a possible replacement for the Rankine cycle in both nuclear and solar-power cycles. This problem will analyze the performance of a supercritical carbon dioxide cycle in a manner that includes the effects of the heat exchangers. The cycle is shown in Figure 8.B-7. Carbon dioxide steadily enters the compressor at state 4 with a volumetric flow rate of $\dot{V} = 0.10\ \text{m}^3/\text{s}$ and pressure of $P_4 = 7.5$ MPa after exiting the pre-cooler (the low temperature heat exchanger). The carbon dioxide is compressed adiabatically with a pressure ratio PR to state 1 where PR is defined as P_1/P_4. At this point, the carbon dioxide is heated in the primary heat exchanger by a hot fluid that has an entering temperature of $T_{H,in} = 600°\text{C}$ and capacitance rate of $\dot{C}_H = 18{,}000$ W/K. The heat transfer effectiveness of the primary heat exchanger is $\varepsilon_{phx} = 0.90$. The heated carbon dioxide is expanded in an adiabatic turbine to state 3. Energy is rejected in the pre-cooler to a cooling stream that enters the heat exchanger at $T_{C,in} = 25°\text{C}$ with a capacitance rate of $\dot{C}_C = 78{,}000$ W/K. The heat exchanger effectiveness of the pre-cooler is $\varepsilon_{pc} = 0.95$. Assume that the pressure losses in the heat exchangers are negligible. Also, in analyzing the heat exchangers, assume that the specific heat capacity of the carbon dioxide is constant throughout the heat exchanger and equal to the value that it has at the heat exchanger inlet. Because of the high pressures, carbon dioxide will not obey the ideal gas law. Therefore, use the fluid 'CarbonDioxide' rather than 'CO2' in your EES program.

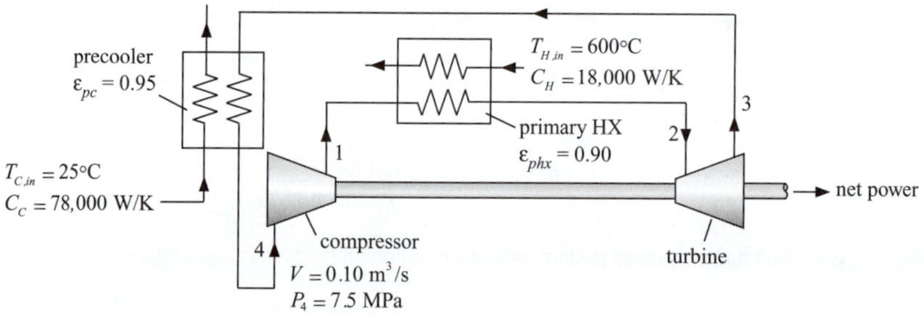

Figure 8.B-7: Supercritical carbon dioxide gas turbine cycle.

a) Assume that the compressor and turbine operate adiabatically and reversibly. Vary the pressure ratio between 1 and 25 and plot the net power output versus the pressure ratio, the cycle efficiency versus the pressure ratio, and the net power output versus the cycle efficiency
b) Repeat part (a), but assume that the turbine and compressor both have isentropic efficiencies of $\eta_t = \eta_c = 0.90$.
c) Summarize the results from your plots. What conclusions can you draw from these results?

8.B-8 A more detailed analysis of the supercritical carbon dioxide gas turbine considered in Problem 8.B-7 is needed. The cooling water is discharged to the environment (at T_{amb} = 25°C and P_{atm} = 1 atm) after leaving the pre-cooler. The hot fluid is returned to a storage unit after leaving the primary heat exchanger. Assume that the isentropic efficiencies of the compressor and turbine are both $\eta_c = \eta_t = 0.90$ and that both components operate adiabatically.

a) Identify the pressure ratio that provides the maximum power for this cycle. What is the maximum power?
b) Determine the specific exergy of the carbon dioxide at all states for the cycle determined in (a).
c) Determine the exergy destruction rate occurring in each system component for this cycle. Please put this information in a table showing a rank ordering of the destruction rates.
d) Determine the First and Second Law efficiencies for this cycle.
e) Based on your results, please provide a suggestion that will improve the performance of this cycle. Model your suggestion and determine the increase in the 2nd Law efficiency.

C: Reciprocating Engines

8.C-3 A supercharged engine uses a compressor to raise the pressure of the gas that is in the cylinder at the start of the cycle. The air-fuel ratio of an internal combustion engine is approximately constant. Therefore, increasing the amount of air in the cylinder increases the amount of fuel and thus the engine power. This purpose of this problem is to determine the power and efficiency behavior of a supercharged engine. The engine considered in this problem is the one described in Example 8.4-1 in which a polytropic model is used to represent the compression and expansion processes and the residual gas in the cylinder is considered. The polytropic exponents for compression and expansion are n_c = 1.35 and n_e = 1.40, respectively. The engine has N_{cyl} = 4 cylinders and operates at N = 3600 rev/min with a compression ratio of CR = 8.3 and an air-fuel ratio of AF = 16. The heat of combustion of the fuel is HC = 44 MJ/kg. The ambient air conditions are T_{amb} = 32°C and P_{atm} = 100 kPa. The supercharger is a compressor that is driven by the engine. It operates adiabatically with an isentropic efficiency of η_c = 0.74. The engine exhausts to atmospheric pressure. Assume that an aftercooler is not employed so that the air entering the engine through the intake valves is at the same pressure and temperature as the air exiting the compressor.

a) Prepare a plot of the power and efficiency of the supercharged engine as a function of the supercharger pressure ratio (defined as the ratio of the outlet to inlet pressures of the supercharger) for a range between PR = 1 (no supercharger) to PR = 5.

8.C-5 Approximating the combustion process in spark-ignition (Otto) or compression-ignition (Diesel) internal combustion engine cycles as a constant-volume or constant-pressure heat-addition process, respectively, is overly simplistic. A better (but slightly more complex) approach would be to model the combustion process in both the Otto and Diesel engines as a combination of two heat-transfer processes, one a constant volume and the other at constant pressure. The ideal cycle based on this concept is called the *dual cycle*. The purpose of this problem is to investigate the power efficiency tradeoffs for the dual cycle, which consists of the following five ideal processes with pure air as the working fluid:

1–2 adiabatic compression
2–3 constant volume heat addition
3–4 isobaric heat addition
4–5 adiabatic expansion
5–1 constant volume heat rejection

Another simplistic assumption in the traditional analyses of internal combustion engines is that the compression and expansion processes are isentropic. In your analyses, assume that the compression process (1-2) and expansion process (4-5) are adiabatic with isentropic efficiencies of $\eta_c = n_e = 0.80$. State 1 may be assumed to be air at $T_{amb} = 300$ K and $P_{atm} = 100$ kPa. The maximum temperature in the cycle is $T_{max} = 2350$ K. The compression ratio is defined as $CR = V_1/V_2$ and the pressure ratio is defined as $PR = P_3/P_2$. The maximum practical compression ratio is $CR_{max} = 30$. Develop a thermodynamic analysis of this cycle. You may ignore combustion processes and the mass of fuel in your cycle, but please consider the variation of the thermodynamic properties of air with temperature.

a) Determine the pressure ratio that will produce the maximum work per cycle for compression ratios ranging between 5 and 30.
b) The maximum work per cycle as a function of compression ratio.
c) The optimum pressure ratio as a function of compression ratio.
d) The maximum work per cycle as a function of the cycle thermal efficiency.

D: Power-Efficiency Tradeoffs

8.D-6 A continuous power source is needed for an application outside of Earth's atmosphere. A solar-driven Carnot cycle has been proposed for this purpose. The Carnot cycle operates between the collector and radiator temperatures. The rate of solar energy input to the collector, $\dot{Q}_c$, is governed by the Stefan-Boltzmann law. Assuming black surfaces,

$$\dot{Q}_C = \sigma \, A_C \, F_{C,S} \left(T_S^4 - T_C^4\right)$$

where $A_C = 2.5$ m^2 is the collector area, $F_{C,S} = 0.75$ is the collector-to-sun view factor, $T_S = 5760$ K is the equivalent blackbody temperature of the sun, T_C is the collector temperature and $\sigma = 5.67 \times 10^{-8}$ W/m^2-K^4 is the Stefan-Boltzmann constant. Similarly, the rate of heat rejection from the radiator, $\dot{Q}_R$, is given by:

$$\dot{Q}_R = \sigma \, A_R F_{R,E} \left(T_R^4 - T_E^4\right)$$

where $A_R = 5$ m^2 is the radiator area, $F_{R,E} = 1$ is the radiator-to-space view factor, T_R is the radiator temperature, and $T_E = 4$ K is the equivalent tempreature of space.

a) What is the maximum efficiency of this engine?
b) What is the maximum power output?
c) What is the efficiency at maximum power output?

8.D-8 An internally-reversible heat engine steadily operates on energy supplied from a gas stream having that has an entering temperature of $T_{H,in} = 1{,}000$°C and capacitance rate of $\dot{C}_H = 5{,}000$ W/K as shown in Figure 8.D-8. The engine rejects energy to a cooling stream which enters the heat exchanger at $T_{C,in} = 30°$C with a capacitance rate of $\dot{C}_C = 10{,}000$ W/K. The high temperature heat exchanger has an effectiveness of $\varepsilon_H = 0.70$ and the low temperature heat exchanger effectiveness is $\varepsilon_C = 0.80$. The heat transfer processes to the ideal gas working fluid in the engine occur at constant pressure, just as in the ideal Brayton cycle, as indicated on the temperature-entropy diagram. The product of the mass flow rate and specific heat of the working fluid is $\dot{C}_{wf} = 20{,}000$ W/K.

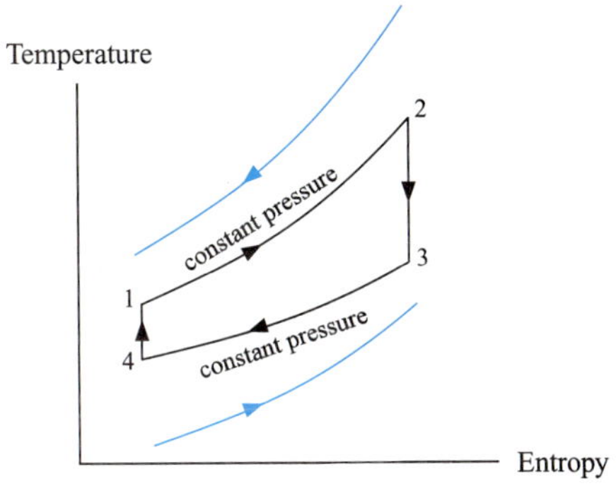

Figure 8.D-8: Temperature-entropy diagram for a heat engine operating as an ideal Brayton cycle.

a) Prepare a plot of the power output of this heat engine as a function of the cycle efficiency.
b) Calculate and plot the power output of a Carnot cycle (with the same external streams and heat exchangers) on the same axes.
c) What can you conclude from these plots?

8.D-9 This problem considers two Carnot cycles operated in series as shown in Figure 8.D-9. The energy source is a hot stream that enters the high temperature heat exchanger at $T_{H,in} = 1{,}000°$C at a capacitance rate (mass flow rate specific heat product) of $\dot{C}_H = 5{,}000$ W/K. The heat exchanger effectiveness is $\varepsilon_H = 0.70$ for each of the two cycles. A second heat exchanger is used for heat rejection. The external cooling stream enters this heat exchanger at $T_{C,in} = 30°$C with a capacitance rate of $\dot{C}_C = 10{,}000$ W/K. The effectiveness of the low temperature heat exchanger is $\varepsilon_C = 0.80$ for each cycle.

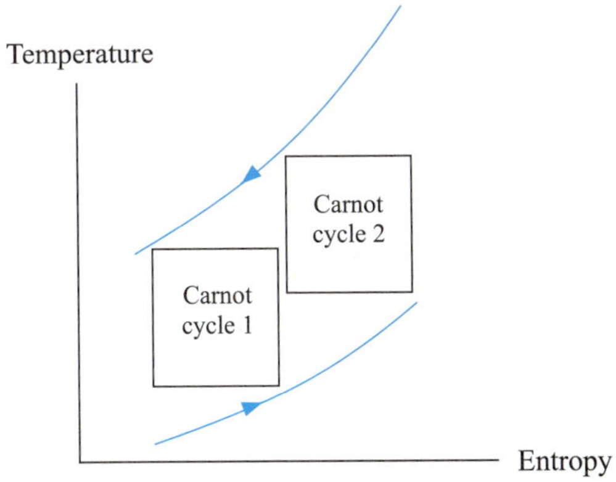

Figure 8.D-9: Power plant consisting of two Carnot cycles.

a) Determine the power output and efficiency of each stage and the overall system efficiency at the conditions in which the combined work of both cycles is maximized.
b) Compare the maximum power output and corresponding efficiency with the maximum power output and efficiency of a single stage Carnot cycle operated under the same conditions.

9 Refrigeration and Heat Pump Cycles

The Second Law states that heat transfers spontaneously in the direction of reduced temperature. However, it is possible to transfer thermal energy from lower to higher temperatures provided that an input of high quality energy (i.e., exergy – most often mechanical power but occasionally high temperature thermal energy) is also provided. This chapter discusses a variety of thermodynamic cycles that accomplish a transfer of thermal energy from low temperature to high temperature; these cycles are referred to as *refrigeration* or *heat pump* cycles depending on whether the intended application is cooling or heating, respectively. This chapter also provides a review of *refrigerants*, the working fluids that are used in refrigeration cycles. Many of the best refrigerants from a thermodynamic perspective have been banned because they were found to cause ozone degradation or other environmental problems when released into the atmosphere.

9.1 The Carnot Cycle

It is appropriate to begin a discussion of refrigeration and heat pump cycles with the Carnot cycle, just as we began the discussion of power cycles in Chapter 8. In theory, the Carnot power cycle can continuously convert heat from a constant temperature source into work with the highest possible efficiency. Operating in reverse, the Carnot cycle is the most efficient refrigeration/heat pump cycle operating between constant temperature thermal reservoirs. However, the same issues that prevent the Carnot power cycle from being practical are also present in the Carnot refrigeration cycle.

The Carnot refrigeration cycle is the reverse of a Carnot power cycle and consists of the four processes summarized in Table 9-1. A refrigeration system is shown Figure 9-1(a) and the Carnot refrigeration cycle is shown on a temperature-specific entropy plot in Figure 9-1(b). The working fluid in the cycle absorbs thermal energy from a source at temperature T_C as it is expanded isothermally from state 1 to state 2. This process provides the desired refrigeration effect in a refrigeration application. During the reversible, adiabatic compression process (from state 2 to state 3) the temperature of the working fluid increases from T_C to T_H. The working fluid transfers heat to the sink at T_H as it is compressed isothermally at temperature T_H from state 3 to state 4. For a heat pumping application, this high temperature thermal energy is the desired product.

Table 9-1: Processes in the Carnot refrigeration cycle.

1-2	Reversible, isothermal expansion at temperature T_C
2-3	Reversible, adiabatic compression from T_C to T_H
3-4	Reversible, isothermal compression at temperature T_H
4-1	Reversible, adiabatic expansion from T_H to T_C

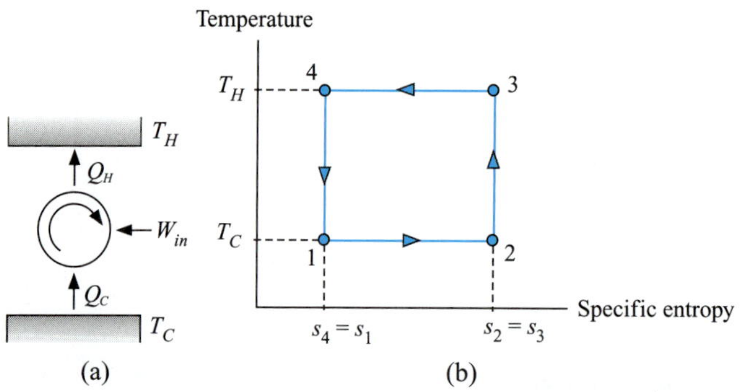

Figure 9-1: (a) Refrigeration system and (b) temperature-specific entropy diagram for a refrigeration Carnot cycle.

The working fluid is then expanded isentropically from T_H to T_C from state 4 to state 1 in order to complete the cycle.

The heat absorbed from the reservoir at T_C during process 1-2 is equal to:

$$Q_C = Q_{1-2} = T_C\, m\, (s_2 - s_1) \tag{9-1}$$

where m is the mass of the working fluid. The energy absorbed per mass of working fluid, $Q_{1\text{-}2}/m$, is the area under the line 1-2 in Figure 9-1(b). The heat rejected to the hot reservoir during process 3-4 is:

$$Q_H = Q_{3-4} = T_H\, m\, (s_3 - s_4) \tag{9-2}$$

The quantity $Q_{3\text{-}4}/m$ is the area under line 3-4 in Figure 9-1(b). An energy balance is used to determine W_{in}, the net work that must be supplied for one complete cycle:

$$W_{in} = Q_H - Q_C \tag{9-3}$$

Therefore, the work required per mass of fluid, W_{in}/m, is the area enclosed within the temperature-specific entropy diagram in Figure 9-1(b).

The index of performance for a refrigeration cycle is called its *Coefficient of Performance* (COP_R) and it is defined as the ratio of the cooling provided to the net work required.

$$COP_R = \frac{Q_C}{W_{in}} \tag{9-4}$$

For the Carnot refrigeration cycle considered here, the coefficient of performance for refrigeration can be obtained by substituting Eqs. (9-1) through (9-3) into Eq. (9-4).

$$COP_R = \frac{Q_C}{W_{in}} = \frac{Q_C}{Q_H - Q_C} = \frac{T_C\,(s_2 - s_1)}{T_H\,(s_3 - s_4) - T_C\,(s_2 - s_1)} \tag{9-5}$$

Recognizing that $(s_2 - s_1)$ must be equal to $(s_3 - s_4)$ leads to:

$$COP_R = \frac{T_C}{T_H - T_C} \tag{9-6}$$

The Carnot refrigeration cycle is reversible because it is composed of only reversible processes. Therefore, Eq. (9-6) provides the highest possible coefficient of performance that can be achieved for a refrigerator operating between two fixed temperatures T_C and T_H. Note that the coefficient of performance can and often does exceed unity. The coefficient of performance decreases as the temperature of the low temperature source decreases (i.e., as you provide refrigeration at lower temperatures) and as the difference between the high and low temperature thermal reservoirs increases. This behavior is generally observed for all refrigeration and heat pump cycles.

In the United States, where mixed units are used, the coefficient of performance for refrigeration is often presented as the *Energy Efficiency Rating* (EER), which is defined in the same manner as the coefficient of performance but with units of Btu/W-hr. The *Seasonal Energy Efficiency Rating* ($SEER$) is often provided for air-conditioning systems. The $SEER$ is measured under carefully specified test conditions that are consistent with the anticipated operating conditions that will be experienced by an air conditioner over a cooling season.

If the desired application is heating, rather than refrigeration, then the index of performance is the coefficient of performance for heating (COP_H), defined as:

$$COP_H = \frac{Q_H}{W_{in}} \tag{9-7}$$

Substituting Eqs. (9-1) through (9-3) into Eq. (9-7) provides the coefficient of performance for a Carnot heat pumping cycle:

$$COP_H = \frac{Q_H}{W_{in}} = \frac{Q_H}{Q_H - Q_C} = \frac{T_H\,(s_3 - s_4)}{T_H\,(s_3 - s_4) - T_C\,(s_2 - s_1)} = \frac{T_H}{T_H - T_C} \tag{9-8}$$

Comparing Eqs. (9-5) and (9-8) shows that COP_H will always be higher than COP_R because the heat transfer to the hot reservoir is always higher than the heat transfer from the cold reservoir by an amount that is equal to the work input to the cycle, as shown by Eq. (9-3). In the United States, the heating coefficient of performance is expressed in units of Btu/W-hr and it is called the heating performance factor (*HPF*) or heating season performance factor (*HSPF*). Often, the subscripts *R* and *H* on *COP* are not used because the meaning is clear by context.

Refrigeration was recognized centuries ago as a method for preserving food. However, the common refrigerator appliance is a relatively new invention that achieved widespread use in homes starting in the 1920's. Before the use of mechanical refrigeration cycles, people depended on ice to provide refrigeration. Ice was harvested from lakes during the winter and kept under sawdust insulation in order to keep it from melting. Stored in this manner, ice was available throughout the year in most locations. The ice was delivered to homes by an 'iceman' and used to keep foods cold in an insulated box called the 'icebox', which was the forerunner of the modern refrigerator. Customers paid for their ice by weight. This history has resulted in the definition of the *ton* as a unit of refrigeration capacity in the English system. If the iceman provided one ton (2000 lb_m or 907 kg) of ice and if that ice melted in 24 hours, the average rate of refrigeration, which is called the *cooling capacity*, would be 12,000 Btu/hr (3616 W). In refrigeration and heating applications, a *ton* is formally defined as 12,000 Btu/hr. The ton unit remains in widespread use today for expressing the heating and cooling capacity of refrigeration, air conditioning, and heat pump systems.

The Carnot refrigeration cycle shown in Figure 9-1(b) achieves the highest possible coefficient of performance. However, the Carnot refrigeration cycle is not practical for the same reasons that the Carnot power cycle studied in Section 8.1 is not practical. The Carnot refrigeration cycle requires an isothermal expansion and compression that will, for a single phase fluid, involve both heat and work simultaneously; it is difficult to efficiently accomplish these processes within the same component. An easy way to implement the constant temperature heat transfer processes that are required in the Carnot refrigeration cycle is to accomplish these processes under two-phase conditions. A Carnot cycle operating with ammonia as the working fluid is shown on a temperature–specific entropy diagram in Figure 9-2. The cycle accepts heat from a low temperature

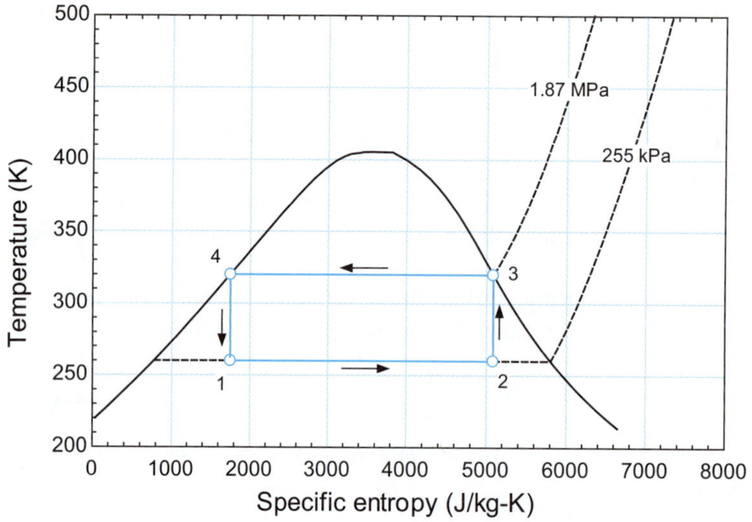

Figure 9-2: Carnot cycle operating under two-phase conditions using ammonia as the working fluid.

reservoir at $T_C = 260$ K and rejects heat to a high temperature reservoir at $T_H = 320$ K.

The heat addition and heat rejection processes (1-2 and 3-4) for the cycle shown in Figure 9-2 can be accomplished rather easily since the temperature of a pure two-phase fluid remains constant if the pressure is constant. However, process 2-3 is a compression process that involves both liquid and vapor phases and process 4-1 is a work-producing expansion process that also involves both liquid and vapor. From a practical perspective, it is difficult to design equipment that can efficiently handle significant amounts of both liquid and vapor phases during compression or expansion processes. The vapor compression cycle provides one method of dealing with these implementation problems while retaining many of the characteristics of the Carnot cycle shown in Figure 9-2. The vapor compression cycle moves the compression process out of the vapor dome into the superheated vapor region and replaces the isentropic expansion with an isenthalpic throttling process.

9.2 The Vapor Compression Cycle

The vapor compression cycle is the refrigeration cycle that is used for the vast majority of refrigeration and air-conditioning applications. Essentially every domestic, commercial, or industrial cooling application employs some derivative of the vapor compression refrigeration cycle discussed in this section.

9.2.1 The Ideal Vapor Compression Cycle

The simple vapor compression cycle retains many of the characteristics of the Carnot refrigeration cycle. However, the two-phase compression process and the two-phase, work-producing expansion processes shown in Figure 9-2 (processes 2-3 and 4-1, respectively) are impractical. Therefore, the vapor compression cycle moves the compression process out of the vapor dome. The need to compress a mixture of liquid and vapor in process 2-3 is eliminated by adding additional heat to the working fluid until state 2 is saturated vapor (or even superheated vapor). The work-producing expansion of a

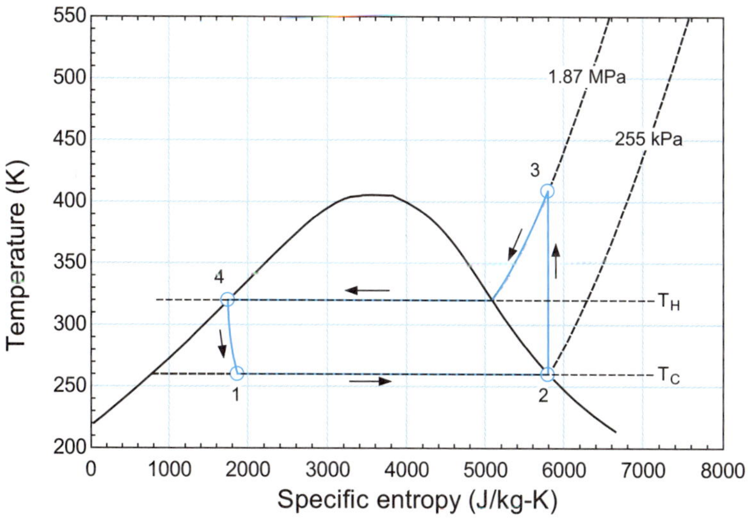

Figure 9-3: Ideal vapor compression cycle with ammonia as the working fluid operating between the same temperatures as the Carnot cycle shown in Figure 9-2; notice that the compression process (2-3) is moved out of the vapor dome and the expansion process (4-1) is isenthalpic rather than isentropic.

mixture of liquid and vapor in process 4-1 is removed and replaced by an isenthalpic throttling process. Figure 9-3 illustrates an ideal vapor compression cycle drawn to scale on a temperature-specific entropy diagram using ammonia as the working fluid that operates between the same temperatures as the Carnot cycle shown in Figure 9-2.

The vapor compression cycle is a closed, steady flow cycle in which the working fluid changes state as it circulates through various pieces of equipment. Figure 9-4 illustrates a schematic of the simple vapor compression cycle that is providing refrigeration to (i.e., receiving heat from) a low temperature reservoir at T_C while rejecting heat to a high temperature reservoir at T_H.

The liquid leaving the condenser at state 4 is saturated (or slightly sub-cooled). This liquid is expanded isenthalpically in a throttling device to state 1. The reduction in pressure leads to a reduction in temperature since the process occurs within the vapor dome. The two-phase mixture leaving the throttle enters the evaporator. Heat is transferred to the fluid from the refrigerated space (i.e., the temperature reservoir at T_C)

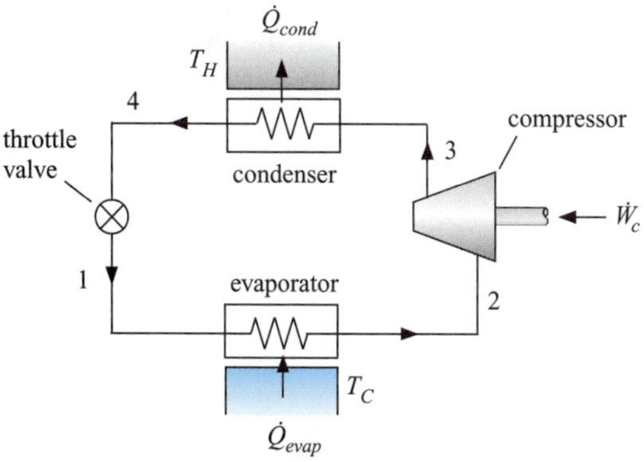

Figure 9-4: The simple vapor compression cycle.

at nearly constant pressure until the fluid completely evaporates. The saturation temperature of the refrigerant at the evaporating pressure will necessarily be lower than T_C. The rate of energy transfer to the refrigerant in the evaporator, $\dot{Q}_{evap}$, is the refrigeration capacity of the cycle. Saturated (or slightly superheated) vapor leaves the evaporator at state 2 and enters the compressor. The compressor increases the pressure of the refrigerant to the condensing pressure. The saturation temperature of the refrigerant at the condensing pressure will necessarily be greater than T_H. The vapor leaving the compressor at state 3 enters the condenser where heat is transferred from the fluid at nearly constant pressure until the refrigerant has completely condensed. The rate of heat transfer from the refrigerant in the condenser is $\dot{Q}_{cond}$. The refrigerant leaves the condenser at state 4, completing the cycle.

Unlike the Carnot cycle shown in Figure 9-2, the vapor compression cycle in Figure 9-3 is fairly easy to implement. Using a pure refrigerant, processes 1-2 and 3-4 are phase change heat transfer processes that occur at constant pressure and therefore they can be implemented with conventional heat exchangers. An added benefit of the vapor compression cycle is that the heat transfer coefficient for a fluid undergoing phase change is typically high and therefore the evaporator and condenser can be relatively small and inexpensive. The compression process in Figure 9-3 is entirely outside of the vapor dome and therefore it can be accomplished using a conventional compressor. The throttling process can be accomplished using any convenient type of flow resistance; for example, a capillary tube, porous plug, thermostatic expansion valve, or any other valve design that provides a resistance to the flow of refrigerant causing a reduction in its pressure.

The ideal vapor compression cycle is the simple cycle shown in Figure 9-4 operating with "perfect" components (i.e., components that individually have the highest possible performance). In order to specify the ideal vapor compression cycle, it is necessary to select the working fluid and the temperatures of the refrigerated space (T_C) and heat rejection sink (T_H).

```
$UnitSystem SI Radian Mass J kg K
T_H=320 [K]                          "high temperature reservoir"
T_C=260 [K]                          "low temperature reservoir"
R$='Ammonia'                         "refrigerant"
```

The condenser in an ideal vapor compression cycle is a perfect heat exchanger. The term ideal implies that there is no pressure loss associated with the flow through the device. It also implies that the temperature of the refrigerant leaving the condenser will be equal to the temperature of the hot reservoir; i.e., the approach temperature difference of the heat exchanger will be zero. State 4, the state of the fluid leaving the condenser, is fixed by the temperature, $T_4 = T_H$, and quality, $x_4 = 0$. The remaining properties (P_4, h_4, and s_4) can be obtained with this information.

```
"State 4, Condenser exit/Throttle inlet"
T[4]=T_H                                 "temperature"
x[4]=0 [-]                               "quality"
s[4]=entropy(R$,T=T[4],x=x[4])           "specific entropy"
P[4]=pressure(R$,T=T[4],x=x[4])          "pressure"
h[4]=enthalpy(R$,T=T[4],x=x[4])          "specific enthalpy"
```

The evaporator in the ideal vapor compression cycle is also a perfect heat exchanger. Therefore, the saturated vapor leaving the evaporator is at the temperature of the cold reservoir; again, the approach temperature difference is zero. State 2, the state of the fluid leaving the evaporator, is fixed by the quality and temperature, $x_2 = 1$ and $T_2 = T_C$. The remaining properties (h_2, s_2, v_2 and P_2) are determined knowing the quality and temperature at state 2.

```
"State 2, Evaporator exit/Compressor inlet"
T[2]=T_C                                  "temperature"
x[2]=1 [-]                                "quality"
s[2]=entropy(R$,T=T[2],x=x[2])            "specific entropy"
P[2]=pressure(R$,T=T[2],x=x[2])           "pressure"
h[2]=enthalpy(R$,T=T[2],x=x[2])           "specific enthalpy"
v[2]=volume(R$,T=T[2],x=x[2])             "specific volume"
```

The compressor in the ideal vapor compression cycle is a reversible and adiabatic device. Therefore, an entropy balance on the compressor is:

$$s_3 = s_2 \tag{9-9}$$

State 3 is fixed by the specific entropy and pressure, $P_3 = P_4$ (recall that there is no pressure loss across the ideal condenser). The remaining properties at state 3 are determined (h_3 and T_3).

```
"State 3, Compressor exit/Condenser inlet"
s[3]=s[2]                                 "specific entropy"
P[3]=P[4]                                 "pressure"
h[3]=enthalpy(R$,s=s[3],P=P[3])           "specific enthalpy"
T[3]=temperature(R$,s=s[3],P=P[3])        "temperature"
```

The throttle is an isenthalpic device. An energy balance on the throttle is:

$$h_1 = h_4 \tag{9-10}$$

State 1 is specified by the specific enthalpy and pressure, $P_1 = P_2$ (since there is no pressure loss across the ideal evaporator). The remaining properties at state 1 are determined (s_1 and T_1).

```
"State 1, Throttle exit/Evaporator inlet"
h[1]=h[4]                                 "specific enthalpy"
P[1]=P[2]                                 "pressure"
s[1]=entropy(R$,h=h[1],P=P[1])            "specific entropy"
T[1]=temperature(R$,h=h[1],P=P[1])        "temperature"
```

Energy balances on the condenser, evaporator, and compressor are:

$$\frac{\dot{Q}_{cond}}{\dot{m}} = h_3 - h_4 \tag{9-11}$$

$$\frac{\dot{Q}_{evap}}{\dot{m}} = h_2 - h_1 \tag{9-12}$$

$$\frac{\dot{W}_c}{\dot{m}} = h_3 - h_2 \tag{9-13}$$

```
"Energy balances"
Q_dot_cond\m_dot=h[3]-h[4]        "condenser"
Q_dot_evap\m_dot=h[2]-h[1]        "evaporator"
W_dot_c\m_dot=h[3]-h[2]           "compressor"
```

Note that the backlash character (\) is part of EES variable name so, for example, Q_dot_cond\m_dot is a variable that represents the heat transfer rate in the condenser divided by the mass flow rate of refrigerant. The mass flow rate can be selected in order to provide a required rate of refrigeration, $\dot{Q}_{evap}$:

$$\dot{Q}_{evap} = \left(\frac{\dot{Q}_{evap}}{\dot{m}}\right)\dot{m} \tag{9-14}$$

The rate of heat rejection in the condenser is determined:

$$\dot{Q}_{cond} = \left(\frac{\dot{Q}_{cond}}{\dot{m}}\right)\dot{m} \tag{9-15}$$

The product of the compressor work transfer per mass and the mass flow rate provides the rate of power consumed by the compressor.

$$\dot{W}_c = \left(\frac{\dot{W}_c}{\dot{m}}\right)\dot{m} \tag{9-16}$$

For a 10 ton system, the mass flow rate and energy transfer rates are calculated according to:

```
Q_dot_evap_ton=10 [ton]                      "refrigeration capacity, in ton"
Q_dot_evap=Q_dot_evap_ton*convert(ton,W)     "refrigeration capacity"
Q_dot_evap=Q_dot_evap\m_dot*m_dot            "mass flow rate"
Q_dot_cond=Q_dot_cond\m_dot*m_dot            "condenser heat transfer"
W_dot_c=W_dot_c\m_dot*m_dot                  "compressor power"
```

which results in $\dot{m} = 0.0344$ kg/s and $\dot{W}_c = 10.43$ kW. The coefficient of performance for refrigeration is:

$$COP_R = \frac{\dot{Q}_{evap}}{\dot{W}_c} \tag{9-17}$$

The *EER* for the cycle is obtained from the coefficient of performance using a unit conversion.

```
COP=Q_dot_evap/W_dot_c              "coefficient of performance"
EER=COP*convert(-,Btu/hr-W)         "energy efficiency rating"
```

Solving these equations leads to $COP = 3.371$ and $EER = 11.5$ Btu/hr-W. Entropy balances can be written for the throttle, condenser, evaporator, and compressor:

$$\dot{S}_{gen,v} = \dot{m}\,(s_1 - s_4) \tag{9-18}$$

$$\dot{S}_{gen,cond} = \frac{\dot{Q}_{cond}}{T_H} + \dot{m}\,(s_4 - s_3) \tag{9-19}$$

$$\frac{\dot{Q}_{evap}}{T_C} + \dot{S}_{gen,evap} = \dot{m}\,(s_2 - s_1) \tag{9-20}$$

$$\dot{S}_{gen,c} = \dot{m}\,(s_3 - s_2) \tag{9-21}$$

```
"Entropy balances"
S_dot_gen_v=m_dot*(s[1]-s[4])                       "valve"
S_dot_gen_cond=Q_dot_cond/T_H+m_dot*(s[4]-s[3])     "condenser"
Q_dot_evap/T_C+S_dot_gen_evap=m_dot*(s[2]-s[1])     "evaporator"
S_dot_gen_c=m_dot*(s[3]-s[2])                       "compressor"
```

which show that $\dot{S}_{gen,v} = 4.15$ W/K and $\dot{S}_{gen,cond} = 3.09$ W/K. The entropy generation in the evaporator and compressor are zero (to within the precision of the properties and the calculation) for the ideal vapor compression cycle. The perfect compressor is a reversible device. The evaporator transfers heat from a thermal reservoir at T_C to a refrigerant that is also always at T_C; therefore, the evaporator in the ideal vapor compression cycle is a reversible device. Entropy is generated in the valve, as is evident in Figure 9-3 by the increase in specific entropy from state 4 to state 1. Entropy is generated in the condenser due to the so-called "superheat horn" that is also evident in Figure 9-3. The vapor leaving the compressor is superheated and therefore heat transfer in the condenser occurs initially from a refrigerant with a temperature higher than T_H to a thermal reservoir at T_H. This heat transfer through a temperature gradient generates entropy. The operation of an ideal vapor compression cycle will inevitably result in the generation of entropy; the cycle is not reversible even though all of the components are perfect. The coefficient of performance for the reversible Carnot refrigeration cycle operating between the same temperatures is obtained from Eq. (9-5), repeated below:

$$COP_{Carnot} = \frac{T_C}{T_H - T_C} \tag{9-22}$$

```
COP_Carnot=T_C/(T_H-T_C)          "COP of a reversible Carnot refrigeration cycle"
```

and provides $COP_{Carnot} = 4.33$, which is significantly higher than the COP of the ideal vapor compression cycle.

The entropy generation that occurs within a refrigeration cycle is associated with "lost work" or exergy destruction, just as it is for a power cycle. In this case, the "lost power" corresponds to power required to run the cycle that is above and beyond the power that would be required to run a reversible refrigeration cycle that provides the same rate of refrigeration and operates between the same temperatures. The theoretical power required by the reversible Carnot cycle is given by:

$$\dot{W}_{rev} = \frac{\dot{Q}_{evap}}{COP_{Carnot}} \tag{9-23}$$

```
W_dot_rev=Q_dot_evap/COP_Carnot     "power required by a reversible refrigeration cycle"
```

which results in $\dot{W}_{rev} = 8.1$ kW. The "lost power" or exergy destruction rate is therefore:

$$\dot{W}_{lost} = \dot{W}_c - \dot{W}_{rev} \tag{9-24}$$

```
W_dot_lost=W_dot_c-W_dot_rev        "lost power"
```

which results in $\dot{W}_{lost} = 2.3$ kW. The product of the dead state temperature (which is T_H, the temperature of the surroundings to which heat is rejected) and the rate of entropy generation must be equal to the "lost power":

$$\dot{W}_{lost} = T_H\,\dot{S}_{gen} \tag{9-25}$$

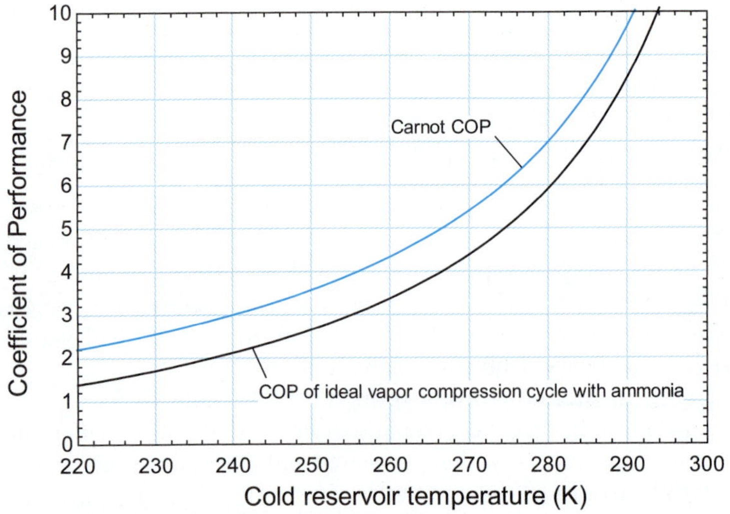

Figure 9-5: Coefficient of performance of the ideal vapor compression cycle using ammonia as a function of the cold reservoir temperature for $T_H = 320$ K. Also shown is the Carnot COP.

where $\dot{S}_{gen}$ is the total rate of entropy generation in the cycle:

$$\dot{S}_{gen} = \dot{S}_{gen,v} + \dot{S}_{gen,cond} + \dot{S}_{gen,evap} + \dot{S}_{gen,c} \tag{9-26}$$

```
S_dot_gen=S_dot_gen_v+S_dot_gen_cond+S_dot_gen_evap+S_dot_gen_c
                                          "total rate of entropy generation"
W_dot_lost_2=S_dot_gen*T_H                "lost power – check"
```

which also results in $\dot{W}_{lost} = 2.3$ kW.

Effect of Refrigeration Temperature

The coefficient of performance for a Carnot refrigeration system, Eq. (9-22), decreases as the refrigeration temperature, T_C, decreases. The coefficient of performance for any real refrigeration system will also decrease as T_C is reduced, usually by more than the Carnot efficiency is reduced. Figure 9-5 illustrates the COP as a function of T_C for the ideal vapor compression cycle using ammonia with $T_H = 320$ K. Also shown is the Carnot COP, predicted using Eq. (9-5).

Another negative effect of operating a vapor compression cycle at low refrigeration temperature is related to the specific volume of the vapor entering the compressor. Most refrigeration compressors are volumetric devices; that is, they process a given volume of refrigerant per unit of time. The size of the compressor dictates the volume processed per cycle and the speed of the compressor dictates the time per cycle. The displacement rate of the compressor, $\dot{V}_{disp}$, is a geometric quantity that can be computed based on the details of the compressor. For example, in a reciprocating compressor, the displacement rate is the ratio of the swept volume of the compressor cylinder to the time per cycle. In the absence of any leakage or other inefficiencies, the displacement rate is equal to the volumetric flow rate produced at the suction port of the compressor (i.e., the volumetric flow rate at state 2 in Figure 9-4). The effects of leakage, pressure drop across intake and exhaust valves, clearance volume, and other inefficiencies reduce the suction volumetric

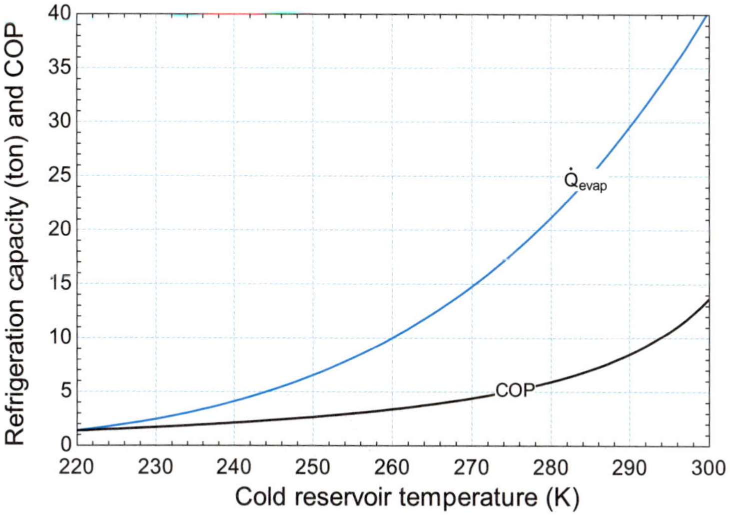

Figure 9-6: Refrigeration capacity and *COP* of the ideal vapor compression cycle using ammonia as a function of the cold reservoir temperature for $T_H = 320$ K, $\eta_{vol} = 0.8$, and $\dot{V}_{disp} = 42.9$ cfm.

flow rate relative to the displacement rate. The volumetric efficiency of the compressor, η_{vol}, is defined as:

$$\eta_{vol} = \frac{\dot{V}_{suction}}{\dot{V}_{disp}} \tag{9-27}$$

where $\dot{V}_{suction}$ is the actual suction volumetric flow rate produced by the compressor. The mass flow rate in the refrigeration cycle is therefore:

$$\dot{m} = \frac{\eta_{vol}\,\dot{V}_{disp}}{v_2} \tag{9-28}$$

where v_2 is the specific volume of the refrigerant at the inlet to the compressor. If the volumetric efficiency of the compressor is known, then it is possible to determine the displacement rate of the compressor that is required by the 10 ton ammonia system modeled in the previous section.

```
eta_vol=0.8 [-]                                    "volumetric efficiency"
m_dot=eta_vol*V_dot_disp/v[2]                      "compressor displacement"
V_dot_disp_cfm=V_dot_disp*convert(m^3/s,cfm)       "in cfm"
```

Solving these equations leads to $\dot{V}_{disp} = 0.0203$ m^3/s (42.9 ft^3/min). A more realistic model of a specific refrigeration system would fix the displacement rate of the compressor rather than the refrigeration capacity:

```
{Q_dot_evap_ton=10 [ton]}                 "refrigeration capacity, in ton"
V_dot_disp_cfm=42.9 [cfm]                 "fixed displacement compressor"
```

Figure 9-6 illustrates the refrigeration capacity of the ideal vapor compression cycle as a function of the refrigeration temperature using ammonia with $T_H = 320$ K, $\eta_{vol} = 0.8$, and $\dot{V}_{disp} = 42.9$ cfm (i.e., a fixed compressor that is suitable for providing $\dot{Q}_{evap} = 10$ ton of refrigeration at $T_C = 260$ K).

Figure 9-7: Household refrigerator (picture courtesy of Embraco, visit http://www.embraco.com/ingles/lancamentos.htm to see an animation of the refrigeration circuit).

Notice that the refrigeration capacity decreases with refrigeration temperature more quickly than the *COP* of the cycle, which is also shown in Figure 9-6. For example, as T_C is reduced from 300 K to 220 K the refrigeration capacity is reduced by a factor of approximately 37 whereas the coefficient of performance is reduced by a factor of approximately 10. The reason that the refrigeration capacity decreases so dramatically is related to the increase in the compressor inlet specific volume. The specific volume of the saturated vapor entering the compressor can be estimated using the ideal gas law:

$$v_2 \approx \frac{R\,T_{sat}}{P_{sat}} \tag{9-29}$$

where R is the ideal gas constant and T_{sat} and P_{sat} are the saturation temperature and pressure, respectively. The saturation pressure decreases exponentially with temperature and therefore the denominator of Eq. (9-29) decreases much more quickly than the numerator. This situation leads to a rapid increase in the specific volume at the compressor inlet as the refrigeration temperature is reduced. As a result, a system operating with a fixed compressor size (i.e., a fixed $\dot{V}_{disp}$) will rapidly lose refrigeration capacity as the refrigeration temperature is reduced because the refrigerant mass flow rate is reduced. A low vapor specific volume (i.e., high density) at low temperatures is a desirable property for a refrigerant.

9.2.2 The Non-Ideal Vapor Compression Cycle

The vapor compression cycle is employed in essentially all of the refrigeration and air-conditioning equipment that is used today. Consider, for example, your refrigerator (Figure 9-7). The compressor is usually located at the bottom and it is visible from behind the unit. All modern refrigerators employ a hermetic design in which the compressor and the electric motor used to energize the compressor are both contained within a black metal canister that is welded shut. The hermetic design eliminates the need for seals around the shaft between the compressor and the motor and the refrigerant is also used as the motor coolant. The refrigeration cycle can be charged at the factory. Leaks seldom occur and the result is a very reliable appliance that rarely, if ever, requires servicing.

The location of the condenser depends on the age of the refrigerator. Older refrigerators may employ a free-convection heat exchanger located at the back of the refrigerator. This design depends on the buoyancy-induced flow of room air past the condenser in order to cool the refrigerant. Newer refrigerators employ a fan-cooled condenser that is usually located at the base of the refrigerator. The fan provides a relatively high flow rate of air past the condenser heat exchanger, resulting in a large heat transfer coefficient and as a consequence, a lower condensing pressure. The reduced condensing pressure leads to improved performance compared to a free-convection condenser. The evaporator in most refrigerators is located in the freezer. You can usually see the refrigerant flow channels at the bottom of the freezer compartment. Although the temperature within the refrigerator is generally 4°C (about 39°F), all of the cooling in most refrigerators is provided at the temperature required for the freezer, which is generally about −18°C (0°F). When the refrigerator requires cooling, air is exchanged between the freezer and refrigerator until the temperature in the refrigerator reaches the desired value. The refrigeration cycle is engaged when the freezer temperature exceeds its thermostat setting. Surprisingly, modern refrigerators employ electric heaters to provide automatic defrost capability in the freezer and also to prevent condensation and ice build-up around the doors of the freezer.

Almost all modern automobiles include air conditioning units as standard equipment. The air conditioning compressor is driven by the engine with a belt. The belt is continuously spinning the pulley wheel that is connected to the compressor. A magnetic clutch between the spinning pulley and the compressor is engaged when you turn the air conditioning system on. The condenser resembles the car radiator and is located in front of the radiator where it can benefit from the lowest outdoor air temperatures when the vehicle is moving. A fan is included to provide air flow through the condenser and radiator when the vehicle is stationary. The evaporator is located in the air ducts behind the dashboard. A fan forces air past the evaporator so that cool air blows out of the dashboard vents and into the cabin of the vehicle.

A residence equipped with a central air-conditioning unit will often place the condenser and compressor in an outdoor unit that is located adjacent to the house. Refrigerant lines carry the refrigerant to an evaporator that is located in the air ducts within the building. A window air-conditioner includes the same components collocated in a single, window-sized enclosure. The compressor and condenser are located in the portion of the air-conditioner that extends outside of the window and a fan blows outdoor air across the condenser unit. The evaporator is located in the portion of the air-conditioner that extends within the room and a blower pulls room air across the evaporator.

Industrial and commercial refrigeration and air-conditioning systems tend to be very large, custom-designed systems. The condensers for such systems are often located on the roof of the facility; the compressors are located in a separate equipment room, and the evaporators are located throughout the building at various points where refrigeration is required.

Useful analyses of any of these real refrigeration systems will require the components to be modeled in greater detail than is described in Section 9.2.1. For example, the compressor will not be isentropic and the heat exchangers must be characterized by both pressure loss and approach temperature differences. Also, the refrigerant leaving the evaporator is usually slightly superheated in order to ensure that no liquid enters the compressor. The difference between the temperature of the refrigerant leaving the evaporator and the saturation temperature at the evaporating pressure is referred to as the superheat or degree of superheat, ΔT_{sh}. The refrigerant leaving the condenser may also be slightly subcooled; the difference between the saturation temperature at the condensing pressure and the temperature of the refrigerant leaving the condenser is referred to as the subcooling or the degree of subcooling, ΔT_{sc}. The analysis of a non-ideal vapor compression is illustrated in Example 9.2-1.

EXAMPLE 9.2-1: INDUSTRIAL FREEZER

The air temperature within a freezer room must be maintained at $T_C = -10^{\circ}$F using the simple vapor compression cycle shown in Figure 9-4 with ammonia as the working fluid. The evaporator approach temperature difference is $\Delta T_{evap} = 5^{\circ}$F and the refrigerant leaving the evaporator has a $\Delta T_{sh} = 4^{\circ}$F degree of superheat. The refrigeration cycle rejects heat to ambient air at $T_H = 78^{\circ}$F. The condenser approach temperature difference is $\Delta T_{cond} = 10^{\circ}$F and the refrigerant leaving the condenser has a $\Delta T_{sc} = 4^{\circ}$F degree of subcooling. The compressor has an isentropic efficiency of $\eta_c = 0.78$, a volumetric efficiency of $\eta_{vol} = 0.75$, and a displacement rate of $\dot{V}_{disp} = 35$ cfm. You may neglect the pressure drop in the heat exchangers.

a) Prepare temperature-specific entropy and pressure-specific enthalpy plots for the cycle.

States are numbered as indicated in Figure 9-4. The inputs are entered in EES:

```
$UnitSystem SI K Pa J Mass Radian
T_C=converttemp(F,K,-10 [F])                         "freezer air temperature"
DT_evap=5 [F]*convert(F,K)                           "evaporator approach temperature difference"
DT_sh=4 [F]*convert(F,K)                             "degree of superheat"
T_H=converttemp(F,K,78 [F])                          "outdoor air temperature"
DT_cond=10 [F]*convert(F,K)                          "condenser approach temperature difference"
DT_sc=4 [F]*convert(F,K)                             "degree of subcooling"
eta_c=0.78 [-]                                       "compressor efficiency"
V_dot_disp_cfm=35 [cfm]                              "compressor displacement, in cfm"
V_dot_disp=V_dot_disp_cfm*convert(cfm,m^3/s)         "compressor displacement"
eta_vol=0.75 [-]                                     "compressor volumetric efficiency"
R$='Ammonia'                                         "refrigerant"
```

The temperature of the refrigerant leaving the condenser is specified by the condenser approach temperature difference:

$$T_4 = T_H + \Delta T_{cond}$$

The condensing temperature (i.e., the saturation temperature at the condensing pressure) is specified by the known degree of subcooling:

$$T_{cond} = T_4 + \Delta T_{sc}$$

The condensing pressure (P_4) is the saturation pressure at the condensing temperature. State 4 is fixed by the temperature and pressure. The specific enthalpy and specific entropy (h_4 and s_4) are computed.

```
"State 4"
T[4]=T_H+DT_cond                    "temperature"
T_cond=T[4]+DT_sc                   "condensing temperature"
P[4]=P_sat(R$,T=T_cond)             "pressure"
s[4]=entropy(R$,T=T[4],P=P[4])      "specific entropy"
h[4]=enthalpy(R$,T=T[4],P=P[4])     "specific enthalpy"
```

The temperature of the refrigerant leaving the evaporator is specified by the evaporator approach temperature difference:

$$T_2 = T_C - \Delta T_{evap}$$

The evaporating temperature is specified by the degree of superheat:

$$T_{evap} = T_2 - \Delta T_{sh}$$

The evaporating pressure (P_2) is the saturation pressure at the evaporating temperature. State 2 is fixed by the temperature and pressure. The specific enthalpy, specific entropy, and specific volume (h_2, s_2, and v_2) are computed.

```
"State 2"
T[2]=T_C-DT_evap                    "temperature"
T_evap=T[2]-DT_sh                   "evaporating temperature"
P[2]=P_sat(R$,T=T_evap)             "pressure"
s[2]=entropy(R$,T=T[2],P=P[2])      "specific entropy"
h[2]=enthalpy(R$,T=T[2],P=P[2])     "specific enthalpy"
v[2]=volume(R$,T=T[2],P=P[2])       "specific volume"
```

There is no pressure loss in the condenser, so $P_3 = P_4$. The specific enthalpy leaving a reversible compressor operating under the same conditions ($h_{s,3}$) is fixed by the specific entropy ($s_{s,3} = s_2$) and pressure. The work required by a reversible compressor per unit mass is:

$$\frac{\dot{W}_{s,c}}{\dot{m}} = h_{s,3} - h_2$$

The work required by the actual compressor per unit mass is:

$$\frac{\dot{W}_c}{\dot{m}} = \frac{1}{\eta_c} \frac{\dot{W}_{s,c}}{\dot{m}}$$

The specific enthalpy leaving the actual compressor is obtained from an energy balance:

$$h_3 = h_2 + \frac{\dot{W}_c}{\dot{m}}$$

State 3 is fixed by the pressure and specific enthalpy. The temperature and specific entropy (T_3 and s_3) are determined.

```
"State 3"
P[3]=P[4]                                   "pressure"
h_s[3]=enthalpy(R$,P=P[3],s=s[2])           "specific enthalpy leaving reversible compressor"
W_dot_s_c\m_dot=h_s[3]-h[2]                 "work per mass required by reversible compressor"
W_dot_c\m_dot=W_dot_s_c\m_dot/eta_c         "work per mass required by actual compressor"
h[3]=h[2]+W_dot_c\m_dot                     "specific enthalpy leaving actual compressor"
s[3]=entropy(R$,P=P[3],h=h[3])              "specific entropy"
T[3]=temperature(R$,P=P[3],h=h[3])          "temperature"
```

There is no pressure loss across the evaporator; therefore, $P_1 = P_2$. A steady-state energy balance on the throttle valve results in:

$$h_1 = h_4$$

State 1 is fixed by the pressure and specific enthalpy. The temperature and specific entropy (T_1 and s_1) are determined.

```
"State 1"
P[1]=P[2]                                   "pressure"
h[1]=h[4]                                   "specific enthalpy"
s[1]=entropy(R$,P=P[1],h=h[1])              "specific entropy"
T[1]=temperature(R$,P=P[1],h=h[1])          "temperature"
```

Figure 1(a) illustrates a temperature-specific entropy diagram for the cycle and Figure 1(b) illustrates a pressure-specific enthalpy diagram for the cycle.

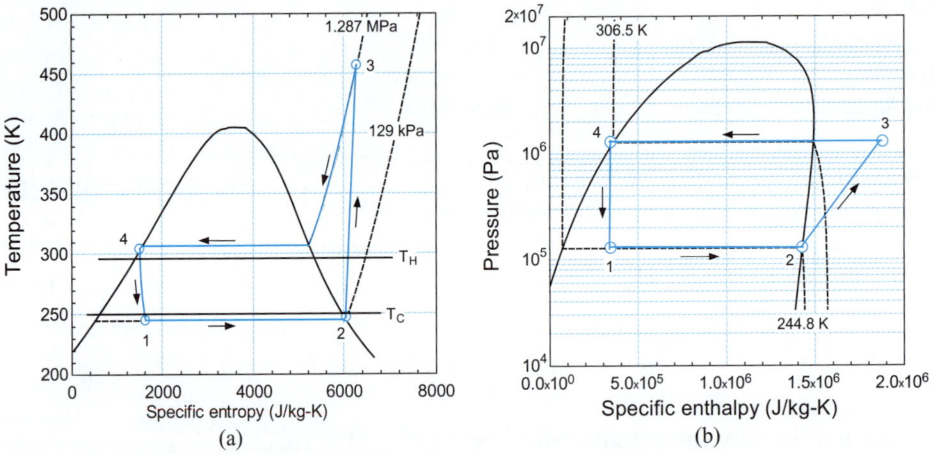

Figure 1: (a) Temperature-specific entropy and (b) pressure-specific enthalpy diagrams for the cycle.

b) Determine the rate of refrigeration (i.e., the refrigeration capacity) provided to the freezer and the coefficient of performance of the system.

The mass flow rate provided by the compressor is determined from the displacement rate and volumetric efficiency:

$$\dot{m} = \frac{\dot{V}_{disp}\,\eta_{vol}}{v_2}$$

The power consumed by the compressor is:

$$\dot{W}_c = \dot{m}\left(h_3 - h_2\right)$$

The rates of heat transfer to the evaporator and from the condenser are computed according to:

$$\dot{Q}_{evap} = \dot{m}\left(h_2 - h_1\right)$$

$$\dot{Q}_{cond} = \dot{m}\left(h_3 - h_4\right)$$

The cycle coefficient of performance is:

$$COP = \frac{\dot{Q}_{evap}}{\dot{W}_c}$$

```
m_dot=V_dot_disp*eta_vol/v[2]                 "mass flow rate"
W_dot_c=m_dot*(h[3]-h[2])                     "compressor power"
Q_dot_evap=m_dot*(h[2]-h[1])                  "evaporator heat transfer"
Q_dot_evap_ton=Q_dot_evap*convert(W,tons)     "in ton"
Q_dot_cond=m_dot*(h[3]-h[4])                  "condenser heat transfer"
COP=Q_dot_evap/W_dot_c                        "coefficient of performance"
```

which results in $\dot{Q}_{evap} = 14.9$ kW (4.23 ton) and $COP = 2.393$.

EXAMPLE 9.2-2: INDUSTRIAL FREEZER DESIGN

A vapor compression system for an industrial freezer, similar to the one described in Example 9.2-1, must be designed to meet a refrigeration load of $\dot{Q}_{evap} = 7.2$ tons at $T_C = -10^{\circ}$F and $T_H = 78^{\circ}$F. The degree of superheat, degree of subcooling, and compressor isentropic and volumetric efficiencies are the same as in Example 9.2-1. The cost of the compressor required by the system scales with the compressor displacement rate according to:

$$Cost_c = 92 \times 10^3 \left[\frac{\text{\$-s}}{\text{m}^3}\right] \dot{V}_{disp} \tag{1}$$

The costs of the condenser and evaporator scale with their conductances (UA_{cond} and UA_{evap}, respectively) according to:

$$Cost_{cond} = 1.5 \left[\frac{\text{\$-K}}{\text{W}}\right] UA_{cond} \tag{2}$$

$$Cost_{evap} = 1.25 \left[\frac{\text{\$-K}}{\text{W}}\right] UA_{evap} \tag{3}$$

The conductances of the condenser and evaporator are estimated according to:

$$UA_{cond} = \frac{\dot{Q}_{cond}}{(T_{cond} - T_H)} \tag{4}$$

$$UA_{evap} = \frac{\dot{Q}_{evap}}{(T_C - T_{evap})} \tag{5}$$

The cost of the electricity required to run the refrigeration system is ec = 0.10 \$/kW-hr and the system runs continuously.

a) Determine the compressor displacement rate, condenser conductance, and evaporator conductance that minimize the overall cost of owning and operating the equipment for *time* = 5 years while providing the required refrigeration rate at the given conditions. Neglect the time value of money for this problem.

The solution from Example 9.2-1 is used as the starting point for this problem. The system design in Example 9.2-1 provided $\dot{Q}_{evap}$ = 4.23 tons; therefore, the size of the compressor must be increased in order to obtain the required refrigeration rate (7.2 tons according to the problem statement). The previously specified compressor displacement is commented out and the refrigeration rate is specified:

```
{V_dot_disp_cfm=35 [cfm]}               "compressor displacement, in cfm"
Q_dot_evap_ton=7.2 [tons]               "required refrigeration capacity"
```

The cost of the compressor is computed using Eq. (1). The conductance of the evaporator and the conductance of the condenser are computed using Eqs. (4) and (5) and the cost of these components are computed using Eqs. (2) and (3). The total cost of the system is:

$$Cost_{system} = Cost_c + Cost_{cond} + Cost_{evap}$$

```
Cost_c=92e3 [$-s/m^3]*V_dot_disp                 "cost of the compressor"
UA_cond=Q_dot_cond/(T_cond-T_H)                  "conductance of condenser"
UA_evap=Q_dot_evap/(T_C-T_evap)                  "conductance of evaporator"
Cost_cond=1.5 [$-K/W]*UA_cond                    "cost of the condenser"
Cost_evap=1.25 [$-K/W]*UA_evap                   "cost of the evaporator"
Cost_system=Cost_c+Cost_cond+Cost_evap           "total system cost"
```

which leads to $Cost_{system}$ = \$15,845. The operating cost associated with running the system is:

$$Cost_{operating} = ec\,\dot{W}_c\,time$$

```
ec=0.10 [$/kW-hr]*convert($/kW-hr,$/J)           "electricity cost"
time=5 [year]*convert(year,s)                    "operating time"
Cost_operating=ec*W_dot_c*time                   "cost of electricity"
```

which results in $Cost_{operating}$ = \$46,340. The total cost of owning and operating the system is:

$$Cost_{total} = Cost_{system} + Cost_{operating}$$

```
Cost_total=Cost_system+Cost_operating       "total cost of owning and operating system"
```

which results in $Cost_{total}$ = \$62,190.

Because the compressor displacement must be selected based on the required refrigeration capacity, there are only two free design parameters: the conductance of the evaporator and the conductance of the condenser. Figure 1 illustrates the total cost, operating cost, system cost, and evaporator cost as a function of the evaporator approach temperature difference (with the condenser approach temperature difference fixed at $\Delta T_{cond} = 10°F$). Notice that there exists an optimal evaporator approach temperature difference that minimizes the total cost. This optimum occurs because of two competing effects. As the evaporator approach temperature difference becomes small, the performance of the system improves and therefore the operating cost is reduced. However, a smaller approach temperature difference requires a physically large evaporator; therefore, the evaporator and system cost tend to increase. The opposite happens when ΔT_{evap} is large: the operating cost is high but the evaporator cost is small. The optimal evaporator approach temperature difference balances these effects.

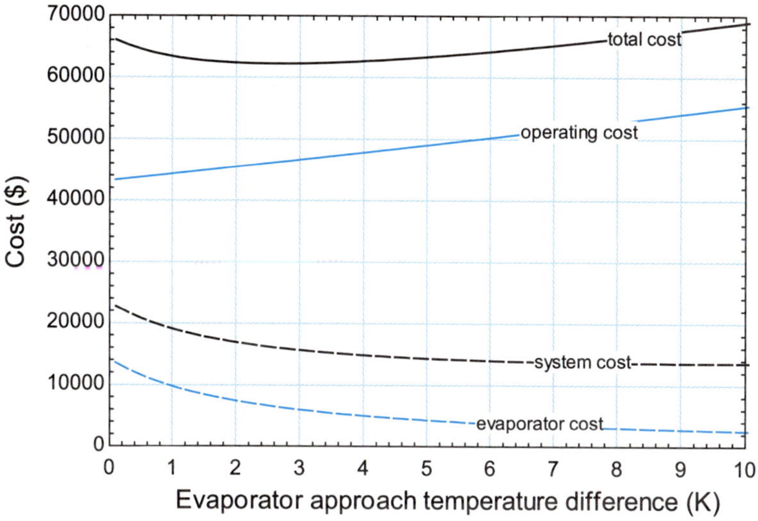

Figure 1: Total cost, operating cost, system cost, and evaporator cost as a function of the evaporator approach temperature difference (with $\Delta T_{cond} = 10°F$).

Both the evaporator and condenser approach temperature differences can be varied in this problem; therefore, we must carry out a multidimensional optimization. For this two-dimensional optimization, it is possible to prepare a contour plot showing the variation of total cost in the parameter space of ΔT_{evap} and ΔT_{cond}. In order to prepare a contour plot using EES, it is necessary to setup a Parametric table in which both of the parameters of interest are varied over a specified range. Create a new Parametric table that includes the two independent variables (the variables DT_cond and DT_evap) as well as the dependent variable of interest (the variable Cost_total). In order to run the simulation for 20 values of ΔT_{evap} and 20 values of ΔT_{cond}, $20 \times 20 = 400$ runs must be included in the table. (Add runs using the Insert/Delete Runs option from the Tables menu.)

It is necessary to set the values of DT_cond and DT_evap in the table. It is possible to vary DT_cond from 1 K to 10 K in 20 equal increments using the "Repeat pattern every" option in the Alter Values dialog that appears by right-clicking on the DT_cond column, as shown in Figure 2(a). In order to completely cover the parameter space, it is necessary to evaluate the solution over a range of DT_evap at each unique value of DT_cond. This pattern of values can be entered using the "Apply pattern every" option in the Alter Values dialog for the DT_evap column of the table; see Figure 2(b).

Comment out the specified values of the variables DT_cond and DT_evap in the Equations window. Run the Parametric table using the Solve Table command from the Calculate menu (F3); 400 values of Cost_total are calculated, one for each

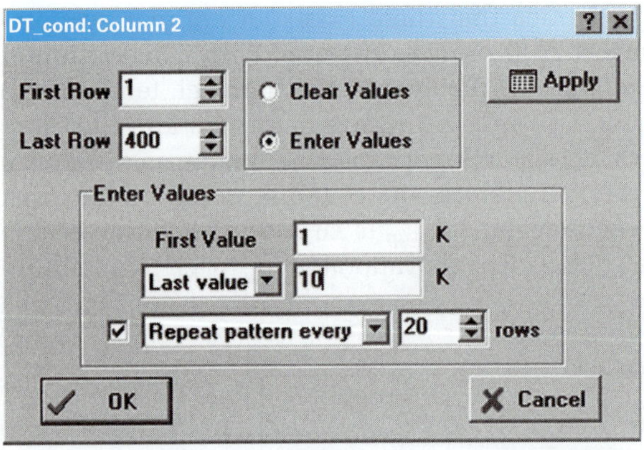

(a)

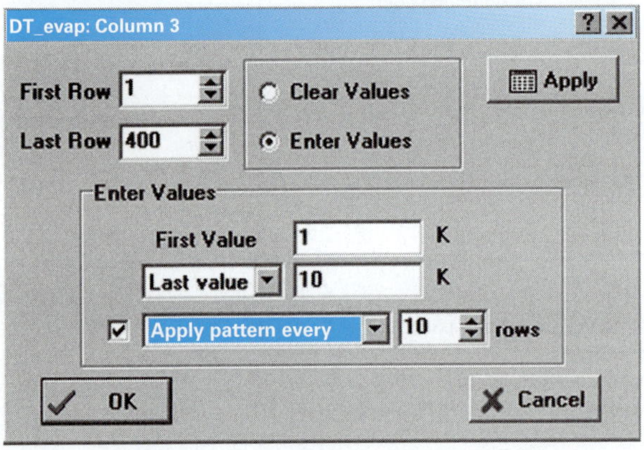

(b)

Figure 2: (a) Vary ΔT_{cond} from 1 K to 10 K 20 times and (b) vary ΔT_{evap} from 1 K to 10 K with 20 runs for each of the 20 values.

combination of ΔT_{cond} and ΔT_{evap} set in the Parametric table. To generate a contour plot, select X-Y-Z Plot from the New Plot Window option in the Plots menu. Select DT_cond as the variable on the x-axis, DT_evap as the variable on the y-axis and Cost_total as the contour variable. The appearance of the resulting contour plot can be adjusted by altering the resolution, smoothing, color options, and the type of function used for interpolation. A contour plot generated using isometric lines is shown in Figure 3 and it shows that the optimal design, corresponding to the minimum total cost, is approximately $\Delta T_{evap} = 2.5$ K and $\Delta T_{cond} = 5$ K.

It is possible to more precisely determine the optimal design using the optimization capabilities provided in EES. With the variables DT_cond and DT_evap commented out, select Min/Max from the Calculate menu. Select Cost_total as the dependent variable and select Minimize (see Figure 4). EES will detect that there are two free parameters for the problem; select the variables DT_cond and DT_evap as the independent variables. Set a reasonable guess and upper and lower bounds for each of the independent variables using the Bounds button. Select an optimization method (e.g., Variable metric method) and then select OK to activate the optimization. EES will identify the optimal design: $\Delta T_{cond} = 5.198$ K and $\Delta T_{evap} = 2.571$ K (consistent with Figure 3), which corresponds to $UA_{cond} = 4821$ W/K and

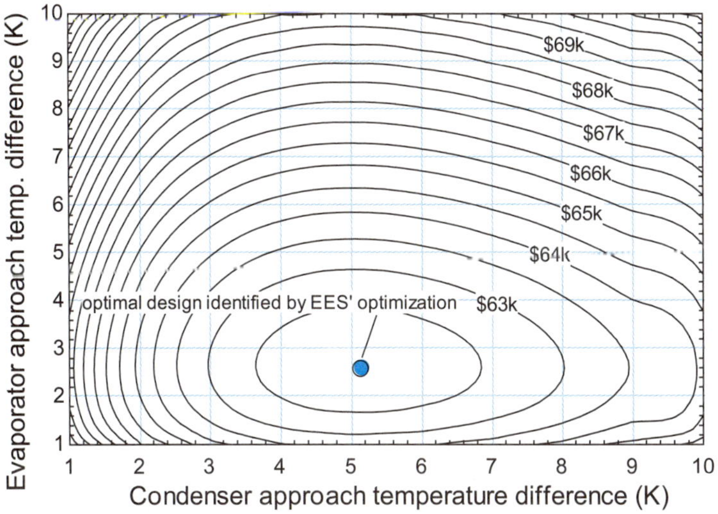

Figure 3: Contours of constant total cost in the parameter space of evaporator and condenser approach temperature differences.

UA_{evap} = 5283 W/K and a total cost of 62,156$. Depending on which optimization method you choose, the answers may be slightly different.

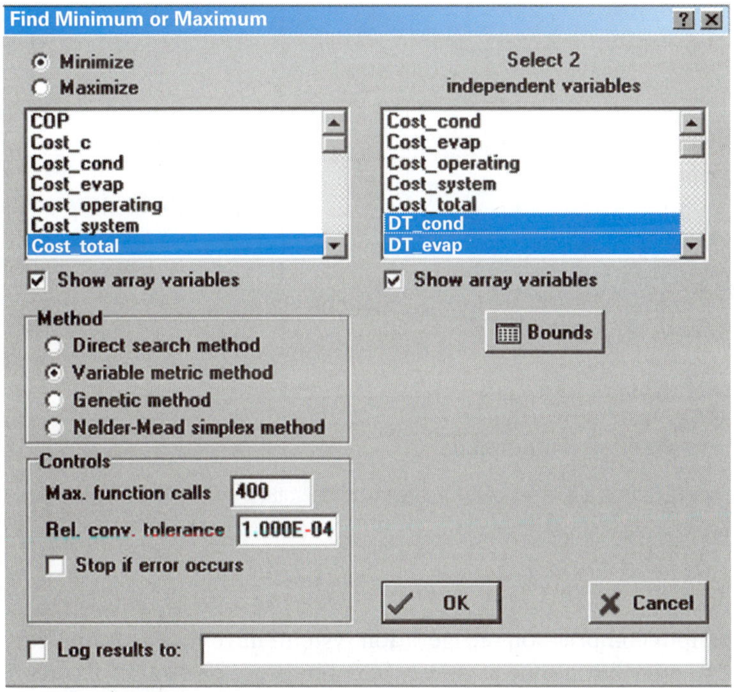

Figure 4: Find Minimum or Maximum dialog.

b) Predict the refrigeration capacity and *COP* of the refrigeration plant that utilizes the fixed components identified in part (a) as a function of the refrigeration temperature.

In order to simulate the refrigeration system with a fixed compressor, evaporator, and condenser, the values of $\dot{V}_{disp}$, UA_{evap}, and UA_{cond} are each fixed. The refrigeration capacity and approach temperature differences are predicted based on the optimal, fixed component sizes identified in (a).

```
{DT_evap=5 [F]*convert(F,K)}          "evaporator approach temperature difference"
{DT_cond=10 [F]*convert(F,K)}         "condenser approach temperature difference"
{Q_dot_evap_ton=7.2 [tons]}           "required refrigeration capacity"

"Optimally sized components"
UA_evap=5283 [W/K]                    "evaporator conductance"
UA_cond=4821 [W/K]                    "condenser conductance"
V_dot_disp_cfm=58.98 [cfm]            "compressor displacement"
```

Figure 5 illustrates the refrigeration capacity and COP as a function of T_C. The refrigeration capacity increases significantly as the freezer temperature increases. However, the actual refrigeration load required to maintain the freezer at its set temperature would likely decrease as the freezer temperature increases, resulting in the refrigeration equipment increasingly having more cooling capacity than is required. One way to control the refrigeration equipment when it has more capacity than is required is to turn the compressor off and on, as determined by a thermostat controller in the freezer.

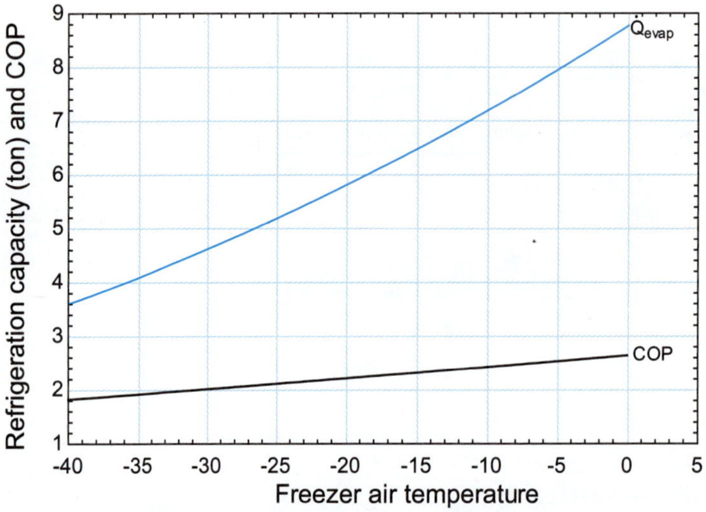

Figure 5: Refrigeration capacity and COP as a function of the freezer air temperature.

9.2.3 Refrigerants

The refrigerants used in vapor compression refrigeration systems have changed considerably over the past century and will continue to change in the future. This section discusses the characteristics that make a refrigerant desirable and presents a brief overview of the evolution of refrigerants over the past several decades.

Desirable Refrigerant Properties

Section 9.2.1 presented the ideal vapor compression cycle and showed that even if each of the components used to implement the cycle are perfect (e.g., the heat exchangers have zero approach temperature differences and the compressor is isentropic) the *COP* of the vapor compression cycle does not achieve the *COP* of a reversible Carnot refrigeration cycle. There is unavoidable entropy generation in the valve and the condenser. The magnitude of this entropy generation, and therefore the *COP* of the ideal vapor compression cycle, depends on the thermodynamic properties of the refrigerant that is used to implement the cycle. Table 9-2 lists the *COP* and other characteristics of various refrigerants used in an ideal vapor compression refrigeration cycle operating between $T_C = 5^\circ$C and $T_H = 30^\circ$C.

The choice of the refrigerant used in a vapor compression cycle is of critical importance. It is not sufficient for the thermodynamic properties to be suitable for the intended application. There are several other environmental and practical considerations that are equally important. Some of the characteristics of a good refrigerant are discussed below.

Positive Evaporator Gage Pressure

The evaporator is the location of the lowest pressure in the cycle. In the event that the system develops a leak, it is desirable to have the refrigerant leak out of the system rather than to have air leak into the system. If the system becomes contaminated with air then the cycle must be evacuated and recharged after the leak is repaired. For these reasons, it is desirable to have the pressure in the evaporator be greater than atmospheric pressure (i.e., the evaporator gage pressure should be positive). Table 9-2 shows the saturation temperature at atmospheric pressure for several refrigerants. Refrigerant R134a, for example, is used in home refrigerators and car air conditioning units. This refrigerant has a saturation pressure that is above atmospheric pressure for evaporator temperatures that are above -26°C. Ammonia provides even better performance in this regard; it has a saturation pressure that is above atmospheric pressure for evaporator temperatures above -33°C. Water could be used as a refrigerant, but it would need to be at a pressure much lower than atmospheric pressure in order to provide refrigeration at any temperature.

Moderate Condensing Pressure

The saturation pressure of a fluid increases exponentially with temperature. Some refrigerants therefore require extremely high condensing pressures in order to reject heat at reasonable temperatures. A high condensing pressure complicates the design of the compressor and requires that the equipment be made of heavy duty materials that are able to withstand high pressure. These requirements increase the weight and cost of the equipment. Table 9-2 lists the saturation pressure associated with a condensing temperature of 30°C, which is a typical condensing temperature for heat rejection to an ambient environment. Notice that carbon dioxide has a particularly high condensing pressure. Carbon dioxide has a critical temperature of about 31°C and therefore refrigeration cycles using carbon dioxide that reject heat to the environment will operate very close to the critical point and even in the supercritical region.

Appropriate Triple Point and Critical Point Temperatures

In order to utilize a vapor compression cycle, it is necessary that the refrigerant remain in a two phase state. Therefore, the operating temperature range must lie between the triple point temperature (below which the refrigerant solidifies) and the critical point temperature (above which the refrigerant ceases to condense and evaporate). Table 9-2 provides the triple point and critical point of each refrigerant. Note that the high triple point of water prevents it from being used below 0°C.

Table 9-2: Characteristics of some refrigerants used in an ideal vapor compression cycle operating between $T_C = 5^\circ$C and $T_H = 30^\circ$C.

Refrigerant	*COP*	T_{sat} at 1 atm	P_{sat} at 30°C	Triple point temp.	Critical temp.	ρ of saturated vapor at 5°C	Enthalpy of vaporization at 5°C
R11	10.37	23.8°C	125.2 kPa	−110.5°C	198.0°C	3.01 kg/m^3	187.1 kJ/kg
Ammonia	10.04	−33.3°C	1167 kPa	−77.7°C	132.3°C	4.13 kg/m^3	1244 kJ/kg
Isobutane	9.92	−11.7°C	404.5 kPa	−159.6°C	134.7°C	5.03 kg/m^3	349.9 kJ/kg
R12	9.87	−29.8°C	744.3 kPa	−157.1°C	112.0°C	21.1 kg/m^3	149.0 kJ/kg
R134a	9.77	−26.1°C	770.6 kPa	−104.3°C	101.0°C	17.1 kg/m^3	194.8 kJ/kg
R22	9.75	−40.8°C	1192 kPa	−157.4°C	96.1°C	24.6 kg/m^3	199.8 kJ/kg
Propane	9.62	−42.1°C	1079 kPa	−187.7°C	96.7°C	12.0 kg/m^3	367.4 kJ/kg
Water	9.60	100.0°C	4.25 kPa	0°C	374.0°C	0.0068 kg/m^3	2489 kJ/kg
Carbon dioxide	5.64	–	7214 kPa	−56.6°C	31.0°C	114.6 kg/m^3	215.0 kJ/kg

High Density/Low Specific Volume at the Compressor Inlet
The saturation pressure of the refrigerant at the evaporator temperature must be sufficiently high in order to provide a reasonable vapor density entering the compressor. The density of the saturated vapor leaving the evaporator and entering the compressor is approximately given by the ideal gas law.

$$\rho = \frac{1}{v} = \frac{P}{RT} \tag{9-30}$$

The temperature is fixed in Eq. (9-30) by the required evaporating temperature. However, the gas constant (on a mass basis) and, more importantly, the saturation pressure depend on the refrigerant. According to Eq. (9-28), the required displacement rate of the compressor (and therefore its physical size and cost) for a given mass flow rate (and therefore refrigeration capacity) is inversely proportional to the density of the saturated vapor leaving the evaporator. Table 9-2 lists the density of saturated vapor at 5°C. Notice that water is a particularly poor choice of refrigerant from this standpoint. The very low density of water vapor compared to other refrigerants will require extremely large compressors. On the other hand, carbon dioxide is attractive from this standpoint.

High Latent Heat (Specific Enthalpy Change) of Vaporization
The refrigeration effect in the vapor compression cycle is the result of the change in phase of the refrigerant from liquid to vapor in the evaporator. The latent heat of vaporization is the difference between the specific enthalpies of the saturated vapor and saturated liquid. Larger values of latent heat result in lower refrigerant mass flow rates required to provide a specific refrigeration capacity and therefore smaller equipment and smaller pressure losses. Table 9-2 provides the latent heat of vaporization for several refrigerants at 5°C. Notice that the latent heat of vaporization of R134a at 5°C is 194.8 kJ/kg whereas ammonia has a latent heat of vaporization of 1244 kJ/kg. Water has an even higher latent heat (2489 kJ/kg) but it is not an ideal refrigerant because of its high triple point and low density of saturated vapor.

High Dielectric Strength
High dielectric strength is only required for hermetically sealed refrigeration cycles that contain the compressor and electric motor within the same enclosure. In these systems, the refrigerant that enters the compressor also cools the motor. Therefore, high dielectric strength is required to ensure that the refrigerant does not interfere with the operation of the electric motor as it passes by the motor windings. The refrigerant and the lubrication oil must also be compatible with the materials used in the motor. R134a is often used in hermetic systems. Ammonia is normally not used in such systems because it reacts with copper, zinc, and bronze, materials that are typically used in electric motors.

Compatibility with Lubricants
Compressors have moving parts that require lubrication in order to reduce frictional effects. Consequently, lubricant must be used in the compressor and the lubricant is in contact with the refrigerant as it flows through the compressor. There are two ways to deal with the lubricant-refrigerant interaction. Most refrigerants employ an oil lubricant that is completely soluble in the refrigerant. The oil moves through the cycle along with the refrigerant and it may tend to accumulate in the coldest part of the cycle, i.e., in the evaporator. Methods to return the oil from the evaporator to the compressor must be included in the system design. An alternative is to use a lubricant that is immiscible with the refrigerant; this is often the case with ammonia refrigerant systems. After leaving the compressor, the oil/refrigerant mixture is directed into a separator where the oil-rich

liquid is allowed to separate from the vapor. The oil is pumped back to the compressor while the refrigerant vapor proceeds to the condenser.

Non-Toxic

The refrigerant is normally contained within the refrigeration cycle; however, refrigerant leaks can occur. Early refrigeration system used methyl formate, sulfur dioxide, methyl chloride, ethyl ether, ammonia, and other toxic substances. The compressors used in these systems were large and noisy, so they were often installed far from the intended refrigeration application; the result was long refrigerant lines that were more prone to leakage and large refrigerant inventories so that leaks could release large amounts of refrigerant. The toxicity of these early refrigerants provided one of the main motivations for the development of non-toxic chlorofluorocarbon refrigerants such as R11 and R12 that were widely used until the mid-1990's. Toxicity remains a major factor in the choice of a refrigerant.

Non-Flammable

Refrigerant flammability is an interesting issue. Obviously, it would be best to use a refrigerant that is non-flammable. A non-flammable refrigerant is particularly important for car air-conditioning systems where the possibility of a crash exists. During an automobile crash, the refrigerant that passes through the passenger cabin might be ignited by hot engine parts. The advantage of a non-flammable refrigerant for a domestic refrigerator is less clear. First, there is relatively little refrigerant contained in the cycle and the possibility of a leak is small. Second, many kitchens already have a much larger supply of other flammable materials, such as natural gas and propane for stoves, so the use of a non-flammable refrigerant does not significantly contribute to improved safety. Nevertheless, non-flammable refrigerants in domestic refrigerators and air-conditioning equipment are currently required by code in the United States. Flammable refrigerants, such as propane and isobutane, are used in refrigerators, vending machines, and other small refrigeration equipment in countries that have less restrictive codes.

Inertness and Stability

Refrigerants should ideally be stable and chemically inert so that they maintain the same chemical composition during their life that they have when they are first used to charge the refrigeration cycle. Decomposition or reactions with trace amounts of air, water, or other materials may generate gaseous products that reduce the effectiveness of the refrigeration cycle. Acids that form from reactions between the refrigerant and water may cause corrosion. The inertness and chemical stability of the chlorofluorocarbon refrigerants R11 and R12 was a major contributor to their success. Unfortunately, the stability of these refrigerants also contributed to their ozone-depletion potential, which is discussed in a subsequent section.

Refrigerant Naming Convention

Some of the more attractive refrigerants in Table 9-2 are no longer used; a particularly notable example is R11, which has the highest theoretical COP of any of the refrigerants listed. In order to understand why some refrigerants are no longer widely used, it is first necessary to understand the refrigerant naming convention and discuss how the refrigerant name corresponds to its chemical composition.

It is common to refer to a refrigerant as FREON, which is the widely-recognized trade name used for a class of refrigerants that are produced by the DuPont chemical company. However, there are many different FREON refrigerants and it is inappropriate to use a trade name to refer to a generic product. Instead, the refrigerant industry uses a

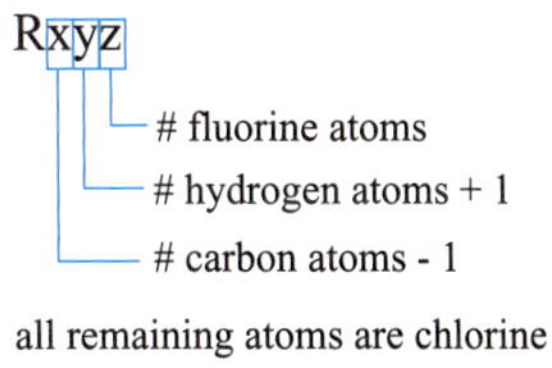

Figure 9-8: Refrigerant naming convention for organic refrigerants.

naming convention for identifying different refrigerants. The refrigerant name is Rxyz, where x, y, and z are numbers that identify the chemical structure of the refrigerant. The code used to assign the numbers x, y, and z depends on the type of refrigerant.

Many refrigerants are composed of small organic molecules consisting of carbon, hydrogen, fluorine, and chlorine. The naming convention for this class of refrigerants is shown in Figure 9-8. If x is zero (i.e., if there is a single carbon atom) then it is normally dropped from the name (thus R11 is really R011). For example, the most common refrigerant used in domestic refrigerators and car air conditioners before 1996 was R12. According to Figure 9-8, R12 has two fluorine atoms, no hydrogen atoms, and a single carbon atom. Carbon forms four covalent bonds. The naming convention assumes that any bonds that are not specifically allocated to fluorine and hydrogen are bonds to chlorine. Thus, R12 is di-fluoro-di-chloro methane and it has the chemical structure shown in Figure 9-9(a). The refrigerant R12 contains chlorine, fluorine, and carbon; it is a ChloroFluoroCarbon (CFC) and it is no longer produced due to its high ozone depletion potential, discussed in the subsequent section.

The refrigerant R22 is commonly used today. According to the naming convention shown in Figure 9-8, R22 (R022) has 2 fluorine atoms, 1 hydrogen atom, and one carbon atom. The single carbon atom has four bonds and the remaining atom must be chlorine. The chemical structure of R22 is shown in Figure 9-9(b). R22 contains hydrogen, chlorine, fluorine, and carbon; it is a HydroChloroFluoroCarbon (HCFC). Regulations that went into effect on January 1, 2010 prohibit the use of R22 in new equipment and production of R22 is scheduled to be terminated by January 1, 2020.

Refrigerant R134a is a common replacement for R12. The naming convention in Figure 9-8 results in the chemical structure for R134a that is shown in Figure 9-9(c). R134a contains hydrogen, fluorine, and carbon; it is a HydroFluoroCarbon (HFC). Notice that R134a does not contain chlorine and, as a result, it has far less ozone depletion potential than either CFCs or HCFCs. The 'a' appearing at the end of R134a requires further explanation. R134a has four fluorine atoms per molecule with three fluorine atoms bonded to one carbon and one bonded to the second carbon. Refrigerant R134 (without the 'a') is a more symmetric molecule with two fluorine atoms bonded to each carbon, as shown in Figure 9-9(d). R134 and R134a, although chemically very similar, have different thermodynamic properties.

Propane, which has the chemical structure shown in Figure 9-9(e), is also a refrigerant. Application of the naming convention shown in Figure 9-8 to propane results in it being designated as R290. Propane has excellent thermodynamic properties for refrigeration, as shown in Table 9-2. Because propane has no chlorine, it has no ozone depletion

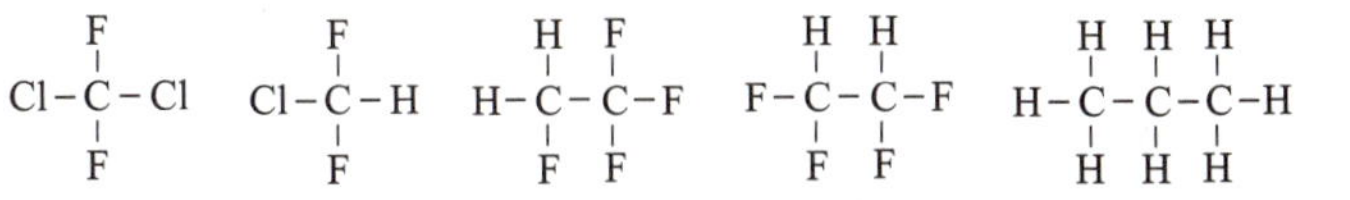

R12 (a) | R22 (b) | R134a (c) | R134 (d) | propane (R290) (e)

Figure 9-9: Chemical structure of (a) R12, (b) R22, (c) R134a, (d) R134, and (e) R290 (propane).

concerns. Propane is non-toxic but it is, of course, flammable. N-butane and isobutane (both C_4H_{10}) are chemically similar to propane and they are also used as refrigerants. However, these refrigerants have 10 hydrogen atoms and the naming convention in Figure 9-8 does not work; therefore, n-butane and isobutane are referred to as R600 and R600a, respectively.

Inorganic refrigerants such as ammonia do not use the naming convention shown in Figure 9-8. Instead, the refrigerant name is determined by adding 700 to the molar mass of the refrigerant. Ammonia, with a molar mass of 17, is therefore designated as refrigerant R717. Carbon dioxide is R744. Water is a common refrigerant for some absorption refrigeration cycles, as described in Section 9.4. Perhaps you can impress your friends by telling them that you often drink refrigerant R718.

Some refrigerants are actually mixtures of several chemicals. The naming convention in Figure 9-8 also does not work for these mixtures. There are two classes of mixtures. *Azeotropic* mixtures behave very much like a pure refrigerant in that their saturation temperature depends only on pressure. Azeotropic mixtures are assigned generic refrigeration names that begin with 500. For example, R507A is a mixture of 50% R125 and 50% R143a (by mass). The letter, e.g., A that follows the number is used to designate different mass fractions for the same chemicals. Non-azeotropic (or *zeotropic*) mixtures do not behave like pure substances in that their saturation temperature is not only a function of pressure but also depends on quality. Zeotropic mixtures exhibit what is referred to as a temperature glide during evaporation and condensation; that is, the saturation temperature changes as the phase change progresses at constant pressure. Zeotropic mixtures are given names that begin with 400. For example, a substitute for R22 is R410A, which is 50% R32 and 50% R125 (by mass).

Ozone Depletion and Global Warming Potential

In order to understand issues related to ozone depletion and global warming, it is first necessary to understand how the earth receives and emits radiation. The earth receives radiation from the sun over a range of wavelengths, λ (i.e., over a spectrum), as shown in Figure 9-10. The peak in the spectrum associated with the radiation emitted by the sun corresponds to visible light, 0.38 μm (violet) $< \lambda <$ 0.78μm (red). Radiation emitted at wavelengths below 0.38 μm is referred to as ultraviolet (UV) radiation and radiation emitted at wavelengths above 0.78 μm is referred to as infrared (IR) radiation. Ultraviolet radiation is further divided into UV-A, UV-B, and UV-C radiation. The UV-B radiation can damage the DNA of plants and animals and cause skin cancer in humans. However, UV-B radiation is strongly absorbed by the ozone (O_3) that resides in the stratosphere, which exists approximately 25 miles above the earth's surface.

The refrigerants R11, R12 and other CFCs are so stable that they remain chemically unchanged even after being released to the environment. Eventually, these components diffuse into the stratosphere where they catalyze reactions that tend to deplete the ozone layer. As a result, an increased amount of UV-B radiation is able to pass through the atmosphere and reach the earth's surface. Atmospheric chemists have proven that the chlorine in the CFC molecules is primarily responsible for ozone depletion, as detailed by ASHRAE (1989). By international agreement, CFC refrigerants are no longer produced because of this environmental problem.

The majority of the radiation that the earth receives from the sun is absorbed by the earth's surface and then re-emitted to space. Because the earth's temperature is so much lower than the sun's temperature, the spectral distribution of the radiation emitted by the earth is very different from the spectral distribution of the radiation that is received from the sun. The earth tends to emit radiation at much longer wavelengths that lie in the infrared range, as shown in Figure 9-10. There are strong absorption bands in the

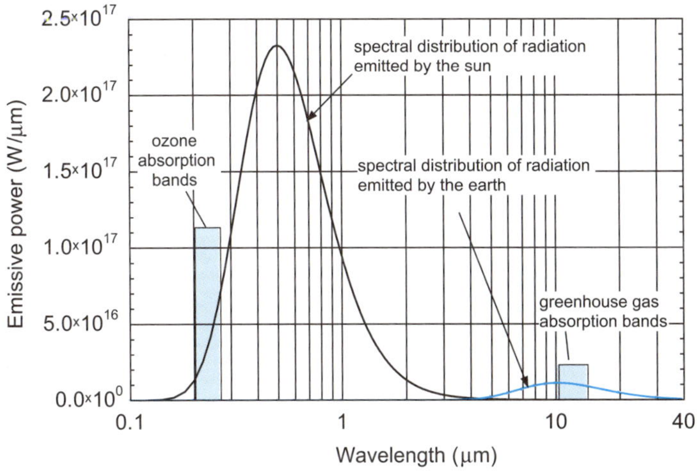

Figure 9-10: Spectrum of radiation received by the earth and emitted by the earth.

infrared range associated with carbon dioxide and other gases in the atmosphere that tend to preferentially "trap" the energy emitted by the earth so that some of it does not make it through the atmosphere. This selective absorption of radiation in specific wavelength bands is referred to as the "greenhouse effect". Global warming is associated with the buildup of carbon dioxide and other so-called greenhouse gases in the atmosphere due to human activity. Some refrigerants are greenhouse gases in that they tend to build up in the atmosphere if they are released and contribute to the global warming effect by absorbing high wavelength radiation.

The concerns discussed above have led to the classification of potential refrigerants in terms of their Ozone Depletion Potential (ODP) and Global Warming Potential (GWP). Ozone depletion potential is a measure of the damage that the substance does to the ozone layer in the atmosphere, defined relative to R11 (i.e., an ODP of 1 indicates that the substance is equivalent to R11). Global warming potential is a measure of the contribution of the substance to the atmosphere's heat retention, also defined relative to R11. Figure 9-11 illustrates the ODP and GWP of various CFCs, HCFCs, and HFCs. Notice that the ODP of CFCs is quite high due to their chlorine content and lack of hydrogen. The ODP of HCFCs is lower, but still significant as these refrigerants also contain chlorine. The ODP of HFCs is zero as they contain no chlorine. However, the HFCs do have some GWP and are considered greenhouse gases.

9.2.4 Vapor Compression Cycle Modifications

As noted in Section 9.2.3, the production of many common CFC refrigerants was banned in 1996 due to environmental concerns and the popular refrigerant R22 has been banned from use in new equipment starting in 2010. Alternative refrigerants have been identified. However, in many cases the alternative refrigerants do not provide cycle performance that is as good as the now-banned original refrigerants. There are other factors that require manufacturers of refrigeration equipment to improve the efficiency of their equipment. For example, the energy performance of new refrigerators and freezers must comply with the increasingly high energy conservation standards that are dictated by the U.S. Department of Energy (DOE, 2009). Consequently, there is interest in enhancements and modifications that improve the performance of refrigeration cycles; some of these common cycle modifications are discussed in this section.

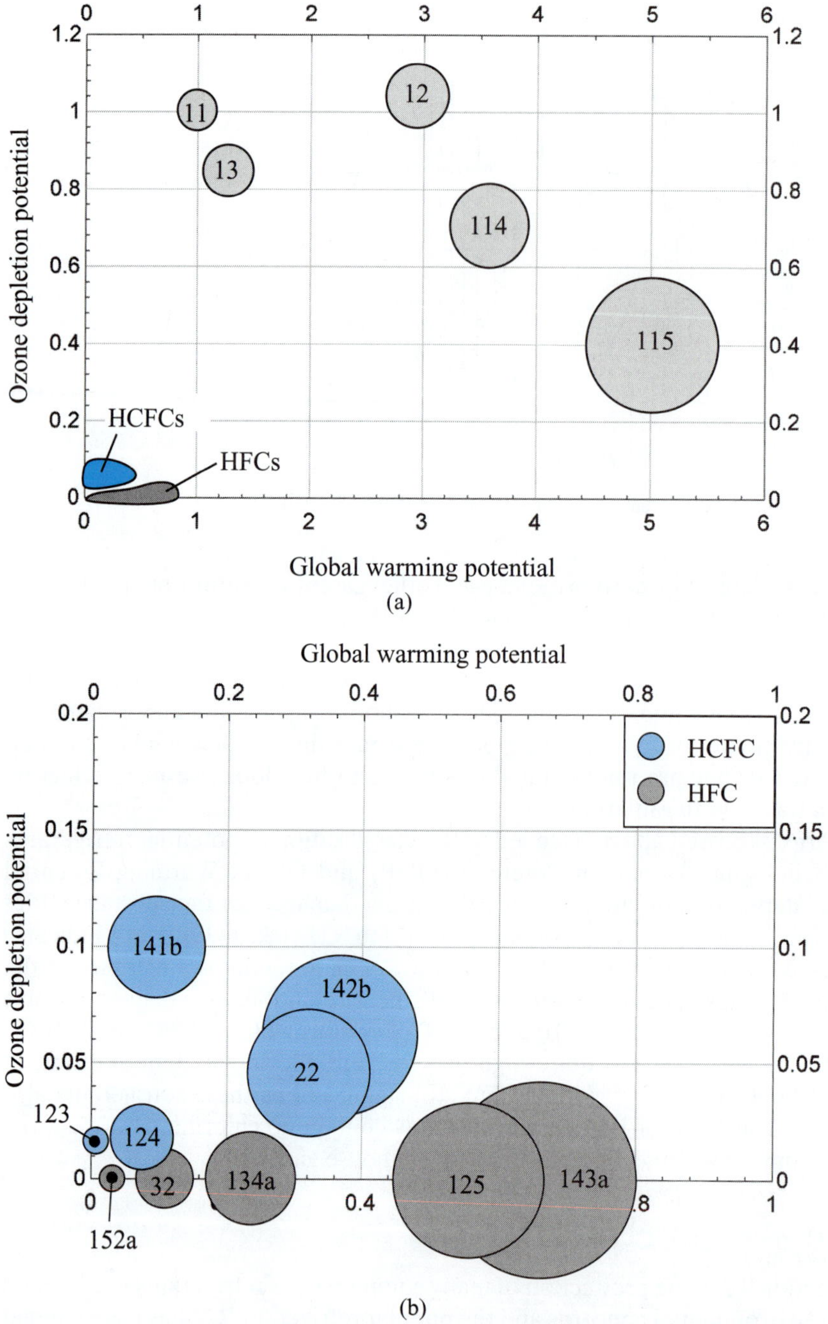

Figure 9-11: ODP and GWP of various (a) CFCs and (b) HCFCs and HFCs. Note that the size of the points are proportional to the atmospheric lifetime of the chemical; for reference, the atmospheric lifetimes of R115, R11, and R152a are 1700 years, 45 years, and 1.4 years, respectively. The source of these data is the World Meteorological Organization (http://www.wmo.int/pages/prog/arep/gaw/ozone/index.html).

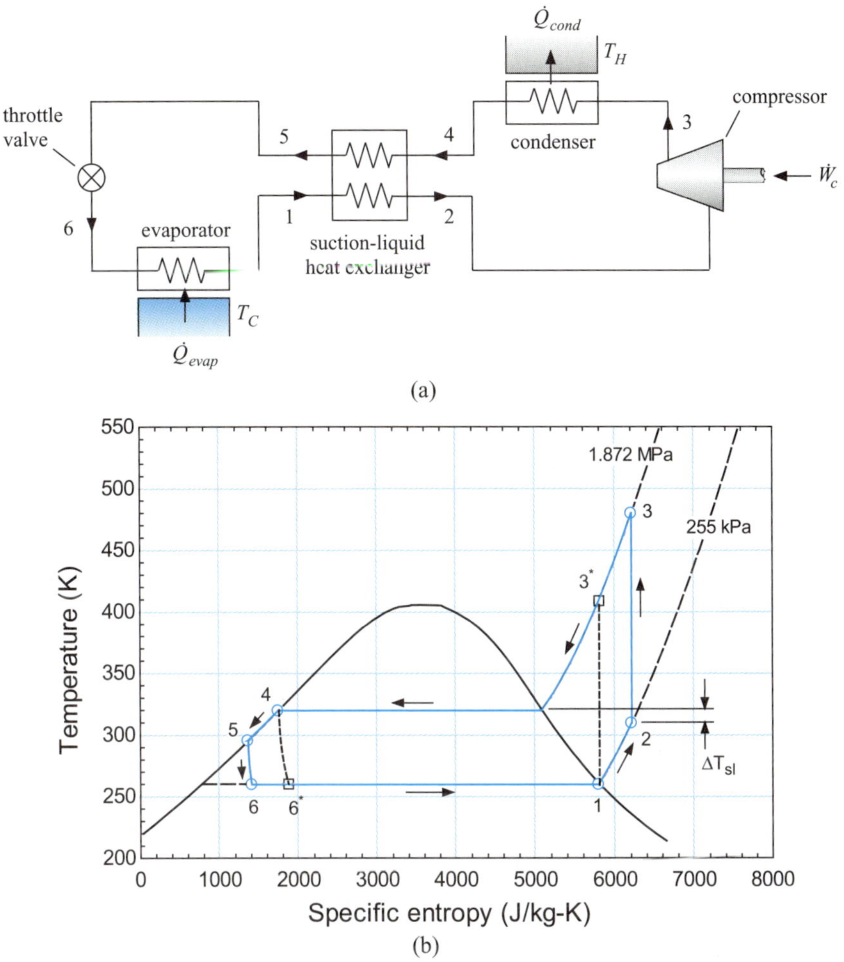

Figure 9-12: (a) Schematic of a refrigerant cycle with a liquid-suction heat exchanger and (b) the cycle shown on a temperature-specific entropy diagram.

Liquid-Suction Heat Exchanger

One possible method of improving refrigeration cycle performance is to use a recuperative heat exchanger (referred to as a liquid-suction heat exchanger), as shown in Figure 9-12(a). The liquid-suction heat exchanger uses the relatively cool refrigerant vapor exiting the evaporator at state 1 in order to precool the relatively warm saturated liquid leaving the condenser at state 4. The result of the insertion of the liquid-suction heat exchanger is that superheated vapor enters the compressor at state 2 while subcooled liquid enters the valve at state 5.

The liquid-suction heat exchanger is typically assumed to operate at steady state with negligible jacket losses (i.e., it is externally adiabatic) and its performance is characterized by either a heat exchanger effectiveness or an approach temperature difference, as discussed in Section 6.6.6. The mass flow rates of the two streams in the suction line heat exchanger are equal, but the vapor stream (entering at state 1 and leaving at state 2) normally has a smaller specific heat capacity than the liquid stream (entering at state 4 and leaving at state 5). Therefore, the pinch point will typically occur at the warm end of the liquid-suction heat exchanger (i.e., where state 4 enters and state 2 leaves). The liquid-suction heat exchanger approach temperature difference (ΔT_{sl}) is indicated on the temperature-specific entropy diagram in Figure 9-12(b).

The addition of a liquid-suction heat exchanger has both a positive and a negative effect on the cycle performance. Subcooled liquid enters the valve and therefore the refrigerant entering the evaporator at state 6 has a lower quality and a lower specific enthalpy than it would have without the liquid-suction heat exchanger, as indicated by state 6^* in Figure 9-12(b). The entropy generation associated with flow through the valve is reduced and the refrigerant effect associated with the phase change in the evaporator is increased. Balancing this positive effect is a negative one. Superheated vapor enters the compressor at state 2 as compared to saturated vapor in the absence of the liquid-suction heat exchanger. This superheated vapor has higher specific volume than saturated vapor at the same pressure and therefore the size of the compressor must increase and the power consumed by the compressor per mass flow rate of refrigerant also increases. Further, the temperature of the refrigerant leaving the compressor at state 3 is higher than it would have been without the liquid-suction heat exchanger, indicated by state 3^* in Figure 9-12(b). The elevated temperature results in a larger rate of entropy generation in the condenser. Depending on the refrigerant, the addition of a liquid-suction heat exchanger can either increase or reduce the performance of the system, as discussed by Klein et al. (2000); these competing effects are explored in Example 9.2-3.

EXAMPLE 9.2-3: REFRIGERATION CYCLE WITH A LIQUID-SUCTION HEAT EXCHANGER

The liquid-suction heat exchanger modification is proposed for the freezer room application described in Example 9.2-1. The air temperature within the freezer room must be maintained at $T_C = -10°\text{F}$ using the cycle shown in Figure 9-12(a) with ammonia as the working fluid. The evaporator approach temperature difference is $\Delta T_{evap} = 5°\text{F}$ and the refrigerant leaving the evaporator has a $\Delta T_{sh} = 4°\text{F}$ degree of superheat. The refrigeration cycle rejects heat to ambient air at $T_H = 78°\text{F}$. The condenser approach temperature difference is $\Delta T_{cond} = 10°\text{F}$ and the refrigerant leaving the condenser has a $\Delta T_{sc} = 4°\text{F}$ degree of subcooling. The compressor has an isentropic efficiency of $\eta_c = 0.78$, a volumetric efficiency of $\eta_{vol} = 0.75$, and a displacement rate of $\dot{V}_{disp} = 35$ cfm. The liquid-suction heat exchanger has an approach temperature difference of $\Delta T_{sl} = 10°\text{F}$. Neglect the pressure drop in the heat exchangers.

a) Determine the *COP* of the system.

The inputs are entered in EES:

```
$UnitSystem SI K Pa J Mass Radian
T_C=converttemp(F,K,-10 [F])                 "freezer air temperature"
DT_evap=5 [F]*convert(F,K)                   "evaporator approach temperature difference"
DT_sh=4 [F]*convert(F,K)                     "degree of superheat"
T_H=converttemp(F,K,78 [F])                  "outdoor air temperature"
DT_cond=10 [F]*convert(F,K)                  "condenser approach temperature difference"
DT_sc=4 [F]*convert(F,K)                     "degree of subcooling"
eta_c=0.78 [-]                               "compressor efficiency"
V_dot_disp_cfm=35 [cfm]                      "compressor displacement rate, in cfm"
V_dot_disp=V_dot_disp_cfm*convert(cfm,m^3/s) "compressor displacement rate"
eta_vol=0.75 [-]                             "compressor volumetric efficiency"
DT_ls=10 [F]*convert(F,K)                    "liquid-suction HX approach temp. difference"
R$='Ammonia'                                 "refrigerant"
```

The temperature of the refrigerant leaving the condenser is specified by the condenser approach temperature difference:

$$T_4 = T_H + \Delta T_{cond}$$

The condensing temperature (i.e., the saturation temperature) is specified by the degree of subcooling:

$$T_{cond} = T_4 + \Delta T_{sc}$$

The condenser pressure (P_4) is the saturation pressure at the condensing temperature. State 4 is fixed by the temperature and pressure. The specific enthalpy and specific entropy (h_4 and s_4) are computed.

```
"State 4"
T[4]=T_H+DT_cond                        "temperature"
T_cond=T[4]+DT_sc                       "condensing temperature"
P[4]=P_sat(R$,T=T_cond)                 "pressure"
s[4]=entropy(R$,T=T[4],P=P[4])          "specific entropy"
h[4]=enthalpy(R$,T=T[4],P=P[4])         "specific enthalpy"
```

The temperature of the refrigerant leaving the evaporator (in a superheated state) is specified by the evaporator approach temperature difference:

$$T_1 = T_C - \Delta T_{evap}$$

The saturation temperature in the evaporator is specified by the degree of superheat:

$$T_{evap} = T_1 - \Delta T_{sh}$$

The evaporating pressure (P_1) is the saturation pressure at the evaporating temperature. State 1 is fixed by the temperature and pressure. The specific enthalpy and specific entropy (h_1 and s_1) are computed.

```
"State 1"
T[1]=T_C-DT_evap                        "temperature"
T_evap=T[1]-DT_sh                       "evaporating temperature"
P[1]=P_sat(R$,T=T_evap)                 "pressure"
s[1]=entropy(R$,T=T[1],P=P[1])          "specific entropy"
h[1]=enthalpy(R$,T=T[1],P=P[1])         "specific enthalpy"
```

There is no pressure loss on the vapor side of the liquid-suction heat exchanger; therefore $P_2 = P_1$. The temperature of the vapor leaving the liquid-suction heat exchanger is obtained from the approach temperature difference:

$$T_2 = T_4 - \Delta T_{ls}$$

State 2 is fixed by the temperature and pressure. The specific enthalpy, specific entropy, and specific volume (h_2, s_2, and v_2) are determined.

```
"State 2"
T[2]=T[4]-DT_ls                         "temperature"
P[2]=P[1]                               "pressure"
s[2]=entropy(R$,T=T[2],P=P[2])          "specific entropy"
h[2]=enthalpy(R$,T=T[2],P=P[2])         "specific enthalpy"
v[2]=volume(R$,T=T[2],P=P[2])           "specific volume"
```

There is no pressure loss on the liquid side of the liquid-suction heat exchanger; therefore $P_5 = P_4$. An energy balance on the liquid-suction heat exchanger provides:

$$h_5 = h_4 + h_1 - h_2$$

State 5 is fixed by the specific enthalpy and pressure. The temperature and specific entropy (T_5 and s_5) are computed.

```
"State 5"
h[5]=h[1]+h[4]-h[2]                    "specific enthalpy"
P[5]=P[4]                              "pressure"
T[5]=temperature(R$,h=h[5],P=P[5])     "temperature"
s[5]=entropy(R$,h=h[5],P=P[5])         "specific entropy"
```

There is no pressure loss in the condenser; therefore $P_3 = P_4$. The specific enthalpy leaving a reversible compressor operating under the same conditions ($h_{s,3}$) is specified by the specific entropy and pressure. The work required by a reversible compressor per unit mass is:

$$\frac{\dot{W}_{s,c}}{\dot{m}} = h_{s,3} - h_2$$

The work required by the actual compressor per unit mass is:

$$\frac{\dot{W}_c}{\dot{m}} = \frac{1}{\eta_c} \frac{\dot{W}_{s,c}}{\dot{m}}$$

The specific enthalpy leaving the actual compressor is obtained from an energy balance:

$$h_3 = h_2 + \frac{\dot{W}_c}{\dot{m}}$$

State 3 is fixed by the pressure and specific enthalpy. The temperature and specific entropy (T_3 and s_3) are determined.

```
"State 3"
P[3]=P[4]                                  "pressure"
h_s[3]=enthalpy(R$,P=P[3],s=s[2])          "specific enthalpy leaving reversible compressor"
W_dot_s_c\m_dot=h_s[3]-h[2]                "work per mass required by reversible compressor"
W_dot_c\m_dot=W_dot_s_c\m_dot/eta_c        "work per mass required by actual compressor"
h[3]=h[2]+W_dot_c\m_dot                    "specific enthalpy leaving actual compressor"
s[3]=entropy(R$,P=P[3],h=h[3])             "specific entropy"
T[3]=temperature(R$,P=P[3],h=h[3])         "temperature"
```

There is no pressure loss across the evaporator; therefore, $P_6 = P_1$. A steady-state energy balance on the throttle valve results in:

$$h_6 = h_5$$

State 6 is fixed by the pressure and specific enthalpy. The temperature and specific entropy (T_6 and s_6) are determined.

```
"State 6"
P[6]=P[1]                                   "pressure"
h[6]=h[5]                                   "specific enthalpy"
s[6]=entropy(R$,P=P[6],h=h[6])              "specific entropy"
T[6]=temperature(R$,P=P[6],h=h[6])          "temperature"
```

The mass flow rate provided by the compressor is determined from the displacement rate and volumetric efficiency:

$$\dot{m} = \frac{\dot{V}_{disp}\,\eta_{vol}}{v_2}$$

The power consumed by the adiabatic compressor is:

$$\dot{W}_c = \dot{m}\left(h_3 - h_2\right)$$

The rates of heat transfer to the evaporator and from the condenser are computed according to:

$$\dot{Q}_{evap} = \dot{m}\left(h_1 - h_6\right)$$

$$\dot{Q}_{cond} = \dot{m}\left(h_3 - h_4\right)$$

The cycle coefficient of performance is:

$$COP = \frac{\dot{Q}_{evap}}{\dot{W}_c}$$

```
m_dot=V_dot_disp*eta_vol/v[2]                     "mass flow rate"
W_dot_c=m_dot*(h[3]-h[2])                         "compressor power"
Q_dot_evap=m_dot*(h[1]-h[6])                      "evaporator heat transfer"
Q_dot_evap_ton=Q_dot_evap*convert(W,tons)         "in ton"
Q_dot_cond=m_dot*(h[3]-h[4])                      "condenser heat transfer"
COP=Q_dot_evap/W_dot_c                            "coefficient of performance"
```

which results in $COP = 2.172$.

b) What is the effectiveness of the liquid-suction heat exchanger?

The actual rate of heat transfer from the liquid to the vapor in the liquid-suction heat exchanger is:

$$\dot{Q}_{ls} = \dot{m}\left(h_2 - h_1\right)$$

The pinch point of the liquid-suction heat exchanger occurs at the warm end. Therefore, the maximum possible temperature of the vapor leaving the liquid-suction heat exchanger at state 2 is T_4. The value of the maximum possible specific enthalpy of the vapor leaving the heat exchanger, $h_{2,max}$, is evaluated at T_4 and P_2. The maximum possible rate of heat transfer in the liquid-suction heat exchanger is:

$$\dot{Q}_{ls,max} = \dot{m}\left(h_{2,max} - h_1\right)$$

and the effectiveness of the liquid-suction heat exchanger is:

$$\varepsilon_{ls} = \frac{\dot{Q}_{ls}}{\dot{Q}_{ls,max}}$$

```
Q_dot_ls=m_dot*(h[2]-h[1])                "heat transfer rate in the liquid-suction heat exchanger"
h_2_max=enthalpy(R$,T=T[4],P=P[2])        "maximum possible specific enthalpy of vapor leaving the HX"
Q_dot_ls_max=m_dot*(h_2_max-h[1])         "maximum possible heat transfer rate"
eff_sl=Q_dot_ls/Q_dot_ls_max              "effectiveness of the liquid-suction heat exchanger"
```

which leads to $\varepsilon_{ls} = 0.905$.

c) Plot the *COP* of the system as a function of the liquid-suction heat exchanger effectiveness. Overlay on your plot the results for refrigerants R22 and R134a in the same system.

A Parametric table is generated that includes ΔT_{sl}, ε_{sl}, and *COP*. The value of ΔT_{sl} is varied from 0 K (corresponding to $\varepsilon_{sl} = 1.0$) to a high value, which corresponds to a low value of ε_{sl}. The calculations are more likely to converge if the approach temperature difference is specified and the effectiveness calculated as opposed to specifying the effectiveness and calculating the approach temperature difference. Figure 1 illustrates the *COP* as a function of effectiveness for ammonia, R22, and R134a. The results are interesting. The *COP* of the ammonia system is reduced by increasing the liquid-suction heat exchanger effectiveness, whereas the *COP* of the R134a system is significantly increased. The R22 system is relatively insensitive to the liquid-suction heat exchanger effectiveness.

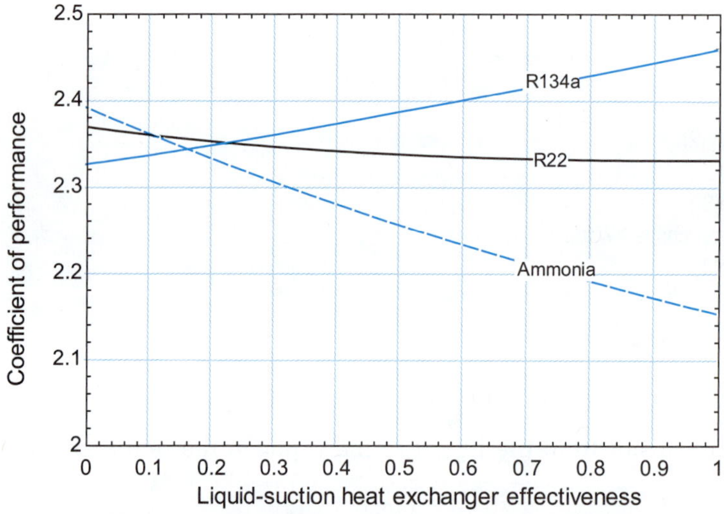

Figure 1: COP as a function of the liquid-suction heat exchanger effectiveness.

Liquid Overfed Evaporator

In a simple vapor compression cycle, the refrigerant leaves the evaporator as saturated vapor, as shown in Figure 9-3, or possibly as slightly superheated vapor, as shown by Figure 1(a) in Example 9.2-1. It is necessary to ensure that the fluid entering the compressor is a vapor since liquids are nearly incompressible and therefore may cause damage to the compressor. However, the performance of the evaporator is reduced by this requirement. The heat transfer coefficient is an indication of the thermal communication between the refrigerant and the inner surface of the evaporator passages and it characterizes the efficiency with which convective heat transfer can occur, as discussed in Section 3.4. A high value of the heat transfer coefficient will lead to a lower approach temperature difference and higher effectiveness for the evaporator. Also, a high value of the heat transfer

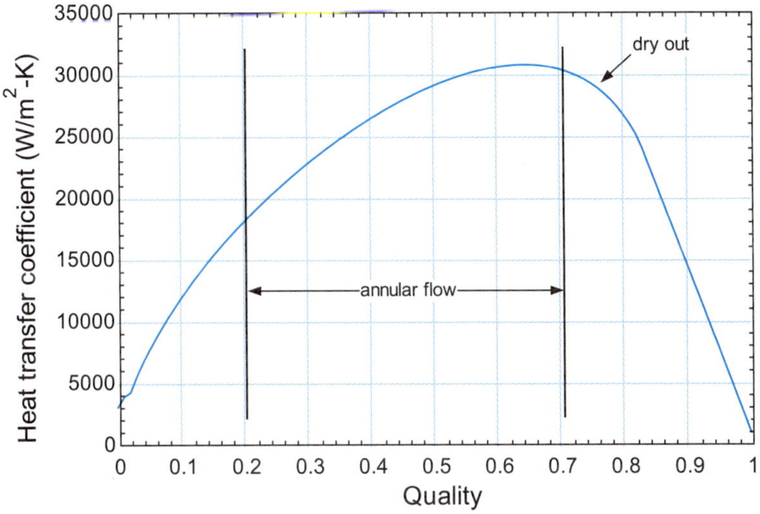

Figure 9-13: Heat transfer coefficient as a function of quality.

coefficient will allow a smaller and less expensive evaporator to be used in the system. Figure 9-13 illustrates the heat transfer coefficient for two-phase ammonia evaporating in a tube at -10°C as a function of quality.

The study of two-phase heat transfer is an important and broad research area that is beyond the scope of this book. Figure 9-13 was generated using a correlation published by Shah (1982) that is based on experimental data; the correlation is one of many that are programmed in EES and it can be accessed from the Heat Transfer function information, as shown in Figure 9-14. Notice in Figure 9-13 that the heat transfer coefficient is very high when the quality of the refrigerant is in the range of 0.2 to 0.7. In this range, the two phase flow has an annular structure in which the liquid forms a thin film on the wall with

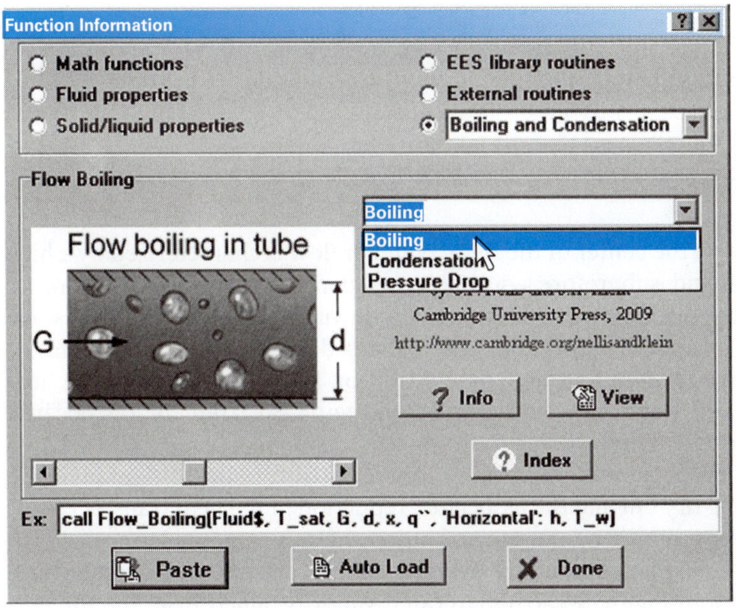

Figure 9-14: Function information window showing the Boiling and Condensation heat transfer functions.

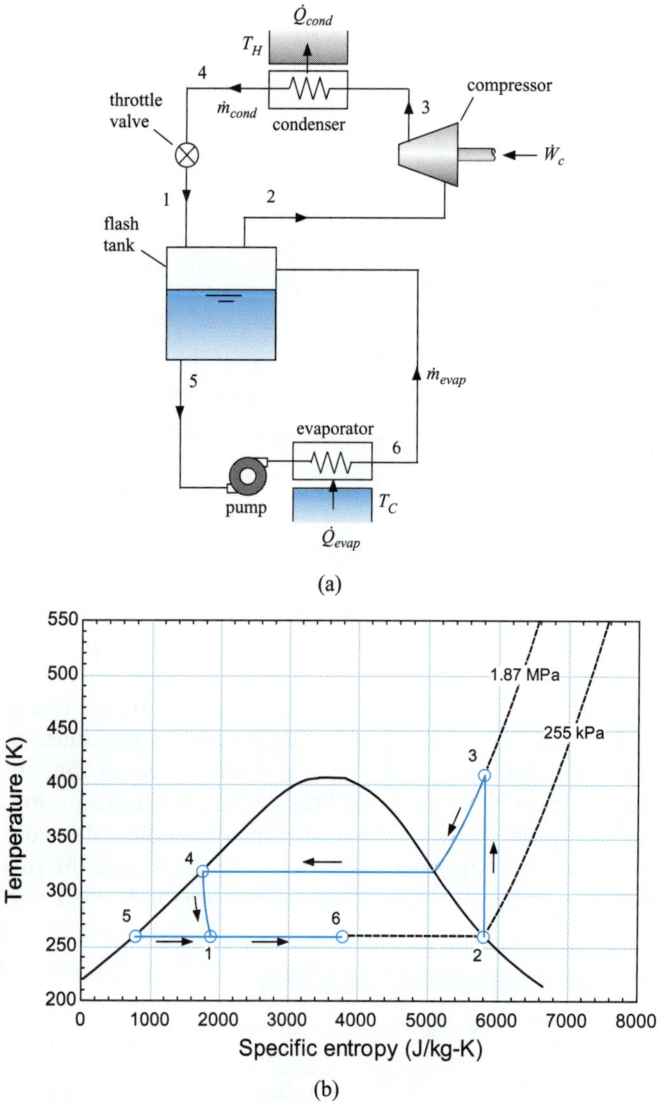

Figure 9-15: (a) Schematic of a liquid overfed evaporator system and (b) the cycle on a *T*-*s* diagram.

the vapor flowing through the center of the tube. Annular flow is characterized by a high heat transfer coefficient and is therefore a desirable flow condition within an evaporator. At high values of quality, the annular structure breaks down and the walls of the passage dry out. This dry out process is associated with a sharp decrease in the heat transfer coefficient and therefore is not a desirable flow condition for an evaporator. The heat transfer coefficient at $x = 1$ and into the superheated region is extremely low because vapor has a very low conductivity.

The liquid overfed evaporator modification keeps the quality of the refrigerant in the evaporator low, avoiding dry out and the associated reduction in the heat transfer coefficient, by placing the evaporator in a separate and independently pumped loop, as shown schematically in Figure 9-15(a). The liquid overfed cycle is shown on a temperature-specific entropy diagram in Figure 9-15(b). Saturated vapor is pulled from the top of a flash tank at state 2. The vapor is compressed to state 3 and then condensed to state 4.

The high pressure liquid leaving the condenser is expanded to state 1 through a throttle valve. States 1 through 4 are consistent with the simple vapor compression cycle shown in Figure 9-3. However, rather than sending the refrigerant at state 1 directly to the evaporator, it is instead collected in a flash tank where the saturated liquid is separated from the saturated vapor; the liquid and vapor naturally separate in the flash tank since their densities are so different. The vapor is pulled from the top of the flash tank and fed to the compressor with mass flow rate $\dot{m}_{cond}$. The liquid is pulled from the bottom of the flash tank at state 5 and fed to the evaporator with mass flow rate $\dot{m}_{evap}$. Because the evaporator is independently pumped, it is possible to adjust the mass flow rate through the evaporator until the quality of the refrigerant leaving the evaporator at state 6 is sufficiently low that dry out is avoided. An energy balance on the flash tank provides the ratio of the mass flow rates:

$$\dot{m}_{cond}\,(h_1 - h_2) + \dot{m}_{evap}\,(h_6 - h_5) = 0 \tag{9-31}$$

Intercooled Cycle

Another method of improving refrigeration cycle performance is to reduce the required compressor power by using *staged compression* and *intercooling*. This process is very similar to the intercooled gas turbine cycle that is discussed in Section 8.3.2. The use of intercooling is advantageous because it requires less work to compress a fluid when it has a low specific volume. The work input per unit mass of refrigerant required by an isentropic compressor was previously derived in Eq. (8-51):

$$\frac{\dot{W}_{s,c}}{\dot{m}} = \Delta h_s = \int_{P_{in}}^{P_{out}} v\,dP \tag{9-32}$$

where the inlet and outlet pressures in Eq. (9-32) correspond to the evaporating and condensing pressures, respectively, for a vapor compression system. As the refrigerant is adiabatically compressed, its temperature increases, which increases the specific volume and thus the required work per unit mass. With the intercooling modification, the refrigerant is compressed in multiple stages. Between each stage, the refrigerant is cooled in order to reduce its specific volume before it enters the next stage. A schematic of a two-stage intercooled process is shown in Figure 9-16(a) and the intercooled cycle is shown on a *T-s* diagram in Figure 9-16(b). Notice that the specific enthalpy change associated with the compression of the refrigerant in the second stage, $h_5 - h_4$ in Figure 9-16(b), is smaller than it would have been if the refrigerant were compressed continuously (i.e., without intercooling) in a single stage from state 3 to the pressure at state 5, which is shown as $h_5^* - h_3$ in Figure 9-16(b).

There are some problems with the intercooling approach illustrated in Figure 9-16(a). The intercooler is a relatively expensive refrigerant-to-air heat exchanger, like the condenser and evaporator already required by the vapor compression cycle. Also, the temperature at the outlet of the first compressor, state 3, may be much higher than it was at the compressor inlet, state 2, but it is likely to be only slightly higher than or even below the ambient temperature, T_H in Figure 9-16(b). In this case, it is not possible to reject heat to ambient and there is no other external temperature reservoir that can serve as a sink to cool the refrigerant between states 3 and 4. There are several intercooling options that overcome these limitations.

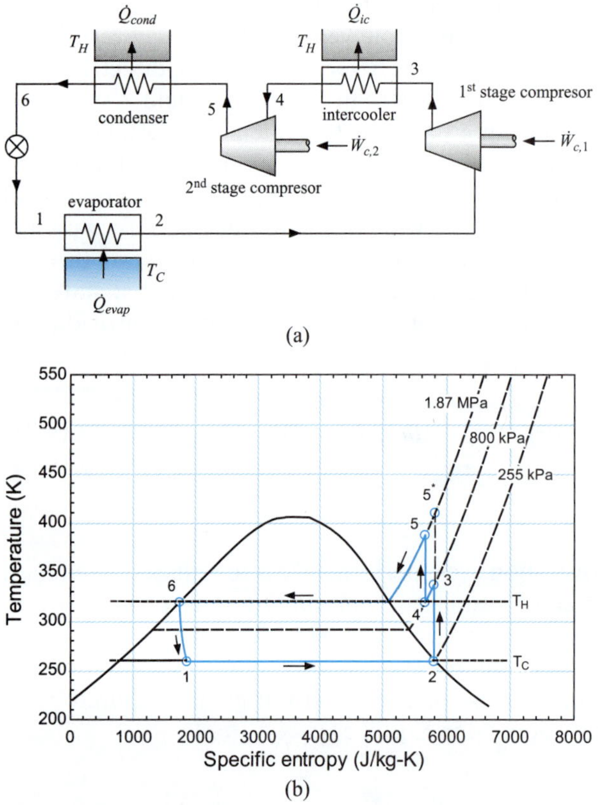

Figure 9-16: (a) Schematic of a vapor compression cycle using a two-stage intercooled compression process and (b) the temperature-specific entropy diagram for the process.

Economized Cycle

An alternative method of providing intercooling is to partially throttle the refrigerant leaving the condenser to an intermediate pressure between the stages using a *flash intercooler* or *flash tank*, as shown in Figure 9-17(a). The economized cycle is shown on a *T-s* diagram in Figure 9-17(b). The liquid exiting the condenser at state 6 is throttled in expansion valve 1 to an intermediate pressure at state 7 where it enters the flash tank. Some of the refrigerant vaporizes (flashes) and saturated vapor at state 9 is removed from the flash tank and mixed with the hot vapor leaving the first stage compressor (sometimes referred to as the booster compressor) at state 3 in order to provide some intercooling for the second stage compressor. The amount of cold vapor that is available for intercooling depends on the quality of the refrigerant at state 7. The saturated liquid in the flash tank is drawn from the bottom at state 8 and throttled to the evaporator pressure through expansion valve 2 where it enters the evaporator and provides the refrigeration load. The vapor exiting the evaporator at state 2 is compressed to state 3. Here, it is mixed with the vapor at state 9 before being compressed in the second stage compressor.

The parameter f is defined as the ratio of the mass flow rate of flash gas leaving the flash tank at state 9 in Figure 9-17(a) to the mass flow rate through the condenser:

$$f = \frac{\dot{m}_{fg}}{\dot{m}_{cond}} \tag{9-33}$$

The value of f is found from an energy balance on the flash tank:

$$h_7 = f\, h_9 + (1 - f)\, h_8 \tag{9-34}$$

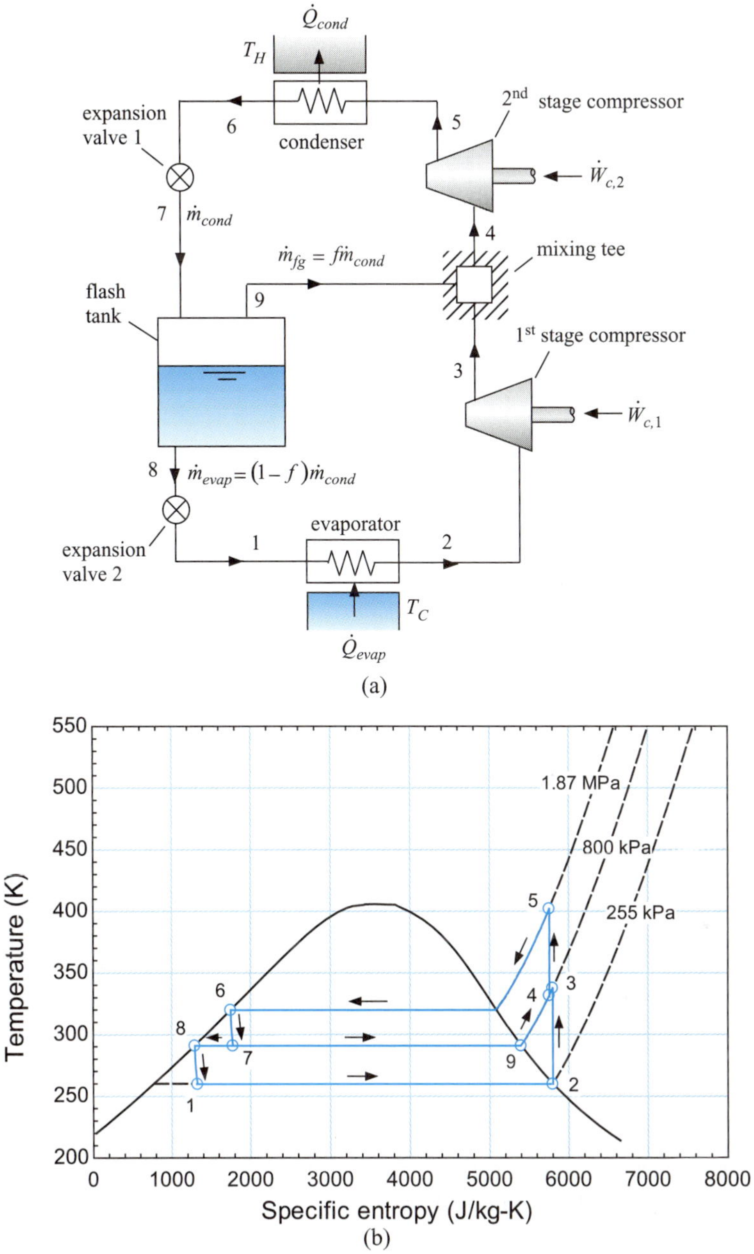

Figure 9-17: (a) Schematic of a flash intercooled (economized) cycle and (b) the cycle on a *T-s* diagram.

Rearranging Eq. (9-34) leads to:

$$f = \frac{h_7 - \overbrace{h_8}^{h_f}}{\underbrace{h_9 - h_8}_{h_g - h_f}} \tag{9-35}$$

Recognizing that $h_8 = h_f$ (the specific enthalpy of saturated liquid) and $h_9 = h_g$ (the specific enthalpy of saturated vapor) shows that f is equal to the quality of the refrigerant leaving expansion valve 1. The vapor entering the second stage compressor at state

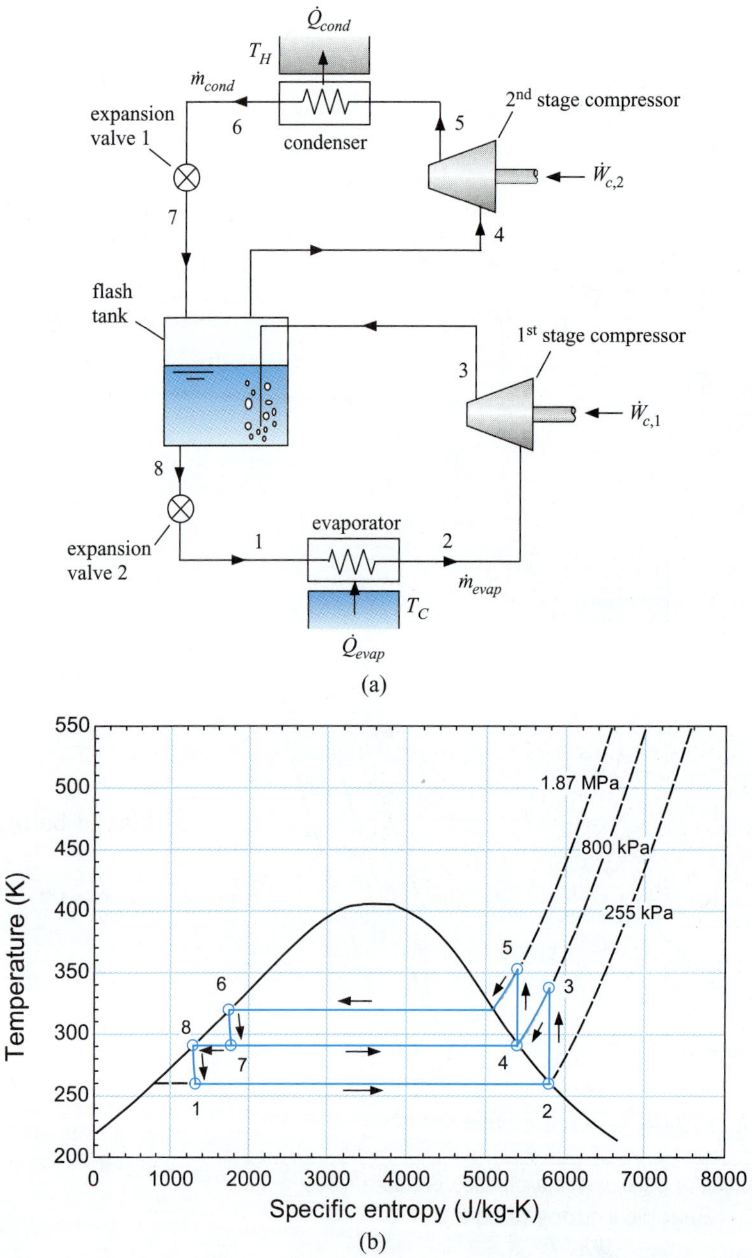

Figure 9-18: (a) Schematic of a flash-intercooled cycle and (b) the cycle on a temperature-specific entropy diagram.

4 is at a temperature that is between the temperatures at state 3 and state 9. The specific enthalpy of the refrigerant at state 4 is found from an energy balance on the mixing tee.

$$h_4 = f\,h_9 + (1 - f)\,h_3 \tag{9-36}$$

The economized cycle is most effective for refrigerants with large throttling losses and therefore a large amount of flash gas generation during the throttling process (i.e., a high value of f). If f is small (as is generally the case for ammonia), then Eq. (9-36) shows that h_4 is very nearly the same as h_3 and very little intercooling is accomplished. The flash intercooled cycle described in the next section provides a method for obtaining a larger amount of intercooling using flash gas even if the refrigerant has a small throttling loss and therefore the quality of the refrigerant leaving the expansion valve is small.

Flash-Intercooled Cycle

The flash-intercooled cycle is shown in Figure 9-18(a). The gas leaving the 1st stage compressor at state 3 is directed to the flash tank. The warm gas causes the generation of additional flash gas and therefore this cycle modification can be used even for refrigerants with low throttling loss and therefore low quality leaving expansion valve 1 at state 7. Saturated vapor is drawn from the flash tank at state 4 and directed into the 2nd stage compressor. The cycle is shown on a *T-s* diagram in Figure 9-18(b); notice that the flash-intercooled cycle results in a large amount of intercooling, much larger than either the intercooled or economized cycles. This extent of intercooling is evident in Figure 9-18(b) because the temperature of the vapor entering the 2nd stage compressor (state 4) is lower than it is in either Figure 9-16(b) or Figure 9-17(b).

The mass flow rate of refrigerant through the condenser and the evaporator are different for the flash intercooled cycle. The ratio of these mass flow rates can be obtained from an energy balance on the flash tank:

$$\dot{m}_{cond}\, h_7 + \dot{m}_{evap}\, h_3 = \dot{m}_{evap}\, h_8 + \dot{m}_{cond}\, h_4 \tag{9-37}$$

EXAMPLE 9.2-4: FLASH INTERCOOLED CYCLE FOR A BLAST FREEZER

Blast freezing (or flash freezing) is the process where food items are quickly frozen by subjecting them to very low temperatures. This process maintains the quality of many food items because it causes water to freeze quickly, avoiding the formation of large crystals. The flash intercooled system shown in Figure 9-18(a) is being considered for the initial design of a blast freezing system.

The flash intercooled system utilizes ammonia. The condenser rejects heat to air at $T_H = 90°\text{F}$ and has an approach temperature difference of $\Delta T_{cond} = 4$ K. The evaporator provides refrigeration to air at $T_C = -50°\text{F}$ and has an approach temperature difference of $\Delta T_{evap} = 5$ K. The isentropic efficiencies of both compressors are $\eta_c = 0.72$. Pressure losses in the heat exchangers are small and may be neglected in this analysis. Also, assume that the degree of superheat in the evaporator and the degree of subcooling in the condenser are negligible.

a) Initially, assume that the intermediate pressure (i.e., the pressure at the discharge of the 1st stage compressor) is $P_{int} = 800$ kPa. Analyze the flash intercooled cycle and determine the *COP* and the refrigeration provided per unit of compressor suction flow rate (the total of both compressors).

The inputs are entered in EES:

```
$UnitSystem SI Radian Mass J kg K
T_H=converttemp(F,K,90 [F])                "outdoor air temperature"
T_C_F=-50 [F]                              "freezer air temperature, in F"
T_C=converttemp(F,K,T_C_F)                 "freezer air temperature"
DT_cond=4 [K]                              "condenser approach temperature difference"
DT_evap=5 [K]                              "evaporator approach temperature difference"
eta_c=0.72 [-]                             "compressor efficiency"
R$='Ammonia'                               "refrigerant"
P_int=800 [kPa]*convert(kPa,Pa)            "intermediate pressure"
```

The temperature of the refrigerant leaving the condenser is specified by the condenser approach temperature difference:

$$T_6 = T_H + \Delta T_{cond}$$

In the absence of any subcooling, saturated liquid refrigerant leaves the condenser. The state of the refrigerant is fixed by the temperature and quality, $x_6 = 0$. The specific entropy, pressure, and specific enthalpy (s_6, P_6, and h_6) are determined.

```
"State 6, Condenser exit/Valve 1 inlet"
T[6]=T_H+DT_cond                          "temperature"
x[6]=0 [-]                                "quality"
s[6]=entropy(R$,T=T[6],x=x[6])            "specific entropy"
P[6]=pressure(R$,T=T[6],x=x[6])           "pressure"
h[6]=enthalpy(R$,T=T[6],x=x[6])           "specific enthalpy"
```

The temperature of the refrigerant leaving the evaporator is specified by the evaporator approach temperature difference:

$$T_2 = T_C - \Delta T_{evap}$$

In the absence of any superheating, saturated vapor refrigerant leaves the evaporator. The state of the refrigerant is fixed by the temperature and quality, $x_2 = 1$. The specific entropy, pressure, specific enthalpy, and specific volume (s_2, P_2, h_2, and v_2) are determined.

```
"State 2, Evaporator exit/1st stage Compressor inlet"
T[2]=T_C-DT_evap                              "temperature"
x[2]=1 [-]                                    "quality"
s[2]=entropy(R$,T=T[2],x=x[2])                "specific entropy"
P[2]=pressure(R$,T=T[2],x=x[2])               "pressure"
h[2]=enthalpy(R$,T=T[2],x=x[2])               "specific enthalpy"
v[2]=volume(R$,T=T[2],x=x[2])                 "specific volume"
```

The pressure leaving the 1st stage compressor is specified by the intermediate pressure, $P_3 = P_{int}$. The specific entropy at the exit of a reversible compressor operating under the same conditions is $s_{s,3} = s_2$. The specific enthalpy at the exit of a reversible compressor, $h_{s,3}$, is fixed by the specific entropy and pressure. The power per evaporator mass flow rate for a reversible compressor is:

$$\frac{\dot{W}_{s,c,1}}{\dot{m}_{evap}} = h_{s,3} - h_2$$

The actual power is determined from the compressor efficiency:

$$\frac{\dot{W}_{c,1}}{\dot{m}_{evap}} = \frac{1}{\eta_c}\frac{\dot{W}_{s,c,1}}{\dot{m}_{evap}}$$

The specific enthalpy leaving the compressor is obtained from an energy balance:

$$h_3 = h_2 + \frac{\dot{W}_{c,1}}{\dot{m}_{evap}}$$

The state of the refrigerant leaving the compressor is fixed by the specific enthalpy and pressure. The specific entropy and temperature (s_3 and T_3) are determined:

```
"State 3, 1st stage Compressor exit/Mixing T inlet"
P[3]=P_int                                "pressure"
s_s[3]=s[2]                               "specific entropy leaving reversible compressor"
h_s[3]=enthalpy(R$,s=s_s[3],P=P[3])       "specific enthalpy leaving reversible compressor"
W_dot_s_c_1\m_dot_evap=h_s[3]-h[2]        "power per evaporator mass flow for reversible compressor"
W_dot_c_1\m_dot_evap=W_dot_s_c_1\m_dot_evap/eta_c
                                          "power per evaporator mass flow"
h[3]=h[2]+W_dot_c_1\m_dot_evap            "specific enthalpy"
T[3]=temperature(R$,h=h[3],P=P[3])        "temperature"
s[3]=entropy(R$,h=h[3],P=P[3])            "specific entropy"
```

The specific enthalpy of the refrigerant leaving the upper stage throttle is specified, $h_7 = h_6$. The pressure leaving the throttle is the intermediate pressure, $P_7 = P_{int}$. State 7 is fixed by the pressure and specific enthalpy, which allows the specific entropy, temperature, and quality (s_7, T_7, and x_7) to be determined.

```
"State 7, Throttle 1 exit/Flash chamber inlet"
h[7]=h[6]                                 "specific enthalpy"
P[7]=P_int                                "pressure"
s[7]=entropy(R$,h=h[7],P=P[7])            "specific entropy"
T[7]=temperature(R$,h=h[7],P=P[7])        "temperature"
x[7]=quality(R$,h=h[7],P=P[7])            "quality"
```

Saturated liquid leaves the flash tank at state 8, $x_8 = 0$. The pressure at state 8 is the intermediate pressure, $P_8 = P_{int}$. Knowing the pressure and quality allows all other properties to be determined.

```
"State 8,Flash chamber exit, Throttle 2 inlet"
x[8]=0 [-]                                "quality"
P[8]=P_int                                "pressure"
s[8]=entropy(R$,P=P[8],x=x[8])            "specific entropy"
T[8]=temperature(R$,x=x[8],P=P[8])        "temperature"
h[8]=enthalpy(R$,P=P[8],x=x[8])           "specific enthalpy"
```

Saturated vapor leaves the flash tank at state 4, $x_4 = 1$. The pressure at state 4 is the intermediate pressure, $P_4 = P_{int}$. The temperature, specific entropy, specific enthalpy, and specific volume (T_4, s_4, h_4, and v_4) are determined.

```
"State 4,Flash chamber exit, 2nd stage compressor inlet"
x[4]=1 [-]                                "quality"
P[4]=P_int                                "pressure"
s[4]=entropy(R$,P=P[4],x=x[4])            "specific entropy"
T[4]=temperature(R$,x=x[4],P=P[4])        "temperature"
h[4]=enthalpy(R$,P=P[4],x=x[4])           "specific enthalpy"
v[4]=volume(R$,P=P[4],x=x[4])             "specific volume"
```

The pressure leaving the 2nd stage compressor is $P_5 = P_6$. The specific entropy leaving a reversible compressor operating under the same conditions is $s_{s,5} = s_4$. The specific enthalpy leaving a reversible compressor, $h_{s,5}$, is fixed by the specific entropy and pressure. The power per condenser mass flow rate for a reversible compressor is:

$$\frac{\dot{W}_{s,c,2}}{\dot{m}_{cond}} = h_{s,5} - h_4$$

The actual power is determined from the compressor efficiency:

$$\frac{\dot{W}_{c,2}}{\dot{m}_{cond}} = \frac{1}{\eta_c}\frac{\dot{W}_{s,c,2}}{\dot{m}_{cond}}$$

The specific enthalpy leaving the compressor is obtained from an energy balance:

$$h_5 = h_4 + \frac{\dot{W}_{c,2}}{\dot{m}_{cond}}$$

The state of the refrigerant leaving the compressor is fixed by the specific enthalpy and pressure, which allows s_5 and T_5 to be determined:

```
"State 5, 2nd stage compressor exit, Condenser inlet"
P[5]=P[6]                                   "pressure"
s_s[5]=s[4]                                 "specific entropy leaving reversible compressor"
h_s[5]=enthalpy(R$,s=s_s[5],P=P[5])         "specific enthalpy leaving reversible compressor"
W_dot_s_c_2\m_dot_cond=h_s[5]-h[4]          "power per condenser mass flow for reversible compressor"
W_dot_c_2\m_dot_cond=W_dot_s_c_2\m_dot_cond/eta_c   "power per condenser mass flow"
h[5]=h[4]+W_dot_c_2\m_dot_cond              "specific enthalpy"
T[5]=temperature(R$,h=h[5],P=P[5])          "temperature"
s[5]=entropy(R$,h=h[5],P=P[5])              "specific entropy"
```

The specific enthalpy of the refrigerant leaving the lower stage throttle is fixed by an energy balance on the throttle, $h_1 = h_8$. The pressure leaving the throttle is the evaporator pressure, $P_1 = P_2$. State 1 is fixed by the pressure and specific enthalpy, therefore s_1 and T_1 can be determined.

```
"State 1,Throttle 2 exit, Evaporator inlet"
h[1]=h[8]                                   "specific enthalpy"
P[1]=P[2]                                   "pressure"
s[1]=entropy(R$,P=P[1],h=h[1])              "specific entropy"
T[1]=temperature(R$,P=P[1],h=h[1])          "temperature"
```

An energy balance on the flash tank provides:

$$\dot{m}_{cond}\, h_7 + \dot{m}_{evap}\, h_3 = \dot{m}_{evap}\, h_8 + \dot{m}_{cond}\, h_4$$

Dividing through by the mass flow rate in the condenser,

$$h_7 + \frac{\dot{m}_{evap}}{\dot{m}_{cond}}\, h_3 = \frac{\dot{m}_{evap}}{\dot{m}_{cond}}\, h_8 + h_4$$

provides the ratio of the evaporator mass flow rate to the condenser mass flow rate. The power required by the 1st stage compressor per unit of condenser mass flow rate is:

$$\frac{\dot{W}_{c,1}}{\dot{m}_{cond}} = \frac{\dot{W}_{c,1}}{\dot{m}_{evap}}\frac{\dot{m}_{evap}}{\dot{m}_{cond}}$$

The refrigeration provided in the evaporator per unit of condenser mass flow rate is:

$$\frac{\dot{Q}_{evap}}{\dot{m}_{cond}} = \frac{\dot{m}_{evap}}{\dot{m}_{cond}} (h_2 - h_1)$$

```
h[7]+m_dot_evap\m_dot_cond*h[3]=m_dot_evap\m_dot_cond*h[8]+h[4]
                    "energy balance on the flash tank"
W_dot_c_1\m_dot_cond=W_dot_c_1\m_dot_evap*m_dot_evap\m_dot_cond
                    "compressor 1 power per condenser mass flow rate"
Q_dot_evap\m_dot_cond=m_dot_evap\m_dot_cond*(h[2]-h[1])
                    "evaporator heat transfer per condenser mass flow rate"
```

The *COP* of the system is:

$$COP = \frac{\dfrac{\dot{Q}_{evap}}{\dot{m}_{cond}}}{\dfrac{\dot{W}_{C,1}}{\dot{m}_{cond}} + \dfrac{\dot{W}_{C,2}}{\dot{m}_{cond}}}$$

The refrigeration provided per unit of suction flow rate (the sum of both compressors) is:

$$\frac{\dot{Q}_{evap}}{\dot{V}_C} = \frac{\dot{Q}_{evap}}{\dot{m}_{cond}\, v_4 + \dot{m}_{evap}\, v_2} = \frac{\dot{Q}_{evap}}{\dot{m}_{cond}\left(v_4 + \dfrac{\dot{m}_{evap}}{\dot{m}_{cond}}\, v_2\right)}$$

```
COP=Q_dot_evap\m_dot_cond/(W_dot_c_1\m_dot_cond+W_dot_c_2\m_dot_cond)
                    "Coefficient of Performance"
Q_dot_evap\V_dot_c=Q_dot_evap\m_dot_cond/(v[4]+m_dot_evap\m_dot_cond*v[2])
                    "refrigeration per comp. suction volumetric flow rate"
Q_dot_evap\V_dot_c_tonpcfm=Q_dot_evap\V_dot_c*convert(J/m^3,ton/cfm)
                    "in ton/cfm"
```

which leads to $COP = 1.36$ and $\dot{Q}_{evap}/\dot{V}_C = 373.2$ kJ/m^3 (0.050 ton/cfm).

b) Determine the intermediate pressure that maximizes the *COP* of the system. Plot the *COP* and refrigeration per unit of suction flow rate as a function of T_C; make sure that the intermediate pressure is optimized at every value of T_C.

Update the guess values, since these results will provide a useful starting point for the following optimization. The intermediate pressure is commented out:

```
{P_int=800 [kPa]*convert(kPa,Pa)}          "intermediate pressure"
```

and the Min/Max option is selected from the Calculate menu. The value of the variable COP is maximized by adjusting the variable P_int. Set appropriate bounds for the intermediate pressure and EES will identify that the optimal intermediate pressure is $P_{int} = 281.3$ kPa, which results in a $COP = 1.445$. In order to generate the requested plot, it is necessary to set up a Parametric table that includes *COP*, P_{int},

$\dot{Q}_{evap}/\dot{V}_c$, and T_C. Adjust T_C from $-60°$F to $0°$F, as shown in Figure 1(a). Comment out the value of T_C that is specified in the Equations window:

```
{T_C_F=-50 [F]}          "freezer air temperature, in F"
```

and select Min/Max Table from the Calculate menu. Maximize *COP* by varying P_{int}, as shown in Figure 1(b). EES will fill out the Parametric table. The optimum *COP* and $\dot{Q}_{evap}/\dot{V}_c$ as a function of T_C are shown in Figure 2.

1..11	1 COP [-]	2 P_{int} [Pa]	3 $T_{C,F}$ [F]	4 $Q_{evap\backslash Vdot,c,ton}$ [ton/cfm]
Run 1			-60	
Run 2			-54	
Run 3			-48	
Run 4			-42	
Run 5			-36	
Run 6			-30	
Run 7			-24	
Run 8			-18	
Run 9			-12	
Run 10			-6	
Run 11			0	

(a)

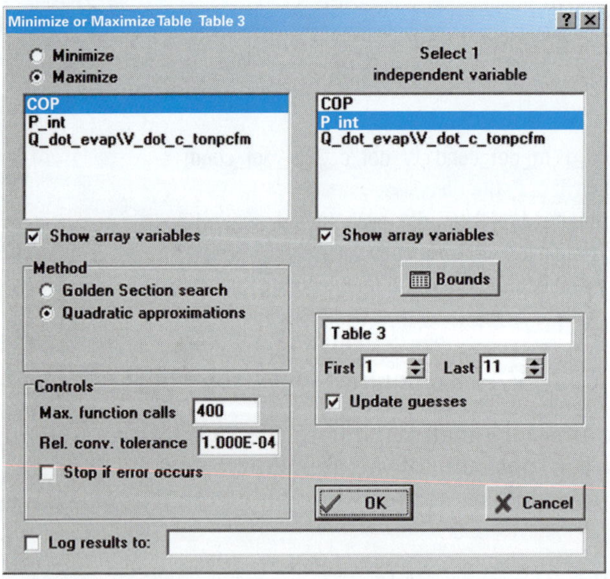

(b)

Figure 1: (a) Parametric table and (b) Min/Max Table dialog.

c) The freezer requires $\dot{Q}_{evap} = 1.5$ ton of refrigeration and runs 24 hours a day all year long. Energy costs $ec = 0.12$ \$/kW-hr. The capital cost of the compressors is directly proportional to the total suction flow rate; the constant of proportionality is $cc = 495{,}000$ \$-s/m^3. Plot the operating, capital, and total cost (ignore all capital costs except the compressor cost) for an economic time frame of *time* = 5 year. Ignore the time value of money.

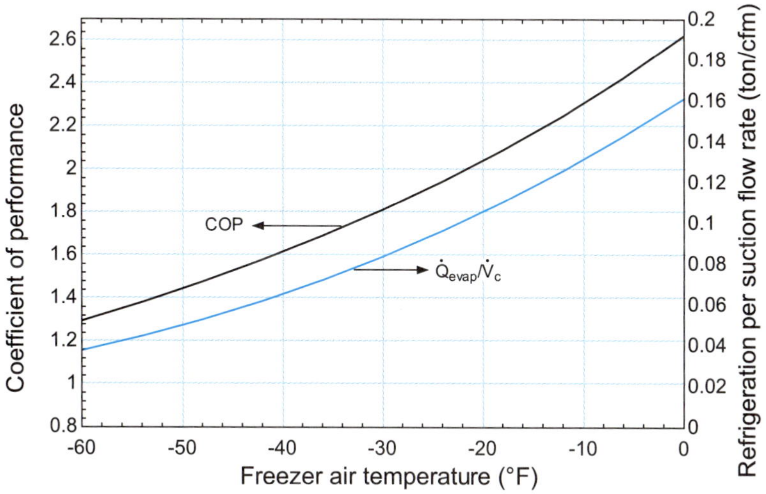

Figure 2: *COP* and refrigeration per unit of compressor suction flow rate as a function of T_C.

The additional inputs are entered in EES:

```
Q_dot_evap=1.5 [ton]*convert(ton,W)            "required freezer load"
ec=0.12 [$/kW-hr]*convert($/kW-hr,$/J)         "electricity cost"
time=5 [year]*convert(year,s)                  "time of operation"
cc=495000 [$-s/m^3]                            "cost of compressor per unit of suction flow rate"
```

The total compressor power is obtained from the *COP*:

$$\dot{W}_c = \frac{\dot{Q}_{evap}}{COP}$$

The operating cost is obtained from:

$$OC = \dot{W}_c\, ec\, time$$

```
W_dot_c=Q_dot_evap/COP                         "compressor power"
OC=W_dot_c*time*ec                             "operating cost"
```

The total suction volumetric flow rate is obtained from:

$$\dot{Q}_{evap} = \frac{\dot{Q}_{evap}}{\dot{V}_c} \dot{V}_c$$

The compressor cost is:

$$CompC = \dot{V}_c\, cc$$

The total cost of owning and operating the system is:

$$TotalC = CompC + OC$$

```
Q_dot_evap=Q_dot_evap\V_dot_c*V_dot_c          "volumetric flow rate for compressors"
CompC=V_dot_c*cc                               "compressor cost"
TotalC=CompC+OC                                "total cost"
```

Three additional columns for the variables OC, CompC, and TotalC are added to the Parametric table from (b). The Min/Max Table option is again selected and used to populate the table. The operating cost, compressor cost, and total cost as a function of the freezer temperature are shown in Figure 3.

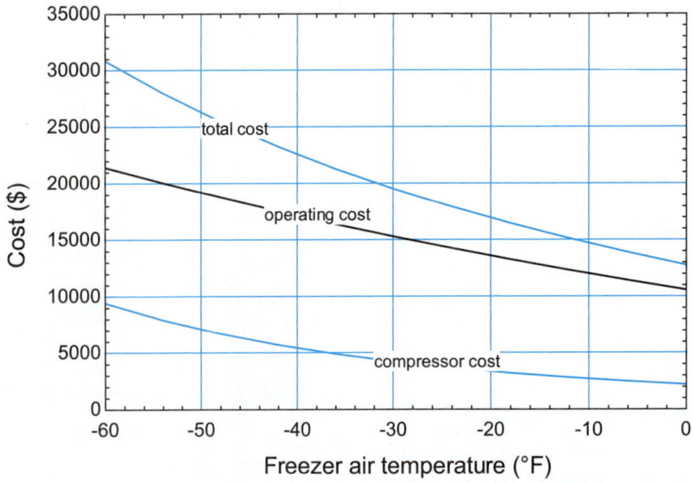

Figure 3: Total cost, operating cost, and compressor cost as a function of the freezer air temperature.

Figure 3 shows that the total cost of owning and operating the flash intercooled ammonia system increases dramatically as the freezer temperature decreases. This behavior occurs because the *COP* decreases with decreasing temperature and also because the specific volume of saturated vapor ammonia becomes very large as the temperature is reduced. Larger specific volumes require larger compressors that become quite expensive, as shown in Figure 3. In order to reduce the cost of the compressors required by the blast freezing system, a cascade refrigeration system is considered in Example 9.2-5.

EXAMPLE 9.2-5: CASCADE CYCLE FOR A BLAST FREEZER

The cascade system shown in Figure 1 is being considered as an alternative to the flash intercooled system analyzed in Example 9.2-4.

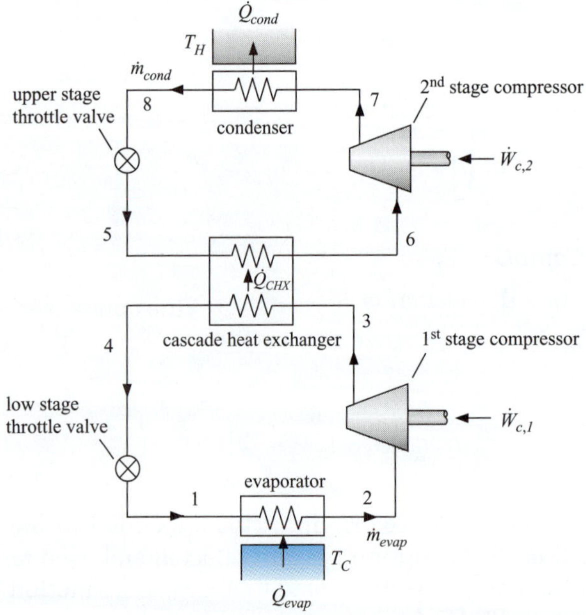

Figure 1: Cascade system.

The cascade system is essentially two separate vapor compression cycles that are integrated by a heat exchanger. The upper stage cycle utilizes ammonia and rejects heat to the ambient air at T_H. The lower stage cycle uses carbon dioxide, which has a much lower specific volume than ammonia and therefore requires a smaller compressor, to provide the refrigeration to the freezer. The lower stage rejects heat to the upper stage in the cascade heat exchanger.

a) Initially, assume that the intermediate temperature (i.e., the temperature of the ammonia leaving the upper stage throttle valve at state 5) is T_{int} = 270 K. Analyze the cascade cycle in order to determine the *COP* and the refrigeration provided per unit of compressor suction flow rate (total of both compressors). Assume that the conditions, costs, and component performance parameters are identical to the flash intercooled cycle analyzed in Example 9.2-4. The approach temperature difference of the cascade heat exchanger is $\Delta T_{CHX} = 6$ K.

A new EES file is created. The inputs are entered:

```
$UnitSystem SI Radian Mass J kg K
T_H=converttemp(F,K,90 [F])                   "outdoor air temperature"
T_C_F=-50 [F]                                 "freezer air temperature, in F"
T_C=converttemp(F,K,T_C_F)                    "freezer air temperature"
DT_cond=4 [K]                                 "condenser approach temperature difference"
DT_evap=5 [K]                                 "evaporator approach temperature difference"
DT_CHX=6 [K]                                  "cascade heat exchanger approach temperature difference"
eta_c=0.72 [-]                                "compressor efficiency"
T_int=270 [K]                                 "temperature of refrigerant entering cascade HX"
R_u$='Ammonia'                                "refrigerant for upper stage"
R_l$='CarbonDioxide'                          "refrigerant for lower stage"
Q_dot_evap=1.5 [ton]*convert(ton,W)           "required freezer load"
ec=0.12 [$/kW-hr]*convert($/kW-hr,$/J)        "electricity cost"
time=5 [year]*convert(year,s)                 "time of operation"
cc=495000 [$-s/m^3]                           "cost of compressor per unit of suction flow rate"
```

The temperature of the refrigerant leaving the condenser is specified by the condenser approach temperature difference:

$$T_8 = T_H + \Delta T_{cond}$$

In the absence of any subcooling, saturated liquid refrigerant leaves the condenser. The state of the refrigerant is fixed by the temperature and quality, $x_8 = 0$, allowing s_8, P_8, and h_8 to be determined.

```
"State 8, Condenser exit/Valve 1 inlet"
T[8]=T_H+DT_cond                              "temperature"
x[8]=0 [-]                                    "quality"
s[8]=entropy(R_u$,T=T[8],x=x[8])              "specific entropy"
P[8]=pressure(R_u$,T=T[8],x=x[8])             "pressure"
h[8]=enthalpy(R_u$,T=T[8],x=x[8])             "specific enthalpy"
```

The temperature of the refrigerant leaving the evaporator is specified by the evaporator approach temperature difference:

$$T_2 = T_C - \Delta T_{evap}$$

In the absence of any superheating, saturated vapor refrigerant leaves the evaporator. The state of the refrigerant is fixed by the temperature and quality, $x_2 = 1$, allowing s_2, P_2, h_2, and v_2 to be determined.

```
"State 2, Evaporator exit/1st stage Compressor inlet"
T[2]=T_C-DT_evap                                   "temperature"
x[2]=1 [-]                                         "quality"
s[2]=entropy(R_l$,T=T[2],x=x[2])                   "specific entropy"
P[2]=pressure(R_l$,T=T[2],x=x[2])                  "pressure"
h[2]=enthalpy(R_l$,T=T[2],x=x[2])                  "specific enthalpy"
v[2]=volume(R_l$,T=T[2],x=x[2])                    "specific volume"
```

The temperature of the upper stage refrigerant entering the cascade heat exchanger is specified by the intermediate temperature, $T_5 = T_{int}$. According to an energy balance, the specific enthalpy of the refrigerant passing through the upper stage throttle valve does not change, i.e., $h_5 = h_8$. State 5 is fixed by the specific enthalpy and temperature allowing P_5 and s_5 to be determined.

```
"State 5, Cascade HX inlet/Valve 1 exit"
h[5]=h[8]                                          "specific enthalpy"
T[5]=T_int                                         "temperature"
P[5]=pressure(R_u$,h=h[5],T=T[5])                  "pressure"
s[5]=entropy(R_u$,h=h[5],T=T[5])                   "specific entropy"
```

Assuming negligible pressure losses, the pressure of the upper stage refrigerant passing through the cascade heat exchanger does not change, $P_6 = P_5$. In the absence of any superheating, saturated vapor refrigerant leaves the cascade heat exchanger. The state of the refrigerant is fixed by the temperature and quality, $x_6 = 1$ allowing s_6, P_6, h_6, and v_6 to be determined.

```
"State 6, Cascade HX exit/2nd stage Compressor inlet"
P[6]=P[5]                                          "pressure"
x[6]=1 [-]                                         "quality"
s[6]=entropy(R_u$,P=P[6],x=x[6])                   "specific entropy"
T[6]=temperature(R_u$,P=P[6],x=x[6])               "temperature"
h[6]=enthalpy(R_u$,P=P[6],x=x[6])                  "specific enthalpy"
v[6]=volume(R_u$,P=P[6],x=x[6])                    "specific volume"
```

The 2nd stage compressor is analyzed as in Example 9.2-4:

```
"State 7, 2nd stage compressor exit/condenser inlet"
P[7]=P[8]                                     "pressure"
s_s[7]=s[6]                                   "specific entropy leaving reversible compressor"
h_s[7]=enthalpy(R_u$,s=s_s[7],P=P[7])         "specific enthalpy leaving reversible compressor"
W_dot_s_c_2\m_dot_cond=h_s[7]-h[6]            "power per condenser mass flow for reversible compressor"
W_dot_c_2\m_dot_cond=W_dot_s_c_2\m_dot_cond/eta_c
                                              "power per condenser mass flow for compressor"
h[7]=h[6]+W_dot_c_2\m_dot_cond                "specific enthalpy"
T[7]=temperature(R_u$,h=h[7],P=P[7])          "temperature"
s[7]=entropy(R_u$,h=h[7],P=P[7])              "specific entropy"
```

which provides the properties at state 7 and the power required by the 2nd stage compressor per unit of condenser mass flow rate, $\dot{W}_{c,2}/\dot{m}_{cond}$. The temperature of the lower stage refrigerant leaving the cascade heat exchanger is specified by the cascade heat exchanger approach temperature difference:

$$T_4 = T_5 + \Delta T_{CHX}$$

In the absence of any subcooling, saturated liquid refrigerant leaves the cascade heat exchanger. The state of the refrigerant is fixed by the temperature and quality, $x_4 = 0$, allowing s_4, P_4, and h_4 to be determined.

```
"State 4, Cascade HX exit/Valve 2 inlet"
T[4]=T[5]+DT_CHX                              "temperature"
x[4]=0 [-]                                    "quality"
s[4]=entropy(R_l$,T=T[4],x=x[4])              "specific entropy"
P[4]=pressure(R_l$,T=T[4],x=x[4])             "pressure"
h[4]=enthalpy(R_l$,T=T[4],x=x[4])             "specific enthalpy"
```

The specific enthalpy of the refrigerant leaving the lower stage throttle is fixed by an energy balance, $h_1 = h_4$. The pressure leaving the throttle is the evaporator pressure, $P_1 = P_2$. State 1 is fixed by the pressure and specific enthalpy. The specific entropy, temperature, and quality (s_1, T_1, and x_1) are determined.

```
"State 1, Throttle 2 exit/Evaporator inlet"
h[1]=h[4]                                     "specific enthalpy"
P[1]=P[2]                                     "pressure"
s[1]=entropy(R_l$,h=h[1],P=P[1])              "specific entropy"
T[1]=temperature(R_l$,h=h[1],P=P[1])          "temperature"
x[1]=quality(R_l$,h=h[1],P=P[1])              "quality"
```

The 1st stage compressor is analyzed as in Example 9.2-4,

```
"State 3, 1st stage compressor exit/cascade HX inlet"
P[3]=P[4]                                  "pressure"
s_s[3]=s[2]                                "specific entropy leaving reversible compressor"
h_s[3]=enthalpy(R_I$,s=s_s[3],P=P[3])      "specific enthalpy leaving reversible compressor"
W_dot_s_c_1\m_dot_evap=h_s[3]-h[2]         "power per evaporator mass flow for reversible compressor"
W_dot_c_1\m_dot_evap=W_dot_s_c_1\m_dot_evap/eta_c
                                           "power per evaporator mass flow rate for compressor"
h[3]=h[2]+W_dot_c_1\m_dot_evap             "specific enthalpy"
T[3]=temperature(R_I$,h=h[3],P=P[3])       "temperature"
s[3]=entropy(R_I$,h=h[3],P=P[3])           "specific entropy"
```

which provides the properties at state 3 and the power required by the 1st stage compressor per unit of evaporator mass flow rate, $\dot{W}_{c,1}/\dot{m}_{evap}$. An energy balance on the cascade heat exchanger provides the ratio of the evaporator to the condenser mass flow rates:

$$h_5 + \frac{\dot{m}_{evap}}{\dot{m}_{cond}} h_3 = h_6 + \frac{\dot{m}_{evap}}{\dot{m}_{cond}} h_4$$

The power required by the 1st stage compressor per unit of condenser mass flow rate is:

$$\frac{\dot{W}_{c,1}}{\dot{m}_{cond}} = \frac{\dot{W}_{c,1}}{\dot{m}_{evap}} \frac{\dot{m}_{evap}}{\dot{m}_{cond}}$$

The refrigeration provided in the evaporator per unit of condenser mass flow rate is:

$$\frac{\dot{Q}_{evap}}{\dot{m}_{cond}} = \frac{\dot{m}_{evap}}{\dot{m}_{cond}} (h_2 - h_1)$$

```
h[5]+m_dot_evap\m_dot_cond*h[3]=h[6]+m_dot_evap\m_dot_cond*h[4]
                         "energy balance on cascade HX"
W_dot_c_1\m_dot_cond=W_dot_c_1\m_dot_evap*m_dot_evap\m_dot_cond
                         "compressor 1 power per condenser mass flow rate"
Q_dot_evap\m_dot_cond=m_dot_evap\m_dot_cond*(h[2]-h[1])
                         "evaporator heat transfer rate per condenser mass flow rate"
```

The *COP* of the system is:

$$COP = \frac{\dfrac{\dot{Q}_{evap}}{\dot{m}_{cond}}}{\dfrac{\dot{W}_{c,1}}{\dot{m}_{cond}} + \dfrac{\dot{W}_{c,2}}{\dot{m}_{cond}}}$$

The refrigeration provided per unit of suction flow rate is:

$$\frac{\dot{Q}_{evap}}{\dot{V}_c} = \frac{\dot{Q}_{evap}}{\dot{m}_{cond}\, v_6 + \dot{m}_{evap}\, v_2} = \frac{\dot{Q}_{evap}}{\dot{m}_{cond} \left(v_6 + \dfrac{\dot{m}_{evap}}{\dot{m}_{cond}} v_2\right)}$$

```
COP=Q_dot_evap\m_dot_cond/(W_dot_c_1\m_dot_cond+W_dot_c_2\m_dot_cond)
                              "Coefficient of Performance"
Q_dot_evap\V_dot_c=Q_dot_evap\m_dot_cond/(v[6]+m_dot_evap\m_dot_cond*v[2])
                              "refrigeration per unit of compressor suction flow rate"
Q_dot_evap\V_dot_c_tonpcfm=Q_dot_evap\V_dot_c*convert(J/m^3,ton/cfm)
                              "in ton/cfm"
```

which leads to $COP = 1.18$ and $\dot{Q}_{evap}/\dot{V}_c = 1435$ kJ/m^3 (0.193 ton/cfm).

b) Determine the intermediate temperature that maximizes the *COP* of the system. Plot the *COP* and refrigeration per unit of suction flow rate as a function of T_C; the intermediate temperature should be optimized for every value of T_C.

The intermediate temperature and freezer air temperature are commented out:

```
{T_int=270 [K]}        "temperature of refrigerant entering cascade HX"
{T_C_F=-50 [F]}        "freezer air temperature, in F"
```

A Parametric table that includes *COP*, T_{int}, $\dot{Q}_{evap}/\dot{V}_c$, and T_C is set up. The value of T_C is adjusted from -60°F to 0°F and the Min/Max Table option is selected from the Calculate menu. The *COP* is maximized by varying T_{int}. The *COP* and $\dot{Q}_{evap}/\dot{V}_c$ as a function of T_C are shown in Figure 2. Comparing Figure 2 for the cascade system to Figure 2 in Example 9.2-4 for the flash intercooled cycle shows that the cascade cycle has a slightly lower *COP* but provides a much larger value of $\dot{Q}_{evap}/\dot{V}_c$, particularly at low temperatures. This result indicates that a smaller (and therefore less expensive) low stage compressor could be employed in the cascade system.

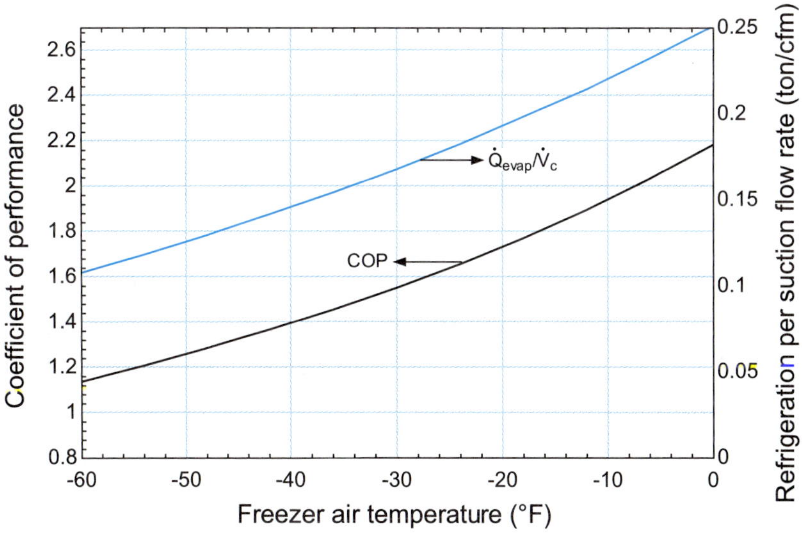

Figure 2: *COP* and refrigeration per compressor suction flow rate as a function of T_C for the cascade system.

c) Overlay on a single plot the operating, compressor, and total cost associated with the cascade and flash intercooled systems. What is the break-even temperature, below which it would make economic sense to consider the cascade cycle?

The operating, compressor, and total costs are calculated as in Example 9.2-4.

```
W_dot_c=Q_dot_evap/COP                              "compressor power"
OC=W_dot_c*time*ec                                  "operating cost"
Q_dot_evap=Q_dot_evap\V_dot_c*V_dot_c               "volumetric flow rate for compressors"
CompC=V_dot_c*cc                                    "compressor cost"
TotalC=CompC+OC                                     "total cost"
```

Figure 3 illustrates the operating, compressor, and total costs associated with the flash intercooled system (from Example 9.2-4) and the cascade system. Notice that the operating cost associated with the cascade system is higher due to its lower *COP*. However, the compressor cost is substantially lower. The advantage associated with having smaller compressors becomes substantial at low temperatures. At temperatures below approximately -35°F, the total cost of owning the cascade system is less than the flash intercooled system for the economic parameters assumed in this problem.

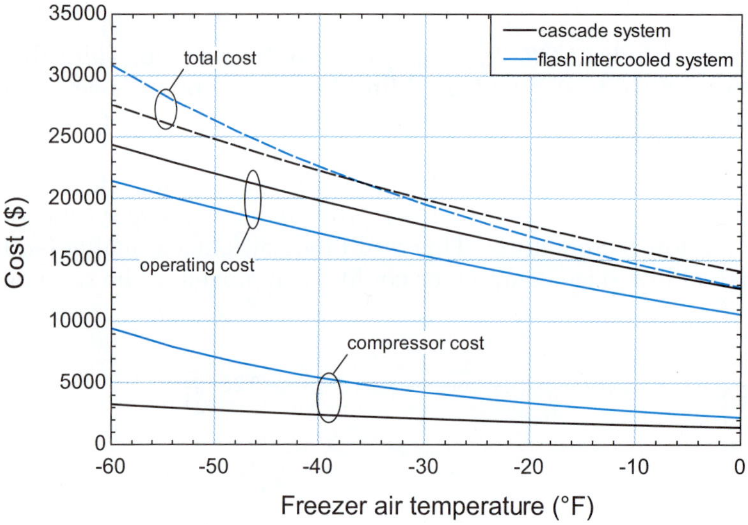

Figure 3: Operating, compressor, and total costs for the flash intercooled and cascade systems.

9.3 Heat Pumps

The vapor compression cycle discussed in Section 9.2 can be used to provide heating as well as cooling. Figure 9-19(a) illustrates a vapor compression system that is utilized to provide cooling to a building during the cooling season when the outdoor air temperature is high. The evaporator is located within the building (typically in the air supply ducts) and the condenser is located outside the building in order to reject heat to the ambient air.

By modifying the flow direction, the same hardware can be used to provide space heating (i.e., heat to maintain the building at a temperature that is above the outdoor temperature) rather than cooling to the building; in this case, the system is referred to as a heat pump. Figure 9-19(b) illustrates the air conditioning system shown in Figure 9-19(a) re-plumbed so that it operates as a heat pump. The high pressure flow leaving the compressor is directed to the heat exchanger within the building where it condenses, rejecting heat to the building air. The flow leaves the condenser and expands through a throttle valve before entering the heat exchanger located outside the building where

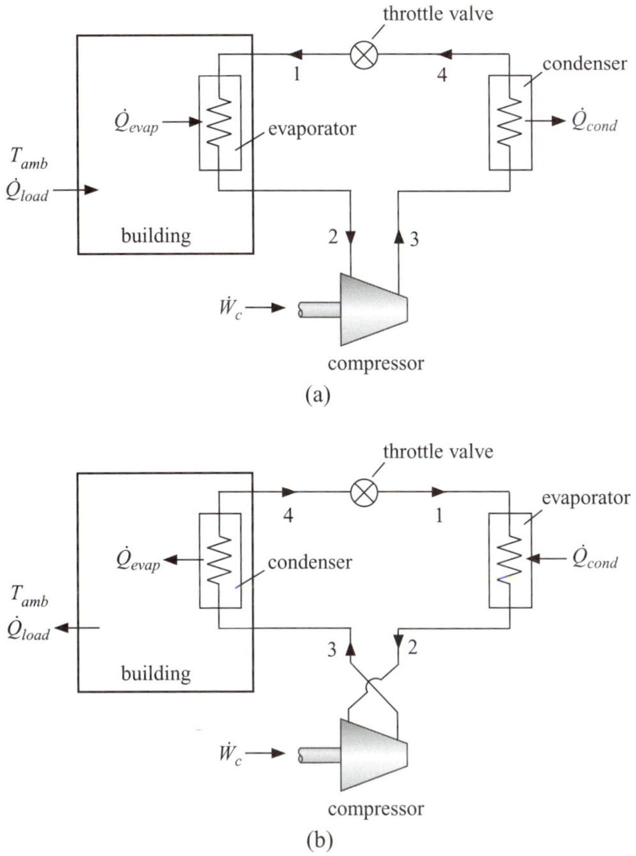

Figure 9-19: Vapor compression cycle providing (a) cooling to a building during the cooling season and (b) heating to a building during the heating season.

it evaporates, accepting heat from the outdoor air. The low pressure vapor leaving the evaporator is directed back to the compressor. The thermodynamic cycle for the heat pump shown in Figure 9-19(b) is the same as the vapor compression cycle providing refrigeration, shown in Figure 9-19(a). You may have noticed that hot air is discharged from the backside of a window air conditioner. One way to think of a heat pump is as an air conditioner that is trying, futilely, to cool the outdoors. (Notice that the evaporator in Figure 9-19(b) is located outdoors.) The outdoor temperature is, of course, not affected by the operation of the heat pump. However, the heat pump rejects heat to the building since the condenser is located indoors and therefore the building temperature rises. The system in Figure 9-19(b) is referred to as an air-source heat pump because it extracts energy from the outdoor air in the evaporator.

The efficiency of a heat pump is defined in terms of its coefficient of performance, which is the ratio of what you want (i.e., $\dot{Q}_{cond}$ – the rate of heat transfer to the building from the condenser) to what you pay for (i.e., $\dot{W}_c$ – the power supplied to the compressor):

$$COP_H = \frac{\dot{Q}_{cond}}{\dot{W}_c} \tag{9-38}$$

If electric resistance heating is used for space heating, then the efficiency of the heating system will be very close to 1. Electrical resistance heaters convert electrical

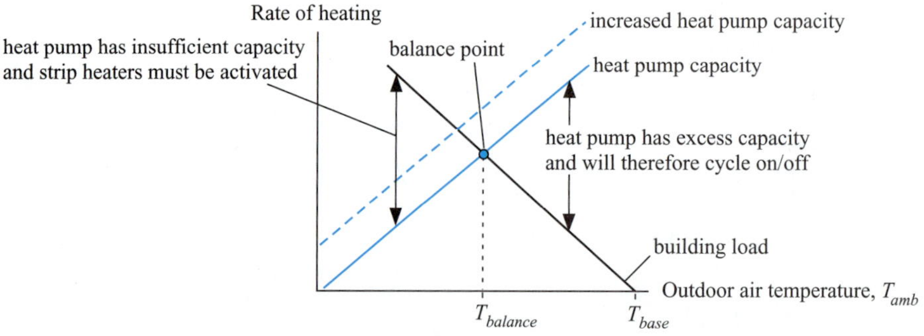

Figure 9-20: Building load and air-source heat pump capacity as a function of outdoor temperature

energy to heat with a First Law efficiency of 100%; the only reason that the efficiency of an electrical heater may be less than unity is due to thermal losses from the jacket of the heater. The major advantage of a heat pump is that the value of the COP_H, as defined in Eq. (9-38), can be significantly greater than 1, which indicates that the First Law efficiency with which the heat pump is converting work to heat is much higher than 100%. Therefore, the use of a heat pump can reduce the expense of heating a building relative to using electric heaters. The use of a heat pump can also be more economical than the alternative of using a furnace that combusts a fuel such as natural gas. In addition, the same vapor compression cycle that provides space heat can also be used to provide air conditioning, as shown in Figure 9-19. Many heating/air conditioning systems provide a four-way valve that effectively interchanges the operation of the indoor and outdoor heat exchangers (i.e., the evaporator and condenser) for summer and winter operation.

The analysis of a vapor compression cycle operating in the heat pump mode is identical to that of the vapor compression refrigeration cycle discussed in Section 9.2. However, there are some additional considerations that become important in heating applications. One consideration is the mismatch between the building load and the heating capacity of an air-source heat pump, which is illustrated in Figure 9-20.

The rate at which energy must be provided to a building ($\dot{Q}_{load}$) to maintain it at a comfortable indoor temperature is nearly a linear function of outdoor temperature, T_{amb}, as indicated by Eq. (9-39):

$$\dot{Q}_{load} = UA\,(T_{base} - T_{amb}) \tag{9-39}$$

where UA is the building heat loss coefficient that depends on the size of the building and the quality of its construction and T_{base} is the outdoor temperature at which the building no longer requires heating. Typically, T_{base} is about 5°C lower than the desired building indoor temperature because energy sources such as lights, appliances, and occupants within the building contribute to the heating process. The black line in Figure 9-20 illustrates the qualitative variation of the building heating load as a function of the outdoor air temperature.

The capacity of an air-source heat pump that utilizes a single-speed compressor is shown qualitatively by the solid blue line in Figure 9-20. The heating capacity of an air source heat pump is a very strong increasing function of the outdoor temperature; the capacity of the air source heat pump shown in Figure 9-19(b) will change much more dramatically during the heating season than the cooling capacity of the air conditioner shown in Figure 9-19(a). This sensitivity arises because the outdoor air temperature typically changes by a much larger amount during the heating season than it does during the cooling season. Also, the state of the refrigerant leaving the indoor evaporator and entering the compressor, state 2 in the air conditioning system shown in, Figure 9-19(a) does not change substantially during the cooling season because the indoor air temperature is relatively constant. As a result, the specific volume of the refrigerant entering the compressor remains constant and the mass flow rate provided by the compressor is not strongly affected by the outdoor air temperature. In contrast, the state of the refrigerant leaving the outdoor evaporator and entering the compressor, state 2 in the heat pump system shown in Figure 9-19(b), changes dramatically as the outdoor air temperature changes. As the outdoor air temperature decreases, the specific volume of the vapor entering the compressor is increased (the density is reduced) leading to a corresponding reduction in the mass flow rate provided by the compressor and the heating capacity provided by the heat pump. This same effect causes the dramatic reduction in the refrigeration capacity of a vapor compression system as the refrigeration temperature is lowered, as discussed in Section 9.2.1 under the subsection *Effect of Refrigeration Temperature*.

Figure 9-20 shows that during much of the heating season there may be a large mismatch between the building heating load and the heating capacity of a heat pump system. At relatively warm outdoor temperatures, the capacity of an air-source heat pump is high but there is only a small building load; therefore, the heat pump has excess capacity. If the heat pump is not equipped with a variable speed motor controller then the compressor must be cycled on and off in order to provide, on average, the heating that is needed by the building. Operating a heat pump in this cyclic, on/off manner tends to lower its effective *COP* and therefore the cost of providing the heat increases. The *COP* is lowered by cycling because the hot and cold temperatures and the high and low pressures in the condenser and evaporator, respectively, must be re-established each time that the unit is activated.

As the outdoor temperature decreases, the building load increases and the heat pump capacity decreases. There is an outdoor temperature that is referred to as the *balance point* where the heat pump capacity just equals the rate at which energy is needed by the building. When the outdoor temperature is equal to the balance point, the heat pump can provide the required heating load by operating continuously. However, as the outdoor temperature is reduced below the balance point, the heat pump has insufficient capacity to meet the load. It is not acceptable to allow the building temperature to drop and therefore auxiliary heating equipment is required in order to maintain the building at its set point temperature. Most heat pump systems also provide electrical resistance heaters called *strip heaters* to provide the additional heating capacity that is required at very low outdoor temperatures. However, the efficiency of the strip heaters (which is, at best, 100%) is much lower than the *COP* of the heat pump and therefore the cost of providing heat increases. Increasing the size of the heat pump increases its capacity, as shown by the dotted blue line in Figure 9-20. The balance point for the larger heat pump will be reduced, causing it to cycle more frequently at higher temperatures and therefore reduce its *COP* at these conditions. The larger heat pump is also more expensive.

The discussion above shows that the outdoor air temperature experienced during the heating season and the characteristics of the building being heated both tend to affect the performance of the system. In the U.S., these effects are captured by the heating season performance factor (*HSPF*) which is defined as a heat pump's estimated seasonal heating output (in Btu) divided by the amount of electrical energy that it requires over the heating season (in Watt-hours). The electrical energy includes the energy that is needed to operate the fans and operate the strip heaters during times that the heat pump capacity is insufficient to meet load. The calculation of the *HSPF* for a particular application is illustrated in Example 9.3-1.

EXAMPLE 9.3-1: HEATING SEASON PERFORMANCE FACTOR

The purpose of this problem is to determine the heating system performance factor (*HSPF*) of an air-source vapor compression heat pump operating in the heating mode using refrigerant R410A. A schematic of the heat pump system is shown in Figure 1.

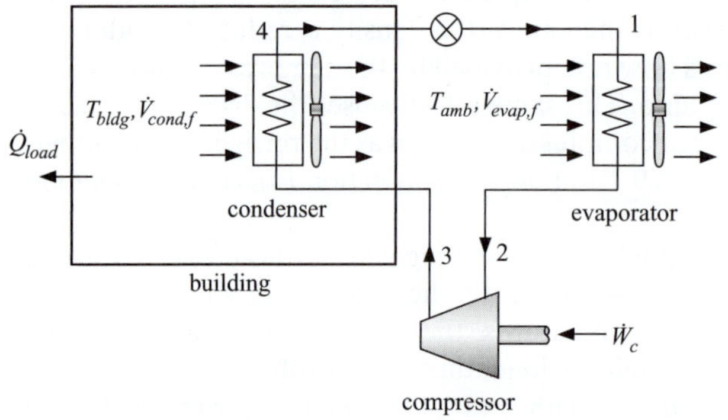

Figure 1: Heat pump system.

The heat pump system utilizes a fixed speed reciprocating compressor. For a reciprocating (i.e., piston-cylinder) compressor with a single cylinder, the displacement rate of the compressor is equal to the ratio of the volume swept by the piston to the cycle time. The pressure-volume processes for a piston-cylinder compressor are shown qualitatively in Figure 2.

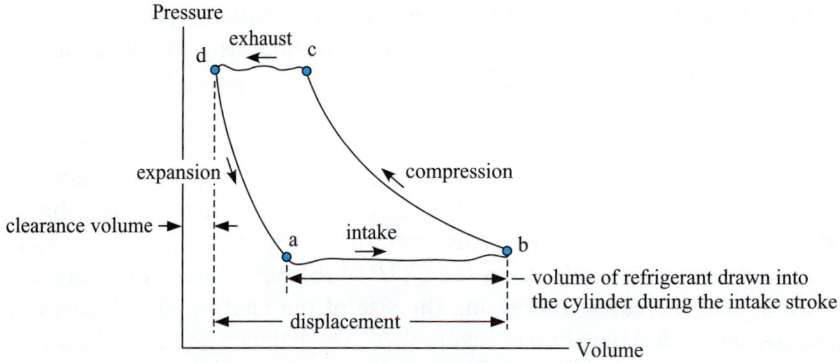

Figure 2: Pressure-volume processes for a piston-cylinder compressor.

Refrigerant enters the compressor through the intake valve during the intake process, from point a to point b in Figure 2. The intake valve closes at point b and the piston compresses the vapor to the pressure indicated by point c. At this point, the exhaust valve lifts and high pressure refrigerant is expelled until the piston reaches the end of its stroke at point d. The volume of the cylinder at point d is called the clearance volume. Some clearance volume is required in order to ensure that the piston does not touch the top of the cylinder. The refrigerant that remains in the cylinder at point d expands to point a as the intake stroke begins;

the refrigerant remaining in the cylinder reduces the amount of new refrigerant that enters the cylinder during the intake stroke. This effect as well as any leakage is quantified in terms of the volumetric efficiency, η_{vol}, which is defined as the ratio of the actual volumetric flow rate of the refrigerant at the compressor suction port to the displacement rate. The volumetric efficiency for a reciprocating compressor can be estimated using Eq. (1):

$$\eta_{vol} = 1 - c\left(\frac{v_2}{v_3} - 1\right) \tag{1}$$

where c is the ratio of the clearance volume to the displaced volume in the compressor and v_2 and v_3 are the refrigerant specific volumes at the compressor inlet and outlet, respectively. The specific compressor used in the heat pump system has a displacement rate $\dot{V}_{disp} = 4.25$ cfm and a clearance volume ratio $c = 0.025$. Compressor test results indicate that the compressor isentropic efficiency can be expressed approximately as:

$$\eta_c = 0.74 - 0.0074\left[\frac{1}{\text{K}}\right]\left(T_{cond} - T_{evap}\right) \tag{2}$$

where T_{cond} and T_{evap} are the condensing and evaporating temperatures. The effectiveness of the condenser is $\varepsilon_{cond} = 0.72$. The condenser fan draws $\dot{V}_{cond,f} = 2.0$ m^3/s of building air at $T_{bldg} = 22°$C across the condenser coil and consumes $\dot{W}_{cond,f} = 250$ W of electrical power. The rate of heat transfer from the refrigerant flowing through the condenser to the air in the building is given by a heat exchanger effectiveness relation, as described in Section 6.6:

$$\dot{Q}_{cond} = \varepsilon_{cond}\,\dot{m}_{a,cond}\,c_{P,a,cond}\left(T_{cond} - T_{bldg}\right) \tag{3}$$

where $\dot{m}_{a,cond}$ is the mass flow rate of air in the condenser and $c_{P,a,cond}$ is the constant pressure specific heat capacity of the building air. The effectiveness of the evaporator is $\varepsilon_{evap} = 0.81$. The evaporator fan draws $\dot{V}_{evap,f} = 2.5$ m^3/s of ambient air through the evaporator and consumes $\dot{W}_{evap,f} = 340$ W of electrical power. The rate of heat transfer between the outdoor air and the refrigerant in the evaporator is given by:

$$\dot{Q}_{evap} = \varepsilon_{evap}\,\dot{m}_{a,evap}\,c_{P,a,evap}\left(T_{amb} - T_{evap}\right) \tag{4}$$

where $\dot{m}_{a,evap}$ is the mass flow rate of air in the evaporator and $c_{P,a,evap}$ is the constant pressure specific heat capacity of the ambient air. Neglect any pressure loss in the system and neglect any superheating or subcooling at the evaporator and condenser exits, respectively.

a) Calculate the heating capacity and the *COP* of the heat pump system when the outdoor air temperature is $T_{amb} = -5°$C.

The inputs are entered in EES:

```
$UnitSystem SI Mass Radian J K Pa
V_dot_disp=4.25 [cfm]*convert(cfm,m^3/s)        "compressor displacement rate"
c=0.025 [-]                                     "clearance volume ratio"
T_bldg=converttemp(C,K,22 [C])                  "building air temperature"
V_dot_cond_f=2.0 [m^3/s]                        "condenser fan volumetric flow rate"
W_dot_cond_f=250 [W]                            "condenser fan power"
eff_cond=0.72 [-]                               "condenser effectiveness"
T_amb_C= -5 [C]                                 "outdoor air temperature, in C"
T_amb=converttemp(C,K,T_amb_C)                  "outdoor air temperature"
V_dot_evap_f=2.5 [m^3/s]                        "evaporator fan volumetric flow rate"
W_dot_evap_f=340 [W]                            "evaporator fan power"
eff_evap=0.81 [-]                               "evaporator effectiveness"
R$='R410a'                                      "working fluid"
```

The effectiveness of the evaporator and condenser are specified rather than the approach temperature differences. Therefore, the evaporating and condensing temperatures are not known. Their values will eventually be determined by simultaneously solving the energy balances and heat exchanger relations for the evaporator and condenser. In order to allow the calculations to proceed in a sequential manner we will assume values for the condensing and evaporating temperatures and then remove these assumed values when the heat exchanger relations are entered.

```
T_cond=T_bldg+5 [K]                             "condensing temperature, guess"
T_evap=T_amb-5 [K]                              "evaporating temperature, guess"
```

In the absence of any subcooling, the temperature of the refrigerant leaving the condenser is the condensing temperature, $T_4 = T_{cond}$, and the quality is $x_4 = 0$. The pressure, specific enthalpy, and specific entropy (P_4, h_4, and s_4) are determined.

```
"State 4, Condenser exit/throttle inlet"
T[4]=T_cond                                     "temperature"
x[4]=0 [-]                                      "quality"
P[4]=pressure(R$,T=T[4],x=x[4])                 "pressure"
h[4]=enthalpy(R$,T=T[4],x=x[4])                 "specific enthalpy"
s[4]=entropy(R$,T=T[4],x=x[4])                  "specific entropy"
```

In the absence of any superheating, the temperature of the refrigerant leaving the evaporator is the evaporating temperature, $T_2 = T_{evap}$, and the quality is $x_2 = 1$ which allows the pressure, specific enthalpy, specific entropy, and specific volume (P_2, h_2, s_2, and v_2) to be determined.

```
"State 2, Evaporator exit/compressor inlet"
T[2]=T_evap                                     "temperature"
x[2]=1 [-]                                      "quality"
P[2]=pressure(R$,T=T[2],x=x[2])                 "pressure"
h[2]=enthalpy(R$,T=T[2],x=x[2])                 "specific enthalpy"
s[2]=entropy(R$,T=T[2],x=x[2])                  "specific entropy"
v[2]=volume(R$,T=T[2],x=x[2])                   "specific volume"
```

Pressure loss in the evaporator is neglected, therefore $P_1 = P_2$. The valve is isenthalpic, therefore $h_1 = h_4$. The temperature and specific entropy leaving the valve (T_1 and s_1) are determined.

```
"State 1, Throttle exit/evaporator inlet"
P[1]=P[2]                                  "pressure"
h[1]=h[4]                                  "specific enthalpy"
T[1]=temperature(R$,P=P[1],h=h[1])         "temperature"
s[1]=entropy(R$,P=P[1],h=h[1])             "specific entropy"
```

Pressure loss in the condenser is neglected, therefore $P_3 = P_4$. The compressor isentropic efficiency is determined using Eq. (2). The specific entropy leaving a reversible compressor operating under the same conditions is $s_{s,3} = s_2$. The specific enthalpy leaving a reversible compressor ($h_{s,3}$) is determined from the specific entropy and pressure. The power per unit mass flow rate required by a reversible compressor is:

$$\left(\frac{\dot{W}_{s,c}}{\dot{m}}\right) = h_{s,3} - h_2$$

The power per unit mass flow rate required by the actual compressor is:

$$\left(\frac{\dot{W}_{c}}{\dot{m}}\right) = \frac{1}{\eta_c}\left(\frac{\dot{W}_{s,c}}{\dot{m}}\right)$$

The specific enthalpy leaving the actual compressor is calculated using an energy balance:

$$h_3 = h_2 + \left(\frac{\dot{W}_{c}}{\dot{m}}\right)$$

State 3 is fixed by the specific enthalpy and pressure allowing the specific entropy, temperature, and specific volume (s_3, T_3, and v_3) to be determined.

```
"State 3, Compressor exit/condenser inlet"
P[3]=P[4]                                        "pressure"
eta_c=0.79 [-] – 0.0039 [1/K]*(T_cond-T_evap)    "isentropic efficiency"
s_s[3]=s[2]                                      "specific entropy leaving isentropic compressor"
h_s[3]=enthalpy(R$,s=s_s[3],P=P[3])              "specific enthalpy leaving isentropic compressor"
W_dot_s_c\m_dot=h_s[3]-h[2]                      "work per mass required by isentropic compressor"
W_dot_c\m_dot=W_dot_s_c\m_dot/eta_c              "work per mass required by compressor"
h[3]=h[2]+W_dot_c\m_dot                          "specific enthalpy"
s[3]=entropy(R$,h=h[3],P=P[3])                   "specific entropy"
T[3]=temperature(R$,h=h[3],P=P[3])               "temperature"
v[3]=volume(R$,h=h[3],P=P[3])                    "specific volume"
```

The volumetric efficiency of the compressor is computed using Eq. (1), which allows the mass flow rate of refrigerant to be determined:

$$\dot{m} = \frac{\eta_{vol}\,\dot{V}_{disp}}{v_2}$$

The power required by the compressor is:

$$\dot{W}_c = \left(\frac{\dot{W}_c}{\dot{m}}\right)\dot{m}$$

```
eta_vol=1-c*(v[2]/v[3]-1)                "volumetric efficiency"
m_dot=eta_vol*V_dot_disp/v[2]            "mass flow rate"
W_dot_c=W_dot_c\m_dot*m_dot              "compressor power"
```

The condenser heat transfer rate is computed using an energy balance on the refrigerant:

$$\dot{Q}_{cond} = \dot{m}\left(h_3 - h_4\right)$$

The specific heat capacity at constant pressure and specific volume of the building air passing through the condenser ($c_{P,a,cond}$ and $v_{a,cond}$) are evaluated at the building air temperature. The mass flow rate of air passing through the condenser is computed from:

$$\dot{m}_{a,cond} = \frac{\dot{V}_{cond,f}}{v_{a,cond}}$$

```
Q_dot_cond=m_dot*(h[3]-h[4])                                "rate of condenser heat transfer"
cP_a_cond=cP(Air,T=T_bldg)                                  "specific heat of the air in the condenser"
v_a_cond=volume(Air,T=T_bldg,P=1 [atm]*convert(atm,Pa))     "specific volume of air in the condenser"
m_dot_a_cond=V_dot_cond_f/v_a_cond                          "mass flow rate of air in the condenser"
```

At this point, the condensing temperature can be computed using Eq. (3); however, this would over-specify the problem. Update the guess values by selecting Update Guesses from the Calculate menu. Comment out the assumed value of T_{cond} and enter Eq. (3).

```
{T_cond=T_bldg+5 [K]}                                       "condensing temperature, guess"
Q_dot_cond=eff_cond*m_dot_a_cond*cP_a_cond*(T_cond-T_bldg)
                                                            "effectiveness relation-condenser"
```

EES will determine the value of the condensing temperature that satisfies both the heat exchanger relationship, Eq. (3), and the energy balance. A similar process is carried out for the evaporator. The evaporator heat transfer rate is computed using an energy balance on the refrigerant:

$$\dot{Q}_{evap} = \dot{m}\left(h_2 - h_1\right)$$

The specific heat capacity at constant pressure and specific volume of the outdoor air passing through the evaporator ($c_{P,a,evap}$ and $v_{a,evap}$) are evaluated at the outdoor air temperature. The mass flow rate of air passing through the evaporator is computed from:

$$\dot{m}_{a,evap} = \frac{\dot{V}_{evap,f}}{v_{a,evap}}$$

```
Q_dot_evap=m_dot*(h[2]-h[1])                              "rate of evaporator heat transfer"
cP_a_evap=cP(Air,T=T_amb)                                 "specific heat capacity of evaporator air"
v_a_evap=volume(Air,T=T_amb,P=1 [atm]*convert(atm,Pa))    "specific volume of the evaporator air"
m_dot_a_evap=V_dot_evap_f/v_a_evap                        "mass flow rate of air in the evaporator"
```

At this point, the evaporating temperature can be computed using Eq. (4). Update the guess values by selecting Update Guesses from the Calculate menu. Comment out the assumed value of T_{evap} and enter Eq. (4).

```
{T_evap=T_amb-5 [K]}                                      "evaporating temperature, guess"
Q_dot_evap=eff_evap*m_dot_a_evap*cP_a_evap*(T_amb-T_evap)
                                                          "effectiveness relation-evaporator"
```

The *COP* of the heat pump system is the ratio of the rate of heat transfer from the condenser to the total electrical power required by the compressor and the fans:

$$COP = \frac{\dot{Q}_{cond}}{\dot{W}_c + \dot{W}_{cond,f} + \dot{W}_{evap,f}}$$

```
COP=Q_dot_cond/(W_dot_c+W_dot_cond_f+W_dot_evap_f)        "COP for heat pump"
```

which leads to $\dot{Q}_{cond} = 10.0$ kW and $COP = 3.81$. Using this heat pump will provide thermal energy for heating the house at an electricity cost that is 3.81 times less than using electrical resistance heaters when the outdoor temperature is -5°C.

b) Plot the heating capacity of the heat pump and the *COP* of the heat pump system as a function of outdoor temperature.

Figure 3 illustrates the heating capacity of the heat pump as a function of the outdoor air temperature. Note that the heating capacity decreases significantly with decreasing outdoor air temperature because the specific volume at the compressor inlet increases with reduced evaporator temperature.

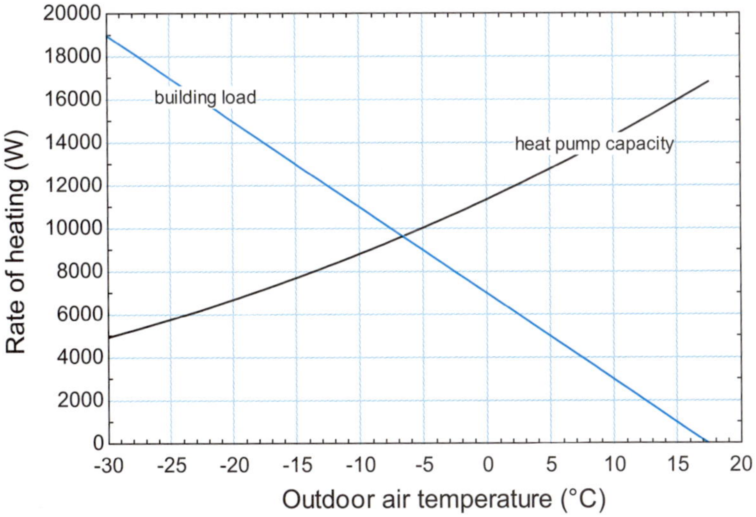

Figure 3: Heating capacity of the heat pump as a function of the outdoor air temperature. Also shown is the building load.

The black line in Figure 4 illustrates the *COP* of the heat pump system as a function of the outdoor air temperature, assuming that the heat pump is operating continuously.

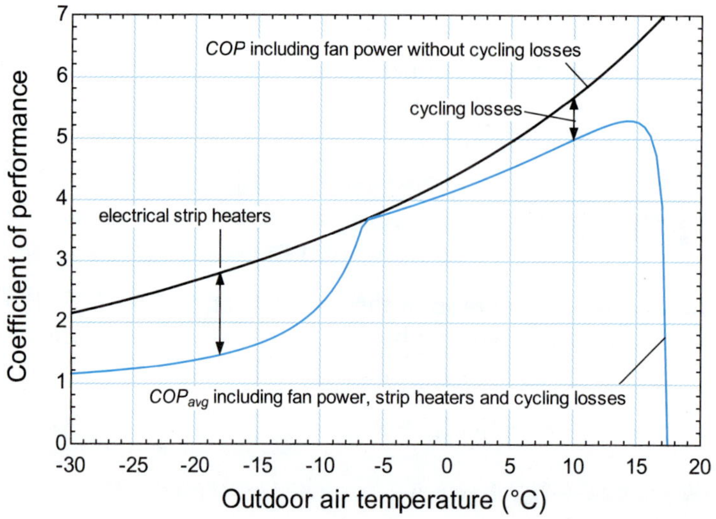

Figure 4: *COP* of the heat pump system operating continuously as a function of the outdoor air temperature. Also shown is the average *COP* of the entire heating system meeting the building load as a function of the outdoor air temperature.

c) The heat pump system is installed in a building with a heat loss coefficient $UA = 400$ W/K and a base temperature $T_{base} = 17.5°$C. Overlay on your plot from (b) the heating load required by the building as a function of the outdoor air temperature.

Equation (9-39) is used to determine the building heating load.

```
UA=400 [W/K]                          "building heat loss coefficient"
T_base=converttemp(C,K,17.5 [C])      "base temperature"
Q_dot_bldg=UA*(T_base-T_amb)          "building load"
```

Figure 3 illustrates the building heating load as a function of the outdoor air temperature. Notice that the balance point for the heat pump system installed in the building is approximately $-7.0°$C; this is the outdoor air temperature at which the heat pump can exactly meet the building load by running continuously.

d) At temperatures below the balance point, the heat pump runs continuously but it has insufficient capacity and therefore additional heat will be provided using electrical resistance heaters. At temperatures above the balance point, the heat pump has more than sufficient capacity and therefore it must cycle on and off in order to provide the required amount of heating, averaged over time. The fraction of time that the heat pump operates, f, is estimated as the ratio of the building load to the heating capacity of the heat pump. Note that the parameter f must range between 0 and 1:

$$f = \min\left(1, \frac{\dot{Q}_{bldg}}{\dot{Q}_{cond}}\right) \tag{5}$$

Inefficiencies resulting from the cycling process cause the actual *COP* of the heat pump to be lower than the value that it would have if it were operated continuously at the same outdoor temperature. The ratio of the actual *COP* (COP_{act}) to the steady state or continuous *COP* (COP_{ss}) is the part load factor, *PLF*:

$$PLF = \frac{COP_{act}}{COP_{ss}} = \frac{\overline{\dot{W}}_{c,ss}}{\overline{\dot{W}}_{c,act}} \tag{6}$$

where $\overline{\dot{W}}_{c,ss}$ is the average compressor power assuming that it is operating steadily and $\overline{\dot{W}}_{c,act}$ is the actual average compressor power. Since the rate of heat provided is the same for both the actual and steady state *COP*, the part load factor is also the ratio of steady-state to the actual integrated average compressor power. If the part load factor is 1 then there is no performance penalty due to cycling. For the system considered here, the part load factor as a function of *f* is shown in Figure 5 and represented by:

$$PLF = \frac{f}{0.01 + 1.18f - 0.24f^2 + 0.05f^3} \tag{7}$$

The difference between 1 and *PLF* is a measure of the cycling losses and this difference increases as *f* is reduced, as seen in Figure 5. Note that the part load factor only includes the part-load performance of the compressor. The electrical energy needed by the fans is also affected by the fraction of time that the heat pump operates.

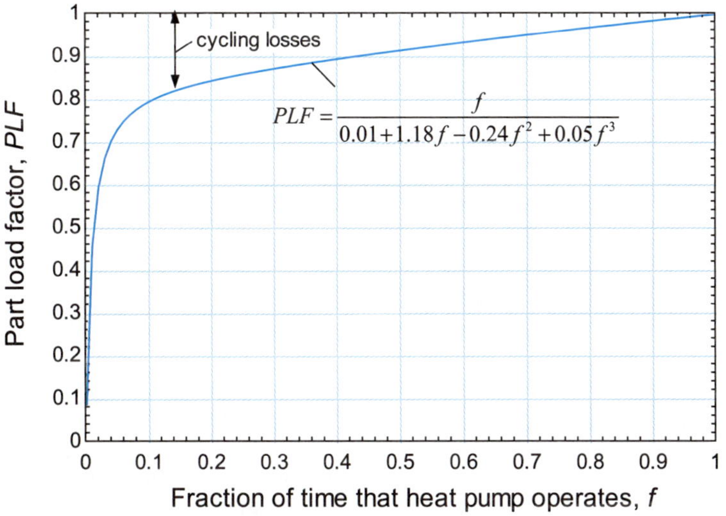

Figure 5: Part load factor as a function of *f*.

Calculate the average electrical power needed by the system (compressor, fan, and strip heater power, if required). Overlay on your plot from (b) the average *COP* of the heating system meeting the building load as a function of outdoor air temperature.

The fraction of time that the heat pump must operate in order to meet load is computed using Eq. (5) with the min function in EES.

```
f=min(1, Q_dot_bldg/Q_dot_cond)     "fraction of time that heat pump must operate to meet load"
```

If the building load exceeds the steady-state heat transfer rate from the condenser, then the strip heaters must be activated in order to meet the additional heating needs of the building; otherwise the strip heaters are not activated. The logic needed to implement this control decision can be implemented with the max function.

```
W_dot_strip=max(0, Q_dot_bldg-Q_dot_cond)        "power required by strip heaters to meet load"
```

If f is less than one, then the building energy load is smaller than the heat pump capacity and therefore the compressor will cycle on and off in order to provide the amount of heating that is needed to meet the load, on average. The average steady-state compressor power is the product of f and the compressor power. The part load factor is computed according to Eq. (7). The average compressor power is calculated according to Eq. (6).

```
PLF=f/(0.01+1.18*f-0.24*f^2+0.05*f^3)      "part-load factor"
W_dot_c_ss=f*W_dot_c                      "average steady-state compressor power"
W_dot_c_avg=W_dot_c_ss/PLF                "average power needed to operate compressor"
```

The steady state fan power required is:

$$\dot{W}_f = \dot{W}_{cond,f} + \dot{W}_{evap,f}$$

We are assuming that there is no cycling loss associated with turning the fans on and off. Therefore, the fan power is equal to the product of the steady state fan power and the fraction of time that the heat pump operates.

```
W_dot_f=W_dot_cond_f+W_dot_evap_f          "steady state fan power"
W_dot_f_avg= f*W_dot_f                     "average fan power"
```

The total electrical power at any outdoor temperature is the sum of the power needed by the strip heaters, the fans, and the compressor:

$$\dot{W}_{elec,avg} = \dot{W}_{cond,f,avg} + \dot{W}_{strip} + \dot{W}_{c,avg}$$

and the average COP of the heating system (including fan power, strip heaters, and cycling losses) is:

$$COP_{avg} = \frac{\dot{Q}_{bldg}}{\dot{W}_{elec,avg}}$$

```
W_dot_elec_avg=W_dot_c_avg+W_dot_f_avg+W_dot_strip   "average rate of electricity consumption"
COP_avg=Q_dot_bldg/W_dot_elec_avg                    "average COP of system"
```

Figure 4 illustrates the COP of the heating system as a function of outdoor air temperature both with and without consideration of the strip heaters and cycling losses. The system COP values are the same at the balance point where the system is operating continuously and the strip heaters are not on. Below the balance point, the use of strip heaters reduces the COP. At very low values of T_{amb}, the system COP approaches unity because most of the heat is being provided by strip heaters. Above the balance point the cycling losses tends to reduce the COP relative to what would be expected under continuous operation.

e) Data that provide the number of hours that the outdoor temperature is within specified intervals during a heating season are referred to as *bin temperature data*. Bin temperature data for Madison, WI are provided in Table 1 at intervals of 5°C. Use these data to compute the heating season performance factor ($HSPF$) for this building-heat pump system located in Madison.

Table 1: Bin temperature data for Madison, WI

Outdoor air temperature (°C)	Number of hours
−30	18
−25	38
−20	105
−15	187
−10	237
−5	476
0	867
5	534
10	518
15	828

The *HSPF* is calculated by integrating the total power required and the heating load provided over the entire heating season. This integration process is simplified by using bin temperature data. Bin temperature data provide the number of hours during the heating season that the outdoor temperature is within a specified bin. For example, the data in Table 1 indicate that there are 476 hours in Madison for which the outdoor temperature is, on average, −5°C (i.e., the temperature is between −7.5°C and −2.5°C). For each bin, the total amount of heating energy provided to the building is:

$$Q_{bldg} = \dot{Q}_{bldg}\, hours$$

where *hours* is the number of hours associated with that bin. The total electrical power consumed is:

$$W_{elec} = \dot{W}_{elec,avg}\, hours$$

```
Q_bldg=Q_dot_bldg*hours*convert(hr,s)            "total heat required for a specific bin"
W_elec=W_dot_elec_avg*hours*convert(hr,s)        "total electricity required for a specific bin"
```

The data in Table 1 are entered into a Parametric table. Select New Parametric Table from the Tables menu and create a Parametric table that includes the variables T_amb_C, hours, W_elec, and Q_bldg. Name the table 'HSPF' by right-clicking in the table header. Enter the data into the Parametric table, as shown in Figure 6.

The *HSPF* is the ratio of the total building heat provided (i.e., the sum of the Q_bldg values in the Parametric table) to the total electrical energy consumed (i.e., the sum of the W_elec values in the Parametric table). These sums can be obtained using the SumParametric function; the first argument of this function is the name of the table and the second is the name of the column to be summed. The third and fourth arguments are the rows at which the summation starts and stops, respectively. Note that TableRun# is a built-in variable that indicates the row in the Parametric table for which calculations are in progress. The value of *HSPF* is converted from

Table 1 | HSPF | Table 3

1..10	$T_{amb,C}$ [C]	hours [hr]	W_{elec} [J]	Q_{bldg} [J]
Run 1	-30	18		
Run 2	-25	38		
Run 3	-20	105		
Run 4	-15	187		
Run 5	-10	237		
Run 6	-5	476		
Run 7	0	867		
Run 8	5	534		
Run 9	10	518		
Run 10	15	828		

Figure 6: Parametric table containing the bin temperature data.

a dimensionless quantity to Btu/W-hr in order to be consistent with *HSPF* values used in the U.S.

```
HSPF=sumParametric('HSPF', 'Q_bldg', 1,TableRun#)/sumParametric('HSPF','W_elec',&
   1,TableRun#)*convert(1,Btu/W-hr)        "heating season performance factor"
```

The calculated value is $HSPF = 9.54$ Btu/W-hr, which is somewhat higher than the current minimum efficiency standard value of 7.7 Btu/W-hr but within the 8-10 range that the U.S. Department of Energy classifies as an "efficient heat pump".

9.4 The Absorption Cycle

The vapor compression refrigeration cycle is, by far, the most widely used refrigeration cycle; however, it does have some disadvantages. The refrigerants that are typically used in vapor compression equipment are associated with environmental concerns, as discussed in Section 9.2.3, despite the great deal of research that has been invested in the identification of alternative refrigerants. In addition, the vapor compression cycle requires significant mechanical power input to operate the compressor. The mechanical power is typically obtained from an electrical motor and the electrical power required to operate vapor compression air conditioning equipment is largely responsible for the peak electrical energy use that occurs in much of the U.S. during the summer.

The absorption cooling cycle is an alternative refrigeration cycle that uses moderate temperature thermal energy, rather than mechanical power, as the primary energy input. The required work input to the cycle is greatly reduced because the refrigerant pressure is increased while it is absorbed in a liquid. The specific volume of a liquid is much smaller than that of a vapor. Therefore, it requires much less work to pump a liquid than it does to compress a vapor, as is required by a vapor compression system. The absorption cycle takes advantage of the temperature dependence of the solubility of a gas in liquid.

9.4.1 The Basic Absorption Cycle

A schematic of the basic absorption cycle is shown in Figure 9-21. Note that the absorption cycle employs a condenser, throttle valve and evaporator; these components are also

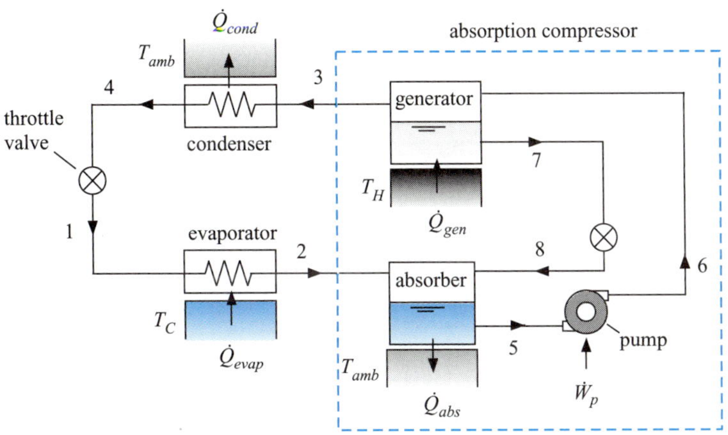

Figure 9-21: Schematic of the basic absorption cooling cycle.

used in the basic vapor compression refrigeration cycle shown in Figure 9-4. The major difference between the vapor compression and the absorption cycles is in the method that is used to compress the refrigerant vapor from the low evaporator pressure at state 2 to the high condenser pressure at state 3.

The compressor used in the vapor compression cycle is replaced by the absorber and generator in the absorption cycle, as shown in Figure 9-21. The absorption cycle uses what could be called an 'absorption compressor' that requires a mixture of two or more chemicals to operate. Gases are absorbed to a greater extent in liquids as the temperature of the liquid solution is lowered. You are probably familiar with this principle. For example, the carbon dioxide gas that is used to create the fizz in soda is absorbed more completely in cold water than it is in warm water. Therefore, if you heat the soda, the carbon dioxide will be driven out causing it to become 'flat'. Trout prefer colder water because it can absorb more oxygen than warm water.

The low pressure, low temperature refrigerant vapor exiting the evaporator (state 2 in Figure 9-21) enters a heat and mass exchanger called the absorber. In the absorber, the refrigerant vapor is absorbed into a liquid solution (the absorbent); because the temperature is low, the solubility of the refrigerant vapor in the absorbent liquid is relatively high. During the absorption process, the refrigerant vapor is converted from vapor to liquid and it therefore releases energy. The energy released per unit mass of refrigerant is comparable to the enthalpy of vaporization of the refrigerant and therefore the absorber must be cooled externally by a heat transfer with the ambient temperature ($\dot{Q}_{abs}$ at T_{amb}). The liquid absorbent is sprayed or distributed on the outside surface of cooling tubes in order to expose a large liquid surface area to the vapor and increase the rate of absorption of refrigerant vapor into the liquid.

The liquid solution leaving the absorber at state 5 is relatively dilute in the absorbent and strong in the refrigerant. The solution is pumped from the evaporator pressure (at state 5) to the condenser pressure (at state 6). The work input to the pump ($\dot{W}_p$) is very small, much less than the power that would be required by a compressor in a vapor compression cycle to increase the pressure of the refrigerant vapor. The small power input associated with the pump occurs because the specific volume of the liquid solution is orders of magnitude lower than the specific volume of the refrigerant vapor. The pump work is so small that it is actually possible to eliminate the pump altogether and just use gravity forces and clever engineering in some types of absorption cycles, as described by Herold et al. (1996).

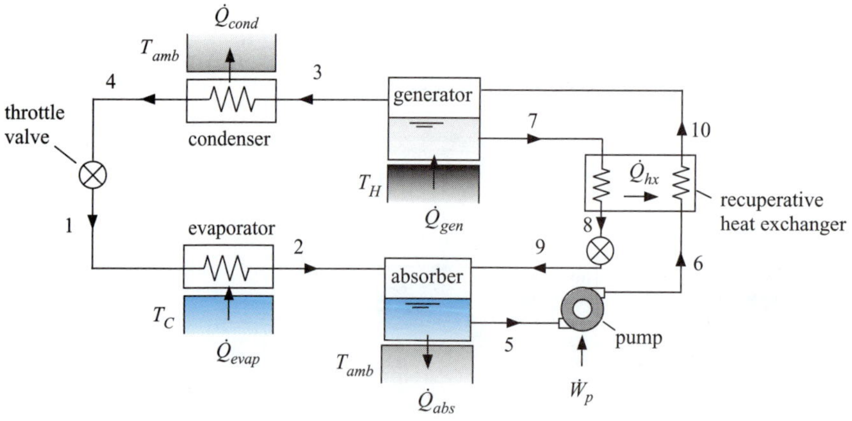

Figure 9-22: Schematic of the absorption cooling cycle with a recuperative heat exchanger.

The high pressure solution at state 6 enters the generator where the heat transfer rate ($\dot{Q}_{gen}$) is provided at temperature T_H in order to raise the temperature of the solution. The value of T_H is typically in the range of 80°C to 150°C. Since vapor is less soluble in liquids at high temperatures, some of the refrigerant vapor is driven out of solution at the generator pressure. This relatively high pressure refrigerant vapor leaves at state 3 and circulates through the condenser, the throttle valve and the evaporator, just as it would in the vapor compression cycle. The hot liquid solution at state 7 is strong in absorbent and weak in refrigerant. This hot, high pressure solution is throttled to the evaporator pressure (state 8) so that it can be re-introduced into the absorber and used to absorb additional refrigerant vapor, continuing the cycle.

In most absorption cycles, a recuperative heat exchanger is placed between the absorber and the generator so that the hot solution leaving the generator at state 7 in Figure 9-21 can be used to preheat the cold solution leaving the pump at state 6. This modification is shown in Figure 9-22 and it reduces the rate of heat transfer required by the generator and the rate of cooling required by the absorber.

The major advantage of the absorption refrigeration cycle compared to the vapor compression cycle is that it requires much less mechanical power. Instead, the absorption cycle requires a heat input to the generator. The heat input must come from a temperature that is greater than the ambient temperature used to cool the absorber. However, the heat input can occur at relatively low temperatures; T_H can be less that 150°C for single-stage machines. The coefficient of performance of an absorption cycle is usually defined as the ratio of the evaporator cooling capacity ($\dot{Q}_{evap}$) to the rate at which thermal energy is supplied to the generator ($\dot{Q}_{gen}$):

$$COP_{abs} = \frac{\dot{Q}_{evap}}{\dot{Q}_{gen}} \tag{9-40}$$

The *COP* defined by Eq. (9-40) differs from the *COP* used for vapor compression cycles since heat, rather than mechanical power, appears in the denominator. The *COP* of the single-stage absorption cycle shown in Figure 9-22 is usually between 0.6 to 0.75; this is much lower than the *COP* of a vapor compression cycle operating at the same conditions. However, the low temperature thermal energy supplied to the absorption cycle can be provided from solar energy collectors or from waste heat sources (e.g., the waste heat from a diesel generator). The major disadvantage of the absorption cycle is that the additional heat exchangers required to operate the system, the absorber, generator, and recuperative heat exchanger in Figure 9-22, cause the required equipment to be physically larger (and more expensive) than a vapor compression cycle of the same capacity. Also, the low *COP* leads to a relatively large rate of heat transfer from the condenser and therefore the condenser in an absorption cycle will need to be physically larger than a comparable condenser in a vapor compression system.

9.4.2 Absorption Cycle Working Fluids

This extended section can be found online at www.cambridge.org/kleinandnellis. This section discusses common absorbent/refrigerant pairs and discusses the property routines that can be used in EES to model absorption systems. Example 9.4-1 is included in this section and illustrates the use of these routines.

Table 9-3: Triple point temperature and critical temperature of some refrigerants used in vapor compression cycles near room temperature.

Refrigerant	Triple point temperature	Critical temperature
Ammonia	195.5 K	405.4 K
R134a	168.9 K	374.2 K
R22	115.7 K	369.3 K
R410A	118.2 K	345.3

9.5 Recuperative Cryogenic Cooling Cycles

The vapor compression and absorption cooling cycles discussed in Sections 9.2 and 9.4, respectively, are capable of providing refrigeration near room temperature but they cannot continue to operate at very low (i.e., cryogenic) temperatures. The refrigerants used in these cycles are selected so that their vapor dome extends over the entire temperature range encountered by the cycle (i.e., the critical temperature is above the condenser temperature and the triple point temperature, or freezing temperature, is below the evaporator temperature). It becomes more difficult to find appropriate working fluids as the refrigeration temperature is reduced while the condenser temperature remains fixed at near room temperature. Table 9-3 lists the triple point temperature and critical temperature for a few common refrigerants used in vapor compression systems that operate near room temperature. Notice that the triple point temperatures of these refrigerants are above 100 K, which is above the desired refrigeration temperature in many cryogenic systems. Even if the working fluid remains liquid at the temperatures of interest, the vapor compression cycle requires the introduction of saturated (or near saturated) refrigerant vapor at the evaporator temperature into the compressor. As the refrigeration temperature is reduced, this requirement leads to a host of practical problems. For example, the specific volume of the refrigerant increases to the point where the compressor becomes impractically large and the temperature decreases to the point where the oil used to lubricate the compressor freezes.

Table 9-4 illustrates the triple point temperature and critical temperature of some potential working fluids for low temperature refrigeration cycles. These fluids have suitably low freezing temperatures; however, they all share the characteristic of having a critical temperature that is far below room temperature. Therefore, in order to provide refrigeration at cryogenic temperatures it is necessary to utilize cycles that operate outside of (i.e., above) the vapor dome. These cycles allow the use of gaseous working fluids

Table 9-4: Triple point temperature and critical temperature of some potential working fluids for cryogenic refrigeration cycles.

Fluid	Triple point temperature	Critical temperature
Argon	83.8 K	150.7 K
Nitrogen	63.2 K	126.2 K
Neon	24.6 K	44.5 K
Hydrogen	14.0 K	33.2 K
Helium	N/A	5.2 K

with very low freezing points and employ compressors that operate at room temperature. The recuperative cycles discussed in this section and the regenerative cycles discussed in Section 9.6 operate outside the vapor dome and therefore these cycles are commonly used to provide cryogenic refrigeration; machines that utilize these cycles are sometimes referred to as *cryocoolers*. Cryogenic temperatures are required by many applications including liquefaction, separation, cryo-pumping, superconductors, cryosurgery, and low temperature detectors.

The basic recuperative cycle is shown in Figure 9-25. Low pressure gas enters the compressor near room temperature at state 1. The gas is compressed using a room temperature compressor. The high pressure, warm gas leaving the compressor at state 2 is cooled to near room temperature (state 3) in an aftercooler. The gas then passes through a recuperative heat exchanger where it is "recuperated"; that is, the fluid is pre-cooled by the low temperature fluid returning from the cold end of the cycle.

The high pressure, cold fluid leaving the recuperator at state 4 undergoes an expansion process where its pressure is reduced. If the expansion process is accomplished using a work extraction device (e.g., a turbine or piston/cylinder apparatus), then the cycle is referred to as a reverse Brayton cycle. Reverse Brayton cycles are discussed in Section 9.5.1. If the expansion process is accomplished using a simple throttle valve, then the cycle is referred to as a Joule-Thomson cycle, discussed in Section 9.5.2. In either case, the adiabatic reduction in the pressure from state 4 to state 5 produces a corresponding reduction in the temperature of the fluid. The low pressure, low temperature gas at state 5 enters the load heat exchanger where it is heated by the device or process being cooled. The heating process results in a temperature rise from state 5 to state 6. It is important to note that the temperature at state 6 must be lower than the temperature at state 4 in order for the recuperative heat exchanger to operate (i.e., heat transfer must occur from hot to cold); the temperature reduction induced by the expansion must be greater than the temperature rise associated with the load heat exchanger. Indeed, the temperature of the high pressure fluid must be greater than the temperature of the low pressure fluid at all locations in the recuperator and the performance of a recuperative cooling cycle is strongly dependent on the effectiveness of the recuperator. The low pressure gas at state 6 is used in the recuperative heat exchanger to precool the high pressure gas, completing the cycle.

The system shown in Figure 9-25 is very simple; the cycles used in industry are often much more complex. The typical recuperative cycle will utilize multiple stages with various configurations of the heat exchangers and expansion devices. A discussion of these systems can be found in Flynn (2005).

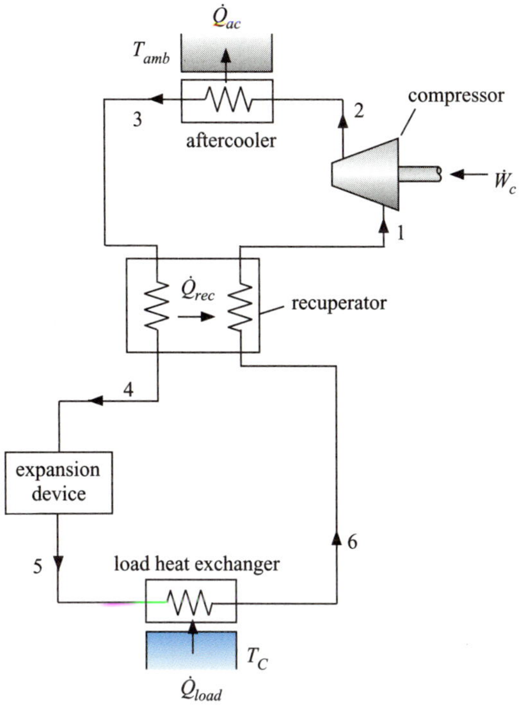

Figure 9-25: Basic recuperative cooling cycle.

9.5.1 The Reverse Brayton Cycle

When a work extraction device is used as the expansion device in a recuperative refrigeration cycle it is referred to as a reverse Brayton cycle. A simple reverse Brayton cycle is shown in Figure 9-26. Notice that the flow is the reverse of the recuperated Brayton power cycle discusssed in Section 8.3.2 in the sense that the flow leaving the compressor is cooled (rather than heated, as it is in the power cycle) and the flow leaving the turbine is heated (rather than cooled or discharged). Reverse Brayton cycles are used in cryogenic refrigeration cycles due to their relatively high efficiency. These cycles are also sometimes used for providing cooling in near room temperature applications where large quantities of compressed air are readily available as, for example, in an airplane where the air can be extracted from the jet engine.

In this section, the simple reverse Brayton cryocooler shown in Figure 9-26 will be analyzed. The cryocooler provides refrigeration at a load temperature $T_C = 60$ K using neon as the working fluid. The cycle employs a cryogenic turbine as the expansion device. The aftercooler rejects heat to ambient at $T_{amb} = 22^{\circ}$C and has an approach temperature difference of $\Delta T_{ac} = 4$ K. The load heat exchanger has an approach temperature difference of $\Delta T_{LHX} = 1$ K. The compressor has an isentropic efficiency of $\eta_c = 0.72$, a displacement rate of $\dot{V}_{disp} = 45$ liter/min, and a volumetric efficiency of $\eta_{vol} = 0.75$. The suction pressure is $P_{low} = 2$ atm and the pressure ratio is $PR = 1.9$. Pressure losses in the recuperator are neglected. The turbine efficiency is $\eta_t = 0.68$. The approach temperature difference for the recuperator is $\Delta T_{rec} = 5$ K and the recuperator pinch point will occur at its warm end for this system. These parameters are entered into EES.

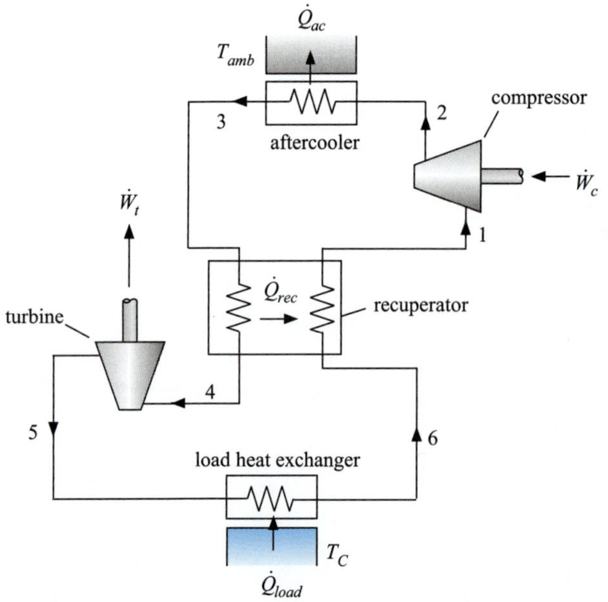

Figure 9-26: Reverse Brayton cycle.

```
$UnitSystem SI Mass Radian J K Pa
V_dot_disp=45 [liter/min]*convert(liter/min,m^3/s)     "displacement of compressor"
eta_vol=0.75 [-]                                        "volumetric efficiency of compressor"
eta_c=0.72 [-]                                          "compressor efficiency"
T_amb=converttemp(C,K, 22[C])                           "ambient temperature"
DT_ac=4 [K]                                             "aftercooler approach temperature difference"
T_C=60 [K]                                              "load temperature"
DT_LHX=1 [K]                                            "load HX approach temperature difference"
DT_rec=5 [K]                                            "recuperator approach temperature difference"
P_low=2 [atm]*convert(atm,Pa)                           "low pressure"
PR=1.9 [-]                                              "pressure ratio"
R$='Neon'                                               "working fluid"
eta_t=0.68 [-]                                          "turbine efficiency"
```

The temperature at state 3 is specified by the aftercooler approach temperature difference:

$$T_3 = T_{amb} + \Delta T_{ac} \tag{9-41}$$

The pressure at state 3 is the compressor discharge pressure, $P_3 = P_{low}\ PR$. State 3 is fixed by the temperature and pressure. The specific enthalpy and specific entropy (h_3 and s_3) are determined.

```
"State 3, Aftercooler exit/Recuperator inlet"
T[3]=T_amb+DT_ac                                  "temperature"
P[3]=P_low*PR                                     "pressure"
s[3]=entropy(R$,T=T[3],P=P[3])                    "specific entropy"
h[3]=enthalpy(R$,T=T[3],P=P[3])                   "specific enthalpy"
```

The temperature at state 6 is specified by the load heat exchanger approach temperature difference:

$$T_6 = T_C - \Delta T_{LHX} \tag{9-42}$$

Since the pressure loss in the heat exchangers is neglected, the pressure at state 6 is the compressor suction pressure, $P_6 = P_{low}$. State 6 is fixed by the temperature and pressure. The specific enthalpy and specific entropy (h_6 and s_6) are determined.

```
"State 6, Load HX exit/Recuperator inlet"
T[6]=T_C-DT_LHX                        "temperature"
P[6]=P_low                             "pressure"
s[6]=entropy(R$,T=T[6],P=P[6])         "specific entropy"
h[6]=enthalpy(R$,T=T[6],P=P[6])        "specific enthalpy"
```

Because the pinch point for the recuperator is at the warm end, the temperature at state 1 is given by:

$$T_1 = T_3 - \Delta T_{rec} \tag{9-43}$$

The pressure at the compressor inlet (state 1) is the suction pressure, $P_1 = P_{low}$. State 1 is fixed by the temperature and pressure. The specific enthalpy, specific entropy, and specific volume (h_1, s_1, and v_1) are determined.

```
"State 1, Compressor inlet/Recuperator outlet"
T[1]=T[3]-DT_rec                       "temperature"
P[1]=P_low                             "pressure"
s[1]=entropy(R$,T=T[1],P=P[1])         "specific entropy"
h[1]=enthalpy(R$,T=T[1],P=P[1])        "specific enthalpy"
v[1]=volume(R$,T=T[1],P=P[1])          "specific volume"
```

The mass flow rate is the same everywhere in the system during steady-state operation. Therefore, an energy balance on the recuperative heat exchanger provides the specific enthalpy at state 4:

$$h_6 + h_3 = h_1 + h_4 \tag{9-44}$$

The pressure at state 4 is the discharge pressure, $P_4 = P_{low}\, PR$. State 4 is fixed by the specific enthalpy and pressure. The temperature and specific entropy (T_4 and s_4) are determined.

```
"State 4, Turbine inlet/Recuperator exit"
h[6]+h[3]=h[1]+h[4]                    "specific enthalpy from recuperator energy balance"
P[4]=P_low*PR                          "pressure"
T[4]=temperature(R$,h=h[4],P=P[4])     "temperature"
s[4]=entropy(R$,h=h[4],P=P[4])         "specific entropy"
```

The pressure at state 2 is the discharge pressure, $P_2 = P_{low}\, PR$. The specific entropy leaving a reversible compressor operating under the same conditions is $s_{s,2} = s_1$. The specific entropy and pressure fix the state of the neon leaving a reversible compressor, allowing the specific enthalpy ($h_{s,2}$) to be determined. The work per mass required by a reversible compressor is:

$$\frac{\dot{W}_{s,c}}{\dot{m}} = h_{s,2} - h_1 \tag{9-45}$$

The work per mass required by the actual compressor is:

$$\frac{\dot{W}_c}{\dot{m}} = \frac{1}{\eta_c}\frac{\dot{W}_{s,c}}{\dot{m}} \tag{9-46}$$

The specific enthalpy leaving the actual compressor is:

$$h_2 = h_1 + \frac{\dot{W}_c}{\dot{m}} \tag{9-47}$$

State 2 is fixed by the specific enthalpy and pressure. The temperature and specific entropy (T_2 and s_2) are determined.

```
"State 2, Compressor exit/Aftercooler inlet"
P[2]=P_low*PR                          "pressure"
s_s[2]=s[1]                            "specific entropy leaving reversible compressor"
h_s[2]=enthalpy(R$,P=P[2],s=s_s[2])    "specific enthalpy leaving reversible compressor"
W_dot_s_c\m_dot=h_s[2]-h[1]            "power per mass flow rate required by reversible compressor"
W_dot_c\m_dot=W_dot_s_c\m_dot/eta_c    "power per mass flow rate required by compressor"
h[2]=h[1]+W_dot_c\m_dot                "specific enthalpy"
T[2]=temperature(R$,h=h[2],P=P[2])     "temperature"
s[2]=entropy(R$,h=h[2],P=P[2])         "specific entropy"
```

The pressure at state 5 is the suction pressure, $P_5 = P_{low}$. The specific entropy leaving a reversible turbine is $s_{s,5} = s_4$. The specific entropy and pressure fix the state of the neon leaving a reversible turbine, allowing the specific enthalpy ($h_{s,5}$) to be determined. The work per mass provided by a reversible turbine is:

$$\frac{\dot{W}_{s,t}}{\dot{m}} = h_4 - h_{s,5} \tag{9-48}$$

The work per mass provided by the actual turbine is:

$$\frac{\dot{W}_t}{\dot{m}} = \eta_t \frac{\dot{W}_{s,t}}{\dot{m}} \tag{9-49}$$

The specific enthalpy leaving the actual turbine is:

$$h_5 = h_4 - \frac{\dot{W}_t}{\dot{m}} \tag{9-50}$$

State 5 is fixed by the specific enthalpy and pressure. The temperature and specific entropy (T_5 and s_5) are determined.

```
"State 5, Turbine exit/Load heat exchanger inlet"
P[5]=P_low                               "pressure"
s_s[5]=s[4]                              "specific entropy leaving reversible turbine"
h_s[5]=enthalpy(R$,P=P[5],s=s_s[5])      "specific enthalpy leaving reversible turbine"
W_dot_s_t\m_dot=h[4]-h_s[5]              "power per mass flow rate produced by reversible turbine"
W_dot_t\m_dot=W_dot_s_t\m_dot*eta_t      "power per mass flow rate produced by turbine"
h[5]=h[4]-W_dot_t\m_dot                  "specific enthalpy"
T[5]=temperature(R$,h=h[5],P=P[5])       "temperature"
s[5]=entropy(R$,h=h[5],P=P[5])           "specific entropy"
```

The mass flow rate of neon is obtained from the compressor displacement and volumetric efficiency:

$$\dot{m} = \frac{\dot{V}_{disp}\,\eta_{vol}}{v_1} \tag{9-51}$$

The power required by the compressor is:

$$\dot{W}_c = \frac{\dot{W}_c}{\dot{m}}\dot{m} \tag{9-52}$$

and the power produced by the turbine is:

$$\dot{W}_t = \frac{\dot{W}_t}{\dot{m}}\dot{m} \tag{9-53}$$

The refrigeration load is obtained from an energy balance on the load heat exchanger:

$$\dot{Q}_{load} = \dot{m}\,(h_6 - h_5) \tag{9-54}$$

The rate of heat rejection in the aftercooler is obtained from an energy balance on the aftercooler:

$$\dot{Q}_{ac} = \dot{m}\,(h_2 - h_3) \tag{9-55}$$

```
m_dot=V_dot_disp*eta_vol/v[1]          "mass flow rate"
W_dot_c=W_dot_c\m_dot*m_dot            "compressor power"
W_dot_t=W_dot_t\m_dot*m_dot            "turbine power"
Q_dot_load=m_dot*(h[6]-h[5])           "refrigeration load"
Q_dot_ac=m_dot*(h[2]-h[1])             "rate of aftercooler heat transfer"
```

Solving these equations results in $\dot{W}_c = 115.9$ W, $\dot{Q}_{load} = 4.56$ W, $\dot{W}_t = 9.46$ W, and $\dot{Q}_{ac} = 111.0$ W. The *COP* of the refrigeration cycle is the ratio of the cooling provided ($\dot{Q}_{load}$) to the power required. If the power produced by the turbine is utilized to help run the compressor then the *COP* should be calculated according to:

$$COP = \frac{\dot{Q}_{load}}{\dot{W}_c - \dot{W}_t} \tag{9-56}$$

However, the turbine power is small in most cryogenic applications in comparison to the compressor power. Under the conditions associated with this example, the power produced by the turbine is only 8% of the compressor power. Therefore, the turbine power is often "thrown away"; i.e., it is dissipated in a bank of electrical resistors or in some other way. In this case, the *COP* is:

$$COP_{nt} = \frac{\dot{Q}_{load}}{\dot{W}_c} \tag{9-57}$$

```
COP=Q_dot_load/(W_dot_c-W_dot_t)       "COP, assuming that the turbine power is utilized"
COP_nt=Q_dot_load/W_dot_c              "COP if the turbine power is not used"
```

which leads to $COP = 0.043$ and $COP_{nt} = 0.039$. This seems like very poor performance compared to the near room temperature vapor compression systems analyzed in Section 9.2 or even the absorption cycle examined in Section 9.4. However, the Second Law shows that the maximum possible COP of a cryogenic refrigerator is quite low. The COP of a Carnot refrigerator was determined in Eq. (9-6) of Section 9.1:

$$COP_{Carnot} = \frac{T_C}{T_{amb} - T_C} \tag{9-58}$$

```
COP_Carnot=T_C/(T_amb T_C)          "COP of reversible refrigeration system"
```

which leads to $COP_{Carnot} = 0.255$; therefore, the performance of the reverse Brayton cycle is a significant fraction of the Carnot COP.

The reverse Brayton cycle is illustrated in Figure 9-27(a) on a T-s diagram. Notice that the cycle operates completely outside of the vapor dome and it therefore can use a refrigerant (e.g., neon) that has a very low critical temperature and a very low triple point temperature.

Figure 9-27(b) illustrates the low temperature portion of the T-s diagram. Notice that the temperature rise across the load heat exchanger (from state 5 to state 6) is less than the temperature drop induced by the turbine (from state 4 to state 5) because some temperature difference is required by the recuperator in order to transfer heat from the high pressure neon that is being cooled (from state 3 to state 4) to the low pressure neon being heated (from state 6 to state 1). The temperature difference required by the recuperator is indicated in Figure 9-27(b) and has a very significant effect on the cycle performance. A recuperative refrigeration cycle requires a high effectiveness recuperator in order to operate. The rate of heat transfer in the recuperator is obtained from an energy balance on the low pressure side:

$$\dot{Q}_{rec} = \dot{m}\,(h_1 - h_6) \tag{9-59}$$

The pinch point in the recuperator occurs at its warm end. Therefore, the maximum temperature that the low pressure neon leaving the recuperator can reach is $T_{1,max} = T_3$. The maximum specific enthalpy of the low pressure neon leaving the recuperator ($h_{1,max}$) is evaluated at $T_{1,max}$ and P_1. The maximum possible rate of heat transfer in the recuperator is:

$$\dot{Q}_{rec,max} = \dot{m}\,(h_{1,max} - h_6) \tag{9-60}$$

and the effectiveness of the recuperator is:

$$\varepsilon_{rec} = \frac{\dot{Q}_{rec}}{\dot{Q}_{rec,max}} \tag{9-61}$$

```
Q_dot_rec=m_dot*(h[1]-h[6])                  "heat transfer rate in the recuperator"
T_1_max=T[3]                                 "maximum temp. leaving low pressure side of recuperator"
h_1_max=enthalpy(R$,T=T_1_max,P=P[1])        "maximum enthalpy leaving low pressure side of recup."
Q_dot_rec_max=m_dot*(h_1_max-h[6])           "maximum possible rate of heat transfer in recuperator"
eff_rec=Q_dot_rec/Q_dot_rec_max              "recuperator effectiveness"
```

which shows that $\varepsilon_{rec} = 0.979$. A Parametric table is created that includes ΔT_{rec}, ε_{rec}, and $\dot{Q}_{load}$. The values of ΔT_{rec} are varied from 0 K (i.e., $\varepsilon_{rec} = 1$) to 12 K. Figure 9-28 shows the refrigeration load and the recuperator approach temperature difference as a function of the recuperator effectiveness. Notice that the system produces no refrigeration if the recuperator effectiveness drops below approximately 0.956. This system requires a very high value of effectiveness that can only be achieved using relatively massive, carefully designed, and expensive heat exchangers.

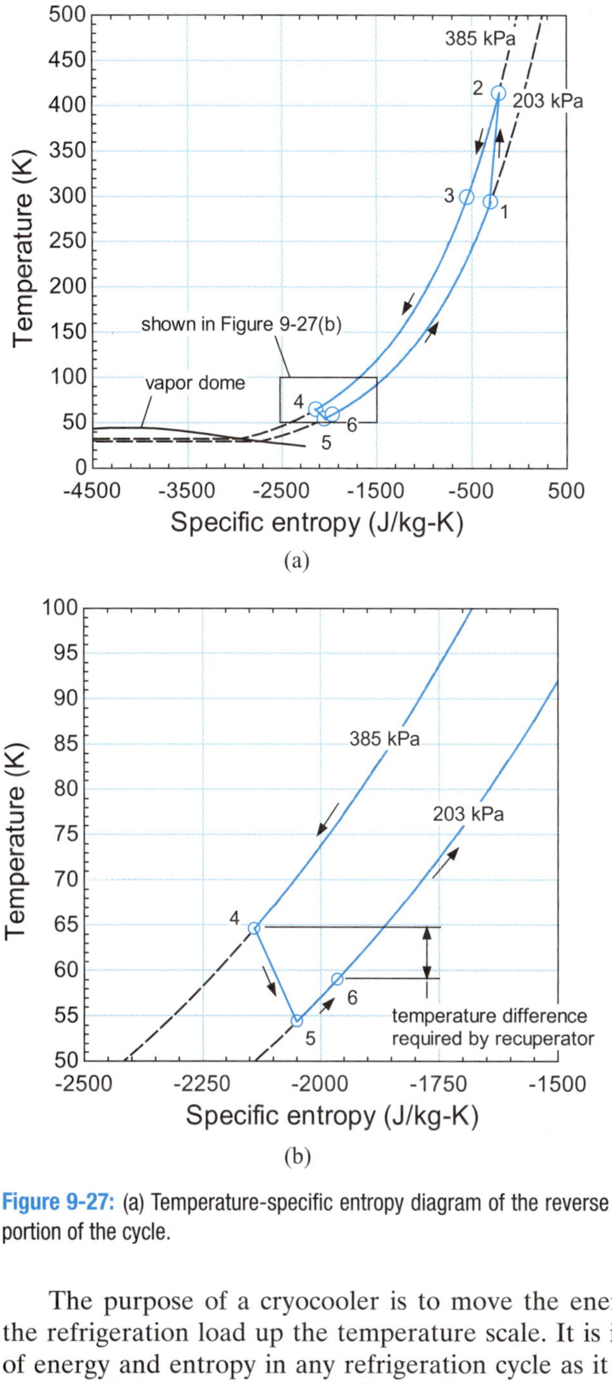

Figure 9-27: (a) Temperature-specific entropy diagram of the reverse Brayton cycle and (b) the low temperature portion of the cycle.

The purpose of a cryocooler is to move the energy and entropy associated with the refrigeration load up the temperature scale. It is interesting to follow the transfer of energy and entropy in any refrigeration cycle as it moves from low temperature to room temperature. In a reverse Brayton system, the energy transfer associated with the refrigeration load is transferred to room temperature as the turbine power. To more clearly see this energy transfer process, a control volume is defined that encompasses the recuperator, the load heat exchanger and the turbine. An energy balance on this control volume is shown in Figure 9-29(a) and it can be written as:

$$\dot{Q}_{load} + \dot{m}\, h_3 = \dot{W}_t + \dot{m}\, h_1 \tag{9-62}$$

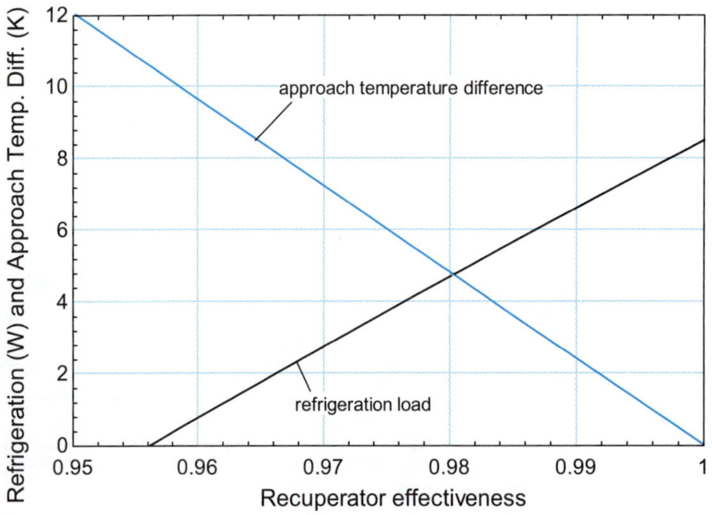

Figure 9-28: Refrigeration load and recuperator approach temperature difference as a function of the recuperator effectiveness.

Equation (9-62) can be rearranged:

$$\dot{Q}_{load} = \dot{W}_t + \underbrace{\dot{m}\,(h_1 - h_3)}_{\dot{H}_{rec}} \tag{9-63}$$

Equation (9-63) shows that the refrigeration load is equal to the sum of the turbine power and the energy transferred through the recuperator. The second term on the right side of Eq. (9-63), $\dot{H}_{rec}$, is sometimes referred to as the recuperator enthalpy flux. The recuperator enthalpy flux is the difference between the enthalpy flow rates associated with the low pressure neon flowing up the recuperator (i.e., towards the hot end) and the flow of high pressure neon flowing down the recuperator (i.e., towards the cold end). In a reverse Brayton cycle, the value of $\dot{H}_{rec}$ will typically be negative, indicating that the energy transfer in the recuperator is actually from the hot end towards the cold end. The enthalpy flux is computed according to:

```
H_dot_rec=m_dot*(h[1]-h[3])          "rate of energy transfer out of the hot end of the recuperator"
```

which results in $\dot{H}_{rec}$ = -4.90 W. The refrigeration power ($\dot{Q}_{load}$ = 4.56 W) is the sum of the turbine power ($\dot{W}_t$ = 9.46 W) and the recuperator enthalpy flux, in this case a negative number that reduces the refrigeration. Therefore, for a reverse-Brayton cycle, the energy associated with the refrigeration load moves up the temperature scale as turbine power and the enthalpy flux represents energy transfer down the temperature scale.

The entropy associated with the refrigeration load is transferred to room temperature as an entropy flux in the recuperator. To show this result, an entropy balance on the control volume that encompasses the recuperator, load heat exchanger, and turbine is shown in Figure 9-29(b) and it can be written as:

$$\frac{\dot{Q}_{load}}{T_C} + \dot{m}\,s_3 + \dot{S}_{gen} = \dot{m}\,s_1 \tag{9-64}$$

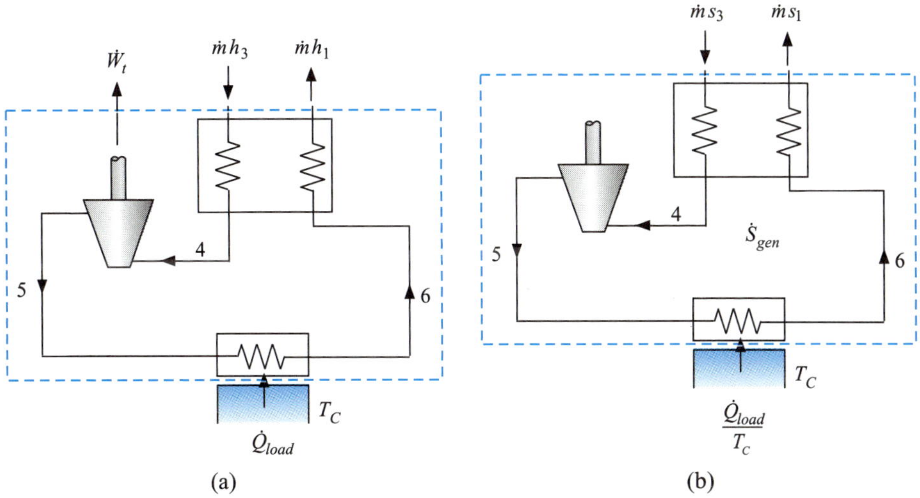

Figure 9-29: (a) Energy balance and (b) entropy balance on a control volume that encompasses the cryogenic components of the reverse Brayton cycle.

Equation (9-64) can be rearranged:

$$\frac{\dot{Q}_{load}}{T_C} = \underbrace{\dot{m}\,(s_1 - s_3)}_{\dot{S}_{rec}} - \dot{S}_{gen} \tag{9-65}$$

where $\dot{S}_{rec}$ is the entropy flux associated with the recuperator. The terms in Eq. (9-65) are evaluated:

```
Q_dot_load\T_C=Q_dot_load/T_C          "rate of entropy transfer into the cooler"
S_dot_rec=m_dot*(s[1]-s[3])            "rate of entropy transfer out of the hot end of the recuperator"
Q_dot_load\T_C=S_dot_rec-S_dot_gen     "entropy generation rate within control volume"
```

which leads to $\dot{Q}_{load}/T_C = 0.076$ W/K, $\dot{S}_{rec} = 0.232$ W/K, and $\dot{S}_{gen} = 0.156$ W/K. The specific entropy of the low pressure neon at state 1 is higher than the specific entropy of the high pressure neon at state 3 because of the pressure difference; therefore, $\dot{S}_{rec}$ is positive and this is how the entropy associated with the refrigeration load is transferred up the temperature scale.

9.5.2 The Joule-Thomson Cycle

When a simple throttle valve is used as the expansion device in a recuperative cycle then the cycle is referred to as a Joule-Thomson cycle, shown in Figure 9-30. The overwhelming advantage of this cycle is its mechanical simplicity; the removal of the work extraction device required by the reverse Brayton cycle allows the system to be compact, cheap, and reliable.

The disadvantage of the Joule-Thomson is that it has a lower efficiency and is much more sensitive to the properties of the working fluid than the reverse Brayton cycle. The work extraction associated with the turbine in the reverse Brayton shown in Figure 9-26 causes the specific enthalpy of the working fluid to decrease, which leads directly to the temperature drop between states 5 and 6 shown in Figure 9-27(b). In a Joule-Thomson

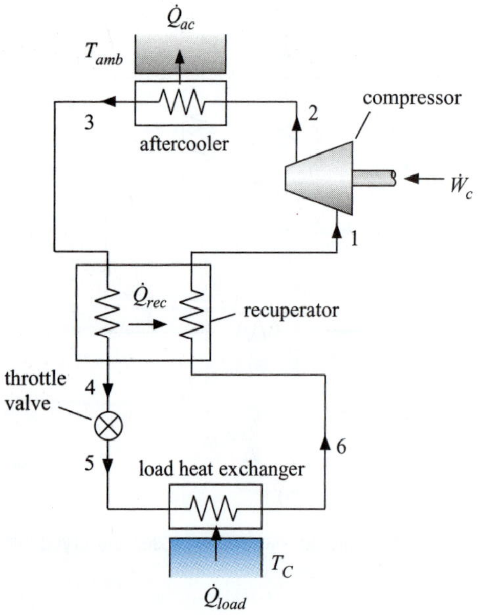

Figure 9-30: Joule-Thomson refrigeration cycle.

system, the specific enthalpy of the working fluid remains constant as it flows through the valve:

$$h_5 = h_4 \tag{9-66}$$

The temperature of a fluid exiting a throttle may be lower, higher, or the same as the throttle inlet temperature depending on the type of fluid and its inlet state. The relationship between the temperature and the pressure of a gas undergoing an isenthalpic process is described in terms of the *Joule-Thomson coefficient*, defined as the partial derivative of temperature with respect to pressure at constant specific enthalpy:

$$\mu_{JT} = \left(\frac{\partial T}{\partial P}\right)_h \tag{9-67}$$

The Joule-Thomson cycle only operates correctly if the flow through the valve induces a temperature decrease. As a result, the Joule-Thomson cycle must use fluids that have a positive Joule-Thomson coefficient at the conditions of interest; examination of Eq. (9-67) shows that a positive value of μ_{JT} leads to reduction in temperature if the pressure decreases for an isenthalpic process.

A fluid that is at a temperature and pressure for which it behaves as an ideal gas will have a zero Joule-Thomson coefficient. The specific enthalpy of an ideal gas is discussed in Section 4.2.3 and provided by integration of:

$$dh = c_P\, dT \quad \text{for an ideal gas} \tag{9-68}$$

where c_P is the specific heat capacity at constant pressure. Equation (9-68) implies that specific enthalpy is only a function of temperature and vice versa. Therefore, a process that occurs at a constant value of specific enthalpy must also occur at constant temperature for an ideal gas; the partial derivative of temperature with respect to pressure at constant enthalpy (μ_{JT}) is zero. A fluid that behaves as an incompressible liquid will have a small and negative Joule-Thomson coefficient. The specific enthalpy of an incompressible substance is discussed in Section 4.2.4 and provided by integration of:

$$dh = c\, dT + v\, dP \quad \text{for an incompressible substance} \tag{9-69}$$

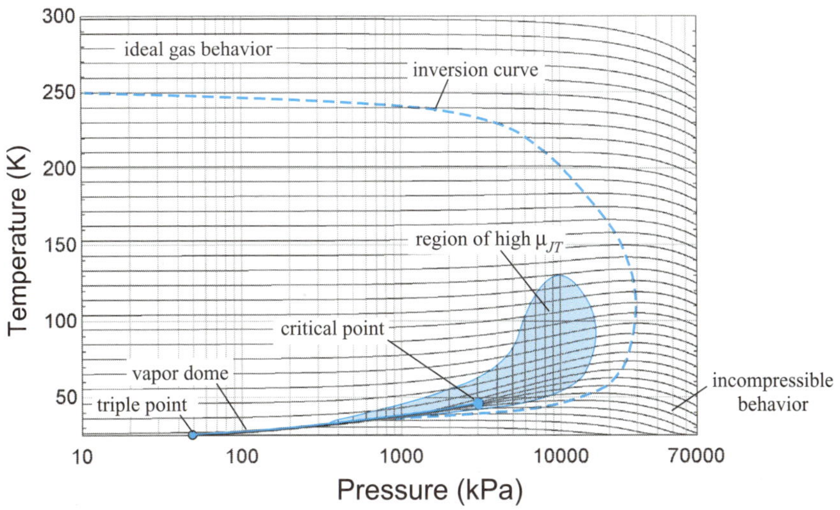

Figure 9-31: Temperature as a function of pressure for various values of specific enthalpy for neon. The vapor dome and inversion curve are also shown.

where c is the specific heat capacity and v is the specific volume of the liquid. Setting $dh = 0$ in Eq. (9-69) and rearranging leads to:

$$\mu_{JT} = \left(\frac{\partial T}{\partial P}\right)_h = -\frac{v}{c} \quad \text{for an incompressible substance} \tag{9-70}$$

Both c and v must be positive and therefore μ_{JT} must be negative for an incompressible substance.

The working fluid for a Joule-Thomson cycle cannot behave as either an ideal gas (with its molecules spaced so far apart that they exert no forces on each other) or an incompressible substance (with its molecules packed so closely together that they repel each other). The Joule-Thomson cycle requires a fluid that is in an intermediate state between these extremes where the molecules experience substantial intermolecular forces and therefore exhibit non-ideal behavior. Figure 9-31 illustrates the temperature of neon as a function of pressure for various values of specific enthalpy. The slope of these lines is equivalent to the Joule-Thomson coefficient and therefore Figure 9-31 shows that, while there are regions where neon has a positive value of μ_{JT} and therefore could be used in a Joule-Thomson cycle, there are also large regions where the value of μ_{JT} is either negative or very close to zero. At high temperatures and low pressures neon behaves as an ideal gas and therefore the lines become flat, consistent with $\mu_{JT} = 0$. At high pressures and low temperatures neon behaves as an incompressible liquid and therefore $\mu_{JT} < 0$, as evident by the negative slope of the lines. Only in a relatively small region, which is also shown in Figure 9-31, is the value of $\mu_{JT} > 0$, as evident by the positive slope of the lines. The inversion curve is defined by the peaks in the lines in Figure 9-31, which corresponds to the point at each specific enthalpy value where $\mu_{JT} = 0$. Inside of the inversion curve, the Joule-Thomson coefficient is positive and outside it is negative. Notice that a Joule-Thomson cycle using neon could not operate under the conditions that were used in Section 9.5.1 to simulate a reverse-Brayton cycle (i.e., 54 K $< T <$ 300 K and 203 kPa $< P <$ 385 kPa) because these conditions extend outside of the inversion curve and are far from the region of high Joule-Thomson coefficient shown in Figure 9-31.

Table 9-5: Maximum inversion temperature of various fluids, from Barron (1985)

Fluid	Maximum inversion temperature
Helium	45 K
Hydrogen	205 K
Neon	250 K
Nitrogen	621 K
Air	603 K
Oxygen	761 K
Argon	794 K

Table 9-5 summarizes the *maximum inversion temperature* (i.e., the maximum temperature at which the Joule-Thomson coefficient is positive) for various fluids. It is interesting to note that the simple Joule-Thomson cycle shown in Figure 9-30 can only operate if the *entire* cycle lies within the inversion curve. Therefore, the maximum inversion temperature associated with the fluid must lie above the temperature to which heat is rejected. Examination of Table 9-5 shows that a simple Joule-Thomson system can be used to produce cooling starting at room temperature with gases such as argon or oxygen but not helium, hydrogen or neon. In fact, argon is often used to energize very simple and compact Joule-Thomson systems for applications such as cryosurgery.

9.5.3 Liquefaction Cycles

This extended section can be found online at www.cambridge.org/kleinandnellis and discusses cryogenic cycles that are used to liquefy gases. Example 9.5-1 in this section presents an analysis of a pre-cooled Linde-Hampson liquefaction cycle.

9.6 Regenerative Cryogenic Cooling Cycles

This extended section can be found online at www.cambridge.org/kleinandnellis and discusses regenerative cryogenic cycles which, like the recuperative cycles discussed in Section 9.5, operate above the vapor dome. As a result, regenerative cooling cycles also overcome the temperature limitations of vapor compression cycles and can produce very low temperature refrigeration.

REFERENCES

Barron, R.F., *Cryogenic Systems, 2nd Edition*, Oxford University Press, New York, ISBN: 0-19-503567-4, (1985).

CFC's: Time of Transition, American Society of Heating, Refrigerating and Air-Conditioning Engineering, Inc., ISBN 0-910110-58-1, (1989).

U.S. Department of Energy, *Energy Efficiency and Renewable Energy*, Building Technologies Program, (2009), http://www1.eere.energy.gov/buildings/appliance_standards/residential/refrigerators_freezers.html

Flynn, T.M., *Cryogenic Engineering, 2nd Edition*, Marcell Dekker, New York, ISBN: 0-8247-5367-4, (2005).

Klein, S.A., Reindl, D.T., and Brownell, K., "Refrigeration System Performance using Liquid-Suction Heat Exchangers," *International Journal of Refrigeration*, Vol. 23, Part 8, pp. 588–596 (2000).
Herold, K.E., Radermacher, R., and Klein, S.A., *Absorption Chillers and Heat Pumps*, CRC Press, ISBN 0-8493-9427-9, (1996).
Shah, M.M., "Chart correlation for saturated boiling heat transfer: equations and further study," *ASHRAE Transactions*, Vol. 88, pp. 185–196, (1982).

Problems

The problems included here have been selected from a larger set of problems that are available the website associated with this book (www.cambridge.org/kleinandnellis).

A: Vapor Compression Problems

9.A-1 Dedicated subcooling is a novel modification for frozen food refrigeration in supermarkets. With the dedicated subcooling modification, liquid refrigerant leaving the condenser is further cooled at constant pressure to an intermediate temperature, T_4, as shown in Figure 9.A-1. The cooling needed for this purpose is provided by another, smaller refrigeration cycle. The overall coefficient of performance is defined as the ratio of the heat removal from the food cases to the total work input for both refrigeration cycles.

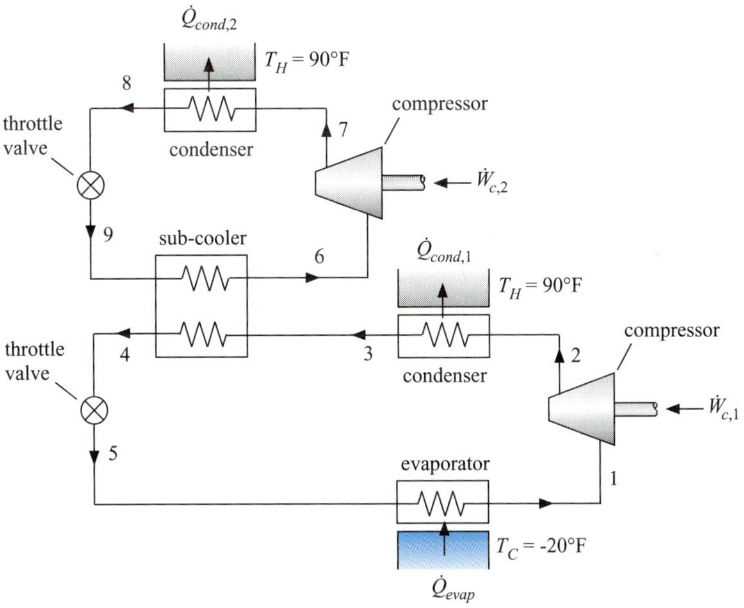

Figure 9.A-1: Dedicated subcooling system for a supermarket application.

Consider a refrigeration cycle designed to maintain food products at $T_C = -20°$F. Air-cooled condensers in the refrigeration cycles reject heat to outdoors at a design temperature of $T_H = 90°$F. At these design conditions, all heat exchange equipment is sized such that there is a $\Delta T_{hx} = 12°$F difference between the entering cold stream temperature and the exiting hot stream temperature (e.g., the saturation temperature in the evaporator of the refrigerated cases is $-32°$F and the temperature at state 6 is 12°F lower than the temperature at state 4). The isentropic efficiency of the compressors is $\eta_c = 0.7$. Assume that the refrigerant leaving the evaporator is saturated vapor and the refrigerant leaving the condensers is saturated liquid. Neglect pressure losses in this analysis. The refrigerant is R22.

a) Determine the value of T_4 that maximizes the COP of the combined cycle.
b) Compare the COP of the optimized dedicated subcooling cycle to that of a simple vapor compression cycle that does not use subcooling.
c) Calculate the Second-Law efficiencies of the optimized dedicated subcooling cycle and a cycle that does not use subcooling.
d) What do you see as the advantages and disadvantages of the subcooling modification?

9.A-6 A $\dot{Q}_{evap}$ = 3-ton single-stage vapor compression refrigeration system is being designed for an application in which the condensing temperature is $T_{cond} = 90°$F and the evaporator temperature is T_{evap} = 0°F. Refrigerants R134a, R22, R717 (ammonia), R290 (propane) and R410A are being considered for the system. The compressor efficiency may be assumed to be $\eta_c = 0.8$ for all refrigerants. Assume that the pressure losses in the heat exchangers and piping are negligible. Assume steady-state operation.

a) Calculate and compare the refrigerant mass flow rate, the compressor power, the cycle COP, and the required volumetric flow rate at the compressor inlet for each refrigerant. Which refrigerant would you recommend based on these comparisons?
b) If a reversible expansion device were used in place of a throttle valve, some power could be produced which would offset the power needed to operate the compressor. In addition, a greater refrigeration effect would occur. Calculate the "lost" power and the reduction in the refrigeration effect resulting from throttling rather than isentropic expansion for each of the refrigerants.

9.A-7 You are considering installation of the conventional vapor compression refrigeration system shown in Figure 9.A-7(a). The system must provide cooling to a space at $T_C = -10°$F and reject heat to ambient air at $T_H = 85°$F.

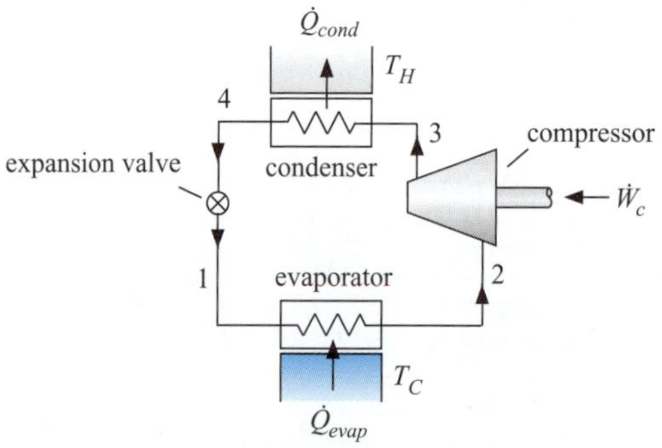

Figure 9.A-7(a): Conventional vapor compression system.

The efficiency of the compressor is η_c = 0.60, the volumetric efficiency of the compressor is η_{vol} = 0.65, and the displacement rate of the compressor is $\dot{V}_{disp}$ = 50 cfm. The condenser has an approach temperature difference ΔT_{cond} = 5 K and the pressure drop associated with the flow of refrigerant through the condenser is ΔP_{cond} = 50 kPa. The refrigerant leaves the condenser with a ΔT_{sc} = 2 K degree of subcool. The evaporator has an approach temperature difference of ΔT_{evap} = 6 K and the pressure drop associated with the flow of refrigerant through the evaporator is $\Delta P_{evap} = 15$ kPa. The refrigerant leaves the evaporator with a superheat of $\Delta T_{sh} = 5$ K. The refrigerant is R134a.

a) Analyze the cycle. Print out the Arrays table containing the state point information and generate a temperature-entropy diagram that shows all states. What is the rate of refrigeration provided and rate of power consumed by the cycle?
b) Check your answer by carrying out an overall energy balance.
c) Determine the rate of entropy generation in each of the components. Check your answer by carrying out an overall entropy balance.
d) Determine the minimum power required to provide the same amount of refrigeration at T_C while rejecting heat to T_H (i.e., the power required by a reversible refrigeration system operating between the same two temperatures). The difference between the actual compressor power and this minimum power is the lost work associated with entropy generation in the cycle and therefore should be equal to the product of the rate of entropy generation and the ambient temperature. Show that this is so.
e) What is the Coefficient of Performance of the cycle?

The liquid overfed system shown in Figure 9.A-7(b) may provide some advantages relative to the conventional system shown in Figure 9.A-7(a). The refrigerant leaving the condenser is expanded into a tank. Saturated liquid is pumped from the tank into the evaporator at state 1. By controlling the flow rate provided by the pump (which is independent of the flow rate provided by the compressor) it is possible to keep the quality of the refrigerant everywhere in the evaporator low; this dramatically improves the thermal performance of the evaporator. Saturated vapor is pulled from the tank by the compressor at state 4.

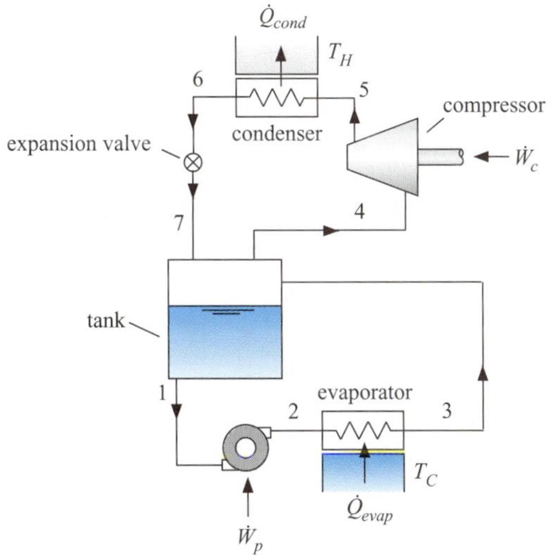

Figure 9.A-7(b): Liquid overfed vapor compression system.

The characteristics of the compressor are the same. The efficiency of the compressor is $\eta_c = 0.60$, the volumetric efficiency of the compressor is $\eta_{vol} = 0.65$, and the displacement of the compressor is $\dot{V}_{disp} = 50$ cfm. The condenser characteristics are also unchanged. The condenser has an approach temperature difference $\Delta T_{cond} = 5$ K and the pressure drop associated with the flow of refrigerant through the condenser is $\Delta P_{cond} = 50$ kPa. The refrigerant leaves the condenser with a $\Delta T_{sc} = 2$ K degree of subcool. The pump has an efficiency of

$\eta_p = 0.5$ and the pump is controlled so that the quality of the refrigerant leaving the evaporator is $x_{evap,out} = 0.5$. The performance of the evaporator improves with the liquid overfed modification; therefore, the evaporator approach temperature difference is reduced to $\Delta T_{evap} = 3$ K. The pressure drop associated with the flow of refrigerant through the evaporator is $\Delta P_{evap} = 15$ kPa. The refrigerant is R134a. The system must provide cooling at $T_C = -10°$F and reject heat to ambient air at $T_H = 85°$F. There is no pressure loss associated with the tank and the tank is insulated.

f) Analyze the liquid overfed system. Print out the Arrays table containing the state point information and generate a temperature-entropy diagram that shows all states. What is the rate of refrigeration provided and rate of power consumed by the cycle?

g) What is the *COP* of the liquid overfed cycle? Compare this value to the *COP* of the conventional system calculated in part (e).

9.A-9 A heat-powered refrigeration cycle is shown in Figure 9.A-9. A Rankine cycle is used to produce the power that drives the compressor in the vapor compression refrigeration cycle. The system operates at steady-state with refrigerant R134a as the working fluid in both cycles. The isentropic efficiencies of the turbine, compressor, and pump are $\eta_t = 0.78$, $\eta_c = 0.72$, and $\eta_p = 0.48$, respectively. At the design condition, the evaporator must provide $\dot{Q}_{evap} = 54$ kW of cooling at an evaporating temperature of $T_{evap} = -10°$C. Saturated vapor leaves the evaporator exit. The condensing temperature is $T_{cond} = 40°$C and saturated liquid leaves the condenser. Saturated vapor leaves the boiler $T_1 = 95°$C. The turbine, compressor, pump, and valve operate adiabatically. Neglect pressure losses in the heat exchangers and piping.

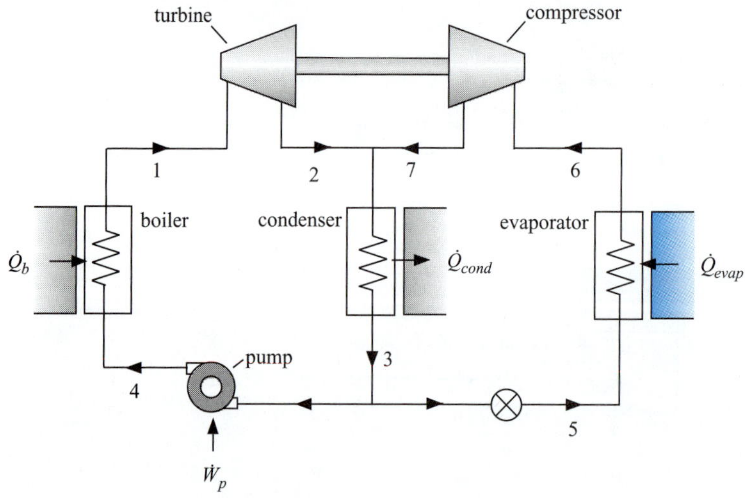

Figure 9.A-9: Heat powered refrigeration cycle.

a) Determine the mass flow rate of R134a through the evaporator.

b) Determine the mass flow rate of R134a through the boiler.

c) The coefficient of performance for this cycle is the ratio of the refrigeration heat input to boiler heat input. What is the *COP* at design conditions?

d) Compare the *COP* determined in part (c) to the maximum possible *COP* for a heat-powered refrigeration system operating within the same temperature limits.

9.A-11 Cooling at two different temperatures is needed in many situations, such as in household refrigerators and in supermarket refrigeration systems. Figure 9.A-11 shows a schematic diagram for a vapor compression refrigeration system in a supermarket that has a single compressor, a single condenser and two evaporators providing cooling capacity at two different temperatures. The compressor efficiency is $\eta_c = 0.78$. Saturated liquid exits the condenser at $T_{cond} = 105^\circ$F. The low temperature evaporator provides a cooling capacity of $\dot{Q}_{evap,1} = 3$ tons at a saturation temperature $T_{evap,1} = -5^\circ$F which is used to maintain a freezer at $T_{C,1} = 5^\circ$F. The higher temperature evaporator provides a cooling capacity of $\dot{Q}_{evap,2} = 2$ tons at a saturation temperature $T_{evap,2} = 25^\circ$F, which is used to maintain a dairy case at $T_{C,2} = 35^\circ$F. The refrigerant is R134a. Assume saturated vapor exists at the evaporator exits and neglect pressure losses in the piping and heat exchangers.

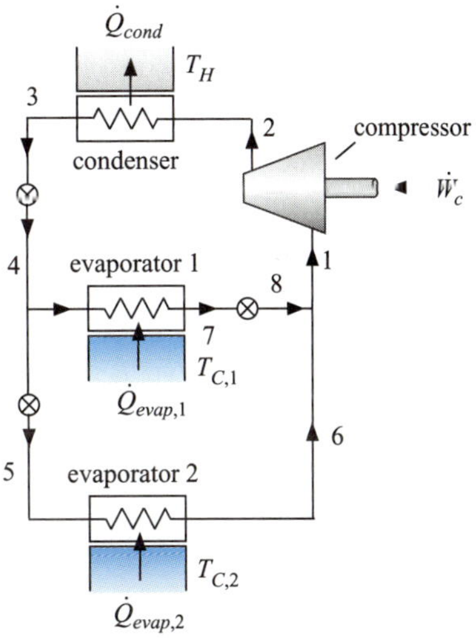

Figure 9.A-11: Refrigeration cycle with two evaporators.

a) Determine the properties at all of the states in the cycle.
b) Determine the power required to operate the cycle at steady-state.
c) Indicate how you would define a *COP* for this cycle and determine its value.
d) Determine the total rate of entropy generation for steady-state operation.
e) Define a Second-Law efficiency and determine its value.

9.A-13 It has been suggested that the refrigeration equipment in an old household refrigerator that is no longer needed, could be modified to serve as a heat pump for heating water in residence. The purpose of this problem is to evaluate this suggestion. The refrigerator uses R134a. The compressor is driven by an electric motor that spins at $N = 1750$ revolutions per minute. The compressor has a displacement of $V_{disp} = 4.5$ in^3, an efficiency of $\eta_c = 0.78$, and a volumetric efficiency of $\eta_{vol} = 0.70$ when operated at condensing temperature of $T_{cond} = 110^\circ$F and an evaporating temperature of $T_{evap} = 0^\circ$F.

a) Determine the *COP* and cooling capacity (in tons) of the refrigerator when it is operated as designed.

The air-cooled condenser is replaced with a water-cooled condenser in order to heat water for domestic purposes. The evaporator temperature increases to $T_{evap} = 60^\circ$F since energy is supplied to the evaporator from indoors. The condenser must operate at $T_{cond} = 140^\circ$C to ensure that the water is heated to $T_{w,out} = 130^\circ$F. The volumetric efficiency of the compressor is reduced to $\eta_{vol} = 0.60$ at these conditions. Assume that the efficiency of the compressor is unaffected by the change in conditions.

b) Determine the number of gallons per hour of water that could be heated from $T_{w,in} = 50°F$ to $T_{w,out}$ with this system. What is the COP of the system.

c) What is your evaluation of the suggestion to use the refrigerator as a water heater?

B: Absorption, Recuperative and Regenerative Cycles

9.B-2 A solar air-conditioning system uses a lithium bromide-water absorption chiller having the configuration shown in Figure 9.B-2. The solar collectors provide thermal energy at a rate sufficient to maintain the solution in the generator at $T_{gen} = 200°F$ (states 4 and 7). The temperature of the strong solution entering the generator at state 3 is $T_3 = 160°F$, while the temperature of the solution leaving the absorber (state 1) is $T_{abs} = 100°F$. The evaporation temperature (states 9 and 10) is $T_{evap} = 40°F$ and the temperature of the saturated liquid water at state 8 is $T_{cond} = 100°F$. The evaporator provides $\dot{Q}_{evap} = 2$ tons of cooling capacity. Assume that saturated vapor leaves the evaporator and neglect the pressure losses in the piping. The pump power should be small so it is sufficient to assume that it operates isentropically.

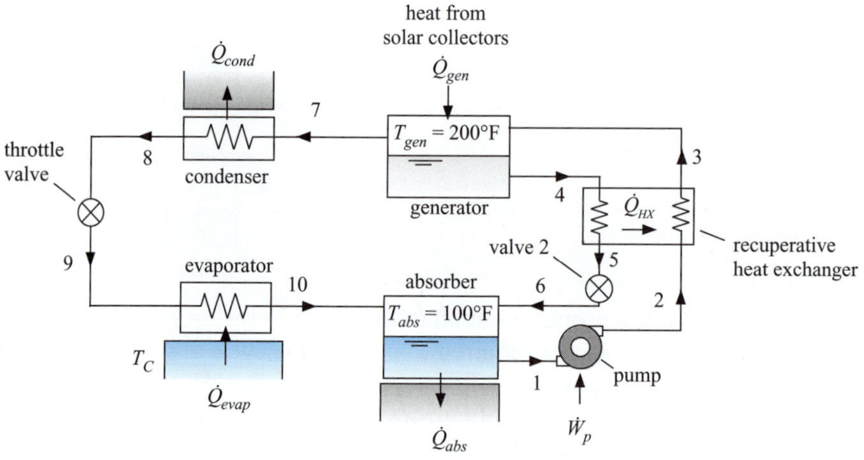

Figure 9.B-2: Solar air conditioning system.

a) Determine the system coefficient of performance at these operating conditions.

b) Determine the required collector area if the solar radiation is 950 W/m^2 and the collector efficiency under these conditions is 0.30.

9.B-4 Figure 9.B-4 illustrates a cryogenic refrigerator that is used to maintain a detector at 70 K. The refrigerator utilizes a reverse-Brayton cycle with neon as the working fluid. Neon enters the compressor at state 1 with pressure $P_{low} = 152$ kPa and mass flow rate $\dot{m} = 0.0006$ kg/s. The isentropic efficiency of the compressor is $\eta_c = 0.78$. The neon leaves the compressor at state 2 with pressure $P_{high} = 532$ kPa. The neon leaving the compressor is quite hot and is therefore cooled in an aftercooler. The aftercooler rejects heat at rate $\dot{Q}_H$ to the ambient temperature at $T_H = 20°C$ with an approach temperature difference $\Delta T_{ac} = 10$ K. Neon flows through the recuperative heat exchanger where it is cooled by heat transfer with the cold neon returning from the cold end of the device. The neon exits the turbine at state 5 with pressure P_{low}. The turbine operates with an isentropic efficiency of

$\eta_t = 0.84$. The neon leaving the turbine enters the load heat exchanger where it is warmed by a heat transfer from the detector which is at temperature $T_C = 70$ K. The approach temperature difference for the load heat exchanger is $\Delta T_{LHX} =$ 2 K. Finally, the neon leaving the load heat exchanger at state 6 passes through the recuperator where it is warmed by heat transfer with the high pressure neon flowing in the opposite direction. The recuperator has an effectiveness of $\varepsilon = 0.92$. Neglect pressure losses in the heat exchangers and assume neon behaves as a real fluid for the conditions in this cycle.

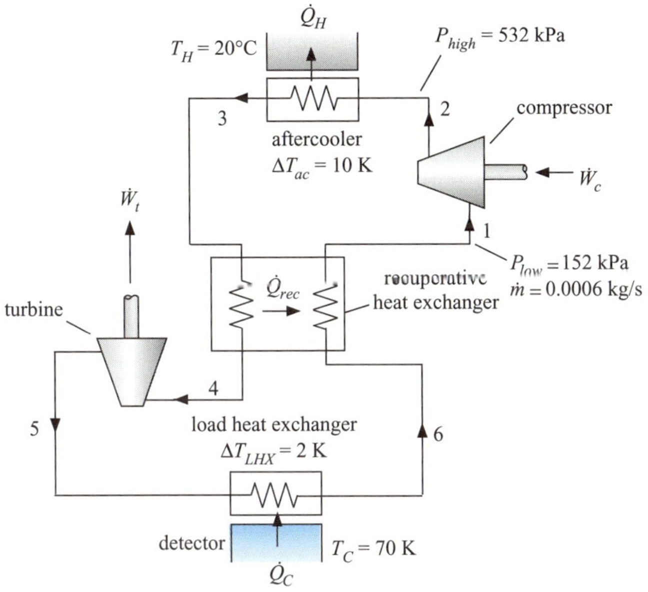

Figure 9.B-4: Reverse-Brayton refrigeration cycle.

a) Determine the temperature and pressure at each of the six numbered states in Figure 9.B-4.
b) Determine the power required by the compressor ($\dot{W}_c$), the power from the turbine ($\dot{W}_t$), and the rate of heat transfer to the neon in the load heat exchanger ($\dot{Q}_C$).
c) Determine the Coefficient of Performance (COP) of the cryocooler.
d) Conduct numerical experiments with your model to determine the importance of the isentropic efficiencies of the compressor and turbine on the COP of this cycle.

9.B-7 Figure 9.B-7 illustrates a system that is used to provide short term cooling for a cryogenic detector on a spacecraft. A bottle of argon is stored onboard the spacecraft. The initial pressure of the bottle is $P_{bottle} = 3500$ psi and the volume of the bottle is $V_{bottle} = 10$ liter. Heat transfer from the ambient environment onboard the spacecraft maintains temperature of the contents of the bottle at $T_{amb} = 20°$C. When cooling is desired, a valve on the bottle is opened and argon passes through a pressure regulator that maintains an exit pressure of $P_{reg} = 2000$ psi. The temperature of the argon may change as it passes through the regulator; therefore an aftercooler is placed downstream of the regulator so that the argon is

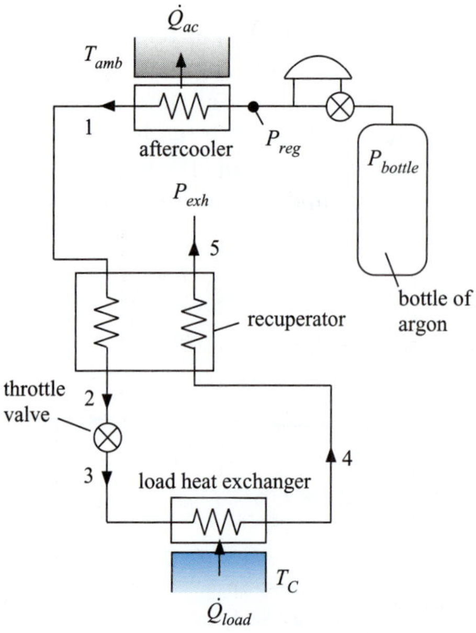

Figure 9.B-7: Joule-Thomson system for detector cooling.

returned to ambient temperature before it enters the recuperative heat exchanger at state 1. The recuperator has an approach temperature difference of $\Delta T_{rec} = 5$ K and the pinch point is at the warm end. The argon leaving the high pressure side of the recuperator at state 2 is throttled to the exhaust pressure, $P_{exh} = 0.25$ atm at state 3 before being heated in a load heat exchanger to state 4 and passing through the low pressure side of the recuperator where it is exhausted to space at state 5. Pressure losses in the recuperator are negligible for both the low and high pressure streams. This problem will ignore the initial cool down transient for the detector and consider only its steady state operation at $T_C = 140$ K. The approach temperature difference associated with the load heat exchanger is $\Delta T_{LHX} = 1$ K. The total refrigeration load required to maintain the temperature of the detector at T_C is $\dot{Q}_{load} = 15$ W. The system is deactivated when the bottle pressure reaches the regulator pressure.

a) Determine the time that the system can maintain the detector at its operating temperature.
b) Plot the operating time as a function of the regulator pressure and determine the optimal regulator pressure that maximizes the operating time of the system.

C: Advanced Problems

9.C-1 Carbon dioxide is of interest as a refrigerant because it is inexpensive, non-toxic, non-flammable, and it has excellent thermodynamic and transport properties. However, carbon dioxide has a critical temperature of about 31°C. Since the ambient temperature is often this high or higher, the carbon-dioxide refrigeration cycle must operate with the high temperature cooler (which serves the function of a condenser in a typical vapor compressor cycle) operating at supercritical pressures. The purpose of this problem is to design a supercritical carbon-dioxide air conditioning cycle to provide $\dot{Q}_{evap} = 10$ kW of cooling with an evaporating temperature $T_{evap} = 5$°C. Assume that the refrigerant leaving the cooler is at

$T_4 = 40°$ C. The mass flow rate of refrigerant can be represented with the following relation:

$$\dot{m} = \left[1 + C - C\left(\frac{p_{discharge}}{p_{suction}}\right)^{\frac{1}{n}}\right]\frac{V\,N}{v_b}$$

where $C = 0.025$ is the clearance volume ratio, n is the polytropic index, V is the compressor cylinder displacement, $N = 3500$ rpm is the compressor speed, and v_b is the specific volume of the refrigerant at the compressor inlet. The polytropic index can be assumed to be equal to the value corresponding to an isentropic process (i.e., $n = c_P/c_v$) evaluated at the compressor inlet conditions. The compressor power can be computed by defining a combined compressor and motor efficiency that is the ratio of the isentropic to the actual power; the combined efficiency is $\eta_c = 0.62$. Compressors are often assumed to be adiabatic. However experience has shown that small compressors have significant heat loss. In this case, the ratio of the heat transfer rate to the compressor power input is $f_H = 0.25$. It will be necessary to determine the optimum compressor discharge pressure, the associated discharge temperature, the refrigerant mass flow rate, the compressor displacement, and the compressor power. A schematic of the proposed refrigeration cycle is shown in Figure 9.C-1. Note that this figure includes a liquid-suction heat exchanger. At design conditions, the refrigerant exits the evaporator as saturated vapor and enters the cold end of the liquid-suction heat exchanger where it is heated so that it leaves at a superheated state. The advantage of the liquid-suction heat exchanger should be investigated in your analysis. Neglect pressure losses in the heat exchangers and piping.

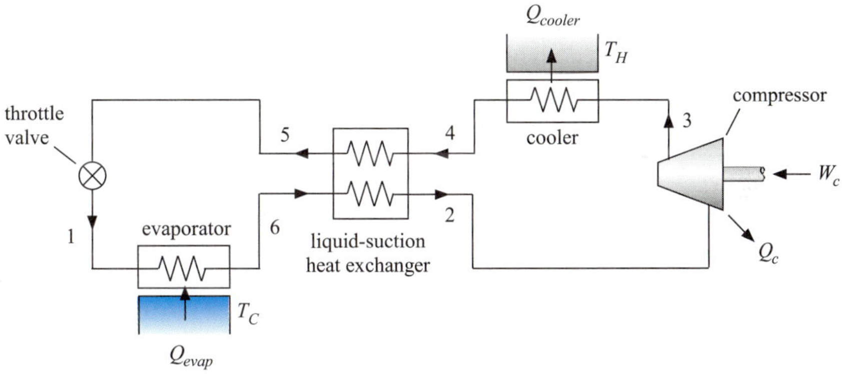

Figure 9.C-1: Carbon dioxide refrigeration cycle with liquid-suction heat exchanger.

a) Assume that the liquid-suction heat exchanger is not present (or alternatively, it has a small heat exchanger effectiveness) and that the discharge pressure of the compressor is $P_{discharge} = 95$ bar. Determine the *COP* of this cycle and plot the state point information on temperature-specific entropy and pressure-specific enthalpy diagrams.

b) Determine optimal compressor discharge pressure that maximizes the *COP* of this cycle for the conditions or part (a). If you find that there is an optimum discharge pressure, please explain why it exists.

c) Prepare a plot of the optimum *COP* as a function of the liquid-to-suction heat exchanger effectiveness. (You will need to determine the optimum discharge pressure at each heat exchanger effectiveness value.) What is your conclusion regarding the benefits of the liquid-to-suction heat exchanger?

d) How does carbon dioxide compare to other refrigerants under similar operating conditions? What is your overall assessment of carbon dioxide as a refrigerant?

9.C-2 The purpose of this problem is to determine the performance of residential vapor compression heat pump operating in the heating mode using refrigerant R134a. The heat pump system has a fixed speed reciprocating compressor that provides a constant displacement rate ($\dot{V}_{disp}$) when it is operating. The volumetric flow rate of the refrigerant entering the compressor is given by:

$$\dot{V}_2 = \dot{V}_{disp}\left[1 - C\left(\frac{v_2}{v_3} - 1\right)\right]$$

where $\dot{V}_{disp} = 0.0055$ m^3/s is the compressor displacement rate, $C = 0.025$ is the clearance volume ratio, and v_2 and v_3 are the specific volumes of the refrigerant at the compressor inlet and outlet, respectively. The compressor isentropic efficiency can be expressed approximately as:

$$\eta_c = 0.84 - 0.0075\ (T_{cond} - T_{evap})$$

where T_{cond} and T_{evap} are the saturation temperatures (in °C) in the condenser and evaporator, respectively. The heat exchange rates in the condenser and evaporator can both be modeled with effectiveness relations of the form:

$$\dot{Q} = \varepsilon\ \dot{C}_{min}\ \Delta T_{max}$$

where $\varepsilon = 1 - \exp(-NTU)$ is the effectiveness, NTU is the number of transfer units, $\dot{C}_{min}$ is the minimum capacitance rate, and ΔT_{max} is the maximum temperature difference between the entering air and the saturation temperature of the refrigerant. Air is supplied to the condenser at $T_{cond,a,in} = 25$°C and $\dot{m}_{a,cond} = 1$ kg/s. The outdoor evaporator coil sees an air supply at the outdoor temperature and $\dot{m}_{a,evap} = 1$ kg/s. The overall heat transfer coefficients for the condenser and evaporator are both $U = 40$ W/m^2-K. Assume that the evaporator outlet is saturated vapor and the condenser outlet is saturated liquid. Pressure drops in the heat exchangers can be neglected.

a) Assume the outdoor air temperature is $T_{a,out} = -5$°C and the heat transfer surface area of condenser and evaporator are each $A = 35$ m^2. Calculate the heating capacity (kW) and COP at this condition and plot the cycle on a pressure-specific enthalpy diagram.

b) Calculate and plot the heating capacity and COP as a function of the outdoor temperature for values between -30°C$< T_{a,out} < 20$°C. Explain the behavior you observe in the plot.

9.C-3 This problem examines the heat pump system analyzed in Problem 9.C-2 installed in a particular building in Madison, WI. The heating season performance factor ($HSPF$) of a heat pump is defined as the estimated seasonal heating output in Btu divided by the amount of electrical energy that it consumes in kW-hr. The heating season is defined as the hours during the year that the outdoor air temperature is less than $T_{a,out} = 15$°C. Table 9.C-3 provides bin temperature information for Madison WI. The first column shows the building load, i.e., the rate of heat transfer required to maintain the building at a comfortable temperature when the outdoor temperature is at the value indicated in the second column. The third column provides the number of hours during the heating season that the temperature is within 2.5°C of the temperature shown in the second column. For example, there are 136 hours for which the ambient temperature is between -22.5°C and -17.5°C.

Table 9.C-3: Heating load and number of hours at each bin temperature in Madison, WI.

Heating load (W)	$T_{a,out}$ (°C)	Hours (hr)
16,438	−30	12
14,737	−25	55
13,037	−20	136
11,336	−15	341
9,636	−10	364
7,935	−5	784
6,234	0	1,487
4,534	5	961
2,833	10	1,058
1,133	15	1,315

a) Determine the heating season performance factor (*HSPF*) for operation of the heat pump described in Problem 9.C-2 installed in a building located in Madison, WI. Include the fan power and auxiliary energy in your calculated *HSPF* value. When operating, the condenser and evaporator fans consume a combined power of $\dot{W}_{fans} = 68$ W. If the heat pump capacity is less than the building load, then auxiliary electrical heating is provided with a $COP = 1$. If the heat pump capacity exceeds the building load then the heat pump is cycled off and on as necessary to supply the load. Neglect any cycling losses for this analysis. Compare your calculated value with recommendation provided by the U.S. DOE that the value of *HSPF* should be greater than 7.7.

9.C-4 You own and operate a natural gas-fired gas turbine engine that provides electrical power to the grid. The gas turbine engine is shown in Figure 9.C-4(a).

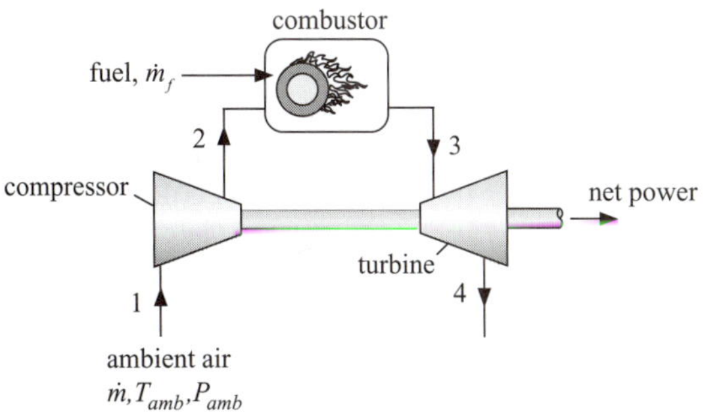

Figure 9.C-4(a): Gas turbine engine.

Both the ambient air temperature and the price that you can sell your electricity vary dramatically. During the day, the ambient air temperature is high and the resulting air-conditioning load causes a high demand for power and therefore leads to high electricity prices. During the night, the ambient air temperature is low and the electricity demand is also low. As a result, you cannot sell your electricity for very much money. For this problem, we will divide the day into two periods. The daytime period has duration $time_{day} = 10$ hr/day and ambient temperature $T_{day} = 96°$F. During the daytime period you can sell the electricity that you produce at a rate $ev_{day} = 0.25$ \$/kW-hr. The nighttime period has duration $time_{night} = 14$ hr/day and ambient temperature $T_{night} = 72°$F. During the nighttime period you can sell the electricity that you produce at a rate $ev_{night} = 0.05$ \$/kW-hr.

The compressor takes in air at ambient conditions, T_{amb} (equal to T_{day} or T_{night}, depending on the time of day) and $P_{amb} = 1$ atm. The compressor pressure ratio is $PR = 5$ and the compressor efficiency is $\eta_c = 0.80$. The compressor is a fixed volume device – it is designed to process a specific volumetric flow rate of air. The volumetric flow rate at the inlet to the compressor is $\dot{V}_{c,in} = 15500$ cfm. The air leaving the compressor at state 2 enters the combustor where it is mixed with natural gas and combusted. The heat of combustion for natural gas is $HC = 50$ MJ/kg and the cost of natural gas is $ngc = 0.25$ \$/kg. The air-fuel ratio is adjusted so that the combustion products leaving the combustor are at $T_{t,in} = 900°$C. The combustion products leaving the turbine are exhausted to atmosphere. The efficiency of the turbine is $\eta_t = 0.84$. Net power produced by the turbine is provided to a generator and converted to electricity (assume that the generator is 100% efficient). The combustion products can be modeled as air and air can be modeled as an ideal gas. Neglect pressure drop through the combustor.

a) Model the cycle for the daytime period. Prepare a T–s diagram and overlay the states on it. Print out an arrays table that includes each of the states and their entropy, enthalpy, temperature, and pressure. Determine the efficiency, power output, and the air-fuel ratio.
b) Determine the profit that you make each day when you run the power plant during the daytime period.
c) Determine the profit that you make each day when you run the power plant during the nighttime period.

Your analysis should have shown that you lose money when you operate at night. The value of the electricity that you produce is not sufficient to pay for the fuel required because you get paid so little for electricity produced during off-peak times. You are considering installing the system shown in Figure 9.C-4(b). The air entering the gas turbine engine is precooled using a cold storage system. The cold storage system is an ice system, except that the temperature of the ice can be adjusted by changing the concentration of antifreeze in the frozen solution. The gas turbine engine is operated only during the daytime period. Therefore, each day the ice in the cold storage system is melted as it cools the air entering the compressor in the precooler. Because the air entering the compressor is colder, it has higher density and therefore the mass flow rate processed by the gas turbine is larger and its power output higher. The result is more daytime profits. The cold storage system is recharged (i.e., the ice is re-solidified) during the nighttime period when the gas turbine is off. The flash intercooled refrigeration system shown in Figure 9.C-4(b) is operated at night in order to re-solidify the ice. This is advantageous because the cost of electricity at night is less than during the day and because the temperature for heat rejection is much lower at night, increasing the COP of the refrigeration system. During the daytime period, ambient air at $T_{day} = 96°$F and $P_{amb} = 1$ atm is drawn into the precooler at state 1. The temperature of the ice in the cold storage system is $T_{ics} = -5°$C and the approach temperature difference for the precooler is $\Delta T_{pc} = 5$ K. Air at $T_2 = T_{ics} + \Delta T_{pc}$ enters the compressor. The remaining parameters for the gas turbine are the same ($\eta_t = 0.84$, $\eta_c = 0.80$,

$T_{t,in} = 900°C$, $PR = 5$, $HC = 50$ MJ/kg, $\dot{V}_{c,in} = 15500$ cfm). There is no pressure loss in the precooler.

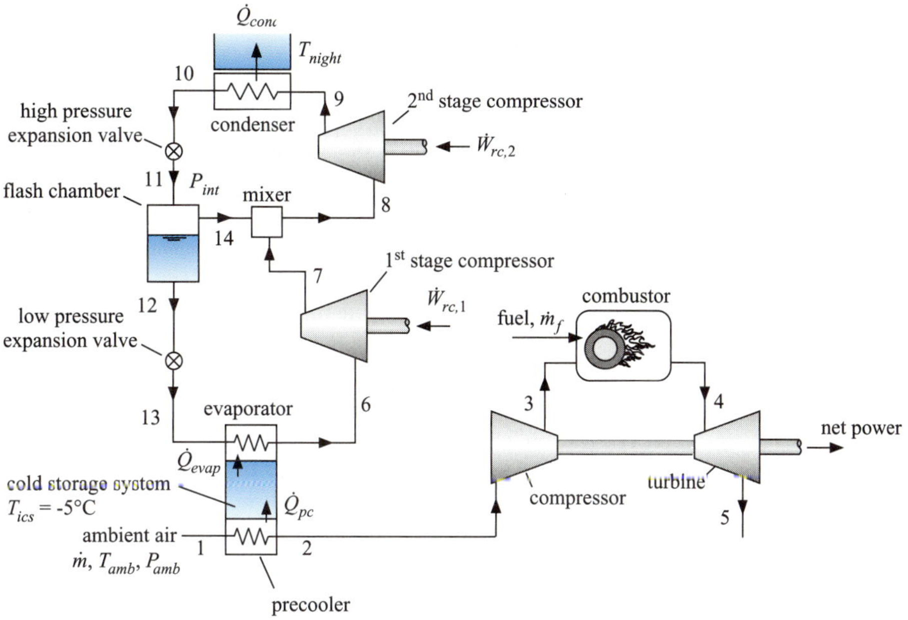

Figure 9.C-4(b): Gas turbine engine with precooling.

d) Model the gas turbine portion of the cycle shown in Figure 9.C-4(b) for the daytime period. Prepare a *T-s* diagram and overlay the states on it. Print out an arrays table that includes each of the states and their entropy, enthalpy, temperature, and pressure. Determine the efficiency, power output, and the air-fuel ratio of the system. Determine the profit that you make each day when you run the power plant during the daytime period, neglecting for now the cost associated with purchasing and running the cold storage system.

The flash intercooled refrigeration system is used to resolidify the ice in the cold storage system during the night. The condenser rejects heat to the ambient air at $T_{night} = 72°F$. The condenser has an approach temperature difference of $\Delta T_{cond} = 4$ K and the refrigerant leaving the condenser is subcooled by $\Delta T_{sc} = 0.5$ K. The evaporator accepts heat from the ice at T_{ics} and has an approach temperature difference of $\Delta T_{evap} = 5$ K. The refrigerant leaving the evaporator is superheated by $\Delta T_{sh} = 5$ K. The isentropic efficiency of the first stage compressor is $\eta_{rc,1} = 0.65$ and the isentropic efficiency of the second stage compressor is $\eta_{rc,2} = 0.70$. The volumetric efficiency of both compressors is $\eta_{vol} = 0.65$. The working fluid in the refrigeration system is R134a. Neglect pressure loss in all of the heat exchangers. The intercooling pressure (P_{int}) is the average of the evaporating and condensing pressures. The flash chamber is a large insulated tank. Saturated liquid is pulled from the bottom of the tank at state 12 and sent to the low pressure expansion valve and evaporator. Saturated vapor is pulled from the top at state 14 and used to provide intercooling by mixing with the superheated vapor leaving the first stage compressor at state 7.

e) Model the refrigeration cycle. Prepare a $T-s$ diagram and overlay the states on it. Print out an arrays table that includes each of the states and their specific entropy, specific enthalpy, temperature, and pressure. Determine the refrigeration provided by the system (in tons). Determine the coefficient of performance of the cycle and the power required by the refrigerant compressors.

The cost of electricity purchased during nighttime hours is $ec_{night} = 0.12$ \$/kW-hr (the cost of electricity to you is higher than what you could sell it for because the power company wants to strongly discourage you from producing power during the night). You have determined that the cost of the refrigeration system can be broken down into the cost of the compressors, condenser, and cold storage system. The cold storage system cost is $Cost_{ics} = \$50{,}000$. The cost of the condenser is given by:

$$Cost_{cond} = 1.5\left[\frac{\$\text{-K}}{\text{W}}\right]\frac{\dot{Q}_{cond}}{\left(T_{cond} - T_{night}\right)}$$

where T_{cond} is the saturation temperature at the condenser pressure. The cost of the compressors is given by:

$$Cost_{comp} = 92000\left[\frac{\$\text{-s}}{\text{m}^3}\right] Displacement$$

where *Displacement* is the total displacement rate of the installed compressors. The displacement rate of the compressor is the volumetric flow rate at the suction port of the compressor divided by the volumetric efficiency of the compressor.

f) Determine the net profit associated with purchasing the refrigeration system and operating the precooled power plant for $Time_{ec} = 1$ year.

As a system designer you have a few free parameters that you can use to optimize your system. The temperature of the cold storage system, T_{ics}, the approach temperature difference in the condenser, ΔT_{cond}, and the intercooling pressure, P_{int}.

g) Plot the net profit as a function of T_{ics}. Explain why an optimal cold storage temperature exists.
h) Plot the net profit as a function of ΔT_{cond}. Explain why an optimal condenser approach temperature difference exits.
i) Optimize your system – determine the optimal values of T_{ics}, ΔT_{cond}, and P_{int} and the associated maximum value of the net profit.

9.C-12 A single-stage refrigeration system provides $\dot{Q}_{evap} = 10$ kW of refrigeration from a space that is maintained at $T_C = -10°$C, while rejecting heat to the surroundings at $T_H = 35°$C. The coefficient of performance for this cycle is $COP = 2.2$ at steady-state operating conditions. The heat transfer rates to and from the refrigeration cycle can be assumed to be linear functions of the temperature differences so that: $\dot{Q}_{evap} = \alpha\,(T_C - T_{evap})$ and $\dot{Q}_H = \beta\,(T_{cond} - T_H)$ where T_{evap} is the evaporator temperature and T_{cond} is the condensing temperature.

a) Determine the steady-state power required to operate this refrigeration cycle.
b) Determine the steady-state rate of exergy destruction resulting from operation of this refrigeration cycle.
c) The evaporating temperature is measured to be $T_{evap} = -30°$C. Assuming that all of the exergy destruction in this refrigeration cycle occurs in the heat exchangers, what are the values of the heat exchanger coefficients, α and β?
d) The sum of the parameters α and β is proportional to the total cost of the heat exchangers. Determine the values of α and β that will minimize this cost while maintaining a *COP* of 2.2 and a refrigeration capacity of 10 kW.

10 Property Relations for Pure Fluids

Thermodynamics is an elegant engineering science because of its simplicity and its generality. All energy transformations are governed by the First and Second Laws. These laws apply to all systems with no exceptions. They apply to systems composed of pure substances and mixtures as well as to chemically reacting systems. Because of this generality, thermodynamics has been successfully applied to biology, cosmology, and other areas that are very different from the steam engines from which thermodynamics originated. But there is a price to pay for this generality. Since the First and Second Laws are independent of substance type, they provide no information related to the properties (internal energy, enthalpy, and entropy) that are needed to apply them. These properties have to be obtained from other information.

We have seen that the property relations for substances that behave as an ideal gas are simple. The specific internal energy and specific enthalpy of a substance that behaves as an ideal gas are functions only of temperature and they can be determined knowing only the specific heat capacity of the substance, which is a measurable property, as shown in Chapters 3 and 4. The specific entropy of an ideal gas depends on temperature and pressure in a simple manner and it can be determined if the specific heat capacity is known, as discussed in Chapter 6. However, the determination of properties for fluids that do not obey the ideal gas law is more complicated. So far, we have managed to avoid these complications by using property tables or the property functions in EES. However, we have not seen how these properties, which are not directly measurable, are determined. That is the focus of this chapter. You may not ever need to determine property data for a substance, because property data for most substances of engineering interest are already available. However, your understanding of thermodynamics will improve by learning how the properties that are required to apply the First and Second Laws of thermodynamics are obtained.

10.1 Equations of State for Pressure, Volume, and Temperature

The phase rule that was introduced in Eq. (2-1) provides the number of degrees of freedom (F) for a system, i.e., the number of intensive properties that must be specified in order to fix the intensive state of a system:

$$F = C - \Pi + 2 \tag{10-1}$$

where C is the number of chemical species and Π is the number of phases. This chapter is concerned with methods to determine the thermodynamic properties of pure substances ($C = 1$). We will initially focus on single-phase systems ($\Pi = 1$) for which the number of degrees of freedom must be $F = 2$ according to Eq. (10-1). Mixtures are considered in Chapter 11.

The easiest properties to consider are the pressure, temperature and specific volume since these properties are directly measurable. However, because $F = 2$ it is only necessary to know two of these three properties in order to determine the third. A mathematical relationship between P, T, and v must therefore exist for any pure single phase substance; this relationship is referred to as an *equation of state*. An equation of state will typically express pressure as a function of temperature and specific volume:

$$P = f(T, v) \tag{10-2}$$

It is shown in this chapter that if an accurate equation of state is known for a pure substance and its specific heat capacity at constant pressure as a function of temperature is also known at one pressure (typically at low pressure, where the substance behaves as an ideal gas) then all other thermodynamic properties (e.g., specific enthalpy and specific entropy) can be derived. The use of measurable properties (i.e., T, P, v, and c_P) to derive very useful but not directly measurable properties (e.g., h and s) is very powerful. The equation of state must be differentiable and the derivatives must accurately represent the shape of the P-v-T surface in order for this process to be successful.

The simplest equation of state is the ideal gas law, which was first introduced in Section 2.5. On a mass basis, the ideal gas law is:

$$P = \frac{R\,T}{v} \tag{10-3}$$

where R is the ideal gas constant. On a molar basis, the ideal gas law is:

$$P = \frac{R_{univ}\,T}{\bar{v}} \tag{10-4}$$

where R_{univ} is the universal gas constant and $\bar{v}$ is the molar specific volume. The ideal gas law is applicable for low density gases for which the molecules are relatively far apart and therefore intermolecular forces are weak. This equation of state works well for many common gases at room temperature and pressures below 5 to 10 atm. However, the ideal gas law becomes increasingly inaccurate as the density of the gas increases. One purpose of this section is to determine whether or not the ideal gas law is applicable at a particular state and how to proceed if it is not.

10.1.1 Compressibility Factor and Reduced Properties

The deviation from ideal gas behavior is quantified in terms of the *compressibility factor*, Z, which is defined as:

$$\boxed{Z = \frac{P\,v}{R\,T} = \frac{P\,\bar{v}}{R_{univ}\,T}} \tag{10-5}$$

The compressibility factor is a non-dimensional index that is equal to unity if the substance obeys the ideal gas law. The compressibility factor can be less than or greater than unity in regions where the substance deviates from the ideal gas law. Figure 10-1 illustrates the force between two molecules as a function of the distance between their centers. Repulsive forces occur when the distance between molecules is small (i.e., the density is high); repulsive forces result in compressibility factors greater than unity. Compressibility factors values less than unity are the result of attractive forces between molecules and tend to occur at lower densities. In either case, the difference between the compressibility factor and unity is a measure of the error that results from using the ideal gas law.

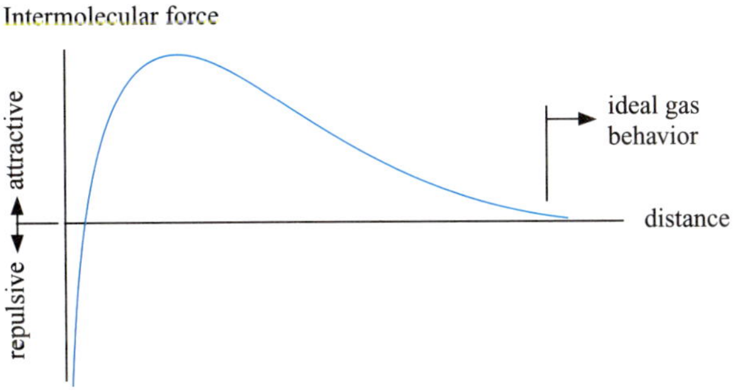

Figure 10-1: Force between two molecules as a function of the distance between their centers.

The *reduced pressure* and *reduced temperature* are non-dimensional parameters defined as the ratio of the absolute pressure and temperature to their critical values (P_{crit} and T_{crit}, respectively):

$$P_r = \frac{P}{P_{crit}} \tag{10-6}$$

$$T_r = \frac{T}{T_{crit}} \tag{10-7}$$

The compressibility factor for nitrogen is shown in Figure 10-2 as a function of reduced pressure for various values of reduced temperature. The *reduced specific volume* is defined as the ratio of the specific volume to the critical specific volume:

$$v_r = \frac{v}{v_{crit}} = \frac{\bar{v}}{\bar{v}_{crit}} \tag{10-8}$$

The critical properties for a number of common fluids are provided in Table 10-1. EES provides the critical temperature, pressure, and specific volume for any substance in

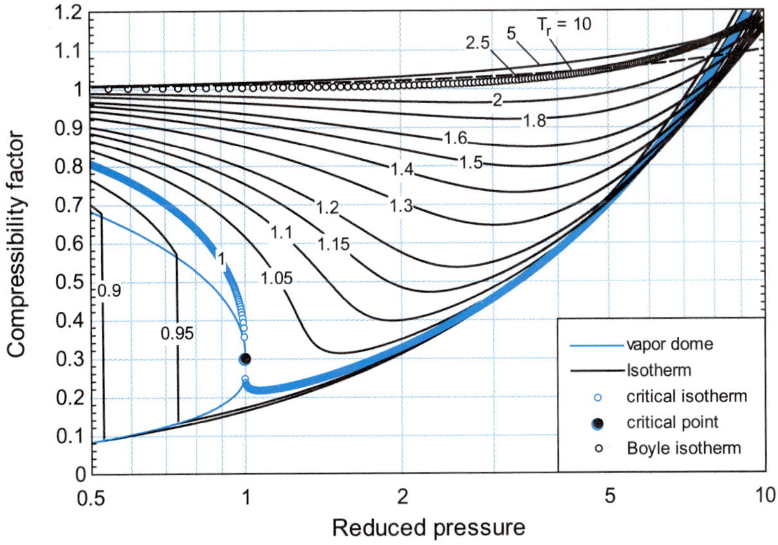

Figure 10-2: Compressibility factor for nitrogen as a function of reduced pressure for several values of reduced temperature.

Table 10-1: Critical Properties of Common Fluids

Substance	*MW* (kg/kmol)	T_{crit} (K)	P_{crit} (MPa)	$\bar{v}_{crit}$ (m^3/kmol)	Z_{crit}	ω
Ammonia	17.03	405.4	11.33	0.0757	0.2545	0.256
Argon	39.95	150.7	4.86	0.0746	0.2895	−0.002
Carbon Monoxide	28.00	132.8	3.49	0.0922	0.2916	0.051
Carbon Dioxide	44.01	304.1	7.38	0.0941	0.2746	0.225
Cyclohexane	84.16	553.6	4.08	0.3083	0.2729	0.209
Deuterium	4.03	38.3	1.67	0.0577	0.3015	−0.175
Ethane	30.07	305.3	4.87	0.1458	0.2799	0.100
Ethanol	46.07	513.9	6.15	0.1669	0.2402	0.644
Fluorine	38.00	144.4	5.24	0.0641	0.2797	0.050
Helium	4.00	5.2	0.23	0.0575	0.3027	−0.385
Hydrogen	2.02	33.2	1.32	0.0670	0.3191	−0.214
Hydrogen Sulfide	34.08	373.4	8.96	0.0980	0.2831	0.096
Isobutane	58.12	407.8	3.64	0.2591	0.2781	0.185
Krypton	83.80	209.4	5.51	0.0923	0.2921	−0.002
Methane	16.04	190.6	4.60	0.0986	0.2863	0.011
Methanol	32.04	513.4	8.10	0.1138	0.2161	0.565
Oxygen	32.00	154.6	5.04	0.0734	0.2879	0.022
n-Butane	58.12	425.1	3.80	0.2551	0.2740	0.200
n-Decane	142.28	617.7	2.10	0.6098	0.2497	0.490
n-Dodecane	170.33	658.1	1.82	0.7519	0.2497	0.574
n-Heptane	100.20	540.1	2.73	0.4319	0.2623	0.348
n-Hexane	86.17	507.9	3.06	0.3682	0.2667	0.312
n-Nonane	128.26	594.5	2.28	0.5525	0.2549	0.447
n-Octane	114.23	569.3	2.50	0.4868	0.2568	0.393
n-Pentane	72.15	469.7	3.36	0.3110	0.2679	0.250
Neon	20.18	44.5	2.68	0.0419	0.3034	−0.037
Nitrogen	28.01	126.2	3.40	0.0895	0.2897	0.037
Propane	44.10	369.8	4.25	0.2018	0.2787	0.152
Propylene	42.08	365.6	4.67	0.1884	0.2891	0.141
R134a	102.03	374.2	4.06	0.2009	0.2620	0.327
R22	86.47	369.3	4.99	0.1663	0.2702	0.221
Sulfur Hexafluoride	146.05	318.7	3.76	0.1968	0.2789	0.210
Water	18.02	647.1	22.06	0.0560	0.2295	0.344
Xenon	131.30	289.7	5.84	0.1195	0.2896	0.004

its property data library using the T_crit, P_crit and v_crit functions, respectively. Note that the critical properties are unique for each fluid and therefore only the fluid name must be specified in order to use these critical point functions. For example, the critical temperature and critical pressure of nitrogen are obtained according to:

```
$UnitSystem SI Mass Radian J K Pa
T_crit=T_crit(Nitrogen)          "Critical temperature for nitrogen"
P_crit=P_crit(Nitrogen)          "Critical pressure for nitrogen"
```

which leads to $T_{crit} = 126.2$ K and $P_{crit} = 3.396 \times 10^6$ Pa (3.396 MPa).

The general behavior of all pure fluids is similar when plotted in terms of reduced coordinates (i.e., T_r, P_r, and v_r or Z); that is, Figure 10-2 would appear similar if it were generated for a fluid other than nitrogen. Notice in Figure 10-2 that the vapor dome is evident for reduced pressures less than unity. The critical point occurs at the intersection of the critical isotherm ($T_r = 1$) and the critical isobar ($P_r = 1$). The value of the compressibility factor at the critical point is called the *critical compressibility factor*, Z_{crit}.

$$Z_{crit} = \frac{P_{crit}\, v_{crit}}{R\, T_{crit}} \tag{10-9}$$

Figure 10-2 shows that the critical compressibility factor for nitrogen is about 0.29. The critical compressibility factor can range from approximately 0.2 to 0.3 depending on the fluid, as seen in Table 10-1. This observation shows that a general relation for reduced volume (or compressibility factor) as a function only of reduced temperature and pressure does not exist. Some substance specific information is needed to accurately determine the equation of state, particularly in the vicinity of the critical point.

10.1.2 Characteristics of the Equation of State

Figure 10-2 shows that the compressibility factor is a complicated function of reduced temperature and reduced pressure. However, the behavior of the compressibility factor must be accurately predicted by an equation of state in order for it to be of engineering use. Careful examination of Figure 10-2 shows that there are some characteristics that all equations of state must have.

Limiting Ideal Gas Behavior

As the temperature is raised or the pressure is reduced, the substance becomes less dense. Eventually, the intermolecular spacing becomes sufficiently large that the substance behaves according to the ideal gas law. Therefore, the compressibility factor must approach unity as the reduced pressure decreases at all values of reduced temperature:

$$\lim_{P_r \to 0} Z = 1 \quad \text{for all } T_r \tag{10-10}$$

The compressibility factor must approach unity as the reduced temperature increases at all values of reduced pressure:

$$\lim_{T_r \to \infty} Z = 1 \quad \text{for all } P_r \tag{10-11}$$

The Boyle Isotherm

The slope of one isotherm, referred to as the *Boyle isotherm*, approaches zero at low pressure:

$$\lim_{P_r \to 0} \left(\frac{\partial Z}{\partial P_r}\right)_{T_r = T_{r,B}} = 0 \tag{10-12}$$

where $T_{r,B}$ is the reduced Boyle temperature. The reduced Boyle temperature is the ratio of the Boyle temperature (T_B) to the critical temperature. The reduced Boyle

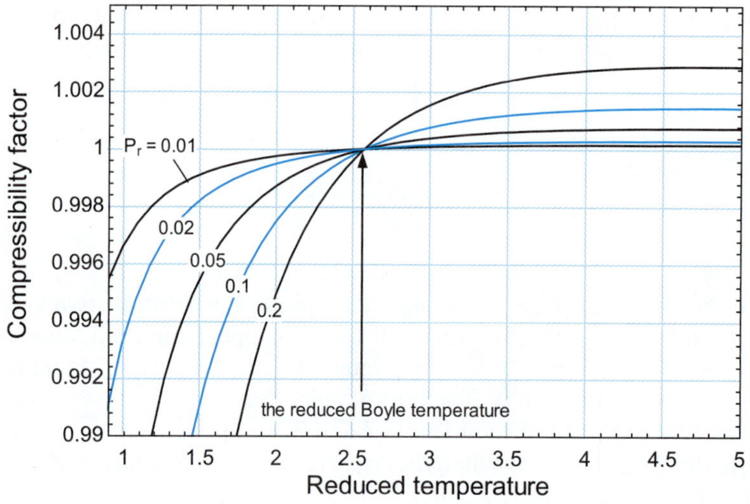

Figure 10-3: Compressibility factor of nitrogen as a function of reduced temperature at several, low values of reduced pressure.

temperature for most substances is approximately 2.5. The Boyle isotherm is shown in Figure 10-2. It is difficult to see the behavior of the Boyle isotherm indicated by Eq. (10-12) in Figure 10-2 because of the scale involved with the figure and the density of the isotherms. The Boyle isotherm can be seen more clearly in Figure 10-3, which illustrates the compressibility factor of nitrogen as a function of reduced temperature for a few, low values of reduced pressure. Notice in Figure 10-3 that the compressibility factor tends to increase as the reduced pressure decreases towards zero at low values of reduced temperature (below the reduced Boyle temperature); therefore:

$$\lim_{P_r \to 0} \left(\frac{\partial Z}{\partial P_r} \right)_{T_r < T_{r,B}} < 0 \tag{10-13}$$

At values of reduced temperature above the reduced Boyle temperature, the compressibility factor decreases as the reduced pressure decreases; therefore:

$$\lim_{P_r \to 0} \left(\frac{\partial Z}{\partial P_r} \right)_{T_r > T_{r,B}} > 0 \tag{10-14}$$

The reduced Boyle temperature for nitrogen is approximately 2.5 and Figure 10-3 shows that the compressibility factor is essentially independent of the reduced pressure at this value of reduced temperature, as indicated by Eq. (10-12).

The compressibility factor is greater than unity for temperatures greater than the Boyle temperature at all pressures:

$$Z \geq 1 \text{ at all } P \qquad \text{for } T \geq T_B \tag{10-15}$$

Equation (10-15) can also be stated in terms of reduced coordinates:

$$Z \geq 1 \text{ at all } P_r \qquad \text{for } T_r \geq T_{r,B} \tag{10-16}$$

Critical Point Behavior

In the two-phase region, the pressure of a pure substance is a function only of temperature and it is therefore independent of specific volume. Specific volume varies with the quality at the saturation temperature but the saturation pressure does not. The critical

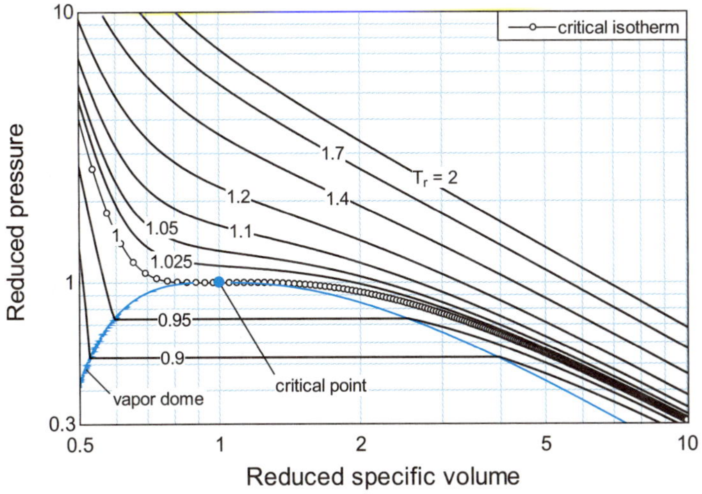

Figure 10-4: Reduced pressure as a function of reduced specific volume for several values of reduced temperature for nitrogen. Note the inflection point associated with the critical isotherm, $T_r = 1$, that occurs at the critical point ($P_r = v_r = 1$).

point is located on the boundary of the two-phase region. Consequently, pressure must be independent of the specific volume at the critical point. Mathematically, this characteristic is stated as:

$$\left(\frac{\partial P}{\partial v}\right)_T = 0 \qquad \text{at the critical point} \tag{10-17}$$

Equation (10-17) can also be stated in terms of reduced coordinates.

$$\left(\frac{\partial P_r}{\partial v_r}\right)_{T_r} = 0 \qquad \text{at the critical point} \tag{10-18}$$

Figure 10-4 illustrates the reduced pressure as a function of reduced specific volume at several values of the reduced temperature for nitrogen. The critical point occurs at $T_r = 1$, $P_r = 1$, and $v_r = 1$; notice that the slope of the reduced pressure as a function of reduced specific volume is zero at the critical point, as evident by the inflection in the critical isotherm at the critical point. This behavior is more clearly seen in Figure 10-5, which shows the partial derivative of reduced pressure with respect to reduced specific volume at various values of the reduced temperature for nitrogen. The partial derivative is zero at the critical point.

Thermodynamic properties exhibit unusual behavior in the vicinity of the critical point. The specific heat capacity at constant pressure, c_P, is defined as:

$$c_P = \left(\frac{\partial h}{\partial T}\right)_P \tag{10-19}$$

The specific heat capacity at constant pressure for a pure fluid in a two-phase state must be infinite because, at constant pressure, the specific enthalpy will increase with quality while the temperature remains constant. As with all two-phase states, the specific heat capacity at constant pressure, c_P, will also be infinity at the critical point. Figure 10-6 illustrates the specific heat capacity of nitrogen as a function of reduced temperature for various values of reduced pressure. Notice the large peak in c_P that occurs in the vicinity of the critical point.

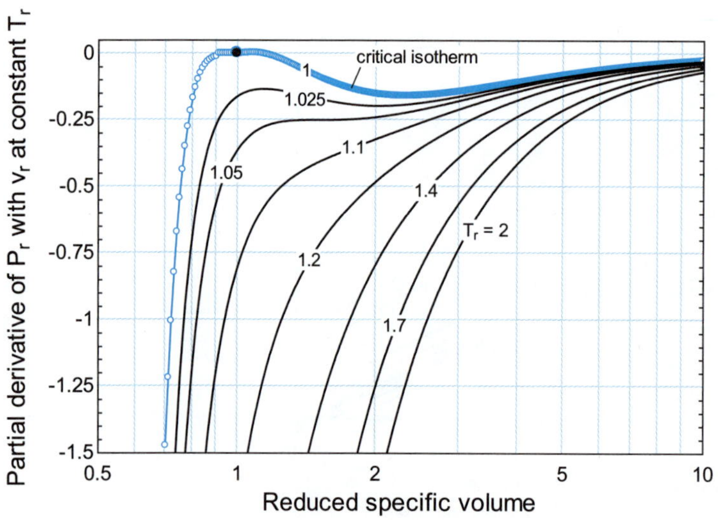

Figure 10-5: Partial derivative of reduced pressure with respect to reduced specific volume at constant reduced temperature for several values of reduced temperature for nitrogen. Note that the partial derivative is zero for the critical isotherm ($T_r = 1$) at the critical point ($P_r = v_r = 1$).

The *isothermal compressibility*, K_T, of a fluid is defined as

$$K_T = -\frac{1}{v}\left(\frac{\partial v}{\partial P}\right)_T \tag{10-20}$$

Equation (10-17) indicates that the isothermal compressibility must also approach infinity at the critical point. Transport properties such as viscosity and thermal conductivity also exhibit enhanced values in the neighborhood of the critical point, as discussed by Sengers and Sengers (1968).

A final characteristic of all fluids is evident in Figure 10-5. Not only is the partial derivative of the reduced pressure with respect to reduced specific volume at constant reduced pressure equal to zero at the critical point, the value of this partial derivative also reaches a maximum value at the critical point. This means that there is an inflection

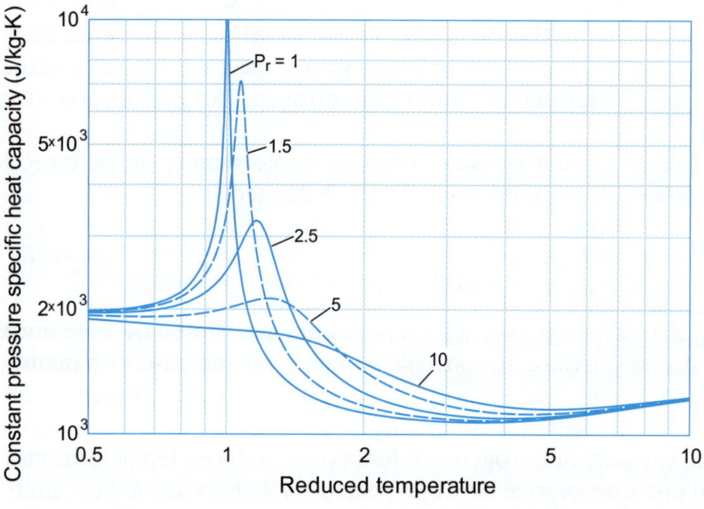

Figure 10-6: Specific heat capacity at constant pressure of nitrogen as a function of reduced temperature for various values of reduced pressure. Note that c_P approaches infinity at the critical point.

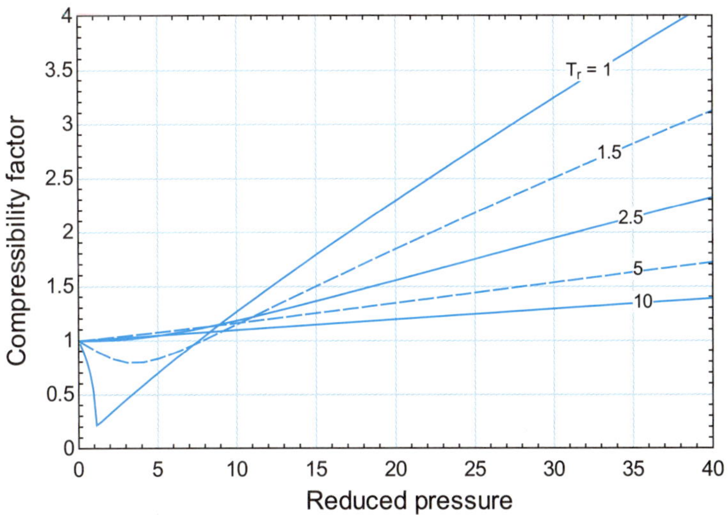

Figure 10-7: Compressibility factor as a function of reduced pressure for several values of reduced temperature for nitrogen.

point in the plot of pressure versus specific volume at the critical point; this inflection is evident in Figure 10-4 and can be expressed mathematically as:

$$\left(\frac{\partial^2 P}{\partial v^2}\right)_T = 0 \quad \text{and} \quad \left(\frac{\partial^2 P_r}{\partial v_r^2}\right)_{T_r} = 0 \quad \text{at the critical point} \tag{10-21}$$

Any equation of state that attempts to accurately represent the relation between pressure, specific volume and temperature for a pure fluid must exhibit the characteristics discussed in this section.

10.1.3 Two-Parameter Equations of State

The compressibility factor predicted by the ideal gas law is always unity. The ideal gas law provides a remarkably accurate equation of state for low pressure gases that are far from their critical point. Figure 10-7 shows the compressibility factor as a function of reduced pressure for various values of reduced temperature for nitrogen. Notice that the ideal gas law becomes inaccurate at high pressures and low temperatures. The ideal gas law is, of course, useless for liquids. This section discusses several equations of state that are more accurate than the ideal gas law and can be used for fluids at high reduced pressure.

The van der Waals Equation of State

The first equation of state to show significant improvement over the ideal gas law was proposed by Dutch scientist Johannes van der Waals in 1873. Based on available experimental data at the time and insightful reasoning, van der Waals proposed the following equation of state:

$$P = \frac{RT}{v - b} - \frac{a}{v^2} \tag{10-22}$$

where a and b are substance-dependent constants. The reasoning that leads from the ideal gas law to the van der Waals equation of state is based on the existence of molecules of finite size that exert intermolecular forces. The ideal gas law assumes that molecules occupy no volume and that there are no forces exerted between molecules. Notice that if the parameters a and b are both zero, then the van der Waals equation of state, Eq. (10-22), reduces to the ideal gas law:

$$P = \frac{RT}{v} \tag{10-23}$$

The parameter b represents the effect of the finite size of the molecules. The importance of b increases as the specific volume, v, is reduced. The specific volume of a real substance cannot be reduced to zero because the molecules occupy a finite volume. The van der Waals equation of state limits the specific volume to a nonzero "molecular" specific volume, b. Any attempt to reduce the specific volume to b will require an infinite pressure. The second term in the van der Waals equation of state accounts for the net effect of attractive intermolecular forces between molecules, which is proportional to the number of gas molecules squared and therefore is inversely proportional to the specific volume squared. These intermolecular forces become much more important as the specific volume is reduced.

The van der Waals equation of state is classified as a two-parameter equation of state because only two parameters, a and b, are required for implementation. The constants a and b can be determined by regressing experimental data to the form of Eq. (10-22). In the absence of experimental data, values of a and b can also be estimated by forcing the van der Waals equation of state to exhibit the critical isotherm behavior that is required by Eqs. (10-17) and (10-21). Taking the first derivative of Eq. (10-22) with respect to specific volume at constant temperature leads to:

$$\left(\frac{\partial P}{\partial v}\right)_T = -\frac{RT}{(v-b)^2} + 2\frac{a}{v^3} \tag{10-24}$$

Substituting Eq. (10-24) into Eq. (10-17) results in:

$$-\frac{R\,T_{crit}}{(v_{crit}-b)^2} + 2\frac{a}{v_{crit}^3} = 0 \tag{10-25}$$

Taking the derivative of Eq. (10-24) with respect to specific volume at constant temperature leads to:

$$\left(\frac{\partial^2 P}{\partial v^2}\right)_T = 2\frac{RT}{(v-b)^3} - 6\frac{a}{v^4} \tag{10-26}$$

Substituting Eq. (10-26) into Eq. (10-21) leads to:

$$2\frac{R\,T_{crit}}{(v_{crit}-b)^3} - 6\frac{a}{v_{crit}^4} = 0 \tag{10-27}$$

Assuming that the critical properties have been measured, Eqs. (10-25) and (10-27) can be solved simultaneously to determine a and b. This algebra can be accomplished most conveniently using Maple. An introduction to the use of Maple can be found in Appendix H. Maple is extremely useful for accomplishing the symbolic manipulations required to solve the algebraic equations that arise in this chapter. Equations (10-25) and (10-27) are entered in Maple as the variables Eq1 and Eq2.

```
> restart;
> Eq1:=-R*T_crit/(v_crit-b)^2+2*a/v_crit^3=0;
```

$$Eq1 := -\frac{RT\_crit}{(v\_crit-b)^2} + \frac{2a}{v\_crit^3} = 0$$

```
> Eq2:=2*R*T_crit/(v_crit-b)^3-6*a/v_crit^4=0;
```

$$Eq2 := \frac{2\,R\,T\_crit}{(v\_crit - b)^3} - \frac{6\,a}{v\_crit^4} = 0$$

The algebra required to obtain a and b is accomplished using the solve command in Maple. The first argument of the solve command is the equation or set of equations to be solved and the second is the variable or group of variables to solve for. In this case, the equations are Eq1 and Eq2 (entered as a list, in curly brackets) and the variables are a and b (also entered as a list using curly brackets).

```
> coef:=solve({Eq1,Eq2},{a,b});
```

$$coef := \left\{a = \frac{9\,R\,T\_crit\;v\_crit}{8},\; b = \frac{v\_crit}{3}\right\}$$

Therefore:

$$a = \frac{9\,R\,T_{crit}\,v_{crit}}{8} \tag{10-28}$$

$$b = \frac{v_{crit}}{3} \tag{10-29}$$

Equations (10-28) and (10-29) provide the coefficients that are required to implement the van der Waals equation of state in terms of the critical temperature and critical specific volume. The critical temperature is relatively easy to measure; however, the critical specific volume is not easily measured since the fluid is very compressible at the critical point and therefore even small forces such as gravity affect the specific volume. As a result, a more useful set of equations for a and b would be expressed in terms of T_{crit} and the critical pressure, P_{crit}, which is also relatively easy to measure. The equation of state, Eq. (10-22), is written at the critical point:

$$P_{crit} = \frac{R\,T_{crit}}{v_{crit} - b} - \frac{a}{v_{crit}^2} \tag{10-30}$$

and used to eliminate v_{crit} in Eqs. (10-28) and (10-29). Equation (10-30) is entered in Maple:

```
> EOS:=P_crit=R*T_crit/(v_crit-b)-a/v_crit^2;
```

$$EOS := P\_crit = \frac{R\,T\_crit}{v\_crit - b} - \frac{a}{v\_crit^2}$$

The equation EOS together with the two elements of the equation set coef (i.e., coef[1] and coef[2], that contain Eqs. (10-28) and (10-29), respectively) are three equations in the variables a, b, T_{crit}, P_{crit}, v_{crit}, and R. These equations can be combined to eliminate any one of these variables, resulting in two equations in the remaining variables. This process is accomplished using the eliminate command in Maple. The first argument required by the eliminate command is the set of equations and the second is the variable (or list of variables) to be eliminated. The eliminate command returns two outputs; the first is an expression for the variable(s) that have been eliminated and the second is the set of equations that remain once the elimination process is complete. The following command eliminates the variable v_crit from the set of equations {coef[1], coef[2], EOS}.

```
> Eqset2:=eliminate({coef[1],coef[2],EOS},v_crit);
```

$$Eqset2 := [\{v\_crit = 3\,b\}, \{-8\,a + 27\,R\,T\_crit\,b,\ -18\,P\_crit\,b^2 + 9\,R\,T\_crit\,b - 2\,a\}]$$

Therefore, the two equations that remain once v_{crit} is eliminated are:

$$-8\,a + 27\,R\,T_{crit} = 0 \tag{10-31}$$

$$-18\,P_{crit}\,b^2 + 9\,R\,T_{crit}\,b - 2\,a = 0 \tag{10-32}$$

Equations (10-31) and (10-32) can be solved simultaneously for a and b in terms of T_{crit} and P_{crit} using the solve command. Notice that the set of two equations are the 2nd element of the list Eqset2, which is the output of the eliminate command.

```
> solve(Eqset2[2],{a,b});
```

$$\{a = 0,\ b = 0\},\ \left\{a = \frac{27\,R^2\,T\_crit^2}{64\,P\_crit},\ b = \frac{R\,T\_crit}{8\,P\_crit}\right\}$$

Therefore, a and b can be expressed in terms of T_{crit} and P_{crit}:

$$a = \frac{27\,R^2\,T_{crit}^2}{64\,P_{crit}} \tag{10-33}$$

$$b = \frac{R\,T_{crit}}{8\,P_{crit}} \tag{10-34}$$

Equations (10-33) and (10-34) together with Eq. (10-22) provides an equation of state that can be applied to any substance provided that the critical temperature and critical pressure have been measured. The van der Waals equation of state is much more accurate than the ideal gas equation of state when applied to low temperature and high pressure regions. However, it is not very accurate near the critical point. Table 10-1 indicates that the critical compressibility factor (Z_{crit}) for most fluids is in the range of 0.2 to 0.3. Substituting Eqs. (10-28) and (10-29) into Eq. (10-30) leads to:

$$P_{crit} = \frac{R\,T_{crit}}{\left(v_{crit} - \frac{v_{crit}}{3}\right)} - \frac{9\,R\,T_{crit}\,v_{crit}}{8}\frac{1}{v_{crit}^2} \tag{10-35}$$

Equation (10-35) can be simplified to:

$$P_{crit} = \frac{3\,R\,T_{crit}}{2\,v_{crit}} - \frac{9\,R\,T_{crit}}{8\,v_{crit}} \tag{10-36}$$

or

$$\frac{P_{crit}\,v_{crit}}{R\,T_{crit}} = \frac{3}{2} - \frac{9}{8} \tag{10-37}$$

Substituting Eq. (10-9) into Eq. (10-37) leads to:

$$Z_{crit} = 0.375 \tag{10-38}$$

which is approximately 50% in error compared to the actual critical compressibility factor for most fluids. The application of the van der Waals equation of state to nitrogen is illustrated in Example 10.1-1.

EXAMPLE 10.1-1: APPLICATION OF THE VAN DER WAALS EQUATION OF STATE

EES properties for nitrogen are determined from a high accuracy reduced Helmholtz equation of state developed by Span et al. (2000) which represents the most accurate property information that is available for nitrogen.

a) Compare the EES correlation for the properties of nitrogen with the van der Waals equation of state where a and b are evaluated using the critical temperature and pressure. Prepare a plot showing the compressibility factor as a function of reduced pressure for various values of reduced temperature and comment on the regions where the van der Waals equation of state is not accurate.

The ideal gas constant for nitrogen is evaluated according to:

$$R = \frac{R_{univ}}{MW}$$

where MW is the molar mass, evaluated using EES' MolarMass function. The critical pressure and temperature, P_{crit} and T_{crit}, are evaluated using the P_crit and T_crit functions, respectively. The coefficients a and b of the van der Waals equation of state are evaluated using Eqs. (10-33) and (10-34).

```
$UnitSystem SI Mass Radian J K Pa
F$='Nitrogen'                      "fluid"
MW=MolarMass(F$)                   "molar mass"
R=R#/MW                            "ideal gas constant"
P_crit=P_crit(F$)                  "critical pressure"
T_crit=T_crit(F$)                  "critical temperature"
a=27*R^2*T_crit^2/(64*P_crit)      "coefficient a for van der Waals EOS"
b=R*T_crit/(8*P_crit)              "coefficient b for van der Waals EOS"
```

The reduced pressure and temperature, P_r and T_r, are set to arbitrary values and used to evaluate the pressure and temperature (P and T) according to:

$$P = P_r\, P_{crit}$$

$$T = T_r\, T_{crit}$$

```
P_r=1.2 [-]                        "reduced pressure"
T_r=1.2 [-]                        "reduced temperature"
P=P_r*P_crit                       "pressure"
T=T_r*T_crit                       "temperature"
```

Equation (10-22) is used to implement the van der Waals equation of state and compute the specific volume (v). The compressibility factor (Z) is obtained from Eq. (10-5). The specific volume is also evaluated using EES' internal property correlation (v_{EES}) and the associated compressibility factor (Z_{EES}) is computed.

```
P=R*T/(v-b)-a/v^2                  "specific volume from van der Waals EOS"
Z=P*v/(R*T)                        "compressibility factor from van der Waals EOS"
v_EES=volume(F$,T=T,P=P)           "specific volume from EES"
Z_EES=P*v_EES/(R*T)                "compressibility factor from EES"
```

Figure 1 illustrates the compressibility factor predicted by the van der Waals equation of state and also the value obtained using the property correlation as a function of reduced pressure for various values of reduced temperature. Figure 1(a) focuses on lower values of P_r whereas Figure 1(b) extends over a wider range of P_r.

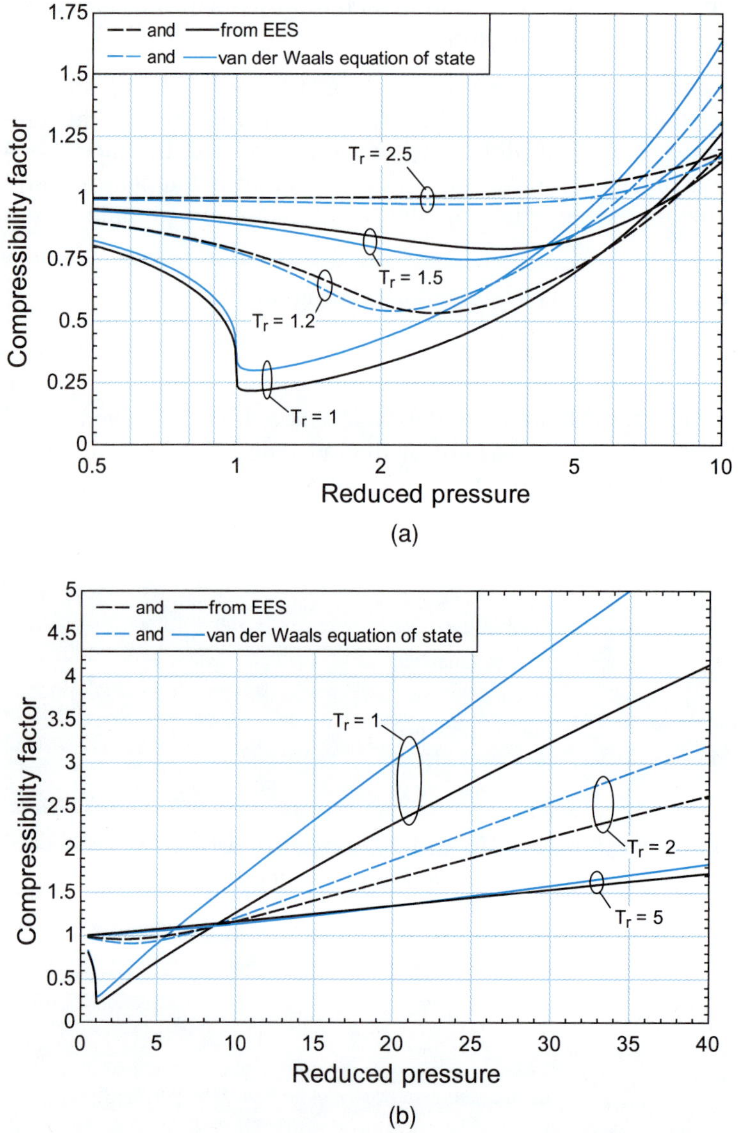

Figure 1: Compressibility factor for nitrogen predicted by the van der Waals equation of state and using the property correlation in EES as a function of reduced pressure for various values of reduced temperature; (a) focuses on reduced pressures near the critical point while (b) extends over a broader range.

The van der Waals equation of state is a great improvement over the ideal gas law (which predicts $Z = 1$ regardless of P_r and T_r) and Figure 1(b) shows that it is quite accurate at a reduced temperature of 5. However, the van der Waals equation of state becomes increasing inaccurate as the reduced temperature is decreased and Figure 1(a) illustrates that it is also inaccurate near the critical point. The van der Waals equation of state is of historic importance, but there are better two-parameter equations of state.

All of the two parameter equations of state that are presented in this section are cubic functions of specific volume. For example, the van der Waals equation of state:

$$P = \frac{R\,T}{v - b} - \frac{a}{v^2} \tag{10-39}$$

can be expressed as a cubic function of v by multiplying both sides of Eq. (10-39) by v^2 $(v - b)$:

$$P\,v^2\,(v - b) = R\,T\,v^2 - a\,(v - b) \tag{10-40}$$

Dividing through by P and collecting like powers of v provides:

$$v^3 - \left(\frac{R\,T}{P} + b\right) v^2 + \left(\frac{a}{P}\right) v - \left(\frac{a\,b}{P}\right) = 0 \tag{10-41}$$

The equation of state can be expressed in terms of reduced coordinates by substituting the definitions for reduced pressure, reduced temperature, and reduced specific volume into Eq. (10-41). Equations(10-33) and (10-34) for a and b, respectively, are also substituted into Eq. (10-41):

$$\begin{aligned} v_r^3\,v_{crit}^3 &- \left(\frac{R\,T_{crit}}{P_{crit}}\frac{T_r}{P_r} + \frac{R\,T_{crit}}{8\,P_{crit}}\right) v_r^?\,v_{crit}^? \\ &+ \frac{27\,R^2\,T_{crit}^2}{64\,P_{crit}}\frac{v_{crit}}{P_{crit}}\frac{v_r}{P_r} - \frac{27\,R^2\,T_{crit}^2}{64\,P_{crit}}\frac{R\,T_{crit}}{8\,P_{crit}}\frac{1}{P_{crit}\,P_r} = 0 \end{aligned} \tag{10-42}$$

Equation (10-37) relates v_{crit}, P_{crit}, and T_{crit} for the van der Waals equation of state:

$$v_{crit} = \frac{3\,R\,T_{crit}}{8\,P_{crit}} \tag{10-43}$$

Equation (10-43) is substituted into Eq. (10-42):

$$\begin{aligned} v_r^3\left(\frac{3\,R\,T_{crit}}{8\,P_{crit}}\right)^3 &- \left(\frac{R\,T_{crit}}{P_{crit}}\frac{T_r}{P_r} + \frac{R\,T_{crit}}{8\,P_{crit}}\right) v_r^2 \left(\frac{3\,R\,T_{crit}}{8\,P_{crit}}\right)^2 \\ &+ \frac{27\,R^2\,T_{crit}^2}{64\,P_{crit}^2}\left(\frac{3\,R\,T_{crit}}{8\,P_{crit}}\right)\frac{v_r}{P_r} - \frac{27\,R^2\,T_{crit}^2}{64\,P_{crit}}\frac{R\,T_{crit}}{8\,P_{crit}}\frac{1}{P_{crit}\,P_r} = 0 \end{aligned} \tag{10-44}$$

Dividing Eq. (10-44) through by $\left(\frac{3\,R\,T_{crit}}{8\,P_{crit}}\right)^3$ results in:

$$v_r^3 - \left(\frac{8}{3}\frac{T_r}{P_r} + \frac{1}{3}\right) v_r^2 + 3\frac{v_r}{P_r} - \frac{1}{P_r} = 0 \tag{10-45}$$

Given temperature and pressure, Eq. (10-41) has three roots for the specific volume (or, given reduced temperature and reduced pressure, Eq. (10-45) has three roots for reduced specific volume). Only one of these roots will be physically meaningful for temperatures above the critical temperature. To demonstrate this result, Eq. (10-45) is entered in EES:

```
v_r^3-(8*T_r/(3*P_r)+1/3)*v_r^2+3*v_r/P_r-1/P_r=0          "van der Waals equation of state"
```

and used to generate Figure 10-8, which shows the predicted reduced pressure as a function of reduced specific volume for several values of reduced temperature.

For reduced temperatures that are greater than 1.0 (e.g., point d), there is only one real value of the reduced specific volume at a given reduced pressure; the other two roots of the van der Waals equation of state are imaginary in this region. However, for

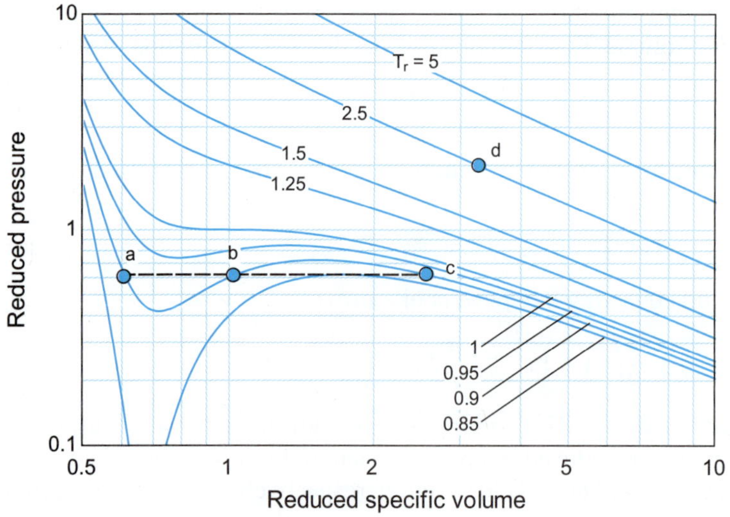

Figure 10-8: Reduced pressure as a function of reduced specific volume for various values of reduced temperature predicted by the van der Waals equation of state, which is a cubic equation of state.

reduced temperatures that are less than 1.0, there are three real roots to Eq. (10-45). This situation implies that there are three possible values of reduced specific volume at a given reduced pressure and reduced temperature. For example, points a, b, and c in Figure 10-8 all lie on the $T_r = 0.9$ isotherm and the $P_r = 0.61$ isobar. Two of the three points have physical significance. Point a corresponds to a saturated liquid state and point c corresponds to a saturated vapor state. The slope of P_r with respect to v_r is negative for both of these states; therefore these are physically possible states. Increasing the pressure exerted on any real substance at constant temperature will cause its specific volume to decrease, corresponding to a negative value of the partial derivative of reduced pressure with respect to reduced specific volume at constant reduced temperature. However, the slope of P_r with respect to v_r is positive at point b. A positive slope is impossible, as it suggests that the specific volume of the substance will somehow increase as the pressure is increased. Point b is an aberration of a cubic equation of state and should be disregarded. Although supercooled gas and superheated liquid can exist in meta-stable states, the trajectory of the isotherm predicted by the equation of state as it moves between the saturated liquid and saturated vapor states (points a and c) is not physical.

When solving a cubic equation of state, it is sometimes difficult to identify the specific root of interest. EES can be used to determine one of the roots, but it is not always clear which one. Manipulation of the guess values and bounds associated with the variables is required to ensure that the solution obtained by EES is, for example, the root corresponding to saturated liquid in Figure 10-8 (i.e., point a rather than points b or c). To facilitate the process of working with cubic equations of state, EES provides the procedure RealCubicRoots. To access information for this procedure, select Function Information from the Options menu and then select EES library routines, as shown in Figure 10-9. The RealCubicRoots procedure returns the real roots of a cubic equation and it is called according to:

```
CALL RealCubicRoots(a_2,a_1,a_0: z_1,z_2,z_3)
```

where the inputs a_2, a_1, and a_0 are the coefficients of the equation, expressed as:

$$z^3 + a_2\, z^2 + a_1\, z + a_0 = 0 \tag{10-46}$$

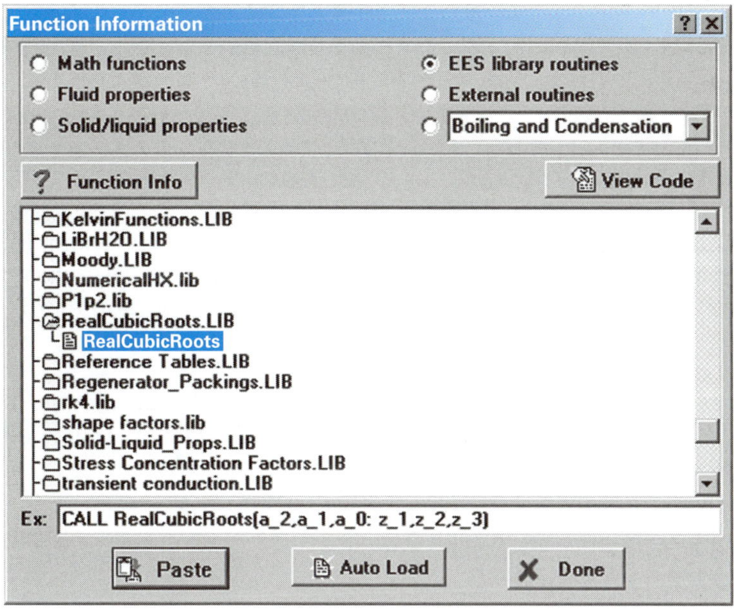

Figure 10-9: Function information dialog for the RealCubicRoots procedure.

and the outputs z_1, z_2, and z_3 are the real roots of the equation; note that if there is only one real root then the values of z_2 and z_3 are set to zero. For example, the procedure RealCubicRoots can be used to determine the reduced specific volume predicted by the van der Waals equation of state at $P_r = 2$ and $T_r = 2.5$ (point d in Figure 10-8). The coefficients a_2, a_1, and a_0 are obtained by examination of Eq. (10-45):

$$a_2 = -\left(\frac{8}{3}\frac{T_r}{P_r} + \frac{1}{3}\right) \tag{10-47}$$

$$a_1 = \frac{3}{P_r} \tag{10-48}$$

$$a_0 = -\frac{1}{P_r} \tag{10-49}$$

```
T_r=2.5 [-]                                      "reduced temperature"
P_r=2 [-]                                        "reduced pressure"
a_2=-(8*T_r/(3*P_r)+1/3)                         "coefficient on v_r^2"
a_1=3/P_r                                        "coefficient on v_r^1"
a_0=-1/P_r                                       "coefficient on v_r^0"
CALL RealCubicRoots(a_2,a_1,a_0: v_r_1,v_r_2,v_r_3)   "obtain real roots"
```

which results in $v_{r,1} = 3.253$ (consistent with point d in Figure 10-9) and $v_{r,2} = v_{r,3} = 0$ because there is only one root to Eq. (10-45) above the critical isotherm. The procedure RealCubicRoots can also be used to determine all three of the values of reduced volume predicted by the van der Waals equation of state at $P_r = 0.61$ and $T_r = 0.90$ (points a, b, and c in Figure 10-8).

```
T_r=0.9 [-]                                           "reduced temperature"
P_r=0.61 [-]                                          "reduced pressure"
a_2=-(8*T_r/(3*P_r)+1/3)                              "coefficient on v_r^2"
a_1=3/P_r                                             "coefficient on v_r^1"
a_0=-1/P_r                                            "coefficient on v_r^0"
CALL RealCubicRoots(a_2,a_1,a_0: v_r_1,v_r_2,v_r_3)   "obtain real roots"
```

Solving leads to $v_{r,1} = 2.64$ (point c), $v_{r,2} = 0.6105$ (point a), and $v_{r,3} = 1.017$ (point b). Note that it is possible to use the cubic equation of state to determine the relationship between saturation pressure and temperature. This topic is discussed in Section 10.5.3.

The Dieterici Equation of State

Hundreds of two-parameter equations of state have been proposed that attempt to improve upon the van der Waals equation of state. One example is the Dieterici equation of state, proposed in 1899:

$$P = \frac{R\,T}{(v-b)} \exp\left(-\frac{a}{R\,T\,v}\right) \tag{10-50}$$

where the two parameters a and b can either be determined from regression of experimental data or estimated by enforcing the behavior at the critical point indicated in Eqs. (10-17) and (10-21). Example 10.1-2 shows how the Dieterici equation of state can be implemented and how it compares with the van der Waals equation of state and the property correlations in EES.

EXAMPLE 10.1-2: DIETERICI EQUATION OF STATE

a) Develop equations for the coefficients a and b required by the Dieterici equation of state in terms of the critical pressure and temperature.

The required algebraic manipulations are carried out using Maple. The Dieterici equation of state, Eq. (10-50), is entered in Maple:

```
> restart;
> P:=R*T*exp(-a/(R*T*v))/(v-b);
```

$$P := \frac{R\,T\,\mathbf{e}^{\left(-\frac{a}{R\,T\,v}\right)}}{v-b}$$

In order to enforce the required behavior at the critical point, the first and second derivatives of pressure with respect to specific volume at constant temperature are obtained using the diff command in Maple.

```
> dPdv:=diff(P,v);
```

$$dP\,dv := \frac{a\,\varepsilon^{\left(-\frac{a}{R\,T\,v}\right)}}{v^2(v-b)} - \frac{R\,T\,\mathbf{e}^{\left(-\frac{a}{R\,T\,v}\right)}}{(v-b)^2}$$

```
> d2Pdv2:=diff(dPdv,v);
```

$$d2P\,dv2 := -\frac{2a\,\mathbf{e}^{\left(-\frac{a}{R\,T\,v}\right)}}{v^3(v-b)} + \frac{a^2\mathbf{e}^{\left(-\frac{a}{R\,T\,v}\right)}}{v^4 R\,T(v-b)} - \frac{2a\,\mathbf{e}^{\left(-\frac{a}{R\,T\,v}\right)}}{v^2(v-b)^2} + \frac{2\,R\,T\,\mathbf{e}^{\left(-\frac{a}{R\,T\,v}\right)}}{(v-b)^3}$$

The two equations required to determine a and b, Eqs. (10-17) and (10-21), are obtained by using the subs command to substitute the critical point properties into the first and second derivatives obtained by Maple.

```
> Eq1:=subs({v=v_crit,T=T_crit},dPdv)=0;
```

$$Eq1 := \frac{a\,\mathbf{e}^{\left(-\frac{a}{R\,T\_crit\;v\_crit}\right)}}{v\_crit^2\,(v\_crit-b)} - \frac{R\,T\_crit\,\mathbf{e}^{\left(-\frac{a}{R\,T\_crit\;v\_crit}\right)}}{(v\_crit-b)^2} = 0$$

```
> Eq2:=subs({v=v_crit,T=T_crit},d2Pdv2)=0;
```

$$Eq2 := -\frac{2\,a\,\mathbf{e}^{\left(-\frac{a}{R\,T\_crit\;v\_crit}\right)}}{v\_crit^3\,(v\_crit-b)} + \frac{a^2\,\mathbf{e}^{\left(-\frac{a}{R\,T\_crit\;v\_crit}\right)}}{v\_crit^4\;R\,T\_crit(v\_crit-b)} - \frac{2\,a\,\mathbf{e}^{\left(-\frac{a}{R\,T\_crit\;v\_crit}\right)}}{v\_crit^2\,(v\_crit-b)^2} + \frac{2\,R\,T\_crit\,\mathbf{e}^{\left(-\frac{a}{R\,T\_crit\;v\_crit}\right)}}{(v\_crit-b)^3} = 0$$

The two equations that result are solved simultaneously using the solve command in order to obtain the coefficients a and b.

```
> coef:=solve({Eq1,Eq2},{a,b});
```

$$coef := \left\{a = 2\,v\_crit\,R\,T\_crit,\;\; b = \frac{v\_crit}{2}\right\}$$

Therefore, the coefficients are:

$$b = \frac{v_{crit}}{2} \tag{1}$$

$$a = 2\,v_{crit}\,R\,T_{crit} \tag{2}$$

In order to obtain an expression for a and b in terms of P_{crit} and T_{crit} it is necessary to combine Eqs. (1) and (2) with Eq. (10-50) evaluated at the critical point.

```
> EOS:=P_crit=R*T_crit*exp(-a/(R*T_crit*v_crit))/(v_crit-b);
```

$$EOS := P\_crit = \frac{R T\_crit\,\mathbf{e}^{\left(-\frac{a}{RT\_crit\,v\_crit}\right)}}{v\_crit-b}$$

The eliminate command is used to develop two equations that do not include v_{crit}.

```
> Eqset2:=eliminate({coef[1],coef[2],EOS},v_crit);
```

$$Eqset2 := \left[\{v\_crit = 2\,b\},\; \left\{-P\_crit\,b + R\,T\_crit\,\mathbf{e}^{\left(-\frac{a}{2\,R\,T\_crit\,b}\right)},\; 4\,b\,R\,T\_crit - a\right\}\right]$$

The two equations are solved for a and b using the solve command.

```
> solve(Eqset2[2],{a,b});
```

$$\left\{a = \frac{4\,R^2\,T\_crit^2}{P\_crit\,\mathbf{e}^2},\quad b = \frac{R\,T\_crit}{P\_crit\,\mathbf{e}^2}\right\}$$

which leads to:

$$b = \frac{R\,T_{crit}}{P_{crit}\exp(2)} \tag{3}$$

$$a = \frac{4\,R^2\,T_{crit}^2}{P_{crit}\exp(2)} \tag{4}$$

b) Determine the critical compressibility factor predicted by the Dieterici equation of state.

Equations (1) and (2) are substituted into Eq. (10-50) evaluated at the critical point:

$$P_{crit} = \frac{R\,T_{crit}}{\left(v_{crit} - \frac{v_{crit}}{2}\right)}\exp\left(-\frac{2\,v_{crit}\,R\,T_{crit}}{R\,T_{crit}\,v_{crit}}\right) \tag{5}$$

Equation (5) is rearranged:

$$\frac{P_{crit}\,v_{crit}}{R\,T_{crit}} = 2\exp(-2) = 0.271$$

The critical compressibility factor $Z_{crit} = 0.271$ is similar to the values observed for most fluids, as shown in Table 10-1, suggesting that the Dieterici equation of state is more accurate in the vicinity of the critical point than the van der Waals equation of state.

c) Plot the compressibility factor associated with nitrogen obtained from the property correlations in EES as well as the value predicted by the van der Waals and Dieterici equations of state as a function of reduced pressure for the critical isotherm ($T_r = 1$).

The required EES code is appended to the code developed in Example 10.1-1. The coefficients a and b are determined using Eqs. (3) and (4). Equation (10-50) is used to determine the specific volume, v_D, and the compressibility factor, Z_D, is computed.

```
a_D=4*R^2*T_crit^2/(P_crit*exp(2))        "coefficient a for Dieterici equation of state"
b_D=R*T_crit/(P_crit*exp(2))              "coefficient b for Dieterici equation of state"
P=R*T*exp(-a_D/(R*T*v_D))/(v_D-b_D)       "specific volume from Dieterici equation of state"
Z_D=P*v_D/(R*T)                           "compressibility factor from Dieterici equation of state"
```

Figure 1 illustrates the compressibility factor predicted by the van der Waals and Dieterici equations of state as well as the value obtained from the property correlations in EES as a function of reduced pressure for the critical isotherm. Notice that the Dieterici equation of state is significantly more accurate than the van der Waals equation of state near the critical point but is no better at high values of reduced pressure.

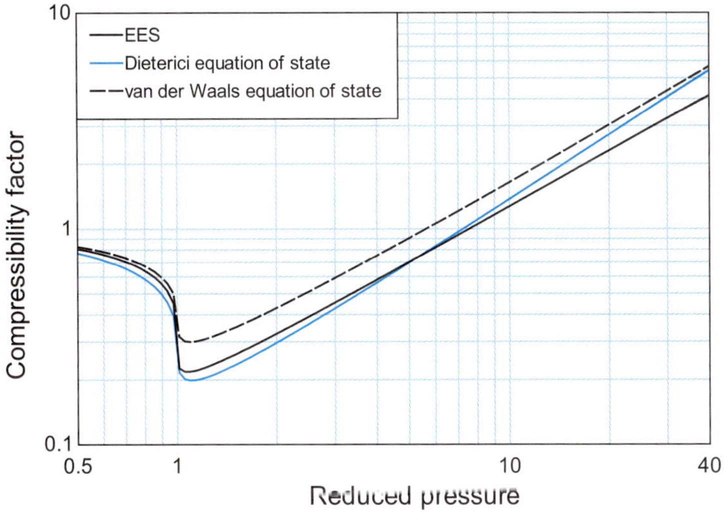

Figure 1: Compressibility factor of nitrogen predicted by the van der Waals and Dieterici equations of state as well as the value obtained from the property correlations in EES as a function of reduced pressure for the critical isotherm.

The Redlich-Kwong Equation of State

The publication of the Redlich-Kwong equation of state in 1949 represented a major improvement in the capability of two-parameter equations of state. The Redlich-Kwong equation of state is:

$$P = \frac{R\,T}{(v - b)} - \frac{a}{v\,(v + b)\,\sqrt{T}} \tag{10-51}$$

The major difference between the Redlich-Kwong and the van der Waals equation of state is the improvement of the second term, which accounts for intermolecular forces. As with the van der Waals and Dieterici equations of state, the parameters a and b can be estimated by requiring that Eqs. (10-17) and (10-21) are satisfied at the critical point, which results in:

$$a = 0.42748 \frac{R^2\,T_{crit}^{5/2}}{P_{crit}} \tag{10-52}$$

$$b = 0.08664 \frac{R\,T_{crit}}{P_{crit}} \tag{10-53}$$

The improvement in the Redlich-Kwong equation of state over the van der Waals and Dieterici equations of state is evident in Figure 10-10, which shows the compressibility factor for nitrogen obtained from the property correlations in EES and using the Redlich-Kwong equation of state as a function of P_r for various values of T_r. Figure 10-10(a) focuses on low values of P_r, near the critical point, while Figure 10-10(b) shows a broader range of P_r.

The success of the Redlich-Kwong equation of state has led to a number of related equations of state that improve the ability of the equation of state in the liquid region. The

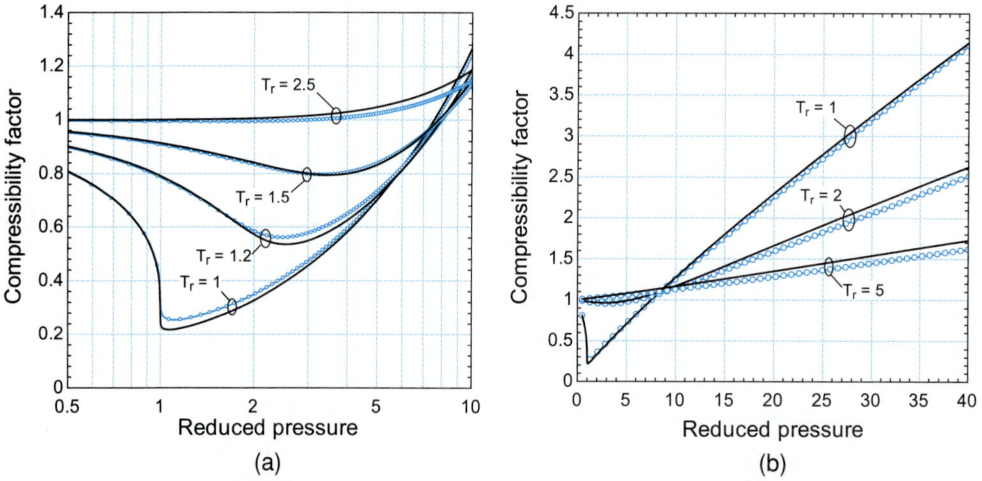

Figure 10-10: The compressibility factor of nitrogen determined using the Redlich-Kwong (1949) equation of state (black line) and the correlations in EES (blue line with symbols) as a function of reduced pressure for various values of reduced temperature; (a) focuses on the region near the critical point while (b) extends over a broad range of reduced pressure.

two most well-known examples are the Soave modification (1972) of the Redlich-Kwong equation of state (often referred to as the RKS equation of state) and the Peng-Robinson (1976) equation of state (referred to as the PR equation of state).

The Redlich-Kwong-Soave (RKS) Equation of State

The Soave modification of the Redlich-Kwong equation of state alters the intermolecular force term in the Redlich-Kwong equation of state, the second term in Eq. (10-51). The RKS equation of state is given by:

$$P = \frac{R\,T}{(v - b)} - \frac{a}{v\,(v + b)} \tag{10-54}$$

The parameter b in Eq. (10-54) is the same as the Redlich-Kwong equation of state:

$$b = 0.08664\frac{R\,T_{crit}}{P_{crit}} \tag{10-55}$$

However, parameter a in the RKS equation of state is a function of temperature, given by:

$$a = 0.42748\,\alpha\,\frac{R^2\,T_{crit}^2}{P_{crit}} \tag{10-56}$$

where α depends on reduced temperature according to:

$$\sqrt{\alpha} = 1 + m\left(1 - \sqrt{T_r}\right) \tag{10-57}$$

The parameter m is defined as:

$$m = 0.480 + 1.574\,\omega - 0.176\,\omega^2 \tag{10-58}$$

where ω is the *acentric factor*. The acentric factor is an easily measured property that depends on the substance and is related to the reduced saturation pressure at a reduced temperature of $T_r = 0.70$ according to:

$$\omega = -1 - \log_{10}\left(P_{r,sat,T_r=0.7}\right) \tag{10-59}$$

where $P_{r,sat,T_r=0.7}$ is the ratio of the saturation pressure at a temperature that is 0.7 T_{crit} to the critical pressure. The function AcentricFactor in EES provides the acentric factor for any substance in the EES fluid database.

The Peng-Robinson (PR) Equation of State

Like the Soave modification to the Redlich-Kwong equation of state, the Peng-Robinson (PR) equation of state (1976) also modifies the intermolecular force term in order to improve its ability to predict the behavior of the liquid phase. The PR equation of state works particularly well for light hydrocarbon substances. The PR equation of state is:

$$P = \frac{R\,T}{(v-b)} - \frac{a}{v\,(v+b) + b\,(v-b)} \tag{10-60}$$

where:

$$a = 0.45724\frac{R^2\,T_{crit}^2}{P_{crit}}\alpha \tag{10-61}$$

$$b = 0.07780\frac{R\,T_{crit}}{P_{crit}} \tag{10-62}$$

$$\sqrt{\alpha} = 1 + m\left(1 - \sqrt{T_r}\right) \tag{10-63}$$

$$m = 0.37464 + 1.54226\,\omega - 0.26992\,\omega^2 \tag{10-64}$$

The acentric factor, ω, appearing in Eq. (10-64) is the same factor used in the RKS equation of state, defined in Eq. (10-59). The Peng-Robinson equation of state can be rearranged to provide a cubic equation for specific volume. The denominators of the fractions on the right side of Eq. (10-60) are cleared.

$$\left[P = \frac{R\,T}{(v-b)} - \frac{a}{v\,(v+b) + b\,(v-b)}\right](v-b)\left[v\,(v+b) + b\,(v-b)\right] \tag{10-65}$$

leading to:

$$P\,(v-b)\left[v\,(v+b) + b\,(v-b)\right] = R\,T\left[v\,(v+b) + b\,(v-b)\right] - a\,(v-b) \tag{10-66}$$

Expanding Eq. (10-66) and collecting like powers of specific volume results in:

$$v^3 + \left(b - \frac{R\,T}{P}\right)v^2 + \left(\frac{a}{P} - 3\,b^2 - \frac{2\,R\,T\,b}{P}\right)v + \left(b^3 + \frac{R\,T\,b^2}{P} - \frac{a\,b}{P}\right) = 0 \tag{10-67}$$

With the appropriate substitutions, the equation of state can also be expressed as a cubic equation for reduced specific volume or compressibility factor. For example, Eq. (10-67) can be written in terms of the compressibility factor, Z. The equation is multiplied by $P^3/(R^3\,T^3)$, leading to:

$$\left[v^3 + \left(b - \frac{R\,T}{P}\right) v^2 + \left(\frac{a}{P} - 3\,b^2 - \frac{2\,R\,T\,b}{P}\right) v + \left(b^3 + \frac{R\,T\,b^2}{P} - \frac{a\,b}{P}\right) = 0\right] \frac{P^3}{R^3\,T^3} \tag{10-68}$$

which results in:

$$Z^3 - (1 - B)\,Z^2 + \left(A - 3\,B^2 - 2\,B\right)\,Z - \left(A\,B - B^2 - B^3\right) = 0 \tag{10-69}$$

where A and B are dimensionless parameters:

$$A = \frac{a\,P}{R^2\,T^2} \tag{10-70}$$

$$B = \frac{b\,P}{R\,T} \tag{10-71}$$

EES provides an external library that facilitates the use of the Peng-Robinson equation of state for pure fluids and mixtures. Information on the use of the Peng-Robinson library can be viewed from the Function Information menu command in the Options menu by clicking the External routines button and selecting Peng-Robinson.DLL, as shown in Figure 10-11.

The procedure AB_PR returns the constants A and B defined by Eqs. (10-70) and (10-71) and required by Eq. (10-69). The functions Z_L_PR and Z_G_PR return the smallest and largest roots of Eq. (10-69), which correspond to the compressibility factors for liquid and vapor, respectively, in a two-phase state. In the supercritical region there is a single physical root to the cubic equation and the functions Z_L_PR and Z_G_PR will both return this value. Use of the Peng-Robinson equation of state is illustrated in Example 10.1-3.

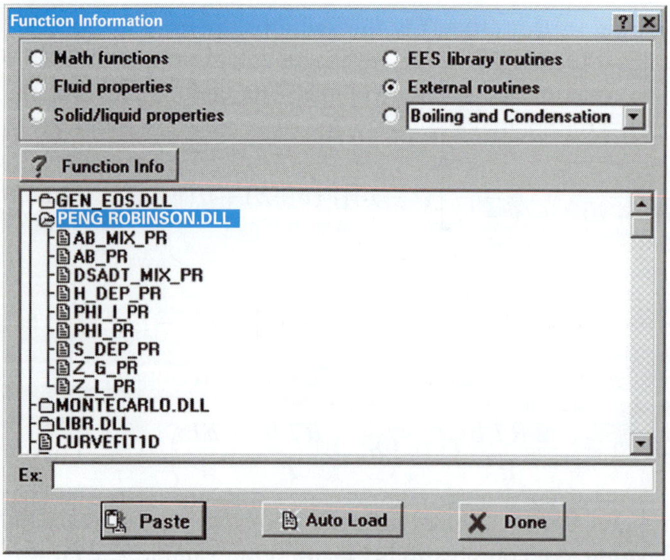

Figure 10-11: Function information dialog for the Peng-Robinson EES library.

EXAMPLE 10.1-3: PENG-ROBINSON EQUATION OF STATE

a) Use the Peng-Robinson equation of state to prepare a plot of the compressibility factor of carbon dioxide as a function of reduced pressure for various values of reduced temperature. Restrict your plot to values of T_r that lie above the critical isotherm.

It is convenient to place the equations that calculate the compressibility factor using the Peng-Robinson equation of state in an EES subprogram. A subprogram in EES solves an independent set of equations using local variables; unlike functions, subprograms consist of the typical EES code that is entered in the Equations window and therefore implicit sets of equations can be solved. Information is passed to and from the subprogram in the argument list. Information can also be passed to the subprogram with the $Common directive. Subprograms must appear at the top of the EES Equations window before any of the equations in the main EES program.

First, specify the unit system. Only one unit system can be specified; the same unit system will be used in the subprogram and the main EES program.

```
$UnitSystem SI Mass Radian J K Pa
```

A subprogram begins with the Subprogram keyword, followed by the name of the subprogram and the argument list. The inputs to the subprogram are the reduced temperature (T_r), reduced pressure (P_r), and the fluid (in the string variable F$). A colon can be optionally used to separate the inputs from the outputs. There is only one output in this case, the compressibility factor (Z).

```
Subprogram PR_EOS(T_r,P_r,F$:Z)
   "This subprogram calculates the compressibility factor using the Peng Robinson equation of state"
   "Inputs:"
   "T_r – reduced temperature"
   "P_r – reduced pressure"
   "F$ – fluid"

   "Output:"
   "Z – compressibility factor"
```

The critical temperature (T_{crit}), critical pressure (P_{crit}), molecular mass (MW), and acentric factor (ω) are determined. The ideal gas constant is computed according to:

$$R = \frac{R_{univ}}{MW}$$

```
   T_crit=T_crit(F$)              "critical temperature"
   P_crit=P_crit(F$)              "critical pressure"
   MW=MolarMass(F$)               "molar mass"
   omega=AcentricFactor(F$)       "acentric factor"
   R=R#/MW                        "gas constant"
```

The parameters m and α are computed using Eqs. (10-64) and (10-63). The parameters a and b are computed using Eqs. (10-61) and (10-62).

EXAMPLE 10.1-3: PENG-ROBINSON EQUATION OF STATE

```
m=0.37464+1.54226*omega-0.26992*omega^2    "parameter m for PR equation of state"
sqrt(alpha)=1+m*(1-sqrt(T_r))               "parameter alpha for PR equation of state"
a=0.45724*R^2*T_crit^2*alpha/P_crit         "paramater a for PR equation of state"
b=0.07780*R*T_crit/P_crit                   "parameter b for PR equation of state"
```

The pressure and temperature are computed according to:

$$P = P_r\, P_{crit}$$

$$T = T_r\, T_{crit}$$

The dimensionless coefficients A and B are computed according to Eqs. (10-70) and (10-71).

```
P=P_r*P_crit                  "pressure"
T=T_r*T_crit                  "temperature"
A_nd=a*P/(R^2*T^2)            "nondimensional coefficient"
B_nd=b*P/(R*T)                "nondimensional coefficient"
```

The function RealCubicRoots is used to find the roots of Eq. (10-69). The coefficients of the cubic equation are computed according to:

$$a_2 = -(1 - B)$$

$$a_1 = A - 3\,B^2 - 2\,B$$

$$a_0 = -\left(A\,B - B^2 - B^3\right)$$

The RealCubicRoots function may return non-physical values for Z that satisfy Eq. (10-69). In the supercritical region, the maximum root of Eq. (10-69) is the physical one; the Max function is used to select the physical root.

```
a_2=-(1-B_nd)                                   "coefficient of Z^2"
a_1=(A_nd-3*B_nd^2-2*B_nd)                      "coefficient of Z^1"
a_0=-(A_nd*B_nd-B_nd^2-B_nd^3)                  "coefficient of Z^0"
CALL RealCubicRoots(a_2,a_1,a_0: Z_1,Z_2,Z_3)   "find real roots of PR equation of state"
Z=MAX(Z_1,Z_2,Z_3)                              "choose the physical root"
end
```

The reduced pressure, reduced temperature, and fluid are set in the Equations window and the subprogram PR_EOS is called in order to determine the compressibility factor.

```
P_r=1.5 [-]                   "reduced pressure"
T_r=2.5 [-]                   "reduced temperature"
F$='CarbonDioxide'            "fluid"
Call PR_EOS(T_r,P_r,F$:Z)     "call subprogram"
```

A Parametric table is generated that includes P_r and Z and the results are used to prepare Figure 1, which shows the compressibility factor as a function of reduced pressure for several values of reduced temperature.

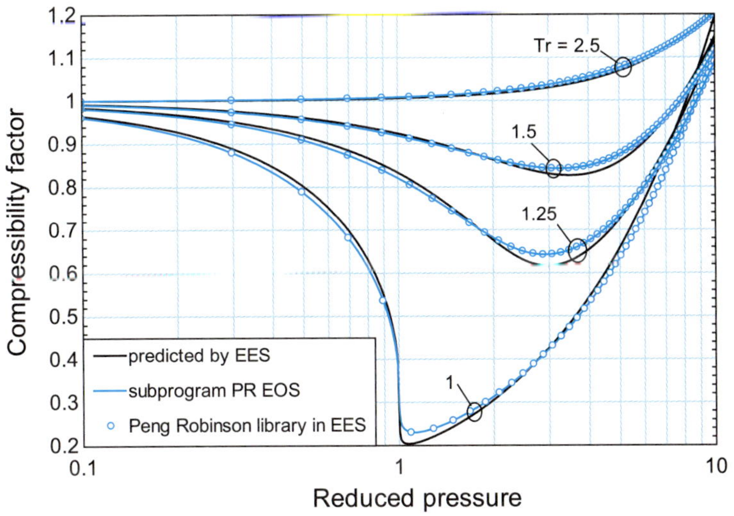

Figure 1: Compressibility factor for carbon dioxide as a function of reduced pressure predicted using the subprogram PR_EOS, using the property correlations in EES, and using the Peng_Robinson library in EES.

b) Overlay on your plot from (a) the compressibility factor predicted using the property correlations in EES.

The required calculations are accomplished in the Equations window. The critical temperature (T_{crit}), critical pressure (P_{crit}), and molecular mass (MW) are determined. The ideal gas constant is computed according to:

$$R = \frac{R_{univ}}{MW}$$

```
T_crit=T_crit(F$)          "critical temperature"
P_crit=P_crit(F$)          "critical pressure"
MW=MolarMass(F$)           "molar mass"
R=R#/MW                    "gas constant"
```

The pressure and temperature are computed according to:

$$P = P_r\, P_{crit}$$

$$T = T_r\, T_{crit}$$

The specific volume (v_{EES}) is obtained using the volume function in EES. The compressibility factor predicted by EES is computed according to:

$$Z_{EES} = \frac{P\, v_{EES}}{R\, T}$$

```
T=T_r*T_crit               "temperature"
P=P_r*P_crit               "pressure"
v_EES=volume(F$,T=T,P=P)   "specific volume from EES"
Z_EES=P*v_EES/(R*T)        "compressibility factor from EES"
```

The compressibility factor predicted by EES is shown in Figure 1 and agrees quite well with the Peng-Robinson correlation, even near the critical point.

c) Use the Peng_Robinson library in EES to predict the compressibility factor and overlay these results on your plot from (a).

The acentric factor (ω) is determined using the AcentricFactor function. The procedure AB_PR is used to determine the dimensionless coefficients A and B. These coefficients are provided to the function Z_G_PR in order to compute the compressibility factor (Z_{PR}).

```
omega=AcentricFactor(F$)                  "acentric factor"
Call AB_PR(T_r, P_r, omega: A_nd, B_nd)   "get coefficients"
Z_PR=Z_G_PR(A_nd,B_nd)                    "compressibility factor from PR library"
```

The value of Z_{PR} is shown in Figure 1 and agrees exactly with the compressibility factor predicted by the subprogram PR_EOS.

10.1.4 Multiple Parameter Equations of State

The accuracy of the Redlich-Kwong-Soave and Peng-Robinson equations of state is remarkable. However, it is not possible to describe the complex behavior of fluids to within the accuracy that they can be measured using equations having only two adjustable terms, particularly in the liquid and high-pressure regions. For example, Figure 10-12 compares the specific volume of water at 100°C calculated using the Peng-Robinson equation of state with the specific volume of water provided using the property correlations in EES, which are based on the international standard presented by Wagner and Pruss (1993) (and accessed with the fluid 'Steam_IAPWS'). Notice that significant differences are evident; the saturated liquid volume at atmospheric pressure predicted by the PR equation of state is about 20% too high and this discrepancy increases with pressure.

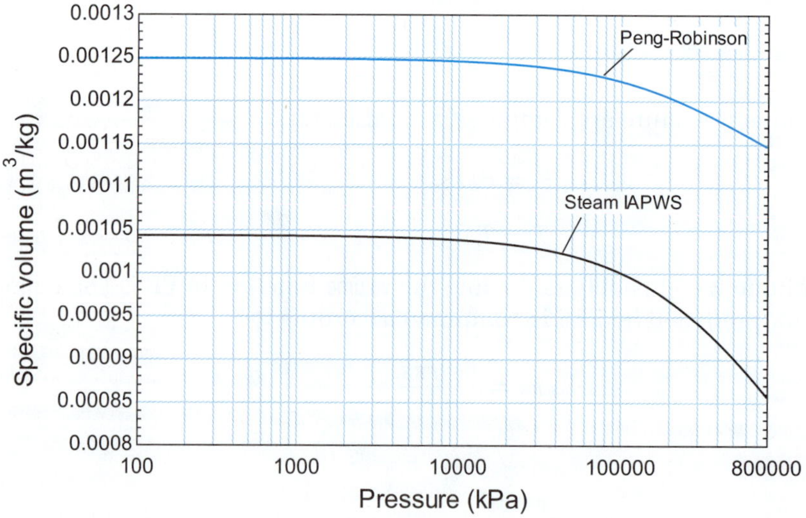

Figure 10-12: The specific volume of water at 100°C as a function of pressure. The value predicted using the Peng-Robinson equation of state is shown together with the value obtained using the EES correlations with the fluid Steam_IAPWS.

One approach to overcome the limitations of two-parameter equations of state is to develop equations that are specifically intended to predict the specific volume in the liquid region where two-parameter equations of state are less accurate. The Rackett (1970) correlation, for example, was developed to predict the specific volume of saturated liquid, v_f:

$$v_f = v_{crit}\, Z_{crit}^{(1-T_r)^{0.2857}} \tag{10-72}$$

Application of Eq. (10-72) for water at 100°C results in a value of $v_f = 0.982\ \text{m}^3/\text{kg}$, which is 6% lower than the accepted value. More elaborate equations for saturated liquid are available with higher accuracy; see, for example, Poling et al. (2000).

Equations of state that employ more than two parameters have been developed in an effort to provide accurate estimates of specific volume over a large range of pressure and temperature. Most of these equations are explicit in pressure and implicit in specific volume, similar to the two-parameter equations of state. Examples are the five parameter Beattie-Bridgeman (1928) equation of state:

$$P = \frac{RT\left(1 - \frac{c}{v\,T^3}\right)}{v^2}\left[v + B_0\left(1 - \frac{b}{v}\right)\right] - \frac{A_0}{v^2}\left(1 - \frac{a}{v}\right) \tag{10-73}$$

as well as the eight parameter Benedict-Webb-Rubin (1940) equation of state:

$$P = \frac{RT}{v} + \left(B_0 RT - A_0 - \frac{C_0}{T^2}\right)\left(\frac{1}{v^2}\right) + \left(\frac{bRT - a}{v^3}\right) + \frac{a\alpha}{v^6} + \frac{c\left(1 + (\gamma/v^2)\right)}{T^2 v^3} e^{-(\gamma/v^2)} \tag{10-74}$$

and the eleven parameter improved Martin-Hou (1955, 1959) equation of state:

$$P = \frac{RT}{v-b} + \frac{A_2 + B_2 T + C_2 exp\left(-k\,T_r\right)}{(v-b)^2} + \frac{A_3 + B_3 T + C_3 exp\left(-k\,T_r\right)}{(v-b)^3} + \frac{A_4}{(v-b)^4} + \frac{A_5 + B_5 T + C_5 exp\left(-k\,T_r\right)}{(v-b)^5} \tag{10-75}$$

The Martin-Hou equation of state was initially used to provide property data in EES. However, EES currently determines much of its property information using reduced Helmholtz free energy equations, as described by Span (2000). The use of a reduced Helmholtz free energy equation is illustrated in Example 10.2-2. One problem with the use of multiple-parameter equations of state, such as those provided by Eqs. (10-73)–(10-75), is that a lot of high accuracy experimental data are needed to determine the parameters for each fluid. In general, the accuracy with which an equation of state can represent experimental data increases as the number of parameters used in the equation of state increases. However, the amount of experimental data needed to determine the parameters also increases. Multiple-parameter equations of state are obviously not useful for hand calculations or for substances for which few data are available. These equations of state are useful for interpolating, extrapolating, and determining the consistency of the experimental data.

10.2 Application of Fundamental Property Relations

Pressure, temperature, and specific volume are all measurable properties. Equation (10-1) indicates that only two of these three properties are needed to fix the state of a single phase pure fluid. The equations of state presented in Section 10.1 provide a means to determine the third property when any two of the three are known. However, other properties such as specific internal energy, specific enthalpy, and specific entropy are required in order to apply the laws of thermodynamics. These properties are not directly measurable, but their values can be determined using fundamental property relations provided that the specific heat capacity and an equation of state are known for the substance, as described in Section 10.3.

10.2.1 The Fundamental Property Relations

The starting point for relating the specific internal energy, specific enthalpy, and specific entropy to measurable properties is the *fundamental property relation*. It is shown in Section 6.2 that the First and Second Laws can be combined in a manner that eliminates path-dependent quantities, such as heat and work, and leads to the fundamental property relation that can be applied to a pure substance:

$$\boxed{du = T\,ds - P\,dv} \tag{10-76}$$

Specific enthalpy is defined in Section 4.1:

$$h = u + P\,v \tag{10-77}$$

The total derivative of Eq. (10-77) is:

$$dh = du + P\,dv + v\,dP \tag{10-78}$$

Substituting Eq. (10-78) into Eq. (10-76) allows the fundamental property relation to be expressed in terms of specific enthalpy rather than specific internal energy:

$$\boxed{dh = T\,ds + v\,dP} \tag{10-79}$$

The fundamental property relation can also be expressed in terms of the *specific Helmholtz free energy* (a) and the *specific Gibbs free energy* (g). These thermodynamic properties have not yet been discussed and are defined according to:

$$a = u - T\,s \tag{10-80}$$

$$g = h - T\,s \tag{10-81}$$

The properties a and g are useful for problems involving phase and chemical equilibrium, as discussed in Section 10.5 and in Chapters 11 and 14. The total derivative of the specific Helmholtz free energy, Eq. (10-80), is:

$$da = du - T\,ds - s\,dT \tag{10-82}$$

Substituting Eq. (10-82) into Eq. (10-76) allows the fundamental property relation to be expressed in terms of the specific Helmholtz free energy:

$$\boxed{da = -s\,dT - P\,dv} \tag{10-83}$$

Repeating this process for the specific Gibbs free energy results in the fundamental property relation expressed in terms of g:

$$\boxed{dg = -s\,dT + v\,dP} \tag{10-84}$$

10.2.2 Complete Equations of State

The fundamental property relation for a pure substance can be expressed in the four different forms that appear in Eqs. (10-76), (10-79), (10-83) and (10-84). These four equations are equivalent in that they express the same fundamental relation in terms of different sets of variables. Each of the four differential fundamental property relations is the differential of a variable, represented as f, that is a function of two other variables, represented as x and y:

$$f = f(x, y) \tag{10-85}$$

The total differential of f in Eq. (10-85) is:

$$df = \left(\frac{\partial f}{\partial x}\right)_y dx + \left(\frac{\partial f}{\partial y}\right)_x dy \tag{10-86}$$

For example, the fundamental property relation expressed in terms of specific internal energy, Eq. (10-76), is the total differential of u as a function of s and v. Comparing Eq. (10-76) with Eq. (10-86) shows that these equations are identical in form if $f = u$, $x = s$, and $y = v$. That is, Eq. (10-76) is the total derivative of the function $u(s, v)$. According to Eq. (10-86), the total derivative of this function is:

$$du = \underbrace{\left(\frac{\partial u}{\partial s}\right)_v}_{T} ds + \underbrace{\left(\frac{\partial u}{\partial v}\right)_s}_{-P} dv \tag{10-87}$$

Comparing Eq. (10-87) with Eq. (10-76) shows that:

$$\left(\frac{\partial u}{\partial s}\right)_v = T \tag{10-88}$$

and

$$\left(\frac{\partial u}{\partial v}\right)_s = -P \tag{10-89}$$

Similarly, Eq. (10-79) is the total derivative of $h(s, P)$, Eq. (10-83) is the total derivative of $a(T, v)$, and Eq. (10-84) is the total derivative of $g(T, P)$.

Suppose that you were given an equation that provides specific internal energy (u) as a function of specific entropy (s) and specific volume (v). This information provides a complete set of thermodynamic property information. You could of course, determine u at any values of s and v, using the function itself. In addition, the derivatives of the equation could be used to provide temperature and pressure according to Eqs. (10-88) and (10-89). Therefore, the equation relates the variables T, P, u, s, and v. Given any two of these variables any of the others could, in theory, be computed. Other properties (e.g., h, a, or g) could be computed from this set using Eqs. (10-77), (10-80), and (10-81). Since h and T are known, the value of c_P could be found by differentiation. Since u and T are known, the value of c_v could also be found by differentiation. In summary, if we had an equation that accurately provided u as a function of s and v, we could use this equation to determine all other thermodynamic properties. Therefore, the function $u = u(s, v)$ is referred to as a *complete equation of state*. The other fundamental property relations, Eqs. (10-79), (10-83), and (10-84), provide the functional form for other complete equations of state. Complete thermodynamic property information would be contained in any of the functions $h(s, P)$, $a(T, v)$, or $g(T, P)$. Much of the thermodynamic property data provided by EES for real substances relies on a Helmholtz equation of state of the form $a(T, v)$.

EXAMPLE 10.2-1: USING A COMPLETE EQUATION OF STATE

a) Treat the enthalpy function in EES as the complete equation of state, $h(s, P)$. Using only the enthalpy function, determine the temperature and specific volume of steam at $s = 8000$ J/kg-K and $P = 1$ MPa. Compare your answers to those obtained using the temperature and volume functions in EES.

The inputs are entered in EES:

```
$UnitSystem SI Mass Radian J K Pa
F$='Steam'                              "fluid"
s=8000 [J/kg-K]                         "specific entropy"
P=1 [MPa]*convert(MPa,Pa)               "pressure"
```

The specific enthalpy (h) at the given entropy and pressure is computed using the enthalpy function in EES.

```
h=enthalpy(F$,s=s,P=P)                  "specific enthalpy"
```

The total differential of $h = h(s, P)$ is:

$$dh = \left(\frac{\partial h}{\partial s}\right)_P ds + \left(\frac{\partial h}{\partial P}\right)_s dP \tag{1}$$

Comparing the first term on the right side of Eq. (1) with Eq. (10-79) shows that

$$\left(\frac{\partial h}{\partial s}\right)_P = T \tag{2}$$

The partial derivative in Eq. (2) is estimated numerically:

$$T = \left(\frac{\partial h}{\partial s}\right)_P \approx \frac{h_{s+ds,P} - h_{s-ds,P}}{2\,ds}$$

where ds is a small, but finite, change in the specific entropy. Here, ds is defined as being 0.1% of the given value of specific entropy. The temperature is also estimated using the temperature function in EES (T_{EES}).

```
ds=0.001*s                              "differential change in specific entropy"
h_plus_s=enthalpy(F$,s=s+ds,P=P)        "specific enthalpy at s+ds"
h_minus_s=enthalpy(F$,s=s-ds,P=P)       "specific enthalpy at s-ds"
T=(h_plus_s-h_minus_s)/(2*ds)           "temperature, estimated using h(s,P)"
T_EES=temperature(F$,s=s,P=P)           "temperature, estimated using EES function"
```

Solving these equations results in $T = 861.8$ K. The temperature function in EES returns $T_{EES} = 861.8$ K. Comparing the second term on the right side of Eq. (1) with Eq. (10-79) shows that:

$$\left(\frac{\partial h}{\partial P}\right)_s = v \tag{3}$$

We can estimate the partial derivative in Eq. (3) numerically according to:

$$\left(\frac{\partial h}{\partial P}\right)_s \approx \frac{h_{s,P+dP} - h_{s,P-dP}}{2\,dP}$$

where dP is a small, but finite, change in the pressure. Here, dP is defined as being 0.1% of the specified pressure. The specific volume is also estimated using the volume function in EES (v_{EES}).

```
dP=0.001*P                                  "differential change in pressure"
h_plus_P=enthalpy(F$,s=s,P=P+dP)            "specific enthalpy at P+dP"
h_minus_P=enthalpy(F$,s=s,P=P-dP)           "specific enthalpy at P-dP"
v=(h_plus_P-h_minus_P)/(2*dP)               "specific volume, estimated using h(s,P)"
v_EES=volume(F$,s=s,P=P)                    "specific volume, estimated using EES function"
```

Solving these equations results in $v = 0.3958$ m^3/kg and $v_{EES} = 0.3958$ m^3/kg. The agreement between v and v_{EES} and between T and T_{EES} is not surprising since, internally, EES uses a fundamental property relation to determine its property information.

EXAMPLE 10.2-2: THE REDUCED HELMHOLTZ EQUATION OF STATE

Most of the thermodynamic property data in EES are provided in terms of a reduced Helmholtz free energy function, α:

$$\alpha(\tau,\delta) = \frac{a}{R\,T} = \frac{(u - T\,s)}{R\,T} \tag{1}$$

The reduced Helmholtz free energy can be expressed as a function of the reduced density (δ) and the inverse reduced temperature (τ):

$$\delta = \frac{\rho}{\rho_{crit}} \tag{2}$$

$$\tau = \frac{T_{crit}}{T} \tag{3}$$

The reduced Helmholtz equation is a complete equation of state, equivalent to $a(T, v)$, so that all thermodynamic properties can be determined from an equation in this form. The reduced Helmholtz function is ordinarily represented as the sum of an ideal gas and residual component (α_{IG} and α_{res}, respectively):

$$\alpha(\tau,\delta) = \alpha_{IG}(\tau,\delta) + \alpha_{res}(\tau,\delta) \tag{4}$$

For carbon dioxide, the ideal gas and residual components have been correlated by Span and Wagner (2003) in the following forms:

$$\alpha_{IG}(\tau,\delta) = \ln(\delta) + a_1 + a_2\,\tau + a_3\,\ln(\tau) + \sum_{i=4}^{8} a_i \ln\left[1 - \exp(-\tau\,\theta_i)\right] \tag{5}$$

$$\begin{aligned}\alpha_{res}(\tau,\delta) = {} & n_1\,\delta\,\tau^{0.25} + n_2\,\delta\,\tau^{1.25} + n_3\,\delta\,\tau^{1.50} + n_4\,\delta^3\,\tau^{0.25} + n_5\,\delta^7\,\tau^{0.875}\\ & + n_6\,\delta\,\tau^{2.375}\exp(-\delta) + n_7\,\delta^2\,\tau^2\exp(-\delta) + n_8\,\delta^5\,\tau^{2.125}\exp(-\delta)\\ & + n_9\,\delta\,\tau^{3.5}\exp\left(-\delta^2\right) + n_{10}\,\delta\,\tau^{6.5}\exp\left(-\delta^2\right) + n_{11}\,\delta^4\,\tau^{4.75}\exp\left(-\delta^2\right)\\ & + n_{12}\,\delta^2\,\tau^{12.5}\exp\left(-\delta^3\right)\end{aligned} \tag{6}$$

where the coefficients a_i, θ_i, and n_i required to implement Eqs. (5) and (6) are provided in Table 1. Equations (5) and (6) are similar to, but not the same as, the equations used in EES for carbon dioxide.

Table 1: Coefficients for Eqs. (5) and (6) for Carbon Dioxide from Span and Wagner (2003).

Index, i	a_i	θ_i	n_i
1	8.373045		0.8987511
2	−3.704543		−2.128199
3	2.500000		-6.819032×10^{-2}
4	1.994270	3.151630	7.635531×10^{-2}
5	0.6210525	6.111900	2.205325×10^{-4}
6	0.4119529	6.777080	0.4154182
7	1.040289	11.32384	0.7133566
8	8.327678×10^{-2}	27.08792	3.035423×10^{-4}
9			−0.3664314
10			-1.440778×10^{-3}
11			-8.916671×10^{-2}
12			-2.369989×10^{-2}

a) Derive an expression for the pressure as a function of α and its partial derivatives with respect to τ and δ.

The fundamental property relation involving the Helmholtz Free energy is introduced in Eq. (10-83):

$$da = -s\,dT - P\,dv \tag{7}$$

Examination of Eq. (7) shows that pressure is the partial derivative of the specific Helmholtz free energy with respect to specific volume at constant temperature.

$$P = -\left(\frac{\partial a}{\partial v}\right)_T \tag{8}$$

According to the definition of α from Eq. (1):

$$a = R\,T\,\alpha\,(\tau, \delta) \tag{9}$$

The partial derivative of a with respect to v can be determined by applying the chain rule.

$$\left(\frac{\partial a}{\partial v}\right)_T = \left(\frac{\partial a}{\partial \delta}\right)_T \left(\frac{\partial \delta}{\partial v}\right)_T \tag{10}$$

The derivative of Eq. (9) with respect to δ is:

$$\left(\frac{\partial a}{\partial \delta}\right)_T = R\,T\left(\frac{\partial \alpha}{\partial \delta}\right)_T \tag{11}$$

The reduced density can be expressed in terms of specific volume:

$$\delta = \frac{\rho}{\rho_{crit}} = \frac{v_{crit}}{v} \tag{12}$$

The derivative of Eq. (12) with respect to v is:

$$\left(\frac{\partial \delta}{\partial v}\right)_T = -\frac{v_{crit}}{v^2} \tag{13}$$

Equation (13) can be written as:

$$\left(\frac{\partial \delta}{\partial v}\right)_T = -\frac{\rho^2}{\rho_{crit}} \tag{14}$$

Substituting Eq. (2) into Eq. (14) leads to:

$$\left(\frac{\partial \delta}{\partial v}\right)_T = -\delta\,\rho \tag{15}$$

Substituting Eqs. (10), (11), and (15) into Eq. (8) results in:

$$P = R\,T\,\delta\,\rho\left(\frac{\partial \alpha}{\partial \delta}\right)_T \tag{16}$$

Substituting Eq. (4) into Eq. (16) provides:

$$P = R\,T\,\delta\,\rho\left[\left(\frac{\partial \alpha_{IG}}{\partial \delta}\right)_\tau + \left(\frac{\partial \alpha_{res}}{\partial \delta}\right)_\tau\right] \tag{17}$$

Note that holding T constant is equivalent to holding τ constant. Taking the derivative of α_{IG}, given by Eq. (5), with respect to δ at constant τ is straightforward:

$$\left(\frac{\partial \alpha_{IG}}{\partial \delta}\right)_\tau = \frac{1}{\delta} \tag{18}$$

Substituting Eq. (18) into Eq. (17) leads to:

$$P = R\,T\,\rho\left[1 + \delta\left(\frac{\partial \alpha_{res}}{\partial \delta}\right)_\tau\right] \tag{19}$$

b) Implement your solution in EES in order to estimate the pressure at $v = 0.01$ m^3/kg and $T = 350$ K. Compare your estimate with the pressure obtained using the pressure function in EES.

The inputs are entered in EES:

```
$UnitSystem SI Radian Mass J K Pa
F$='CarbonDioxide'                     "fluid"
T=350 [K]                              "temperature"
v=0.01 [m^3/kg]                        "specific volume"
```

The critical temperature, critical specific volume, and molar mass (T_{crit}, v_{crit}, and MW) are obtained. The gas constant is computed according to:

$$R = \frac{R_{univ}}{MW}$$

```
T_crit=T_crit(F$)                      "critical temperature"
v_crit=v_crit(F$)                      "critical specific volume"
MW=MolarMass(F$)                       "molar mass"
R=R#/MW                                "ideal gas constant"
```

The density is computed as the inverse of the specific volume. The reduced density is obtained from Eq. (12) and the inverse reduced temperature from Eq. (3).

```
rho=1/v                                "density"
delta=v_crit/v                         "reduced density"
tau=T_crit/T                           "inverse reduced temperature"
```

The coefficients in Table 1 are entered in the EES Equation window as arrays.

```
"coefficients for the equation of state for carbon dioxide"
a[1..8]=[8.37304456, -3.70454304, 2.5, 1.99427042, 0.62105248, 0.41195293,&
   1.04028922, 0.08327678]
theta[4..8]=[3.15163, 6.11190, 6.77708, 11.32384, 27.08792]
n[1..12]=[0.89875108, -0.21281985E1, -0.68190320E-1, 0.76355306E-1,&
   0.22053253E-3, 0.41541823, 0.71335657, 0.30354234E-3,&
     -0.36643143, -0.14407781E-2, -0.89166707E-1, -0.23699887E-1]
```

The derivative of α_{res} with respect to δ is obtained using Maple. Equation (6) is entered in Maple.

```
> restart;
> alpha_res:=n[1]*delta*tau^0.25+n[2]*delta*tau^1.25+n[3]*delta*tau^1.5+n[4]*
  delta^3*tau^0.25+n[5]*delta^7*tau^0.875+n[6]*delta*tau^2.375*exp(-delta)+n[7]*
  delta^2*tau^2*exp(-delta)+n[8]*delta^5*tau^2.125*exp(-delta)+n[9]*
  delta*tau^3.5*exp(-delta^2)+n[10]*delta*tau^6.5*exp(-delta^2)+n[11]*delta^4*tau^4.75*exp(-delta^2)+
  n[12]*delta^2*tau^12.5*exp(-delta^3);
```

$$alpha\_res := n_1\,\delta\,\tau^{0.25} + n_2\,\delta\,\tau^{1.25} + n_3\,\delta\,\tau^{1.5} + n_4\,\delta^{3}\,\tau^{0.25} + n_5\,\delta^{7}\,\tau^{0.875} + n_6\,\delta\,\tau^{2.375}\,\mathbf{e}^{(-\delta)} + n_7\,\delta^{2}\,\tau^{2}\,\mathbf{e}^{(-\delta)} + n_8\,\delta^{5}\,\tau^{2.125}\,\mathbf{e}^{(-\delta)} + n_9\,\delta\,\tau^{3.5}\,\mathbf{e}^{(-\delta^2)} + n_{10}\,\delta\,\tau^{6.5}\,\mathbf{e}^{(-\delta^2)} + n_{11}\,\delta^{4}\,\tau^{4.75}\,\mathbf{e}^{(-\delta^2)} + n_{12}\,\delta^{2}\,\tau^{12.5}\,\mathbf{e}^{(-\delta^3)}$$

The derivative is obtained using the diff command in Maple.

```
> dalpha_resddelta_tau:=diff(alpha_res,delta);
```

$$dalpha\_resddelta\_tau := n_1\,\tau^{0.25} + n_2\,\tau^{1.25} + n_3\,\tau^{1.5} + 3\,n_4\,\delta^{2}\,\tau^{0.25} + 7\,n_5\,\delta^{6}\,\tau^{0.875} + n_6\,\tau^{2.375}\,\mathbf{e}^{(-\delta)} - n_6\,\delta\,\tau^{2.375}\,\mathbf{e}^{(-\delta)} + 2\,n_7\,\delta\,\tau^{2}\,\mathbf{e}^{(-\delta)} - n_7\,\delta^{2}\,\tau^{2}\,\mathbf{e}^{(-\delta)} + 5n_8\,\delta^{4}\,\tau^{2.125}\,\mathbf{e}^{(-\delta)} - n_8\,\delta^{5}\,\tau^{2.125}\,\mathbf{e}^{(-\delta)} + n_9\,\tau^{3.5}\,\mathbf{e}^{(-\delta^2)} - 2\,n_9\,\delta^{2}\,\tau^{3.5}\,\mathbf{e}^{(-\delta^2)} + n_{10}\,\tau^{6.5}\,\mathbf{e}^{(-\delta^2)} - 2\,n_{10}\,\delta^{2}\,\tau^{6.5}\,\mathbf{e}^{(-\delta^2)} + 4\,n_{11}\,\delta^{3}\,\tau^{4.75}\,\mathbf{e}^{(-\delta^2)} - 2\,n_{11}\,\delta^{5}\,\tau^{4.75}\,\mathbf{e}^{(-\delta^2)} + 2\,n_{12}\,\delta\,\tau^{12.5}\,\mathbf{e}^{(-\delta^3)} - 3\,n_{12}\,\delta^{4}\,\tau^{12.5}\,\mathbf{e}^{(-\delta^3)}$$

The Maple result is copied and pasted into EES; the only modification required is to replace the := from Maple with = in EES.

```
"partial derivative of alpha with respect to delta at constant tau, from Maple"
dalpha_resddelta_tau = n[1]*tau^.25+n[2]*tau^1.25+n[3]*tau^1.5+3*n[4]*delta^2*tau^.25&
   +7*n[5]*delta^6*tau^.875+n[6]*tau^2.375*exp(-delta)-n[6]*delta*tau^2.375*exp(-delta)&
   +2*n[7]*delta*tau^2*exp(-delta)-n[7]*delta^2*tau^2*exp(-delta)&
   +5*n[8]*delta^4*tau^2.125*exp(-delta)-n[8]*delta^5*tau^2.125*exp(-delta)&
   +n[9]*tau^3.5*exp(-delta^2)-2*n[9]*delta^2*tau^3.5*exp(-delta^2)&
   +n[10]*tau^6.5*exp(-delta^2)-2*n[10]*delta^2*tau^6.5*exp(-delta^2)&
   +4*n[11]*delta^3*tau^4.75*exp(-delta^2)-2*n[11]*delta^5*tau^4.75*exp(-delta^2)&
   +2*n[12]*delta*tau^12.5*exp(-delta^3)-3*n[12]*delta^4*tau^12.5*exp(-delta^3)
```

Equation (19) is used to evaluate the pressure (P). The pressure is also obtained using the pressure function in EES (P_{EES}).

```
P=R*T*rho*(1+delta*dalpha_resddelta_tau)        "pressure, from the reduced Helmholtz eos"
P_kPa=P*convert(Pa,kPa)                         "in kPa"
P_EES=pressure(F$,T=T,v=v)                      "pressure, from EES correlation"
P_EES_kPa=P_EES*convert(Pa,kPa)                 "in kPa"
```

which results in $P = 5.472$ MPa and $P_{EES} = 5.472$ MPa.

c) Plot the pressure as a function of specific volume for several values of temperature. Also show the value obtained using the EES property correlations.

Figure 1 illustrates a P-v diagram for carbon dioxide. The results obtained from the pressure function in EES and by using the Helmholtz equation of state are shown. The pressures returned by the EES property relation are nearly identical to the values determined from the reduced Helmholtz equation for these conditions.

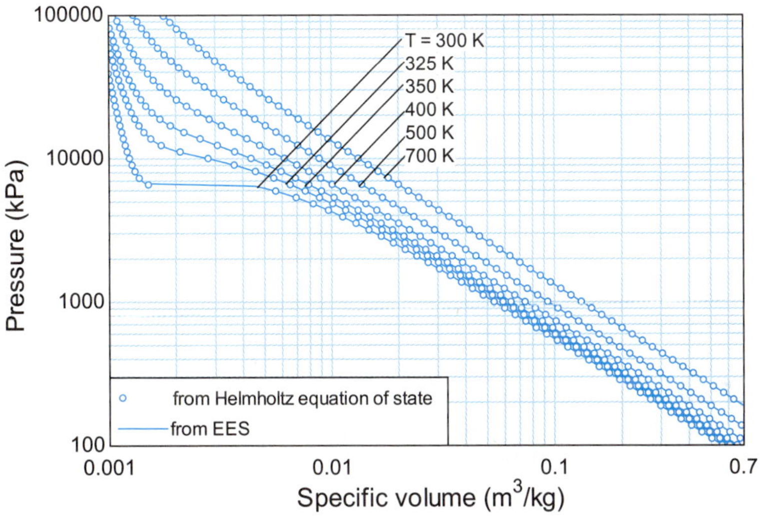

Figure 1: Pressure as a function of specific volume predicted using the Helmholtz equation of state and from EES' property correlations.

d) Derive an expression for the specific internal energy as a function of α and its partial derivatives with respect to τ and δ.

The definition of the reduced Helmholtz free energy, Eq. (1), is:

$$\alpha = \frac{u}{R\,T} - \frac{s}{R} \tag{20}$$

To obtain u, take the derivative of α with respect to T at constant v:

$$\left(\frac{\partial \alpha}{\partial T}\right)_v = \frac{1}{R\,T}\underbrace{\left(\frac{\partial u}{\partial T}\right)_v}_{c_v} - \frac{u}{R\,T^2} - \frac{1}{R}\left(\frac{\partial s}{\partial T}\right)_v \tag{21}$$

The partial derivative of specific entropy with respect to temperature at constant volume is c_v/T, as shown in Chapter 6. To see this, rearrange the fundamental property relation, Eq. (10-76):

$$ds = \frac{1}{T}du + \frac{P}{T}dv \tag{22}$$

According to Eq. (22), the partial derivative of specific entropy with respect to temperature at constant specific volume is:

$$\left(\frac{\partial s}{\partial T}\right)_v = \frac{1}{T}\underbrace{\left(\frac{\partial u}{\partial T}\right)_v}_{c_v} = \frac{c_v}{T} \tag{23}$$

Substituting Eq. (23) into Eq. (21) leads to:

$$\left(\frac{\partial \alpha}{\partial T}\right)_v = \frac{c_v}{R\,T} - \frac{u}{R\,T^2} - \frac{c_v}{R\,T} = -\frac{u}{R\,T^2} \tag{24}$$

Eq. (24) can be rearranged to provide a relation for u:

$$u = -R\,T^2\left(\frac{\partial \alpha}{\partial T}\right)_v \tag{25}$$

Using the chain rule,

$$\left(\frac{\partial \alpha}{\partial T}\right)_v = \left(\frac{\partial \alpha}{\partial \tau}\right)_v\left(\frac{\partial \tau}{\partial T}\right)_v \tag{26}$$

The partial derivative of the inverse reduced temperature, Eq. (3), with respect to temperature leads is:

$$\left(\frac{\partial \tau}{\partial T}\right)_v = -\frac{T_{crit}}{T^2} \tag{27}$$

Substituting Eqs. (26) and (27) into Eq. (25) leads to:

$$u = R\,T_{crit}\left(\frac{\partial \alpha}{\partial \tau}\right)_\delta \tag{28}$$

e) Implement your solution in EES in order to estimate the specific internal energy at $v = 0.01$ m^3/kg and $T = 350$ K. Compare your estimate with the specific internal energy obtained using the intenergy function in EES.

According to Eq. (4):

$$\left(\frac{\partial \alpha}{\partial \tau}\right)_\delta = \left(\frac{\partial \alpha_{IG}}{\partial \tau}\right)_\delta + \left(\frac{\partial \alpha_{res}}{\partial \tau}\right)_\delta \tag{29}$$

These partial derivatives are obtained using Maple. The expression for the ideal gas component, Eq. (5), is entered in Maple:

```
> alpha_IG:=ln(delta)+a[1]+a[2]*tau+a[3]*ln(tau)+a[4]*ln(1-exp(-tau*theta[4]))
+a[5]*ln(1-exp(-tau*theta[5]))+a[6]*ln(1-exp(-tau*theta[6]))+
a[7]*ln(1-exp(-tau*theta[7]))+a[8]*ln(1-exp(-tau*theta[8]));
```

$$alphaIG := \ln(\delta) + a_1 + a_2\,\tau + a_3\,\ln(\tau) + a_4\,\ln(1 - e^{(-\tau\,\theta_4)}) + a_5\,\ln(1 - e^{(-\tau\,\theta_5)}) + a_6\,\ln(1 - e^{(-\tau\,\theta_6)}) + a_7\,\ln(1 - e^{(-\tau\,\theta_7)}) + a_8\,\ln(1 - e^{(-\tau\,\theta_8)})$$

The derivatives of α_{IG} and α_{res} with respect to τ at constant δ are obtained with the diff command.

```
> dalpha_IGdtau_delta:=diff(alpha_IG,tau);
```

$$dalpha\_IGdtau\_delta := a_2 + \frac{a_3}{\tau} + \frac{a_4\,\theta_4\,\mathbf{e}^{(-\tau\,\theta_4)}}{1-\mathbf{e}^{(-\tau\,\theta_4)}} + \frac{a_5\,\theta_5\mathbf{e}^{(-\tau\,\theta_5)}}{1-\mathbf{e}^{(-\tau\,\theta_5)}} + \frac{a_6\,\theta_6\mathbf{e}^{(-\tau\,\theta_6)}}{1-\mathbf{e}^{(-\tau\,\theta_6)}} + \frac{a_7\,\theta_7\mathbf{e}^{(-\tau\,\theta_7)}}{1-\mathbf{e}^{(-\tau\,\theta_7)}} + \frac{a_8\,\theta_8\mathbf{e}^{(-\tau\,\theta_8)}}{1-\mathbf{e}^{(-\tau\,\theta_8)}}$$

```
> dalpha_resdtau_delta:=diff(alpha_res,tau);
```

$$dalpha\_resdtau\_delta := \frac{0.25\,n_1\,\delta}{\tau^{0.75}} + 1.25\,n_2\,\delta\,\tau^{0.25} + 1.5\,n_3\,\delta\,\tau^{0.5} + \frac{0.25\,n_4\,\delta^3}{\tau^{0.75}} + \frac{0.875\,n_5\,\delta^7}{\tau^{0.125}} + 2.375\,n_6\,\delta\,\tau^{1.375}\,\mathbf{e}^{(-\delta)} + 2\,n_7\,\delta^2\,\tau\,\mathbf{e}^{(-\delta)} + 2.125\,n_8\,\delta^5\,\tau^{1.125}\,\mathbf{e}^{(-\delta)} + 3.5\,n_9\,\delta\,\tau^{2.5}\,\mathbf{e}^{(-\delta^2)} + 6.5\,n_{10}\,\delta\,\tau^{5.5}\,\mathbf{e}^{(-\delta^2)} + 4.75\,n_{11}\,\delta^4\,\tau^{3.75}\,\mathbf{e}^{(-\delta^2)} + 12.5\,n_{12}\,\delta^2\,\tau^{11.5}\,\mathbf{e}^{(-\delta^3)}$$

These expressions are copied and pasted into EES:

```
"partial derivative of alpha_IG with respect to tau at constant delta, from Maple"
dalpha_IGdtau_delta = a[2]+a[3]/tau+a[4]*theta[4]*exp(-tau*theta[4])/(1-exp(-tau*theta[4]))&
   +a[5]*theta[5]*exp(-tau*theta[5])/(1-exp(-tau*theta[5]))&
   +a[6]*theta[6]*exp(-tau*theta[6])/(1-exp(-tau*theta[6]))&
   +a[7]*theta[7]*exp(-tau*theta[7])/(1-exp(-tau*theta[7]))&
   +a[8]*theta[8]*exp(-tau*theta[8])/(1-exp(-tau*theta[8]))

"partial derivative of alpha_res with respect to tau at constant delta, from Maple"
dalpha_resdtau_delta =.25*n[1]*delta/tau^.75+1.25*n[2]*delta*tau^.25+1.5*n[3]*delta*tau^.5&
   +.25*n[4]*delta^3/tau^.75+.875*n[5]*delta^7/tau^.125+2.375*n[6]*delta*tau^1.375*exp(-delta)&
   +2*n[7]*delta^2*tau*exp(-delta)+2.125*n[8]*delta^5*tau^1.125*exp(-delta)&
   +3.5*n[9]*delta*tau^2.5*exp(-delta^2)+6.5*n[10]*delta*tau^5.5*exp(-delta^2)&
   +4.75*n[11]*delta^4*tau^3.75*exp(-delta^2)+12.5*n[12]*delta^2*tau^11.5*exp(-delta^3)
```

Equations (29) and (28) are used to determine the specific internal energy (u). The specific internal energy is also obtained using the intenergy command in EES.

```
"partial derivative of alpha with respect to tau at constant delta"
dalphadtau_delta=dalpha_IGdtau_delta+dalpha_resdtau_delta
u=R*T_crit*dalphadtau_delta        "specific internal energy from the reduced Helmholtz eos"
u_EES=intenergy(F$,v=v,T=T)        "specific internal energy, from EES correlation"
```

Solving results in $u = -50793$ J/kg and $u_{EES} = -50742$ J/kg. The values are not identical because EES uses a different formulation of the reduced Helmholtz free energy for carbon dioxide that includes more terms. However, the values are quite close.

f) Plot the specific internal energy as a function of pressure for several values of temperature. Also show the value obtained using the EES property correlations.

Figure 2 illustrates the specific internal energy of carbon dioxide as a function of pressure for various values of temperature; both the results from the intenergy function in EES and from using the Helmholtz equation of state are shown.

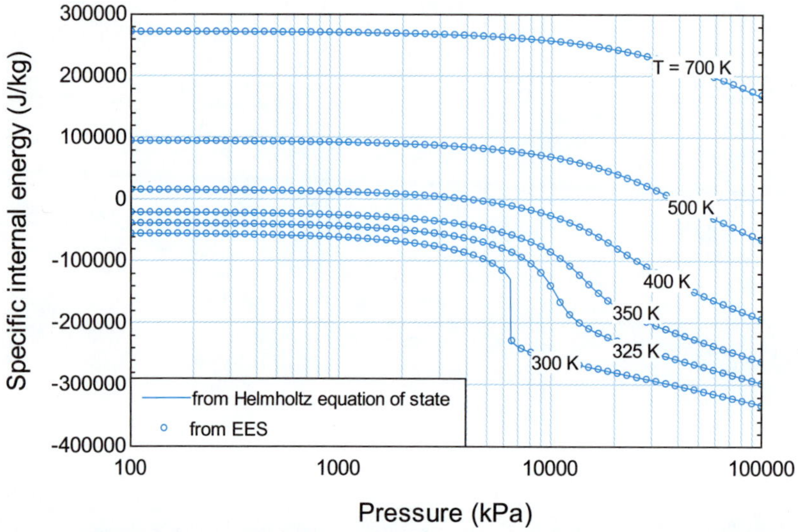

Figure 2: Specific internal energy as a function of pressure predicted using the Helmholtz equation of state and from EES' internal correlations.

g) Derive an expression for the specific entropy as a function of α and its partial derivatives with respect to τ and δ.

Eq. (20) can be solved for s:

$$s = \frac{u}{T} - \alpha R \tag{30}$$

Substituting Eq. (28) into Eq. (30) leads to:

$$s = \frac{R\, T_{crit}}{T}\left(\frac{\partial \alpha}{\partial \tau}\right)_{\delta} - \alpha R \tag{31}$$

or

$$s = R\,\tau\left(\frac{\partial \alpha}{\partial \tau}\right)_{\delta} - \alpha R \tag{32}$$

h) Implement your solution in EES in order to estimate the specific entropy at $v = 0.01\ \text{m}^3/\text{kg}$ and $T = 350$ K. Compare your estimate with the specific entropy obtained using the entropy function in EES.

The expressions for α_{IG} and α_{res} are copied from Maple and pasted into EES. Equation (4) is used to compute α and Eq. (32) is used to compute s. The EES function entropy is also used to estimate the specific entropy (s_{EES}).

```
"ideal gas component of alpha"
alpha_IG = ln(delta)+a[1]+a[2]*tau+a[3]*ln(tau)+a[4]*ln(1-exp(-tau*theta[4]))&
    +a[5]*ln(1-exp(-tau*theta[5]))&
    +a[6]*ln(1-exp(-tau*theta[6]))+a[7]*ln(1-exp(-tau*theta[7]))+a[8]*ln(1-exp(-tau*theta[8]))

"residual component of alpha"
alpha_res = n[1]*delta*tau^.25+n[2]*delta*tau^1.25+n[3]*delta*tau^1.5+n[4]*delta^3*tau^.25&
    +n[5]*delta^7*tau^.875+n[6]*delta*tau^2.375*exp(-delta)+n[7]*delta^2*tau^2*exp(-delta)&
    +n[8]*delta^5*tau^2.125*exp(-delta)+n[9]*delta*tau^3.5*exp(-delta^2)&
    +n[10]*delta*tau^6.5*exp(-delta^2)+n[11]*delta^4*tau^4.75*exp(-delta^2)&
    +n[12]*delta^2*tau^12.5*exp(-delta^3)

alpha=alpha_IG+alpha_res                "reduced Helmholtz free energy"
s=R*tau*dalphadtau_delta-alpha*R        "specific entropy from reduced Helmholtz equation of state"
s_EES=entropy(F$,v=v,T=T)               "specific entropy, from EES correlation"
```

which results in $s = -700.6$ J/kg-K and $s_{EES} = -700.6$ J/kg-K.

i) Plot the specific entropy as a function of pressure for several values of temperature. Also show the value obtained using the EES property correlations.

Figure 3 illustrates the specific entropy of carbon dioxide as a function of pressure for various values of temperature; both the results from the entropy function in EES and from using the Helmholtz equation of state are shown.

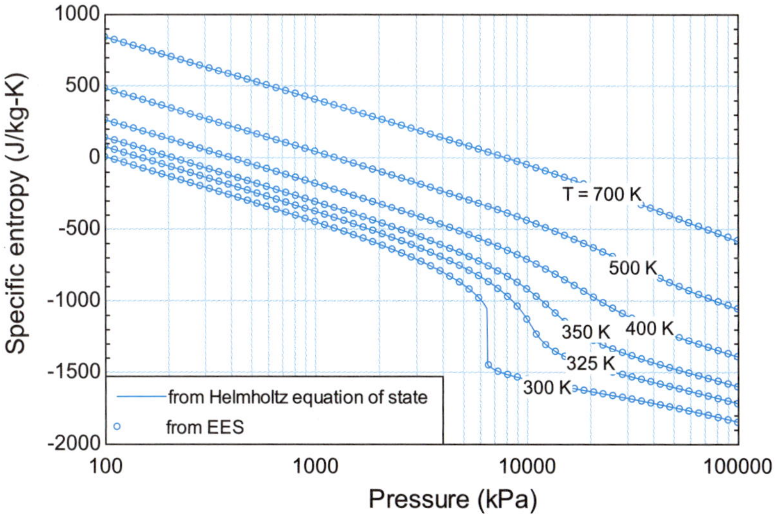

Figure 3: Specific entropy as a function of pressure predicted using the Helmholtz equation of state and from EES' property correlations.

10.3 Derived Thermodynamic Properties

Section 10.2 discusses the fundamental property relation and shows that it can be expressed in four equivalent ways for a pure fluid. Table 10-2 summarizes these four fundamental property relations. Each fundamental property relation is the total differential of a function that is a complete equation of state. These functions are listed in the second column of Table 10-2. If any one of these complete equations of state is known, then all thermodynamic property information can be determined. However, all four of the complete equations of state involve at least one thermodynamic property that is not directly measurable. For example, to determine the function $u = u(s, v)$, we would need values of u, s, and v; only v is directly measurable. This section discusses methods for deriving the values of properties such as u, h, and s that are not measurable from properties such as P, T, and v that are measurable.

Table 10-2: Fundamental equations of state and the corresponding Maxwell relations

Fundamental property relation	Functional form of complete equation of state	Maxwell relation
$du = T\,ds - P\,dv$	$u = u(s, v)$	$\left(\frac{\partial T}{\partial v}\right)_s = -\left(\frac{\partial P}{\partial s}\right)_v$
$dh = T\,ds + v\,dP$	$h = h(s, P)$	$\left(\frac{\partial T}{\partial P}\right)_s = \left(\frac{\partial v}{\partial s}\right)_P$
$da = -s\,dT - P\,dv$	$a = a(T, v)$	$\left(\frac{\partial s}{\partial v}\right)_T = \left(\frac{\partial P}{\partial T}\right)_v$
$dg = -s\,dT + v\,dP$	$g = g(T, P)$	$\left(\frac{\partial s}{\partial P}\right)_T = -\left(\frac{\partial v}{\partial T}\right)_P$

10.3.1 Maxwell's Relations

The key to determining the thermodynamic properties that are not directly measurable is the recognition that the functions in Table 10-2 (u, h, a, and g) are *exact differentials*. If a variable $f(x, y)$ is an exact differential of the variables x and y then f has a unique value at given values of x and y. This characteristic is entirely consistent with the concept of a state property. For example, we know that u has a unique value at any specified values of s and v. The cyclic integral of an exact differential must be zero and all of the thermodynamic properties exhibit this characteristic. For example, if we change s and v in some arbitrary manner and then change them back to their original values along some different path, the property u will have the same value at the end of the process that it had at the beginning. The change in the value of an exact differential is independent of the path taken during the change. Thermodynamic properties also exhibit this behavior. We know that the change in specific internal energy, for example, depends only on the initial and final values of two specific properties (i.e., on the initial and final states), but not on the path taken as it passes from the initial state to the final state. Therefore, the mathematical characteristics of an exact differential are entirely consistent with our understanding of how a thermodynamic property must behave.

A necessary and sufficient requirement for an exact differential is that the order in which the function is differentiated is irrelevant. For example, if we determine the second derivative of f with respect to x and y, it does not matter if we first differentiate with respect to x and then y or first with respect to y and then x.

$$\frac{\partial}{\partial x}\left[\left(\frac{\partial f}{\partial y}\right)_x\right]_y = \frac{\partial}{\partial y}\left[\left(\frac{\partial f}{\partial x}\right)_y\right]_x \tag{10-90}$$

We can apply this requirement to each of the fundamental property relations in Table 10-2, since they are all exact differentials. For example, the functional form for the complete equation of state associated with specific internal energy is $u = u(s, v)$ and the corresponding fundamental property relation is:

$$du = T\,ds - P\,dv \tag{10-91}$$

Examination of Eq. (10-91) shows that the first partial differential of u with respect to s at constant v is T:

$$\left(\frac{\partial u}{\partial s}\right)_v = T \tag{10-92}$$

The second partial derivative of u with respect first to s and then to v is therefore:

$$\frac{\partial}{\partial v}\left[\underbrace{\left(\frac{\partial u}{\partial s}\right)_v}_{T}\right]_s = \left(\frac{\partial T}{\partial v}\right)_s \tag{10-93}$$

The order in which we determine the second derivatives of u is irrelevant because u is an exact differential of s and v. Therefore, we could have taken the first derivative with respect to v and then the second derivative with respect to s. Examination of Eq. (10-91) shows that the first partial derivative of u with respect to v at constant s is $-P$:

$$\left(\frac{\partial u}{\partial v}\right)_s = -P \tag{10-94}$$

The second partial derivative of u with respect to first v and then s is:

$$\frac{\partial}{\partial s}\left[\underbrace{\left(\frac{\partial u}{\partial v}\right)_s}_{-P}\right]_v = -\left(\frac{\partial P}{\partial s}\right)_v \tag{10-95}$$

Since the order of differentiation is irrelevant for exact differentials:

$$\frac{\partial}{\partial v}\left[\left(\frac{\partial u}{\partial s}\right)_v\right]_s = \frac{\partial}{\partial s}\left[\left(\frac{\partial u}{\partial v}\right)_s\right]_v \tag{10-96}$$

Substituting Eqs. (10-93) and (10-95) into Eq. (10-96) leads to:

$$\left(\frac{\partial T}{\partial v}\right)_s = -\left(\frac{\partial P}{\partial s}\right)_v \tag{10-97}$$

This requirement can be applied to each of the functional forms in Table 10-2. The result, collectively, is called *Maxwell's relations* and is provided in the third column of Table 10-2. Note that the most useful Maxwell relations appear in last two rows of the table.

These Maxwell relations express the dependence of specific entropy on specific volume and pressure in terms of the measurable quantities v, T, and P.

$$\left(\frac{\partial s}{\partial v}\right)_T = \left(\frac{\partial P}{\partial T}\right)_v \tag{10-98}$$

$$\left(\frac{\partial s}{\partial P}\right)_T = -\left(\frac{\partial v}{\partial T}\right)_P \tag{10-99}$$

10.3.2 Calculus Relations for Partial Derivatives

Maxwell's relations provide a tool that is useful for determining the properties u, h, and s based on experimental measurements. Calculus provides some additional relations that are also required; these are reviewed in this section.

Partial derivatives are invertible; therefore, if f is a function of x and y then:

$$\left(\frac{\partial f}{\partial x}\right)_y = \frac{1}{\left(\frac{\partial x}{\partial f}\right)_y} \tag{10-100}$$

A relation for a constant potential is often needed. Taking again $f = f(x, y)$, the total derivative of f can be expressed as:

$$df = \left(\frac{\partial f}{\partial x}\right)_y dx + \left(\frac{\partial f}{\partial y}\right)_x dy \tag{10-101}$$

But suppose f is constant. In this case, $df = 0$:

$$\left(\frac{\partial f}{\partial x}\right)_y dx + \left(\frac{\partial f}{\partial y}\right)_x dy = 0 \tag{10-102}$$

Equation (10-102) can be reduced to:

$$\left(\frac{\partial f}{\partial x}\right)_y \left(\frac{\partial x}{\partial y}\right)_f \left(\frac{\partial y}{\partial f}\right)_x = -1 \tag{10-103}$$

The most useful relation for property manipulations is the chain rule, which allows a partial derivative expressed in terms of one set of independent variables to be transformed to a different set of independent variables. For example, suppose that an equation is of the form $f(x, y)$ but x and y are each functions of other variables, a and b, i.e., $x(a, b)$ and $y(a, b)$. You may wish to express the partial derivative of f with respect to a and b rather than x and y. The chain rule allows this change in independent variables:

$$\left(\frac{\partial f}{\partial a}\right)_b = \left(\frac{\partial f}{\partial x}\right)_y \left(\frac{\partial x}{\partial a}\right)_b + \left(\frac{\partial f}{\partial y}\right)_x \left(\frac{\partial y}{\partial a}\right)_b \tag{10-104}$$

The chain rule can also be used to expand the partial derivative to include another variable. For example:

$$\left(\frac{\partial f}{\partial a}\right)_b = \left(\frac{\partial f}{\partial x}\right)_b \left(\frac{\partial x}{\partial a}\right)_b \tag{10-105}$$

10.3.3 Derived Relations for *u*, *h*, and *s*

Now that we have established Maxwell's relations and reviewed some calculus tools, it is time to return to the problem of how the values of u, h, and s can be determined, even though it is not possible to directly measure these quantities.

We can measure pressure, specific volume, and temperature. According to the phase rule in Eq. (10-1), we only need to know two properties in order to fix the state of a pure single-phase substance. Temperature and pressure are perhaps the two most convenient properties that can be used to fix the state of a substance. Suppose that we want to determine specific enthalpy as a function of temperature and pressure. Mathematically, we are looking for a function of the form:

$$h = h(T, P) \tag{10-106}$$

The fundamental property relation for specific enthalpy in Table 10-2 is $h(s, P)$ and the fundamental property relation associated with specific enthalpy is:

$$dh = T\,ds + v\,dP \tag{10-107}$$

However, this relation is not convenient because specific entropy is not directly measurable. The total derivative of Eq. (10-106) is:

$$dh = \underbrace{\left(\frac{\partial h}{\partial T}\right)_P}_{c_P} dT + \left(\frac{\partial h}{\partial P}\right)_T dP \tag{10-108}$$

The first partial derivative in Eq. (10-108) should be familiar. It is, by definition, the specific heat capacity at constant pressure:

$$c_P = \left(\frac{\partial h}{\partial T}\right)_P \tag{10-109}$$

Because c_P is measurable, the first term on the right side of Eq. (10-108) can be integrated in order to evaluate the change in specific enthalpy for processes that occur at constant pressure. We need to find a method for evaluating the second term on the right side of Eq. (10-108), the partial derivative of specific enthalpy with respect to pressure at constant temperature. Equation (10-107) can be used to determine the partial derivatives of h with respect to s and P. In order to obtain $h(T, P)$ it is necessary to change from independent variables s and P to independent variables T and P. This change in independent variables is provided by the chain rule in Eq. (10-104). In this case, the function $f = h$ and the original independent variables are $x = s$ and $y = P$. We need to transform to the new set of independent variables $a = P$, $b = T$. Applying the chain rule provides:

$$\left(\frac{\partial h}{\partial P}\right)_T = \underbrace{\left(\frac{\partial h}{\partial s}\right)_P}_{T} \left(\frac{\partial s}{\partial P}\right)_T + \underbrace{\left(\frac{\partial h}{\partial P}\right)_s}_{v} \left(\frac{\partial P}{\partial P}\right)_T \tag{10-110}$$

It may appear that we have complicated the problem rather than simplified it. When we started we only needed to evaluate one partial derivative whereas Eq. (10-110) involves four. Fortunately, the four partial derivatives that appear on the right side of Eq. (10-110) can easily be determined. The reason for using the fundamental property relation, $h(s, P)$, in order to begin the application of the chain rule is that we know the partial derivatives that result from this relation. Examination of Eq. (10-107) shows that:

$$\left(\frac{\partial h}{\partial s}\right)_P = T \tag{10-111}$$

and

$$\left(\frac{\partial h}{\partial P}\right)_s = v \tag{10-112}$$

These partial derivatives can be substituted into Eq. (10-110):

$$\left(\frac{\partial h}{\partial P}\right)_T = T\underbrace{\left(\frac{\partial s}{\partial P}\right)_T}_{-\left(\frac{\partial v}{\partial T}\right)_P} + v\underbrace{\left(\frac{\partial P}{\partial P}\right)_T}_{1} \tag{10-113}$$

The first partial derivative on the right side of Eq. (10-113) is one of the four Maxwell relations in Table 10-2:

$$\left(\frac{\partial s}{\partial P}\right)_T = -\left(\frac{\partial v}{\partial T}\right)_P \tag{10-114}$$

The second partial derivative on the right side of Eq. (10-113) must be unity. Therefore, Eq. (10-113) reduces to

$$\left(\frac{\partial h}{\partial P}\right)_T = -T\left(\frac{\partial v}{\partial T}\right)_P + v \tag{10-115}$$

Substituting Eqs. (10-109) and (10-115) into Eq. (10-108) provides the final relation for the differential of $h(T, P)$:

$$\boxed{dh = c_P\, dT + \left[v - T\left(\frac{\partial v}{\partial T}\right)_P\right] dP} \tag{10-116}$$

Note that all of the quantities on the right side of Eq. (10-116) are measurable. The derivative of specific volume with respect to temperature at constant pressure can be evaluated using an equation of state. Integration of Eq. (10-116) with respect to temperature and pressure provides $h(T, P)$; this process is discussed in Section 10.4.

We can use the same procedure to develop an expression that provides specific entropy as a function of temperature and pressure, $s = s(T, P)$. The total derivative of this function is:

$$ds = \left(\frac{\partial s}{\partial T}\right)_P dT + \left(\frac{\partial s}{\partial P}\right)_T dP \tag{10-117}$$

The fundamental property relation expresses specific entropy as a function of specific enthalpy and pressure, $s(h, P)$. Rearranging Eq. (10-107) leads to:

$$ds = \frac{1}{T}dh - \frac{v}{T}dP \tag{10-118}$$

The first partial derivative on the right side of Eq. (10-117) is determined by applying the chain rule in Eq. (10-104) to change independent variables from $s(h, P)$ to $s(T, P)$; therefore $f = s$, $x = h$ and $y = P$, $a = T$ and $b = P$. The chain rule:

$$\left(\frac{\partial f}{\partial a}\right)_b = \left(\frac{\partial f}{\partial x}\right)_y \left(\frac{\partial x}{\partial a}\right)_b + \left(\frac{\partial f}{\partial y}\right)_x \left(\frac{\partial y}{\partial a}\right)_b \tag{10-119}$$

with these substitutions becomes:

$$\left(\frac{\partial s}{\partial T}\right)_P = \underbrace{\left(\frac{\partial s}{\partial h}\right)_s}_{1/T} \underbrace{\left(\frac{\partial h}{\partial T}\right)_P}_{c_P} + \underbrace{\left(\frac{\partial s}{\partial P}\right)_h}_{-v/T} \underbrace{\left(\frac{\partial P}{\partial T}\right)_P}_{0} \tag{10-120}$$

The first and third partial derivatives on the right side of Eq. (10-120) are found by inspection of Eq. (10-118). The second partial derivative is, by definition, the specific heat capacity at constant pressure. The last partial derivative asks how does pressure change with temperature when pressure is held constant; this partial derivative must be identically zero. As a result, Eq. (10-120) reduces to:

$$\left(\frac{\partial s}{\partial T}\right)_P = \frac{c_P}{T} \tag{10-121}$$

The second partial derivative on the right side of Eq. (10-117) is a Maxwell relation, Eq. (10-114). Substituting Eqs. (10-121) and (10-114) into Eq. (10-117) results in

$$\boxed{ds = \frac{c_P}{T} dT - \left(\frac{\partial v}{\partial T}\right)_P dP} \tag{10-122}$$

All of the variables appearing the right side of Eq. (10-122) are measurable. Therefore, the value of s at any specified values of T and P can be found by integration, as discussed in Section 10.4. If we apply these same mathematical operations to determine $u = u\,(T, v)$ and $s = s\,(T, v)$ the results are:

$$\boxed{du = c_v\, dT + \left[T\left(\frac{\partial P}{\partial T}\right)_v - P\right] dv} \tag{10-123}$$

and:

$$\boxed{ds = \frac{c_v}{T} dT + \left(\frac{\partial P}{\partial T}\right)_v dv} \tag{10-124}$$

As a simple example, we can apply Eqs. (10-116) and (10-122) to the case where a substance obeys the ideal gas law in order to determine its specific enthalpy and specific entropy. The ideal gas law is:

$$v = \frac{R\,T}{P} \tag{10-125}$$

The partial derivative of the specific volume predicted by the ideal gas law with respect to temperature at constant pressure is:

$$\left(\frac{\partial v}{\partial T}\right)_P = \frac{R}{P} \tag{10-126}$$

Substituting Eqs. (10-125) and (10-126) into Eq. (10-115) leads to:

$$\left(\frac{\partial h}{\partial P}\right)_T = -T\left(\frac{\partial v}{\partial T}\right)_P + v = -T\frac{R}{P} + \frac{R\,T}{P} = 0 \quad \text{for an ideal gas} \tag{10-127}$$

The partial derivative of specific enthalpy with respect to pressure for an ideal gas is zero. The specific enthalpy of an ideal gas only depends upon temperature; this conclusion is consistent with the discussion in Section 4.2.3. Substituting this result into Eq. (10-116) simplifies it to:

$$dh = c_P\, dT \quad \text{for an ideal gas} \tag{10-128}$$

Substituting Eq. (10-126) into Eq. (10-122) provides:

$$ds = \frac{c_P}{T} dT - \frac{R}{P} dP \quad \text{for an ideal gas} \tag{10-129}$$

Equations (10-128) and (10-129) are identical to Eqs. (4-31) and (6-27), respectively, which are used in Chapter 4 and Chapter 6 in order to determine the specific enthalpy and specific entropy of an ideal gas.

EXAMPLE 10.3-1: ISOTHERMAL COMPRESSION PROCESS

During a test of a refrigeration compressor, ammonia is compressed isothermally, as shown in Figure 1.

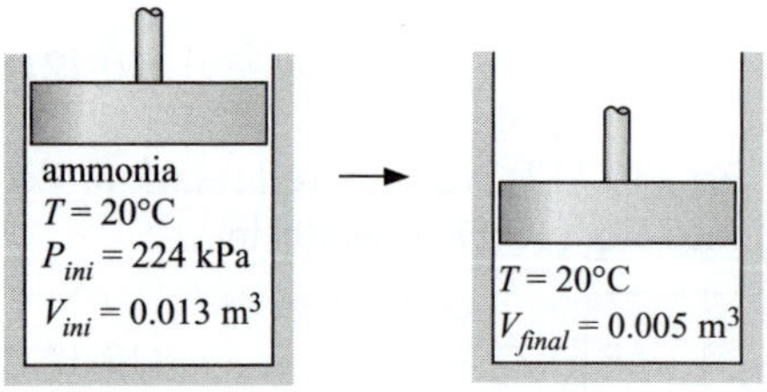

Figure 1: Isothermal compression of ammonia.

The ammonia is contained in a piston cylinder with initial volume $V_{ini} = 0.013$ m^3 and final volume $V_{final} = 0.005$ m^3. The temperature of the ammonia is maintained at $T = 20^\circ$C during this process. The initial pressure of the ammonia is $P_{ini} = 224$ kPa.

a) Use the Redlich-Kwong-Soave equation of state to estimate the heat transfer from the ammonia and the work done on the ammonia assuming that the compression process is reversible.

We need to determine the relationship between pressure, volume and temperature for ammonia many times in order to evaluate the property changes occurring in this process. These calculations will be done in a subprogram called RKS that implements the RKS equation of state using the relations provided by Eqs. (10-54) to (10-58) in Section 10.1.3.

```
$UnitSystem SI Radian Mass J K Pa
Subprogram RKS(F$,P,T:v)                            "implements the RKS equation of state"
    "Inputs:"
    "F$: fluid name"
    "P: pressure (Pa)"
    "T: temperature (K)"

    "Outputs:"
    "v: specific volume (m^3/kg)"
    R=R#/MolarMass(F$)                              "ideal gas constant"
    T_crit=T_crit(F$)                               "critical temperature"
    P_crit=P_crit(F$)                               "critical pressure"
    b=0.08664*R*T_crit/P_crit                       "parameter b in equation of state"
    omega=AcentricFactor(F$)                        "acentric factor"
    m=0.480+1.574*omega-0.176*omega^2               "parameter m used in alpha"
    T_r=T/T_crit                                    "reduced temperature"
    sqrt(alpha)=1+m*(1-sqrt(T_r))                   "parameter alpha used in a"
    a=0.42747*alpha*R^2*T_crit^2/P_crit             "parameter a in equation of state"
    P=R*T/(v-b)-a/(v*(v+b))                         "RKS equation of state"
end
```

The inputs are entered in EES.

```
Vol_ini=0.013 [m^3]                          "initial volume"
T=converttemp(C,K,20 [C])                    "temperature"
P_ini=224 [kPa]*convert(kPa,Pa)              "initial pressure"
Vol_final=0.005 [m^3]                        "final volume"
F$='Ammonia'                                 "fluid"
```

The initial pressure and temperature of the ammonia are known. The subprogram RKS is used to determine the initial specific volume (v_{ini}). The mass of ammonia is computed:

$$m = \frac{V_{ini}}{v_{ini}}$$

```
Call RKS(F$,P_ini,T:v_ini)                   "initial specific volume"
m=Vol_ini/v_ini                              "mass"
```

which results in $m = 0.02073$ kg. The final specific volume of the ammonia is computed according to:

$$v_{final} = \frac{V_{final}}{m}$$

The final temperature and specific volume are used with the subprogram RKS to determine the final pressure (P_{final}).

```
v_final=Vol_final/m                          "final specific volume"
Call RKS(F$,P_final,T:v_final)               "final pressure"
```

which results in $P_{final} = 564.9$ kPa. An entropy balance on the ammonia assuming a reversible process (and therefore, no entropy generation) is:

$$0 = \frac{Q_{out}}{T} + m\,\Delta s \tag{1}$$

where Q_{out} is the heat transfer out of the ammonia and Δs is the change in the specific entropy of the ammonia ($s_{final} - s_{ini}$). An energy balance on the ammonia is:

$$W_{in} = Q_{out} + m\,\Delta u \tag{2}$$

where W_{in} is the work done on the ammonia and Δu is change in the specific internal energy of the ammonia ($u_{final} - u_{ini}$). In order to evaluate Eqs. (1) and (2), it is necessary to compute the change in specific entropy and the change in specific internal energy of the ammonia. The change in specific internal energy is computed according to:

$$\Delta u = \int_{u_{ini}}^{u_{final}} du = \int_{T_{ini}}^{T_{final}} \left(\frac{\partial u}{\partial T}\right)_v dT + \int_{v_{ini}}^{v_{final}} \left(\frac{\partial u}{\partial v}\right)_T dv \tag{3}$$

Because the temperature of the ammonia does not change, Eq. (3) reduces to:

$$\Delta u = \int_{v_{ini}}^{v_{final}} \left(\frac{\partial u}{\partial v}\right)_T dv \tag{4}$$

The change in specific entropy is computed according to:

$$\Delta s = \int_{s_{ini}}^{s_{final}} ds = \int_{T_{ini}}^{T_{final}} \left(\frac{\partial s}{\partial T}\right)_v dT + \int_{v_{ini}}^{v_{final}} \left(\frac{\partial s}{\partial v}\right)_T dv \tag{5}$$

Because the temperature of the ammonia does not change, Eq. (5) reduces to:

$$\Delta s = \int_{v_{ini}}^{v_{final}} \left(\frac{\partial s}{\partial v}\right)_T dv \tag{6}$$

The RKS equation of state is used in conjunction with Eqs. (10-123) and (10-124) in order to evaluate the integrals in Eqs. (4) and (6). According to Eq. (10-123):

$$\left(\frac{\partial u}{\partial v}\right)_T = \left[T\left(\frac{\partial P}{\partial T}\right)_v - P\right] \tag{7}$$

and according to Eq. (10-124):

$$\left(\frac{\partial s}{\partial v}\right)_T = \left(\frac{\partial P}{\partial T}\right)_v \tag{8}$$

Equations (7) and (8) show that it will be necessary to evaluate the partial derivative of pressure with respect to temperature at constant specific volume in order to carry out the integrations in Eqs. (4) and (6). The value of this partial derivative will change with specific volume and the specific volume will vary between v_{ini} and v_{final} during the integrations in Eqs. (4) and (6). To start the process, set the specific volume (v) equal to an arbitrary value; we will remove this assignment when we do the integration. The subprogram RKS is used to determine the pressure at v and T.

```
v=v_ini                    "value of the integration variable – remove for integration"
Call RKS(F$,P,T:v)         "determine pressure"
```

The partial derivative of pressure with respect to temperature at constant specific volume is estimated numerically according to:

$$\left(\frac{\partial P}{\partial T}\right)_v \approx \frac{P_{T+dT,v} - P_{T-dT,v}}{2\,dT}$$

where dT is a small, but finite, change in the temperature. Here, dT is defined as being 0.1 K. Equations (7) and (8) are used to compute the integrands of Eqs. (4) and (6).

```
dT=0.1 [K]                     "differential value of temperature change"
Call RKS(F$,P_plus,T+dT:v)     "pressure at T+dT"
Call RKS(F$,P_minus,T-dT:v)    "pressure at T-dT"
dPdT_v=(P_plus-P_minus)/(2*dT) "partial differential of P with respect to T at constant v"
dsdv_T=dPdT_v                  "partial differential of s with respect to v at constant T"
dudv_T=T*dPdT_v-P              "partial differential of u with respect to v at constant T"
```

The specified value of the integration variable, v, is commented out and the Integral command in EES is used to evaluate the integrals in Eqs. (4) and (6). Use of the Integral command is discussed in Appendix G.5. The integration algorithm in EES provides several options that can be set using the Preferences dialog in the Options menu or with the $IntegralAutoStep directive. The directive included below uses a fixed step size with 200 steps and turns off the Richardson extrapolation method.

```
{v=v_ini}                                          "value of the integration variable – remove for integration"
Deltau=Integral(dudv_T,v,v_ini,v_final)            "change in specific internal energy"
Deltas=Integral(dsdv_T,v,v_ini,v_final)            "change in specific entropy"
$IntegralAutoStep Fixed=200 Richardson=Off         "control the numerical integration"
```

The values of Δu and Δs are substituted into Eqs. (1) and (2) in order to determine the heat and work.

```
0=Q_out/T+m*Deltas                                 "heat transfer"
W_in=m*Deltau+Q_out                                "work transfer"
```

which results in $Q_{out} = 2912$ J and $W_{in} = 2747$ J.

b) Assume that ammonia behaves as an ideal gas. Estimate the heat transfer from the ammonia and the work done on the ammonia during the process.

The ideal gas constant is computed according to:

$$R = \frac{R_{univ}}{MW}$$

where MW is the molecular weight, determined using the MolarMass function. The initial specific volume is:

$$v_{ini} = \frac{R\,T}{P_{ini}}$$

and the mass of ammonia is:

$$m = \frac{V_{ini}}{v_{ini}}$$

The final specific volume is:

$$v_{final} = \frac{V_{final}}{m}$$

```
"Calculations assuming ideal gas"
R=R#/MolarMass(F$)                                 "ideal gas constant"
v_ini_IG=R*T/P_ini                                 "initial specific volume"
m_IG=Vol_ini/v_ini_IG                              "mass"
v_final_IG=Vol_final/m_IG                          "final specific volume"
```

The specific internal energy of an ideal gas is only a function of temperature. Therefore, the change in the specific internal energy is zero, $\Delta u = 0$. The change in the specific entropy is calculated from:

$$\Delta s = R \ln\left(\frac{v_{final}}{v_{ini}}\right)$$

Equations (1) and (2) are used to compute the heat and work.

```
Deltau_IG=0 [J/kg]                                 "change in specific internal energy"
Deltas_IG=R*ln(v_final_IG/v_ini_IG)                "change in specific entropy"
0=Q_out_IG/T+m_IG*Deltas_IG                        "heat transfer"
W_in_IG=m_IG*Deltau_IG+Q_out_IG                    "work transfer"
```

which leads to $Q_{out} = 2782$ J and $W_{in} = 2782$ J; these answers are 4.5% and 1.3%, respectively, different from those computed in part (a) using the RKS equation of state.

c) Use the internal property correlations in EES in order to estimate the heat transfer from the ammonia and the work transfer to the ammonia during the process.

The initial state of the ammonia is specified by the temperature and pressure. The initial specific volume, specific internal energy, and specific entropy (v_{ini}, u_{ini}, and s_{ini}) are computed. The mass of ammonia is computed according to:

$$m = \frac{V_{ini}}{v_{ini}}$$

```
v_ini_EES=volume(F$,T=T,P=P_ini)          "initial specific volume"
u_ini_EES=intenergy(F$,T=T,P=P_ini)       "initial specific internal energy"
s_ini_EES=entropy(F$,T=T,P=P_ini)         "initial specific entropy"
m_EES=Vol_ini/v_ini_EES                   "mass"
```

The final specific volume is:

$$v_{final} = \frac{V_{final}}{m}$$

The final state of the ammonia is specified by the specific volume and temperature. The final specific internal energy and specific entropy (u_{final} and s_{final}) are determined and used to compute the change in the specific internal energy and specific entropy.

$$\Delta u = u_{final} - u_{ini}$$

$$\Delta s = s_{final} - s_{ini}$$

Equations (1) and (2) are used to compute the heat transfer and work transfer.

```
v_final_EES=Vol_final/m_EES                      "final specific volume"
u_final_EES=intenergy(F$,T=T,v=v_final_EES)      "final specific internal energy"
s_final_EES=entropy(F$,T=T,v=v_final_EES)        "final specific entropy"
Deltau_EES=u_final_EES-u_ini_EES                 "change in specific internal energy"
Deltas_EES=s_final_EES-s_ini_EES                 "change in specific entropy"
Q_out_EES=-T*m_EES*Deltas_EES                    "heat transfer"
W_in_EES=m_EES*Deltau_EES+Q_out_EES              "work transfer"
```

which solves to $Q_{out} = 3107$ J and $W_{in} = 2734$ J; these answers are 6.7% and 0.5%, respectively, different from those computed in part (a) using the RKS equation of state.

The heat and work calculated with the RKS equation of state agree more closely with the EES values than do the results obtained with the ideal gas law. Recalculating part (a) with a reduced step size does not change the answers (to within 5 significant figures), so integration error is apparently not the cause of the discrepancy. The discrepancy is due to a limitation in the RKS equation of state for ammonia. Note that it is much more difficult to obtain accurate values for specific internal energy and specific entropy from the equation of state than it is to obtain accurate values of specific volume. The calculations for u and s require an accurate estimate of partial derivative of the pressure with respect to temperature at constant specific volume, in addition to an accurate relation between pressure, specific volume and temperature. The need for high accuracy derivatives is the primary reason that higher order equations of state are needed.

10.3.4 Derived Relations for other Thermodynamic Quantities

Using the fundamental property relation and calculus, we were able to derive the following relationships in Section 10.3.3:

$$du = c_v\, dT + \left[T\left(\frac{\partial P}{\partial T}\right)_v - P\right] dv \tag{10-130}$$

$$dh = c_P\, dT + \left[v - T\left(\frac{\partial v}{\partial T}\right)_P\right] dP \tag{10-131}$$

$$ds = \frac{c_v}{T} dT + \left(\frac{\partial P}{\partial T}\right)_v dv \tag{10-132}$$

and

$$ds = \frac{c_P}{T} dT - \left(\frac{\partial v}{\partial T}\right)_P dP \tag{10-133}$$

These relations are all very useful because they allow the properties specific internal energy, specific enthalpy, and specific entropy, which are not measurable, to be calculated through the integration of measurable properties. There are many other useful properties that can be obtained. For example, the *Joule-Thomson coefficient*, indicates the magnitude of the temperature change that is induced by an *isenthalpic* (i.e., constant enthalpy) process such as flow through a valve or restriction. The Joule-Thomson coefficient is defined as the partial derivative of temperature with respect to pressure at constant specific enthalpy.

$$\mu_{JT} = \left(\frac{\partial T}{\partial P}\right)_h \tag{10-134}$$

The Joule-Thomson coefficient was introduced in Section 9.5.2 because the Joule-Thomson refrigeration cycle can only operate with fluids that have a positive value of μ_{JT}. The flow of such a fluid through a valve or other flow restriction that induces a drop in pressure will result in a temperature drop.

It is possible to determine a relation for the Joule-Thomson coefficient, or any thermodynamic property, in terms of measurable properties. In general, the property manipulations needed to arrive at the intended result are not obvious and occasionally it may seem that you are going around in circles. However, the process is simplified when it is done using the procedure for manipulating property relations that is summarized in Table 10-3.

Table 10-3: Rules for manipulating property relations.

Step 1:	If any of the properties *u*, *a*, *h*, or *g* appear as a constraint on the derivative then use Eq. (10-103) to bring it inside the derivative.
Step 2:	If the derivative contains *u*, *a*, *h*, or *g*, eliminate it using a fundamental property relationship or by applying the definitions of c_P and c_v.
Step 3:	(All references to *u*, *a*, *h*, or *g* have been eliminated by Steps 1 and 2) Eliminate references to *s* using Eq. (10-103), the Maxwell's relations in Table 10-2, and/or the definitions of c_P and c_v. Note that if *s* appears alone then it cannot be removed.

We can apply the rules in Table 10-3 to identify an expression for the Joule-Thomson coefficient in terms of measurable quantities. Step 1 utilizes Eq. (10-103) to remove specific enthalpy as a constraint.

$$\mu_{JT} = \left(\frac{\partial T}{\partial P}\right)_h = -\left(\frac{\partial h}{\partial P}\right)_T \left(\frac{\partial T}{\partial h}\right)_P \tag{10-135}$$

Step 2 eliminates references to h in the partial derivatives on the right side of Eq. (10-135). Equation (10-131) provides an expression for the partial derivative of specific enthalpy with respect to pressure at constant temperature. The second partial derivative on the right side of Eq. (10-135) is by definition the inverse of the specific heat capacity, c_P. With these substitutions, Eq. (10-135) becomes:

$$\mu_{JT} = -\frac{\left[v - T\left(\frac{\partial v}{\partial T}\right)_P\right]}{c_P} \tag{10-136}$$

Step 3 is not needed because specific entropy does not appear on the right side of Eq. (10-136). Equation (10-136) only involves measurable properties and their derivatives. Therefore, the value of the Joule-Thomson coefficient can be determined if the specific heat capacity at constant pressure and an equation of state are known for the substance.

EXAMPLE 10.3-2: SPEED OF SOUND OF CARBON DIOXIDE

When a pressure disturbance occurs in a compressible fluid, the velocity of the disturbance wave that results is a thermodynamic property referred to as the speed of sound (c). The speed of sound is an easily measured thermodynamic quantity that is discussed in Section 16.1. The speed of sound is defined according to:

$$c = \sqrt{\left(\frac{\partial P}{\partial \rho}\right)_s} \tag{1}$$

where P is pressure, ρ is the fluid density, and s is specific entropy.

a) Derive an equation for the speed of sound in terms of measurable properties.

We will start by expressing the speed of sound in terms of specific volume rather than density. The partial derivative of pressure with respect to density at constant specific entropy can be written as:

$$\left(\frac{\partial P}{\partial \rho}\right)_s = \left(\frac{\partial P}{\partial v}\right)_s \left(\frac{\partial v}{\partial \rho}\right)_s \tag{2}$$

The derivative of specific volume with respect to density is:

$$\frac{dv}{d\rho} = -\frac{1}{\rho^2} = -v^2 \tag{3}$$

Substituting Eq. (3) into Eq. (2) leads to:

$$\left(\frac{\partial P}{\partial \rho}\right)_s = -v^2 \left(\frac{\partial P}{\partial v}\right)_s \tag{4}$$

The partial derivative in Eq. (4) does not involve u, a, h, or g; therefore, we can skip Steps 1 and 2 in Table 10-3. Partial derivatives of P with respect to v are more conveniently manipulated when they are expressed in terms of temperature. This transformation is accomplished using the chain rule in Eq. (10-105).

$$\left(\frac{\partial P}{\partial v}\right)_s = \left(\frac{\partial P}{\partial T}\right)_s \left(\frac{\partial T}{\partial v}\right)_s \tag{5}$$

Substituting Eq. (5) into Eq. (4) provides:

$$\left(\frac{\partial P}{\partial \rho}\right)_s = -v^2 \left(\frac{\partial P}{\partial T}\right)_s \left(\frac{\partial T}{\partial v}\right)_s \tag{6}$$

Equation (10-103) is applied to the first partial derivative on the right side of Eq. (6):

$$\left(\frac{\partial P}{\partial T}\right)_s = -\underbrace{\left(\frac{\partial s}{\partial T}\right)_P}_{\frac{c_P}{T}} \underbrace{\left(\frac{\partial P}{\partial s}\right)_T}_{-\left(\frac{\partial T}{\partial v}\right)_P} \tag{7}$$

The partial derivative of s with respect to T at constant P is shown to be c_P/T in Eq. (10-121). The partial derivative of P with respect to s at constant T is the inverse of the final Maxwell relation in Table 10-2. With these substitutions, Eq. (7) becomes:

$$\left(\frac{\partial P}{\partial T}\right)_s = \frac{c_P}{T} \left(\frac{\partial T}{\partial v}\right)_P \tag{8}$$

Equation (10-103) is also applied to the second partial derivative on the right side of Eq. (6):

$$\left(\frac{\partial T}{\partial v}\right)_s = -\underbrace{\left(\frac{\partial s}{\partial v}\right)_T}_{\left(\frac{\partial P}{\partial T}\right)_v} \underbrace{\left(\frac{\partial T}{\partial s}\right)_v}_{T/c_v} \tag{9}$$

The first partial derivative in Eq. (9) is the third Maxwell equation in Table 10-2. The second partial derivative in Eq. (9) is evident by inspection of Eq. (10-132). With these substitutions, Eq. (9) becomes:

$$\left(\frac{\partial T}{\partial v}\right)_s = -\left(\frac{\partial P}{\partial T}\right)_v \frac{T}{c_v} \tag{10}$$

Substituting Eqs. (8) and (10) into Eq. (6) leads to:

$$\left(\frac{\partial P}{\partial \rho}\right)_s = v^2 \frac{c_P}{c_v} \left(\frac{\partial T}{\partial v}\right)_P \left(\frac{\partial P}{\partial T}\right)_v \tag{11}$$

Applying Eq. (10-103) to Eq. (11) allows it to be simplified to:

$$\left(\frac{\partial P}{\partial \rho}\right)_s = -v^2 \frac{c_P}{c_v} \left(\frac{\partial P}{\partial v}\right)_T \tag{12}$$

Substituting Eq. (12) into Eq. (1) provides the final result:

$$c = \sqrt{-v^2 \frac{c_p}{c_v} \left(\frac{\partial P}{\partial v}\right)_T} \tag{13}$$

b) Use the RKS equation of state to develop an expression for the speed of sound.

The RKS equation of state appears in Eq. (10-54).

$$P = \frac{R\,T}{v-b} - \frac{a}{v\,(v+b)} \tag{14}$$

The partial derivative of P with respect to v at constant T is

$$\left(\frac{\partial P}{\partial v}\right)_T = -\frac{R\,T}{(v-b)^2} + \frac{a}{v^2(v+b)} + \frac{a}{v\,(v+b)^2} \tag{15}$$

Equation (15) is substituted into Eq. (13) to provide the sound speed predicted by the RKS equation of state:

$$c = \sqrt{-v^2 \frac{c_P}{c_v}\left[-\frac{R\,T}{(v-b)^2} + \frac{a}{v^2(v+b)} + \frac{a}{v\,(v+b)^2}\right]} \tag{16}$$

c) Use the RKS equation of state to predict the speed of sound for carbon dioxide at $T = 350$ K and $P = 100$ kPa. Compare your answer with the value obtained using the SoundSpeed function in EES. Obtain the specific heat capacity values from EES property functions.

The inputs are entered in EES.

```
$UnitSystem SI Radian Mass J K Pa
F$='CarbonDioxide'                 "fluid"
T=350 [K]                          "temperature"
P_kPa=100 [kPa]                    "pressure, in kPa"
P=P_kPa*convert(kPa,Pa)            "pressure"
```

The gas constant for carbon dioxide is obtained according to:

$$R = \frac{R_{univ}}{MW}$$

where MW is the molar mass, obtained using the MolarMass function in EES. The critical temperature, critical pressure, and acentric factor (T_{crit}, P_{crit}, and ω) are obtained.

```
R=R#/molarmass(F$)                 "gas constant for fluid F$"
T_crit=T_crit(F$)                  "critical temperature"
P_crit=P_crit(F$)                  "critical pressure"
omega=acentricFactor(F$)           "acentric factor"
```

The RKS equation of state is implemented using Eqs. (10-54) through (10-58).

```
T_r=T/T_crit                              "reduced temperature"
m=0.480+1.574*omega-0.176*omega^2         "parameter m used in alpha"
alpha^0.5=1+m*(1-sqrt(T_r))               "parameter alpha used in a"
a=0.42747*alpha*R^2*T_crit^2/P_crit       "parameter a in equation of state"
b=0.08664*R*T_crit/P_crit                 "parameter b in equation of state"
P=R*T/(v-b)-a/(v*(v+b))                   "RKS equation of state"
```

The specific heat capacities (c_P and c_v) are obtained using the functions cP and cv in EES.

```
c_P=cP(F$,T=T,P=P)                              "c_P from EES"
c_v=cv(F$,T=T,P=P)                              "c_v from EES"
```

Equation (16) is used to obtain the speed of sound predicted by the RKS equation of state (c). The EES function SoundSpeed is also used to predict the speed of sound (c_{EES}).

```
c=sqrt(-v^2*c_P*(-R*T/(v-b)^2+a/(v^2*(v+b))+a/(v*(v+b)^2))/c_v)   "speed of sound from RKS eq. of state"
c_EES=SoundSpeed(F$,T=T,P=P)                                     "speed of sound from EES"
```

These calculations result in $c = 289.1$ m/s and $cs_{EES} = 289.1$ m/s.

d) Plot the speed of sound for carbon dioxide as a function of pressure for several values of temperature. Include the value predicted by the RKS equation of state and by the SoundSpeed function in EES.

Figure 1 illustrates c and c_{EES} as a function of P for several values of T. Note that the difference between the sound speed values obtained from the RKS equation of state and EES increases with increasing pressure. Since sound speed is easily and accurately measured, it can be used, along with other data, to establish an accurate equation of state.

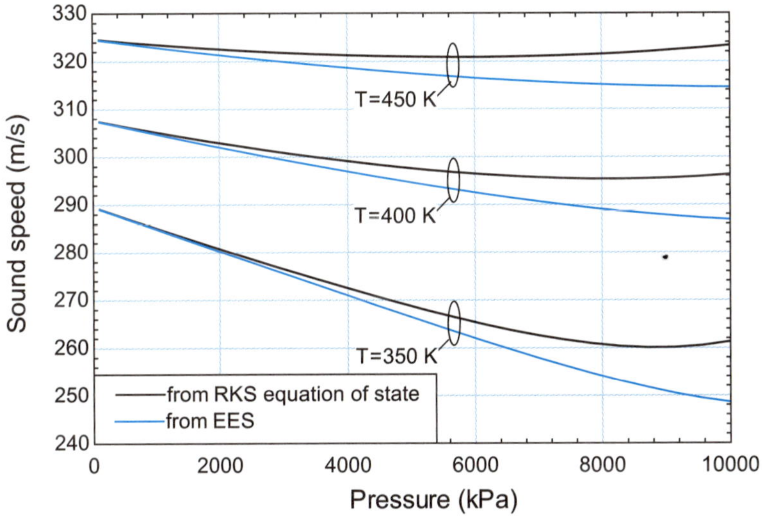

Figure 1: Sound speed of carbon dioxide as a function of pressure for various values of temperature predicted using the RKS equation of state and the SoundSpeed function in EES.

10.3.5 Relations Involving Specific Heat Capacity

The specific heat capacities at constant pressure and constant volume, c_P and c_v, are measurable thermodynamic properties. Their values can be measured at fixed pressure or fixed volume as a function of temperature using a calorimeter. Like other properties, the specific heat capacities are subject to the phase rule, Eq. (10-1). Therefore, two intensive properties fix the value of the specific heat capacity for a pure substance. The two specific heat capacities, c_P and c_v, are related. If one is known, the other can be determined using an equation of state. The pressure dependence of specific heat capacity can also be determined with an equation of state.

It was shown in Section 10.3.3 that specific enthalpy can be expressed as a function of temperature and pressure. The total derivative of the function $h(T, P)$ is given by Eq. (10-116), which is repeated below:

$$dh = c_P\,dT + \left[v - T\left(\frac{\partial v}{\partial T}\right)_P\right]dP \tag{10-137}$$

Specific enthalpy is an exact differential of temperature and pressure and so the mathematical requirements of an exact differential can be applied to Eq. (10-137). The Maxwell relations provided in Table 10-2 were determined by applying Eq. (10-90), which states that the order in which an exact differential is differentiated does not matter. We can also apply Eq. (10-90) to Eq. (10-137):

$$\frac{\partial}{\partial P}\left[\left(\frac{\partial h}{\partial T}\right)_P\right]_T = \left(\frac{\partial c_P}{\partial P}\right)_T \tag{10-138}$$

$$\frac{\partial}{\partial T}\left[\left(\frac{\partial h}{\partial P}\right)_T\right]_P = \frac{\partial}{\partial T}\left[v - T\left(\frac{\partial v}{\partial T}\right)_P\right]_P \tag{10-139}$$

Equating Eq. (10-138) and Eq. (10-139) leads to:

$$\left(\frac{\partial c_P}{\partial P}\right)_T = \frac{\partial}{\partial T}\left[v - T\left(\frac{\partial v}{\partial T}\right)_P\right]_P \tag{10-140}$$

Carrying out the differentiation of the right hand side of Eq. (10-140) provides information on how c_P varies with pressure:

$$\left(\frac{\partial c_P}{\partial P}\right)_T = \left(\frac{\partial v}{\partial T}\right)_P - T\left(\frac{\partial^2 v}{\partial T^2}\right)_P - \left(\frac{\partial v}{\partial T}\right)_P \tag{10-141}$$

or

$$\left(\frac{\partial c_P}{\partial P}\right)_T = -T\left(\frac{\partial^2 v}{\partial T^2}\right)_P \tag{10-142}$$

Application of Eq. (10-142) shows that c_P is not a function of pressure for an ideal gas. The ideal gas law is:

$$v = \frac{RT}{P} \tag{10-143}$$

The first derivative of Eq. (10-143) with respect to temperature is:

$$\left(\frac{\partial v}{\partial T}\right)_P = \frac{R}{P} \tag{10-144}$$

The second derivative of Eq. (10-143) with respect to temperature is:

$$\left(\frac{\partial^2 v}{\partial T^2}\right)_P = 0 \tag{10-145}$$

Substituting Eq. (10-145) into Eq. (10-142) leads to:

$$\left(\frac{\partial c_P}{\partial P}\right)_T = 0 \qquad \text{for an ideal gas} \tag{10-146}$$

Equation (10-141) eliminates the need to measure c_P as a function of pressure for real fluids. If c_P is known at a specific temperature, T_1, and pressure, P_1, then its value can be determined at any pressure (at that same temperature, T_1) by integration.

$$c_P(P, T_1) = c_P(P_1, T_1) + \int_{P_1}^{P} \left(\frac{\partial c_P}{\partial P}\right)_T dP = c_P(P_1, T_1) - T \int_{P_1}^{P} \left(\frac{\partial^2 v}{\partial T^2}\right)_P dP \qquad (10\text{-}147)$$

In practice, it is most convenient to measure c_P at a low pressure where the fluid behaves according to the ideal gas law and the specific heat capacity is independent of pressure. The ideal gas specific heat capacity at constant pressure is designated c_P^o. The specific heat capacity at constant pressure can then be found at any pressure by integrating from the ideal gas condition at low pressure (P approaching zero) to the actual pressure.

$$c_P(P, T_1) = c_P^o(P, T_1) - T \int_{0}^{P} \left(\frac{\partial^2 v}{\partial T^2}\right)_P dP \qquad (10\text{-}148)$$

It is not necessary to measure both c_P and c_v because the difference between these specific heat capacities can be determined from the equation of state. Equation (10-122) shows that:

$$c_P = T\left(\frac{\partial s}{\partial T}\right)_P \qquad (10\text{-}149)$$

and Eq. (10-124) shows that:

$$c_v = T\left(\frac{\partial s}{\partial T}\right)_v \qquad (10\text{-}150)$$

These relations can also be derived using the chain rule in Eq. (10-105) and the fundamental property relations. The fundamental property relation for specific enthalpy, Eq. (10-79), is:

$$dh = T\,ds + v\,dP \qquad (10\text{-}151)$$

The definition of c_P is:

$$c_P = \left(\frac{\partial h}{\partial T}\right)_P \qquad (10\text{-}152)$$

Applying the chain rule to Eq. (10-152) leads to:

$$c_P = \underbrace{\left(\frac{\partial h}{\partial s}\right)_P}_{T} \left(\frac{\partial s}{\partial T}\right)_P \qquad (10\text{-}153)$$

According to Eq. (10-151), the partial derivative of specific enthalpy with respect to specific entropy at constant pressure is temperature; therefore:

$$c_P = T\left(\frac{\partial s}{\partial T}\right)_P \qquad (10\text{-}154)$$

The fundamental property relation for specific internal energy, Eq. (10-76), is:

$$du = T\,ds - P\,dv \qquad (10\text{-}155)$$

The definition of c_v is:

$$c_v = \left(\frac{\partial u}{\partial T}\right)_v \qquad (10\text{-}156)$$

Applying the chain rule to Eq. (10-156) leads to:

$$c_v = \underbrace{\left(\frac{\partial u}{\partial s}\right)_v}_{T} \left(\frac{\partial s}{\partial T}\right)_v \tag{10-157}$$

According to Eq. (10-155), the partial derivative of specific internal energy with respect to specific entropy at constant pressure is temperature; therefore:

$$c_v = T\left(\frac{\partial s}{\partial T}\right)_v \tag{10-158}$$

According to Eqs. (10-154) and (10-158):

$$c_P - c_v = T\left(\frac{\partial s}{\partial T}\right)_P - T\left(\frac{\partial s}{\partial T}\right)_v \tag{10-159}$$

The right side of Eq. (10-159) involves *s*, *T*, *P*, and *v*. We can eliminate references to *s* using the chain rule in Eq. (10-104).

$$\left(\frac{\partial s}{\partial T}\right)_P = \left(\frac{\partial s}{\partial T}\right)_v \underbrace{\left(\frac{\partial T}{\partial T}\right)_P}_{1} + \underbrace{\left(\frac{\partial s}{\partial v}\right)_T}_{\left(\frac{\partial P}{\partial T}\right)_v} \left(\frac{\partial v}{\partial T}\right)_P \tag{10-160}$$

The third partial derivative on the right side of Eq. (10-160) is a Maxwell relation, Eq. (10-98). With this substitution, Eq. (10-160) becomes:

$$\left(\frac{\partial s}{\partial T}\right)_P = \left(\frac{\partial s}{\partial T}\right)_v + \left(\frac{\partial P}{\partial T}\right)_v \left(\frac{\partial v}{\partial T}\right)_P \tag{10-161}$$

Substituting Eq. (10-161) into Eq. (10-159) provides:

$$c_P - c_v = T\left[\left(\frac{\partial s}{\partial T}\right)_v + \left(\frac{\partial P}{\partial T}\right)_v \left(\frac{\partial v}{\partial T}\right)_P\right] - T\left(\frac{\partial s}{\partial T}\right)_v \tag{10-162}$$

which can be simplified to:

$$c_P - c_v = T\left(\frac{\partial P}{\partial T}\right)_v \left(\frac{\partial v}{\partial T}\right)_P \tag{10-163}$$

Equation (10-163) shows that if c_P is known then c_v can be obtained using only an equation of state. Notice that if the fluid obeys the ideal gas law then:

$$c_P^o - c_v^o = T\underbrace{\left(\frac{\partial P}{\partial T}\right)_v}_{R/v} \underbrace{\left(\frac{\partial v}{\partial T}\right)_P}_{R/P} = \frac{R^2 T}{P v} = R\underbrace{\left(\frac{RT}{Pv}\right)}_{=1} = R \quad \text{for an ideal gas} \tag{10-164}$$

10.4 Methodology for Calculating *u*, *h*, and *s*

The major results of Section 10.3 are the general relations for *u*, *h*, and *s* for a pure single phase fluid expressed in terms of measurable properties. These are repeated below:

$$du = c_v\, dT + \left[T\left(\frac{\partial P}{\partial T}\right)_v - P\right] dv \tag{10-165}$$

$$dh = c_P\, dT + \left[v - T\left(\frac{\partial v}{\partial T}\right)_P\right] dP \tag{10-166}$$

and

$$ds = \begin{cases} \dfrac{c_v}{T} dT + \left(\dfrac{\partial P}{\partial T}\right)_v dv \\ \dfrac{c_P}{T} dT - \left(\dfrac{\partial v}{\partial T}\right)_P dP \end{cases} \tag{10-167}$$

In principle, it should be easy to evaluate Eqs. (10-165) through (10-167) in order to determine values of u, h, and s (relative to their values at some reference state). However, the details of the process are not intuitive. For example, suppose that you wish to determine the change in the value of specific enthalpy between state 2 (defined by T_2 and P_2) and state 1 (defined by T_1 and P_1). Integration of Eq. (10-166) results in:

$$h_2 - h_1 = \int_{T_1}^{T_2} c_P \, dT + \int_{P_1}^{P_2} \left[v - T \left(\frac{\partial v}{\partial T}\right)_P \right] dP \tag{10-168}$$

The evaluation of the integral in Eq. (10-168) is not directly possible. The integrand of the temperature integral involves c_P, which depends on pressure. What pressure should be used here? The pressure varies between P_1 and P_2. The pressure integral involves T, but T varies between T_1 and T_2; which temperature should be used?

The key to evaluating properties is to recognize that the properties are exact differentials and therefore the path of the integration between states 1 and 2 does not matter. Consequently, we can choose a path that starts at state 1 and ends at state 2 and makes Eq. (10-168) easy to integrate. The most convenient path is shown in Figure 10-13. From state 1, the path proceeds isothermally at T_1 to state a, which is at a sufficiently low pressure (P^o) so that the fluid obeys the ideal gas law. Next, the temperature is changed from T_1 to T_2 at constant pressure, P^o, for the portion of the path that extends from state a to state b; the fluid obeys the ideal gas law for this temperature change. The final leg of the path, from state b to state 2, increases the pressure isothermally at T_2 from P^o to P_2. The change in specific enthalpy between states 1 and 2 for the path indicated in Figure 10-13 is:

$$h_2 - h_1 = (h_2 - h_b) + (h_b - h_a) + (h_a - h_1) \tag{10-169}$$

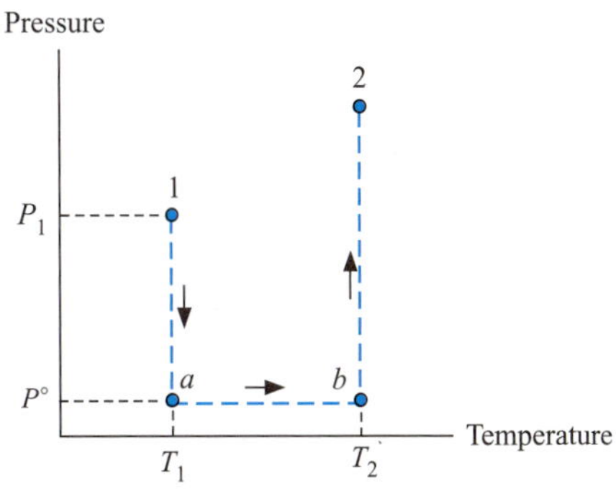

Figure 10-13: Convenient path for evaluation of thermodynamic property changes.

Equation (10-166) is integrated along each of the three legs of the path shown in Figure 10-13. Because the path is designated, the integration is now possible. Also, because of the way that the path has been selected, the integration is simplified. The integration along the leg of the path from state b to state 2 is:

$$h_2 - h_b = \cancel{\int_{T_2}^{T_2} c_P dT} + \int_{P^o}^{P_2} \left[v - T_2 \left(\frac{\partial v}{\partial T} \right)_P \right] dP \tag{10-170}$$

Since this leg of the path is isothermal, the temperature integral must be zero and the integrand of the pressure integral must be evaluated at T_2. Similarly, the integration along the leg of the path from state 1 to state a is:

$$h_a - h_1 = \cancel{\int_{T_1}^{T_1} c_P dT} + \int_{P_1}^{P^o} \left[v - T_1 \left(\frac{\partial v}{\partial T} \right)_P \right] dP \tag{10-171}$$

The portion of the path from state a to state b is isobaric and the pressure, P^o, has been selected so that it is sufficiently low that the fluid behaves as an ideal gas. As a result, the specific heat capacity that appears in the temperature integral is independent of pressure and equal to the ideal gas specific heat capacity, c_P^o.

$$h_b - h_a = \int_{T_1}^{T_2} c_P^o \, dT + \cancel{\int_{p^o}^{p^o} \left[v - T \left(\frac{\partial v}{\partial T} \right)_P \right] dT} \tag{10-172}$$

Substituting Eqs. (10-170) through (10-172) into Eq. (10-169) leads to:

$$h_2 - h_1 = \int_{P^o}^{P_2} \underbrace{\left[v - T_2 \left(\frac{\partial v}{\partial T} \right)_P \right]}_{\text{evaluated at } T_2} dP + \int_{T_1}^{T_2} c_P^o \, dT + \int_{P_1}^{P^o} \underbrace{\left[v - T_1 \left(\frac{\partial v}{\partial T} \right)_P \right]}_{\text{evaluated at } T_1} dP \tag{10-173}$$

All that is needed to evaluate the specific enthalpy change in Eq. (10-173) are the specific heat capacity at constant pressure as a function of temperature measured under ideal gas conditions and an accurate equation of state.

The change in specific entropy between states 1 and 2 can also be evaluated along the same path shown in Figure 10-13.

$$s_2 - s_1 = (s_2 - s_b) + (s_b - s_a) + (s_a - s_1) \tag{10-174}$$

Applying Eq. (10-167) to Eq. (10-174) allows the specific entropy change to be written as:

$$s_2 - s_1 = -\int_{P^o}^{P_2} \underbrace{\left(\frac{\partial v}{\partial T} \right)_P}_{\text{evaluated at } T_2} dP + \int_{T_1}^{T_2} \frac{c_P^o}{T} dT - \int_{P_1}^{P^o} \underbrace{\left(\frac{\partial v}{\partial T} \right)_P}_{\text{evaluated at } T_1} dP \tag{10-175}$$

Integration of Eqs. (10-173) and (10-175) provide the specific enthalpy and entropy at any other desired state. Tables of h and s can be developed by setting state 1 to be a reference state with specified reference values of specific enthalpy and specific entropy. There are, however, some strategies that can be used to simplify the integration process if it is done numerically, as is usually the case. The integrands of the pressure integrals in Eq. (10-173) are well-behaved in that they approach zero as the pressure approaches zero. Numerical integration is easy for this situation. However, the integrands in the pressure integrals in Eq. (10-175) approach infinity as pressure goes to zero; that is:

$$\lim_{P\to 0}\left(\frac{\partial v}{\partial T}\right)_P = \infty \tag{10-176}$$

Numerical problems can be avoided during the integration by making the following substitution for each of the pressure integrals:

$$-\int_{P^\circ}^{P_2}\underbrace{\left(\frac{\partial v}{\partial T}\right)_P}_{\text{evaluated at } T_2} dP = -\int_{P^\circ}^{P_2}\underbrace{\left[\left(\frac{\partial v}{\partial T}\right)_P - \frac{R}{P}\right]}_{\text{evaluated at } T_1} dP - R\ln\left(\frac{P_2}{P^\circ}\right) \tag{10-177}$$

$$-\int_{P_1}^{P^\circ}\underbrace{\left(\frac{\partial v}{\partial T}\right)_P}_{\text{evaluated at } T_1} dP = -\int_{P_1}^{P^\circ}\underbrace{\left[\left(\frac{\partial v}{\partial T}\right)_P - \frac{R}{P}\right]}_{\text{evaluated at } T_1} dP - R\ln\left(\frac{P^\circ}{P_1}\right) \tag{10-178}$$

The quantity $\int\left[\left(\frac{\partial v}{\partial T}\right)_P - \frac{R}{P}\right]dP$ is called the *entropy departure* because it is related to the difference between the specific entropy of the actual fluid and the specific entropy of an ideal gas under the same conditions. Substituting Eqs. (10-177) and (10-178) into Eq. (10-175) leads to:

$$s_2 - s_1 = -\int_{P^\circ}^{P_2}\underbrace{\left[\left(\frac{\partial v}{\partial T}\right)_P - \frac{R}{P}\right]}_{\text{evaluated at } T_2} dP + \int_{T_1}^{T_2}\frac{c_P}{T}dT$$
$$-\int_{P_1}^{P^\circ}\underbrace{\left[\left(\frac{\partial v}{\partial T}\right)_P - \frac{R}{P}\right]}_{\text{evaluated at } T_1} dP - R\ln\left(\frac{P_2}{P_1}\right) \tag{10-179}$$

Numerical integration is easier with this transformation because:

$$\lim_{P\to 0}\left[\left(\frac{\partial v}{\partial T}\right)_P - \frac{R}{P}\right] = 0 \tag{10-180}$$

and so the pressure integrals in Eq. (10-179) are well-behaved.

EXAMPLE 10.4-1: CALCULATING THE PROPERTIES OF ISOBUTANE

The ideal gas specific heat capacity at constant pressure for isobutane is approximately represented by:

$$c_P^o = R\left(d + e\,T\right) \tag{1}$$

where $d = 2.285$, $e = 0.03157\ \text{K}^{-1}$, R is the ideal gas constant (in the same units as c_P^o), and T is the temperature in K.

a) Write a computer program that can calculate the specific volume, specific internal energy, specific enthalpy, and specific entropy of isobutane using the Peng-Robinson equation of state and the c_P^o function given by Eq. (1). Use a reference state, $T_{ref} = 300$ K and $P_{ref} = 10$ kPa and assign the reference specific enthalpy and specific entropy to be $h_{ref} = 605345$ J/kg and $s_{ref} = 2862.8$ J/kg-K, respectively. (These values have been chosen to be consistent with the reference conditions used by EES for isobutane.)

A subprogram PR is written in order to implement the PR equation of state using Eqs. (10-60) through (10-64).

```
$UnitSystem SI Radian Mass J K Pa
Subprogram PR(F$,P,T:v)                              "implements the PR equation of state"
    "Inputs:"
    "F$: fluid name"
    "P: pressure (Pa)"
    "T: temperature (K)"

    "Outputs:"
    "v: specific volume (m^3/kg)"

    R=R#/MolarMass(F$)                               "gas constant for fluid F$"
    T_crit=T_crit(F$)                                "critical temperature"
    P_crit=P_crit(F$)                                "critical pressure"
    b=0.07780*R*T_crit/P_crit                        "parameter b in equation of state"
    omega=AcentricFactor(F$)                         "acentric factor"
    m=0.37464+1.54226*omega-0.26992*omega^2          "parameter m used in alpha"
    T_r=T/T_crit                                     "reduced temperature"
    sqrt(alpha)=1+m*(1-sqrt(T_r))                    "parameter alpha used in a"
    a=0.45724*alpha*R^2*T_crit^2/P_crit              "parameter a in equation of state"
    P=R*T/(v-b)-a/(v*(v+b)+b*(v-b))                  "PR equation of state"
end
```

The reference conditions and other inputs are entered:

```
T_ref=300 [K]                                        "reference temperature"
P_ref=10 [kPa]*convert(kPa,Pa)                       "reference pressure"
h_ref=605345 [J/kg]                                  "reference specific enthalpy"
s_ref=2862.8 [J/kg-K]                                "reference specific entropy"
F$='Isobutane'                                       "fluid"
d=2.285 [-]                                          "constant for cP_o function"
e=0.03157 [1/K]                                      "constant for cP_o function"
```

The subprogram PR is used to determine the specific volume at the reference state, v_{ref}. The ideal gas constant is computed according to:

$$R = \frac{R_{univ}}{MW}$$

where MW is the molar mass of isobutane. The compressibility factor at the reference state is given by:

$$Z_{ref} = \frac{P_{ref}\, v_{ref}}{R\, T_{ref}}$$

```
Call PR(F$,P_ref,T_ref:v_ref)          "reference specific volume"
R=R#/MolarMass(F$)                     "ideal gas constant"
Z_ref=P_ref*v_ref/(R*T_ref)            "reference compressibility factor"
```

which results in $Z_{ref} = 0.9975$. The compressibility factor at the reference state is extremely close to unity, indicating that the reference state is essentially an ideal gas state. This situation simplifies the integration required to determine the change in properties from the reference state because P_{ref} is sufficiently low that it can be taken to be equal to P°. The integration path is therefore given by Figure 1.

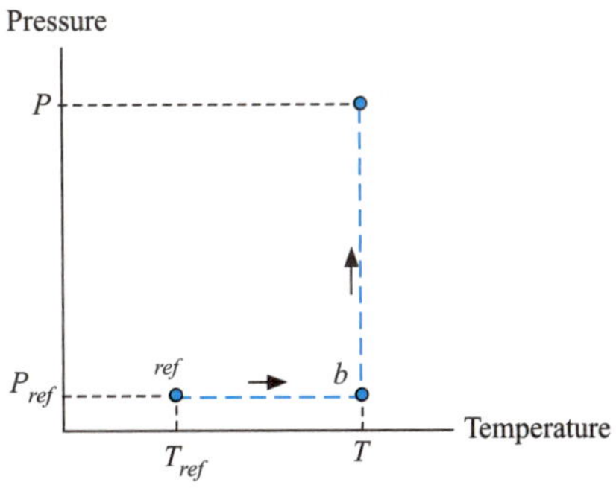

Figure 1: Integration path.

In order to obtain the specific enthalpy and specific entropy (h and s) at an arbitrary value of T and P, it is necessary to carry out the integrations specified by Eqs. (10-173) and (10-179) with $T_2 = T$, $P_2 = P$, $P_1 = P_{ref}$, and $T_1 = T_{ref}$. Also, $P^{\circ} = P_{ref}$ which eliminates the integration from P_1 to P° that is shown in Figure 10-13.

$$h = h_{ref} + \int_{P_{ref}}^{P} \underbrace{\left[v - T\left(\frac{\partial v}{\partial T}\right)_P\right]}_{\text{evaluated at } T} dP + \int_{T_{ref}}^{T} c_P^{\circ}\, dT \tag{2}$$

and

$$s = s_{ref} - \int_{P_{ref}}^{P} \underbrace{\left[\left(\frac{\partial v}{\partial T}\right)_P - \frac{R}{P}\right]}_{\text{evaluated at } T} dP + \int_{T_{ref}}^{T} \frac{c_P}{T} dT - R \ln\left(\frac{P}{P_{ref}}\right) \tag{3}$$

Substituting Eq. (1) into Eqs. (2) and (3) and carrying out the resulting temperature integrations leads to:

$$h = h_{ref} + \underbrace{\int_{P_{ref}}^{P} \underbrace{\left[v - T\left(\frac{\partial v}{\partial T}\right)_P\right]}_{\text{Integrand 1}} dP}_{\text{Integral 1}} + R\, d\,(T - T_{ref}) + \frac{R\, e}{2}\left(T^2 - T_{ref}^2\right) \tag{4}$$

and

$$s = s_{ref} - \underbrace{\int_{P_{ref}}^{P} \underbrace{\left[\left(\frac{\partial v}{\partial T} \right)_P - \frac{R}{P} \right]}_{\text{Integrand 2}} dP}_{\text{Integral 2}} + R\, d \ln\left(\frac{T}{T_{ref}}\right) + R\, e\, (T - T_{ref}) - R \ln\left(\frac{P}{P_{ref}}\right) \tag{5}$$

The pressure integrations in Eqs. (4) and (5) are accomplished numerically. Arbitrary values of T and P are set:

```
T=450 [K]                              "temperature at which to evaluate h and s"
P=10 [MPa]*convert(MPa,Pa)             "pressure at which to evaluate h and s"
```

Before carrying out the numerical integrations, it is a good idea to test the code by evaluating the integrands, Integrand 1 and Integrand 2 in Eqs. (4) and (5), respectively, at some arbitrary value of the integration variable, P_{int}. An arbitrary value of P_{int} is specified and the PR subprogram is used to obtain the specific volume, v_{int}:

```
P_int=P_ref                            "arbitrary value of P for the integration"
Call PR(F$,P_int,T:v_int)              "get specific volume"
```

The partial derivative of specific volume with respect to temperature at constant pressure is estimated numerically:

$$\left(\frac{\partial v}{\partial T} \right)_P \approx \frac{v_{T+dT,P_{int}} - v_{T-dT,P_{int}}}{2\, dT}$$

where dT is a small but finite perturbation in the temperature.

```
dT=1 [K]                                      "differential value of temperature change"
Call PR(F$,P_int,T+dT:v_int_plus)             "get specific volume at T+dT"
Call PR(F$,P_int,T-dT:v_int_minus)            "get specific volume at T-dT"
dvdT_P_int=(v_int_plus-v_int_minus)/(2*dT)    "partial derivative of v with respect to T at constant P"
```

The integrand in the specific enthalpy equation, Integrand 1 in Eq. (4), is computed:

```
Integrand1=v_int-T*dvdT_P_int        "Integrand of pressure integral in specific enthalpy equation"
```

and the integrand in the specific entropy equation, Integrand 2 in Eq. (5), is computed:

```
Integrand2=dvdT_P_int-R/P_int        "Integrand of pressure integral in specific entropy equation"
```

The assumed value of the integration pressure is commented out and the two pressure integrals, Integral 1 and Integral 2 in Eqs. (4) and (5), respectively, are computed using the Integral command.

```
{P_int=P_ref}                                   "arbitrary value of P for the integration"
Integral1=Integral(Integrand1,P_int,P_ref,P)    "pressure integral for specific enthalpy"
Integral2=Integral(Integrand2,P_int,P_ref,P)    "pressure integral for specific entropy"
```

When the EES code is solved at this point it is likely that the error shown in Figure 2 will be encountered.

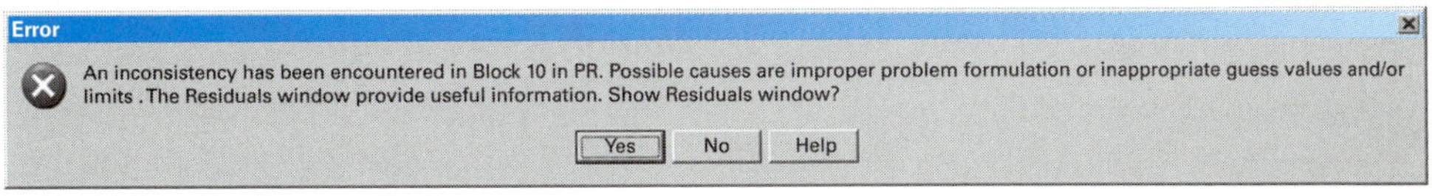

Figure 2: Error message.

This error occurs because EES is having difficulty solving for specific volume at some state along the integration path, given the temperature and pressure. The convergence problem can be eliminated by restricting the value of the specific volume in the subprogram PR to positive values. Select Variable Information from the Options menu. Bring up the variables for Subprogram PR by selecting this program segment from the drop-down list at the top of the Variable Information dialog. Change the lower limit for variable v to 0, as shown in Figure 3. Click the OK button.

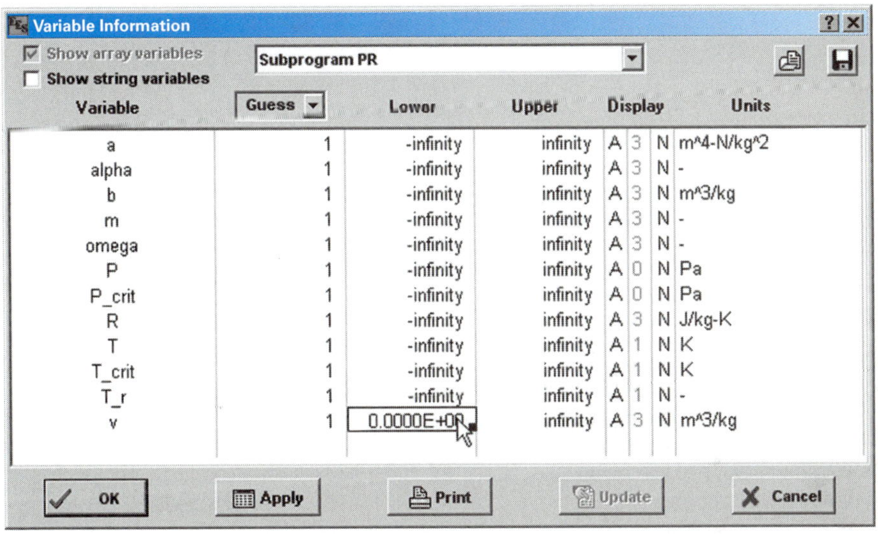

Figure 3: Variable information window with limit set for specific volume.

With the limit set on specific volume it should be possible to evaluate each of the integrals numerically. Specific enthalpy and specific entropy can be evaluated using Eqs. (4) and (5), respectively.

```
h=h_ref+Integral1+R*d*(T-T_ref)+R*e*(T^2-T_ref^2)/2          "specific enthalpy"
s=s_ref-Integral2+R*d*ln(T/T_ref)+R*e*(T-T_ref)-R*ln(P/P_ref) "specific entropy"
```

The specific internal energy can be obtained from the definition of enthalpy:

$$u = h - P\,v$$

where v is obtained from the subprogram PR.

```
Call PR(F$,P,T:v)          "get specific volume"
u=h-P*v                    "specific internal energy"
```

b) Use the program to calculate the specific enthalpy along the isotherm $T = 450$ K for pressures ranging between 10 kPa and 10 MPa. Note that 450 K is above the critical temperature, $T_{crit} = 407.8$ K for isobutane. Prepare a plot showing the specific enthalpy and specific entropy predicted by your program and also the values obtained from the EES functions enthalpy and entropy.

The properties predicted by EES (h_{EES}, s_{EES}, and u_{EES}) are obtained.

```
h_EES=enthalpy(F$,T=T,P=P)            "specific enthalpy from EES"
s_EES=entropy(F$,T=T,P=P)             "specific entropy from EES"
u_EES=intenergy(F$,T=T,P=P)           "specific internal energy from EES"
```

A Parametric table is setup that contains P, h, s, h_{EES}, and s_{EES}. The pressure is varied from 10 kPa to 10 MPa and the result is used to generate Figure 4, which shows the specific enthalpy and specific entropy predicted using the Peng-Robinson equation of state with Eq. (1) as well as the same values predicted using the EES property correlations.

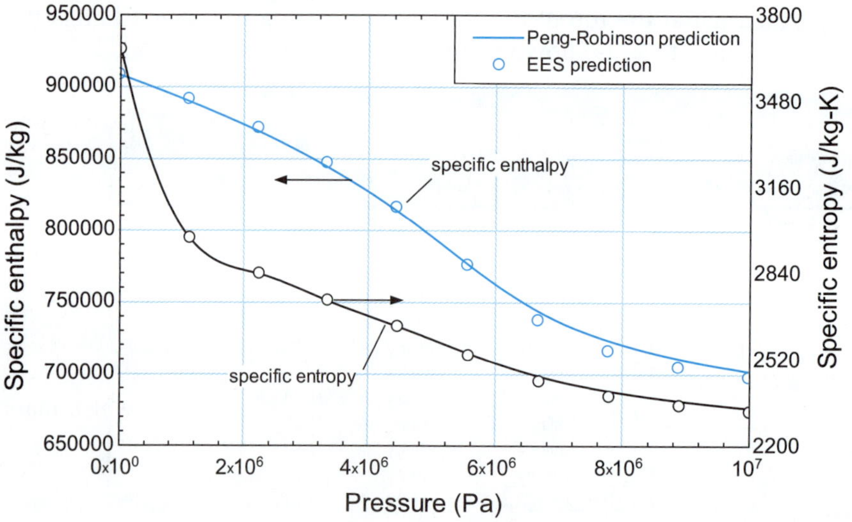

Figure 4: The specific enthalpy and specific entropy of isobutane at $T = 450$ K calculated with the Peng-Robinson equation of state as a function pressure. Also shown are the values obtained from the EES property correlations.

Note that the method used to generate Figure 4 was more computationally expensive than is necessary because each of the points along the isotherm required the same integration with pressure. A more expedient approach would be to utilize an Integral table that records the specific enthalpy and specific entropy along the isotherm as a function of the integration pressure. The modified code required to accomplish this is shown below.

```
h=h_ref+Integral1+R*d*(T-T_ref)+R*e*(T^2-T_ref^2)/2                 "specific enthalpy"
s=s_ref-Integral2+R*d*ln(T/T_ref)+R*e*(T-T_ref)-R*ln(P_int/P_ref)   "specific entropy"
u=h-P_int*v_int                                                     "specific internal energy"

h_EES=enthalpy(F$,T=T,P=P_int)                                      "specific enthalpy from EES"
s_EES=entropy(F$,T=T,P=P_int)                                       "specific entropy from EES"
u_EES=intenergy(F$,T=T,P=P_int)                                     "specific internal energy from EES"

$IntegralTable P_int:100000,h,h_EES,s,s_EES,u,u_EES
```

With these modifications, the code need only be run for P = 10 MPa and the properties at every pressure from P_{ref} to P will be tabulated in the Integral table.

10.5 Phase Equilibria for Pure Fluids

The property relations presented in this chapter have, up to this point, been applicable to single-phase, single-component systems. This section discusses multi-phase, single-component systems; for example a pure fluid that consists of liquid and vapor phases or solid and liquid phases. Multi-component phase equilibrium is discussed in Section 11.5. The phase rule in Eq. (10-1) indicates that there is only one degree of freedom for a pure substance that is coexisting in two-phases. Therefore, specifying one intensive variable fixes the intensive state of both phases. For example, specifying the pressure of water in which both liquid and vapor phases are present fixes the saturation temperature. The pressure also fixes the specific volumes, specific enthalpies and specific entropies of both phases. This section reviews the relationships between thermodynamic properties in a two-phase system.

10.5.1 Criterion for Phase Equilibrium

Suppose we have a system consisting of liquid and vapor in equilibrium, such as the system contained within the piston-cylinder apparatus shown in Figure 10-14. We would like to establish the thermodynamic relationship that exists between saturation temperature and saturation pressure.

The pure fluid in the piston-cylinder device shown in Figure 10-14 exists at a fixed pressure and therefore at a fixed temperature. After reaching an equilibrium condition, no observable changes occur on a macroscopic scale. However, there is a great deal of activity occurring on a microscopic scale. If we could observe the microscopic events, we would see mass leaving the liquid surface and becoming part of the vapor phase. Simultaneously, mass from the vapor phase hits the liquid surface and becomes part of the liquid phase. At a fixed temperature (or pressure) the average rate at which mass from the liquid phase becomes vapor is equal to the average rate at which mass from the vapor phase becomes liquid; the system is in a state of dynamic equilibrium.

Figure 10-14 shows a small amount of mass, *dm*, leaving the liquid phase to become part of the vapor. We can calculate the changes in thermodynamic properties that occur as a result of this process. It should be understood that the mass, *dm*, is small relative to the mass of liquid and vapor in each phase so that the system remains in a two-phase condition. The pressure is constant during the process, as required by a force balance on the free-floating piston:

$$dP = 0 \tag{10-181}$$

The phase rule indicates that there is one degree of freedom. If the pressure is fixed, all other intensive properties for each phase remain unchanged during the process in which *dm* transfers from liquid to vapor. Therefore, the temperature is unchanged:

$$dT = 0 \tag{10-182}$$

as are the specific volumes (v_f and v_g), specific enthalpies (h_f and h_g), and specific entropies (s_f and s_g) of both the liquid and vapor phases. However, the quality of the

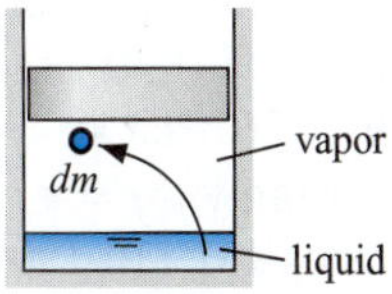

Figure 10-14: Two phase liquid-vapor system at equilibrium showing a small mass moving from liquid to vapor phase.

system does change due to the vaporization of dm and that results in changes in the extensive properties of the system. The change in the volume of the system that accompanies the process is given by:

$$dV = dm\,(v_g - v_f) > 0 \tag{10-183}$$

The enthalpy and entropy of the system also change:

$$dH = dm\,(h_g - h_f) > 0 \tag{10-184}$$

$$dS = dm\,(s_g - s_f) > 0 \tag{10-185}$$

The Gibbs free energy function, G, was introduced in Section 10.2. Gibbs free energy is defined as:

$$G = H - TS = mg = m(h - Ts) \tag{10-186}$$

where g in Eq. (10-186) is the specific Gibbs free energy. So far, Gibbs free energy has played a very minor role in our understanding of thermodynamic property relations. It was introduced in Section 10-2 in order to derive one of the more useful Maxwell relations listed in Table 10-2. Gibbs free energy plays a much more important role in phase and chemical equilibria, as shown in this section and again in Chapters 11 and 14.

What is the change in Gibbs free energy resulting from the transfer of dm from the liquid to vapor phase? The Gibbs free energy is an extensive property, like enthalpy and entropy, and so the change in G for this process can be written in a manner that is analogous to the changes in the other extensive properties given by Eqs. (10-183) through (10-185):

$$dG = dm\,(g_g - g_f) \tag{10-187}$$

where g_f and g_g are the specific Gibbs free energies associated with the liquid and vapor phases, respectively. As shown in Eq. (10-84), the fundamental property relation written in terms of the Gibbs free energy is:

$$dG = -S\,dT + V\,dP \tag{10-188}$$

However, both T and P are constant for this phase change process; therefore, the criterion for phase equilibrium is that the Gibbs free energy of the system must not change:

$$\boxed{dG = 0 \quad \text{criterion for phase equilibrium}} \tag{10-189}$$

Since $dm > 0$, Eq. (10-187) and Eq. (10-189) together require that:

$$g_f = g_g \tag{10-190}$$

The specific Gibbs free energy of the liquid and vapor phases must be equal in order for these phases to be in equilibrium:

$$h_f - Ts_f = h_g - Ts_g \tag{10-191}$$

Equation (10-190) or Eq. (10-191) is a requirement of phase equilibrium for a pure fluid. This relation must be obeyed for a system in two-phase equilibrium, which helps to determine property values. For example, for saturated water at 100°C, $h_f = 419.2$ kJ/kg and $s_f = 1.307$ kJ/kg-K; therefore $g_f = h_f - Ts_f = -68.62$ kJ/kg. For the vapor phase, $h_g = 2676$ kJ/kg and $s_g = 7.354$ kJ/kg-K; therefore $g_g \equiv h_g - Ts_g = -68.62$ kJ/kg. Not surprisingly, $g_f = g_g$ for this two-phase state because the property relations were developed using Eq. (10-190).

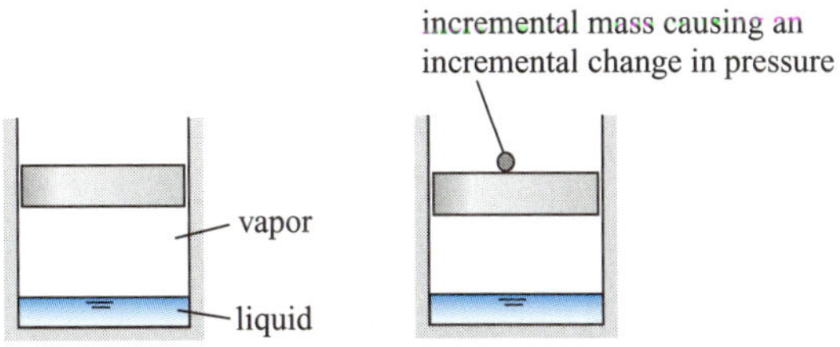

Figure 10-15: (a) A liquid-vapor system in equilibrium and (b) the equilibrium state disturbed by the addition of a small mass on the piston which causes a small increase in the pressure.

10.5.2 Relations between Properties during a Phase Change

We know that a single component two-phase system has one degree of freedom and that the specific Gibbs free energies of each phase must have the same value. This statement can be used to relate the pressure and temperature for a pure two-phase system. Figure 10-15(a) shows a two-phase system that is initially in an equilibrium state. At this state,

$$g_f = g_g \tag{10-192}$$

However, we now slightly disturb this equilibrium state by placing a small mass on the piston, as shown in Figure 10-15(b). The additional mass on the piston causes a slight increase in the pressure of the liquid and vapor in the cylinder. Provided that the change is sufficiently small that the system remains in a two-phase state, there is only one degree of freedom and the increased pressure results in a change in all of the other intensive properties associated with the system, including the temperature. The total derivative of Eq. (10-192) is:

$$dg_f = dg_g \tag{10-193}$$

Substituting the fundamental property relation from Eq. (10-84) into Eq. (10-193) leads to:

$$-s_f\,dT + v_f\,dP = -s_g\,dT + v_g\,dP \tag{10-194}$$

The relationship between the change in the saturation pressure and the change in the saturation temperature for this two-phase system is obtained by rearranging Eq. (10-194).

$$\left.\frac{dP}{dT}\right|_{sat} = \frac{(s_g - s_f)}{(v_g - v_f)} \tag{10-195}$$

One form of the fundamental property relation is given in terms of specific enthalpy, as shown in Eq. (10-79):

$$dh = T\,ds + v\,dP \tag{10-196}$$

This equation can be integrated for a system that is changing from saturated liquid to saturated vapor:

$$\int_{h_f}^{h_g} dh = \int_{s_f}^{s_g} T\,ds + \int_{P_f}^{P_g} v\,dP \tag{10-197}$$

The temperature and pressure remain constant during a phase change for a single-component substance; therefore Eq. (10-197) can be simplified to:

$$h_g - h_f = T \int_{s_f}^{s_g} ds + \cancel{\int_{P_f}^{P_g} v\,dP} \tag{10-198}$$

or

$$h_g - h_f = T\,(s_g - s_f) \tag{10-199}$$

Substituting Eq. (10-199) into Eq. (10-195) results in:

$$\boxed{\left.\frac{dP}{dT}\right|_{sat} = \frac{(h_g - h_f)}{T\,(v_g - v_f)}} \tag{10-200}$$

Equation (10-200) is called the Clapeyron equation in honor of Benoît Clapeyron, an early 19th century French physicist. The Clapeyron equation indicates how saturation pressure and saturation temperature are related in two-phase systems. Although this equation is written in terms of saturated liquid (*f*) and saturated vapor (*g*), Eq. (10-200) is general and can be applied to any phase change process.

Several approximations can be applied in order to simplify the integration of Eq. (10-200) for liquid-vapor (or solid-vapor) systems. At states far from the critical point, the specific volume of the liquid is ordinarily much smaller than the specific volume of the vapor, so that

Approximation 1: $$(v_g - v_f) \approx v_g \tag{10-201}$$

The vapor pressure is often low so that the specific volume of the saturated vapor can be approximated by the ideal gas law:

Approximation 2: $$v_g \approx \frac{R\,T_{sat}}{P_{sat}} \tag{10-202}$$

For a small change in conditions, the specific enthalpy change associated with vaporization (Δh_{fg}) may be assumed to be constant:

Approximation 3: $$(h_g - h_f) = \Delta h_{fg} \approx \text{constant} \tag{10-203}$$

Substituting Eqs. (10-201) through (10-203) into Eq. (10-200) results in:

$$\left.\frac{dP}{dT}\right|_{sat} \approx \frac{\Delta h_{fg}\,P_{sat}}{T_{sat}^2\,R} \tag{10-204}$$

Equation (10-204) can be separated and integrated between any two saturation states:

$$\int_{P_{sat,1}}^{P_{sat,2}} \frac{dP_{sat}}{P_{sat}} = \frac{\Delta h_{fg}}{R} \int_{T_{sat,1}}^{T_{sat,2}} \frac{dT_{sat}}{T_{sat}^2} \tag{10-205}$$

which results in:

$$\ln\left(\frac{P_{sat,2}}{P_{sat,1}}\right) = -\frac{\Delta h_{fg}}{R}\left(\frac{1}{T_{sat,2}} - \frac{1}{T_{sat,1}}\right) \tag{10-206}$$

Equation (10-206) can be rearranged to provide the saturation pressure:

$$P_{sat,2} = P_{sat,1} \exp\left[-\frac{\Delta h_{fg}}{R}\left(\frac{1}{T_{sat,2}} - \frac{1}{T_{sat,1}}\right)\right] \tag{10-207}$$

Equation (10-207) is called the Clausius-Clapeyron equation and it is an approximate relationship between the saturation pressure and the saturation temperature. The utility of the Clausius-Clapeyron equation is based on the accuracy of the three approximations in Eqs. (10-201) through (10-203).

EXAMPLE 10.5-1: EVALUATING A NEW REFRIGERANT

It is necessary to provide a preliminary evaluation of a recently synthesized organic fluid for use in the simple vapor compression refrigeration shown in Figure 1.

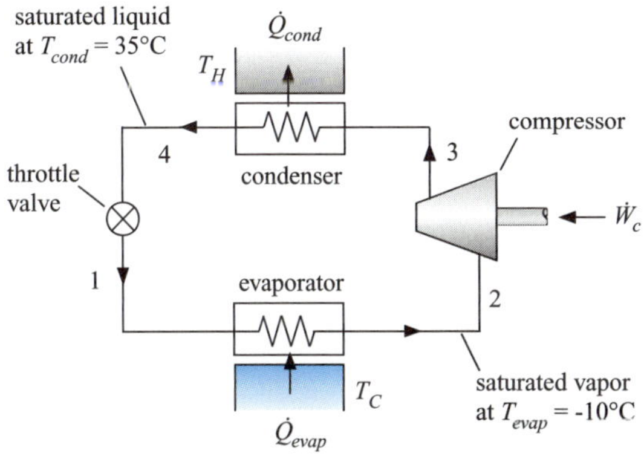

Figure 1: Vapor compression refrigeration system.

A limited set of measurements is currently available. The molecular weight of the fluid is $MW = 117$ kg/kgmol. The critical temperature and critical pressure are $T_{crit} = 477$ K and $P_{crit} = 4250$ kPa, respectively. The specific heat capacity and specific volume of the liquid phase have been measured; for the range of conditions of interest for the refrigeration cycle it is appropriate to model the liquid phase as being incompressible with $c_f = 1050$ J/kg-K and $v_f = 0.00077$ m^3/kg. The saturation pressure (i.e., the vapor pressure) of the refrigerant has been measured as a function of saturation temperature and the data have been used to fit the equation:

$$\ln\left(\frac{P_{sat}}{a_1}\right) = \frac{a_2}{T_{sat}} + a_3\, T_{sat} + a_4\left(1 - \frac{T_{sat}}{T_{crit}}\right)^{1.5} \tag{1}$$

where $a_1 = 4.000\times10^{10}$ Pa, $a_2 = -4206$ K, $a_3 = -0.0006$ K^{-1}, and $a_4 = 4.979$.

a) Use the property data listed above to estimate the evaporator heat transfer per mass of refrigerant in a vapor compression cycle in which saturated liquid exits the condenser at $T_{cond} = 35°$ and saturated vapor leaves the evaporator at $T_{evap} = -10°$C.

The inputs are entered in EES.

```
$UnitSystem SI Mass J Radian K Pa
MW=117 [kg/kgmol]                          "molecular weight"
P_crit=4250 [kPa]*convert(kPa,Pa)          "critical pressure"
T_crit=477 [K]                             "critical temperature"
c_f=1050 [J/kg-K]                          "specific heat capacity of liquid"
v_f=0.00077 [m^3/kg]                       "specific volume of liquid"
a1=4.0003e10 [Pa]                          "constants for saturation pressure fit"
a2= -4206.35 [K]
a3= -0.00060 [1/K]
a4= 4.9789 [-]
T_cond=converttemp(C,K,35 [C])             "condensing temperature"
T_evap=converttemp(C,K,-10 [C])            "evaporating temperature"
```

The temperature at state 4 is the condensing temperature, $T_4 = T_{cond}$. The pressure at state 4 is the saturation pressure, estimated using Eq. (1):

$$\ln\left(\frac{P_4}{a_1}\right) = \frac{a_2}{T_4} + a_3\,T_4 + a_4\left(1 - \frac{T_4}{T_{crit}}\right)^{1.5}$$

The temperature at state 1 is the evaporating temperature, $T_1 = T_{evap}$. The pressure at state 1 is the saturation pressure, estimated using Eq. (1):

$$\ln\left(\frac{P_1}{a_1}\right) = \frac{a_2}{T_1} + a_3\,T_1 + a_4\left(1 - \frac{T_1}{T_{crit}}\right)^{1.5}$$

```
T_4=T_cond                                        "temperature at state 4"
ln(P_4/a1)=a2/T_4+a3*T_4+a4*(1-T_4/T_crit)^1.5    "saturation pressure at state 4"
T_1=T_evap                                        "temperature at state 1"
ln(P_1/a1)=a2/T_1+a3*T_1+a4*(1-T_1/T_crit)^1.5    "saturation pressure at state 1"
```

An energy balance on the evaporator leads to:

$$\frac{\dot{Q}_{evap}}{\dot{m}} = h_2 - h_1 \tag{2}$$

An energy balance on the expansion valve provides:

$$h_1 = h_4 \tag{3}$$

Substituting Eq. (3) into Eq. (2) leads to:

$$\frac{\dot{Q}_{evap}}{\dot{m}} = h_2 - h_4 \tag{4}$$

The specific enthalpy at state 4 can be estimated assuming that the saturated liquid behaves as an incompressible substance with a constant specific heat capacity using the relation provided in Eq. (4-43):

$$h_4 = h_{f,T_1} + c_f\,(T_4 - T_1) + v_f\,(P_4 - P_1) \tag{5}$$

where h_{f,T_1} is the specific enthalpy of saturated liquid at T_1. The temperature at state 2 is equal to the temperature at state 1. The specific enthalpy at state 2 is given by:

$$h_2 = h_{f,T_1} + \Delta h_{fg,T_1} \tag{6}$$

where $\Delta h_{fg,T_1}$ is the latent heat of vaporization associated with the refrigerant at T_1. Substituting Eqs. (5) and (6) into Eq. (4) leads to:

$$\frac{\dot{Q}_{evap}}{\dot{m}} = \Delta h_{fg,T_1} - c_f\,(T_4 - T_1) - v_f\,(P_4 - P_1) \tag{7}$$

The latent heat of vaporization is estimated using Eq. (10-200):

$$\left.\frac{dP}{dT}\right|_{sat} = \frac{\Delta h_{fg}}{T\,(v_g - v_f)} \tag{8}$$

The specific volume of the saturated vapor at state 2 is estimated using the ideal gas law:

$$v_g = \frac{R\,T_1}{P_1}$$

where R is the gas constant, computed from:

$$R = \frac{R_{univ}}{MW}$$

```
R=R#/MW                         "gas constant"
v_g=R*T_1/P_1                   "specific volume of saturated vapor"
```

The derivative of the saturation pressure with respect to temperature is obtained by taking the derivative of Eq. (1):

$$\frac{d}{dT}\left[\ln\left(\frac{P_{sat}}{a_1}\right)\right] = -\frac{a_2}{T^2} + a_3 - \frac{1.5\,a_4}{T_{crit}}\left(1 - \frac{T}{T_{crit}}\right)^{0.5}$$

which provides:

$$\frac{1}{P_{sat}}\frac{dP_{sat}}{dT} = -\frac{a_2}{T^2} + a_3 - \frac{1.5\,a_4}{T_{crit}}\left(1 - \frac{T}{T_{crit}}\right)^{0.5} \tag{9}$$

Equation (9) is used to evaluate the derivative of the saturation pressure with respect to temperature at T_1 and the latent heat of vaporization is estimated using Eq. (8). Equation (7) is used to estimate the evaporator heat transfer per unit mass.

```
dPsatdT_1/P_1=-a2/T_1^2+a3-1.5*a4*(1-T_1/T_crit)/T_crit
                                "partial derivative of P_sat w/respect to T at 1"
dPsatdT_1=Dh_fg_1/(T_1*(v_g-v_f))   "latent heat at state 1"
Q_dot_evap\m_dot=Dh_fg_1-c_f*(T_4-T_1)-v_f*(P_4-P_1)
                                "evaporator heat transfer per mass"
```

which results in $\dot{Q}_{evap}/\dot{m} = 213.9$ kJ/kg.

10.5.3 Estimating Saturation Properties using an Equation of State

This extended section can be found online at www.cambridge.org/kleinandnellis. This section discusses the use of an equation of state to estimate saturation pressure and saturation temperature.

10.6 Fugacity

This section introduces a thermodynamic property called *fugacity*. Fugacity is not really needed for pure single-phase substances, but it is very useful for the multi-component phase and chemical equilibrium calculations that are discussed in Chapters 11 and 14. The fugacity of a component within a mixture can often be approximated as being equal to the fugacity of the pure substance. Methods of estimating the fugacity of a pure substance are presented here.

Specific Gibbs free energy is defined as:

$$g = h - T s \tag{10-211}$$

When the definition for the specific Gibbs free energy is substituted into the fundamental property relation, the differential of Gibbs free energy can be expressed as:

$$dg = -s\,dT + v\,dP \tag{10-212}$$

Now, consider a process that occurs at constant temperature between pressures P_1 and P_2. The change in the specific Gibbs free energy for this isothermal process is:

$$\Delta g_T = \int_{P_1}^{P_2} v\,dP \tag{10-213}$$

where the subscript T is used to indicate an isothermal process; therefore, Δg_T is the change in specific Gibbs free energy for an isothermal process. The change in specific Gibbs free energy is equal to the minimum work per unit mass that is required to isothermally compress the fluid from P_1 to P_2. To see this result, consider the system shown in Figure 10-17 in which a flow stream enters at state 1 and proceeds reversibly and isothermally to state 2. Figure 10-17(a) illustrates an entropy balance on a reversible system:

$$\dot{m} s_1 = \frac{\dot{Q}_{out}}{T} + \dot{m} s_2 \tag{10-214}$$

where T is the temperature of the flow. Figure 10-17(b) illustrates an energy balance on the system:

$$\dot{m} h_1 + \dot{W}_{in} = \dot{Q}_{out} + \dot{m} h_2 \tag{10-215}$$

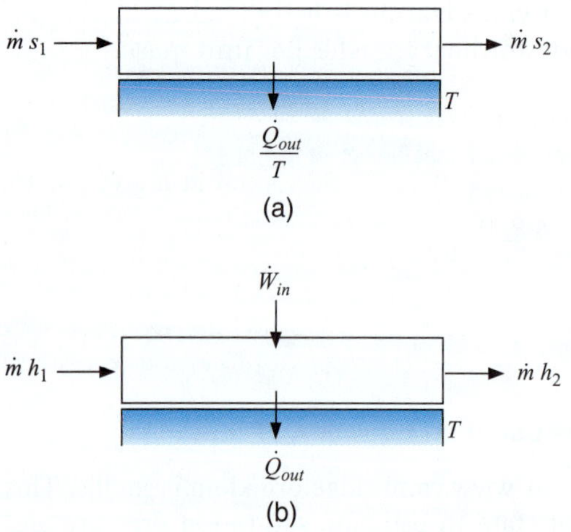

Figure 10-17: A reversible flow system undergoing an isothermal process; (a) an entropy balance and (b) an energy balance.

Solving Eq. (10-214) for $\dot{Q}_{out}$ and substituting the result into Eq. (10-215) leads to:

$$\dot{W}_{in} = \dot{m}\left[\underbrace{(h_2 - T\,s_2)}_{g_2} - \underbrace{(h_1 - T\,s_1)}_{g_1}\right] = \dot{m}\,\Delta g_T \tag{10-216}$$

If the substance involved in this process obeys the ideal gas law, then the integration of Eq. (10-213) is accomplished by substituting the ideal gas law:

$$\Delta g_T = \int_{P_1}^{P_2} v\,dP = R\,T\int_{P_1}^{P_2}\frac{dP}{P} \quad \text{for an ideal gas} \tag{10-217}$$

which results in:

$$\Delta g_T = R\,T\ln\left(\frac{P_2}{P_1}\right) \quad \text{for an ideal gas} \tag{10-218}$$

If the substance does not obey the ideal gas law then it is more difficult to integrate Eq. (10-213) because the specific volume is related to pressure through an equation of state. We could select an equation of state and complete the integration numerically, but there is a simpler alternative. The property *fugacity* (f) is defined such that:

$$\boxed{dg_T = RT\,d\ln(f)} \tag{10-219}$$

Integration of Eq. (10-219) between two states at the same temperature, T, results in:

$$\Delta g_T = \int_{g_1}^{g_2} dg_T = R\,T\int_{g_1}^{g_2} d\ln(f) \tag{10-220}$$

which leads to:

$$\Delta g_T = R\,T\ln\left(\frac{f_2}{f_1}\right) \tag{10-221}$$

Comparing Eqs. (10-218) and (10-221) suggests that the fugacity must be equal to the absolute pressure in the limit that the substance behaves as an ideal gas:

$$\lim_{P\to 0} f = P \tag{10-222}$$

Fugacity is a pseudo-pressure. It has the same units as pressure. The difference between fugacity and pressure is an indirect indication of the degree of non-ideal gas behavior of a substance. Fugacity is used for problems in multi-component phase and chemical equilibrium involving substances that do not obey the ideal gas law.

The function Fugacity in EES provides the fugacity of a pure fluid at a given state, specified in the usual way. The function Fugacity behaves in the same way as other EES property functions (e.g., volume or pressure). Figure 10-18 illustrates the ratio of the fugacity to the pressure, which is called the *fugacity coefficient* (ϕ) for water as a function of the specific volume at various values of the temperature. Notice that the fugacity coefficient approaches one in regions where the specific volume is large and therefore water behaves according to the ideal gas law.

Fugacity can be determined by substituting Eq. (10-213) into Eq. (10-221) to obtain:

$$R\,T\ln\left(\frac{f_2}{f_1}\right) = \int_{P_1}^{P_2} v\,dP \tag{10-223}$$

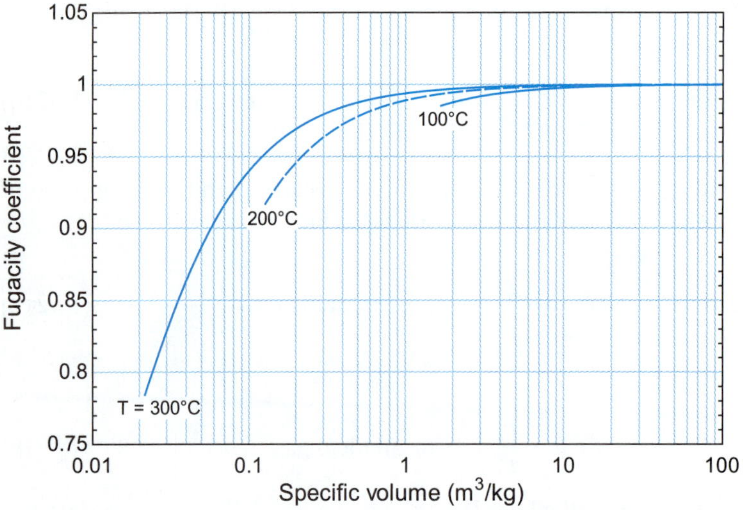

Figure 10-18: The fugacity coefficient as a function of specific volume for water at various values of temperature.

Equation (10-223) can be evaluated by numerically integrating specific volume with respect to pressure using an equation of state.

10.6.1 The Fugacity of Gases

It is convenient to set the pressure at state 1 in Eq. (10-223) to a pressure that is sufficiently low such that the ideal gas law is satisfied (P^o). In this case, $f_1 = P^o$ and $f_2 = f$ is the fugacity at temperature T and pressure P; with these substitutions Eq. (10-223) becomes:

$$\ln\left(\frac{f}{P^o}\right) = \frac{1}{RT}\int_{P^o}^{P} v\,dP \tag{10-224}$$

The integration required in Eq. (10-224) is not easy because most equations of state do not provide an explicit expression for specific volume. For example, the RKS equation of state from Eq. (10-54) is:

$$P = \frac{RT}{v-b} - \frac{a}{v\,(v+b)} \tag{10-225}$$

Also, at the lower limit of integration (P°, a pressure approaching zero), the integrand in Eq. (10-224) (v) becomes infinitely large. This issue can be resolved by adding and subtracting the integral of dP/P to the right side of Eq. (10-224):

$$\ln\left(\frac{f}{P^o}\right) = \frac{1}{RT}\int_{P^o}^{P} v\,dP - \int_{P^o}^{P}\frac{dP}{P} + \int_{P^o}^{P}\frac{dP}{P} \tag{10-226}$$

The first and second integrals on the right side of Eq. (10-226) are combined:

$$\ln\left(\frac{f}{P^o}\right) = \int_{P^o}^{P}\left(\frac{v}{RT} - \frac{1}{P}\right)dP + \int_{P^o}^{P}\frac{dP}{P} \tag{10-227}$$

The integrand of the first integral in Eq. (10-227) now approaches zero as the pressure approaches P° which makes numerical integration more tractable. The final integral in Eq. (10-227) is evaluated, leading to:

$$\ln\left(\frac{f}{P^o}\right) = \int_{P^o}^{P} \left(\frac{v}{RT} - \frac{1}{P}\right) dP + \ln(P) - \ln(P^\circ) \tag{10-228}$$

which provides:

$$\ln\left(\frac{f}{P}\right) = \int_{P^o}^{P} \left(\frac{v}{RT} - \frac{1}{P}\right) dP \tag{10-229}$$

Equation (10-229) is appropriate for numerical integration because the integrand is well-behaved. However, we still have the problem that most equations of state are not explicit for v but rather are written in terms of P. The integration of specific volume with respect to pressure in Eq. (10-229) can be transformed to an integration of pressure with respect to specific volume by applying integration by parts. Integration by parts utilizes the chain rule to rearrange an integral that involves the product of two variables; for example, the differential of the product of arbitrary variables y and z can be written as:

$$d(y\,z) = y\,dz + z\,dy \tag{10-230}$$

Integrating both sides of Eq. (10-230) leads to:

$$\int_{y_1\,z_1}^{y_2\,z_2} d(y\,z) = \int_{z_1}^{z_2} y\,dz + \int_{y_1}^{y_2} z\,dy \tag{10-231}$$

The term on the left side of Eq. (10-231) is the integral of an exact differential, therefore:

$$y_2\,z_2 - y_1\,z_1 = \int_{z_1}^{z_2} y\,dz + \int_{y_1}^{y_2} z\,dy \tag{10-232}$$

Equation (10-232) can be used to recast an integral of y with respect to z into an integration of z with respect to y:

$$\int_{z_1}^{z_2} y\,dz = y_2\,z_2 - y_1\,z_1 - \int_{y_1}^{y_2} z\,dy \tag{10-233}$$

For our purposes, we would like to use Eq. (10-233) in order to recast the integral of v with respect to P in Eq. (10-229), which is rearranged below:

$$\ln\left(\frac{f}{P}\right) = \frac{1}{RT}\int_{P^o}^{P} v\,dP - \int_{P^o}^{P} \frac{dP}{P} \tag{10-234}$$

into one that involves P with respect to v. Applying Eq. (10-233) with $y = v$ and $z = P$ leads to:

$$\int_{P^o}^{P} v\,dP = P\,v - \underbrace{P^o\,v^o}_{RT} - \int_{v^o}^{v} P\,dv \tag{10-235}$$

where v^o and v are the specific volumes at temperature T and pressures P° and P, respectively. The pressure P^o can be any pressure that is sufficiently low so that the substance obeys the ideal gas law. In the following analyses, P^o is chosen to be vanishingly small in order to ensure that the ideal gas law is observed. At this ideal gas state, $P^o v^o = R\,T$ and therefore Eq. (10-235) becomes:

$$\int_{P^o}^{P} v\,dP = P\,v - R\,T - \int_{v^o}^{v} P\,dv \tag{10-236}$$

Substituting Eq. (10-236) into Eq. (10-234) leads to:

$$\ln\left(\frac{f}{P}\right) = \frac{1}{RT}\left[P\,v - R\,T - \int_{v^o}^{v} P\,dv\right] - \int_{P^o}^{P} \frac{dP}{P} \tag{10-237}$$

or

$$\ln\left(\frac{f}{P}\right) = Z - 1 - \frac{1}{RT}\int_{v^o}^{v} P\,dv - \int_{P^o}^{P} \frac{dP}{P} \tag{10-238}$$

where Z is the compressibility factor. Equation (10-238) can be written in terms of the fugacity coefficient, the ratio of fugacity to pressure:

$$\ln(\phi) = Z - 1 - \frac{1}{RT}\int_{v^o}^{v} P\,dv - \int_{P^o}^{P} \frac{dP}{P} \tag{10-239}$$

It is possible to substitute a pressure-explicit equation of state directly into Eq. (10-238) and analytically carry out the integration.

Calculating Fugacity using the RKS and PR Equations of State

This extended section can be found online at www.cambridge.org/kleinandnellis. This section shows the steps required to substitute the RKS and PR equations of state into Eq. (10-238) in order to estimate the fugacity of a gas. After some algebra, the RKS equation of state provides:

$$\ln\left(\frac{f}{P}\right) = Z - 1 - \ln\left[Z - \frac{b\,P}{RT}\right] - \frac{a}{b\,R\,T}\ln\left[\frac{Z + \frac{b\,P}{RT}}{Z}\right] \tag{10-250}$$

where a and b are defined in Eqs. (10-56) and (10-55). The PR equation of state provides:

$$\ln\left(\frac{f}{P}\right) = Z - 1 - \ln[Z - B] - \frac{A}{2\sqrt{2}B}\ln\left[\frac{Z + 2.414B}{Z - 0.414B}\right] \tag{10-251}$$

where A and B are defined in Eqs. (10-70) and (10-71).

10.6.2 The Fugacity of Liquids

The specific volume of most liquids is a weak function of pressure; that is, liquids are typically very nearly incompressible. We can take advantage of this behavior in order to determine the difference in fugacity between a subcooled liquid at temperature T and pressure P and a liquid at its saturation pressure at temperature T. Applying Eq. (10-223) between these states leads to:

$$RT\ln\left(\frac{f_{T,P}}{f_{T,P_{sat}}}\right) = \int_{P_{sat}}^{P} v\,dP \tag{10-252}$$

Making the approximation that v is constant provides:

$$RT\ln\left(\frac{f_{T,P}}{f_{T,P_{sat}}}\right) \approx v\,(P - P_{sat}) \tag{10-253}$$

The fugacity of a subcooled liquid is then given by:

$$f_{T,P} = f_{T,P_{sat}}\exp\left(\frac{v\,(P - P_{sat})}{R\,T}\right) \tag{10-254}$$

The value of the exponential in Eq. (10-254) is ordinarily close to unity unless very large pressures are involved. Therefore, the fugacity of a subcooled liquid will be close to the fugacity of the saturated liquid. For example, consider water at $T = 25°$C and $P = 1$ MPa.

```
$UnitSystem SI Mass Radian J Pa K
F$='Water'                              "fluid"
T=convertTemp(C,K, 25 [C])              "temperature"
P=1 [MPa]*convert(MPa, Pa)              "pressure"
```

The specific volume (v), saturation pressure (P_{sat}), and gas constant (R) are computed:

```
v=volume(F$,T=T,P=P)                    "specific volume"
P_sat=P_sat(F$,T=T)                     "saturation pressure"
R=R#/MolarMass(F$)                      "gas constant"
```

The exponential in Eq. (10-254) is computed:

```
expfactor=exp(v*(P-P_sat)/(R*T))        "exponential factor"
```

and found to be 1.007; therefore, the fugacity of water at 25°C and 1 MPa is only 0.7% different from the fugacity of saturated liquid water at the same temperature. This change in fugacity can usually be ignored and the fugacity of a subcooled liquid is often approximated using the fugacity of saturated liquid, i.e.:

$$f_{T,P} \approx f_{T,P_{sat}} = f_f \tag{10-255}$$

The change in the specific Gibbs free energy between a saturated liquid and a saturated vapor can be found by integrating Eq. (10-219) for a phase change process:

$$\int_{g_f}^{g_g} dg_T = g_g - g_f = \int_{f_f}^{f_g} RT\,d\ln(f) = RT\,\ln\left(\frac{f_g}{f_f}\right) \tag{10-256}$$

where f_f and f_g are the fugacities for saturated liquid and saturated vapor, respectively, at temperature T. The specific Gibbs free energy for saturated liquid and vapor are equal, as indicated in Eq. (10-190). Substituting this result into Eq. (10-256) results in:

$$f_f = f_g \tag{10-257}$$

The fugacity of the coexisting liquid and vapor phases are equal. Equations (10-255) and (10-257) show that the fugacity of a subcooled liquid is approximately the same as the fugacity of saturated vapor at the same temperature:

$$f_{T,P} \approx f_{T,P_{sat}} = f_f = f_g \tag{10-258}$$

Equation (10-258) is useful for estimating the fugacity of substances in liquid solutions.

REFERENCES

Beattie, J.A. and Bridgeman, O.C., "A New Equation of State For Fluids. II. Application To Helium, Neon, Argon, Hydrogen, Nitrogen, Oxygen, Air And Methane," *Journal of the American Chemical Society*, Vol. 50, pp. 3133–3138, (1928).

Benedict, M., Webb, G. B., and Rubin, L. C., "An Empirical Equation for Thermodynamic Properties of Light Hydrocarbons and Their Mixtures: I. Methane, Ethane, Propane, and n-Butane," *J. Chem. Phys*., Vol. 8, No.4, pp. 334–345, (1940).

Dieterici, C., *Ann. Phys. Chem. Wiedemanns*, Vol. 69, pp. 685, (1899).

Jacobsen, R.T., Stewart, R.B., and Jahangiri, M., "Thermodynamic Properties of Nitrogen from the Freezing Line to 2000 K at Pressures to 1000 MPa," *J. Phys, Chem, Ref. Data*, Vol. 15, No. 2, (1986).

Martin, J.J. and Hou, Y.C., "Development of an Equation of State for Gases," *A.I.Ch.E. Journal*, Vol. 1, pp. 142, (1955).

Martin, J.J., Kapoor, R.M., and De Nevers, N., "An Improved Equation of State for Gases," *A.I.Ch.E. Journal*, Vol. 5, No. 2, (1959).

Rackett, J., *Chem Eng. Data*, Vol. 15, pp. 514, (1970).

Redlich, O., Kwong, J.N.S, *Chem. Rev*., Vol. 44, pp. 233–244, (1949).

Peng, Ding-Yu, and Robinson, D.B, "A New Two-Constant Equation of State," *Ind. Eng. Chem. Fundam*., Vol. 15, No. 1, pp. 59, (1976).

Poling, B.E., Prausnitz, J.M. and O'Connell, J., *The Properties of Gases and Liquids*, McGraw-Hill, 5th Edition, ISBN 9780070116825,(2000).

Sengers, J.V. and Levelt Sengers, J.M.H., "The critical region," *Chem. & Engr. News*, Vol. 46, No. 25, pp. 104–118, (1968).

Soave, G., "Equilibrium Constants from a Modified Redlich-Kwong Equation of State," *Chemical Engineering Science*, Vol. 27, pp. 1197–1203, (1972).

Span, R., *Multiparameter Equations of State*, Springer-Verlag, Berlin, ISBN 3-540-67311-3, (2000).

Span, R., Lemmon, E.W,, Jacobsen, R.T., Wagner, W., and Yokozeki, A., "A Reference Equation of State for the Thermodynamic Properties of Nitrogen for Temperatures from 63.151 to 1000 K and Pressures to 2200 MPa," *J. Phys. Chem. Ref. Data*, Vol. 29, No. 6, (2000).

Span, R. and Wagner, W., "Equations of State for Technical Applications. III. Results for Polar Fluids," *International Journal of Thermophysics*, Vol. 24, No. 1, (2003).

van der Waals, J.D. *On the Continuity of the Gaseous and Liquid States*, Doctoral Dissertation, Universiteit Leiden, (1873).

van der Waals, J.D., "The Equation of State for Gases and Liquids," Nobel Lecture, Dec. 12, (1910).

Wagner and Pruss, *J. Phys. Chem. Ref. Data*, Vol. 22, pp. 783, (1993).

Problems

The problems included here have been selected from a larger set of problems that are available from the website associated with this book (www.cambridge.org/kleinandnellis).

A: Equations of State

10.A-1 One of the first advanced equations of state was the Beattie-Bridgeman equation of state originally proposed in 1928, which has the following form:

$$P = \frac{R_{univ}\, T\left(1 - \frac{c}{v\,T^3}\right)}{v^2}\left[v + B_0\left(1 - \frac{b}{v}\right)\right] - \frac{A_0}{v^2}\left(1 - \frac{a}{v}\right)$$

where A_0, B_0, a, b, and c, are substance-dependent coefficients. Values of these coefficients for many gases can found in the literature. For carbon dioxide, the gas of interest in this problem, the coefficients have the values provided in Table 10.A-1.

Table 10.A-1: Coefficients for the Beattie-Bridgeman equation of state for carbon dioxide (from Cravalho and Smith, Jr., *Engineering Thermodynamics*, Pitman, Boston, 1981).

Coefficient	Value
a	1.62129×10^3 m^3/kg
A_0	262.07 N-m^4/kg^2
b	1.6444×10^3 m^3/kg
B_0	2.3811×10^3 m^3/kg
c	1.4997×10^4 m$^{3-1}$K^3/kg

a) Test the Beattie-Bridgeman equation's ability to accurately predict the specific volume of carbon dioxide by comparing the specific volumes and compressibility factors obtained from the equation of state to a reliable source for isotherms at 250 K, 304.1 K and 350 K and pressures ranging from atmospheric to 200 bar. What is your assessment of the accuracy of the Beattie-Bridgeman equation of state for carbon dioxide?

b) It is difficult to accurately measure the critical specific volume, so there is some incentive to obtain it from the equation of state. One way to do so is to plot the partial derivative of pressure with respect to specific volume at constant temperature along the critical isotherm in order to determine the specific volume at which it is zero. Can you obtain an accurate estimate of the critical volume from the Beattie-Bridgeman equation of state in this manner?

10.A-2 The Benedict-Webb-Rubin (BWR) equation of state, originally proposed in 1940, has the following form:

$$P = \frac{R\,T}{\bar{v}} + \frac{B_o R\,T - A_o - \frac{C_o}{T^2}}{\bar{v}^2} + \frac{b\,R\,T - a}{\bar{v}^3} + \frac{a\,\alpha}{\bar{v}^6} + \frac{c}{\bar{v}^3\,T^2}\left(1 + \frac{\gamma}{\bar{v}^2}\right)\exp\left(-\frac{\gamma}{\bar{v}^2}\right)$$

where A_o, B_o, C_o, a, b, c, α and γ are substance-dependent coefficients. Values of these coefficients can be found in the literature. For methane, the gas of interest in this problem, the coefficients have the values indicated in Table 10.A-2 with P in Pa, $\bar{v}$ in m^3/kmol and T in K.

Table 10.A-2: Coefficients for the BWR equation of state for methane (from Van Wylen, G. and Sonntag, R., *Fundamentals of Classical Thermodynamics*, 3rd edition, Wiley, New York, 1986).

Coefficient	Value	Coefficient	Value
a	5000 Pa-m^3/kmol^3	A_o	187.91×10^3 Pa-m^6/kmol^2
b	0.003380 m^6/kmol^2	B_o	0.04260 m^3/kmol
c	2.578×10^8 K^2-Pa-m^9/kmol^3	C_o	2.287×10^9 K^2-Pa-m^6/kmol^2
α	1.244×10^{-4} m^9/kmol^3	γ	0.0060 m^6/kmol^2

a) Test the BWR equation's ability to accurately predict the molar specific volume of methane in the superheated, subcooled and saturated regimes by comparing the molar specific volumes obtained from the equation to a reliable source. Summarize your results in a table and indicate the percentage error.

b) It is much more difficult to accurately measure the critical molar specific volume than it is to measure critical temperature and critical pressure. Assuming that the BWR constants provided above are correct, determine the critical molar specific volume at which the critical isotherm exhibits a slope of 0. How does your result compare to the accepted value?

10.A-3 One proposed solution to the environmental problems caused by transportation vehicles is to use fuel cells powered by hydrogen fuel. In place of the gasoline tank, a pressurized hydrogen tank would be needed. Tank pressures as high as 800 atm have been suggested. Use the Guggenheim equation of state (Guggenheim, E.A., *Molecular Physics*, Vol. 9, pp. 199–200, 1965) to do your calculations and compare the results that it provides with a reliable source. The Guggenheim equation of state is given by:

$$z = \frac{-a}{R\,T\,v} + \frac{1}{\left(1 - \frac{b}{4\,v}\right)^4}$$

a) Determine the constants, a and b, for hydrogen by forcing Eqs. (10-17) and (10-21) to be satisfied. Critical property data for hydrogen are provided in EES.

b) Determine the hydrogen storage volume required to provide a vehicle range equivalent to 15 gallons of gasoline for storage pressures between 100 to 800 atm. Note that the density of gasoline on a mass basis is 70% of the density of water at 300 K and 1 atm. The energy content of one kg of gasoline is about 44.4 MJ/kg whereas the energy content of hydrogen is 119.95 MJ/kg.

c) Compare the critical compressibility factor predicted by the Guggenheim equation to the value determined from EES property functions or other reported critical point data.

10.A-4 It has been suggested that the two-parameter MMM equation of state proposed by Mohsen-Nia et al. (Mohsen-Nia, M., Moddaress, H., and Mansoori, G.A., "A Simple Cubic Equation of State for Hydrocarbons and Other Compounds," SPE Paper #26667, *Proceedings of the 1993 Annual Technical Conference and Exhibition of the Society of Petroleum Engineers*, Houston, TX, 1993) is more accurate than the PR or the RKS equations. The MMM equation can be expressed in the following form:

$$Z = \frac{\bar{v} + 1.319b}{\bar{v} - b} - \frac{a}{R_{univ}\ T^{3/2}\,(\bar{v} + b)}$$

where Z is the compressibility factor. The parameter a is determined from critical point information as follows:

$$a_{crit} = 0.486989\frac{R_{univ}^2\ T_{crit}^{2.5}}{P_{crit}},\ \alpha_1 = -0.036139 + 0.14167\,\omega,\ a = a_{crit}\frac{(1 + \alpha_1/T_r)^3}{(1 + \alpha_1)^3}$$

where T_{crit} and P_{crit} are the critical temperature and critical pressure, respectively, ω is the acentric factor, and T_r is the reduced temperature. The parameter b is determined as follows:

$$b_{crit} = 0.064662\frac{R_{univ}\ T_{crit}}{P_{crit}},\ \beta_1 = 0.0634 - 0.18769\,\omega,\ b = b_{crit}\frac{(1 + \beta_1/T_r)^3}{(1 + \beta_1)^3}$$

a) Test the MMM equation's ability to accurately predict the molar specific volume of carbon dioxide in the superheated, subcooled and saturated regimes by plotting the 240 K, 280 K, and 320 K isotherms over a range of pressures between atmospheric and 275 bar. Compare your predictions with a reliable source and prepare a short summary of your results indicating the accuracy of the equation in the different regimes.

b) Apply the requirements that the equation of state must have zero first and second derivatives of P with respect to v at the critical point and use this information to determine parameters a_{crit} and b_{crit}. Compare your results with the values provided above.

10.A-7 A heating plant for a building complex uses natural gas (methane) as a fuel. A concern has been raised regarding a possible interruption in service. The plant manager has suggested that they stockpile cylinders of methane to be used in such an emergency. Each cylinder has a volume of $V = 2\ \text{ft}^3$ and, when fully filled, the pressure is $P_s = 3000$ psia at a temperature of $T_{amb} = 70°\text{F}$. When combusted in the plant equipment, methane provides $HC = 18{,}060\ \text{Btu/lb}_m$.

a) Estimate the number of cylinders of methane needed to supply a heating load of $\dot{Q} = 500{,}000$ Btu/hr for a 24 hour period using the Peng-Robinson equation of state.

b) Compare the result in part (a) with the value obtained using EES property data.

10.A-8 The Carnahan-Starling-DeSantis (CSD) equation of state (Morrison, G and McLinden, M.O., *NBS Technical Note 1226*, August, 1986) has the following form:

$$Z = \frac{1 + \left(\frac{b}{4\bar{v}}\right) + \left(\frac{b}{4\bar{v}}\right)^2 + \left(\frac{b}{4\bar{v}}\right)^3}{\left[1 - \left(\frac{b}{4\bar{v}}\right)\right]^3} - \frac{a}{R_{univ}\ T\ (\bar{v} + b)}$$

where Z is the compressibility factor. It has been claimed that this equation equation of state provides accurate predictions of liquid density without the complex parameter fitting procedures required by more elaborate equations. The equation of state can also be used in refrigerant mixture calculations. When used for a pure fluid without additional data available, the values of the parameters a and b must be determined by requiring that the critical isotherm have a slope of zero and an inflection point at the critical point.

Like most equations of state, the CSD equation is explicit in pressure but implicit in specific volume. A numerical method (e.g., Newton's method) is needed to solve for molar specific volume at specified temperature and pressure. For some conditions, there may be three real solutions for the molar specific volume. In this case, the smallest and largest solutions are the estimated molar specific volumes of saturated liquid and vapor, respectively. The intermediate solution corresponds to an unstable state which is physically unrealizable.

In this problem, we will test the CSD equation's ability to accurately predict the behavior of R1234yf, which is a new refrigerant that has properties that are similar to R134a. Critical property data for R1234yf can be obtained with the P_crit, T_crit and v_crit property routines in EES. EES also provides preliminary property data for this fluid so that you can compare the results of the CSD equation of state with EES.

a) Using the critical temperature, pressure, and volume provided by EES, calculate the values of a and b (on a molar basis) for the CSD equation of state and determine the critical compressibility factor.
b) Use the CSD equation of state to determine the critical molar specific volume at the critical pressure and critical temperature provided by EES. Note that the values of a and b are needed to calculate the critical molar specific volume. These values should be obtained by repeating part (a), but using the critical molar specific volume obtained from the CSD equation of state in place of the critical molar specific volume supplied by EES. Compare the values of a and b and the critical compressibility factor with the values obtained in part (a).
c) Test the CSD equation of state by calculating the molar specific volume as a function of pressure for pressures ranging from 1 bar to 50 bar at temperatures of 500 K, 400 K, and 300 K. Compare the results with EES. (Note that at conditions below the critical temperature you will need to select the correct root for the CSD equation. One way to do this is to use the EES molar specific volume value as the guess for the CSD molar specific volume.)

10.A-9 Properties are commonly formulated in terms of a reduced Helmholtz free energy function, α, that can be represented according to:

$$\alpha(T, v) = \frac{(u - Ts)}{RT}$$

The reduced Helmholtz free energy function is a complete equation of state so that all thermodynamic properties can be determined from an equation in this form. The reduced Helmholtz function is ordinarily represented as the sum of an ideal gas and residual component (α_{IG} and α_{res}, respectively):

$$\alpha(T, v) = \alpha_{IG}(T, v) + \alpha_{res}(T, v)$$

This problem illustrates that a pressure explicit equation of state can be converted into the reduced Helmholtz form.

a) Show that:

$$\alpha_{res}(T, v) = \frac{1}{RT}\int_{v=\infty}^{v}\left(\frac{RT}{v} - P\right)_T dv$$

b) Use the Peng-Robinson equation of state to show that:

$$\alpha_{res}(T, v) = \ln\left(\frac{v}{v-b}\right) + \frac{a}{2RT\sqrt{2}b}\ln\left[\frac{v + \left(1-\sqrt{2}\right)b}{v + \left(1+\sqrt{2}\right)b}\right]$$

B. Evaluation of Properties

10.B-1 The isentropic index, k, is defined as the coefficient relating pressure and volume such that during an isentropic process, $P v^k$ = constant. For an ideal gas, k is the ratio of the specific heat capacities, c_P/c_v.

a) Derive a relation for k that is applicable for non-ideal gas behavior. Your relation should involve only the specific heat ratio and expressions involving pressure, specific volume, and temperature.

10.B-2 The specific heat at constant volume of a gas is $c_v = 825$ J/kg-K at 800 K. The molar mass of the gas is $MW = 30$ kg/kmol. The gas is known to obey the Berthelot equation of state at 800 K for pressures ranging from atmospheric to 100 MPa. The Berthelot equation of state is:

$$P = \frac{RT}{v-b} - \frac{a}{T v^2}$$

where $a = 21{,}420$ N-m^4-K/kg^2 and $b = 0.00126$ m^3/kg.

a) Prepare a plot of the constant volume and constant pressure specific heat capacities as a function of pressure for pressures between atmospheric and 100 MPa.

10.B-4 Consider a piece of rubber as a thermodynamic system. The differential work done on the rubber is given by $\delta W = -F\,dl$, where F is the force exerted on the rubber when it is extended to a length of l.

a) Assuming a reversible process, show that the fundamental property relation for this system is $dU = T\,dS + F\,dl$.

b) Develop an expression for the differential change in entropy of the system, dS, in terms of independent variables T and l. Your result should involve only T, F, l, and C_l. The parameter C_l is the heat capacity at constant length, defined as: $C_l = \left(\frac{\partial U}{\partial T}\right)_l$.

c) Derive an equation that can determine the temperature as a function of length for a reversible adiabatic stretching process, given the following equation of state for the rubber: $F = bT(l - l_o)$ where b is a positive-valued constant, l_o is the unstretched length at temperature T_o. Assume C_l to be constant.

d) Use the relationship determined in part (c) to determine whether the temperature of the rubber increases or decreases when it is adiabatically stretched.

10.B-6 Propane at $P = 50$ atm is heated at constant pressure from $T_1 = 500$ K to $T_2 = 600$ K in a constant volume container. Assume that the constant pressure specific heat of propane under ideal gas conditions is $c_P^o = 125$ J/gmol-K and that propane obeys the van der Waals equation of state:

$$P = \frac{R_{univ}\,T}{\bar{v} - b} - \frac{a}{\bar{v}^2}$$

where $a = 9.255$ liters^2-atm/gmol^2 and $b = 0.09033$ liters/gmol.

a) Determine the heat transfer and entropy change for this process per mole of propane.

10.B-7 The purpose of this problem is to determine the specific volume, specific enthalpy, and specific entropy of carbon dioxide using the Beattie-Bridgeman equation of state described in Problem 10.A-1. The ideal gas specific heat capacity of carbon dioxide is given by:

$$c_P^o = -3.7357\left[\frac{\text{kJ}}{\text{kmol-K}}\right] + 30.529\left[\frac{\text{kJ}}{\text{kmol-K}}\right]\theta^{0.5} - 4.1034\left[\frac{\text{kJ}}{\text{kmol-K}}\right]\theta + 0.024198\left[\frac{\text{kJ}}{\text{kmol-K}}\right]\theta^2$$

where θ is a dimensionless temperature, $\theta = T/100$ [K].

a) Write a program to calculate and plot the compressibility factor, specific enthalpy, and specific entropy of carbon dioxide as a function of pressure for 100 Pa $< P <$ 10e7 Pa for isotherms of 310 K and 350 K. Refer your values of h and s to reference values of $h_{ref} = 9.211$ kJ/kg and $s_{ref} = 0.03123$ kJ/kg-K, respectively, at $T_{ref} = 310$ K and $P_{ref} = 101.3$ kPa. These reference conditions will result in your values having the same reference states as used in EES. Compare your results with values from EES. At what conditions do significant errors occur and what is the major cause of these errors?

10.B-11 A superheated organic vapor flowing through an insulated pipeline passes through a restriction. Upstream of the restriction, the pressure is $P_{in} = 35$ bar and the temperature is $T_{in} = 230°$C. Downstream of the restriction, the pressure is $P_{out} = 31.5$ bar and the temperature is $T_{out} = 225.5°$C. Pressure, specific volume, and temperature data for this substance are provided in Table 10.B-11.

Table 10.B-11: Specific volume (m^3/kg) over an array of pressure and temperature for a superheated organic vapor.

	Temperature		
Pressure	220° C	230° C	240° C
30 bar	0.02302	0.02431	0.0255
35 bar	0.01839	0.01971	0.02089
40 bar	0.01466	0.01611	0.01733

a) Estimate an average value for the constant pressure specific heat capacity at the conditions encountered at the pipeline restriction.

b) Estimate the change in specific entropy of the vapor as it passes through the restriction.

c) It has been reported that in some sections of the pipeline, the temperature increases as it passes through a restriction. Is this possible? Would the entropy change be positive in this case? Explain.

10.B-13 When a pressure disturbance occurs in a compressible fluid, the disturbance travels with a velocity that depends on the state of the fluid. A sound wave is a very small pressure disturbance which can be approximated as an isentropic process. The speed of sound, c, is an easily measured thermodynamic quantity defined by:

$$c = \sqrt{\left(\frac{\partial P}{\partial \rho}\right)_s}$$

where P is pressure, ρ is the fluid density, and s is specific entropy.

a) Derive an equation for the speed of sound through a fluid that is described by the MMM equation of state proposed by Mohsen-Nia et al., which is described in Problem 10.A-4. Your equation should involve only of P, v, T, c_P/c_v and derivatives involving these properties. Use your equation to calculate and plot the speed of sound through carbon dioxide at $T = 350$, 300, and 250 K as a function of reduced pressures for values between $P_r =$ 0.1 and 1.0. Use the EES value of c_P/c_v in your evaluations.

b) Compare your results with the SoundSpeed function results provided by EES. Provide an explanation for differences between your results and the accepted values.

10.B-15 A gas obeys the equation of state:

$$P = \frac{R_{univ}\, T}{\bar{v} - b} - \frac{a}{\sqrt{T}\,\bar{v}^2}$$

where $a = 1.426\text{e}4$ kJ-m^3-K$^{1/2}$/kmol^2 and $b = 0.0211$ m^3/kmol. The specific heat capacity at constant volume under ideal gas conditions is $c_v^o = 30$ kJ/kmol-K. The molar specific volume of the gas is $\bar{v} = 0.423$ m^3/kmol at $T_1 = 400$ K. The gas is heated at constant volume to $T_2 = 800$ K.

a) Determine the initial and final pressures

b) Determine the change in internal energy per kmol of gas.

C. Phase Equilibrium

10.C-1 A $m = 180$ lbm person is planning to ice skate on blades that have a total area in contact with the ice of $A = 0.012$ in^2. The ice temperature is $T_{ice} = 28°$F. Data for water at its triple point are provided in Table 10.C-1.

Table 10.C-1: Triple point data for water.

Pressure	611.7 Pa
Temperature	273.16 K
Liquid specific volume	0.001000 m^3/kg
Solid specific volume	0.001091 m^3/kg
Liquid specific enthalpy	23.26 J/kg
Solid specific enthalpy	−333,316 J/kg

a) Will the ice melt under the blades?

10.C-2 The critical temperature and pressure of n-butane are $T_{crit} = 425.2$ K and $P_{crit} = 3{,}796$ kPa, respectively. Its molar mass is $MW = 58.12$ kg/kmol. In the temperature range between 300 K and 400 K, the saturation pressure of n-butane is represented by:

$$\ln(P_{sat}) = 21.54 - \frac{2,722.08}{T_{sat}}$$

where P_{sat} is in Pa and T_{sat} is in K. The specific volume of liquid n-butane in this temperature range is given approximately by:

$$v_f = 0.00535\left[\frac{\text{m}^3}{\text{kg}}\right] - 2.577\times10^{-5}\left[\frac{\text{m}^3}{\text{kg-K}}\right]T + 4.608\times10^{-8}\left[\frac{\text{m}^3}{\text{kg-K}^2}\right]T^2$$

a) Using these data, estimate the specific enthalpy of vaporization of n-butane liquid at 340 K using the Clausius-Clapeyron equation
b) Using these data, estimate the specific enthalpy of vaporization of n-butane liquid at 340 K using the Clapeyron equation
c) Compare the results to each other and to the value obtained from EES.

10.C-3 Shown in Figure 10.C-3 is a plot of pressure versus specific volume along an isotherm of 90°C for R134a calculated with the Peng-Robinson equation of state.

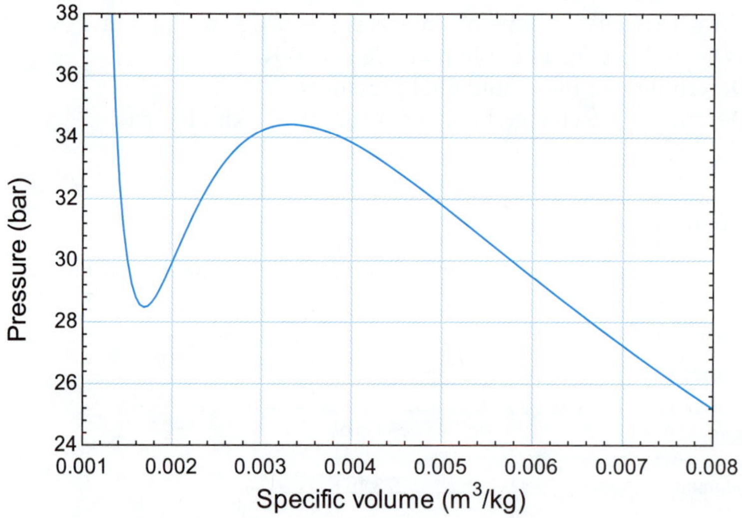

Figure 10.C-3 Pressure versus specific volume at 90°C for R134a determined with the Peng Robinson equation of state.

a) Using only this plot, estimate the vapor pressure of R134a at 90°C.
b) Compare your result with the vapor pressure provided by EES.

10.C-4 The saturation pressure, liquid density and vapor density of a refrigerant have been measured as a function of temperature. These data are reported in Table 10.C-4.

Table 10.C-4: Measured saturation data for a refrigerant.

Temperature (°C)	Pressure (bar)	Liquid density (kg/m³)	Vapor density (kg/m³)
50	2.346	1,414	12.88
55	2.707	1,402	14.75
60	3.11	1,389	16.82
65	3.557	1,375	19.11
70	4.05	1,362	21.64
75	4.594	1,348	24.42
80	5.191	1,334	27.48
85	5.844	1,320	30.84
90	6.556	1,306	34.52
95	7.332	1,291	38.54
100	8.174	1,276	42.94

a) Calculate and plot the specific enthalpy change of vaporization and the specific entropy change of vaporization as a function of temperature.
b) Estimate the normal boiling point for this refrigerant.

10.C-7 Refrigerant R134a at $T = 300$ K is isothermally compressed from an initial pressure of $P_1 = 100$ kPa until it reaches a condition of saturation. Use the Peng Robinson equation of state to solve this problem. You are welcome to use the Peng-Robinson library in EES. Compare your values with the R134a property data.

a) Determine the saturation pressure at T by equating the fugacities of the liquid and vapor.
b) Determine the change in specific volume.
c) Determine the change in specific enthalpy.
d) Determine the change in specific entropy.

10.C-8 The vapor pressure of water can be represented by the equation:

$$\ln(P_{sat}) = 32.99 - \frac{5301}{T_{sat}} - 1.2236 \ln(T_{sat})$$

where P_{sat} is the saturation pressure in Pa and T_{sat} is the saturation temperature in K.

a) Using only the information given above, prepare a plot of the specific enthalpy of vaporization of water as a function of temperature for temperatures between 300 K and 635 K. Compare the results with the steam table data.
b) Use the specific volume data provided in the steam tables to obtain a more correct result.

10.C-10 Using the Peng Robinson equation of state, estimate the saturation vapor pressure of carbon dioxide at 250 K and the corresponding specific volumes of saturated liquid and vapor by:

a) using the fundamental property relation involving Gibbs free energy, and by
b) equating the fugacities of saturated liquid and vapor.
c) Compare these estimates with values from a respected source.

11 Mixtures and Multi-Component Phase Equilibrium

In Section 2.1 it was stated that the number of intensive internal properties needed to fix the state of a non-reacting thermodynamic system, F, is given by:

$$F = C - \Pi + 2 \tag{11-1}$$

where C is the number of distinguishable chemical species in the system and Π is the number of phases (i.e., solid, liquid, or vapor) that are present. Up to this point, all of the thermodynamic systems that have been considered consist of a pure fluid or a *pseudo-pure fluid* for which $C = 1$. A pseudo-pure fluid is a mixture with a constant composition so that it can be treated in the same manner as a pure fluid. Dry air has been treated as a pseudo-pure fluid thus far in this text. However, gas mixtures arise in many common applications. For example, *psychrometric* processes involve air-water vapor mixtures and are discussed in Chapter 12. Chemical reaction processes involving mixtures are discussed in Chapters 13 and 14. Equation (11-1) indicates that the number of degrees of freedom increases with the number of components in a mixture; this increases the complexity associated with determining thermodynamic properties for mixtures. Unlike pure substances, tables of property information are rarely available for mixtures because there are too many degrees of freedom to make this approach practical. Instead, it is necessary to estimate the mixture properties in terms of the properties of the pure components that form the mixture. This process is considerably simplified if the components in a mixture obey the ideal gas law. This chapter discusses the properties of ideal gas mixtures as well as mixtures in which the components do not obey the ideal gas law. Two-phase liquid-vapor mixtures are also examined.

11.1 *P-v-T* Relations for Ideal Gas Mixtures

11.1.1 Composition Relations

The relative amounts of the different species in a mixture can be expressed on either a mass or molar basis. The total mass of the mixture (m) is the sum of the mass of each substance in the mixture (m_i, where i identifies the particular component):

$$m = \sum_{i=1}^{C} m_i \tag{11-2}$$

where C is the number of substances that are present in the mixture. The mass fraction of component i in the mixture (mf_i) is the ratio of the mass of component i to the total mass of the mixture:

$$mf_i = \frac{m_i}{m} \tag{11-3}$$

It is often more convenient to specify composition and specific thermodynamic properties on a molar basis rather than on a mass basis. A mole of substance i is defined as the amount of mass that is equal to the molar mass of substance i (MW_i). Therefore, the mass units need to be specified when specifying a mole. For example, if the mass unit is chosen to be kg then the corresponding mole is called a kmol (or kgmol). The molar mass of helium is 4 and therefore a kmol of helium has a mass of 4 kg. A pound mole (lbmol) of helium has a mass of 4 lb_m. A gram mole (gmol) of helium has a mass of 4 grams, and so on. Occasionally, the term "mole" is expressed without reference to mass units; in this case, mole usually refers to a gmol. The number of moles and the mass of substance i are related by:

$$n_i = \frac{m_i}{MW_i} \tag{11-4}$$

The total number of moles in a mixture (n) is the sum of the number of moles of each substance (n_i):

$$n = \sum_{i=1}^{C} n_i \tag{11-5}$$

The mole fraction of substance i (y_i) is the number of moles of that substance divided by the total number of moles of mixture:

$$y_i = \frac{n_i}{n} \tag{11-6}$$

Note that the mass (m) and the number of moles (n) of a mixture are related by MW, the molar mass of the mixture:

$$\frac{m}{n} = MW \tag{11-7}$$

Substituting Eq. (11-4) into Eq. (11-2) leads to:

$$m = \sum_{i=1}^{C} MW_i\, n_i \tag{11-8}$$

Dividing Eq. (11-8) by n, the total number of moles of mixture, provides:

$$\frac{m}{n} = \sum_{i=1}^{C} MW_i \frac{n_i}{n} \tag{11-9}$$

or

$$\frac{m}{n} = \sum_{i=1}^{C} MW_i\, y_i \tag{11-10}$$

Comparing Eq. (11-7) with Eq. (11-10) shows that:

$$MW = \sum_{i=1}^{C} y_i\, MW_i \tag{11-11}$$

The mass fractions and mole fractions are related by the molar mass. Substituting Eqs. (11-7) and (11-4) into Eq. (11-3) leads to:

$$mf_i = \frac{n_i\, MW_i}{n\, MW} \tag{11-12}$$

or

$$mf_i = y_i \frac{MW_i}{MW} \tag{11-13}$$

Consider dry air, which is a gas mixture consisting approximately of nitrogen ($y_1 = 0.781$), oxygen ($y_2 = 0.210$) and argon ($y_3 = 0.009$). The names of the components and mole fractions are assigned using the arrays C$ and y, respectively, in the EES code below.

```
$UnitSystem SI Mass Radian J K Pa
C=3                                  "Number of components"
C$[1..C]=['N2','02','Ar']            "Components"
y[1..C]=[0.781, 0.21, 0.009]         "Mole fractions"
```

Note that the assignments for an array can be accomplished on a single line using the protocol:

```
x[1..N]=[x1,x2,...,xN]
```

where the 1..N notation indicates the elements 1 through N of the array will be assigned using a comma-delimited (for US keyboard configurations) list enclosed in square brackets. An array containing the molar mass of each component is obtained using the duplicate command and the MolarMass function.

```
duplicate i=1,C
   MW[i]=MolarMass(C$[i])            "molar mass of component i"
end
```

Dry air can be modeled as a pseudo-pure fluid with an equivalent molar mass obtained using Eq. (11-11):

```
MW=sum(y[i]*MW[i],i=1,3)             "molar mass of the mixture"
```

which results in $MW = 28.96$ kg/kgmol. The mass fraction of each component can be obtained from Eq. (11-13).

```
duplicate i=1,C
   mf[i]=y[i]*MolarMass(C$[i])/MW    "mass fraction of each substance"
end
```

The mass fractions of nitrogen, oxygen, and argon in dry air are $mf_1 = 0.756$, $mf_2 = 0.232$, and $mf_3 = 0.0124$, respectively.

11.1.2 Mixture Rules for Ideal Gas Mixtures

For a pure gas, the phase rule indicates that there are two degrees of freedom. If the pressure and temperature are specified, for example, then the specific volume of a pure gas is fixed. For a gas mixture consisting of C components, the phase rule indicates that there are C+1 degrees of freedom. The additional degrees of freedom require specification of the mixture composition, which can be accomplished by providing C-1 mole fractions. (The last mole fraction can be obtained by requiring that the sum of the mole fractions equal unity.) For example, in a two component mixture ($C = 2$) there are three degrees of freedom. Therefore, in addition to specifying two intensive properties such as temperature and pressure it is also necessary to specify the mixture composition by setting one of the two mole fractions.

The focus in this section is on determining the relationship between pressure, specific volume, and temperature for an ideal gas mixture of specified composition. The ideal gas model provides a fairly accurate description of the *P-v-T* relationship for most low pressure gas mixtures existing at room temperature or higher and therefore it is a very

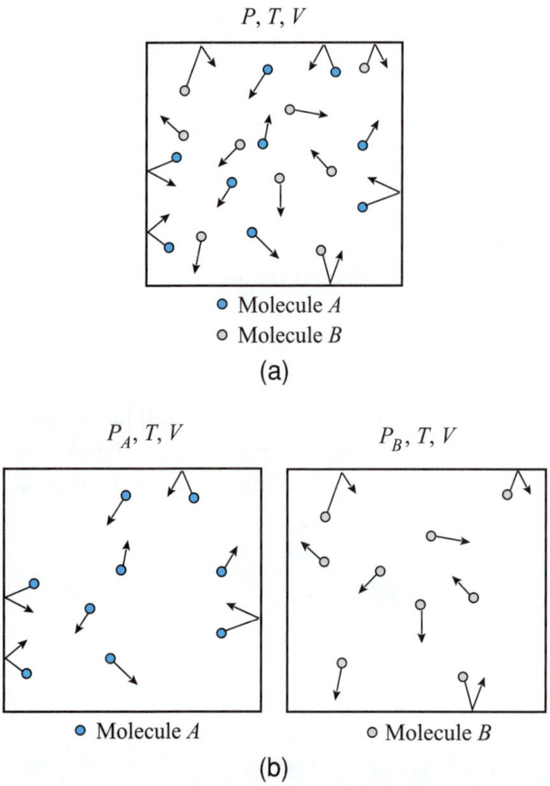

Figure 11-1: A conceptual representation of (a) a two-component mixture of molecules A and B and (b) the same molecules alone occupying the same volume at the same temperature.

useful model for engineering calculations in many applications. The ideal gas model assumes that each gas in the mixture behaves just as it would if the other gases were not present. Consider the two component mixture shown in Figure 11-1(a). The mixture consists of gases A and B. However, because the mixture obeys the ideal gas law, the P-v-T relationship for the mixture can be modeled with two systems, as shown in Figure 11-1(b). The two systems consist of the molecules of gas A and the molecules of gas B, each existing alone in the same volume at the same temperature as the original mixture shown in Figure 11-1(a).

The conceptual representation of an ideal gas mixture that is shown in Figure 11(b) is referred to as the the *rule of additive pressures*, which is also called *Dalton's rule*. Dalton's rule, approximates the pressure of a gas mixture, P, as the sum of the *partial pressures* of each of the component gases in the mixture:

$$\boxed{P = \sum_{i=1}^{C} P_i} \tag{11-14}$$

The partial pressure of a component in the mixture is the pressure that the component would exert if it were alone in the vessel at the same temperature and volume as the mixture, $P_i = P(\overline{v}_i, T)$, where $\overline{v}_i$ is the molar specific volume of component i. Physically, Eq. (11-14) makes sense provided that the different gases do not exhibit any intermolecular forces, as assumed by the ideal gas law. Pressure is the manifestation of the momentum associated with many individual molecules impacting the walls of the vessel. The pressure associated with the mixture is therefore made up of the partial pressures due to collisions associated with each of the components. For the two component mixture shown in Figure 11-1(a), the total pressure is the sum of the partial pressure of gas A and the partial pressure of gas B:

$$P = P_A + P_B \tag{11-15}$$

The partial pressure of each component is the pressure that the component would exert if it alone occupied the entire volume at the same temperature. This definition is entirely consistent with the idea of an ideal gas mixture, which is that each gas in the mixture behaves just as it would if the other gases were not present. Using the rule of additive pressures, each component behaves as if it occupies the entire volume at the same temperature. For the two component mixture shown in Figure 11-1(a), the total pressure is the sum of the pressures that would be exerted by the two systems shown in Figure 11-1(b). The pressure due to the molecules of component A existing alone is P_A, the partial pressure of component A, and the pressure due to the molecules of component B existing alone is P_B, the partial pressure of component B. Each gas in an ideal gas mixture obeys the ideal gas law. Therefore, the partial pressure of component A is:

$$P_A = \frac{n_A\, R_{univ}\, T}{V} \tag{11-16}$$

and the partial pressure of component B is:

$$P_B = \frac{n_B\, R_{univ}\, T}{V} \tag{11-17}$$

where n_A and n_B are the number of molecules of gases A and B, respectively, V is the volume of the container, and T is the temperature of the mixture. The total pressure experienced by the mixture is obtained by substituting Eqs. (11-16) and (11-17) into Eq. (11-15):

$$P = \frac{n_A\, R_{univ}\, T}{V} + \frac{n_B\, R_{univ}\, T}{V} \tag{11-18}$$

Rearranging Eq. (11-18) reveals that an ideal gas mixture itself obeys the ideal gas law:

$$P = \frac{(n_A + n_B)\; R_{univ}\, T}{V} \tag{11-19}$$

or

$$P = \frac{n\, R_{univ}\, T}{V} = \frac{R_{univ}\, T}{\overline{v}} \tag{11-20}$$

where $\overline{v}$ is the molar specific volume of the mixture. Dividing the partial pressure of component A, Eq. (11-16), by the total pressure of the mixture, Eq. (11-20), leads to:

$$\frac{P_A}{P} = \frac{n_A\, R_{univ}\, T}{V}\,\frac{V}{n\, R_{univ}\, T} = \frac{n_A}{n} = y_A \tag{11-21}$$

That is, the partial pressure of a component is equal to the product of its mole fraction and the total pressure:

$$P_i = y_i\, P \tag{11-22}$$

Equation (11-22) is physically intuitive. The fraction of the total pressure exerted on the walls of the container that can be attributed to component i must be related to the fraction of the total number of molecules that are present which are component i. Therefore, the partial pressure of component i should be directly related to its mole fraction. The concept of a partial pressure is broadly useful and it will be applied often in psychrometric and chemical reaction applications in Chapters 12 to 14.

There is an alternative representation of ideal gas mixtures that is referred to as the *rule of additive volumes*, also called *Amagat's rule* and the *ideal solution model*. Equation (11-5) requires that the total number of moles be equal to the sum of the number of moles of each component:

$$n = \sum_{i=1}^{C} n_i \tag{11-23}$$

Multiplying both sides of Eq. (11-23) by $R_{univ}\, T/P$ leads to:

$$\underbrace{\frac{n\, R_{univ}\, T}{P}}_{V} = \sum_{i=1}^{C} \underbrace{\frac{n_i\, R_{univ}\, T}{P}}_{\substack{V_i,\ partial\ volume \\ of\ component\ i}} \tag{11-24}$$

According to the ideal gas law, Eq. (11-20), the quantity on the left side of Eq. (11-24) must be the total volume of the mixture. The quantities in the summation on the right side of Eq. (11-24) are referred to as the *partial volumes* of each of the component gases. For an ideal gas mixture, the partial volume of component i is defined as:

$$V_i = \frac{n_i\, R_{univ}\, T}{P} \tag{11-25}$$

The partial volume of component i is equal to the volume that the component would occupy, alone, if it were at the same temperature and pressure as the mixture, $V_i = n_i\, \overline{v}_i\,(T, P)$. Substituting Eq. (11-25) into Eq. (11-24) provides a relation for the total volume that is referred to as Amagat's rule:

$$\boxed{V = \sum_{i=1}^{C} V_i} \tag{11-26}$$

Dalton's and Amagat's rules result in identical pressure-volume-temperature relationships for ideal gas mixtures. However, if one or more of the gases in the mixture do not behave according to the ideal gas law then Dalton's rule, Eq. (11-14), and Amagat's rule, Eq. (11-26), will lead to different results and neither is entirely correct, as discussed in Section 11.3.

11.2 Energy, Enthalpy, and Entropy for Ideal Gas Mixtures

Each component of an ideal gas mixture acts as if it occupies the entire volume of the mixture by itself at its partial pressure and the mixture temperature, as shown conceptually in Figure 11-1. Therefore, any extensive property of the mixture can be obtained by adding together that property for each of the components of the mixture. For example, the internal energy of an ideal gas mixture can be expressed as:

$$U = U_1 + U_2 + \cdots = \sum_{i=1}^{C} U_i \tag{11-27}$$

where U_i is the partial internal energy of component i. The partial molar specific internal energy of each component, $\overline{u}_i$, can be determined using the property relations for a pure component evaluated at the partial pressure of component i and the mixture temperature:

$$U = \sum_{i=1}^{C} n_i \, \overline{u}_i \,(T, P_i) \tag{11-28}$$

The molar specific internal energy of an ideal gas is independent of pressure and depends only on temperature. Therefore, the partial molar specific internal energy of gas i in an ideal gas mixture is identical to the molar specific internal energy of the pure gas i evaluated at the temperature of the mixture:

$$U = \sum_{i=1}^{C} n_i \, \overline{u}_i \,(T) \tag{11-29}$$

The molar specific internal energy for the mixture, $\overline{u}$, is defined by dividing Eq. (11-28) by the number of moles of gas mixture.

$$\overline{u} = \sum_{i=1}^{C} y_i \, \overline{u}_i \,(T) \tag{11-30}$$

The molar specific heat capacity at constant volume is defined as:

$$\overline{c}_v = \left(\frac{\partial \overline{u}}{\partial T}\right)_v \tag{11-31}$$

Substituting Eq. (11-30) into Eq. (11-31) leads to:

$$\overline{c}_v = \frac{\partial}{\partial T}\left[\sum_{i=1}^{C} y_i \, \overline{u}_i \,(T)\right]_v = \sum_{i=1}^{C} y_i \left(\frac{\partial \overline{u}_i}{\partial T}\right)_v \tag{11-32}$$

Therefore, the molar specific heat capacity at constant volume is equal to the mole fraction weighted average of the molar specific heat capacity at constant volume of the constituents:

$$\overline{c}_v = \sum_{i=1}^{C} y_i \, \overline{c}_{v,i} \tag{11-33}$$

Analogous relations can be derived for the enthalpy of a mixture; note that the molar specific enthalpy of an ideal gas is also only a function of temperature.

$$H = H_1 + H_2 + \cdots = \sum_{i=1}^{C} H_i = \sum_{i=1}^{C} n_i \, \overline{h}_i \,(T) \tag{11-34}$$

$$\overline{h} = \sum_{i=1}^{C} y_i \, \overline{h}_i \,(T) \tag{11-35}$$

The molar specific heat capacity at constant pressure is defined as:

$$\overline{c}_P = \left(\frac{\partial \overline{h}}{\partial T}\right)_P \tag{11-36}$$

Substituting Eq. (11-35) into Eq. (11-36) leads to:

$$\overline{c}_P = \sum_{i=1}^{C} y_i \, \overline{c}_{P,i} \tag{11-37}$$

In problems where the concentration of the mixture remains constant and the values of $\bar{c}_{v,i}$ and $\bar{c}_{P,i}$ for each component can be assumed to be nearly constant, it is useful to use Eqs. (11-33) and (11-37) to compute a mixture $\bar{c}_v$ and $\bar{c}_P$ and then treat the mixture as if it were a pure ideal gas. Equation (6-26) shows that the gas constant for a pure gas is equal to the difference between the specific heat capacites at constant pressure and consant volume. On a molar basis, the ideal gas constant for every component is equal to R_{univ} and therefore:

$$\bar{c}_P = \bar{c}_v + R_{univ} \tag{11-38}$$

Unlike specific internal energy and specific enthalpy, the specific entropy of an ideal gas depends on both temperature and pressure (or specific volume), as discussed in Section 6.3.3. The entropy of an ideal gas mixture is the sum of the entropy of each of the components:

$$S = S_1 + S_2 + \cdots = \sum_{i=1}^{C} S_i \tag{11-39}$$

The entropy of each component can be determined using the molar specific entropy evaluated at its partial pressure and temperature:

$$S = \sum_{i=1}^{C} n_i\, \bar{s}_i\,(T, P_i) \tag{11-40}$$

The molar specific entropy of the mixture is obtained by dividing Eq. (11-40) by the number of moles of the mixture:

$$\bar{s} = \sum_{i=1}^{C} y_i\, \bar{s}_i\,(T, P_i) \tag{11-41}$$

11.2.1 Changes in Properties for Ideal Gas Mixtures with Fixed Composition

The change in molar specific internal energy for a mixture undergoing a process with fixed composition can be obtained by integrating the molar specific heat capacity at constant volume of the mixture, defined according to Eq. (11-33):

$$\bar{u}_2 - \bar{u}_1 = \int_{T_1}^{T_2} \bar{c}_v\, dT \quad \text{for a fixed composition mixture} \tag{11-42}$$

In the limit that the molar specific heat capacity at constant volume of the mixture can be assumed to be constant over the temperature range of interest, Eq. (11-42) can be simplified to:

$$\bar{u}_2 - \bar{u}_1 = \bar{c}_v\,(T_2 - T_1) \quad \text{for a fixed composition mixture with constant } \bar{c}_v \tag{11-43}$$

Similarly, the change in the molar specific enthalpy of a mixture can be evaluated according to:

$$\bar{h}_2 - \bar{h}_1 = \int_{T_1}^{T_2} \bar{c}_P\, dT \quad \text{for a fixed composition mixture} \tag{11-44}$$

where $\overline{c}_P$ is the molar specific heat capacity at constant pressure for the mixture, defined according to Eq. (11-37). If $\overline{c}_P$ can be assumed to be constant, then Eq. (11-44) becomes:

$$\overline{h}_2 - \overline{h}_1 = \overline{c}_P (T_2 - T_1) \quad \text{for a fixed composition mixture with constant } \overline{c}_P \quad (11\text{-}45)$$

The change in the molar specific entropy of a mixture undergoing a constant composition process is obtained using Eq. (11-41) where the molar specific entropy of component *i* in an ideal gas mixture can be evaluated as if gas *i* is pure and exists at the temperature of the mixture and at a pressure that is equal to the partial pressure that it exhibits in the gas mixture:

$$\overline{s}_2 - \overline{s}_1 = \sum_{i=1}^{C} y_i \left[\overline{s}_i (T_2, P_{i,2}) - \overline{s}_i (T_1, P_{i,1})\right] \quad (11\text{-}46)$$

Substituting the expression for the specific entropy change of an ideal gas, Eq. (6-28), into Eq. (11-46) provides:

$$\overline{s}_2 - \overline{s}_1 = \sum_{i=1}^{C} y_i \left[\int_{T_1}^{T_2} \frac{\overline{c}_{P,i}}{T} dT - R_{univ} \ln\left(\frac{P_{i,2}}{P_{i,1}}\right)\right] \quad (11\text{-}47)$$

Substituting Eqs. (11-22) and (11-37) into Eq. (11-47) results in a relation for the molar specific entropy difference for a mixture of fixed composition:

$$\overline{s}_2 - \overline{s}_1 = \int_{T_1}^{T_2} \frac{\overline{c}_P}{T} dT - R_{univ} \ln\left(\frac{P_2}{P_1}\right) \quad \text{for a fixed composition mixture} \quad (11\text{-}48)$$

In the limit that $\overline{c}_P$ can be assumed to be constant, Eq. (11-48) becomes:

$$\overline{s}_2 - \overline{s}_1 = \overline{c}_P \ln\left(\frac{T_2}{T_1}\right) - R_{univ} \ln\left(\frac{P_2}{P_1}\right) \quad \text{for a fixed composition mixture with constant } \overline{c}_P \quad (11\text{-}49)$$

In applications involving an ideal gas mixture with a constant composition, it is often easiest to define a mixture $\overline{c}_P$ and $\overline{c}_v$ and treat the mixture as a pseudo-pure ideal gas. Indeed, this is how we have dealt with dry air to this point in the text.

11.2.2 Enthalpy and Entropy Change of Mixing

It is often convenient to evaluate the property change that occurs when a mixture is formed by mixing pure gases that are each at the same temperature and pressure as the mixture. For example, the *molar specific enthalpy change of mixing* is given by:

$$\Delta\overline{h}_{mix} = \overline{h} - \sum_{i=1}^{C} y_i \overline{h}_i(T, P) \quad (11\text{-}50)$$

where $\overline{h}$ is the molar specific enthalpy of the mixture and $\overline{h}_i(T, P)$ is the molar specific enthalpy of each pure component evaluated at the mixture temperature and total pressure. Substituting $\overline{h}$ from Eq. (11-35) into Eq. (11-50) shows that the molar specific enthalpy change of mixing for an ideal gas mixture is zero:

$$\Delta\overline{h}_{mix} = \sum_{i=1}^{C} y_i\,\overline{h}_i\,(T) - \sum_{i=1}^{C} y_i\,\overline{h}_i(T) = 0 \quad \text{for an ideal gas mixture} \tag{11-51}$$

The molar specific volume change of mixing ($\Delta\overline{v}_{mix}$) and the molar specific internal energy change of mixing ($\Delta\overline{u}_{mix}$) for an ideal gas mixture are also zero. The *molar specific entropy change of mixing* is given by:

$$\Delta\overline{s}_{mix} = \overline{s} - \sum_{i=1}^{C} y_i\,\overline{s}_i\,(T, P) \tag{11-52}$$

where $\overline{s}_i(T, P)$ is the molar specific entropy of each component evaluated at the mixture temperature and total pressure. The molar specific entropy of an ideal gas mixture is given in Eq. (11-41); substituting Eq. (11-41) into Eq. (11-52) provides:

$$\Delta\overline{s}_{mix} = \sum_{i=1}^{C} y_i\,\overline{s}_i\,(T, P_i) - \sum_{i=1}^{C} y_i\,\overline{s}_i\,(T, P) \tag{11-53}$$

or

$$\Delta\overline{s}_{mix} = \sum_{i=1}^{C} y_i\,[\overline{s}_i\,(T, P_i) - \overline{s}_i\,(T, P)] \tag{11-54}$$

Substituting the expression for the change in the molar specific entropy of an ideal gas, Eq. (6-28), into Eq. (11-54) leads to:

$$\Delta\overline{s}_{mix} = \sum_{i=1}^{C} y_i \left[\int_{T}^{T} \frac{\overline{c}_{P,i}}{T} dT - R_{univ}\,\ln\left(\frac{P_i}{P}\right)\right] \tag{11-55}$$

The first integral in Eq. (11-55) is zero:

$$\Delta\overline{s}_{mix} = -\sum_{i=1}^{C} y_i\,R_{univ}\,\ln\left(\frac{P_i}{P}\right) \tag{11-56}$$

The partial pressure of a component in an ideal gas mixture is the product of the total pressure and the mole fraction, therefore Eq. (11-56) can be written as:

$$\Delta\overline{s}_{mix} = -\sum_{i=1}^{C} y_i\,R_{univ}\,\ln\left(y_i\right) \tag{11-57}$$

Equation (11-57) shows that the molar specific entropy change of mixing is always greater than zero for an ideal gas mixture because the pressure of gas i is reduced from the total pressure, P, to its partial pressure, P_i, during the mixing process. This pressure reduction results in an entropy increase for each component during the process in which the pure gases are mixed isothermally.

EXAMPLE 11.2-1: POWER AND EFFICIENCY OF A GAS TURBINE

A volumetric analysis of a gas mixture resulting from a combustion process shows that it consists of CO_2 ($y_1 = 0.036$), O_2 ($y_2 = 0.131$), H_2O ($y_3 = 0.072$), and N_2. The gas mixture enters a gas turbine at $T_{in} = 1204°C$ and $P_{in} = 11.4$ bar with a volumetric flow rate of $\dot{V}_{in} = 1.7\ m^3/s$. The gases exit the turbine at $T_{out} = 710°C$ and atmospheric pressure, $P_{out} = 1$ atm. Model the mixture as an ideal gas mixture.

a) Determine the mass flow rate of the gas mixture.

The inputs are entered in EES:

```
$UnitSystem Radian SI K Pa J Molar
C=4                                  "number of components"
C$[1..4]=['CO2', 'O2', 'H2O', 'N2']  "mixture components"
y[1..3]=[0.036, 0.131, 0.072]        "mole fractions of first 3 components"
T_in=converttemp(C,K,1204 [C])       "temperature of mixture entering turbine"
P_in=11.4 [bar]*convert(bar,Pa)      "pressure of mixture entering turbine"
T_out=converttemp(C,K,710 [C])       "temperature of mixture leaving turbine"
P_out=1 [atm]*convert(atm,Pa)        "pressure of mixture leaving turbine"
V_dot_in=1.7 [m^3/s]                 "volumetric flow rate entering turbine"
```

The mole fraction of nitrogen, y_4, is determined by requiring that the mole fractions sum to unity:

$$y_C = 1 - \sum_{i=1}^{C-1} y_i$$

```
Cminus1=C-1                          "number of components, minus 1"
y[C]=1-sum(y[1..Cminus1])            "mole fraction of final component"
```

The molar specific volume at the inlet to the turbine is determined with the ideal gas law:

$$\bar{v} = \frac{R_{univ}\, T_{in}}{P_{in}}$$

and the molar flow rate is obtained from:

$$\dot{n} = \frac{\dot{V}_{in}}{\bar{v}_{in}}$$

The molar mass of the mixture is obtained from Eq. (11-11):

$$MW = \sum_{i=1}^{C} y_i\, MW_i$$

The mass flow rate is related to the molar flow rate according to:

$$\dot{m} = \dot{n}\, MW$$

```
v_bar_in=R#*T_in/P_in                   "molar specific volume at turbine inlet"
n_dot=V_dot_in/v_bar_in                 "molar flow rate"
MW=sum(y[i]*MolarMass(C$[i]),i=1,C)     "molar mass of mixture"
m_dot=n_dot*MW                          "mass flow rate entering turbine"
```

which leads to $\dot{m} = 4.48$ kg/s.

b) Estimate the power output of the turbine assuming that it operates adiabatically.

The molar specific enthalpies of the mixture entering and leaving the turbine are calculated using Eq. (11-35):

$$\bar{h}_{in} = \sum_{i=1}^{C} y_i \, \bar{h}_i \left(T_{in}\right)$$

$$\bar{h}_{out} = \sum_{i=1}^{C} y_i \, \bar{h}_i \left(T_{out}\right)$$

The power is determined by an energy balance on the turbine, which is assumed to operate adiabatically at steady-state with negligible kinetic energy changes:

$$\dot{W}_t = \dot{n}\left(\bar{h}_{in} - \bar{h}_{out}\right)$$

```
h_bar_in=sum(y[i]*enthalpy(C$[i],T=T_in),i=1,C)       "molar specific enthalpy entering turbine"
h_bar_out=sum(y[i]*enthalpy(C$[i],T=T_out),i=1,C)     "molar specific enthalpy leaving turbine"
W_dot_t=n_dot*(h_bar_in-h_bar_out)                    "energy balance"
```

which results in $\dot{W}_t = 2.771 \times 10^6$ W (2.771 MW).

c) Determine the rate of entropy generation in the turbine.

The molar specific entropy of the gas mixture entering the turbine is evaluated from Eq. (11-41), where the molar specific entropy of each gas in the mixture is determined as if it were a pure gas at the inlet temperature and the partial pressure that it has at the turbine inlet, evaluated using Eq. (11-22):

$$\bar{s}_{in} = \sum_{i=1}^{C} y_i \, \bar{s}_i \left(T_{in}, \, y_i \, P_{in}\right)$$

The molar specific entropy of the mixture at the outlet is evaluated in the same way:

$$\bar{s}_{out} = \sum_{i=1}^{C} y_i \, \bar{s}_i \left(T_{out}, \, y_i \, P_{out}\right)$$

An entropy balance on the adiabatic turbine is:

$$\dot{S}_{gen} = \dot{n}\left(\bar{s}_{out} - \bar{s}_{in}\right)$$

```
s_bar_in=sum(y[i]*entropy(C$[i],T=T_in,P=P_in*y[i]),i=1,C)      "molar specific entropy entering turbine"
s_bar_out=sum(y[i]*entropy(C$[i],T=T_out,P=P_out*y[i]),i=1,C)   "molar specific entropy entering turbine"
S_dot_gen=n_dot*(s_bar_out-s_bar_in)                            "rate of entropy generation"
```

which leads to $\dot{S}_{gen} = 898.3$ W/K.

d) Estimate the isentropic efficiency of this turbine.

The isentropic efficiency of the turbine is the ratio of the actual power to the power that would be produced by a reversible turbine operating under the same conditions. An entropy balance on the reversible turbine requires that the molar specific entropy at the turbine inlet be equal to the molar specific entropy at the turbine outlet. However, the temperature of the gas mixture at the exit of the reversible turbine ($T_{s,out}$) is not known. Enter a guess for this temperature so that a value for the specific entropy at the exit of the turbine can be calculated.

```
T_s_out=1000 [K]          "guess for isentropic turbine exit temperature"
```

The molar specific entropy of the gas mixture leaving the isentropic turbine is computed according to:

$$\bar{s}_{s,out} = \sum_{i=1}^{C} y_i \, \bar{s}_i \left(T_{s,out}, y_i \, P_{out}\right)$$

```
s_bar_s_out=sum(y[i]*entropy(C$[i],T=T_s_out,P=P_out*y[i]),i=1,C)
          "molar specific entropy leaving isentropic turbine"
```

Solve and update the guess values. Now comment out the guess for the outlet temperature and in its place, specify that the molar specific entropy at the outlet must equal the molar specific entropy at the inlet:

$$\bar{s}_{s,out} = \bar{s}_{in}$$

```
{T_s_out=1000 [K]}        "guess for isentropic turbine exit temperature"
s_bar_s_out=s_bar_in      "entropy balance on isentropic turbine"
```

Solving should result in an outlet temperature for the isentropic turbine of $T_{s,out} =$ 830.2 K. Note that EES would not have found the correct value for T_s_out if its guess value were not first set to a reasonable value. The equation that defines s_bar_s_out is implicit in the temperature T_s_out when s_bar_s_out is set to a known value. EES must solve this equation iteratively. It will start this process using the guess value provided in the Variable Information dialog (which is 1 K by default) unless the guess value T_s_out is set to some other value. This equation has multiple solutions since the specific heat of the gases that are used to calculate the specific entropies depend upon temperature in a non-linear manner. If you do not change the guess value for T_s_out in the Variable Information dialog or set its value by using the Update Guess Values option, as was done in the above code, then EES will find the solution for $T_{s,out} = 87.8$ K, which is obviously not possible. You should always check the results of a calculation to ensure that the results are within a range that you expect.

With $T_{s,out}$ known, the molar specific enthalpy at the outlet of the reversible turbine can be directly calculated:

$$\bar{h}_{s,out} = \sum_{i=1}^{C} y_i \, \bar{h}_i \left(T_{s,out}\right)$$

which allows the power output for the reversible turbine to be determined:

$$\dot{W}_{s,t} = \dot{n}\left(\bar{h}_{in} - \bar{h}_{s,out}\right)$$

The isentropic turbine efficiency is calculated according to:

$$\eta_t = \frac{\dot{W}_t}{\dot{W}_{s,t}}$$

```
h_bar_s_out=sum(y[i]*enthalpy(C$[i],T=T_s_out),i=1,C)
                                          "molar specific enthalpy leaving isentropic turbine"
W_dot_s_t=n_dot*(h_bar_in-h_bar_s_out)    "energy balance on isentropic turbine"
eta_t=W_dot_t/W_dot_s_t                   "turbine isentropic efficiency"
```

which shows that the isentropic efficiency is $\eta_t = 0.7732$ (77.32%).

EXAMPLE 11.2-2: SEPARATING CO_2 FROM THE ATMOSPHERE

Figure 1 illustrates a proposed method for mitigating global climate change caused by carbon dioxide in the atmosphere.

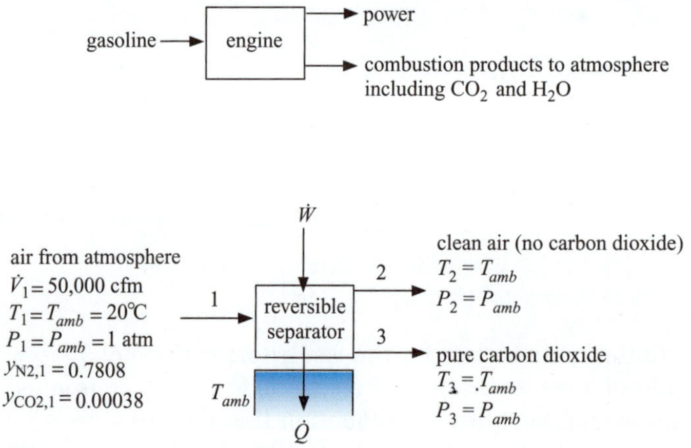

Figure 1: Burning gasoline in an engine and removing the resulting CO_2 from the atmosphere using a reversible separator.

Oil extracted from deep wells is refined to produce gasoline, which is eventually combusted in engines. The primary result of the combustion process is CO_2 and H_2O, which enter the atmosphere. The buildup of CO_2 in the atmosphere raises concerns related to climate change. Therefore, it has been suggested that oil-rich nations may want to invest in large CO_2-scrubbers that separate the CO_2 from the air and sequester it. The logic is that, by removing CO_2 from the atmosphere, it will be possible to continue to use oil-based fuels for a longer time period and therefore preserve the value of the world's oil reserves. This problem determines the thermodynamic lower limit to the amount of work that is required to accomplish this separation process. The overall practicality of the idea can be assessed based on this analysis.

A separator device draws in ambient air with a volumetric flow rate of $\dot{V}_1$ = 50,000 cfm at T_{amb} = 20°C and P_{amb} = 1 atm. The molar composition of the entering air is $y_{N2,1} = 0.7808$, $y_{CO2,1} = 0.00038$, and the balance of the incoming air is modeled as being oxygen. Assume that the separator removes all of the CO_2 from the air that leaves at state 2 and the pure CO_2 leaves the separator at state 3 to be stored. The clean air at state 2 and the pure CO_2 at state 3 are both at T_{amb} and

P_{amb}. In order to evaluate the minimum possible power required, assume that the separator is reversible. The separator operates at steady state and exchanges heat with the ambient at T_{amb}.

a) Determine the minimum power that will be required by the separator.

The inputs are entered in EES.

```
$UnitSystem SI Molar Radian K Pa J
T_amb=converttemp(C,K,20[C])                    "ambient temperature"
P_amb=1 [atm]*convert(atm,Pa)                   "ambient pressure"
V_dot[1]=50000 [cfm]*convert(cfm,m^3/s)         "volumetric flow rate entering separator"
y_N2[1]=0.7808 [-]                              "mole fraction of nitrogen entering separator"
y_CO2[1]=0.00038 [-]                            "mole fraction of CO2 entering separator"
```

The temperature and pressure at state 1 are given, $T_1 = T_{amb}$ and $P_1 = P_{amb}$. The mole fraction of oxygen at state 1 is computed according to:

$$y_{O2,1} = 1 - y_{N2,1} - y_{CO2}$$

The molar flow rate of nitrogen at state 1 is:

$$\dot{n}_{N2} = \frac{\dot{V}_1}{\bar{v}_{N2,1}} \tag{1}$$

where $\bar{v}_{N2,1}$ is the molar specific volume of nitrogen at the inlet, calculated using the ideal gas law:

$$\bar{v}_{N2,1} = \frac{R_{univ}\, T_1}{P_{N2,1}} \tag{2}$$

The partial pressure of nitrogen at the inlet is computed according to:

$$P_{N2,1} = y_{N2,1}\, P_1 \tag{3}$$

Combining Eqs. (1) through (3) provides the molar flow rate of nitrogen:

$$\dot{n}_{N2} = \frac{\dot{V}_1\, y_{N2,1}\, P_1}{R_{univ}\, T_1}$$

The molar flow rates of oxygen and carbon dioxide are computed similarly:

$$\dot{n}_{O2} = \frac{\dot{V}_1\, y_{O2,1}\, P_1}{R_{univ}\, T_1}$$

$$\dot{n}_{CO2} = \frac{\dot{V}_1\, y_{CO2,1}\, P_1}{R_{univ}\, T_1}$$

```
"State 1"
T[1]=T_amb                                      "temperature"
P[1]=P_amb                                      "pressure"
y_O2[1]=1-y_N2[1]-y_CO2[1]                      "mole fraction of O2 entering separator"
n_dot_N2=V_dot[1]*y_N2[1]*P[1]/(R#*T[1])        "molar flow rate of N2"
n_dot_O2=V_dot[1]*y_O2[1]*P[1]/(R#*T[1])        "molar flow rate of O2"
n_dot_CO2=V_dot[1]*y_CO2[1]*P[1]/(R#*T[1])      "molar flow rate of CO2"
```

The temperature and pressure at state 2 are given, $T_2 = T_{amb}$ and $P_2 = P_{amb}$. All of the carbon dioxide is removed from the air so that the mole fractions of nitrogen and oxygen that exit at state 2 are:

$$y_{N2,2} = \frac{\dot{n}_{N2}}{\dot{n}_{O2} + \dot{n}_{N2}}$$

and

$$y_{O2,2} = 1 - y_{N2,2}$$

```
"State 2"
T[2]=T_amb                                  "temperature"
P[2]=P_amb                                  "pressure"
y_N2[2]=n_dot_N2/(n_dot_N2+n_dot_02)        "mole fraction of nitrogen"
y_02[2]=1-y_N2[2]                           "mole fraction of oxygen"
```

The pressure and temperature at state 3 are given, $T_3 = T_{amb}$ and $P_3 = P_{amb}$. State 3 consists of pure carbon dioxide.

```
"State 3"
T[3]=T_amb                                  "temperature"
P[3]=P_amb                                  "pressure"
```

An entropy balance on the reversible separator leads to:

$$\dot{n}_{N2}\,\overline{s}_{N2,1} + \dot{n}_{O2}\,\overline{s}_{O2,1} + \dot{n}_{CO2}\,\overline{s}_{CO2,1} = \frac{\dot{Q}}{T_{amb}} + \dot{n}_{N2}\,\overline{s}_{N2,2} + \dot{n}_{O2}\,\overline{s}_{O2,2} + \dot{n}_{CO2}\,\overline{s}_{CO2,3}$$

which can be rearranged:

$$0 = \frac{\dot{Q}}{T_{amb}} + \dot{n}_{N2}\left(\overline{s}_{N2,2} - \overline{s}_{N2,1}\right) + \dot{n}_{O2}\left(\overline{s}_{O2,2} - \overline{s}_{O2,1}\right) + \dot{n}_{CO2}\left(\overline{s}_{CO2,3} - \overline{s}_{CO2,1}\right) \quad (4)$$

The expressions for the isothermal changes in the molar specific entropy of an ideal gas are substituted into Eq. (4):

$$0 = \frac{\dot{Q}}{T_{amb}} + \dot{n}_{N2}\left[-R_{univ}\ln\left(\frac{P_{N2,2}}{P_{N2,1}}\right)\right] + \dot{n}_{O2}\left[-R_{univ}\ln\left(\frac{P_{O2,2}}{P_{O2,1}}\right)\right] + \dot{n}_{CO2}\left[-R_{univ}\ln\left(\frac{P_{CO2,3}}{P_{CO2,1}}\right)\right] \quad (5)$$

The partial pressure of each component is written as the product of the mole fraction and total pressure. The atmospheric pressure cancels from all of the logarithmic terms:

$$0 = \frac{\dot{Q}}{T_{amb}} + \dot{n}_{N2}\left[-R_{univ}\ln\left(\frac{y_{N2,2}}{y_{N2,1}}\right)\right] + \dot{n}_{O2}\left[-R_{univ}\ln\left(\frac{y_{O2,2}}{y_{O2,1}}\right)\right] + \dot{n}_{CO2}\left[-R_{univ}\ln\left(\frac{1}{y_{CO2,1}}\right)\right] \quad (6)$$

Solving Eq. (6) for the rate of heat transfer leads to:

$$\dot{Q} = T_{amb} R_{univ} \left[\dot{n}_{N2} \ln\left(\frac{y_{N2,2}}{y_{N2,1}}\right) + \dot{n}_{O2} \ln\left(\frac{y_{O2,2}}{y_{O2,1}}\right) + \dot{n}_{CO2} \ln\left(\frac{1}{y_{CO2,1}}\right) \right]$$

```
0=Q_dot/T_amb-R#*n_dot_O2*ln(y_O2[2]/y_O2[1])-R#*n_dot_N2*ln(y_N2[2]/y_N2[1])&
 -R#*n_dot_CO2*ln(1/y_CO2[1])                    "entropy balance on reversible separator"
```

which results in $\dot{Q} = 8064$ W. An energy balance on the separator is:

$$\dot{W} + \dot{n}_{N2}\,\overline{h}_{N2,1} + \dot{n}_{O2}\,\overline{h}_{O2,1} + \dot{n}_{CO2}\,\overline{h}_{CO2,1} = \dot{Q} + \dot{n}_{N2}\,\overline{h}_{N2,2} + \dot{n}_{O2}\,\overline{h}_{O2,2} + \dot{n}_{CO2}\,\overline{h}_{CO2,3} \tag{7}$$

Rearranging Eq. (7) provides:

$$\dot{W} = \dot{Q} + \dot{n}_{N2}\left(\overline{h}_{N2,2} - \overline{h}_{N2,1}\right) + \dot{n}_{O2}\left(\overline{h}_{O2,2} - \overline{h}_{O2,1}\right) + \dot{n}_{CO2}\left(\overline{h}_{CO2,3} - \overline{h}_{CO2,1}\right)$$

Recognizing that the molar specific enthalpy of an ideal gas is only a function of temperature leads to:

$$\dot{W} = \dot{Q}$$

```
W_dot=Q_dot    "energy balance on reversible separator"
```

which provides $\dot{W} = 8064$ W.

b) Determine the minimum work per mole of carbon dioxide required to separate the CO_2 from the atmosphere. Compare this value to the power produced per mole of CO_2 emitted from an engine. Assume that the engine has an efficiency of $\eta_{engine} = 0.35$ and burns gasoline with chemical composition C_8H_{18}, molar mass $MW_{C_8H_{18}} = 114.2$ kg/kgmol, and heat of combustion $HC_{C_8H_{18}} = 44.4$ MJ/kg.

The inputs related to the engine are entered in EES:

```
HC_C8H18=44.4 [MJ/kg]*convert(MJ/kg,J/kg)    "heat of combustion of gasoline"
MW_C8H18=114.2 [kg/kmol]                     "molar mass of gasoline"
eta_engine=0.35 [-]                          "efficiency of engine"
```

The power required by the separator per mole of carbon dioxide is computed according to:

$$\frac{W_{separator}}{n_{CO2}} = \frac{\dot{W}}{\dot{n}_{CO2}}$$

```
W_separator\n_CO2=W_dot/n_dot_CO2    "work required per kgmol of CO2"
```

which results in $W_{separator}/n_{CO2} = 2.163\times10^7$ J/kmol. The heat provided to the engine per mole of gasoline is given by:

$$\frac{Q_{engine}}{n_{C8H18}} = HC_{C_8H_{18}}\, MW_{C_8H_{18}}$$

The power produced by the engine per mole of gasoline is the product of the heat released and the engine efficiency:

$$\frac{W_{engine}}{n_{C8H18}} = HC_{C_8H_{18}} MW_{C_8H_{18}} \eta_{engine}$$

Each mole of gasoline that is burned produces 8 moles of carbon dioxide. Therefore:

$$\frac{W_{engine}}{n_{CO2}} = \frac{HC_{C_8H_{18}} MW_{C_8H_{18}} \eta_{engine}}{8}$$

```
W_engine\n_CO2=HC_C8H18*MW_C8H18*eta_engine/8
            "work produced per kgmol of CO2 produced by burning gasoline in an engine"
```

which results in $W_{engine}/n_{CO2} = 2.218 \times 10^8$ J/kgmol. Therefore, about 10% of the work potential of the gasoline must be used to remove the carbon dioxide that is produced from the atmosphere. Of course, the fraction is actually much larger than this value because any real separator will not operate near the reversible limit.

11.3 *P-v-T* Relations for Non-Ideal Gas Mixtures

Gas mixtures at sufficiently high pressures or low temperatures may not be accurately described by the ideal gas law. Liquid mixtures obviously do not obey the ideal gas law. Equations of state that provide relations between pressure, specific volume, and temperature for pure fluids were reviewed in Section 10.1. This section shows how the Redlich-Kwong-Soave (Soave, 1972) and Peng-Robinson (Peng and Robinson, 1976) equations of state can be applied to determine the pressure-specific volume-temperature relations of mixtures. Although the methodology is applicable to both gases and liquids, the RKS and PR equations of state are more accurate for gases.

The additional information needed for application of an equation of state to a gas mixture (beyond that required for pure gases) is the compositional dependence of the terms in the equation of state. Dalton's and Amagat's rules were introduced for ideal gas mixtures in Section 11.1. Both of these rules can be extended to describe mixtures of non-ideal gases.

11.3.1 Dalton's Rule

Dalton's rule, shown in Eq. (11-14), assumes that the partial pressures of pure substances that form the mixture are additive:

$$P = \sum_{i=1}^{C} P_i \tag{11-58}$$

The partial pressure of each pure gas, P_i, is the pressure that the gas would have if it occupied the entire volume alone at the mixture temperature. Therefore, the partial pressure is fixed by the molar specific volume and temperature according to the equation of state for the pure fluid:

$$P_i = P_i\left(\overline{v}_i, T\right) \tag{11-59}$$

where $\overline{v}_i$ is the molar specific volume of gas i:

$$\overline{v}_i = \frac{V}{n_i} = \frac{V}{y_i\, n} = \frac{\overline{v}}{y_i} \tag{11-60}$$

The mixture specific volume is estimated using Dalton's rule by forcing Eqs. (11-58) through (11-60) to be satisfied. Given a temperature, pressure, and molar composition, the mixture molar specific volume, $\overline{v}$, is assumed and used to compute each of the component molar specific volumes using Eq. (11-60). These component molar specific volumes are used in the pure fluid equations of state to determine the partial pressure of each gas, according to Eq. (11-59). Finally, the partial pressures are added in order to determine the total pressure, P, as shown in Eq. (11-58). The assumed value of the mixture molar specific volume, $\overline{v}$, must be adjusted in order to obtain the correct value of P. Therefore, some iteration is required if the known mixture properties are temperature, pressure and composition.

If the mixture is an ideal gas mixture, then Eq. (11-59) becomes:

$$P_i = \frac{R_{univ}\,T}{\overline{v}_i} = \frac{n_i\,R_{univ}\,T}{V} \tag{11-61}$$

For a non-ideal gas, the partial pressure in Eq. (11-59) can be expressed in terms of the compressibility factor, Z_i, defined in Section 10.1.1:

$$P_i = \frac{R_{univ}\,T\,Z_i\,(\overline{v}_i, T)}{\overline{v}_i} = \frac{n_i\,R_{univ}\,T\,Z_i\,(\overline{v}_i, T)}{V} \tag{11-62}$$

The compressibility factor can be estimated using an equation of state. Using Dalton's rule, the compressibility factor of each gas must be determined at the temperature of the mixture and the molar specific volume of the gas in the mixture. If gas i obeys the ideal gas law, then $Z_i = 1$.

11.3.2 Amagat's Rule

Amagat's rule is shown in Eq. (11-26) and assumes that the partial volumes of the pure components that form the mixture are additive:

$$V = \sum_{i=1}^{C} V_i \tag{11-63}$$

The partial volume of gas i is the volume that the pure gas would have if it existed alone at the temperature and total pressure of the mixture (T and P):

$$V_i = n_i\,\overline{v}_i\,(T, P) \tag{11-64}$$

where $\overline{v}_i$ is the molar specific volume of pure gas i evaluated at T and P. Substituting Eq. (11-64) into Eq. (11-63) provides:

$$V = \sum_{i=1}^{C} n_i\,\overline{v}_i\,(T, P) \tag{11-65}$$

Dividing Eq. (11-65) through by n, the total number of moles of mixture, leads to:

$$\overline{v} = \sum_{i=1}^{C} y_i\,\overline{v}_i\,(T, P) \tag{11-66}$$

The mixture molar specific volume is estimated using Amagat's rule directly using Eq. (11-66). Given a temperature, pressure, and molar composition, the molar specific volume of each component, $\overline{v}_i$, is calculated using the equations of state associated with each pure fluid evaluated at the mixture temperature and total pressure. The molar specific volume of the mixture is the mole fraction weighted average of the molar specific volume of each component, as given by Eq. (11-66). No iteration is required when the known mixture properties are temperature, pressure and composition.

If the mixture is an ideal gas mixture then the molar specific volume of each component in Eq. (11-66) is given by:

$$\overline{v}_i = \frac{R_{univ}\, T}{P} \tag{11-67}$$

For a non-ideal gas mixture, the molar specific volume of each component in Eq. (11-66) can be expressed in terms of the compressibility factor,

$$\overline{v}_i = \frac{R_{univ}\, T\, Z_i\,(T, P)}{P} \tag{11-68}$$

where the compressibility factor for each gas in the mixture, Z_i, is determined at the temperature and total pressure of the mixture.

Dalton's and Amagat's rules may appear to be identical, but closer inspection shows that they are not because the molar specific volume (or compressibility factors) for each gas in the mixture are not determined in the same manner. In Dalton's rule, the compressibility factor for gas i is determined at the mixture temperature and the molar specific volume that gas i would have if it were alone and occupying the entire volume. In Amagat's rule, the compressibility factor for gas i is determined at the mixture temperature and the total pressure. Therefore Dalton's and Amagat's rules do not result in the same pressure-specific volume-temperature behavior for a non-ideal gas mixture and neither rule is completely correct, as shown in Example 11.3-1. Mixtures that obey Amagat's rule form what is called an *ideal solution*.

11.3.3 Empirical Mixing Rules

A variety of empirical mixing rules have been proposed that can provide estimates of the pressure-specific volume-temperature behavior of gas mixtures; these mixing rules are more complex but also more accurate than either Dalton's or Amagat's rules.

Kay's Rule

The simplest mixing rule is Kay's rule, as described by Poling et al. (2000), which introduces pseudo-critical and pseudo-reduced properties. The pseudo-critical properties, referred to here as the effective critical properties, are simply the mole fraction weighted values of the critical properties:

$$T_{crit,eff} = \sum_{i=1}^{C} y_i\, T_{crit,i} \tag{11-69}$$

$$P_{crit,eff} = \sum_{i=1}^{C} y_i\, P_{crit,i} \tag{11-70}$$

$$v_{crit,eff} = \sum_{i=1}^{C} y_i\, v_{crit,i} \tag{11-71}$$

An averaged or effective acentric factor can be defined in a similar manner, if the acentric factor is needed in an equation of state.

$$\omega_{eff} = \sum y_i\, \omega_i \tag{11-72}$$

The pseudo-reduced temperature, or effective reduced temperature, is the ratio of the mixture temperature to the pseudo-critical temperature. Similar definitions apply for effective reduced pressure and specific volume.

$$T_{r,eff} = \frac{T}{T_{crit,eff}} \tag{11-73}$$

$$P_{r,eff} = \frac{P}{P_{crit,eff}} \tag{11-74}$$

$$v_{r,eff} = \frac{v}{v_{crit,eff}} \tag{11-75}$$

The pseudo-reduced properties of a mixture can be used in an equation of state, just as if they were reduced properties for a pure fluid as shown in Example 11.3-1. Other, more elaborate mixing rules exist and are also described by Poling et al. (2000).

Mixing Rules

Perhaps the best alternative for estimating the *P-v-T* relation for gas mixtures is to use mixing rules that have been developed for the Redlich-Kwong-Soave (RKS) and Peng-Robinson (PR) equations of state. The RKS and PR equations of state for pure fluids both rely on two parameters, a and b. Definitions of a and b for the RKS and PR equations of state are provided in Section 10.1.3. Soave (1972) has proposed the following mixing rules to determine the coefficients, a and b, that can be used to apply the RKS equation of state to a mixture:

$$a = \left(\sum_{i=1}^{C} y_i \sqrt{a_i}\right)^2 \tag{11-76}$$

$$b = \sum_{i=1}^{C} y_i \, b_i \tag{11-77}$$

where a_i and b_i are the parameters evaluated for pure component i. Peng and Robinson (1976) recommend the same mixing rule for the parameter b required by the RKS equation of state. However, they suggest that the parameter a in the PR equation of state be calculated according to:

$$a = \sum_{i=1}^{C}\sum_{j=1}^{C} y_i \, y_j \, a_{ij} \tag{11-78}$$

where

$$a_{ij} = (1 - k_{ij}) \sqrt{a_i}\sqrt{a_j} \tag{11-79}$$

and k_{ij} is a binary mixing parameter that is zero if $i = j$. If k_{ij} is not known for a set of components then it can be estimated using Eq. (11-80), as proposed by Poling et al. (2000):

$$k_{ij} = 1 - \left[\frac{2\left(v_{crit,i} v_{crit,j}\right)^{1/6}}{\left(v_{crit,i}^{1/3} + v_{crit,j}^{1/3}\right)}\right]^3 \tag{11-80}$$

where $v_{crit,i}$ and $v_{crit,j}$ are the critical specific volumes of gases i and j. Note that the relations for parameter a in Eqs. (11-76) and (11-78) are the same if the binary mixing parameters are zero.

EXAMPLE 11.3-1: SPECIFIC VOLUME OF A GAS MIXTURE

The molar composition of a two-component mixture is $y_1 = 0.8$ where component 1 is carbon dioxide and $y_2 = 0.2$ where component 2 is ammonia. The mixture is at $T = 150°C$ and $P = 1000$ kPa.

a) Use Dalton's rule with the RKS equation of state to estimate the molar specific volume and compressibility factor of the mixture.

The inputs are entered in EES:

```
$UnitSystem SI Molar J K Pa
T=convertTemp(C,K,150 [C])                 "temperature of the mixture"
P_kPa=10000 [kPa]                          "pressure, in kPa"
P=P_kPa*convert(kPa,Pa)                    "pressure"
C=2 [-]                                    "number of components"
G$[1..C]=['CarbonDioxide','Ammonia']       "names of the gases in the mixture"
y[1..C]=[0.20,0.80]                        "mole fractions of the gases in the mixture"
```

The critical temperature, critical pressure and the acentric factor for each pure gas ($T_{crit,i}$, $P_{crit,i}$, and ω_i) are obtained from the EES data base. These properties are used to determine the a and b parameters in the RKS equation of state for each pure component (a_i and b_i), as discussed in Section 10.1.1.

```
duplicate i=1,C
  T_crit[i]=T_crit(G$[i])                          "critical temperature of each gas"
  P_crit[i]=P_crit(G$[i])                          "critical pressure of each gas"
  v_bar_crit[i]=v_crit(G$[i])                      "critical molar specific volume of each gas"
  omega[i]=AcentricFactor(G$[i])                   "acentric factor of each gas"
  m[i]=0.480+1.574*omega[i]-0.176*omega[i]^2       "parameter m used in alpha for each gas"
  T_r[i]=T/T_crit[i]                               "reduced temperature of each gas"
  alpha[i]^0.5=1+m[i]*(1-sqrt(T_r[i]))             "parameter alpha used in a for each gas"
  a[i]=0.42747*alpha[i]*R#^2*T_crit[i]^2/P_crit[i] "parameter a in equation of state for each gas"
  b[i]=0.08664*R#*T_crit[i]/P_crit[i]              "parameter b in equation of state for each gas"
end
```

In order to implement Dalton's rule, we start by guessing the mixture molar specific volume, $\overline{v}_D$. Here, the ideal gas law is used to establish a reasonable guess.

```
"Dalton's Law"
v_bar_D=R#*T/P                             "initial guess for total molar specific volume"
```

Equation (11-60) is used to determine the molar specific volume of each component in the mixture ($\overline{v}_{D,i}$), based on this initial guess. The subscript D is used to indicate a result relating to Dalton's rule.

$$\overline{v}_{D,i} = \frac{\overline{v}_D}{y_i} \quad \text{for} \quad i = 1..C$$

```
duplicate i=1,C
  v_bar_D[i]=v_bar_D/y[i]                  "molar specific volume of gas i"
end
```

Dalton's rule evaluates the partial pressure of each component at the molar specific volume of the component and the total temperature of the mixture, as given by Eq. (11-59). Here, the RKS equation of state is utilized for this purpose:

$$P_{D,i} = \frac{R_{univ}\,T}{\left(\overline{v}_{D,i} - b_i\right)} - \frac{a_i}{\overline{v}_{D,i}\left(\overline{v}_{D,i} + b_i\right)} \quad \text{for} \quad i = 1..C$$

```
duplicate i=1,C
  P_D[i]=R#*T/(v_bar_D[i]-b[i])-a[i]/(v_bar_D[i]*(v_bar_D[i]+b[i]))   "partial pressure of gas i"
end
```

At this point, the assumed value of the molar specific volume of the mixture must be adjusted so that Dalton's rule, Eq. (11-58), is satisfied. The guess values for the problem are updated (select Update Guesses from the Calculate menu) and the assumed value of $\overline{v}_D$ is commented out. In its place, Dalton's rule, Eq. (11-58), is enforced:

$$P = \sum_{i=1}^{C} P_{D,i}$$

```
{v_bar_D=R#*T/P}    "initial guess for total molar specific volume"
P=sum(P_D[1..C])    "Dalton's law"
```

which requires that $\overline{v}_D = 0.2852$ m^3/kgmol. The compressibility factor estimated using Dalton's rule is:

$$Z_D = \frac{P\,\overline{v}_D}{R_{univ}\,T}$$

```
Z_D=P*v_bar_D/(R#*T)    "Dalton's law estimate of the compressibility factor"
```

which provides $Z_D = 0.8107$.

b) Use Amagat's rule with the RKS equation of state to estimate the molar specific volume and compressibility factor of the mixture.

Amagat's rule is easier to implement for this problem than Dalton's rule. The molar specific volume of each gas is calculated at the mixture temperature and total pressure. Using the RKS equation of state, the molar specific volumes are computed according to:

$$P = \frac{R_{univ}\,T}{\left(\overline{v}_{A,i} - b_i\right)} - \frac{a_i}{\overline{v}_{A,i}\left(\overline{v}_{A,i} + b_i\right)} \quad \text{for} \quad i = 1..C$$

where the subscript A indicates a result based on Amagat's rule.

```
"Amagat's Law"
duplicate i=1,C
  P=R#*T/(v_bar_A[i]-b[i])-a[i]/(v_bar_A[i]*(v_bar_A[i]+b[i]))   "molar specific volume of gas i"
end
```

The molar specific volume of the mixture estimated using Amagat's rule ($\overline{v}_A$) is obtained from Eq. (11-66):

$$\overline{v}_A = \sum_{i=1}^{C} y_i\,\overline{v}_{A,i}\,(T, P)$$

The compressibility factor is:

$$Z_A = \frac{P\,\overline{v}_A}{R_{univ}\,T}$$

```
v_bar_A=sum(y[i]*v_bar_A[i],i=1,C)     "Amagat's law"
Z_A=P*v_bar_A/(R#*T)                   "Amagat's law estimate of the compressibility factor"
```

which leads to $\overline{v}_A = 0.2526$ m^3/kgmol and $Z_A = 0.7179$.

c) Use Kay's method with the RKS equation of state to estimate the molar specific volume and compressibility factor of the mixture.

Kay's method is also easy to implement for this problem. The pseudo-critical temperature, pressure, and molar specific volume of the mixture ($T_{crit,eff}$, $P_{crit,eff}$, and $\overline{v}_{crit,eff}$) are calculated according to Eqs. (11-69) through (11-71) and the effective acentric factor of the mixture (ω_{eff}) is calculated according to Eq. (11-72).

```
"Kay's Rule"
T_crit_eff=sum(y[i]*T_crit[i],i=1,C)             "pseudo-critical temperature"
P_crit_eff=sum(y[i]*P_crit[i],i=1,C)             "pseudo-critical pressure"
v_bar_crit_eff=sum(y[i]*v_bar_crit[i],i=1,C)     "pseudo-critical molar specific volume"
omega_eff=sum(y[i]*omega[i],i=1,C)               "effective acentric factor"
```

The pseudo-reduced temperature is calculated using Eq. (11-73) and used to determine the effective RKS parameters that characterize the mixture, a and b.

```
m_eff=0.480+1.574*omega_eff-0.176*omega_eff^2      "effective parameter m used in alpha"
T_r_eff=T/T_crit_eff                               "effective reduced temperature"
alpha_eff^0.5=1+m_eff*(1-sqrt(T_r_eff))            "effective parameter alpha"
a=0.42747*alpha_eff*R#^2*T_crit_eff^2/P_crit_eff   "parameter a in equation of state"
b=0.08664*R#*T_crit_eff/P_crit_eff                 "parameter b in equation of state"
```

The molar specific volume of the mixture is determined by Kay's rule ($\overline{v}_K$) using the RKS equation of state implemented with the effective RKS parameters:

$$P = \frac{R_{univ}\,T}{(\overline{v}_K - b)} - \frac{a}{\overline{v}_K\,(\overline{v}_K + b)}$$

where the subscript K indicates a result determined using Kay's rule. The compressibility factor estimated using Kay's rule is:

$$Z_K = \frac{P\,\overline{v}_K}{R_{univ}\,T}$$

```
P=R#*T/(v_bar_K-b)-a/(v_bar_K*(v_bar_K+b))     "molar specific volume of gas mixture"
Z_K=P*v_bar_K/(R#*T)                           "Kay's rule estimate of the compressibility factor"
```

which results in $\bar{v}_K = 0.2572\ \mathrm{m^3/kgmol}$ and $Z_K = 0.7310$.

d) Use the mixing rules suggested by Soave (1972) with the RKS equation of state to estimate the molar specific volume and compressibility factor of the mixture.

The mixing rules suggested by Soave (1972) are used to compute the values of the RKS parameters, a and b, with Eqs. (11-76) and (11-77):

$$a = \left(\sum_{i=1}^{C} y_i \sqrt{a_i}\right)^2$$

$$b = \sum_{i=1}^{C} y_i\, b_i$$

```
"Mixing Rules"
a=sum(y[i]*sqrt(a[i]),i=1,C)^2          "value of a determined from mixing rule"
b=sum(y[i]*b[i],i=1,C)                  "value of b determined from mixing rule"
```

The molar specific volume of the mixture using the mixing rule ($\bar{v}_{MR}$) is determined with the RKS equation of state implemented with the RKS parameters:

$$P = \frac{R_{univ}\, T}{(\bar{v}_{MR} - b)} - \frac{a}{\bar{v}_{MR}(\bar{v}_{MR} + b)}$$

The compressibility factor estimated using the mixing rule is:

$$Z_{MR} = \frac{P\, \bar{v}_{MR}}{R_{univ}\, T}$$

```
P=R#*T/(v_bar_MR-b)-a/(v_bar_MR*(v_bar_MR+b))    "molar specific volume of gas mixture"
Z_MR=P*v_bar_MR/(R#*T)                           "mixing rule estimate of compressibility factor"
```

which results in $\bar{v}_{MR} = 0.2583\ \mathrm{m^3/kgmol}$ and $Z_{MR} = 0.7340$.

e) Plot the compressibility factors predicted using Dalton's rule, Amagat's rule, Kay's rule, and the mixing rules as a function of pressure.

The requested plot is shown in Figure 1. Unfortunately, the actual compressibility factor is not known for this mixture, but it is likely to be close to the result obtained

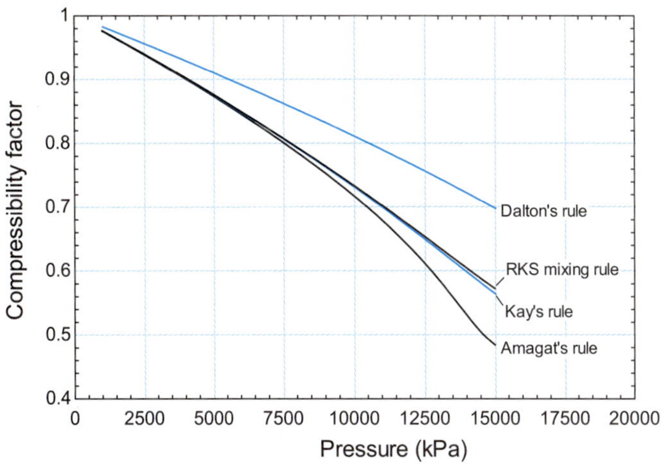

Figure 1: Compressibility factors predicted using Dalton's rule, Amagat's rule, Kay's rule, and the RKS mixing rule as a function of pressure for a mixture that is 20% CO_2 and 80% ammonia (by mole) at 150°C.

with the RKS mixing rules or Kay's method, which agree very well. Dalton's rule provides good results at low pressures, but it becomes increasingly inaccurate as the pressure increases and is not recommended for non-ideal gas mixtures. Amagat's rule agrees with Kay's rule and the RKS mixing rules for pressures up to about 100 atm (10000 kPa) in this case.

The methods described in this section for non-ideal gas mixtures could, in principle, be used for liquid mixtures as well. In particular, Amagat's rule of additive volumes provides reasonable estimates of liquid mixture specific volumes for mixtures that have chemically similar components. Mixtures that obey Amagat's law are called ideal solutions. The Redlich-Kwong-Soave and Peng-Robinson equations of state can be used for liquids as well as non-ideal gases, but the accuracy of these equations of state when they are applied to liquids is low compared to gases. The properties of liquid mixtures are often described in terms of empirical data presented in the form of activity coefficients, which are beyond the scope of this book. See Kyle (1999) or Smith et al. (2001) for more information on liquid mixtures.

11.4 Energy and Entropy for Non-Ideal Gas Mixtures

Property information for the specific volume, specific enthalpy and specific entropy as a function of pressure and temperature are generally available for pure fluids. Even when the pure fluid property relations are not available, they can be readily estimated using the methods discussed in Chapter 10. However, mixture property data are rarely available and therefore it is advantageous to be able to estimate these data based on information that is available for the pure components that make up the mixture.

11.4.1 Enthalpy and Entropy Changes of Mixing

The change in a property that occurs upon mixing can be defined in order to relate a mixture property to the properties of the pure components that compose the mixture at a specified temperature and pressure (T and P). For example, the volume change of mixing at a fixed temperature and pressure is the difference between the volume of the mixture (V) and the sum of the volumes of the pure components from which the mixture is formed (V_i):

$$\Delta V_{mix} = V - \sum_{i=1}^{C} V_i \tag{11-81}$$

Dividing Eq. (11-81) through by n, the total number of moles of mixture, leads to:

$$\frac{\Delta V_{mix}}{n} = \frac{V}{n} - \sum_{i=1}^{C} \frac{V_i}{n} \tag{11-82}$$

which can be rearranged to provide:

$$\underbrace{\frac{\Delta V_{mix}}{n}}_{\Delta \overline{v}_{mix}} = \underbrace{\frac{V}{n}}_{\overline{v}_{mix}} - \sum_{i=1}^{C} \underbrace{\frac{V_i}{n_i}}_{\overline{v}_i} \underbrace{\frac{n_i}{n}}_{y_i} \tag{11-83}$$

Therefore, the molar specific volume change of mixing is given by:

$$\Delta\overline{v}_{mix} = \overline{v} - \sum_{i=1}^{C} y_i\,\overline{v}_i\,(T, P) \tag{11-84}$$

In Eq. (11-84), $\overline{v}$ is the molar specific volume of the mixture and $\overline{v}_i\,(T, P)$ is the molar specific volume of pure component i evaluated at the mixture temperature and total pressure. Analogous relations can be written for the molar specific enthalpy of mixing and the molar specific entropy of mixing:

$$\Delta\overline{h}_{mix} = \overline{h} - \sum_{i=1}^{C} y_i\,\overline{h}_i\,(T, P) \tag{11-85}$$

$$\Delta\overline{s}_{mix} = \overline{s} - \sum_{i=1}^{C} y_i\,\overline{s}_i\,(T, P) \tag{11-86}$$

where $\overline{h}$ and $\overline{s}$ are the molar specific enthalpy and molar specific entropy, respectively, for the mixture. The quantities $\overline{h}_i\,(T, P)$ and $\overline{s}_i\,(T, P)$ are the molar specific enthalpy and molar specific entropy, respectively, of pure component i evaluated at the mixture temperature and total pressure. Methods for estimating $\overline{v}$ were presented in Section 11.3. For example, Amagat's rule in Eq. (11-66) results in:

$$\overline{v} = \sum_{i=1}^{C} y_i\,\overline{v}_i\,(T, P) \tag{11-87}$$

which, by inspection of Eq. (11-84) requires that the molar specific volume change of mixing, $\Delta\overline{v}_{mix}$, must be zero. Solutions for which $\Delta\overline{v}_{mix} = 0$ are called *ideal solutions*. The molar specific volume change of mixing therefore indicates the degree to which Amagat's rule is not satisfied (i.e., the deviation of the behavior of the mixture from an ideal solution).

Our concern in this section is with methods for estimating $\overline{h}$ and $\overline{s}$, the molar specific enthalpy and molar specific entropy of a mixture. If the composition is known and the specific properties of the pure components can be determined, as is generally the case, then $\overline{h}$ and $\overline{s}$ can be determined using Eqs. (11-85) and (11-86) provided that the values of $\Delta\overline{h}_{mix}$ and $\Delta\overline{s}_{mix}$ can be determined.

If the mixture obeys the ideal gas law then the molar specific enthalpy change of mixing at constant temperature must be zero. This was shown in Section 11.2.2 and the result is repeated below:

$$\Delta\overline{h}^{o}_{mix} = \overline{h}^{o} - \sum_{i=1}^{C} y_i\,\overline{h}^{o}_i\,(T) = 0 \tag{11-88}$$

where the supercript o is used to designate quantities that apply to ideal gas conditions. Equation (11-88) can be subtracted from Eq. (11-85) to provide:

$$\Delta\overline{h}_{mix} = \underbrace{\left(\overline{h} - \overline{h}^{o}\right)}_{-\overline{h}_{dep}} - \sum_{i=1}^{C} y_i \underbrace{\left[\overline{h}_i\,(T, P) - \overline{h}^{o}_i\,(T)\right]}_{-\overline{h}_{dep,i}} \tag{11-89}$$

The first term on the right side of Eq. (11-89) is the difference between the molar specific enthalpy of the mixture and the molar specific enthalpy that the mixture would have at the same temperature and pressure if it obeyed the ideal gas law. This term is related to the *molar specific enthalpy departure* for the mixture:

$$\text{molar specific enthalpy departure of the mixture} = \overline{h}_{dep} = \left(\overline{h}^{\circ} - \overline{h}\right) \quad (11\text{-}90)$$

The term $\overline{h}_i^{\circ}(T) - \overline{h}_i(T, P)$ in Eq. (11-89) is the molar specific enthalpy departure for pure gas i:

$$\text{molar specific enthalpy departure for pure gas } i$$
$$= \overline{h}_{dep,i} = \left[\overline{h}_i^{\circ}(T) - \overline{h}_i(T, P)\right] \quad (11\text{-}91)$$

Equation (11-89) can be rewritten in terms of the molar specific enthalpy departures of the mixture and of the pure components:

$$\boxed{\Delta\overline{h}_{mix} = \sum_{i=1}^{C} y_i\, \overline{h}_{dep,i} - \overline{h}_{dep}} \quad (11\text{-}92)$$

The molar specific enthalpy departure can be evaluated for the mixture if an equation of state is available for the mixture, as will be shown in Section 11.4.2. Molar specific enthalpy departures for each of the pure gases can be evaluated using an equation of state for the pure gases, as explained in Section 10.4. Equation (11-89) shows that knowledge of these molar specific enthalpy departures is sufficient to compute the molar specific enthalpy change of mixing, which allows the calculation of the molar specific enthalpy of the mixture according to Eq. (11-85).

The molar specific entropy change of mixing for an ideal gas mixture is presented in Section 11.2.2 and given by:

$$\Delta\overline{s}^{o}_{mix} = \overline{s}^{o}(T, P) - \sum_{i=1}^{C} y_i\, \overline{s}_i^{o}(T, P) = -R_{univ} \sum_{i=1}^{C} y_i \ln(y_i) \quad (11\text{-}93)$$

Subtracting Eq. (11-93) from Eq. (11-86) allows the molar specific entropy change of mixing to be expressed in terms of molar specific entropy departures:

$$\Delta\overline{s}_{mix} = \underbrace{(\overline{s} - \overline{s}^{\circ})}_{-\overline{s}_{dep}} - \sum_{i=1}^{C} y_i \underbrace{\left[\overline{s}_i(T, P) - \overline{s}_i^{\circ}(T, P)\right]}_{-\overline{s}_{dep,i}} - R_{univ} \sum_{i=1}^{C} y_i \ln(y_i) \quad (11\text{-}94)$$

The first term on the right side of Eq. (11-94) is the difference between the molar specific entropy of the mixture and the molar specific entropy that the mixture would have at the same temperature and pressure if it obeyed the ideal gas law. This term is related to the *molar specific entropy departure* for the mixture:

$$\text{molar specific entropy departure of the mixture} = \overline{s}_{dep} = (\overline{s}^{\circ} - \overline{s}) \quad (11\text{-}95)$$

The term $\overline{s}_i^{\circ}(T, P) - \overline{s}_i(T, P)$ in Eq. (11-94) is the molar specific entropy departure for pure gas i.

$$\text{molar specific entropy departure for pure gas } i = \overline{s}_{dep,i} = \left[\overline{s}_i^{\circ}(T, P) - \overline{s}_i(T, P)\right] \quad (11\text{-}96)$$

Equation (11-94) can be written in terms of molar specific entropy departures:

$$\boxed{\Delta\overline{s}_{mix} = \sum_{i=1}^{C} y_i\, \overline{s}_{dep,i} - \overline{s}_{dep} - R_{univ} \sum_{i=1}^{C} y_i \ln(y_i)} \quad (11\text{-}97)$$

The molar specific entropy departures can be evaluated if an equation of state is available for the mixture and for the pure gases. Equation (11-94) shows that knowledge of these molar specific entropy departures is sufficient to compute the molar specific entropy change of mixing, which allows the calculation of the molar specific entropy of the mixture according to Eq. (11-86).

11.4.2 Enthalpy and Entropy Departures

This section discusses the use of an equation of state to evaluate the molar specific enthalpy departure that is defined in Eqs. (11-90) and (11-91) for a mixture and for a pure gas, respectively and the molar specific entropy departure that is defined in Eqs. (11-95) and (11-96) for a mixture and a pure gas, respectively.

A relation for the molar specific enthalpy as a function of temperature and pressure is developed in Section 10.3.3 and given by Eq. (10-116); this relation is expressed on a molar basis as:

$$d\bar{h} = \bar{c}_P\, dT + \left[\bar{v} - T\left(\frac{\partial \bar{v}}{\partial T}\right)_P\right] dP \tag{11-98}$$

Equation (11-98) can be used to evaluate the molar specific enthalpy departure associated with both a pure gas and a mixture with a known composition. The molar specific enthalpy departure is the difference between the molar specific enthalpy that the fluid would have if it behaved as an ideal gas and the actual molar specific enthalpy at the same temperature and pressure. In order to evaluate the molar specific enthalpy departure, the molar specific enthalpy that an ideal gas would have at T and P is evaluated; this is $\bar{h}^{\circ}$ in Eq. (11-90). The value of $\bar{h}^{\circ}$ must be determined relative to the molar specific enthalpy of the real fluid at the same T and P. The real fluid will be an ideal gas at temperature T and pressure P° where P° is a sufficiently small pressure. The molar specific enthalpy of the real fluid at T and P° is $\bar{h}(T, P^{\circ})$. The difference between $\bar{h}^{\circ}$ and $\bar{h}(T, P^{\circ})$ is obtained by integrating Eq. (11-98) along a line of constant temperature from P° to P using Eq. (11-98), with the assumption that the substance behaves as an ideal gas:

$$\bar{h}^{\circ} - \bar{h}(T, P^{\circ}) = \int_{P^{\circ}}^{P} \underbrace{\left[\bar{v} - T\left(\frac{\partial \bar{v}}{\partial T}\right)_P\right]}_{\text{assuming ideal gas behavior}} dP \tag{11-99}$$

For an ideal gas the integrand of Eq. (11-99) is zero and therefore:

$$\bar{h}^{\circ} - \bar{h}(T, P^{\circ}) = 0 \tag{11-100}$$

It is next necessary to obtain the molar specific enthalpy that the actual mixture has at state T and P (i.e., $\bar{h}$ in Eq. (11-90)) relative to the molar specific enthalpy that it has at an ideal gas state at the same temperature (i.e., $\bar{h}(T, P^{\circ})$). This value is obtained by integrating Eq. (11-98) along the same line of constant temperature from P° to P, but without the assumption that the substance behaves as an ideal gas:

$$\bar{h} - \bar{h}(T, P^{\circ}) = \int_{P^{\circ}}^{P} \underbrace{\left[\bar{v} - T\left(\frac{\partial \bar{v}}{\partial T}\right)_P\right]}_{\text{assuming real gas behavior}} dP \tag{11-101}$$

Subtracting Eq. (11-101) from Eq. (11-100) provides the molar specific enthalpy departure of the mixture:

$$\overline{h}_{dep} = \left(\overline{h}^{\circ} - \overline{h}\right) = -\int_{P^{\circ}}^{P} \left[\overline{v} - T\left(\frac{\partial \overline{v}}{\partial T}\right)_P\right] dP \tag{11-102}$$

In order to evaluate Eq. (11-102) it is necessary to have an equation of state that provides the molar specific volume of the mixture, $\overline{v}$, in terms of temperature and pressure. Several equations of state for mixtures are presented in Section 11.3. In like manner, the molar specific enthalpy departure for each pure gas in the mixture can be evaluated according to:

$$\overline{h}_{dep,i} = \left(\overline{h}_i^{\circ} - \overline{h}_i\right) = -\int_{P^{\circ}}^{P} \left[\overline{v}_i - T\left(\frac{\partial \overline{v}_i}{\partial T}\right)_P\right] dP \tag{11-103}$$

where $\overline{v}_i$ is the molar specific volume of the pure gas.

A general relation for specific entropy as a function of temperature and pressure is provided in Eq. (10-122) and is expressed on a molar basis as:

$$d\overline{s} = \frac{\overline{c}_P}{T} dT - \left(\frac{\partial \overline{v}}{\partial T}\right)_P dP \tag{11-104}$$

The molar specific entropy departure is the difference between the molar specific entropy that the mixture would have if it behaved as an ideal gas and the actual molar specific entropy of the mixture at the same temperature and pressure. In order to evaluate the molar specific entropy departure it is first necessary to obtain the molar specific entropy that an ideal gas would have at state T and P (i.e., $\overline{s}^{\circ}$ in Eq. (11-95)). The real gas will behave as an ideal gas at temperature T and pressure P°, where P° is a sufficiently small pressure. The molar specific entropy of the real gas at T and P° is $\overline{s}\,(T, P^{\circ})$. The difference between $\overline{s}^{\circ}$ and $\overline{s}\,(T, P^{\circ})$ is obtained by integrating Eq. (11-104) along a line of constant temperature (T) from P° to P using Eq. (11-104) with the assumption that the substance behaves as an ideal gas:

$$\overline{s}^{\circ} - \overline{s}\,(T, P^{\circ}) = -\int_{P^{\circ}}^{P} \underbrace{\left(\frac{\partial \overline{v}}{\partial T}\right)_P}_{\substack{\text{assuming ideal}\\ \text{gas behavior}}} dP \tag{11-105}$$

For an ideal gas, the partial derivative of molar specific volume with respect to temperature at constant pressure is R_{univ}/P:

$$\overline{s}^{\circ} - \overline{s}\,(T, P^{\circ}) = -\int_{P^{\circ}}^{P} \frac{R_{univ}}{P} dP \tag{11-106}$$

Carrying out the integration in Eq. (11-106) leads to:

$$\overline{s}^{\circ} - \overline{s}\,(T, P^{\circ}) = -R_{univ} \ln\left(\frac{P}{P^{\circ}}\right) \tag{11-107}$$

Next, it is necessary to obtain the molar specific entropy that the actual mixture has at state T and P (i.e., $\overline{s}$ in Eq. (11-95)) relative to the molar specific entropy that the mixture has at pressure P° and the same temperature (i.e., $\overline{s}\,(T, P^{\circ})$, the state where the mixture behaves as an ideal gas). This difference is obtained by integrating Eq. (11-104) along a line of constant temperature (T) from P° to P, but this time without the assumption that the substance behaves as an ideal gas:

$$\overline{s} - \overline{s}(T, P^\circ) = -\int_{P^\circ}^{P} \underbrace{\left(\frac{\partial \overline{v}}{\partial T}\right)_P}_{\substack{\text{assuming real}\\ \text{gas behavior}}} dP \tag{11-108}$$

Subtracting Eq. (11-108) from Eq. (11-106) provides the molar specific entropy departure of the mixture:

$$\boxed{\overline{s}_{dep} = (\overline{s}^o - \overline{s}) = \int_{P^\circ}^{P} \left[\left(\frac{\partial \overline{v}}{\partial T}\right)_P - \frac{R_{univ}}{P}\right] dP} \tag{11-109}$$

In like manner, the molar specific entropy departure for pure gas i is:

$$\boxed{\overline{s}_{dep,i} = (s_i^o - s_i) = \int_{P^\circ}^{P} \left[\left(\frac{\partial \overline{v}_i}{\partial T}\right)_P - \frac{R_{univ}}{P}\right] dP} \tag{11-110}$$

Molar Specific Enthalpy and Entropy Departures from a Two-Parameter Equation of State

This extended section can be found online at www.cambridge.org/kleinandnellis. Two-parameter equations of state are presented in Section 10.1.3 for a pure fluid and extended to mixtures in Section 11.3 using Dalton's rule, Amagat's rule, and various mixing rules. Equations (11-102), (11-103), (11-109), and (11-110) (which are needed to calculate the enthalpy and entropy change of mixing) are inconvenient for use with a two-parameter equation of state because they involve molar specific volume and its derivatives whereas the equations of state are explicit in pressure and implicit in molar specific volume. In this section the algebraic manipulations required to obtain the specific enthalpy and entropy changes of mixing using a pressure explicit equation of state are presented. The final results appear in Eqs. (11-125) and (11-130):

$$\begin{aligned}\Delta\overline{h}_{mix} &= \left(\overline{h} - \overline{h}^\circ\right) - \sum_{i=1}^{C} y_i \left[\overline{h}_i(T,P) - \overline{h}_i^\circ(T)\right]\\ &= -\int_{\overline{v}^\circ}^{\overline{v}} \left[P - T\left(\frac{\partial P}{\partial T}\right)_{\overline{v}}\right] d\overline{v} + P\overline{v}\\ &\quad - \sum_{i=1}^{C} y_i \left[-\int_{\overline{v}_i^\circ}^{\overline{v}_i} \left[P - T\left(\frac{\partial P}{\partial T}\right)_{\overline{v}}\right] d\overline{v}_i + P\overline{v}_i\right]\end{aligned} \tag{11-125}$$

$$\begin{aligned}\Delta\overline{s}_{mix} &= (\overline{s} - \overline{s}^o) - \sum_{i=1}^{C} y_i(\overline{s}_i - \overline{s}_i^o) - R_{univ}\sum_{i=1}^{C} y_i \ln(y_i)\\ &= \int_{\overline{v}^\circ}^{\overline{v}} \left(\frac{\partial P}{\partial T}\right)_{\overline{v}} d\overline{v} - \sum_{i=1}^{C} y_i \left[\int_{\overline{v}_i^\circ}^{\overline{v}_i} \left(\frac{\partial P}{\partial T}\right)_{\overline{v}} d\overline{v}\right] - R_{univ}\sum_{i=1}^{C} y_i \ln(y_i)\end{aligned} \tag{11-130}$$

11.4.3 Enthalpy and Entropy for Ideal Solutions

A mixture that obeys Amagat's rule of additive volumes given by Eq. (11-63) is called an ideal solution. In this case:

$$\overline{v} - \sum_{i=1}^{C} y_i\,\overline{v}_i = 0 \tag{11-131}$$

In order to evaluate the molar specific enthalpy of mixing using Eq. (11-92) or the molar specific entropy of mixing using Eq. (11-97), it is necessary to determine the partial derivative of molar specific volume with respect to temperature at constant pressure. For an ideal solution, the partial derivative of molar specific volume can be obtained from Eq. (11-131):

$$\left(\frac{\partial \overline{v}}{dT}\right)_P - \sum_{i=1}^{C} y_i \left(\frac{\partial \overline{v}_i}{dT}\right)_P = 0 \tag{11-132}$$

Substituting Eqs. (11-131) and (11-132) into Eqs. (11-92), (11-102), and (11-103) leads to:

$$\Delta\overline{h}_{mix} = \int_{P^\circ}^{P} \left[\overline{v} - T\sum_{i=1}^{C} y_i \left(\frac{\partial \overline{v}_i}{dT}\right)_P\right] dP - \sum_{i=1}^{C} y_i \int_{P^\circ}^{P} \left[\overline{v}_i - T\left(\frac{\partial \overline{v}_i}{\partial T}\right)_P\right] dP \tag{11-133}$$

which can be rearranged to provide:

$$\Delta\overline{h}_{mix} = \int_{P^\circ}^{P} \overline{v}\, dP - \sum_{i=1}^{C} y_i \int_{P^\circ}^{P} T\left(\frac{\partial \overline{v}_i}{dT}\right)_P dP$$
$$- \underbrace{\sum_{i=1}^{C} y_i \int_{P^\circ}^{P} \overline{v}_i\, dP}_{=\int_{P^\circ}^{P} \overline{v}\, dP} + \sum_{i=1}^{C} y_i \int_{P^\circ}^{P} T\left(\frac{\partial \overline{v}_i}{\partial T}\right)_P dP \tag{11-134}$$

Notice that all of the terms in Eq. (11-134) cancel, resulting in:

$$\Delta\overline{h}_{mix} = 0 \quad \text{for an ideal solution} \tag{11-135}$$

As a result, Eq. (11-85) indicates that the molar specific enthalpy of a mixture that forms an ideal solution is given by:

$$\boxed{\overline{h} = \sum_{i=1}^{C} y_i\,\overline{h}_i \quad \text{for an ideal solution}} \tag{11-136}$$

Equation (11-136) works reasonably well for many gas mixtures. It may also be applicable for liquid mixtures when the components of the mixture are chemically similar.

Substituting Eq. (11-132) into Eqs. (11-97), (11-109), and (11-110) results in:

$$\Delta\overline{s}_{mix} = \int_{P}^{P^o} \left[\sum_{i=1}^{C} y_i \left(\frac{\partial \overline{v}_i}{dT}\right)_P - \frac{R_{univ}}{P}\right] dP$$
$$- \sum_{i=1}^{C} y_i \int_{P}^{P^o} \left[\left(\frac{\partial \overline{v}_i}{\partial T}\right)_P - \frac{R_{univ}}{P}\right] dP - R_{univ}\sum_{i=1}^{C} y_i \ln(y_i) \tag{11-137}$$

which can be rearranged to provide:

$$\Delta\overline{s}_{mix} = \sum_{i=1}^{C} y_i \int_{P}^{P^o} \left(\frac{\partial \overline{v}_i}{dT}\right)_P - \int_{P}^{P^o} \frac{R_{univ}}{P} dP - \sum_{i=1}^{C} y_i \int_{P}^{P^o} \left(\frac{\partial \overline{v}_i}{\partial T}\right)_P dP$$
$$+ \sum_{i=1}^{C} y_i \int_{P}^{P^o} \frac{R_{univ}}{P} dP - R_{univ} \sum_{i=1}^{C} y_i \ln(y_i) \quad (11\text{-}138)$$

Notice that all of the terms in Eq. (11-138) cancel except for the last one, resulting in:

$$\Delta\overline{s}_{mix} = -R_{univ} \sum_{i=1}^{C} y_i \ln(y_i) \quad \text{for an ideal solution} \quad (11\text{-}139)$$

Equation (11-139) is identical to the molar specific entropy change of mixing for an ideal gas, given by Eq. (11-57). Substituting Eq. (11-139) into Eq. (11-86) shows that the molar specific entropy of a mixture that forms an ideal solution is:

$$\boxed{\overline{s} = \sum_{i=1}^{C} y_i \overline{s}_i(T, P) - R_{univ} \sum_{i=1}^{C} y_i \ln(y_i) \quad \text{for an ideal solution}} \quad (11\text{-}140)$$

11.4.4 Enthalpy and Entropy using a Two-Parameter Equation of State

Equations (11-125) and (11-130) can be used in conjunction with any pressure-explicit equation of state that is applicable for mixtures in order to estimate the molar specific enthalpy and entropy departures that are required to compute $\Delta\overline{h}_{mix}$ and $\Delta\overline{s}_{mix}$.

The RKS Equation of State

This extended section can be found online at www.cambridge.org/kleinandnellis. Shown below are the molar specific departures that result from applying the RKS equation of state to the general departure relations in Eqs. (11-102) and (11-103) for enthalpy and Eqs. (11-109) and (11-110) for entropy:

$$\overline{h}_{dep} = \left(\overline{h}^{\circ} - \overline{h}\right) = \left(\frac{a}{b} - \frac{T}{b}\frac{da}{dT}\right) \ln\left(\frac{\overline{v} + b}{\overline{v}}\right) + R_{univ} T - P\overline{v} \quad (11\text{-}147)$$

$$\overline{h}_{dep,i} = \left(\overline{h}_i^{\circ} - \overline{h}_i\right) = \left(\frac{a_i}{b_i} - \frac{T}{b_i}\frac{da_i}{dT}\right) \ln\left(\frac{\overline{v}_i + b_i}{\overline{v}_i}\right) + R_{univ} T - P\overline{v}_i \quad (11\text{-}148)$$

$$\overline{s}_{dep} = (\overline{s}^{\circ} - \overline{s}) = R_{univ} \ln\left[\frac{\overline{v}}{Z(\overline{v} - b)}\right] + \frac{da}{dT}\frac{1}{b} \ln\left(\frac{\overline{v}}{\overline{v} + b}\right) \quad (11\text{-}155)$$

$$\overline{s}_{dep,i} = (\overline{s}_i^{\circ} - \overline{s}_i) = R_{univ} \ln\left[\frac{\overline{v}_i}{Z_i(\overline{v}_i - b_i)}\right] + \frac{1}{b_i}\frac{da_i}{dT} \ln\left(\frac{\overline{v}_i}{\overline{v}_i + b_i}\right) \quad (11\text{-}156)$$

where Z is the compressibility factor for the mixture and Z_i is the compressibility factor for the pure gas.

The Peng-Robinson Equation of State

The Peng-Robinson (PR) equation of state is presented in Eq. (10-60). The PR equation of state for a mixture is:

$$P = \frac{R_{univ}\,T}{(\overline{v} - b)} - \frac{a}{\overline{v}\,(\overline{v} + b) + b\,(\overline{v} - b)} \tag{11-158}$$

where a and b are the mixture parameters defined in Eqs. (11-77) and (11-78). The molar specific enthalpy departure and molar specific entropy departure can be determined using the steps that are discussed in the previous section for the RKS equation of state. The molar specific enthalpy departure for the mixture using the Peng Robinson equation of state is:

$$\overline{h}_{dep} = \left(\overline{h}^{o} - \overline{h}\right) = R_{univ}\,T\,(1 - Z) + \left(\frac{a - T\dfrac{da}{dT}}{2\sqrt{2}\,b}\right)\ln\left(\frac{Z + 2.414B}{Z - 0.414B}\right) \tag{11-159}$$

where

$$B = \frac{b\,P}{R_{univ}\,T} \tag{11-160}$$

The molar specific enthalpy departure for pure gas i has the same form:

$$\overline{h}_{dep,i} = \left(\overline{h}_i^{o} - \overline{h}_i\right) = R_{univ}\,T\,(1 - Z_i) + \left(\frac{a_i - T\dfrac{da_i}{dT}}{2\sqrt{2}\,b_i}\right)\ln\left(\frac{Z_i + 2.414B_i}{Z_i - 0.414B_i}\right) \tag{11-161}$$

where

$$B_i = \frac{b_i\,P}{R_{univ}\,T} \tag{11-162}$$

The molar specific entropy departure for the mixture based on the Peng Robinson equation of state is:

$$\overline{s}_{dep} = (\overline{s}^{o} - \overline{s}) = -R_{univ}\ln(Z - B) - \frac{1}{2\sqrt{2}\,b}\frac{da}{dT}\ln\left(\frac{Z + 2.414B}{Z - 0.414B}\right) \tag{11-163}$$

The molar specific entropy departure for gas i is provided by the analogous relation:

$$\overline{s}_{dep,i} = (\overline{s}_i^{o} - \overline{s}_i) = -R_{univ}\ln(Z_i - B_i) - \frac{1}{2\sqrt{2}\,b_i}\frac{da_i}{dT}\ln\left(\frac{Z_i + 2.414B_i}{Z_i - 0.414B_i}\right) \tag{11-164}$$

EXAMPLE 11.4-1

EXAMPLE 11.4-1: ANALYSIS OF A COMPRESSOR WITH A GAS MIXTURE

A gaseous mixture with molar composition 50% methane ($y_1 = 0.5$), 20% nitrogen ($y_2 = 0.2$), and 30% ethane ($y_3 = 0.3$) enters a compressor at $P_{in} = 10$ atm and $T_{in} = 0^{\circ}$C with a mass flow rate of $\dot{m} = 10$ kg/min. The exit pressure is $P_{out} = 100$ atm.

a) If the compression process is isothermal and reversible, estimate the power required by the compressor. Assume that the mixture behaves as an ideal solution. Plot the power required by the compressor as a function of the exit pressure.

The inputs are entered in EES.

```
$UnitSystem SI Molar Radian J K Pa
C=3 [-]                                       "number of components"
G$[1..C]=['Methane','Nitrogen','Ethane']      "names of components"
y[1]=0.5 [-]                                  "mole fraction of methane"
y[2]=0.2 [-]                                  "mole fraction of nitrogen"
y[3]=0.3 [-]                                  "mole fraction of ethane"
P_in=10 [atm]*convert(atm,Pa)                 "inlet pressure"
T_in=converttemp(C,K,0[C])                    "inlet temperature"
P_out_atm=100 [atm]                           "exit pressure, in atm"
P_out=P_out_atm*convert(atm,Pa)               "exit pressure"
m_dot=10 [kg/min]*convert(kg/min,kg/s)        "mass flow rate"
```

The molar mass of the mixture is determined:

$$MW = \sum_{i=1}^{C} y_i \, MW_i$$

and the molar flow rate is:

$$\dot{n} = \frac{\dot{m}}{MW}$$

```
MW=sum(y[i]*MolarMass(G$[i]),i=1,C)           "molar mass of mixture"
n_dot=m_dot/MW                                "molar flow rate"
```

The molar specific enthalpy and molar specific entropy of each pure substance at the inlet state ($\overline{h}_{in,i}$ and $\overline{s}_{in,i}$) are determined using the known inlet temperature and total inlet pressure.

```
duplicate i=1,C
   h_bar_in[i]=enthalpy(G$[i],T=T_in,P=P_in)   "specific enthalpy of pure substance"
   s_bar_in[i]=entropy(G$[i],T=T_in,P=P_in)    "specific entropy of pure substance"
end
```

The molar specific enthalpy of mixing for an ideal solution is $\Delta\overline{h}_{mix,in} = 0$. The molar specific entropy of mixing for an ideal solution is computed according to Eq. (11-139):

$$\Delta\overline{s}_{mix,in} = -R_{univ} \sum_{i=1}^{C} y_i \ln\left(y_i\right)$$

```
Deltah_bar_mix_in=0 [J/kmol]                        "ideal solution molar specific enthalpy of mixing"
Deltas_bar_mix_in=-R#*sum(y[i]*ln(y[i]),i=1,C)      "ideal solution molar specific entropy of mixing"
```

The molar specific enthalpy of the mixture at the inlet state is computed using Eq. (11-85):

$$\overline{h}_{in} = \Delta\overline{h}_{mix,in} + \sum_{i=1}^{C} y_i \, \overline{h}_{in,i} \qquad (1)$$

and the molar specific entropy of the mixture at the inlet state is computed using Eq. (11-86):

$$\bar{s}_{in} = \Delta\bar{s}_{mix,in} + \sum_{i=1}^{C} y_i\,\bar{s}_{in,i} \tag{2}$$

```
h_bar_in=sum(y[i]*h_bar_in[i],i=1,C)+Deltah_bar_mix_in     "molar specific enthalpy"
s_bar_in=sum(y[i]*s_bar_in[i],i=1,C)+Deltas_bar_mix_in     "molar specific entropy"
```

For an isothermal compression process, the temperature of the exit state is given by $T_{out} = T_{in}$. The molar specific enthalpy and molar specific entropy of each pure substance at the outlet state ($\bar{h}_{out,i}$ and $\bar{s}_{out,i}$) are determined using the outlet temperature and total outlet pressure.

```
T_out=T_in                                          "outlet temperature"
duplicate i=1,C
  h_bar_out[i]=enthalpy(G$[i],T=T_out,P=P_out)      "molar specific enthalpy of pure substance"
  s_bar_out[i]=entropy(G$[i],T=T_out,P=P_out)       "molar specific entropy of pure substance"
end
```

The molar specific enthalpy of mixing for an ideal solution is $\Delta\bar{h}_{mix,out} = 0$. The molar specific entropy of mixing for an ideal solution is computed according to Eq. (11-139):

$$\Delta\bar{s}_{mix,out} = -R_{univ}\sum_{i=1}^{C} y_i \ln(y_i)$$

```
Deltah_bar_mix_out=0 [J/kmol]                     "ideal solution molar specific enthalpy of mixing"
Deltas_bar_mix_out=-R#*sum(y[i]*ln(y[i]),i=1,C)   "ideal solution molar specific entropy of mixing"
```

The molar specific enthalpy of the mixture at the outlet state is computed using Eq. (11-85):

$$\bar{h}_{out} = \Delta\bar{h}_{mix,out} + \sum_{i=1}^{C} y_i\,\bar{h}_{out,i}$$

and the molar specific entropy of the mixture at the outlet state is computed using Eq. (11-86):

$$\bar{s}_{out} = \Delta\bar{s}_{mix,out} + \sum_{i=1}^{C} y_i\,\bar{s}_{out,i}$$

```
h_bar_out=sum(y[i]*h_bar_out[i],i=1,C)+Deltah_bar_mix_out     "molar specific enthalpy"
s_bar_out=sum(y[i]*s_bar_out[i],i=1,C)+Deltas_bar_mix_out     "molar specific entropy"
```

An entropy balance on the isothermal reversible compressor requires that:

$$\dot{n}\,\bar{s}_{in} = \frac{\dot{Q}_{out}}{T_{in}} + \dot{n}\,\bar{s}_{out}$$

An energy balance on the compressor provides:

$$\dot{n}\,\bar{h}_{in} + \dot{W} = \dot{Q}_{out} + \dot{n}\,\bar{h}_{out}$$

```
n_dot*s_bar_in=Q_dot_out/T_in+n_dot*s_bar_out        "entropy balance"
n_dot*h_bar_in+W_dot=Q_dot_out+n_dot*h_bar_out       "energy balance"
```

which leads to $\dot{W} = 29.8$ kW. Figure 1 illustrates the power required by the compressor as a function of the outlet pressure.

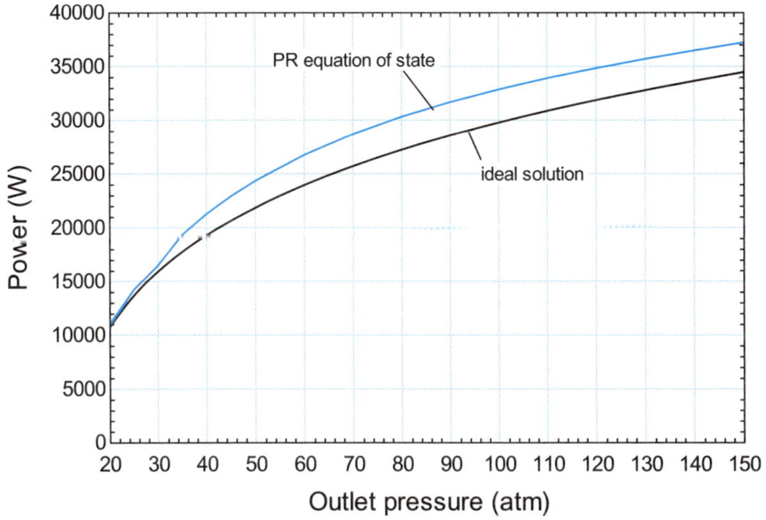

Figure 1: Power required by a reversible, isothermal compressor as a function of P_{out} estimated by assuming that the mixture behaves as an ideal solution and also by using the Peng-Robinson equation of state.

b) Estimate the power required by the reversible, isothermal compression process using the Peng-Robinson equation of state to determine the molar specific enthalpy departures and the molar specific entropy departures. Use Eq. (11-80) to determine the interaction parameters. Overlay on your plot from (b) the power required by the compressor as a function of the outlet pressure.

The molar specific enthalpy and molar specific entropy departures predicted using the ideal solution model are commented out.

```
{Deltah_bar_mix_in=0 [J/kgmol]                      "ideal solution molar specific enthalpy of mixing"
Deltas_bar_mix_in=-R#*sum(y[i]*ln(y[i]),i=1,C)       "ideal solution molar specific entropy of mixing"}
{Deltah_bar_mix_out=0 [J/kgmol]                     "ideal solution molar specific enthalpy of mixing"
Deltas_bar_mix_out=-R#*sum(y[i]*ln(y[i]),i=1,C)      "ideal solution molar specific entropy of mixing"}
```

It will be necessary to determine the molar specific enthalpy and molar specific entropy of mixing using Eqs. (11-125) and (11-130). The molar specific enthalpy and molar specific entropy departures in these equations are evaluated by applying the Peng-Robinson equation of state. The critical properties ($T_{crit,i}$, $P_{crit,i}$, and $v_{crit,i}$) and acentric factors (ω_i) of each of the pure components are computed:

```
duplicate i=1,C
   T_crit[i]=T_crit(G$[i])                    "critical temperature"
   P_crit[i]=P_crit(G$[i])                    "critical pressure"
   v_bar_crit[i]=v_crit(G$[i])                "critical molar specific volume"
   omega[i]=acentricFactor(G$[i])             "acentric factor"
end
```

It is convenient to use subprograms to implement the solution as the same calculations are needed for each fluid. Subprogram ab determines Peng-Robinson parameters a, b, and the derivative of a with respect to temperature for each pure substance. The subprogram takes as inputs the temperature, critical temperature, critical pressure, and acentric factor. The reduced temperature is computed:

$$T_r = \frac{T}{T_{crit}}$$

The parameters m, α, a, and b required by the Peng-Robinson equation of state are computed as discussed in Section 10.1.3.

$$\sqrt{\alpha} = 1 + m\left(1 - \sqrt{T_r}\right) \tag{3}$$

$$m = 0.37464 + 1.54226\,\omega - 0.26992\,\omega^2$$

$$a = 0.45724\frac{R^2\,T_{crit}^2}{P_{crit}}\alpha \tag{4}$$

$$b = 0.07780\frac{R\,T_{crit}}{P_{crit}}$$

```
Subprogram ab(T,T_crit, P_crit, omega: a, b, da\dT)
  T_r=T/T_crit                                 "reduced temperature"
  m=0.37464+1.54226*omega-0.26992*omega^2      "parameter m used in alpha"
  alpha=(1+m*(1-sqrt(T_r)))^2                  "parameter alpha used in a"
  a=0.45724*R#^2*T_crit^2/P_crit*alpha         "parameter a in equation of state"
  b=0.07780*R#*T_crit/P_crit                   "parameter b in equation of state"
```

The derivative of α, Eq. (3), with respect to T is:

$$\frac{1}{2}\alpha^{-1/2}\frac{d\alpha}{dT} = -\frac{1}{2}m\left(\frac{T}{T_{crit}}\right)^{-1/2}\frac{1}{T_{crit}} \tag{5}$$

Equation (5) can be rearranged to provide:

$$\frac{d\alpha}{dT} = -\frac{m}{T_{crit}}\sqrt{\frac{\alpha}{T_r}} \tag{6}$$

```
  dalphadT=-m*sqrt(alpha/T_r)/T_crit      "derivative of alpha with respect to T"
```

The derivative of a with respect to T is obtained by substituting Eq. (6) into the derivative of Eq. (4).

$$\frac{da}{dT} = 0.45724\frac{R^2\,T_{crit}^2}{P_{crit}}\frac{d\alpha}{dT}$$

```
  da\dT=0.45724*R#^2*T_crit^2*dalphadT/P_crit    "derivative of a with respect to T"
end
```

Subprogram PR implements the Peng Robinson equation of state for the pure fluids and determines the molar specific enthalpy and molar specific entropy departures at specified values of temperature and pressure. The subprogram takes as inputs the parameters a, b, and the derivative of a with respect to T as well as the temperature and pressure. The dimensionless parameters A and B are computed:

$$A = \frac{a\,P}{R^2\,T^2}$$

$$B = \frac{b\,P}{R\,T}$$

```
Subprogram PR(a, b, da\dT,T,P : Z, h_bar_dep, s_bar_dep)
  AA=a*P/(R#^2*T^2)                          "A parameter"
  BB=b*P/(R#*T)                              "B parameter"
  Z_Low=1.01*BB                              "lower limit for Z"
```

The dimensionless PR equation of state (from Eq. 10-69) is:

$$Z^3 - (1 - B)\,Z^2 + \left(A - 3\,B^2 - 2\,B\right)\,Z - \left(A\,B - B^2 - B^3\right) = 0$$

where Z is the compressibility factor.

```
  Z^3-(1-BB)*Z^2+(AA-3*BB^2-2*BB)*Z-(AA*BB-BB^2-BB^3)=0     "PR equation of state"
```

The cubic equation for the compressibility factor can return undesired roots. If the value of Z becomes less than B during the iteration process, numerical problems will occur in the evaluation of the entropy departure. To prevent this possibility, define Z_Low to be 1.01*B and set Z_Low to be the lower limit on variable Z in the Variable Information dialog, as shown in Figure 2.

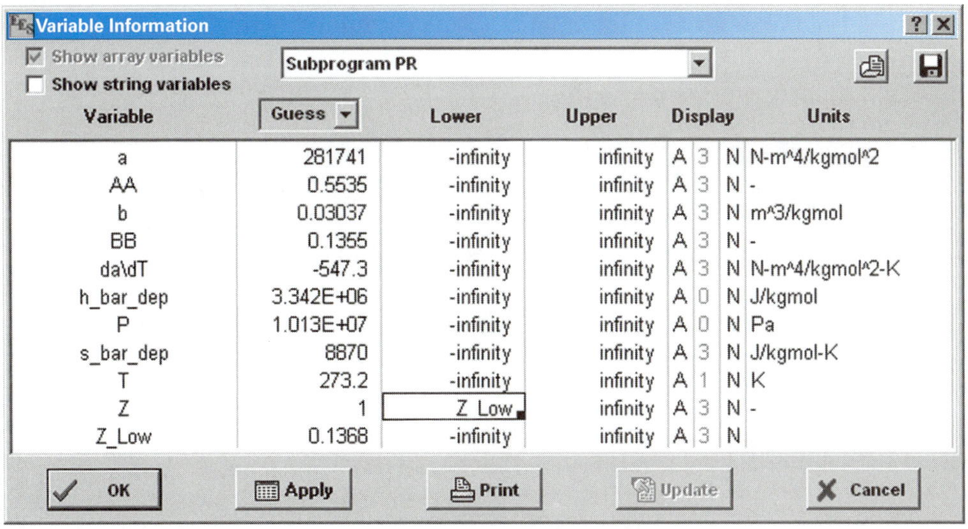
Variable Information
Show array variables
Show string variables
Subprogram PR
Variable Guess Lower Upper Display Units
a 281741 -infinity infinity A 3 N N-m^4/kgmol^2
AA 0.5535 -infinity infinity A 3 N -
b 0.03037 -infinity infinity A 3 N m^3/kgmol
BB 0.1355 -infinity infinity A 3 N -
da\dT -547.3 -infinity infinity A 3 N N-m^4/kgmol^2-K
h_bar_dep 3.342E+06 -infinity infinity A 0 N J/kgmol
P 1.013E+07 -infinity infinity A 0 N Pa
s_bar_dep 8870 -infinity infinity A 3 N J/kgmol-K
T 273.2 -infinity infinity A 1 N K
Z 1 Z_Low infinity A 3 N -
Z_Low 0.1368 -infinity infinity A 3 N
OK
Apply
Print
Update
Cancel

Figure 2: Variable Information for Subroutine PR showing lower limit on Z.

The molar specific enthalpy departure is computed using Eq. (11-161) and the molar specific entropy departure is computed using Eq. (11-164).

$$\overline{h}_{dep,i} = R_{univ}\,T\,(1 - Z_i) + \left(\frac{a_i - T\dfrac{d\,a_i}{dT}}{2\sqrt{2}\,b_i}\right)\ln\left(\frac{Z_i + 2.414B_i}{Z_i - 0.414B_i}\right)$$

$$\overline{s}_{dep,i} = -R_{univ}\ln(Z_i - B_i) - \frac{1}{2\sqrt{2}\,b_i}\frac{d\,a_i}{dT}\ln\left(\frac{Z_i + 2.414B_i}{Z_i - 0.414B_i}\right)$$

```
  h_bar_dep=R#*T*(1-Z)+(a-T*da\dT)/(2*sqrt(2)*b)*ln((Z+2.414*BB)/(Z-0.414*BB))
       "molar specific enthalpy departure"
  s_bar_dep=-R#*ln(Z-BB)-1/(2*sqrt(2)*b)*da\dT*ln((Z+2.414*BB)/(Z-0.414*BB))
       "molar specific entropy departure"
end
```

Subprogram ab is called to determine the PR equation of state parameters at the inlet state for each component, $a_{in,i}$, $b_{in,i}$, and $(\frac{da}{dT})_{in,i}$.

```
duplicate i=1,C
  Call ab(T_in,T_crit[i], P_crit[i], omega[i]: a_in[i], b_in[i], da\dT_in[i])   "Peng-Robinson parameters"
end
```

Subprogram PR is called to determine the compressibility factor ($Z_{in,i}$), molar specific enthalpy departure ($\overline{h}_{dep,in,i}$), and molar specific entropy departure ($\overline{s}_{dep,in,i}$) at the inlet state.

```
duplicate i=1,C
  Call PR(a_in[i], b_in[i], da\dT_in[i],T_in,P_in: Z_in[i],h_bar_dep_in[i], s_bar_dep_in[i])
                                                                "PR Equation of state"
end
```

The molar specific enthalpy and molar specific entropy departure of the mixture at the conditions associated with the inlet state must be computed next. The effective b parameter of the mixture at the inlet, $b_{eff,in}$, is computed using Eq. (11-77):

$$b_{eff,in} = \sum_{i=1}^{C} y_i\,b_{in,i}$$

```
b_eff_in=sum(y[i]*b_in[i],i=1,C)    "b for mixture at inlet"
```

The interaction parameters are computed according to Eq. (11-80):

$$k_{ij} = 1 - \left[\frac{2\left(v_{crit,i}v_{crit,j}\right)^{1/6}}{\left(v_{crit,i}^{1/3} + v_{crit,j}^{1/3}\right)}\right]^3$$

and the effective a parameter of the mixture, $a_{eff,in}$, is computed according to Eqs. (11-78) and (11-79):

$$a_{eff,in} = \sum_{i=1}^{C}\sum_{j=1}^{C} y_i\,y_j\,a_{in,ij} \tag{7}$$

where

$$a_{in,ij} = \left(1 - k_{ij}\right) \sqrt{a_{in,i}} \sqrt{a_{in,j}} \tag{8}$$

```
duplicate i=1,C
  duplicate j=1,C
    k[i,j]=1-(2*(v_bar_crit[i]*v_bar_crit[j])^(1/6)/(v_bar_crit[i]^(1/3)+v_bar_crit[j]^(1/3)))^3
  end
end
duplicate i=1,C
  duplicate j=1,C
    a_in[i,j]=(1-k[i,j])*sqrt(a_in[i])*sqrt(a_in[j])
  end
end
duplicate i=1,C
  a_eff_in[i]=sum(y[i]*y[j]*a_in[i,j],j=1,C)
end
a_eff_in=sum(a_eff_in[i],i=1,C)
```

The dimensionless parameter B_{in} is computed for the mixture at the inlet state according to Eq. (11-160):

$$B_{in} = \frac{b_{eff,in}\, P_{in}}{R_{univ}\, T_{in}}$$

```
BB_in=b_eff_in*P_in/(R#*T_in)    "B parameter for mixture at inlet"
```

The derivative of $a_{eff,in}$ with respect to T is computed by substituting Eq. (8) into Eq. (7) and taking the derivative:

$$\frac{d\,a_{eff,in}}{dT} = \sum_{i=1}^{C} \sum_{j=1}^{C} y_i\, y_j\, (1 - k_{ij}) \left(-\frac{1}{2} a_{in,i}^{-1/2} a_{in,j}^{1/2} \frac{d\,a_{in,i}}{dT} - \frac{1}{2} a_{in,j}^{-1/2} a_{in,i}^{1/2} \frac{d\,a_{in,j}}{dT} \right)$$

which can be rearranged:

$$\frac{d\,a_{eff,in}}{dT} = \sum_{i=1}^{C} \sum_{j=1}^{C} \frac{y_i\, y_j\, (1 - k_{ij})}{2} \left(\sqrt{\frac{a_{in,j}}{a_{in,i}}} \frac{d\,a_{in,i}}{dT} + \sqrt{\frac{a_{in,i}}{a_{in,j}}} \frac{d\,a_{in,j}}{dT} \right)$$

```
duplicate i=1,C
  duplicate j=1,C
    da\dT_in[i,j]=y[i]*y[j]*(1-k[i,j])*(sqrt(a_in[j]/a_in[i])*da\dT_in[i]+sqrt(a_in[i]/a_in[j])*da\dT_in[j])/2
  end
end
duplicate i=1,C
  da\dT_eff_in[i]=sum(da\dT_in[i,j],j=1,C)
end
da\dT_eff_in=sum(da\dT_eff_in[i],i=1,C)
```

The Peng-Robinson equation of state is applied to the mixture at the inlet state using the parameters $a_{eff,in}$ and $b_{eff,in}$. The molar specific enthalpy departure and molar specific entropy departure of the mixture at the inlet ($\bar{h}_{dep,in}$ and $\bar{s}_{dep,in}$) are computed using Eqs. (11-159) and (11-163).

$$\bar{h}_{dep,in} = R_{univ}\,T_{in}\,(1 - Z_{in}) + \left(\frac{a_{eff,in} - T_{in}\dfrac{d\,a_{eff,in}}{dT}}{2\sqrt{2}\,b_{eff,in}}\right)\ln\left(\frac{Z_{in} + 2.414B_{in}}{Z_{in} - 0.414B_{in}}\right)$$

$$\bar{s}_{dep,in} = -R_{univ}\ln(Z_{in} - B_{in}) - \frac{1}{2\sqrt{2}\,b_{eff,in}}\frac{d\,a_{eff,in}}{dT}\ln\left(\frac{Z_{in} + 2.414B_{in}}{Z_{in} - 0.414B_{in}}\right)$$

The values of Z_{in}, $\bar{h}_{dep,in}$ and $\bar{s}_{dep,in}$ are all determined by calling the PR subprogram while providing it with the mixture parameters at the inlet state.

```
Call PR(a_eff_in, b_eff_in, da\dT_eff_in,T_in, P_in: Z_in, h_bar_dep_in, s_bar_dep_in)
```

The molar specific volume of the mixture at the inlet state is computed according to:

$$Z_{in} = \frac{P_{in}\,\bar{v}_{in}}{R_{univ}\,T_{in}}$$

```
Z_in=P_in*v_bar_in/(R#*T_in)    "compressibility factor"
```

The molar specific enthalpy of mixing and the molar specific entropy of mixing are computed using Eqs. (11-92) and (11-97).

$$\Delta\bar{h}_{mix,in} = \sum_{i=1}^{C} y_i\,\bar{h}_{dep,in,i} - \bar{h}_{dep,in}$$

$$\Delta\bar{s}_{mix,in} = \sum_{i=1}^{C} y_i\,\bar{s}_{dep,in,i} - \bar{s}_{dep,in} - R_{univ}\sum_{i=1}^{C} y_i\,\ln(y_i)$$

```
Deltah_bar_mix_in=sum(y[i]*h_bar_dep_in[i],i=1,C)-h_bar_dep_in
                                   "molar specific enthalpy of mixing"
Deltas_bar_mix_in=sum(y[i]*s_bar_dep_in[i],i=1,C)-s_bar_dep_in-R#*sum(y[i]*ln(y[i]),i=1,C)
                                   "molar specific entropy of mixing"
```

Now that the molar specific enthalpy change of mixing and the molar specific entropy change of mixing at the inlet conditions are known, the molar specific enthalpy and molar specific entropy of the mixture at the inlet are computed as in part (a), using Eqs. (1) and (2).

```
h_bar_in=sum(y[i]*h_bar_in[i],i=1,C)+Deltah_bar_mix_in    "enthalpy"
s_bar_in=sum(y[i]*s_bar_in[i],i=1,C)+Deltas_bar_mix_in    "entropy"
```

The calculations are repeated in order to determine the molar specific enthalpy and molar specific entropy of mixing at the outlet state.

```
T_out=T_in                                                    "outlet temperature"
duplicate i=1,C
  h_bar_out[i]=enthalpy(G$[i],T=T_out,P=P_out)                "molar specific enthalpy of pure substance"
  s_bar_out[i]=entropy(G$[i],T=T_out,P=P_out)                 "molar specific entropy of pure substance"
end
duplicate i=1,C
  Call ab(T_out,T_crit[i], P_crit[i], omega[i]:a_out[i], b_out[i], da\dT_out[i])
                                                              "Peng-Robinson parameters"
end
duplicate i=1,C
  Call PR(a_out[i], b_out[i], da\dT_out[i],T_out,P_out:Z_out[i], h_bar_dep_out[i], s_bar_dep_out[i])
end
b_eff_out=sum(y[i]*b_out[i],i=1,C)                            "b for mixture at outlet"
duplicate i=1,C
  duplicate j=1,C
      a_out[i,j]=(1-k[i,j])*sqrt(a_out[i])*sqrt(a_out[j])
  end
end
duplicate i=1,C
  a_eff_out[i]=sum(y[i]*y[j]*a_out[i,j],j=1,C)
end
a_eff_out=sum(a_eff_out[i],i=1,C)
BB_out=b_eff_out*P_out/(R#*T_out)                             "B parameter for mixture at outlet"
duplicate i=1,C
  duplicate j=1,C
      da\dT_out[i,j]=y[i]*y[j]*(1-k[i,j])*(sqrt(a_out[j]/a_out[i])*da\dT_out[i]+sqrt(a_out[i]/&
      a_out[j])*da\dT_out[j])/2
  end
end
duplicate i=1,C
  da\dT_eff_out[i]=sum(da\dT_out[i,j],j=1,C)
end
da\dT_eff_out=sum(da\dT_eff_out[i],i=1,C)
Call PR(a_eff_out, b_eff_out, da\dT_eff_out,T_out,P_out: Z_out, h_bar_dep_out, s_bar_dep_out)
Z_out=P_out*v_bar_out/(R#*T_out)                              "compressibility factor"
Deltah_bar_mix_out=sum(y[i]*h_bar_dep_out[i],i=1,C)-h_bar_dep_out
                                                              "molar specific enthalpy of mixing"
Deltas_bar_mix_out=sum(y[i]*s_bar_dep_out[i],i=1,C)-s_bar_dep_out-R#*sum(y[i]*ln(y[i]),i=1,C)
                                                              "molar specific entropy of mixing"
```

The resulting power required by the compressor is $\dot{W} = 32.85$ kW, which is approximately 10% different from the prediction from part (a) using the ideal solution model. Figure 1 illustrates power required by the compressor as a function of the outlet pressure using the Peng-Robinson equation of state. Note that at outlet pressures lower than 30 atm, the Peng-Robinson and ideal solution methods provide nearly the same compressor power values.

c) Estimate the power required by an adiabatic compression process using the Peng-Robinson equation of state to determine the molar specific enthalpy and molar specific entropy departures.

The outlet temperature is not known and it must be determined in the calculations. The outlet temperature is initially set to a reasonable guess value; because the compression process is adiabatic, it is expected that the outlet temperature will be substantially higher than the inlet temperature:

```
T_out=T_in+150 [K]    "guess for the outlet temperature"
```

Solve the equations and then update the guess values. (Select Update Guesses from the Calculate menu.) Then comment out the equation that provides the guess for the outlet temperature:

```
{T_out=T_in+150 [K]}    "outlet temperature"
```

and set the heat transfer rate to zero, $\dot{Q} = 0$.

```
Q_dot_out=0 [W]    "adiabatic compressor"
```

Solve the equation set. The solution should show that $T_{out} = 440.4$ K and $\dot{W} = 47.69$ kW.

11.4.5 Peng-Robinson Library Functions

Example 11.4-1 demonstrates that the use of the Peng-Robinson equation of state for calculating the properties of a mixture can be tedious. An external Peng-Robinson library has been developed for EES that simplifies the process. Access to the library routine and documentation are provided through the Function Info menu in the Options menu. Click the External routines button and then double-click on the Peng_Robinson.DLL item in order to expand the list of routines, which will appear as shown in Figure 10-11. Clicking the Function Info button above the list will provide detailed documentation for the selected item. Some of the routines in this library are useful for pure substances and they are described in Section 10.1.3 and demonstrated in Example 10.1-3. Additional routines in the library that are useful for mixtures are summarized in this section.

The procedure AB_MIX_PR returns the dimensionless parameters A and B associated with a mixture of up to 10 fluids using the mixing rules provided in Eqs. (11-77) through (11-79). The procedure AB_MIX_PR requires the number of fluids, the mole fraction of each fluid, the Peng-Robinson constants for each fluid (obtained from the procedure AB_PR), and the binary interaction coefficient between each of the fluids.

The function H_DEP_PR returns the molar specific enthalpy departure (normalized by $R_{univ}\ T_{crit}$) for a pure fluid using Eq. (11-148). The function S_DEP_PR returns the molar specific entropy departure (normalized by R_{univ}) for a pure fluid using Eq. (11-156). Both functions require the reduced temperature, acentric factor, compressibility factor (obtained from either function Z_L_PR or function Z_G_PR), and the dimensionless parameter B (obtained from procedure AB_PR).

The function DADT_MIX_PR returns the value of $\frac{da}{dT}$ for the mixture, which is required by Eqs. (11-159) and (11-163) in order to compute the molar specific enthalpy and molar specific entropy departure of the mixture. The function DADT_MIX_PR requires the number of fluids, the temperature, the critical temperature, critical pressure, and acentric factors of each fluid, the composition, and the binary interaction coefficient between each of the fluids. Use of these Peng-Robinson routines is demonstrated in Example 11.4-2.

EXAMPLE 11.4-2: ANALYSIS OF A COMPRESSOR WITH A GAS MIXTURE (REVISITED)

In this example, the compressor analysis presented in Example 11.4-1 is repeated using the Peng-Robinson library in EES. A gaseous mixture with molar composition 50% methane ($y_1 = 0.5$), 20% nitrogen ($y_2 = 0.2$), and 30% ethane ($y_3 = 0.3$) enters a compressor at $P_{in} = 10$ atm and $T_{in} = 0°$C with a mass flow rate of $\dot{m} = 10$ kg/min. The exit pressure is $P_{out} = 100$ atm.

a) Estimate the power required by the reversible, isothermal compression process using the Peng-Robinson equation of state to determine the molar specific enthalpy and molar specific entropy departures. Use Eq. (11-80) to determine the interaction parameters.

The inputs are entered in EES.

```
$Constant C#=3                                  "number of components"

$UnitSystem SI Molar Radian J K Pa
$TabStops 0.25 0.5 0.75 3.5 in
"Inputs"
G$[1..C#]=['Methane','Nitrogen','Ethane']       "names of components"
y[1]=0.5 [-]                                    "mole fraction of methane"
y[2]=0.2 [-]                                    "mole fraction of nitrogen"
y[3]=0.3 [-]                                    "mole fraction of ethane"
P_in=10 [atm]*convert(atm,Pa)                   "inlet pressure"
T_in=converttemp(C,K,0[C])                      "inlet temperature"
P_out_atm=100 [atm]                             "exit pressure, in atm"
P_out=P_out_atm*convert(atm,Pa)                 "exit pressure"
m_dot=10 [kg/min]*convert(kg/min,kg/s)          "mass flow rate"
```

Note that the number of components, C, is entered as a constant, i.e., as variable C# using the $Constant directive. The $Constant directive must be the first equation in the Equations window, appearing before the subprograms or any other EES code. EES is not able to accept a normal variable as the limit in the arrays that are passed to the subprograms that will be used in this implementation because the value of a normal variable will not be assigned at the time that the equations are parsed. It would be possible to enter a number for the upper limit of the arrays, e.g., 3, but the use of a constant such as C# makes it easier to change the number of pure components in the mixture and therefore makes the EES code more flexible.

The molar mass of the mixture is determined:

$$MW = \sum_{i=1}^{C} y_i \, MW_i$$

and the molar flow rate is:

$$\dot{n} = \frac{\dot{m}}{MW}$$

```
MW=sum(y[i]*MolarMass(G$[i]),i=1,C#)    "molar mass of mixture"
n_dot=m_dot/MW                          "molar flow rate"
```

The inlet state is specified by the temperature and pressure, $T_1 = T_{in}$ and $P_1 = P_{in}$:

```
T[1]=T_in                               "temperature"
P[1]=P_in                               "pressure"
```

Subprogram PR_Properties calculates the compressibility factor, the molar specific enthalpy, and the molar specific entropy of the mixture (Z_{mix}, $\bar{h}_{mix}$, and $\bar{s}_{mix}$) at a specified temperature, pressure and composition. The subprogram takes as inputs arrays holding the names of the components, their composition, and the temperature and pressure.

```
SubProgram PR_Properties(G$[1..C#],y[1..C#],T,P:Z_mix,h_bar_mix,s_bar_mix)
  "Calculates the properties of a mixture using the PR equation of state"
  "Inputs"
  "G$[1..C#] - array with the name of each component"
  "y[1..C#] - array with the mole fraction of each component"
  "T - temperature"
  "P - pressure"
  "Outputs"
  "Z_mix - compressibility factor"
  "h_bar_mix - molar specific enthalpy"
  "s_bar_mix - molar specific entropy"
```

The critical point information ($T_{crit,i}$, $P_{crit,i}$, and $v_{crit,i}$) and the acentric factor (ω_i) for each pure component are determined.

```
"Parameters for each constituent"
duplicate i=1,C#
  T_crit[i]=T_crit(G$[i])                    "critical temperature"
  P_crit[i]=P_crit(G$[i])                    "critical pressure"
  v_bar_crit[i]=v_crit(G$[i])                "critical molar specific volume"
  omega[i]=acentricFactor(G$[i])             "acentric factors"
end
```

The A and B parameters required by the PR equation of state are computed for each pure component (A_i and B_i) using the procedure AB_PR. The compressibility factor predicted for each component using the PR equation of state (Z_i) is obtained using the Z_G_PR procedure. The molar specific enthalpy departure and the molar specific entropy departure for each pure component ($\bar{h}_{dep,i}$ and $\bar{s}_{dep,i}$) are computed using the procedures H_DEP_PR and S_DEP_PR, respectively. The molar specific enthalpy and molar specific entropy of each pure component ($\bar{h}_i$ and $\bar{s}_i$) are computed using

EES' internal property routines.

```
duplicate i=1,C#
  Call AB_PR(T/T_crit[i], P/P_crit[i], omega[i]: A[i], B[i])       "get PR parameters for pure fluids"
  Z[i]=Z_G_PR(A[i],B[i])                                           "get compressibility factor for pure fluids"
  h_bar_dep[i]=R#*T_crit[i]*H_DEP_PR(T/T_crit[i],omega[i], Z[i], B[i])
                                                     "get molar specific enthalpy departure for pure fluids"
  s_bar_dep[i]=R#*S_DEP_PR(T/T_crit[i],omega[i], Z[i], B[i])
                                                     "get molar specific entropy departure for pure fluids"
  h_bar[i]=enthalpy(G$[i],T=T,P=P)                                 "get molar specific enthalpy of pure fluids"
  s_bar[i]=entropy(G$[i],T=T,P=P)                                  "get molar specific entropy of pure fluids"
end
```

The interaction parameters are computed according to Eq. (11-80):

$$k_{ij} = 1 - \left[\frac{2\left(v_{crit,i} v_{crit,j}\right)^{1/6}}{\left(v_{crit,i}^{1/3} + v_{crit,j}^{1/3}\right)} \right]^3$$

```
"interaction parameters"
duplicate i=1,C#
  duplicate j=1,C#
    k[i,j]=1-(2*(v_bar_crit[i]*v_bar_crit[j])^(1/6)/(v_bar_crit[i]^(1/3)+v_bar_crit[j]^(1/3)))^3
  end
end
```

The A and B parameters for the mixture (A_{mix} and B_{mix}) are determined using the AB_mix_PR procedure. The compressibility factor of the mixture (Z_{mix}) is determined using the Z_G_PR procedure with the values A_{mix} and B_{mix}.

```
Call AB_mix_PR(C#,y[1..C#],A[1..C#],B[1..C#],k[1..C#,1..C#]:AA_mix,BB_mix)
                                               "A and B parameters for mixture"
Z_mix=Z_G_PR(AA_mix,BB_mix)                    "compressibility factor for mixture"
```

The dimensional PR parameters for the mixture (a_{mix} and b_{mix}) are computed.

$$a_{mix} = \frac{R_{univ}^2\, T^2\, A_{mix}}{P}$$

$$b_{mix} = \frac{R_{univ}\, T\, B_{mix}}{P}$$

```
a_mix=R#^2*T^2*AA_mix/P                        "a parameter for mixture"
b_mix=BB_mix*R#*T/P                            "b parameter for mixture"
```

The derivative of a_{mix} with respect to T is determined using the DADT_MIX_PR procedure. The molar specific enthalpy departure and the molar specific entropy departure of the mixture ($\overline{h}_{dep,mix}$ and $\overline{s}_{dep,mix}$) are computed according to:

$$\bar{h}_{dep,mix} = R_{univ}\, T\,(1 - Z_{mix}) + \left(\frac{a_{mix} - T \dfrac{d\,a_{mix}}{dT}}{2\sqrt{2}\, b_{mix}} \right) \ln\left(\frac{Z_{mix} + 2.414 B_{mix}}{Z_{mix} - 0.414 B_{mix}} \right)$$

$$\bar{s}_{dep,mix} = -R_{univ} \ln(Z_{mix} - B_{mix}) - \frac{1}{2\sqrt{2}\, b_{mix}} \frac{d\,a_{mix}}{dT} \ln\left(\frac{Z_{mix} + 2.414 B_{mix}}{Z_{mix} - 0.414 B_{mix}} \right)$$

```
da_mix\dT=DADT_MIX_PR(C#, T, T_crit[1..C#], P_crit[1..C#], omega[1..C#], y[1..C#], k[1..C#,1..C#])
    "dadT parameter for the mixture"
h_bar_dep_mix=R#*T*(1-Z_mix)+(a_mix-T*da_mix\dT)/(2*sqrt(2)*b_mix)&
    *ln((Z_mix+2.414*BB_mix)/(Z_mix-0.414*BB_mix))"enthalpy departure of mixture"
s_bar_dep_mix=-R#*ln(Z_mix-BB_mix)-1/(2*sqrt(2)*b_mix)*da_mix\dT*&
    ln((Z_mix+2.414*BB_mix)/(Z_mix-0.414*BB_mix))"entropy departure of mixture"
```

The molar specific enthalpy of mixing and the molar specific entropy of mixing are computed using Eqs. (11-92) and (11-97).

$$\Delta\bar{h}_{mix} = \sum_{i=1}^{C} y_i\, \bar{h}_{dep,i} - \bar{h}_{dep,mix}$$

$$\Delta\bar{s}_{mix} = \sum_{i=1}^{C} y_i\, \bar{s}_{dep,i} - \bar{s}_{dep,mix} - R_{univ} \sum_{i=1}^{C} y_i \ln(y_i)$$

The molar specific enthalpy of the mixture is computed using Eq. (11-85):

$$\bar{h}_{mix} = \Delta\bar{h}_{mix} + \sum_{i=1}^{C} y_i\, \bar{h}_i$$

and the molar specific entropy of the mixture is computed using Eq. (11-86):

$$\bar{s} = \Delta\bar{s}_{mix} + \sum_{i=1}^{C} y_i\, \bar{s}_i$$

```
  Deltah_bar_mix=sum(y[i]*h_bar_dep[i],i=1,C#)-h_bar_dep_mix
                                                    "molar specific enthalpy of mixing"
  Deltas_bar_mix=sum(y[i]*s_bar_dep[i],i=1,C#)-s_bar_dep_mix-R#*sum(y[i]*ln(y[i]),i=1,C#)
                                                    "molar specific entropy of mixing"
  h_bar_mix=sum(y[i]*h_bar[i],i=1,C#)+Deltah_bar_mix "molar specific enthalpy"
  s_bar_mix=sum(y[i]*s_bar[i],i=1,C#)+Deltas_bar_mix "molar specific entropy"
end
```

Subprogram PR_Properties is called to determine the molar specific enthalpy and molar specific entropy at the inlet state, which is designated as state 1 ($\bar{h}_1$ and $\bar{s}_1$).

```
Call PR_Properties(G$[1..C#],y[1..C#],T[1],P[1]:Z[1],h_bar[1],s_bar[1])
    "get molar specific enthalpy, entropy, and compressibility factor using PR equation of state"
```

The outlet state for the isothermal case is fixed by the temperature and pressure, $T_2 = T_{in}$ and $P_2 = P_{out}$. Subprogram PR_Properties is called to determine the molar specific enthalpy and molar specific entropy at state 2 ($\overline{h}_2$ and $\overline{s}_2$).

```
T[2]=T_in                                   "temperature"
P[2]=P_out                                  "pressure"
Call PR_Properties(G$[1..C#],y[1..C#],T[2],P[2]:Z[2],h_bar[2],s_bar[2])
        "get molar specific enthalpy, entropy, and compressibility factor using PR equation of state"
```

An entropy balance on the compressor leads to:

$$\dot{n}\overline{s}_1 = \frac{\dot{Q}_{out}}{T_{in}} + \dot{n}\overline{s}_2$$

and an energy balance on the compressor leads to:

$$\dot{n}\overline{h}_1 + \dot{W} = \dot{Q}_{out} + \dot{n}\overline{h}_2$$

```
n_dot*s_bar[1]=Q_dot_out/T_in+n_dot*s_bar[2]        "entropy balance"
n_dot*h_bar[1]+W_dot=Q_dot_out+n_dot*h_bar[2]       "energy balance"
```

which leads to $\dot{W} = 32.86$ kW, the same value that was computed in part (b) of Example 11.4-1.

11.5 Multi-Component Phase Equilibrium

The previous sections of this chapter dealt with single-phase mixtures. This section examines the thermodynamic behavior of multi-phase mixtures. An important result of this section is that the mole fraction of a substance in the vapor phase will normally be quite different from its mole fraction in the liquid phase at equilibrium. This behavior occurs even for the simplest case in which the vapor obeys the ideal gas law and the liquid forms an ideal solution.

11.5.1 Criterion of Multi-Component Phase Equilibrium

This extended section can be found at www.cambridge.org/kleinandnellis and shows that the criterion of multi-component phase equilibrium can be expressed as

$$dG_{T,P} = 0 \quad \text{at equilibrium} \tag{11-177}$$

11.5.2 Chemical Potentials

The molar specific Gibbs free energy of a pure substance can be expressed as a function of temperature and pressure:

$$\overline{g} = \overline{g}(T, P) \tag{11-179}$$

The total Gibbs free energy for a pure substance is the product of the molar specific Gibbs free energy and the number of moles of the substance, n:

$$G = n\overline{g} = G(T, P, n) \tag{11-180}$$

The focus of this section is on systems consisting of liquid and vapor phases, although Eq. (11-165) is applicable to all multi-phase systems at equilibrium. In general, the mole fraction of a component in the liquid phase will differ from its mole fraction in the vapor phase. Therefore, it is necessary to distinguish these mole fractions. We will use the symbol x_i to represent the mole fraction of substance i in the liquid phase and y_i to represent the mole fraction of substance i in the vapor phase.

The Gibbs free energy of a system that consists of C components coexisting in the liquid phase (f) and the vapor phase (g) is, by extension of Eq. (11-180), a function of temperature, pressure, and the number of moles of each component in each phase:

$$G = G(T, P, n_{f,1}, n_{f,2}, \ldots, n_{f,C}, n_{g,1}, n_{g,2}, \ldots, n_{g,C}) \tag{11-181}$$

where $n_{f,1}$ is the number of moles of substance 1 in the liquid phase, $n_{g,1}$ is the number of moles of substance 1 in the vapor phase, and so on for the other substances. The total derivative of G is given by:

$$dG = \left(\frac{\partial G}{\partial T}\right)_{\substack{P \\ \text{all}\, n_{f,i} \\ \text{all}\, n_{g,i}}} dT + \left(\frac{\partial G}{\partial P}\right)_{\substack{T \\ \text{all}\, n_{f,i} \\ \text{all}\, n_{g,i}}} dP + \sum_{i=1}^{C} \left(\frac{\partial G}{\partial n_{f,i}}\right)_{\substack{T,P, \\ \text{all}\, n_{f,j}, j\neq i \\ \text{all}\, n_{g,i}}} dn_{f,i} + \sum_{i=1}^{C} \left(\frac{\partial G}{\partial n_{g,i}}\right)_{\substack{T,P, \\ \text{all}\, n_{f,i} \\ \text{all}\, n_{g,j}, j\neq i}} dn_{g,i} \tag{11-182}$$

The partial derivatives with respect to $n_{f,i}$ and $n_{g,i}$ that appear in the summations in Eq. (11-182) are called *chemical potentials*. The chemical potential of substance i in the liquid phase is the partial derivative of the Gibbs free energy with respect to the number of moles of substance i in the liquid phase, holding T, P and the number of moles of all other substances in all phases constant:

$$\mu_{f,i} = \left(\frac{\partial G}{\partial n_{f,i}}\right)_{\substack{T,P, \\ \text{all}\, n_{f,j}, j\neq i \\ \text{all}\, n_{g,i}}} \tag{11-183}$$

The chemical potential of substance i in the vapor phase is the partial derivative of the Gibbs free energy with respect to the number of moles of substance i in the vapor phase holding T, P and the number of moles of all other substances in all phases constant:

$$\mu_{g,i} = \left(\frac{\partial G}{\partial n_{g,i}}\right)_{\substack{T,P, \\ \text{all}\, n_{f,i} \\ \text{all}\, n_{g,j}, j\neq i}} \tag{11-184}$$

Note that the chemical potential of a pure substance is the same as its molar specific Gibbs free energy.

$$\mu_i = \overline{g}_i = \overline{h}_i - T\overline{s}_i \quad \text{for a pure substance} \tag{11-185}$$

Substituting Eqs. (11-183) and (11-184) allows Eq. (11-182) to be written as:

$$dG = \left(\frac{\partial G}{\partial T}\right)_{\substack{P \\ \text{all}\, n_{f,i} \\ \text{all}\, n_{g,i}}} dT + \left(\frac{\partial G}{\partial P}\right)_{\substack{T \\ \text{all}\, n_{f,i} \\ \text{all}\, n_{g,i}}} dP + \sum_{i=1}^{C} \mu_{f,i}\, dn_{f,i} + \sum_{i=1}^{C} \mu_{g,i}\, dn_{g,i} \tag{11-186}$$

Under equilibrium conditions at a fixed temperature and pressure, the first two terms in Eq. (11-186) are zero since T and P are constant. The criterion for phase equilibrium, Eq. (11-165), requires that the left side of Eq. (11-182) be zero.

$$dG_{T,P} = \sum_{i=1}^{C} \mu_{f,i}\, dn_{f,i} + \sum_{i=1}^{C} \mu_{g,i}\, dn_{g,i} = 0 \tag{11-187}$$

If no chemical reactions occur then the number of moles of substance i contained in the liquid and vapor phases must be constant and equal to n_i, the total number of moles of substance i:

$$n_{f,i} + n_{g,i} = n_i \tag{11-188}$$

The derivative of Eq. (11-188) is:

$$dn_{f,i} + dn_{g,i} = 0 \tag{11-189}$$

Substituting Eq. (11-189) into Eq. (11-187) allows the criterion for phase equilibrium to be expressed according to:

$$dG_{T,P} = \sum_{i=1}^{C} (\mu_{f,i} - \mu_{g,i})\, dn_{f,i} = 0 \tag{11-190}$$

The only way Eq. (11-190) can be satisfied in general is if the chemical potential of substance i in the liquid phase is equal to the chemical potential of substance i in the vapor phase:

$$\boxed{\mu_{f,i} = \mu_{g,i} \quad \text{for} \quad i = 1, C} \tag{11-191}$$

Chemical potential differences are the driving forces for mass transfer from phase to phase, just as temperature differences are the driving forces for heat transfer. At equilibrium, the chemical potential of substance i must be the same in all co-existing phases indicating that there is no net transfer of mass of any species from one phase to the other.

11.5.3 Evaluation of Chemical Potentials for Ideal Gas Mixtures

According to Eq. (11-191), the chemical potential of each component in each phase must be the same at equilibrium. In order to apply Eq. (11-191) we need to know how to determine the chemical potentials for each substance in the gas and liquid phases. Determining the chemical potentials is very simple for a mixture in which the vapor phase obeys the ideal gas law. As discussed in Section 11.1, each gas in an ideal gas mixture behaves as if it were a pure substance at the same temperature as the mixture but at a pressure that is equal to its partial pressure. For an ideal gas the partial pressure of substance i is the product of its mole fraction and the total pressure, $P_i = y_i\, P$. This idea can be used to evaluate the chemical potential of each component i in an ideal gas mixture:

$$\mu_{i,mix}(T, P) = \mu_{i,pure}(T, P_i) = \bar{g}_i(T, P_i) = \bar{h}_i - T\bar{s}_i \tag{11-192}$$

Equation (11-192) can be expressed in an equivalent form in which the molar specific enthalpy and molar specific entropy are referred to values at a reference pressure, P^o.

The reference pressure is often chosen to be atmospheric pressure and the gas typically obeys the ideal gas law at this reference pressure. Properties at the reference pressure are indicated with a superscript °. The molar specific enthalpy of an ideal gas is independent of pressure, therefore:

$$\overline{h}_i^o = \overline{h}_i \tag{11-193}$$

The molar specific entropy of an ideal gas is a function of temperature and pressure, as discussed in Section 6.3. At a specified temperature and partial pressure, the molar specific entropy of an ideal gas can be expressed as:

$$\overline{s}_i = \overline{s}_i^o - R_{univ} \ln\left(\frac{P_i}{P^o}\right) \tag{11-194}$$

where $\overline{s}_i^o$ is the molar specific entropy of the pure gas at temperature T and reference pressure P^o. Substituting the definition of partial pressure into Eq. (11-194) provides the molar specific entropy of substance i in a mixture at total pressure P:

$$\overline{s}_i = \overline{s}_i^o - R_{univ} \ln\left(\frac{y_i P}{P^o}\right) \tag{11-195}$$

where y_i is the mole fraction of the gas in the vapor phase of the mixture. Substituting Eqs. (11-193) and (11-195) into Eq. (11-192), provides an expression for the chemical potential of a component in an ideal gas mixture.

$$\mu_i = \overline{g}_i\,(T, P_i) = \overline{h}_i - T\,\overline{s}_i = \overline{h}_i^o - T\left[\overline{s}_i^o - R_{univ} \ln\left(\frac{y_i P}{P^o}\right)\right] \tag{11-196}$$

Equation (11-196) can be expanded:

$$\mu_i = \underbrace{\overline{h}_i^o - T\,\overline{s}_i^o}_{\mu_i^o} + R_{univ}\,T \ln\left(\frac{y_i P}{P^o}\right) \tag{11-197}$$

The first two terms in Eq. (11-197) correspond to μ_i^o, the chemical potential (which is also the molar specific Gibbs free energy) of substance i at temperature T and reference pressure P^o.

$$\boxed{\mu_i = \mu_i^o + R_{univ}\,T \ln\left(\frac{y_i P}{P^o}\right)} \tag{11-198}$$

where

$$\mu_i^o = \overline{g}_i^o = \overline{h}_i^o - T\,\overline{s}_i^o \tag{11-199}$$

11.5.4 and 11.5.5 Evaluation of Chemical Potentials for Ideal Solutions and Liquid Mixtures

These extended sections can be found at www.cambridge.org/kleinandnellis. They show how chemical potentials can be determined for each component in a mixture that conforms to the ideal solution model but may not obey the ideal gas law. The chemical potential of component i with mole fraction y_i in the gas phase for a solution that obeys the ideal solution model is:

$$\mu_{g,i} = \mu_i^o + R_{univ}\,T\,\ln\left(\frac{y_i\,f_{g,i}}{P^o}\right) \quad \text{for an ideal solution} \tag{11-221}$$

where $f_{g,i}$ is the fugacity of pure fluid i in the gaseous state at the temperature and pressure of the mixture and P^o is a pressure that is sufficiently low such that fluid i obeys the ideal gas law.

The chemical potential of component i in the liquid phase for a solution that obeys the ideal solution model is:

$$\mu_{f,i} \approx \mu_{g,i}\left(T, P_{sat,i}\right) = \mu_i^o + R_{univ}\, T \ln\left(\frac{x_i\, f_{sat,i}}{P^o}\right) \quad \text{for an ideal liquid solution} \tag{11-228}$$

where x_i is the mole fraction of substance i in the liquid phase and $f_{sat,i}$ is the fugacity of pure substance i at temperature T and its vapor pressure $P_{sat,i}$.

11.5.6 Applications of Multi-Component Phase Equilibrium

Equation (11-191) states that the chemical potential of each substance in a mixture must have the same value in all coexisting phases. For a system that forms liquid and vapor phases at a specified temperature and pressure:

$$\mu_{f,i} = \mu_{g,i} \quad \text{for } i = 1..C \tag{11-230}$$

Assuming that the liquid and vapor phases both form ideal solutions, the chemical potentials in Eq. (11-230) can be evaluated using Eqs. (11-228) and (11-221):

$$\mu_i^o + R_{univ}\, T \ln\left(\frac{x_i\, f_{sat,i}}{P_0}\right) = \mu_i^o + R_{univ}\, T \ln\left(\frac{y_i\, f_i}{P_0}\right) \quad \text{for } i = 1..C \tag{11-231}$$

Equation (11-231) can be simplified to provide:

$$\boxed{x_i\, f_{sat,i} = y_i\, f_i \text{ for } i = 1..C} \tag{11-232}$$

If the gas phase obeys the ideal gas law then Eq. (11-232) can be further simplified to:

$$\boxed{x_i\, P_{sat,i} = y_i\, P \quad \text{for } i = 1, C} \tag{11-233}$$

Equation (11-233) is known as Raoult's Law. Although Raoult's Law provides a phase equilibrium relation that is very simple, it can be used to predict complex behavior. For example, we can apply Raoult's law to a liquid-vapor mixture of propane and n-butane at a total pressure of $P = 2$ atm.

```
$UnitSystem SI Molar Radian J K Pa
C=2 [-]                                 "number of components"
F$[1..C]=['propane','n-butane']         "components"
P=2 [atm]*convert(atm,Pa)               "pressure"
```

We will initially set the mole fraction of propane in the liquid phase equal to $x_1 = 0.5$. The mole fraction of n-butane in the liquid phase must be given by $x_2 = 1 - x_1$.

```
x[1]=0.5 [-]          "mole fraction of propane in the liquid phase"
x[2]=1-x[1]           "mole fraction of n-butane in the liquid phase"
```

Initially, the temperature of the mixture, T, will be guessed. We will see that there is a unique equilibrium temperature associated with this condition.

```
T=250 [K]                                  "temperature – a guess"
```

The saturation pressure for each of the pure components, $P_{sat,i}$, is determined at the mixture temperature, T. Raoult's Law, Eq. (11-233), is used to determine the mole fraction of each component in the vapor phase, y_i. Raoult's law is assumed to be applicable to this mixture because propane and n-butane are chemically similar and therefore the liquid phase is likely to form an ideal solution. Also, the pressure is sufficiently low that the vapor phase is likely to obey the ideal gas law.

```
duplicate i=1,2
   P_sat[i] = P_sat(F$[i],T=T)             "saturation pressure"
   x[i]*P_sat[i]=y[i]*P                    "criterion for phase equilibrium"
end
```

Solving at this point provides $y_1 = 0.5377$ and $y_2 = 0.09665$. This is clearly not a viable solution because the mole fractions in the vapor phase must add to unity. The temperature of the mixture must be adjusted until this is the case. Update the guess values, comment out the equation that sets the guess value for T and add the constraint that the sum of y_1 and y_2 must be 1.

```
{T=250 [K]}                                "temperature – a guess"
y[1]+y[2]=1                                "mole fractions in vapor phase must add to one"
```

which leads to $y_1 = 0.833$, $y_2 = 0.167$, and $T = 262.5$ K. Create a Parametric table with columns for temperature and the molar concentrations. Comment out the value of the mole fraction of propane in the liquid phase, x_1, and vary its value from 0.001 to 0.999. Figure 11-2(a) illustrates the mole fraction of propane in the vapor (y_1) as a function of the mole fraction of propane in the liquid (x_1). Figure 11-2(b) illustrates the mole fraction of n-butane in the vapor (y_2) as a function of the mole fraction of n-butane in the liquid (x_2). The relation between the liquid and vapor mole fractions is referred to as the equilibrium line. The dotted lines in Figure 11-2(a) and (b) illustrate $y_1 = x_1$ and $y_2 = x_2$, respectively. An important result of phase equilibrium that is evident in Figure 11-2 is that the mole fraction of a substance in the vapor phase will normally be quite different from its mole fraction in the liquid phase. Figure 11-2(a) shows that the vapor will be richer in propane than the liquid and Figure 11-2(b) shows that the vapor will be leaner in n-butane than the liquid.

Figure 11-3 illustrates the calculated temperature at 2 atm as a function of the propane mole fractions in the liquid (x_1) and vapor (y_1). The plot of temperature as a function of vapor mole fraction (y_1) is referred to as the *dew point line* and the plot of the temperature as a function of the liquid mole fraction (x_1) is referred to as the *bubble point line*. For a mixture of propane and n-butane at 2 atm, Figure 11-3 shows that liquid and vapor can coexist only over a limited range of temperatures between approximately 248 K and 292 K. At temperatures below 248 K, the mixture will always be in a liquid phase. At temperatures above 292 K, the mixture will always be in a vapor phase.

Figure 11-3 can be used to help understand the behavior of a two phase mixture. Consider a process in which a mixture with an overall mole fraction of propane that is equal to 0.60 is heated at a constant pressure of 2 atm. Initially the mixture is at 250 K and it is entirely liquid, as indicated by point a in Figure 11-3. The mixture remains a liquid as it is heated until it reaches the bubble point line at a temperature of 259 K,

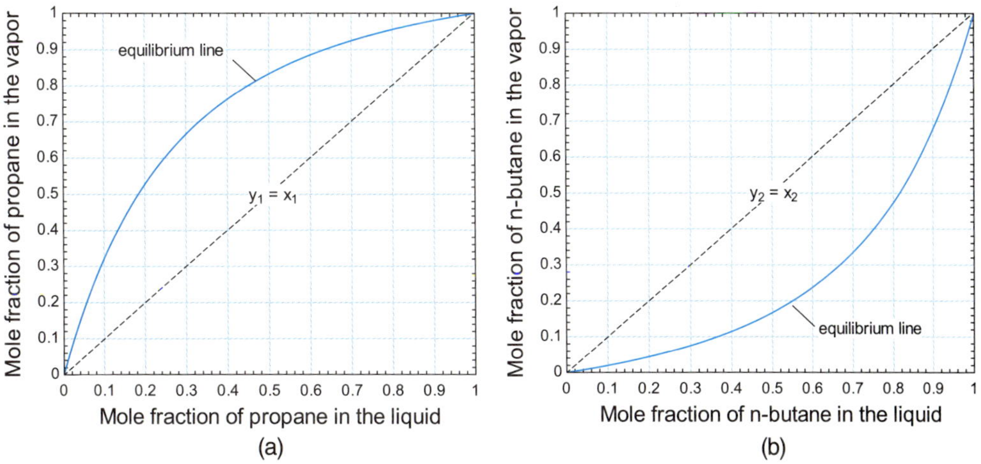

Figure 11-2: (a) Mole fraction of propane in the vapor as a function of the mole fraction of propane in the liquid phase and (b) mole fraction of n-butane in the vapor as a function of the mole fraction of n-butane in the liquid phase for a mixture at 2 atm.

indicated by point b in Figure 11-3. At point b, the first bubble of vapor appears. The propane mole fraction in the liquid is equal to the overall mole fraction of propane, $x_1 = 0.6$, at point b. However, the first bubble of vapor that appears has a propane mole fraction of approximately $y_1 = 0.89$; this mole fraction is found by following the *tie line* from point b to its intersection with the dew point line at point b_g.

As the mixture temperature is increased further, it forms a two-phase mixture. At point c, for example, the temperature is 265 K and both liquid and vapor phases will be present. The propane mole fractions in the liquid and vapor phases are approximately $x_1 = 0.437$ and $y_1 = 0.791$, as indicated by the tie lines from point c to points c_f and c_g, respectively.

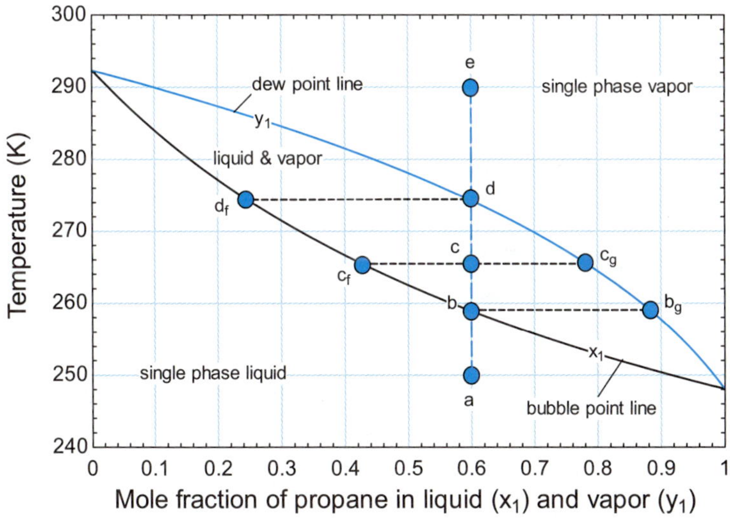

Figure 11-3: Temperature as a function of the mole fraction of propane in the liquid (x_1) and vapor (y_1) phases at 2 atm.

The *molar quality* of the mixture (Q) is defined as the number of moles of vapor (n_g) to the total moles of liquid and vapor ($n = n_f + n_g$):

$$Q = \frac{n_g}{n} \tag{11-234}$$

The molar quality can be found by applying a molar balance on the propane. The total number of moles of propane at state c is the sum of the number of moles in the liquid phase and the number of moles in the vapor phase:

$$n_1 = x_1\, n_f + y_1\, n_g \tag{11-235}$$

Dividing Eq. (11-235) by the total number of moles of mixture leads to:

$$\frac{n_1}{n} = x_1 \underbrace{\frac{n_f}{n}}_{1-Q} + y_1 \underbrace{\frac{n_g}{n}}_{Q} \tag{11-236}$$

The total mole fraction of component i, n_i/n, is sometimes given the symbol z_i in order to differentiate it from the mole fraction of component i in the liquid (x_i) and the mole fraction of component i in the vapor (y_i). Substituting Eq. (11-234) into Eq. (11-236) allows it to be written as:

$$z_1 = x_1\,(1 - Q) + y_1\, Q \tag{11-237}$$

Solving Eq. (11-237) provides the relation for the molar quality:

$$Q = \frac{(z_1 - x_1)}{(y_1 - x_1)} \tag{11-238}$$

For point c in Figure 11-3, the molar quality must be $Q = (0.6 - 0.437)/(0.791 - 0.437) = 0.460$. In general, Eq. (11-237) can be written for each of the components in the mixture:

$$z_i = x_i\,(1 - Q) + y_i\, Q \quad \text{for } i = 1..C \tag{11-239}$$

Continued heating of the mixture will increase the molar quality. At point d, the quality is 1 and the vapor has a propane mole fraction of 0.6. The temperature at this point is about 274 K and the last droplet of liquid remaining has a propane mole fraction of $x_1 = 0.24$, as indicated by the tie line to point d_f. Further heating to point e results in a single phase gas mixture with a 0.6 propane mole fraction.

EXAMPLE 11.5-1

EXAMPLE 11.5-1: USE OF A MIXTURE IN A REFRIGERATION CYCLE

Mixtures of propane and butane are used as a refrigerant in domestic refrigerators in Europe (but not currently in the U.S.). In a particular case, the molar composition of the refrigerant is $z_p = 0.25$ propane and $z_b = 0.75$ n-butane. This refrigerant is used in the vapor compression cycle shown in Figure 1. The refrigerant exits the condenser at state 1 as a saturated liquid at $P_{cond} = 700$ kPa and is then throttled to a pressure of $P_{evap} = 200$ kPa at the evaporator inlet, state 2. The refrigerant exits the evaporator at state 3 as a saturated vapor at P_{evap}.

a) Determine the temperature of the saturated liquid at the condenser outlet assuming that the mixture obeys Raoult's Law.

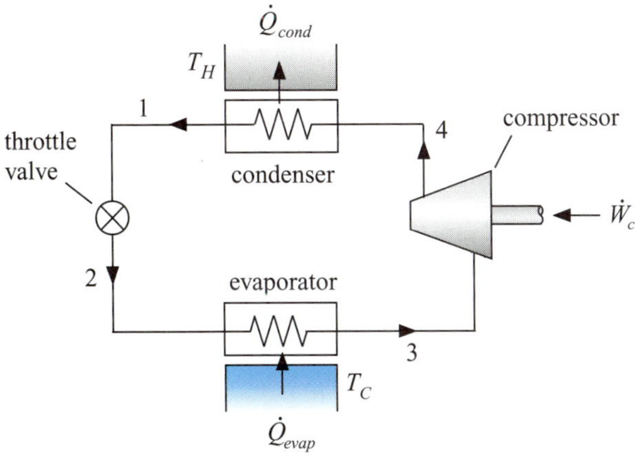

Figure 1: Mixture in a refrigeration cycle.

The inputs are entered in EES.

```
$UnitSystem SI Molar Radian J Pa K
P_cond=700 [kPa]*convert(kPa,Pa)     "condenser pressure"
P_evap=200 [kPa]*convert(kPa,Pa)     "evaporator pressure"
z_p=0.25 [-]                         "total mole fraction of propane"
z_b=0.75 [-]                         "total mole fraction of n-butane"
```

The pressure and molar quality at state 1 are given, $P_1 = P_{cond}$ and $Q_1 = 0$. Initially, the temperature at state 1, T_1, will be guessed in order to proceed with the problem in a linear manner.

```
P[1]=P_cond          "pressure at state 1"
Q[1]=0 [-]           "molar quality at state 1"
T[1]=300 [K]         "guess for the temperature"
```

The saturation pressure of the n-butane and propane ($P_{sat,b,1}$ and $P_{sat,p,1}$) are determined at T_1.

```
P_sat_b[1]=P_sat(n-butane,T=T[1])    "saturation vapor pressure of propane"
P_sat_p[1]=P_sat(propane,T=T[1])     "saturation vapor pressure of n_butane"
```

At state 1, the quality is zero so that all of the mixture is liquid. As a result, the mole fraction of propane in the liquid must be equal to the overall mole fraction of propane and the mole fraction of butane in the liquid must be equal to the overall mole fraction of butane:

$$z_p = x_{p,1} \tag{1}$$

$$z_b = x_{b,1} \tag{2}$$

Equations (1) and (2) can be obtained from Eq. (11-237) by setting $Q = 0$.

```
z_p=x_p[1]           "mole fraction of propane in liquid"
z_b=x_b[1]           "mole fraction of butane in liquid"
```

At state 1, the mixture is saturated liquid. Therefore, the liquid must be in equilibrium with vapor that is just beginning to form. According to Raoult's Law, Eq. (11-233), the mole fraction of propane and butane in the vapor must be given by:

$$x_{p,1}\,P_{sat,p,1} = y_{p,1}\,P_1$$

$$x_{b,1}\,P_{sat,b,1} = y_{b,1}\,P_1$$

```
x_p[1]*P_sat_p[1]=y_p[1]*P[1]          "Raoult's Law for propane"
x_b[1]*P_sat_b[1]=y_b[1]*P[1]          "Raoult's Law for n-butane"
```

Solving now provides $y_{p,1} = 0.3564$ and $y_{b,1} = 0.2765$. Clearly, the guessed value of the temperature at state 1 is not correct because the mole fractions in the vapor phase do not add to unity. The guess values are updated, the guessed temperature is commented out, and the constraint that the sum of $y_{p,1}$ and $y_{b,1}$ must be unity is added.

```
{T[1]=300 [K]}                         "guess for the temperature"
y_p[1]+y_b[1]=1                        "sum of vapor mole fractions must equal unity"
```

The solution provides $T_1 = 317.6$ K.

b) Determine the temperature, liquid and vapor compositions, and molar quality of the refrigerant at state 2, the exit of the throttle.

The molar specific enthalpies of saturated liquid propane and saturated liquid n-butane at T_1 ($\overline{h}_{p,f,1}$ and $\overline{h}_{b,f,1}$) are determined, assuming that the liquid forms an ideal solution. The molar specific enthalpy at state 1 is given by:

$$\overline{h}_1 = x_{p,1}\,\overline{h}_{p,f,1} + x_{b,1}\,\overline{h}_{b,f,1}$$

```
h_p_f[1]=enthalpy(Propane,T=T[1],x=0 [-])    "molar specific enthalpy of saturated liquid propane"
h_b_f[1]=enthalpy(n-butane,T=T[1],x=0 [-])   "molar specific enthalpy of saturated liquid n-butane"
h[1]=x_p[1]*h_p_f[1]+x_b[1]*h_b_f[1]         "molar specific enthalpy"
```

The pressure at state 2 is $P_2 = P_{evap}$. Initially, the temperature at state 2 (T_2) will be assumed in order to allow the solution to proceed.

```
P[2]=P_evap                            "pressure"
T[2]=260 [K]                           "guess for temperature"
```

The saturation pressure of the n-butane and propane ($P_{sat,b,2}$ and $P_{sat,p,2}$) are determined at T_2.

```
P_sat_p[2]=P_sat(propane,T=T[2])       "saturation vapor pressure of n_butane"
P_sat_b[2]=P_sat(n-butane,T=T[2])      "saturation vapor pressure of propane"
```

According to Raoult's Law, the mole fractions of propane and butane in the vapor are related to the mole fractions in the liquid by:

$$x_{p,2}\,P_{sat,p,2} = y_{p,2}\,P_2 \quad (3)$$

$$x_{b,2}\,P_{sat,b,2} = y_{b,2}\,P_2 \quad (4)$$

The mole fractions in the vapor and liquid must add to unity, providing the additional constraints:

$$y_{p,2} + y_{b,2} = 1 \tag{5}$$

$$x_{p,2} + x_{b,2} = 1 \tag{6}$$

Equations (3) through (6) are four equations in the four unknowns $x_{p,2}$, $x_{b,2}$, $y_{p,2}$, and $y_{b,2}$.

```
x_p[2]*P_sat_p[2]=y_p[2]*P[2]          "Raoult's Law for propane"
x_b[2]*P_sat_b[2]=y_b[2]*P[2]          "Raoult's Law for n-butane"
y_p[2]+y_b[2]=1                        "sum of vapor mole fractions must equal unity"
x_p[2]+x_b[2]=1                        "sum of liquid mole fractions must equal unity"
```

Equation (11-239) can be written for any one of the components in order to provide the molar quality:

$$z_p = x_{p,2}\,(1 - Q_2) + y_{p,2}\,Q_2$$

```
z_p=x_p[2]*(1Q[2])+y_p[2]*Q[2]         "molar quality"
```

The molar specific enthalpies of saturated liquid propane and saturated liquid n-butane at T_2 ($\overline{h}_{p,f,2}$ and $\overline{h}_{b,f,2}$) are determined. The molar specific enthalpies of saturated vapor propane and saturated vapor n-butane at T_2 ($\overline{h}_{p,g,2}$ and $\overline{h}_{b,g,2}$) are also determined. Assuming an ideal solution, the molar specific enthalpy of the liquid phase at state 2 is:

$$\overline{h}_{f,2} = x_{p,2}\,\overline{h}_{p,f,2} + x_{b,2}\,\overline{h}_{b,f,2}$$

and the molar specific enthalpy of the vapor phase at state 2 (which is assumed to be an ideal gas mixture) is:

$$\overline{h}_{g,2} = y_{p,2}\,\overline{h}_{p,g,2} + y_{b,2}\,\overline{h}_{b,g,2}$$

The molar specific enthalpy at state 2 is:

$$\overline{h}_2 = \overline{h}_{f,2} + Q_2\left(\overline{h}_{g,2} - \overline{h}_{f,2}\right)$$

```
h_p_f[2]=enthalpy(Propane,T=T[2],x=0 [-])     "molar specific enthalpy of saturated liquid propane"
h_b_f[2]=enthalpy(n-butane,T=T[2],x=0 [-])    "molar specific enthalpy of saturated liquid n-butane"
h_p_g[2]=enthalpy(Propane,T=T[2],x=1 [-])     "molar specific enthalpy of saturated vapor propane"
h_b_g[2]=enthalpy(n-butane,T=T[2],x=1 [-])    "molar specific enthalpy of saturated vapor n-butane"
h_f[2]=x_p[2]*h_p_f[2]+x_b[2]*h_b_f[2]        "molar specific enthalpy of liquid"
h_g[2]=y_p[2]*h_p_g[2]+y_b[2]*h_b_g[2]        "molar specific enthalpy of vapor"
h[2]=h_f[2]+Q[2]*(h_g[2]-h_f[2])              "molar specific enthalpy"
```

An energy balance on the valve provides:

$$\overline{h}_1 = \overline{h}_2 \tag{7}$$

Update the guesses, comment out the assumed value of T_2, and enforce the energy balance from Eq. (7) in order to solve the problem.

```
{T[2]=260 [K]}                                "guess for temperature"
h[2]=h[1]                                     "energy balance on the valve"
```

The solution provides $T_2 = 261.3$ K, $Q_2 = 0.3687$, $x_{p,2} = 0.1368$, $x_{b,2} = 0.8632$, $y_{p,2} = 0.4438$, and $y_{b,2} = 0.5562$.

c) The refrigerant mixture exits the evaporator at P_{evap} as a saturated vapor. Determine the temperature of the refrigerant mixture exiting the evaporator and the evaporator heat transfer per mole of refrigerant.

The pressure and molar quality at state 3 are given, $P_3 = P_{evap}$ and $Q_3 = 1$. Initially, the temperature at state 3, T_3, will be guessed in order to proceed with the problem in a linear manner.

```
P[3]=P_evap                 "pressure"
Q[3]=1 [-]                  "molar quality"
T[3]=260 [K]                "guess for temperature"
```

The saturation pressure of the n-butane and propane ($P_{sat,b,3}$ and $P_{sat,p,3}$) are determined at T_3.

```
P_sat_p[3]=P_sat(propane,T=T[3])     "saturation vapor pressure of n_butane"
P_sat_b[3]=P_sat(n-butane,T=T[3])    "saturation vapor pressure of propane"
```

At state 3, the quality is unity so that all of the mixture is vapor. As a result, the mole fraction of propane in the vapor must be equal to the overall mole fraction of propane and the mole fraction of butane in the vapor must be equal to the overall mole fraction of butane:

$$z_p = y_{p,3} \tag{8}$$

$$z_b = y_{b,3} \tag{9}$$

Equations (8) and (9) can be obtained from Eq. (11-237) by setting $Q = 1$.

```
z_p=y_p[3]                  "mole fraction of propane in vapor"
z_b=y_b[3]                  "mole fraction of butane in vapor"
```

At state 3, the mixture is saturated vapor. Therefore, the vapor must be in equilibrium with liquid that is just beginning to form. Raoult's Law, Eq. (11-233), is again applied in order to determine the mole fractions of propane and butane in the liquid that is in equilibrium with the vapor:

$$x_{p,3}\, P_{sat,p,3} = y_{p,3}\, P_3$$

$$x_{b,3}\, P_{sat,b,3} = y_{b,3}\, P_3$$

```
x_p[3]*P_sat_p[3]=y_p[3]*P[3]     "Raoult's Law for propane"
x_b[3]*P_sat_b[3]=y_b[3]*P[3]     "Raoult's Law for n-butane"
```

Solving now provides $x_{p,3} = 0.08046$ and $x_{b,3} = 1.229$. Clearly, the guessed value of the temperature at state 3 is not correct. The guess values are updated, the guessed temperature is commented out, and the constraint that the sum of $x_{p,3}$ and $x_{b,3}$ must be unity is added.

```
{T[3]=260 [K]}              "guess for temperature"
x_p[3]+x_b[3]=1             "sum of liquid mole fractions must equal unity"
```

which leads to $T_3 = 266.6$ K. The refrigerant mixture enters the evaporator at 261.3 K. Note that, unlike a pure refrigerant, the temperature of a refrigerant mixture is not constant during a constant pressure evaporation process. The so-called *temperature glide* that occurs during constant pressure evaporation can improve the performance of refrigeration cycles that use mixtures. The molar specific enthalpies of saturated vapor propane and saturated vapor n-butane at T_3 ($\bar{h}_{p,g,3}$ and $\bar{h}_{b,g,3}$) are determined. The molar specific enthalpy at state 3 is given by:

$$\bar{h}_3 = y_{p,3}\,\bar{h}_{p,g,3} + y_{b,3}\,\bar{h}_{b,g,3}$$

```
h_p_g[3]=enthalpy(Propane,T=T[3],x=1 [-])     "molar specific enthalpy of saturated vapor propane"
h_b_g[3]=enthalpy(n-butane,T=T[3],x=1 [-])    "molar specific enthalpy of saturated vapor n-butane"
h[3]=y_p[3]*h_p_g[3]+y_b[3]*h_b_g[3]          "molar specific enthalpy"
```

An energy balance on the evaporator provides:

$$\frac{\dot{Q}}{\dot{n}} = \bar{h}_3 - \bar{h}_2$$

```
Q_dot\n_dot=h[3]h[2]                          "refrigeration per unit mole"
```

which leads to a heat transfer per unit mole of refrigerant of 1.437×10^7 J/kmol.

d) Solve the problem without assuming ideal gas behavior (i.e., without using Raoult's Law) but still assuming that the mixture behaves as an ideal solution. That is, use Eq. (11-232) rather than Eq. (11-233) in order to enforce the phase equilibrium. Compare your answers with those obtained using Raoult's Law.

The easiest way to modify the program is to use the fugacity coefficients. The fugacity coefficient is the ratio of the fugacity to the pressure. Written in terms of the fugacity coefficients, Equation (11-232) becomes:

$$x_i\,\phi_{sat,i}\,P_{sat,i} = y_i\,\phi_i\,P \text{ for } i = 1..C$$

where $\phi_{sat,i}$ is the fugacity coefficient of component i in its saturated state. The fugacity coefficient of saturated propane at each state is computed according to:

$$\phi_{sat,p,j} = \frac{f_{sat,p,j}}{P_{sat,p,j}} \quad \text{for } j = 1..3$$

where $f_{sat,p,j}$ is the fugacity of saturated propane at state j, computed using the Fugacity function. The fugacity coefficient of saturated n-butane at each state is computed according to:

$$\phi_{sat,b,j} = \frac{f_{sat,b,j}}{P_{sat,b,j}} \quad \text{for } j = 1..3$$

```
"Fugacity coefficients in saturated state"
duplicate i=1,3
  phi_sat_p[i]=Fugacity(propane,T=T[i],x=1 [-])/P_sat_p[i]
  phi_sat_b[i]=Fugacity(n-butane,T=T[i],x=1 [-])/P_sat_b[i]
end
```

The fugacity coefficient of each component at the temperature and total pressure of the mixture must be determined. First, the phase of each pure fluid at the temperature and pressure of the mixture at each state is determined using the Phase$ function.

```
"Determine the phase of each fluid at each state"
duplicate i=1,3
   Phase_p$[i]=Phase$(Propane,T=T[i],P=P[i])       "phase of propane"
   Phase_b$[i]=Phase$(n-butane,T=T[i],P=P[i])      "phase of n-butane"
end
```

Examination of the arrays table shows that pure propane is a vapor while pure n-butane is a liquid at the temperature and total pressure of the mixture at all three states. The fugacity coefficient for propane can be determined directly according to:

$$\phi_{p,j} = \frac{f_{p,j}}{P_j} \quad \text{for } j = 1..3$$

where $f_{p,j}$ is the fugacity of pure propane at state j evaluated at T_j and P_j.

```
"Fugacity coefficient for propane in gas phase"
duplicate i=1,3
  phi_p[i]=Fugacity(Propane,T=T[i],P=P[i])/P[i]
end
```

Because n-butane is a liquid at the temperature and total pressure of the mixture, it is necessary to use the approximation provided in Eq. (11-222) to estimate the fugacity coefficient of n-butane.

$$\phi_{b,j} = \phi_{sat,b,j} \quad \text{for } j = 1..3$$

```
"Fugacity coefficient for n-butane in gas phase"
duplicate i=1,3
  phi_b[i]=phi_sat_b[i]
end
```

In order to facilitate switching between Raoult's law and Eq. (11-232), a string variable Idealgas$ is defined. The string variable is used as the condition in a $IF directive to switch between the two equilibrium criteria; Eq. (11-233) is used if the variable Idealgas$ is set to 'yes' and Eq. (11-232) is used otherwise.

```
Idealgas$='no'                  "string variable indicating whether to assume ideal gas behavior"
$IF Idealgas$='yes'
   x_p[1]*P_sat_p[1]=y_p[1]*P[1]                          "Raoult's Law for propane"
   x_b[1]*P_sat_b[1]=y_b[1]*P[1]                          "Raoult's Law for n-butane"
$ELSE
   x_p[1]*phi_sat_p[1]*P_sat_p[1]=y_p[1]*phi_p[1]*P[1]    "equilibrium constraint for propane"
   x_b[1]*phi_sat_b[1]*P_sat_b[1]=y_b[1]*phi_b[1]*P[1]    "equilibrium constraint for n-butane"
$ENDIF
```

```
$IF Idealgas$='yes'
  x_p[2]*P_sat_p[2]=y_p[2]*P[2]                        "Raoult's Law for propane"
  x_b[2]*P_sat_b[2]=y_b[2]*P[2]                        "Raoult's Law for n-butane"
$ELSE
  x_p[2]*phi_sat_p[2]*P_sat_p[2]=y_p[2]*phi_p[2]*P[2]  "equilibrium constraint for propane"
  x_b[2]*phi_sat_b[2]*P_sat_b[2]=y_b[2]*phi_b[2]*P[2]  "equilibrium constraint for n-butane"
$ENDIF
$IF Idealgas$='yes'
  x_p[3]*P_sat_p[3]=y_p[3]*P[3]                        "Raoult's Law for propane"
  x_b[3]*P_sat_b[3]=y_b[3]*P[3]                        "Raoult's Law for n-butane"
$ELSE
  x_p[3]*phi_sat_p[3]*P_sat_p[3]=y_p[3]*phi_p[3]*P[3]  "equilibrium constraint for propane"
  x_b[3]*phi_sat_b[3]*P_sat_b[3]=y_b[3]*phi_b[3]*P[3]  "equilibrium constraint for n-butane"
$ENDIF
```

Solving these equations leads to $T_1 = 320.5$ K, $T_2 = 259.2$ K, and $T_3 = 266.7$ K with an evaporator heat transfer rate per mole of refrigerant of 1.397x10^7 J/kmol. These results are quite close to the results obtained with Raoult's Law.

11.6 The Phase Rule

The Phase Rule states that the number of intensive internal properties needed to fix the state of a non-reacting thermodynamic system, F, is given by:

$$F = C - \Pi + 2 \tag{11-240}$$

where C is the number of distinguishable chemical species in the system and Π is the number of phases (i.e., solid, liquid, or vapor) that are present. The Phase Rule was first presented in Section 2.1 and it is presented again in Eq. (11-1). The Phase Rule, first proposed by Gibbs in 1875, has been applied throughout this text. The basic postulate behind this rule is that the intensive state of a system is fixed if the temperature, pressure and mole fraction of each component in each phase is known. The reasoning behind this postulate is based on the experimental observation that the specific volume of a phase cannot be independently changed if the temperature, pressure and mole fraction of each component in the phase are fixed.

With this postulate there are $(2+C)\,\Pi$ variables that must be specified in order to fix the state of a system than has C components and Π phases. The mole fractions of each phase must sum to unity, leading to Π equations:

$$\sum_{i=1}^{C} x_{i,\text{phase } j} = 1 \quad \text{for } j = 1..\Pi \tag{11-241}$$

At equilibrium, the temperature of each co-existing phase must be equal, leading to Π – 1 additional equations:

$$T_{\text{phase } 1} = T_{\text{phase } 2} = \cdots = T_{\text{phase}\,\Pi} \tag{11-242}$$

The pressure of each co-existing phase must also be equal, providing Π –1 additional equations:

$$P_{\text{phase } 1} = P_{\text{phase } 2} = \cdots = P_{\text{phase}\,\Pi} \tag{11-243}$$

A final requirement is that, at equilibrium, the chemical potential of substance i must have the same value in all coexisting phases, as indicated in Eq. (11-191). This requirement provides $C\,(\Pi - 1)$ additional equations:

$$\mu_{i,\text{phase 1}} = \mu_{i,\text{phase 2}} = \cdots = \mu_{i,\text{phase }\Pi} \quad \text{for } i = 1..C \tag{11-244}$$

The number of degrees of freedom, F, is the difference between the number of variables that must be specified and the number of equations that relate these variables, provided by Eqs. (11-241) through (11-244).

$$\begin{aligned} F &= (2 + C)\,\Pi - [\Pi + 2\,(\Pi - 1) + C\,(\Pi - 1)] \\ &= C - \Pi + 2 \end{aligned} \tag{11-245}$$

Equation (11-245) is the Phase Rule.

REFERENCES

Jacobsen, R.T., Penoncello, S.G., and Lemmon, E.W., *Thermodynamic Properties of Cryogenic Fluids*, Plenum Press, NY, ISBN 0-306-45522-6, (1997).

Kyle, B.G., *Chemical and Process Thermodynamics*, 3rd edition, Prentice-Hall, ISBN 0-13-081244-7, (1999).

Peng, Ding-Yu and Robinson, D.B, "A New Two-Constant Equation of State," *Ind. Eng. Chem. Fundamentals*, Vol. 15, No. 1, p. 59, (1976).

Poling, B.E., Prausnitz, J.M. and O'Connell, J., *The Properties of Gases and Liquids*, 5th edition, McGraw-Hill, ISBN 9780070116825, (2000).

Smith, J.M., Van Ness, H.C., and Abbott, M.M., *Chemical Engineering Thermodynamics*, 6th edition, McGraw-Hill, ISBN 0-07-240296-2, (2001).

Soave, G., "Equilibrium Constants from a Modified Redlich-Kwong Equation of State," *Chemical Engineering Science*, Vol. 27, pp. 1197–1203, (1972).

Problems

The problems included here have been selected from a larger set of problems that are available from the website associated with this book (www.cambridge.org/kleinandnellis).

A. Ideal Gas Mixtures

11.A-1 It is often necessary to prepare a gas mixture with accurately known composition in order to calibrate gas analysis instruments. In this problem, a mixture with of 80% ethylene (C_2H_4) and 20% carbon dioxide (CO_2), on a molar basis, is prepared in the following manner. Two uninsulated tanks, each of $V = 4$ liter volume, are connected together. One tank contains ethylene gas at $P_{C2H4,1} = 100$ atm and $T = 25°$C. The other contains carbon dioxide at $P_{CO2,1}$ and the same temperature. The valve connecting the tanks is opened and it remains open for a long time until a homogeneous mixture at the desired proportions is obtained. The surroundings are at $T_{amb} = 25°$C and $P_{atm} = 1$ atm. Answer the following questions assuming that the gas mixture obeys the ideal gas law. Note that, in EES, ideal gas ethylene is represented with substance C_2H_4 and ideal gas carbon dioxide is represented with substance CO_2.

a) What is the initial CO_2 pressure, $P_{CO2,1}$?
b) What is the final pressure of the gas mixture?
c) What is the heat transfer for this process?
d) What is the total exergy destruction for this process?

11.A-2 A mixture of 80% methane and 20% ethane on a molar basis is contained in an insulated tank with volume $V = 0.5\ \text{m}^3$ at $P_1 = 6$ bar and $T_{amb} = 30°\text{C}$. The value is opened accidentally, and the pressure quickly drops to $P_2 = 2$ bar before the valve is closed. Assume that the mixture behaves as an ideal gas.

a) Calculate the mass of gas mixture that escapes.
b) Eventually, the gas mixture returns to the temperature of the surroundings, T_{amb}. Estimate the tank pressure at this time.
c) Calculate the heat transferred to the gas mixture in the tank during the process.

11.A-3 A mixture of helium and ethane is prepared using the device shown in Figure 11.A-3. There are separate supply manifolds for helium and ethane. The well-insulated mixing tank is first evacuated to a very low pressure. Helium from a pipeline at $T_{He,s} = 100°\text{F}$ and $P_{He,s} = 10$ atm is admitted very rapidly until the pressure in the tank reaches $P_1 = 2$ atm. The helium supply valve is then closed. Thirty minutes later, the ethane supply valve is opened to allow ethane from a pipeline at $T_{Ethane,s} = 100°\text{F}$ and $P_{Ethane,s} = 5$ atm to flow into the tank. The valve is closed when the tank pressure reaches $P_2 = 3$ atm. A day later, the gas mixture is drawn off through valve C. The tank is cylindrical with an inner diameter of $D = 1$ ft and a height of $H = 1$ ft. The walls are made of aluminum with a uniform thickness of $th = 0.25$ in. At these low pressures, assume helium and ethane behave as ideal gases.

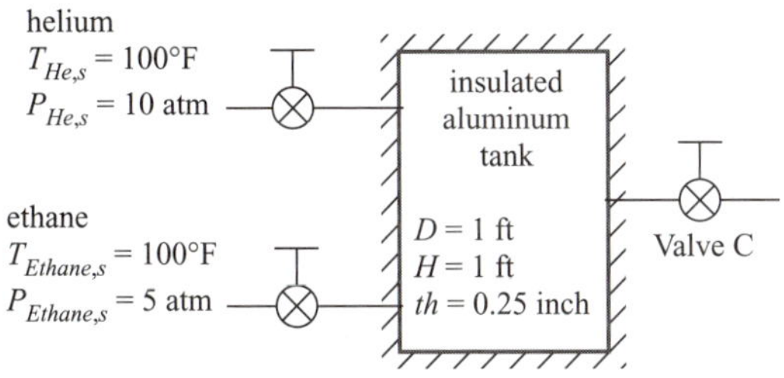

Figure 11.A-3: Preparation of a gas mixture of helium and ethane.

a) What is the temperature of the helium in the tank when valve A is closed?
b) What is the temperature of the gas mixture in the tank when valve B is closed?
c) What is your best engineering estimate of the composition of the mixture removed through valve C?
d) What is the total entropy generation for this process?

11.A-4 A mixture of 70% methane and 30% nitrogen on a molar basis enters a compressor at $T_{in} = -40°\text{C}$, and $P_{in} = 10$ atm with a mass flow rate of $\dot{m} = 10$ kg/min. Estimate the minimum power required to compress the mixture to $P_{out} = 100$ atm, assuming ideal gas behavior.

11.A-5 A gas mixture is prepared by steadily mixing carbon dioxide and ethylene, both at $T = 100°\text{C}$ and $P = 175.8$ atm, in molar proportions of 69.5% carbon dioxide and 30.5% ethylene. The mixture exits at the same temperature and pressure. Assuming that the mixture obeys the ideal gas law,

a) Determine the molar specific volume of the mixture.
b) Determine the required heat transfer per mole of mixture.
c) Determine the entropy generation per mole of mixture.

11.A-6 A binary gas mixture can be prepared by charging one component into a tank of known volume that contains the other gas. The pressure of the gas that is initially in the tank must be set so that the desired composition results when the second gas is charged. In a particular case, a $V = 40$ liter tank is available containing nitrogen at $T_1 = 25^{\circ}$C. The tank is connected to a fill line carrying ethane gas at $T_s = 25^{\circ}$C and $P_s = 84$ bar, as shown in Figure 11.A-6. The valve is opened to allow ethane to enter the tank. The valve remains open until the gas mixture within the tank returns to T_1. Assume ideal gas behavior.

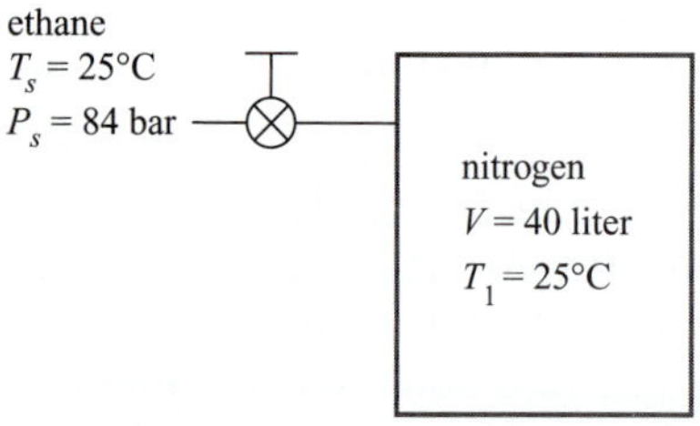

Figure 11.A-6: Schematic of the charging process.

a) Determine the pressure that the nitrogen should initially have in order to reach an equimolar mixture when the charging process has completed.
b) Determine the heat transfer for this process.
c) Determine the entropy generation for this process.

B. Real Fluid Mixtures

11.B-1 The gases in Problem 11.A-1 may not obey the ideal gas law at the pressures occurring in the problem. However, the gas mixture is still likely to behave as an ideal solution.

a) Answer the questions posed in Problem 11.A-1 using the pure gas properties in EES with substances ethylene and carbondioxide and ideal solution theory.
b) Compare your results with the results obtained assuming ideal gas behavior in Problem 11.A-1.

11.B-4 Resolve Problem 11.A-5 assuming the mixture obeys:

a) the ideal solution assumption, and
b) the Peng-Robinson equation of state.
c) Compare the results with each other and with the results from the ideal gas law.

11.B-5 Refrigerant 410A is a mixture of 50% Refrigerant 32 and 50% Refrigerant 125, where the percentages are on a mass basis. Property information for R32 and R125 are available in EES. Use the pure fluid property data and the Peng-Robinson equation of state for a mixture.

a) Calculate and plot the specific volume, specific enthalpy and specific entropy of R410A as a function of temperature between 40°C and 100°C for pressures of 5, 10, and 20 bar.
b) Compare your results with the property data in EES for R410A. Note that the specific enthalpy and specific entropy reference states for R410A are not consistent with the reference states for R32 and R125. Therefore, an offset between the EES values and those calculated with the Peng-Robinson equation of state can be expected.

11.B-6 Nitrogen and hydrogen are mixed at $P = 300$ atm and $T = 25^{\circ}$C as the first step in the production of ammonia. Using the Peng-Robinson equation of state, prepare the following plots as a function of the mole fraction of hydrogen, ranging from 0 to 1:

a) the mixture specific volume,
b) the specific enthalpy change of mixing,
c) and the specific entropy change of mixing.
d) Indicate how the plots would look if the mixture were assumed to be an ideal solution following Amagat's rule.

11.B-7 A natural gas has the following composition on a weight basis: 85% methane, 5% ethane and 10% nitrogen. The gas is compressed to $P = 4$ MPa for transport though a pipeline at $T = 10^\circ$C. The power needed to transport the gas is a function of its density. Estimate the density of this gas mixture with the following methods and compare the results.
a) the ideal gas law
b) the ideal solution of real gases
c) Kay's approximation
d) the Redlich-Kwong-Soave equation of state
e) the Peng-Robinson equation of state

11.B-8 Solve Problem 11.A-6 using the Peng-Robinson equation of state for the properties of the mixture. The EES property data base can be used for the pure fluids.

11.B-9 A gas mixture consisting of 70% methane and 30% nitrogen, by mole, at $P_1 =$ 100 atm and $T_1 = 300$ K expands through an insulated nozzle to a final pressure of $P_2 = 35$ atm. The temperature at the nozzle exit is reported to be $T_2 = 255$ K. Using the Peng-Robinson equation of state, determine:
a) the velocity at the nozzle exit, and
b) the rate of entropy generation for this process.

11.B-11 A gas mixture used in a Joule-Thomson expansion cycle consists of 50% R23 and 50% methane, on a molar basis. This mixture enters an compressor at $P_1 = 2.5$ atm and $T_1 = -40^\circ$C with a mass flow rate of $\dot{m} = 0.25$ kg/s where it is compressed to $P_2 = 92$ atm.
a) Using the ideal solution method, estimate the molar specific volume at the compressor exit, the minimum compressor power, and the compressor exit temperature for both an adiabatic and an isothermal compressor.
b) Estimate the same quantities using Kay's method.
c) Estimate the same quantities using the Peng-Robinson equation of state.
d) You may use the GEN_EOS and Peng-Robinson libraries in EES to simplify the calculations, if you wish. The interaction coefficient between these gases can be estimated using Eq. (11-80).

11.B-12 Compare the molar specific volumes of an equimolar mixture of propane and n-butane at $T = 450$ K and $P = 25{,}000$ kPa using:
a) Dalton's rule,
b) Amagat's rule, and
c) Kay's rule.

C. Multi-component Phase Equilibrium

11.C-1 A mixture of oxygen and helium is prepared as shown in Figure 11.C-1. Helium gas at $T_{in} = 25^\circ$C and pressure P is bubbled through a tank containing liquid oxygen at $T_{tank} = 125$ K. The pressure in the tank is maintained at P with a pressure regulator. The temperature of the liquid oxygen is maintained at T_{tank}

by heat transfer to a liquid nitrogen coil. The gas mixture is eventually heated to $T_{out} = 25°\text{C}$ and throttled to $P_{out} = 1$ bar with a volumetric flow rate of $\dot{V}_{out} = 0.034\ \text{m}^3/\text{s}$. The helium has negligible solubility in the liquid oxygen.

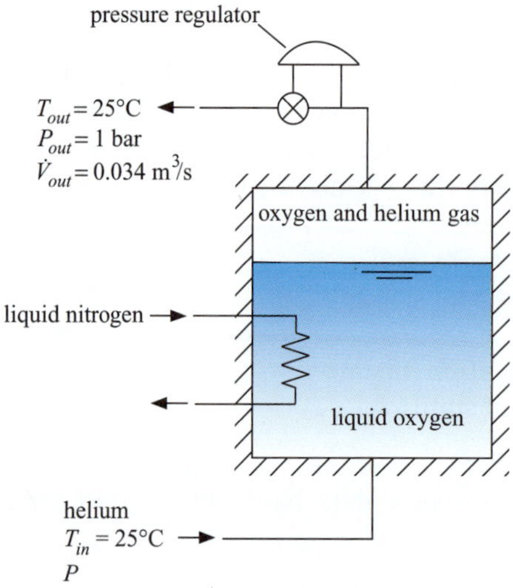

Figure 11.C-1: System to generate mixture of helium and oxygen.

Assuming that the gas mixture behaves as an ideal solution. Calculate and plot the mole fraction of the oxygen and the required heat transfer rate to the liquid nitrogen as a function of the pressure in the tank for pressures ranging from the saturation pressure of oxygen at 125 K to 120 bar.

11.C-2 A mixture of refrigerants R12 and R114 enters the evaporator of a vapor compression refrigeration system at $T_{in} = -25°\text{C}$ and $P_{in} = 60$ kPa. The quality (on a mass basis) of the entering mixture is 35%. R12 and R114 are chemically similar and therefore it can be assumed that the mixture obeys Raoult's law and that the specific enthalpy of the liquid and vapor phases can be determined assuming that they form ideal solutions. Property data for pure refrigerant R12 and R114 are available in EES.

a) Calculate the entering mole fractions for the R12 and R114 in the liquid and vapor phases.
b) Determine the overall mass fraction of R114.
c) Determine the temperature of the refrigerant mixture exiting the evaporator as a saturated vapor at $P_{out} = P_{in}$.
d) Estimate the heat transfer in the evaporator per mass of entering refrigerant for these conditions.

11.C-3 The "propane" tanks that are used for heating and cooking actually contain a mixture of propane and butane. Some suppliers will tailor the composition of the fuel based on the local weather conditions. In cold weather, a higher percentage of propane is used in order to achieve a higher delivery pressure. During the summer, the propane percentage is reduced. Assume that liquid and vapor phases of propane and n-butane form ideal solutions due to their chemical similarity. Property data for the pure fluids are available in EES using the substances 'Propane' and 'n-Butane', respectively. The lower heating values of propane and n-butane are 46,327 kJ/kg and 45,348 kJ/kg, respectively.

a) Suppose that a $V = 0.42\ \text{m}^3$ tank is charged on a hot summer day with a saturated mixture of propane and n-butane to 900 kPa at 40°C. The volume fraction of the liquid is 95% after charging. What are the total mass of fuel in the tank and the overall mole fraction of propane?

b) What is the pressure in the tank if its contents are later cooled to –30°C without any of the fuel having been used?

c) What will be the mass of fuel and the overall mole fraction of propane if the tank is charged with a mixture of propane and n-butane at –30°C, 115 kPa so that 80% of the volume is liquid?

d) If the tank contents determined in part (c) are later heated to 40°C without any of the fuel having been used, what will the pressure be?

e) Estimate the percent difference in energy content for a fully charged tank for cases (a) and (c).

11.C-4 A saturated liquid mixture of propane and n-butane at $T_{in} = 40°C$ and $P_{in} = 900$ kPa is throttled to $P_{out} = 125$ kPa as it passes from an LPG tank through a pipe and enters a heating appliance. Assuming that the throttling process is adiabatic, estimate the temperature, the quality (on a mass basis), and the molar composition of the liquid and vapor streams of the fuel just downstream of the throttle. You may assume Raoult's law is applicable for this mixture. Further, at the low pressure downstream of the throttle, the vapor phase may be assumed to obey the ideal gas law.

11.C-5 The performance of standard vapor compression refrigeration cycles is normally estimated assuming that the refrigerant is a pure substance. In most refrigeration cycles, however, the refrigerant is mixed with oil. The presence of oil in the liquid refrigerant lowers the vapor pressure exerted by the refrigerant at a given temperature, leading to reduced refrigeration capacity or higher evaporator temperatures than would otherwise be expected based on the properties of pure refrigerants. Consider a specific situation in which refrigerant R22 with 3% (by mass) mineral oil exits the condenser of a refrigeration cycle as a liquid solution at 35°C and 1500 kPa. This liquid is adiabatically throttled into the evaporator. The refrigerant oil mixture exits the evaporator as a two-phase system at –10°C and 300 kPa. The oil-rich liquid phase returns oil to the compressor crankcase. The compressor is a reciprocating constant-displacement device that produces a constant volumetric flow. Assume that the oil and R22 form an ideal solution and that Raoult's law is applicable for phase equilibrium. The molecular weight of the oil is 390 kg/kmol. The oil specific heat capacity is $c = 1.75$ kJ/kg-K. The saturation pressure of the oil can be represented by:

$$\ln(P_{sat}) = 24.56 - \frac{10{,}510}{T_{sat}}$$

where P_{sat} is in [mm Hg] and T_{sat} is in [K].

a) What is the quality (on a mass basis) of the refrigerant-oil mixture exiting the evaporator at 300 kPa and -10°C?

b) What is the mass fraction of oil in the liquid phase that exits the evaporator?

c) What is the evaporator heat transfer per mass of refrigerant-oil mixture passing through the evaporator? Compare this value with the heat transfer per mass that would occur if there were no oil in the refrigerant and the system operated at the same temperatures and pressures.

d) One problem resulting from the oil is that some of the refrigerant exits the evaporator as a liquid and it cannot enter the compressor in this condition. Assuming the pressure remains at 300 kPa, what temperature must the refrigerant-oil mixture be heated to so that the fraction of refrigerant remaining in the liquid phase is less than 0.001 of the total mass of R22?

11.C-6 Determine the liquid and vapor composition of a mixture that has an overall molar composition of 60% methane, 20% ethane, with the remainder being propane. The mixture is at –120°F and 250 psia. Compare the results determined in the following ways:

a) using Raoult's Law,
b) assuming the liquid and vapor form ideal solutions, and
c) using the Peng-Robinson library to determine the fugacity of each fluid in the gaseous solution. Note that the fugacity of each component in the solution is the product of the mole fraction in the gas, the partial fugacity coefficient returned by function PHI_I_PR and the total pressure. Assume that the liquid phase forms an ideal solution, as in part (b).

11.C-7 Prepare a temperature composition diagram for a mixture of propane and n-butane at fixed pressures of 1, 5, and 10 bar assuming Raoult's law is applicable for these conditions. Label the dew and bubble point curves. Superimpose all of the plots on one set of axes. Explain the trends observed in the plots.

12 Psychrometrics

This chapter discusses moist gas mixtures, i.e., mixtures in which one component can condense. The moist gas mixture that is of primary engineering importance is dry air and water vapor. *Psychrometrics* is the term used to describe the study of air-water vapor mixtures. Psychrometrics is important for human comfort, drying foods, product packaging, and industrial drying processes. A significant fraction of the energy consumed in air conditioning processes is used to lower the humidity of air rather than reduce its temperature.

12.1 Psychrometric Definitions

Consider the air-water vapor mixture that might be present in a typical classroom. The temperature in the room is $T_1 = 20°\text{C}$ and the total pressure is $P = 1$ atm. A typical value for the mole fraction of water vapor in the air is $y_{v,1} = 0.01$; that is, 1% of the gas in the room (by mole) is water vapor and the remaining gas is *dry air*. Dry air is air (itself a mixture of nitrogen, oxygen, etc.) that contains no water vapor. You likely have never been exposed to dry air because some water vapor is normally present in air even in the driest climates. Assuming that the air-water vapor mixture behaves as an ideal gas mixture, the partial pressure of the water vapor is $P_{v,1} = y_{v,1}\,P = 1013$ Pa. Therefore, the state of the water vapor in the mixture (state v,1) is fixed by $P_{v,1} = 1013$ Pa and $T_1 = 20°\text{C}$. State v,1 is located on a (very qualitative) temperature-specific entropy diagram for water in Figure 12-1. The partial pressure of the dry air in the mixture is $P_{a,1} = P - P_{v,1} = 100312$ Pa. The state of the dry air in the classroom is fixed by $P_{a,1} =$ 100312 and $T_1 = 20°\text{C}$.

The most direct indicator of the amount of water in the air is the *humidity ratio*, ω, which is defined as the ratio of the mass of water vapor (m_v) to the mass of dry air (m_a):

$$\omega = \frac{\text{mass of water vapor}}{\text{mass of dry air}} = \frac{m_v}{m_a} \tag{12-1}$$

Substituting the ideal gas law into Eq. (12-1) leads to:

$$\omega = \frac{P_v\,V}{R_v\,T}\,\frac{R_a\,T}{P_a\,V} \tag{12-2}$$

where V is the volume, T is the temperature, and R_a and R_v are the ideal gas constants for dry air and water vapor, provided by:

$$R_a = \frac{R_{univ}}{MW_a} \tag{12-3}$$

$$R_v = \frac{R_{univ}}{MW_v} \tag{12-4}$$

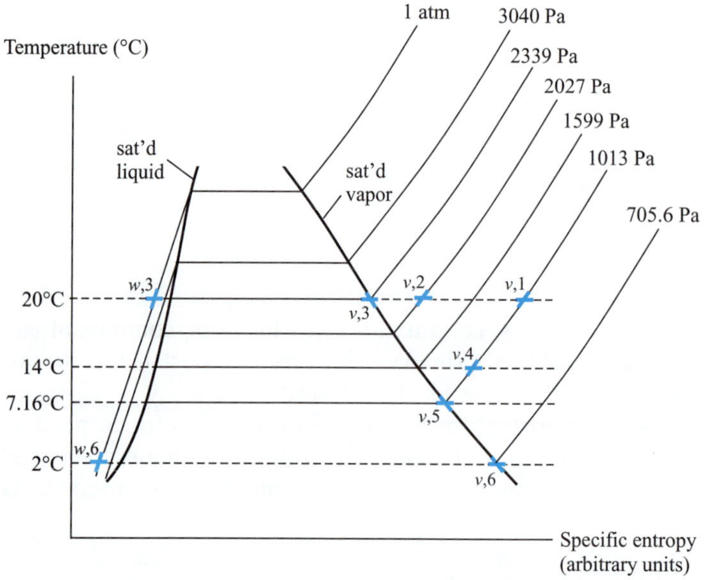

Figure 12-1: States of the water in the air-water vapor mixture located on a qualitative temperature-specific entropy diagram.

where $MW_a = 28.97$ kg/kmol and $MW_v = 18.0$ kg/kmol are the molar masses of air and water vapor, respectively. Substituting Eqs. (12-3) and (12-4) into Eq. (12-2) leads to:

$$\omega = \frac{P_v \, MW_v}{P_a \, MW_a} \tag{12-5}$$

The partial pressure of air is the total pressure (P) less the partial pressure of water vapor; therefore, Eq. (12-5) can be written as:

$$\omega = 0.622 \frac{P_v}{(P - P_v)} \tag{12-6}$$

where 0.622 is the ratio of the molar masses. Substituting $P_{v,1} = 1013$ Pa and $P = 101325$ Pa into Eq. (12-6) provides $\omega_1 = 0.006284$ kg_v/kg_a (i.e., the air-water vapor mixture contains 6.284 g of vapor for every kg of dry air).

What happens to the air-water vapor mixture in the classroom if water vapor is added? Perhaps a humidifier is allowed to inject water vapor in the air or there are people in the room who release water by sweating and breathing. We will assume that the temperature in the room remains the same, $T_2 = T_1$, and the total pressure is unchanged. However, the humidification causes the mole fraction of water vapor to increase, perhaps to $y_{v,2} = 0.02$ (i.e., the mole fraction of water vapor doubles so that 2% of the mixture is water vapor, by mole). We can carry out the same calculations at state 2 in order to identify the partial pressure of the water vapor, $P_{v,2} = y_{v,2}\, P =$ 2027 Pa. The state of the water vapor is fixed by $P_{v,2}$ and T_2 in Figure 12-1. Substituting

$P_{v,2} = 2027$ Pa into Eq. (12-6) provides $\omega_2 = 0.0126$ kg$_v$/kg$_a$ (12.7 g of vapor for every kg of air).

Can the humidification process continue until $y_{v,3} = 0.03$ (i.e., 3% of the mixture is water vapor, by mole) while the room remains at the same temperature and pressure? The partial pressure of the water vapor is computed by $P_{v,3} = y_{v,3}\,P = 3040$ Pa. However, the saturation pressure of water at 20°C is 2339 Pa. Examination of Figure 12-1 shows that water vapor cannot exist at $P_{v,3} = 3040$ Pa and $T_3 = 20$°C; this state is liquid water, not water vapor.

So what actually happens during the continued humidification process? The air-water vapor mixture in the room becomes *saturated*; that is, the moist air contains the maximum amount of water vapor that it possibly can have at the given temperature and total pressure. The partial pressure of the water vapor will be equal to the saturation pressure in this case, $P_{v,3} = P_{sat,3}$. State v,3 is located in Figure 12-1 and it lies on the vapor dome. At 20°C, the largest partial pressure of water vapor that can be achieved is 2339 Pa. Any additional water in the room cannot evaporate into the air and must instead exist as liquid water at 20°C and 1 atm, as shown by state w,3 in Figure 12-1.

The individual gases in an ideal gas mixture behave as if the other gases were not present. We know that pure water will exist as a gas at a fixed temperature and at a suitably low pressure. However, if the pressure is slowly increased while holding the temperature constant then a two-phase mixture of liquid and vapor will eventually appear. The pressure at which liquid and vapor coexist is called the saturation vapor pressure, P_{sat}. Raising the pressure of a pure water sample above the saturation vapor pressure while holding the temperature fixed will cause the water vapor to condense to liquid. Because an ideal gas mixture behaves as if the other gases are not present, the *saturation vapor pressure* is also the maximum partial pressure that water vapor can have in an air-water vapor mixture. If the partial pressure of water vapor in an air-water vapor mixture is at its saturation vapor pressure, then the mixture is said to be *saturated*, i.e., it contains the maximum possible amount of water vapor. The humidity ratio for a saturated mixture is given by Eq. (12-6) with P_v replaced by P_{sat}:

$$\omega_{sat} = \frac{0.622\,P_{sat}}{(P - P_{sat})} \tag{12-7}$$

Equation (12-7) indicates that the maximum amount of water vapor that can exist in air depends strongly on the temperature of the air because the saturation pressure is an exponential function of temperature. At the saturated state v,3 in Figure 12-1, the humidity ratio is $\omega_3 = 0.0147$ kg$_v$/kg$_a$; that is, at 20°C, the maximum amount of water vapor that can be held in an air-water vapor mixture at 1 atm is 14.7 g of water vapor per kg of air.

The most familiar psychrometric index is the relative humidity, ϕ. Relative humidity is an indirect measure of the amount of water vapor in air that is related to how close the air-water mixture is to being saturated. Relative humidity is defined as the ratio of the mass of water vapor in the air to the maximum possible (or saturation) mass of water vapor at the same temperature:

$$\phi = \frac{m_v}{m_{sat}} \tag{12-8}$$

Substituting the ideal gas law into Eq. (12-8) results in:

$$\phi = \frac{P_v\,V}{R_v\,T}\frac{R_v\,T}{P_{sat}\,V} \tag{12-9}$$

which can be simplified to:

$$\phi = \frac{P_v}{P_{sat}} \tag{12-10}$$

The relative humidity by itself does not provide an indication of the amount of water in the air. For example, the relative humidity of cold outdoor air in winter may be nearly 100%, but the amount of water vapor in the air is low because the temperature (and thus the saturation vapor pressure) is low.

The relative humidity at state v,1 in Figure 12-1 is $\phi_1 = P_{v,1}/P_{sat,1} = 1013/2339 = 0.4332$ (43.32% relative humidity). State v,2 in Figure 12-1 is closer to being saturated and therefore has a higher relative humidity, $\phi_2 = P_{v,2}/P_{sat,2} = 0.8666$ (86.66%). The relative humidity of state v,3 in Figure 12-1 is 1.0 (100%).

Let's return to state v,1 in Figure 12-1 and imagine a different process. Instead of adding water to the air using a humidifier, we could reduce the temperature of the air-water vapor mixture. Suppose that the temperature in the room is reduced to $T_4 = 14°\text{C}$ at constant pressure. Provided that no water condenses, the amount of water vapor in the air remains unchanged; therefore $y_{v,4} = y_{v,1} = 0.01$ and the partial pressure of the water vapor in the air remains at $P_{v,4} = P_{v,1} = 1013$ Pa. State v,4 is specified by $P_{v,4}$ and T_4, as shown in Figure 12-1. The humidity ratio at state 4 does not change, $\omega_4 = \omega_1 = 0.006284$ kg$_v$/kg$_a$. However, the relative humidity at state 4 is substantially higher than it is at state 1 because the saturation pressure at 14°C ($P_{sat,4} = 1599$ Pa) is much less than it was at state 1 and therefore the moist air is closer to being saturated. The relative humidity at state 4 is $\phi_4 = P_{v,4}/P_{sat,4} = 1013/1599 = 0.634$ (63.4%).

The *dew point temperature* is the temperature at which condensation occurs when moist air is cooled at constant pressure. The dew point temperature is the saturation temperature associated with the partial pressure of the water vapor. Examination of Figure 12-1 shows that the dew point temperature associated with the air in the room (at state 1) is 7.16°C. If the air in the room is cooled to $T_5 = 7.16°\text{C}$ then it will be saturated and have a relative humidity of $\phi_5 = 1$ (100%), as shown by state v,5 in Figure 12-1.

What happens if we continue to cool the air-water mixture, for example to $T_6 = 2°\text{C}$? The air-water mixture at state 6 will contain the maximum amount of water vapor possible. Therefore, the partial pressure of the water vapor will be equal to the saturation pressure at T_6, $P_{v,6} = P_{sat,6} = 705.6$ Pa, as shown in Figure 12-1. Notice that the amount of water vapor in the air must have decreased during the cooling process from state 5 to state 6. At state 5, the humidity ratio is 0.006284 kg$_v$/kg$_a$. At state 6, the humidity ratio is obtained by substituting $P_v = 705.6$ Pa into Eq. (12-6) which leads to $\omega_6 = 0.00436$ kg$_v$/kg$_a$. The reduction in the water content occurs because water will condense out of the air-water mixture onto the surfaces in the room. The state of the liquid water that forms is shown as state w,6 in Figure 12-1. The dew (water droplets) that is seen on the grass in the morning during much of the year forms because the temperature at night falls below the dew point temperature of the air. The condensate persists into the morning because evaporation is a mass transfer process that takes time to occur. Liquid droplets form on the external surface of a cold beverage or on the heat transfer surfaces of an evaporator because the temperature adjacent to the surface is lower than the dew point temperature.

EXAMPLE 12.1-1: BUILDING AIR CONDITIONING SYSTEM

Figure 1 illustrates an air-conditioning system installed in a large building.

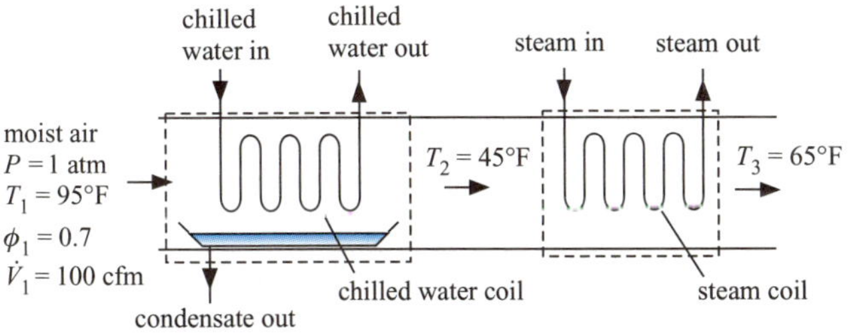

Figure 1: Building air-conditioning system.

Hot, humid outdoor air is drawn into the system with a volumetric flow rate $\dot{V}_1 = 100$ cfm. The temperature of the outdoor air is $T_1 = 95°F$ and it has relative humidity $\phi_1 = 0.7$. In order to provide a comfortable environment, the air must be both cooled and dehumidified. The air passes over a chilled water coil in order to dehumidify it. The air exits the chilled water coil at $T_2 = 45°F$. The air then passes over a steam coil in order to reheat it to a comfortable temperature. The air exits the steam coil at $T_3 = 65°F$.

a) What is the humidity ratio of the air entering the chilled water coil?

The inputs are entered in EES.

```
$UnitSystem SI Mass Radian J K Pa
V_dot_1=100 [cfm]*convert(cfm,m^3/s)     "volumetric flow rate entering chilled water coil"
T_1=converttemp(F,K,95 [F])              "temperature of outdoor air"
P=1 [atm]*convert(atm,Pa)                "pressure"
phi_1=0.70 [-]                           "relative humidity of outdoor air"
T_2=converttemp(F,K,45 [F])              "temperature of air leaving chilled water coil"
T_3=converttemp(F,K,65 [F])              "temperature of air leaving steam coil"
```

The saturation pressure of water at the inlet temperature ($P_{sat,1}$) can be determined using the P_sat function in EES. The partial pressure of water vapor at the inlet is obtained from:

$$P_{v,1} = \phi_1 P_{sat,1}$$

and the humidity ratio is calculated according to:

$$\omega_1 = 0.622 \frac{P_{v,1}}{(P - P_{v,1})}$$

```
P_sat_1=P_sat(Water,T=T_1)               "saturation pressure at state 1"
P_v_1=phi_1*P_sat_1                      "partial pressure of water vapor at state 1"
omega_1=P_v_1*0.622/(P-P_v_1)            "humidity ratio at state 1"
```

which results in $\omega_1 = 0.02516$ kg_v/kg_a.

b) What are the mass flow rates of dry air and water vapor entering the chilled water coil?

The partial pressure of dry air at state 1 is obtained from:

$$P_{a,1} = P - P_{v,1}$$

The specific volume of the dry air at state 1 is obtained from the ideal gas law using $P_{a,1}$ and T_1:

$$v_{a,1} = \frac{R_a \, T_1}{P_{a,1}}$$

where $R_a = R_{univ} \,/\, MW_a$ is the ideal gas constant for air. The mass flow rate of dry air is obtained from:

$$\dot{m}_a = \frac{\dot{V}_1}{v_{a,1}}$$

The mass flow rate of water vapor is obtained from the humidity ratio:

$$\dot{m}_{v,1} = \omega_1 \, \dot{m}_a$$

```
P_a_1=P-P_v_1                  "partial pressure of air at state 1"
R_a=R#/MolarMass(Air)          "ideal gas constant for dry air"
v_a_1=R_a*T_1/P_a_1            "specific volume of air at state 1"
m_dot_a=V_dot_1/v_a_1          "mass flow rate of dry air at state 1"
m_dot_v_1=m_dot_a*omega_1      "mass flow rate of water vapor at state 1"
```

which results in $\dot{m}_a = 0.05196$ kg$_a$/s and $\dot{m}_{v,1} = 0.001307$ kg$_v$/s.

c) At what temperature does condensation begin as the air flows through the chilled water coil?

The temperature of the air reaches its dew point temperature before condensation occurs. The dew point temperature is the temperature at which the partial pressure of the water vapor becomes equal to the saturation pressure; at this point, the air is saturated and water will begin to condense on the chilled water coil upon further cooling. The dew point temperature ($T_{dp,1}$) is equal to the saturation temperature at $P_{v,1}$ and can be obtained using the T_sat function in EES.

```
T_dp=T_sat(Water,P=P_v_1)        "dew point temperature"
T_dp_F=converttemp(K,F,T_dp)     "in F"
```

The solution provides $T_{dp} = 301.9$ K (83.66°F).

d) What is the humidity ratio of the air leaving the chilled water coil?

The temperature of the air leaving the coil, T_2, is less than the dew point temperature calculated in part (c). Therefore, the air leaving the chilled water coil will be saturated (i.e., it will have a relative humidity of 100%). The partial pressure of water vapor at state 2 will be equal to the saturation pressure evaluated at the exit temperature ($P_{sat,2}$).

$$P_{v,2} = P_{sat,2}$$

The humidity ratio at state 2 is obtained from:

$$\omega_2 = \frac{0.622 \, P_{v,2}}{(P - P_{v,2})}$$

```
P_v_2=P_sat(Water,T=T_2)           "partial pressure of water vapor at state 2"
omega_2=P_v_2*0.622/(P-P_v_2)      "humidity ratio at state 2"
```

Solving provides $\omega_2 = 0.006309$ kg_v/kg_a; notice that the water content of the air has been reduced by the condensation process on the coil. Indeed, this is the primary purpose of the chilled water coil – to dehumidify the entering air.

e) What is the volumetric flow rate of condensate that must be drained from the chilled water coil?

The mass flow rate of air leaving the coil at state 2 is the same as the mass flow rate of air that enters at state 1, $\dot{m}_a$. The mass flow rate of water vapor leaving the coil is:

$$\dot{m}_{v,2} = \omega_2 \, \dot{m}_a$$

A water mass balance on the coil provides:

$$\dot{m}_{v,1} = \dot{m}_c + \dot{m}_{v,2}$$

The specific volume of the condensate (v_c) is obtained assuming that it leaves the coil at T_2 and P. The volumetric flow rate of condensate is:

$$\dot{V}_c = \dot{m}_c \, v_c$$

```
m_dot_v_2=m_dot_a*omega_2                      "mass flow rate of water vapor at state 2"
m_dot_v_1=m_dot_c+m_dot_v_2                    "mass flow rate of condensate"
v_c=volume(Water,T=T_2,P=P)                    "specific volume of condensate"
V_dot_c=m_dot_c*v_c                            "volumetric flow rate of condensate"
V_dot_c_gph=V_dot_c*convert(m^3/s,gal/hr)      "in gallon/hr"
```

which provides $\dot{V}_c = 0.932$ gal/hr.

f) What is the relative humidity of the air leaving the steam coil that is provided to the conditioned space?

The process of heating the air from state 2 to state 3 moves the state of the water vapor in the mixture away from the vapor dome and therefore does not change the amount of water vapor in the air. As a result, the humidity ratio and partial pressure of water vapor in the air at state 3 are the same as at state 2:

$$P_{v,3} = P_{v,2}$$

The saturation pressure at the temperature leaving the steam coil ($P_{sat,3}$) is calculated and the relative humidity is determined according to:

$$\phi_3 = \frac{P_{v,3}}{P_{sat,3}}$$

```
P_v_3=P_v_2                                    "vapor pressure at state 3"
P_sat_3=P_sat(Water,T=T_3)                     "saturation pressure at state 3"
phi_3=P_v_3/P_sat_3                            "relative humidity at state 3"
```

Solving provides $\phi_3 = 0.4827$ (48.27%). This is a comfortable relative humidity range for a building. However, to achieve this level of dehumidification it was necessary to cool the air to an uncomfortably low temperature ($T_2 = 45°F$) and then reheat the air to a comfortable temperature ($T_3 = 65°F$).

g) What are the rates of heat transfer to the chilled water in the chilled water coil and from the steam in the steam coil?

An energy balance on the chilled water coil is:

$$\dot{m}_a\, h_{a,1} + \dot{m}_{v,1}\, h_{v,1} = \dot{Q}_{cwc} + \dot{m}_a\, h_{a,2} + \dot{m}_{v,2}\, h_{v,2} + \dot{m}_c\, h_c \qquad (1)$$

where $h_{a,1}$ and $h_{a,2}$ are the specific enthalpies of dry air at states 1 and 2, respectively, $h_{v,1}$ and $h_{v,2}$ are the specific enthalpies of water vapor at states 1 and 2, respectively, and h_c is the specific enthalpy of the liquid condensate.

```
h_a_1=enthalpy(Air,T=T_1)                   "specific enthalpy of dry air at state 1"
h_v_1=enthalpy(Water,T=T_1,P=P_v_1)  "specific enthalpy of water vapor at state 1"
h_c=enthalpy(Water,T=T_2,P=P)               "specific enthalpy of condensate"
h_a_2=enthalpy(Air,T=T_2)                   "specific enthalpy of dry air at state 2"
h_v_2=enthalpy(Water,T=T_2,x=1 [-])   "specific enthalpy of water vapor at state 2"
m_dot_a*h_a_1+m_dot_v_1*h_v_1=Q_dot_cwc+m_dot_c*h_c+m_dot_a*h_a_2+m_dot_v_2*h_v_2
                                            "energy balance on chilled water coil"
```

The solution provides $\dot{Q}_{cwc} = 3.949$ kW. In the above EES code, notice that the substance 'Water' is used to evaluate h_c, $h_{v,1}$, and $h_{v,2}$, the specific enthalpies of the condensate and water vapor. This is done even though the water vapor in the air-water mixture is modeled as an ideal gas which would suggest that the substance 'H2O' is the appropriate choice for this situation. However, the condensate specific enthalpy must be evaluated using the substance 'Water' and the property correlations for 'H2O' and 'Water' are based on different reference states. Also notice that the specific enthalpy of water vapor at state 2 is evaluated using temperature and quality rather than temperature and pressure. The water vapor at state 2 is saturated vapor and therefore the state cannot be specified by temperature and pressure. Finally, the energy contribution associated with the condensate is small and therefore it is often neglected. Removing this term from Eq. (1) leads to:

$$\dot{m}_a\, h_{a,1} + \dot{m}_{v,1}\, h_{v,1} = \dot{Q}_{cwc} + \dot{m}_a\, h_{a,2} + \dot{m}_{v,2}\, h_{v,2} \qquad (2)$$

```
m_dot_a*h_a_1+m_dot_v_1*h_v_1=Q_dot_cwc+{m_dot_c*h_c+}m_dot_a*h_a_2+m_dot_v_2*h_v_2
                                            "energy balance on chilled water coil"
```

which provides $\dot{Q}_{cwc} = 3.979$ kW, a difference of less than 1% from the result obtained with Eq. (1).

An energy balance on the steam coil is:

$$\dot{m}_a\, h_{a,2} + \dot{m}_{v,2}\, h_{v,2} + \dot{Q}_{stm} = \dot{m}_a\, h_{a,3} + \dot{m}_{v,2}\, h_{v,3}$$

where $h_{a,3}$ and $h_{v,3}$ are the specific enthalpies of dry air and water vapor at state 3, respectively.

```
h_a_3=enthalpy(Air,T=T_3)                   "specific enthalpy of dry air at state 3"
h_v_3=enthalpy(Water,T=T_3,P=P_v_3)   "specific enthalpy of water vapor at state 3"
m_dot_a*h_a_2+m_dot_v_2*h_v_2+Q_dot_stm=m_dot_a*h_a_3+m_dot_v_2*h_v_3
                                            "energy balance on steam coil"
```

The solution provides $\dot{Q}_{stm} = 586.6$ W.

12.2 Wet Bulb and Adiabatic Saturation Temperatures

The *dry bulb temperature* is simply the temperature of the air as measured by an ordinary thermometer. It is sometimes referred to as the dry bulb temperature in order to distinguish it from the wet bulb temperature. The *wet bulb temperature* is the temperature of an air-water vapor mixture that is measured with a thermometer that has a wetted measurement point. An instrument that measures wet bulb temperature is called a *psychrometer*. A simple psychrometer consists of two ordinary thermometers, as shown in Figure 12-2; the bulb surface of one of the thermometers (the dry bulb thermometer) is dry while the other thermometer bulb (the wet bulb thermometer) is wetted by covering it with a wick made of cotton or another material that has been soaked in water. The thermometer bulbs are then exposed to a high velocity flow of ambient air, either by swinging the thermometers or by blowing air over them with a small fan. The dry thermometer measures the normal air temperature and the thermometer exposed to liquid water measures the wet bulb temperature.

Initially, both the wet bulb and dry bulb thermometers measure the same temperature. However, liquid water is touching the wet bulb thermometer and therefore the surface temperature of the wet bulb thermometer will decrease due to the evaporation of water that occurs if the surrounding air is at a relative humidity less than 100%. The temperature reduction will continue until the energy required by evaporation of water from the surface of the bulb is entirely provided by convection heat transfer from the surrounding air. Thus, the wet bulb temperature is a result of simultaneous heat and mass transfer. The wet bulb temperature measurement is useful because it can be used to determine the moisture content of the air. The wet bulb temperature closely approximates the adiabatic saturation temperature (sometimes referred to as the thermodynamic wet bulb temperature). The *adiabatic saturation temperature* is the temperature that an air-water vapor mixture attains if it is adiabatically humidified until it becomes saturated (i.e., it achieves a relative humidity of 100%) at constant pressure. This process could be accomplished with a device similar to the one shown in Figure 12-3. The adiabatic saturation temperature is a thermodynamic property of an air-water vapor mixture. Therefore, the adiabatic saturation temperature (measured approximately using a wet bulb temperature sensor) together with the dry bulb temperature can be used to determine the relative humidity and humidity ratio of the air.

We can analyze the adiabatic saturation process shown in Figure 12-3 for the particular case where air is drawn into the device at $T_1 = 68°\text{F}$ (20°C). The pressure, P, is nearly atmospheric throughout. As the air passes through the device, it comes into contact with liquid water. When the relative humidity of the air is less than 100%, some liquid water will evaporate into the air which will cause the humidity ratio of the air to increase. The device provides sufficient heat and mass transfer surface area so that the air leaves saturated with water (i.e., with a humidity ratio $\phi_2 = 1$) at an adiabatic saturation temperature of $T_2 = 50°\text{F}$ (10°C) (in this example). Liquid water is continuously

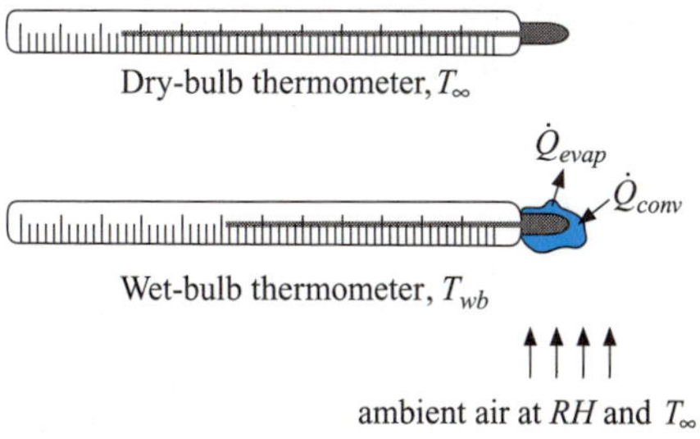

Figure 12-2: Psychrometer consisting of a wet bulb and a dry bulb thermometer.

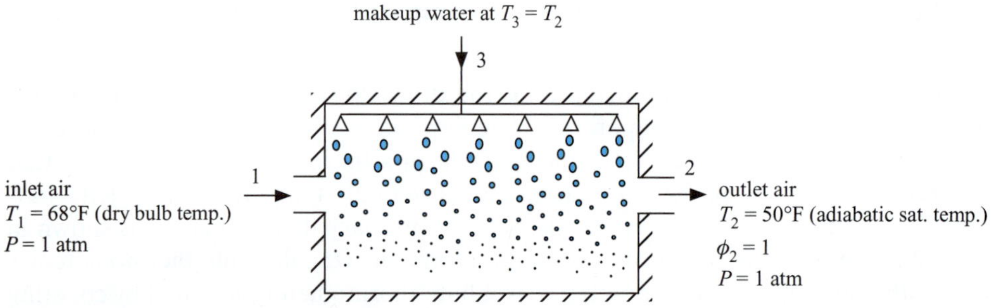

Figure 12-3: Adiabatic saturation process.

provided to the device at state 3 in order to make up for the water that is evaporated. Water is transformed from a lower to a higher energy state in the evaporation process. The energy needed for this process must be provided by the air since the device is insulated to prevent heat transfer from the surroundings. Consequently, the adiabatic saturation temperature of the air at the exit of the device, T_2, will be lower than the dry bulb temperature that it has as it enters, T_1, and the humidity ratio at the exit, ω_2, will be higher than ω_1.

Measurements of the inlet dry bulb temperature (T_1) and the adiabatic saturation temperature (T_2) are sufficient to determine the inlet relative humidity. These quantities and the pressure are entered in EES.

```
$UnitSystem SI Radian Mass J K Pa
T_1=converttemp(F,K,68 [F])          "inlet air dry bulb temperature"
T_2=converttemp(F,K,50 [F])          "adiabatic saturation temperature"
P=1 [atm]*convert(atm,Pa)            "pressure"
```

A reliable way to solve this problem is to initially guess the inlet relative humidity (ϕ_1) so that the problem can be solved in a linear manner.

```
phi_1=0.5 [-]                        "guess for inlet air relative humidity"
```

The saturation pressure at the inlet temperature ($P_{sat,1}$) is determined. The vapor pressure at the inlet is calculated from:

$$P_{v,1} = \phi_1 \, P_{sat,1} \tag{12-11}$$

and the humidity ratio at the inlet is calculated using Eq. (12-6):

$$\omega_1 = \frac{0.622 \, P_{v,1}}{(P - P_{v,1})} \tag{12-12}$$

```
P_sat_1=P_sat(Water,T=T_1)           "saturation pressure at state 1"
P_v_1=phi_1*P_sat_1                  "vapor pressure at state 1"
omega_1=0.622*P_v_1/(P-P_v_1)        "humidity ratio at state 1"
```

The specific enthalpy of the dry air and water vapor at the inlet state ($h_{a,1}$ and $h_{v,1}$) are determined.

```
h_a_1=enthalpy(Air,T=T_1)            "specific enthalpy of dry air at state 1"
h_v_1=enthalpy(Water,T=T_1,P=P_v_1)  "specific enthalpy of water vapor at state 1"
```

The saturation pressure at the exit state ($P_{sat,2}$) is computed. The air is saturated at the exit of the device, $P_{v,2} = P_{sat,2}$. The humidity ratio at the exit state is obtained from Eq. (12-6):

$$\omega_2 = \frac{0.622\, P_{v,2}}{(P - P_{v,2})} \tag{12-13}$$

```
P_sat_2=P_sat(Water,T=T_2)        "saturation pressure at state 2"
P_v_2=P_sat_2                     "vapor pressure at state 2"
omega_2=0.622*P_v_2/(P-P_v_2)     "humidity ratio at state 2"
```

The specific enthalpies of the dry air and water vapor at the exit state are computed ($h_{a,2}$ and $h_{v,2}$). The makeup water is assumed to be provided at the adiabatic saturation temperature (i.e., at the exit temperature); we will see that the makeup water temperature has a very small effect on the calculated relative humidity. The specific enthalpy of the makeup water (h_3) is computed.

```
h_a_2=enthalpy(Air,T=T_2)              "specific enthalpy of dry air at state 2"
h_v_2=enthalpy(Water,T=T_2,x=1 [-])    "specific enthalpy of water vapor at state 2"
h_3=enthalpy(Water,T=T_2,P=P)          "specific enthalpy of makeup water at state 3"
```

A water mass balance for the device is:

$$\dot{m}_{v,1} + \dot{m}_3 = \dot{m}_{v,2} \tag{12-14}$$

Substituting the definition of the humidity ratio into Eq. (12-14) provides:

$$\dot{m}_3 = \dot{m}_a\,(\omega_2 - \omega_1) \tag{12-15}$$

An energy balance on the device is:

$$\dot{m}_{v,1}\, h_{v,1} + \dot{m}_a\, h_{a,1} + \dot{m}_3\, h_3 = \dot{m}_{v,2}\, h_{v,2} + \dot{m}_a\, h_{a,2} \tag{12-16}$$

Substituting the definition of the humidity ratio into Eq. (12-16) provides:

$$\dot{m}_a\, \omega_1\, h_{v,1} + \dot{m}_a\, h_{a,1} + \dot{m}_3\, h_3 = \dot{m}_a\, \omega_2\, h_{v,2} + \dot{m}_a\, h_{a,2} \tag{12-17}$$

Substituting Eq. (12-15) into Eq. (12-17) and dividing by the mass flow rate of air provides:

$$\omega_1\, h_{v,1} + h_{a,1} + (\omega_2 - \omega_1)\, h_3 = \omega_2\, h_{v,2} + h_{a,2} \tag{12-18}$$

The guess values are updated, the assumed value of the inlet relative humidity is commented out, and Eq. (12-18) is entered.

```
{phi_1=0.5 [-]}                                        "guess for inlet air relative humidity"
h_a_1+omega_1*h_v_1+(omega_2-omega_1)*h_3=h_a_2+omega_2*h_v_2      "energy balance"
```

The solution provides $\phi_1 = 0.246$ (24.6% relative humidity). With these calculations, the measurements of the dry bulb temperature (T_1) and the adiabatic saturation temperature (T_2) are sufficient to determine the inlet relative humidity (ϕ_1).

The wet bulb temperature closely approximates the adiabatic saturation temperature and therefore the wet bulb temperature is typically used in place of the adiabatic saturation temperature in psychrometric calculations. Figure 12-4 illustrates the relative humidity as a function of the *wet bulb temperature depression*, which is defined as the difference between the dry bulb temperature and the wet bulb temperature.

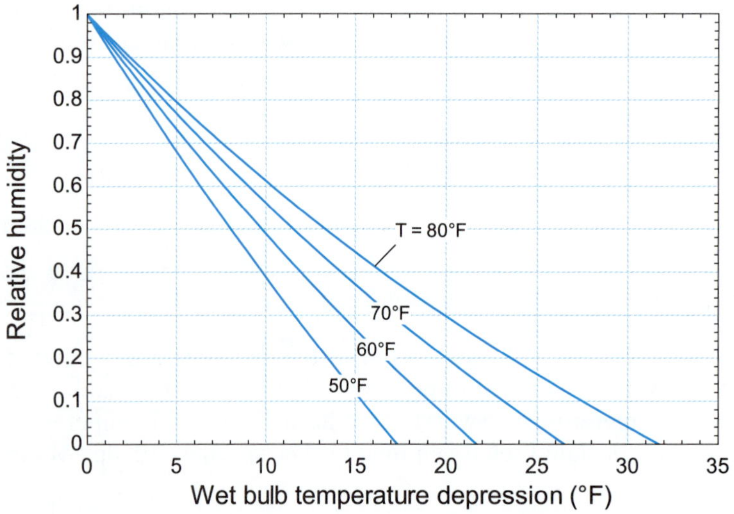

Figure 12-4: Relative humidity as a function of the wet bulb temperature depression (the dry bulb less the wet bulb temperature) for various values of the dry bulb temperature.

12.3 The Psychrometric Chart and EES' Psychrometric Functions

The methods that are discussed in Sections 12.1 and 12.2 are valid for any moist gas mixture in which the vapor phase can be treated as an ideal gas mixture. This section introduces psychrometric properties that have been defined in order to facilitate solving problems involving air-water vapor mixtures.

12.3.1 Psychrometric Properties

Psychrometric properties have been defined in a way that makes psychrometric problems easy to solve. In a typical problem, the quantities that are required include the total flow of enthalpy, entropy, and volume associated with an air-water vapor mixture. We have seen in Sections 12.1 and 12.2 that these flows can be evaluated by separately considering the water vapor and the dry air portions of the flow. The psychrometric functions are defined as the ratio of the total quantity of interest (e.g., enthalpy or volume) per unit mass of dry air. Defining properties on a per unit mass of dry air basis is convenient because the mass (or mass flow rate) of dry air does not change as water vapor is added or removed and therefore this approach eliminates the need to separately calculate the properties of dry air and water vapor.

Consider, for example, the device shown in Figure 12-5, which processes an air-water mixture. The temperature, pressure, relative humidity, and volumetric flow rate (T, P, ϕ, and $\dot{V}$) of the air-water vapor mixture entering the device have been measured. It is necessary to carry out an air mass balance, a water mass balance, an energy balance, and (perhaps) an entropy balance in order to solve the problem. Therefore, the air mass flow rate, water mass flow rate, total enthalpy flow rate, and total entropy flow rate associated with the air-water mixture must be computed.

The volume of the air-water vapor mixture per mass of dry air, v_{av}, is one of the psychrometric properties that can be read from a psychrometric chart or obtained from EES' psychrometric functions, as discussed in Sections 12.3.2 and 12.3.3:

$$v_{av} = \frac{V}{m_a} \left[\frac{\text{m}^3}{\text{kg}_\text{a}}\right] \tag{12-19}$$

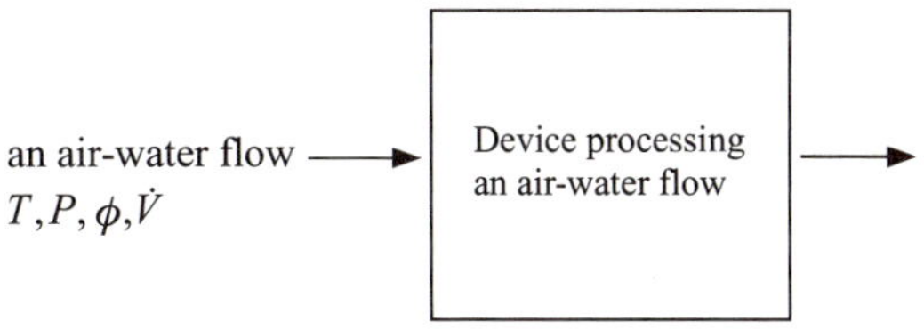

Figure 12-5: Device processing an air-water mixture.

The subscript *av* is used in this text to indicate a quantity that is defined per mass of dry air. The units of v_{av} are m^3/kg$_a$. The mass flow rate of air entering the device is computed from:

$$\dot{m}_a = \frac{\dot{V}}{v_{av}} \left[\frac{\text{m}^3}{\text{s}} \middle| \frac{\text{kg}_\text{a}}{\text{m}^3} = \frac{\text{kg}_\text{a}}{\text{s}} \right] \tag{12-20}$$

The total enthalpy of the air-water vapor mixture per mass of dry air, h_{av}, is another psychrometric property:

$$h_{av} = \frac{H}{m_a} \left[\frac{\text{J}}{\text{kg}_\text{a}} \right] \tag{12-21}$$

In many texts and references, the total enthalpy of an air-water vapor mixture per mass of dry air is given the symbol h. However, it is important to understand the basis for the psychrometric property and therefore this text utilizes the subscript *av* to reinforce the fact that the quantity is defined per mass of dry air rather than per mass of mixture. The total flow rate of enthalpy into the device is:

$$\dot{H} = \dot{m}_a\, h_{av} \left[\frac{\text{kg}_\text{a}}{\text{s}} \middle| \frac{\text{J}}{\text{kg}_\text{a}} = \text{W} \right] \tag{12-22}$$

The mass of water per mass of dry air has already been introduced as the humidity ratio, ω:

$$\omega = \frac{m_v}{m_a} \left[\frac{\text{kg}_\text{v}}{\text{kg}_\text{a}} \right] \tag{12-23}$$

The mass flow rate of water entering the device with the dry air is:

$$\dot{m}_v = \dot{m}_a\, \omega \left[\frac{\text{kg}_\text{a}}{\text{s}} \middle| \frac{\text{kg}_\text{v}}{\text{kg}_\text{a}} = \frac{\text{kg}_\text{v}}{\text{s}} \right] \tag{12-24}$$

Finally, the total entropy of the air-water mixture per mass of dry air, s_{av}, is a psychrometric property that may be useful in some problems:

$$s_{av} = \frac{S}{m_a} \left[\frac{\text{J}}{\text{K-kg}_\text{a}} \right] \tag{12-25}$$

The total flow rate of entropy into the device is:

$$\dot{S} = \dot{m}_a\, s_{av} \left[\frac{\text{kg}_\text{a}}{\text{s}} \middle| \frac{\text{J}}{\text{K-kg}_\text{a}} = \frac{\text{W}}{\text{K}} \right] \tag{12-26}$$

The psychrometric quantities v_{av}, h_{av}, and s_{av} (which are properties of an ideal gas mixture) can all be computed using the techniques discussed in Sections 11.1 and 11.2. The dry air specific volume is calculated according to:

$$v_{av} = \frac{R_a\, T}{P_a} \tag{12-27}$$

where R_a is the gas constant for dry air and P_a is the partial pressure of dry air. The humidity ratio is given by:

$$\omega = \frac{0.622\, P_v}{(P - P_v)} \tag{12-28}$$

The dry air specific enthalpy is:

$$h_{av} = \frac{H}{m_a} = \frac{m_a\, h_a + m_v\, h_v}{m_a} = h_a + \omega\, h_v \tag{12-29}$$

and the dry air specific entropy can be calculated according to:

$$s_{av} = \frac{S}{m_a} = \frac{m_a\, s_a + m_v\, s_v}{m_a} = s_a + \omega\, s_v \tag{12-30}$$

When the psychrometric functions are used to directly compute v_{av}, ω, h_{av}, and s_{av} it is no longer necessary to separately keep track of the states of the dry air and the vapor and compute their individual properties. The mixture properties of dry air and water vapor can be combined according to Eqs. (12-27) through (12-30). The psychrometric charts and functions save time during the solution process.

12.3.2 The Psychrometric Chart

The thermodynamic properties that are most often needed to do mass and energy balances on psychrometric systems include:

T	dry bulb temperature
T_{wb}	wet bulb temperature
T_{dp}	dew point temperature
ω	humidity ratio
ϕ	relative humidity
h_{av}	total enthalpy per unit mass dry air
v_{av}	total volume per unit mass dry air

All of these properties are graphically provided on a psychrometric chart. The psychrometric chart is widely used in Heating, Ventilating and Air-Conditioning (HVAC) calculations because it allows psychrometric properties to be quickly determined with an appropriate accuracy. Throughout this book, we have been advocating the use of computer tools for obtaining property information and it is true that EES can also provide all of the psychrometric properties listed above as well as others (e.g., entropy per unit mass of dry air and various transport properties), as discussed in Section 12.3.3. However, the psychrometric chart remains useful for quick calculations as well as for easy visualization of psychrometric processes.

Figure 12-6 illustrates the psychrometric chart prepared for standard atmospheric pressure, 101.3 kPa. The psychrometric chart is basically a plot of humidity ratio on the ordinate as a function of dry bulb temperature on the abscissa for a fixed total pressure. Many companies and professional societies provide psychrometric charts and there are minor differences in the manner in which they present information. For example, the chart shown in Figure 12-6 provides the humidity ratio in units of g of water vapor per kg of dry air. Other psychrometric charts are presented in English units and may provide the humidity ratio in units of grains of water vapor per pound mass of dry air. (There are 7,000 grains per pound mass.)

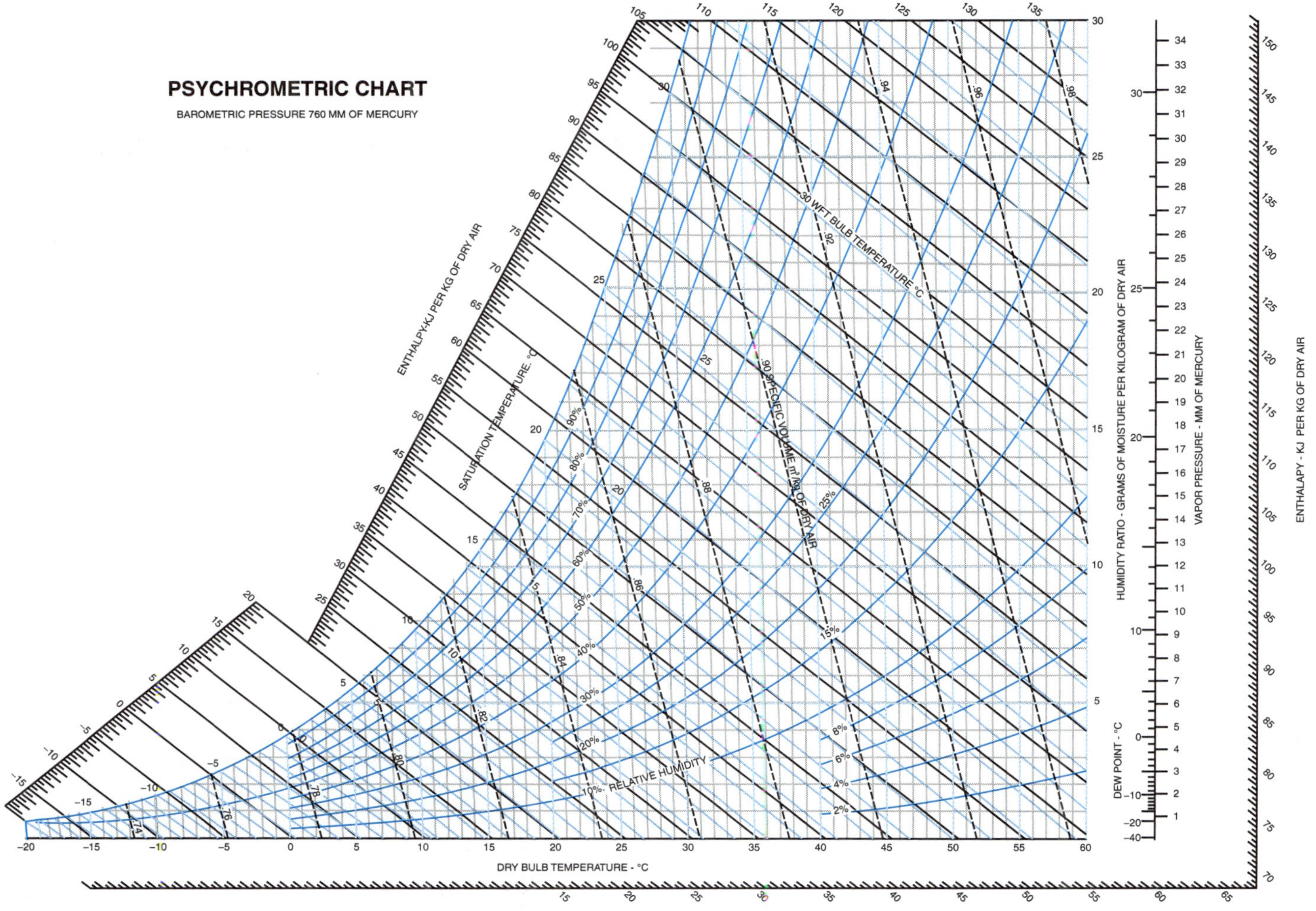

Figure 12-6: Psychrometric chart in SI units for P = 101.3 kPa. (From Universal Industrial Gases Inc., http://www.uigi.com/UIGI_SI.PDF with permission)

The maximum humidity ratio at any temperature is provided by the saturation curve, the humidity ratio as a function of temperature associated with a relative humidity of 100%. Lines of constant relative humidity run below the saturation curve.

The dry air specific enthalpy provided on the psychrometric chart, h_{av}, is given by Eq. (12-29) and shown in the psychrometric chart in Figure 12-6. Note that the quantity h_{av} is labeled simply as enthalpy in Figure 12-6, as it is in most psychrometric charts. Lines of constant h_{av} appear in black and lie at an angle of approximately 135° in Figure 12-6; notice that h_{av} is an increasing function of both temperature and humidity ratio. The scale for dry air specific enthalpy is usually provided on both sides of the chart. Accurate values of dry air specific enthalpy can be read from the chart by using a transparent straight edge to align the point of interest with the scales on both the left and right sides of the chart.

Lines of constant wet bulb temperature (actually adiabatic saturation temperature) are shown with blue lines in Figure 12-6 and are nearly parallel to the lines of constant dry air specific enthalpy. The reason that these lines are nearly parallel can be seen by examining the energy balance for the adiabatic saturation device, Eq. (12-18), repeated below:

$$\omega_1 h_{v,1} + h_{a,1} + \underbrace{(\omega_2 - \omega_1)\, h_3}_{\approx 0} = \omega_2 h_{v,2} + h_{a,2} \tag{12-31}$$

If the relatively small contribution associated with enthalpy of the makeup water is ignored then Eq. (12-31) becomes:

$$\omega_1 h_{v,1} + h_{a,1} \approx \omega_2 h_{v,2} + h_{a,2} \tag{12-32}$$

Substituting Eq. (12-29) into Eq. (12-32) simplifies to:

$$h_{av,1} \approx h_{av,2} \tag{12-33}$$

which shows that the adiabatic saturation process approximately follows a line of constant dry air specific enthalpy. A scale for wet bulb temperature is provided along the saturation curve.

The dry air specific volume provided on the psychrometric chart, v_{av}, is the volume of the mixture per mass of dry air; this quantity is calculated according to Eq. (12-27).

The dew point is the temperature at which condensation will occur when the air-water vapor mixture is cooled at constant pressure. The psychrometric chart is drawn for a constant pressure. Therefore, the dew point can be read from the chart by following a horizontal line (i.e., a line of constant humidity ratio) from the point of interest to the left until it intersects the saturation curve.

In order to demonstrate the use of the psychrometric chart, let's determine the properties of an air-water vapor mixture that is at $T = 25°\text{C}$, $P = 101.3$ kPa, and $\phi = 0.4$ (40%). Note that three intensive properties must be provided to fix the state of an air-water vapor mixture according to the phase rule, Eq. (11-1). The psychrometric chart in Figure 12-6 has been developed for standard atmospheric pressure. Psychrometric properties are not strongly dependent upon pressure; however, charts are available for other pressures in order to accommodate locations, such as Denver, Colorado, that have significantly lower barometric pressure because of their elevation. Two additional properties (other than pressure) must be specified in order to determine the state of an air-water vapor mixture on a psychrometric chart. In this case, the state of the air-water mixture is specified by the point of intersection of a line of constant temperature, $T = 25°\text{C}$, and a line of constant relative humidity, $\phi = 40\%$. The humidity ratio is seen to be $\omega = 8\ \text{g}_v/\text{kg}_a$ from the scale at the right of the chart. Following a horizontal line from the intersection point to the saturation curve at the left locates the dew point, which is about $T_{dp} = 10.5°C$. Following a line of constant wet bulb temperature to the saturation curve shows a wet bulb temperature of about $T_{wb} = 16°C$. A straight edge is needed to accurately read the dry air specific enthalpy, which is about $h_{av} = 45$ kJ/kg$_a$. Finally, the dry air specific volume can be read to be about $v_{av} = 0.855$ m^3/kg$_a$ by visual interpolation.

We can compare the values read from the psychrometric chart with those calculated manually. The conditions are entered in EES.

```
$UnitSystem SI Mass J K Pa
T=converttemp(C,K,25 [C])          "dry bulb temperature"
P=1 [atm]*convert(atm,Pa)          "pressure"
phi=0.4 [-]                        "relative humidity"
```

The saturation pressure (P_{sat}) is obtained and used to determine the vapor pressure:

$$P_v = \phi \, P_{sat} \tag{12-34}$$

The humidity ratio is computed:

$$\omega = \frac{0.622 \, P_v}{(P - P_v)} \tag{12-35}$$

```
P_sat=P_sat(Water,T=T)             "saturation pressure"
P_v=phi*P_sat                      "vapor pressure"
omega=0.622*P_v/(P-P_v)            "humidity ratio"
```

which results in 7.88 g$_v$/kg$_a$, close to the value read off of the chart (8 g$_v$/kg$_a$). The dew point is obtained by determining the saturation temperature at the vapor pressure of the water:

```
T_dp=T_sat(Water,P=P_v)            "dew point temperature"
T_dp_C=converttemp(K,C,T_dp)       "in C"
```

which leads to $T_{dp} = 10.47°C$, close to the value read off of the chart (10.5°C). The partial pressure of dry air is obtained from:

$$P_a = P - P_v \tag{12-36}$$

and the dry air specific volume is computed from:

$$v_{av} = \frac{R_a \, T}{P_a} \tag{12-37}$$

```
R_a=R#/MolarMass(Air)              "gas constant for air"
P_a=P-P_v                          "partial pressure of air"
v_av=R_a*T/P_a                     "dry air specific volume"
```

which results in $v_{av} = 0.8553$ m^3/kg, close to the value read off of the chart (0.855 m^3/kg). The specific enthalpy of the dry air and water vapor (h_a and h_v) are determined. The dry air specific enthalpy is obtained from:

$$h_{av} = h_a + \omega \, h_v \tag{12-38}$$

```
h_a=enthalpy(Air,T=T)              "enthalpy of dry air"
h_v=enthalpy(Water,T=T,P=P_v)      "enthalpy of water"
h_av=h_a+omega*h_v                 "dry air specific enthalpy"
```

which results in $h_{av} = 318.6$ kJ/kg$_a$. This value is very different from the value $h_{av} = 45$ kJ/kg$_a$ that is read from the psychrometric chart. The explanation for this apparent discrepancy is that the reference state used to calculate the specific enthalpy of dry air is different for EES and the psychrometric chart. The specific enthalpy of dry air used in preparing the psychrometric chart shown in Figure 12-6 is based on a reference state of 0°C whereas the specific enthalpy of air provided by the EES function enthalpy is based on a reference state of 0 K. For this reason, you should avoid doing problems using some property values that are read from the psychrometric chart and others obtained from EES. Because the reference states may not be consistent, the results determined using property values obtained from different sources in an energy balance may not be correct. In fact, you should also be careful not to use results read from different psychrometric charts or read from a psychrometric chart and obtained from hand calculations together because they also may not use the same reference state for specific enthalpy.

EXAMPLE 12.3-1: BUILDING AIR CONDITIONING SYSTEM (REVISITED)

Example 12.1-1 examined a building air conditioning system in which hot, humid outdoor air is drawn into the system with a volumetric flow rate $\dot{V}_1 = 100$ cfm (0.0472 m^3/s). The temperature of the outdoor air is $T_1 = 95$°F (35°C) and the air has relative humidity $\phi_1 = 0.70$. In order to provide a comfortable environment, the air must be both cooled and dehumidified. The air passes over a chilled water coil in order to dehumidify it. The air exits the chilled water coil at $T_2 = 45$°F (7.2°C). The air then passes over a steam coil in order to reheat it to a comfortable temperature. The air exits the steam coil at $T_3 = 65$°F (18.3°C). Here we will solve the problem using the psychrometric chart shown in Figure 1.

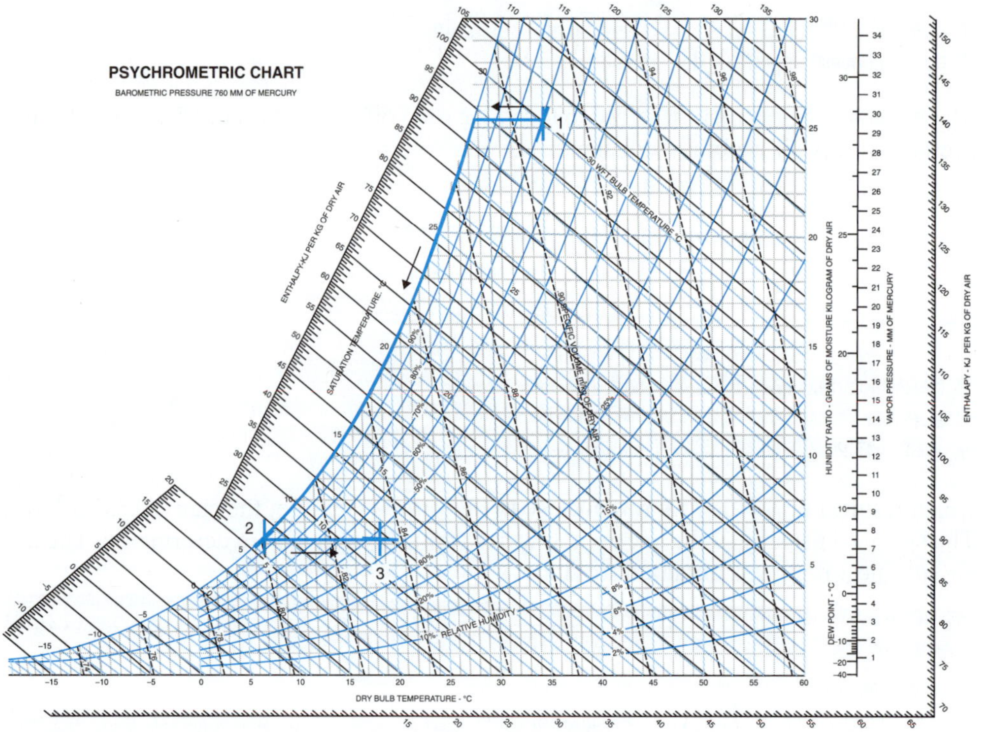

Figure 1: Psychrometric chart.

a) What is the humidity ratio of the air entering the chilled water coil?

State 1 is specified by the temperature and relative humidity and located on the psychrometric chart in Figure 1 as the intersection between the line of constant relative humidity $\phi_1 = 70\%$ and the line of constant temperature $T_1 = 35°\text{C}$. The other properties at state 1 can be read off of the psychrometric chart: $v_{av,1} = 0.908\ \text{m}^3/\text{kg}_\text{a}$, $\omega_1 = 25.3\ \text{g}_v/\text{kg}_\text{a}$, and $h_{av,1} = 100\ \text{kJ/kg}_\text{a}$.

b) What is the mass flow rate of dry air entering the chilled water coil?

The mass flow rate of dry air is computed according to:

$$\dot{m}_a = \frac{\dot{V}_1}{v_{av,1}} = \frac{0.0472\ \text{m}^3}{\text{s}}\left|\frac{\text{kg}_\text{a}}{0.908\ \text{m}^3}\right. = 0.0520\frac{\text{kg}_\text{a}}{\text{s}}$$

c) At what temperature does condensation begin as the air flows through the chilled water coil?

The temperature of the air can be reduced to its dew point temperature before condensation occurs. This cooling process corresponds to a line of constant humidity ratio that extends to the saturation curve, as shown in Figure 1. The dew point temperature is the temperature at which the horizontal line intersects the saturation curve, $T_{dp} = 28.6°\text{C}$ (83.5°F).

d) What is the humidity ratio of the air leaving the chilled water coil?

The temperature of the air leaving the coil, T_2, is less than the dew point temperature calculated in part (c). Therefore, the air leaving the chilled water coil will be saturated. State 2 is fixed by the intersection of the saturation curve and the line of constant temperature $T_2 = 7.2°\text{C}$, as shown in Figure 1. The remaining properties at state 2 can be read off of the psychrometric chart: $\omega_2 = 6.3\ \text{g}_v/\text{kg}_\text{a}$ and $h_{av,2} = 23\ \text{kJ/kg}_\text{a}$.

e) What is the volumetric flow rate of condensate that forms on the chilled water coil?

A water mass balance on the chilled water coil provides:

$$\dot{m}_c = \dot{m}_a\,(\omega_1 - \omega_2) = \frac{0.0520\ \text{kg}_\text{a}}{\text{s}}\left|\frac{(25.3 - 6.3)\,\text{g}_\text{v}}{\text{kg}_\text{a}}\right. = 0.988\frac{\text{g}}{\text{s}}$$

The specific volume of the condensate (v_c) is obtained from the water tables in Appendix B, $v_c = 0.001000\ \text{m}^3/\text{kg}$, and used to compute the volumetric flow rate of condensate:

$$\dot{V}_c = \dot{m}_c\,v_c = \frac{0.988\ \text{g}}{\text{s}}\left|\frac{0.001000\ \text{m}^3}{\text{kg}}\right\|\frac{\text{kg}}{1000\ \text{g}}\left|\frac{\text{gal}}{3.785\times 10^{-3}\ \text{m}^3}\right|\frac{3600\ \text{s}}{\text{hr}} = 0.940\frac{\text{gal}}{\text{hr}}$$

f) What is the relative humidity of the air leaving the steam coil that is provided to the conditioned space?

State 3 is located on the psychrometric chart as the intersection of a line of constant humidity ratio, $\omega_3 = \omega_2$, and a line of constant temperature, $T_3 = 18.3°\text{C}$, as shown in Figure 1. The remaining properties can be read from the psychrometric chart: $\phi_3 = 49\%$, $h_{av,3} = 34\ \text{kJ/kg}_\text{a}$.

g) What are the rates of heat transfer to the chilled water in the chilled water coil and from the steam in the steam coil?

An energy balance on the chilled water coil, neglecting the enthalpy associated with the condensate (which was demonstrated in Example 12.1-1 to be small), is:

$$\dot{Q}_{cwc} = \dot{m}_a \left(h_{av,1} - h_{av,2}\right) = \frac{0.0520\ \text{kg}_\text{a}}{\text{s}} \left| \frac{(100-23)\ \text{kJ}}{\text{kg}_\text{a}} \right. = 4.00\ \text{kW} \tag{1}$$

An energy balance on the steam coil leads to:

$$\dot{Q}_{stm} = \dot{m}_a \left(h_{av,3} - h_{av,2}\right) = \frac{0.0520\ \text{kg}_\text{a}}{\text{s}} \left| \frac{(34-23)\ \text{kJ}}{\text{kg}_\text{a}} \right. = 0.57\ \text{kW}$$

The answers obtained using the psychrometric chart are close to those obtained in Example 12.1-1.

12.3.3 Psychrometric Properties in EES

The EES property functions will return psychrometric property information if the substance AirH2O (i.e., an air-water mixture) is specified as the fluid name. Most of the property functions that have been used for pure fluids (e.g., enthalpy, intenergy, volume, temperature, and entropy) can be used with the substance AirH2O. The only exception is the pressure function, which is not implemented for the substance AirH2O. In addition, there are several new property functions that can only be used with the substance AirH2O. These include wetbulb, dewpoint, humrat, and relhum, which return the thermodynamic wet bulb temperature (i.e., the adiabatic saturation temperature), dew point temperature, humidity ratio, and relative humidity, respectively. Note that those property routines that return a specific property (e.g., enthalpy and entropy) will return the specific property on a per unit mass of dry air basis when used with the substance AirH2O.

Three properties must be included in order to fix the state when using any of the property functions with the substance AirH2O, rather than the two properties that are required to fix the state of a pure fluid. One of the three properties must be the pressure. The properties used to fix the state are identified in the usual way, with a single case-insensitive letter followed by an equal sign. The one letter indicators that are recognized in psychrometric functions and their meaning are listed in Table 12-1; note that some of these indicators are only applicable to the substance AirH2O.

In order to demonstrate the use of the psychrometric properties in EES, let's determine (again) the properties of an air-water vapor mixture that is at $T = 25°\text{C}$, $P = 101.3\ \text{kPa}$, and $\phi = 0.4\ (40\%)$.

Table 12-1: One letter indicators for EES property functions.

Indicator	Thermodynamic Property
B	thermodynamic wet bulb temperature (only for substance 'AirH2O')
D	dew point temperature (only for substance 'AirH2O')
H	dry air specific enthalpy
P	pressure
R	relative humidity (only for substance 'AirH2O')
S	dry air specific entropy
T	temperature
U	dry air specific internal energy
V	dry air specific volume
W	humidity ratio (only for substance 'AirH2O')
X	quality (not valid for substance 'AirH2O')

```
$UnitSystem SI Mass Radian J K Pa
T=converttemp(C,K,25 [C])                 "dry bulb temperature"
P=101.3 [kPa]*convert(kPa,Pa)             "total pressure"
phi=0.40 [-]                              "relative humidity"
```

The functions humrat, dewpoint, wetbulb, enthalpy, and volume are used to determine the humidity ratio, dew point temperature, thermodynamic wet bulb temperature, dry air specific enthalpy, and dry air specific volume.

```
omega=humrat(AirH2O,T=T,P=P,R=phi)        "humidity ratio"
T_dp=dewpoint(AirH2O,T=T,P=P,R=phi)       "dew point temperature"
T_dp_C=converttemp(K,C,T_dp)              "in C"
T_wb=wetbulb(AirH2O,T=T,P=P,R=phi)        "wet bulb temperature"
T_wb_C=converttemp(K,C,T_wb)              "in C"
h_av=enthalpy(AirH2O,T=T,P=P,R=phi)       "dry air specific enthalpy"
v_av=volume(AirH2O,T=T,P=P,R=phi)         "dry air specific volume"
```

Solving these equations results in $\omega = 0.007882$ kg$_v$/kg$_a$ (compared to 8 g$_v$/kg$_a$ read off of the psychrometric chart in Section 12.3.2), $T_{dp} = 10.48°C$ (compared to 10.5°C), $T_{wb} = 16.21°C$ (compared to 16°C), $h_{av} = 45.24$ kJ/kg$_a$ (compared to 45 kJ/kg$_a$), and $v_{av} = 0.8555$ m^3/kg$_a$ (compared to 0.855 m^3/kg$_a$). It is important to remember that the specific enthalpy and specific volume values provided by the EES psychrometric functions are on a per unit mass of dry air basis, as defined in Eqs. (12-21) and (12-19), respectively. These functions are defined to be consistent with the values provided in the psychrometric chart. The values calculated using the EES code above agree with the values determined using the psychrometric chart for the same state in Section 12.3.2 because the reference state used to define the specific enthalpy for dry air is 0°C, which is consistent with the psychrometric chart provided in Figure 12-6.

Note that it is not necessary to use relative humidity as an input for each property function. For example, the dry air specific enthalpy could have been obtained in any of the following ways and they all would have returned the same value.

```
h_av_2=enthalpy(AirH2O,T=T,P=P,w=omega)   "dry air specific enthalpy with humidity ratio as an input"
h_av_3=enthalpy(AirH2O,T=T,P=P,D=T_dp)    "dry air specific enthalpy with dew point as an input"
h_av_4=enthalpy(AirH2O,T=T,P=P,B=T_wb)    "dry air specific enthalpy with wet bulb as an input"
```

EXAMPLE 12.3-2: BUILDING AIR CONDITIONING SYSTEM (REVISITED AGAIN)

Examples 12.1-1 and 12.1-2 examined a building air conditioning system in which hot, humid outdoor air is drawn into the system with a volumetric flow rate $\dot{V}_1 =$ 100 cfm (0.0472 m^3/s). The temperature of the outdoor air is $T_1 = 95°F$ (35°C) and it has relative humidity $\phi_1 = 0.70$. In order to provide a comfortable environment, the air must be both cooled and dehumidified. The air passes over a chilled water coil in order to dehumidify it. The air exits the chilled water coil at $T_2 = 45°F$ (7.2°C). The air then passes over a steam coil in order to reheat it to a comfortable temperature. The air exits the steam coil at $T_3 = 65°F$ (18.3°C). Here we will solve the problem using EES' psychrometric functions.

a) What is the humidity ratio of the air entering the chilled water coil?

The inputs are entered in EES.

```
$UnitSystem SI Mass Radian J K Pa
V_dot[1]=100 [cfm]*convert(cfm,m^3/s)       "inlet volumetric flow rate of air"
T[1]=converttemp(F,K,95 [F])                "inlet temperature of air"
phi[1]=0.70 [-]                             "inlet relative humidity of air"
T[2]=converttemp(F,K,45 [F])                "temperature of air leaving chilled water coil"
T[3]=converttemp(F,K,65 [F])                "temperature of air leaving steam coil"
P=1 [atm]*convert(atm,Pa)                   "pressure"
```

State 1 is specified by the pressure, temperature, and relative humidity. The properties at state 1 (ω_1, $v_{av,1}$, and $h_{av,1}$) are obtained using the EES functions humrat, volume, and enthalpy.

```
omega[1]=humrat(AirH2O,P=P,T=T[1],R=phi[1])      "relative humidity"
v_av[1]=volume(AirH2O,P=P,T=T[1],R=phi[1])       "dry air specific volume"
h_av[1]=enthalpy(AirH2O,P=P,T=T[1],R=phi[1])     "dry air specific enthalpy"
```

Solving provides $\omega_1 = 0.02516$ kg$_v$/kg$_a$.

b) What is the mass flow rate of dry air entering the chilled water coil?

The mass flow rate of dry air is computed according to:

$$\dot{m}_a = \frac{\dot{V}_1}{v_{av,1}}$$

```
m_dot_a=V_dot[1]/v_av[1]                         "mass flow rate of dry air"
```

which results in $\dot{m}_a = 0.05196$ kg$_a$/s.

c) At what temperature does condensation begin as the air flows through the chilled water coil?

The dew point temperature of the incoming air (T_{dp}) is obtained using the dewpoint function.

```
T_dp=dewpoint(AirH2O,P=P,T=T[1],R=phi[1])        "dew point temperature"
T_dp_C=converttemp(K,C,T_dp)                     "in C"
```

Solving results in $T_{dp} = 28.70°C$.

d) What is the humidity ratio of the air leaving the chilled water coil?

The state of the air leaving the chilled water coil is specified by the pressure, temperature, and relative humidity ($\phi_2 = 1$). The properties at state 2 (ω_2 and $h_{av,2}$) are obtained using the EES functions humrat and enthalpy:

```
phi[2]=1 [-]                                        "relative humidity of air leaving chilled water coil"
omega[2]=humrat(AirH2O,P=P,T=T[2],R=phi[2])         "humidity ratio"
h_av[2]=enthalpy(AirH2O,P=P,T=T[2],R=phi[2])        "dry air specific enthalpy"
```

which shows that $\omega_2 =$ kg_v/kg_a.

e) What is the volumetric flow rate of condensate that forms on the chilled water coil?

A water mass balance on the chilled water coil provides:

$$\dot{m}_c = \dot{m}_a (\omega_1 - \omega_2)$$

The state of the condensate is fixed by the pressure and temperature. The specific volume of the condensate (v_c) is obtained and used to compute the volumetric flow rate of condensate:

$$\dot{V}_c = \dot{m}_c \, v_c$$

```
m_dot_c=m_dot_a*(omega[1]-omega[2])             "mass flow rate of condensate"
v_c=volume(Water,P=P,T=T[2])                    "specific volume of condensate"
V_dot_c=m_dot_c*v_c                             "volumetric flow rate of condensate"
V_dot_c_gph=V_dot_c*convert(m^3/s,gal/hr)       "in gph"
```

which leads to $\dot{V}_c = 0.932$ gal/hr.

f) What is the relative humidity of the air leaving the steam coil that is provided to the conditioned space?

State 3 is fixed by the pressure, temprature, and humidity ratio ($\omega_3 = \omega_2$). The properties at state 3 (ϕ_3 and $h_{av,3}$) are obtained using the EES functions relhum and enthalpy:

```
omega[3]=omega[2]                                   "humidity ratio leaving steam coil"
phi[3]=relhum(AirH2O,P=P,T=T[3],w=omega[3])         "relative humidity"
h_av[3]=enthalpy(AirH2O,P=P,T=T[3],w=omega[3])      "dry air specific enthalpy"
```

which provides $\phi_3 = 0.4821$ (48.21%)

g) What are the rates of heat transfer to the chilled water in the chilled water coil and from the steam in the steam coil?

An energy balance on the chilled water coil, neglecting the enthalpy associated with the condensate, is:

$$\dot{Q}_{cwc} = \dot{m}_a \left(h_{av,1} - h_{av,2}\right) \tag{1}$$

An energy balance on the steam coil leads to:

$$\dot{Q}_{stm} = \dot{m}_a \left(h_{av,3} - h_{av,2}\right)$$

```
Q_dot_cwc=m_dot_a*(h_av[1]-h_av[2])    "heat transfer to chilled water"
Q_dot_stm=m_dot_a*(h_av[3]-h_av[2])    "heat transfer from steam"
```

which results in $\dot{Q}_{cwc} = 3.984$ kW and $\dot{Q}_{stm} = 587.1$ W. The answers obtained using the EES psychrometric functions are identical to those obtained in Example 12.1-1 and close to those obtained with the psychrometric chart in Example 12.3-1. EES can be used to generate a psychrometric chart that shows these results. Select Property Plot from the Plots menu and then select the substance AirH2O. Overlay your states onto the resulting psychrometric plot by selecting Overlay Plot from the Plots menu. The result should look similar to Figure 1.

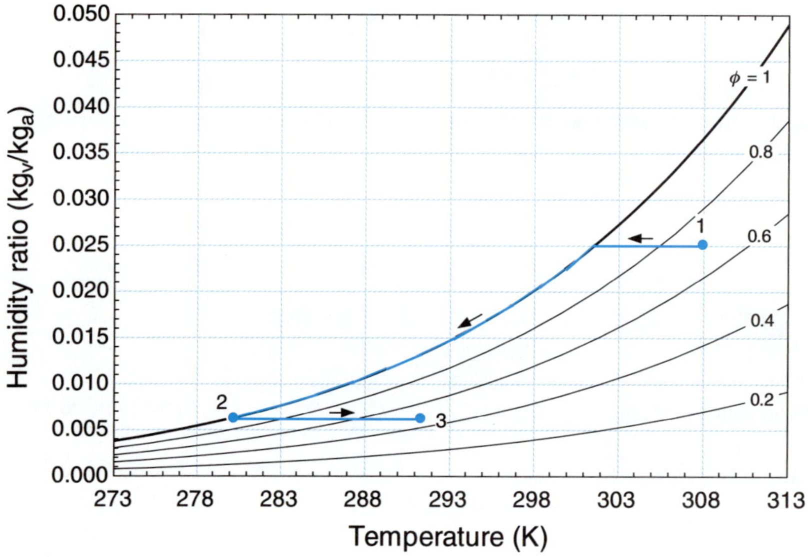

Figure 1: Psychrometric chart produced by EES.

12.4 Psychrometric Processes for Comfort Conditioning

One of the accomplishments of modern society is the ability to condition living spaces to a state that humans find comfortable, regardless of the outdoor environment. Human comfort depends strongly on humidity as well as temperature. The ranges of temperature and humidity that most people find to be comfortable have been the object of considerable study and depend on how the person is dressed, the work activity he or she is involved in, as well as individual preference. Shown in Figure 12-7 are the summer and winter comfort conditions recommended by the ASHRAE (the American Society of Heating, Refrigeration and Air-Conditioning Engineers, 2001).

This section reviews a number of common psychrometric processes that are used to provide comfortable indoor conditions. All of these psychrometric processes can be analyzed using the methods discussed in Sections 12.1 through 12.3. The processes

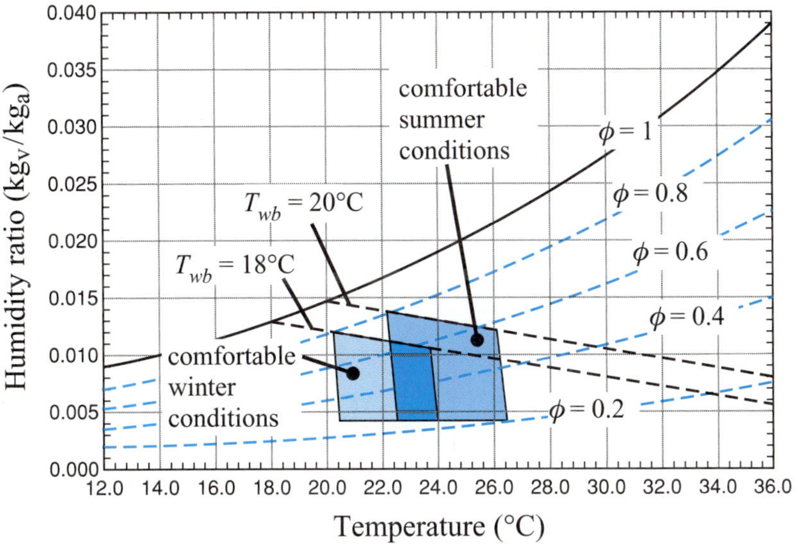

Figure 12-7: Comfortable indoor conditions during the winter and summer based on the ASHRAE Handbook of Fundamentals, Chapter 8, (2001).

occur at nearly constant pressure and can therefore be conveniently represented on a psychrometric chart.

12.4.1 Humidification Processes

Buildings located in cold climates must be heated. A large portion of the energy consumed by a building results from the conditioning of outdoor air that infiltrates into the building. In a typical residence, all of the air in the building may be replaced with outdoor air every hour. Both the temperature and the humidity of the outdoor air that enters the building must be adjusted in order to maintain comfortable indoor air conditions.

Consider a typical situation. A building is in a location where the outdoor air is at $T_{out} = -12°\text{C}$ with a relative humidity of $\phi_{out} = 85\%$. In the psychrometric chart shown in Figure 12-8, the outdoor air condition is state 1. According to Figure 12-7, in order to modify the outdoor air state so that it is comfortable for the building occupants, both the moisture content and the temperature of the air must be increased; that is, the state of the air must move both to the right (as it is heated) and up (as it is humidified) on the psychrometric chart. If the outdoor air were just heated to the indoor temperature, $T_{in} = 22°\text{C}$, then the relative humidity would be only about 5%, as shown by state 2 in Figure 12-8. This low humidity condition would cause dry skin and generate static electricity; it would be uncomfortable for most people.

There are a number of ways to both heat and humidify the air. For example, the outdoor air might be initially heated to a relatively high temperature, $T_3 = 38°\text{C}$, and then adiabatically humidified to the desired comfort condition, labeled state 4 in Figure 12-8. The adiabatic humidification process follows a line of constant adiabatic saturation temperature on the psychrometric chart. Alternatively, non-adiabatic humidification might be employed in which hot water (provided from, for example, steam) is injected into the air stream.

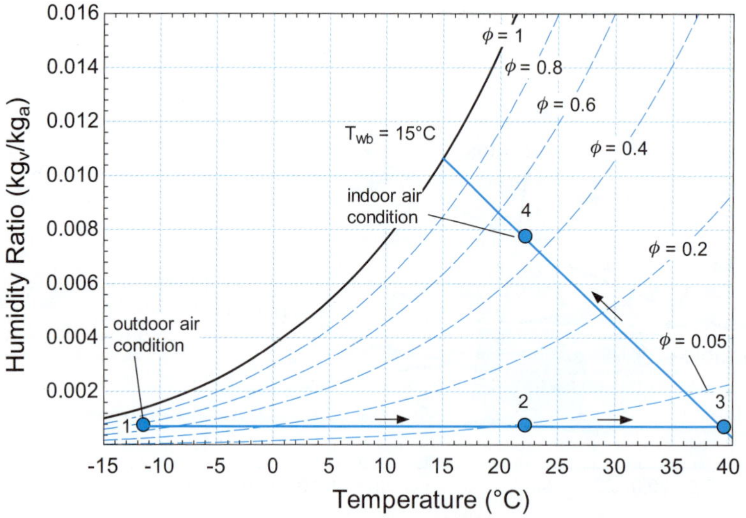

Figure 12-8: Psychrometric chart showing the outdoor (state 1) and desired indoor (state 4) air conditions. The process 1-2 corresponds to heating without humidification. The processes 1-3 and 3-4 correspond to heating followed by adiabatic humidification.

EXAMPLE 12.4-1: HEATING/HUMIDIFICATION SYSTEM

Figure 1 illustrates a heating/humidification system for a building.

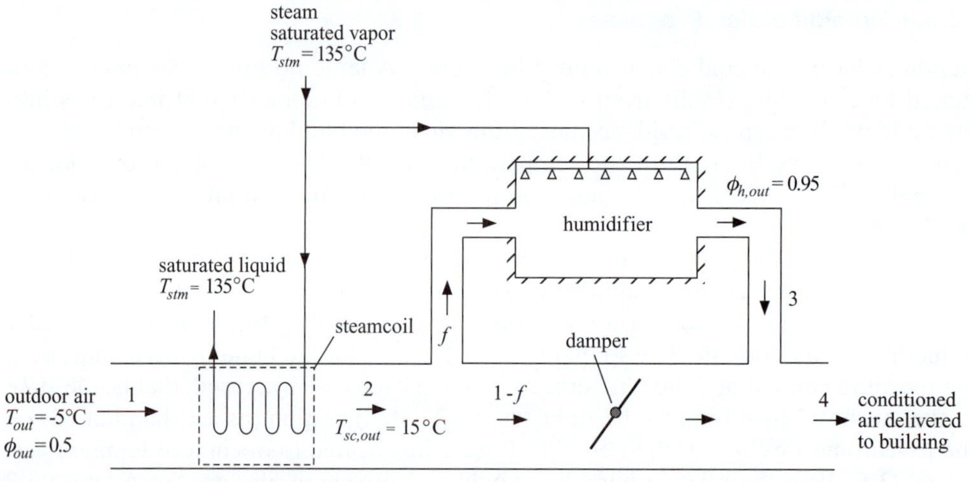

Figure 1: Heating/humidification system.

Outdoor air with temperature $T_{out} = -5°\text{C}$ and relative humidity $\phi_{out} = 0.5$ is heated to $T_{sc,out} = 15°\text{C}$ by a steam coil (i.e., a heat exchanger that transfers heat from the steam to the air without mixing). A fraction of the air leaving the steam coil, $f = 0.2$, is diverted to a humidifier where it is mixed with a flow of steam. The air leaving the humidifier has relative humidity $\phi_{h,out} = 0.95$. The steam provided to the humidifier and the steam coil is saturated vapor at $T_{stm} = 135°\text{C}$. The steam leaves the steam coil as saturated liquid at the same temperature. The pressure of the air throughout the device is atmospheric. The humidified air leaving the humidifier at state 3 is mixed with the main flow of air and provided to the building at state 4.

a) Determine the temperature and relative humidity of the air provided to the building at state 4.

The inputs are entered in EES.

```
$UnitSystem SI Mass J K Pa
T_out_C= -5 [C]                          "outdoor air temperature, in C"
T_out=converttemp(C,K,T_out_C)           "outdoor air temperature"
phi_out=0.5 [-]                          "outdoor air relative humidity"
T_sc_out_C=15 [C]                        "furnace exit temperature, in C"
T_sc_out=converttemp(C,K,T_sc_out_C)     "furnace exit temperature"
f=0.2 [-]                                "fraction of flow that passes through humidifier"
phi_h_out=0.95 [-]                       "relative humidity of air leaving humidifier"
T_stm=converttemp(C,K,135 [C])           "temperature of saturated steam"
P=1 [atm]*convert(atm,Pa)                "pressure"
```

State 1 is fixed by the temperature, $T_1 = T_{out}$, and relative humidity, $\phi_1 = \phi_{out}$. The humidity ratio and dry air specific enthalpy (ω_1 and $h_{av,1}$) are computed.

```
"State 1"
T[1]=T_out                                     "temperature"
phi[1]=phi_out                                 "relative humidity"
omega[1]=humrat(AirH2O,P=P,T=T[1],R=phi[1])    "humidity ratio"
h_av[1]=enthalpy(AirH2O,P=P,T=T[1],R=phi[1])   "dry air specific enthalpy"
```

State 2 is fixed by the temperature, $T_2 = T_{sc,out}$, and humidity ratio, $\omega_2 = \omega_1$. The relative humidity and dry air specific enthalpy (ϕ_2 and $h_{av,2}$) are computed.

```
"State 2"
T[2]=T_sc_out                                       "furnace exit temperature"
omega[2]=omega[1]                                   "humidity ratio"
phi[2]=relhum(AirH2O,P=P,T=T[2],w=omega[2])         "relative humidity"
h_av[2]=enthalpy(AirH2O,P=P,T=T[2],w=omega[2])      "dry air specific enthalpy"
```

The relative humidity of the air at state 3 is given by $\phi_3 = \phi_{h,out}$. The temperature of the air leaving the humidifier (T_3) is initially guessed. This guess will be removed in order to enforce an energy balance on the humidifier. The temperature and relative humidity are used to compute the humidity ratio and dry air specific enthalpy (ω_3 and $h_{av,3}$). The state of the steam provided to the humidifier is fixed by the temperature and quality. The specific enthalpy of the steam (h_{stm}) is computed.

```
"State 3"
T[3]=converttemp(C,K,50 [C])                   "guess for humidifier exit temperature"
phi[3]=phi_h_out                               "relative humidity"
omega[3]=humrat(AirH2O,P=P,T=T[3],R=phi[3])"humidity ratio"
h_av[3]=enthalpy(AirH2O,P=P,T=T[3],R=phi[3]) "dry air specific enthalpy"
h_stm=enthalpy(Water,T=T_stm,x=1 [-])          "specific enthalpy of saturated steam entering humidifier"
```

A water mass balance on the humidifier provides:

$$\dot{m}_a\, f\, \omega_2 + \dot{m}_{stm,h} = \dot{m}_a\, f\, \omega_3 \tag{1}$$

where $\dot{m}_a$ is the mass flow rate of dry air entering the steam coil and $\dot{m}_{stm,h}$ is the mass flow rate of steam provided to the humidifier. Solving Eq. (1) for the mass flow rate of steam:

$$\dot{m}_{stm,h} = \dot{m}_a f \left(\omega_3 - \omega_2\right) \tag{2}$$

An energy balance on the humidifier provides:

$$\dot{m}_a f \, h_{av,2} + \dot{m}_{stm,h} \, h_{stm} = \dot{m}_a f \, h_{av,3} \tag{3}$$

Substituting Eq. (2) into Eq. (3) and dividing by $\dot{m}_a f$ leads to:

$$h_{av,2} + \left(\omega_3 - \omega_2\right) h_{stm} = h_{av,3} \tag{4}$$

The guess values are updated, the guessed value of the temperature is commented out and Eq. (4) is entered in its place.

```
{T[3]=converttemp(C,K,50 [C])}                  "guess for humidifier exit temperature"
h_av[2]+(omega[3]-omega[2])*h_stm=h_av[3]       "energy balance"
```

A water mass balance on the mixing process provides:

$$\dot{m}_a f \, \omega_3 + \dot{m}_a \left(1 - f\right) \omega_2 = \dot{m}_a \, \omega_4 \tag{5}$$

Dividing Eq. (5) through by $\dot{m}_a$ leads to an expression for ω_4:

$$f \, \omega_3 + \left(1 - f\right) \omega_2 = \omega_4$$

An energy balance on the mixing process provides:

$$\dot{m}_a f \, h_{av,3} + \dot{m}_a \left(1 - f\right) h_{av,2} = \dot{m}_a \, h_{av,4} \tag{6}$$

Dividing Eq. (6) through by $\dot{m}_a$ leads to an expression for $h_{av,4}$:

$$f \, h_{av,3} + \left(1 - f\right) h_{av,2} = h_{av,4}$$

```
"State 4"
f*omega[3]+(1-f)*omega[2]=omega[4]              "water mass balance"
f*h_av[3]+(1-f)*h_av[2]=h_av[4]                 "energy balance"
```

State 4 is fixed by the humidity ratio and the dry air specific enthalpy. The temperature, relative humidity, and dry air specific volume (T_4, ϕ_4, and $v_{av,4}$) are computed.

```
T[4]=temperature(AirH2O,P=P,w=omega[4],h=h_av[4])   "temperature"
T_4_C=converttemp(K,C,T[4])                         "in C"
phi[4]=relhum(AirH2O,P=P,w=omega[4],h=h_av[4])      "relative humidity"
v_av[4]=volume(AirH2O,P=P,w=omega[4],T=T[4])        "dry air specific volume"
```

Solving results in $T_4 = 15.4°\text{C}$ and $\phi_4 = 0.3044$ (30.44%).

b) The desired temperature and relative humidity provided to the building are $T_{in} = 23°\text{C}$ and $\phi_{in} = 0.5$, respectively. Determine the values of $T_{sc,out}$ and f that should be used in order to obtain these exit conditions.

The guess values are updated and the specified values of f and $T_{sc,out}$ are commented out. The values of T_4 and ϕ_4 are set to the desired indoor air conditions.

```
{T_sc_out_C=15 [C]}                "furnace exit temperature, in C"
{f=0.2 [-]}                        "fraction of flow that passes through humidifier"
T_in=converttemp(C,K,23 [C])       "indoor air temperature"
phi_in=0.5 [-]                     "indoor air relative humidity"
T[4]=T_in                          "required indoor air temperature"
phi[4]=phi_in                      "required indoor air relative humidity"
```

Solving provides $f = 0.4282$ and $T_{sc,out} = 21.63$°C. Figure 2 illustrates a psychrometric chart with the states overlaid onto it.

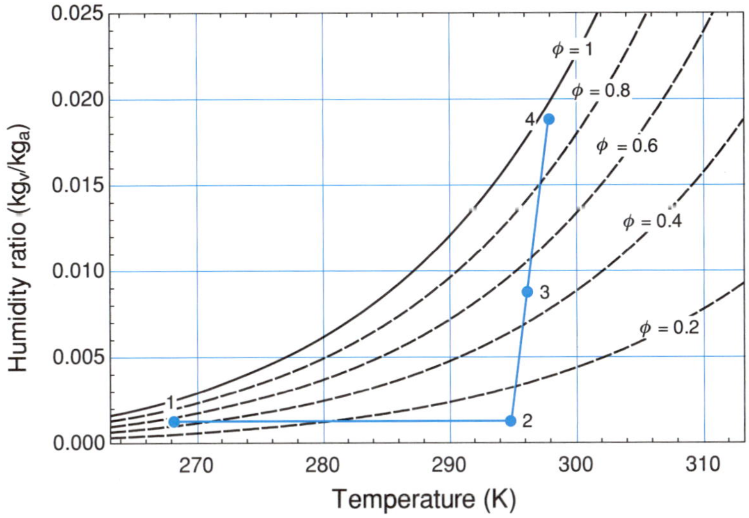

Figure 2: Psychrometric chart with the states indicated.

You should always review your psychrometric chart to ensure that it makes sense. Air at state 1 is heated to state 2 without adding water vapor; therefore, the line connecting these states is horizontal (i.e., the humidity ratio is constant). The steam added to the humidifier increases both the energy and moisture content of the air. Therefore, state 4 lies at a higher dry air specific enthalpy and at a higher humidity ratio than state 2. State 3 is at 23°C and 50% relative humidity. States 2 and 4 are mixed adiabatically, which results in state 3 lying on a straight line that connects these points.

c) The building has a total volume of $V_{bld} = 590$ m^3 and it experiences an infiltration rate of $ac = 0.80$ air changes per hour. Determine the mass flow rate of steam required by the steam coil and the humidifier. Determine the total mass flow rate of steam required by the system.

The volumetric flow rate of air provided to the building is:

$$\dot{V}_4 = V_{bld}\, ac$$

The mass flow rate of dry air provided to the building is:

$$\dot{m}_a = \frac{\dot{V}_4}{v_{av,4}}$$

```
V_bld=590 [m^3]                        "building volume"
ach=0.8 [1/hr]*convert(1/hr,1/s)       "rate of air changes"
V_dot[4]=V_bld*ach                     "volumetric flow rate at exit"
m_dot_a=V_dot[4]/v_av[4]               "mass flow rate of dry air"
```

The mass flow rate of steam provided to the humidifier is obtained from Eq. (2):

```
m_dot_stm_hum=m_dot_a*f*(omega[3]-omega[2])   "mass flow rate of steam consumed by humidifier"
```

which results in $\dot{m}_{stm,h} = 0.001158$ kg/s. An energy balance on the air-side of the steam coil provides:

$$\dot{m}_a\, h_{av,1} + \dot{Q}_{sc} = \dot{m}_a\, h_{av,2}$$

where $\dot{Q}_{sc}$ is the rate of heat transfer from the steam to the air. An energy balance on the steam-side of the steam coil provides:

$$\dot{m}_{stm,sc}\, h_{stm} = \dot{Q}_{sc} + \dot{m}_{stm,sc}\, h_{stm,out}$$

where $\dot{m}_{stm,sc}$ is the mass flow rate of steam required by the steam coil and $h_{stm,out}$ is the specific enthalpy of the steam leaving the coil, obtained from the temperature and quality.

```
Q_dot_sc=m_dot_a*(h_av[2]-h_av[1])           "steam coil heat transfer rate"
h_stm_out=enthalpy(Water,T=T_stm,x=0 [-])    "specific enthalpy of steam leaving steam coil"
m_dot_stm_sc*(h_stm-h_stm_out)=Q_dot_sc      "mass flow rate of steam consumed by steam coil"
m_dot_stm=m_dot_stm_h+m_dot_stm_sc           "total mass flow rate of steam consumed by the system"
```

Solving these equations results in $\dot{m}_{stm,sc} = 0.001917$ kg/s. The total mass flow rate of steam consumed by the system is:

$$\dot{m}_{stm} = \dot{m}_{stm,sc} + \dot{m}_{stm,h}$$

```
m_dot_stm=m_dot_stm_h+m_dot_stm_sc
                    "total mass flow rate of steam consumed by the system"
m_dot_stm_kgphr=m_dot_stm*convert(kg/s,kg/hr)                    "in kg/hr"
```

which provides $\dot{m}_{stm} = 0.003075$ kg/s (11.07 kg/hr).

d) Plot the fraction of flow diverted through the humidifier (f) and the total mass flow rate of steam consumed by the system as a function of the outdoor air temperature for various values of outdoor air relative humidity.

Figure 3 illustrates f as a function of T_{out} for various values of ϕ_{out}. Notice that the value of f is independent of T_{out} if the outdoor air is completely dry ($\phi_{out} = 0$) but it decreases as the water content of the outdoor air increases, corresponding to either increased T_{out} or ϕ_{out}.

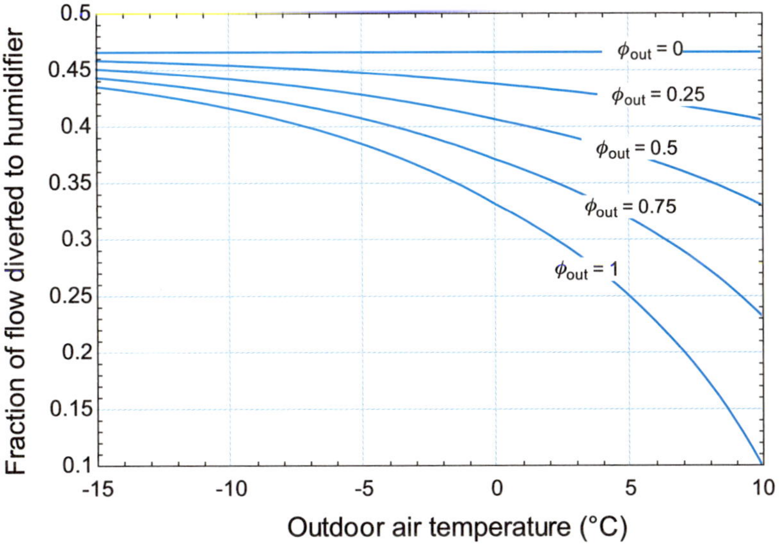

Figure 3: Fraction of air diverted to humidifier as a function of the outdoor air temperature for various values of the outdoor air relative humidity.

Figure 4 illustrates $\dot{m}_{stm}$ as a function of T_{out} for various values of ϕ_{out}. Steam is used for both heating and humidification in this system. Therefore, the mass flow rate of steam increases as either T_{out} or ϕ_{out} is reduced. At very low outdoor air temperatures, the impact of ϕ_{out} is small because the water content of the outdoor air is small regardless of ϕ_{out}. As the temperature increases, the outdoor air can hold a significant amount of water.

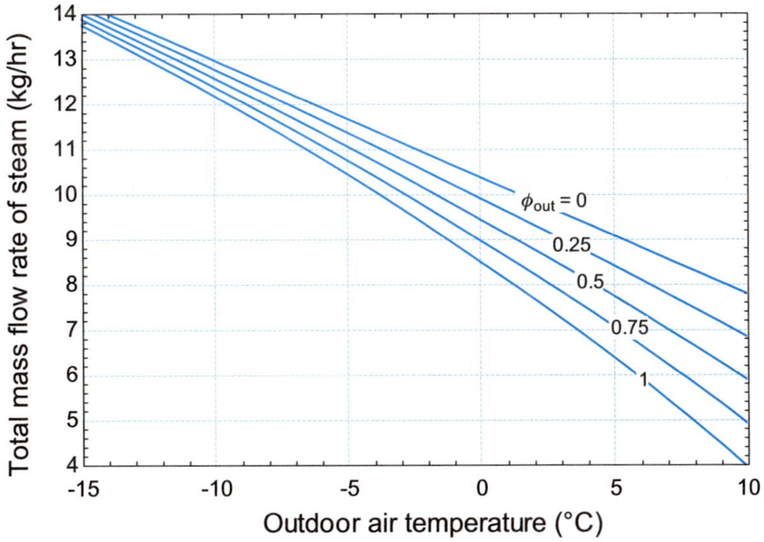

Figure 4: Total mass flow rate of steam as a function of the outdoor air temperature for various values of the outdoor air relative humidity.

12.4.2 Dehumidification Processes

The discomfort that you feel on a hot summer day is, in part, a result of high humidity. For example, an outdoor condition of 28°C and 60% relative humidity would be quite uncomfortable for most people, even though the temperature itself is not extremely high. Figure 12-7 shows that a comfortable indoor summer condition would be 25°C and 40% relative humidity. These two states are shown in Figure 12-9 as states 1 and 2, respectively. It would be great if we could devise a process in which outdoor air enters at state 1 and directly proceeds to the desired indoor air condition, state 2, along the dashed blue line shown in Figure 12-9. However, in order to proceed from state 1 to state 2, the outdoor air must be dehumidified. Dehumidification at constant pressure is possible only by cooling the air or by using desiccants, as described in Section 12.4.4.

The most common way to dehumidify air is to cool it to below its dew point temperature so that water vapor condenses out of the air-water mixture. This method of dehumidification was studied in Examples 12.1-1, 12.3-1, and 12.3-2. The dehumidification process is illustrated again in Figure 12-9. The outdoor air is cooled until it reaches its dew point temperature, indicated by state a. The dew point temperature for the chosen conditions is about 20°C. Further cooling, to state b, causes water vapor to condense out of the air-water mixture resulting in its dehumidification. Figure 12-9 shows that the air must be cooled to approximately 10.5°C in order to achieve the desired humidity ratio. However, the air at state b is now too cold for comfort and therefore it must be reheated in order to bring it to the desired indoor air condition at state 2. The conditioning process shown by 1-a-b-2 in Figure 12-9 is referred to as the cooling-reheat process. A disadvantage of this process is that the air must be cooled to a temperature that is lower than desired and then reheated; these processes result in increased cooling and heating costs. However, the cooling-reheat process allows the temperature and humidity to be independently controlled.

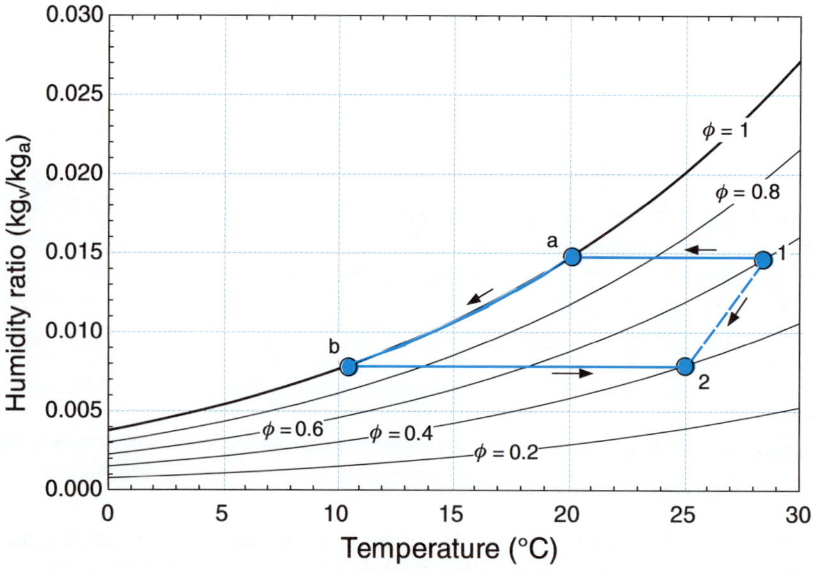

Figure 12-9: Psychrometric chart showing a hypothetical process that transforms outdoor air at state 1 directly to a comfortable condition at state 2. Also shown is the cooling-reheat process, states 1-a-b-2.

EXAMPLE 12.4-2: AIR CONDITIONING SYSTEM

Figure 1 illustrates the air conditioning system used to condition a large building on a hot and humid day.

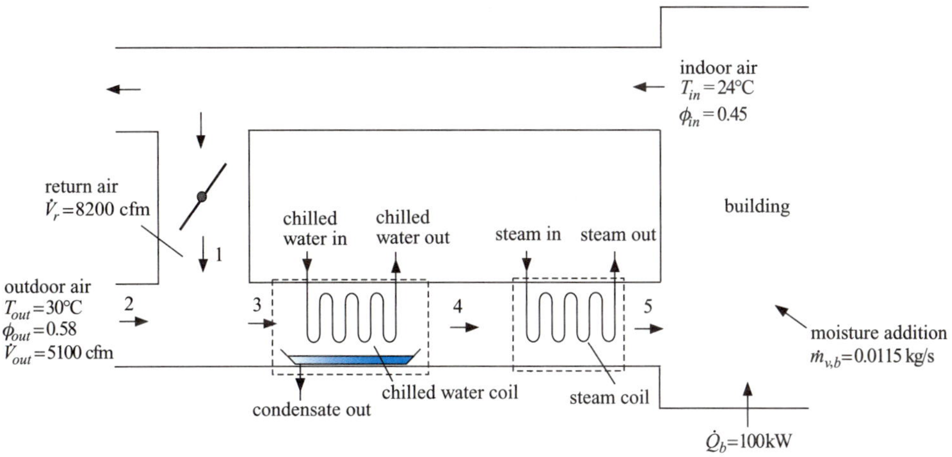

Figure 1: Air conditioning system for a large building.

The indoor air is conditioned to $T_{in} = 24°\text{C}$ and $\phi_{in} = 0.45$. The building is subjected to a heat load from solar gain, conduction, and air infiltration as well as from internal loads associated with various sources. The total rate of heat transfer to the building is $\dot{Q}_b = 100$ kW. The building air is also subjected to moisture addition from the occupants and other sources. The total rate of moisture addition to the building air is $\dot{m}_{v,b} = 0.0115$ kg/s. Outdoor air at $T_{out} = 30°\text{C}$ and $\phi_{out} = 0.58$ is drawn into the system with flow rate $\dot{V}_{out} = 5100$ cfm. Some of the exhaust air drawn from the building is diverted and mixed with the outdoor air. The volumetric flow rate of the return air flow is $\dot{V}_r = 8200$ cfm. The mixed air flow passes through a chilled water coil where it is cooled and dehumidified and then through a steam coil where it is reheated. The pressure is atmospheric throughout the system.

a) Determine the temperature and relative humidity of the air that is provided to the building at state 5.

The inputs are entered in EES.

```
$UnitSystem SI Mass Radian J K Pa
T_out=converttemp(C,K,30 [C])                     "outdoor air temperature"
phi_out=0.58 [-]                                  "outdoor air relative humidity"
V_dot_out=5100 [cfm]*convert(cfm,m^3/s)           "volumetric flow rate of outdoor air"
T_in=converttemp(C,K,24[C])                       "indoor air temperature"
phi_in=0.45 [-]                                   "indoor air relative humidity"
V_dot_r_cfm=8200 [cfm]                            "volumetric flow rate of return air, in cfm"
V_dot_r=V_dot_r_cfm*convert(cfm,m^3/s)            "volumetric flow rate of return air"
P=1 [atm]*convert(atm,Pa)                         "pressure"
Q_dot_b=100 [kW]*convert(kW,W)                    "building heat load"
m_dot_v_b=0.0115 [kg/s]                           "building water load"
```

The state of the air leaving the building, state 1, is fixed by the temperature, $T_1 = T_{in}$ and relative humidity, $\phi_1 = \phi_{in}$. The humidity ratio, dry air specific enthalpy, and dry air specific volume (ω_1, $h_{av,1}$, and $v_{av,1}$) are determined. The mass flow rate of return air that is used to mix with the outdoor air is computed:

$$\dot{m}_{a,r} = \frac{\dot{V}_r}{v_{av,1}}$$

```
"State 1"
T[1]=T_in                                       "temperature"
phi[1]=phi_in                                   "relative humidity"
omega[1]=humrat(AirH2O,P=P,T=T[1],R=phi[1])     "humidity ratio"
h_av[1]=enthalpy(AirH2O,P=P,T=T[1],R=phi[1])    "dry air specific enthalpy"
v_av[1]=volume(AirH2O,P=P,T=T[1],R=phi[1])      "dry air specific volume"
m_dot_a_r=V_dot_r/v_av[1]                       "mass flow rate of return air"
```

The state of the air drawn from outdoors, state 2, is fixed by the temperature, $T_2 = T_{out}$ and relative humidity, $\phi_2 = \phi_{out}$. The humidity ratio, dry air specific enthalpy, and dry air specific volume (ω_2, $h_{av,2}$, and $v_{av,2}$) are determined. The mass flow rate of outdoor air is computed.

$$\dot{m}_{a,out} = \frac{\dot{V}_{out}}{v_{av,2}}$$

```
"State 2"
T[2]=T_out                                      "temperature"
phi[2]=phi_out                                  "relative humidity"
omega[2]=humrat(AirH2O,P=P,T=T[2],R=phi[2])     "humidity ratio"
h_av[2]=enthalpy(AirH2O,P=P,T=T[2],R=phi[2])    "dry air specific enthalpy"
v_av[2]=volume(AirH2O,P=P,T=T[2],R=phi[2])      "dry air specific volume"
m_dot_a_out=V_dot_out/v_av[2]                   "mass flow rate of outdoor air"
```

A steady state, dry air mass balance on the mixing process is:

$$\dot{m}_{a,r} + \dot{m}_{a,out} = \dot{m}_a$$

where $\dot{m}_a$ is the mass flow rate of dry air passing through the chilled water and steam coils. A water mass balance on the mixing process:

$$\dot{m}_{a,r}\,\omega_1 + \dot{m}_{a,out}\,\omega_2 = \dot{m}_a\,\omega_3$$

provides the humidity ratio of the air entering the chilled water coil. A steady state energy balance on the adiabatic mixing process:

$$\dot{m}_{a,r}\,h_{av,1} + \dot{m}_{a,out}\,h_{av,2} = \dot{m}_a\,h_{av,3}$$

provides the dry air specific enthalpy entering the chilled water coil. State 3 is fixed by ω_3 and $h_{av,3}$. The temperature and relative humidity (T_3 and ϕ_3) are computed.

```
m_dot_a_r+m_dot_a_out=m_dot_a                               "mass balance on dry air"
m_dot_a_r*omega[1]+m_dot_a_out*omega[2]=m_dot_a*omega[3]    "mass balance on water"
m_dot_a_r*h_av[1]+m_dot_a_out*h_av[2]=m_dot_a*h_av[3]       "energy balance"
T[3]=temperature(AirH2O,P=P,h=h_av[3],w=omega[3])           "temperature"
phi[3]=relhum(AirH2O,P=P,h=h_av[3],w=omega[3])              "relative humidity"
```

A water mass balance on the building is:

$$\dot{m}_a\,\omega_5 + \dot{m}_{v,b} = \dot{m}_a\,\omega_1$$

which provides the humidity ratio of the air entering the building. An energy balance on the building leads to:

$$\dot{m}_a\, h_{av,5} + \dot{Q}_b = \dot{m}_a\, h_{av,1}$$

which provides the dry air specific enthalpy entering the building. State 5 is fixed by ω_5 and $h_{av,5}$. The temperature and relative humidity (T_5 and ϕ_5) are computed.

```
"State 5"
m_dot_a*omega[5]+m_dot_v_b=m_dot_a*omega[1]          "water mass balance on building"
m_dot_a*h_av[5]+Q_dot_b=m_dot_a*h_av[1]              "building energy balance"
T[5]=temperature(AirH2O,P=P,h=h_av[5],w=omega[5])    "temperature"
T_5_C=converttemp(K,C,T[5])                          "in C"
phi[5]=relhum(AirH2O,P=P,h=h_av[5],w=omega[5])       "relative humidity"
```

Solving these equations results in $T_5 = 14.46°\text{C}$ and $\phi_5 = 0.663$.

b) Determine the rate of heat transfer from the air to the chilled water in the chilled water coil (in tons) and the rate of condensate formation (in gal/hr). Also determine the rate of heat transfer from the steam to the air in the steam coil.

The humidity ratio at state 4 must be equal to the humidity ratio at state 5 as there is no moisture added or removed in the steam coil, $\omega_4 = \omega_5$. Because the humidity ratio has decreased between states 3 and 4, the air must have been cooled below the dew point temperature at state 3 and it will emerge from the chilled water coil in a saturated state, $\phi_4 = 1$. State 4 is fixed by ω_4 and ϕ_4. The temperature and the dry air specific enthalpy (T_4 and $h_{av,4}$) are determined.

```
"State 4"
omega[4]=omega[5]                                    "humidity ratio"
phi[4]=1 [-]                                         "relative humidity"
T[4]=temperature(AirH2O,P=P,w=omega[4],R=phi[4])     "temperature"
h_av[4]=enthalpy(AirH2O,P=P,w=omega[4],R=phi[4])     "dry air specific enthalpy"
```

A water mass balance on the chilled water coil provides:

$$\dot{m}_a\, \omega_3 = \dot{m}_c + \dot{m}_a\, \omega_4$$

where $\dot{m}_c$ is the mass flow rate of condensate. The specific volume of the condensate, v_c, is computed assuming that it leaves the coil at T_4 and atmospheric pressure. The volumetric flow rate of condensate is:

$$\dot{V}_c = \dot{m}_c\, v_c$$

```
m_dot_a*omega[3]=m_dot_c+m_dot_a*omega[4]    "water mass balance on chilled water coil"
v_c=volume(Water,T=T[4],P=P)                 "specific volume of condensate"
V_dot_c=m_dot_c*v_c                          "volumetric flow rate of condensate"
V_dot_c_gph=V_dot_c*convert(m^3/s,gal/hr)    "in gph"
```

which leads to $\dot{V}_c = 29.5$ gal/hr. An energy balance on the chilled water coil is:

$$\dot{m}_a\, h_{av,3} = \dot{m}_c\, h_c + \dot{Q}_{cwc} + \dot{m}_a\, h_{av,4}$$

where h_c is the specific enthalpy of the condensate (evaluated at T_4 and atmospheric pressure).

```
h_c=enthalpy(Water,T=T[4],P=P)                        "specific enthalpy of condensate"
m_dot_a*h_av[3]=m_dot_c*h_c+Q_dot_cwc+m_dot_a*h_av[4]  "energy balance on chilled water coil"
Q_dot_cwc_ton=Q_dot_cwc*convert(W,ton)                 "cooling coil heat transfer, in ton"
```

Solving these equations leads to $\dot{Q}_{cwc} = 60.15$ ton. An energy balance on the steam coil provides:

$$\dot{m}_a\, h_{av,4} + \dot{Q}_{sc} = \dot{m}_a\, h_{av,5}$$

```
m_dot_a*h_av[4]+Q_dot_sc=m_dot_a*h_av[5]               "energy balance on steam coil"
```

which results in $\dot{Q}_{sc} = 45.94$ kW. Figure 2 illustrates a psychrometric chart with the states in the system indicated.

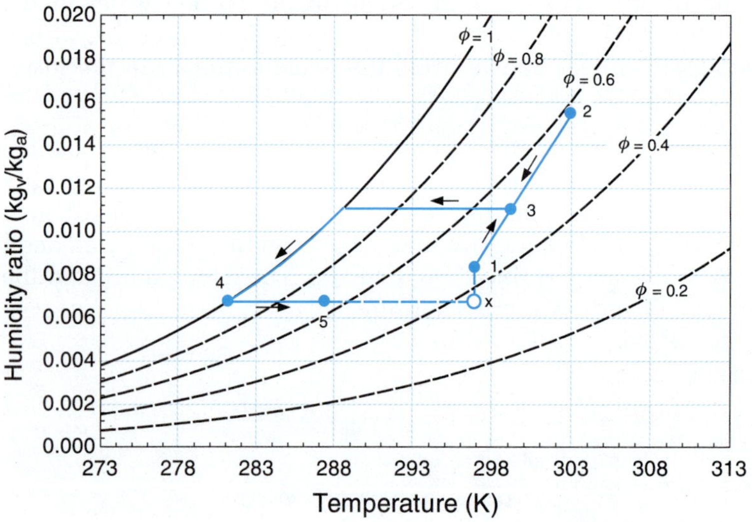

Figure 2: Psychrometric chart with the states in the system overlaid. Also shown is state x, which corresponds to warming the air entering the building at state 5 to the building air temperature (T_1) without changing its humidity ratio.

c) Determine the sensible heat ratio associated with the building. The sensible heat ratio (*SHR*) is defined as the ratio of the energy required to warm the air entering the building to the building air temperature without dehumidification (i.e., at constant humidity ratio) to the total energy change that the air undergoes as it passes through the building.

Air is provided to the building at state 5 and it is returned from the building hotter and more humid at state 1 due to the heat and moisture addition. The energy that is required to condition the air from state 5 to state 1 can be viewed as the sum of two parts. The first part, called the sensible term, is the energy required to heat the air from T_5 to T_1 at constant humidity ratio. This hypothetical process is represented by the dotted line between state 5 and state x in Figure 2. The second part, called the latent term, is the energy required to humidify the air from ω_5 to ω_1 at constant temperature. This process is shown by the dotted line between state x and state 1.

State x is fixed by the temperature, $T_x = T_1$, and humidity ratio, $\omega_x = \omega_5$. The dry air specific enthalpy ($h_{av,x}$) is determined. The sensible heat ratio is the ratio of the sensible term to the sum of the sensible and latent terms:

$$SHR = \frac{(h_{av,x} - h_{av,5})}{(h_{av,1} - h_{av,5})}$$

```
"calculate sensible heat ratio"
T_x=T[1]                                          "temperature of state x"
omega_x=omega[5]                                  "humidity ratio at state x"
h_av_x=enthalpy(AirH2O,P=P,T=T_x,w=omega_x)       "dry air specific enthalpy at state x"
SHR=(h_av_x-h_av[5])/(h_av[1]-h_av[5])            "sensible heat ratio"
```

The sensible heat ratio is $SHR = 0.71$ for this building.

d) Plot the cooling required by the chilled water coil, the heating provided by the steam coil, and the sensible heat ratio as a function of the moisture load associated with the building, $\dot{m}_{v,b}$.

Figure 3 illustrates $\dot{Q}_{cwc}$, $\dot{Q}_{sc}$, and SHR as a function of $\dot{m}_{v,b}$. Notice that both the cooling and heating energy required increases dramatically as the moisture addition in the building increases. The sensible heat ratio tends to be reduced as the moisture addition increases, indicating that more of the energy required by the building is related to dehumidification as compared to cooling at constant humidity ratio.

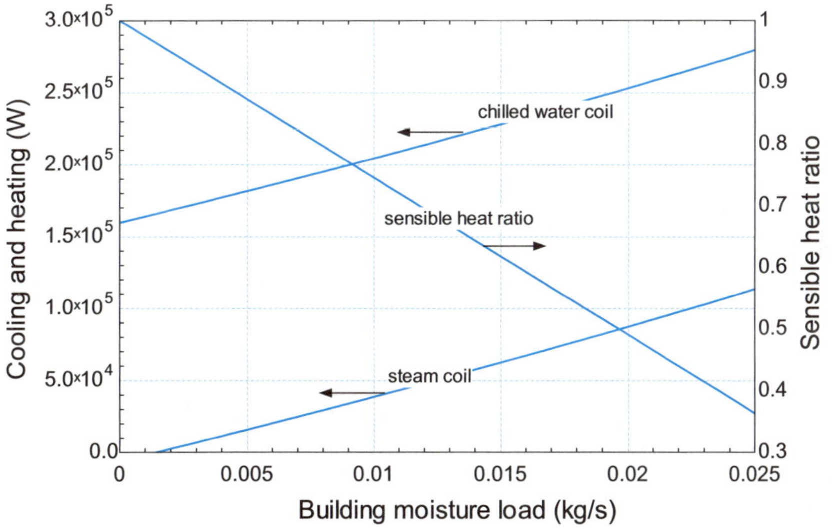

Figure 3: Cooling, heating, and sensible heat ratio as a function of the building moisture load.

12.4.3 Evaporative Cooling

The temperature of air that is not saturated can be reduced by humidifying the air in an adiabatic process, as described in Section 12.2. This process is particularly effective for providing air conditioning in dry locations, such as in the southwestern region of the United States. The device used for this purpose is called an *evaporative cooler* which is affectionately referred to as a "swamp cooler". Figure 12-10 illustrates the evaporative cooling process for air entering an evaporative cooler at state 1 with $T_1 = 35°C$ and

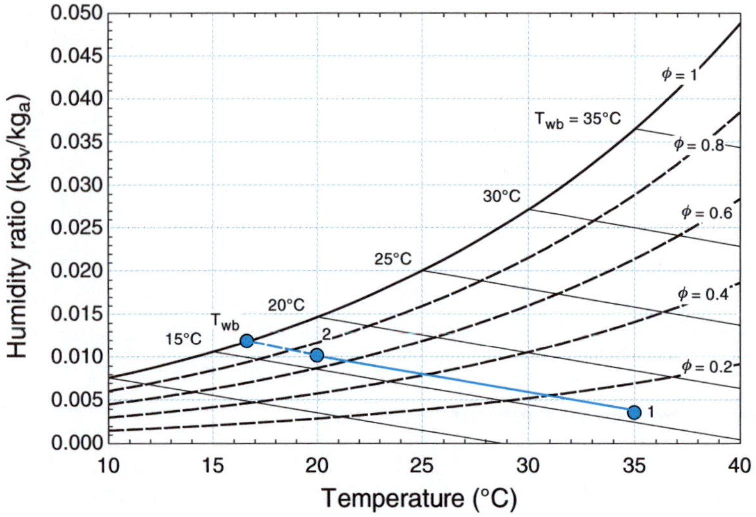

Figure 12-10: Evaporative cooling process.

$\phi_1 = 10\%$ relative humidity on a psychrometric chart. The air state approximately follows a line of constant wet bulb temperature as it passes through the evaporative cooler. If the air were humidified to saturation, it would emerge saturated at the thermodynamic wet bulb temperature of the inlet air, which in this case is about $T_{wb} = 16°\text{C}$. In an actual application, the air would likely not be brought to complete saturation. For example, Figure 12-10 shows an outlet state with $T_2 = 20°\text{C}$. The performance of an evaporative cooler can be quantified in terms of its effectiveness, defined as:

$$\varepsilon_{ec} = \frac{T_1 - T_2}{T_1 - T_{wb}} \tag{12-39}$$

where T_1 and T_2 are the inlet and outlet dry bulb temperatures, respectively, and T_{wb} is the wet bulb temperature. The effectiveness of the evaporative cooling process shown in Figure 12-10 is about $\varepsilon_{ec} = 0.79$.

The advantage of evaporative cooling is that it eliminates the significant expense required to operate a vapor compression air conditioning unit. However, evaporative cooling becomes less effective as the outdoor relative humidity increases (i.e., at state 1 in Figure 12-10 moves vertically up). One way to extend the applicability of evaporative cooling to outdoor conditions with higher relative humidity is to use the *indirect two-stage evaporative cooling process* that is shown in Figure 12-11.

In the indirect two-stage evaporative cooling process, the outdoor air is split into two streams. One stream is cooled and humidified as it passes through an evaporative cooler. This stream is then used as the cooling source in a heat exchanger in order to lower the temperature of the stream that is provided to the building. The advantage of this process is that the air entering the system at state 1 can be cooled without increasing its humidity ratio. The indirect two-stage evaporative cooling process is shown on a psychrometric chart in Figure 12-12.

Outdoor air at $T_1 = 30°\text{C}$ and $\phi_1 = 30\%$ relative humidity is split into two equal mass flow streams. One stream passes through an evaporative cooler operating with an effectiveness of $\varepsilon_{ec} = 0.9$ where it is cooled to $T_2 = 19°\text{C}$ at state 2. The cooled and outdoor air streams enter a heat exchanger having a heat exchanger effectiveness of $\varepsilon_{hx} = 0.70$ where energy, but not mass, is exchanged. The outdoor air entering the heat exchanger at state 1 is cooled from 30°C to $T_4 = 22.4°\text{C}$ at constant humidity ratio. The

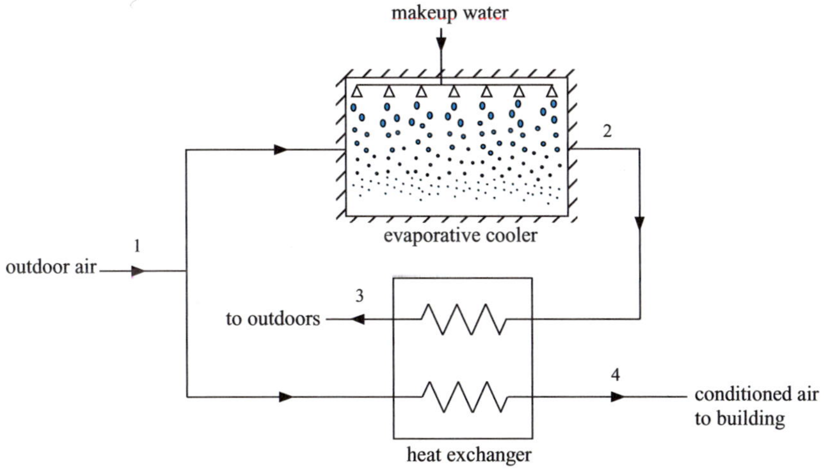

Figure 12-11: Schematic of the indirect two-stage evaporative cooling process.

relative humidity of the air leaving the heat exchanger at state 4 is $\phi_4 = 47\%$. State 4 is within the summer time comfort region shown in Figure 12-7 and therefore it can be used to condition the building. If the outdoor air were evaporatively cooled to the same temperature (i.e., from 30°C to 22.4°C), as shown by state ec in Figure 12-12, then it would have a relative humidity of about $\phi_{ec} = 65\%$ which is too high to be used for air conditioning. The disadvantage of the indirect two-stage evaporative cooling is that more air flow, and thus more fan power, is required than for the single-stage direct evaporative cooling process.

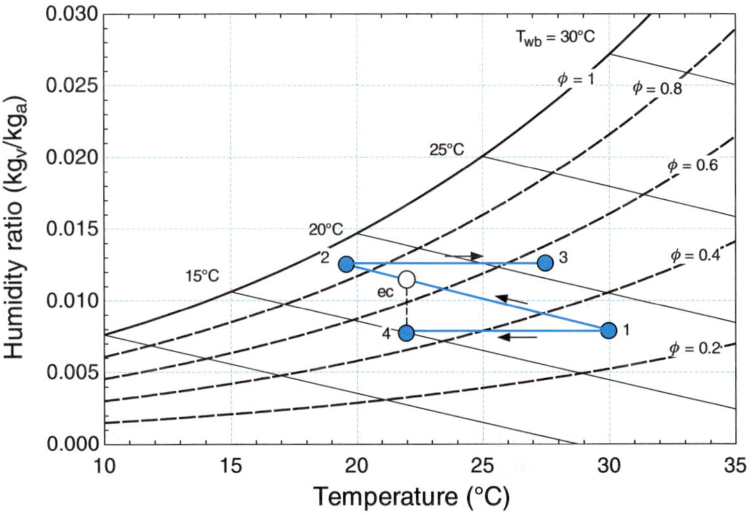

Figure 12-12: Indirect two-stage evaporative cooling process.

12.4.4 Desiccants

This extended section, which can be found online at www.cambridge.org/kleinandnellis, describes dehumidification using desiccant systems.

12.5 Cooling Towers

The performance of both power and refrigeration cycles improves as the temperature to which heat is rejected is lowered. For example, the maximum (Carnot) efficiency of a heat engine, identified in Section 5.3, is:

$$\eta_{max} = 1 - \frac{T_C}{T_H} \tag{12-40}$$

where T_H is the temperature of heat source and T_C is the temperature to which the cycle rejects heat. Equation (12-40) shows that reducing T_C will increase the maximum efficiency that the cycle can attain. Although actual power cycles do not operate at the Carnot efficiency, their performance also improves as the thermal sink temperature is reduced. For example, in the Rankine power cycle (discussed in Section 8.2) lowering the temperature of the cooling water provided to the condenser results in a reduction in the condensing temperature and therefore a reduction in the condensing pressure. The larger pressure difference applied across the turbine increases both the power and the efficiency of the cycle. A similar trend is observed for refrigeration cycles. The maximum coefficient of performance for a refrigeration cycle is derived in Section 9.1:

$$COP_{max} = \frac{T_C}{T_H - T_C} \tag{12-41}$$

where T_C is the refrigeration temperature and T_H is the heat sink temperature. Equation (12-41) shows that reducing the heat sink temperature increases the maximum attainable *COP* of the cycle. In the vapor compression cycle (discussed in Section 9.2) lowering the temperature of the external stream provided to the condenser causes a reduction in the condensing pressure, which reduces the pressure rise that must be produced by the compressor. As a consequence, the compressor power decreases which improves the *COP*.

Large power and refrigeration plants may use cooling water, rather than air, as the external fluid in the condenser for several reasons. First, the convection heat transfer coefficient is much higher for water than for air, which reduces the heat transfer resistance between the cycle working fluid and the coolant and therefore decreases the required size of the heat exchange equipment. Second, water is generally available at a lower temperature than ambient air. For example, the temperature of water in rivers and lakes is generally lower than the air temperature during the summer when the demand for refrigeration and power cycles is highest. However, power and refrigerant plants are often not located near a water source and even when they are located near a lake or river, they are often restricted from using the water due to environmental concerns.

Cooling towers are simple heat and mass exchangers that allow the outdoor air to interact with a stream of water which, in turn, acts as the heat sink for the cycle. Figure 12-15 illustrates a basic Rankine cycle using a cooling tower. The condenser rejects heat to a stream of water that is subsequently cooled in the cooling tower.

The water entering the cooling tower at state 1 is distributed or sprayed over a packing of some type so that it comes into intimate contact with the air that is blown through the tower entering at state 3. Some of the water evaporates into the air, which results in an energy transfer from the remaining water and a reduction in its temperature. The water leaves at state 2 at a lower temperature and returns to the condenser. The air leaves the cooling tower at state 4 with increased moisture content; ideally, this air is nearly saturated. Because some of the water from the condenser evaporates within the cooling tower, it is necessary to provide a relatively small flow of makeup water to the system at state 5. Additional water is added periodically to reduce the concentration of the salts that are dissolved in the water.

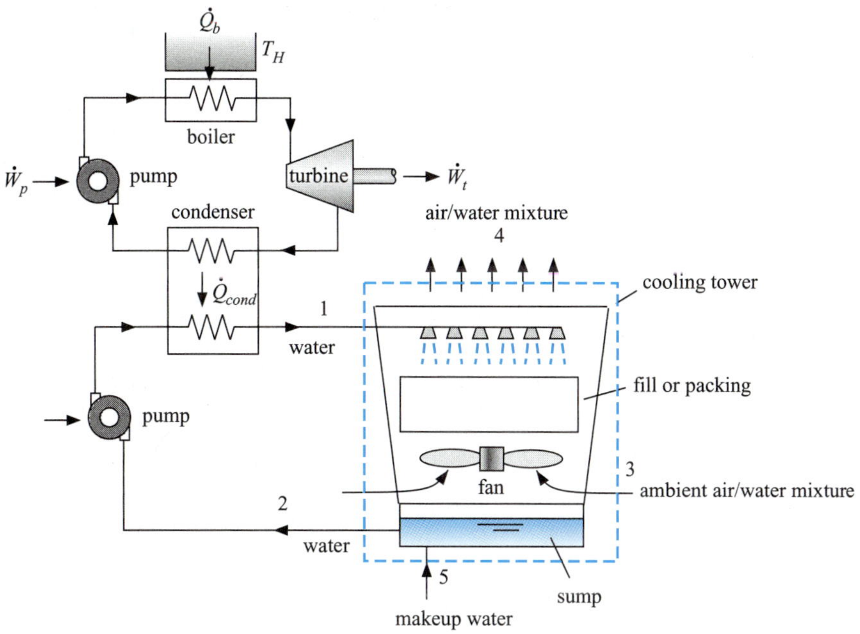

Figure 12-15: Cooling tower integrated with a Rankine power cycle.

Cooling towers are widely employed in large power and refrigeration cycles because they can cool water below the ambient temperature due to the evaporative cooling effect. Ideally, the water could be cooled to near the wet bulb temperature of the entering air. In addition, the mass transfer that occurs in a cooling tower increases the heat transfer coefficient between the water and the air so that cooling towers can be considerably smaller and less expensive than ordinary heat exchangers for the same heat transfer rate.

12.5.1 Cooling Tower Nomenclature

Cooling towers are classified as mechanical draft or natural draft. Mechanical draft cooling towers employ a fan to move air through the device; forced draft designs provide the fan at the top of the cooling tower whereas induced draft designs provide a fan at the bottom, as shown in Figure 12-15. A photograph of a mechanical, forced draft cooling tower is shown in Figure 12-16(a). Natural draft cooling towers eliminate the need for a fan by providing a nozzle shape that promotes air motion due to the buoyancy force that is caused by the reduced density of the heated air. These cooling towers must be quite tall in order to be effective and therefore they are only used in very large installations. A picture of the natural draft cooling tower in use at the Three Mile Island power plant in Pennsylvania is shown in Figure 12-16(b).

Cooling towers consist of a *fill* or *packing* that provides surface area for the water and air streams to interact. The packing can be any type of porous structure ranging from treated wooden planks to the fiberglass packing material that is used in modern cooling towers, which has an aerodynamic shape in order to reduce the required fan power. A cooling tower may consist of one or more *cells*. Each cell is equipped with its own fan so that it can be operated individually. Water enters the cooling tower at the top and is sprayed or evenly distributed over the packing material. Some of the water evaporates into the air that is blown from the bottom as the water moves down through the packing due to gravity. The water that is not evaporated collects in a reservoir at the bottom of

(a) (b)

Figure 12-16: (a) Forced draft cooling tower at University of Wisconsin, Madison and (b) natural draft cooling towers at the Three Mile Island power plant in Harrisburg, Pennsylvania.

the tower called the *sump*. The warm, humid conditions in the sump are conducive to the undesired growth of biological materials and so cooling tower water must be treated with chlorine, ozone, or other oxidants.

Water loss occurs by evaporation and by drift and therefore makeup water is required. *Drift* is the loss of small droplets of liquid water that are entrained in the air. Cooling towers are generally equipped with *drift eliminators*, which are screens that restrict the movement of liquid droplets and thereby reduce the water loss.

The *range* of a cooling tower is defined as the difference between the inlet and outlet temperatures of the water; i.e., the range for the cooling tower in Figure 12-15 is $(T_1 - T_2)$. The lowest possible temperature of the water leaving at state 2 is the wet bulb temperature of the ambient air entering at state 3. The *approach* is defined as the difference between the outlet water temperature and the inlet air wet bulb temperature; i.e., the approach for the cooling tower in Figure 12-15 is $T_2 - T_{wb,3}$. The cooling tower *load* is the rate of energy transfer from the water to the air.

12.5.2 Cooling Tower Analysis

Cooling towers operate in much the same way as an evaporative cooler, discussed in Section 12.4.3, and therefore they are analyzed in a similar manner. Cooling towers normally operate at steady state. Taking the entire cooling tower as a system (see the dashed line shown in Figure 12-15) leads to a steady-state mass balance on dry air:

$$\dot{m}_{a,3} = \dot{m}_{a,4} = \dot{m}_a \tag{12-42}$$

Makeup water is provided so that the mass flow rate of water returned to the plant ($\dot{m}_2$) is the same as the mass flow rate of water provided to the cooling tower ($\dot{m}_1$). A steady state balance on the water determines the rate at which makeup water must be provided.

$$\dot{m}_5 = \dot{m}_a\left(\omega_4 - \omega_3\right) \tag{12-43}$$

Neglecting fan power and assuming that the tower operates adiabatically, a steady-state energy balance on the tower is:

$$\dot{m}_1\,h_1 + \dot{m}_a\,h_{av,3} + \dot{m}_5\,h_5 = \dot{m}_2\,h_2 + \dot{m}_a\,h_{av,4} \tag{12-44}$$

Fan power can be significant. However, it can be determined in a separate analysis as the fan power does not significantly affect the specific enthalpy of the air or water streams. The pressure drop depends on the type of packing, the geometry of the tower, and the air flow rate.

The purpose of a thermodynamic analysis of a cooling tower is normally to determine the temperature of the cooled water returning to the plant and the rate at which makeup water must be supplied. The mass flow rates of air and water and the state of the inlet air (state 3) and the entering water (state 1) are usually specified. Equations (12-42) through (12-44) are not sufficient to solve this problem unless more information, such as the state of the outlet air, is provided. This situation is similar to the analysis of a heat exchanger, presented in Section 6.6.6, where it was found that a heat transfer rate equation is needed in addition to mass and energy balances. In Section 6.6.6, we found that the performance of a heat exchanger can be expressed in terms of a heat exchanger effectiveness. Cooling towers are heat and mass exchangers and their performance with regard to each of these phenomena can also be expressed in terms of an effectiveness. The cooling tower heat transfer effectiveness ($\varepsilon_{ct,Q}$) is defined as the ratio of the actual rate of energy transfer to the air to the maximum possible rate of energy transfer to the air, as proposed by Braun et al. (1989):

$$\varepsilon_{ct,Q} = \frac{\dot{Q}}{\dot{Q}_{max}} \tag{12-45}$$

The actual rate of heat transfer can be obtained from an energy balance on the air:

$$\dot{Q} = \dot{m}_a\left(h_{av,4} - h_{av,3}\right) \tag{12-46}$$

The maximum possible rate of energy transfer results if the air at state 4 were to achieve a saturated state at the temperature of the inlet water (i.e., if $\phi_4 = 1$ and $T_4 = T_1$).

$$\dot{Q}_{max} = \dot{m}_a\left(h_{av,T=T_1,\phi=1} - h_{av,3}\right) \tag{12-47}$$

Substituting Eqs. (12-46) and (12-47) into Eq. (12-45) results in:

$$\varepsilon_{ct,Q} = \frac{\left(h_{av,4} - h_{av,3}\right)}{\left(h_{av,T=T_1,\phi=1} - h_{av,3}\right)} \tag{12-48}$$

The cooling tower effectiveness depends on the packing surface area and the air flow rate. Application of Eq. (12-48) determines the dry air specific enthalpy at state 4 provided that the effectiveness is known.

The cooling tower mass transfer effectiveness ($\varepsilon_{ct,m}$) is defined as the ratio of the actual rate of evaporation of water into the air to the maximum possible rate of evaporation:

$$\varepsilon_{ct,m} = \frac{\dot{m}_{evap}}{\dot{m}_{evap,max}} \tag{12-49}$$

The actual rate of evaporation can be obtained from a water mass balance on the air-water mixture flowing through the cooling tower:

$$\dot{m}_{evap} = \dot{m}_a\left(\omega_4 - \omega_3\right) \tag{12-50}$$

The maximum possible rate of evaporation results if the air leaving at state 4 were to achieve a saturated state at the temperature of the inlet water (i.e., if $\phi_4 = 1$ and $T_4 = T_1$):

$$\dot{m}_{evap,max} = \dot{m}_a\left(\omega_{T=T_1,\phi=1} - \omega_3\right) \tag{12-51}$$

Substituting Eqs. (12-50) and (12-51) into Eq. (12-49) results in:

$$\varepsilon_{ct,m} = \frac{\left(\omega_4 - \omega_3\right)}{\left(\omega_{T=T_1,\phi=1} - \omega_3\right)} \tag{12-52}$$

The analogous behavior of heat and mass transfer suggests that the effectiveness values defined by Eqs. (12-48) and (12-52) should have approximately the same value, $\varepsilon_{ct,Q} \approx \varepsilon_{ct,m} \approx \varepsilon_{ct}$. State 4 is fixed by the dry air specific enthalpy, calculated using Eq. (12-48), and the humidity ratio, calculated using Eq. (12-52).

EXAMPLE 12.5-1: ANALYSIS OF A COOLING TOWER

A cooling tower is shown in Figure 1.

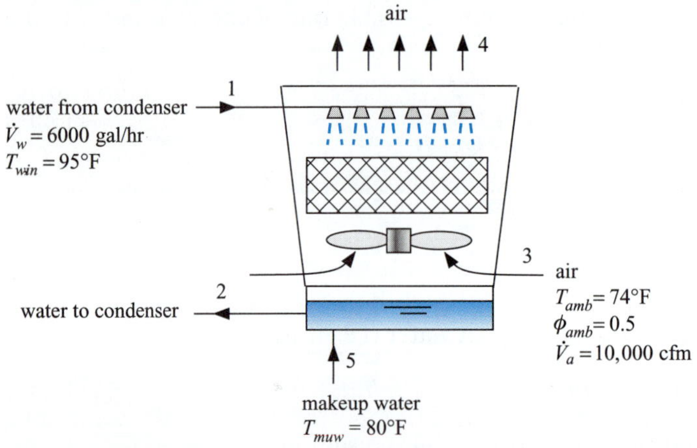

Figure 1: Cooling tower.

The cooling tower is designed to cool water from the condenser of a large air conditioning system that enters with a flow rate $\dot{V}_w = 6{,}000$ gallons per hour with temperature $T_{w,in} = 95°\text{F}$. Outdoor air at $T_{amb} = 74°\text{F}$ and $\phi_{amb} = 50\%$ relative humidity is forced through the tower at a rate of $\dot{V}_a = 10{,}000$ cfm. Makeup water is provided at $T_{muw} = 80°\text{F}$. The effectiveness of the cooling tower is $\varepsilon_{ct} = 0.7$.

a) Determine the load, range, approach, and makeup water flow rate.

The inputs are entered in EES.

```
$UnitSystem SI Mass Radian J K Pa
T_w_in=converttemp(F,K,95 [F])                      "water inlet temperature"
V_dot_w=6000 [gal/hr]*convert(gal/hr,m^3/s)         "water inlet volumetric flow rate"
T_amb=converttemp(F,K,74 [F])                       "ambient air temperature"
phi_amb=0.5 [-]                                     "ambient air relative humidity"
P=1 [atm]*convert(atm,Pa)                           "ambient pressure"
V_dot_a=10000 [cfm]*convert(cfm,m^3/s)              "volumetric flow rate of air entering tower"
T_muw=converttemp(F,K,80 [F])                       "temperature of makeup water"
eff_ct=0.7 [-]                                      "cooling tower effectiveness"
```

State 3 is fixed by the temperature, $T_3 = T_{amb}$, and relative humidity, $\phi_3 = \phi_{amb}$. The dry air specific volume, dry air specific enthalpy, humidity ratio, and thermodynamic wet bulb temperature ($v_{av,3}$, $h_{av,3}$, ω_3, and $T_{wb,3}$) are determined. The mass flow rate of dry air is given by:

$$\dot{m}_a = \frac{\dot{V}_a}{v_{av,3}}$$

```
"State 3"
T[3]=T_amb                                          "temperature"
phi[3]=phi_amb                                      "relative humidity"
v_av[3]=volume(AirH2O,P=P,T=T[3],R=phi[3])          "dry air specific volume"
h_av[3]=enthalpy(AirH2O,P=P,T=T[3],R=phi[3])        "dry air specific enthalpy"
omega[3]=humrat(AirH2O,P=P,T=T[3],R=phi[3])         "humidity ratio"
T_wb[3]=wetbulb(AirH2O,P=P,T=T[3],R=phi[3])         "thermodynamic wet bulb temperature"
m_dot_a=V_dot_a/v_av[3]                             "mass flow rate of dry air"
```

The temperature at state 1 is given, $T_1 = T_{w,in}$. The pressure is assumed to be atmospheric. Note that pressure has very little effect on the properties of liquid water. The specific volume and specific enthalpy (v_1 and h_1) are determined. The mass flow rate of water is:

$$\dot{m}_w = \frac{\dot{V}_w}{v_1}$$

```
"State 1"
T[1]=T_w_in                                         "temperature"
v[1]=volume(Water,T=T[1],P=P)                       "specific volume"
h[1]=enthalpy(Water,T=T[1],P=P)                     "specific enthalpy"
m_dot_w=V_dot_w/v[1]                                "mass flow rate of water"
```

The maximum possible dry air specific enthalpy leaving the cooling tower at state 4 ($h_{av,4,max}$) occurs if the air reaches $T_{w,in}$ and 100% relative humidity. Equation (12-48) is used to determine the dry air specific enthalpy at state 4 based on the cooling tower effectiveness:

$$\varepsilon_{ct} = \frac{(h_{av,4} - h_{av,3})}{(h_{av,T=T_1,\phi=1} - h_{av,3})}$$

The maximum possible humidity ratio leaving the cooling tower at state 4 ($\omega_{4,max}$) also occurs if the air reaches $T_{w,in}$ and 100% relative humidity. Equation (12-52) is used to determine the humidity ratio at state 4 based on the cooling tower effectiveness:

$$\varepsilon_{ct} = \frac{(\omega_4 - \omega_3)}{\left(\omega_{T=T_1,\phi=1} - \omega_3\right)}$$

State 4 is fixed by $h_{av,4}$ and ω_4. The temperature and relative humidity (T_4 and ϕ_4) are determined.

```
"State 4"
h_av_4_max=enthalpy(AirH2O,P=P,T=T_w_in,R=1 [-])      "maximum dry air specific enthalpy at state 4"
eff_ct=(h_av[4]-h_av[3])/(h_av_4_max-h_av[3])         "dry air specific enthalpy at state 4"
omega_4_max=humrat(AirH2O,P=P,T=T_w_in,R=1 [-])       "maximum possible humidity ratio at state 4"
eff_ct=(omega[4]-omega[3])/(omega_4_max-omega[3])     "humidity ratio at state 4"
T[4]=temperature(AirH2O,P=P,h=h_av[4],w=omega[4])     "temperature"
T_a_out_F=converttemp(K,F,T[4])                       "in F"
phi[4]=relhum(AirH2O,P=P,h=h_av[4],w=omega[4])        "relative humidity"
```

State 5 is fixed by the temperature, $T_5 = T_{muw}$, and pressure, which is assumed to be atmospheric. The enthalpy at state 5 (h_5) is determined. The mass flow rate of makeup water is calculated from a water mass balance on the cooling tower:

$$\dot{m}_{muw} = \dot{m}_a\left(\omega_4 - \omega_3\right)$$

```
"State 5"
T[5]=T_muw                                            "temperature"
h[5]=enthalpy(Water,T=T[5],P=P)                       "enthalpy"
m_dot_muw=m_dot_a*(omega[4]-omega[3])                 "mass flow rate of makeup water"
```

which results in $\dot{m}_{muw} = 0.107$ kg/s. An energy balance on the cooling tower provides the specific enthalpy of the water at state 2.

$$\dot{m}_w\, h_1 + \dot{m}_{muw}\, h_5 + \dot{m}_a\, h_{av,3} = \dot{m}_w\, h_2 + \dot{m}_a\, h_{av,4}$$

The specific enthalpy at state 2 is used to determine the temperature of the water returning to the condenser (T_2).

```
"State 2"
m_dot_w*h[1]+m_dot_muw*h[5]+m_dot_a*h_av[3]=m_dot_a*h_av[4]+m_dot_w*h[2]
                                                      "energy balance"
T[2]=temperature(Water,P=P,h=h[2])                    "temperature"
T_w_out_F=converttemp(K,F,T[2])                       "in F"
```

which results in $T_2 = 296.4$ K (73.77°F). The cooling tower load is obtained from an energy balance on the water:

$$\dot{Q}_{ct} = \dot{m}_w\left(h_1 - h_2\right)$$

The range and approach are computed according to:

$$Range = T_1 - T_2$$

$$Approach = T_2 - T_{wb,3}$$

```
Q_dot_ct=m_dot_w*(h[1]-h[2])                "load"
Q_dot_ct_ton=Q_dot_ct*convert(W,ton)        "in ton"
Range=T[1]-T[2]                             "range"
Approach=T[2]-T_wb[3]                       "approach"
```

which leads to $\dot{Q}_{ct} = 309.5$ kW (88 ton), $Range = 11.8$ K, and $Approach = 6.681$ K. Figure 2 illustrates the state of the air entering (state 3) and leaving (state 4) on a psychrometric chart.

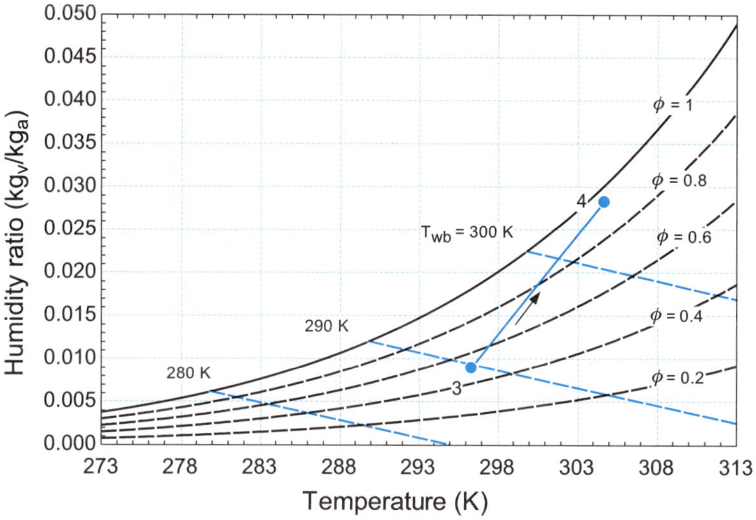

Figure 2: Entering and leaving air states on a psychrometric chart.

b) Plot the load, range, approach, and mass flow rate of makeup water as a function of the cooling tower effectiveness.

Figure 3 illustrates the approach and the range of the cooling tower as a function of the cooling tower effectiveness. Figure 4 illustrates the load and cooling tower makeup water flow rate as a function of the cooling tower effectiveness.

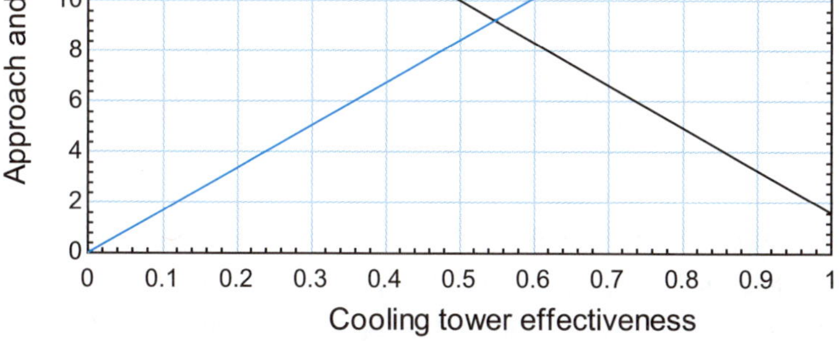

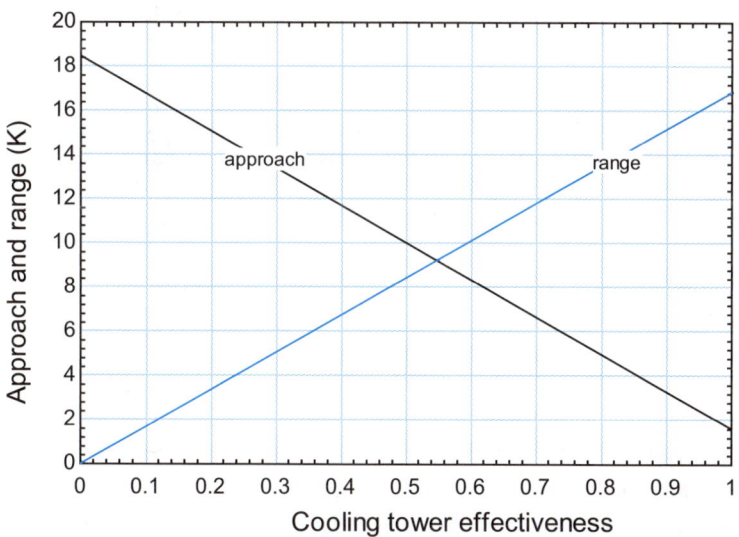

Figure 3: Approach and range of the cooling tower as a function of the cooling tower effectiveness.

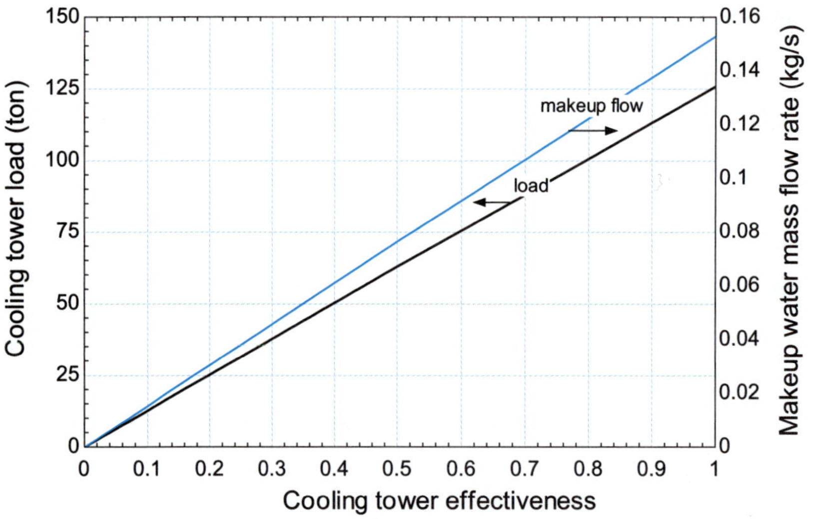

Figure 4: Cooling tower load and makeup flow rate as a function of the cooling tower effectiveness.

12.6 Entropy for Psychrometric Mixtures

This extended section, which can be found online at www.cambridge.org/kleinandnellis, indicates how to calculate the specific entropy of an air-water vapor mixture and demonstrates the use of entropy calculations in psychrometric applications.

REFERENCES

ASHRAE Handbook of Fundamentals, American Society of Heating, Refrigeration and Air-Conditioning Engineers, Chapter 8, Figure 5 (2001).

Braun, J.E., Klein, S.A. and Mitchell, J.W., "Effectiveness Models for Cooling Towers and Cooling Coils," *ASHRAE Transactions*, ISSN 0001-2505, Vol. 95, Pt. 2, p. 164–174 (1989).

Duffie, J.A. and Beckman, W.A., *Solar Engineering of Thermal Processes*, 3rd Edition, Wiley, (2006).

Problems

The problems included here have been selected from a much larger set of problems that are available on the website associated with this book, www.cambridge.org/kleinandnellis.

A: Psychrometric Definitions

12.A-1 A mixture of helium and water vapor is flowing through a pipe at $T = 90°C$ and $P = 150$ kPa. The mole fraction of helium is $y_{He} = 0.80$.

a) What is the relative humidity of the mixture?

b) What is the humidity ratio of the mixture?

c) What is the dew point of the mixture?

12.A-2 Figure 12.A-2 shows a humidifier that takes in an air-water vapor mixture with relative humidity $\phi_1 = 0.1$ (10%), temperature $T_1 = 18°C$, and pressure $P_1 =$ 1 atm and mixes it with a source of steam at $P_2 = 50$ psia in order to generate an air-water mixture that has relative humidity $\phi_3 = 0.7$ (70%), temperature $T_3 = 20°\text{C}$, and pressure $P_3 = 1$ atm. The mass flow rate of the inlet air-water mixture is $\dot{m}_1 = 0.25$ kg/s.

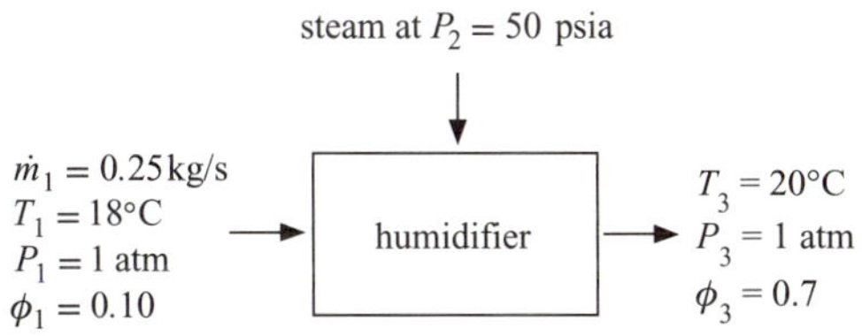

Figure 12.A-2: Humidifier.

The humidifier is at steady-state and is externally adiabatic.

a) Determine the humidity ratio of the air-water vapor mixture entering the humidifier at state 1 (ω_1) and the humidity ratio at state 3 (ω_3).
b) Determine the mass flow rates of air and the mass flow rates of water vapor entering the humidifier at state 1 and leaving at state 3.
c) Determine the mass flow rate of steam entering the humidifier at state 2.
d) What is the temperature of the steam entering the humidifier at state 2, assuming that the humidifier operates adiabatically?

12.A-3 Figure 12.A-3 shows an air compressor system that is used in a factory. At state 1, the compressor draws in ambient air (which is actually an air/water vapor mixture) with $T_{amb} = 20°\text{C}$ and $P_{atm} = 1$ atm. The relative humidity of the ambient air/water mixture is $\phi_1 = 0.5$ (50%). The mass flow rate of the incoming air/water mixture is $\dot{m}_1 = 0.1$ kg/s. The rate of heat transfer from the compressor to the atmosphere is $\dot{Q}_c = 50$ kW. The air/water mixture leaves the compressor at $T_2 = 175°\text{C}$ and $P_2 = 90$ psig. The hot air/water mixture is cooled in an aftercooler that rejects heat to ambient. The air/water mixture leaving the aftercooler is at $T_3 = T_{amb}$ and $P_3 = P_2$. The aftercooler also serves as a filter and drier. Condensate (i.e., liquid water condensed out of the air) leaves the aftercooler at state 4, $T_4 = T_{amb}$ and $P_4 = P_{atm}$.

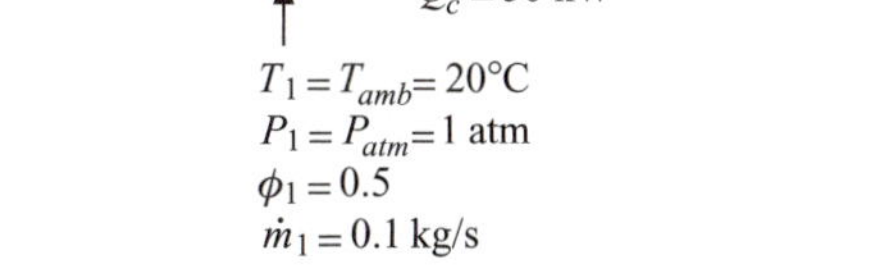

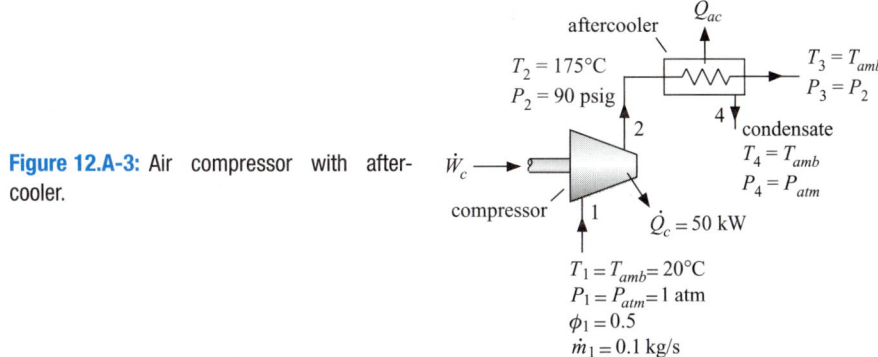

Figure 12.A-3: Air compressor with aftercooler.

a) Determine the humidity ratio of the air at state 1. What are the mass flow rates of dry air and water vapor entering the compressor ($\dot{m}_{a,1}$ and $\dot{m}_{v,1}$)?
b) Assume that no condensation of water occurs in the compressor. Determine the relative humidity and the humidity ratio at state 2. Is the assumption correct? How do you know?
c) Determine the power consumed by the compressor ($\dot{W}_c$) and the rate of entropy generation within the compressor ($\dot{S}_{gen,c}$).

d) Determine the humidity ratio of the air at state 3.

e) What are the mass flow rates of dry air and water vapor leaving the aftercooler ($\dot{m}_{a,3}$ and $\dot{m}_{v,3}$)? What is the mass flow rate of condensate leaving the aftercooler ($\dot{m}_4$)? What is volumetric flow rate of condensate (in gallons/day)?

f) What is the rate of heat transfer from the aftercooler to the environment ($\dot{Q}_{ac}$)?

g) Use EES to draw a *T-s* diagram for the states of the water in the problem (i.e., the water vapor at states 1, 2, and 3 and the water leaving at state 4). Label the states.

12.A-4 During the winter, indoor air with relative humidity $\phi_1 = 0.7$ (70%), temperature $T_1 = 22°C$, and volumetric flow rate $\dot{V}_1 = 0.04\ m^3/s$ is used to pre-heat air drawn from the outdoors at relative humidity $\phi_3 = 0.2$ (20%), temperature $T_3 = 10°C$, and volumetric flow rate $\dot{V}_3 = 0.06\ m^3/s$ in a heat exchanger, as shown in Figure 12.A-4. The outdoor air leaving the heat exchanger is heated to $T_4 = 20°C$. The process occurs at atmospheric pressure and the heat exchanger is externally adiabatic.

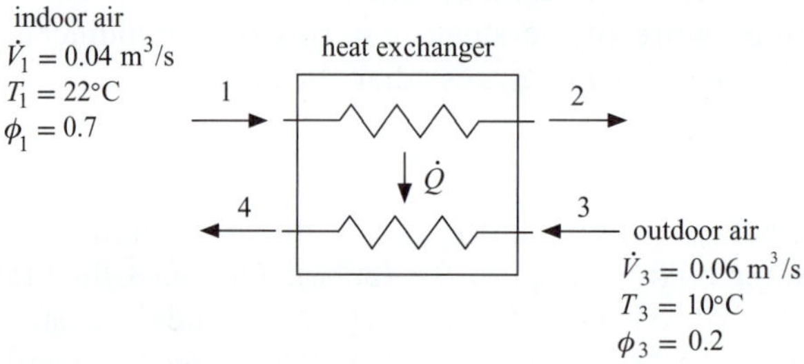

Figure 12.A-4: Heat exchanger.

a) What is the relative humidity of the air leaving the heat exchanger at state 4?

b) What is the rate of heat transfer between the air streams in the heat exchanger, $\dot{Q}$?

c) Will there be any condensation as the indoor air flows through the heat exchanger? Justify your answer.

d) Assume that your answer to (c) is yes. At what temperature will the indoor air begin to condense?

e) What is the exit temperature of the indoor air (T_2)? What is the mass flow rate of the condensate from the indoor air? You may neglect the enthalpy carried by the condensate to answer this question.

f) Locate and label all four states on the psychrometric chart.

12.A-5 Two flows of air/water vapor mixtures enter a steady flow device, as shown in Figure 12.A-5.

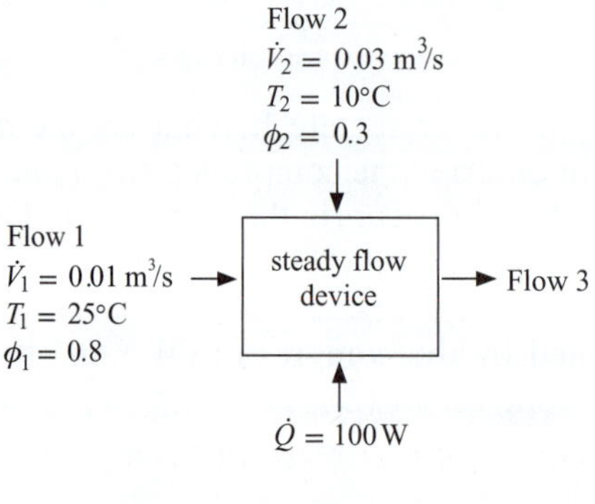

Figure 12.A-5: Steady flow device.

Flow 1 enters the device with temperature $T_1 = 25°C$ and relative humidity $\phi_1 = 0.8$. The volumetric flow rate at state 1 is $\dot{V}_1 = 0.01\ m^3/s$. Flow 2 enters the device with temperature $T_2 = 10°C$ and relative humidity $\phi_2 = 0.3$. The volumetric flow rate at state 2 is $\dot{V}_2 = 0.03\ m^3/s$. Heat is transferred to the device at a rate of $\dot{Q} = 100$ W. A single flow of air/water vapor mixture leaves at state 3 and there is no condensation in the device. The pressure of each stream (entering and leaving) is $P = 1$ atm.

a) Determine the temperature and relative humidity of the air leaving the device.
b) Indicate each of the states on the psychrometric chart.

12.A-9 A building having a volume of $V = 140\ m^3$ is to be maintained at $T_b = 25°C$ and $\phi_b = 50\%$ relative humidity on a winter day. Ventilation air is provided at a rate of $\dot{V}_v = 0.02\ m^3/s$, $T_v = 5°C$, and $\phi_v = 30\%$ relative humidity. Heat losses from the building space to outdoors occur at a rate of $\dot{Q} = 1.5$ kW. A humidifier provides the necessary moisture to the building space in the form of saturated water vapor at $T_s = 100°C$. Space heaters within the building space maintain the required room temperature. Assume that the air in the room is fully mixed.

a) Determine the rate at which moisture must be added.
b) Determine the rate at which heating must be provided in the building space.
c) Determine the mass of water vapor in the room at any time.
d) Determine the temperature of the walls in the room that would produce condensation.

12.A-11 Two tanks, A and B, are connected by a pipe and a closed valve. Tank A contains $m_A = 4$ kg of butane (C_8H_{18}) at $P_{A,1} = 3.5$ bar and $T_{A,1} = 175°C$. Tank B contains $m_B = 1.4$ kg of water vapor at $P_{B,1} = 0.5$ bar and $T_{B,1} = 95°C$. The valve is opened and, after some time, the contents of the tanks completely mix. A temperature measurement indicates that the mixture is at $T_2 = 126°C$.

a) What is the final pressure of the gas mixture?
b) What is the humidity ratio of the gas mixture?
c) Determine the value of the relative humidity of the final gas mixture.

12.A-13 A rigid tank with volume $V = 1\ m^3$ initially contains pure refrigerant R134a vapor at $P_1 = 100$ kPa and $T_1 = T_{amb} = 20°C$. The tank is connected to a source of high pressure nitrogen at $P_s = 500$ kPa and $T_s = 200°C$, as shown in Figure 12.A-13.

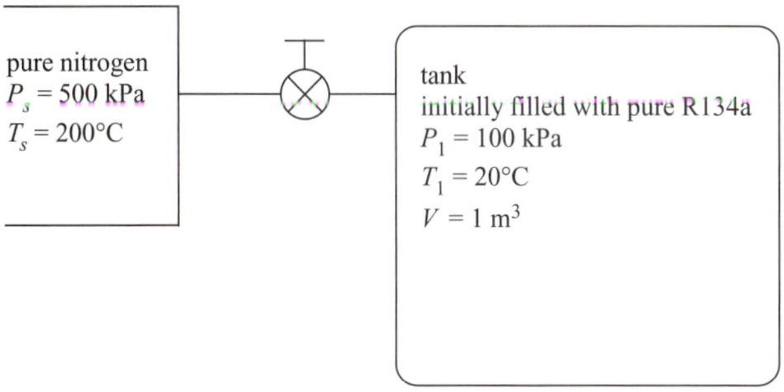

Figure 12.A-13: Tank containing refrigerant vapor connected to a source of nitrogen.

The valve is opened, allowing pure nitrogen to flow into the tank. At the conclusion of the flow process, the pressure in the tank is $P_2 = P_s$ and the temperature in the tank is $T_2 = 100°C$. Model both the nitrogen and the refrigerant as an ideal gas and assume that they form an ideal gas mixture. The properties of the refrigerant should be evaluated using $R_r = 81.5$ J/kg-K, $c_{P,r} = 840$ J/kg-K, and $c_{v,r} = 758.5$ J/kg-K. The properties of nitrogen should be evaluated using $R_{N_2} = 297$ J/kg-K, $c_{P,N_2} = 1040$ J/kg-K, and $c_{v,N_2} = 743$ J/kg-K.

a) Determine the amount of heat transferred from the tank to the surroundings at T_{amb} during the filling process.
b) Determine the entropy generated by the filling process.
c) At the conclusion of the filling process, the valve is closed and the temperature of the tank is gradually reduced. At what temperature will R134a begin to condense? Note that your text includes tables for the properties of saturated R134a.

12.A-14 A closed rigid tank having a volume of $V = 3$ m^3 contains air at $T_1 = 100$°C, $P_1 = 4.4$ bar, and $\phi_1 = 40\%$ relative humidity. The tank contents cool to $T_2 = 20$°C as a result of heat transfer with the surroundings. The gas mixture may be assumed to behave as an ideal gas.

a) Determine the mass of water in the tank.
b) Determine the temperature at which condensation begins.
c) Determine the mass of liquid water in the tank at 20° C.

B: Psychrometric Processes

12.B-1 In an air conditioning system that operates at steady-state, return air at $\dot{V}_1 = 35$ m^3/min, $T_1 = 22$°C, $\phi_1 = 54\%$ relative humidity is mixed with outdoor ventilation air at $\dot{V}_2 = 12$ m^3/min, $T_2 = 30$°C, $\phi_2 = 55\%$ relative humidity, as shown in Figure 12.B-1. The mixed air at state 3 passes over a cooling coil and exits in a saturated condition at state 4. Some water vapor is condensed in the cooling coil and leaves at state 6. The saturated moist air and condensate streams exit the dehumidifier at the same temperature, $T_4 = T_6$. The moist air at state 4 then passes through a heating coil. The air is exhausted to the building at $T_5 = 24$°C and $\phi_5 = 40\%$ relative humidity. The entire process occurs at atmospheric pressure.

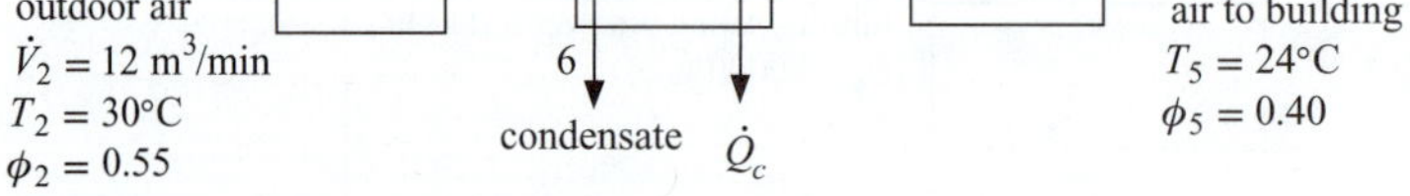

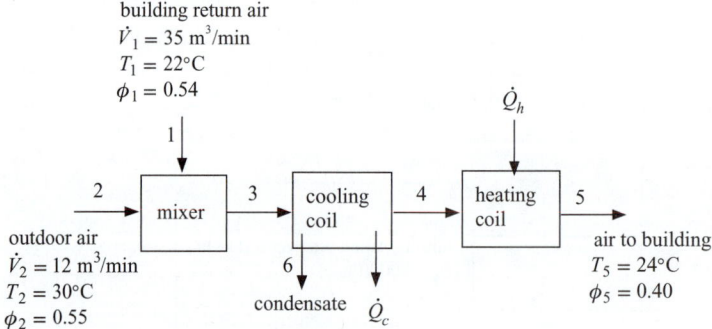

Figure 12.B-1: Air conditioning system.

a) Determine the temperature and relative humidity of the air at state 3.
b) Determine the temperature of the moist air exiting the cooling coil at state 4.
c) Determine the rate at which cooling must be provided in the cooling coil in kW and in tons. Also, determine the mass flow rate of condensate leaving the cooling coil.
d) Plot the cooling required as a function of the outdoor air relative humidity.

12.B-2 One type of residential humidifier is shown in Figure 12.B-2. The humidifier is directly connected to the furnace. In a particular case, air from the building at $T_1 = 20°\text{C}$ and $\phi_1 = 30\%$ relative humidity (state 1) is heated in the furnace to $T_2 = 95°\text{C}$. A fraction of furnace outlet air (f) passes through a humidifier; it exits as saturated air at $T_3 = 50°\text{C}$. The humidified air mixes with the air from the furnace and the mixture re-enters the building at state 4. The entire process occurs at atmospheric pressure.

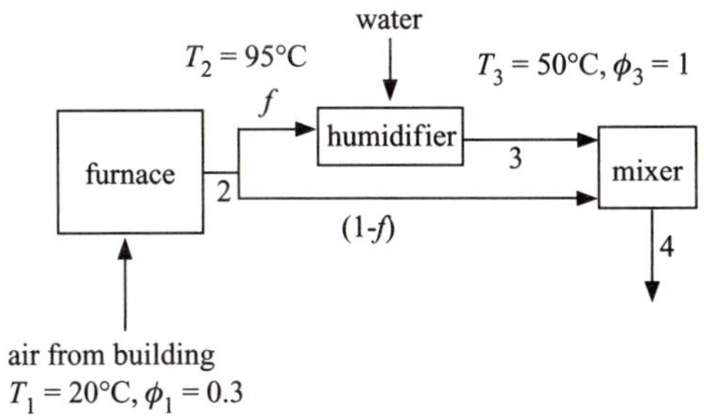

Figure 12.B-2: Humidifier.

a) What is the humidity ratio of the air in the building at state 1?
b) What is the relative humidity of the air at state 2?
c) What is the humidity ratio of the air at state 3?
d) What is the ratio of the mass flow rate of liquid water that must be added to the humidifier to the mass flow rate of the air entering the humidifier?
e) Determine the temperature and relative humidity of the air leaving the mixer that is supplied to the building if $f = 0.5$ (i.e., 50% of the flow passes through the humidifier).
f) Plot the temperature and relative humidity of the air returning to the building as a function of the fraction of the flow diverted through the humidifier, f.

12.B-3 Your cabin has no air conditioning system. Therefore you have installed a swamp cooler, as shown in Figure 12.B-3. The cooler draws in air at a rate of $\dot{V}_1 = 0.035\ \text{m}^3/\text{s}$ from outdoors. Your cabin is located in a hot but dry climate; the outdoor air is at $T_1 = 30°\text{C}$ and $\phi_1 = 0.2$. The air is forced to flow through

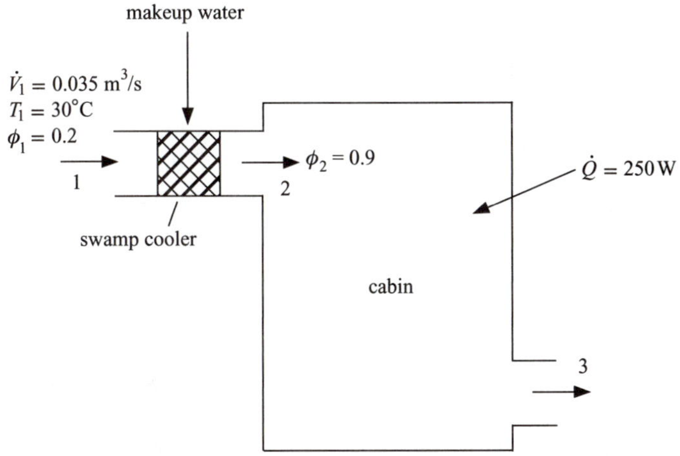

Figure 12.B-3: Swamp cooler.

a chamber that is filled with a porous structure that is saturated with liquid water. The chamber is adiabatic and the air exits from the chamber at $\phi_2 = 0.90$. The air/water mixture passing through the chamber is being adiabatically saturated; therefore, it follows a line of constant adiabatic saturation (or wet bulb) temperature. The air leaving the chamber is directed to your cabin. The cabin has no significant moisture gain, but does experience heating at a rate of $\dot{Q} = 250$ W. Air leaves the cabin at temperature T_3, the indoor air temperature of the cabin.

a) Determine the humidity ratio of the outdoor air.
b) Determine the mass flow rate of dry air passing through the system.
c) Determine the humidity ratio and temperature of the air leaving the chamber.
d) Determine the volume of makeup water (in gallon) that must be provided to the chamber each day.
e) Determine the temperature and relative humidity of the air leaving the cabin.
f) Sketch the three states on a psychrometric chart.
g) Assume that you want to keep your cabin at the temperature that you calculated in part (e) and that the outdoor air temperature remains at 30°C. Estimate the outdoor air relative humidity at which the swamp cooler would no longer provide any useful cooling to the cabin (assume the air leaving the swamp cooler will always be at $\phi_2 = 0.9$).

12.B-4 Figure 12.B-4 illustrates a system used to humidify building air.

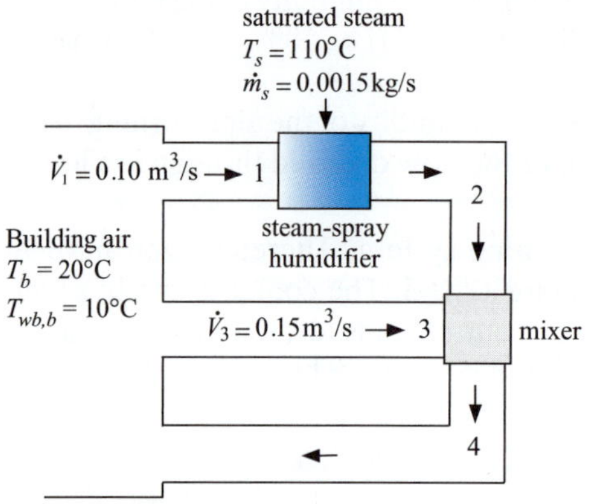

Figure 12.B-4: Steady flow device.

Two streams of air are removed from the building space. The building air has temperature (i.e., dry bulb temperature) $T_b = 20°C$ and adiabatic saturation temperature (i.e., wet bulb temperature) $T_{wb,b} = 10°C$. The stream of air at state 1 has volumetric flow rate $\dot{V}_1 = 0.1\ m^3/s$ and enters a steam-spray humidifier where it is humidified by the injection of saturated steam at $T_s = 110°C$ with mass flow rate $\dot{m}_s = 0.0015$ kg/s. A separate flow of air is pulled from the building with volumetric flow rate $\dot{V}_3 = 0.15\ m^3/s$ and mixes with the air leaving the humidifier. The mixed air flow at state 4 re-enters the building. The pressure of each air stream is at ambient pressure.

a) Determine the temperature and relative humidity of the air leaving the steam spray humidifier at state 2.
b) Determine the temperature and relative humidity of the air that is provided to the building at state 4.

12.B-7 Supermarkets have unusual cooling needs during the summer. Because the freezer cases are often left uncovered, frost accumulates on the frozen food products making them unsightly. The frost also reduces the efficiency of the frozen food cases. To reduce this problem, the humidity in the store can be lowered. A cooling system designed for this purpose is shown in Figure 12.B-7.

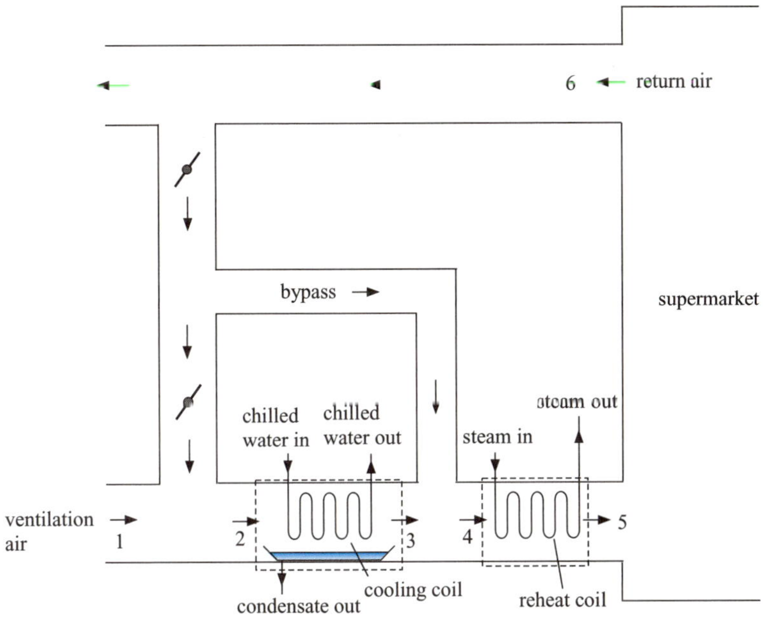

Figure 12.B-7: Supermarket refrigeration system.

Air with flow rate $\dot{V}_5 = 4000$ cfm at $T_5 = 62°$F and $\phi_5 = 55\%$ relative humidity is supplied to the supermarket at state 5 in order to maintain comfort conditions in the supermarket. Air returns from the supermarket at state 6 with $T_6 = 74°$F and $\phi_6 = 54\%$. 15% of the return air (on a mass basis) is exhausted and replaced with outdoor ventilation air at state 1 with $T_1 = 82°$F and $\phi_1 = 48\%$ relative humidity. Some of the re-circulated air is mixed with the ventilation air to achieve state 2. This flow is passed through the cooling coil from which it emerges at state 3 at saturated conditions. The rest of the re-circulated air bypasses the cooling coil, mixes with the cooling coil outlet air and then enters the reheat coil at state 4. Calculate and plot the following quantities as a function of the fraction of the recirculation air that bypasses the cooling coil.

a) The required heat transfer rate in the cooling coil.
b) The cooling coil outlet air temperature.
c) The rate of heat transfer in the reheat coil
d) What is the largest value of the bypass that will allow the air at state 3 to remain above the freezing temperature?

12.B-8 A cooling tower operates with a condenser water mass flow rate $\dot{m}_{cw} = 50{,}000$ lb_m/hr that enters the tower at $T_{cw,in} = 95°$F. The ambient air used to cool the water is at a dry bulb temperature of $T_{a,in} = 70°$F and a relative humidity of $\phi_{a,in} = 50$ %. The air flow rate is $\dot{V}_{a,in} = 10{,}000$ cfm. The performance of a cooling tower can be estimated in terms of a cooling tower effectiveness defined in Eq. (12-48). Assume that air exits the tower at a relative humidity of $\phi_{a,out} = 0.98$ at all effectiveness values.

a) Plot cooling tower water outlet temperature and the required rate of water replacement as a function of the cooling tower effectiveness for values between 0.2 and 0.8.
b) Plot the cooling tower Range and Approach parameters as a function of the cooling tower effectiveness for values between 0.2 and 0.8.
c) Determine the heat transfer rate that would result for the same effectiveness and air and water flow rates if a 'dry' counterflow heat exchanger were used instead of a cooling tower.

12.B-9 Figure 12.B-9 shows a schematic of an indirect/direct evaporative cooling system. Outdoor air at $T_1 = 26°C$ and $\phi_1 = 40\%$ relative humidity enters the equipment and is split into two flow streams. One stream passes through an evaporative cooler that lowers the temperature of the air to T_2. This cooled air is then used as the cold flow in a counter flow heat exchanger in order to reduce the temperature of the other ambient air stream to state 4. The counter flow heat exchanger effectiveness is $\varepsilon_{hx} = 0.85$. The air exiting the heat exchanger at state 4 then passes through a second evaporative cooler in order to emerge at the final conditioned state.

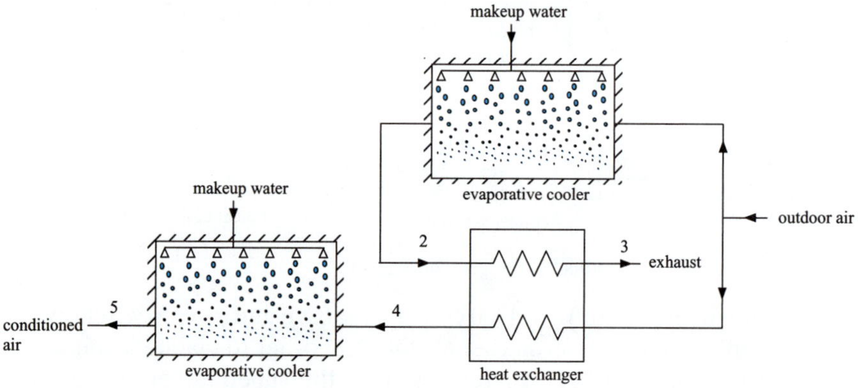

Figure 12.B-9: Indirect/direct evaporative cooling system.

a) Assume that the effectiveness of both evaporative coolers is $\varepsilon_{ec} = 0.75$ at any air flow rate. Calculate and plot the energy rate at which conditioned air at state 5 is provided (relative to state 1) as a function of the ratio of the mass flow rates at states 1 and 3.
b) If the effectiveness of one of the evaporative coolers could be increased to 0.85 while the other remained at 0.75, which evaporative cooler would you select to have the higher effectiveness?
c) Indicate the advantages and disadvantages of the system in Figure 12.B-9 compared to a traditional evaporative cooling system.

12.B-12 Figure 12.B-12 shows a novel air conditioning system. It is similar to a conventional air conditioning system, but it includes an additional heat exchanger. Air at $T_1 = 28°C$ and $\phi_1 = 45\%$ relative humidity (state 1) is cooled to state 2 (without condensation) in the heat exchanger. The air then passes through a conventional cooling coil where cooling and condensation both occur. The pinch point for the cooling coil occurs at state 3, where the air leaving the cooling coil is $\Delta T_{cc} = 10°C$ warmer than the evaporating refrigerant. The air exiting the cooling cool passes through the heat exchanger as the cold stream and then leaves through the reheat coil where heat is provided so that air exits at at state 5 with $T_5 = 18°C$ and $\phi_5 = 40\%$ relative humidity.

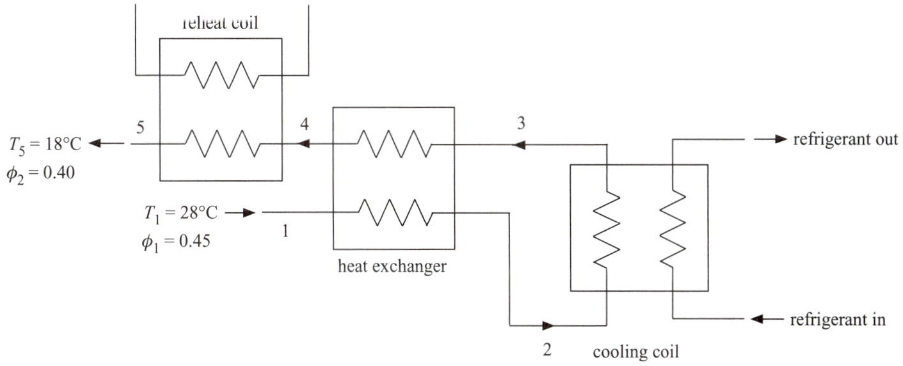

Figure 12.B-12: Schematic of a novel air-conditioning system.

The purpose of this problem is to determine the benefit of the additional heat exchanger.

a) Determine the maximum value of the heat exchanger effectiveness that will cool the air at state 2 without condensation occurring.
b) Plot the required rates of cooling and reheat per mass of dry air for effectiveness values between 0 and the value you determined in part (a).
c) Discuss the benefits (if any) and disadvantages of the proposed system.

12.B-13 An air-conditioning system for a supermarket uses two cooling coils, as shown in Figure 12.B-13. The cooling coil in the outdoor air duct cools and dehumidifies the ventilation air so that air exiting this coil is at $T_2 = 40°F$ with a humidity ratio of $\omega_2 = 0.005\ kg_v/kg_a$. The return air cooling coil operates at higher temperature and does not provide any condensation. The amount of cooling and the air flows for each cooling coil are controlled so as to provide supply

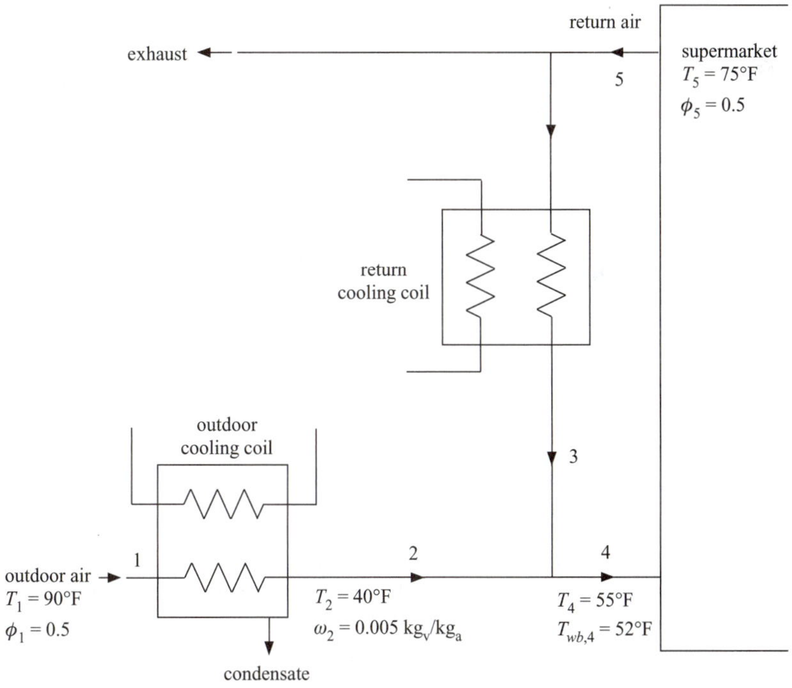

Figure 12.B-13: Supermarket cooling system with two cooling coils.

air at $T_4 = 55°F$ with a wet bulb temperature of $T_{wb,4} = 52°F$. The return air taken from the supermarket is maintained at $T_5 = 75°F$ and $\phi_5 = 50\%$ relative humidity. The total cooling load for the supermarket is $\dot{Q}_b = 25$ tons when the outdoor conditions are $T_1 = 95°F$ with $\phi_1 = 50\%$ relative humidity. Determine:
a) the dry air flow rate associated with the supply (state 1).
b) the dry air flow rates through each cooling coil.
c) the cooling load for each cooling coil.
d) the operating cost, in \$/hr, if the *COP* of the system used to provide cooling to the outdoor cooling coil is 2.8, the *COP* of the system used to provide cooling to the return air cooling coil is 3.6, and electricity is purchased at \$0.15/kwhr. (Neglect the cost to operate the fans.)

12.B-17 A cooling tower is shown in Figure 12.B-17.

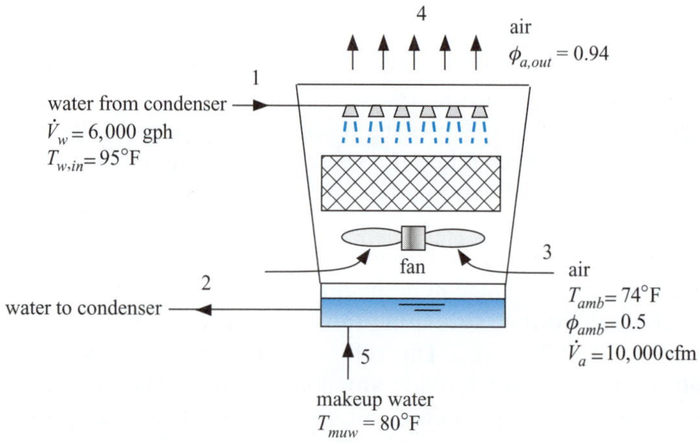

Figure 12.B-17: Cooling tower.

The cooling tower is designed to cool water entering with a flow rate $\dot{V}_w = 6{,}000$ gallons per hour and temperature $T_{w,in} = 95°F$ from the condenser of a large air conditioning system. Outdoor air at $T_{amb} = 74°F$ and $\phi_{amb} = 50\%$ relative humidity is forced through the tower at a rate of $\dot{V}_a = 10{,}000$ cfm. Makeup water is provided at $T_{muw} = 80°F$. The temperature of the air leaving the cooling tower at state 4 is $T_{a,out} = 89°F$ and the relative humidity of the air leaving at state 4 is $\phi_{a,out} = 0.94$.
a) Determine the rate of makeup water (in gal/hr) that must be provided to the cooling tower.
b) Determine the temperature of the water leaving the cooling tower at state 2.
c) Determine the rate of cooling that is provided to the condenser using the cooling tower (in ton).
d) Clearly indicate states 3 and 4 on a psychrometric chart. Note that one of the reasons that the cooling tower works so well is that the air passing through the tower is not only heated, it is also humidified. Therefore, both its temperature and humidity ratio are increased. What fraction of the total enthalpy change of the air is due to it temperature change only (i.e., what fraction of the enthalpy change of the air would be achieved if its humidity ratio did not change but it left at the same temperature, $T_{a,out}$)? This fraction is referred to as the sensible heat ratio of the device.

12.B-18 Figure 12.B-18 illustrates a building that is cooled using an earth tube system.

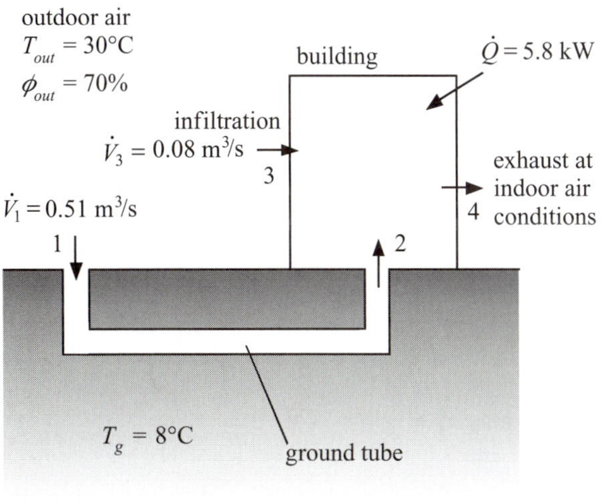

Figure 12.B-18: Earth tube system.

Outdoor air at $\phi_{out} = 70\%$ and $T_{out} = 30°\mathrm{C}$ is drawn into tubes at state 1 with volumetric flow rate $\dot{V}_1 = 0.51\ \mathrm{m^3/s}$. The tubes are buried in the ground. The ground temperature is $T_g = 8°\mathrm{C}$ and the air leaving the tubes and entering the building at state 2 has been cooled to within $\Delta T_g = 3°\mathrm{C}$ of the ground temperature. There is an infiltration of outdoor air directly into the building at state 3. The volumetric flow rate of outdoor air that directly enters the building is $\dot{V}_3 = 0.08\ \mathrm{m^3/s}$. The building experiences a heat transfer of $\dot{Q} = 5.8$ kW. Assume that no moisture is added to the air as it passes through the building. Indoor air leaves the building through the exhaust duct at state 4. Assume that the pressure of the air is atmospheric throughout the system.

a) Is there any condensation within the ground tube? If so then determine the volumetric flow rate of condensation draining into the ground in the ground tube (in gallons/day).
b) Determine the temperature and relative humidity of the indoor air.

C: Advanced Problems

12.C-1 Consider a summer day with outdoor conditions at $T_1 = 27°\mathrm{C}$ and $\phi_1 = 40\%$ relative humidity. An outdoor air flow is split into two (possibly unequal) streams, as shown in Figure 12.C-1. One stream is passed through an evaporative cooler with an effectiveness of $\varepsilon_{ec} = 1$. The two streams then enter a device that uses the temperature difference between the streams to produce power. The work producing device may be adiabatic (i.e., $\dot{Q}$ in Figure 12.C-1 may be zero) and the streams do not mix.

a) Assuming that power can be produced in this manner. Explain why this power production scheme does not violate the Kelvin-Planck statement of the Second Law: "*It is impossible for any device that operates continuously to receive heat from a single reservoir and produce a net amount of work*".
b) How would you define the efficiency of this work producing process? What is the upper bound for this efficiency?

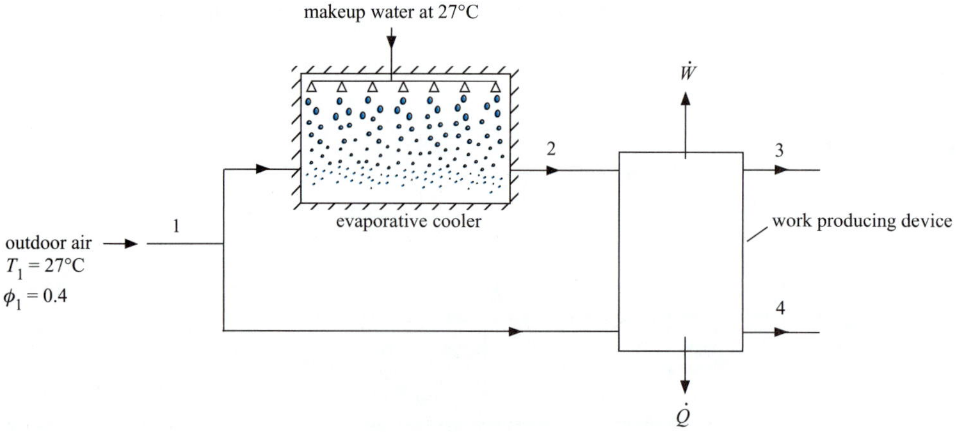

Figure 12.C-1: Power production from an evaporative cooler.

c) Estimate the maximum possible work per mass of processed air that can be produced if the two streams have the same flow rate.
d) What is the fraction of the entering air that should be passed through the evaporative cooler in order to result in the largest power?
e) What do you see as practical limitations that prevent this work production method from becoming commercially successful? What improvements could be made?

12.C-2 An auditorium is to be maintained at 25°C dry-bulb temperature and 40% relative humidity when the outdoor conditions are 32°C dry-bulb and 25°C wet bulb. The space will be conditioned by the system shown in Figure 12.C-2. Outdoor air enters the duct at state 5 with flow rate 0.25 m^3/s. The air exiting the cooling coil at state 7 is saturated at 7°C. Air is blown into the space at state 1 with flow rate 0.90 m^3/s. Pressure losses in the ducts and fan power are

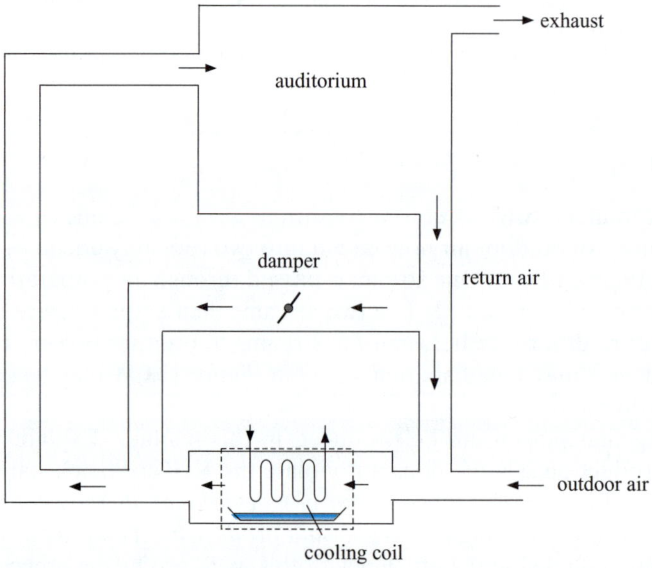

Figure 12.C-2: HVAC system for an auditorium.

negligible. As shown in Figure 12.C-2, the system provides a bypass damper that controls how much of the re-circulated air passes through the cooling coil. Unless otherwise specified, assume that the bypass damper is positioned such that the 50% of the re-circulated air at state 3 passes through the cooling coil.

a) Determine the temperature, humidity ratio, relative humidity and mass flow rate at all states.
b) Determine the rate of condensate removal in the cooling coil.
c) Determine the sensible heat ratio, defined as the ratio of the sensible to the total cooling load.
d) Determine the cost to condition this space assuming that the cooling coil is served by a cooling system that has a system COP of 3.5 for the conditions of part (a).
e) Vary the bypass damper position between 0 and 1, where 0 indicates that all of the air passes through the cooling coil and 1 indicates that only the outdoor air that enters at state 5 passes through the cooling coil. Prepare a plot of the total load and operation cost as a function of the bypass damper position. Explain the results you see in the plot.

12.C-4 In hot dry climates, air conditioning can be achieved simply by evaporative cooling. Water is sprayed into the dry air and subsequently evaporates with a resulting decrease in the temperature of the exit air stream. The evaporative cooler design is characterized by an effectiveness that is defined as:

$$\varepsilon = \frac{T_{in} - T_{out}}{T_{in} - T_{wb}}$$

where T_{in} is the entering air temperature, T_{out} is the exiting air temperature, and T_{wb} is the entering wet bulb temperature. In a particular case, outdoor air at $T_{in} = 35°\text{C}$ and $\phi_{in} = 20\%$ relative humidity enters an evaporative cooler at $\dot{V} = 6.0\ \text{m}^3/\text{sec}$ and is evaporatively cooled with liquid water that enters at $T_{muw} = 35°\text{C}$. Calculate and plot the following quantities as a function of the evaporator cooler effectiveness:

a) the outlet air temperature,
b) the rate of exergy destruction, and
c) the Second Law efficiency.
d) Is the Second Law efficiency of this process equal to one when the effectiveness of the evaporative cooler is equal to one? If not, explain why.
e) What do you see as the advantages and disadvantages of evaporative cooling?

13 Combustion

Combustion is the reaction of a fuel with oxygen. It is a subset of the more general subject of chemical equilibrium, which is considered in Chapter 14. Combustion is treated as a separate topic because combustion reactions tend to progress until the fuel is completely consumed. Combustion of various hydrocarbon fuels is the major source of useful energy (i.e., exergy) for transportation, electrical generation, space conditioning, water heating, and industrial processes.

13.1 Introduction to Combustion

Table 13-1 summarizes the sources of the energy that are consumed in the United States. The nation consumes about 102 quads (1.02×10^{17} Btu or 1.08×10^{11} GJ) of useful energy each year, which represents about 25% of the world energy consumption. The majority of this energy is provided by combustion. There are two major concerns with this current situation. First, nearly all combustible fuels contain carbon, which results in the generation of carbon dioxide during combustion, as shown in this chapter. Combustion in the U.S. alone resulted in the production of 5,990 million metric tons (5.9×10^{12} kg) of carbon dioxide in 2007 according to the U.S. Department of Energy (2009). Carbon dioxide in the atmosphere absorbs a portion of the thermal energy that is re-radiated from the earth and thereby contributes to global warming. Second, reserves of petroleum and natural gas, which account for over 60% of our useful energy supply, are finite. Although the extent of these reserves is a subject of debate, it is likely that they will become depleted in less than 50 years at the present rate of use. There are hundreds of years of coal reserves, but coal generates more carbon dioxide per unit energy than natural gas and coal also produces other contaminants when combusted. The sustainability of our energy supply and the associated problem of global warming are perhaps the most serious problems that the human race has ever faced.

Combustion necessarily involves a mixture of two or more substances. Two questions arise whenever a mixture is formed:

1) What is the chemical composition of the equilibrium state?
2) How fast will the reaction proceed to equilibrium?

The first question can be answered by thermodynamics. It is not generally true that a chemical reaction will continue to occur until all of the reactants are consumed. Instead, the equilibrium composition may consist of both reactants and products in amounts that depend upon the temperature and pressure as well as the composition of the reactants. The chemical composition of the equilibrium state can be determined by application of the First and Second Laws of Thermodynamics, as shown in Chapter 14. However, combustion reactions tend to proceed to near completion (i.e., until all of the fuel is consumed) and therefore the determination of the equilibrium state is not typically a concern for combustion reactions.

Table 13-1: Source of useful energy (i.e., exergy) consumed in the U.S. in 2007 from DOE/EIA-0383 (2009)

Exergy Consumption	Quads[1]
Liquid fuels and other petroleum	40.75
Natural gas	23.70
Coal	22.74
Nuclear power	8.41
Hydropower	2.46
Biomass	2.62
Other renewable energy	0.97
Other	0.23
Total	**101.89**

[1] 1 quad = 10^{15} Btu = 1.05×10^{9} GJ

The study of the rate of chemical reactions is called chemical kinetics and is not considered in this book. However, chemical kinetics is generally not a concern for combustion reactions because, once initiated, they tend to reach a chemical equilibrium state very rapidly.

We can examine the simple combustion reaction of hydrogen (a fuel) with oxygen. The reaction of these chemicals forms water according to:

$$H_2 + \frac{1}{2}O_2 \rightarrow H_2O \tag{13-1}$$

Consider an experiment in which hydrogen and oxygen in the proportions indicated by the reactants in Eq. (13-1) are mixed together at 25°C and 1 atm. A chemical equilibrium analysis for this situation could be conducted following the methods presented in Chapter 14 and it would show that the equilibrium state is nearly all water with only trace amounts of hydrogen and oxygen. However, when these chemicals are mixed at 25°C and 1 atm, you would not observe the formation of any water because the rate at which the reaction occurs at 25°C is imperceptibly small. The rate of a chemical reaction can be altered by introducing a catalyst, which is an intermediate substance that participates in the reaction mechanism, but is not consumed. A small piece of platinum metal would serve as catalyst for the reaction shown in Eq. (13-1). Introduction of the platinum into the mixture would cause it to react to equilibrium in an explosively rapid manner. A spark would also initiate the reaction and provide the same result. Once started, the reaction will precede until essentially all of the hydrogen is gone with an energy release of about 240,000 kJ per kmol of hydrogen. Combustion reactions, such as the one shown in Eq. (13-1), proceed nearly to completion very rapidly.

The fundamental balances of mass, energy, and entropy introduced in earlier chapters for non-reacting systems are applied in this chapter in order to analyze combustion reactions. Mass balances for reacting systems are complicated by the fact that atoms can be rearranged to form different molecules as the reaction proceeds. That is, the substances that are present at the end of the reaction will not be the same as those that were present initially. Mass balances are ensured by requiring that the molar amounts of each of the elements remain constant during the reaction. Atoms must be conserved even though the mass of each substance involved in the reaction may change.

Energy must also be conserved in a reacting system. However, energy balances are complicated by the fact that chemical bonds between atoms are associated with the storage of energy. Therefore, the breaking and forming of chemical bonds that occurs during the chemical reaction must be specifically considered in the energy balances. Up to this point, the reference state used to define the internal energy of a specific substance has not mattered because energy balances have only involved differences in the energy of a particular substance at two states. In combustion problems, it will be necessary to establish a consistent reference state that will allow the energy of different substances to be directly compared.

13.2 Balancing Chemical Reactions

Combustion reactions are written with the reactants appearing on the left side of the reaction symbol (→) and the products on the right, as indicated in Eq. (13-2):

$$\text{Reactants} \rightarrow \text{Products} \tag{13-2}$$

A chemical equation is a statement of conservation of mass, expressed on a molar basis. The reactants consist of fuel and an oxidizer (usually oxygen). The fuel in a combustion reaction can be gaseous (e.g., natural gas or propane), liquid (e.g., gasoline or ethane) or solid (e.g., coal or wood). Nearly all combustible fuels are hydrocarbons that may also include some oxygen; therefore, the fuel can in general be represented by the chemical formula $C_xH_yO_z$ where x, y, and z are related to the relative amounts of carbon, hydrogen, and oxygen in the fuel. The products associated with the combustion of a hydrocarbon fuel include carbon dioxide and water as well as products of incomplete combustion.

The first task for any combustion problem is to balance the reaction, i.e., to ensure that the number of atoms of each element is the same for the reactants and the products. Balancing a reaction ensures that a mass balance will be satisfied. Equation (13-3) provides the chemical reaction for the combustion of methane (CH_4) with oxygen (methane is the fuel and oxygen is the oxidizer). The products of complete combustion of a hydrocarbon are carbon dioxide and water; therefore, we can place these chemical species on the right of the reaction equation:

$$a\,CH_4 + b\,O_2 \rightarrow c\,CO_2 + d\,H_2O \tag{13-3}$$

Balancing the reaction corresponds to selecting the coefficients a, b, c, and d in Eq. (13-3). The process begins with the specification of a *basis*. A basis is a convenient amount of reactant (or product) to use when writing the chemical reaction. Here, we will use one mole of CH_4 as the basis; therefore, the coefficient a in Eq. (13-3) is equal to unity:

$$CH_4 + b\,O_2 \rightarrow c\,CO_2 + d\,H_2O \tag{13-4}$$

The *stoichiometric coefficients* must be determined; these are the values of the coefficients in Eq. (13-4). These coefficients are determined so that the number of moles of each element (i.e., each atom) is the same on both sides of the chemical equation. All that is needed to balance the reaction is to set up algebraic equations that enforce the conservation of each element. There are three stoichiometric coefficients and three elements (C, H, and O) in Eq. (13-4). Therefore, the balances are simple in this case. One mole of carbon atoms appears in the reactants in Eq. (13-4). In order for the carbon atoms to balance, there can only be one mole of carbon dioxide in the products; therefore $c = 1$. Four moles of hydrogen atoms appear in the reactants. In order for hydrogen atoms to balance, there can only be two moles of water (each with two hydrogen atoms) in the products; therefore, $d = 2$:

$$CH_4 + b\,O_2 \rightarrow CO_2 + 2\,H_2O \tag{13-5}$$

There are four moles of oxygen atoms in the products in Eq. (13-5). In order for oxygen atoms to balance, there can only be two moles of oxygen (each oxygen molecule has two oxygen atoms) in the reactants; therefore, $b = 2$:

$$CH_4 + 2\,O_2 \rightarrow CO_2 + 2\,H_2O \tag{13-6}$$

Equation (13-6) is a properly balanced reaction for the complete combustion of methane. A properly balanced reaction will inherently conserve mass because atoms are conserved. For example, the mass of the reactants in Eq. (13-6) can be computed according to:

$$m_{reactant} = n_{CH_4}\,MW_{CH_4} + n_{O_2}\,MW_{O_2} = \frac{1\,\text{kmol}}{}\left|\frac{16\,\text{kg}}{\text{kmol}}\right. + \frac{2\,\text{kmol}}{}\left|\frac{32\,\text{kg}}{\text{kmol}}\right. = 80\,\text{kg} \tag{13-7}$$

and the mass of the products in Eq. (13-6) can be computed according to:

$$m_{product} = n_{CO_2}\,MW_{CO_2} + n_{H_2O}\,MW_{H_2O} = \frac{1\,\text{kmol}}{}\left|\frac{44\,\text{kg}}{\text{kmol}}\right. + \frac{2\,\text{kmol}}{}\left|\frac{18\,\text{kg}}{\text{kmol}}\right. = 80\,\text{kg} \tag{13-8}$$

Equation (13-6) is called a *stoichiometric* or *theoretical* reaction because the exact amount of oxygen that is needed to completely oxidize the fuel is provided. Therefore, both the oxygen and the fuel are entirely consumed by the reaction and there is no excess oxygen or unburned fuel appearing in the products. The stoichiometric coefficient for oxygen in Eq. (13-6) is called the *stoichiometric* or *theoretical amount* of oxygen.

13.2.1 Air as an Oxidizer

Pure oxygen is not naturally available and therefore the oxygen needed for combustion reactions is normally provided from atmospheric air. Dry air is a mixture of many gases, but the primary constituents of dry air are nitrogen and oxygen. Although nitrogen may react to a small extent with oxygen at high temperatures, we will assume that nitrogen is inert in this chapter. To simplify the chemical description of air, the small amount of argon, carbon dioxide and other minor inert substances present in air are typically lumped together with the nitrogen. With this assumption, dry air can be modeled as a mixture of 79% (by mole) nitrogen and 21% (by mole) oxygen. One mole of air consists of 0.79 moles of nitrogen and 0.21 moles of oxygen. Therefore one mole of air can be represented in a chemical reaction as (0.21 O_2 + 0.79 N_2). The balanced complete combustion reaction of 1 mole of methane with a stoichiometric amount of air is given by:

$$CH_4 + 9.524\,(0.21\,O_2 + 0.79\,N_2) \rightarrow CO_2 + 2\,H_2O + 7.524\,N_2 \tag{13-9}$$

Equation (13-9) indicates that 9.524 moles of air are required per mole of methane. Note that 9.524 moles of dry air provide exactly 2 moles of oxygen ($9.524 \times 0.21 = 2$), as required by the balanced stoichiometric reaction in Eq. (13-6). The additional nitrogen that is carried along with the air ($9.524 \times 0.79 = 7.524$ moles of nitrogen per mole of methane) does not participate in the combustion reaction and therefore appears in the products in Eq. (13-9). Equation (13-9) may appear to be more complicated than is necessary since the amount of nitrogen that appears in the reactants is the same as the amount of nitrogen present in the products. Mathematically, nitrogen could be eliminated completely from the chemical equation. However, as shown in Section 13.3, the nitrogen in the products will generally be at a different temperature than the nitrogen in the reactants and therefore nitrogen must be included in energy and entropy balances on the combustion process. The procedure of specifically showing all of the reactants and all of the products in the chemical equation, regardless of whether they are inert, will be necessary for combustion analyses.

13.2.2 Methods for Quantifying Excess Air

Equation (13-9) is the stoichiometric or theoretical reaction of methane with air because no excess oxygen is provided; the amount of oxygen available in the reactants is exactly the right amount required for complete combustion. However, it is often the case that more air is provided to a combustion reaction than is needed to completely oxidize the fuel. The excess air may be used to control the temperature of the products and also to ensure that the fuel is entirely oxidized. For example, if 15 moles of air per mole of methane were provided in the reactants rather than the 9.524 moles of air that is required by the stoichiometric reaction, Eq. (13-9), then the balanced equation for complete combustion of methane would be:

$$CH_4 + 15\,(0.21\,O_2 + 0.79\,N_2) \rightarrow CO_2 + 2\,H_2O + 1.15\,O_2 + 11.85\,N_2 \quad (13\text{-}10)$$

Notice that the excess oxygen provided to the reaction shows up in the products of Eq. (13-10). The excess air that is provided to a combustion reaction can be described by the % stoichiometric air, the % excess air, or the air fuel ratio. The *% stoichiometric air* (also called the *% theoretical air*) is the ratio of the molar amount of air that is actually provided to the reaction to the amount of air that is needed for a stoichiometric reaction, expressed as a percentage. For example, Eq. (13-10) is the chemical equation for complete combustion of methane with 157.5% stoichiometric air because 15 moles of air were provided and 9.524 moles are required for complete combustion (15/9.524 = 157.5%). The *% excess air* is the % stoichiometric air minus 100%; therefore, Eq. (13-10) is written for 57.5% excess air.

The *air fuel ratio* is normally expressed on a mass, rather than on a molar, basis. It is defined as the ratio of the mass of air in the reactants to the mass of fuel in the reactants. The mass of air can be determined using the equivalent molar mass of dry air, $MW_a =$ 28.96 kg/kmol, as shown in Section 11.1. The air fuel ratio for the reaction given by Eq. (13-10) is calculated according to:

$$AF = \frac{m_a}{m_f} = \frac{n_a\,MW_a}{n_{CH_4}\,MW_{CH_4}} = \frac{15\,\text{kmol}}{} \left| \frac{28.96\,\text{kg}}{\text{kmol}} \right| \frac{}{1\,\text{kmol}} \left| \frac{\text{kmol}}{16\,\text{kg}} \right. = 27.15 \quad (13\text{-}11)$$

Air fuel ratios for hydrocarbon combustion reactions usually range between 15 and 50. A low air fuel ratio corresponds to a *rich* combustion condition for which the amount of air provided is close to the stoichiometric amount (i.e., the mixture is rich in fuel). Spark ignition automobile engines tend to run rich. A high air fuel ratio corresponds to a lean condition with a large amount of excess air (i.e., the mixture is lean in fuel). Gas turbines tend to run lean in order to keep the temperature of the combustion products below the maximum operating temperature of the turbine.

Other methods of expressing the amount of excess air may also be used. For example, the fuel air ratio is the inverse of the air fuel ratio. The equivalence ratio, which is used in the automotive industry, is the ratio of the actual fuel air ratio to the fuel air ratio at stoichiometric conditions. The many ways used to express the amount of air provided in a combustion reaction are an indication of its importance.

13.2.3 Psychrometric Issues

Water is formed whenever a hydrocarbon fuel is reacted with oxygen. Water may also enter with the reactants; both the fuel and the air may contain some amount of water. Condensation will occur if the combustion products are cooled below the *dew point* temperature of the gas mixture.

The method for determining the dew point temperature of a combustion gas mixture follows the principles that are discussed in Chapter 12 and used to determine the dew point of mixtures of dry air and water vapor. However, the psychrometric chart and the corresponding DewPoint function in EES cannot be used for this purpose because the mixture that results from a combustion process has a composition that differs from air. Assuming that the combustion mixture obeys the ideal gas law, water will condense when the temperature of the products reaches the saturation temperature corresponding to the partial pressure of the water vapor in the mixture:

$$P_v = y_v\, P \tag{13-12}$$

where y_v is the mole fraction of water vapor and P is the total pressure of the products. Assuming that no condensation has occurred, the mole fraction of water vapor is:

$$y_v = \frac{n_{H_2O}}{n_{dry} + n_{H_2O}} \tag{13-13}$$

where n_{H_2O} is number of moles of water in the products and n_{dry} is the sum of the number of moles of all of the other gases in the products.

Consider the reaction in Eq. (13-10). Assuming that no condensation has occurred, the water vapor mole fraction is given by:

$$y_v = \frac{n_{H_2O}}{n_{dry} + n_{H_2O}} = \frac{2}{1 + 1.15 + 11.85 + 2} = 0.125 \tag{13-14}$$

Assuming that the total pressure of the products is 1 atm (101.3 kPa), the partial pressure of the water vapor in the products is:

$$P_v = y_v\, P = 0.125\,(101.3\,\text{kPa}) = 12.66\,\text{kPa} \tag{13-15}$$

The saturation temperature of water at 12.66 kPa is 50.48°C; therefore, the dew point temperature of the combustion products in Eq. (13-10) is 50.48°C. Liquid water will appear if the combustion products are cooled below 50.48°C.

It is sometimes necessary to determine how much condensate will be formed (relative to the chosen basis of the reaction) if the combustion products are cooled below the dew point temperature. If condensation occurs, then the products will be saturated and therefore the partial pressure of the remaining water vapor will be equal to the saturation pressure. The mole fraction of the water that remains in a gas phase is then given by:

$$y_v = \frac{P_v}{P} = \frac{P_{sat}(T)}{P} = \frac{n_{H_2O,g}}{n_{dry} + n_{H_2O,g}} \tag{13-16}$$

where $P_{sat}(T)$ is the saturation pressure of water at temperature T and $n_{H_2O,g}$ is the number of moles of water vapor that remain in gas phase. The number of moles of water that have condensed, $n_{H_2O,f}$, can then be determined from the difference between the total number of moles of water in the products, n_{H_2O}, and $n_{H_2O,g}$.

$$n_{H_2O,f} = n_{H_2O} - n_{H_2O,g} \tag{13-17}$$

Note that the moles of condensate computed using Eq. (13-17) are relative to the basis chosen for the reaction.

If, for example, the products of the combustion reaction shown in Eq. (13-10) are cooled to 25°C then the saturation pressure of the water vapor is $P_{sat} = 3.169$ kPa and the mole fraction of water vapor must be given by:

$$y_v = \frac{P_{sat}(T)}{P} = \frac{3.169\,\text{kPa}}{101.3\,\text{kPa}} = 0.03128 \tag{13-18}$$

The number of moles of water vapor can be obtained by rearranging Eq. (13-16):

$$y_v\, n_{dry} = n_{H_2O,g}\,(1 - y_v) \tag{13-19}$$

or

$$n_{H_2O,g} = \frac{y_v}{(1 - y_v)} n_{dry} = \frac{0.03128}{(1 - 0.03128)}(1 + 1.15 + 11.85)\,\text{kmol} = 0.4521\,\text{kmol} \tag{13-20}$$

Therefore, the number of moles of water that have condensed is given by:

$$n_{H_2O,f} = n_{H_2O} - n_{H_2O,g} = 2\,\text{kmol} - 0.4521\,\text{kmol} = 1.548\,\text{kmol} \tag{13-21}$$

These calculations show that 1.548 kmol of water will condense per kmol of fuel that is burned. The mass of condensate per mass of fuel can be computed according to:

$$\frac{m_{H_2O,f}}{m_f} = \frac{n_{H_2O,f}\, MW_{H_2O}}{n_{CH_4}\, MW_{CH_4}} = \frac{1.548\,\text{kmol}}{}\left|\frac{18\,\text{kg}}{\text{kmol}}\right|\frac{}{1\,\text{kmol}}\left|\frac{\text{kmol}}{16\,\text{kg}}\right. = 1.743 \tag{13-22}$$

Therefore, 1.743 kg of water will condense per kg of fuel that is burned.

EXAMPLE 13.2-1: COMBUSTION OF A PRODUCER GAS

The gas that is driven off when low-grade coal is burned with insufficient air for complete combustion is known as producer gas. A particular producer gas has been analyzed and it has the volumetric (i.e., molar) composition summarized in Table 1.

Table 1: Molar analysis of a producer gas

Constituent	Mole fraction
Methane, CH_4	0.038
Ethane, C_2H_6	0.001
Carbon dioxide, CO_2	0.048
Hydrogen, H_2	0.117
Oxygen, O_2	0.006
Carbon monoxide, CO	0.232
Nitrogen, N_2	balance

The producer gas is combusted with 150% stoichiometric dry air.

a) Determine the air fuel ratio for the combustion reaction.

The known information is entered in EES.

```
$UnitSystem SI Mass Radian J K Pa
y_CH4=0.038 [-]                "mole fraction of methane"
y_C2H6=0.001 [-]               "mole fraction of ethane"
y_CO2=0.048 [-]                "mole fraction of carbon dioxide"
y_H2=0.117 [-]                 "mole fraction of hydrogen"
y_O2=0.006 [-]                 "mole fraction of oxygen"
y_CO=0.232 [-]                 "mole fraction of carbon monoxide"
StoichiometricAir=1.5 [-]      "stoichiometric air"
```

The first step is to choose a basis. A convenient basis for this problem is one mole of producer gas. With this basis, the stoichiometric reaction can be written as:

$$\left(y_{CH_4}CH_4 + y_{C_2H_6}C_2H_6 + y_{CO_2}CO_2 + y_{H_2}H_2 + y_{O_2}O_2 + y_{CO}CO + y_{N_2}N_2\right) + a_s\,(0.21\,O_2 + 0.79N_2) \rightarrow b_s\,CO_2 + c_s\,H_2O + d_s\,N_2 \quad (1)$$

where the coefficients a_s, b_s, c_s, and d_s are the coefficients for the stoichiometric reaction. The parameters y_{CH_4}, $y_{C_2H_6}$, etc. indicate the mole fractions of the components in the producer gas, as listed in Table 1. The mole fraction of nitrogen in the producer gas is obtained from:

$$y_{N_2} = 1 - y_{CH_4} - y_{C_2H_6} - y_{CO_2} - y_{H_2} - y_{O_2} - y_{CO}$$

```
y_N2=1-y_CH4-y_C2H6-y_CO2-y_H2-y_O2-y_CO      "mole fraction of nitrogen"
```

The stoichiometric reaction, Eq. (1), is balanced by conserving the individual atoms. A balance of carbon atoms provides:

$$y_{CH_4} + 2\,y_{C_2H_6} + y_{CO_2} + y_{CO} = b_s$$

A balance of hydrogen atoms provides:

$$4\,y_{CH_4} + 6\,y_{C_2H_6} + 2\,y_{H_2} = 2\,c_s$$

A balance of oxygen atoms provides:

$$2\,y_{CO_2} + 2\,y_{O_2} + y_{CO} + 2\,(0.21)\;a_s = 2\,b_s + c_s$$

A balance of nitrogen atoms provides:

$$2\,y_{N_2} + 2\,(0.79)\,a_s = 2\,d_s$$

```
"Balance stoichiometric reaction"
y_CH4+2*y_C2H6+y_CO2+y_CO=b_s                "carbon atom balance"
4*y_CH4+6*y_C2H6+2*y_H2=2*c_s                "hydrogen atom balance"
2*y_CO2+2*y_O2+y_CO+a_s*0.21*2=2*b_s+c_s     "oxygen atom balance"
2*y_N2+a_s*0.79*2=2*d_s                      "nitrogen atom balance"
```

Solving these equations results in $a_s = 1.181$, $b_s = 0.32$, $c_s = 0.196$, and $d_s = 1.491$. Therefore, the balanced stoichiometric reaction, Eq. (1), is:

$$\left(y_{CH_4}CH_4 + y_{C_2H_6}C_2H_6 + y_{CO_2}CO_2 + y_{H_2}H_2 + y_{O_2}O_2 + y_{CO}CO + y_{N_2}N_2\right) + 1.181\,(0.21\,O_2 + 0.79N_2) \rightarrow 0.32\,CO_2 + 0.196\,H_2O + 1.491\,N_2 \quad (2)$$

The actual reaction is carried out with 150% stoichiometric air. Therefore the actual number of moles of air is calculated according to:

$$a = StoichiometricAir\,a_s$$

where $StoichiometricAir = 1.5$ (150%). The actual reaction can be written as:

$$\begin{aligned}(y_{CH_4}CH_4 + y_{C_2H_6}C_2H_6 + y_{CO_2}CO_2 + y_{H_2}H_2 + y_{O_2}O_2 + y_{CO}CO + y_{N_2}N_2)\\ +a(0.21\,O_2 + 0.79N_2) \rightarrow b\,CO_2 + c\,H_2O + d\,N_2 + e\,O_2\end{aligned} \quad (3)$$

The actual reaction, Eq. (3), is also balanced by conserving the individual atoms. A balance of carbon atoms provides:

$$y_{CH_4} + 2\,y_{C_2H_6} + y_{CO_2} + y_{CO} = b$$

A balance of hydrogen atoms provides:

$$4\,y_{CH_4} + 6\,y_{C_2H_6} + 2\,y_{H_2} = 2\,c$$

A balance of oxygen atoms provides:

$$2\,y_{CO_2} + 2\,y_{O_2} + y_{CO} + 2\,(0.21)\,a = 2\,b + c + 2\,e$$

A balance of nitrogen atoms provides:

$$2\,y_{N_2} + 2\,(0.79)\,a = 2\,d$$

```
"Balance actual reaction"
a=StoichiometricAir*a_s                         "air"
y_CH4+2*y_C2H6+y_CO2+y_CO=b                     "carbon atom balance"
4*y_CH4+6*y_C2H6+2*y_H2=2*c                     "hydrogen atom balance"
2*y_CO2+2*y_O2+y_CO+a*0.21*2=2*b+c+2*e          "oxygen atom balance"
2*y_N2+a*0.79*2=2*d                             "nitrogen atom balance"
```

which results in $a = 1.771$, $b = 0.32$, $c = 0.196$, $d = 1.957$, and $e = 0.124$. Therefore, the balanced reaction of the producer gas with 150% stoichiometric air is:

$$\begin{aligned}(y_{CH_4}CH_4 + y_{C_2H_6}C_2H_6 + y_{CO_2}CO_2 + y_{H_2}H_2 + y_{O_2}O_2 + y_{CO}CO + y_{N_2}N_2)\\ +1.771\,(0.21\,O_2 + 0.79\,N_2) \rightarrow 0.32\,CO_2 + 0.196\,H_2O + 1.957\,N_2 + 0.124\,O_2\end{aligned} \quad (4)$$

The molar mass of the producer gas is obtained according to:

$$\begin{aligned}MW_f = y_{CH_4}MW_{CH_4} + y_{C_2H_6}MW_{C_2H_6} + y_{CO_2}MW_{CO_2} + y_{H_2}MW_{H_2} + y_{O_2}MW_{O_2}\\ + y_{CO}MW_{CO} + y_{N_2}MW_{N_2}\end{aligned} \quad (5)$$

The air fuel ratio is found using Eq. (13-11):

$$AF = \frac{a\,MW_a}{MW_f}$$

```
MW_f=y_CH4*MolarMass(CH4)+y_C2H6*MolarMass(C2H6)+y_CO2*MolarMass(CO2)&
   +y_H2*MolarMass(H2)+y_O2*MolarMass(O2)+y_CO*MolarMass(CO)&
   +y_N2*MolarMass(N2)                    "molar mass of fuel"
AF=a*MolarMass(Air)/MW_f                  "air fuel ratio"
```

which leads to $AF = 2.027$ kg of air per kg of fuel. This air fuel ratio is low, relative to the values expected for complete combustion of a hydrocarbon fuel, because oxygen is provided in the producer gas and also because much of the producer gas consists of inert or partially oxidized constituents.

b) Determine the dew point of the products of the reaction at 1 atm.

The mole fraction of water vapor in the products, assuming that no condensation occurs, is given by Eq. (13-13):

$$y_v = \frac{c}{b + d + e + c}$$

where b, c, d, and e are the coefficients in Eq. (3). The partial pressure of water vapor is:

$$P_v = y_v \, P$$

The dew point temperature (T_{dp}) is the saturation temperature of water at the partial pressure, obtained using the T_sat function in EES.

```
P=1 [atm]*convert(atm,Pa)          "pressure"
y_v=c/(c+b+d+e)                    "mole fraction of water vapor"
P_v=y_v*P                          "partial pressure of water vapor"
T_dp=T_sat(Water,P=P_v)            "dew point temperature"
T_dp_C=converttemp(K,C,T_dp)       "in C"
```

Solving provides $T_{dp} = 313.8$ K (40.66°C).

c) Determine the mass of condensate per unit mass of producer gas that forms when the products are cooled to $T_2 = 25°$C.

Since the products are cooled below their dew point temperature, determined in part (b), condensation will occur. The products are saturated and therefore the partial pressure of the water will be the saturation pressure at T_2 ($P_{sat,2}$), obtained using the P_sat function in EES. The mole fraction of water vapor in the products is given by:

$$y_{v,2} = \frac{P_{sat,2}}{P}$$

The number of moles of water vapor per mole of producer gas is given by:

$$n_{H_2O,g} = \frac{y_{v,2}}{(1 - y_{v,2})} (b + d + e)$$

The number of moles of condensate per mole of producer gas is given by:

$$n_{H_2O,f} = n_{H_2O} - n_{H_2O,g}$$

The ratio of the mass of condensate to the mass of producer gas is computed according to:

$$r = \frac{n_{H_2O,f} \, MW_{H_2O}}{MW_f}$$

```
T_2=converttemp(C,K,25 [C])               "temperature"
P_sat_2=P_sat(Water,T=T_2)                "saturation pressure"
y_v_2=P_sat_2/P                           "mole fraction of water vapor"
n_H2O_g=y_v_2*(b+d+e)/(1-y_v_2)           "number of moles of water vapor per mole of fuel"
n_H2O_f=c-n_H2O_g                         "number of moles of condensate per mole of fuel"
r=n_H2O_f*MolarMass(Water)/MW_f           "mass of condensate per mass of fuel"
```

which results in $r = 0.0843$ kg of condensate per kg of producer gas.

d) Repeat parts (b) and (c) assuming that the air used for combustion is at 25°C and 1 atm with a relative humidity of 50%.

The inputs and the equations used to balance the stoichiometric reaction, Eq. (1), in order to obtain the coefficients a_s, b_s, c_s, and d_s remain the same as in part (a).

```
$UnitSystem SI Mass Radian J K Pa
y_CH4=0.038 [-]                                  "mole fraction of methane"
y_C2H6=0.001 [-]                                 "mole fraction of ethane"
y_CO2=0.048 [-]                                  "mole fraction of carbon dioxide"
y_H2=0.117 [-]                                   "mole fraction of hydrogen"
y_O2=0.006 [-]                                   "mole fraction of oxygen"
y_CO=0.232 [-]                                   "mole fraction of carbon monoxide"
StoichiometricAir=1.5 [-]                        "stoichiometric air"
T=converttemp(C,K,25 [C])                        "temperature"
P=1 [atm]*convert(atm,Pa)                        "pressure"
phi=0.5 [-]                                      "relative humidity"
y_N2=1-y_CH4-y_C2H6-y_CO2-y_H2-y_O2-y_CO         "mole fraction of nitrogen"
"Balance stoichiometric reaction"
y_CH4+2*y_C2H6+y_CO2+y_CO=b_s                    "carbon atom balance"
4*y_CH4+6*y_C2H6+2*y_H2=2*c_s                    "hydrogen atom balance"
2*y_CO2+2*y_O2+y_CO+a_s*0.21*2=2*b_s+c_s         "oxygen atom balance"
2*y_N2+a_s*0.79*2=2*d_s                          "nitrogen atom balance"
```

The number of moles of air is given by:

$$a = StoichiometricAir\, a_s$$

where $StoichiometricAir = 1.5$. The actual reaction includes some water on the reactants side that is associated with the moist air:

$$\left(y_{CH_4}CH_4 + y_{C_2H_6}C_2H_6 + y_{CO_2}CO_2 + y_{H_2}H_2 + y_{O_2}O_2 + y_{CO}CO + y_{N_2}N_2\right) + a\,(0.21\,O_2 + 0.79N_2) + f\,H_2O \rightarrow b\,CO_2 + c\,H_2O + d\,N_2 + e\,O_2 \tag{6}$$

The coefficient f in Eq. (6) depends on the ratio of the number of moles of water vapor in the air to the number of moles of dry air. This quantity is related to the humidity ratio introduced in Chapter 12. The humidity ratio of combustion air (ω) is obtained using the humrat function in EES. The humidity ratio is defined as:

$$\omega = \frac{m_v}{m_a} \tag{7}$$

Equation (7) is written in terms of the number of moles of water and the number of moles of air:

$$\omega = \frac{n_v\, MW_{H_2O}}{n_a\, MW_a} \tag{8}$$

According to Eq. (6), the number of moles of dry air is a and the number of moles of water vapor in the reactants is f. Therefore, Eq. (8) can be written as:

$$\omega = \frac{f\, MW_{H_2O}}{a\, MW_a} \tag{9}$$

Solving Eq. (9) for the coefficient f provides:

$$f = \frac{a\, MW_a\, \omega}{MW_{H_2O}}$$

```
"Balance actual reaction"
a=StoichiometricAir*a_s                              "air"
omega=humrat(AirH2O,T=T,P=P,R=phi)                   "humidity ratio"
f=a*MolarMass(Air)*omega/MolarMass(Water)            "number of moles of water in reactants per mole of fuel"
```

The actual reaction with moist air, Eq. (6), is balanced by conserving the individual atoms. A balance of carbon atoms provides:

$$y_{CH_4} + 2\, y_{C_2H_6} + y_{CO_2} + y_{CO} = b$$

A balance of hydrogen atoms provides:

$$4\, y_{CH_4} + 6\, y_{C_2H_6} + 2\, y_{H_2} + 2\, f = 2\, c$$

A balance of oxygen atoms provides:

$$2\, y_{CO_2} + 2\, y_{O_2} + y_{CO} + 2\,(0.21)\; a + f = 2\, b + c + 2\, e$$

A balance of nitrogen atoms provides:

$$2\, y_{N_2} + 2\,(0.79)\, a = 2\, d$$

```
y_CH4+2*y_C2H6+y_CO2+y_CO=b                          "carbon atom balance"
4*y_CH4+6*y_C2H6+2*y_H2+2*f=2*c                      "hydrogen atom balance"
2*y_CO2+2*y_O2+y_CO+a*0.21*2+f=2*b+c+2*e             "oxygen atom balance"
2*y_N2+a*0.79*2=2*d                                  "nitrogen atom balance"
```

The molar mass of the producer gas is obtained from Eq. (5).

```
MW_f=y_CH4*MolarMass(CH4)+y_C2H6*MolarMass(C2H6)+y_CO2*MolarMass(CO2)&
   +y_H2*MolarMass(H2)+y_O2*MolarMass(O2)+y_CO*MolarMass(CO)&
   +y_N2*MolarMass(N2)                               "molar mass of fuel"
```

The process of determining the dew point proceeds as discussed in part (b).

```
y_v=c/(c+b+d+e)                                      "mole fraction of water vapor"
P_v=y_v*P                                            "partial pressure of water vapor"
T_dp=T_sat(Water,P=P_v)                              "dew point temperature"
T_dp_C=converttemp(K,C,T_dp)                         "in C"
```

Solving these equations shows that $T_{dp} = 316.1$ K (42.98°C); the dew point temperature has increased by approximately 2.3°C due to the presence of water vapor in the air. The process of determining the mass of condensate per mass of producer gas proceeds as discussed in part (c).

```
T_2=converttemp(C,K,25 [C])                          "temperature"
P_sat_2=P_sat(Water,T=T_2)                           "saturation pressure"
y_v_2=P_sat_2/P                                      "mole fraction of water vapor"
n_H2O_g=y_v_2*(b+d+e)/(1-y_v_2)                      "number of moles of water vapor per mole of fuel"
n_H2O_f=c-n_H2O_g                                    "number of moles of condensate per mole of fuel"
r=n_H2O_f*MolarMass(Water)/MW_f                      "mass of condensate per mass of fuel"
```

Solving these equations results in $r = 0.1044$ kg of condensate per kg of producer gas (increased from 0.0843 kg/kg for dry air).

13.3 Energy Considerations

This section discusses the application of the First Law of Thermodynamics for systems involving combustion reactions.

13.3.1 Enthalpy of Formation

An energy balance on a system that is undergoing a chemical reaction is conducted in same manner as for a non-reacting system. The first step is to choose the system. Then, the general energy balance presented in Section 4.2 as Eq. (4-22) can be simplified as appropriate for the situation.

$$\sum_{i=1}^{\#inlets} \dot{m}_{in,i}\left(h_{in,i}+\frac{1}{2}\tilde{V}_{in,i}^2+g\,z_{in,i}\right)+\sum_{i=1}^{\#heat\ terms}\dot{Q}_i$$
$$=\sum_{i=1}^{\#outlets}\dot{m}_{out,i}\left(h_{out,i}+\frac{1}{2}\tilde{V}_{out,i}^2+g\,z_{out,i}\right)+\sum_{i=1}^{\#work\ terms}\dot{W}_i+\frac{dU}{dt}+\frac{dKE}{dt}+\frac{dPE}{dt} \quad (13\text{-}23)$$

In order to apply Eq. (13-23) to a problem involving chemical reactions it is necessary to consider the energy stored in the chemical bonds of the reactants and products. Consider the steady-state reaction of carbon with a stoichiometric amount of oxygen. The carbon and oxygen enter a reactor at $T_{in} = 25°C$ and $P_{in} = 1$ atm and undergo a complete combustion reaction to form carbon dioxide that leaves at $T_{out} = 25°C$ and $P_{out} = 1$ atm, as shown in Figure 13-1.

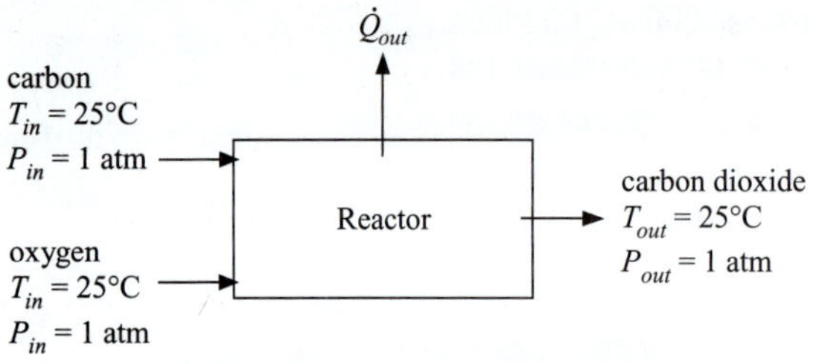

Figure 13-1: Steady state reaction of carbon and oxygen in a reactor.

The balanced chemical reaction for the situation shown in Figure 13-1 is:

$$C + O_2 \rightarrow CO_2 \quad (13\text{-}24)$$

Applying Eq. (13-23), written on a molar basis, to a steady-state analysis of the reactor leads to:

$$\dot{n}_{C,in}\,\bar{h}_{C,in} + \dot{n}_{O_2,in}\,\bar{h}_{O_2,in} = \dot{Q}_{out} + \dot{n}_{CO_2,out}\,\bar{h}_{CO_2,out} \quad (13\text{-}25)$$

where $\bar{h}_{C,in}$ and $\bar{h}_{O_2,in}$ are the molar specific enthalpies of the carbon and oxygen entering the reactor, $\bar{h}_{CO_2,out}$ is the molar specific enthalpy of the carbon dioxide leaving the reactor, and $\dot{n}_{C,in}$, $\dot{n}_{O_2,in}$, and $\dot{n}_{CO_2,out}$ are the molar flow rates of carbon, oxygen, and carbon dioxide, respectively. It is important to notice in Eq. (13-25) that we are using the absolute molar specific enthalpy at a specific state rather than the difference in the molar specific enthalpies of the same substance evaluated at two different states. For example, there is no way to rearrange Eq. (13-25) in order to express it in terms of the difference in the molar specific enthalpy of carbon at the inlet state and the molar specific enthalpy of carbon at the exit state. Rather, Eq. (13-25) requires that we know the absolute value of the molar specific enthalpy of carbon at the inlet state and, more importantly, how it compares with the molar specific enthalpy of oxygen at the inlet state and the molar specific enthalpy of carbon dioxide at the outlet state. This is the first time in this text that we have needed to know the absolute value of the specific enthalpy of a substance because it is the first time that a problem has allowed a substance (i.e., a molecule) to break apart and reform into a different substance.

Enthalpy, like position in a gravitational field, is a thermodynamic property that does not have an absolute zero point. To establish tables of specific enthalpy values, we arbitrarily select a reference state at which the specific enthalpy is assigned a reference value (typically zero). Values of specific enthalpy at any other state can then be determined relative to the reference value by knowing the specific heat capacity of the substance and its equation of state (i.e., the relation between pressure, volume, and temperature). Methods for calculating changes in the specific enthalpy of a substance are discussed in Section 10.4. The reference condition and the associated reference specific enthalpy are completely arbitrary for substances that do not chemically react. For example, the reference state that is used to construct the water tables contained in Appendix B is saturated liquid at the triple point of water (273.16 K, 611.6 Pa) and the reference specific enthalpy at that condition is chosen to be zero. However, a consistent set of reference states must be chosen for all substances that participate in a chemical reaction. To demonstrate this requirement, we can arbitrarily set the reference state for each of the substances involved in the reaction to be 25°C and 1 atm and set the reference specific enthalpy for each substance to be zero at these conditions. Doing so would lead to $\bar{h}_{C,in} = \bar{h}_{O_2,in} = \bar{h}_{CO_2,out} = 0$ in Eq. (13-25) and therefore $\dot{Q}_{out} = 0$. This result is contrary to the experimental observation that the combustion of carbon actually releases a substantial amount of heat.

Clearly, a consistent reference state is needed for all of the substances appearing in the chemical equation. The convention that has been established is to choose a reference state of $T_{ref} = 25°\text{C}$ and $P_{ref} = 1$ atm. The specific enthalpy of all elements appearing in their preferred form at this reference state is set to zero. For example, the preferred state of oxygen at T_{ref} and P_{ref} is the O_2 molecule; therefore, the specific enthalpy of O_2 (but not monatomic oxygen) is defined as being zero at T_{ref} and P_{ref}. Since all compounds necessarily consist of elements, the specific enthalpy of a compound at T_{ref} and P_{ref} is then determined relative to the reference specific enthalpy of its elements. The specific enthalpy of a compound at the reference state is referred to as the *specific enthalpy of formation*. The specific enthalpy of formation is the difference between the specific enthalpy of the compound at the reference state and the specific enthalpy of the elements from which it is formed.

In theory, the specific enthalpy of formation could be measured by calorimetry. The energy balance provided in Eq. (13-25) can be expressed as:

$$n_{C,in}\,\bar{h}_{C,in} + n_{O_2,in}\,\bar{h}_{O_2,in} = Q_{out} + n_{CO_2,out}\,\bar{h}_{CO_2,out} \tag{13-26}$$

The inlet condition is equal to the reference state and both carbon and O_2 are the preferred forms of these elements; therefore $\bar{h}_{C,in} = \bar{h}_{O_2,in} = 0$ and so Eq. (13-26) can be simplified to:

$$0 = Q_{out} + n_{CO_2,out}\,\bar{h}_{CO_2,out} \tag{13-27}$$

Solving Eq. (13-27) for $\bar{h}_{CO_2,out}$ shows that:

$$\bar{h}_{CO_2,out} = -\frac{Q_{out}}{n_{CO_2,out}} \tag{13-28}$$

For the reaction of carbon and oxygen at 25°C, a calorimetric measurement would show that Q_{out} is 393,486 kJ per kmol of carbon dioxide. Therefore, the molar specific enthalpy of formation of carbon dioxide must be $\bar{h}_{form,CO_2} = -393{,}486$ kJ/kmol. The specific enthalpy of formation of most substances can be determined more readily using statistical thermodynamics than calorimetry. Values of the molar specific enthalpy of formation at 25°C for many substances are provided in Table 13-2. These values and others can also be determined using EES, as discussed in Section 13.3.4.

The specific enthalpy of a substance, defined using the reference state discussed in this section, is referred to as the *standardized specific enthalpy*. The standardized specific enthalpy can be used directly in an energy balance for a chemical reaction because it correctly accounts for the energy associated with the chemical bonds required to form the substance from its stable elements. The standardized molar specific enthalpy is given by:

$$\bar{h}_{std}(T, P) = \bar{h}_{form} + \left[\bar{h}(T, P) - \bar{h}\left(T_{ref}, P_{ref}\right)\right] \tag{13-29}$$

Notice that the standardized molar specific enthalpy at the reference state is simply the molar specific enthalpy of formation. The change in the molar specific enthalpy of the substance relative to its value at the reference state, shown in square brackets in Eq. (13-29), can be evaluated using any of the techniques discussed in earlier chapters.

13.3.2 Heating Values

Combustion reactions are exothermic. The heat released during a combustion reaction is appropriately called the *heat of combustion*. However, if the temperature of the reactants and the products are both equal to the reference temperature, $T_{ref} = 25$°C, and the products are fully oxidized then the heat of combustion is also called the *heating value*. The heating value of a fuel is a measure of the thermal energy that can reasonably be obtained from a fuel. Heating values are often used to define the efficiency of systems that operate based on the combustion of a fuel. For example, the efficiency of a house heating furnace is defined as the ratio of the heat provided to the house to the heating value of the fuel combusted in the furnace. The efficiency of an automotive engine is defined as the ratio of the mechanical power that the engine produces to the heating value of the fuel that it consumes.

The heating value of a fuel can be determined from a steady-state energy balance on the combustion reaction of 1 mole of fuel in which the products and reactants are both at T_{ref}. Kinetic and potential energy terms are not considered and power is not involved in the heating value calculation; therefore, the heating value, HV, is given by:

$$HV = H_R - H_P \tag{13-30}$$

Table 13-2: Molar specific enthalpy of formation of some common substances at 25°C (77°F)

Substance	Phase	Chemical formula	Molar mass (kg/kmol)	Molar specific enthalpy of formation ($\bar{h}_{form}$) (kJ/kmol)	(Btu/lbmol)
Carbon	solid	C	12.00	0	0
Hydrogen	gas	H_2	2.016	0	0
Oxygen	gas	O_2	32.00	0	0
Nitrogen	gas	N_2	28.01	0	0
Carbon Monoxide	gas	CO	28.01	−110,528	−47,519
Carbon Dioxide	gas	CO_2	44.01	−393,486	−169,168
Nitrogen Oxide	gas	NO	30.01	91266	39,237
Nitrogen Dioxide	gas	NO_2	46.01	34,191	14,699
Sulfur Dioxide	gas	SO_2	64.06	−296,792	−127,597
Water	liquid	H_2O	18.02	−285,813	−122,877
Water	gas	H_2O	18.02	−241,811	−103,960
Ammonia	gas	NH_3	17.03	−45,937	−19,749
Methane	gas	CH_4	16.04	−74,595	−32,070
Ethane	gas	C_2H_6	30.07	−83,846	−36047
Ethylene	gas	C_2H_4	28.04	52,497	22,570
Acetylene	gas	C_2H_2	26.04	228,186	98,102
Propane	gas	C_3H_8	44.10	−104,674	−45,002
n−Butane	liquid	C_4H_8	58.12	−147,379	−63,362
n−Butane	gas	C_4H_8	58.12	−125,782	−54,077
n−Pentane	liquid	C_5H_{12}	72.15	−173,419	−74,557
n−Pentane	gas	C_5H_{12}	72.15	−146,751	−63,092
n−Hexane	liquid	C_6H_{14}	86.17	−198,135	−85,183
n−Hexane	gas	C_6H_{14}	86.17	−166,910	−71,758
n−Octane	liquid	C_8H_{18}	114.20	−250,302	−107,611
n−Octane	gas	C_8H_{18}	114.20	−208,737	−89,741
Methanol	liquid	CH_3OH	32.04	−239,004	−102,753
Methanol	gas	CH_3OH	32.04	−200,928	−86,383
Ethanol	liquid	C_2H_5OH	46.07	−277,402	−119,261
Ethanol	gas	C_2H_5OH	46.07	−234,936	−101,004

Source: McBride, B.J., Zehe, M.J. and Gordon, S., "NASA Glenn Coefficients for Calculating Thermodynamic Properties of Individual Species", NASA/TP–2002-211556, Sept. 2002, http://www.lerc.nasa.gov/WWW/CEAWeb/

where H_R is the sum of the standardized enthalpies of the reactants and H_P is the sum of the standardized enthalpies of the products, both expressed on a 1 mole of fuel basis. Notice that the inlet and outlet states are both equal to the reference state and therefore, according to Eq. (13-29), the standardized molar specific enthalpy of each reactant and product must be equal to its molar specific enthalpy of formation.

To determine the heating value of propane (C_3H_8), for example, we begin by writing the balanced chemical equation for the complete combustion of one mole of C_3H_8 with stoichiometric oxygen:

$$C_3H_8 + 5\,O_2 \rightarrow 3\,CO_2 + 4\,H_2O \tag{13-31}$$

Note that we could also write the reaction using a stoichiometric amount of air (instead of pure oxygen) or even with some excess air. However, the heating value that we calculate would not change because the unreacted oxygen and nitrogen that appear in the products are at the same state and therefore will have the same enthalpy that they do in the reactants. Consequently, these terms will cancel from an energy balance.

The total standardized enthalpy of the reactants for this reaction, per mole of propane, is:

$$H_R = \bar{h}_{std,C_3H_8} + 5\,\bar{h}_{std,O_2} \tag{13-32}$$

The temperature of the inlet state is equal to the reference temperature. However, the pressure of propane and oxygen in the reactants is not necessarily equal to the reference pressure. Both propane and oxygen can be modeled as ideal gases at the inlet condition as well as at the reference condition. Therefore, the enthalpy of these substances is only a function of temperature and the standardized molar specific enthalpies in Eq. (13-32) are simply equal to the molar specific enthalpy of formation for each substance. These enthalpies of formation can be obtained from Table 13-2.

$$H_R = \bar{h}_{form,C_3H_8} + 5\,\bar{h}_{form,O_2} = -104,674\frac{\text{kJ}}{\text{kmol}} + 5\,(0) = -104,674\frac{\text{kJ}}{\text{kmol}} \tag{13-33}$$

The enthalpy of the products of the reaction per mole of propane is:

$$H_P = 3\,\bar{h}_{std,CO_2} + 4\,\bar{h}_{std,H_2O} \tag{13-34}$$

The outlet temperature is equal to the reference temperature. The carbon dioxide in the products can be considered to be an ideal gas at both the reference state and the outlet state; therefore, the standardized molar specific enthalpy of the carbon dioxide in the products must be equal to its molar specific enthalpy of formation. Notice that there are two choices in Table 13-2 for the enthalpy of formation of water, corresponding to liquid and vapor phases. At the reference state, $T_{ref} = 25°\text{C}$ and $P_{ref} = 1$ atm, water is a liquid and therefore, technically, the enthalpy of formation for water correctly corresponds to the value associated with liquid water in Table 13-2. If the water in the products of the combustion reaction in Eq. (13-31) is in liquid phase, then the molar specific standardized enthalpy of water in the products is the molar specific enthalpy of formation of liquid water, found in Table 13-2 to be $\bar{h}_{form,H_2O,liquid} = -285{,}813$ kJ/kmol. In this case, the enthalpy of the products is given by:

$$\begin{aligned} H_P &= 3\,\bar{h}_{form,CO_2} + 4\,\bar{h}_{form,H_2O,liquid} \\ &= 3\,(-393,486)\,\frac{\text{kJ}}{\text{kmol}} + 4\,(-285,813)\,\frac{\text{kJ}}{\text{kmol}} = -2.324\times 10^{6}\,\frac{\text{kJ}}{\text{kmol}} \end{aligned} \tag{13-35}$$

and the heating value of propane is determined according to Eq. (13-30):

$$HV = H_R - H_P = -104,674\frac{\text{kJ}}{\text{kmol}} - \left(-2.324\times 10^6\right)\frac{\text{kJ}}{\text{kmol}} = 2.219\times 10^6\frac{\text{kJ}}{\text{kmol}} \quad (13\text{-}36)$$

However, water may exist in the products as either liquid or vapor depending on the pressure. If the temperature of the outlet state is below the dew point temperature, then it is likely that water exists in the products in both the liquid and vapor phases. We could use the techniques discussed in Section 13.2 in order to determine exactly how much water exists in each phase. However, a simpler approach is usually adopted and that is to assume that any water that is formed is either all liquid or all vapor. If all of the water in the products is liquid, then the heating value that is calculated is referred to as the *higher heating value* because the enthalpy of liquid water is lower than that of vapor and therefore the heat release will be larger. The higher heating value of propane is $HHV = 2.219\text{x}10^6$ kJ/kmol, as calculated by Eq. (13-36). Using the molar mass of propane, the higher heating value can also be expressed on a per mass basis:

$$HHV = \frac{2.219\times 10^6\,\text{kJ}}{\text{kmol}}\left|\frac{\text{kmol}}{44.1\text{kg}}\right. = 50,321\,\frac{\text{kJ}}{\text{kg}} \quad (13\text{-}37)$$

If all of the water appearing in the products is vapor, then the standardized molar specific enthalpy of water in the products is not equal to the molar specific enthalpy of formation of liquid water. Rather, according to Eq. (13-29), the standardized enthalpy of water vapor is equal to the enthalpy of formation of liquid water plus the molar specific enthalpy change associated with converting the liquid to vapor:

$$\bar{h}_{std,H_2O} = \bar{h}_{form,H_2O,liquid} + \underbrace{\bar{h}_g\left(T_{ref}\right) - \bar{h}_f\left(T_{ref}\right)}_{h_{fg}} = \bar{h}_{form,H_2O,vapor} \quad (13\text{-}38)$$

The standardized molar specific enthalpy of water vapor is listed in Table 13-2 as the molar specific enthalpy of formation for water vapor. If the water in the products is all vapor, then the enthalpy of the products should be computed according to:

$$\begin{aligned} H_P &= 3\,\bar{h}_{form,CO_2} + 4\,\bar{h}_{form,H_2O,vapor} \\ &= 3\left(-393,486\right)\frac{\text{kJ}}{\text{kmol}} + 4\left(-241,811\right)\frac{\text{kJ}}{\text{kmol}} = -2.148\times 10^6\frac{\text{kJ}}{\text{kmol}} \end{aligned} \quad (13\text{-}39)$$

and the heating value of propane is determined according to Eq. (13-30):

$$HV = H_R - H_P = -104,674\frac{\text{kJ}}{\text{kmol}} - \left(-2.148\times 10^6\right)\frac{\text{kJ}}{\text{kmol}} = 2.043\times 10^6\frac{\text{kJ}}{\text{kmol}} \quad (13\text{-}40)$$

The heating value calculated assuming that all of the water in the products is vapor is referred to as the *lower heating value*. The lower heating value of propane is $LHV = 2.043\times 10^6$ kJ/kmol (46,330 kJ/kg). For propane, the lower heating value is about 8% less than the higher heating value.

There are then two heating values that are reported when water appears in the products of a combustion reaction. The heating value corresponding to the case where all of the water in the products is liquid is larger and it is called the *higher heating value, HHV*. The heating value corresponding to the case where all of the water in the products is vapor is called the *lower heating value, LHV*. Liquid water in the products produces a higher heating value because the heat from the reaction includes the enthalpy change associated with the condensation of water from vapor to liquid in addition to the enthalpy change of combustion. Table 13-3 lists both heating values of some common fuels.

Table 13-3: The heating values for some common fuels

Substance	Phase	Chemical Formula	Molar Mass (kg/kmol)	Lower heating value, *LHV*		Higher heating value, *HHV*	
				kJ/kg	Btu/lb_m	kJ/kg	Btu/lb_m
Carbon	solid	C	12.00	32,790	14,097	32,790	14,097
Hydrogen	gas	H_2	2.016	119,946	51,567	141,764	60,947
Carbon Monoxide	gas	CO	28.01	10,102	43,43	10,102	4,343
Ammonia	gas	NH_3	17.03	16,594	7,134	20,468	8,800
Methane	gas	CH_4	16.04	50,032	21,510	55,516	23,868
Ethane	gas	C_2H_6	30.07	47,508	20,425	51,896	22,311
Ethylene	gas	C_2H_4	28.04	47,186	20,286	50,323	21,635
Acetylene	gas	C_2H_2	26.04	48,271	20,753	49,960	21,479
Propane	gas	C_3H_8	44.10	46,330	19,917	50,321	21,632
n-Butane	liquid	C_4H_8	58.12	45,348	19,496	49,132	21,123
n-Butane	gas	C_4H_8	58.12	45,719	19,656	49,503	21,283
n-Pentane	liquid	C_5H_{12}	72.15	44,974	19,335	48,632	20,908
n-Pentane	gas	C_5H_{12}	72.15	45,344	19,494	49,001	21,067
n-Hexane	liquid	C_6H_{14}	86.17	44,742	19,236	48,316	20,772
n-Hexane	gas	C_6H_{14}	86.17	45,105	19,392	48,678	20,928
n-Octane	liquid	C_8H_{18}	114.20	44,430	19,101	47,896	20,592
n-Octane	gas	C_8H_{18}	114.20	44,794	19,258	48,260	20,748
Methanol	liquid	CH_3OH	32.04	19,916	8,562	22,661	9,743
Methanol	gas	CH_3OH	32.04	21,104	9,073	23,850	10,254
Ethanol	liquid	C_2H_5OH	46.07	26,807	11,525	29,671	12,756
Ethanol	gas	C_2H_5OH	46.07	27,729	11,921	30,593	13,153

EXAMPLE 13.3-1: HEATING VALUE OF A PRODUCER GAS

a) Determine the lower heating value of the producer gas described in Example 13.2-1.

The molar composition of the producer gas is entered in EES:

```
$UnitSystem SI Mass Radian kJ K Pa
y_CH4=0.038 [-]                          "mole fraction of mcthanc"
y_C2H6=0.001 [-]                         "mole fraction of ethane"
y_CO2=0.048 [-]                          "mole fraction of carbon dioxide"
y_H2=0.117 [-]                           "mole fraction of hydrogen"
y_O2=0.006 [-]                           "mole fraction of oxygen"
y_CO=0.232 [-]                           "mole fraction of carbon monoxide"
y_N2=1-y_CH4-y_C2H6-y_CO2-y_H2-y_O2-y_CO "mole fraction of nitrogen"
```

Following Example 13.2-1, a basis of one mole of producer gas is chosen and the combustion reaction is balanced for a stoichiometric amount of air:

$$\left(y_{CH_4}\mathrm{CH_4} + y_{C_2H_6}\mathrm{C_2H_6} + y_{CO_2}\mathrm{CO_2} + y_{H_2}\mathrm{H_2} + y_{O_2}\mathrm{O_2} + y_{CO}\mathrm{CO} + y_{N_2}\mathrm{N_2}\right) + a_s\left(0.21\,\mathrm{O_2} + 0.79\mathrm{N_2}\right) \rightarrow b_s\,\mathrm{CO_2} + c_s\,\mathrm{H_2O} + d_s\,\mathrm{N_2} \tag{1}$$

where the parameters a_s, b_s, c_s, and d_s are the coefficients for the stoichiometric reaction. The parameters y_{CH_4}, $y_{C_2H_6}$, etc. indicate the mole fractions of the components in the producer gas, as provided in Table 1 from Example 13.2-1. Note that it does not matter if stoichiometric or excess air is used in the balance as the enthalpy associated with the excess oxygen and nitrogen in the products at 25° will cancel with the enthalpy of the oxygen and nitrogen in the reactants at 25°C. It is simplest to use stoichiometric air. The stoichiometric reaction, Eq. (1), is balanced by conserving the individual atoms. A balance of carbon atoms provides:

$$y_{CH_4} + 2\,y_{C_2H_6} + y_{CO_2} + y_{CO} = b_s$$

A balance of hydrogen atoms provides:

$$4\,y_{CH_4} + 6\,y_{C_2H_6} + 2\,y_{H_2} = 2\,c_s$$

A balance of oxygen atoms provides:

$$2\,y_{CO_2} + 2\,y_{O_2} + y_{CO} + 2\,(0.21)\,a_s = 2\,b_s + c_s$$

A balance of nitrogen atoms provides:

$$2\,y_{N_2} + 2\,(0.79)\,a_s = 2\,d_s$$

```
"Balance stoichiometric reaction"
y_CH4+2*y_C2H6+y_CO2+y_CO=b_s             "carbon atom balance"
4*y_CH4+6*y_C2H6+2*y_H2=2*c_s             "hydrogen atom balance"
2*y_CO2+2*y_O2+y_CO+a_s*0.21*2=2*b_s+c_s  "oxygen atom balance"
2*y_N2+a_s*0.79*2=2*d_s                   "nitrogen atom balance"
```

The lower heating value is determined using Eq. (13-30):

$$LHV = H_R - H_P \tag{2}$$

where the enthalpy of the products, H_P, is evaluated assuming that all of the water in the products is in vapor form. The enthalpy of the reactants in Eq. (1) is given by:

$$H_R = y_{CH_4}\,\bar{h}_{form,CH_4} + y_{C_2H_6}\,\bar{h}_{form,C_2H_6} + y_{CO_2}\,\bar{h}_{form,CO_2} + y_{H_2}\bar{h}_{form,H_2} + y_{O_2}\,\bar{h}_{form,O_2}$$
$$+ y_{CO}\,\bar{h}_{form,CO} + y_{N_2}\,\bar{h}_{form,N_2} + a_s\,(0.21)\,\bar{h}_{form,O_2} + a_s\,(0.79)\,\bar{h}_{form,N_2}$$

where the molar specific enthalpy of formations are obtained from Table 13-2.

```
H_R=y_CH4*(-74595 [kJ/kmol])+y_C2H6*(-83846 [kJ/kmol])+y_CO2*(-393486 [kJ/kmol])&
   +y_H2*(0 [kJ/kmol])+y_O2*(0 [kJ/kmol])+y_CO*(-110528 [kJ/kmol])+y_N2*(0 [kJ/kmol])&
   +a_s*0.21*(0 [kJ/kmol])+a_s*0.79*(0 [kJ/kmol]) "enthalpy of reactants per mole of producer gas"
```

The enthalpy of the products in Eq. (2) is given by:

$$H_P = b_s\,\bar{h}_{form,CO_2} + c_s\,\bar{h}_{form,H_2O,vapor} + d_s\,\bar{h}_{form,N_2}$$

where the molar specific enthalpy of formation of vapor water is used because we are computing the lower heating value.

```
H_P=b_s*(-393486 [kJ/kmol])+c_s*(-241811 [kJ/kmol])+d_s*(0 [kJ/kmol])
                     "enthalpy of products per mole of producer gas, assuming water is vapor"
```

The lower heating value is computed according to Eq. (2):

```
LHV=H_R-H_P                "lower heating value"
```

which leads to $LHV = 125{,}862$ kJ/kmol of producer gas. The LHV can be expressed on a mass basis using the molar mass of the producer gas, calculated according to:

$$MW_f = y_{CH_4}MW_{CH_4} + y_{C_2H_6}MW_{C_2H_6} + y_{CO_2}MW_{CO_4} + y_{H_2}MW_{H_2} + y_{O_2}MW_{O_2}$$
$$+ y_{CO}MW_{CO} + y_{N_2}MW_{N_2}$$

The lower heating value on a mass basis is obtained by dividing the molar specific heating value by the molar mass.

```
MW_f=y_CH4*MolarMass(CH4)+y_C2H6*MolarMass(C2H6)+y_CO2*MolarMass(CO2)&
   +y_H2*MolarMass(H2)+y_O2*MolarMass(O2)+y_CO*MolarMass(CO)&
   +y_N2*MolarMass(N2)      "molar mass of fuel"
LHV_m=LHV/MW_f              "lower heating value on a mass basis"
```

Solving these equations provides $LHV = 4{,}973$ kJ/kg of producer gas.

b) Determine the higher heating value of the producer gas.

The higher heating value is also computed according to Eq. (2), except that the enthalpy of the products is calculated assuming that all of the water is in liquid form:

$$H_P = b_s\,\bar{h}_{form,CO_2} + c_s\,\bar{h}_{form,H_2O,liquid} + d_s\,\bar{h}_{form,N_2}$$

```
H_P_hhv=b_s*(-393486 [kJ/kmol])+c_s*(-285813 [kJ/kmol])+d_s*(0 [kJ/kmol])
                          "enthalpy of products per mole of producer gas, assuming water is liquid"
HHV=H_R-H_P_hhv           "higher heating value"
HHV_m=HHV/MW_f            "higher heating value on a mass basis"
```

which results in $HHV = 134{,}487$ kJ/kmol (5,314 kJ/kg).

13.3.3 Enthalpy and Internal Energy as a Function of Temperature

Table 13-2 provides the molar specific enthalpy of formation of many substances. The values in this table correspond to the molar specific standardized enthalpy of these substances at $T_{ref} = 25°C$. Standardized enthalpies are based on a common reference state for all of the substances; that is, the elements having zero enthalpy at T_{ref}. However, the reactants and products of a combustion reaction will generally not be at T_{ref} and therefore it is necessary to estimate their values at other temperatures in order to complete an energy balance for a combustion process.

In general, the specific enthalpy of a pure substance depends on two properties, e.g., temperature and pressure. However, most of the substances that are involved in combustion reactions are well-represented by the ideal gas law (e.g., O_2, N_2, CO_2) or they are nearly incompressible (e.g, carbon, liquid C_8H_{18}, and liquid water). The specific enthalpy of an ideal gas is only a function of temperature, as discussed in Section 4.2.3. The standardized molar specific enthalpy of an ideal gas in a mixture is the same as the standardized molar specific enthalpy of the pure gas and it can be evaluated according to:

$$\bar{h}_{std} = \bar{h}_{form} + \int_{T_{ref}}^{T} \bar{c}_P \, dT \text{ for an ideal gas} \tag{13-41}$$

where $\bar{h}_{form}$ is the molar specific enthalpy of formation and $\bar{c}_p$ is the molar specific heat capacity at constant pressure. The specific enthalpy of an incompressible substance is a function of both temperature and pressure, as discussed in Section 4.2.4. Incompressible solids and liquids that form their own separate phase (e.g., solid carbon) are essentially pure substances. The standardized molar specific enthalpy of an incompressible substance at an arbitrary temperature and pressure can then be evaluated as:

$$\bar{h}_{std} = \bar{h}_{form} + \int_{T_{ref}}^{T} \bar{c} \, dT + \bar{v} \left(P - P_{ref}\right) \text{ for an incompressible substance} \tag{13-42}$$

where $\bar{c}$ is the molar specific heat capacity and $\bar{v}$ is the molar specific volume. The magnitude of the last term in Eq. (13-42) is very small under most conditions and therefore the last term is usually neglected so that:

$$\bar{h}_{std} \approx \bar{h}_{form} + \int_{T_{ref}}^{T} \bar{c} \, dT \text{ for an incompressible substance} \tag{13-43}$$

The standardized molar specific internal energy is related to the standardized molar specific enthalpy according to:

$$\bar{u}_{std} = \bar{h}_{std} - P\,\bar{v} \tag{13-44}$$

The ideal gas law states that:

$$P\,\bar{v} = R_{univ}\,T \tag{13-45}$$

Substituting Eq. (13-45) into Eq. (13-44) leads to Eq. (13-46) for an ideal gas:

$$\boxed{\bar{u}_{std} = \bar{h}_{std} - R_{univ}\,T \text{ for an ideal gas}} \tag{13-46}$$

The product of pressure and molar specific volume in Eq. (13-44) will generally be negligible for an incompressible substance; therefore:

$$\boxed{\bar{u}_{std} \approx \bar{h}_{std} \text{ for an incompressible substance}} \tag{13-47}$$

Using Eqs. (13-46) and (13-47), the standardized specific internal energy is based on a consistent reference state across all substances and therefore can be used in energy balances involving chemical reactions.

The molar specific heat capacity at constant pressure, $\bar{c}_P$, appearing in Eq. (13-41) is a measurable property that can be determined by calorimetry. The molar specific heat capacity can also be determined using statistical thermodynamics. For most substances, $\bar{c}_P$ is a function of temperature. The constant pressure specific heat capacity values for several substances at 20°C are provided in Appendix D. If the temperature of a substance is close to room temperature, then the standardized molar specific enthalpy equation given in Eq. (13-42) for an ideal gas can be approximated according to:

$$\bar{h}_{std} \approx \bar{h}_{form} + \bar{c}_P\,(T - T_{ref}) \text{ for an ideal gas} \tag{13-48}$$

where the $\bar{c}_P$ value is obtained from Appendix D. Similarly, the standardized molar specific enthalpy of an incompressible substance given by Eq. (13-43) can be approximated near room temperature according to:

$$\bar{h}_{std} \approx \bar{h}_{form} + \bar{c}\,(T - T_{ref}) \text{ for an incompressible substance} \tag{13-49}$$

where $\bar{c}$ is the molar specific heat capacity evaluated at a temperature near room temperature.

The temperature changes occurring in combustion reactions are often large and therefore it is usually best to carry out the integrations indicated in Eqs. (13-41) or (13-43) rather than assume a constant specific heat capacity. The computational effort involved in repeatedly evaluating specific enthalpy with Eq. (13-41) is significant and no one will want to do this calculation by hand. Appendix F provides the standardized specific internal energy and standardized specific enthalpy for several ideal gases as a function of temperature, determined using Eq. (13-41). The specific enthalpy values provided by EES for substances modeled as an ideal gas (e.g., '02' and 'H20') are also determined using Eq. (13-41). The use of EES for evaluating the properties of substances involved in combustion reactions is discussed in the next section.

EXAMPLE 13.3-2: PROPANE HEATER

Figure 1 illustrates a space heater in which propane (C_3H_8) is burned with 150% theoretical air.

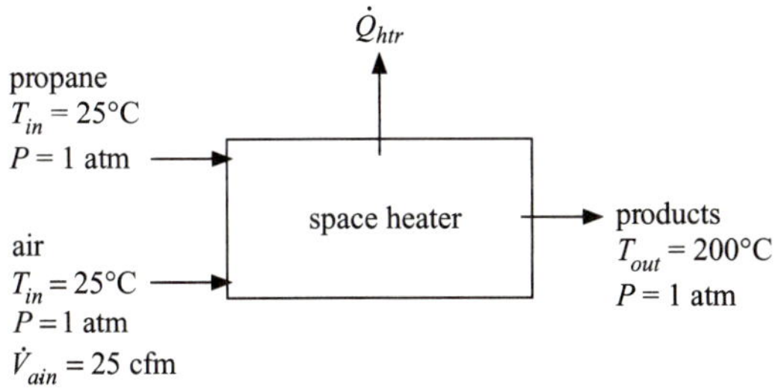

Figure 1: Space heater.

The air enters the heater at $T_{in} = 25°\text{C}$ and $P = 1$ atm with volumetric flow rate $\dot{V}_{a,in} = 25$ cfm. The propane is also supplied to the heater at $T_{in} = 25°\text{C}$. The combustion products leave the heater at $T_{out} = 200°\text{C}$ and $P = 1$ atm.

a) Determine the rate of heat transfer from the heater to the surroundings.

The inputs are entered in EES.

```
$UnitSystem SI Molar Radian J K Pa
T_in=converttemp(C,K,25 [C])                "inlet temperature"
T_out=converttemp(C,K,200 [C])              "exit temperature"
P=1 [atm]*convert(atm,Pa)                   "pressure"
V_dot_a_in=25 [cfm]*convert(cfm,m^3/s)      "volumetric flow rate of entering air"
```

The actual reaction must be defined relative to a stoichiometric reaction. The stoichiometric reaction is written below using a basis of 1 mole of fuel:

$$C_3H_8 + a_s\,(0.21\,O_2 + 0.79\,N_2) \rightarrow b_s\,CO_2 + c_s\,H_2\,O + d_s\,N_2 \tag{1}$$

Balances on the carbon, hydrogen, oxygen, and nitrogen atoms in Eq. (1) provide:

$$3 = b_s \tag{2}$$

$$8 = 2\,c_s \tag{3}$$

$$0.21\,(2)\,a_s = 2\,b_s + c_s \tag{4}$$

$$0.79\,(2)\,a_s = 2\,d_s \tag{5}$$

Equations (2) through (5) can either be solved by inspection or by entering them into EES.

```
"Balance stoichiometric reaction"
3=b_s                      "carbon atom balance"
8=2*c_s                    "hydrogen atom balance"
0.21*2*a_s=2*b_s+c_s       "oxygen atom balance"
0.79*2*a_s=2*d_s           "nitrogen atom balance"
```

The actual reaction is:

$$C_3H_8 + a\,(0.21\,O_2 + 0.79\,N_2) \rightarrow b\,CO_2 + c\,H_2O + d\,N_2 + e\,O_2 \tag{6}$$

where the amount of air is 150% of that required by the stoichiometric reaction:

$$a = 1.5\,a_s \tag{7}$$

Balances on the carbon, hydrogen, oxygen, and nitrogen atoms for the actual reaction indicated in Eq. (6) provide:

$$3 = b \tag{8}$$

$$8 = 2\,c \tag{9}$$

$$0.21\,(2)\,a = 2\,b + c + 2\,e \tag{10}$$

$$0.79\,(2)\,a = 2\,d \tag{11}$$

Equations (7) through (11) are entered in EES in order to define the actual reaction.

```
"Balance actual reaction"
a=1.5*a_s                     "150% excess air"
3=b                           "carbon atom balance"
8=2*c                         "hydrogen atom balance"
0.21*2*a=2*b+c+2*e            "oxygen atom balance"
0.79*2*a=2*d                  "nitrogen atom balance"
```

The molar specific volume of the air entering the heater ($\bar{v}_{a,in}$) is obtained using the volume function in EES. The molar flow rate of air entering the heater is computed according to:

$$\dot{n}_{a,in} = \frac{\dot{V}_{a,in}}{\bar{v}_{a,in}}$$

```
v_bar_a_in=volume(Air,T=T_in,P=P)      "molar specific volume of air entering"
n_dot_a_in=V_dot_a_in/v_bar_a_in       "molar flow rate of air entering"
```

The molar flow rates of oxygen and nitrogen entering the heater are computed according to:

$$\dot{n}_{O_2,in} = 0.21\,\dot{n}_{a,in}$$

$$\dot{n}_{N_2,in} = 0.79\,\dot{n}_{a,in}$$

```
n_dot_02_in=0.21*n_dot_a_in            "molar flow rate of oxygen entering"
n_dot_N2_in=0.79*n_dot_a_in            "molar flow rate of nitrogen entering"
```

The balanced chemical reaction, Eq. (6), determines the molar flow rates of each substance relative to the molar flow rate of air. Therefore, the molar flow rate of the propane entering the heater is:

$$\dot{n}_{C_3H_8,in} = \frac{\dot{n}_{a,in}}{a}$$

and the molar flow rates of carbon dioxide, water, oxygen, and nitrogen leaving the heater are:

$$\dot{n}_{CO_2,out} = \frac{b\,\dot{n}_{a,in}}{a}$$

$$\dot{n}_{H_2O,out} = \frac{c\,\dot{n}_{a,in}}{a}$$

$$\dot{n}_{N_2,out} = \frac{d\,\dot{n}_{a,in}}{a}$$

$$\dot{n}_{O_2,out} = \frac{e\,\dot{n}_{a,in}}{a}$$

```
n_dot_C3H8_in=n_dot_a_in/a          "molar flow rate of propane entering"
n_dot_CO2_out=b*n_dot_a_in/a        "molar flow rate of carbon dioxide leaving"
n_dot_H2O_out=c*n_dot_a_in/a        "molar flow rate of water leaving"
n_dot_N2_out=d*n_dot_a_in/a         "molar flow rate of nitrogen leaving"
n_dot_O2_out=e*n_dot_a_in/a         "molar flow rate of oxygen leaving"
```

An energy balance on the heater provides:

$$\dot{H}_R = \dot{H}_P + \dot{Q}_{htr} \tag{12}$$

where $\dot{H}_R$ and $\dot{H}_P$ are the total flow rates of enthalpy associated with the reactants and the products, respectively.

$$\dot{H}_R = \dot{n}_{C_3H_8,in}\,\bar{h}_{std,C_3H_8,in} + \dot{n}_{O_2,in}\,\bar{h}_{std,O_2,in} + \dot{n}_{N_2,in}\,\bar{h}_{std,N_2,in} \tag{13}$$

$$\dot{H}_P = \dot{n}_{CO_2,out}\,\bar{h}_{std,CO_2,out} + \dot{n}_{H_2O,out}\,\bar{h}_{std,H_2O,out} + \dot{n}_{N_2,out}\,\bar{h}_{std,N_2,out} + \dot{n}_{O_2,out}\,\bar{h}_{std,O_2,out} \tag{14}$$

Notice that standardized molar enthalpies are used in Eqs. (13) and (14). According to Eq. (13-29), the standardized enthalpies of the oxygen and nitrogen at the inlet condition are zero because these substances are at an inlet temperature of T_{ref} and the specific enthalpy of formation for each of these stable substances is zero (see Table 13-2). The standardized molar specific enthalpy of propane at the inlet is equal to the molar specific enthalpy of formation of propane, again because the inlet temperature is equal to T_{ref}.

```
h_std_O2_in=0 [kJ/kmol]             "standardized molar specific enthalpy of oxygen at inlet"
h_std_N2_in=0 [kJ/kmol]             "standardized molar specific enthalpy of nitrogen at inlet"
h_std_C3H8_in=-104674 [kJ/kmol]     "standardized molar specific enthalpy of propane at inlet"
```

The standardized molar specific enthalpy of carbon dioxide at the outlet condition is obtained from Eq. (13-29):

$$\bar{h}_{std,CO_2,out} = \bar{h}_{form,CO_2} + \int_{T_{ref}}^{T_{out}} \bar{c}_{P,CO_2}\, dT$$

The value of $\bar{h}_{std,CO_2,out}$ can be obtained from the CO_2 table in Appendix F. The temperature at the outlet of the heater is 200°C or 473.15 K. Values of the standardized molar specific enthalpy of CO_2 are available in Appendix F at

460 K where $\bar{h}_{std,CO_2,460\,K} = -386{,}936$ kJ/kmol and 480 K where $\bar{h}_{std,CO_2,480\,K} = -386{,}060$ kJ/kmol. Interpolation is applied between these table values:

$$\frac{\bar{h}_{std,CO_2,out} - (-386,936)}{(-386060) - (-386,936)} = \frac{473.15 - 460}{480 - 460}$$

```
(h_std_CO2_out-(-386936 [kJ/kmol]))/(-386060[kJ/kmol]-(-386936[kJ/kmol]))=&
   (T_out-460[K])/(480 [K]-460 [K])
                         "interpolate using the entries Appendix F for standardized
                            specific enthalpy of CO2 at outlet"
```

Because the molar specific enthalpies of formation of nitrogen and oxygen are both zero, the standardized molar specific enthalpies of oxygen and nitrogen at the heater outlet are obtained according to:

$$\bar{h}_{std,O_2,out} = \int_{T_{ref}}^{T_{out}} \bar{c}_{P,O_2}\, dT$$

$$\bar{h}_{std,N_2,out} = \int_{T_{ref}}^{T_{out}} \bar{c}_{P,N_2}\, dT$$

Again, the specific enthalpy values in Appendix F are employed with interpolation.

```
(h_std_O2_out-4861[kJ/kmol])/(5481[kJ/kmol]-4861[kJ/kmol])=(T_out-460 [K])/(480 [K]-460 [K])
                          "standardized molar specific enthalpy of O2 at outlet"
(h_std_N2_out-4727[kJ/kmol])/(5318[kJ/kmol]-4727[kJ/kmol])=(T_out-460 [K])/(480 [K]-460 [K])
                          "standardized molar specific enthalpy of N2 at outlet"
```

In order to evaluate the standardized molar specific enthalpy of water at the outlet it is necessary to determine the state of the water. Assuming that no condensation occurs, the partial pressure of water vapor at the outlet is:

$$P_{v,out} = \frac{c\,P}{(b + c + d + e)}$$

The dew point temperature of the combustion products (T_{dp}) is the saturation temperature of water evaluated at $P_{v,out}$:

```
P_v_out=c*P/(b+c+d+e)            "vapor pressure of water leaving"
T_dp=T_sat(Water,P=P_v_out)      "dew point of combustion products"
T_dp_C=converttemp(K,C,T_dp)     "in C"
```

which leads to $T_{dp} = 47.21$°C. Therefore, the water in the products will be entirely vapor. The standardized molar specific enthalpy of water in the products can be obtained by interpolation of the ideal gas water vapor tables in Appendix F:

```
(h_std_H2O_out-(-236290 [kJ/kmol]))/(-235594 [kJ/kmol]-(-236290 [kJ/kmol]))=&
   (T_out-460 [K])/(480 [K]-460 [K])
                  "interpolate Appendix F for standardized enthalpy of H2O at outlet"
```

The enthalpy flow rate of the reactants and products are computed according to Eqs. (13) and (14) and the rate of heat transfer from the heater is obtained from Eq. (12).

```
H_dot_R=n_dot_C3H8_in*h_std_C3H8_in+n_dot_N2_in*h_std_N2_in+n_dot_02_in*h_std_02_in
                          "enthalpy flow rate associated with reactants"
H_dot_P=n_dot_C02_out*h_std_C02_out+n_dot_H20_out*h_std_H20_out&
   +n_dot_02_out*h_std_02_out+n_dot_N2_out*h_std_N2_out
                          "enthalpy flow rate associated with products"
H_dot_R=H_dot_P+Q_dot_htr "energy balance on the heater"
```

Solving these equations leads to $\dot{Q}_{htr} = 24.85$ kW.

Reading and interpolating property tables is time-consuming and likely to lead to error. The following section will demonstrate how EES can directly provide the property information needed to solve combustion problems.

13.3.4 Use of EES for Determining Properties

At this point, you are likely familiar with the EES property functions. However, for problems involving chemical reactions it is important to ensure that EES returns values of the standardized specific internal energy and the standardized specific enthalpy.

You will almost always want to determine thermodynamic properties on a molar (rather than on a mass) basis when dealing with chemical reactions. There are two ways to instruct EES to return properties on a molar basis. The first way is to specify Molar Basis for specific properties in the Unit System dialog (select Unit System from the Options menu), as indicated in Figure 13-2. The alternative to using the Unit System dialog is to use a $UnitSystem directive. The following directive placed anywhere within an EES program will set the units in the same way as indicated in Figure 13-2.

```
$UnitSystem SI K kJ Molar Pa Deg
```

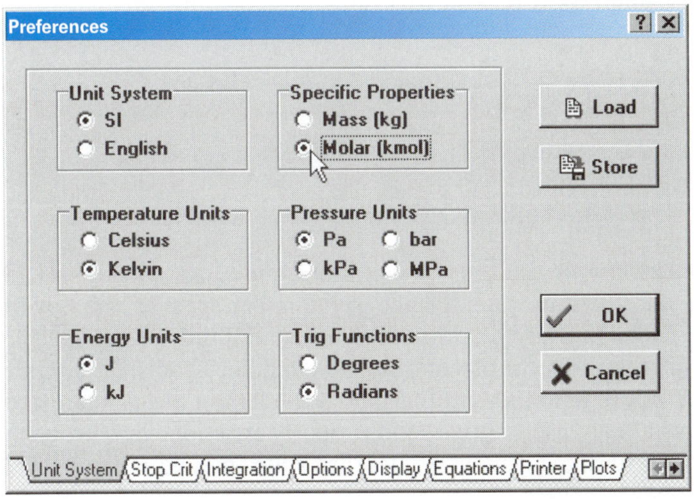

Figure 13-2: Unit System dialog in EES showing specific properties set for molar basis.

Note that the keyword 'Molar' in this directive will set the specific properties to be on a molar basis. It does not matter what order the unit specifications appear in the $UnitSystem directive.

The intEnergy and enthalpy functions in EES will return the standardized specific internal energy and standardized specific enthalpy for all of the ideal gas substances (except air). Note that ideal gases in EES are identified by having a chemical symbol notation rather than a spelled out name. For example, the enthalpy function in EES called with the substances H2O, N2, and CO2 will provide standardized specific enthalpy, consistent with Eq. (13-41). The enthalpy function in EES called with the substances Water, Nitrogen, and CarbonDioxide will provide real fluid property data that are based on arbitrary reference states. Real fluid property data cannot be directly used in energy and entropy balances involving chemical reactions unless the reference values for specific enthalpy and specific entropy values are shifted.

To illustrate these differences, enter the following equations into EES in order to calculate the specific enthalpy of water, nitrogen, and carbon dioxide at $T = 25°\text{C}$ and $P = 1$ atm, assuming that these substances can be modeled as ideal gases.

```
$UnitSystem SI K kJ Molar Pa Deg
T=convertTemp(C,K,25 [C])          "temperature"
P=1 [atm]*convert(atm,Pa)          "pressure"
h_bar_H2O=enthalpy(H2O,T=T)        "molar specific enthalpy of water, modeled as an ideal gas"
h_bar_N2=enthalpy(N2,T=T)          "molar specific enthalpy of nitrogen, modeled as an ideal gas"
h_bar_CO2=enthalpy(CO2,T=T)        "molar specific enthalpy of carbon dioxide, modeled as an ideal gas"
```

The solution is $\bar{h}_{H_2O} = -241{,}811$ kJ/kmol, $\bar{h}_{N_2} = 0$ kJ/kmol, and $\bar{h}_{CO_2} = -393{,}486$ kJ/kmol. These values are the same as the values of the molar specific enthalpy of formation for these substances appearing in Table 13-2. Change the temperature to $T = 1000$ K and solve again. You should see that the molar specific enthalpy values provided by EES agree exactly with the standardized molar specific enthalpy values for these substances provided in Appendix F at this temperature.

If we were to use real fluid designations (e.g., Nitrogen instead of N2) in the calls to the function enthalpy then it is necessary to specify a second property, e.g., pressure, since the state of a real fluid is not fixed by temperature alone. The molar specific enthalpies of the real fluids at $T = 25°\text{C}$ and $P = 101.3$ kPa are evaluated according to:

```
h_bar_water=enthalpy(Water,T=T,P=P)
                              "molar specific enthalpy of water, modeled as a real fluid"
h_bar_nitrogen=enthalpy(Nitrogen,T=T,P=P)
                              "molar specific enthalpy of nitrogen, modeled as a real fluid"
h_bar_carbondioxide=enthalpy(CarbonDioxide,T=T,P=P)
                              "molar specific enthalpy of carbon dioxide, modeled as a real fluid"
```

The solution is $\bar{h}_{water} = 1{,}889$ kJ/kmol, $\bar{h}_{nitrogen} = 8{,}664$ kJ/kmol, and $\bar{h}_{carbondioxide} = -41.28$ kJ/kmol. Note that these values bear no resemblance to the molar specific enthalpy of formation values listed in Table 13-2 because the reference state used for the specific enthalpy of real fluids is arbitrary and inconsistent from fluid to fluid. That is, the specific enthalpy returned by the enthalpy function when a real fluid is used is not the standardized specific enthalpy. If you mistakenly use the real fluid property functions in a problem involving chemical reactions without shifting the reference state then your results will be completely incorrect.

The ideal gas properties in EES provide standardized molar specific enthalpy using a reference state that is based on the molar specific enthalpy of formation. However, there may be problems where you need to determine the standardized molar specific enthalpy of a liquid involved in a chemical reaction. In this case, you will need to adjust the reference state using Eq. (13-29), restated below:

$$\bar{h}_{std}(T, P) = \bar{h}_{form} + \bar{h}(T, P) - \bar{h}\left(T_{ref}, P_{ref}\right) \tag{13-50}$$

For example, a chemical reaction may involve methanol at $T = 50°\text{C}$ and $P = 1$ atm. Methanol is a liquid under these conditions and therefore the real fluid substance methanol (rather than the ideal gas substance CH3OH) should be used to determine the specific enthalpy. However, the reference state for the substance methanol is not consistent with the definition of the standardized molar specific enthalpy and therefore Eq. (13-50) must be used to shift the reference state. The molar specific enthalpy of formation for liquid methanol is listed in Table 13-2 and the change in the molar specific enthalpy relative to the reference state can be computed using the enthalpy function called with the substance methanol. Because the difference between the last two terms in Eq. (13-50) corresponds to a specific enthalpy difference, the reference state cancels out.

```
$UnitSystem SI Molar kJ Pa K
P=1 [atm]*convert(atm,Pa)                 "pressure"
T=converttemp(C,K,50 [C])                 "temperature"
T_ref=converttemp(C,K,25 [C])             "reference temperature"
P_ref=1 [atm]*convert(atm,Pa)             "reference pressure"
h_bar_form=-239004 [kJ/kmol]              "molar specific enthalpy of formation for liquid methanol"
h_bar_std=h_bar_form+enthalpy(methanol,T=T,P=P)-enthalpy(methanol,T=T_ref,P=P_ref)
                                          "standardized molar specific enthalpy"
```

Solving results in $\bar{h}_{std} = -236{,}906$ kJ/kmol.

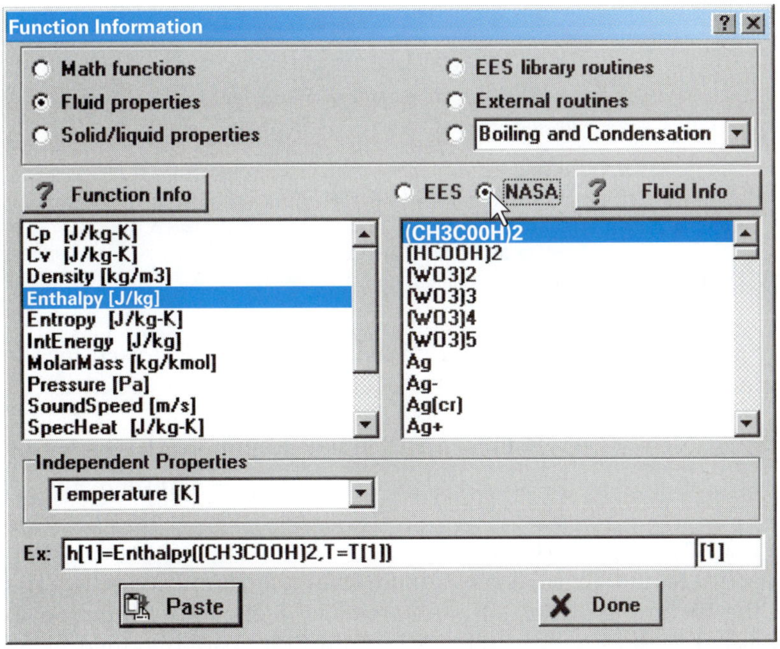

Figure 13-3: Select the NASA option in order to access fluids that are part of the NASA database.

EES provides property data with its built-in property functions for most of the substances that are involved in combustion reactions. EES also integrates directly to the NASA external library. The NASA library provides the specific heat capacity, specific enthalpy, and specific entropy of more than 1200 ideal gas substances, as described by McBride et al. (2002). In order to access these substances in EES, select the NASA radio button from the Fluid Properties dialog as shown in Figure 13-3. Note that the specific enthalpy that is returned for each of the substances in the NASA database is the standardized specific enthalpy; that is, the reference state is consistent with stable elements at $T_{ref} = 25°C$ and $P_{ref} = 1$ atm.

EXAMPLE 13.3-3: FURNACE EFFICIENCY

The purpose of this problem is to compare the efficiencies of alternative residential furnaces. The fuel used in each furnace is natural gas, which enters the furnace at $T_{ng,in} = 10°C$ and $P = 1$ atm. A volumetric analysis of the natural gas shows that it is composed of methane, nitrogen, and water vapor with composition $y_{ng,CH_4} = 0.94$, $y_{ng,N_2} = 0.05$, and $y_{ng,H_2O} = 0.01$. The natural gas is combusted with 50% excess air that enters at $T_{a,in} = 25°C$, $P = 1$ atm, and $\phi_{\text{a,in}} = 60\%$ relative humidity.

a) First consider a conventional furnace. The gas flow rate is adjusted so that the combustion products leave the furnace at $T_{out} = 175°C$ in order to promote exhaust gas removal by buoyancy and avoid condensation in the chimney. The exit pressure is $P = 1$ atm. Determine the efficiency of this furnace, defined relative to the higher heating value of the fuel.

The inputs are entered in EES:

```
$UnitSystem SI Molar kJ Pa K
P=1 [atm]*convert(atm,Pa)                "pressure"
T_ng_in=converttemp(C,K,10 [C])          "temperature of natural gas entering"
y_ng_CH4=0.94 [-]                        "mole fraction of methane in natural gas"
y_ng_N2=0.05 [-]                         "mole fraction of nitrogen in natural gas"
y_ng_H2O=0.01 [-]                        "mole fraction of water in natural gas"
T_a_in=converttemp(C,K,25 [C])           "temperature of air entering"
phi_a_in=0.6 [-]                         "inlet relative humidity"
ExcessAir=0.5 [-]                        "excess air used in furnace"
T_out=converttemp(C,K,175 [C])           "furnace exit temperature"
```

The stoichiometric combustion reaction is provided below; the basis used for the reaction is one mole of natural gas:

$$\left(y_{ng,CH_4}\,\mathrm{CH_4} + y_{ng,N_2}\,\mathrm{N_2} + y_{ng,H_2O}\,\mathrm{H_2O}\right) + a_s\left(0.21\,\mathrm{O_2} + 0.79\,\mathrm{N_2} + w\,\mathrm{H_2O}\right) \rightarrow b_s\,\mathrm{CO_2} + c_s\,\mathrm{H_2O} + d_s\,\mathrm{N_2} \quad (1)$$

The coefficient w in Eq. (1) is equal to the number of moles of water vapor associated with one mole of dry air and it can be obtained from the humidity ratio. The humidity ratio of the incoming air ($\omega_{a,in}$) is obtained with the humrat function in EES. The molar mass of air and water vapor (MW_a and MW_v) are obtained using the MolarMass function.

```
omega_a_in=humrat(AirH2O,P=P,T=T_a_in,R=phi_a_in)      "inlet humidity ratio"
MW_a=MolarMass(Air)                                    "molar mass of air"
MW_v=MolarMass(H2O)                                    "molar mass of water"
```

The humidity ratio is defined as the ratio of the mass of water vapor to the mass of dry air:

$$\omega_{a,in} = \frac{m_v}{m_a} \quad (2)$$

Substituting the molar mass of air and water into Eq. (2) leads to:

$$\omega_{a,in} = \frac{n_v \, MW_v}{n_a \, MW_a} \quad (3)$$

Rearranging Eq. (3) provides the coefficient w in Eq. (1):

$$w = \frac{n_v}{n_a} = \omega_{a,in} \frac{MW_a}{MW_v}$$

```
"balance stoichiometric reaction"
w=omega_a_in*MW_a/MW_v          "number of moles of water vapor per mole of dry air"
```

Carbon, hydrogen, oxygen, and nitrogen atom balances for Eq. (1) provide:

$$y_{ng,CH_4} = b_s$$

$$4\, y_{ng,CH_4} + 2\, y_{ng,H_2O} + a_s\, 2\, w = 2\, c_s$$

$$y_{ng,H_2O} + a_s\, 0.21\, (2) + a_s\, w = 2\, b_s + c_s$$

$$2\, y_{ng,N_2} + a_s\, 0.79\, (2) = d_s\, 2$$

```
y_ng_CH4=b_s                               "carbon atom balance"
4*y_ng_CH4+2*y_ng_H2O+a_s*2*w=2*c_s        "hydrogen atom balance"
y_ng_H2O+0.21*2*a_s+a_s*w=2*b_s+c_s        "oxygen atom balance"
2*y_ng_N2+2*a_s*0.79=2*d_s                 "nitrogen atom balance"
```

The actual reaction occurs with 50% excess air. Therefore, the chemical reaction for the actual reaction is given by:

$$\left(y_{ng,CH_4}\, CH_4 + y_{ng,N_2}\, N_2 + y_{ng,H_2O}\, H_2O\right) + a\, (0.21\, O_2 + 0.79\, N_2 + w\, H_2O) \rightarrow b\, CO_2 + c\, H_2O + d\, N_2 + e\, O_2 \quad (4)$$

where

$$a = (1 + ExcessAir)\, a_s$$

and $ExcessAir = 0.5$ (i.e., 50% excess air).

```
"balance actual reaction"
a=(1+ExcessAir)*a_s          "amount of air"
```

Carbon, hydrogen, oxygen, and nitrogen atom balances on Eq. (4) provide:

$$y_{ng,CH_4} = b$$

$$4\, y_{ng,CH_4} + 2\, y_{ng,H_2O} + a\, 2\, w = 2\, c$$

$$y_{ng,H_2O} + a\,0.21\,(2) + a\,w = 2\,b + c + 2\,e$$

$$2\,y_{ng,N_2} + a\,0.79\,(2) = 2\,d$$

```
y_ng_CH4=b                              "carbon atom balance"
4*y_ng_CH4+2*y_ng_H2O+2*a*w=2*c         "hydrogen atom balance"
y_ng_H2O+0.21*2*a+a*w=2*b+c+2*e         "oxygen atom balance"
2*y_ng_N2+2*a*0.79=2*d                  "nitrogen atom balance"
```

The standardized molar specific enthalpies of the methane, nitrogen, and water vapor in the natural gas at the inlet state ($\bar{h}_{CH_4,ng,in}$, $\bar{h}_{N_2,ng,in}$, and $\bar{h}_{H_2O,ng,in}$) are computed using EES, assuming that these substances behave as ideal gases (i.e., using the substances CH4, N2, and H2O, respectively). Note that EES directly returns the standardized molar specific enthalpies for these subtances provided that the unit system has been set to return values on a molar basis.

```
h_bar_CH4_ng_in=enthalpy(CH4,T=T_ng_in)
                    "molar specific enthalpy of the CH4 entering with ng"
h_bar_N2_ng_in=enthalpy(N2,T=T_ng_in)
                    "molar specific enthalpy of the N2 entering with ng"
h_bar_H2O_ng_in=enthalpy(H2O,T=T_ng_in)
                    "molar specific enthalpy of the water vapor entering with ng"
```

The standardized molar specific enthalpies of the oxygen, nitrogen, and water vapor entering the furnace with the air at the inlet ($\bar{h}_{O_2,a,in}$, $\bar{h}_{N_2,a,in}$, and $\bar{h}_{H_2O,a,in}$) are also computed.

```
h_bar_O2_a_in=enthalpy(O2,T=T_a_in)
                    "molar specific enthalpy of the O2 entering with air"
h_bar_N2_a_in=enthalpy(N2,T=T_a_in)
                    "molar specific enthalpy of the N2 entering with air"
h_bar_H2O_a_in=enthalpy(H2O,T=T_a_in)
                    "molar specific enthalpy of the water vapor entering with air"
```

The total enthalpy of the reactants per mole of natural gas is computed according to:

$$H_R = y_{ng,CH_4}\,\bar{h}_{CH_4,ng,in} + y_{ng,N_2}\,\bar{h}_{N_2,ng,in} + y_{ng,H_2O}\,\bar{h}_{H_2O,ng,in} + 0.21\,a\,\bar{h}_{O_2,a,in} + 0.79\,a\,\bar{h}_{N_2,a,in} + a\,w\,\bar{h}_{H_2O,a,in}$$

```
H_R=y_ng_CH4*h_bar_CH4_ng_in+y_ng_N2*h_bar_N2_ng_in+y_ng_H2O*h_bar_H2O_ng_in&
   +a*0.21*h_bar_O2_a_in+a*0.79*h_bar_N2_a_in+a*w*h_bar_H2O_a_in
                    "enthalpy of reactants per mole of natural gas"
```

We next turn our attention to the products. In order to calculate the molar specific enthalpy of the water in the products it is necessary to determine whether the water is in a vapor or liquid state. Assuming no condensation, the molar coefficients provided in Eq. (4) can be used to determine the partial pressure of the water at the outlet:

$$P_{v,out} = \frac{c}{(b + c + d + e)}P$$

The dew point temperature of the combustion products ($T_{dp,out}$) is the saturation temperature of water at $P_{v,out}$, determined using the T_sat function in EES.

```
P_v_out=c*P/(b+c+d+e)                  "partial pressure of water in products"
T_dp_out=T_sat(Water,P=P_v_out)        "dew point temperature of products"
T_dp_out_C=converttemp(K,C,T_dp_out)   "in C"
```

Solving these equations provides $T_{dp,out} = 53.69°\text{C}$. Because $T_{dp,out}$ is less than $T_{out} = 175°\text{C}$, the water is vapor in the combustion products and the standardized molar specific enthalpy ($\bar{h}_{H_2O,g,out}$) can be obtained using the substance H2O in EES.

```
h_bar_H2O_g_out=enthalpy(H2O,T=T_out)     "molar specific enthalpy of H2O leaving furnace"
```

The standardized molar specific enthalpies of the carbon dioxide, nitrogen, and oxygen leaving the furnace ($\bar{h}_{CO_2,out}$, $\bar{h}_{N_2,out}$, and $\bar{h}_{O_2,out}$) are computed and the enthalpy of the products per mole of natural gas is obtained from:

$$H_P = b\,\bar{h}_{CO_2,out} + c\,\bar{h}_{H_2O,g,out} + d\,\bar{h}_{N_2,out} + e\,\bar{h}_{O_2,out}$$

```
h_bar_CO2_out=enthalpy(CO2,T=T_out)  "molar specific enthalpy of CO2 leaving furnace"
h_bar_N2_out=enthalpy(N2,T=T_out)    "molar specific enthalpy of N2 leaving furnace"
h_bar_O2_out=enthalpy(O2,T=T_out)    "molar specific enthalpy of O2 leaving furnace"
H_P=b*h_bar_CO2_out+c*h_bar_H2O_g_out+d*h_bar_N2_out+e*h_bar_O2_out
                                     "enthalpy of products per mole of natural gas"
```

An energy balance on the furnace (on a per mole of natural gas basis) provides:

$$H_R = Q_{out,conv} + H_P$$

```
H_R=Q_out_conv+H_P                   "energy balance on the conventional furnace"
```

which results in $Q_{out,conv} = 686.3$ MJ/kmol of natural gas. In order to compute the efficiency of the furnace, it is necessary to compute the higher heating value of the natural gas. By convention, the heating value is determined by calculating the heat that would result for complete combustion of the gas at $T_{ref} = 25°\text{C}$, despite the fact that the gas actually enters the furnace at a slightly different temperature. The chemical reaction for the complete combustion of one mole of natural gas with a stoichiometric amount of dry air is:

$$\left(y_{ng,CH_4}\text{CH}_4 + y_{ng,N_2}\text{N}_2 + y_{ng,H_2O}\text{H}_2\text{O}\right) + a_{s,da}\left(0.21\,\text{O}_2 + 0.79\text{N}_2\right) \rightarrow b_{s,da}\text{CO}_2 + c_{s,da}\text{H}_2\text{O} + d_{s,da}\text{N}_2 \tag{5}$$

Carbon, hydrogen, oxygen, and nitrogen atom balances on Eq. (5) provide:

$$y_{ng,CH_4} = b_{s,da}$$

$$4\,y_{ng,CH_4} + 2\,y_{ng,H_2O} = 2\,c_{s,da}$$

$$y_{ng,H_2O} + a_{s,da}\,0.21\,(2) = 2\,b_{s,da} + c_{s,da}$$

$$2\,y_{ng,N_2} + a_{s,da}\,0.79\,(2) = d_{s,da}\,2$$

```
"balance stoichiometric reaction with dry air"
y_ng_CH4=b_s_da                               "carbon atom balance"
4*y_ng_CH4+2*y_ng_H2O=2*c_s_da                "hydrogen atom balance"
y_ng_H2O+0.21*2*a_s_da=2*b_s_da+c_s_da        "oxygen atom balance"
2*y_ng_N2+2*a_s_da*0.79=2*d_s_da              "nitrogen atom balance"
```

The standardized molar specific enthalpy of water vapor at T_{ref} ($\bar{h}_{H_2O,g,ref}$) is evaluated using the substance H2O. The standardized molar specific enthalpy of the liquid water leaving at T_{ref} ($\bar{h}_{H_2O,f,ref}$) is evaluated using the molar specific enthalpy of formation provided in Table 13-2.

```
T_ref=converttemp(C,K,25 [C])            "reference temperature"
h_bar_H2O_g_ref=enthalpy(H2O,T=T_ref)    "molar specific enthalpy of vapor H2O at ref. temperature"
h_bar_form_H2O_f= -285813 [kJ/kmol]      "molar specific enthalpy of formation for liquid water"
h_bar_H2O_f_ref=h_bar_form_H2O_f         "molar specific enthalpy of liquid H2O at reference condition"
```

The standardized molar specific enthalpies of the other substances in the reaction are evaluated at T_{ref} ($\bar{h}_{CH_4,ref}$, $\bar{h}_{N_2,ref}$, $\bar{h}_{O_2,ref}$, and $\bar{h}_{CO_2,ref}$).

```
h_bar_CH4_ref=enthalpy(CH4,T=T_ref)     "molar specific enthalpy of CH4 at reference condition"
h_bar_N2_ref=enthalpy(N2,T=T_ref)       "molar specific enthalpy of N2 at reference condition"
h_bar_O2_ref=enthalpy(O2,T=T_ref)       "molar specific enthalpy of O2 at reference condition"
h_bar_CO2_ref=enthalpy(CO2,T=T_ref)     "molar specific enthalpy of CO2 at reference condition"
```

The enthalpy of the reactants at reference conditions per mole of natural gas is evaluated according to:

$$H_{R,ref} = y_{ng,CH_4}\,\bar{h}_{CH_4,ref} + \left(y_{ng,N_2} + 0.79\,a_{s,da}\right)\bar{h}_{N_2,ref} + y_{ng,H_2O}\,\bar{h}_{H_2O,g,ref} + 0.21\,a_{s,da}\,\bar{h}_{O_2,a,in}$$

The enthalpy of the products at reference conditions per mole of natural gas is evaluated according to:

$$H_{P,ref} = b_{s,da}\,\bar{h}_{CO_2,ref} + c_{s,da}\,\bar{h}_{H_2O,f,ref} + d_{s,da}\,\bar{h}_{N_2,ref}$$

The higher heating value is obtained from:

$$H_{R,ref} = HHV + H_{P,ref}$$

and the efficiency of the conventional furnace is given by:

$$\eta_{conv} = \frac{Q_{out,conv}}{HHV}$$

```
H_R_ref=y_ng_CH4*h_bar_CH4_ref+(y_ng_N2+a_s_da*0.79)*h_bar_N2_ref&
   +y_ng_H2O*h_bar_H2O_g_ref
                    "enthalpy of reactants at reference conditions per mole of natural gas"
H_P_ref=b_s_da*h_bar_CO2_ref+c_s_da*h_bar_H2O_f_ref+d_s_da*h_bar_N2_ref
                    "enthalpy of products at reference conditions per mole of natural gas"
H_R_ref=HHV+H_P_ref        "higher heating value"
eta_conv=Q_out_conv/HHV    "efficiency of conventional furnace"
```

which results in $HHV = 837.5$ MJ/kmol of natural gas and $\eta_{conv} = 0.8194$ (81.94%).

b) An advanced furnace uses pulse combustion to force combustion gases out of the furnace, rather than relying on a buoyancy effect as in a conventional furnace. As a result, combustion gases can exit the furnace at a lower temperature; in this case, they exit at $T_{out,adv} = 42°C$. A drain is provided for the condensate that may form due to the reduced product temperature. What is the efficiency of this pulse combustion furnace based on the higher heating value of the fuel?

The actual combustion reaction remains as given by Eq. (4); however, the outlet temperature is now below the dew point temperature computed in part (a), $T_{dp,out} = 53.69°C$. Therefore, some of the water vapor in the combustion products will condense and this effect must be accounted for in the energy balance. The combustion reaction in Eq. (4) is written again; this time, the water in the products is broken into two components corresponding to water vapor (c_g) and liquid water (c_f):

$$\begin{aligned}&\left(y_{ng,CH_4}CH_4 + y_{ng,N_2}N_2 + y_{ng,H_2O}H_2O\right) + a\left(0.21\,O_2 + 0.79N_2 + w\,H_2O\right)\\&\rightarrow b\,CO_2 + c_g H_2O_g + c_f\,H_2O_f + d N_2 + e\,O_2\end{aligned} \tag{6}$$

Note that the coefficients c_f and c_g must add to the original coefficient c in Eq. (4):

$$c_f + c_g = c$$

The coefficient c_f represents the number of moles of liquid water per mole of natural gas and c_g represents the number of moles of water vapor per mole of natural gas. Because the outlet temperature is below the dew point temperature, the combustion products will be saturated and the partial pressure of water vapor in the products will be equal to the saturation vapor pressure at 42°C ($P_{v,out,adv}$), determined using the P_sat function in EES. The mole fraction of the water that remains in vapor form is the ratio of the partial pressure of water vapor to atmospheric pressure:

$$y_{v,out,adv} = \frac{P_{v,out,adv}}{P}$$

According to Eq. (6), the mole fraction of water vapor in the products that exist in the vapor phase can be written as:

$$y_{v,out,adv} = \frac{c_g}{b + c_g + d + e}$$

```
"Advanced furnace"
T_out_adv=converttemp(C,K,42 [C])   "exit temperature of advanced furnace"
P_v_out_adv=P_sat(Water,T=T_out_adv)
                                     "partial pressure of water vapor in combustion products"
y_v_out_adv=P_v_out_adv/P            "mole fraction of water vapor in the vapor products"
y_v_out_adv=c_g/(b+c_g+d+e)          "mole fraction based on coefficients of vapor products"
c_g+c_f=c                            "coefficients of liquid and vapor water must add to total water"
```

The enthalpy of the reactants per mole of natural gas, H_R, is the same as was calculated in part (a). The standardized molar specific enthalpy of the liquid water leaving at $T_{out,adv}$ is obtained from Eq. (13-50):

$$\bar{h}_{H_2O,f,out,adv} = \bar{h}_{form,H_2O,f} + \bar{h}_{H_2O}\left(T_{out,adv}, P\right) - \bar{h}\left(T_{ref}, P_{ref}\right)$$

```
h_bar_H2O_f_out_adv=h_bar_form_H2O_f+enthalpy(Water,T=T_out,P=P)-enthalpy(Water,T=T_ref,P=P)
                          "molar specific enthalpy of liquid H2O at T_out"
```

The molar specific enthalpies of the other constituents of the products leaving the advanced furnace ($\bar{h}_{CO_2,out,adv}$, $\bar{h}_{N_2,out,adv}$, $\bar{h}_{H_2O,g,out,adv}$ and $\bar{h}_{O_2,out,adv}$) are evaluated using the enthalpy property function for the ideal gas substances in EES.

```
h_bar_CO2_out_adv=enthalpy(CO2,T=T_out_adv)      "molar specific enthalpy of CO2 leaving furnace"
h_bar_H2O_g_out_adv=enthalpy(H2O,T=T_out_adv)    "molar specific enthalpy of H2O vapor furnace"
h_bar_N2_out_adv=enthalpy(N2,T=T_out_adv)        "molar specific enthalpy of N2 leaving furnace"
h_bar_O2_out_adv=enthalpy(O2,T=T_out_adv)        "molar specific enthalpy of O2 leaving furnace"
```

The enthalpy of the products per mole of natural gas is evaluated according to:

$$H_{P,adv} = b\,\bar{h}_{CO_2,out,adv} + c_g\,\bar{h}_{H_2O,g,out,adv} + c_f\,\bar{h}_{H_2O,f,out,adv} + d\,\bar{h}_{N_2,out,adv} + e\,\bar{h}_{O_2,out,adv}$$

An energy balance on the advanced furnace provides:

$$H_R = Q_{out,adv} + H_{P,adv}$$

```
H_P_adv=b*h_bar_CO2_out_adv+c_g*h_bar_H2O_g_out_adv+c_f*h_bar_H2O_f_out_adv&
   +d*h_bar_N2_out_adv+e*h_bar_O2_out_adv
                       "enthalpy of products per mole of natural gas"
H_R=Q_out_adv+H_P_adv  "energy balance on advanced furnace"
```

which provides $Q_{out,adv} = 791.4$ kJ/kmol of natural gas. The efficiency of the advanced furnace is computed:

$$\eta_{adv} = \frac{Q_{out,adv}}{HHV}$$

```
eta_adv=Q_out_adv/HHV     "efficiency of advanced furnace"
```

which results in $\eta_{adv} = 0.9449$ (94.49%). This efficiency is significantly higher than the furnace efficiency determined for the conventional furnace in part (a). The pulse combustion furnace will likely be more expensive, but it may pay for itself in fuel cost savings.

c) Determine the volumetric flow rate of the natural gas required to provide $\dot{Q} = 23.5$ kW of heat using the pulse combustion furnace. What is the volumetric flow rate at which condensate is formed?

The molar flow rate of natural gas can be obtained from:

$$\dot{Q} = \dot{n}_{ng}\, Q_{out,adv}$$

where $Q_{out,adv}$ is the heat provided per mole of natural gas in the advanced furnace, computed in part (b). The molar specific volume of the natural gas is obtained from the ideal gas law on a molar basis:

$$\bar{v}_{ng} = \frac{R_{univ}\, T_{ng,in}}{P}$$

The volumetric flow rate of natural gas is:

$$\dot{V}_{ng} = \bar{v}_{ng}\, \dot{n}_{ng}$$

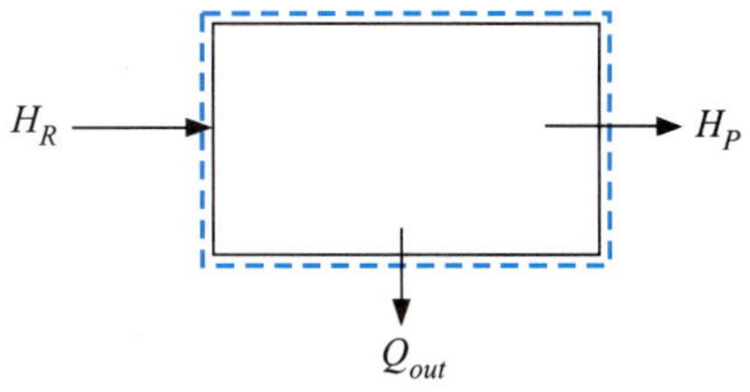

Figure 13-4: Open combustion system.

```
Q_dot=23.5 [kW]        "required heating capacity"
Q_dot=n_dot_ng*Q_out_adv
                       "required molar flow rate"
v_bar_ng=R#*convert(kJ/kmol-K,J/kmol-K)*T_ng_in/P
                       "molar specific volume of natural gas entering"
V_dot_ng=n_dot_ng*v_bar_ng
                       "volumetric flow rate of natural gas entering"
V_dot_ng_cfm=V_dot_ng*convert(m^3/s,cfm)
                       "in cfm"
```

which provides $\dot{V}_{ng} = 1.462$ cfm. The molar flow rate of condensate is computed according to:

$$\dot{n}_c = c_f\, \dot{n}_{ng}$$

The molar specific volume of the condensate ($\bar{v}_c$) is computed using the volume function in EES. The volumetric flow rate of condensate is obtained according to:

$$\dot{V}_c = \bar{v}_c\, \dot{n}_c$$

```
n_dot_c=c_f*n_dot_ng                           "molar flow rate of condensate"
v_bar_c=volume(Water,T=T_out_adv,P=P)          "molar specific volume of condensate"
V_dot_c=n_dot_c*v_bar_c                        "volumetric flow rate of condensate"
V_dot_c_gph=V_dot_c*convert(m^3/s,gal/hr)      "in gal/hr"
```

which leads to $\dot{V}_c = 0.5347$ gal/hr.

13.3.5 Adiabatic Reactions

Large energy changes occur in combustion reactions. These energy changes can be used to provide heat; alternatively, the energy changes can be used to raise the temperature of the combustion products. Figure 13-4 illustrates an open combustion system in which the fuel and oxidizer continuously flow in and the products of combustion continuously flow out. A steady-state energy balance on the combustor provides:

$$H_R = Q_{out} + H_P \tag{13-51}$$

where H_R is the enthalpy of the reactants, H_P is the enthalpy of the products, and Q_{out} is the heat transfer from the combustor. The products leaving the combustor will achieve their highest temperature when the reaction occurs adiabatically. In this case, the energy balance reduces to:

$$H_R = H_P \tag{13-52}$$

The temperature of the products that satisfies Eq. (13-52) is called the *adiabatic combustion temperature* (or the *adiabatic flame temperature*). The adiabatic combustion temperature can be quite high. It will be highest when a stoichiometric amount of air is provided to the reaction (assuming that all of the fuel is combusted) since any excess air in the products increases the heat capacity of the products and therefore lowers the product temperature. Even higher adiabatic combustion temperatures can be achieved if pure oxygen rather than air is used as the oxidizer in the combustion process because then there is no nitrogen present and the heat capacity of the products is lower.

Acetylene (C_2H_2) is commonly used for welding and cutting torches because of the high temperatures that can result during its combustion. We can calculate the adiabatic combustion temperature associated with the combustion of acetylene with stoichiometric air entering at $T_{in} = 25°C$. The combustion reaction, using a basis of 1 mole of acetylene, is:

$$C_2H_2 + a_s\,(0.21\,O_2 + 0.79\,N_2) \rightarrow b_s\,CO_2 + c_s\,H_2O + d_s\,N_2 \tag{13-53}$$

Balances on the carbon, hydrogen, oxygen and nitrogen atoms are:

$$2 = b_s \tag{13-54}$$

$$2 = 2\,c_s \tag{13-55}$$

$$(2)0.21\,a_s = 2b_s + c_s \tag{13-56}$$

$$(2)\,0.79\,a_s = 2\,d_s \tag{13-57}$$

```
$UnitSystem SI Molar J K Pa
T_in=converttemp(C,K,25 [C])              "inlet temperature"

"balance reaction"
2=b_s                                     "carbon atoms"
2=2*c_s                                   "hydrogen atoms"
2*0.21*a_s=2*b_s+c_s                      "oxygen atoms"
2*0.79*a_s=2*d_s                          "nitrogen atoms"
```

The enthalpy of the reactants per mole of acetylene is provided by:

$$H_R = \bar{h}_{C_2H_2,in} + 0.21\,a_s\,\bar{h}_{O_2,in} + 0.79\,a_s\,\bar{h}_{N_2,in} \tag{13-58}$$

where the standardized molar specific enthalpies of acetylene, oxygen, and nitrogen ($\bar{h}_{C_2H_2,in}$, $\bar{h}_{O_2,in}$, and $\bar{h}_{N_2,in}$) entering the combustor are evaluated using the EES enthalpy function with ideal gas substances.

```
h_bar_C2H2_in=enthalpy(C2H2,T=T_in)  "molar specific enthalpy of acetylene"
h_bar_02_in=enthalpy(02,T=T_in)      "molar specific enthalpy of oxygen"
h_bar_N2_in=enthalpy(N2,T=T_in)      "molar specific enthalpy of nitrogen"
H_R=h_bar_C2H2_in+0.21*a_s*h_bar_02_in+0.79*a_s*h_bar_N2_in
                                     "enthalpy of reactants"
```

Assuming an adiabatic constant pressure reaction with negligible kinetic and potential energy effects, the enthalpy of the products must be equal to the enthalpy of the reactants, as indicated by Eq. (13-52). We will further assume that the reaction is complete and that the products of the reaction are those included in Eq. (13-53). In the actual case, additional reactions occur at the high temperatures that are encountered during adiabatic combustion. For example, oxygen and nitrogen may react to some extent and hydrogen and carbon dioxide may dissociate. However, the energy effects of these reactions are small and we will neglect them here. These effects will be examined in more detail in Chapter 14. The enthalpy of the products per mole of acetylene is calculated according to:

$$H_P = b_s\,\bar{h}_{CO_2,out} + c_s\,\bar{h}_{H_2O,out} + d_s\,\bar{h}_{N_2,out} \tag{13-59}$$

Equation (13-59) is difficult to solve by hand because the specific enthalpies of the products depend on the outlet temperature (i.e., the adiabatic combustion temperature) and this temperature is not known. One way to solve the equation is to prepare a plot of the difference between H_R and H_P as a function of outlet temperature and use the plot to identify the outlet temperature that satisfies Eq. (13-52). For example, assume that the products are at $T_{out} = 1000$ K. The molar specific enthalpies of the products exiting the reactor ($\bar{h}_{CO_2,out}$, $\bar{h}_{H_2O,out}$, and $\bar{h}_{N_2,out}$) are evaluated using the EES **enthalpy** function. Note that the high temperatures associated with an adiabatic reaction will likely preclude the possibility that any water in the products condenses. This assumption can be checked once the problem is completed by computing the dew point temperature of the products.

```
T_out=1000 [K]                              "guess for the outlet temperature"
h_bar_CO2_out=enthalpy(CO2,T=T_out)         "molar specific enthalpy of carbon dioxide"
h_bar_H2O_out=enthalpy(H2O,T=T_out)         "molar specific enthalpy of water vapor"
h_bar_N2_out=enthalpy(N2,T=T_out)           "molar specific enthalpy of nitrogen"
H_P=b_s*h_bar_CO2_out+c_s*h_bar_H2O_out+d_s*h_bar_N2_out
                                            "enthalpy of products"
```

The error in the energy balance, Eq. (13-52), is calculated according to:

$$err = H_R - H_P \tag{13-60}$$

```
err=H_R-H_P                                 "error in energy balance"
```

For the assumed outlet temperature $T_{out} = 1000$ K the error in the energy balance is $err = 9.623 \times 10^8$ J/kmol. Figure 13-5 illustrates the error in the energy balance, calculated according to Eq. (13-60), as a function of the assumed outlet temperature.

If the problem were completed by hand, it would be necessary to guess a few outlet temperatures and then interpolate or extrapolate based on the energy balance errors computed at these temperatures in order to estimate the correct temperature (i.e., the temperature where $err = 0$ J/kmol). Based on Figure 13-5, the adiabatic combustion temperature must be approximately 2850 K. This is a tedious process if done by hand; it is much easier to solve this problem with EES. Comment out the guessed value for T_{out}:

```
{T_out=1000 [K]}                            "guess for the outlet temperature"
```

and specify that the error in the energy balance must be zero.

```
err=0 [J/kgmol]                             "set error to zero"
```

The program will solve readily, but look at the solution. The calculated value of T_{out} is 0.61 K! The problem here is that the specific heat capacities are non-linear functions of temperature and therefore Eq. (13-52) has multiple solutions. EES solves equations in an iterative manner and it will often find a solution that is closest to the guess value for the variables that are being calculated. Unless you specify otherwise, the guess value for

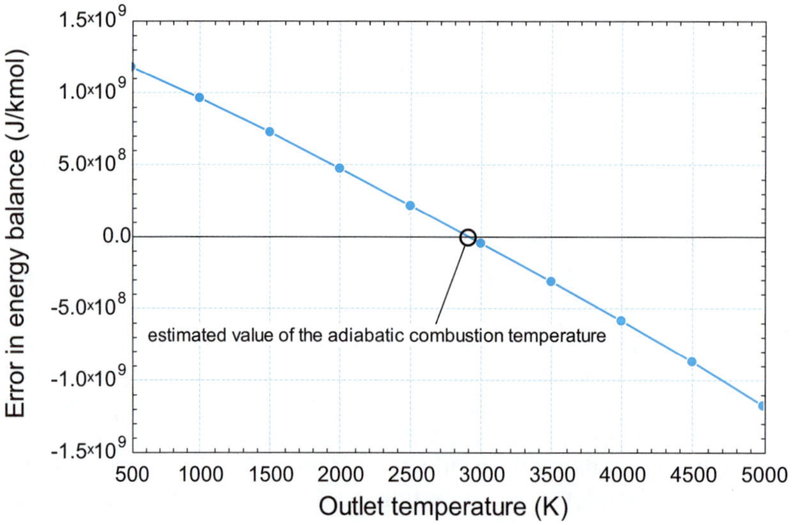

Figure 13-5: Error in energy balance as a function of the outlet temperature.

every variable is 1.0. Use the Variable Information dialog (selected from the Options menu) to change the guess value of T_out to 1000 K, as shown in Figure 13-6. Solve again and you should find that the correct value of the adiabatic combustion temperature is obtained, $T_{out} = 2909$ K. Note that the correct value of T_{out} will be predicted by EES for any reasonable value of the guess used for T_{out}.

The adiabatic combustion temperature will be much higher if pure oxygen, rather than air, is employed as the oxidizer. We can determine the adiabatic combustion temperature with pure oxygen by simply commenting out the enthalpy contribution of the nitrogen in the reactants and the products.

```
H_R=h_bar_C2H2_in+0.21*a_s*h_bar_02_in{+0.79*a_s*h_bar_N2_in}  "enthalpy of reactants"
H_P=b_s*h_bar_C02_out+c_s*h_bar_H20_out{+d_s*h_bar_N2_out}     "enthalpy of products"
```

Solving shows that the adiabatic combustion temperature using pure oxygen is $T_{out} =$ 6877 K. Note that this temperature is higher than the upper limit of the property correlations used in EES and therefore warnings will be generated to indicate that this result may not be accurate. In addition, the actual adiabatic combustion temperature will be lower than this value due to heat transfer and also due to the energy effects

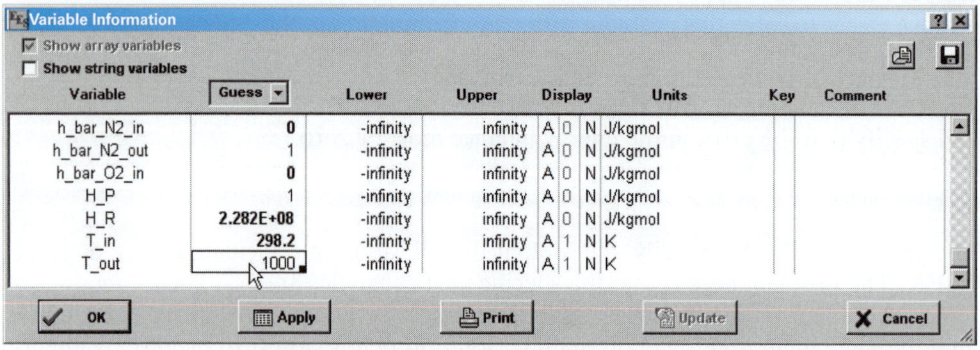

Figure 13-6: Variable Information dialog used to set the guess value of T_out to 1000 K.

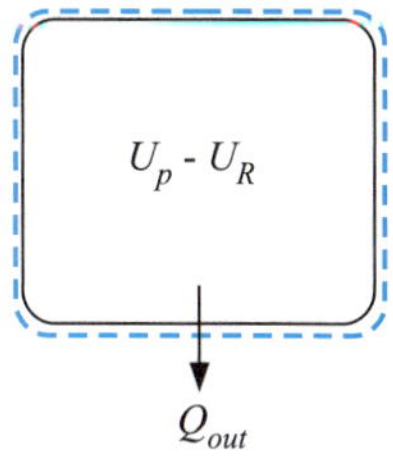

Figure 13-7: Closed combustion system.

associated with dissociation reactions, which will tend to become more important as the temperature is increased.

Open system combustion reactions of the type depicted in Figure 13-4 commonly occur in an open system such as in the combustor of a gas turbine engine, described in Section 8.3. Figure 13-7 illustrates a closed combustion system. The reactants are initially placed within a rigid container and ignited, causing the combustion process to occur. At the conclusion of the combustion process the products exist within the container. This situation is similar to the process that occurs in the cylinder of a four-stroke internal combustion engine, as discussed in Section 8.4.

Both the temperature and the pressure of the reaction products may significantly increase in a closed combustion system. An energy balance on the combustor provides:

$$0 = Q_{out} + U_P - U_R \tag{13-61}$$

where Q_{out} is the heat transfer from the combustor. The pressure and temperature within the container will achieve their highest value when the reaction occurs adiabatically with no power output and negligible changes in kinetic and potential energy. In this case, the energy balance reduces to:

$$U_R = U_P \tag{13-62}$$

where U_R is the internal energy of the reactants and U_P is the internal energy of the products. The pressure that is achieved if the combustion process occurs adiabatically is sometimes referred to as the *adiabatic explosion pressure*. Enthalpy is defined as:

$$H = U + P\,V \tag{13-63}$$

Substituting the ideal gas law into Eq. (13-63) provides:

$$H = U + n_g\,R_{univ}\,T \tag{13-64}$$

where n_g is the number of moles of gas. Equation (13-64) can be used to evaluate the internal energy of both the reactants and products.

$$U_R = H_R - n_{g,R}\,R_{univ}\,T_R \tag{13-65}$$

$$U_P = H_P - n_{g,P}\,R_{univ}\,T_P \tag{13-66}$$

where T_R and T_P are the temperatures of the reactants and products, respectively.

EXAMPLE 13.3-4: DETERMINATION OF THE EXPLOSION PRESSURE OF METHANE

An experimental study on the pressure change that occurs during closed vessel adiabatic reactions was reported by Razus et al. (2006). The investigators studied the influence of initial pressure, fuel type, air fuel ratio, ignition source, vessel geometry and heat losses on the explosion pressure for many different hydrocarbon fuels. A spherical vessel having an internal diameter of $D_v = 10$ cm was used in some of their experiments. During one experiment, an explosion pressure $P_{exp} = 8.25$ bar was measured for $y_{CH_4} = 9\%$ methane combusted in air at $T_R = 25°C$ and $P_R = 1$ bar.

a) Determine the mass of methane initially contained in the vessel

The inputs are entered in EES.

```
$UnitSystem SI K Pa J molar
D_v=10 [cm]*convert(cm,m)              "vessel diameter"
T_R=convertTemp(C,K,25 [C])            "temperature of reactants"
P_R=1 [bar]*convert(bar, Pa)           "pressure of reactants"
y_CH4=0.09 [-]                         "mole fraction of methane in the reactants"
```

The volume contained in the vessel is computed according to:

$$V = \frac{4}{3}\pi\left(\frac{D_v}{2}\right)^3$$

The total number of moles of reactants is computed using the ideal gas law:

$$n_R = \frac{P_R\, V}{R_{univ}\, T_R}$$

```
Vol=4/3*pi*(D_v/2)^3                   "volume of vessel"
n_R=(P_R*Vol)/(R#*T_R)                 "total moles of reactants"
```

The number of moles of methane in the reactants is given by:

$$n_{CH_4} = y_{CH_4}\, n_R$$

The number of moles of air in the reactants is:

$$n_{air} = n_R - n_{CH_4}$$

```
n_CH4=y_CH4*n_R                        "moles of CH4 in vessel"
n_air=n_R-n_CH4                        "moles of air in vessel"
```

The mass of methane in the vessel is obtained according to:

$$m_{CH_4} = MW_{CH_4}\, n_{CH_4}$$

where MW_{CH_4} is the molar mass of methane, obtained using the MolarMass function in EES.

```
m_CH4=n_CH4*MolarMass(CH4)             "mass of methane in vessel"
m_CH4_g=m_CH4*convert(kg,g)            "in g"
```

Solving results in $m_{CH_4} = 0.0305$ g.

b) Determine the % theoretical air associated with the experiment.

The stoichiometric reaction of air with methane can be written as:

$$CH_4 + a_s\,(0.21\,O_2 + 0.79\,N_2) \rightarrow b_s CO_2 + c_s H_2O + d_s N_2 \tag{1}$$

Balancing carbon, hydrogen, oxygen, and nitrogen atoms in Eq. (1) produces the following equations:

$$1 = b_s$$

$$4 = 2\,c_s$$

$$a_s\,2\,(0.21) = 2\,b_s + c_s$$

$$a_s\,2\,(0.79) = 2\,d_s$$

```
"balance stoichiometric reaction"
1=b_s                          "carbon atom balance"
4=2*c_s                        "hydrogen atom balance"
a_s*2*0.21=2*b_s+c_s           "oxygen atom balance"
a_s*2*0.79=2*d_s               "nitrogen atom balance"
```

The solution of these equations provides the coefficients for the balanced, stoichiometric chemical reaction shown in Eq. (2).

$$CH_4 + 9.524\,(0.21\,O_2 + 0.79\,N_2) \rightarrow 1\,CO_2 + 2\,H_2O + 7.524\,N_2 \tag{2}$$

The actual reaction can be written as:

$$CH_4 + a\,(0.21\,O_2 + 0.79\,N_2) \rightarrow b\,CO_2 + c\,H_2O + d\,N_2 + e\,O_2 \tag{3}$$

The coefficient a in Eq. (3) is the ratio of the number of moles of air to the number of moles of methane in the reactants:

$$a = \frac{n_{air}}{n_{CH_4}}$$

```
"balance actual reaction"
a=n_air/n_CH4                  "coefficient on air in the chemical reaction"
```

which results in $a = 10.11$. The % theoretical air is obtained from:

$$\%\ \text{theoretical air} = 100\,\frac{a}{a_s}$$

```
percent_theoretical_air=(a/a_s)*convert(-,%)        "percent theoretical air"
```

which results in 106.2% theoretical air. Balancing carbon, hydrogen, oxygen, and nitrogen atoms in the actual reaction, Eq. (3), provide the following equations:

$$1 = b$$

$$4 = 2\,c$$

$$a\,2\,(0.21) = 2\,b + c + 2\,e$$

$$a\,2\,(0.79) = 2\,d$$

```
1=b                       "carbon atom balance"
4=2*c                     "hydrogen atom balance"
a*2*0.21=2*b+c+2*e        "oxygen atom balance"
a*2*0.79=2*d              "nitrogen atom balance"
```

The solution provides the coefficients for the balanced reaction in Eq. (4).

$$CH_4 + 10.11\,(0.21\,O_2 + 0.79\,N_2) \rightarrow 1\,CO_2 + 2\,H_2O + 7.988\,N_2 + 0.123\,O_2 \quad (4)$$

c) Calculate the adiabatic explosion pressure and the corresponding temperature of the products for this experiment.

The reaction is assumed to be adiabatic. An energy balance indicates that the internal energy of the reactants is equal to the internal energy of the products:

$$U_R = U_P \quad (5)$$

The molar specific enthalpy of the reactants ($\bar{h}_{CH_4,R}$, $\bar{h}_{N_2,R}$, and $\bar{h}_{O_2,R}$) are obtained using the enthalpy function in EES; note that the enthalpy function returns the standardized specific enthalpy for any ideal gas substance. The total enthalpy of the reactants is obtained according to:

$$H_R = n_{CH_4}\left(\bar{h}_{CH_4,R} + a\,0.21\,\bar{h}_{N_2,R} + a\,0.79\,\bar{h}_{O_2,R}\right)$$

The internal energy of the reactants is obtained according to Eq. (13-65):

$$U_R = H_R - n_R\,R_{univ}\,T_R$$

```
h_bar_CH4_R=enthalpy(CH4,T=T_R) "molar specific enthalpy of CH4 in reactants"
h_bar_N2_R=enthalpy(N2,T=T_R)   "molar specific enthalpy of N2 in reactants"
h_bar_O2_R=enthalpy(O2,T=T_R)   "molar specific enthalpy of O2 in reactants"
H_R=n_CH4*(h_bar_CH4_R+a*0.21*h_bar_O2_R+a*0.79*h_bar_N2_R)
                                "enthalpy of the reactants"
U_R=H_R-n_R*R#*T_R              "internal energy of the reactants"
```

Note that the internal energy of the reactants could alternatively have been obtained according to:

$$U_R = n_{CH_4}\left(\bar{u}_{CH_4,R} + a\,0.21\,\bar{u}_{N_2,R} + a\,0.79\,\bar{u}_{O_2,R}\right)$$

where $\bar{u}_{CH_4,R}$, $\bar{u}_{N_2,R}$, and $\bar{u}_{O_2,R}$ are the molar specific internal energies obtained using the IntEnergy function in EES.

```
u_bar_CH4_R=IntEnergy(CH4,T=T_R) "molar specific internal energy of CH4 in reactants"
u_bar_N2_R=IntEnergy(N2,T=T_R)   "molar specific internal energy of N2 in reactants"
u_bar_O2_R=IntEnergy(O2,T=T_R)   "molar specific internal energy of O2 in reactants"
U_R_check=n_CH4*(u_bar_CH4_R+a*0.21*u_bar_O2_R+a*0.79*u_bar_N2_R)
                                 "alternative to find U_R"
```

The temperature of the products, T_P, is not yet known. Enter a reasonable value for this temperature so that the calculations can proceed without iteration. The molar specific enthalpy of the products ($\bar{h}_{CO_2,P}$, $\bar{h}_{H_2O,P}$, $\bar{h}_{N_2,P}$, and $\bar{h}_{O_2,P}$) are evaluated at the assumed value of T_P. The water in the products is assumed to be in vapor phase; this situation is likely given the high temperatures that result from the combustion reaction, but we can check to see that this is true at the conclusion of the problem. The total enthalpy of the reactants is obtained according to:

$$H_P = n_{CH_4}\left(b\,\bar{h}_{CO_2,P} + c\,\bar{h}_{H_2O,P} + d\,\bar{h}_{N_2,P} + e\,\bar{h}_{O_2,P}\right)$$

The internal energy of the reactants is obtained according to Eq. (13-65):

$$U_P = H_P - n_P\, R_{univ}\, T_P$$

where n_P is the total number of moles of products (we are assuming that the water is entirely in vapor phase), calculated according to:

$$n_P = n_{CH_4}\,(b + c + d + e)$$

```
T_P=2000 [K]                           "guess for the products temperature"
h_bar_CO2_P=enthalpy(CO2,T=T_P)        "molar specific enthalpy of CO2 in products"
h_bar_H2O_P=enthalpy(H2O,T=T_P)        "molar specific enthalpy of H2O in products"
h_bar_N2_P=enthalpy(N2,T=T_P)          "molar specific enthalpy of N2 in products"
h_bar_O2_P=enthalpy(O2,T=T_P)          "molar specific enthalpy of O2 in products"
H_P=n_CH4*(b*h_bar_CO2_P+c*h_bar_H2O_P+d*h_bar_N2_P+e*h_bar_O2_P)
                                       "enthalpy of the products"
n_P=n_CH4*(b+c+d+e)                    "number of moles of product"
U_P=H_P-n_P*R#*T_P                     "internal energy of the products"
```

Solve the problem and then update the guess values. The internal energy of the reactants must equal the internal energy of products. Comment out the guess that was made for the product temperature and replace this equation with the energy balance, Eq. (5).

```
{T_P=2000 [K]}                         "guess for the products temperature"
U_R=U_P                                "energy balance"
```

The resulting temperature of the products is $T_P = 2715$ K. The adiabatic explosion pressure is obtained by applying the ideal gas law to the products:

$$P_P = \frac{n_P\, R_{univ}\, T_P}{V}$$

```
P_P=n_P*R#*T_P/Vol                     "adiabatic explosion pressure"
P_P_bar=P_P*convert(Pa,bar)            "in bar"
```

The calculated adiabatic explosion pressure is $P_P = 9.106$ bar whereas the experimentally measured value is 8.25 bar. The calculated temperature of the products is far above the critical point for water; clearly, any water in the products will be in the vapor phase.

d) Explain why the calculated result differs from the experimental value.

There are many possible reasons for the discrepancy with the experimental value. The most likely explanation is heat transfer between the hot products and the cold vessel walls. Further, the reaction may not have been complete. At the high temperatures in the vessel additional reactions may occur; for example, the reaction of oxygen and nitrogen to form nitrous oxides. These chemical equilibrium considerations are the subject of Chapter 14.

13.4 Entropy Considerations

Entropy is used in combustion applications to determine the maximum efficiency of the conversion of thermal energy into mechanical power and to quantify the entropy generation in real processes. Entropy is also used to determine the exergy of fuels, as discussed in Section 13.5.

Like internal energy and enthalpy, entropy does not have a natural absolute value. For pure fluids that do not participate in a chemical reaction, we arbitrarily set the specific entropy to zero at a convenient reference state and then determine the values of entropy at all other states relative to this reference state. For example, the specific entropy of the substance 'Water' in EES is arbitrarily set to zero for water in the liquid state at the triple point (273.16 K, 611.6 Pa). However, when chemical reactions occur it is necessary to specify a reference state that is consistent for each of the various substances that are involved in the reaction. We encountered a similar situation for enthalpy in Section 13.3.1; this led to the use of a standardized specific enthalpy that is defined relative to the specific enthalpy of the elements at $T_{ref} = 25°\text{C}$. The standardized specific enthalpy of all other substances is found using the enthalpy change of formation for that particular substance from the elements. This process does not work for entropy since we have no direct way to determine the entropy change of formation of a substance.

Experimental data on the behavior of matter at low temperatures suggest that the specific entropy of a pure substance approaches a constant value as temperature approaches absolute zero. This observation is referred to as the *Third Law of Thermodynamics*. It is also called *Nernst's theorem*, after Walther Nernst who proposed this postulate in the early 1900's. The Third Law of Thermodynamics sets the entropy of all *pure crystalline* substances to zero at 0 K. The Third Law applies to both elements and compounds. The word *pure* in the statement of the Third Law implies that the substance is not in a mixture. A mixture will have an entropy value that is greater than zero, even at 0 K, due to the entropy change of mixing discussed in Chapter 11. Most substances are in a solid phase near 0 K. Substances in a solid phase can be arranged in an organized manner referred to as a crystalline structure. It is this organized state at 0 K that is assigned a zero entropy value. A substance may be able to exist in several different crystalline forms at 0 K; however, all of these forms would have an entropy value of zero at 0 K. The entropy of a substance will be greater than zero at 0 K if it exists in a non-crystalline or glass state.

Like the First and Second Laws, the Third Law of Thermodynamics cannot be proven. It is supported by experimental data at low temperatures and by statistical thermodynamics. The specific entropy of a pure substance is a function of temperature and pressure. At temperatures approaching absolute zero, experiments show that the specific heat capacity at constant pressure, c_P, approaches zero and that the quantity c_P/T also approaches zero. This observation suggests that entropy becomes independent of temperature at 0 K since the partial derivative of specific entropy with respect to temperature at constant pressure, derived in Section 10.3.3, is:

$$\left(\frac{\partial \bar{s}}{\partial T}\right)_P = \frac{\bar{c}_P}{T} \tag{13-67}$$

Experiments also show that the partial derivative of specific volume with respect to pressure at constant temperature approaches zero as the temperature approaches absolute zero. According to one of the Maxwell's relations derived in Section 10.3.1:

$$\left(\frac{\partial \bar{s}}{\partial P}\right)_T = -\left(\frac{\partial \bar{v}}{\partial T}\right)_P \tag{13-68}$$

Therefore, at temperatures approaching absolute zero, entropy becomes independent of pressure. If entropy is independent of both temperature and pressure at absolute zero then it must have a constant value. The Third Law of Thermodynamics assigns this constant to be zero.

With a reference state set by the Third Law, the specific entropy of any pure substance can be obtained from experimental measurements. For example, since the specific entropy at absolute zero is independent of pressure, we can set the specific entropy to be zero at $T = 0$ K and $P = 1$ atm. The molar specific entropy of a gas, e.g. oxygen, at any temperature T and $P = 1$ atm could, in principle, be obtained by the integration indicated in Eq. (13-69):

$$\bar{s}(T, P = 1\,\text{atm}) = \underbrace{\bar{s}\,(T = 0\,\text{K}, P = 1\,\text{atm})}_{=0\text{ by Third Law of Thermodynamics}} + \underbrace{\int_{0K}^{T_m} \frac{\bar{c}_{P,s}}{T} dT}_{\substack{\text{entropy change}\\ \text{of solid with } T}} + \underbrace{\frac{\Delta \bar{h}_{fs}}{T_m}}_{\substack{\text{entropy change}\\ \text{due to melting}}}$$

$$+ \underbrace{\int_{T_m}^{T_v} \frac{\bar{c}_{P,f}}{T} dT}_{\substack{\text{entropy change}\\ \text{of liquid with } T}} + \underbrace{\frac{\Delta \bar{h}_{fg}}{T_v}}_{\substack{\text{entropy change}\\ \text{due to vaporization}}} + \underbrace{\int_{T_v}^{T} \frac{\bar{c}_{P,g}}{T} dT}_{\substack{\text{entropy change}\\ \text{of vapor with } T}} \tag{13-69}$$

In Eq. (13-69), $\bar{c}_{P,s}$, $\bar{c}_{P,f}$ and $\bar{c}_{P,g}$ are the specific heat capacities at constant pressure of the solid, liquid and vapor phases as a function of temperature at 1 atm, $\Delta \bar{h}_{fs}$ and $\Delta \bar{h}_{fg}$ are the molar specific enthalpy changes associated with melting and vaporization at 1 atm, respectively, and T_m and T_v are the melting and vaporization temperatures. In reality, evaluation of specific entropy using Eq. (13-69) is difficult. The specific entropy of an ideal gas can be determined with much less effort using statistical thermodynamics, as shown in Chapter 15.

Values of entropy that are referred to the Third Law reference state are called *absolute entropies*. Absolute entropy values are always greater than zero. Absolute molar specific entropy values at 25°C and 1 atm are provided in Table 13-4 for substances that are commonly involved in combustion reactions. The molar specific entropy of the substances in Table 13-4 that can be modeled as ideal gases can be found at other temperatures and pressures using the ideal gas entropy relations:

$$\bar{s}\,(T, P) = \bar{s}\,(T = 298.15\,\text{K}, P = 1\,\text{atm}) + \int_{298.15\,\text{K}}^{T} \frac{\bar{c}_P}{T} dT - R_{univ} \ln\left(\frac{P}{1\,\text{atm}}\right) \quad \text{for ideal gases} \tag{13-70}$$

The pressure dependence of entropy is small for condensed substances (liquids and solids) and it is generally neglected. Therefore, the entropy of substances in Table 13-4 that can be modeled as being incompressible can be estimated according to:

$$\bar{s}\,(T, P) \approx \bar{s}\,(T = 298.15\,\text{K}, P = 1\,\text{atm}) + \int_{298.15\,\text{K}}^{T} \frac{\bar{c}}{T} dT \quad \text{for incompressible substance} \tag{13-71}$$

Table 13-4: Absolute molar specific entropy values at 298.15 K (25°C or 77°F) and 1 atm for some common substances

Substance	Phase	Chemical Formula	Molar Mass	Absolute Entropy at 1 atm and 25°C (77°F) (kJ/kmol-K)	(Btu/lbmol-R)
Carbon	solid	C	12.00	5.734	1.369
Hydrogen	gas	H_2	2.016	130.673	31.211
Oxygen	gas	O_2	32.00	205.137	48.996
Nitrogen	gas	N_2	28.01	191.598	45.762
Carbon Monoxide	gas	CO	28.01	197.648	47.207
Carbon Dioxide	gas	CO_2	44.01	213.774	51.059
Nitrogen Oxide	gas	NO	30.01	210.735	50.333
Nitrogen Dioxide	gas	NO_2	46.01	240.156	57.360
Sulfur Dioxide	gas	SO_2	64.06	248.207	59.283
Water	liquid	H_2O	18.02	69.938	16.704
Water	gas	H_2O	18.02	188.818	45.098
Ammonia	gas	NH_3	17.03	192.758	46.040
Methane	gas	CH_4	16.04	186.360	44.511
Ethane	gas	C_2H_6	30.07	229.207	54.745
Ethylene	gas	C_2H_4	28.04	219.309	52.381
Acetylene	gas	C_2H_2	26.04	200.904	47.985
Propane	gas	C_3H_8	44.10	270.298	64.560
n-Butane	liquid	C_4H_8	58.12	230.632	55.085
n-Butane	gas	C_4H_8	58.12	309.862	74.009
n-Pentane	liquid	C_5H_{12}	72.15	263.420	62.917
n-Pentane	gas	C_5H_{12}	72.15	349.539	83.486
n-Hexane	liquid	C_6H_{14}	86.17	297.366	71.025
n-Hexane	gas	C_6H_{14}	86.17	388.826	92.870
n-Octane	liquid	C_8H_{18}	114.20	360.833	86.183
n-Octane	gas	C_8H_{18}	114.20	467.321	111.618
Methanol	liquid	CH_3OH	32.04	126.976	30.328
Methanol	gas	CH_3OH	32.04	239.795	57.274
Ethanol	liquid	C_2H_5OH	46.07	159.206	38.026
Ethanol	gas	C_2H_5OH	46.07	280.576	67.014

Source: McBride, B.J., Zehe, M.J. and Gordon, S.,"NASA Glenn Coefficients for Calculating Thermodynamic Properties of Individual Species", NASA/TP—2002-211556, Sept. 2002, http://www.lerc.nasa.gov/WWW/CEAWeb/

Equation (13-70) is applicable to pure substances in gaseous form that can be modeled with the ideal gas law. Gaseous substances that are involved in a chemical reaction are, of course, not pure. However, if the mixture obeys the ideal gas law then the specific entropy of a substance in a gaseous mixture is the same as the specific entropy of the pure substance evaluated at the mixture temperature and its partial pressure in the mixture, as discussed in Section 11.2. EES evaluates the specific entropy of ideal gases using Eq. (13-70) and provides absolute entropy values for these substances that are appropriate for use in a combustion problem. The absolute specific entropy of a gas in a mixture that obeys the ideal gas law can be obtained using the entropy function evaluated using the temperature and partial pressure.

If EES is not available, then Eq. (13-70) can be used to evaluate the absolute entropy of a gas in an ideal gas mixture. However, the specific heat capacity at constant pressure is a function of temperature for most substances and that complicates the evaluation of the integral of c_P with respect to T. The integral in Eq. (13-70) has been evaluated and tabulated as a function of temperature at a pressure of 1 atm for several common gases in the property tables provided in Appendix F. Alternatively, if the temperature of the gas is close to 25°C then it may be sufficiently accurate to approximate $\bar{c}_P$ as being constant, so that:

$$\int_{298.15\,\text{K}}^{T} \frac{\bar{c}_P}{T}\, dT \approx \bar{c}_P \ln\left(\frac{T}{298.15\,\text{K}}\right) \tag{13-72}$$

EXAMPLE 13.4-1: PERFORMANCE OF A GAS TURBINE ENGINE

A stationary gas turbine engine is used for generating electricity during times of peak demand, as shown in Figure 1.

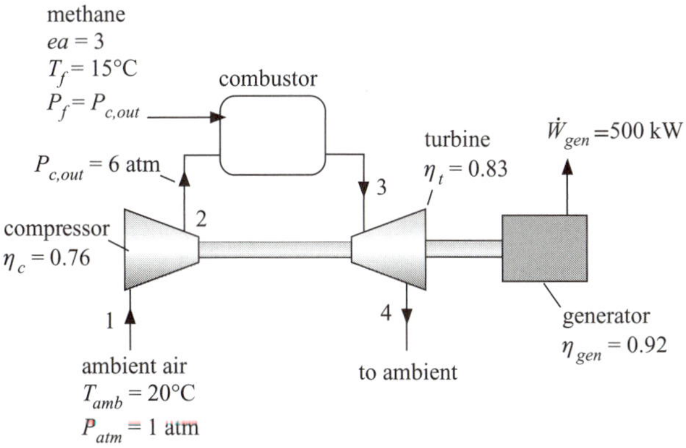

Figure 1: Gas turbine engine.

Dry air at $T_{amb} = 20°\text{C}$ and $P_{atm} = 1$ atm enters the compressor. The pressure of the air is increased to $P_{c,out} = 6$ atm in the compressor. The isentropic efficiency of the compressor is $\eta_c = 0.76$. The air leaving the combustor is mixed with methane that enters at $T_f = 15°\text{C}$ and $P_f = P_{c,out}$ and the mixture is combusted. The methane flow rate is adjusted so that the combustion reaction uses 300% excess air, $ea = 3$ (i.e., 400% theoretical air). The combustion products are expanded in a turbine with an isentropic efficiency of $\eta_t = 0.83$. A generator mounted on the same shaft as the turbine and the compressor produces $\dot{W}_{gen} = 500$ kW of electrical power with a generator efficiency of $\eta_{gen} = 0.92$. Pressure and thermal losses in the piping can be neglected in this analysis.

a) Determine the temperature of the air entering the combustor at state 2.

The inputs are entered in EES.

```
$UnitSystem SI Molar J K Pa
T_amb=converttemp(C,K,20[C])              "ambient temperature"
P_amb=1 [atm]*convert(atm,Pa)             "ambient pressure"
eta_c=0.76 [-]                            "compressor isentropic efficiency"
T_f=converttemp(C,K,15 [C])               "fuel inlet temperature"
P_c_out=6 [atm]*convert(atm,Pa)           "compressor exit pressure"
ea=3 [-]                                  "excess air"
eta_t=0.83 [-]                            "turbine isentropic efficiency"
W_dot_gen=500 [kW]*convert(kW,W)          "generator electrical power"
eta_gen=0.92 [-]                          "generator efficiency"
```

The compressor is operating with dry air and can therefore be analyzed using the substance Air in EES. The temperature and pressure of the air entering the compressor are specified; $T_1 = T_{amb}$ and $P_1 = P_{atm}$. The molar specific enthalpy, molar specific entropy and molar specific volume ($\bar{h}_1$, $\bar{s}_1$, and $\bar{v}_1$) are obtained.

```
"State 1"
T_1=T_amb                                 "temperature"
P_1=P_amb                                 "total pressure"
h_bar_1=enthalpy(Air,T=T_1)               "molar specific enthalpy"
s_bar_1=entropy(Air,T=T_1,P=P_1)          "molar specific entropy"
v_bar_1=volume(Air,T=T_1,P=P_1)           "molar specific volume"
```

The compressor outlet pressure is specified, $P_2 = P_{c,out}$. The molar specific enthalpy leaving a reversible compressor ($\bar{h}_{s,2}$) is specified by the exit pressure and inlet molar specific entropy. The molar specific enthalpy leaving the actual compressor is obtained using the definition of the compressor isentropic efficiency:

$$\bar{h}_2 = \bar{h}_1 + \frac{\left(\bar{h}_{s,2} - \bar{h}_1\right)}{\eta_c}$$

The state of the air entering the combustor is fixed by P_2 and $\bar{h}_2$. The temperature of the air entering the combustor (T_2) is obtained.

```
"State 2"
P_2=P_c_out                                  "pressure"
h_bar_s_2=enthalpy(Air,P=P_2,s=s_bar_1)      "molar specific enthalpy leaving reversible compressor"
h_bar_2=h_bar_1+(h_bar_s_2-h_bar_1)/eta_c    "molar specific enthalpy"
T_2=temperature(Air,h=h_bar_2)               "temperature"
T_2_C=converttemp(K,C,T_2)                   "in C"
```

Solving these equations leads to $T_2 = 547.6$ K (274.4°C).

b) Determine the temperature of the combustion products leaving the combustor and entering the turbine at state 3.

The stoichiometric reaction of air with methane can be written as:

$$CH_4 + a_s\,(0.21\,O_2 + 0.79N_2) \rightarrow b_sCO_2 + c_sH_2O + d_sN_2 \tag{1}$$

The following equations provide conservation of carbon, hydrogen, oxygen, and nitrogen atoms for Eq. (1):

$$1 = b_s$$

$$4 = 2\,c_s$$

$$a_s\,2\,(0.21) = 2\,b_s + c_s$$

$$a_s\,2\,(0.79) = 2\,d_s$$

```
"balance stoichiometric reaction"
1=b_s                       "carbon atom balance"
4=2*c_s                     "hydrogen atom balance"
a_s*2*0.21=2*b_s+c_s        "oxygen atom balance"
a_s*2*0.79=2*d_s            "nitrogen atom balance"
```

The actual reaction with excess air can be written as:

$$CH_4 + a\,(0.21\,O_2 + 0.79\,N_2) \rightarrow b\,CO_2 + c\,H_2O + d\,N_2 + e\,O_2 \tag{2}$$

where the number of moles of air per mole of methane, a, is obtained from the excess air:

$$a = (1 + ea)\,a_s$$

Balancing carbon, hydrogen, oxygen, and nitrogen atoms for the actual reaction, Eq. (2), provides:

$$1 = b$$

$$4 = 2\,c$$

$$a\,2\,(0.21) = 2\,b + c + 2\,e$$

$$a\,2\,(0.79) = 2\,d$$

```
"balance actual reaction"
a=(1+ea)*a_s                "air"
1=b                         "carbon atom balance"
4=2*c                       "hydrogen atom balance"
a*2*0.21=2*b+c+2*e          "oxygen atom balance"
a*2*0.79=2*d                "nitrogen atom balance"
```

The molar specific enthalpy of the methane entering the combustor ($\bar{h}_{CH_4,R}$) is computed at T_f. The molar specific enthalpies of the oxygen and nitrogen entering the combustor ($\bar{h}_{O_2,R}$ and $\bar{h}_{N_2,R}$) are computed at T_2. The enthalpy of the reactants entering the combustor per mole of fuel is computed according to:

$$H_R = \bar{h}_{CH_4,R} + 0.21\,a\,\bar{h}_{O_2,R} + 0.79\,a\,\bar{h}_{N_2,R}$$

```
h_bar_CH4_R=enthalpy(CH4,T=T_f)   "molar specific enthalpy of methane in reactants"
h_bar_N2_R=enthalpy(N2,T=T_2)     "molar specific enthalpy of nitrogen in reactants"
h_bar_O2_R=enthalpy(O2,T=T_2)     "molar specific enthalpy of oxygen in reactants"
H_R=h_bar_CH4_R+0.21*a*h_bar_O2_R+0.79*a*h_bar_N2_R
                                  "enthalpy of reactants entering combustor per mole of fuel"
```

The temperature of the combustion products entering the turbine, T_3, is initially guessed.

```
T_3=3000 [K]                     "guess for combustor exit temperature"
```

The molar specific enthalpies of the products leaving the combustor ($\bar{h}_{CO_2,P}$, $\bar{h}_{H_2O,P}$, $\bar{h}_{N_2,P}$ and $\bar{h}_{O_2,P}$) are computed at T_3. The enthalpy of the products leaving the combustor per mole of fuel is computed according to:

$$H_P = b\,\bar{h}_{CO_2,P} + c\,\bar{h}_{H_2O,P} + d\,\bar{h}_{N_2,P} + e\,\bar{h}_{O_2,P}$$

```
h_bar_CO2_P=enthalpy(CO2,T=T_3)  "molar specific enthalpy of CO2 in products"
h_bar_H2O_P=enthalpy(H2O,T=T_3)  "molar specific enthalpy of H2O in products"
h_bar_N2_P=enthalpy(N2,T=T_3)    "molar specific enthalpy of N2 in products"
h_bar_O2_P=enthalpy(O2,T=T_3)    "molar specific enthalpy of O2 in products"
H_P=b*h_bar_CO2_P+c*h_bar_H2O_P+d*h_bar_N2_P+e*h_bar_O2_P
                                 "enthalpy of products leaving combustor per mole of fuel"
```

An energy balance on the combustor requires:

$$H_R = H_P \tag{3}$$

The guess values for the problem are updated and the assumed value of T_3 is commented out. Equation (3) is implemented in its place.

```
{T_3=3000 [K]}                   "guess for combustor exit temperature"
H_P=H_R                          "energy balance on combustor"
T_3_C=converttemp(K,C,T_3)       "combustor exit temperature, in C"
```

The solution provides $T_3 = 1161$ K (887.4°C).

c) Determine the volumetric flow rate of the air entering the compressor.

It will be necessary to determine the turbine power output in order to calculate the molar flow rate of air and therefore the volumetric flow rate of the air. The mole fractions of carbon dioxide, water, oxygen, and nitrogen in the products ($y_{CO_2,P}$, $y_{H_2O,P}$, $y_{O_2,P}$, and $y_{N_2,P}$, respectively) are computed using Eq. (2):

$$y_{CO_2,P} = \frac{b}{b+c+d+e}$$

$$y_{H_2O,P} = \frac{c}{b+c+d+e}$$

$$y_{N_2,P} = \frac{d}{b+c+d+e}$$

$$y_{O_2,P} = \frac{e}{b+c+d+e}$$

There is assumed to be no pressure loss in the combustor; therefore $P_3 = P_2$. The molar specific entropy of the carbon dioxide, water, nitrogen, and oxygen entering the turbine ($\bar{s}_{CO_2,P}$, $\bar{s}_{H_2O,P}$, $\bar{s}_{N_2,P}$ and $\bar{s}_{O_2,P}$) are computed at T_3 and the partial pressures of each component (i.e., the product of the mole fraction of the component and the total pressure at state 3). The entropy of the products entering the turbine per mole of fuel is determined according to:

$$S_P = b\,\bar{s}_{CO_2,P} + c\,\bar{s}_{H_2O,P} + d\,\bar{s}_{N_2,P} + e\,\bar{s}_{O_2,P}$$

```
y_CO2_P=b/(b+c+d+e)            "mole fraction of CO2 in the combustion products"
y_H2O_P=c/(b+c+d+e)            "mole fraction of H2O in the combustion products"
y_N2_P=d/(b+c+d+e)             "mole fraction of N2 in the combustion products"
y_O2_P=e/(b+c+d+e)             "mole fraction of O2 in the combustion products"
P_3=P_2                        "pressure at combustor outlet"
s_bar_CO2_P=entropy(CO2,T=T_3,P=y_CO2_P*P_3)
                               "molar specific entropy of CO2 in the products"
s_bar_H2O_P=entropy(H2O,T=T_3,P=y_H2O_P*P_3)
                               "molar specific entropy of H2O in the products"
s_bar_N2_P=entropy(N2,T=T_3,P=y_N2_P*P_3)
                               "molar specific entropy of N2 in the products"
s_bar_O2_P=entropy(O2,T=T_3,P=y_N2_P*P_3)
                               "molar specific entropy of O2 in the products"
S_P=b*s_bar_CO2_P+c*s_bar_H2O_P+d*s_bar_N2_P+e*s_bar_O2_P
                               "entropy of products leaving combustor per mole of fuel"
```

The pressure leaving the turbine is atmospheric, $P_4 = P_{atm}$. The temperature leaving a reversible turbine, $T_{s,4}$, is initially assumed. Note that the guess for $T_{s,4}$ does not need to be accurate, but it should be physically reasonable.

```
P_4=P_atm                      "pressure at the exit of the turbine"
T_s_4=800 [K]                  "guess for temperature leaving reversible turbine"
```

The molar specific entropies of the carbon dioxide, water, nitrogen, and oxygen leaving a reversible turbine ($\bar{s}_{CO_2,s,4}$, $\bar{s}_{H_2O,s,4}$, $\bar{s}_{N_2,s,4}$ and $\bar{s}_{O_2,s,4}$) are computed at $T_{s,4}$ and the partial pressures of each component. The entropy of the products leaving a reversible turbine per mole of fuel is determined according to:

$$S_{s,4} = b\,\bar{s}_{CO_2,s,4} + c\,\bar{s}_{H_2O,s,4} + d\,\bar{s}_{N_2,s,4} + e\,\bar{s}_{O_2,s,4}$$

```
s_bar_CO2_s_4=entropy(CO2,T=T_s_4,P=y_CO2_P*P_4)
                               "molar specific entropy of CO2 leaving rev. turbine"
s_bar_H2O_s_4=entropy(H2O,T=T_s_4,P=y_H2O_P*P_4)
                               "molar specific entropy of H2O leaving rev. turbine"
s_bar_N2_s_4=entropy(N2,T=T_s_4,P=y_N2_P*P_4)
                               "molar specific entropy of N2 leaving rev. turbine"
s_bar_O2_s_4=entropy(O2,T=T_s_4,P=y_O2_P*P_4)
                               "molar specific entropy of N2 leaving rev. turbine"
S_s_4=b*s_bar_CO2_s_4+c*s_bar_H2O_s_4+d*s_bar_N2_s_4+e*s_bar_O2_s_4
                               "entropy of products leaving reversible turbine per mole of fuel"
```

A steady-state entropy balance on the reversible turbine provides:

$$S_P = S_{s,4} \tag{4}$$

The guess values for the problems are updated and the assumed value of $T_{s,4}$ is commented out. Equation (4) is implemented in its place.

```
{T_s_4=800 [K]}        "guess for temperature leaving isentropic turbine"
S_s_4=S_P              "entropy balance on reversible turbine"
```

The molar specific enthalpies of the carbon dioxide, water, nitrogen, and oxygen leaving a reversible turbine ($\bar{h}_{CO_2,s,4}$, $\bar{h}_{H_2O,s,4}$, $\bar{h}_{N_2,s,4}$ and $\bar{h}_{O_2,s,4}$) are computed at $T_{s,4}$. The enthalpy of the products leaving a reversible turbine per mole of fuel is computed according to:

$$H_{s,4} = b\,\bar{h}_{CO_2,s,4} + c\,\bar{h}_{H_2O,s,4} + d\,\bar{h}_{N_2,s,4} + e\,\bar{h}_{O_2,s,4}$$

The enthalpy of the products leaving the actual turbine per mole of fuel is computed using the turbine efficiency:

$$H_4 = H_P - \eta_t\,(H_P - H_{s,4})$$

```
h_bar_CO2_s_4=enthalpy(CO2,T=T_s_4)
                    "molar specific enthalpy of CO2 leaving isentropic turbine"
h_bar_H2O_s_4=enthalpy(H2O,T=T_s_4)
                    "molar specific enthalpy of H2O leaving isentropic turbine"
h_bar_N2_s_4=enthalpy(N2,T=T_s_4)
                    "molar specific enthalpy of N2 leaving isentropic turbine"
h_bar_O2_s_4=enthalpy(O2,T=T_s_4)
                    "molar specific enthalpy of O2 leaving isentropic turbine"
H_s_4=b*h_bar_CO2_s_4+c*h_bar_H2O_s_4+d*h_bar_N2_s_4+e*h_bar_O2_s_4
                    "enthalpy of products leaving isentropic turbine per mole of fuel"
H_4=H_P-(H_P-H_s_4)*eta_t   "enthalpy of products leaving actual turbine per mole of fuel"
```

The power generated by the engine is given by:

$$\dot{W}_{gen} = \left(\dot{W}_t - \dot{W}_c\right)\eta_{gen} \tag{5}$$

The power required by the compressor is:

$$\dot{W}_c = \dot{n}_{air}\left(\bar{h}_2 - \bar{h}_1\right) \tag{6}$$

The molar specific enthalpies $\bar{h}_1$ and $\bar{h}_2$ are defined on a per mole of air basis and therefore they are multiplied by the molar flow rate of air in Eq. (6). The power generated by the turbine is:

$$\dot{W}_t = \dot{n}_{CH_4}\,(H_P - H_4) \tag{7}$$

Note that the quantities H_P and H_4 correspond to the enthalpy of the combustion products per mole of fuel and therefore they are multiplied by the molar flow rate of methane in Eq. (7). Substituting Eqs. (6) and (7) into Eq. (5) provides:

$$\dot{W}_{gen} = \left[\dot{n}_{CH_4}\,(H_P - H_4) - \dot{n}_{air}\left(\bar{h}_2 - \bar{h}_1\right)\right]\eta_{gen} \tag{8}$$

According to Eq. (2), the ratio of the molar flow rate of air to the molar flow rate of methane is given by the coefficient a. Therefore, Eq. (8) can be written as:

$$\dot{W}_{gen} = \dot{n}_{CH_4}\left[(H_P - H_4) - a\left(\bar{h}_2 - \bar{h}_1\right)\right]\eta_{gen} \tag{9}$$

Equation (9) can be used to determine the molar flow rate of fuel that is required to produce the required generator power. The molar flow rate of air is computed from:

$$\dot{n}_{air} = a\,\dot{n}_{CH_4}$$

and the volumetric flow rate of air entering the compressor is given by:

$$\dot{V}_1 = \dot{n}_{air}\, \bar{v}_1$$

```
W_dot_gen=n_dot_CH4*((H_P-H_4)-a*(h_bar_2-h_bar_1))*eta_gen
                              "fuel flow rate is set to achieve required generator power"
n_dot_a=a*n_dot_CH4           "molar flow rate of air"
V_dot_a_1=n_dot_a*v_bar_1     "volumetric flow rate entering compressor"
```

which provides $\dot{V}_1 = 2.299\ \text{m}^3/\text{s}$.

d) Determine the efficiency of the gas turbine engine relative to the lower heating value of the methane.

The lower heating value of methane is provided in Table 13-3; however, it is easy to calculate its value. The heating value is the difference between the enthalpy of the reactants and the enthalpy of the products for the complete combustion of one mole of methane at $T_{ref} = 25°\text{C}$. In order to obtain the lower heating value, all of the water in the products is assumed to be in a vapor phase. The standardized specific enthalpy of elements such as oxygen and nitrogen are zero at 25°C. Therefore, only the specific enthalpy of the methane, carbon dioxide and water vapor ($\bar{h}_{CH_4,ref}$, $\bar{h}_{CO_2,ref}$, and $\bar{h}_{H_2O,g,ref}$) must be considered. The lower heating value is calculated according to:

$$LHV = \bar{h}_{CH_4,ref} - 2\,\bar{h}_{H_2O,g,ref} - \bar{h}_{CO_2,ref}$$

The efficiency of the gas turbine engine is:

$$\eta = \frac{\dot{W}_{gen}}{\dot{n}_{CH_4}\, LHV}$$

```
T_ref=converttemp(C,K,25[C])                   "reference temperature"
h_bar_CH4_ref=enthalpy(CH4,T=T_ref)            "molar specific enthalpy of methane at T_ref"
h_bar_H2O_g_ref=enthalpy(H2O,T=T_ref)          "molar specific enthalpy of water vapor at T_ref"
h_bar_CO2_ref=enthalpy(CO2,T=T_ref)            "molar specific enthalpy of carbon dioxide at T_ref"
LHV=h_bar_CH4_ref-2*h_bar_H2O_g_ref-h_bar_CO2_ref
                                               "LHV of methane on a molar basis"
eta=W_dot_gen/(n_dot_CH4*LHV)                  "efficiency based on the lower heating value"
```

which results in $\eta = 0.2483(24.83\%)$.

13.5 Exergy of Fuels

This extended section can be found online at www.cambridge.org/kleinandnellis and presents a methodology for determining the exergy of a fuel and the Second-Law efficiency of processes that employ the combustion of fuels.

REFERENCES

Annual Energy Outlook, U.S. DOE, Table A1, page 109 and Figure 91, p. 84, DOE/EIA-0383, http://www.eia.doe.gov/oiaf/aeo/pdf/0383 (2009).pdf, (2009).

Lemmon, E. W., Jacobsen, R. T, Penoncello, S. G., and Friend, D., "Thermodynamic Properties of Air and Mixtures of Nitrogen, Argon, and Oxygen from 60 to 2000 K at Pressures to 2000 MPa," *J. Phys. Chem. Ref. Data*, Vol. 29, No. 3, 2000.

McBride, B.J., Zehe, M.J. and Gordon, S., "NASA Glenn Coefficients for Calculating Thermodynamic Properties of Individual Species", NASA/TP—2002-211556, http://www.lerc.nasa.gov/WWW/CEAWeb/, (2002).

Razus, D. Movileanu, C., Brinzea, V., and Oancea, D., "Explosion pressures of hydrocarbon-air mixtures in closed vessels," *Journal of Hazardous Materials*, B135, 58-65, (2006).

Problems

The problems included here have been selected from a larger set of problems that are available on the website associated with this book (www.cambridge.org/kleinandnellis).

A: Stoichiometry

13.A-2 A small air-cooled gasoline engine is tested and the output is found to be $\dot{W} = 1.34$ hp. The fuel is liquid octane (C_8H_{18}) and it is provided to the engine at a mass flow rate of $\dot{m}_f = 0.15$ g/s. The fuel and air both enter the engine at $T_{in} = 25°C$. The relative humidity of the entering air is $\phi_{in} = 42\%$. The products are analyzed with the results in Table 13.A-2, reported on a dry volumetric basis:

Table 13.A-2: Combustion gas analysis

Chemical species	Mole fraction
CO_2	11.40%
O_2	1.47%
CO	2.90%
Inert gases	84.23%

a) Determine the air fuel ratio.
b) Determine the percent of theoretical air used in this combustion process.
c) The combustion products cool as they pass through the exhaust system. At what temperature will condensation initiate?
d) If the exhaust products exit the exhaust pipe at $T_{out} = 32°C$, at what rate will liquid water need to be removed?

13.A-3 An analysis of a hydrocarbon fuel indicates that it consists of 83% carbon and 17% hydrogen on a mass basis. This fuel is completely combusted with 50% excess dry air at 25°C and 1 atm pressure. The combustion products are cooled to 35°C at 1 atm.
a) What is the dew point temperature of the combustion products?
b) What is the mass of condensate resulting from the combustion of 1 kg of the fuel?
c) How are your answers to parts (a) and (b) affected if the air used for combustion has a relative humidity of 100%?

13.A-5 An experimental automotive fuel consists of 50% (by mole) liquid ethanol (C_2H_5OH) and 50% (by mole) liquid octane (C_8H_{18}). This mixture is steadily combusted with 120% theoretical dry air at 1 atm (101.3 kPa).
a) Determine the balanced chemical reaction equation for complete combustion of 1 mole of fuel mixture with 120% theoretical air.
b) Determine the air fuel ratio for these conditions.
c) What is the dew point of the combustion products at 1 atm pressure?

d) How does the dew point calculated in part (c) compare to the dew point that would result if the fuel were pure liquid octane?

13.A-6 Butane (C_8H_{18}) is combusted with air that is at $T_{in} = 18°C$, $P_{in} = 1$ atm, and $\phi_{in} = 44\%$ relative humidity. The combustion products are eventually cooled to $T_{out} = 18°C$.

a) Calculate and plot the dew point temperature and the mass of condensate per mass of butane as a function of the percent excess air for values ranging from 0% to 500%.
b) How would the plot changed if dry air were used in the combustion process?

13.A-7 The life of an automobile's exhaust system can be related to the amount of liquid water that is condensed from the combustion products. A company interested in manufacturing replacement exhaust systems would like you to develop estimates for exhaust system life expectancy. The model engine operates using octane (C_8H_{18}) as the fuel with an average air fuel ratio of 20. The products are cooled to $T_{out} = 30°C$ prior to exiting the exhaust system. Empirical data suggest that the life expectancy (in months) of the exhaust is given by:

$$Life = 48\,[\text{month}] - 8\,[\text{month}]\; m_w$$

where m_w is the mass of water condensed per mass of fuel burned.

a) Determine the percent excess air being used for this model engine.
b) What is the dew point of the exhaust products?
c) Assuming that the empirical relation is correct, what is the life expectancy of the exhaust system?

13.A-8 A fuel gas mixture used in steel production is delivered at $T_{in} = 700$ K and $P_{in} = 250$ kPa. This gas mixture is combusted with 200% theoretical dry air in a boiler. The composition of the gas mixture is provided in Table 13.A-8 on a volumetric basis. Complete combustion occurs and combustion products exit the boiler at $T_{out} = 500$ K and $P_{out} = 100$ kPa.

Table 13.A-8: Volume percentages of fuel gas mixture

Chemical species	Mole fraction
H_2	2.4%
CH_4	5.1%
CO	23.3%
CO_2	9.4%
Inert gases	59.8%

a) What is the apparent molar mass of the fuel gas mixture? (Assume the inert gases to be primarily nitrogen.)
b) What is the air fuel ratio for the combustion process?
c) What is the dew point of the combustion products?

13.A-10 Liquid methyl alcohol (CH_3OH) is combusted with air. The air and methanol enter a combustion chamber separately at $T_{in} = 25°C$ and $P_{in} = 100$ kPa. The combustion products are cooled to $T_{out} = 44°C$ and exhausted to the surroundings at $P_{out} = 100$ kPa. A volumetric analysis of the dried combustion products resulted in the following composition: 1.2% CO, 10.7% CO_2, 4.8% O_2, with the remainder being inert gases.

a) What is the air fuel ratio, assuming the air is complete dry?
b) Calculate the ratio of the mass of water that condenses to the mass of liquid methanol.
c) Repeat part (b) assuming that the air used for combustion is saturated with water.

B: Energy Considerations

13.B-1 Propylene (C_3H_6) is burned with 10% excess air in a combustor. The molar specific enthalpy of formation of propylene at 25°C is $\bar{h}_{form,C_3H_6} = 20{,}410$ kJ/kmol. The flow rate of air entering the combustor is $\dot{V}_{air} = 2.4$ cfm (0.00113 m^3/s). The air and fuel are drawn into the combustor at $T_{in} = 25$°C and $P_{in} = 1$ atm. The products leave at $T_{out} = 450$°C and $P_{out} = 1$ atm. Model the gases as ideal gases with constant specific heat capacity at constant pressure: $\bar{c}_{P,N_2} = 29.18$ kJ/kmol-K, $\bar{c}_{P,O_2} = 29.63$ kJ/kmol-K, $\bar{c}_{P,CO_2} = 39.31$ kJ/kmol-K, and $\bar{c}_{P,H_2O} = 33.87$ kJ/kmol-K.

a) What is the air fuel ratio for the reaction?
b) Determine the rate of heat transfer from the combustor.

13.B-2 Figure 13.B-2 illustrates a simple, un-recuperated gas turbine engine.

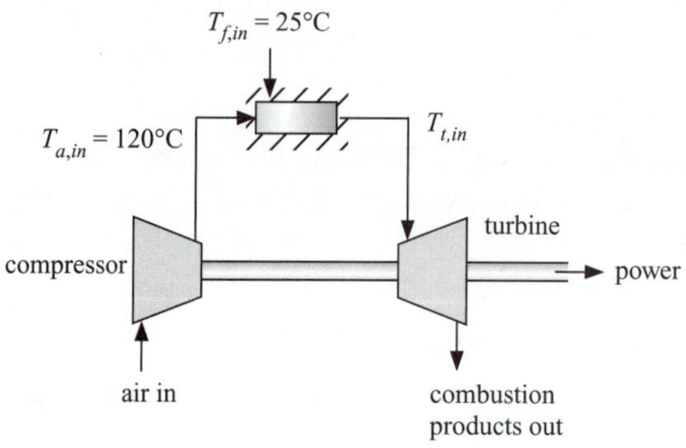

Figure 13.B-2: Un-recuperated gas turbine engine.

The air leaves the compressor and enters the combustor at $T_{a,in} = 120$°C where it is mixed with jet fuel that enters at $T_{f,in} = 25$°C. Assume that jet fuel has composition C_8H_{18}. The air flow rate is selected so that there is 50% excess air. The combustor is adiabatic. Model the mixture as an ideal gas, but do not assume constant specific heat capacities.

a) Determine the temperature of the combustion products leaving the combustor and entering the turbine, $T_{t,in}$.
b) Plot the temperature of the gas entering the turbine as a function of the percent excess air.
c) Plot the temperature of the gas entering the turbine as a function of the air fuel ratio.

13.B-3 Pentane (C_5H_{12}) is burned with 50% excess air.

a) Write the chemical reaction for the combustion.

b) What is the air fuel ratio (on a mass basis) for the reaction in part (a)? The molecular weight of pentane is 72.151 kg/kmol.
c) What is the dew point temperature of the products of the reaction in part (a) at 1 atm?
d) Determine the higher heating value (in kJ/kg) of pentane. The enthalpy of formation of pentane is -146,440 kJ/kmol.

13.B-4 Ethane (C_2H_6) is burned with a stoichiometric amount of air in a combustor, as shown in Figure 13.B-4.

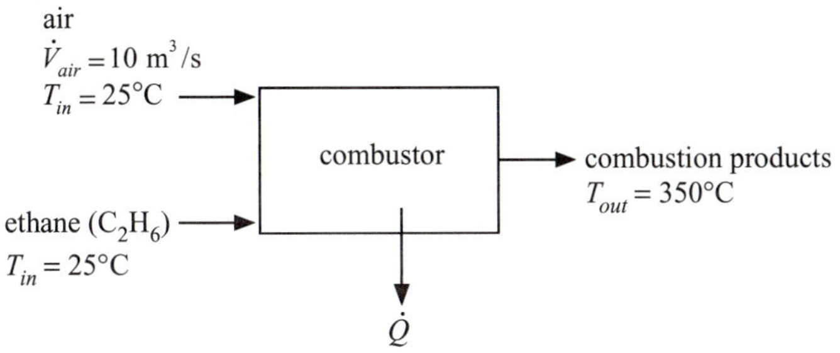

Figure 13.B-4: Ethane burned in a combustor.

The fuel and air enter the combustor at $T_{in} = 25°C$ and combustion products leave at $T_{out} = 350°C$. The inlet and exit pressure is $P = 1$ atm. The volumetric flow rate of air entering the combustor is $\dot{V}_{air} = 1\ m^3/s$. Assume that the specific heat capacities of carbon dioxide, water vapor, nitrogen, and oxygen are constant and equal to $\bar{c}_{P,CO_2} = 37.0$ kJ/kmol-K, $\bar{c}_{P,H_2O} = 33.7$ kJ/kmol-K, $\bar{c}_{P,N_2} = 29.2$ kJ/kmol-K, and $\bar{c}_{P,O_2} = 29.5$ kJ/kmol-K, respectively.
a) Determine the rate of heat transfer from the combustor, $\dot{Q}$.

13.B-5 Fuel oil is used in a modern condensing furnace. The fuel oil consists of a mixture of hydrocarbons. An elemental analysis shows that the fuel oil consists of 84.28% carbon (by mass), 15.22% hydrogen, and 0.5% sulfur. The lower heating value of this fuel oil is 42.5 MJ/kg. The furnace, which provides residential space heat, uses 50% excess air at 25°C, 60% relative humidity. Fuel oil enters the furnace at 25°C. Combustion gases and condensate exit at 43°C, 1 atm.
a) Determine the dew point temperature of the combustion products.
b) Calculate the ratio of condensate to fuel oil on a mass basis.
c) Determine the higher heating value of the fuel oil.
d) Determine the efficiency of the furnace based on the lower and higher heating values.

13.B-8 A solid oxide fuel cell operates on an equimolar mixture of carbon monoxide (CO) and hydrogen (H_2) produced from a reformation process. The fuel gas mixture enters at 25°C and reacts with dry air at 25°C within the fuel cell stack. The fuel cell produces 8 kJ of electrical energy for each gram of fuel mixture that enters. The fuel cell operation is nearly adiabatic and products exit at 800°C. The high temperatures and catalytic processes within the fuel cell ensure complete combustion. Determine:
a) the % of excess air that must be provided.
b) the efficiency of the fuel cell at these conditions. (Indicate how you are defining efficiency.)

13.B-9 Hydrogen has been proposed as a alternative to natural gas as a fuel for residential use. Hydrogen can be produced by electrolysis of water. In a particular process, water at 50°F and 1 atm is supplied to an electrolysis cell. The electrical current supplied to the cell decomposes the water to produce steady streams of hydrogen and oxygen at 100°F. For each pound of hydrogen produced, 45,000 Btu of heat are transferred from the cell to the surroundings in a steady operating condition.

a) Determine the efficiency of the electrolysis process.
b) The collected hydrogen is then pumped through a pipeline where it is cooled to the environmental temperature (77°F) and later used to heat water from 50°F to 140°F in a water heater by combusting it with 80% excess air at 75°F. The combustion products exit the water heater at 150°F. Determine the efficiency of the water heating process.
c) A figure of merit for the overall process can be obtained by calculating the volume (in gallons) of water heated per kW-hr of electricity expended in the electrolysis process. Calculate this figure of merit and compare it with the corresponding value that would be expected from a common electrical water heater. Would you recommend hydrogen as a substitute for natural gas if it were produced and used as described? Why or why not?

13.B-13 Combustion fuels are seldom pure chemicals, but rather a mixture of many chemicals. In a particular case, the lower heating value of a fuel mixture is determined to be 21,300 kJ/kg and a chemical analysis indicates that it contains 12.5% H, 37.5% C, and 50% O on a mass basis.

a) Plot the adiabatic combustion temperature as a function of the % theoretical air for values between 100% and 400%. Assume complete combustion.

13.B-14 The purpose of this problem is to compare methane (CH_4), propane (C_3H_8) and hydrogen (H_2) as possible fuels for a residential furnace. 100% excess air is used for all fuels. Assume that the fuel and air enter the furnace at 25°C. For each fuel, calculate and plot the following quantities for flue gas temperatures ranging between 30°C and 80°C:

a) the furnace efficiencies based on the lower and higher heating values of the fuel,
b) the mass flow rate of fuel required to supply 10 kW of thermal energy, and
c) the mass flow rate of condensate.

13.B-15 A mixture of octane (C_8H_{18}) and 20% excess air at $T_1 = 25$°C and $P_1 = 1$ atm occupies $V_1 = 0.575$ liters. The mixture is adiabatically compressed with a compression ratio of 8 in one cylinder of an engine. Combustion is then initiated at constant volume and proceeds to complete combustion in an adiabatic process.

a) What is the temperature and pressure of the mixture of octane and air after the adiabatic compression process has concluded and before combustion is initiated?
b) What is the work required to compress the mixture?
c) The reaction of octane and air is initiated and occurs adiabatically at constant volume. What is the temperature and pressure of the combustion products?
d) The combustion products expand isentropically until the volume is again 0.575 liters. Determine the work produced in this process.
e) The engine operates at 3000 rpm using a 4-stroke process (i.e., one combustion process for every 2 revolutions of the crank shaft). What is the average power generated from the one cylinder?

f) The efficiency of the engine is the net work divided by the lower heating value of the octane. What is the efficiency for this engine?

13.B-17 A furnace combusts propane with air. The air enters at the outdoor temperature of $T_{out} = 10°F$ and may be considered to be dry. The propane is taken from an outdoor tank that also is at T_{out}. A volumetric analysis of the combustion products on a dry basis (i.e., after removing the water) results in the composition listed in Table 13.B-17. The flow rate of the propane is 10 scfm (note that scfm corresponds to a mass flow rate that would lead to a volumetric flow rate in ft^3/min at standard conditions: 77°F, 1 atm). Combustion products exit the furnace at $T_{out} = 95°F$.

Table 13.B-17: Volumetric analysis of combustion products

Chemical Species	Mole Fraction
CO_2	5.5%
CO	1.1%
O_2	11.0%
Inert gases and N_2	balance

a) Determine the percent excess air used in this furnace.
a) At what rate must condensate be removed (in gallons/hr)?
b) Determine the rate at which heat is provided to the building.
c) What is the furnace efficiency, based on the higher heating value of propane?

13.B-19 A proton exchange membrane (PEM) fuel cell involves electrochemical reactions between hydrogen and the oxygen that is in air. In a particular case, air and hydrogen are provided at $T_{in} = 25°C$ and atmospheric pressure in separate streams, as shown in Figure 13.B-19. Two exit streams exhaust from the PEM at $T_{out} = 92°C$ and atmospheric pressure. The fuel cell generates $\dot{W} = 3.2$ kW. The hydrogen and air flow rates are carefully measured to be $\dot{m}_{H_2} = 0.0564$ g/s and $\dot{m}_{air} = 2.9$ g/s, respectively. It is known that 5% of the hydrogen exits the fuel cell unreacted with the water vapor.

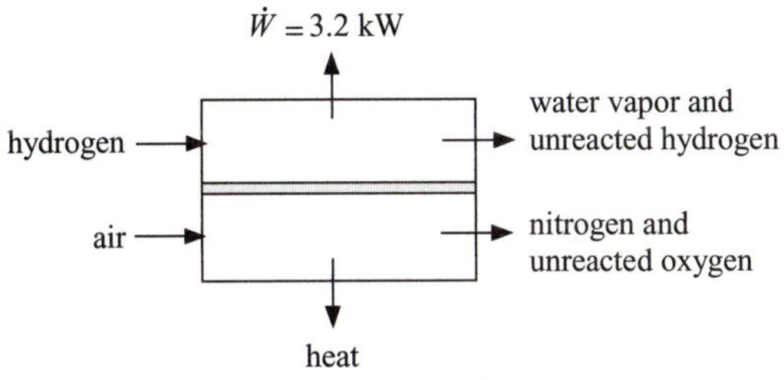

Figure 13.B-19: Schematic of a hydrogen fuel cell.

a) What is the percent excess air that is provided relative to that needed for complete combustion?
b) What is the heat transfer rate from the cell?
c) What is the efficiency of the fuel cell at this operating condition?

13.B-21 Liquid hydrogen peroxide (H_2O_2) is used as the oxidizer with methane fuel in a rocket engine. The methane and hydrogen peroxide steadily enter the engine in separate streams at $T_{in} = 25°C$ and $P_{in} = 1$ atm at low velocity. To ensure complete combustion, 25% excess oxidizer is supplied. Combustion products (which are all gaseous) exit the engine at $T_{out} = 1200$ K and $P_{out} = 1$ atm through a nozzle at high velocity. Note: The enthalpy of vaporization of liquid H_2O_2 at 25°C is $\Delta h_{fg} = 1535$ kJ/kg. Properties for gaseous H_2O_2 are available in EES.

a) Estimate the velocity of the products leaving the nozzle.
b) Estimate the Mach number, defined as the ratio of the velocity of the products to the sound speed of the products.

13.B-22 A space heating application uses liquid ethanol (C_2H_5OH) as a fuel and it employs a combustion air preheater, as shown in Figure 13.B-22. Liquid methanol enters the combustion chamber at $T_{f,in} = 15°C$ and 1 atm. At the design condition, outdoor air enters the preheater at $T_1 = -10°C$ and it is heated to $T_2 = 34°C$ before entering the furnace. Building air at atmospheric pressure and $T_5 = 15°C$ is circulated through a heat exchanger within the furnace at a rate of $\dot{m}_{ba} = 0.88$ kg/s; the building air exits the furnace at $T_6 = 38°C$. Combustion products exit the furnace at $T_3 = 185°C$. Thermal losses from the jackets of the boiler and preheater are negligible. An exhaust gas analysis of the combustion gases resulted in the following volumetric percentages, expressed on a dry basis: 1.2% CO; 10.7% CO_2; 4.8% O_2; 83.3% N_2. The specific enthalpy of liquid methanol at 25°C is -277.69 MJ/kmol.

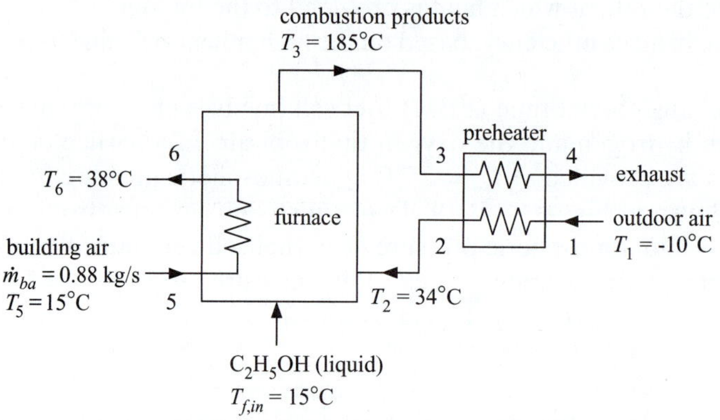

Figure 13.B-22: Furnace with preheater that uses ethanol as the fuel.

a) What is the % stoichiometric air?
b) Determine the rate at which ethanol must be provided.
c) Will condensation occur in the preheater? Why or why not?
d) What is the effectiveness of the preheater?
e) What is the 2nd Law efficiency of this process, assuming that the exergy of the ethanol is equal to its lower heating value?

13.B-25 You are probably aware of the global warming concerns. Carbon dioxide is one of many 'greenhouse' gases that is mostly transparent to visible radiation, but absorbent in certain bands within the infrared region. The current U.S. administration has stated a goal of capping and later reducing carbon dioxide production. A logical question is how best to do this, since all hydrocarbon fuels produce carbon dioxide as an end product. Table 13.B-25 provides the molar specific enthalpy of formation, the absolute molar specific entropy, and the molar specific heat capacity at 25°C and 1 atm for four common fuels. Rank these fuels with respect to their contribution to global warming. Explain the methodology behind your ranking.

Table 13.B-25: Characteristics of some common fuels

		Properties at 25°C and 1 atm		
Fuel	Chemical formula	Molar specific enthalpy of formation (kJ/kmol)	Absolute molar specific entropy (kJ/kmol-K)	Molar specific heat capacity (kJ/kmol-K)
Ethanol (liquid)	C_2H_5OH	−277,402	38.0	115.4
Coal (solid)	C	0	1.369	8.53
Natural gas	CH_4	−32,211	186.3	36.06
Octane (liquid)	C_8H_{18}	−250	360.8	254.3
	CO_2	−393,500	213.7	37.0
	H_2O (g)	−241,810	188.7	33.6
	O_2	0	205.0	29.2

C: Advanced Problems

13.C-1 The purpose of this problem is to provide a thermodynamic analysis of the solid oxide fuel cell (SOFC) system shown in Figure 13.C-1.

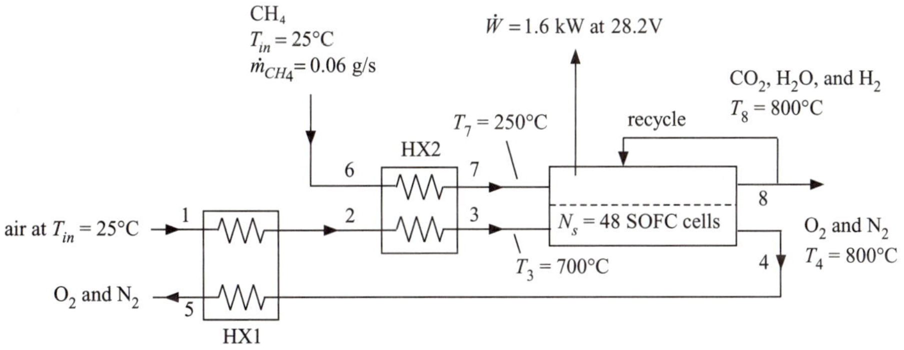

Figure 13.C-1: Schematic of solid oxide fuel cell system.

Inside of the fuel cell, methane reacts with water that is recycled from the anode of the fuel cell. Four moles of hydrogen are produced per mole of methane reacted according to the reaction:

$$CH_4 + 2\,H_2O \rightarrow 4\,H_2 + CO_2$$

All of the methane entering the cell is reacted. The hydrogen that is produced is subsequently reacted with oxygen in the fuel cell according to:

$$4\,H_2 + a\,(O_2{+}3.76N_2) \rightarrow 4\,U\,H_2O + 4(1-U)\,H_2 + a\,3.76\,N_2 + b\,O_2$$

According to Faraday's Law, each mole of hydrogen that is reacted produces two moles of hydrogen protons, which results in a charge transfer of 2 F where F is 9.64867x10^7 coulomb/kmol. Thus, the current is directly proportional to the rate at which hydrogen is converted into water. Note that not all of the hydrogen is reacted. The fraction of the available hydrogen (from reformed methane) that is reacted is U, the fuel utilization. It is necessary to having some unreacted hydrogen exit the stack in order to maintain the voltage in the fuel cell. Consequently, the current through each cell is:

$$I = \frac{8\,F\,U\,\dot{n}_{CH_4}}{N_s}$$

where $\dot{n}_{CH_4}$ is the molar flow rate of methane and $N_s = 48$ is the number of cells in series. Air and methane enter at $T_{in} = 25°\text{C}$ and $P_{in} = 101.3$ kPa at states 1 and 6, respectively. The methane flow rate is $\dot{m}_{CH_4} = 0.06$ g/sec. Water vapor, carbon dioxide and some unreacted hydrogen at $T_8 = 800°\text{C}$ emerge from the stack at state 8. Some of this gas is recycled back to the stack so that additional water is not required. Unreacted oxygen and nitrogen at $T_4 = 800°\text{C}$ exit the stack at state 4 and are used to preheat the incoming air. The temperature at state 5 is measured and found to be $T_5 = 108°\text{C}$. A second heat exchanger uses the preheated air to preheat the methane to $T_7 = 250°\text{C}$. The air enters the fuel cell (state 3) at $T_3 = 700°\text{C}$. The fuel cell stack produces $\dot{W} = 1.6$ kW of electrical power at $V = 28.2$ V; since there are 48 cells in series, each cell operates at 0.5875 V. Heat losses from the jackets of the heat exchangers are negligible, but there is heat loss from the SOFC stack.

a) Calculate the efficiency of the fuel cell system at these conditions (not considering parasitic losses). Please be sure to indicate how you are defining efficiency.
b) Calculate the fuel utilization, U.
c) Calculate the % stoichiometric air supplied.
d) Calculate the rate of heat loss from the SOFC stack.

13.C-5 The purpose of this problem is to evaluate the feasibility of injecting liquid water into an engine in order to increase power output. Consider an Otto cycle in which n-octane (C_8H_{18} in a vapor state) is combusted with 5% excess air that enters the cylinder at 25°C and 1 atm. The temperature and pressure after the compression stroke is completed is 335°C and 17.2 bar. At this point, liquid water at 25°C is quickly injected in order to produce a saturated mixture with no liquid droplets. The spark ignites the mixture and the combustion process proceeds adiabatically at constant volume. The combustion products adiabatically expand to 1 atm, thereby producing mechanical power. You may assume complete combustion and neglect friction and other irreversible processes in your analysis.

a) What is the compression ratio for these conditions?
b) What is the temperature and pressure of the fuel-air mixture immediately after the water is adiabatically injected and the corresponding mole fraction of water vapor, assuming complete mixing and no change in volume during the injection process?
c) Calculate the work required to compress the fuel-air mixture per mole of n-octane, assuming an adiabatic compression process.
d) Determine the maximum temperature and pressure of the combustion products and compare the results with those that would result if there were no water injection.
e) A concern has been raised regarding the effect of the water on the exhaust system. Compare the dew point temperatures of the exiting exhaust products with and without water injection.

13.C-6 This problem is a continuation of Problem 13.C-5. Please prepare an analysis to answer the following questions.

a) Determine the exhaust temperature with and without the water injection assuming that the expansion process is adiabatic and internally reversible.
b) Calculate and compare the net work per mole of fuel with and without the water injection.
c) Calculate the First Law efficiency of the engine with and without the water injection.
d) Calculate the mass of water required per mass of n-octane consumed. Is this water injection process feasible in passenger vehicles? Indicate the advantages and disadvantages of water injection and whether you think it is a useful idea.

13.C-8 A small stationary gas turbine plant is used for provide $\dot{W}_{out} = 50$ kW of electrical power during times of peak demand using a generator that has an efficiency of $\eta_{gen} = 0.95$. A schematic of the plant is shown in Figure 13.C-8.

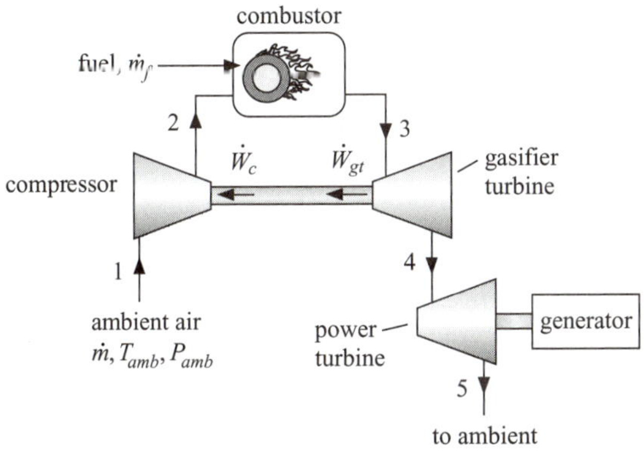

Figure 13.C-8: Stationary gas turbine system.

Dry air enters the system at $T_1 = 25°C$ and $P_1 = 1$ atm. The fuel is methane, which enters the combustor at 25°C. The turbine inlet temperature is $T_3 = 1450$ K. The turbine isentropic efficiencies are both $\eta_t = 0.84$ and the compressor isentropic efficiency is $\eta_c = 0.82$. Assume complete combustion and neglect the pressure loss in the combustor and between components. Prepare plots of following quantities as a function of the pressure ratio $PR = P_2/P_1$ for a range of $1.5 < PR < 15$:

a) the air fuel ratio,
b) the required flow rate of methane in kmol/s,
c) the efficiency of the cycle, and
d) the temperature at state 5.

13.C-9 Although fuel cells are receiving a great deal of developmental attention, a major impediment to their widespread adoption is that they use hydrogen as a fuel and the low density of hydrogen makes it difficult to store and transport. Consequently, alternative ways of providing storing or providing a hydrogen source are being investigated. One alternative is to use liquid ammonia. Liquid anhydrous ammonia has a relatively high heating value, and products of combustion, in at least hydrogen transferring fuel cells, are only water and elemental nitrogen. Ammonia of course has several major drawbacks, but it is worth investigating. The molar mass of ammonia is $MW = 17.03$. The enthalpy of formation is $\bar{h}_{form} = -45{,}937$ kJ/kmol and the absolute entropy at 25°C is $\bar{s} = 192.8$ kJ/K-kmol.

a) What are the lower and higher heating values of liquid ammonia, in units of kJ/kg?
b) What pressure would the ammonia need to be maintained at in the fuel tank? Provide a short description of how you arrive at this answer.
c) What is the ratio of the volume of the ammonia tank to volume of a gasoline tank that is necessary to provide the same theoretical driving distance?

13.C-11 One method to improve the performance of a stationary gas turbine system is to inject steam into the combustion products before they enter the turbine, as shown in Figure 13.C-11. The steam is produced by a heat exchange process using the exhaust of the gas turbine as the heat source.

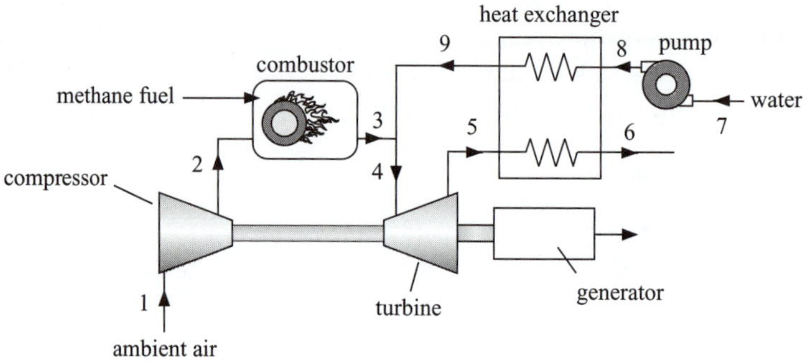

Figure 13.C-11: Gas turbine system with steam injection.

In a particular case, ambient air at $T_1 = 20°C$ and $P_1 = 1$ atm (state 1) enters the compressor and is compressed to 12 atm (state 2). The isentropic efficiency of the compressor is $\eta_c = 0.62$. The air then enters the combustor, which is fueled with methane that enters at 20°C. To ensure complete combustion, the fuel flow rate is adjusted to achieve 75% excess air. The combustion products exit at state 3. Metallurgical considerations dictate that the temperature of the gas entering the turbine at state 4 must be no higher than $T_4 = 1200°C$. The turbine isentropic efficiency is $\eta_t = 0.74$. Steam at $T_9 = 365°C$ and $P_9 = 12$ bar (state 9) is injected into the combustion products at a rate sufficient to maintain this temperature. The steam is generated from entering water at $T_7 = 20°C$ and $P_7 = 1$ bar (state 7) by heat exchange with the combustion products exiting the turbine at state 5.

a) What is the ratio of the mass flow rate of steam to the mass flow rate of air entering the compressor that is required to keep the turbine inlet stream at state 4 from exceeding 1200°C?
b) What are the temperature and the dew point temperature of the exhaust stream at state 6?
c) What is the First Law efficiency of the system with the steam injection and the corresponding back-work ratio?
d) If steam injection were not used, what would the % excess air have to be in order to maintain a combustor outlet temperature of 1200°C?
e) What is the First Law efficiency of the system without steam injection and the corresponding back-work ratio?
f) Provide a short explanation for the results you found in parts (e) and (f).

13.C-15 A simplified analysis of the processes occurring in a spark-ignition internal combustion engine is provided by the air-standard Otto cycle, as presented in Chapter 8. The purpose of this problem is to develop a more realistic model which considers combustion and heat transfer processes.

a) Calculate and plot the thermal efficiency and mean effective pressure for an air standard Otto cycle for compression ratios between 6 and 12 with a maximum cycle temperature of 3,000 K. Assume that the state of the air before compression is 25°C, 1 atm.
b) The working fluid entering the engine is not air, but rather a mixture of propane (C_3H_8) and stoichiometric air. The maximum cycle temperature is the temperature that results during the adiabatic constant volume combustion process. Compute the maximum cycle temperature and plot the engine thermal efficiency (based on the lower heating value of the fuel) and the mean effective pressure as a function of the compression ratio for compression ratios between 6 and 12 using the same assumptions that are employed in the air-standard Otto cycle, i.e., isentropic compression and expansion with combustion occurring at constant volume.
c) Some of the energy released in the combustion process is transferred to the 'cold' engine walls which are maintained at $T_{wall} = 105°C$. Assume that the amount of heat transfer to the engine walls during the combustion process is given by:

$$q_{wall} = K\left(\bar{T} - T_{wall}\right)$$

where $K = 628$ kJ/K-kmol propane and $\bar{T}$ is the average temperature occurring during the combustion process, e.g, $(T_2+T_3)/2$. Calculate and plot the thermal efficiency and mean effective pressure as a function of the compression ratio including this heat transfer consideration.

13.C-20 A 1.5 ft^3 rigid container contains 0.025 lb$_m$ of liquid methyl alcohol (CH_3OH) with a stoichiometric amount of air. The container is well-insulated and equipped with a spark plug to initiate the reaction. The methanol and air are initially at 77°F. The enthalpy of formation of liquid methanol at 77°F is –102,605 Btu/lbmol.
a) What is the pressure in the container before ignition?
b) Assuming that complete combustion occurs, estimate the temperature and pressure of the combustion products directly after the reaction occurs. (Assume ideal gas behavior.)
c) As the combustion gas is cooled, at what temperature will condensation begin?

13.C-22 Diesel engines are used in many communities for generating electricity. A particular system is shown schematically in Figure 13.C-22. The engine produces 385 kW of electrical power. Cooling water enters at 25°C and exits at 95°C and 1 atm. Fuel at a mass flow rate of 0.055 kg/sec with 10% excess air enters the engine at 25°C. Exhaust products exit at 425°C. (Assume complete combustion.) Depending on the cost of fuel relative to the cost of capital equipment, opportunities exist for recovering exergy from the cooling water, from the exhaust products, or from both. The Diesel fuel has a composition of 85% carbon and 15% hydrogen on a mass basis. The lower heating value of the fuel is 42,600 kJ/kg. The reference environment consists of a mixture of the following gases at 25°C and a total pressure of 1.0 atm: 20.4% oxygen, 0.04% carbon dioxide, 78.68% nitrogen, and 0.88% water vapor (on a molar basis). Assume that the generator efficiency is 1.0.
a) What is the efficiency of the engine (assuming that the cooling water and exhaust products are discarded)?
b) Estimate the enthalpy of formation of the Diesel fuel at 25°C in J per kg.

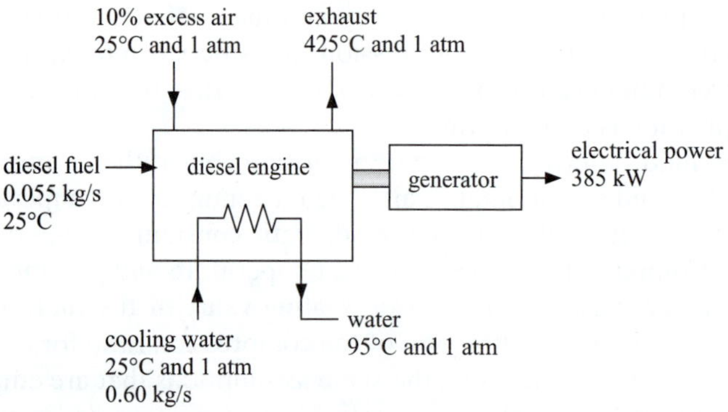

Figure 13.C-22: Schematic of the Diesel generator system.

c) What is the rate at which cooling water must be provided?
d) What is the rate of exergy associated with the exhaust products exiting the engine?
e) What is the rate of exergy associated with the exiting cooling water stream?
f) Estimate the rate of exergy destruction in the engine, assuming that the specific exergy of the fuel is equal to its lower heating value.

13.C-25 Figure 13.C-25 shows a schematic of a residential heating system that uses an engine-driven heat pump. The fuel to the engine is propane (C_3H_8). During normal operation, the engine drives the compressor of the heat pump, which uses R134a as the refrigerant. The condenser of the heat pump is located in one of two coils in the furnace ductwork and provides heat to the building. In addition, a glycol coolant is circulated through the engine and then through a heat exchanger that recovers some of the energy in the engine exhaust gas. The heated glycol solution then proceeds through a second coil in the furnace providing additional heat to the building.

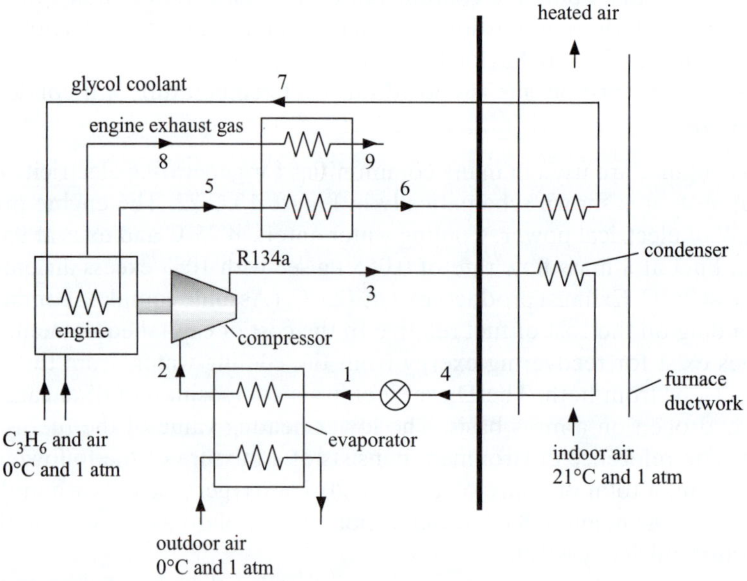

Figure 13.C-25: Schematic of a gas-fired heat pump system.

Test data for the engine driven heat pump are as follows. The engine provides 3.59 kW of power to the compressor. The isentropic efficiency of the compressor is known to be 0.72 under these operating conditions. The saturation temperatures in the evaporator and condenser are −12°C and 36°C, respectively. (Assume states 2 and 4 to be saturated vapor and liquid, respectively and neglect pressure losses in the piping and heat exchangers.) The glycol solution, which has a specific heat capacity of 3.43 kJ/kg-K, is circulated at a rate of 0.062 kg/s. The measured temperatures of the glycol solution at states 5, 6, and 7 are 76°C, 91°C, and 58°C. The exhaust gas exits at state 9 at 124°C. A combustion gas analysis of the indicated the following volume percentages of the exhaust gases, expressed on a dry basis: 8.3% CO_2, 1.5% O_2, 0.6% CO, remainder inert gases (primarily N_2).

a) What is the mass flow rate of the R134a?
b) What is the percent theoretical air and the corresponding air fuel ratio for the engine?
c) What is the efficiency of the engine based on the lower heating value of the fuel?
d) The exhaust gas recovery heat exchanger uses a cross-flow design with both fluids unmixed. Estimate the heat transfer rate and the effectiveness of this heat exchanger.
e) Determine the total rate of heat transfer to the building. Also, indicate the fractional contributions of this heat transfer from the heat pump, the recovery of energy from the engine coolant, and the recovery of energy from the exhaust gas.
f) Determine the efficiency of this heating system based on the lower heating value of the fuel defined in the same manner as for a conventional furnace.

13.C-28 The advent of fuel cell technology has increased interest in hydrogen. One way that hydrogen can be produced is by electrolysis of water. In a particular process, water at 25°C and 1 atm together with electrical energy are supplied to an electrolysis cell. The final products from this process are hydrogen and oxygen in separate tanks, each at 100°C and 5 atm. The necessary electrical energy is 500,000 kJ per kmol of hydrogen. Atmospheric conditions are 25°C, 1 atm, 40% relative humidity, with 21% oxygen by mole.

a) What is the specific molar exergy of the hydrogen produced in this process?
b) What is the specific molar exergy of the oxygen produced in this process?
c) Determine the required heat transfer per kmol of hydrogen.
d) What is the Second Law efficiency of the overall process?

14 Chemical Equilibrium

Many devices of interest to mechanical engineers involve chemical reactions. Examples are internal combustion engines that produce work and furnaces and boilers that provide heat. Therefore, it is useful for a mechanical engineer to have an understanding of the principles governing chemical reacting systems. An initial examination of combustion reactions is presented in Chapter 13. However, in Chapter 13 it is assumed that the combustion reactions proceed to completion; that is, the reaction continues until one or more of the reactants are consumed. Although this assumption is approximately correct for many combustion reactions, it is not generally true. At high temperatures, dissociation and reverse reactions occur and result in the generation of pollutants such as carbon monoxide, hydrocarbon fragments, and oxides of nitrogen. These pollutants are usually only generated in small quantities, but they can have significant environmental consequences. The assumption of complete reactions is also inappropriate for many reformation reactions in which fuel of one chemical form is converted to a different chemical form. In general, the equilibrium state achieved by a reaction will include both reactants and products at a condition of chemical equilibrium. This chapter presents methods for determining the equilibrium state of a system that can chemically react in one or more ways.

14.1 Criterion for Chemical Equilibrium

Suppose we place two or more different substances, A_1, A_2, A_3, etc., each with known mass into a container and maintain the pressure and temperature at constant values, as shown in Figure 14-1. The contents of the container will eventually reach an equilibrium state. The equilibrium state could be multi-phase (e.g., liquid and vapor), as considered in Chapter 11; however for now we will assume that the equilibrium condition is gas phase. The substances in the container may participate in one or more chemical reactions. We are interested in determining the amount of each substance that exists at the equilibrium state.

Any change that occurs as the process proceeds towards equilibrium must obey the laws of thermodynamics. The system shown in Figure 14-1 is the contents of the cylinder. This is a closed system for which there are no significant changes in kinetic or potential energy. An energy balance on this system results in:

$$\dot{Q}_{in} = \dot{W}_{out} + \frac{dU}{dt} \tag{14-1}$$

The pressure is constant in this process. Therefore the rate of work out of the system is given by:

$$\dot{W}_{out} = P\frac{dV}{dt} \tag{14-2}$$

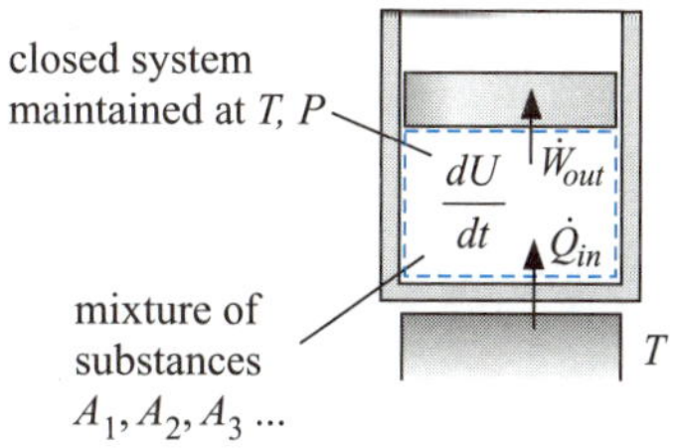

Figure 14-1: Closed system containing reactive substances maintained at constant temperature and pressure.

where V is the volume of the system. Substituting Eq. (14-2) into Eq. (14-1) leads to:

$$\dot{Q}_{in} = P\frac{dV}{dt} + \frac{dU}{dt} \tag{14-3}$$

Recognizing that pressure is constant for this process allows Eq. (14-3) to be written as:

$$\dot{Q}_{in} = \frac{dU}{dt} + \frac{d(PV)}{dt} \tag{14-4}$$

Substituting the definition of enthalpy ($H = U + PV$) into Eq. (14-4) provides:

$$\dot{Q}_{in} = \frac{dH}{dt} \quad \text{at constant } P \tag{14-5}$$

An entropy balance on the closed system shown in Figure 14-1 is:

$$\frac{\dot{Q}_{in}}{T} + \dot{S}_{gen} = \frac{dS}{dt} \tag{14-6}$$

where the entropy generation rate, $\dot{S}_{gen}$, must be greater than zero if any spontaneous change in the system occurs. Equation (14-6) can be solved for the rate of heat transfer into the system.

$$\dot{Q}_{in} = T\frac{dS}{dt} - T\dot{S}_{gen} \tag{14-7}$$

Substituting Eq. (14-5) into Eq. (14-7) provides:

$$\frac{dH}{dt} = T\frac{dS}{dt} - T\dot{S}_{gen} \quad \text{at constant } P \tag{14-8}$$

which can be rearranged to provide:

$$\frac{dH}{dt} - T\frac{dS}{dt} = \underbrace{-T\dot{S}_{gen}}_{\text{must be}<0} \quad \text{at constant } P \tag{14-9}$$

Note the both the absolute temperature (T) and the entropy generation rate ($\dot{S}_{gen}$) must be positive. Therefore, Eq. (14-9) indicates that if any change within the system occurs, it does so subject to the constraint:

$$\frac{dH}{dt} - T\frac{dS}{dt} < 0 \quad \text{at constant } P \tag{14-10}$$

The Gibbs free energy (G) is defined in Eq. (10-81) as:

$$G = H - TS \tag{14-11}$$

The total differential of Gibbs free energy with respect to time is:

$$\frac{dG}{dt} = \frac{dH}{dt} - T\frac{dS}{dt} - S\frac{dT}{dt} \tag{14-12}$$

If temperature is constant, then the last term in Eq. (14-12) must be zero:

$$\frac{dG}{dt} = \frac{dH}{dt} - T\frac{dS}{dt} \qquad \text{at constant } T \tag{14-13}$$

Comparing Eqs. (14-10) and (14-13) suggests that:

$$\frac{dG}{dt} < 0 \qquad \text{at constant } T \text{ and } P \tag{14-14}$$

The importance of Gibbs free energy can be understood by considering Eq. (14-14). Any spontaneous changes that occur in a system while the temperature and pressure of the system are held constant must result in a reduction of the system's Gibbs free energy.

After some time, the system will achieve an equilibrium state and there will be no further changes in the thermodynamic properties of the system. The time required for equilibrium to occur may be a few seconds or a few centuries. Thermodynamics does not provide any information concerning the rates at which the reactions proceed. Thermodynamics only allows us to determine the equilibrium condition. Gibbs free energy is a thermodynamic property. Therefore, once the system reaches equilibrium:

$$\frac{dG}{dt} = 0 \qquad \text{at equilibrium} \tag{14-15}$$

Equation (14-15) is the criterion for chemical equilibrium. It is also the criterion for phase equilibrium, as discussed in Section 11.5.1. Equation (14-15) indicates that the chemical reactions and phase changes occurring in a system will spontaneously result in a reduction of its Gibbs free energy. At the equilibrium point, the Gibbs free energy at a specified temperature and pressure will reach its minimum value. The entropy generation will be maximized at the equilibrium point.

Note that we are only interested in the equilibrium state and the equilibrium state will be characterized by some temperature and pressure. The path that the process takes to get the equilibrium state is irrelevant. Thus, there really is no need to be concerned with a path and the concept of keeping the temperature and pressure constant to arrive at Eq. (14-15) is a convenient, but unnecessary, mental crutch. The Gibbs free energy at a specified temperature and pressure will reach a minimum at equilibrium regardless of the process that is followed to reach the equilibrium condition.

14.2 Reaction Coordinates

Balancing combustion reactions that are assumed to react to completion, as assumed in Chapter 13, is straightforward. The situation becomes more complicated when dealing with one or more chemical reactions that may not proceed to completion. This section introduces the concept of a *reaction coordinate* (also called the *degree of reaction*), ε, that simplifies balances in chemical reacting systems.

Any chemical reaction can be written in algebraic form as:

$$\nu_1 A_1 + \nu_2 A_2 + \cdots . = \sum_{i=1}^{C} \nu_i A_i = 0 \tag{14-16}$$

where ν_i are *stoichiometric coefficients*, A_i represents chemical substance i, and C is the total number of substances. By convention, ν_i is considered to be a positive value for products and a negative value for reactants. For example, consider the stoichiometric reaction of hydrogen with oxygen:

$$2\,H_2 + O_2 \leftrightharpoons 2H_2O \tag{14-17}$$

Table 14-1: Substances and stoichiometric coefficients for the reaction given by Eq. (14-17).

i	A_i	ν_i
1	H_2	-2
2	O_2	-1
3	H_2O	2

This chemical reaction can be written in the form of Eq. (14-16), $\sum_{i=1}^{C} \nu_i A_i = 0$, with the stoichiometric coefficients listed in Table 14-1.

Although the reaction in Eq. (14-17) is written with hydrogen and oxygen as the reactants (and therefore with negative stoichiometric coefficients), the reaction may not actually proceed in this direction. For example, at high temperatures water can decompose into hydrogen and oxygen. Therefore, under some conditions water could be the reactant and hydrogen and oxygen the products. The use of the $\leftrightharpoons$ symbol rather than the $\rightarrow$ symbol, as in Chapter 13, reinforces this point. However, the reaction can be written in the form of Eq. (14-17) regardless of which way it actually proceeds. The purpose of the chemical equilibrium analyses provided in this chapter is to determine which direction the reaction will proceed and to what extent.

The stoichiometric coefficients relate the changes in the molar amounts of each substance. For example, in Eq. (14-17), the change in number of moles of water ($dn_{H_2O} = dn_3$) is related to the change in the number of moles of hydrogen ($dn_{H_2} = dn_1$) according to:

$$\frac{dn_{H_2O}}{dn_{H_2}} = \frac{dn_3}{dn_1} = \frac{\nu_3}{\nu_1} = \frac{2}{-2} = -1 \quad \text{or} \quad \frac{dn_3}{\nu_3} = \frac{dn_1}{\nu_1} \tag{14-18}$$

The change in the number of moles of water and the change in the number of moles of oxygen are related according to:

$$\frac{dn_{H_2O}}{dn_{O_2}} = \frac{dn_3}{dn_2} = \frac{\nu_3}{\nu_2} = \frac{2}{-1} = -2 \quad \text{or} \quad \frac{dn_3}{\nu_3} = \frac{dn_2}{\nu_2} \tag{14-19}$$

In general, the ratio of the change in the number of moles of a substance to the associated stoichiometric coefficient must be the same for all substances involved in a single reaction:

$$\frac{dn_1}{\nu_1} = \frac{dn_2}{\nu_2} = \frac{dn_3}{\nu_2} = \cdots = \frac{dn_C}{\nu_C} \qquad \text{(14-20}$$

Equation (14-20) can be written in a general manner that is applicable for any single chemical reaction by introducing the *reaction coordinate* (ε, also called the *degree of reaction*):

$$\frac{dn_1}{\nu_1} = \cdots = \frac{dn_i}{\nu_i} = \cdots = \frac{dn_C}{\nu_C} = d\varepsilon \tag{14-21}$$

The reaction coordinate relates the molar amounts of all substances involved in one chemical reaction. Since it is defined as a differential, a boundary condition must be specified to determine its value. Normally, the reaction coordinate is set equal to $\varepsilon_0 = 0$ at the start of the reaction. Note that the stoichiometric coefficients are generally considered to be dimensionless quantities, which requires the reaction coordinate, ε, to have the same molar units as n_i. Also, the value of ε is bounded by stoichiometry in that it is not possible to have negative amounts of any substance. However, even with this limitation the value of ε is not necessarily constrained to be positive or to be less than unity.

Equation (14-21) can be integrated for each substance in the chemical reaction:

$$\int_{n_{0,i}}^{n_i} \frac{dn_i}{\nu_i} = \int_{\varepsilon_0}^{\varepsilon} d\varepsilon \tag{14-22}$$

where ε_0 is the reaction coordinate at the start of the reaction and $n_{0,i}$ is the number of moles of substance i at the start of the reaction. Carrying out the integration in Eq. (14-22) provides:

$$\varepsilon = \varepsilon_0 + \frac{(n_i - n_{0,i})}{\nu_i} \quad \text{for } i = 1..C \tag{14-23}$$

Solving Eq. (14-23) for the number of moles of substance i provides:

$$n_i = n_{0,i} + \nu_i\,(\varepsilon - \varepsilon_0) \quad \text{for } i = 1..C \tag{14-24}$$

If the reaction coordinate is set to zero at the start of the reaction ($\varepsilon_0 = 0$) then Eq. (14-23) can be written as:

$$\varepsilon = \frac{(n_i - n_{0,i})}{\nu_i} \quad \text{for } i = 1..C \tag{14-25}$$

and Eq. (14-24) can be written as:

$$\boxed{n_i = n_{0,i} + \nu_i\,\varepsilon \quad \text{for } i = 1..C} \tag{14-26}$$

Applying Eq. (14-25) to the hydrogen-oxygen reaction in Eq. (14-17) provides:

$$\varepsilon = \frac{(n_{H_2} - n_{0,H_2})}{-2} = \frac{(n_{O_2} - n_{0,O_2})}{-1} = \frac{(n_{H_2O} - n_{0,H_2O})}{2} \tag{14-27}$$

where n_{0,H_2}, n_{0,O_2}, and n_{0,H_2O} are the initial amounts of hydrogen, oxygen, and water, respectively. The reaction coordinate is the single quantity that defines the number of moles of every substance involved in the reaction, given their initial values. In Eq. (14-27), if the number of moles of hydrogen, oxygen, or water at any point in the reaction is known, then the reaction coordinate can be computed and used to determine the number of moles of the other substances that are involved in the chemical reaction.

Often, two or more reactions occur simultaneously in the same system. In this case, a reaction coordinate must be defined for each independent chemical reaction (ε_j, where j represents a specific reaction). The number of moles of any substance participating in the reactions can be determined by expanding Eq. (14-24) so that all of the r independent reactions are considered:

$$n_i = n_{0,i} + \sum_{j=1}^{r} \nu_{i,j}(\varepsilon_j - \varepsilon_{0,j}) \quad \text{for } i = 1..C \tag{14-28}$$

where $\nu_{i,j}$ is the stoichiometric coefficient for substance i in reaction j and $\varepsilon_{0,j}$ is the initial value of the reaction coordinate for reaction j. If substance i does not appear in reaction j, then the stoichiometric coefficient $\nu_{i,j}$ is equal to zero. Assuming that the initial value of the reaction coordinate for each reaction is specified to be zero, $\varepsilon_{0,j} = 0$, allows Eq. (14-28) to be written as:

$$n_i = n_{0,i} + \sum_{j=1}^{r} \nu_{i,j}\,\varepsilon_j \quad \text{for } i = 1..C \tag{14-29}$$

EXAMPLE 14.2-1: SIMULTANEOUS CHEMICAL REACTIONS

The following two chemical reactions occur simultaneously during the reformation of methane to produce hydrogen:

$$\text{Reaction 1: } CH_4 + H_2O \leftrightharpoons CO + 3\,H_2 \tag{1}$$

$$\text{Reaction 2: } H_2 + CO_2 \leftrightharpoons H_2O + CO \tag{2}$$

Initially, there are $n_{0,CH_4} = 2$ kmol of CH_4, $n_{0,H_2O} = 3$ kmol of H_2O, and $n_{0,CO_2} = 1$ kmol of CO_2 present in the reaction vessel.

a) Develop general relationships for the mole fraction of each species in terms of the reaction coordinates for the $r = 2$ simultaneous reactions, ε_1 and ε_2.

The inputs are entered in EES.

```
$UnitSystem SI Molar J K Pa
n_0_CH4=2 [kmol]          "initial amount of CH4"
n_0_H2O=3 [kmol]          "initial amount of H2O"
n_0_CO2=1 [kmol]          "initial amount of CO2"
C=5 [-]                   "number of substances"
r=2 [-]                   "number of reactions"
```

The matrix of stoichiometric coefficients is setup based on the reactions given by Eqs. (1) and (2). The substances are numbered according to $1 = CH_4$, $2 = H_2O$, $3 = CO$, $4 = H_2$, and $5 = CO_2$.

```
CH4=1; H2O=2; CO=3; H2=4; CO2=5;     "substance numbering system"
```

Examination of Eq. (1) suggests that: $\nu_{1,1} = -1$, $\nu_{2,1} = -1$, $\nu_{3,1} = 1$, $\nu_{4,1} = 3$, and $\nu_{5,1} = 0$. (Note that the first subscript identifies the substance and the second subscript corresponds to the reaction number.) Examination of Eq. (2) provides: $\nu_{1,2} = 0$, $\nu_{2,2} = 1$, $\nu_{3,2} = 1$, $\nu_{4,2} = -1$, and $\nu_{5,2} = -1$. The matrix of stoichiometric coefficients is defined in EES.

```
"Stoichiometric coefficient matrix"
"Reaction 1"
nu[CH4,1]= -1 [-]         "CH4 in reaction 1"
nu[H2O,1]= -1 [-]         "H2O in reaction 1"
nu[CO,1]= 1 [-]           "CO in reaction 1"
nu[H2,1]= 3 [-]           "H2 in reaction 1"
nu[CO2,1]= 0 [-]          "CO2 in reaction 1"

"Reaction 2"
nu[CH4,2]= 0 [-]          "CH4 in reaction 2"
nu[H2O,2]= 1 [-]          "H2O in reaction 2"
nu[CO,2]= 1 [-]           "CO in reaction 2"
nu[H2,2]= -1 [-]          "H2 in reaction 2"
nu[CO2,2]= -1 [-]         "CO2 in reaction 2"
```

The initial quantities of each of the five substances are assigned.

$$n_{0,1} = n_{0,CH_4}$$

$$n_{0,2} = n_{0,H_2O}$$

$$n_{0,3} = 0$$

$$n_{0,4} = 0$$

$$n_{0,5} = n_{0,CO_2}$$

```
"Initial quantities of each substance"
n_0[CH4]=n_0_CH4                   "CH4"
n_0[H2O]=n_0_H2O                   "H2O"
n_0[CO]=0 [kmol]                   "CO2"
n_0[H2]=0 [kmol]                   "H2"
n_0[CO2]=n_0_CO2                   "CO2"
```

The reaction coordinates, ε_1 and ε_2, are initially assigned arbitrary values.

```
e[1]=0 [kmol]                      "reaction coordinate for reaction 1"
e[2]=0 [kmol]                      "reaction coordinate for reaction 2"
```

Equation (14-29) is used to determine the number of moles of each of the substances based on the degrees of reaction:

$$n_i = n_{0,i} + \sum_{j=1}^{r} \nu_{i,j}\, \varepsilon_j \quad \text{for } i = 1..C$$

```
duplicate i=1,C
   n[i]=n_0[i]+sum(nu[i,j]*e[j], j=1,r)   "number of moles of each substance"
end
```

The total number of moles is obtained from:

$$n = \sum_{i=1}^{C} n_i$$

and the mole fraction of each component is given by:

$$y_i = \frac{n_i}{n} \quad \text{for } i = 1..C$$

```
n=sum(n[i],i=1,C)                  "total number of moles"
duplicate i=1,C
    y[i]=n[i]/n                    "mole fraction"
end
```

b) Determine the maximum possible mole fraction of hydrogen.

This question can be examined using a contour plot. In order to generate a contour plot, a parametric table with 400 rows is generated that includes ε_1, ε_2, y_1, and y_4. The values of the reaction coordinates are each varied between 0 kmol and 3 kmol. However, the values of ε_1 are varied from 0 kmol to 3 kmol repeating the pattern every 20 rows, as shown in Figure 1(a), while the values of ε_2 are varied from 0 kmol to 3 kmol applying the pattern every 20 rows, as shown in Figure 1(b). The result is a parametric table containing 400 unique combinations of ε_1 and ε_2 that generate a grid over the parameter space 0 kmol $< \varepsilon_1 <$ 3 kmol and 0 kmol $< \varepsilon_2 <$ 3 kmol.

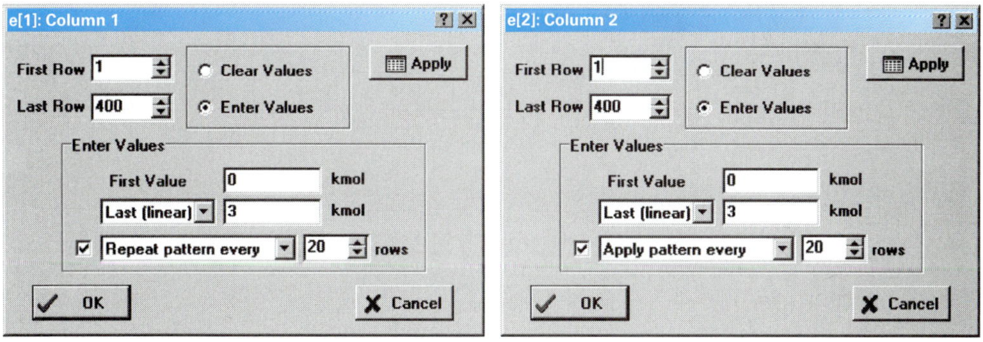

(a) (b)

Figure 1: Setting the values for (a) ε_1 and (b) ε_2 in order to generate a contour plot.

The specified values of ε_1 and ε_2 in the EES code are commented out

```
{e[1]=0 [kmol]    "reaction coordinate for reaction 1"
 e[2]=0 [kmol]    "reaction coordinate for reaction 2"}
```

and the Parametric table is solved. The table is used to generate the contour plots of the mole fraction of hydrogen (y_4) and the mole fraction of methane (y_1) in the parameter space of ε_1 and ε_2 that are shown in Figures 2(a) and 2(b), respectively. Select New Plot Window from the Plots menu and then select X-Y-Z Plot in order to generate these plots.

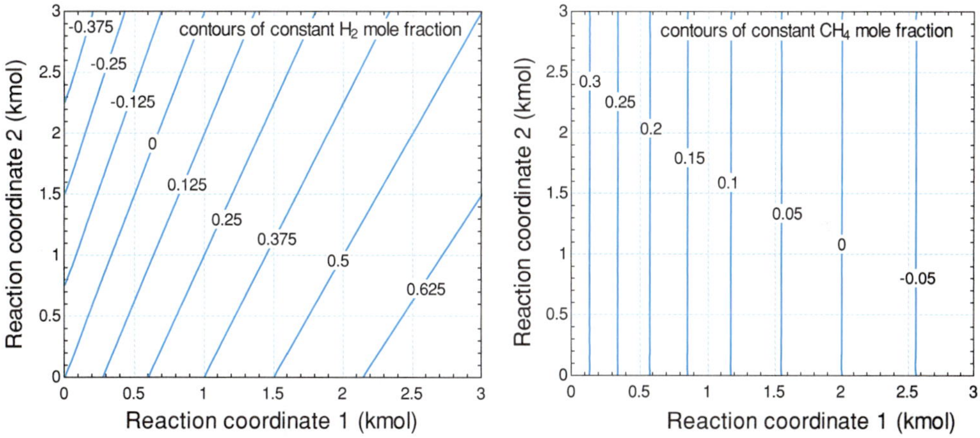

(a) (b)

Figure 2: Contour plot of (a) mole fraction of hydrogen (y_4) and (b) mole fraction of methane (y_1) as a function of ε_1 and ε_2.

Figure 2(a) shows that the amount of hydrogen increases as ε_1 increases and decreases as ε_2 increases; examination of Eqs. (1) and (2) shows that reaction 1 tends to produce hydrogen while reaction 2 consumes hydrogen. Figure 2(a) also shows that there are some combinations of reaction coordinates that are not possible. For example, the region to the left of the hydrogen mole fraction contour $y_4 = 0$ in Figure 2(a) is not possible because the mole fraction of any substance cannot go below 0 or above 1. Figure 2(b) shows that the mole fraction of methane is only a function of ε_1. Further, the value of ε_1 must be bounded by 2 since y_1 cannot be less than zero.

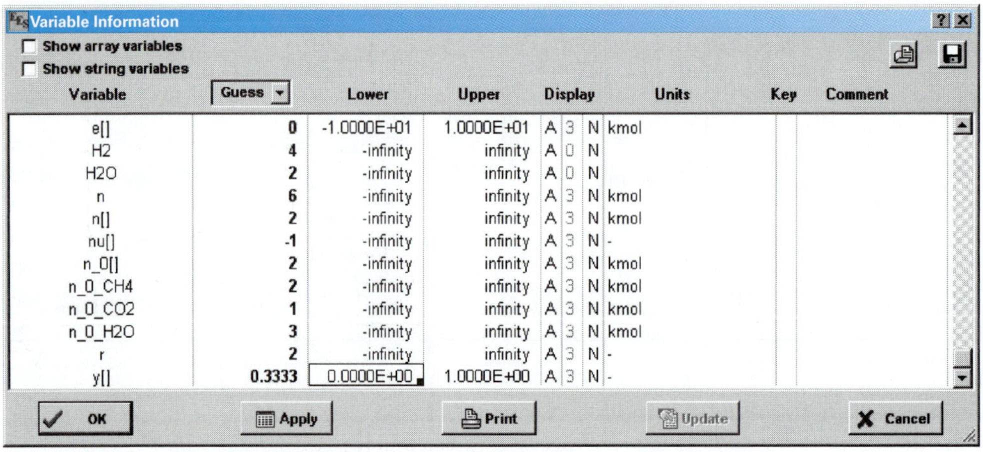

Figure 3: Variable Information dialog showing the limits on the array of mole fractions.

We are asked to determine the maximum value of the hydrogen mole fraction (y_4) subject to mass balance constraints. This problem is an example of a constrained optimization problem. Analytical optimization methods, such as the Lagrange Method of Undetermined Multipliers discussed in Section 14.4.2, can be applied to this type of problem. However, this problem can also be solved numerically in EES. First, set constraints for each of the mole fractions by selecting Variable Info from the Options menu. In the Variable Information dialog, Figure 3, deselect the Show array variables and scroll to the mole fraction array, y[ ]. Set the lower and upper bounds on the array to 0 and 1, respectively.

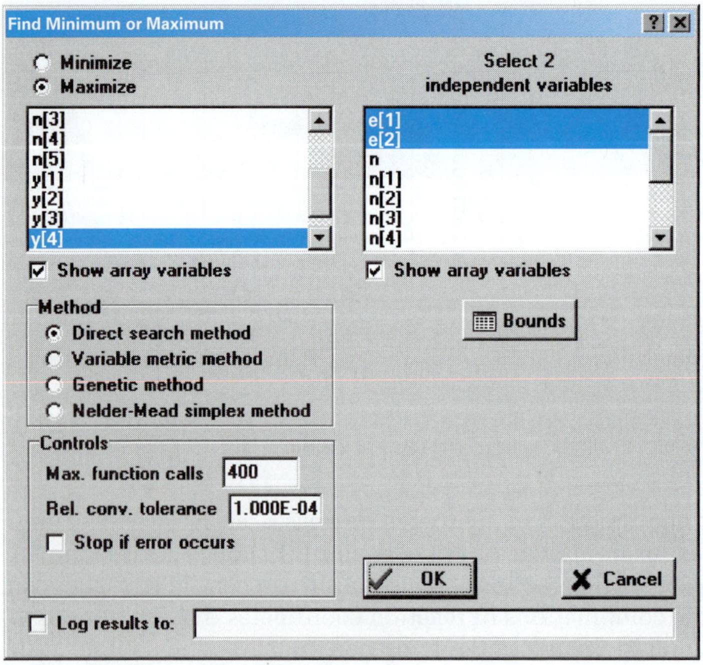

Figure 4: Find Minimum or Maximum dialog.

With the values of ε_1 and ε_2 commented out, select Min/Max from the Calculate menu. In the Find Minimum or Maximum dialog, Figure 4, select Maximize and indicate that the dependent variable is y_4 and choose ε_1 and ε_2 to be the independent variables.

Select the Bounds button and set reasonable bounds for the reaction coordinates. Note that the constraints on the mole fractions will constrain these variables such that no mole fraction exceeds its physical limits; therefore, the lower and upper bounds on ε_1 and ε_2 need only be set to very low and very high values, respectively. Finally, select the OK button and EES will identify that the maximum possible value of $y_4 = 0.7$ and this value occurs when $\varepsilon_1 = 2$ kmol and $\varepsilon_2 = -1$ kmol. Note that this is the maximum possible mole fraction of hydrogen that will not result in negative amounts of other substances. However, this mole fraction may not be attainable due to chemical equilibrium limitations.

14.3 The Law of Mass Action

The criterion for chemical equilibrium is that the Gibbs free energy of the system must reach a minimum value, as discussed in Section 14.1; the minimum value of the Gibbs free energy defines the equilibrium state. Therefore, it is necessary to identify the relation between Gibbs free energy and composition in order to determine the equilibrium composition of a system.

14.3.1 The Criterion of Equilibrium in terms of Chemical Potentials

The molar specific Gibbs free energy for a pure substance, $\overline{g}$, is a function of two intensive properties; these are most conveniently chosen to be T and P.

$$\overline{g} = \overline{h} - T\overline{s} = \overline{g}(T, P) \quad \text{for a pure substance} \tag{14-30}$$

By definition, the molar specific Gibbs free energy for a pure substance is the total Gibbs free energy (G) divided by the number of moles (n) of the substance. The total Gibbs free energy is then a function of T, P, and n:

$$G = n\overline{g} = G(T, P, n) \quad \text{for a pure substance} \tag{14-31}$$

In a single phase chemical reaction, the system necessarily consists of a mixture of substances with molar amounts $n_1, n_2, \ldots, n_C$ where C is the number of components. The total Gibbs free energy for the mixture is a function of the temperature, pressure, and the number of moles of each substance:

$$G = n\overline{g} = G(T, P, n_1, n_2, \cdots, n_C) \quad \text{for a mixture} \tag{14-32}$$

The total derivative of the Gibbs free energy function for the mixture is:

$$dG = \left(\frac{\partial G}{\partial T}\right)_{P,n_i} dT + \left(\frac{\partial G}{\partial P}\right)_{T,n_i} dP + \sum_{i=1}^{C}\left(\frac{\partial G}{\partial n_i}\right)_{T,P,n_{j,j\neq i}} dn_i \tag{14-33}$$

The partial derivative within the summation in Eq. (14-33) appears often in multicomponent phase and chemical equilibrium problems. The partial derivative of total Gibbs free energy with respect to the number of moles of component i is called the *chemical potential* of component i (μ_i); chemical potential was discussed in Section 11.5.2.

$$\mu_i = \left(\frac{\partial G}{\partial n_i}\right)_{T,P,n_{j,j\neq i}} \tag{14-34}$$

Equation (14-15) can be written in terms of the chemical potentials of each substance.

$$dG = \left(\frac{\partial G}{\partial T}\right)_{P,n_i} dT + \left(\frac{\partial G}{\partial P}\right)_{T,n_i} dP + \sum_{i=1}^{C} \mu_i \, dn_i \tag{14-35}$$

For a process that occurs at constant temperature and pressure, Eq. (14-35) can be written as:

$$dG = \sum_{i=1}^{C} \mu_i \, dn_i \text{ at constant } T \text{ and } P \tag{14-36}$$

The criteria for chemical equilibrium, Eq. (14-15), requires that the change in Gibbs free energy must be equal to zero:

$$dG = \sum_{i=1}^{C} \mu_i \, dn_i = 0 \quad \text{at equilibrium at constant } T \text{ and } P \tag{14-37}$$

For a single reaction, the change in the number of moles of substance i can be expressed in terms of the reaction coordinate and the stoichiometric coefficient for substance i using Eq. (14-21):

$$dn_i = \nu_i \, d\varepsilon \quad \text{for a single reaction} \tag{14-38}$$

Equation (14-38) can be substituted into Eq. (14-37) in order to provide a relation in which the reaction coordinate is the only independent variable:

$$dG = \sum_{i=1}^{C} \mu_i \, \nu_i \, d\varepsilon = 0 \text{ for a single reaction at equilibrium at constant } T \text{ and } P \tag{14-39}$$

Rearranging Eq. (14-39) provides:

$$\frac{dG}{d\varepsilon} = \sum_{i=1}^{C} \mu_i \, \nu_i = 0 \quad \text{for a single reaction at equilibrium at constant } T \text{ and } P \tag{14-40}$$

For two or more simultaneous reactions, the change in the number of moles of substance i can be expressed in terms of the reaction coordinates for each of the reactions:

$$dn_i = \sum_{j=1}^{r} \nu_{i,j} \, d\varepsilon_j \quad \text{for } r \text{ simultaneous reactions} \tag{14-41}$$

Substituting Eq. (14-41) into Eq. (14-37) provides:

$$dG = \sum_{i=1}^{C} \sum_{j=1}^{r} \nu_{i,j} \, \mu_i \, d\varepsilon_j = 0 \quad \text{at equilibrium at constant } T \text{ and } P \tag{14-42}$$

The only general way to ensure that the change in Gibbs free energy will always be equal to zero is to set the coefficients that multiply the change in each of the reaction coordinates, $d\varepsilon_j$ in Eq. (14-42), equal to zero. This process results in the following criteria for equilibrium:

$$\boxed{\sum_{i=1}^{C} \nu_{i,j} \, \mu_i = 0 \quad \text{for } j = 1..r \quad \text{at equilibrium}} \tag{14-43}$$

14.3.2 Chemical Potentials for an Ideal Gas Mixture

In order to use Eq. (14-43) it is necessary to evaluate the chemical potential of each substance in the mixture. This process is simple for a mixture that obeys the ideal gas law. As shown in Section 11.1, each gas in an ideal gas mixture behaves as if it were a pure substance at the same temperature as the mixture, but at a pressure that is equal to its partial pressure, P_i, which is the product of its mole fraction and the total pressure:

$$\mu_i\,(T,P) = \mu_{i,pure}\,(T,P_i) = \overline{g}_i\,(T,P_i) = \overline{h}_i - T\,\overline{s}_i \quad \text{for an ideal gas mixture} \qquad (14\text{-}44)$$

It is shown in Section 11.5.3 that the chemical potential for an ideal gas mixture, which appears in Eq. (14-44), can be equivalently expressed as:

$$\mu_i = \mu_i^o + R_{univ}\,T \ln\left(\frac{y_i\,P}{P^o}\right) \quad \text{for an ideal gas mixture} \qquad (14\text{-}45)$$

where μ_i^o is the chemical potential (which is also the molar specific Gibbs free energy, $\overline{g}_i^o$) of pure substance i at temperature T and a reference pressure P^o (sometimes referred to as the standard state pressure). The reference pressure is normally chosen to be 1 atm or 100 kPa for gas-phase reactions.

$$\mu_i^o = \overline{g}_i^o = \overline{h}_i^o - T\,\overline{s}_i^o \qquad (14\text{-}46)$$

14.3.3 Equilibrium Constant and the Law of Mass Action for Ideal Gas Mixtures

The criterion of equilibrium for a system in which r simultaneous chemical reactions occur is provided in Eq. (14-43), which is repeated below:

$$\sum_{i=1}^{C} \nu_{i,j}\,\mu_i = 0 \quad \text{for } j = 1..r \text{ at equilibrium} \qquad (14\text{-}47)$$

Substituting the chemical potential relation for ideal gases, given by Eq. (14-45), into Eq. (14-47) provides:

$$\sum_{i=1}^{C} \nu_{i,j}\left[\overline{g}_i^o + R_{univ}\,T \ln\left(\frac{y_i\,P}{P^o}\right)\right] = 0 \quad \begin{array}{l}\text{for } j = 1..r \text{ at equilibrium}\\ \text{for an ideal gas mixture}\end{array} \qquad (14\text{-}48)$$

Equation (14-48) can be rearranged:

$$\sum_{i=1}^{C} \nu_{i,j}\,\overline{g}_i^o = -R_{univ}\,T \sum_{i=1}^{C} \ln\left(\frac{y_i\,P}{P^o}\right)^{\nu_{i,j}} \quad \begin{array}{l}\text{for } j = 1..r \text{ at equilibrium}\\ \text{for an ideal gas mixture}\end{array} \qquad (14\text{-}49)$$

The left side of Eq. (14-49) is defined as the *standard state Gibbs free energy change of reaction* for reaction j, ΔG_j^o:

$$\boxed{\Delta G_j^o = \sum_{i=1}^{C} \nu_{i,j}\,\overline{g}_i^o \quad \text{for } j = 1..r} \qquad (14\text{-}50)$$

The value of ΔG_j^o is only a function of temperature since the pressure used to compute the reference molar specific Gibbs free energies of each substance ($\overline{g}_i^o$ for $i = 1..C$) is the reference pressure P^o and there is no composition dependence in Eq. (14-50).

The right side of Eq. (14-49) involves a summation of the logarithm of terms. Summing the logarithms of a series of values is equivalent to taking the logarithm of the product of the values:

$$\sum_{i=1}^{C} \ln\left(\frac{y_i P}{P^o}\right)^{\nu_{i,j}} = \ln\left[\prod_{i=1}^{C}\left(\frac{y_i P}{P^o}\right)^{\nu_{i,j}}\right] \quad \text{for } j = 1..r \tag{14-51}$$

The product sign (Π) in Eq. (14-51) is similar to the summation sign; however, it indicates that the individual terms in the series are multiplied by one another rather than summed together. Substituting Eqs. (14-50) and (14-51) into Eq. (14-49) leads to:

$$\Delta G_j^o = -R_{univ}\, T \ln\left[\underbrace{\prod_{i=1}^{C}\left(\frac{y_i\, P}{P^o}\right)^{\nu_{i,j}}}_{K_j}\right] \quad \begin{array}{l}\text{for } j = 1..r \text{ at equilibrium}\\ \text{for an ideal gas mixture}\end{array} \tag{14-52}$$

The term inside the brackets of Eq. (14-52) is called the *equilibrium constant* for reaction j and given the symbol K_j. The standard state Gibbs free energy change of reaction, ΔG_j^o, is only a function of temperature. Therefore, according to Eq. (14-52), the equilibrium constant must also be only a function of temperature, even though it appears to depend on the mole fractions of the gases and the pressure:

$$K_j = \prod_{i=1}^{C}\left(\frac{y_i\, P}{P^o}\right)^{\nu_{i,j}} \quad \text{for } j = 1..r \quad \text{for an ideal gas mixture} \tag{14-53}$$

Substituting Eq. (14-53) into Eq. (14-52) results in the *Law of Mass Action*, which describes the equilibrium state of a reacting system:

$$\boxed{\Delta G_j^o = -R_{univ}\, T \ln\left(K_j\right) \quad \text{for } j = 1..r} \tag{14-54}$$

At a particular temperature, the values of ΔG_j^o and K_j can be evaluated for each reaction using Eqs. (14-50) and (14-54). Each reaction is characterized by a reaction coordinate, as discussed in Section 14.2, and the molar composition of the mixture is completely specified by the values of the reaction coordinates for each independent reaction together with the initial composition of the mixture. Therefore, at a specified pressure and temperature, the values of the reaction coordinates must be adjusted until Eq. (14-54) is satisfied in order to determine the equilibrium composition.

EXAMPLE 14.3-1: REFORMATION OF METHANE

The following two chemical reactions occur simultaneously during the reformation of methane:

$$\text{Reaction 1: } CH_4 + \mathrm{H_2O} \leftrightharpoons \mathrm{CO} + 3\,\mathrm{H_2} \tag{1}$$

$$\text{Reaction 2: } H_2 + \mathrm{CO_2} \leftrightharpoons \mathrm{H_2O} + \mathrm{CO} \tag{2}$$

These reactions are being studied in an attempt to improve the yield of hydrogen. Initially, the reaction vessel contains methane, water, and carbon dioxide in molar ratios of 2 to 3 to 1 and there is no carbon monoxide or hydrogen.

a) Prepare a plot that shows the equilibrium hydrogen mole fraction as a function of temperature at various pressures.

The ratios of the initial amounts of methane, water, and carbon dioxide are specified. Therefore, this problem will be done on a per unit mole of CO_2 basis: $n_{0,CH_4} = 2$ kmol, $n_{0,H_2O} = 3$ kmol, and $n_{0,CO_2} = 1$ kmol. The inputs are entered in EES and an arbitrary temperature and pressure are initially specified in order to develop the model.

```
$UnitSystem SI Molar J K Pa
n_0_CH4=2 [kmol]                "initial amount of CH4"
n_0_H2O=3 [kmol]                "initial amount of H2O"
n_0_CO2=1 [kmol]                "initial amount of CO2"
C=5 [-]                         "number of substances"
r=2 [-]                         "number of reactions"
T=1100 [K]                      "temperature"
P_bar=10 [bar]                  "pressure, in bar"
P=P_bar*convert(bar,Pa)         "pressure"
```

The components and component numbering scheme are setup as in Example 14.2-1.

```
C$[1..C]=['CH4','H2O','CO','H2','CO2']   "Components"
CH4=1; H2O=2; CO=3; H2=4; CO2=5;         "substance numbering system"
```

The matrix of stoichiometric coefficients for the reactions is setup as in Example 14.2-1.

```
"Stoichiometric coefficient matrix"
"Reaction 1"
nu[CH4,1]= -1 [-]               "CH4 in reaction 1"
nu[H2O,1]= -1 [-]               "H2O in reaction 1"
nu[CO,1]= 1 [-]                 "CO in reaction 1"
nu[H2,1]= 3 [-]                 "H2 in reaction 1"
nu[CO2,1]= 0 [-]                "CO2 in reaction 1"

"Reaction 2"
nu[CH4,2]= 0 [-]                "CH4 in reaction 2"
nu[H2O,2]= 1 [-]                "H2O in reaction 2"
nu[CO,2]= 1 [-]                 "CO in reaction 2"
nu[H2,2]= -1 [-]                "H2 in reaction 2"
nu[CO2,2]= -1 [-]               "CO2 in reaction 2"
```

The array of initial number of moles for each substance is setup as in Example 14.2-1.

```
"Initial quantities of each substance"
n_0[CH4]=n_0_CH4                          "CH4"
n_0[H2O]=n_0_H2O                          "H2O"
n_0[CO]=0 [kmol]                          "CO2"
n_0[H2]=0 [kmol]                          "H2"
n_0[CO2]=n_0_CO2                          "CO2"
```

The reaction coordinates will eventually be determined based on equilibrium considerations. Initially, a guess will be implemented for each of the reaction coordinates.

```
"Initial guess for the reaction coordinates"
duplicate j=1,r
   e[j]=1 [kmol]                          "guess for reaction coordinate"
end
```

Equation (14-29) is used to determine the number of moles of each of the components based on the reaction coordinates:

$$n_i = n_{0,i} + \sum_{j=1}^{r} \nu_{i,j}\,\varepsilon_j \quad \text{for } i = 1..C$$

The total number of moles is obtained from:

$$n = \sum_{i=1}^{C} n_i$$

and the mole fraction of each component is given by:

$$y_i = \frac{n_i}{n} \quad \text{for } i = 1..C$$

```
duplicate i=1,C
   n[i]=n_0[i]+sum(nu[i,j]*e[j],j=1,r)    "number of moles of each substance"
end
n=sum(n[i],i=1,C)                         "total number of moles"
duplicate i=1,C
   y[i]=n[i]/n                            "mole fraction"
end
```

The standard state molar specific Gibbs free energy is computed for each substance using the temperature and reference pressure with Eq. (14-46); note that the reference pressure is set to 1 atm according to convention.

$$\overline{g}_i^{\,o} = \overline{h}_i^{\,o} - T\,\overline{s}_i^{\,o} \quad \text{for } i = 1..C$$

```
P_ref=1 [atm]*convert(atm,Pa)             "reference pressure"
"Standard Gibbs free energy for each substance"
duplicate i=1,C
   g_bar_o[i]=enthalpy(C$[i],T=T)-T*entropy(C$[i],T=T,P=P_ref)
end
```

The standard state Gibbs free energy change is calculated for each reaction using Eq. (14-50):

$$\Delta G_j^{\text{o}} = \sum_{i=1}^{C} \nu_{i,j}\, \overline{g}_i^{\text{o}} \quad \text{for } j = 1..r$$

```
"standard state Gibbs free energy change for each reaction"
duplicate j=1,r
   DELTAG_o[j]=sum(nu[i,j]*g_bar_o[i],i=1,C)
end
```

The equilibrium constant for each reaction is obtained from Eq. (14-54):

$$\Delta G_j^{\text{o}} = -R_{univ}\, T \ln\left(K_j\right) \quad \text{for } j = 1..r$$

```
"get equilibrium constant for each reaction"
duplicate j=1,r
   DELTAG_o[j]=-R#*T*ln(K[j])
end
```

The assumed values of the reaction coordinates must be removed and the equilibrium value of the reaction coordinates selected based on Eq. (14-53):

$$K_j = \prod_{i=1}^{C} \left(\frac{y_i P}{P^{\text{o}}}\right)^{\nu_{i,j}} \quad \text{for } j = 1..r \qquad (3)$$

The system of equations that must be solved in order implement Eq. (3) is highly nonlinear and therefore the guess values should be updated by selecting Update Guesses from the Calculate menu. Also, the limits on the mole fractions of each component should be set so that they are bounded between 0 and 1. The lower limit on both the equilibrium constants and the number of moles for each component should be set to 0. These constraints can be implemented in the Variable Information Window, as shown in Figure 1.

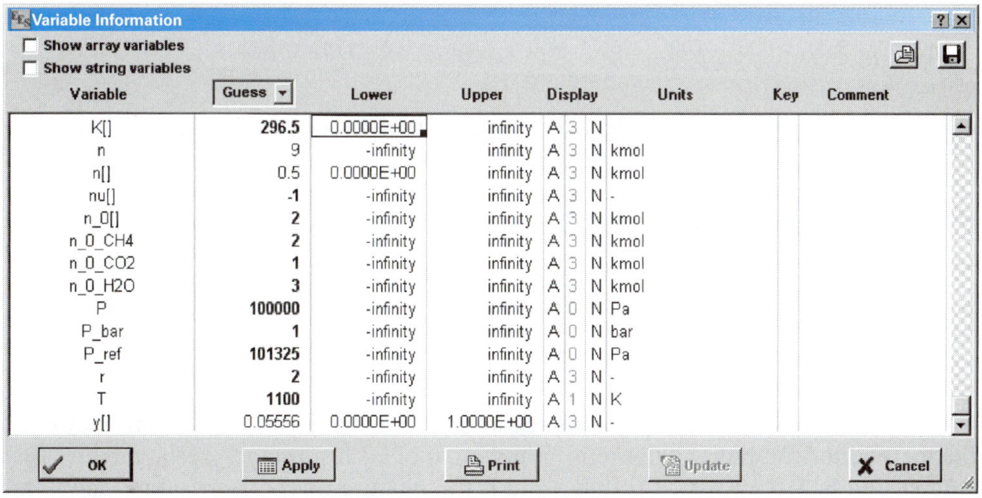

Variable	Guess	Lower	Upper	Display	Units	Key	Comment
K[]	296.5	0.0000E+00	infinity	A 3 N			
n	9	-infinity	infinity	A 3 N	kmol		
n[]	0.5	0.0000E+00	infinity	A 3 N	kmol		
nu[]	-1	-infinity	infinity	A 3 N	-		
n_0[]	2	-infinity	infinity	A 3 N	kmol		
n_0_CH4	2	-infinity	infinity	A 3 N	kmol		
n_0_CO2	1	-infinity	infinity	A 3 N	kmol		
n_0_H2O	3	-infinity	infinity	A 3 N	kmol		
P	100000	-infinity	infinity	A 0 N	Pa		
P_bar	1	-infinity	infinity	A 0 N	bar		
P_ref	101325	-infinity	infinity	A 0 N	Pa		
r	2	-infinity	infinity	A 3 N	-		
T	1100	-infinity	infinity	A 1 N	K		
y[]	0.05556	0.0000E+00	1.0000E+00	A 3 N	-		

Figure 1: Variables Information window with constraints on y[ ], K[ ], and n[ ] arrays.

Comment out the assumed values of the reaction coordinates:

```
{"Initial guess for the reaction coordinates"
duplicate j=1,r
   e[j]=1 [kmol]                    "reaction coordinate"
end}
```

and instead implement Eq. (3). Note the use of the Product function in EES, which multiplies each of the arguments provided to it, but otherwise is used in the same way as the Sum function.

```
"enforce the equilibrium constraint"
duplicate j=1,r
   K[j]=Product((y[i]*P/P_ref)^nu[i,j],i=1,C)
end
```

Solving the EES code leads to $y_4 = 0.4899$; this is the equilibrium mole fraction of hydrogen at the specified temperature and pressure. Figure 2 illustrates the equilibrium mole fraction of hydrogen as a function of temperature for various values of pressure.

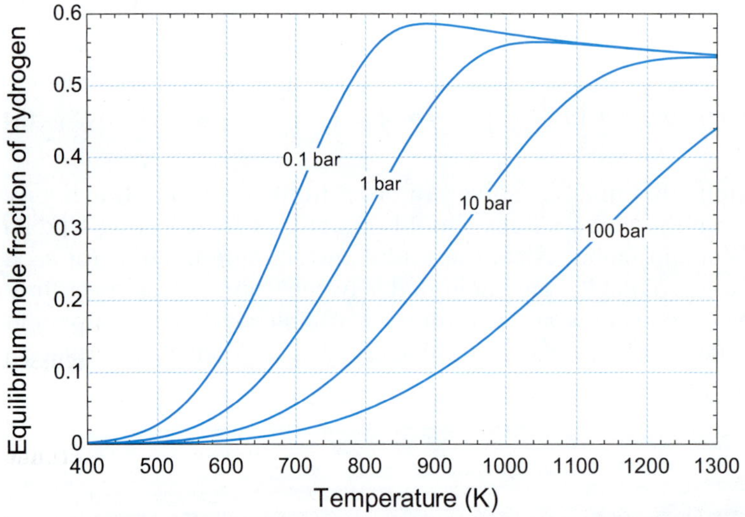

Figure 2: Equilibrium mole fraction of hydrogen as a function of temperature for various pressures.

14.3.4 Equilibrium Constant and the Law of Mass Action for an Ideal Solution

The criterion for equilibrium given by Eq. (14-43):

$$\sum_{i=1}^{C} \nu_{i,j}\,\mu_i = 0 \quad \text{for } j = 1..r \quad \text{at equilibrium} \tag{14-55}$$

was derived without any assumptions regarding the behavior of the system. However, the method for computing the chemical potential of each substance in the mixture that is discussed in Section 14.3.2 is only valid if the mixture obeys the ideal gas law. This assumption can be relaxed without adding much additional complexity, as shown in this section.

In an ideal gas mixture, each substance behaves as if it were pure (i.e., as if it were alone at the same temperature and volume as the mixture) and each substance obeys the ideal gas law. The degree to which the behavior of a gas mixture can be predicted by the ideal gas law decreases as the density of the mixture increases. However, many mixtures, particularly gas mixtures, behave as an *ideal solution* even under conditions in which they do not obey the ideal gas law. An ideal solution obeys Amagat's rule of additive volumes, presented in Section 11.1. The substances in an ideal solution behave as if they were pure substances at the temperature and pressure of the mixture, but they do not necessarily obey the ideal gas law. The formal requirement of an ideal solution is that there is no observed change in volume resulting from mixing the pure substances to form the mixture.

The chemical potential for a gas in ideal solution is shown in Section 11.5.4 to be:

$$\mu_{g,i} = \mu_i^o + R_{univ}\, T\, \ln\left(\frac{y_i\, f_{g,i}}{P^o}\right) \quad \text{for an ideal solution} \tag{14-56}$$

where μ_i^o is the chemical potential (which is also the molar specific Gibbs free energy, $\overline{g}_i^o$) of substance i at the standard pressure (P^o) and $f_{g,i}$ is the fugacity of pure substance i in the gaseous state at the temperature and pressure of the mixture. Equation (14-56) assumes that gas i obeys the ideal gas law at pressure P^o. Note the similarity of Eq. (14-56) to Eq. (14-45). The fugacity of a gas is equal to its pressure if the gas obeys the ideal gas law. The Law of Mass Action, Eq. (14-54), can be applied to mixtures that behave as an ideal solution simply by replacing the partial pressure in Eq. (14-53), $y_i\, P$, with the partial fugacity defined as the product of the mole fraction and the fugacity of pure gas i, $y_i\, f_{g,i}$. The equilibrium constant for reaction j in a set of r independent reactions then becomes:

$$\boxed{K_j = \prod_{i=1}^{C} \left(\frac{y_i\, f_{g,i}}{P^o}\right)^{\nu_{i,j}} \quad \text{for } j = 1..r} \tag{14-57}$$

Using Eq. (14-57) in place of Eq. (14-53) extends the range of applicability of the Law of Mass Action in Eq. (14-54) to reactive systems that may not obey the ideal gas law.

The steps required to solve the problem are essentially the same as those discussed in Section 14.3.3. At a particular temperature, the values of ΔG_j^o and K_j can be evaluated for each reaction using Eqs. (14-50) and (14-54). At a specified pressure, the values of the reaction coordinates are adjusted until Eq. (14-57) is satisfied in order to determine the equilibrium composition.

EXAMPLE 14.3-2: AMMONIA SYNTHESIS

Ammonia (NH_3) is used extensively as a fertilizer. It does not occur naturally and must be made by reacting hydrogen and nitrogen according to the reaction:

$$0.5\, N_2 + 1.5\, H_2 \leftrightharpoons NH_3 \tag{1}$$

Higher yields of ammonia are obtained at high pressures; therefore, this reaction is typically conducted at $P = 300$ atm. The purpose of this problem is to determine the maximum equilibrium yield of ammonia at this pressure and over a range of temperature when nitrogen and hydrogen are supplied in stoichiometric proportions.

a) Plot the equilibrium yield of ammonia as a function of temperature assuming that the reacting gas mixture obeys the ideal gas law.

The inputs are entered in EES.

```
$UnitSystem SI Molar J K Pa
P=300 [atm]*convert(atm,Pa)          "pressure"
T_C=500 [C]                          "temperature, in C"
T=converttemp(C,K,T_C)               "temperature"
```

The number of components and numbering system are set up; nitrogen is designated as substance 1, hydrogen as 2, and ammonia as 3. Notice that the component names in the array C_IG$ are set to their ideal gas values.

```
C=3 [-]                              "number of substances"
C_IG$[1..C]=['N2','H2','NH3']        "components - ideal gas"
N2=1; H2=2; NH3=3;                   "substance numbering system"
```

The stoichiometric coefficients are entered in the matrix nu based on inspection of Eq. (1); $\nu_1 = -0.5$, $\nu_2 = -1.5$, and $\nu_3 = 1$.

```
"Stoichiometric coefficients"
nu[N2]= -0.5 [-]                     "N2"
nu[H2]= -1.5 [-]                     "H2"
nu[NH3]= 1 [-]                       "NH3"
```

The initial quantities of each substance are entered into the array n_0. These quantities are specified based on the statement that the initial amounts of nitrogen and hydrogen are provided in stoichiometric amounts; $n_{0,1} = 0.5$ kmol, $n_{0,2} = 1.5$ kmol, $n_{0,3} = 0$ kmol.

```
"Initial quantities of each substance"
n_0[N2]=n_0_N2                       "N2"
n_0[H2]=n_0_H2                       "H2"
n_0[NH3]=0 [kmol]                    "NH3"
```

An initial guess is made for the reaction coordinate, ε, and the number of moles of each substance is calculated using Eq. (14-26).

$$n_i = n_{0,i} + \nu_i \varepsilon \quad \text{for } i = 1..C$$

The total number of moles is computed according to:

$$n = \sum_{i=1}^{C} n_i$$

The mole fractions are computed according to:

$$y_i = \frac{n_i}{n} \quad \text{for } i = 1..C$$

```
e=0.5 [kmol]                         "initial guess for reaction coordinate"
duplicate i=1,C
   n[i]=n_0[i]+nu[i]*e               "number of moles of each substance"
end
n=sum(n[i],i=1,C)                    "total number of moles"
duplicate i=1,C
   y[i]=n[i]/n                       "mole fraction"
end
```

The standard state molar specific Gibbs free energy is computed for each substance using the temperature and reference pressure with Eq. (14-46); note that the reference pressure is set to 1 atm according to convention:

$$\overline{g}_i^o = \overline{h}_i^o - T\,\overline{s}_i^o \quad \text{for } i = 1..C$$

```
P_ref=1 [atm]*convert(atm,Pa)                "reference pressure"
"Standard Gibbs free energy for each substance"
duplicate i=1,C
   g_bar_o[i]=enthalpy(C_IG$[i],T=T)-T*entropy(C_IG$[i],T=T,P=P_ref)
end
```

The standard state Gibbs free energy change of the reaction is computed:

$$\Delta G^o = \sum_{i=1}^{C} \nu_i\,\overline{g}_i^o$$

```
DELTAG_o=sum(nu[i]*g_bar_o[i],i=1,C)    "standard state Gibbs free energy change"
```

The equilibrium constant for the reaction is obtained.

$$\Delta G^o = -R_{univ}\,T\ln(K)$$

```
DELTAG_o=−R#*T*ln(K)                    "determine equilibrium constant"
```

The assumed value of the reaction coordinate must be removed since the equilibrium value of the reaction coordinate will be determined by Eq. (14-53).

$$K = \prod_{i=1}^{C}\left(\frac{y_i\,P}{P^o}\right)^{\nu_i} \tag{2}$$

Update the guess values by selecting Update Guesses from the Calculate menu. Also, the limits on the mole fractions of each component should be set so that they are bounded between 0 and 1. The lower limit on both the equilibrium constant and the number of moles for each component should be set to 0. Comment out the assumed values of the reaction coordinate:

```
{e=0.5 [kmol]}                          "initial guess for reaction coordinate"
```

and instead implement Eq. (2):

```
K=Product((y[i]*P/P_ref)^nu[i],i=1,C)   "enforce the equilibrium constraint"
```

Figure 1 illustrates the equilibrium mole fraction of hydrogen as a function of temperature; the curve labeled ideal gas corresponds to the predictions from part (a).

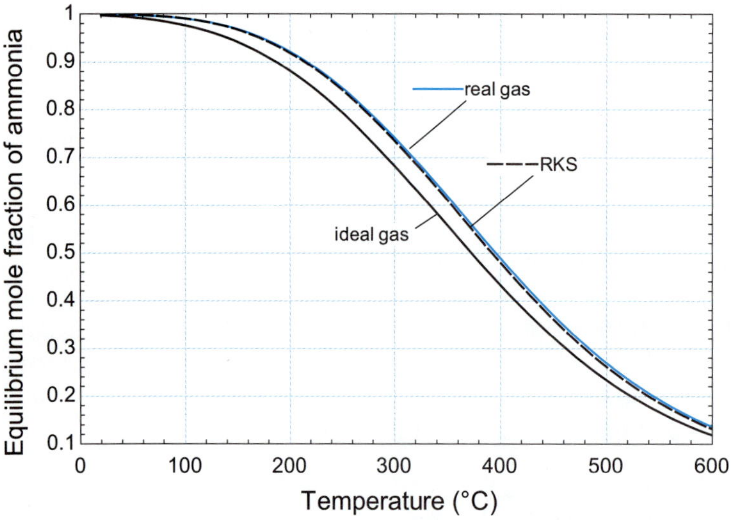

Figure 1: Equilibrium mole fraction of ammonia as a function of temperature for the ideal gas, ideal solution with real gas models, and ideal solution with RKS models.

Notice that the equilibrium mole fraction of ammonia increases as the temperature is reduced. However, it is likely that the chemical kinetics are such that the reaction requires more time to reach equilibrium at lower temperatures. The rate of the reaction is not considered by these equilibrium calculations.

b) Since this reaction takes place at high pressure, the mixture may not obey the ideal gas law. Repeat the calculation in part (a) with the assumption that the gas mixture is an ideal solution but not necessarily an ideal gas. Compare the results with those from part (a).

The calculation of the equilibrium constant using the ideal gas assumption, Eq. (2), is commented out.

```
{K=Product((y[i]*P/P_ref)^nu[i],i=1,C)     "enforce the equilibrium constraint"}
```

A vector containing the real gas names of the components is created and the fugacity of each component at the temperature and pressure of the mixture is determined using the Fugacity function in EES. The equilibrium constant is calculated assuming ideal solution according to Eq. (14-57):

$$K_j = \prod_{i=1}^{C} \left(\frac{y_i f_{g,i}}{P^o} \right)^{\nu_{i,j}} \quad \text{for } j = 1..r \tag{3}$$

```
C_RG$[1..C]=['Nitrogen', 'Hydrogen', 'Ammonia']     "components - real gas"
duplicate i=1,C
   f[i]=fugacity(C_RG$[i],T=T,P=P)                    "fugacity of each component"
end
K=Product((y[i]*f[i]/P_ref)^nu[i],i=1,C)             "enforce the equilibrium constraint"
```

Figure 1 illustrates the equilibrium mole fraction of ammonia as a function of temperature using the real gas properties and the ideal solution model.

The temperatures involved in this problem exceed the recommended upper limit for the applicability of the equation of state for ammonia that is programmed in EES. Therefore, you may see warning messages stating "Temperature out of range for Ammonia" depending on whether warning messages are enabled. An alternative way to solve this problem is to use the Redlich-Kwong-Soave (RKS) equation of state to estimate the fugacities, as discussed in Section 10.6. The calculation of fugacity using the RKS equation of state is provided in the FugCoef function, which is one of several non-ideal gas functions provided in the GEN_EOS external library for EES. Information on these functions can be viewed from the Function Information dialog by clicking the External routines button, selecting the GEN_EOS.DLL folder and then clicking the Function Info button. The FugCoef function returns the fugacity coefficient, which is the ratio of the fugacity to pressure, as a function of the reduced temperature, reduced pressure, and acentric factor of the substance. The previous code used to compute the equilibrium constant is commented out.

```
{duplicate i=1,C
   f[i]=fugacity(C_RG$[i],T=T,P=P)        "fugacity of nitrogen"
end
K=Product((y[i]*f[i]/P_ref)^nu[i],i=1,C)   "enforce the equilibrium constraint"}
```

The critical temperature, critical pressure, and acentric factor are determined for each of the components ($T_{crit,i}$, $P_{crit,i}$, and ω_i). The fugacity of each of the components is estimated as the product of the fugacity coefficient, calculated using the FugCoef function, and the pressure. Equation (3) is used to compute the equilibrium constant.

```
duplicate i=1,C
   T_crit[i]=T_crit(C_RG$[i])                            "critical temperature"
   P_crit[i]=P_crit(C_RG$[i])                            "critical pressure"
   omega[i]=acentricFactor(C_RG$[i])                     "acentric factor"
   f[i]=FUGCOEF(T/T_crit[i],P/P_crit[i],omega[i])*P      "fugacity"
end
K=Product((y[i]*f[i]/P_ref)^nu[i],i=1,C)                 "enforce the equilibrium constraint"
```

Figure 1 illustrates the equilibrium mole fraction of ammonia predicted assuming an ideal solution with the RKS equation of state; the results are nearly identical to those obtained using the real gas functions in EES.

14.4 Alternative Methods for Chemical Equilibrium Problems

There are several disadvantages to using the Law of Mass Action given by Eq. (14-54) to determine the equilibrium composition of a chemically reactive system. One disadvantage is that the equations involved are non-linear and therefore they can be difficult to solve, particularly when more than one chemical reaction is occurring simultaneously. The major disadvantage, however, is that it is necessary to know exactly what chemical reactions are occurring in order to apply Eq. (14-54). Two additional methods are presented in this section for determining the equilibrium composition of a reactive mixture. These methods only require information concerning the various substances that can be present in the reactive mixture and not what chemical reactions are occurring. Both of these methods are better suited than the Law of Mass Action for determining the equilibrium state when there is more than one reaction occurring.

14.4.1 Direct Minimization of Gibbs Free Energy

Section 14.1 showed that the equilibrium state of a chemically reactive system corresponds to the conditions at which its Gibbs free energy is minimized at a specified temperature and pressure. The Law of Mass Action presented in Section 14.3 is an analytical method that can be used to determine the state for which the total derivative of the Gibbs free energy is zero, given a set of reactions. However, an alternative way to determine the equilibrium state is to compute the total Gibbs free energy of the system (G) and then adjust the molar amounts of each substance, subject to element balances, so as to minimize G. Multi-dimensional optimization algorithms can be used for this purpose.

The problem can be stated mathematically as follows. The Gibbs free energy of the system at a specified temperature and pressure is the sum of the product of the number of moles of each substance and the chemical potential of the substance in the mixture, μ_i:

$$G = \sum_{i=1}^{C} n_i \, \mu_i \tag{14-58}$$

If the mixture obeys the ideal gas law, then the chemical potential for substance i is identical to the molar specific Gibbs free energy of the substance evaluated at the mixture temperature and the partial pressure of the substance in the mixture ($y_i \, P$). The chemical potential of an ideal gas in a mixture is provided in Eq. (14-45), which is repeated below:

$$\mu_i = \mu_i^{\text{o}} + R_{univ} \, T \ln\left(\frac{y_i \, P}{P^{\text{o}}}\right) \quad \text{for ideal gas mixture} \tag{14-59}$$

where μ_i^{o} is the chemical potential (which is also the molar specific Gibbs free energy, $\overline{g}_i^{o}$) of substance i at temperature T and reference pressure P^{o} (sometimes referred to as the standard state pressure). The reference pressure is normally chosen to be 1 atm or 100 kPa for gas-phase reactions.

$$\mu_i^{\text{o}} = \overline{g}_i^{\text{o}} = \overline{h}_i^{\text{o}} - T \, \overline{s}_i^{\text{o}} \tag{14-60}$$

More generally, if the mixture behaves as an ideal solution in gas phase, then the chemical potential of substance i is expressed in terms of the fugacity, as indicated in Eq. (14-56) which is repeated below:

$$\mu_i = \mu_i^{\text{o}} + R_{univ} \, T \, \ln\left(\frac{y_i \, f_{g,i}}{P^{\text{o}}}\right) \quad \text{for an ideal solution} \tag{14-61}$$

where $f_{g,i}$ is the fugacity of pure substance i in the gaseous state at the temperature and pressure of the mixture.

The initial composition of the reacting mixture is known. Therefore, the number of moles of each element (atom) that is present (E_j) can be determined. The molar amount of each element remains constant as the reaction progresses. The initial number of moles of each element is given by:

$$E_{0,j} = \sum_{i=1}^{C} n_{0,i} \, e_{i,j} \quad \text{for } j = 1..E \tag{14-62}$$

where $n_{0,i}$ is the initial number of moles of each substance, $e_{i,j}$ is the number of moles of element j in a mole of substance i, and E is the total number of elements appearing in all of the substances. As the reaction progresses, the number of moles of each element is given by:

$$E_j = \sum_{i=1}^{C} n_i\, e_{i,j} \quad \text{for } j = 1..E \tag{14-63}$$

The element balances require that:

$$E_j = E_{0,j} \quad \text{for } j = 1..E \tag{14-64}$$

The equilibrium state is found by minimizing G, calculated using Eq. (14-58), subject to the element balances provided by Eqs. (14-62) through (14-64). The method is illustrated in Example 14.4-1.

EXAMPLE 14.4-1: REFORMATION OF METHANE (REVISITED)

Example 14.3-1 determined the equilibrium state for the simultaneous reactions involving CH_4, H_2, H_2O, CO and CO_2 that describe the reformation of methane by applying the Law of Mass Action. At the start, the reaction vessel contains methane, water and carbon dioxide in molar ratios of 2 to 3 to 1 with no carbon monoxide or hydrogen. Assume that the reacting mixture behaves as an ideal gas mixture.

a) Solve the problem by directly minimizing the Gibbs free energy. Plot the equilibrium hydrogen mole fraction at 1 bar as a function of temperature and compare the result to the prediction from Example 14.3-1.

The ratios of the initial amounts of methane, water, and carbon dioxide are specified. Therefore, this problem will be done on a per unit mole of CO_2 basis: $n_{0,CH_4} = 2$ kmol, $n_{0,H_2O} = 3$ kmol, and $n_{0,CO_2} = 1$ kmol. Initially, we will assume a value of temperature and pressure although these properties will eventually be varied in order to generate the plot requested in the problem statement.

```
$UnitSystem SI Molar K Pa J
n_0_CH4=2 [kmol]                    "initial amount of CH4"
n_0_H2O=3 [kmol]                    "initial amount of H2O"
n_0_CO2=1 [kmol]                    "initial amount of CO2"
T=1100 [K]                          "temperature"
P_bar=1 [bar]                       "pressure, in bar"
P=P_bar*convert(bar,Pa)             "pressure"
```

It is not necessary to know what reactions may be taking place. All we need to know is what substances are expected to be present in the reaction vessel at equilibrium. In this case, we know that the reaction vessel contains $C = 5$ substances: CH_4, H_2, H_2O, CO and CO_2. There are $E = 3$ elements represented in these substances, C, H, and O. For convenience when working with summations, an index number is assigned to each substance and each element. The substance names are stored in the array, C$.

```
N_C=5 [-]                                    "number of components"
N_E=3 [-]                                    "number of elements"
C=1; H=2; O=3;                               "numbering system for elements"
CH4=1; H2O=2; CO=3; H2=4; CO2=5;             "substance numbering system"
C$[1..N_C]=['CH4','H2O','CO','H2','CO2']     "components"
```

The initial molar amount of each substance is assigned in array n_0.

```
"Initial quantities of each substance"
n_0[CH4]=n_0_CH4                    "CH4"
n_0[H2O]=n_0_H2O                    "H2O"
n_0[CO]=0 [kmol]                    "CO2"
n_0[H2]=0 [kmol]                    "H2"
n_0[CO2]=n_0_CO2                    "CO2"
```

The array $e_{i,j}$ is set up where $e_{i,j}$ corresponds to the number of moles of element j that is present per mole of substance i. This is done by inspection of the chemical formula of the substances.

```
"Element indices"
e[CH4,C]=1; e[CH4,H]=4; e[CH4,O]=0;    "CH4"
e[H2O,C]=0; e[H2O,H]=2; e[H2O,O]=1;    "H2O"
e[CO,C]=1; e[CO,H]=0; e[CO,O]=1;       "CO"
e[H2,C]=0; e[H2,H]=2; e[H2,O]=0;       "H2"
e[CO2,C]=1; e[CO2,H]=0; e[CO2,O]=2;    "CO2"
```

The number of moles of each element that are initially present in the reaction vessel is determined by applying Eq. (14-63) to the number of moles of each substance that is initially present ($n_{0,i}$):

$$E_{0,j} = \sum_{i=1}^{C} n_{0,i}\, e_{i,j} \quad \text{for } j = 1..E \tag{1}$$

```
"Total amount of each element initially present"
duplicate j=1,N_E
   E_s_0[j]=sum(n_0[i]*e[i,j],i=1,N_C)
end
```

The mole fraction of all but one of the components is initially assumed; the last mole fraction is computed from:

$$y_C = 1 - \sum_{i=1}^{(C-1)} y_i$$

The final number of moles (n) is also initially assumed.

```
"Initial guess for the mole fraction of each gas and number of moles"
y[CH4]=0.1 [-]
y[H2O]=0.2 [-]
y[CO]=0.2 [-]
y[H2]=0.2 [-]
y[CO2]=1-sum(y[i],i=1,(N_C-1))
n=10 [kmol]
```

The number of moles of each of the components can be computed according to:

$$n_i = y_i\, n \quad \text{for } i = 1..C$$

```
duplicate i=1,N_C
   n[i]=n*y[i]
end
```

The total number of moles of each element that is present is computed using Eq. (14-63):

$$E_j = \sum_{i=1}^{C} n_i\, e_{i,j} \quad \text{for } j = 1..E \tag{2}$$

```
"Total amount of each element present"
duplicate j=1,N_E
   E_s[j]=sum(n[i]*e[i,j],i=1,N_C)
end
```

The element balances require that the number of elements present in the reaction, given by Eq. (2), must be equal to the number of elements that are initially present, given by Eq. (1):

$$E_j = E_{0,j} \quad \text{for } j = 1..E \tag{3}$$

Equation (3) provides E equations (one for each element) in the C unknowns (y_1 through $y_{C\text{-}1}$ and n). The guess values are updated. The assumed mole fractions of carbon monoxide and hydrogen (y_{CO} and y_{H_2}) and the assumed value of the number of moles (n) are commented out.

```
{y[CO]=0.2 [-]
y[H2]=0.2 [-]}
{n=10 [kmol]}
```

In their place, Eq. (3) is used to enforce element balances.

```
"Enforce element balances"
duplicate j=1,N_E
   E_s[j]=E_s_0[j]
end
```

The resulting values of y_{CO}, y_{H_2}, and n satisfy element mass balances given the assumed values of y_{CH_4} and y_{H_2O}. Once element balances are enforced, there are generally $C - E$ remaining free parameters (in this case, there are 5 substances -3 elements $= 2$ remaining mole fractions); these free parameters must be adjusted in order to minimize the Gibbs free energy of the mixture. The value of $\mu_i^{o} = \overline{g}_i^{o}$ is computed for each of the substances using Eq. (14-60):

$$\mu_i^{o} = \overline{g}_i^{o} = \overline{h}_i^{o} - T\,\overline{s}_i^{o} \quad \text{for } i = 1..C$$

The chemical potential of each substance is computed using Eq. (14-59):

$$\mu_i = \mu_i^{o} + R_{univ}\, T \ln\left(\frac{y_i P}{P^{o}}\right) \quad \text{for } i = 1..C$$

The total Gibbs free energy is computed according to Eq. (14-58):

$$G = \sum_{i=1}^{C} n_i\, \mu_i$$

```
P_ref=1 [atm]*convert(atm,Pa)                                    "reference pressure"
duplicate i=1,N_C
   g_bar_o[i]=enthalpy(C$[i],T=T)-T*entropy(C$[i],T=T,P=P_ref)   "chemical potential at T and P_ref"
   mu[i]=g_bar_o[i]+R#*T*ln(y[i]*P/P_ref)                        "chemical potential"
end
G=sum(n[i]*mu[i],i=1,N_C)                                        "total Gibbs free energy of the system"
```

We need to minimize G by adjusting the two degrees of freedom that remain in this optimization problem, y_{CH_4} and y_{H_2O}. Select Variable Info from the Options menu and constrain the values of each of the mole fractions in the vector y to be between 0 and 1. Comment out the assumed values of y_{CH_4} and y_{H_2O}.

```
{y[CH4]=0.1 [-]
y[H2O]=0.2 [-]}
```

Select the Min/Max menu item in the Calculate menu in order to start the optimization. Minimize G with independent variables y[1] and y[2]. Uncheck the Stop if error occurs box; we do not want the optimization to stop if, for example, the program reaches a negative mole fraction since that surely is not the optimum solution. Note that there are several optimization algorithms to choose from. Your version may not have the Genetic method as an option, as that is provided only in the Professional version. The Direct search method works fine in this problem. The problem should solve to show an optimum with the hydrogen mole fraction $y_4 = 0.5579$. The value found for the equilibrium hydrogen mole fraction in Example 14.3-1 under the same conditions is also 0.5579. The accuracy of the result obtained from the optimization can be increased by reducing the relative convergence tolerance in the Min/Max dialog. You may also need to increase the maximum allowable number of function calls.

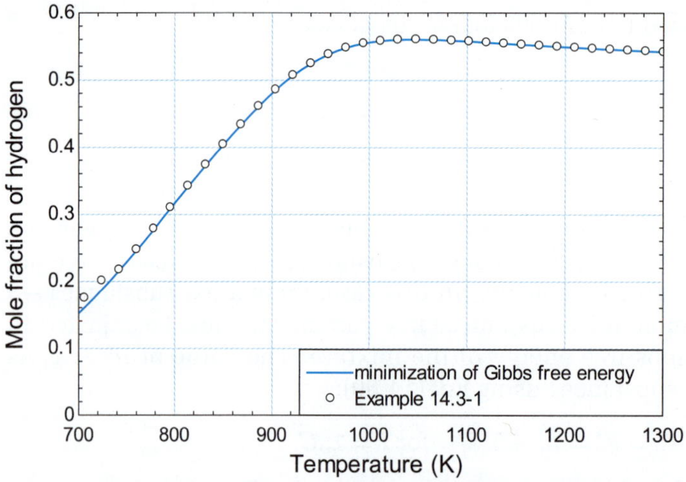

Figure 1: Equilibrium mole fraction of hydrogen as a function of temperature at 1 bar predicted by minimizing the Gibbs free energy of the mixture. Also shown is the result from Example 14.3-1.

In order to generate a plot of the equilibrium mole fraction of hydrogen as a function of temperature it is necessary to construct a Parametric Table that includes columns for the variable you want to change (T), the variable that should be optimized (G), the independent variables (y_1 and y_2) and any other variables that you are interested in (y_4). Vary the temperature from 700 K to 1300 K in the Parametric

Table. We need to find the optimal concentration for every row of this table. Comment out the specified value of temperature in the Equations window and select Min/Max Table from the Calculate menu. In the resulting dialog, specify that G should be maximized by varying y_1 and y_2. Set the Maximum function calls to 1000 since convergence may not occur with 200 iterations at low temperatures. Click OK to initiate the optimization for all rows in the table. Figure 1 illustrates the equilibrium concentration of hydrogen predicted by minimizing the Gibbs free energy as a function of temperature. Also shown in Figure 1 is the equilibrium concentration of hydrogen predicted in Example 14.3-1.

14.4.2 Lagrange Method of Undetermined Multipliers

The method described in Section 14.4.1 for finding the equilibrium state by minimizing the Gibbs free energy of the system offers the advantage of not needing to know what reactions are ongoing. However, the optimization process is computationally intensive and not guaranteed to converge. For example, in Example 14.4-1 the method of minimizing the method of Gibbs free energy will not converge well for temperatures below about 700 K without adjusting the convergence tolerance. These problems become more difficult as the number of substances involved, and thus the number of degrees of freedom for the optimization problem, increases.

The general problem that needs to be solved is classified as a constrained optimization problem. In this case, the optimization is constrained by the element balances. There is a very powerful analytical solution method for such problems called the Lagrange Method of Undetermined Multipliers. This method can be applied to the solution of chemical equilibrium problems.

The Gibbs free energy of the chemically reacting system at a specified temperature and pressure is a function of the number of moles of each substance, n_i:

$$G = \sum_{i=1}^{C} n_i \, \mu_i \tag{14-65}$$

We want to find the number of moles of each substance that result in the minimum value for G. The total derivative of G at a given temperature and pressure must be zero at equilibrium:

$$dG = \sum_{i=1}^{C} \mu_i \, dn_i = 0 \text{ at equilibrium at constant } T \text{ and } P \tag{14-66}$$

If there were no constraints, we could find the minimum value of G by setting the chemical potential of each substance to zero. However, the moles of each substance, n_i, are not independent variables because they are constrained by the fact that the number of moles of each element must remain constant. The element balances provided by Eqs. (14-62) through (14-64) can be represented according to:

$$\psi_j = \sum_{i=1}^{C} n_i \, e_{i,j} - E_{0,j} = 0 \quad \text{for } j = 1..E \tag{14-67}$$

where $E_{0,j}$ is the number of moles of each element that is initially in the mixture. The quantity ψ_j is a function that evaluates to zero when the element balance j is observed. Because the constraint equation must be equal to zero, the total derivative of Eq. (14-67) must be zero. The function ψ_j depends only on the number of moles of each substance; therefore, the total derivative of ψ_j is given by:

$$d\psi_j = \sum_{i=1}^{C} \left(\frac{\partial \psi_j}{\partial n_i}\right)_{n_j, j\neq i} dn_i = 0 \quad \text{for } j = 1..E \tag{14-68}$$

where, according to Eq. (14-67), the partial derivative of ψ_j with respect to n_i must be $e_{i,j}$, the number of moles of element j in substance i:

$$\left(\frac{\partial \psi_j}{\partial n_i}\right)_{n_j, j\neq i} = e_{i,j} \tag{14-69}$$

In the Lagrange method, the total derivative of each constraint equation is multiplied by an underdetermined multiplier, λ_j, and added to the total derivative of G. Since the total derivative of each constraint equation is zero, adding the total derivative of each constraint to the derivative of G, with or without multiplying it by a value, should not change the value of the derivative of G:

$$dG + \sum_{j=1}^{E} \lambda_j \, d\psi_j = 0 \tag{14-70}$$

The total derivative of G from Eq. (14-66) and the constraint equations in Eq. (14-68) are substituted into Eq. (14-70):

$$dG = \sum_{i=1}^{C} \mu_i \, dn_i + \sum_{j=1}^{E} \lambda_j \sum_{i=1}^{C} \left(\frac{\partial \psi_j}{\partial n_i}\right)_{n_j, j\neq i} dn_i = 0 \tag{14-71}$$

Equation (14-71) can be rearranged:

$$dG = \sum_{i=1}^{C} \left[\mu_i + \sum_{j=1}^{E} \lambda_j \left(\frac{\partial \psi_j}{\partial n_i}\right)_{n_j, j\neq i} \right] dn_i = 0 \tag{14-72}$$

The only way to ensure that the total derivative, dG, is always zero is to set the value in the brackets for each term in the summation equal to zero:

$$\mu_i + \sum_{j=1}^{E} \lambda_j \left(\frac{\partial \psi_j}{\partial n_i}\right)_{n_j, j\neq i} = 0 \quad \text{for } i = 1..C \tag{14-73}$$

Substituting Eq. (14-69) into Eq. (14-73) leads to:

$$\boxed{\mu_i + \sum_{j=1}^{E} \lambda_j \, e_{i,j} = 0 \quad \text{for } i = 1..C} \tag{14-74}$$

We started with C unknowns, corresponding to the number of moles of each substance. In the Lagrange method, we introduced E additional unknowns, the undetermined multipliers, λ_j. Therefore, there are now a total of $C + E$ unknown parameters. However, there are C equations provided by Eq. (14-74) and E equations provided by the constraints in Eq. (14-67). Thus, there are an equal number of equations and unknowns and the problem can be solved without requiring optimization in order to determine the moles of each substance at equilibrium.

EXAMPLE 14.4-2: REFORMATION OF METHANE (REVISITED AGAIN)

Example 14.4-1 determined the equilibrium concentration of a reacting system consisting of CH_4, H_2, H_2O, CO and CO_2 by directly minimizing the Gibbs free energy of the system, subject to element constraints. The equation set in Example 14.4-1 has two degrees of freedom, meaning that there are two fewer equations than variables. The problem was solved using an optimization algorithm that varies two selected variables in an organized manner so as to locate the values that minimize the Gibbs free energy. This optimization algorithm requires significant computational effort and may not always converge. In this example, the same problem will be solved using the Lagrange Method of Undetermined Multipliers.

a) Plot the equilibrium hydrogen mole fraction at 1 bar as a function of temperature and compare the results to those obtained in Example 14.3-1.

The initial portion of the EES code is identical to Example 14.4-1.

```
$UnitSystem SI Molar K Pa J
n_0_CH4=2 [kmol]                          "initial amount of CH4"
n_0_H2O=3 [kmol]                          "initial amount of H2O"
n_0_CO2=1 [kmol]                          "initial amount of CO2"
T=1100 [K]                                "temperature"
P_bar=100 [bar]                           "pressure, in bar"
P=P_bar*convert(bar,Pa)                   "pressure"

N_C=5 [-]                                 "number of components"
N_E=3 [-]                                 "number of elements"
C=1; H=2; O=3;                            "numbering system for elements"
CH4=1; H2O=2; CO=3; H2=4; CO2=5;          "substance numbering system"
C$[1..N_C]=['CH4','H2O','CO','H2','CO2']  "components"

"Initial quantities of each substance"
n_0[CH4]=n_0_CH4                          "CH4"
n_0[H2O]=n_0_H2O                          "H2O"
n_0[CO]=0 [kmol]                          "CO2"
n_0[H2]=0 [kmol]                          "H2"
n_0[CO2]=n_0_CO2                          "CO2"

"Element indices"
e[CH4,C]=1; e[CH4,H]=4; e[CH4,O]=0;       "CH4"
e[H2O,C]=0; e[H2O,H]=2; e[H2O,O]=1;       "H2O"
e[CO,C]=1; e[CO,H]=0; e[CO,O]=1;          "CO"
e[H2,C]=0; e[H2,H]=2; e[H2,O]=0;          "H2"
e[CO2,C]=1; e[CO2,H]=0; e[CO2,O]=2;       "CO2"
```

The number of moles of each element that are initially present in the reaction vessel are determined by applying Eq. (14-63) to the number of moles of each substance that is initially present ($n_{0,i}$):

$$E_{0,j} = \sum_{i=1}^{C} n_{0,i}\, e_{i,j} \quad \text{for } j = 1..E$$

```
"Total amount of each element initially present"
duplicate j=1,N_E
   E_s_0[j]=sum(n_0[i]*e[i,j],i=1,N_C)
end
```

The equilibrium number of moles of each of the components is initially guessed.

```
"initial guesses for the unknowns"
duplicate i=1,N_C
   n[i]=1 [kmol]
end
```

The total number of moles is computed according to:

$$n = \sum_{i=1}^{C} n_i$$

and the mole fraction of each component is obtained from:

$$y_i = \frac{n_i}{n} \quad \text{for } i = 1..C$$

```
n=sum(n[i],i=1,N_C)    "total number of moles"
duplicate i=1,N_C
   y[i]=n[i]/n         "mole fraction"
end
```

The value of $\mu_i^o = \overline{g}_i^o$ is computed for each of the substances using Eq. (14-60):

$$\mu_i^o = \overline{g}_i^o = \overline{h}_i^o - T\,\overline{s}_i^o \quad \text{for } i = 1..C$$

The chemical potential of each substance is computed using Eq. (14-59):

$$\mu_i = \mu_i^o + R_{univ}\,T\ln\left(\frac{y_i P}{P^o}\right) \quad \text{for } i = 1..C$$

```
P_ref=1 [atm]*convert(atm,Pa)                                   "reference pressure"
duplicate i=1,N_C
   g_bar_o[i]=enthalpy(C$[i],T=T)-T*entropy(C$[i],T=T,P=P_ref)  "chemical potential at T and P_ref"
   mu[i]=g_bar_o[i]+R#*T*ln(y[i]*P/P_ref)                       "chemical potential"
end
```

The guess values are updated and the assumed number of moles for each substance are commented out.

```
{"initial guesses for the unknowns"
duplicate i=1,N_C
   n[i]=1 [kmol]
end}
```

The equations required to implement the Lagrange Method of Undetermined Multipliers, Eqs. (14-67) and (14-74):

$$\psi_j = \sum_{i=1}^{C} n_i\,e_{i,j} - E_{0,j} = 0 \quad \text{for } j = 1..E$$

$$\mu_i + \sum_{j=1}^{E} \lambda_j \, e_{i,j} = 0 \quad \text{for } i = 1..C$$

are entered in EES.

```
"Lagrange method of undetermined multipliers"
duplicate j=1,N_E
   sum(n[i]*e[i,j],i=1,N_C)-E_s_0[j]=0
end
duplicate i=1,N_C
   mu[i]+sum(lambda[j]*e[i,j],j=1,N_E)=0
end
```

Solving the problem leads to $y_4 = 0.5579$, which is the same answer obtained in Examples 14.3-1 and 14.4-1. Figure 1 illustrates the equilibrium mole fraction of hydrogen as a function of temperature for various values of pressure. Also shown in Figure 1 is the result obtained in Example 14.3-1.

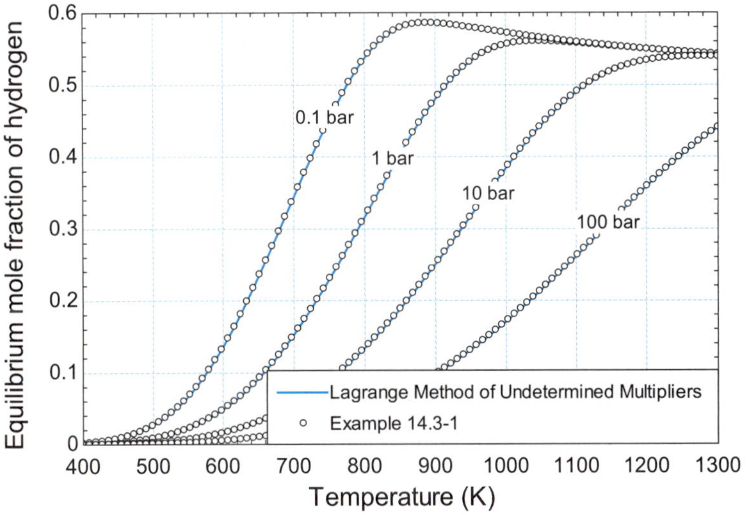

Figure 1: Equilibrium mole fraction of hydrogen as a function of temperature for various values of pressure predicted by the Lagrange Method of Undetermined Multipliers and from Example 14.3-1.

Notice that the Lagrange Method of Undetermined Multipliers does not require optimization and therefore is computationally efficient and continues to converge well under all of the conditions encountered in Figure 1. Further, the method does not require that the reactions are specified; it is only necessary to know what chemicals are present at equilibrium.

14.5 Heterogeneous Reactions

This extended section can be found online at www.cambridge.org/kleinandnellis and extends the chemical equilibrium treatment to reactions that takes place between substances in more than one phase.

14.6 Adiabatic Reactions

The methods presented in this chapter to determine an equilibrium state for a chemically reactive system have so far assumed that the temperature of the reaction is known. However, this is not always the case. The reactions that take place to arrive at an equilibrium condition have energy implications that will affect the temperature of the mixture in an adiabatic system. In order to determine the equilibrium temperature and composition for an adiabatic reaction, it is necessary to apply an energy balance as well as the chemical equilibrium condition. No new concepts are needed to solve problems of this type; however, the implementation of the solution is more difficult because of the coupling that exists between the energy balance and the chemical equilibrium constraint. Direct minimization of Gibbs free energy will not lead to the correct result because each value of Gibbs free energy is determined at a different temperature, which confounds the results. However, the Law of Mass Action and the Lagrange Method of Undetermined Multipliers can be applied to these problems, as illustrated in Examples 14.6-1 and 14.6-2.

EXAMPLE 14.6-1: ADIABATIC COMBUSTION OF HYDROGEN

A mixture of hydrogen and oxygen at $T_1 = 1000$ K and $P = 10$ bar is ignited by a spark according to the reaction:

$$H_2 + 0.5O_2 \leftrightharpoons H_2O \tag{1}$$

and proceeds to equilibrium in a constant pressure adiabatic process. Initially there is twice as much oxygen as hydrogen (by mole) present and no water. Assume that the gases behave according to the ideal gas law.

a) Determine the temperature and composition of the equilibrium mixture using the Law of Mass Action. Plot the temperature and mole fraction of hydrogen in the products as a function of pressure.

The inputs are entered in EES. The initial amounts of hydrogen and oxygen are not specified. However, we know the ratio of the number of moles of each gas. Therefore, a basis of 1 kmol of hydrogen is chosen for the calculation.

```
$UnitSystem SI Molar J K Pa
T_1=1000 [K]                  "initial temperature"
P_bar=10 [bar]                "pressure, in bar"
P=P_bar*convert(bar,Pa)       "pressure"
n_H2_0=1 [kmol]               "initial amount of hydrogen"
n_O2_0=2 [kmol]               "initial amount of oxygen"
```

The numbering system for the components is selected and an array containing the name of each component is specified.

```
N_C=3                         "number of components"
C$[1..N_C]=['H2','O2','H2O']  "components"
H2=1; O2=2; H2O=3;            "component index designations"
```

An array containing the initial number of moles of each substance, $n_{0,i}$, is setup.

```
"Initial quantities of each substance"
n_0[H2]=n_H2_0                    "H2"
n_0[O2]=n_O2_0                    "O2"
n_0[H2O]=0 [kmol]                 "H2O"
```

The initial enthalpy of the substances in the reaction is determined according to:

$$H_1 = \sum_{i=1}^{C} n_{0,i}\,\overline{h}_{i,T=T_1}$$

where C is the number of components and $\overline{h}_{i,T=T_1}$ is the molar specific enthalpy of component i at the initial temperature, T_1.

```
H_1=sum(n_0[i]*enthalpy(C$[i],T=T_1),i=1,N_C)    "enthalpy of substances entering reactor"
```

We will first solve the problem using the Law of Mass Action. Since the problem will be solved in different ways, it is convenient to employ a **$If** directive so that the solution method can be selected and changed. The directive uses a string variable, **Method$**, to determine which equations will be solved. The **Method$** variable is set and the code required to implement the solution is placed between a **$If** and a **$EndIf** Directive.

```
Method$='Law of Mass Action'          "select solution method"
$If Method$='Law of Mass Action'      "use Law of Mass Action"
```

The stoichiometric coefficients for the reaction (ν_i) shown in Eq. (1) are specified.

```
"Stoichiometric coefficients"
nu[H2]= -1 [-]                    "H2"
nu[O2]= -0.5 [-]                  "O2"
nu[H2O]= 1 [-]                    "H2O"
```

The reaction coordinate (ε) and final temperature (T_2) are both guessed. The values of these parameters will eventually be chosen in order to satisfy both chemical equilibrium and an energy balance.

```
e=0.5 [kmol]                      "guess for reaction coordinate"
T_2=2000 [K]                      "guess for final temperature"
```

The number of moles of each substance is evaluated using the reaction coordinate according to Eq. (14-26):

$$n_i = n_{0,i} + \nu_i\,\varepsilon \quad \text{for } i = 1..C$$

The total number of moles of gas is computed:

$$n = \sum_{i=1}^{C} n_i$$

and the mole fraction of each gas is obtained:

$$y_i = \frac{n_i}{n} \quad \text{for } i = 1..C$$

```
duplicate i=1,N_C
   n[i]=n_0[i]+nu[i]*e                         "moles of each substance"
end
n=sum(n[i],i=1,N_C)                            "number of moles"
duplicate i=1,N_C
   y[i]=n[i]/n                                 "mole fraction"
end
```

The standard state molar specific Gibbs free energy is computed for each substance using the assumed value of the final temperature and reference pressure, according to Eq. (14-46); note that the reference pressure is set to 1 atm according to convention:

$$\overline{g}_i^{o} = \overline{h}_{i,T=T_2}^{o} - T_2\,\overline{s}_{i,T=T_2}^{o} \quad \text{for } i = 1..C$$

where $\overline{h}_{i,T=T_2}^{o}$ is the molar specific enthalpy of component i evaluated at T_2 and $\overline{s}_{i,T=T_2}^{o}$ is the molar specific enthalpy of component i evaluated at the reference pressure and T_2.

```
P_ref=1 [atm]*convert(atm,Pa)                  "reference pressure"
"Standard Gibbs free energy for each substance"
duplicate i=1,N_C
   g_bar_o[i]=enthalpy(C$[i],T=T_2)-T_2*entropy(C$[i],T=T_2,P=P_ref)
end
```

The standard state Gibbs free energy change is calculated using Eq. (14-50); in this case there is only a single reaction and therefore:

$$\Delta G^{o} = \sum_{i=1}^{C} \nu_i\,\overline{g}_i^{o}$$

```
DELTAG_o=sum(nu[i]*g_bar_o[i],i=1,N_C)     "standard state Gibbs free energy change"
```

The equilibrium constant for each reaction is obtained from Eq. (14-54):

$$\Delta G^{o} = -R_{univ}\,T\ln(K)$$

```
DELTAG_o=-R#*T_2*ln(K)                         "equilibrium constant"
```

Update the guess values for the problem. The assumed value of the reaction coordinate must be removed; the equilibrium value of the reaction coordinate as determined based on Eq. (14-53):

$$K = \prod_{i=1}^{C}\left(\frac{y_i P}{P^{o}}\right)^{\nu_i}$$

```
{e=0.5 [kmol]}                                 "guess for reaction coordinate"
K=Product((y[i]*P/P_ref)^nu[i],i=1,N_C)        "enforce equilibrium constraint"
```

The result is the equilibrium concentration at the assumed value of T_2. Next, it is necessary to adjust T_2 in order to satisfy an energy balance. The enthalpy of the equilibrium mixture is computed according to:

$$H_2 = \sum_{i=1}^{C} n_i \, \overline{h}_{i,T=T_2}$$

where $\overline{h}_{i,T=T_2}$ is the molar specific enthalpy of component i evaluated at T_2.

```
H_2=sum(n[i]*enthalpy(C$[i],T=T_2),i=1,N_C)     "enthalpy of substances leaving reactor"
```

The guess values are updated and the assumed value of T_2 is commented out.

```
{T_2=2000 [K]}                                  "guess for final temperature"
```

The energy balance:

$$H_2 = H_1$$

is implemented.

```
H_1=H_2                                         "enforce energy balance"
$EndIf
```

Solving leads to $T_2 = 3209$ K with composition $y_1 = 0.01351$, $y_2 = 0.6027$, and $y_3 = 0.3838$. Notice that the reaction proceeds very nearly to complete combustion (i.e., nearly all of the hydrogen is consumed). The answer obtained by assuming complete combustion, as we did in Chapter 13, would be very nearly the same as the answer obtained here considering chemical equilibrium.

Figure 1 illustrates the temperature and mole fraction of hydrogen as a function of pressure. Notice that as pressure is increased, the equilibrium condition contains less hydrogen (the combustion is more complete) and therefore more energy is released during the combustion process so the final temperature is higher.

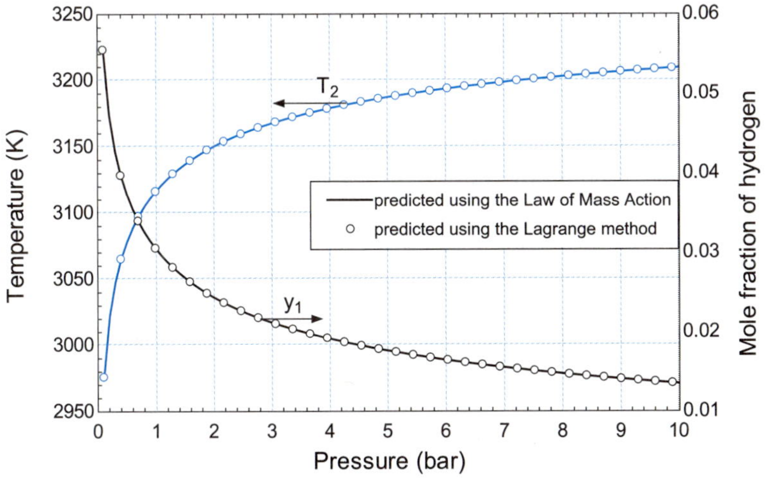

Figure 1: Temperature and mole fraction of hydrogen as a function of pressure predicted using the Law of Mass Action and the Lagrange Method of Undetermined Multipliers.

b) Determine the temperature and composition of the equilibrium mixture using the Lagrange Method of Undetermined Multipliers. Overlay on your plot from

(a) the temperature and mole fraction of hydrogen in the products as a function of pressure.

The LaGrange method described in Section 14.4.2 can also be used for adiabatic reactions. Change the value of the Method$ variable to 'LaGrange'.

```
Method$='LaGrange'                 "select solution method"
```

The code that is placed within the statements:

```
$If Method$='Law of Mass Action'      "use Law of Mass Action"
```

and

```
$EndIf
```

is no longer executed when the value of the variable Method$ is changed. Develop a new $If statement that contains the code that will be activated if the LaGrange Method of Undetermined Multipliers is used.

```
$If Method$='LaGrange'             "use LaGrange Method of Undetermined Multipliers"
```

Set up the index numbering system for the elements that are present.

```
N_E=2                              "number of elements"
H=1; O=2;                          "element numbering system"
```

The indices of the element index matrix, $e_{i,j}$, contains the number of moles of element j in 1 mole of component i.

```
"Element indices"
e[H2,H]=2; e[H2,O]=0;              "H2"
e[O2,H]=0; e[O2,O]=2;              "O2"
e[H2O,H]=2; e[H2O,O]=1;            "H2O"
```

The number of moles of each element that are initially present are computed according to:

$$E_{0,j} = \sum_{i=1}^{C} e_{i,j}\, n_{0,i} \quad \text{for } j = 1..E$$

where E is the total number of elements that are present.

```
duplicate j=1,N_E
   E_s_O[j]=sum(n_O[i]*e[i,j],i=1,N_C)   "number of moles of each element initially present"
end
```

Initially, we will guess the values of the number of moles of each of the components present at equilibrium (n_i for $i = 1..\ C$) as well as the final temperature, T_2. Eventually, these values will be adjusted in order to satisfy element balances and chemical equilibrium (using the Lagrange Method of Undetermined Multipliers) as well as an energy balance.

```
"initial guesses for the unknowns"
duplicate i=1,N_C
   n[i]=1 [kmol]                   "guess for final molar amounts"
end
T_2=2000 [K]                       "guess for final temperature"
```

The total number of moles of gas is computed:

$$n = \sum_{i=1}^{C} n_i$$

and the mole fraction of each gas is obtained from:

$$y_i = \frac{n_i}{n} \quad \text{for } i = 1..C$$

```
n=sum(n[i],i=1,N_C)                    "total number of moles"
duplicate i=1,N_C
   y[i]=n[i]/n                          "mole fraction"
end
```

The value of $\mu_i^o = \overline{g}_i^o$ is computed for each of the substances at T_2 and the reference pressure using Eq. (14-60).

$$\mu_i^o = \overline{g}_i^o = \overline{h}_i^o - T_2\,\overline{s}_i^o \quad \text{for } i = 1..C$$

The chemical potential of each substance is computed using Eq. (14-59):

$$\mu_i = \mu_i^o + R_{univ}\,T_2 \ln\left(\frac{y_i P}{P^o}\right) \quad \text{for } i = 1..C$$

```
P_ref=1 [atm]*convert(atm,Pa)           "reference pressure"
duplicate i=1,N_C
   g_bar_o[i]=enthalpy(C$[i],T=T_2)-T_2*entropy(C$[i],T=T_2,P=P_ref)
                                         "chemical potential at T_2 and P_ref"
   mu[i]=g_bar_o[i]+R#*T_2*ln(y[i]*P/P_ref)  "chemical potential"
end
```

The Lagrange method of undetermined multipliers determines the equilibrium state subject to element constraints by solving the set of equations given by Eq. (14-67) and Eq. (14-74):

$$\sum_{i=1}^{C} n_i\, e_{i,j} - E_{0,j} = 0 \quad \text{for } j = 1..E \tag{2}$$

$$\mu_i + \sum_{j=1}^{E} \lambda_j\, e_{i,j} = 0 \quad \text{for } i = 1..C \tag{3}$$

The guess values for the problem are updated. The assumed values of the molar quantities are commented out:

```
{duplicate i=1,N_C
   n[i]=1 [kmol]                        "guess for final molar amounts"
end}
```

and Eqs. (2) and (3) are added:

```
"Lagrange method of undetermined multipliers"
duplicate j=1,N_E
   sum(n[i]*e[i,j],i=1,N_C)-E_s_0[j]=0
end
duplicate i=1,N_C
   mu[i]+sum(lambda[j]*e[i,j],j=1,N_E)=0
end
```

Solving the EES code provides the equilibrium concentration at the assumed value of T_2. Next, it is necessary to adjust T_2 in order to satisfy an energy balance. The enthalpy of the equilibrium mixture is computed according to:

$$H_2 = \sum_{i=1}^{C} n_i \, \overline{h}_{i,T=T_2}$$

where $\overline{h}_{i,T=T_2}$ is the molar specific enthalpy of component i evaluated at T_2.

```
H_2=sum(n[i]*enthalpy(C$[i],T=T_2),i=1,N_C)     "enthalpy of substances leaving reactor"
```

The guess values are updated and the assumed value of T_2 is commented out.

```
{T_2=2000 [K]}                                  "guess for final temperature"
```

The energy balance:

$$H_2 = H_1$$

is implemented.

```
   H_1=H_2                                      "enforce energy balance"
$EndIf
```

Solving leads to $T_2 = 3209$ K with composition $y_1 = 0.01351$, $y_2 = 0.6027$, and $y_3 = 0.3838$; these answers are identical to those obtained in (a). Figure 1 illustrates the temperature and mole fraction of hydrogen as a function of pressure predicted by the Lagrange method. A more realistic problem that takes advantage of the Lagrange method for solving these types of problems is provided in Example 14.6-2.

EXAMPLE 14.6-2: ADIABATIC COMBUSTION OF ACETYLENE

Acetylene (C_2H_2) gas is combusted with ta = 110% theoretical air in a torch at atmospheric pressure. The acetylene and air are both supplied to the torch at T_{in} = 25°C and atmospheric pressure, P = 1 atm. The torch is adiabatic and the products emerge from the torch at a high temperature, T_{out}, and atmospheric pressure, P = 1 atm. At high temperatures, reactions can occur between the oxygen and nitrogen to form NO and NO_2. In addition, the C_2H_2 could possibly be converted to C_2H_4 or CH_4. The gases that may be present in the products include C_2H_2, C_2H_4, CH_4, CO, CO_2, H_2, H_2O, O_2, N_2, NO, and NO_2.

a) Determine the temperature and composition of the products assuming that the reaction proceeds to completion (i.e., without consideration of chemical equilibrium). Plot the temperature of the mixture leaving the torch as a function of theoretical air from $0.75 < ta < 2$ (75% to 200% theoretical air).

The input information is entered in EES.

```
$UnitSystem SI Molar J K Pa
T_in=converttemp(C,K, 25[C])      "inlet temperature"
P=1 [atm]*convert(atm,Pa)         "atmospheric pressure"
ta=1.1 [-]                        "theoretical air"
```

The stoichiometric reaction of acetylene with air, assuming complete combustion, is:

$$C_2H_2 + a_s\,(0.21\,O_2 + 0.79\,N_2) \rightleftharpoons b_s\,CO_2 + c_s\,H_2O + d_s\,N_2 \tag{1}$$

Balances on the carbon, hydrogen, oxygen, and nitrogen provide the following equations:

$$2 = b_s$$

$$2 = 2\,c_s$$

$$2\,(0.21)\,a_s = 2\,b_s + c_s$$

$$2\,(0.79)\,a_s = 2\,d_s$$

```
"Balance stoichiometric reaction"
2=b_s                             "carbon"
2=2*c_s                           "hydrogen"
a_s*0.21*2=2*b_s+c_s              "oxygen"
a_s*0.79*2=d_s*2                  "nitrogen"
```

The reaction (still assuming complete combustion) is given by:

$$C_2H_2 + a\,(0.21\,O_2 + 0.79\,N_2) \rightleftharpoons b\,CO_2 + c\,H_2O + d\,N_2 + e\,O_2 + f\,C_2H_2 \tag{2}$$

There may be either oxygen or acetylene left in the products, depending on the amount of air supplied with the acetylene; therefore, both of these substances are included in Eq. (2). The amount of air is calculated based on the theoretical air according to:

$$a = ta\,a_s$$

```
a=ta*a_s                          "excess air"
```

If the reacting mixture is rich (i.e., $ta < 1$), then there will be no oxygen remaining in the products:

$$e_{rich} = 0$$

Balances on the oxygen, carbon, hydrogen, and nitrogen in Eq. (2) for a rich mixture lead to:

$$2\,(0.21)\,a = 2\,b_{rich} + c_{rich}$$

$$2 = b_{rich} + 2\,f_{rich}$$

$$2 = 2\,c_{rich} + 2\,f_{rich}$$

$$2\,(0.79)\,a = 2\,d_{rich}$$

```
"Balance actual reaction – rich (ta<1)"
e_rich=0                              "no O2 is left"
0.21*2*a=2*b_rich+c_rich              "oxygen"
2=b_rich+2*f_rich                     "carbon"
2=2*c_rich+2*f_rich                   "hydrogen"
a*(0.79)*2=2*d_rich                   "nitrogen"
```

If the reacting mixture is lean (i.e., $ta > 1$), then there will be no fuel remaining in the products:

$$f_{lean} = 0$$

Balances on the carbon, hydrogen, oxygen, and nitrogen in Eq. (2) for a lean mixture lead to:

$$2 = 2\, b_{lean}$$

$$2 = 2\, c_{lean}$$

$$2\,(0.21)\,a = 2\, b_{lean} + c_{lean} + 2\, e_{lean}$$

$$2\,(0.79)\,a = 2\, d_{lean}$$

```
"Balance actual reaction - lean (ta>1)"
f_lean=0                              "no fuel left"
2=b_lean                              "carbon"
2=2*c_lean                            "hydrogen"
0.21*2*a=2*b_lean+c_lean+2*e_lean     "oxygen"
0.79*2*a=2*d_lean                     "nitrogen"
```

The coefficients b, c, d, e, and f in Eq. (2) are selected using the IF function in EES based on whether the mixture is rich or lean. The IF function requires five arguments and has the general form IF(a, b, x, y, z). If a < b, the function returns the value of x. If a = b, the function returns y and if a > b, the function returns z.

```
"coefficients"
b_a=IF(ta,1,b_rich,b_rich,b_lean)
c_a=IF(ta,1,c_rich,c_rich,c_lean)
d_a=IF(ta,1,d_rich,d_rich,d_lean)
e_a=IF(ta,1,e_rich,e_rich,e_lean)
f_a=IF(ta,1,f_rich,f_rich,f_lean)
```

The number of gases, index information, and the initial amount of each gas ($n_{0,i}$ for $i = 1..C$) are entered; note that these quantities are entered on a per unit mole of acetylene basis using the coefficients in Eq. (2).

```
N_C=11                  "number of gases"
"information for each gas"
C2H2=1;                 C$[C2H2]='C2H2';     n_0[C2H2]=1
C2H4=2;                 C$[C2H4]='C2H4';     n_0[C2H4]=0
CH4=3;                  C$[CH4]='CH4';       n_0[CH4]=0
CO2=4;                  C$[CO2]='CO2';       n_0[CO2]=0
H2O=5;                  C$[H2O]='H2O';       n_0[H2O]=0
O2=6;                   C$[O2]='O2';         n_0[O2]=a*0.21
```

```
N2=7;     C$[N2]='N2';     n_0[N2]=a*0.79
H2=8;     C$[H2]='H2';     n_0[H2]=0
CO=9;     C$[CO]='CO';     n_0[CO]=0
NO=10;    C$[NO]='NO';     n_0[NO]=0
NO2=11;   C$[NO2]='NO2';   n_0[NO2]=0
```

The enthalpy of the reactants per mole of fuel is calculated according to:

$$H_R = \sum_{i=1}^{C} n_{0,i}\, \overline{h}_{i,T=T_{in}}$$

where $\overline{h}_{i,T=T_{in}}$ is the molar specific enthalpy of component i evaluated at T_{in}.

```
H_R=sum(n_0[i]*enthalpy(C$[i],T=T_in),i=1,N_C)   "enthalpy of reactants per mole of fuel"
```

The final amount of each component per mole of fuel (n_i) (assuming complete combustion) can be entered based on inspection of Eq. (2).

```
"assuming complete combustion"
n[C2H2]=f_a
n[C2H4]=0
n[CH4]=0
n[CO2]=b_a
n[H2O]=c_a
n[O2]=e_a
n[N2]=d_a
n[H2]=0
n[CO]=0
n[NO]=0
n[NO2]=0
```

The outlet temperature, T_{out}, is assumed. The enthalpy of the products per mole of fuel is computed according to:

$$H_P = \sum_{i=1}^{C} n_i\, \overline{h}_{i,T=T_{out}}$$

where $\overline{h}_{i,T=T_{out}}$ is the molar specific enthalpy of component i evaluated at T_{out}.

```
T_out=2000 [K]                                     "guess for outlet temperature"
H_P=sum(n[i]*enthalpy(C$[i],T=T_out),i=1,N_C)      "enthalpy of products per mole of fuel"
```

The guess values are updated and the assumed value of T_{out} is commented out. An energy balance on the reaction is entered.

$H_R = H_P$

```
{T_out=2000 [K]}                                   "guess for outlet temperature"
H_R=H_P                                            "energy balance"
```

Solving the equations leads to $T_{out} = 2720$ K. Figure 1 illustrates the outlet temperature assuming complete combustion as a function of the theoretical air.

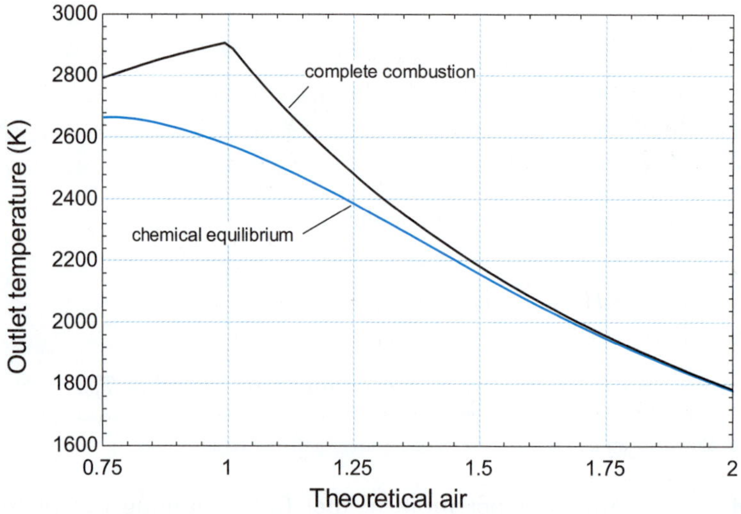

Figure 1: Outlet temperature as a function of the theoretical air computed assuming complete combustion. Also shown is the outlet temperature obtained considering chemical equilibrium.

b) Determine the temperature and composition of the products assuming that the reaction proceeds to the equilibrium condition required by chemical equilibrium.

There are 11 gases composed of 4 elements in the products. The problem is further complicated by the fact that the temperature of the products is not known and must be determined from an energy balance. The Lagrange Method of Undetermined Multipliers (Section 14.4.2) provides the best alternative for solving this problem.

The code used to solve the problem assuming complete combustion in part (a) is commented out.

```
{"assuming complete combustion"
n[C2H2]=f_a
n[C2H4]=0
n[CH4]=0
n[CO2]=b_a
n[H2O]=c_a
n[O2]=e_a
n[N2]=d_a
n[H2]=0
n[CO]=0
n[NO]=0
n[NO2]=0

{T_out=2000 [K]}                                   "guess for outlet temperature"
H_P=sum(n[i]*enthalpy(C$[i],T=T_out),i=1,N_C)      "enthalpy of products per mole of fuel"
H_R=H_P                                            "energy balance"}
```

The matrix of the coefficients $e_{i,j}$ is implemented. Recall that the coefficient $e_{i,j}$ is the number of moles of element j appearing in one mole of substance i. Index numbers are assigned to the four elements to make the equations more readable.

```
"Equilibrium"
N_E=4                                                      "number of elements"
C=1; H=2; O=3; N=4;                                        "index number for elements"
e[C2H2,C]=2;   e[C2H2,H]=2;   e[C2H2,O]=0;   e[C2H2,N]=0;   "element indices for C2H2"
e[C2H4,C]=2;   e[C2H4,H]=4;   e[C2H4,O]=0;   e[C2H4,N]=0;   "element indices for C2H4"
e[CH4,C]=1;    e[CH4,H]=4;    e[CH4,O]=0;    e[CH4,N]=0;    "element indices for CH4"
e[CO2,C]=1;    e[CO2,H]=0;    e[CO2,O]=2;    e[CO2,N]=0;    "element indices for CO2"
e[H2O,C]=0;    e[H2O,H]=2;    e[H2O,O]=1;    e[H2O,N]=0;    "element indices for H2O"
e[O2,C]=0;     e[O2,H]=0;     e[O2,O]=2;     e[O2,N]=0;     "element indices for O2"
e[N2,C]=0;     e[N2,H]=0;     e[N2,O]=0;     e[N2,N]=2;     "element indices for N2"
e[H2,C]=0;     e[H2,H]=2;     e[H2,O]=0;     e[H2,N]=0;     "element indices for H2"
e[CO,C]=1;     e[CO,H]=0;     e[CO,O]=1;     e[CO,N]=0;     "element indices for CO"
e[NO,C]=0;     e[NO,H]=0;     e[NO,O]=1;     e[NO,N]=1;     "element indices for NO"
e[NO2,C]=0;    e[NO2,H]=0;    e[NO2,O]=2;    e[NO2,N]=1;    "element indices for NO2"
```

The number of moles of each element that enter (per mole of fuel) is determined according to:

$$E_{0,j} = \sum_{i=1}^{C} n_{0,j}\, e_{i,j} \quad \text{for } j = 1..E$$

```
duplicate j=1,N_E
   E_s_0[j]=sum(n_0[i]*e[i,j],i=1,N_C)   "number of moles of each element initially present"
end
```

Initially, the number of moles of each component leaving (n_i for $i = 1..C$) and the exit temperature, T_{out}, are assumed.

```
duplicate i=1,N_C
   n[i]=0.01                              "guess for final molar amounts"
end
T_out=2500 [K]                            "guess for outlet temperature"
```

The total number of moles is computed according to:

$$n = \sum_{i=1}^{C} n_i$$

and the mole fraction of each component is calculated:

$$y_i = \frac{n_i}{n}$$

```
n_total=sum(n[i],i=1,N_C)                 "total number of moles"
duplicate i=1,N_C
   y[i]=n[i]/n_total                      "mole fraction"
end
```

The value of $\mu_i^o = \overline{g}_i^o$ is computed for each of the substances at T_{out} and the reference pressure using Eq. (14-60):

$$\mu_i^o = \overline{g}_i^o = \overline{h}_i^o - T_{out}\,\overline{s}_i^o \quad \text{for } i = 1..C$$

The chemical potential of each substance is computed using Eq. (14-59):

$$\mu_i = \mu_i^o + R_{univ}\,T_{out}\ln\left(\frac{y_i P}{P^o}\right) \quad \text{for } i = 1..C$$

```
P_ref=1 [atm]*convert(atm,Pa)                         "reference pressure"
duplicate i=1,N_C
   g_bar_o[i]=enthalpy(C$[i],T=T_out)-T_out*entropy(C$[i],T=T_out,P=P_ref)
                                                      "chemical potential at T_out and P_ref"
   mu[i]=g_bar_o[i]+R#*T_out*ln(y[i]*P/P_ref)         "chemical potential"
end
```

The Lagrange Method of Undetermined Multipliers determines the equilibrium state subject to element constraints by solving the set of equations given by Eq. (14-67) and Eq. (14-74):

$$\sum_{i=1}^{C} n_i\,e_{i,j} - E_{0,j} = 0 \quad \text{for } j = 1..E \tag{3}$$

$$\mu_i + \sum_{j=1}^{E} \lambda_j\,e_{i,j} = 0 \quad \text{for } i = 1..C \tag{4}$$

The guess values for the problem are updated. The assumed values of the molar quantities are commented out:

```
{duplicate i=1,N_C
   n[i]=0.01            "guess for final molar amounts"
end}
```

and Eqs. (3) and (4) are added:

```
"Lagrange method of undetermined multipliers"
duplicate j=1,N_E
   sum(n[i]*e[i,j],i=1,N_C)-E_s_0[j]=0
end
duplicate i=1,N_C
   mu[i]+sum(lambda[j]*e[i,j],j=1,N_E)=0
end
```

The result is the equilibrium concentration at the assumed value of T_{out}. Next, it is necessary to adjust T_{out} in order to satisfy an energy balance. The enthalpy of the equilibrium mixture per mole of fuel is computed according to:

$$H_{out} = \sum_{i=1}^{C} n_i\,\overline{h}_{i,T=T_{out}}$$

where $\overline{h}_{i,T=T_{out}}$ is the molar specific enthalpy of component i evaluated at T_{out}.

```
H_out=sum(n[i]*enthalpy(C$[i],T=T_out),i=1,N_C)   "enthalpy of products per mole of fuel"
```

The guess values are updated and the assumed value of T_{out} is commented out.

```
{T_out=2500 [K]}    "guess for outlet temperature"
```

The energy balance:

$$H_R = H_{out}$$

is implemented.

```
H_R=H_out    "energy balance"
```

Solving provides $T_{out} = 2507$ K, which is substantially less than the value determined in part (a) assuming complete combustion (2720 K). The results show that there will not be any measureable amount of C_2H_2, C_2H_4 or CH_4 remaining at equilibrium. The mole fractions of NO and NO_2 are small, but even small values may be a concern in some applications (e.g., motor vehicles) due to regulations in place to reduce smog. Figure 1 illustrates the outlet temperature as a function of theoretical air predicted using the Lagrange method of undetermined multipliers to determine the equilibrium concentration of products. The difference between the temperatures calculated assuming complete combustion and chemical equilibrium is reduced as the amount of excess air is increased. The excess air results in less carbon monoxide and hydrogen production, which results in a solution that agrees more closely with the result obtained assuming complete combustion. The mole fraction of NO exhibits a maximum at an excess air fraction of about 1.2, but the molar amounts of NO and NO_2 are both too small to have any significant energy implications. Another interesting observation is that the highest temperature of the products can be obtained with theoretical air that is slightly lower than the stoichiometric amount.

REFERENCE

McBride, B.J., Zehe, M.J. and Gordon, S., "NASA Glenn Coefficients for Calculating Thermodynamic Properties of Individual Species", NASA/TP—2002-211556, Sept. 2002, http://www.lerc.nasa.gov/WWW/CEAWeb/

Problems

The problems included here have been selected from a larger set of problems that are available on the website associated with this book (www.cambridge.org/kleinandnellis).

A: Simple Reactions

14.A-1 The proton exchange membrane (PEM) fuel cell requires hydrogen as the fuel. Unfortunately, hydrogen does not naturally occur and it must therefore be generated from another fuel. One way to do this is to reform a hydrocarbon fuel such as methane (CH_4). A preliminary step in the reformation process is to react methane with water in a catalytic converter to produce hydrogen and carbon monoxide according to the reaction: $CH_4 + H_2O \leftrightharpoons 3H_2 + CO$. Methane and water are fed to reactor in molar proportions of 1 mole of methane to 4 moles of water.

a) Prepare a plot of the equilibrium hydrogen mole fraction as a function of temperature for temperatures ranging from 600 K and 1000 K at 1 atm.

b) How will the equilibrium mole fraction of hydrogen be affected if the pressure in the converter is raised to 5 atm?

14.A-3 A mixture of methane and 350% stoichiometric pure oxygen at 25°C and 1 bar is ignited in a constant volume, well-insulated container. The only significant decomposition is assumed to be that of carbon dioxide reacting to produce carbon monoxide and oxygen. Estimate the maximum temperature, pressure and composition assuming no hydrogen or decomposition products other than carbon monoxide are present in the reaction vessel.

14.A-4 Methanol can be mixed with gasoline and used in internal combustion engines. The oxygen provided in methanol can reduce the concentration of pollutants in the exhaust. Methanol can be produced in many ways. One way is to react carbon monoxide with hydrogen according to the reaction: $CO + 2\,H_2 \leftrightharpoons CH_3OH$.

a) Calculate and plot the equilibrium yield of methanol per mole of CO at 125°C as a function of pressure for pressures between 1 and 10 bar. Assume ideal gas behavior and assume that CO and H_2 are provided in stoichiometric proportions.

b) It is possible to provide excess carbon monoxide or excess hydrogen rather than stoichiometric amounts of the two gases. What ratio of CO to H_2 would you recommend if the concentration of methanol is to be maximized at 125°C and 5 bar?

14.A-5 A gas mixture at 400°C consisting of 7.8% SO_2, 10.8% O_2, with the remainder being N_2 (on a molar basis) is passed through a catalytic converter in which the reaction $0.5\,O_2 + SO_2 \leftrightharpoons SO_3$ occurs adiabatically. Assume ideal gas behavior. Prepare a plot of the equilibrium temperature at the reactor outlet as a function of pressure for pressures between 1 atm and 10 atm. Assume ideal gas behavior.

14.A-6 Portable propane stoves are a concern because they can easily be dropped which could cause them to malfunction and result in the generation of carbon monoxide. Also, these stoves are sometimes used in closed spaces (against manufacturer's guidelines). Assume that the products of the combustion of propane consist of CO_2, CO, O_2, N_2 and H_2O. Plot the equilibrium mole fraction of CO as a function of the % theoretical air for a range of 80% and 200% at temperatures of 1000 K, 1500 K and 2000 K. What conditions result in the largest mole fraction of carbon monoxide?

14.A-7 One method of producing hydrogen gas is to react hydrogen sulfide (H_2S) with water vapor in proportions of 1 mole of H_2S to 5 moles of water vapor. The mixture of H_2S and H_2O is provided to the reaction at 1.5 atm and 110°C. The vapor phase reaction occurs according to: $H_2S + 2\,H_2O \leftrightharpoons 3\,H_2 + SO_2$. Calculate and plot the required heat input per mole of H_2 produced and the equilibrium mole fraction of hydrogen associated with this reaction as a function of temperature for temperatures between 1000 K and 1500 K.

14.A-8 The dissociation of the hydrogen molecule (H_2) into elemental hydrogen atoms is being studied at high temperatures and 1 atm total pressure. Prepare a plot of the mole fraction of elemental hydrogen as a function of temperature for temperatures between 3000 K and 4000 K by directly minimizing the Gibbs free energy.

14.A-9 When ammonia dissociates into hydrogen and nitrogen according to $NH_3 \leftrightharpoons 0.5\,N_2 + 1.5\,H_2$, the number of moles of gas are increased. An engineer has proposed that the increased number of moles of gas could increase the pressure and thus the power output of a turbine. As a preliminary evaluation of this concept, assume ammonia enters an isentropic turbine at 300°C and 3.5 bar and is exhausted to 1 bar. Assume ideal gas behavior.

a) Calculate the work per mole of ammonia, assuming dissociation does not occur.

b) Calculate the work per mole of ammonia, assuming chemical equilibrium is achieved. Compare the result with the value from part (a).

14.A-12 The reaction forming hydrogen from methane and water is $CH_4 + H_2O \leftrightharpoons 3H_2 + CO$. One mole of methane and four moles of water are heated to 1000 K.

a) Prepare a plot of the pressure required to obtain 1.5 moles of hydrogen as a function of temperature for temperatures between 800 K and 1200 K, assuming ideal gas behavior.

B. Simultaneous and Heterogeneous Reactions

14.B-1 Furnaces are designed so that the combustion of the fuel occurs with excess air. However, an improperly sized fuel jet or an air restriction can result in insufficient air supply. The major concern in this case is the formation of carbon monoxide gas, which is deadly. In the present case, the fuel is methane. The combustion gas that exits the furnace when 90% of theoretical air is provided is believed to be a mixture of CH_4, CO_2, CO, N_2, NO, H_2O, H_2 and O_2.

a) Prepare a plot of the mole fraction of CO versus temperature for temperatures between 1500 K and 2500 K at atmospheric pressure using the LaGrange method.

b) What is the maximum temperature that the combustion products gas mixture can reach without external heating?

14.B-2 The combustion process that occurs in spark-ignition engines produces small amounts of carbon monoxide (CO), nitrogen oxide (NO) and nitrogen dioxide (NO_2). The amounts of these gases are small. However, the large number of engines and the subsequent atmospheric reactions involving these gases cause them to be a concern. Assume that octane (C_8H_{18}) and air are provided in stoichiometric amounts at atmospheric pressure and 25°C to an engine and that the combustion process occurs adiabatically. Estimate the molar amounts of CO, NO and NO_2 that are produced for each mole of octane combusted. Estimate the adiabatic combustion temperature.

14.B-5 You have been asked to investigate methods to improve the energy efficiency of the ammonia synthesis process. Currently the process is operated at 300 atm and 425°C with the nitrogen and hydrogen provided in stoichiometric proportions at 25°C, 300 atm. One idea that you have is to provide excess nitrogen since nitrogen is inexpensive. Prepare plots of the mole fraction of ammonia and the required heat input per mole of ammonia produced as a function of the ratio of the actual nitrogen provided to the stoichiometric amount. Note that because of the high pressures involved, the ideal gas law behavior is not valid. Assume that the mixture obeys ideal solution theory.

14.B-6 Hydrogen reforming is one of the main processes used for the industrial production of hydrogen. The reformation process is based on the following reaction of methane with water vapor: $CH_4 + H_2O \leftrightharpoons CO + 3\,H_2$. The reaction is endothermic and the equilibrium constant at low temperature results in a very small yield of hydrogen. Thus, the reaction is carried out at 1100°C. This temperature remains constant during the reaction by combusting controlled amounts of methane with oxygen according to: $2\,CH_4 + 3\,O_2 \leftrightharpoons 2CO + 4\,H_2O$. A well-insulated reactor operating at steady conditions is fed with separate streams of 1 kmol/sec of methane, 1 kmol/sec of water vapor and oxygen at a rate which is to be determined. All streams are preheated so that they enter the reactor at 1100°C and all streams are at 1 atm pressure. The oxygen is completely consumed. Leaving the reactor is a mixture of methane, water vapor, carbon monoxide, and hydrogen in chemical equilibrium at 1100°C, 1 atm.

a) Determine the required molar flow rate of oxygen and the mole fraction of hydrogen in the exit stream.

14.B-7 Oxygen and nitrogen in the air supplied to a combustion process can react at sufficient rates at high temperatures. The extent of the reaction is small but the presence of even small amounts of the various oxides of nitrogen in combustion products is an important factor from an air pollution perspective. Consider a mixture consisting of the following basic products of combustion: 11% CO_2, 12% H_2O, 4% O_2 and 73% N_2 (on a molar basis). At the high temperatures and pressures occurring within the cylinder of an engine, both NO and NO_2 may form. It is likely that carbon monoxide will also be formed. Prepare plots showing the equilibrium moles fractions of CO, NO and NO_2 as a function of pressure for pressures between 5 atm and 15 atm at 2000 K.

14.B-9 An evacuated 2 liter cylinder at 25°C contains 1.6 grams of methane (CH_4). The cylinder and its contents are then slowly heated to 1500°C. Methane is expected to dissociate into elemental carbon and hydrogen gas during this heating process according to: $CH_4 \leftrightharpoons C + 2\,H_2$. Prepare plots of the expected pressure in the cylinder and the mole fraction of methane as a function of temperature for temperatures between 50°C and 1500°C. Assume ideal gas behavior.

14.B-10 The vast coal deposits in the U.S. and elsewhere are a likely result of a chemical equilibrium process in which small hydrocarbon molecules dissociate into carbon. A simple model to study this process is provided by the chemical equilibrium of the ethane formation reaction: $2\,C + 3\,H_2 \leftrightharpoons C_2H_6$. In order to study this reaction, a 4 liter volume is filled with pure ethane at 15 bar and 25°C. The ethane is then slowly heated to 1000°C.

a) Determine the pressure and mole fraction of hydrogen, assuming ideal gas behavior.

b) Non-ideal gas behavior can be expected at the high pressures involved in this process. Repeat the calculations requested in part (a) assuming that the gases form an ideal solution.

14.B-11 Coal is increasingly less desirable as a fuel because it contains trace elements (e.g., mercury, sulfur) that form pollutants when combusted and also because of the large amount of CO_2 that is produced per unit energy. A proposed alternative is to react coal (assumed here to be pure carbon) with steam at 825°C and a pressure that is to be determined. The following reactions are believed to occur simultaneously during this process:

$$C + H_2O \leftrightharpoons CO + H_2 \quad CO + H_2O \leftrightharpoons CO_2 + H_2 \quad C + CO_2 \leftrightharpoons 2\,CO$$

Prepare plots of the hydrogen and carbon monoxide mole fractions as a function of pressure for pressures between 1 and 12 atm. Assume ideal gas behavior.

14.B-12 Methane needs to be steam-reformed in order to produce hydrogen for use in a fuel-cell system. The desired reaction is: $CH_4 + 2\,H_2O \rightarrow CO_2 + 4\,H_2$. However, other species may also be present at equilibrium. The equilibrium condition can be considered to be a mixture of CO_2, CO, H_2, H_2O, CH_4 and solid carbon. Molecular oxygen is not one of the products. The presence of CO in the mixture is a problem, since CO is a poison to some types of fuel cell systems.

a) Calculate and plot the equilibrium mole fractions of hydrogen and carbon monoxide for a stoichiometric mixture of methane and water at 1 atm as a function of temperature for temperatures between 500 and 1000 K. Assume that solid carbon does not form. Ideal gas behavior can be assumed.

b) Repeat part (b) but allow for the possibility that solid carbon may be present at equilibrium. Plot the moles of solid carbon per mole of methane that form.

14.B-13 Acetylene (C_2H_2) at 25°C, 1 atm is used a fuel for a cutting torch. The fuel is reacted with air at 25°C, 1 atm resulting in products that include CO_2, CO, H_2O, H, and N_2, all at 1 atm. The proportions of the acetylene and air are controlled by adjustable valves. A student has noticed that smoke appears if he adjusts the torch to run lean, i.e., with insufficient air. He suspects that the smoke could be visible carbon particles that form during the reaction. Is this possible? Assume that the products exit the torch at equilibrium at 800 K. Calculate and plot the moles of solid carbon that form per mole of acetylene for theoretical air percentages between 20% and 100%. Assume ideal gas behavior.

14.B-15 A constant volume bomb is charged with 62 g of pure ethane (C_2H_6) at 24°C, 5 bar. The bomb and its contents are then heated to 500°C and allowed to come to equilibrium. The equilibrium products are expected to be ethane (C_2H_6), methane (CH_4), ethylene (C_2H_4), acetylene (C_2H_2), hydrogen (H_2), and possibly solid carbon.

a) What is the pressure in the bomb and the composition of the bomb contents at the 500°C equilibrium state, assuming that carbon does not form. Assume the gases form an ideal gas mixture.

b) Repeat part (a) assuming carbon may be a product. Note that properties of solid carbon are available from the NASA property library in EES with the species name set to 'C(gr)'.

15 Statistical Thermodynamics

As noted in Chapter 1, there are two very different methods to describe a thermodynamic system corresponding to the microscopic approach and the macroscopic approach. So far, only the macroscopic (or classical) approach has been used in this book. The macroscopic approach is based on empirical laws and it describes the equilibrium state of a system in terms of a relatively small number of properties such as temperature, pressure, internal energy, and entropy. The macroscopic approach can be applied to any system and it is mathematically much simpler than the microscopic approach. At no point in the development of the First or Second Laws is it necessary to know that matter consists of individual particles (molecules).

The disadvantage of the macroscopic approach to thermodynamics is that it does not provide any physical insight into the First and Second Laws upon which it is based. More importantly, classical thermodynamics does not provide any means to directly calculate the thermodynamic properties that are needed to apply the First and Second Laws. For example, we know that the specific heat capacities, c_P and c_v, are needed to evaluate specific enthalpy and specific internal energy. Macroscopic thermodynamics shows us that c_P and c_v are related (e.g., $c_P - c_v = R$ for an ideal gas) and it can also explain how these specific heat capacities vary with pressure at a given temperature, as discussed in Chapter 10. However, macroscopic thermodynamics offers no information on how c_P and c_v vary with temperature at constant pressure. Indeed, it is even difficult to understand what the property temperature refers to using only concepts from macroscopic thermodynamics. Statistical thermodynamics takes a microscopic approach in an attempt to advance our knowledge in these areas.

Statistical thermodynamics provides methods for calculating specific heat capacities and related thermodynamic properties based on spectroscopic data (i.e., absorption of different wavelengths of electromagnetic radiation.) Statistical thermodynamics can be used to provide accurate estimates of thermodynamic properties (e.g., specific heat capacity, specific internal energy and specific entropy), which greatly reduces the need for experimental measurements. Statistical thermodynamics has also yielded useful property information for cryogenic and high temperature systems for which alternative experimental techniques are complicated or impossible. However, since these property data are now known and tabulated, it is unlikely that you will be able to solve many engineering problems using statistical thermodynamics that you cannot also solve (more easily) using classical, macroscopic thermodynamics.

A microscopic approach shows that the laws of thermodynamics are in agreement with the postulates of quantum theory. The physical significance of properties such as temperature, pressure, and entropy becomes clear when these quantities are examined from a microscopic viewpoint. Most importantly, the microscopic approach provides an explanation for the Second Law of Thermodynamics. Statistical thermodynamics shows that entropy is related to the probability of particles distributing themselves among the allowable energy levels.

The goal of this chapter is to present a simple treatment of statistical thermodynamics applied to monatomic ideal gases in order to provide greater understanding of the concept of properties and the origin of the Second Law of Thermodynamics. Statistical thermodynamics provides an explanation of how matter behaves, which will surely expand your understanding of thermodynamic concepts.

15.1 A Brief Review of Quantum Theory History

The microscopic approach requires an understanding of the behavior of elementary particles of matter (atoms and molecules). As we will see, these elementary particles behave in a manner that is non-intuitive and very different from the behavior of the macroscopic objects that we are familiar with. Some history of quantum theory is therefore helpful.

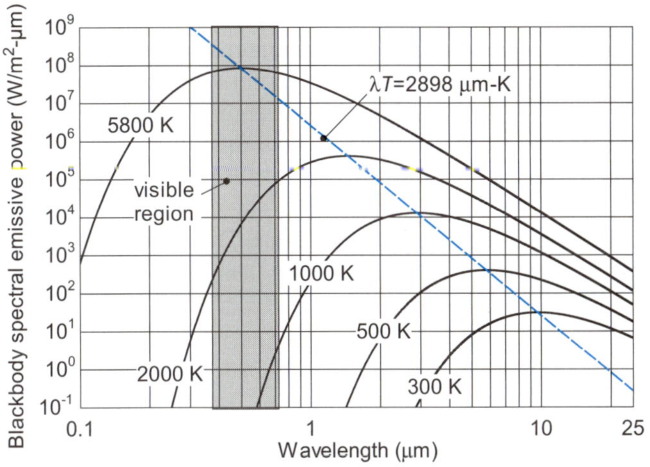

Figure 15-1: Blackbody spectral emissive power, $E_{b,\lambda}$, as a function of wavelength for various values of temperature (from Nellis and Klein, 2009).

15.1.1 Electromagnetic Radiation

The behavior of elementary particles of matter is described by quantum theory, which was developed early in the 20th century. Quantum theory originated with attempts to explain electromagnetic radiation. Stefan experimentally showed in 1879 that the total blackbody energy flux was proportional to the fourth power of absolute temperature. However, the spectral (wavelength dependent) behavior of electromagnetic radiation remained unexplained. The first accurate measurements of the spectral behavior of black body emissive power were published by Lummer and Pringsheim in 1899, for which they received the Nobel Prize in 1911. The blackbody spectral emissive power, $E_{b,\lambda}$, as a function of wavelength (λ) is shown in Figure 15-1 for various temperatures.

Early 20th century physicists attempted to develop a theory that explains the behavior shown in Figure 15-1. Experiments conducted at this time revealed that electromagnetic radiation has both wave-like and particle-like behaviors. Wave-like behavior was evident in experiments by Young in the early 1800s that showed that electromagnetic radiation can be diffracted and can exhibit constructive-destructive interference. Particle-like behavior was demonstrated by the force that is exerted on a surface as a result of incident electromagnetic radiation.

To illustrate the particle-like behavior of radiation, consider a parallel beam of light traveling at $c = 3 \times 10^8$ m/s (which is the speed of light through a vacuum) with an energy flux of 10 W/cm^2 that is impinging on a perfect mirror with an area of 1 cm^2. The rate that energy is incident on the surface of the mirror is given by:

$$\dot{E} = \frac{10\,\text{W}}{\text{cm}^2}\bigg|\frac{1\text{cm}^2}{} = 10\ \text{W} \tag{15-1}$$

If electromagnetic radiation behaves like particles, then this energy transfer must be related to the rate of change of the momentum of the particles. If the radiation is reflected from the mirror, then the force associated with the change in momentum of the particles can be computed according to:

$$F = \frac{2\,\dot{E}}{c} = \frac{2}{}\bigg|\frac{10\,\text{W}}{}\bigg|\frac{\text{s}}{3 \times 10^8\text{m}}\bigg\|\frac{\text{N–m}}{\text{W}} = 6.67 \times 10^{-8}\,\text{N} \tag{15-2}$$

If the surface absorbs (rather than reflects) the radiation then the force is one-half of what is calculated by Eq. (15-2) because the particles come to rest rather than reversing their direction; therefore, the particles experience a smaller momentum change. The force exerted by the particles on the mirror is very small, but it was experimentally determined that electromagnetic radiation exerts a force in conformance with Eq. (15-2) by Nichols and Hull in 1901.

The wave-like behavior of electromagnetic radiation prompted physicists to try to explain the blackbody emission spectrum in terms of a three-dimensional wave equation:

$$\nabla^2 \bar{E} = \frac{1}{c^2}\frac{\partial \bar{E}}{\partial t^2} \tag{15-3}$$

where $\bar{E}$ is the intensity of the electromagnetic radiation and ∇^2 is the Laplacian operator:

$$\nabla^2 = \bar{i}\frac{\partial^2}{\partial x^2} + \bar{j}\frac{\partial^2}{\partial y^2} + \bar{k}\frac{\partial^2}{\partial z^2} \tag{15-4}$$

where $\bar{i}$, $\bar{j}$, and $\bar{k}$ are the unit vectors in the x, y, and z directions, respectively. The solutions resulting from these attempts replicated the behavior of electromagnetic radiation at long wavelengths, but failed to predict the experimentally observed behavior at short wavelengths. This failure was referred to as the 'ultraviolet catastrophe' since the difference between the solutions and experimental data was largest for ultraviolet electromagnetic radiation.

In 1901, Max Planck published an empirical equation that fit the experimental data at both long and short wavelengths. His equation, called Planck's law, predicts the blackbody spectral emissive power according to:

$$E_{b,\lambda} = \frac{C_1}{\lambda^5\left[\exp\left(\dfrac{C_2}{\lambda\,T}\right) - 1\right]} \tag{15-5}$$

where $C_1 = 3.742 \times 10^8$ W-μm^4/m^2 and $C_2 = 14{,}388$ μm-K. Planck then tried to derive his empirical equation using classical theory of wave behavior, but these attempts failed. He was later able to provide a successful derivation using Boltzmann's statistical reasoning. His derivation assumes that a radiating particle can only have discrete amounts of energies, given by:

$$\varepsilon_i = i\,\frac{h\,c}{\lambda} \quad \text{where } i = 0, 1, 2, \ldots \tag{15-6}$$

where the parameter h is Planck's constant (6.625×10^{-34} J-s). Einstein expanded on Planck's theory in 1905 based on his analysis of the photoelectric effect in which he shows that the energy of the *photons* or packets of electromagnetic energy emitted from a surface depends on the frequency rather than the intensity of the incident radation.

The implications of these studies are not intuitive. Electromagnetic radiation has behavior that is characteristic of both waves and particles and the energy of electromagnetic radiation can only occur in discrete or *quantized* amounts. These observations lead to the quantum theory of matter.

15.1.2 Extension to Particles

If a photon of electromagnetic radiation behaves like a particle with mass m traveling at the speed of light c then its kinetic energy is given by:

$$\varepsilon = mc^2 \tag{15-7}$$

where m is the *equivalent mass* of the photon. This famous relation is attributed to Einstein and differs from the normal kinetic energy relation by a factor of 2 because electromagnetic radiation is polarized into two conjugate directions. Planck's reasoning showed that the energy of a photon is an integer multiple of the quantity (hc/λ), as indicated in Eq. (15-6). Combining Eqs. (15-6) and (15-7) provides a relation for the equivalent momentum (p) of a photon:

$$p = mc = \frac{h}{\lambda} \tag{15-8}$$

Although $p = h/\lambda$ was initially applied to photons of electromagnetic radiation, it was later proposed by de Broglie in 1923 that all mass in motion has wave-like properties and thus elementary particles such as gaseous atoms should also obey Eq. (15-8). Support for this idea was provided by experiments that showed that a beam of electrons can exhibit diffraction, which is a characteristic of waves. According to de Broglie, elementary particles of mass m may only have speeds ($\tilde{V}$) and momenta (p) that result in integral numbers of a characteristic wavelength (λ), as indicated in Eq. (15-9):

$$\lambda = \frac{h}{p} = \frac{h}{m\tilde{V}} \tag{15-9}$$

If Eq. (15-9) is applied to a mass of $m = 1$ kg at a velocity of $\tilde{V} = 1$ m/s, the characteristic wavelength is $\lambda = 6.625 \times 10^{-34}$ m. Perhaps the motion of this mass is restricted to incremental movements of this magnitude. If so, we would never notice. However, if Eq. (15-9) is applied to an electron of mass $m = 9.11 \times 10^{-31}$ kg and velocity $\tilde{V} = 2 \times 10^6$ m/s then the characteristic wavelength is $\lambda = 3.64 \times 10^{-10}$ m. This wavelength is large relative to the size of microscopic particles. For example, the radius of a helium atom is about 1.28×10^{-10} m. Therefore, although we do not notice quantum effects in our macroscopic world, these effects are important for microscopic particles.

It would have been obliging on the part of nature if it were possible to attribute to the smallest pieces of matter the same Newtonian properties – in particular definite positions and velocities – as are normally associated with the notion of a particle. Experience has shown, however, that these minute objects do not have such clear-cut properties. In some experiments they seem to have the extended character of a wave.

Denbigh (1981)

15.2 The Wave Equation and Degeneracy for a Monatomic Ideal Gas

Our development of statistical thermodynamics will begin with the simplest case, a monatomic ideal gas in an enclosure of dimension a, b, c. We assume that there are no intermolecular forces and no gravitational, magnetic or other types of external force fields exerted on the gas particles. Collisions between particles are assumed to be perfectly elastic. With these assumptions, the only form of energy that a gas particle can have is its kinetic energy of translation. The kinetic energy of a gas particle in the x-direction is given by:

$$\varepsilon_x = \frac{1}{2} m \tilde{V}_x^2 = \frac{p_x^2}{2m} \tag{15-10}$$

15.2.1 Probability of Finding a Particle

Consider a single particle having a non-zero kinetic energy of ε_x. Where along the x-coordinate is this gas particle most likely to be found within the enclosure? The immediate response most of us would provide to this question is that the particle could be located at any value of x between 0 and a (the x-dimension of the enclosure) as all values of x have an equal likelihood of occurrence. However, this is not true. For example, we know that the velocity of the particle must be zero at the enclosure walls ($x = 0$ and $x = a$) and so there is zero probability of finding the particle with finite kinetic energy ε_x at these locations. What would a plot of the probability (P_x) of locating a particle with kinetic energy ε_x as a function of position x look like? We know that P_x must be zero at $x = 0$ and $x = a$ and also that the integral of any probability distribution must be unity:

$$\int_{x=0}^{x=a} P_x \, dx = 1 \tag{15-11}$$

A probability function that satisfies these requirements is shown in Figure 15-2. This probability function could be viewed as a wave form with a wave length of 2 a. Many other functions could be proposed that also satisfy these requirements.

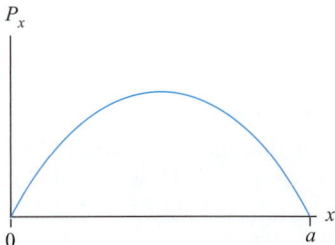

Figure 15-2: One possible probability function corresponding to the probability of finding a particle with kinetic energy ε_x at position x.

15.2.2 Application of a Wave Equation

Experimental evidence shows that particles exhibit wave-like behavior. With that in mind, Schrodinger in 1926 applied the classical differential equation for a wave to de Broglie's particle wave concept. You have likely encountered a wave equation in the description of a vibrating string for a one-dimensional wave in your physics course. This wave equation has the form:

$$\frac{\partial^2 A}{\partial x^2} = \frac{1}{\tilde{V}^2}\left(\frac{\partial^2 A}{\partial t^2}\right) \tag{15-12}$$

where A is the amplitude of the wave at axial position x and $\tilde{V}$ is the phase velocity, i.e., the velocity at which the crest of a wave appears to travel. We will solve Eq. (15-12) using a separation of variables technique in which the amplitude is assumed to be the product of a function of position $f(x)$ and a function of time $g(t)$.

$$A(x, t) = f(x)\, g(t) \tag{15-13}$$

Substituting Eq. (15-13) into Eq. (15-12) provides:

$$g \frac{d^2 f}{dx^2} = \frac{1}{\tilde{V}^2} f \left(\frac{d^2 g}{dt^2} \right) \tag{15-14}$$

Dividing Eq. (15-14) by the product $f\,g$ leads to:

$$\underbrace{\frac{1}{f} \frac{d^2 f}{dx^2}}_{\text{function of } x} = \underbrace{\frac{1}{g\,\tilde{V}^2} \left(\frac{d^2 g}{dt^2} \right)}_{\text{function of } t} \tag{15-15}$$

Notice that the left side of Eq. (15-15) is a function only of x while the right side is a function only of t. Since x and t are independent, the only possible solution is that both sides are equal to a constant, which is chosen here to be $-\alpha^2$:

$$\frac{1}{f} \frac{d^2 f}{dx^2} = \frac{1}{g\,\tilde{V}^2} \left(\frac{d^2 g}{dt^2} \right) = -\alpha^2 \tag{15-16}$$

Equation (15-16) therefore provides two ordinary differential equations:

$$\frac{d^2 f}{dx^2} + \alpha^2 f = 0 \tag{15-17}$$

$$\frac{d^2 g}{dt^2} + \alpha^2 \tilde{V}^2 g = 0 \tag{15-18}$$

The solution to Eq. (15-18) is:

$$g = C_1 \cos\left(\alpha\, \tilde{V}\, t\right) + C_2 \sin\left(\alpha\, \tilde{V}\, t\right) \tag{15-19}$$

where C_1 and C_2 are constants. Equation (15-19) indicates that g is a periodic function of time having a frequency (ν) of:

$$\nu = \frac{\alpha\, \tilde{V}}{2\pi} = \frac{\tilde{V}}{\lambda} \tag{15-20}$$

Solving Eq. (15-20) for the constant α provides:

$$\alpha = \frac{2\pi\, \nu}{\tilde{V}} = \frac{2\pi}{\lambda} \tag{15-21}$$

Substituting Eq. (15-21) into Eq. (15-17) leads to:

$$\frac{d^2 f}{dx^2} + \left(\frac{2\pi}{\lambda} \right)^2 f = 0 \tag{15-22}$$

Equation (15-22) is the equation describing the amplitude of a standing wave with wavelength λ. This equation is now applied to a particle moving in the x direction with velocity $\tilde{V}_x$ and kinetic energy defined by Eq. (15-10). The characteristic wavelength (λ) is given in Eq. (15-9):

$$\left(\frac{2\pi}{\lambda} \right)^2 = \left(\frac{2\pi\, m\, \tilde{V}_x}{h} \right)^2 = \frac{4\pi^2 m^2 \tilde{V}_x^2}{h^2} = \frac{8\pi^2 m\, \varepsilon_x}{h^2} \tag{15-23}$$

With these definitions, Eq. (15-22) can be written as:

$$\frac{d^2 f}{dx^2} + \left(\frac{8\pi^2 m}{h^2}\varepsilon_x\right) f = 0 \tag{15-24}$$

But what exactly is the function f? Schrodinger proposed that f is related to the probability of finding a particle with kinetic energy ε_x at position x. However, f can be positive or negative so the probability is assumed to be the product of f and its complex conjugate. If f is real then this product is simply f^2.

The general solution to Eq. (15-24) is given by:

$$f = C_1 \sin\left(x\sqrt{\frac{8\pi^2 m}{h^2}\varepsilon_x}\right) + C_2 \cos\left(x\sqrt{\frac{8\pi^2 m}{h^2}\varepsilon_x}\right) \tag{15-25}$$

The boundary conditions for the second-order differential equation, Eq. (15-24), are that $f = 0$ at $x = 0$ and also at $x = a$, as shown in Figure 15-2. Substituting the boundary condition that $f = 0$ at $x = 0$ into Eq. (15-25) provides:

$$0 = C_1 \underbrace{\sin\left(0\sqrt{\frac{8\pi^2 m}{h^2}\varepsilon_x}\right)}_{=0} + C_2 \underbrace{\cos\left(0\sqrt{\frac{8\pi^2 m}{h^2}\varepsilon_x}\right)}_{=1} \tag{15-26}$$

which can only be true if $C_2 = 0$.

$$f = C_1 \sin\left(x\sqrt{\frac{8\pi^2 m}{h^2}\varepsilon_x}\right) \tag{15-27}$$

The boundary condition $f = 0$ at $x = a$ requires that:

$$\sin\left(a\sqrt{\frac{8\pi^2 m}{h^2}\varepsilon_x}\right) = 0 \tag{15-28}$$

Equation (15-28) is the *eigencondition* for this problem, which has an infinite number of roots (*eigenvalues*) of the form:

$$a\sqrt{\frac{8\pi^2 m}{h^2}\varepsilon_x} = n_x \pi \tag{15-29}$$

where n_x is an integer that is called a *principle quantum number* in this context. Solving Eq. (15-29) for the kinetic energy provides:

$$\boxed{\varepsilon_x = \frac{1}{2} m \tilde{V}_x^2 = \frac{p_x^2}{2m} = \frac{h^2}{2m\lambda^2} = \left(\frac{n_x^2 h^2}{8ma^2}\right) \text{ for } n_x = 0, 1, 2, 3, \ldots.} \tag{15-30}$$

The discrete values of energy that Planck found necessary in his derivation of the behavior of electromagnetic radiation are understandable in the context that they are eigenvalues.

15.2.3 Degeneracy

Equation (15-30) only considers movement in the x direction. A particle in an enclosure can translate in three directions (x, y, and z) so that the total kinetic energy of a particle in one particular state is given by:

$$\varepsilon_i = \frac{p_x^2 + p_y^2 + p_z^2}{2m} = \left(\frac{h^2}{8m}\right)\left(\frac{n_x^2}{a^2} + \frac{n_y^2}{b^2} + \frac{n_z^2}{c^2}\right) \tag{15-31}$$

where a, b, and c are the spatial extents of the enclosure in the x, y, and z directions, respectively. Specification of quantum numbers in each direction, n_x, n_y, and n_z, determines the kinetic energy of the particle according to Eq. (15-31). If we knew the quantum numbers for every particle in the system then we would know the total energy of the system. This specification of the quantum numbers of every particle is referred to as a *quantum state* or a *microstate*.

Recall that the quantum numbers must be integers. According to Eq. (15-31), all particles having values of n_x, n_y, and n_z such that $n_x^2 + n_y^2 + n_z^2 = constant$ have the same energy, ε_i, for given dimensions a, b, and c. The number of unique combinations of quantum numbers that result in an energy level ε_i, is called the *degeneracy* of that energy level, g_i. For example, consider an enclosure with dimensions $a = b = c$; in this case, Eq. (15-31) can be written as:

$$\varepsilon_i = \frac{h^2}{8\,m\,a^2}\left(n_x^2 + n_y^2 + n_z^2\right) \tag{15-32}$$

One possible energy level for this system is $\varepsilon = h^2/\left(8\,m\,a^2\right)$ which corresponds to $n_x^2 + n_y^2 + n_z^2 = 1$. The degeneracy of this energy level is $g = 3$ since this energy level can be achieved with three combinations of n_x, n_y, and n_z; specifically, (1,0,0), (0,1,0), and (0,0,1). Another possible energy level for the system is $\varepsilon = 9\,h^2/\left(8\,m\,a^2\right)$, which corresponds to $n_x^2 + n_y^2 + n_z^2 = 9$. The degeneracy of this energy level is $g = 6$. The possible quantum number combinations that lead to this energy level are (3,0,0), (0,3,0), (0,0,3), (2,2,1), (2,1,2), (1,2,2). The degeneracy of the energy level $\varepsilon = 66\,h^2/\left(8\,m\,a^2\right)$ corresponding to $n_x^2 + n_y^2 + n_z^2 = 66$ is $g = 12$.

The magnitudes of the quantum numbers needed to represent the energy levels associated with ideal gases are large (on the order of $n_x^2 + n_y^2 + n_z^2 \approx 1 \times 10^{10}$). The degeneracy of an energy level tends to increase with increasing magnitude of $n_x^2 + n_y^2 + n_z^2$. Consequently, the degeneracy of most of the energy levels available to ideal gas particles will be very large. It can be shown that for ideal gases, the degeneracy of an energy level (g_i) will usually be much larger than the number of particles in the system that occupy that energy level (N_i). (See Problem 15-16.) Thus, it is unlikely that more than one ideal gas particle in the system will have the same set of quantum numbers at any instant of time.

15.3 The Equilibrium Distribution

Our focus in this chapter is on a monatomic ideal gas in an enclosure of fixed dimensions that is maintained at a constant temperature. From classical thermodynamics, we know that the internal energy of the ideal gas depends only on its mass and temperature. The internal energy of the gas in the enclosure is therefore fixed if the temperature is held constant. If we knew the quantum numbers of every gas particle in the system, we would then know its microstate. The sum of the energies of the individual particles would be the total internal energy of the gas. However, the number of particles in the enclosure

(N) is huge and therefore it is clear that we can never know the microstate of the system. Instead, we need to apply statistical analyses and probability theory in order to determine the most likely state of the particles.

On a macroscopic level, we observe the gas in the enclosure to be in an unchanging and unique equilibrium state at constant temperature and volume. However, on a microscopic level the situation is dynamic. Particles collide and exchange momentum and energy. As a result, the particles will have a distribution of velocities and an associated distribution of energy levels at any given time. We will try to determine the most likely distribution of energy levels as that is the distribution that is observed most of the time.

Quantum theory requires the energy levels assumed by the particles to be discrete. At any instant of time, there will be a distribution of energy levels. That is, there will be N_1 particles with energy level ε_1, N_2 particles with energy level ε_2, and so on. The total number of particles (N) is fixed:

$$\boxed{N = \sum_{i}^{\#\,\text{levels}} N_i = \text{constant}} \tag{15-33}$$

The total energy of all of the particles in the container is also fixed. Kinetic energy of translation is the only energy storage mode for monatomic ideal gas particles. Therefore, the total energy of the gas is the product of the number of particles in each energy level (N_i) and the energy level ε_i summed for all energy levels:

$$\boxed{U = \sum_{i}^{\#\,\text{levels}} N_i\,\varepsilon_i = \text{constant}} \tag{15-34}$$

15.3.1 Macrostates and Thermodynamic Probability

There are many possible ways for the particles to be arranged among the different possible energy levels that would result in a particular value of total energy, U, for the system. A specification of the number of particles in each energy level (N_1, N_2, ...) that satisfies Eqs. (15-33) and (15-34) is called a *macrostate*. For example, a possible macrostate would be one in which all of the particles except one are stationary and the one non-stationary particle is moving at a very high velocity such that the total energy of all of the particles is U. However, this macrostate does not have a high likelihood of occurrence. Our goal is to determine the most probable macrostate.

Each candidate macrostate can be realized with many different microstates. There are N_i particles in energy level ε_i and a large degeneracy is associated with the energy level (g_i). Therefore, there are many possible microstates just for this one energy level. A fundamental postulate of quantum theory referred to as the *equipartition principle*, states that every microstate having the same energy level has an equal likelihood of being occupied. However, the different macrostates do not have an equal likelihood of occurrence. The number of possible microstates that correspond to a particular macrostate is called the *thermodynamic probability*, Ω. The macrostate that has the largest value of Ω is the most probable macrostate, simply because it has the highest probability of occurrence. It will be shown that the thermodynamic probability of a macrostate is related to entropy.

To proceed, we need to use probability relations to determine the number of microstates (Ω) that correspond to a particular macrostate. The number of microstates is the number of different ways that the N particles can be arranged among the different energy levels, accounting for the degeneracy of each energy level. If the particles are distinguishable, the number of possible ways that the particles can be arranged is given by Eq. (15-35):

$$\Omega_{DIST} = \frac{N!}{\prod_{i=1}^{\#\text{ levels}} N_i!} \prod_{i=1}^{\#\text{ levels}} g_i^{N_i} = N! \prod_{i=1}^{\#\text{ levels}} \frac{g_i^{N_i}}{N_i!} \tag{15-35}$$

where the product sign Π indicates that the individual terms in the series are multiplied by one another. Note that the exclamation sign refers to the factorial operator, which is defined as:

$$N! = (N)(N-1)(N-2)\ldots.(1) \tag{15-36}$$

The monatomic ideal gas particles are not distinguishable to us. Therefore, the number of possible microstates is given by the Bose-Einstein model for indistinguishable particles:

$$\Omega_{BE} = \prod_{i=1}^{\#\text{ levels}} \frac{(g_i + N_i - 1)!}{(g_i - 1)!\ N_i!} \tag{15-37}$$

Finally, we could require that no more than one particle occupy any one microstate. In this case, the number of possible microstates is given by the Fermi-Dirac model:

$$\Omega_{FD} = \prod_{i=1}^{\#\text{levels}} \frac{g_i!}{(g_i - N_i)!\ N_i!} \tag{15-38}$$

To illustrate these probability relations, consider a situation in which there is only one energy level (# levels = 1) having a degeneracy of three ($g_1 = 3$) for a system that consists of two particles ($N_1 = 2$). Applying Eq. (15-35) to the system shows that there are Ω_{DIST} = 9 microstates:

$$\Omega_{DIST} = N! \prod_{i=1}^{\#\text{ levels}} \frac{g_i^{N_i}}{N_i!} = 2! \prod_{i=1}^{1} \frac{g_i^{N_i}}{N_i!} = 2! \frac{3^2}{2!} = 3^2 = 9 \tag{15-39}$$

There are very few microstates for this situation; we can identify every one of them, as indicated in Table 15-1. The two particles, which are distinguishable, are represented as **a** and **b**. The three possible ways that the energy level can be achieved (i.e., the degeneracy of the energy level is 3) are identified as D_1, D_2 and D_3.

If the particles are not distinguishable, then the number of microstates is given by the Bose-Einstein model, Eq. (15-37):

$$\Omega_{BE} = \prod_{i=1}^{\#\text{ levels}} \frac{(g_i + N_i - 1)!}{(g_i - 1)!\ N_i!} = \prod_{i=1}^{1} \frac{(g_i + N_i - 1)!}{(g_i - 1)!\ N_i!} = \frac{(3 + 2 - 1)!}{(3 - 1)!\ (2!)} = \frac{4!}{2!\ 2!} = 6 \tag{15-40}$$

The six possible indistinguishable microstates are arrangements 1 through 6 in Table 15-1. Note that microstates 7, 8, and 9 would be indistinguishable from microstates 4, 5, and 6 if particle **a** were indistinguishable from particle **b**. If we now restrict the microstates so that the particles are indistinguishable and also allow only one particle in any one microstate, then the number of possible microstates is given by the Fermi-Dirac model, Eq. (15-38):

$$\Omega_{FD} = \prod_{i=1}^{\#\text{levels}} \frac{g_i!}{(g_i - N_i)!\ N_i!} = \frac{3!}{(3-2)!\ 2!} = 3 \tag{15-41}$$

Table 15-1: Possible arrangements of 2 distinguishable particles in 1 energy level with a degeneracy of 3. The particles are designated as a and b and the three arrangements of quantum numbers that can provide the energy level are designated as D_1, D_2, and D_3

	Arrangement of quantum numbers		
Microstate	D_1	D_2	D_3
1	a, b		
2		a, b	
3			a, b
4	a	b	
5	a		b
6		a	b
7	b	a	
8	b		a
9		b	a

The three possible microstates associated with the Fermi-Dirac model would correspond to arrangements 4, 5, and 6 in Table 15-1.

For an ideal gas, the number of particles in any energy level is likely to be quite large and the degeneracy of an energy level is therefore much greater than the number of particles in the energy level. Under these conditions, the Bose-Einstein and Fermi-Dirac models, Eqs. (15-37) and (15-38), tend to provide nearly the same result. However, Eq. (15-35), which assumes distinguishable particles, provides a much larger value. For example, for a single energy level with five particles ($N_1 = 5$) and a degeneracy of 1000 ($g_1 = 1000$) the results of Eqs. (15-35), (15-37), and (15-38) are $\Omega_{MB} = 1.00 \times 10^{15}$, $\Omega_{BE} = 8.42 \times 10^{12}$, and $\Omega_{FD} = 8.25 \times 10^{12}$, respectively. In the limit that the degeneracy of the energy levels is much larger than the number of particles in the energy levels (i.e., $g_i \gg N_i$), both the Bose-Einstein and Fermi-Dirac models can be approximated according to:

$$\boxed{\Omega = \prod_{i}^{\#\,\text{levels}} \frac{g_i^{N_i}}{N_i!}} \tag{15-42}$$

The thermodynamic probability predicted by Eq. (15-42) is $\Omega = 8.33 \times 10^{12}$. The approximation associated with Eq. (15-42) improves as N_i and g_i increases. Equation (15-42) will be used in the following analyses because it is algebraically simpler than Eq. (15-37) or (15-38) and it provides essentially the same result as the Bose-Einstein and Fermi-Dirac models for ideal gases.

15.3.2 Identification of the Most Probable Macrostate

The identification of the most probable distribution of particles within the available discrete energy levels can be formulated as an optimization problem. We wish to identify the macrostate (i.e., the number of particles in each energy level) that maximizes the thermodynamic probability Ω subject to the constraints indicated in Eqs. (15-33) and (15-34). We assume that Ω for any macrostate is given by the approximate model for indistinguishable particles, Eq. (15-42).

Since Ω is a very large number, it is more convenient (and equivalent) to maximize the natural logarithm of the thermodynamic probability, $\ln(\Omega)$:

$$\ln(\Omega) = \ln\left(\prod_{i=1}^{\#\text{levels}} \frac{g_i^{N_i}}{N_i!}\right) = \ln\left(\frac{g_1^{N_1}}{N_1!}\right) + \ln\left(\frac{g_2^{N_2}}{N_2!}\right) + \ln\left(\frac{g_3^{N_3}}{N_3!}\right) + \ldots \quad (15\text{-}43)$$

The arguments of the natural logarithms in Eq. (15-43) can be expanded:

$$\ln(\Omega) = [N_1 \ln(g_1) - \ln(N_1!)] + [N_2 \ln(g_2) - \ln(N_2!)] + [N_3 \ln(g_3) - \ln(N_3!)] + \ldots \quad (15\text{-}44)$$

Equation (15-44) can be written in terms of summations:

$$\ln(\Omega) = \sum_{i=1}^{\#\text{ levels}} N_i \ln(g_i) - \sum_{i=1}^{\#\text{ levels}} \ln(N_i!) \quad (15\text{-}45)$$

Equation (15-45) can be simplified using Stirling's approximation, which is applicable for large arguments within a logarithm:

$$\boxed{\ln(x!) \approx [x \ln(x) - x] \text{ for large values of } x} \quad (15\text{-}46)$$

Substituting Eq. (15-46) into Eq. (15-45) provides:

$$\ln(\Omega) = \sum_{i=1}^{\#\text{ levels}} N_i \ln(g_i) - \sum_{i=1}^{\#\text{ levels}} [N_i \ln(N_i) - N_i] \quad (15\text{-}47)$$

Equation (15-47) can be simplified to:

$$\ln(\Omega) = \sum_{i=1}^{\#\text{ levels}} N_i \ln\left(\frac{g_i}{N_i}\right) + N \quad (15\text{-}48)$$

where N is the total number of particles. We need to maximize $\ln(\Omega)$ with respect to N_i subject to the constraints of fixed N and fixed U. The Lagrange Method of Undetermined Multipliers provides the easiest way to determine this maximum. The Lagrange Method of Undetermined Multipliers was introduced in Chapter 14 where it was used to minimize the Gibbs free energy of a chemically reactive system. The LaGrange Method can be summarized as follows. The extremum (minimum or maximum) of an arbitrary function F subject to one or more constraints, given by the equations ψ_j, is found by solving the set of algebraic equations that result from:

$$dF + \sum_{j=1}^{\#\text{constraints}} \lambda_j \, d\psi_j = 0 \quad (15\text{-}49)$$

where λ_j is the undetermined multiplier for constraint j.

In this case, the function to be maximized (F) is $\ln(\Omega)$ and the independent variables are N_i. The total derivative of $\ln(\Omega)$ is obtained by differentiating Eq. (15-48):

$$dF = d[\ln(\Omega)] = \sum_{i=1}^{\#\text{ levels}} d\left(N_i \ln\left(\frac{g_i}{N_i}\right)\right) + \underbrace{dN}_{0} \quad (15\text{-}50)$$

Applying the chain rule to the term within the summation in Eq. (15-50) allows it to be written as:

$$dF = d[\ln(\Omega)] = \sum_{i=1}^{\#\,\text{levels}} N_i\, d\ln\left(\frac{g_i}{N_i}\right) + \sum_{i=1}^{\#\,\text{levels}} \ln\left(\frac{g_i}{N_i}\right) dN_i \tag{15-51}$$

Equation (15-51) can be further reduced by recognizing that:

$$d\ln\left(\frac{g_i}{N_i}\right) = \left(-\frac{N_i}{g_i}\right)\left(\frac{g_i}{N_i^2}\right) dN_i = -\frac{dN_i}{N_i} \tag{15-52}$$

Substituting Eq. (15-52) into Eq. (15-51) produces:

$$dF = d[\ln(\Omega)] = -\underbrace{\sum_{i=1}^{\#\,\text{levels}} dN_i}_{=0} + \sum_{i=1}^{\#\,\text{levels}} \ln\left(\frac{g_i}{N_i}\right) dN_i \tag{15-53}$$

Recognizing that $\sum_{i}^{\#\,\text{levels}} dN_i = 0$ eliminates the first term on the right side of Eq. (15-53):

$$dF = d[\ln(\Omega)] = \sum_{i=1}^{\#\,\text{levels}} \ln\left(\frac{g_i}{N_i}\right) dN_i \tag{15-54}$$

The first constraint on the optimization is that the number of particles in the enclosure is fixed:

$$\psi_1 = \sum_{i=1}^{\#\,\text{levels}} N_i - N = 0 \tag{15-55}$$

The derivative of this constraint is:

$$d\psi_1 = \sum_{i}^{\#\,\text{levels}} dN_i = 0 \tag{15-56}$$

The second constraint is that the energy of the particles is fixed:

$$\psi_2 = \sum_{i}^{\#\,\text{levels}} N_i\, \varepsilon_i - U = 0 \tag{15-57}$$

The derivative of this constraint is:

$$d\psi_2 = \sum_{i=1}^{\#\,\text{levels}} \varepsilon_i\, dN_i + \sum_{i=1}^{\#\,\text{levels}} N_i \left(\frac{\partial \varepsilon_i}{\partial N_i}\right)_{N_j, j\neq i} dN_i = 0 \tag{15-58}$$

Note that the energy per particle associated with each available energy level, ε_i, is independent of the number of particles that occupy the energy level; therefore:

$$\left(\frac{\partial \varepsilon_i}{\partial N_i}\right) = 0 \tag{15-59}$$

Substituting Eq. (15-59) into Eq. (15-58) provides:

$$d\psi_2 = \sum_{i=1}^{\#\,\text{levels}} \varepsilon_i\, dN_i = 0 \tag{15-60}$$

Substituting Eqs. (15-54), (15-55), and (15-60) into Eq. (15-49) results in:

$$\sum_{i=1}^{\#\text{ levels}} \ln\left(\frac{g_i}{N_i}\right) dN_i + \lambda_1 \left(\sum_{i}^{\#\text{ levels}} dN_i\right) + \lambda_2 \left(\sum_{i=1}^{\#\text{ levels}} \varepsilon_i\, dN_i\right) = 0 \tag{15-61}$$

The summations in Eq. (15-61) can be combined in order to provide:

$$\sum_{i=1}^{\#\text{ levels}} \left[\ln\left(\frac{g_i}{N_i}\right) + \lambda_1 + \lambda_2\, \varepsilon_i\right] dN_i = 0 \tag{15-62}$$

Since the number of particles contained in each energy level are independent variables, Equation (15-62) can only be true if each term in the summation is zero:

$$\ln\left(\frac{g_i}{N_i}\right) + \lambda_1 + \lambda_2\, \varepsilon_i = 0 \quad \text{for } i = 1 \ldots \#\text{ levels} \tag{15-63}$$

The multipliers, λ_1 and λ_2, are typically renamed according to:

$$\lambda_1 = \ln(B) \tag{15-64}$$

$$\lambda_2 = -\beta \tag{15-65}$$

where B and β are simply undetermined values at this point, in the same way that the Lagrange multipliers λ_1 and λ_2 were undetermined values in Chapter 14. The significance of these variables will become evident in the following sections. Substituting Eqs. (15-64) and (15-65) into Eq. (15-63) results in:

$$\ln\left(\frac{g_i}{N_i}\right) + \ln(B) - \beta\, \varepsilon_i = 0 \quad \text{for } i = 1 \ldots \#\text{ levels} \tag{15-66}$$

Solving Eq. (15-66) for N_i provides an important result:

$$\boxed{N_i = B\, g_i \exp(-\beta\, \varepsilon_i) \quad \text{for } i = 1 \ldots \#\text{ levels}} \tag{15-67}$$

Equation (15-67) is called the Maxwell-Boltzmann distribution and it defines the most probable macrostate, i.e., the macrostate that has the highest number of possible microstates.

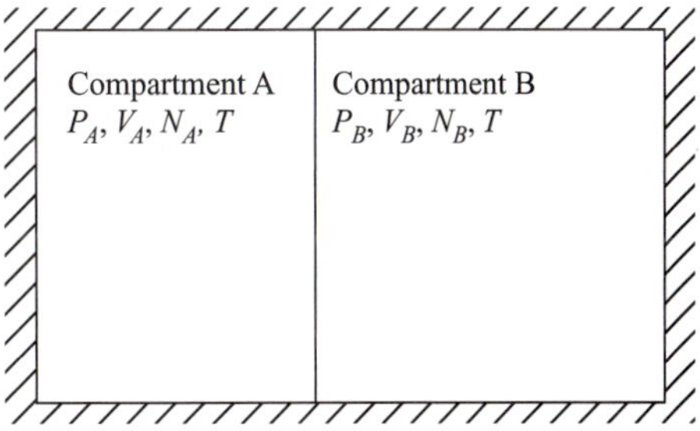

Figure 15-3: System consisting of two compartments at the same temperature; each compartment contains a monatomic ideal gas.

15.3.3 The Significance of β

A system consists of two compartments each containing a monatomic ideal gas, as shown in Figure 15-3. The pressure, volume, and number of particles in the two compartments are different. However, heat transfer can occur between the particles in compartment A and those in compartment B; therefore, the temperature of the two gases is the same after a steady-state condition has been achieved.

The particles in compartment A are independent of the particles in compartment B. The particles in each compartment are arranged within the available energy levels so that the thermodynamic probability of compartment A is Ω_A and the thermodynamic probability of compartment B is Ω_B. The thermodynamic probability of both compartments is the product of the thermodynamic probabilities of the two compartments:

$$\Omega = \Omega_A \, \Omega_B \tag{15-68}$$

or, equivalently:

$$\ln(\Omega) = \ln(\Omega_A) + \ln(\Omega_B) \tag{15-69}$$

The most likely distribution of the particles in compartments A and B is found by maximizing the thermodynamic probability of both compartments, ln(Ω), subject to the constraints that the number of particles in each compartment and the total energy of the entire system (i.e., the particles in both compartments) are constant; these constraints are shown algebraically below:

$$\psi_1 = \sum_{i=1}^{\#\,\text{levels}} N_{A,i} - N_A = 0 \tag{15-70}$$

$$\psi_2 = \sum_{i=1}^{\#\,\text{levels}} N_{B,i} - N_B = 0 \tag{15-71}$$

$$\psi_3 = \sum_{i=1}^{\#\,\text{levels}} \varepsilon_{A,i} \, N_{A,i} + \sum_{i=1}^{\#\,\text{levels}} \varepsilon_{B,i} \, N_{B,i} - U = 0 \tag{15-72}$$

The most likely distribution is found by applying the Lagrange Method of Undetermined Multipliers in order to maximize ln(Ω) subject to the three constraints provided by Eqs. (15-70) to (15-72). Applying Eq. (15-54) to compartments A and B provides:

$$d\left[\ln(\Omega_A)\right] = \sum_{i=1}^{\#\,\text{levels}} \ln\left(\frac{g_{A,i}}{N_{A,i}}\right) dN_{A,i} \tag{15-73}$$

$$d\left[\ln(\Omega_B)\right] = \sum_{i=1}^{\#\,\text{levels}} \ln\left(\frac{g_{B,i}}{N_{B,i}}\right) dN_{B,i} \tag{15-74}$$

Substituting Eqs. (15-73) and (15-74) into the total derivative of Eq. (15-69) leads to:

$$d\left[\ln(\Omega)\right] = d\left[\ln(\Omega_A)\right] + d\left[\ln(\Omega_B)\right] = \sum_{i=1}^{\#\,\text{levels}} \ln\left(\frac{g_{A,i}}{N_{A,i}}\right) dN_{A,i} + \sum_{i=1}^{\#\,\text{levels}} \ln\left(\frac{g_{B,i}}{N_{B,i}}\right) dN_{B,i} \tag{15-75}$$

The derivatives of the three constraint equations, Eqs. (15-70) through (15-72), are

$$d\psi_1 = \sum_{i=1}^{\#\,\text{levels}} dN_{A,i} = 0 \tag{15-76}$$

$$d\psi_2 = \sum_{i=1}^{\#\,\text{levels}} dN_{B,i} = 0 \tag{15-77}$$

$$d\psi_3 = \sum_{i=1}^{\#\text{ levels}} \left[\varepsilon_{A,i}\, dN_{A,i} + \varepsilon_{B,i}\, dN_{B,i}\right] = 0 \tag{15-78}$$

The Lagrange Method of Undetermined Multipliers sums the differential of the quantity to be optimized, Eq. (15-75), and the differentials of the constraint equations, Eqs. (15-76) through (15-78), each multiplied by a undetermined multiplier. Following Section 15.3.2, the multipliers are given the symbols $\ln(B_A)$, $\ln(B_B)$, and $-\beta$, respectively. The result is:

$$\sum_{i=1}^{\#\text{ levels}} \ln\left(\frac{g_{A,i}}{N_{A,i}}\right) dN_{A,i} + \sum_{i=1}^{\#\text{ levels}} \ln\left(\frac{g_{B,i}}{N_{B,i}}\right) dN_{B,i} + \ln(B_A) \sum_{i=1}^{\#\text{ levels}} dN_{A,i}$$
$$+ \ln(B_B) \sum_{i=1}^{\#\text{ levels}} dN_{B,i} - \beta \sum_{i=1}^{\#\text{ levels}} \left[\varepsilon_{A,i}\, dN_{A,i} + \varepsilon_{B,i}\, dN_{B,i}\right] = 0 \tag{15-79}$$

Equation (15-79) can be written as:

$$\sum_{i=1}^{\#\text{ levels}} \left[\ln\left(\frac{g_{A,i}}{N_{A,i}}\right) + \ln(B_A) - \beta\,\varepsilon_{A,i}\right] dN_{A,i}$$
$$+ \sum_{i=1}^{\#\text{ levels}} \left[\ln\left(\frac{g_{B,i}}{N_{B,i}}\right) + \ln(B_B) - \beta\,\varepsilon_{B,i}\right] dN_{B,i} = 0 \tag{15-80}$$

Setting the coefficients for $dN_{A,i}$ and $dN_{B,i}$ equal to zero results in:

$$N_{A,i} = B_A\, g_{A,i} \exp(-\beta\,\varepsilon_{A,i}) \tag{15-81}$$

$$N_{B,i} = B_B\, g_{B,i} \exp(-\beta\,\varepsilon_{B,i}) \tag{15-82}$$

The only property shared by the gases in compartments A and B is the temperature. Therefore, the undetermined multiplier, β, that appears in both Eqs. (15-81) and (15-82) must be related to temperature as it is the only variable in the analysis that is common to both compartments. The exact relationship between β and temperature is identified in the following section.

15.3.4 Boltzmann's Law

From classical thermodynamics, we know that any spontaneous change in the state of a system is accompanied by an increase in entropy. At equilibrium, the entropy of an isolated system reaches a maximum. The entropy of the system in Figure 15-3 is given by:

$$S = S_A + S_B \tag{15-83}$$

where S_A and S_B are the entropy of the particles in compartments A and B, respectively. On a microscopic scale, the equilibrium state has the largest thermodynamic probability. The thermodynamic probability of the system in Figure 15-3 is the product of the thermodynamic probabilities of the two compartments, as indicated in Eq. (15-68). Thus, there is an apparent relationship between the entropy and the thermodynamic probability of a system that must be of the form:

$$\boxed{S = k \ln(\Omega)} \tag{15-84}$$

where k is Boltzmann's constant, which will be shown to be related to the ideal gas constant. Equation (15-84) is Boltzmann's Law and is one of the most famous equations of physical science. This equation appears on Boltzmann's tombstone.

The fundamental property relation from classical thermodynamics, Eq. (10-76), is:

$$dU = T\,dS - P\,dV \tag{15-85}$$

Solving Eq. (15-85) for temperature provides,

$$T = \left(\frac{\partial U}{\partial S}\right)_V \tag{15-86}$$

Temperature is a macroscopic property. However, U and S can be found from statistical thermodynamics; U is defined by:

$$U = \sum_{i=1}^{\#\,\text{levels}} \varepsilon_i\, N_i \tag{15-87}$$

and S is related to the thermodynamic probability, Ω, according to Eq. (15-84). Equation (15-86) provides a link between classical and statistical thermodynamics. The results that we have already derived can be used to determine the nature of this link.

The total derivative of the thermodynamic probability at a fixed volume is given by Eq. (15-54):

$$d\ln(\Omega) = \sum_{i=1}^{\#\,\text{levels}} \ln\left(\frac{g_i}{N_i}\right) dN_i \tag{15-88}$$

The Maxwell-Boltzmann distribution that maximizes the thermodynamic probability is given by Eq. (15-67):

$$N_i = B\,g_i \exp(-\beta\,\varepsilon_i) \quad \text{for } i = 1 \ldots \#\text{ levels} \tag{15-89}$$

Equation (15-89) can be rearranged:

$$\ln\left(\frac{g_i}{N_i}\right) + \ln(B) - \beta\,\varepsilon_i = 0 \quad \text{for } i = 1 \ldots \#\text{ levels} \tag{15-90}$$

Substituting Eq. (15-90) into Eq. (15-88) allows the total derivative of $\ln(\Omega)$ to be written as:

$$d\ln(\Omega) = \sum_{i=1}^{\#\,\text{levels}} [\beta\,\varepsilon_i - \ln(B)]\,dN_i \tag{15-91}$$

The first term in the summation in Eq. (15-91) is related to the total derivative of internal energy dU at a fixed volume. Note that the energy levels depend on the volume as indicated in Eq. (15-31), but the enclosure dimensions are constant. The derivative of the total internal energy, Eq. (15-87), is:

$$dU_V = \sum_{i=1}^{\#\,\text{levels}} \varepsilon_i\,dN_i + \sum_{i=1}^{\#\,\text{levels}} N_i \underbrace{d\varepsilon_i}_{=0} \tag{15-92}$$

The energy levels have fixed values in a enclosure of fixed dimensions; therefore $d\varepsilon_i = 0$ and Eq. (15-92) can be reduced to:

$$dU_V = \sum_{i=1}^{\#\,\text{levels}} \varepsilon_i\,dN_i \tag{15-93}$$

Substituting Eq. (15-93) into Eq. (15-91) leads to:

$$d\ln(\Omega) = \beta\, dU_V - \ln(B) \underbrace{\sum_{i=1}^{\#\text{ levels}} dN_i}_{0} \tag{15-94}$$

The last term in Eq. (15-94) must be zero according to the constraint that the number of particles is fixed.

$$d\ln(\Omega) = \beta\, dU_V \tag{15-95}$$

Solving Boltzmann's Law, Eq. (15-84), for the natural logarithm of the thermodynamic probability:

$$\ln(\Omega) = \frac{S}{k} \tag{15-96}$$

and taking the derivative provides:

$$d\ln(\Omega) = \frac{1}{k} dS \tag{15-97}$$

Equating Eqs. (15-94) and (15-97):

$$\beta\, dU_V = \frac{1}{k} dS \tag{15-98}$$

Equation (15-98) can be rearranged to provide:

$$\left(\frac{\partial U}{\partial S}\right)_V = \frac{1}{\beta\, k} \tag{15-99}$$

Comparing Eq. (15-99) with Eq. (15-86) provides:

$$\left(\frac{\partial U}{\partial S}\right)_V = \frac{1}{\beta\, k} = T \tag{15-100}$$

Therefore:

$$\boxed{\beta = \frac{1}{k\, T}} \tag{15-101}$$

15.4 Properties and the Partition Function

15.4.1 Definition of the Partition Function

Summing the number of particles for all energy levels results in:

$$N = \sum_{i=1}^{\#\text{ levels}} N_i \tag{15-102}$$

The Maxwell-Boltzmann distribution in Eq. (15-67) provides the number of particles in each energy level for the most probable macrostate. Substituting Eq. (15-67) into Eq. (15-102) provides:

$$N = B \sum_{i=1}^{\#\text{ levels}} g_i \exp(-\beta\, \varepsilon_i) \tag{15-103}$$

The multiplier β is related to temperature according to Eq. (15-101):

$$N = B \sum_{i=1}^{\#\,\text{levels}} g_i \exp\left(-\frac{\varepsilon_i}{k\,T}\right) \tag{15-104}$$

The *partition function*, *f*, is defined as:

$$f = \sum_{i=1}^{\#\,\text{levels}} g_i \exp\left(-\frac{\varepsilon_i}{k\,T}\right) \tag{15-105}$$

Substituting Eq. (15-105) into Eq. (15-104) provides:

$$B = \frac{N}{f} \tag{15-106}$$

Substituting Eq. (15-106) into Eq. (15-67) provides:

$$N_i = \frac{N}{f} g_i \exp\left(-\frac{\varepsilon_i}{k\,T}\right) \tag{15-107}$$

Knowledge of the partition function as a function of temperature and volume provides a complete equation of state. All thermodynamic property information is contained in the partition function. The partition function, $f = f(T, V)$, is therefore equivalent to the Helmholtz free energy function, $A = A(T, V)$, defined on a specific basis in Eq. (10-80). However, there is no way of determining the Helmholtz free energy as a function of temperature and volume without empirical data in the form of an equation of state relating pressure, volume and specific heat capacity as a function of temperature. Statistical thermodynamics provides a means for determining the partition function as a function of temperature and volume from basic assumptions concerning the behavior of particles. All of the thermodynamic properties can be determined using the partition function, as shown in the subsequent sections.

15.4.2 Internal Energy from the Partition Function

From a microscopic viewpoint, the internal energy is given by:

$$U = \sum_{i=1}^{\#\,\text{levels}} N_i\, \varepsilon_i \tag{15-108}$$

We would like to determine the internal energy based on the partition function provided as a function of T and V. The partition function is defined in Eq. (15-105).

$$f = \sum_{i=1}^{\#\,\text{levels}} g_i \exp\left(-\frac{\varepsilon_i}{k\,T}\right) \tag{15-109}$$

The partial derivative of the partition function with respect to temperature at constant volume is given by:

$$\left(\frac{\partial f}{\partial T}\right)_V = \frac{1}{k\,T^2} \sum_{i=1}^{\#\,\text{levels}} g_i\, \varepsilon_i \exp\left(-\frac{\varepsilon_i}{k\,T}\right) \tag{15-110}$$

Equation (15-107) can be solved for the degeneracy, g_i:

$$g_i = \frac{f\,N_i}{N} \exp\left(\frac{\varepsilon_i}{k\,T}\right) \tag{15-111}$$

and substituted into Eq. (15-110):

$$\left(\frac{\partial f}{\partial T}\right)_V = \frac{1}{k\,T^2} \sum_{i=1}^{\#\,\text{levels}} \left[\frac{f\,N_i}{N} \exp\left(\frac{\varepsilon_i}{k\,T}\right)\right] \varepsilon_i \exp\left(-\frac{\varepsilon_i}{k\,T}\right) \tag{15-112}$$

Equation (15-112) can be simplified:

$$\left(\frac{\partial f}{\partial T}\right)_V = \frac{f}{N\,k\,T^2} \underbrace{\sum_{i=1}^{\#\,\text{levels}} N_i\,\varepsilon_i}_{U} \tag{15-113}$$

Substituting Eq. (15-108) into Eq. (15-113) provides:

$$\left(\frac{\partial f}{\partial T}\right)_V = \frac{f\,U}{N\,k\,T^2} \tag{15-114}$$

Rearrangement of Eq. (15-114) provides the relation for the internal energy, U, in terms of the partition function:

$$U = \frac{N\,k\,T^2}{f}\left(\frac{\partial f}{\partial T}\right)_V \tag{15-115}$$

or

$$\boxed{U = N\,k\,T^2\left[\frac{\partial \ln(f)}{\partial T}\right]_V} \tag{15-116}$$

15.4.3 Entropy from the Partition Function

The macroscopic property entropy is related to microscopic properties through Boltzmann's Law, Eq. (15-84):

$$S = k \ln(\Omega) \tag{15-117}$$

A relation for $\ln(\Omega)$ was derived in Eq. (15-48).

$$\ln(\Omega) = \sum_{i=1}^{\#\,\text{levels}} N_i \ln\left(\frac{g_i}{N_i}\right) + N \tag{15-118}$$

The degeneracy, g_i, given in Eq. (15-111) can be substituted into Eq. (15-118) to yield:

$$\ln(\Omega) = \sum_{i=1}^{\#\,\text{levels}} N_i \ln\left[\frac{1}{N_i}\frac{f\,N_i}{N} \exp\left(\frac{\varepsilon_i}{k\,T}\right)\right] + N \tag{15-119}$$

Equation (15-119) is simplified to provide:

$$\ln(\Omega) = \sum_{i=1}^{\#\,\text{levels}} N_i \left[\ln\left(\frac{f}{N}\right) + \frac{\varepsilon_i}{k\,T}\right] + N \tag{15-120}$$

Separating the terms in the summation leads to:

$$\ln(\Omega) = \ln\left(\frac{f}{N}\right) \underbrace{\sum_{i=1}^{\#\,\text{levels}} N_i}_{N} + \frac{1}{k\,T} \underbrace{\sum_{i=1}^{\#\,\text{levels}} N_i\,\varepsilon_i}_{U} + N \tag{15-121}$$

or:

$$\ln(\Omega) = N \ln\left(\frac{f}{N}\right) + \frac{U}{k\,T} + N \tag{15-122}$$

Substituting Eq. (15-122) into Boltzmann's Law, Eq. (15-117), provides:

$$S = k\,N \ln\left(\frac{f}{N}\right) + \frac{U}{T} + k\,N \tag{15-123}$$

Substituting Eq. (15-116) into Eq. (15-123) provides the entropy in terms of the partition function:

$$S = k\,N \ln\left(\frac{f}{N}\right) + N k\,T \left[\frac{\partial \ln(f)}{\partial T}\right]_V + k\,N \tag{15-124}$$

Equation (15-124) can be written as:

$$\boxed{S = k\,N\left[\ln\left(\frac{f}{N}\right) + T\left(\frac{\partial \ln(f)}{\partial T}\right)_V + 1\right]} \tag{15-125}$$

15.4.4 Pressure from the Partition Function

Pressure, like temperature, is a macroscopic property. It is related to other thermodynamic properties by the fundamental property relation. The fundamental property relation resulting from the Helmholtz free energy ($A = U - T\,S$) is shown in Eq. (10-83) to be:

$$d A = dU - T\,dS - S\,dT = -S\,dT - P\,d\,V \tag{15-126}$$

From Eq. (15-126), pressure is the partial derivative of Helmholtz free energy with respect to volume at constant temperature:

$$P = -\left(\frac{\partial A}{\partial V}\right)_T = -\left[\frac{\partial\,(U - T\,S)}{\partial V}\right]_T \tag{15-127}$$

The relation for S in Eq. (15-123) can be substituted into Eq. (15-127):

$$P = -\left[\frac{\partial\left(U - T\left[k\,N \ln\left(\frac{f}{N}\right) + \frac{U}{T} + k\,N\right]\right)}{\partial V}\right]_T \tag{15-128}$$

Carrying out the differentiation in Eq. (15-128) leads to:

$$P = -\left(\frac{\partial U}{\partial V}\right)_T + T k\,N\left(\frac{\partial \ln f}{\partial V}\right)_T + \left(\frac{\partial U}{\partial V}\right)_T \tag{15-129}$$

which can be written as:

$$\boxed{P = T k\,N\left[\frac{\partial \ln(f)}{\partial V}\right]_T} \tag{15-130}$$

Knowing the partition function as a function of T and V has provided equations for U, S and P. All other thermodynamic properties can now be determined. Enthalpy, for example, is given by:

$$H = U + P\,V \tag{15-131}$$

15.5 Partition Function for an Monatomic Ideal Gas

It is shown in Section 15.4 that all thermodynamic properties can be determined if the partition function is provided as a function of temperature and volume. In this section, a partition function is derived for an ideal monatomic gas. The partition function is defined in Eq. (15-105) to be:

$$f = \sum_{i=1}^{\#\,\text{levels}} g_i \exp\left(-\frac{\varepsilon_i}{k\,T}\right) \tag{15-132}$$

The summation in Eq. (15-132) is over all discrete energy levels. However, the number of possible microstates for energy level i is the degeneracy of that energy level, g_i. Equation (15-132) can be equivalently expressed as a summation over all possible microstates:

$$f = \sum_{j=1}^{\#\text{microstates}} \exp\left(-\frac{\varepsilon_j}{k\,T}\right) \tag{15-133}$$

A monatomic ideal gas particle can only have kinetic energy associated with translation. A relation for the kinetic energy of a monatomic ideal gas particle was established in Eq. (15-31) in terms of the quantum numbers of the particle.

$$\varepsilon_i = \frac{p_x^2 + p_y^2 + p_z^2}{2m} = \left(\frac{h^2}{8m}\right)\left(\frac{n_x^2}{a^2} + \frac{n_y^2}{b^2} + \frac{n_z^2}{c^2}\right) \tag{15-134}$$

Equation (15-134) can be substituted into the partition function definition in Eq. (15-133) to obtain a summation over all quantum numbers:

$$f = \sum_{n_x}\sum_{n_y}\sum_{n_z} \exp\left[\frac{-h^2}{8m\,k\,T}\left(\frac{n_x^2}{a^2} + \frac{n_y^2}{b^2} + \frac{n_z^2}{c^2}\right)\right] \tag{15-135}$$

The triple summation in Eq. (15-135) can be broken into the product of three summations:

$$f = \sum_{n_x} \exp\left(\frac{-h^2\,n_x^2}{8m\,k\,T\,a^2}\right) \sum_{n_y} \exp\left(\frac{-h^2\,n_y^2}{8m\,k\,T\,b^2}\right) \sum_{n_z} \exp\left(\frac{-h^2\,n_z^2}{8m\,k\,T\,c^2}\right) \tag{15-136}$$

The quantum numbers (n_x, n_y, and n_z) for an ideal gas can be very large. The upper bound on the summations in Eq. (15-136) can be assumed to be essentially infinite since the value of each of the exponentials limits to zero as the quantum number becomes large; therefore, an exact determination of the upper bound on each of the summations is not a concern.

A numerical approximation of an integral can be provided by a summation:

$$\int F(n)\,dn = \sum F(n_i)\,\Delta n_i \tag{15-137}$$

We are going to reverse this process here and approximate the summations in Eq. (15-137) as integrals. For example, the first summation in Eq. (15-136) can be approximated as:

$$\sum_{n_x} \exp\left(\frac{-h^2\,n_x^2}{8m\,k\,T\,a^2}\right) \approx \int_0^{\infty} \exp\left(\frac{-h^2\,n_x^2}{8m\,k\,T\,a^2}\right) dn_x \tag{15-138}$$

Since the quantum numbers are very large, a change in a quantum number by 1 results in a very small change in energy so the integral approximation in Eq. (15-138) is nearly perfect in this case. Substituting Eq. (15-138) into Eq. (15-136) allows it to be approximated as:

$$f \approx \left[\int_0^\infty \exp\left(-\frac{h^2 n_x^2}{8 m k T a^2}\right) dn_x\right]\left[\int_0^\infty \exp\left(-\frac{h^2 n_y^2}{8 m k T b^2}\right) dn_y\right] \times \left[\int_0^\infty \exp\left(-\frac{h^2 n_z^2}{8 m k T c^2}\right) dn_z\right] \tag{15-139}$$

Each of the three integrals in Eq. (15-139) has an analytical solution of the form:

$$\int_0^\infty \exp(-c x^2)\, dx = \frac{1}{2}\sqrt{\frac{\pi}{c}} \tag{15-140}$$

Substituting Eq. (15-140) into Eq. (15-139) results in:

$$f = \left[\frac{a}{2}\sqrt{\frac{8\pi m k T}{h^2}}\right]\left[\frac{b}{2}\sqrt{\frac{8\pi m k T}{h^2}}\right]\left[\frac{c}{2}\sqrt{\frac{8\pi m k T}{h^2}}\right] \tag{15-141}$$

The product of dimensions a, b, and c is the volume (V) of the enclosure. Therefore, the partition function for the monatomic gas is given by:

$$\boxed{f = V\left(\frac{2\pi m k T}{h^2}\right)^{3/2}} \tag{15-142}$$

15.5.1 Pressure for a Monatomic Ideal Gas

Equation (15-130) provides a general relation for the pressure in terms of the partition function:

$$P = k T N \left[\frac{\partial \ln(f)}{\partial V}\right]_T \tag{15-143}$$

The natural logarithm of the partition function for a monatomic ideal gas, given by Eq. (15-142), is:

$$\ln(f) = \ln(V) + \frac{3}{2}\ln(T) + \frac{3}{2}\ln\left(\frac{2\pi m k}{h^2}\right) \tag{15-144}$$

The partial derivative of ln(f) with respect to volume at constant temperature is:

$$\left[\frac{\partial \ln(f)}{\partial V}\right]_T = \frac{1}{V} \tag{15-145}$$

Substituting Eq. (15-145) into Eq. (15-143) shows that

$$P = \frac{k T N}{V} \tag{15-146}$$

The form of Eq. (15-146) agrees with the ideal gas law:

$$P = \frac{R_{univ}\, T n}{V} \tag{15-147}$$

where n is the number of moles of gas. Equations (15-146) and (15-147) show that:

$$k N = n\, R_{univ} \tag{15-148}$$

The ratio of the number of molecules of gas to the number of moles is Avogadro's number.

$$N_A = \frac{N}{n} = 6.023 \times 10^{26} \text{ atoms/kmol} \tag{15-149}$$

Therefore, Boltzmann's constant is the ratio of the universal gas constant to Avogadro's number:

$$\boxed{k = \frac{R_{univ}}{N_A} = 1.3805 \times 10^{-23} \frac{\text{J}}{\text{K}}} \tag{15-150}$$

15.5.2 Internal Energy for a Monatomic Ideal Gas

Equation (15-116) provides the internal energy of a system in terms of the partition function:

$$U = N\, k\, T^2 \left[\frac{\partial \ln(f)}{\partial T}\right]_V \tag{15-151}$$

The natural logarithm of the partition function for a monatomic ideal gas is provided in Eq. (15-144). The partial derivative of ln(*f*) with respect to temperature at constant volume is:

$$\left[\frac{\partial \ln(f)}{\partial T}\right]_V = \frac{3}{2\,T} \tag{15-152}$$

Substituting Eq. (15-152) into Eq. (15-151) provides the internal energy of a monatomic gas:

$$U = \frac{3\, N\, k\, T}{2} \tag{15-153}$$

Substituting Eq. (15-148) into Eq. (15-153) leads to:

$$U = \frac{3}{2}\, R_{univ}\, n\, T \tag{15-154}$$

Equation (15-154) indicates that the molar specific heat capacity at constant volume ($\bar{c}_v$) for a monatomic ideal gas is given by:

$$\bar{c}_v = \left(\frac{\partial \bar{u}}{\partial T}\right)_V = \frac{3}{2} N_A\, k = \frac{3}{2} R_{univ} \tag{15-155}$$

Equation (15-155) agrees with experimental data for the specific heat capacity of monatomic ideal gases. For example, the constant volume specific heat capacity of helium at 300 K and 1 kPa is $\bar{c}_v = 12{,}471$ J/kmol-K. This value is exactly equal to 3/2 R_{univ}. Statistical thermodynamics provides an alternative to measurement for determining specific heat capacity values.

15.5.3 Entropy for a Monatomic Ideal Gas

Equation (15-125) provides the entropy of a system in terms of the partition function:

$$S = k\, N\left[\ln\left(\frac{f}{N}\right) + T\left(\frac{\partial \ln(f)}{\partial T}\right)_V + 1\right] \tag{15-156}$$

The natural logarithm of the partition function and its partial derivative with respect to temperature for a monatomic ideal gas are provided in Eqs. (15-144) and (15-152), respectively. Substituting these two equations into Eq. (15-156) produces:

$$S = k\,N\left[\ln\left(\frac{V}{N}\right) + \frac{3}{2}\ln(T) + \frac{3}{2}\ln\left(\frac{2\,\pi\,m\,k}{h^2}\right) + T\left(\frac{3}{2\,T}\right) + 1\right] \tag{15-157}$$

Equation (15-157) can be simplified:

$$S = k\,N\left[\ln\left(\frac{k\,T}{P}\right) + \ln\left(\left[\frac{2\,\pi\,m\,k\,T}{h^2}\right]^{3/2}\right) + \frac{5}{2}\right] \tag{15-158}$$

The product $k\,N$ is $n\,R_{univ}$, from Eq. (15-148), where n is the number of moles of gas. With this substitution, the molar specific entropy ($\bar{s}$) of a monatomic ideal gas is given by:

$$\boxed{\bar{s} = R_{univ}\left[\ln\left(\frac{k\,T}{P}\right) + \ln\left(\left[\frac{2\,\pi\,m\,k\,T}{h^2}\right]^{3/2}\right) + \frac{5}{2}\right]} \tag{15-159}$$

Equation (15-159) is known as the *Sackur-Tetrode equation* and it is a particularly impressive result of statistical thermodynamics. The specific entropy value obtained from Eq. (15-159) is the absolute entropy, with a reference value of zero at 0 K consistent with the Third Law of Thermodynamics. As explained in Section 13.4, the alternative method for evaluating the Third Law entropy is to solve Eq. (13-69), repeated below,

$$\bar{s}(T, P = 1\,\text{atm}) = \underbrace{\bar{s}\,(T = 0\,\text{K}, P = 1\,\text{atm})}_{=0\text{ by Third Law of Thermodynamics}} + \underbrace{\int_{0K}^{T_m}\frac{\bar{c}_{P,s}}{T}dT}_{\substack{\text{entropy change}\\\text{of solid with }T}} + \underbrace{\frac{\Delta\bar{h}_{fs}}{T_m}}_{\substack{\text{entropy change}\\\text{due to melting}}}$$
$$+ \underbrace{\int_{T_m}^{T_v}\frac{\bar{c}_{P,f}}{T}dT}_{\substack{\text{entropy change}\\\text{of liquid with }T}} + \underbrace{\frac{\Delta\bar{h}_{fg}}{T_v}}_{\substack{\text{entropy change}\\\text{due to vaporization}}} + \underbrace{\int_{T_v}^{T}\frac{\bar{c}_{P,g}}{T}dT}_{\substack{\text{entropy change}\\\text{of vapor with }T}} \tag{15-160}$$

which is tedious and requires extensive experimental data and careful integration. Equation (15-159) provides a much simpler alternative to determining absolute entropy values.

As temperature approaches 0 K, the Sackur-Tetrode Equation, Eq. (15-159), predicts that $\bar{s}$ approaches $-\infty$, rather than zero. Entropy is related to the thermodynamic probability (Ω) by Boltzmann's Law in Eq. (15-84). Boltzmann's Law shows that at 0 K, Ω should go to 1 (with S going to 0). However, the partition function used to derive the Sackur-Tetrode equation is based on the approximate relation in Eq. (15-42) for Ω, which breaks down as T approaches 0 K.

Equation (15-159) is only valid for monatomic ideal gases. However, corresponding relations can be formulated for more complex molecules by including energy storage contributions other than the kinetic energy of translation, such as rotation and vibration, as discussed in Section 15.6.

EXAMPLE 15.5-1: CALCULATION OF ABSOLUTE ENTROPY VALUES

a) Compare the specific entropy of argon, helium and monatomic oxygen calculated using the Sackur-Tetrode equation, Eq. (15-159), with values obtained from the EES fluid database for temperatures ranging between 200 K and 1000 K. Offer reasons for any apparent discrepancy.

The inputs are entered in EES.

```
$UnitSystem SI Molar J K Pa
G$='Ar'                              "gas"
T=300 [K]                            "temperature"
P=1 [atm]*convert(atm,Pa)            "pressure"
```

The mass that appears in Eq. (15-159) is the mass of a particle, which is the ratio of the molar mass of the gas (the mass per mole of gas) to Avogadro's number:

$$m = \frac{MW}{N_A}$$

Note that N_A is a pre-defined constant in EES, NA#. (The constants that are available in EES can be viewed or extended by selecting the Constants menu item in the Options menu.)

```
MW=MolarMass(G$)                     "molar mass"
m=MW/NA#                             "mass of a particle"
```

The Sackur-Tetrode equation is entered in EES in order to evaluate the molar specific entropy. Notice that both k and h are both pre-defined constants in EES (k# and h#). The entropy is also obtained using the built-in functions in EES.

```
s_bar=R#*(ln(k#*T*(2*pi*m*k#*T/(h#^2))^(3/2)/P)+5/2)     "Sackur-Tetrode equation"
s_bar_EES=Entropy(G$,T=T,P=P)                             "check using EES/NASA database"
```

Figure 1 shows the molar specific entropy of the three gases calculated using the Sackur-Tetrode equation and obtained from the fluid database in EES. The specific entropy calculated with the Sackur-Tetrode equation is nearly identical to the specific entropy obtained from EES for argon and helium. However, there is significant deviation between the Sackur-Tetrode and EES results for monatomic oxygen. Oxygen normally occurs in the oxygen doublet molecule. The monatomic oxygen particle does not obey the ideal gas law, since it exhibits relatively strong inter-particle forces. The energy storage model in Eq. (15-134) that is used to obtain the Sackur-Tetrode equation does not account for all of the energy storage modes that are available to monatomic oxygen.

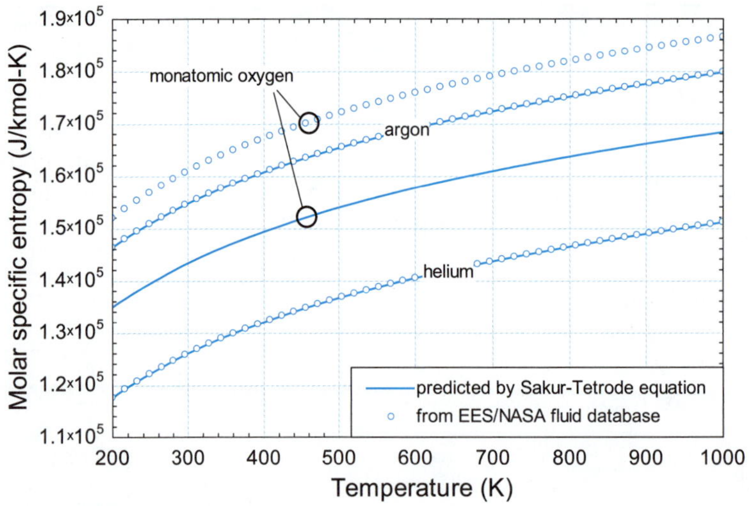

Figure 1: Specific entropy as a function of temperature at 1 atm predicted by the Sackur-Tetrode equation and the EES database.

15.6 Extension to More Complex Particles

Thus far, this chapter has only considered monatomic ideal gas particles. A monatomic ideal gas provides the simplest case to illustrate statistical thermodynamic concepts because the only energy storage is related to the kinetic energy of translation and there are no energy interactions between particles. The translational kinetic energy of a monatomic ideal gas particle was identified in Eq. (15-31) to be:

$$\varepsilon_{trans,i} = \frac{p_x^2 + p_y^2 + p_z^2}{2m} = \left(\frac{h^2}{8m}\right)\left(\frac{n_x^2}{a^2} + \frac{n_y^2}{b^2} + \frac{n_z^2}{c^2}\right) \tag{15-161}$$

In developing this equation, it was assumed that the x, y, and z translational energies are independent; that is:

$$\varepsilon_{trans,i} = \varepsilon_{x,i} + \varepsilon_{y,i} + \varepsilon_{z,i} \tag{15-162}$$

The particle's energy is equally split among the three translational directions. The partition function defined in Eq. (15-105) incorporates the energy model of the particle. For example, the partition function for a monatomic ideal gas, from Eq. (15-136), is given by:

$$f = \sum_{n_x} \exp\left(\frac{-h^2\, n_x^2}{8\,m\,k\,T\,a^2}\right) \sum_{n_y} \exp\left(\frac{-h^2\, n_y^2}{8\,m\,k\,T\,b^2}\right) \sum_{n_z} \exp\left(\frac{-h^2\, n_z^2}{8\,m\,k\,T\,c^2}\right) \tag{15-163}$$

Note that the partition function in Eq. (15-163), which only includes translational energy forms, can be expressed as:

$$f_{trans} = f_x\, f_y\, f_z \tag{15-164}$$

where f_x, f_y, and f_z correspond to each of the summation terms in Eq. (15-163).

To be useful, the statistical thermodynamic concepts should be extended to more complex particles. Diatomic molecules provide the simplest next case. A monatomic gas has nearly all of its mass concentrated at its nucleus and therefore has negligible tumbling motion or rotational energy. However, when a molecule consists of more than one atom, its rotational energy is no longer negligible.

Consider a diatomic molecule consisting of masses m_1 and m_2 rigidly located a distance r_o apart. The *moment of inertia*, *I*, for this molecule is:

$$I = \underbrace{\left(\frac{m_1\,m_2}{m_1+m_2}\right)}_{m_r} r_o^2 = m_r\,r_o^2 \tag{15-165}$$

Rotation can occur around two conjugate directions, so there are two rotational energy modes. Quantum theory dictates that, like translational energy, the rotational energy is also not continuous. Instead, the rotational energy must be some integer multiple of the quantity hc/λ, just as was assumed for translational energy in Section 15.2.2. The energy of a rotating molecule is a function of its angular velocity, ω, according to:

$$\varepsilon_{r,i} = h\,\nu = \frac{1}{2} I\,\omega^2 = \frac{p_{r,i}^2}{2\,m_r} \quad \text{for } i = 1 \text{ or } 2 \tag{15-166}$$

where $p_{r,i}$ is the rotational momentum for rotational mode i and m_r is the quantity in parenthesis in Eq. (15-165). To proceed, the probability of a molecule having a particular rotational energy is assumed to be described as a wave, as was assumed for translation energy forms in Section 15.3. Solving the wave equation for the rotational energy probability results in the eigenvalue solutions:

$$\varepsilon_{r,i} = h\,\nu = \frac{p_{r,i}^2}{2\,m_r} = \frac{1}{2} I\,\omega^2 = \frac{h^2}{8\pi^2 I} j\,(j+1) \quad \text{for } j = 0, 1, 2, \ldots \text{ and } i = 1 \text{ or } 2 \tag{15-167}$$

The degeneracy of rotational energy level j is given by:

$$g_{r,j} = 2j + 1 \tag{15-168}$$

A multi-atom molecule can also store energy in the bonds between atoms in different ways (or modes) such as stretching, bending, and twisting. These energy storage modes are modeled as masses connected with a spring. The energy associated with these energy storage modes is represented as:

$$\varepsilon_{v,l} = \frac{1}{2}\,k_l\,x_l^2 \tag{15-169}$$

where $\varepsilon_{v,l}$ is the energy associated with vibrational mode l, k_l is a spring constant, and x_l is a displacement of the spring. Using these energy modes in the wave equation results in:

$$\varepsilon_{v,l} = \frac{1}{2} k_l\,x_l^2 = \left(k + \frac{1}{2}\right) h\,\nu_l \quad \text{for } k = 0, 1, 2, \ldots. \tag{15-170}$$

The different energy storage modes (e.g., translational, rotational, vibrational) are additive so that the energy of a particle in microstate j is given by:

$$\varepsilon_j = \underbrace{\varepsilon_x + \varepsilon_y + \varepsilon_z}_{\text{translation}} + \underbrace{\varepsilon_{r,1} + \varepsilon_{r,2}}_{\text{rotation}} + \underbrace{\varepsilon_{v,1} + \varepsilon_{v,2} + \ldots}_{\text{vibration}}. \tag{15-171}$$

Equation (15-171) shows that there are three translational energy modes, two rotational energy modes and a variable number of vibrational energy modes. Note that each energy mode can be expressed as:

$$\varepsilon_m = b_m\,p_m^2 \tag{15-172}$$

where b_m is a constant related to energy mode m and p_m is the momentum related to energy mode m. For example, the constant for translational energy in the x direction is

$1/(2\,m)$ from Eq. (15-10) and the constant for rotational energy around axis 1 is $1/(2\,m_r)$ from Eq. (15-166).

The energy of a particle can be entered into the partition function defined in Eq. (15-133):

$$f = \sum_{j=1}^{\#\,\text{microstates}} \exp\left(\frac{-\varepsilon_j}{k\,T}\right) \tag{15-173}$$

Substituting Eq. (15-172) into Eq. (15-173) allows the partition function to be represented as:

$$f = \prod_{m=1}^{M} \sum_{j}^{\infty} \exp\left(-\frac{b_m\,p_{m,j}^2}{k\,T}\right) \tag{15-174}$$

where M is the total number of energy modes and the number of quantum numbers associated with each energy mode has been approximated to be infinite. Each of the summations can be approximated as an integral, as indicated in Eq. (15-137), and the integrals can be evaluated analytically since they all have the same form, as indicated in Eq. (15-140):

$$f \approx \prod_{m=1}^{M} \int_0^{\infty} \exp\left(-\frac{b_m\,p_m^2}{k\,T}\right) d\,p_m = \prod_{m=1}^{M} \frac{1}{2}\sqrt{\frac{k\,T\,\pi}{b_m}} \tag{15-175}$$

The natural logarithm of the partition function is given by:

$$\ln(f) = \ln\left(\prod_{m=1}^{M} \frac{1}{2}\sqrt{\frac{k\,T\,\pi}{b_m}}\right) = M\ln\left(\frac{1}{2}\right) + \frac{M}{2}\ln(k\,T) + \sum_{m=1}^{M} \ln\left(\sqrt{\frac{\pi}{b_m}}\right) \tag{15-176}$$

The internal energy depends on the partial derivative of the natural logarithm of the partition function with respect to temperature at constant volume, as shown in Eq. (15-116).

$$U = \frac{N\,k\,T^2}{f}\left(\frac{\partial f}{\partial T}\right)_V = N\,k\,T^2\left[\frac{\partial \ln(f)}{\partial T}\right]_V \tag{15-177}$$

Taking the partial derivative of the partition function in Eq. (15-176) with respect to temperature at constant volume leads to:

$$\left[\frac{\partial \ln(f)}{\partial T}\right]_V = \frac{M}{2\,T} \tag{15-178}$$

Substituting Eq. (15-178) into Eq. (15-177) provides the equation for the internal energy of the system:

$$U = \frac{N\,M\,k\,T}{2} \tag{15-179}$$

The molar specific internal energy is then:

$$\bar{u} = \frac{U}{n} = \frac{N\,M\,k\,T}{2\,n} = M\,\frac{R_{univ}\,T}{2} \tag{15-180}$$

This analysis indicates that each of the M energy modes contributes $(R_{univ}\,T)/2$ to the molar specific internal energy of the gas. For monatomic gases, $M = 3$ corresponding to translation in the x, y, and z directions. Therefore, $\bar{c}_v$, the molar specific heat capacity at constant volume is $3\,R_{univ}/2$. For diatomic molecules, there are two rotational energy modes in addition to the translational energy modes. If the vibrational energy modes are

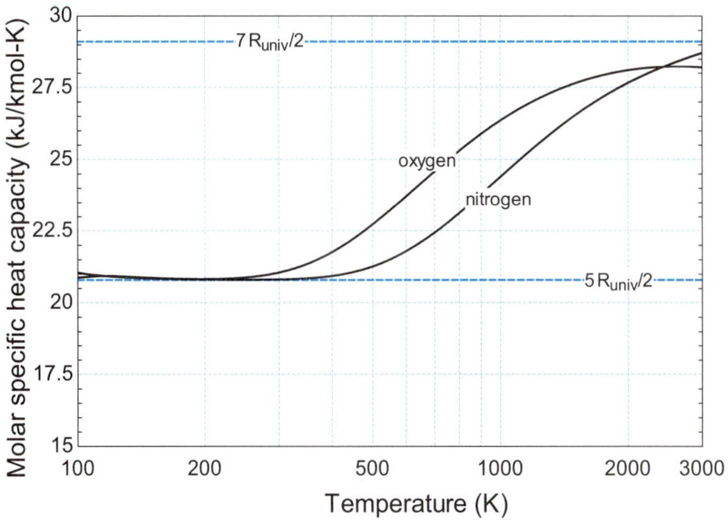

Figure 15-4: The molar specific heat capacity at constant volume as a function of temperature for oxygen and nitrogen.

not significant, then $\bar{c}_v$ for a diatomic ideal gas should be 5 R_{univ}/2 or 20.785 kJ/kmol. This result compares reasonably well with the specific heat capacity values for oxygen (O_2) and nitrogen (N_2) at temperatures between 100 K and 300 K, as seen in Figure 15-4. The vibrational modes become more significant with increasing temperature, which results in the observed temperature dependence of the specific heat capacity of these diatomic gases. More detailed models of the energy per particle, including consideration of the electronic states, have been developed for multi-atom particles that result in improved estimates of the specific heat capacity. There are many excellent books on this subject, e.g., Smith (1982), which provide a more detailed discussion of this topic.

15.7 Heat and Work from a Statistical Thermodynamics Perspective

Heat and work are introduced in Chapter 3 as energy transfer quantities. Heat is defined as an energy transfer, independent of mass transfer, across the boundary of a system as a result of a temperature difference between the surface of the system and its surroundings. Work is defined as the energy transfer, independent of mass transfer, which results from a difference in any potential other than temperature. Heat and work are treated equally in energy balances; therefore it is not necessary to distinguish between heat and work in the development of the general statement of the First Law of Thermodynamics. The difference between heat and work becomes apparent in the development of the Second Law, where heat is associated with an entropy transfer but work is not. However, macroscopic thermodynamics does not provide a clear explanation as to why this is true. Statistical thermodynamics provides an explanation for the difference between heat and work.

Consider a closed system consisting of a compressible gas undergoing a reversible process. A differential energy balance for this system is:

$$\delta Q_{in} = \delta W_{out} + dU \tag{15-181}$$

Substituting the expression for *P-V* work into Eq. (15-181) provides:

$$\delta Q_{in} = P\,dV + dU \tag{15-182}$$

Application of the Second Law to this closed system results in:

$$\frac{\delta Q_{in}}{T} + dS_{gen} = dS \tag{15-183}$$

where the entropy generation is zero because the process is reversible. Substituting Eq. (15-183) into Eq. (15-181) results in the Fundamental Property Relation, first introduced in Section 6.2:

$$dU = T\,dS - P\,dV \tag{15-184}$$

From a statistical thermodynamics viewpoint, internal energy is the sum of the energies of the individual particles. Assuming that there are no energy interactions between the particles, the internal energy can be written as:

$$U = \sum_{i=1}^{\#\,\text{levels}} N_i\,\varepsilon_i \tag{15-185}$$

The total derivative of the internal energy is then:

$$dU = \sum_{i=1}^{\#\,\text{levels}} N_i\,d\varepsilon_i + \sum_{i=1}^{\#\,\text{levels}} \varepsilon_i\,dN_i \tag{15-186}$$

A relation for the energy of a monatomic ideal gas particle that only exhibits translational energy was determined in Eq. (15-134):

$$\varepsilon_i = \frac{p_x^2 + p_y^2 + p_z^2}{2m} = \left(\frac{h^2}{8m}\right)\left(\frac{n_x^2}{a^2} + \frac{n_y^2}{b^2} + \frac{n_z^2}{c^2}\right) \tag{15-187}$$

The presentation is simplified for a cubical enclosure for which $a = b = c = L$. In this case, Eq. (15-187) can be simply written in terms of the volume of the enclosure, V:

$$\varepsilon_i = \left(\frac{h^2}{8m\,L^2}\right)\left(n_x^2 + n_y^2 + n_z^2\right) = \left(\frac{h^2\,V^{-2/3}}{8m}\right)\left(n_x^2 + n_y^2 + n_z^2\right) \tag{15-188}$$

A given energy level is identified by a specified set of quantum numbers and hence specific values of n_x, n_y, and n_z. For a given quantum state (n_x, n_y, and n_z), Eq. (15-188) shows that the energy level depends only on the volume, V. At a given quantum state, the total differential of ε_i is then:

$$d\varepsilon_i = \left(\frac{h^2\left(n_x^2 + n_y^2 + n_z^2\right)}{8m}\right)\left(-\frac{2}{3}\,V^{-5/3}\right)dV \tag{15-189}$$

Substituting Eq. (15-188) into Eq. (15-189) leads to:

$$d\varepsilon_i = -\frac{2}{3}\frac{\varepsilon_i}{V}dV \tag{15-190}$$

Equation (15-190) can be substituted into the first summation in Eq. (15-186):

$$\sum_{i=1}^{\#\,\text{levels}} N_i\,d\varepsilon_i = -\frac{2}{3}\frac{dV}{V}\sum_{i}^{\#\,\text{levels}} N_i\,\varepsilon_i \tag{15-191}$$

Substituting Eq. (15-185) into Eq. (15-191) provides:

$$\sum_{i}^{\#\,levels} N_i\,d\varepsilon_i = -\frac{2}{3}\frac{dV}{V}U \tag{15-192}$$

The internal energy for a monatomic gas, from Eq. (15-154), is given by:

$$U = \frac{3}{2} N k T = \frac{3}{2} n R_{univ} T \tag{15-193}$$

and the pressure for an ideal gas is given by Eq. (15-146):

$$P = \frac{N k T}{V} = \frac{n R_{univ} T}{V} \tag{15-194}$$

Substitution of Eqs. (15-193) and (15-194) into Eq. (15-191) shows that the first term in Eq. (15-186) is

$$\sum_{i}^{\# \text{levels}} N_i d\varepsilon_i = -\frac{2}{3}\frac{dV}{V} U = -\frac{2}{3}\frac{dV}{V}\frac{3}{2} n R_{univ} T = -\underbrace{\frac{n R_{univ} T}{V}}_{P} dV = -P\, dV \tag{15-195}$$

Next we consider the second summation in Eq. (15-186). Boltzmann's Law relates entropy to the thermodynamic probability:

$$S = k \ln(\Omega) \tag{15-196}$$

Substituting Eq. (15-118) into Eq. (15-196) allows the entropy of the gas to be written as:

$$S = k \left[\sum_{i=1}^{\# \text{levels}} N_i \ln\left(\frac{g_i}{N_i}\right) + N \right] \tag{15-197}$$

The total derivative of S is then given by:

$$dS = k \sum_{i=1}^{\# \text{levels}} d\left(N_i \ln\left(\frac{g_i}{N_i}\right)\right) + k \underbrace{dN}_{=0} \tag{15-198}$$

Expanding the differential within the summation of Eq. (15-198) leads to:

$$dS = k \sum_{i=1}^{\# \text{levels}} N_i\, d\left[\ln\left(\frac{g_i}{N_i}\right)\right] + k \sum_{i=1}^{\# \text{levels}} \ln\left(\frac{g_i}{N_i}\right) dN_i \tag{15-199}$$

The first term in Eq. (15-199) is evaluated, noting that the degeneracies are constants:

$$dS = \underbrace{k \sum_{i=1}^{\# \text{levels}} N_i \left(-\frac{1}{N_i}\right) dN_i}_{=0} + k \sum_{i=1}^{\# \text{levels}} \ln\left(\frac{g_i}{N_i}\right) dN_i \tag{15-200}$$

Equation (15-200) is simplified:

$$dS = k \sum_{i=1}^{\# \text{levels}} \ln\left(\frac{g_i}{N_i}\right) dN_i \tag{15-201}$$

The Maxwell-Boltzmann distribution in Eq. (15-107) provides:

$$N_i = \frac{N}{f} g_i \exp\left(-\frac{\varepsilon_i}{k T}\right) \tag{15-202}$$

Solving Eq. (15-202) for g_i/N_i provides:

$$\frac{g_i}{N_i} = \frac{f}{N} \exp\left(\frac{\varepsilon_i}{k T}\right) \tag{15-203}$$

The natural logarithm of Eq. (15-203) is:

$$\ln\left(\frac{g_i}{N_i}\right) = \ln\left(\frac{f}{N}\right) + \frac{\varepsilon_i}{k\,T} \tag{15-204}$$

Substituting Eq. (15-204) into Eq. (15-201) provides:

$$dS = k \sum_{i=1}^{\#\text{ levels}} \left[\ln\left(\frac{f}{N}\right) + \frac{\varepsilon_i}{k\,T}\right] dN_i \tag{15-205}$$

Equation (15-205) is rewritten:

$$dS = k\,\ln\left(\frac{f}{N}\right) \underbrace{\sum_{i=1}^{\#\text{ levels}} dN_i}_{=0} + \sum_{i=1}^{\#\text{ levels}} \frac{\varepsilon_i}{T}\,dN_i \tag{15-206}$$

The first summation in Eq. (15-206) is the differential of the total number of particles and therefore must be zero. Therefore, Eq. (15-206) can be written as:

$$dS = \sum_{i=1}^{\#\text{ levels}} \frac{\varepsilon_i}{T}\,dN_i \tag{15-207}$$

Rearranging Eq. (15-207) shows that the second summation in Eq. (15-186) is:

$$\sum_{i=1}^{\#\text{ levels}} \varepsilon_i\,dN_i = T\,dS \tag{15-208}$$

Substituting Eqs. (15-195) and (15-208) into Eq. (15-186) provides:

$$\begin{aligned} dU &= \sum_{i=1}^{\#\text{levels}} N_i\,d\varepsilon_i + \sum_{i=1}^{\#\text{levels}} \varepsilon_i\,dN_i \\ &= -P\,dV + T\,dS \end{aligned} \tag{15-209}$$

Equation (15-209) is exactly the same as the Fundamental Property Relation in Eq. (15-184). However it also provides some understanding of the difference between the heat and work terms from which the Fundamental Property Relation was derived. The transfer of heat to a system ($T\,dS$) results in more particles appearing in higher energy level states; thus it is related to dN_i, the second term in Eq. (15-209). However, the absorption of heat causes the energy levels themselves to remain unchanged. The transfer of work to a system ($P\,dV$) results in a change in the energy levels themselves; thus it is related to $d\varepsilon_i$, the first term in Eq. (15-209). However, the relative distribution of the particles over the energy states remains unchanged.

REFERENCES

Denbigh, K.G., *Principles of Chemical Equilibrium*, 4th Edition, Cambridge University Press, (1981).

McBride, B.J., Zehe, M.J. and Gordon, S., "NASA Glenn Coefficients for Calculating Thermodynamic Properties of Individual Species," NASA/TP—2002-211556, (2002) http://www.lerc.nasa.gov/WWW/CEAWeb/

Smith, N.O, *Elementary Statistical Thermodynamics*, Plenum Press, New York, ISBN 0-306-41205-5, (1982).

Problems

The problems included here have been selected from a larger set of problems that are available on the website associated with this book (www.cambridge.org/kleinandnellis).

15-1 Assuming that Ω, the number of possible combinations of particle energies in a macrostate, is given by:

$$\Omega = \prod_{i=1}^{\#\,\text{levels}} \frac{g_i^{N_i}}{N_i!}$$

it is shown in Section 15.4 that the most probable distribution of energy levels is the Maxwell-Boltzmann distribution, $N_i = B\, g_i \exp\left(-\beta\, \varepsilon_i\right)$. Derive the corresponding relationship for N_i using the Bose-Einstein model defined by:

$$\Omega_{BE} = \prod_{i=1}^{\#\,\text{levels}} \frac{(g_i + N_i - 1)!}{(g_i - 1)!\;\; N_i!}$$

15-2 Four indistinguishable particles are to be placed in two energy levels, each of which has a degeneracy of three. There is no restriction on the number of particles that can occupy an energy state. Determine the thermodynamic probability of all possible arrangements and identify the most probable arrangement.

15-3 A system consists of eight particles that can occupy five different energy levels. The energy levels are equally spaced and differ by ε units with the value of the first energy level being zero. The degeneracy of each of the five energy levels is provided in Table 15-3.

Table 15-3: Degeneracy of each energy level

Energy level	Degeneracy
0	1
ε	3
2ε	6
3ε	6
4ε	12

The total energy of the system is 24ε. Calculate the total number of macrostates and microstates assuming that:

a) the particles are distinguishable,
b) the particles are indistinguishable, and
c) the particles are indistinguishable and only one particle can occupy a microstate.
d) What is the most likely macrostate?
e) What happens to the number of microstates as the energy of the system is increased, all else being the same?
f) What happens to the number of microstates as the spacing between energy levels increases with the total energy remaining the same?
g) What happens to the number of microstates as the number of particles is increased with the total energy remaining the same?

15-4 Sound travels through a gas as a wave at an average velocity that depends on the temperature. What is the quantum number of a neon molecule traveling at the speed of sound at 300 K in a cubic enclosure that is 10 m on a side?

15-5 Prepare a plot of the velocity of an argon atom as a function of L where L is the length of the cubical enclosure. The three quantum numbers are each 1×10^8. Vary L from 1 cm to 1 m.

15-6 Air at 101 kPa, 298 K circulates freely in a large box that is subdivided into leaky cells, each of volume 1 cm^3. The thermodynamic probability for the distribution of equal numbers of molecules, $N_{eq,}$ in each of the cells should exceed the value for the non-equilibrium situation in which the numbers in the cells differ. Consider a non-equilibrium situation in which in which one cell contains 1% more molecules and a second cell contains 1% fewer molecules than the value N_{eq} with all other cells containing N_{eq} molecules. Calculate and compare:

a) the difference in entropy between the equilibrium and non-equilibrium states using classical thermodynamics, and
b) the difference in the logarithms of the thermodynamic probabilities for the equilibrium and non-equilibrium states.

15-7 The Sackur-Tetrode equation fails to provide an accurate estimate of the specific entropy of monatomic oxygen because this molecule has internal energy modes related to its excited electron state in addition to its translation energy modes. The electronic states of molecule can be determined by solving the wave equation based on a Coulomb potential in spherical polar coordinates. The Pauli exclusion principle must be applied to disallow two electrons in a single atom to simultaneously have an identical set of quantum numbers. The solution for monatomic oxygen results in the energy levels and atomic momentum numbers (J) that are provided in Table 15-7.

Table 15-7: Energy levels and atomic momentum numbers for monatomic oxygen

Energy/(h c) (1/mm)	Atomic momentum number, J
0	2
15.85	1
22.65	0
1586.77	2

The degeneracy for each energy level is $2J + 1$. The partition function, f, is defined in general as $f = \sum_{i=1}^{\#\,\text{levels}} g_i \exp\left(\frac{-\varepsilon_i}{k\,T}\right)$.

a) Using the information in the table above for the degeneracies and energy levels for monatomic oxygen, calculate and plot the specific heat capacity at constant volume and specific entropy for monatomic oxygen for temperatures between 200 K and 2000 K at atmospheric pressure.
b) Compare these values to the values provided in with the NASA external procedure in EES.

15-8 Calculate and plot the specific enthalpy and specific entropy of mercury vapor at a pressure of 5 atm as a function of temperature for temperatures between 300 K and

1000 K based on statistical thermodynamics. Compare the results with the values provided by the NASA external procedure in EES and explain any discrepancies.

15-10 As a first approximation, the quantum-mechanical energy contribution resulting from one-dimensional atomic vibrations in a solid at constant frequency is:

$$\varepsilon_i = \left(i + \frac{1}{2}\right) h\,\nu \quad i = 0,\ 1,\ 2, \ldots.$$

where h is Planck's constant. The partition function corresponding to these vibrations, assuming only one microstate is allowed per energy level, is:

$$f = \frac{\exp\left(-\dfrac{h\,\nu}{2\,k\,T}\right)}{\left[1 - \exp\left(-\dfrac{h\,\nu}{k\,T}\right)\right]}$$

Derive a relation for the constant volume specific heat capacity of a substance that has this partition function.

15-11 A system has available energy levels of 0, 1, and 2 units and each of the energy levels has a degeneracy of 10,000. Consider a thermodynamic state consisting of 2,000 particles having a total energy of 1,000 units.

a) Determine the values of N_0, N_1 and N_2 that will result in the most probable macrostate where N_i is the number of particles with energy level i.
b) Determine the entropy of the system for the conditions of part (a).
c) In words, indicate how the distribution of energy levels will be affected if the total energy of the particles is increased.

15-12 The partition function for rotation of a symmetric diatomic molecule, like N_2, can be represented as:

$$f_{rot} = \sum_{j=0}^{\infty} \left(\frac{2j+1}{2}\right) \exp\left[\frac{-j\,(j+1)\,T_r}{T}\right] \approx \frac{T}{2\,T_r}$$

where T_r is a characteristic temperature that is a constant for each substance; the value of $T_r = 2.89$ K for diatomic nitrogen gas. The approximation at the right of the above equation is valid when $T >> T_r$. Assume that the partition function for diatomic nitrogen can be represented by translational and rotational contributions at 300 K and 100 kPa (i.e., the vibrational contributions are negligible).

a) Prepare a plot of f_{rot} versus the number of terms in the summation. How many terms are needed in order for the approximation for f_{rot} to be accurate?
b) Derive analytical expressions for the specific heat capacity at constant pressure and the specific entropy of nitrogen gas based on principles of statistical thermodynamics.
c) Compare the results of the expressions with the accepted values.

15-13 The sun can be considered a sphere with diameter 1.4×10^9 m with a equivalent blackbody surface temperature of 5780 K. As you know, radiation has both particle and wave-like behavior. The particle-like behavior is exhibited by the pressure exerted on a surface that reflects the radiation. It has been proposed that this radiation pressure could be harnessed and used to propel a vehicle for space travel. This option is explored in this problem. Consider a vehicle having a 'sail' with an effective area of 20 m^2 coated with an aluminized film that has a reflectivity of 0.90. Assume that the sail is always oriented perpendicular to the sun. The vehicle has a mass of 184 kg. The vehicle is initially located (at zero velocity) just outside of earth's atmosphere at a distance of 1.5×10^{11} meters from the sun. The solar 'wind' will be harnessed to propel the vehicle towards Mars which is 142 million miles distant from the Sun.

a) Estimate the solar radiation per unit area incident on the sail at the earth-sun distance (1.5×10^{11} m). This is the so-called 'solar constant' for earth.
b) Calculate the force per unit area exerted on the sail when the vehicle is located at the earth-sun distance.
c) Estimate the solar constant for Mars.
d) Calculate the force per unit area exerted on the sail when the vehicle is located just outside of the Mars' atmosphere.
e) Calculate the distance traveled by the vehicle in 10 years (ignoring relativistic effects). What is the vehicle velocity at this time?
f) Calculate the time required for the vehicle to travel to Mars

15-16 In order to establish a simple algebraic relation for the thermodynamic probability, we assumed that the degeneracy of an energy level is much greater than the number of particles in that energy level. Prove that this is true for Argon gas at 1 atm and 300 K.

16 Compressible Flow

This extended section can be found online at www.cambridge.org/kleinandnellis. Nozzles were first discussed in Section 4.4.4 and the concept of a nozzle efficiency was introduced in Section 6.6.4. The purpose of a nozzle is to increase the velocity of a fluid as its pressure is reduced. Nozzles are used in many applications. They are used in gas and steam turbine engines to allow the conversion of a high pressure gas into a high kinetic energy flow stream that can then be converted into mechanical power by the blades of a turbine. Nozzles are used directly to provide thrust in jet engines and rockets. In addition, the unique behavior of nozzles makes them very useful for flow rate measurement. This chapter provides a more detailed analysis of nozzle behavior than the discussion provided in Section 6.6.4. Very high velocities can be achieved in a properly designed nozzle; it is even possible to achieve velocities that greatly exceed the speed of sound.

Problems

The problems included here have been selected from a larger set of problems that are available on the website associated with this book (www.cambridge.org/kleinandnellis).

16-1 A nozzle is used to determine the mass flow rate of air through a 1.5 inch internal diameter pipe. The air in the line upstream of the meter is at 70°F and 95 psig. The barometric pressure is 14.7 psia. The diameter at the throat of the nozzle is 0.25 in. Assume that the nozzle operates isentropically.

a) Prepare a plot that relates the mass flow rate of the air (in lb_m/min) to the absolute pressure at the nozzle throat (in psia).
b) What is the maximum flow rate that can be measured by this nozzle (in lb_m/min)?

16-2 For the Formula Car challenge, the air inlet to the engine is required to have a minimum area of 0.25 in^2. You have been asked to evaluate the two inlet air design options shown in Figure 16P-2. Design A uses a converging nozzle with a throat area of 0.25 in^2. Design B uses a converging-diverging nozzle with a throat area of 0.25 in^2 and an exit area of 0.415 in^2. The nozzle in both of these designs exhausts to the inlet plenum for a turbo charger. Assume that both nozzles operate isentropically.

a) Compare the maximum air flow rates for the nozzles in Design A and Design B. Which nozzle provides the largest mass flow rate?
b) Prepare a plot of the mass flow rate through the converging nozzle in Design A as a function of the ratio of the plenum pressure to the inlet pressure.
c) Prepare a plot of the mass flow rate through the converging-diverging nozzle in Design B as a function of the ratio of the plenum pressure to the inlet pressure, assuming that flow in the diverging section of the nozzle is subsonic.

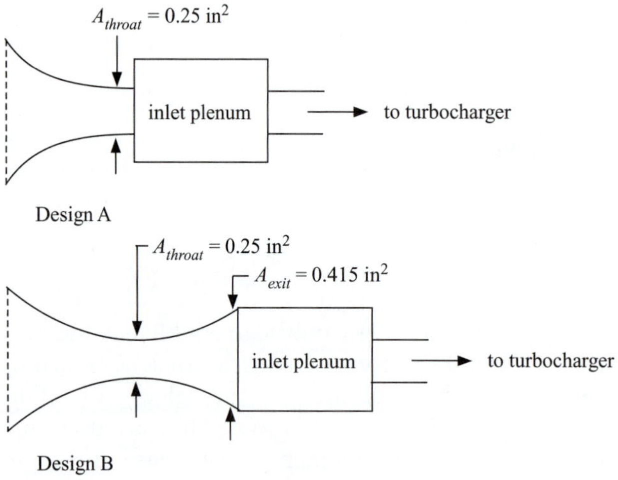

Figure 16P-2: Two inlet air design options.

d) The turbocharger raises the pressure of the air from the pressure in the inlet plenum pressure to the pressure at which air is charged into the cylinders. Explain which nozzle (Design A or Design B) you would you recommend for this purpose and why.

16-3 A converging-diverging nozzle has the geometry shown in Figure 16P-3.

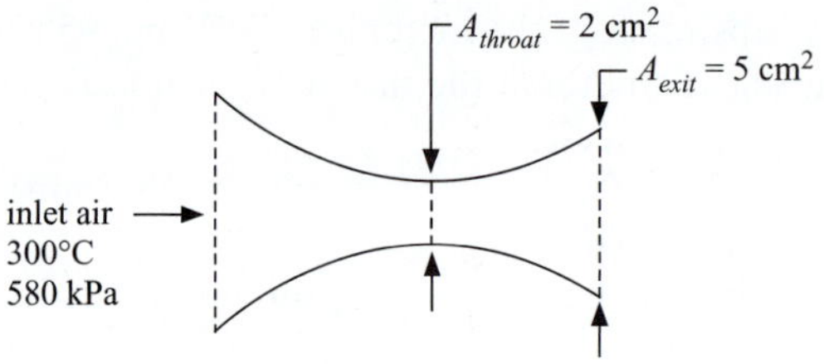

Figure 16P-3: Nozzle Geometry.

a) Assuming isentropic flow conditions otherwise, determine the lowest and highest pressures at the exit plane for which a shock wave is expected to develop within the diverging part of the nozzle.
b) The pressure at the exit of the nozzle is measured to be 400 kPa. Determine the velocity and temperature at the exit.

16-4 A rocket engine generates combustion gas at a pressure of 682 psia, measured by a pressure sensor attached to the body of the motor as shown in Figure 16P-4. The combustion gas consists of 40% carbon dioxide and 60% water vapor (on a molar basis) at 1245°F. The nozzle has a throat diameter of 18 in and it discharges to the atmosphere with a barometric pressure of 14.5 psia. Experimental data verify that

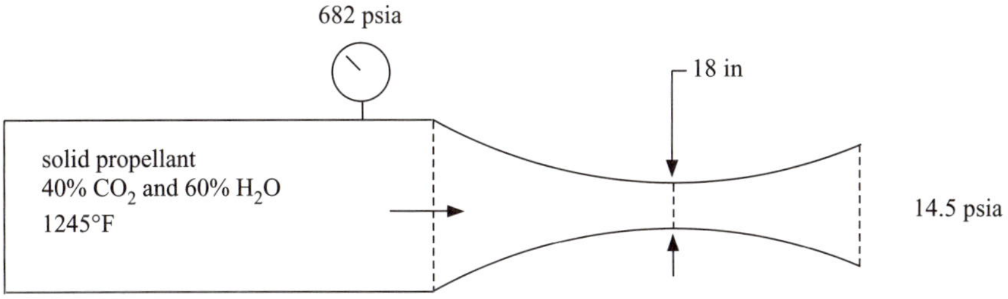

Figure 16P-4: Rocket engine.

the performance of this system can be predicted by one-dimensional isentropic flow theory. Assuming ideal gas behavior, determine:

a) the required exit diameter of the nozzle in order for it to operate at design conditions,
b) the mass flow rate of gas out of the engine, and
c) the thrust produced by the engine (in lb_f).

16-6 An engineer is studying the leakage that occurs from a labyrinth seal in a super-critical carbon dioxide Brayton cycle. He has decided to model the leakage that occurs through the seal as a nozzle in order to estimate a limit on leakage rate and the temperature that may be experienced at the seal. Carbon dioxide enters the seal at 55°C and 75 bar. The exit area is estimated to be 4.5×10^{-6} m^2. Assume that the process can be modeled as an isentropic process and that choked flow occurs at the seal exit.

a) Calculate the leakage rate and the pressure and temperature at the seal exit assuming carbon dioxide can be modeled as an ideal gas.
b) The inlet pressure is above the critical pressure. Repeat part (a), but do not assume that carbon dioxide obeys the ideal gas law.

16-7 A nozzle is used to determine the mass flow rate of methane (CH_4) as it flows through a pipeline. The stagnation conditions of the methane are 30°C, 10 bar. The exit diameter of the nozzle is 5.0 cm. The pressure downstream of the nozzle is 9 bar. The nozzle is adiabatic and, except for a normal shock (if one is present), the flow may be assumed to be reversible.

a) If a converging nozzle is used to make this measurement, what is the mass flow rate of methane?
b) If a converging-diverging nozzle with a throat diameter of 3.5 cm and an exit diameter of 5.0 cm is used to make this measurement, what is the mass flow rate of methane?

16-8 Air is flowing through a converging-diverging nozzle for which the cross-sectional area as a function of position is provided in Table 16P-8. The pressure at the inlet plenum is 550 kPa and the temperature is 42°C. A shock is observed to occur at position 8.

a) Determine the mass flow rate of air through this nozzle.
b) Determine the pressures just upstream and just downstream of the shock wave, assuming that the nozzle otherwise behaves isentropically.
c) Determine the temperature, pressure and velocity of air at the nozzle exit plane.

Table 16P-8: Cross-sectional area at various positions in a nozzle

Position	Area (cm^2)
0 (inlet)	95
1	65
2	45
3	30
4	25
5	28
6	31
7	34
8	37
9	41
10	45
11	49
12 (exit plane)	54

16-9 A gas cylinder having a volume of 0.24 m^3 contains air at 20°C, 24 bar. The cylinder is equipped with a valve. The valve is opened and the air is rapidly vented to the atmosphere at 101.3 kPa. The exit diameter of the valve is 0.025 cm. It is not clear how much thermal interaction there is between the air remaining in the cylinder and the cylinder walls. At one extreme, the venting process can be considered to be adiabatic. At the other extreme, the heat transfer coefficient between the air and the cylinder walls may be high so that the air remains at 20°C throughout the venting process. The valve can be modeled as a converging nozzle which may or may not be choked, depending on the pressure in the cylinder. Assume air to behave as an ideal gas.

a) Assuming that the venting process is adiabatic, calculate and plot the pressure in the tank as a function of time. How much time is needed to reduce the pressure to 110 kPa?
b) Repeat part (a), but assume that the venting process is isothermal at 20°C.

16-10 Hydrogen has been proposed as a fuel for vehicles. This problem is concerned with the time that it will take to fill the fuel tank in a vehicle with high pressure hydrogen. The tank is a carbon-fiber reinforced shell with a volume of 150 liters. The filling station supplies compressed hydrogen at 400 bar and 25°C. A vehicle enters a filling station with its hydrogen fuel tank at 25°C and 20 bar and the tank is filled to a pressure of 375 bar through a converging nozzle that has an exit diameter of 1 mm. (The nozzle is also used for measuring the mass flow rate.) In the actual filling process, the flow through the nozzle will initially be choked and then as the pressure in the tank increases the flow at nozzle exit will become subsonic. Assume ideal gas behavior in your analysis.

a) Prepare a plot of the pressure in the tank as a function of time assuming that choked flow always occurs in the nozzle during the filling process. Based on your plot, determine the time required to fill the tank to 375 bar and the mass of hydrogen delivered.
b) Repeat the calculations for part (a), but in this case, account for subsonic flow when it occurs. Compare the time and mass of hydrogen delivered to the tank with the results from part (a).

Appendix A Unit Conversions and Useful Information

The base SI unit system is completely self-consistent, as shown in Table A-1.

Table A-1: Unit conversions between SI units

$1\ \text{J} = 1\ \text{N-m} = 1\ \text{kg-m}^2/\text{s}^2$
$1\ \text{N} = 1\ \text{kg-m/s}^2$
$1\ \text{Pa} = 1\ \text{N/m}^2 = 1\ \text{kg/m-s}^2$
$1\ \text{W} = 1\ \text{J/s} = 1\ \text{N-m/s} = 1\ \text{kg-m}^2/\text{s}^3$

Table A-2 provides the prefixes used to delineate between orders of magnitude.

Table A-2: Prefixes

Prefix	Symbol	Factor
femto	f	1×10^{-15}
pico	p	1×10^{-12}
nano	n	1×10^{-9}
micro	μ	1×10^{-6}
milli	m	1×10^{-3}
centi	c	1×10^{-2}
deci	d	1×10^{-1}
deka	da	1×10^{1}
kilo	k	1×10^{3}
mega	M	1×10^{6}
giga	G	1×10^{9}

Table A-3 provides formulae to convert between different temperature scales.

Table A-3: Formulae to convert between different temperature scales

from Kelvin	from Celsius
T[°C] = T[K] – 273.15	T[°F] = 1.8 T[°C] + 32
T[°F] = 1.8 T[K] – 459.67	T[K] = T[°C] + 273.15
T[R] = 1.8 T[K]	T[R] = 1.8 (T[°C] + 273.15)
from Fahrenheit	**from Rankine**
T[°C] = (T[°F] – 32)/1.8	T[°C] = (T[R] – 491.67)/1.8
T[K] = (T[°F] + 459.67)/1.8	T[°F] = T[R] – 491.67
T[R] = T[°F] + 459.67	T[K] = T[R]/1.8

Table A-4 provides unit conversions between specified units and their base SI equivalent; they are categorized (alphabetically) by dimension.

Table A-4: Unit conversions between arbitrary and base SI units

Area (m^2)	Mass (kg)	Time (s)
1 acre = 4046.9 m^2	1 amu = 1.6605×10^{-27} kg	1 min = 60 s
1 hectare = 1×10^4 m^2	1 grain = 6.4799×10^{-5} kg	1 hr = 3600 s
Energy (J)	1 lb_m = 0.45359 kg	1 day = 8.6400×10^4 s
1 Btu = 1055.1 J	1 oz = 0.028350 kg	1 week = 6.0480×10^5 s
1 calorie = 4.1868 J	1 slug = 14.594 kg	1 year = 3.1536×10^7 s
1 erg = 1×10^{-7} J	1 stone = 6.3503 kg	**Velocity (m/s)**
1 eV = 1.6021×10^{-19} J	1 ton = 907.18 kg	1 ft/s = 0.30480 m/s
1 kW-hr = 3.6×10^6 J	**Power (W)**	1 in/s = 0.0254 m/s
1 therm = 1.0551×10^8 J	1 Btu/s = 1055.1 W	1 knot = 0.51444 m/s
Force (N)	1 cal/s = 4.1868 W	1 micron/s = 1×10^{-6} m/s
1 lb_f = 4.4482 N	1 erg/s = 1×10^{-7} W	1 mph = 0.44704 m/s
1 dyne = 1×10^{-5} N	1 eV/s = 1.6021×10^{-19} W	1 yard/s = 0.91440 m/s
Length (m)	1 hp = 745.70 W	**Viscosity – dynamic (Pa-s)**
1 Angstrom = 1×10^{-10} m	1 therm/s = 1.0551×10^8 W	1 poise = 0.1 Pa-s
1 fathom = 1.8288 m	1 ton = 3516.9 W	1 Reyn = 6.8947×10^3 Pa-s
1 ft = 0.30480 m	**Pressure (Pa)**	**Viscosity – kinematic (m/s^2)**
1 furlong = 201.17 m	1 atm = 101325 Pa	1 Stoke = 1×10^{-4} m^2/s
1 in = 0.025400	1 bar = 1×10^5 Pa	**Volume (m^3)**
1 league = 5556.0 m	1 inHg = 3386.4 Pa	1 gal = 0.0037854 m^3
1 micron = 1×10^{-6} m	1 inH20 = 249.09 Pa	1 liter = 0.001 m^3
1 mil = 2.5400×10^{-5} m	1 mmH20 = 9.8066 Pa	1 quart = 9.4632×10^{-4} m^3
1 mile = 1609.3 m	1 mmHg = 133.32 Pa	**Volumetric flow rate (m^3/s)**
1 rod = 5.0292 m	1 psi = 6894.8 Pa	1 cfm = 4.7195×10^{-4} m^3/s
1 yard = 0.91440 m	1 torr = 133.32 Pa	1 gpm = 6.3090×10^{-5} m^3/s

Table A-5 lists standard atmospheric pressure in several units.

Table A-5: Standard atmospheric pressure

1 atm
1.01325 bar
14.695 psi
101325 Pa
101.325 kPa
760 mmHg
29.92 inHg

Table A-6 provides the universal gas constant in several units.

Table A-6: Universal Gas Constant

SI Unit System	English Unit System
8.314 kJ/kmol-K	1.987 Btu/lbmol-R
0.08314 bar-m^3/kmol-K	0.730 atm-ft^3/lbmol-R
8314 J/kmol-K	1545 ft-lb$_f$/lbmol-R

Appendix B Property Tables for Water

Tables B-1 and B-2 present data for saturated liquid and saturated vapor water. Table B-1 presents property data at regular intervals of temperature while Table B-2 presents data at regular intervals of pressure. Table B-3 presents data for superheated vapor over a matrix of temperatures and pressures. Table B-4 presents data for compressed liquid over a matrix of temperatures and pressures. These tables were generated using EES with the substance 'Steam_IAPWS' which implements the high accuracy thermodynamic properties of water described in *1995 Formulation for the Thermodynamic Properties of Ordinary Water Substance for General and Scientific Use*, issued by the International Associated for the Properties of Water and Steam (IAPWS). Note that these tables can be printed from the website associated with this text, www.cambridge.org/kleinandnellis, for use during closed book examinations.

Table B-1: Properties of Saturated Water, Presented at Regular Intervals of Temperature

Temp. T (°C)	Pressure P (kPa)	Specific volume (m^3/kg)		Specific internal energy (kJ/kg)		Specific enthalpy (kJ/kg)		Specific entropy (kJ/kg-K)		T (°C)
		$10^3\, v_f$	v_g	u_f	u_g	h_f	h_g	s_f	s_g	
0.01	0.6117	1.0002	206.00	0	2374.9	0.000	2500.9	0	9.1556	0.01
2	0.7060	1.0001	179.78	8.3911	2377.7	8.3918	2504.6	0.03061	9.1027	2
4	0.8135	1.0001	157.14	16.812	2380.4	16.813	2508.2	0.06110	9.0506	4
6	0.9353	1.0001	137.65	25.224	2383.2	25.225	2511.9	0.09134	8.9994	6
8	1.0729	1.0002	120.85	33.626	2385.9	33.627	2515.6	0.12133	8.9492	8
10	1.2281	1.0003	106.32	42.020	2388.7	42.022	2519.2	0.15109	8.8999	10
12	1.4028	1.0006	93.732	50.408	2391.4	50.410	2522.9	0.18061	8.8514	12
14	1.5989	1.0008	82.804	58.791	2394.1	58.793	2526.5	0.20990	8.8038	14
16	1.8187	1.0011	73.295	67.169	2396.9	67.170	2530.2	0.23898	8.7571	16
18	2.0646	1.0015	65.005	75.542	2399.6	75.544	2533.8	0.26784	8.7112	18
20	2.3392	1.0018	57.762	83.913	2402.3	83.915	2537.4	0.29649	8.6661	20
22	2.6452	1.0023	51.422	92.280	2405.1	92.283	2541.1	0.32493	8.6217	22
24	2.9857	1.0028	45.861	100.65	2407.8	100.65	2544.7	0.35318	8.5782	24
26	3.3638	1.0033	40.975	109.01	2410.5	109.01	2548.3	0.38123	8.5354	26

(continued)

Table B-1 *(continued)*

Temp. T (°C)	Pressure P (kPa)	Specific volume (m^3/kg)		Specific internal energy (kJ/kg)		Specific enthalpy (kJ/kg)		Specific entropy (kJ/kg-K)		T (°C)
		$10^3\ v_f$	v_g	u_f	u_g	h_f	h_g	s_f	s_g	
28	3.7830	1.0038	36.673	117.37	2413.2	117.37	2551.9	0.40909	8.4933	28
30	4.2469	1.0044	32.879	125.73	2415.9	125.74	2555.6	0.43676	8.4520	30
32	4.7596	1.0050	29.527	134.09	2418.6	134.10	2559.2	0.46425	8.4114	32
34	5.3251	1.0057	26.560	142.45	2421.3	142.46	2562.8	0.49155	8.3714	34
36	5.9480	1.0064	23.929	150.81	2424.0	150.82	2566.4	0.51868	8.3322	36
38	6.6330	1.0071	21.593	159.17	2426.7	159.18	2569.9	0.54563	8.2935	38
40	7.3851	1.0079	19.515	167.53	2429.4	167.53	2573.5	0.57241	8.2556	40
42	8.2098	1.0087	17.663	175.89	2432.1	175.90	2577.1	0.59902	8.2182	42
44	9.1127	1.0095	16.010	184.25	2434.8	184.26	2580.7	0.62546	8.1815	44
46	10.100	1.0104	14.534	192.61	2437.4	192.62	2584.2	0.65174	8.1454	46
48	11.178	1.0112	13.212	200.97	2440.1	200.98	2587.8	0.67786	8.1098	48
50	12.352	1.0122	12.026	209.33	2442.7	209.34	2591.3	0.70382	8.0748	50
55	15.763	1.0146	9.5639	230.24	2449.3	230.26	2600.1	0.76803	7.9898	55
60	19.947	1.0171	7.6670	251.16	2455.9	251.18	2608.8	0.83130	7.9082	60
65	25.043	1.0199	6.1935	272.09	2462.4	272.12	2617.5	0.89366	7.8296	65
70	31.202	1.0228	5.0396	293.04	2468.9	293.07	2626.1	0.95514	7.7540	70
75	38.597	1.0258	4.1291	313.99	2475.3	314.03	2634.6	1.01578	7.6812	75
80	47.416	1.0291	3.4053	334.97	2481.6	335.02	2643.0	1.07559	7.6111	80
85	57.868	1.0324	2.8260	355.96	2487.8	356.02	2651.4	1.13461	7.5435	85
90	70.183	1.0360	2.3593	376.97	2494.0	377.04	2659.6	1.19288	7.4782	90
95	84.609	1.0396	1.9808	398.00	2500.1	398.09	2667.6	1.25040	7.4151	95
100	101.42	1.0435	1.6720	419.06	2506.0	419.17	2675.6	1.30722	7.3542	100
105	120.90	1.0474	1.4186	440.15	2511.9	440.28	2683.4	1.3634	7.2952	105
110	143.38	1.0516	1.2094	461.27	2517.7	461.42	2691.1	1.4188	7.2382	110
115	169.18	1.0559	1.0360	482.42	2523.3	482.59	2698.6	1.4737	7.1829	115
120	198.67	1.0603	0.89136	503.60	2528.9	503.81	2706.0	1.5279	7.1292	120
125	232.23	1.0649	0.77012	524.83	2534.3	525.07	2713.1	1.5816	7.0771	125
130	270.28	1.0697	0.66808	546.10	2539.5	546.38	2720.1	1.6346	7.0265	130

(continued)

Table B-1 *(continued)*

Temp. T (°C)	Pressure P (kPa)	Specific volume (m^3/kg)		Specific internal energy (kJ/kg)		Specific enthalpy (kJ/kg)		Specific entropy (kJ/kg-K)		T (°C)
		$10^3\ v_f$	v_g	u_f	u_g	h_f	h_g	s_f	s_g	
135	313.22	1.0746	0.58179	567.41	2544.7	567.75	2726.9	1.6872	6.9773	135
140	361.53	1.0798	0.50850	588.77	2549.6	589.16	2733.5	1.7392	6.9294	140
145	415.68	1.0850	0.44600	610.19	2554.4	610.64	2739.8	1.7908	6.8827	145
150	476.16	1.0905	0.39248	631.66	2559.1	632.18	2745.9	1.8418	6.8371	150
160	618.23	1.1020	0.30680	674.79	2567.8	675.47	2757.5	1.9426	6.7492	160
170	792.18	1.1143	0.24260	718.20	2575.7	719.08	2767.9	2.0417	6.6650	170
180	1002.8	1.1274	0.19385	761.92	2582.8	763.05	2777.2	2.1392	6.5841	180
190	1255.2	1.1414	0.15636	806.00	2589.0	807.43	2785.3	2.2355	6.5059	190
200	1554.9	1.1565	0.12721	850.46	2594.2	852.26	2792.0	2.3305	6.4302	200
210	1907.7	1.1727	0.10429	895.38	2598.3	897.61	2797.3	2.4245	6.3563	210
220	2319.6	1.1901	0.086094	940.79	2601.3	943.55	2801.0	2.5176	6.2840	220
230	2797.1	1.2089	0.071505	986.76	2602.9	990.14	2802.9	2.6100	6.2128	230
240	3347.0	1.2294	0.059707	1033.4	2603.1	1037.5	2803.0	2.7018	6.1424	240
250	3976.2	1.2516	0.050085	1080.7	2601.8	1085.7	2801.0	2.7933	6.0721	250
260	4692.3	1.2759	0.042175	1128.9	2598.7	1134.8	2796.6	2.8847	6.0017	260
270	5503.0	1.3028	0.035622	1178.0	2593.7	1185.1	2789.7	2.9762	5.9305	270
280	6416.6	1.3326	0.030153	1228.2	2586.4	1236.7	2779.9	3.0681	5.8579	280
290	7441.8	1.3660	0.025554	1279.7	2576.5	1289.8	2766.7	3.1608	5.7834	290
300	8587.9	1.4038	0.021659	1332.7	2563.6	1344.8	2749.6	3.2548	5.7059	300
310	9865.0	1.4475	0.018333	1387.7	2547.1	1402.0	2727.9	3.3506	5.6243	310
320	11284.3	1.4987	0.015470	1445.1	2526.0	1462.0	2700.6	3.4491	5.5372	320
330	12858.1	1.5604	0.012979	1505.7	2499.2	1525.8	2666.0	3.5516	5.4422	330
340	14600.7	1.6377	0.010783	1570.7	2464.5	1594.6	2622.0	3.6602	5.3358	340
350	16529.3	1.7407	0.008806	1642.4	2418.3	1671.2	2563.9	3.7788	5.2114	350
360	18666.0	1.8950	0.006950	1726.2	2351.9	1761.5	2481.6	3.9165	5.0537	360
370	21043.8	2.2172	0.004953	1844.5	2230.1	1891.2	2334.3	4.1119	4.8009	370
373.95	22064.0	3.1056	0.003106	2015.7	2015.7	2084.3	2084.3	4.4070	4.4070	373.95

Table B-2: Properties of Saturated Water, Presented at Regular Intervals of Pressure

Pressure P (kPa)	Temp. T (°C)	Specific volume (m^3/kg) $10^3\ v_f$	v_g	Specific internal energy (kJ/kg) u_f	u_g	Specific enthalpy (kJ/kg) h_f	h_g	Specific entropy (kJ/kg-K) s_f	s_g	P (kPa)
1	6.9705	1.0001	129.19	29.302	2384.5	29.303	2513.7	0.10593	8.9749	1
1.5	13.0205	1.0007	87.964	54.686	2392.8	54.688	2524.7	0.19558	8.8270	1.5
2	17.4957	1.0014	66.990	73.431	2398.9	73.433	2532.9	0.26058	8.7227	2
2.5	21.0777	1.0021	54.242	88.422	2403.8	88.424	2539.4	0.31184	8.6421	2.5
3	24.0796	1.0028	45.654	100.98	2407.9	100.98	2544.8	0.35430	8.5765	3
4	28.9607	1.0041	34.791	121.39	2414.5	121.39	2553.7	0.42240	8.4734	4
5	32.8743	1.0053	28.185	137.75	2419.8	137.75	2560.7	0.47620	8.3938	5
6	36.1587	1.0065	23.733	151.47	2424.2	151.48	2566.6	0.52082	8.3291	6
7	38.9992	1.0075	20.524	163.34	2428.1	163.35	2571.7	0.55903	8.2745	7
8	41.5082	1.0085	18.099	173.83	2431.4	173.84	2576.2	0.59249	8.2273	8
10	45.8056	1.0103	14.670	191.80	2437.2	191.81	2583.9	0.64919	8.1488	10
12	49.4178	1.0119	12.358	206.90	2442.0	206.91	2590.3	0.69628	8.0850	12
14	52.5458	1.0134	10.691	219.98	2446.1	219.99	2595.8	0.73663	8.0311	14
16	55.3120	1.0147	9.4307	231.55	2449.8	231.57	2600.7	0.77201	7.9847	16
18	57.7971	1.0160	8.4432	241.95	2453.0	241.96	2605.0	0.80354	7.9438	18
20	60.0569	1.0172	7.6481	251.40	2456.0	251.42	2608.9	0.83202	7.9073	20
25	64.9618	1.0199	6.2034	271.93	2462.4	271.96	2617.5	0.89319	7.8302	25
30	69.0942	1.0222	5.2287	289.24	2467.7	289.27	2624.6	0.94407	7.7675	30
40	75.8560	1.0264	3.9933	317.58	2476.3	317.63	2636.1	1.02607	7.6691	40
50	81.3163	1.0299	3.2403	340.49	2483.2	340.54	2645.2	1.09120	7.5931	50
60	85.9255	1.0331	2.7320	359.85	2489.0	359.91	2652.9	1.14545	7.5312	60
70	89.9314	1.0359	2.3650	376.68	2493.9	376.75	2659.4	1.19208	7.4791	70
80	93.4853	1.0385	2.0873	391.63	2498.2	391.71	2665.2	1.23305	7.4340	80
100	99.6059	1.0432	1.6941	417.40	2505.6	417.51	2675.0	1.30277	7.3589	100
101.325	99.9743	1.0434	1.6734	418.95	2506.00	419.06	2675.56	1.30693	7.35451	101.325
120	104.7837	1.0473	1.4285	439.24	2511.7	439.36	2683.1	1.36094	7.2978	120
140	109.2919	1.0510	1.2367	458.27	2516.9	458.42	2690.0	1.41101	7.2461	140
160	113.2977	1.0544	1.0915	475.21	2521.4	475.38	2696.1	1.45507	7.2015	160

(continued)

Table B-2 *(continued)*

Pressure P (kPa)	Temp. T (°C)	Specific volume (m³/kg) $10^3\ v_f$	v_g	Specific internal energy (kJ/kg) u_f	u_g	Specific enthalpy (kJ/kg) h_f	h_g	Specific entropy (kJ/kg-K) s_f	s_g	P (kPa)
180	116.9117	1.0576	0.97759	490.51	2525.5	490.70	2701.4	1.49448	7.1621	180
200	120.2106	1.0605	0.88578	504.50	2529.1	504.71	2706.3	1.53018	7.1270	200
250	127.4120	1.0672	0.71873	535.08	2536.8	535.35	2716.5	1.60724	7.0525	250
300	133.5230	1.0732	0.60582	561.11	2543.2	561.43	2724.9	1.67173	6.9917	300
350	138.8577	1.0786	0.52422	583.89	2548.5	584.26	2732.0	1.72738	6.9402	350
400	143.6089	1.0836	0.46242	604.22	2553.1	604.66	2738.1	1.77646	6.8955	400
500	151.8315	1.0925	0.37483	639.54	2560.7	640.09	2748.1	1.86039	6.8207	500
600	158.8268	1.1006	0.31560	669.72	2566.8	670.38	2756.2	1.93083	6.7593	600
700	164.9464	1.1080	0.27278	696.23	2571.8	697.00	2762.8	1.99178	6.7071	700
800	170.4066	1.1148	0.24035	719.97	2576.0	720.87	2768.3	2.04566	6.6616	800
900	175.3505	1.1212	0.21489	741.55	2579.6	742.56	2773.0	2.09405	6.6213	900
1000	179.88	1.1272	0.19437	761.39	2582.8	762.51	2777.1	2.1381	6.5850	1000
1100	184.06	1.1330	0.17745	779.78	2585.5	781.03	2780.7	2.1785	6.5520	1100
1200	187.96	1.1385	0.16326	796.96	2587.8	798.33	2783.8	2.2159	6.5217	1200
1300	191.60	1.1438	0.15119	813.10	2589.9	814.59	2786.5	2.2508	6.4936	1300
1400	195.04	1.1489	0.14078	828.35	2591.8	829.96	2788.9	2.2835	6.4675	1400
1500	198.29	1.1539	0.13171	842.82	2593.4	844.55	2791.0	2.3143	6.4430	1500
1600	201.37	1.1587	0.12374	856.59	2594.8	858.44	2792.8	2.3435	6.4200	1600
1700	204.31	1.1633	0.11668	869.75	2596.1	871.72	2794.5	2.3711	6.3982	1700
1800	207.11	1.1679	0.11037	882.35	2597.3	884.46	2795.9	2.3975	6.3775	1800
1900	209.80	1.1724	0.10471	894.46	2598.3	896.69	2797.2	2.4226	6.3578	1900
2000	212.38	1.1767	0.099587	906.12	2599.1	908.47	2798.3	2.4467	6.3390	2000
2200	217.25	1.1852	0.090701	928.24	2600.6	930.85	2800.1	2.4921	6.3038	2200
2400	221.79	1.1934	0.083247	948.97	2601.7	951.83	2801.4	2.5342	6.2712	2400
2600	226.05	1.2013	0.076901	968.51	2602.4	971.63	2802.4	2.5735	6.2409	2600
2800	230.06	1.2091	0.071432	987.02	2602.9	990.41	2802.9	2.6105	6.2124	2800
3000	233.85	1.2166	0.066667	1004.6	2603.2	1008.3	2803.2	2.6454	6.1856	3000

(continued)

Table B-2 *(continued)*

Pressure P (kPa)	Temp. T (°C)	Specific volume (m^3/kg)		Specific internal energy (kJ/kg)		Specific enthalpy (kJ/kg)		Specific entropy (kJ/kg-K)		P (kPa)
		$10^3\ v_f$	v_g	u_f	u_g	h_f	h_g	s_f	s_g	
3250	238.33	1.2258	0.061508	1025.6	2603.2	1029.5	2803.1	2.6866	6.1541	3250
3500	242.56	1.2349	0.057061	1045.4	2603.0	1049.7	2802.7	2.7253	6.1244	3500
3750	246.55	1.2437	0.053186	1064.3	2602.5	1069.0	2801.9	2.7618	6.0963	3750
4000	250.35	1.2524	0.049779	1082.4	2601.7	1087.4	2800.8	2.7966	6.0696	4000
4500	257.44	1.2695	0.044061	1116.4	2599.7	1122.1	2798.0	2.8613	6.0198	4500
5000	263.94	1.2862	0.039448	1148.1	2597.0	1154.5	2794.2	2.9207	5.9737	5000
6000	275.59	1.3190	0.032449	1205.8	2589.9	1213.8	2784.6	3.0275	5.8902	6000
7000	285.83	1.3515	0.027378	1258.0	2581.0	1267.5	2772.6	3.1220	5.8148	7000
8000	295.01	1.3843	0.023525	1306.0	2570.5	1317.1	2758.7	3.2077	5.7450	8000
9000	303.35	1.4177	0.020489	1350.9	2558.5	1363.7	2742.9	3.2866	5.6791	9000
10,000	311.00	1.4522	0.018028	1393.3	2545.2	1407.9	2725.5	3.3603	5.6159	10,000
11,000	318.08	1.4881	0.015988	1433.9	2530.4	1450.2	2706.3	3.4299	5.5544	11,000
12,000	324.68	1.5260	0.014264	1473.0	2514.3	1491.3	2685.4	3.4964	5.4939	12,000
13,000	330.85	1.5663	0.012781	1511.1	2496.6	1531.4	2662.7	3.5606	5.4336	13,000
14,000	336.67	1.6097	0.011487	1548.4	2477.1	1571.0	2637.9	3.6232	5.3728	14,000
15,000	342.16	1.6572	0.010341	1585.5	2455.7	1610.3	2610.8	3.6848	5.3108	15,000
16,000	347.36	1.7100	0.009312	1622.6	2432.0	1649.9	2581.0	3.7461	5.2466	16,000
18,000	356.99	1.8402	0.007504	1699.1	2375.0	1732.2	2510.0	3.8720	5.1064	18,000
20,000	365.75	2.0378	0.005862	1785.8	2294.8	1826.6	2412.1	4.0146	4.9310	20,000
22,000	373.71	2.7031	0.003644	1951.7	2092.4	2011.1	2172.6	4.2942	4.5439	22,000
22,064	373.95	3.1056	0.003106	2015.7	2015.7	2084.3	2084.3	4.4067	4.4072	22,064

Table B-3: Properties of Superheated Water Vapor: Pressures from 10 kPa to 400 kPa

P (kPa)		Temperature, T (°C) 50	100	150	200	300	400	500	600	800	1000
10	v (m³/kg)	14.867	17.196	19.513	21.826	26.446	31.063	35.68	40.296	49.527	58.758
	u (kJ/kg)	2443.3	2515.5	2587.9	2661.4	2812.3	2969.3	3132.9	3303.3	3665.4	4055.3
	h (kJ/kg)	2592.0	2687.5	2783.0	2879.6	3076.7	3280.0	3489.7	3706.3	4160.6	4642.8
	s (kJ/kg-K)	8.1741	8.4489	8.6893	8.9049	9.2827	9.6094	9.8998	10.163	10.631	11.043
20	v (m³/kg)		8.5855	9.7486	10.907	13.220	15.530	17.839	20.147	24.763	29.379
	u (kJ/kg)		2514.5	2587.4	2661.0	2812.1	2969.2	3132.8	3303.3	3665.3	4055.2
	h (kJ/kg)		2686.2	2782.3	2879.2	3076.5	3279.8	3489.6	3706.2	4160.6	4642.8
	s (kJ/kg-K)		8.1263	8.3681	8.5843	8.9625	9.2893	9.5798	9.8432	10.311	10.723
40	v (m³/kg)		4.2799	4.8662	5.448	6.6067	7.7628	8.9179	10.073	12.381	14.689
	u (kJ/kg)		2512.5	2586.3	2660.3	2811.7	2969.0	3132.7	3303.2	3665.3	4055.2
	h (kJ/kg)		2683.7	2780.9	2878.2	3076.0	3279.5	3489.4	3706.1	4160.5	4642.7
	s (kJ/kg-K)		7.8011	8.0456	8.2629	8.642.0	8.9691	9.2597	9.5231	9.9913	10.403
60	v (m³/kg)		2.8445	3.2387	3.6283	4.4023	5.1739	5.9444	6.7144	8.2538	9.7927
	u (kJ/kg)		2510.5	2585.2	2659.6	2811.4	2968.8	3132.5	3303.0	3665.2	4055.1
	h (kJ/kg)		2681.1	2779.5	2877.3	3075.5	3279.2	3489.2	3705.9	4160.4	4642.7
	s (kJ/kg-K)		7.6084	7.8559	8.0743	8.4542	8.7816	9.0723	9.3359	9.8041	10.216
80	v (m³/kg)		2.1267	2.4250	2.7184	3.3002	3.8794	4.4576	5.0353	6.1901	7.3444
	u (kJ/kg)		2508.4	2584.1	2658.9	2811.0	2968.5	3132.4	3302.9	3665.1	4055.1
	h (kJ/kg)		2678.5	2778.1	2876.4	3075.0	3278.9	3489.0	3705.7	4160.3	4642.6
	s (kJ/kg-K)		7.4699	7.7204	7.9401	8.3208	8.6485	8.9394	9.2030	9.6712	10.083
100	v (m³/kg)		1.6959	1.9367	2.1724	2.6389	3.1027	3.5655	4.0279	4.9519	5.8755
	u (kJ/kg)		2506.2	2582.9	2658.2	2810.7	2968.3	3132.2	3302.8	3665.0	4055.0
	h (kJ/kg)		2675.8	2776.6	2875.5	3074.5	3278.6	3488.7	3705.6	4160.2	4642.6
	s (kJ/kg-K)		7.3611	7.6148	7.8356	8.2172	8.5452	8.8362	9.0999	9.5682	9.9800
200	v (m³/kg)			0.9599	1.0805	1.3162	1.5493	1.7814	2.0130	2.4755	2.9375
	u (kJ/kg)			2577.1	2654.6	2808.8	2967.2	3131.4	3302.2	3664.7	4054.8
	h (kJ/kg)			2769.1	2870.7	3072.1	3277.0	3487.7	3704.8	4159.8	4642.3
	s (kJ/kg-K)			7.2810	7.5081	7.8941	8.2236	8.5153	8.7793	9.2479	9.6599

(continued)

Table B-3 *(continued)*

P (kPa)		Temperature, T (°C)							
		150	200	300	400	500	600	800	1000
300	v (m³/kg)	0.6340	0.7164	0.8753	1.0315	1.1867	1.3414	1.6500	1.9582
	u (kJ/kg)	2571.0	2651.0	2807.0	2966.0	3130.6	3301.6	3664.3	4054.5
	h (kJ/kg)	2761.2	2865.9	3069.6	3275.5	3486.6	3704.0	4159.3	4642.0
	s (kJ/kg-K)	7.0792	7.3132	7.7037	8.0347	8.3271	8.5915	9.0605	9.4726
400	v (m³/kg)	0.4709	0.5343	0.6549	0.7726	0.8894	1.0056	1.2373	1.4686
	u (kJ/kg)	2564.4	2647.2	2805.1	2964.9	3129.8	3301.0	3663.9	4054.3
	h (kJ/kg)	2752.8	2860.9	3067.1	3273.9	3485.5	3703.3	4158.9	4641.7
	s (kJ/kg-K)	6.9306	7.1723	7.5677	7.9003	8.1933	8.4580	8.9274	9.3396

P (kPa)		Temperature, T (°C)									
		200	250	300	400	500	600	700	800	900	1000
500	v (m³/kg)	0.4250	0.4744	0.5226	0.6173	0.7109	0.8041	0.897	0.9897	1.0823	1.1748
	u (kJ/kg)	2643.3	2723.8	2803.3	2963.7	3129.0	3300.4	3478.6	3663.6	3855.4	4054.0
	h (kJ/kg)	2855.8	2961.0	3064.6	3272.4	3484.5	3702.5	3927.0	4158.4	4396.6	4641.4
	s (kJ/kg-K)	7.0610	7.2725	7.4614	7.7956	8.0893	8.3544	8.5978	8.8240	9.0362	9.2364
600	v (m³/kg)	0.3521	0.3939	0.4344	0.5137	0.5920	0.6698	0.7472	0.8246	0.9018	0.9789
	u (kJ/kg)	2639.4	2721.2	2801.4	2962.5	3128.2	3299.8	3478.1	3663.2	3855.1	4053.8
	h (kJ/kg)	2850.6	2957.6	3062.0	3270.8	3483.4	3701.7	3926.4	4157.9	4396.2	4641.1
	s (kJ/kg-K)	6.9683	7.1833	7.3740	7.7097	8.0041	8.2695	8.5132	8.7395	8.9518	9.1521
700	v (m³/kg)	0.3000	0.3364	0.3714	0.4398	0.5070	0.5738	0.6403	0.7066	0.7729	0.8390
	u (kJ/kg)	2635.3	2718.6	2799.5	2961.4	3127.4	3299.3	3477.6	3662.8	3854.8	4053.5
	h (kJ/kg)	2845.3	2954.0	3059.5	3269.2	3482.3	3700.9	3925.9	4157.5	4395.9	4640.8
	s (kJ/kg-K)	6.8884	7.1070	7.2995	7.6368	7.9319	8.1977	8.4415	8.6681	8.8804	9.0807
800	v (m³/kg)	0.2609	0.2932	0.3242	0.3843	0.4433	0.5019	0.5601	0.6182	0.6762	0.7341
	u (kJ/kg)	2631.1	2715.9	2797.5	2960.2	3126.6	3298.7	3477.2	3662.5	3854.5	4053.3
	h (kJ/kg)	2839.8	2950.4	3056.9	3267.7	3481.3	3700.1	3925.3	4157.0	4395.5	4640.5
	s (kJ/kg-K)	6.8177	7.0402	7.2345	7.5735	7.8692	8.1354	8.3794	8.6061	8.8185	9.0189
900	v (m³/kg)	0.2304	0.2596	0.2874	0.3411	0.3938	0.4459	0.4977	0.5494	0.6010	0.6525
	u (kJ/kg)	2626.7	2713.2	2795.6	2959.1	3125.8	3298.1	3476.7	3662.1	3854.2	4053.0
	h (kJ/kg)	2834.1	2946.8	3054.3	3266.1	3480.2	3699.4	3924.7	4156.6	4395.1	4640.3
	s (kJ/kg-K)	6.7539	6.9805	7.1767	7.5173	7.8138	8.0804	8.3246	8.5514	8.7639	8.9644

(continued)

Table B-3 *(continued)*

P (kPa)		Temperature, T (°C) 200	250	300	400	500	600	700	800	900	1000
1000	v (m³/kg)	0.2060	0.2327	0.2580	0.3066	0.3541	0.4011	0.4478	0.4944	0.5408	0.5872
	u (kJ/kg)	2622.3	2710.4	2793.7	2957.9	3125.0	3297.5	3476.3	3661.7	3853.9	4052.7
	h (kJ/kg)	2828.3	2943.1	3051.6	3264.5	3479.1	3698.6	3924.1	4156.1	4394.8	4640.0
	s (kJ/kg-K)	6.6956	6.9265	7.1246	7.4670	7.7642	8.0311	8.2755	8.5024	8.7150	8.9155
1200	v (m³/kg)	0.1693	0.1924	0.2139	0.2548	0.2946	0.3339	0.3730	0.4118	0.4506	0.4893
	u (kJ/kg)	2612.9	2704.7	2789.7	2955.5	3123.4	3296.3	3475.3	3661.0	3853.3	4052.2
	h (kJ/kg)	2816.1	2935.6	3046.3	3261.3	3477.0	3697.0	3922.9	4155.2	4394.0	4639.4
	s (kJ/kg-K)	6.5909	6.8313	7.0335	7.3793	7.6779	7.9456	8.1904	8.4176	8.6303	8.8310
1400	v (m³/kg)	0.1430	0.1636	0.1823	0.2178	0.2522	0.2860	0.3195	0.3529	0.3861	0.4193
	u (kJ/kg)	2602.7	2698.9	2785.7	2953.1	3121.8	3295.1	3474.4	3660.3	3852.7	4051.7
	h (kJ/kg)	2803.0	2927.9	3040.9	3258.1	3474.8	3695.5	3921.7	4154.3	4393.3	4638.8
	s (kJ/kg-K)	6.4975	6.7488	6.9553	7.3046	7.6047	7.8730	8.1183	8.3458	8.5587	8.7595
1600	v (m³/kg)		0.1419	0.1587	0.1901	0.2203	0.2500	0.2794	0.3087	0.3378	0.3669
	u (kJ/kg)		2692.9	2781.6	2950.8	3120.1	3293.9	3473.5	3659.5	3852.1	4051.2
	h (kJ/kg)		2919.9	3035.4	3254.9	3472.6	3693.9	3920.5	4153.4	4392.6	4638.2
	s (kJ/kg-K)		6.6753	6.8864	7.2394	7.5410	7.8101	8.0558	8.2834	8.4965	8.6974

P (kPa)		Temperature, T (°C) 250	300	350	400	500	600	700	800	900	1000
1800	v (m³/kg)	0.1250	0.1402	0.1546	0.1685	0.1955	0.2220	0.2482	0.2743	0.3002	0.3261
	u (kJ/kg)	2686.7	2777.4	2863.6	2948.3	3118.5	3292.7	3472.6	3658.8	3851.5	4050.7
	h (kJ/kg)	2911.7	3029.9	3141.9	3251.6	3470.4	3692.3	3919.4	4152.4	4391.9	4637.6
	s (kJ/kg-K)	6.6088	6.8246	7.0120	7.1814	7.4845	7.7543	8.0005	8.2284	8.4417	8.6427
2000	v (m³/kg)	0.1115	0.1255	0.1386	0.1512	0.1757	0.1996	0.2233	0.2467	0.2701	0.2934
	u (kJ/kg)	2680.3	2773.2	2860.5	2945.9	3116.9	3291.5	3471.7	3658.0	3850.9	4050.2
	h (kJ/kg)	2903.3	3024.2	3137.7	3248.4	3468.3	3690.7	3918.2	4151.5	4391.1	4637.1
	s (kJ/kg-K)	6.5475	6.7684	6.9583	7.1292	7.4337	7.7043	7.9509	8.1791	8.3925	8.5936
3000	v (m³/kg)	0.07063	0.08118	0.09056	0.09938	0.1162	0.1325	0.1484	0.1642	0.1799	0.1955
	u (kJ/kg)	2644.7	2750.8	2844.4	2933.6	3108.6	3285.5	3467.0	3654.3	3847.9	4047.7
	h (kJ/kg)	2856.5	2994.3	3116.1	3231.7	3457.2	3682.8	3912.2	4146.9	4387.5	4634.2
	s (kJ/kg-K)	6.2893	6.5412	6.7450	6.9235	7.2359	7.5103	7.759	7.9885	8.2028	8.4045

(continued)

Table B-3 *(continued)*

P (kPa)		Temperature, T (°C) 250	300	350	400	500	600	700	800	900	1000
4000	v (m³/kg)		0.05887	0.06647	0.07343	0.08644	0.09886	0.1110	0.1229	0.1348	0.1465
	u (kJ/kg)		2726.2	2827.4	2920.8	3100.3	3279.4	3462.4	3650.6	3844.8	4045.1
	h (kJ/kg)		2961.7	3093.3	3214.5	3446	3674.9	3906.3	4142.3	4383.9	4631.2
	s (kJ/kg-K)		6.3639	6.5843	6.7714	7.0922	7.3706	7.6214	7.8523	8.0675	8.2698
5000	v (m³/kg)		0.04535	0.05197	0.05784	0.06858	0.07870	0.08852	0.09816	0.1077	0.1172
	u (kJ/kg)		2699.0	2809.5	2907.5	3091.8	3273.3	3457.7	3646.9	3841.8	4042.6
	h (kJ/kg)		2925.7	3069.3	3196.7	3434.7	3666.9	3900.3	4137.7	4380.2	4628.3
	s (kJ/kg-K)		6.2111	6.4516	6.6483	6.9781	7.2605	7.5136	7.7458	7.9619	8.1648
6000	v (m³/kg)		0.03619	0.04225	0.04742	0.05667	0.06527	0.07355	0.08165	0.08964	0.09756
	u (kJ/kg)		2668.4	2790.4	2893.7	3083.1	3267.2	3453.0	3643.2	3838.8	4040.1
	h (kJ/kg)		2885.6	3043.9	3178.3	3423.1	3658.8	3894.3	4133.1	4376.6	4625.4
	s (kJ/kg-K)		6.0703	6.3357	6.5432	6.8826	7.1693	7.4247	7.6582	7.8751	8.0786
7000	v (m³/kg)		0.02949	0.03526	0.03996	0.04816	0.05567	0.06285	0.06986	0.07675	0.08357
	u (kJ/kg)		2633.5	2770.1	2879.5	3074.3	3261.0	3448.3	3639.5	3835.7	4037.5
	h (kJ/kg)		2839.9	3016.9	3159.2	3411.4	3650.6	3888.3	4128.5	4373.0	4622.5
	s (kJ/kg-K)		5.9337	6.2305	6.4502	6.8000	7.0910	7.3487	7.5836	7.8014	8.0055
8000	v (m³/kg)		0.02428	0.02997	0.03434	0.04177	0.04846	0.05483	0.06101	0.06708	0.07308
	u (kJ/kg)		2592.3	2748.3	2864.6	3065.4	3254.7	3443.6	3635.7	3832.7	4035.0
	h (kJ/kg)		2786.5	2988.1	3139.4	3399.5	3642.4	3882.2	4123.8	4369.3	4619.6
	s (kJ/kg-K)		5.7937	6.1321	6.3658	6.7266	7.0221	7.2822	7.5185	7.7372	7.9419
9000	v (m³/kg)			0.02582	0.02996	0.03679	0.04286	0.04859	0.05413	0.05956	0.06492
	u (kJ/kg)			2725.0	2849.2	3056.3	3248.4	3438.8	3632.0	3829.6	4032.4
	h (kJ/kg)			2957.3	3118.8	3387.4	3634.1	3876.1	4119.2	4365.7	4616.7
	s (kJ/kg-K)			6.038	6.2876	6.6603	6.9605	7.2229	7.4606	7.6802	7.8855

P (kPa)		Temperature, T (°C) 350	400	500	600	700	800	900	1000	1100	1200
10,000	v (m³/kg)	0.02244	0.02644	0.03281	0.03838	0.04360	0.04863	0.05355	0.05839	0.06318	0.06794
	u (kJ/kg)	2699.6	2833.1	3047.0	3242.0	3434.0	3628.2	3826.5	4029.9	4238.5	4452.4
	h (kJ/kg)	2924.0	3097.5	3375.1	3625.8	3870.0	4114.5	4362.0	4613.8	4870.3	5131.7
	s (kJ/kg-K)	5.946	6.2141	6.5995	6.9045	7.1693	7.4085	7.6290	7.8349	8.0289	8.2126

(continued)

Table B-3 *(continued)*

P (kPa)		Temperature, T (°C)									
		350	400	500	600	700	800	900	1000	1100	1200
12,000	v (m³/kg)	0.01722	0.02111	0.02683	0.03165	0.03611	0.04037	0.04452	0.0486	0.05262	0.05661
	u (kJ/kg)	2641.4	2798.7	3028.1	3229.1	3424.4	3620.7	3820.4	4024.8	4234.2	4448.7
	h (kJ/kg)	2848.1	3052.0	3350.0	3608.9	3857.7	4105.1	4354.7	4608.0	4865.6	5128.0
	s (kJ/kg-K)	5.7609	6.0764	6.4903	6.8054	7.0753	7.3173	7.5396	7.7468	7.9416	8.126
14,000	v (m³/kg)	0.01323	0.01724	0.02254	0.02684	0.03076	0.03448	0.03808	0.04161	0.04508	0.04852
	u (kJ/kg)	2567.8	2761.0	3008.4	3216.0	3414.7	3613.1	3814.3	4019.7	4229.9	4444.9
	h (kJ/kg)	2753.1	3002.3	3324.1	3591.8	3845.4	4095.8	4347.4	4602.1	4861.0	5124.2
	s (kJ/kg-K)	5.5598	5.9460	6.3932	6.7191	6.9941	7.2391	7.4632	7.6716	7.8673	8.0523
16,000	v (m³/kg)	0.009766	0.01428	0.01932	0.02324	0.02675	0.03006	0.03325	0.03636	0.03942	0.04245
	u (kJ/kg)	2460.7	2719.1	2988.1	3202.6	3404.9	3605.4	3808.1	4014.6	4225.5	4441.2
	h (kJ/kg)	2617.0	2947.6	3297.3	3574.4	3832.9	4086.4	4340.1	4596.3	4856.3	5120.4
	s (kJ/kg-K)	5.3045	5.8180	6.3046	6.6421	6.9224	7.1704	7.3964	7.6060	7.8025	7.9882
18,000	v (m³/kg)		0.01192	0.01681	0.02043	0.02363	0.02662	0.02949	0.03228	0.03502	0.03773
	u (kJ/kg)		2671.9	2967.1	3189.1	3395.1	3597.8	3801.9	4009.4	4221.2	4437.5
	h (kJ/kg)		2886.4	3269.7	3556.8	3820.4	4076.9	4332.7	4590.5	4851.6	5116.6
	s (kJ/kg-K)		5.6883	6.2223	6.5721	6.8579	7.1089	7.3368	7.5476	7.7451	7.9313
20,000	v (m³/kg)		0.009950	0.01479	0.01819	0.02113	0.02387	0.02648	0.02902	0.03150	0.03396
	u (kJ/kg)		2617.9	2945.3	3175.3	3385.1	3590.1	3795.7	4004.3	4216.9	4433.8
	h (kJ/kg)		2816.9	3241.2	3539.0	3807.8	4067.5	4325.4	4584.7	4847.0	5112.9
	s (kJ/kg-K)		5.5526	6.1446	6.5075	6.7991	7.0531	7.2829	7.4950	7.6933	7.8802
22,000	v (m³/kg)		0.008256	0.01314	0.01635	0.01909	0.02162	0.02403	0.02635	0.02863	0.03086
	u (kJ/kg)		2554.2	2922.7	3161.4	3375.1	3582.4	3789.5	3999.2	4212.5	4430.1
	h (kJ/kg)		2735.8	3211.8	3521.0	3795.1	4058.0	4318.1	4578.9	4842.3	5109.1
	s (kJ/kg-K)		5.4051	6.0705	6.4474	6.7448	7.0020	7.2337	7.4470	7.6462	7.8337
24,000	v (m³/kg)		0.006732	0.01175	0.01481	0.01739	0.01975	0.02198	0.02413	0.02623	0.02829
	u (kJ/kg)		2475.9	2899.4	3147.2	3365.0	3574.6	3783.3	3994.0	4208.2	4426.4
	h (kJ/kg)		2637.5	3181.4	3502.7	3782.4	4048.5	4310.7	4573.1	4837.7	5105.4
	s (kJ/kg-K)		5.2369	5.9991	6.3908	6.6943	6.9546	7.1883	7.4029	7.6029	7.7911
26,000	v (m³/kg)		0.005285	0.01058	0.01352	0.01595	0.01816	0.02024	0.02225	0.02420	0.02611
	u (kJ/kg)		2373	2875.1	3132.8	3354.8	3566.8	3777.1	3988.9	4203.9	4422.7
	h (kJ/kg)		2510.4	3150.2	3484.3	3769.6	4039	4303.4	4567.3	4833.1	5101.6
	s (kJ/kg-K)		5.0302	5.9298	6.3373	6.6469	6.9105	7.1461	7.362	7.5629	7.7517

Table B-4: Properties of Compressed Liquid Water

P (MPa)		Temperature, T (°C) 0	20	40	60	80	100	120	140	160	180
5	10^3 v (m^3/kg)	0.9977	0.9996	1.0057	1.0149	1.0267	1.0410	1.0576	1.0769	1.0988	1.1240
	u (kJ/kg)	0.0441	83.609	166.92	250.29	333.82	417.65	501.91	586.80	672.55	759.47
	h (kJ/kg)	5.030	88.610	171.95	255.36	338.96	422.85	507.19	592.18	678.04	765.09
	s (kJ/kg-K)	0.00014	0.29543	0.57046	0.82865	1.0723	1.3034	1.5236	1.7344	1.9374	2.1338
10	10^3 v (m^3/kg)	0.9952	0.9973	1.0035	1.0127	1.0244	1.0385	1.0549	1.0738	1.0954	1.1200
	u (kJ/kg)	0.1171	83.308	166.33	249.43	332.69	416.23	500.18	584.72	670.06	756.48
	h (kJ/kg)	10.069	93.281	176.37	259.55	342.94	426.62	510.73	595.45	681.01	767.68
	s (kJ/kg-K)	0.00034	0.29435	0.56852	0.82602	1.0691	1.2996	1.5191	1.7293	1.9316	2.1272
20	10^3 v (m^3/kg)	0.9904	0.9929	0.9992	1.0084	1.0199	1.0337	1.0496	1.0679	1.0886	1.1122
	u (kJ/kg)	0.2257	82.708	165.17	247.75	330.50	413.50	496.85	580.71	665.28	750.78
	h (kJ/kg)	20.033	102.57	185.16	267.92	350.90	434.17	517.84	602.07	687.05	773.02
	s (kJ/kg-K)	0.00047	0.29208	0.56461	0.8208	1.0627	1.2920	1.5105	1.7194	1.9203	2.1143
40	10^3 v (m^3/kg)	0.9811	0.9845	0.9911	1.0001	1.0113	1.0245	1.0397	1.0569	1.0762	1.0980
	u (kJ/kg)	0.3078	81.520	162.96	244.58	326.37	408.36	490.61	573.26	656.43	740.32
	h (kJ/kg)	39.553	120.90	202.60	284.59	366.82	449.34	532.20	615.53	699.48	784.24
	s (kJ/kg-K)	−0.00024	0.28716	0.55676	0.81054	1.0503	1.2775	1.4938	1.7006	1.8990	2.0903
60	10^3 v (m^3/kg)	0.9725	0.9765	0.9833	0.9923	1.0032	1.0159	1.0304	1.0467	1.0649	1.0852
	u (kJ/kg)	0.2357	80.345	160.86	241.62	322.54	403.61	484.88	566.44	648.40	730.90
	h (kJ/kg)	58.584	138.94	219.87	301.16	382.73	464.56	546.70	629.24	712.30	796.01
	s (kJ/kg-K)	−0.00208	0.2818	0.54885	0.80049	1.0383	1.2637	1.4781	1.6829	1.8792	2.0681
80	10^3 v (m^3/kg)	0.9643	0.9690	0.9760	0.9849	0.9956	1.0078	1.0217	1.0372	1.0545	1.0735
	u (kJ/kg)	0.03710	79.182	158.88	238.85	318.96	399.20	479.57	560.16	641.05	722.35
	h (kJ/kg)	77.184	156.70	236.96	317.64	398.61	479.83	561.31	643.14	725.41	808.23
	s (kJ/kg-K)	−0.00489	0.27604	0.54087	0.79062	1.0266	1.2503	1.4631	1.6661	1.8605	2.0474
100	10^3 v (m^3/kg)	0.9567	0.9619	0.9691	0.9779	0.9883	1.0002	1.0136	1.0284	1.0448	1.0628
	u (kJ/kg)	−0.2637	78.031	156.99	236.24	315.62	395.09	474.65	554.36	634.29	714.52
	h (kJ/kg)	95.40	174.22	253.90	334.03	414.46	495.11	576.01	657.20	738.77	820.80
	s (kJ/kg-K)	−0.00851	0.26992	0.53284	0.7809	1.0153	1.2375	1.4487	1.6501	1.8429	2.0280

Appendix C Property Tables for R134a

Tables C-1 and C-2 present thermodynamic property data for saturated liquid and saturated vapor R134a. Table C-1 presents data at regular intervals of temperature while Table C-2 presents data at regular intervals of pressure. Table C-3 presents data for superheated vapor over a matrix of temperatures and pressures. These tables were generated using EES with the substance 'R134a', which implements the fundamental equation of state developed by R. Tillner-Roth and H. D. Baehr, "An International Standard Formulation for the Thermodynamic Properties of 1,1,1,2-Tetrafluoroethane (HFC-134a) for Temperatures from 170 K to 455 K and Pressures up to 70 MPa," *J. Phys. Chem, Ref. Data*, Vol. 23, No. 5, (1994). Note that these tables can be printed from the website associated with this text, www.cambridge.org/kleinandnellis, for use during closed book examinations.

Table C-1: Properties of Saturated R134a, Presented at Regular Intervals of Temperature

Temp. T (°C)	Pressure P (kPa)	Specific volume (m^3/kg)		Specific internal energy (kJ/kg)		Specific enthalpy (kJ/kg)		Specific entropy (kJ/kg-K)		T (°C)
		$10^3\ v_f$	v_g	u_f	u_g	h_f	h_g	s_f	s_g	
−40	51.25	0.7053	0.36064	−0.04	207.38	0.00	225.86	0.0000	0.9687	−40
−35	66.19	0.7126	0.28373	6.25	210.25	6.29	229.03	0.0267	0.9619	−35
−30	84.43	0.7201	0.22577	12.58	213.12	12.64	232.19	0.0530	0.9559	−30
−25	106.5	0.7280	0.18152	18.95	215.99	19.03	235.32	0.0789	0.9505	−25
−20	132.8	0.7361	0.14735	25.37	218.86	25.47	238.43	0.1046	0.9457	−20
−15	164.0	0.7445	0.12066	31.85	221.72	31.97	241.51	0.1299	0.9415	−15
−10	200.7	0.7533	0.099600	38.38	224.56	38.53	244.55	0.1550	0.9378	−10
−5	243.5	0.7625	0.082823	44.96	227.38	45.15	247.55	0.1798	0.9345	−5
0	293.0	0.7722	0.069335	51.61	230.18	51.83	250.50	0.2043	0.9316	0
5	349.9	0.7823	0.058401	58.31	232.96	58.59	253.39	0.2287	0.9290	5
10	414.9	0.7929	0.049466	65.09	235.69	65.42	256.22	0.2528	0.9266	10
15	488.7	0.8041	0.042110	71.93	238.39	72.32	258.97	0.2768	0.9245	15

(continued)

Table C-1 *(continued)*

Temp. T (°C)	Pressure P (kPa)	Specific volume (m^3/kg)		Specific internal energy (kJ/kg)		Specific enthalpy (kJ/kg)		Specific entropy (kJ/kg-K)		T (°C)
		10^3 v_f	v_g	u_f	u_g	h_f	h_g	s_f	s_g	
20	572.1	0.8160	0.036012	78.85	241.04	79.32	261.64	0.3006	0.9225	20
25	665.8	0.8286	0.030922	85.85	243.64	86.40	264.23	0.3243	0.9207	25
30	770.6	0.8421	0.026648	92.93	246.17	93.58	266.71	0.3479	0.9190	30
35	887.5	0.8565	0.023037	100.11	248.63	100.87	269.08	0.3714	0.9173	35
40	1017	0.8720	0.019968	107.39	251.00	108.28	271.31	0.3949	0.9155	40
45	1161	0.8889	0.017344	114.79	253.27	115.82	273.40	0.4184	0.9137	45
50	1319	0.9072	0.015089	122.30	255.42	123.50	275.32	0.4419	0.9117	50
55	1492	0.9274	0.013140	129.96	257.43	131.35	277.03	0.4655	0.9095	55
60	1688	0.9498	0.011444	137.79	259.25	139.38	278.51	0.4893	0.9069	60
65	1891	0.9751	0.0099591	145.80	260.86	147.64	279.69	0.5133	0.9038	65
70	2118	1.0038	0.0086500	154.04	262.20	156.16	280.52	0.5377	0.9000	70
75	2366	1.0372	0.0074858	162.54	263.17	165.00	280.88	0.5625	0.8953	75
80	2635	1.0774	0.0064393	171.43	263.66	174.27	280.63	0.5881	0.8893	80
85	2928	1.1273	0.0054843	180.81	263.45	184.11	279.51	0.6149	0.8812	85
90	3247	1.1938	0.0045914	190.94	262.13	194.82	277.04	0.6435	0.8699	90
95	3594	1.2945	0.0037133	202.49	258.73	207.14	272.08	0.6760	0.8524	95
100	3975	1.5269	0.0026575	218.73	248.46	224.80	259.02	0.7222	0.8139	100
101.03	4059	1.9685	0.0019685	233.90	233.90	241.88	241.88	0.7678	0.7678	101.03

Table C-2: Properties of Saturated R134a, Presented at Regular Intervals of Pressure

Pressure P (kPa)	Temp. T (°C)	Specific volume (m³/kg) $10^3\ v_f$	v_g	Specific internal energy (kJ/kg) u_f	u_g	Specific enthalpy (kJ/kg) (kJ/kg) h_f	h_g	Specific entropy (kJ/kg-K) s_f	s_g	P (kPa)
40	−44.61	0.699	0.45483	−5.79	204.74	−5.76	222.94	−0.0249	0.9757	40
60	−36.95	0.710	0.31108	3.79	209.13	3.84	227.80	0.0163	0.9644	60
80	−31.13	0.718	0.23749	11.14	212.48	11.20	231.47	0.0471	0.9572	80
100	−26.37	0.726	0.19255	17.19	215.21	17.27	234.46	0.0718	0.9519	100
200	−10.09	0.753	0.099951	38.26	224.51	38.41	244.50	0.1545	0.9379	200
300	0.65	0.773	0.067777	52.48	230.55	52.71	250.88	0.2075	0.9312	300
400	8.91	0.791	0.051266	63.61	235.10	63.92	255.61	0.2476	0.9271	400
500	15.71	0.806	0.041168	72.92	238.77	73.32	259.36	0.2802	0.9242	500
600	21.55	0.820	0.034335	81.01	241.86	81.50	262.46	0.3080	0.9220	600
700	26.69	0.833	0.029392	88.24	244.51	88.82	265.08	0.3323	0.9201	700
800	31.31	0.846	0.025645	94.80	246.82	95.48	267.34	0.3541	0.9185	800
900	35.51	0.858	0.022703	100.84	248.88	101.62	269.31	0.3738	0.9171	900
1000	39.37	0.870	0.020330	106.47	250.71	107.34	271.04	0.3920	0.9157	1000
1200	46.29	0.893	0.016728	116.72	253.84	117.79	273.92	0.4245	0.9132	1200
1400	52.40	0.917	0.014185	125.96	256.40	127.25	276.17	0.4532	0.9107	1400
1600	57.88	0.940	0.012134	134.45	258.50	135.96	277.92	0.4792	0.9080	1600
1800	62.87	0.964	0.010568	142.36	260.21	144.09	279.23	0.5030	0.9052	1800
2000	67.45	0.989	0.0092970	149.81	261.56	151.78	280.15	0.5252	0.9020	2000
2200	71.70	1.015	0.0082396	156.90	262.57	159.13	280.70	0.5460	0.8985	2200
2400	75.66	1.042	0.0073419	163.70	263.27	166.20	280.89	0.5658	0.8946	2400
2600	79.37	1.072	0.0065657	170.29	263.63	173.08	280.70	0.5848	0.8901	2600
2800	82.86	1.104	0.0058830	176.73	263.64	179.82	280.11	0.6033	0.8849	2800
3000	86.16	1.141	0.0052722	183.09	263.26	186.51	279.08	0.6213	0.8789	3000
3200	89.29	1.182	0.0047157	189.41	262.41	193.19	277.50	0.6392	0.8718	3200
3400	92.26	1.233	0.0041973	195.91	260.96	200.10	275.23	0.6575	0.8631	3400
3600	95.08	1.297	0.0036987	202.66	258.65	207.32	271.97	0.6765	0.8521	3600
3800	97.76	1.387	0.0031898	210.26	254.87	215.54	266.99	0.6980	0.8367	3800
4000	100.31	1.562	0.0025558	220.43	246.82	226.68	257.05	0.7272	0.8085	4000
4059	101.03	1.9685	0.0019685	233.90	233.90	241.88	241.88	0.7678	0.7678	4059

Table C-3: Properties of Superheated R134a: Pressures from 80 kPa to 400 kPa

P (kPa)		Temperature, *T* (°C)										
		−30	−20	−10	0	10	20	30	40	50	60	70
80	*v* (m³/kg)	0.2388	0.2501	0.2611	0.2720	0.2828	0.2935	0.3041	0.3147	0.3252	0.3357	0.3462
	u (kJ/kg)	213.2	220.2	227.2	234.3	241.6	249.1	256.7	264.5	272.4	280.6	288.8
	h (kJ/kg)	232.4	240.2	248.1	256.1	264.3	272.6	281.0	289.7	298.5	307.4	316.5
	s (kJ/kg-K)	0.9608	0.9922	1.023	1.053	1.082	1.111	1.139	1.167	1.195	1.222	1.249
100	*v* (m³/kg)		0.1984	0.2074	0.2163	0.2251	0.2337	0.2423	0.2509	0.2594	0.2678	0.2763
	u (kJ/kg)		219.7	226.8	234.0	241.3	248.8	256.5	264.3	272.2	280.4	288.7
	h (kJ/kg)		239.5	247.5	255.6	263.8	272.2	280.7	289.4	298.2	307.1	316.3
	s (kJ/kg-K)		0.9721	1.003	1.033	1.063	1.092	1.12	1.149	1.176	1.204	1.231
120	*v* (m³/kg)		0.1639	0.1716	0.1792	0.1866	0.1939	0.2011	0.2083	0.2155	0.2226	0.2296
	u (kJ/kg)		219.2	226.4	233.6	241.0	248.5	256.2	264.0	272.0	280.2	288.5
	h (kJ/kg)		238.9	246.9	255.1	263.4	271.8	280.3	289.0	297.9	306.9	316.0
	s (kJ/kg-K)		0.9553	0.9866	1.017	1.047	1.076	1.105	1.133	1.161	1.188	1.215
140	*v* (m³/kg)			0.1461	0.1526	0.1591	0.1654	0.1717	0.1779	0.1841	0.1903	0.1964
	u (kJ/kg)			225.9	233.2	240.7	248.2	255.9	263.8	271.8	280	288.3
	h (kJ/kg)			246.4	254.6	262.9	271.4	280	288.7	297.6	306.6	315.8
	s (kJ/kg-K)			0.9724	1.003	1.033	1.062	1.091	1.12	1.147	1.175	1.202
160	*v* (m³/kg)			0.1268	0.1327	0.1385	0.1441	0.1496	0.1551	0.1606	0.1660	0.1714
	u (kJ/kg)			225.5	232.9	240.4	248.0	255.7	263.6	271.6	279.8	288.1
	h (kJ/kg)			245.8	254.1	262.5	271.0	279.6	288.4	297.3	306.3	315.5
	s (kJ/kg-K)			0.9599	0.9909	1.021	1.051	1.08	1.108	1.136	1.164	1.191
180	*v* (m³/kg)			0.1119	0.1172	0.1224	0.1275	0.1325	0.1374	0.1423	0.1471	0.152
	u (kJ/kg)			225.0	232.5	240.0	247.7	255.4	263.3	271.4	279.6	287.9
	h (kJ/kg)			245.2	253.6	262.1	270.6	279.3	288.1	297.0	306.1	315.3
	s (kJ/kg-K)			0.9485	0.9799	1.010	1.040	1.069	1.098	1.126	1.153	1.181
200	*v* (m³/kg)			0.09991	0.1048	0.1096	0.1142	0.1187	0.1232	0.1277	0.1321	0.1364
	u (kJ/kg)			224.6	232.1	239.7	247.4	255.2	263.1	271.2	279.4	287.7
	h (kJ/kg)			244.6	253.1	261.6	270.2	278.9	287.7	296.7	305.8	315.0
	s (kJ/kg-K)			0.9381	0.9699	1.001	1.030	1.060	1.088	1.116	1.144	1.171
300	*v* (m³/kg)					0.07093	0.07425	0.07748	0.08063	0.08372	0.08677	0.08978
	u (kJ/kg)					237.9	245.8	253.8	261.9	270.1	278.4	286.8
	h (kJ/kg)					259.2	268.1	277.0	286.1	295.2	304.4	313.8
	s (kJ/kg-K)					0.9611	0.9920	1.022	1.051	1.080	1.108	1.136
400	*v* (m³/kg)					0.05151	0.05421	0.05680	0.05929	0.06172	0.06410	0.06644
	u (kJ/kg)					236.0	244.2	252.4	260.6	268.9	277.3	285.9
	h (kJ/kg)					256.6	265.9	275.1	284.3	293.6	303.0	312.5
	s (kJ/kg-K)					0.9306	0.9628	0.9937	1.024	1.053	1.081	1.109

(continued)

Table C-3 *(continued)*

P (kPa)		Temperature, *T* (°C)										
		20	30	40	50	60	70	80	900	100	110	120
500	*v* (m³/kg)	0.0421	0.0443	0.0465	0.0485	0.0505	0.0524	0.0543	0.0562	0.0580	0.0599	0.0617
	u (kJ/kg)	242.4	250.9	259.3	267.7	276.3	284.9	293.7	302.5	311.5	320.6	329.9
	h (kJ/kg)	263.5	273.0	282.5	292.0	301.5	311.1	320.8	330.6	340.5	350.6	360.8
	s (kJ/kg-K)	0.9384	0.9704	1.001	1.031	1.06	1.088	1.116	1.144	1.171	1.197	1.223
600	*v* (m³/kg)		0.0360	0.0379	0.0397	0.0414	0.0431	0.0447	0.0463	0.0479	0.0494	0.051
	u (kJ/kg)		249.2	257.9	266.5	275.2	283.9	292.7	301.7	310.7	319.9	329.2
	h (kJ/kg)		270.8	280.6	290.3	300.0	309.8	319.6	329.5	339.5	349.6	359.8
	s (kJ/kg-K)		0.95	0.9817	1.012	1.042	1.071	1.099	1.126	1.154	1.18	1.207
700	*v* (m³/kg)		0.0300	0.0317	0.0333	0.0349	0.0364	0.0378	0.0393	0.0406	0.0420	0.0434
	u (kJ/kg)		247.5	256.4	265.2	274.0	282.9	291.8	300.8	310	319.2	328.6
	h (kJ/kg)		268.5	278.6	288.5	298.4	308.3	318.3	328.3	338.4	348.6	358.9
	s (kJ/kg-K)		0.9314	0.9642	0.9955	1.026	1.055	1.084	1.112	1.139	1.166	1.192
800	*v* (m³/kg)			0.0270	0.0286	0.0300	0.0313	0.0327	0.0339	0.0352	0.0364	0.0376
	u (kJ/kg)			254.8	263.9	272.8	281.8	290.9	300.0	309.2	318.5	327.9
	h (kJ/kg)			276.5	286.7	296.8	306.9	317	327.1	337.3	347.6	358.0
	s (kJ/kg-K)			0.9481	0.9803	1.011	1.041	1.07	1.098	1.126	1.153	1.18
900	*v* (m³/kg)			0.0234	0.0248	0.0262	0.0274	0.0286	0.0298	0.0310	0.0321	0.0332
	u (kJ/kg)			253.2	262.5	271.6	280.7	289.9	299.1	308.4	317.7	327.2
	h (kJ/kg)			274.2	284.8	295.1	305.4	315.6	325.9	336.2	346.6	357.0
	s (kJ/kg-K)			0.9328	0.9661	0.9977	1.028	1.057	1.086	1.114	1.141	1.168
1000	*v* (m³/kg)			0.0204	0.0218	0.0231	0.0243	0.0254	0.0265	0.0276	0.0286	0.0296
	u (kJ/kg)			251.3	261.0	270.3	279.6	288.9	298.2	307.5	317.0	326.5
	h (kJ/kg)			271.7	282.8	293.4	303.9	314.3	324.7	335.1	345.5	356.1
	s (kJ/kg-K)			0.918	0.9526	0.9851	1.016	1.046	1.075	1.103	1.131	1.158
1100	*v* (m³/kg)				0.0193	0.0205	0.0217	0.0228	0.0238	0.0248	0.0257	0.0267
	u (kJ/kg)				259.4	269.0	278.4	287.8	297.2	306.7	316.2	325.8
	h (kJ/kg)				280.6	291.6	302.3	312.9	323.4	333.9	344.5	355.1
	s (kJ/kg-K)				0.9396	0.973	1.005	1.035	1.065	1.093	1.121	1.148
1200	*v* (m³/kg)				0.0172	0.0184	0.0195	0.0205	0.0215	0.0224	0.0234	0.0242
	u (kJ/kg)				257.6	267.6	277.2	286.8	296.3	305.8	315.4	325.1
	h (kJ/kg)				278.3	289.7	300.6	311.4	322.1	332.7	343.4	354.1
	s (kJ/kg-K)				0.9268	0.9615	0.9939	1.025	1.055	1.084	1.112	1.139
1300	*v* (m³/kg)				0.0154	0.0166	0.0177	0.0187	0.0196	0.0205	0.0213	0.0222
	u (kJ/kg)				255.8	266.1	276.0	285.7	295.3	304.9	314.6	324.3
	h (kJ/kg)				275.8	287.6	298.9	309.9	320.8	331.5	342.3	353.1
	s (kJ/kg-K)				0.914	0.9501	0.9835	1.015	1.045	1.075	1.103	1.131

(continued)

Table C-3 *(continued)*

		Temperature, T (°C)									
P (kPa)		60	70	80	90	100	110	120	130	140	150
1400	v (m^3/kg)	0.0150	0.0161	0.0170	0.0179	0.0188	0.0196	0.0204	0.0212	0.0219	0.0226
	u (kJ/kg)	264.5	274.6	284.5	294.3	304	313.8	323.6	333.4	343.4	353.4
	h (kJ/kg)	285.5	297.1	308.4	319.4	330.3	341.2	352.1	363	374	385.1
	s (kJ/kg-K)	0.939	0.9734	1.006	1.036	1.066	1.095	1.123	1.15	1.177	1.204
1600	v (m^3/kg)	0.0124	0.0134	0.0144	0.0152	0.0160	0.0168	0.0175	0.0182	0.0189	0.0196
	u (kJ/kg)	260.9	271.8	282.1	292.2	302.2	312.1	322	332	342.1	352.2
	h (kJ/kg)	280.7	293.3	305.1	316.5	327.8	338.9	350	361.1	372.3	383.5
	s (kJ/kg-K)	0.9164	0.9536	0.9875	1.019	1.05	1.08	1.108	1.136	1.163	1.19
1800	v (m^3/kg)		0.0113	0.0123	0.0131	0.0139	0.0146	0.0153	0.0159	0.0165	0.0171
	u (kJ/kg)		268.6	279.5	289.9	300.2	310.3	320.4	330.6	340.7	351
	h (kJ/kg)		288.9	301.5	313.5	325.1	336.6	347.9	359.2	370.5	381.8
	s (kJ/kg-K)		0.9338	0.97	1.003	1.035	1.065	1.094	1.123	1.15	1.178
2000	v (m^3/kg)		0.00957	0.0105	0.0114	0.0121	0.0128	0.0134	0.0141	0.0146	0.0152
	u (kJ/kg)		264.8	276.6	287.5	298.1	308.5	318.8	329.1	339.4	349.7
	h (kJ/kg)		283.9	297.6	310.3	322.3	334.1	345.7	357.2	368.6	380.1
	s (kJ/kg-K)		0.9131	0.9525	0.9877	1.02	1.052	1.081	1.11	1.138	1.166
2200	v (m^3/kg)			0.00909	0.00993	0.0107	0.0113	0.0120	0.0125	0.0131	0.0136
	u (kJ/kg)			273.3	284.9	295.9	306.6	317.1	327.5	338	348.4
	h (kJ/kg)			293.3	306.7	319.3	331.5	343.4	355.1	366.8	378.4
	s (kJ/kg-K)			0.9346	0.9722	1.006	1.039	1.069	1.099	1.127	1.155
2400	v (m^3/kg)			0.00781	0.00870	0.00944	0.0101	0.0107	0.0113	0.0118	0.0123
	u (kJ/kg)			269.4	282	293.5	304.5	315.3	325.9	336.5	347.1
	h (kJ/kg)			288.2	302.9	316.2	328.8	341	353	364.8	376.6
	s (kJ/kg-K)			0.9154	0.9565	0.9926	1.026	1.057	1.087	1.117	1.145
2600	v (m^3/kg)			0.00660	0.00763	0.00839	0.00905	0.00964	0.0102	0.0107	0.0112
	u (kJ/kg)			264.6	278.8	291	302.4	313.4	324.3	335	345.7
	h (kJ/kg)			281.9	298.6	312.8	325.9	338.5	350.8	362.9	374.8
	s (kJ/kg-K)			0.8935	0.9401	0.9787	1.013	1.046	1.077	1.106	1.135
2800	v (m^3/kg)				0.00666	0.00747	0.00813	0.00872	0.00926	0.00976	0.0102
	u (kJ/kg)				275	288.2	300.1	311.5	322.6	333.5	344.3
	h (kJ/kg)				293.6	309.1	322.9	335.9	348.5	360.8	373
	s (kJ/kg-K)				0.9226	0.9645	1.001	1.035	1.066	1.096	1.126
3000	v (m^3/kg)				0.00575	0.00664	0.00733	0.00792	0.00845	0.00894	0.00940
	u (kJ/kg)				270.4	285	297.7	309.4	320.8	331.9	342.9
	h (kJ/kg)				287.7	305	319.6	333.2	346.1	358.7	371.1
	s (kJ/kg-K)				0.9027	0.9497	0.9885	1.023	1.056	1.087	1.116

Appendix D Ideal Gas & Incompressible Substances

The ideal gas characteristics of some common gases are summarized in Table D-1. Note that the specific heat capacity values provided in Table D-1 are evaluated at 20°C using the EES cv and cp property functions. The specific heat capacity values for most gases depend on temperature and will have somewhat different values at temperatures other than 20°C.

Table D-1: Ideal gas characteristics of some common gases; the specific heat capacities are evaluated at 20°C.

Gas	Chemical formula	MW	R (J/kg-K)	c_P (J/kg-K)	c_v (J/kg-K)	$\bar{c}_P$ (J/kmol-K)	$\bar{c}_v$ (J/kmol-K)	k
Air	–	28.967	287.0	1,004.4	717.4	29,094	20,780	1.400
Argon	Ar	39.948	208.1	520.3	312.2	20,786	12,472	1.667
Butane	C_4H_{10}	58.124	143.0	1,662.8	1,519.8	96,650	88,336	1.094
Carbon dioxide	CO_2	44.010	188.9	836.2	647.2	36,799	28,485	1.292
Ethane	C_2H_6	30.070	276.5	1,739.8	1,463.3	52,315	44,001	1.189
Helium	He	4.003	2,077.0	5,192.6	3,115.6	20,786	12,472	1.667
Hydrogen	H_2	2.016	4124.2	14,330	10,207	28,891	20,576	1.404
Krypton	Kr	83.800	99.2	248.0	148.8	20,785	12,471	1.667
Methane	CH_4	16.043	518.3	2,225.3	1,707.1	35,701	27,387	1.304
Neon	Ne	20.179	412.0	1,030.0	618.0	20,785	12,471	1.667
Nitrogen	N_2	28.013	296.8	1,037.7	740.9	29,069	20,755	1.401
Octane	C_8H_{18}	114.23	72.8	1,622.2	1,549.4	185,307	176,993	1.047
Oxygen	O_2	31.999	259.8	912.6	652.8	29,203	20,889	1.398
Propane	C_3H_8	44.097	188.5	1,633.4	1,444.8	72,027	63,713	1.130
Water vapor	H_2O	18.016	461.5	1,867.1	1,405.6	33,638	25,324	1.328

The incompressible characteristics of some substances are summarized in Table D-2. Note that the values provided in Table D-2 are evaluated at 20°C using the EES Solid/Liquid property functions,

Table D-2: Incompressible characteristics of some substances (at 1 atm and 20°C)

Substance	ρ (kg/m^3)	c ($J/kg\text{-}K$)	k ($W/m\text{-}K$)
Solids			
Aluminum	2703	895.8	236.1
Brass	8530	376.3	114.6
Copper	8936	383.0	401.8
Glass	2500	750.0	1.40
Gold	19,306	128.7	317.7
Lead	11,347	128.7	35.4
Plywood	545	1215	0.12
Rock (granite)	2630	775.0	2.79
Silver	10,504	234.3	429.1
304 Stainless steel	7902	471.9	14.74
Liquids			
Engine oil	888.1	1881	0.145
Syltherm 800[1]	935.2	1608	0.135
Therminol 59[1]	975.1	1681	0.121
Water (liquid)	998.2	4183	0.586

[1] Commercially available heat transfer fluids

Appendix E Ideal Gas Properties of Air

Ideal gas properties of air are provided in Table E-1. The specific internal energy provided in Table E-1 is computed by integration of the ideal gas specific heat capacity at constant volume:

$$u = \int_{T_{ref}}^{T} c_v(T)\, dT$$

and the specific enthalpy, h, provided in Table E-1 is computed by integration of the ideal gas specific heat capacity at constant pressure:

$$h = \int_{T_{ref}}^{T} c_P(T)\, dT$$

The data in Table E-1 have been obtained from EES. For temperatures between 100 K and 2000 K, the property routines use the ideal gas specific heat capacity relations given in:

E.W. Lemmon, R.T. Jacobsen, S.G. Penoncello, and D. Friend, "Thermodynamic Properties of Air and Mixtures of Nitrogen, Argon, and Oxygen from 60 to 2000 K at Pressures to 2000 MPa," *J. Phys. Chem. Ref. Data*, Vol. 29, No. 3, (2000).

For temperatures between 2000 K and 3500 K, the thermodynamic properties are based on data from Keenan, Chao, and Kaye, *Gas Tables*, Wiley, (1983). Note that these tables can be printed from the website associated with this text, www.cambridge.org/kleinandnellis, for use during closed book examinations.

Table E-1: Ideal gas properties of air.

Temp. (K)	c_v (kJ/kg-K)	c_P (kJ/kg-K)	u (kJ/kg)	h (kJ/kg)	$\int_{T_{ref}}^{T} \frac{c_P(T)}{T} dT$ (kJ/kg-K)
200	0.7153	1.002	142.7	200.1	5.299
220	0.7155	1.003	157.0	220.2	5.394
240	0.7158	1.003	171.3	240.2	5.481
260	0.7162	1.003	185.6	260.3	5.562
280	0.7168	1.004	200.0	280.3	5.636
300	0.7177	1.005	214.3	300.4	5.705
320	0.7188	1.006	228.7	320.5	5.770
340	0.7202	1.007	243.1	340.7	5.831

(continued)

Table E-1 *(continued)*

Temp. (K)	c_V (kJ/kg-K)	c_P (kJ/kg-K)	u (kJ/kg)	h (kJ/kg)	$\int_{T_{ref}}^{T} \frac{c_P(T)}{T} dT$ (kJ/kg-K)
360	0.7219	1.009	257.5	360.8	5.889
380	0.7239	1.011	272.0	381.0	5.944
400	0.7262	1.013	286.5	401.3	5.995
420	0.7289	1.016	301.0	421.6	6.045
440	0.7318	1.019	315.6	441.9	6.092
460	0.7350	1.022	330.3	462.3	6.137
480	0.7385	1.026	345.0	482.8	6.181
500	0.7423	1.029	359.8	503.3	6.223
520	0.7462	1.033	374.7	524.0	6.263
540	0.7504	1.037	389.7	544.7	6.302
560	0.7547	1.042	404.7	565.5	6.340
580	0.7592	1.046	419.9	586.3	6.377
600	0.7638	1.051	435.1	607.3	6.412
620	0.7685	1.055	450.4	628.4	6.447
640	0.7732	1.060	465.8	649.5	6.480
660	0.7780	1.065	481.3	670.8	6.513
680	0.7828	1.070	497.0	692.1	6.545
700	0.7876	1.075	512.7	713.6	6.576
720	0.7925	1.079	528.5	735.1	6.606
740	0.7973	1.084	544.4	756.8	6.636
760	0.8020	1.089	560.3	778.5	6.665
780	0.8068	1.094	576.4	800.3	6.693
800	0.8114	1.098	592.6	822.2	6.721
820	0.8160	1.103	608.9	844.3	6.748
840	0.8206	1.108	625.3	866.4	6.775
860	0.8250	1.112	641.7	888.6	6.801
880	0.8294	1.116	658.3	910.8	6.827
900	0.8337	1.121	674.9	933.2	6.852

(continued)

Table E-1 *(continued)*

Temp. (K)	c_v (kJ/kg-K)	c_P (kJ/kg-K)	u (kJ/kg)	h (kJ/kg)	$\int_{T_{ref}}^{T} \frac{c_P(T)}{T} dT$ (kJ/kg-K)
920	0.8379	1.125	691.6	955.7	6.876
940	0.8420	1.129	708.4	978.2	6.901
960	0.8460	1.133	725.3	1001	6.924
980	0.8500	1.137	742.2	1024	6.948
1000	0.8538	1.141	759.3	1046	6.971
1020	0.8575	1.145	776.4	1069	6.993
1040	0.8612	1.148	793.6	1092	7.016
1060	0.8648	1.152	810.8	1115	7.038
1080	0.8682	1.155	828.2	1138	7.059
1100	0.8716	1.159	845.6	1161	7.080
1120	0.8749	1.162	863.0	1185	7.101
1140	0.8782	1.165	880.6	1208	7.122
1160	0.8813	1.168	898.2	1231	7.142
1180	0.8843	1.171	915.8	1255	7.162
1200	0.8873	1.174	933.5	1278	7.182
1220	0.8902	1.177	951.3	1301	7.201
1240	0.8930	1.180	969.1	1325	7.220
1260	0.8958	1.183	987.0	1349	7.239
1280	0.8985	1.185	1005	1372	7.258
1300	0.9011	1.188	1023	1396	7.276
1320	0.9036	1.191	1041	1420	7.294
1340	0.9061	1.193	1059	1444	7.312
1360	0.9085	1.196	1077	1468	7.330
1380	0.9109	1.198	1095	1492	7.347
1400	0.9132	1.200	1114	1516	7.365
1420	0.9154	1.202	1132	1540	7.382
1440	0.9176	1.205	1150	1564	7.398
1460	0.9197	1.207	1169	1588	7.415

(continued)

Table E-1 *(continued)*

Temp. (K)	c_v (kJ/kg-K)	c_P (kJ/kg-K)	u (kJ/kg)	h (kJ/kg)	$\int_{T_{ref}}^{T} \frac{c_P(T)}{T} dT$ (kJ/kg-K)
1480	0.9218	1.209	1187	1612	7.432
1500	0.9239	1.211	1206	1636	7.448
1520	0.9259	1.213	1224	1660	7.464
1540	0.9278	1.215	1243	1685	7.480
1560	0.9297	1.217	1261	1709	7.495
1580	0.9316	1.219	1280	1733	7.511
1600	0.9334	1.220	1298	1758	7.526
1620	0.9352	1.222	1317	1782	7.541
1640	0.9369	1.224	1336	1807	7.556
1660	0.9386	1.226	1355	1831	7.571
1680	0.9403	1.227	1373	1856	7.586
1700	0.9419	1.229	1392	1880	7.600
1720	0.9435	1.231	1411	1905	7.615
1740	0.9451	1.232	1430	1929	7.629
1760	0.9466	1.234	1449	1954	7.643
1780	0.9481	1.235	1468	1979	7.657
1800	0.9496	1.237	1487	2003	7.671
1820	0.9511	1.238	1506	2028	7.684
1840	0.9525	1.240	1525	2053	7.698
1860	0.9539	1.241	1544	2078	7.711
1880	0.9553	1.242	1563	2103	7.725
1900	0.9566	1.244	1582	2127	7.738
1920	0.9579	1.245	1601	2152	7.751
1940	0.9592	1.246	1620	2177	7.764
1960	0.9605	1.248	1640	2202	7.776
1980	0.9618	1.249	1659	2227	7.789
2000	0.9630	1.250	1678	2252	7.802
2020	0.9633	1.250	1698	2277	7.814
2040	0.9645	1.252	1717	2303	7.826
2060	0.9656	1.253	1736	2328	7.839

(continued)

Table E-1 *(continued)*

Temp. (K)	c_v (kJ/kg-K)	c_P (kJ/kg-K)	u (kJ/kg)	h (kJ/kg)	$\int_{T_{ref}}^{T} \frac{c_P(T)}{T} dT$ (kJ/kg-K)
2080	0.9668	1.254	1756	2353	7.851
2100	0.9679	1.255	1775	2378	7.863
2120	0.9689	1.256	1794	2403	7.875
2140	0.9700	1.257	1814	2428	7.886
2160	0.9711	1.258	1833	2453	7.898
2180	0.9721	1.259	1853	2478	7.910
2200	0.9731	1.260	1872	2503	7.921
2220	0.9741	1.261	1891	2529	7.933
2240	0.9751	1.262	1911	2554	7.944
2260	0.9761	1.263	1930	2579	7.955
2280	0.9770	1.264	1950	2604	7.966
2300	0.9779	1.265	1970	2630	7.977
2320	0.9789	1.266	1989	2655	7.988
2340	0.9798	1.267	2009	2680	7.999
2360	0.9807	1.268	2028	2706	8.010
2380	0.9815	1.269	2048	2731	8.021
2400	0.9824	1.269	2068	2756	8.031
2420	0.9833	1.270	2087	2782	8.042
2440	0.9841	1.271	2107	2807	8.052
2460	0.9850	1.272	2127	2833	8.063
2480	0.9858	1.273	2146	2858	8.073
2500	0.9866	1.274	2166	2884	8.083
2520	0.9874	1.274	2186	2909	8.093
2540	0.9882	1.275	2206	2935	8.103
2560	0.9890	1.276	2225	2960	8.113
2580	0.9897	1.277	2245	2986	8.123
2600	0.9905	1.278	2265	3011	8.133
2650	0.9924	1.279	2314	3075	8.158
2700	0.9942	1.281	2364	3139	8.182
2750	0.9960	1.283	2414	3203	8.205

(continued)

Table E-1 *(continued)*

Temp. (K)	c_v (kJ/kg-K)	c_P (kJ/kg-K)	u (kJ/kg)	h (kJ/kg)	$\int_{T_{ref}}^{T} \frac{c_P(T)}{T} dT$ (kJ/kg-K)
2800	0.9977	1.285	2464	3267	8.228
2850	0.9994	1.286	2514	3332	8.251
2900	1.001	1.288	2564	3396	8.273
2950	1.003	1.290	2614	3460	8.295
3000	1.004	1.291	2664	3525	8.317
3050	1.006	1.293	2714	3590	8.338
3100	1.007	1.294	2765	3654	8.359
3150	1.009	1.296	2815	3719	8.380
3200	1.010	1.297	2865	3784	8.401
3250	1.012	1.299	2916	3849	8.421
3300	1.013	1.300	2967	3914	8.441
3350	1.015	1.302	3017	3979	8.460
3400	1.016	1.303	3068	4044	8.479
3450	1.017	1.304	3119	4109	8.498
3500	1.019	1.306	3170	4174	8.517

Appendix F Ideal Gas Properties of Common Combustion Gases

The specific internal energy, enthalpy and entropy at 1 atm pressure for common combustion gases are provided as a function of temperature in the following tables.

Table F-1: Ideal gas properties of CO_2

Table F-2: Ideal gas properties of CO

Table F-3: Ideal gas properties of O_2

Table F-4: Ideal gas properties of N_2

Table F-5: Ideal gas properties of H_2O

The data in these tables were obtained from EES. The reference state for specific enthalpy is based on the enthalpy of formation relative to the elements at 25°C. The reference state for specific entropy is based on the Third Law of Thermodynamics. The reference values are from:

Bonnie J. McBride, Michael J. Zehe, and Sanford Gordon

"NASA Glenn Coefficients for CalculatingThermodynamic Properties of Individual Species"

NASA/TP-2002-211556, Sept. 2002

http://www.lerc.nasa.gov/WWW/CEAWeb/

Note that these tables can be printed from the website associated with this text, www.cambridge.org/kleinandnellis, for use during closed book examinations.

Table F-1: Ideal gas properties of CO_2

Temp. (K)	$\bar{u}$ (kJ/kmol)	$\bar{h}$ (kJ/kmol)	$\int_{T_{ref}}^{T} \frac{\bar{c}_p(T)}{T} dT$ (kJ/kmol-K)
280	−396,478	−394,150	211.37
298.15	−395,965	−393,486	213.67
300	−395,912	−393,417	213.90
320	−395,326	−392,666	216.32
340	−394,722	−391,896	218.66
360	−394,101	−391,108	220.91
380	−393,464	−390,304	223.08
400	−392,810	−389,484	225.18
420	−392,141	−388,649	227.22
440	−391,458	−387,799	229.20
460	−390,761	−386,936	231.11
480	−390,050	−386,060	232.98
500	−389,328	−385,171	234.79
520	−388,593	−384,270	236.56
540	−387,847	−383,358	238.28
560	−387,091	−382,435	239.96
580	−386,324	−381,501	241.60
600	−385,546	−380,558	243.20
620	−384,760	−379,605	244.76
640	−383,964	−378,643	246.29
660	−383,159	−377,672	247.78
680	−382,346	−376,692	249.24
700	−381,525	−375,705	250.67
720	−380,696	−374,709	252.07
740	−379,859	−373,706	253.45
760	−379,015	−372,696	254.80
780	−378,164	−371,679	256.12
800	−377,306	−370,655	257.41
820	−376,442	−369,624	258.69
840	−375,571	−368,587	259.94
860	−374,694	−367,544	261.16
880	−373,811	−366,495	262.37
900	−372,923	−365,440	263.55
920	−372,029	−364,380	264.72
940	−371,129	−363,314	265.87
960	−370,225	−362,243	266.99
980	−369,315	−361,167	268.10
1000	−368,400	−360,086	269.19
1050	−366,093	−357,363	271.85
1100	−363,758	−354,612	274.41
1150	−361,398	−351,836	276.88
1200	−359,013	−349,036	279.26
1250	−356,607	−346,214	281.57
1300	−354,180	−343,371	283.80
1350	−351,733	−340,509	285.96
1400	−349,269	−337,629	288.05
1450	−346,787	−334,732	290.08
1500	−344,291	−331,819	292.06
1550	−341,779	−328,892	293.98
1600	−339,255	−325,952	295.84
1650	−336,717	−322,998	297.66
1700	−334,168	−320,034	299.43
1750	−331,608	−317,058	301.16
1800	−329,038	−314,072	302.84
1850	−326,458	−311,077	304.48
1900	−323,870	−308,072	306.08
1950	−321,273	−305,060	307.65
2000	−318,669	−302,040	309.18

(continued)

Table F-1 *(continued)*

Temp. (K)	$\bar{u}$ (kJ/kmol)	$\bar{h}$ (kJ/kmol)	$\int_{T_{ref}}^{T} \frac{\bar{c}_p(T)}{T} dT$ (kJ/kmol-K)
2050	−316,058	−299,014	310.67
2100	−313,440	−295,980	312.13
2150	−310,817	−292,941	313.56
2200	−308,188	−289,896	314.96
2250	−305,554	−286,846	316.34
2300	−302,915	−283,792	317.68
2350	−300,271	−280,733	318.99
2400	−297,624	−277,669	320.28
2450	−294,972	−274,602	321.55
2500	−292,317	−271,531	322.79
2550	−289,659	−268,457	324.01
2600	−286,997	−265,380	325.20
2650	−284,333	−262,300	326.38
2700	−281,665	−259,216	327.53
2750	−278,995	−256,130	328.66
2800	−276,322	−253,042	329.77
2850	−273,646	−249,950	330.87
2900	−270,968	−246,857	331.94
2950	−268,287	−243,760	333.00
3000	−265,604	−240,661	334.04
3050	−262,918	−237,559	335.07
3100	−260,229	−234,454	336.08
3150	−257,537	−231,347	337.07
3200	−254,842	−228,236	338.05

Table F-2: Ideal gas properties of CO

Temp. (K)	$\overline{u}$ (kJ/kmol)	$\overline{h}$ (kJ/kmol)	$\int_{T_{ref}}^{T} \frac{\bar{c}_p(T)}{T} dT$ (kJ/kmol-K)
280	−113,383	−111,055	195.72
298.15	−113,007	−110,528	197.54
300	−112,969	−110,475	197.72
320	−112,554	−109,893	199.60
340	−112,138	−109,311	201.36
360	−111,721	−108,728	203.03
380	−111,302	−108,143	204.61
400	−110,882	−107,557	206.11
420	−110,460	−106,968	207.55
440	−110,036	−106,378	208.92
460	−109,610	−105,785	210.24
480	−109,181	−105,190	211.50
500	−108,750	−104,593	212.72
520	−108,317	−103,994	213.90
540	−107,881	−103,392	215.03
560	−107,443	−102,787	216.13
580	−107,002	−102,180	217.20
600	−106,559	−101,570	218.23
620	−106,113	−100,958	219.24
640	−105,664	−100,343	220.21
660	−105,213	−99,725	221.16
680	−104,759	−99,105	222.09
700	−104,303	−98,483	222.99
720	−103,844	−97,857	223.87
740	−103,382	−97,229	224.73
760	−102,918	−96,599	225.57
780	−102,451	−95,966	226.40
800	−101,981	−95,330	227.20
820	−101,509	−94,692	227.99
840	−101,035	−94,051	228.76
860	−100,558	−93,408	229.52
880	−100,079	−92,762	230.26
900	−99,597	−92,114	230.99
920	−99,113	−91,464	231.70
940	−98,627	−90,811	232.40
960	−98,138	−90,156	233.09
980	−97,647	−89,499	233.77
1000	−97,153	−88,839	234.44
1050	−95,911	−87,181	236.06
1100	−94,655	−85,509	237.61
1150	−93,386	−83,824	239.11
1200	−92,105	−82,128	240.55
1250	−90,812	−80,419	241.95
1300	−89,509	−78,700	243.30
1350	−88,195	−76,971	244.60
1400	−86,872	−75,232	245.87
1450	−85,539	−73,483	247.09
1500	−84,198	−71,726	248.28
1550	−82,848	−69,961	249.44
1600	−81,491	−68,188	250.57
1650	−80,126	−66,408	251.66
1700	−78,755	−64,621	252.73
1750	−77,378	−62,828	253.77
1800	−75,996	−61,030	254.78
1850	−74,608	−59,226	255.77
1900	−73,215	−57,417	256.74
1950	−71,817	−55,604	257.68
2000	−70,416	−53,787	258.60

(continued)

Table F-2 *(continued)*

Temp. (K)	$\bar{u}$ (kJ/kmol)	$\bar{h}$ (kJ/kmol)	$\int_{T_{ref}}^{T} \frac{\bar{c}_P(T)}{T} dT$ (kJ/kmol-K)	Temp. (K)	$\bar{u}$ (kJ/kmol)	$\bar{h}$ (kJ/kmol)	$\int_{T_{ref}}^{T} \frac{\bar{c}_P(T)}{T} dT$ (kJ/kmol-K)
2050	−69,011	−51,967	259.50	2650	−51,962	−29,929	268.92
2100	−67,603	−50,142	260.38	2700	−50,531	−28,083	269.61
2150	−66,191	−48,315	261.24	2750	−49,100	−26,236	270.29
2200	−64,777	−46,485	262.08	2800	−47,668	−24,387	270.96
2250	−63,360	−44,653	262.90	2850	−46,234	−22,538	271.61
2300	−61,941	−42,818	263.71	2900	−44,800	−20,688	272.26
2350	−60,520	−40,981	264.50	2950	−43,364	−18,837	272.89
2400	−59,097	−39,143	265.27	3000	−41,928	−16,985	273.51
2450	−57,673	−37,303	266.03	3050	−40,490	−15,132	274.12
2500	−56,247	−35,461	266.77	3100	−39,051	−13,277	274.73
2550	−54,820	−33,618	267.50	3150	−37,611	−11,421	275.32
2600	−53,391	−31,774	268.22	3200	−36,169	−9,563	275.91

Table F-3: Ideal gas properties of O_2

Temp. (K)	$\overline{u}$ (kJ/kmol)	$\overline{h}$ (kJ/kmol)	$\int_{T_{ref}}^{T} \frac{\overline{c}_p(T)}{T} dT$ (kJ/kmol-K)
280	−2,858	−530	203.20
298.15	−2,479	0	205.03
300	−2,440	54	205.21
320	−2,020	641	207.10
340	−1,595	1,231	208.89
360	−1,167	1,826	210.59
380	−735	2,424	212.21
400	−298	3,027	213.76
420	142	3,634	215.24
440	587	4,246	216.66
460	1,037	4,861	218.03
480	1,490	5,481	219.35
500	1,947	6,104	220.62
520	2,408	6,731	221.85
540	2,873	7,362	223.04
560	3,341	7,997	224.19
580	3,813	8,635	225.31
600	4,288	9,277	226.40
620	4,767	9,922	227.46
640	5,249	10,570	228.49
660	5,734	11,222	229.49
680	6,222	11,876	230.47
700	6,713	12,534	231.42
720	7,207	13,194	232.35
740	7,704	13,857	233.26
760	8,204	14,523	234.15
780	8,706	15,191	235.01
800	9,211	15,862	235.86
820	9,718	16,536	236.69
840	10,228	17,212	237.51
860	10,740	17,890	238.31
880	11,254	18,571	239.09
900	11,771	19,253	239.86
920	12,289	19,939	240.61
940	12,810	20,626	241.35
960	13,333	21,315	242.07
980	13,859	22,007	242.79
1000	14,386	22,700	243.49
1050	15,712	24,442	245.19
1100	17,050	26,195	246.82
1150	18,398	27,960	248.39
1200	19,758	29,735	249.90
1250	21,127	31,519	251.36
1300	22,505	33,314	252.76
1350	23,892	35,117	254.12
1400	25,288	36,928	255.44
1450	26,693	38,748	256.72
1500	28,105	40,576	257.96
1550	29,524	42,412	259.16
1600	30,951	44,254	260.33
1650	32,385	46,104	261.47
1700	33,826	47,960	262.58
1750	35,273	49,823	263.66
1800	36,727	51,692	264.71
1850	38,186	53,568	265.74
1900	39,652	55,449	266.74
1950	41,123	57,336	267.72
2000	42,600	59,228	268.68

(continued)

Table F-3 *(continued)*

Temp. (K)	$\bar{u}$ (kJ/kmol)	$\bar{h}$ (kJ/kmol)	$\int_{T_{ref}}^{T} \frac{\bar{c}_p(T)}{T} dT$ (kJ/kmol-K)
2050	44,082	61,126	269.62
2100	45,570	63,030	270.54
2150	47,063	64,938	271.43
2200	48,561	66,852	272.31
2250	50,064	68,771	273.18
2300	51,572	70,695	274.02
2350	53,084	72,623	274.85
2400	54,602	74,556	275.67
2450	56,124	76,495	276.46
2500	57,652	78,437	277.25
2550	59,183	80,385	278.02
2600	60,720	82,337	278.78
2650	62,261	84,294	279.52
2700	63,807	86,256	280.26
2750	65,357	88,222	280.98
2800	66,913	90,193	281.69
2850	68,473	92,168	282.39
2900	70,037	94,149	283.08
2950	71,607	96,134	283.76
3000	73,181	98,124	284.43
3050	74,760	100,119	285.08
3100	76,344	102,119	285.74
3150	77,933	104,123	286.38
3200	79,528	106,133	287.01

Table F-4: Ideal gas properties of N_2

Temp. (K)	$\bar{u}$ (kJ/kmol)	$\bar{h}$ (kJ/kmol)	$\int_{T_{ref}}^{T} \frac{\bar{c}_p(T)}{T} dT$ (kJ/kmol-K)
280	−2,856	−528	189.67
298.15	−2,479	0	191.49
300	−2,441	54	191.67
320	−2,025	635	193.55
340	−1,610	1,217	195.31
360	−1,194	1,799	196.97
380	−777	2,382	198.55
400	−359	2,967	200.05
420	60	3,552	201.48
440	481	4,139	202.84
460	903	4,727	204.15
480	1,327	5,318	205.41
500	1,752	5,910	206.62
520	2,180	6,504	207.78
540	2,610	7,100	208.91
560	3,042	7,698	209.99
580	3,476	8,298	211.05
600	3,913	8,901	212.07
620	4,351	9,506	213.06
640	4,792	10,114	214.03
660	5,236	10,724	214.97
680	5,682	11,336	215.88
700	6,130	11,951	216.77
720	6,581	12,568	217.64
740	7,035	13,187	218.49
760	7,491	13,809	219.32
780	7,949	14,434	220.13
800	8,410	15,061	220.93
820	8,873	15,691	221.70

Temp. (K)	$\bar{u}$ (kJ/kmol)	$\bar{h}$ (kJ/kmol)	$\int_{T_{ref}}^{T} \frac{\bar{c}_p(T)}{T} dT$ (kJ/kmol-K)
840	9,339	16,323	222.46
860	9,807	16,957	223.21
880	10,277	17,594	223.94
900	10,750	18,233	224.66
920	11,225	18,875	225.37
940	11,703	19,518	226.06
960	12,183	20,165	226.74
980	12,665	20,813	227.41
1000	13,150	21,464	228.06
1050	14,371	23,101	229.66
1100	15,605	24,751	231.19
1150	16,852	26,414	232.67
1200	18,112	28,089	234.10
1250	19,384	29,777	235.47
1300	20,667	31,476	236.81
1350	21,961	33,185	238.10
1400	23,265	34,905	239.35
1450	24,578	36,634	240.56
1500	25,901	38,373	241.74
1550	27,232	40,119	242.88
1600	28,571	41,874	243.99
1650	29,918	43,637	245.08
1700	31,272	45,406	246.14
1750	32,632	47,182	247.16
1800	33,998	48,964	248.17
1850	35,370	50,751	249.15
1900	36,747	52,544	250.10
1950	38,128	54,341	251.04
2000	39,514	56,143	251.95

(continued)

Table F-4 *(continued)*

Temp. (K)	$\bar{u}$ (kJ/kmol)	$\bar{h}$ (kJ/kmol)	$\int_{T_{ref}}^{T} \frac{\bar{c}_p(T)}{T} dT$ (kJ/kmol-K)
2050	40,904	57,949	252.84
2100	42,298	59,758	253.72
2150	43,695	61,571	254.57
2200	45,095	63,387	255.41
2250	46,498	65,205	256.22
2300	47,903	67,026	257.02
2350	49,311	68,849	257.81
2400	50,720	70,675	258.58
2450	52,132	72,502	259.33
2500	53,545	74,331	260.07
2550	54,960	76,162	260.80
2600	56,376	77,994	261.51

Temp. (K)	$\bar{u}$ (kJ/kmol)	$\bar{h}$ (kJ/kmol)	$\int_{T_{ref}}^{T} \frac{\bar{c}_p(T)}{T} dT$ (kJ/kmol-K)
2650	57,794	79,827	262.21
2700	59,214	81,663	262.90
2750	60,635	83,499	263.57
2800	62,057	85,337	264.24
2850	63,480	87,176	264.89
2900	64,906	89,017	265.53
2950	66,332	90,860	266.16
3000	67,761	92,704	266.78
3050	69,191	94,550	267.39
3100	70,623	96,398	267.99
3150	72,058	98,248	268.58
3200	73,494	100,100	269.16

Table F-5: Ideal gas properties of H_2O

Temp. (K)	$\bar{u}$ (kJ/kmol)	$\bar{h}$ (kJ/kmol)	$\int_{T_{ref}}^{T} \frac{\bar{c}_P(T)}{T} dT$ (kJ/kmol-K)
280	−244,749	−242,421	186.60
298.15	−244,290	−241,811	188.71
300	−244,243	−241,749	188.92
320	−243,735	−241,075	191.09
340	−243,226	−240,399	193.14
360	−242,714	−239,721	195.08
380	−242,200	−239,041	196.92
400	−241,684	−238,358	198.67
420	−241,164	−237,672	200.34
440	−240,641	−236,983	201.95
460	−240,115	−236,290	203.49
480	−239,584	−235,594	204.97
500	−239,050	−234,893	206.40
520	−238,512	−234,189	207.78
540	−237,969	−233,480	209.12
560	−237,422	−232,766	210.41
580	−236,871	−232,048	211.67
600	−236,314	−231,326	212.90
620	−235,753	−230,598	214.09
640	−235,187	−229,866	215.25
660	−234,616	−229,129	216.39
680	−234,040	−228,387	217.50
700	−233,459	−227,639	218.58
720	−232,873	−226,887	219.64
740	−232,282	−226,129	220.68
760	−231,685	−225,366	221.69
780	−231,083	−224,598	222.69
800	−230,476	−223,825	223.67
820	−229,864	−223,046	224.63
840	−229,247	−222,263	225.58
860	−228,624	−221,474	226.50
880	−227,996	−220,679	227.42
900	−227,363	−219,880	228.32
920	−226,724	−219,075	229.20
940	−226,081	−218,265	230.07
960	−225,432	−217,450	230.93
980	−224,778	−216,630	231.77
1000	−224,119	−215,804	232.61
1050	−222,448	−213,718	234.64
1100	−220,746	−211,601	236.61
1150	−219,013	−209,452	238.52
1200	−217,249	−207,272	240.38
1250	−215,456	−205,063	242.18
1300	−213,632	−202,824	243.94
1350	−211,781	−200,556	245.65
1400	−209,901	−198,261	247.32
1450	−207,994	−195,938	248.95
1500	−206,060	−193,589	250.54
1550	−204,101	−191,213	252.10
1600	−202,116	−188,813	253.63
1650	−200,107	−186,388	255.12
1700	−198,074	−183,940	256.58
1750	−196,019	−181,469	258.01
1800	−193,941	−178,975	259.42
1850	−191,842	−176,460	260.80
1900	−189,722	−173,924	262.15
1950	−187,581	−171,368	263.48
2000	−185,422	−168,793	264.78

(continued)

Table F-5 *(continued)*

Temp. (K)	$\bar{u}$ (kJ/kmol)	$\bar{h}$ (kJ/kmol)	$\int_{T_{ref}}^{T} \frac{\bar{c}_p(T)}{T} dT$ (kJ/kmol-K)	Temp. (K)	$\bar{u}$ (kJ/kmol)	$\bar{h}$ (kJ/kmol)	$\int_{T_{ref}}^{T} \frac{\bar{c}_p(T)}{T} dT$ (kJ/kmol-K)
2050	−183,244	−166,199	266.06	2650	−155,874	−133,841	279.89
2100	−181,047	−163,587	267.32	2700	−153,508	−131,059	280.93
2150	−178,834	−160,958	268.56	2750	−151,131	−128,267	281.95
2200	−176,603	−158,312	269.77	2800	−148,744	−125,464	282.96
2250	−174,357	−155,649	270.97	2850	−146,348	−122,652	283.96
2300	−172,094	−152,971	272.15	2900	−143,941	−119,830	284.94
2350	−169,817	−150,279	273.31	2950	−141,526	−116,999	285.91
2400	−167,526	−147,572	274.45	3000	−139,103	−114,160	286.86
2450	−165,221	−144,851	275.57	3050	−136,671	−111,312	287.80
2500	−162,903	−142,117	276.67	3100	−134,231	−108,457	288.73
2550	−160,572	−139,370	277.76	3150	−131,784	−105,594	289.65
2600	−158,228	−136,611	278.83	3200	−129,329	−102,723	290.55

Appendix G Numerical Solution to ODEs

This extended section can be found online at www.cambridge.org/kleinandnellis. The mathematical description of many interesting problems in thermodynamics and other areas of engineering involves ordinary differential equations (ODEs). In some cases, the ordinary differential equations are sufficiently simple that an analytical solution can be derived. However, in most cases this is not possible and therefore numerical solutions to the ODEs is required. This appendix provides an introduction to numerical techniques and a discussion of the use of the Integral command in EES.

Appendix H Introduction to Maple

This extended section can be found on the website www.cambridge.org/kleinandnellis. Maple is an application that can be used to analytically solve algebraic and differential equations. The capability to differentiate, integrate and algebraically manipulate mathematical expressions in symbolic form can be a very powerful aid in solving many types of engineering problems, including some thermodynamics problems. Maple also provides a very convenient mathematical reference; if, for example, you've forgotten that the derivative of sine is cosine, it is easy to use Maple to quickly provide this information. Therefore, Maple can replace the numerous mathematical reference books that might otherwise be required to carry out all of the integration, differentiation, simplification, etc. that is required to solve many engineering problems. Maple and EES can be used effectively together; Maple can determine the analytical solution to a problem and these symbolic expressions can subsequently be copied (almost directly) into EES for convenient numerical evaluation and manipulation in the context of a specific application. This appendix summarizes the commands that are the most useful for thermodynamics problems.

INDEX

Note that page numbers starting with Ex indicate that these pages can be found on the extended section x which is available at www.cambridge.org/kleinandnellis. For example, E6 is indicates extended section E6.